Theorems from Vector Calculus

In the following ϕ, ψ, and $\mathbf{A}$ are well-behaved scalar or vector functions, V is a three-dimensional volume with volume element d^3x, S is a closed two-dimensional surface bounding V, with area element da and unit outward normal $\mathbf{n}$ at da.

$$\int_V \boldsymbol{\nabla} \cdot \mathbf{A} \, d^3x = \int_S \mathbf{A} \cdot \mathbf{n} \, da \qquad \text{(Divergence theorem)}$$

$$\int_V \boldsymbol{\nabla}\psi \, d^3x = \int_S \psi \mathbf{n} \, da$$

$$\int_V \boldsymbol{\nabla} \times \mathbf{A} \, d^3x = \int_S \mathbf{n} \times \mathbf{A} \, da$$

$$\int_V (\phi\nabla^2\psi + \boldsymbol{\nabla}\phi \cdot \boldsymbol{\nabla}\psi) \, d^3x = \int_S \phi\mathbf{n} \cdot \boldsymbol{\nabla}\psi \, da \qquad \text{(Green's first identity)}$$

$$\int_V (\phi\nabla^2\psi - \psi\nabla^2\phi) \, d^3x = \int_S (\phi\boldsymbol{\nabla}\psi - \psi\boldsymbol{\nabla}\phi) \cdot \mathbf{n} \, da \qquad \text{(Green's theorem)}$$

In the following S is an open surface and C is the contour bounding it, with line element $d\mathbf{l}$. The normal $\mathbf{n}$ to S is defined by the right-hand-screw rule in relation to the sense of the line integral around C.

$$\int_S (\boldsymbol{\nabla} \times \mathbf{A}) \cdot \mathbf{n} \, da = \oint_C \mathbf{A} \cdot d\mathbf{l} \qquad \text{(Stokes's theorem)}$$

$$\int_S \mathbf{n} \times \boldsymbol{\nabla}\psi \, da = \oint_C \psi \, d\mathbf{l}$$

Classical
Electrodynamics

Classical Electrodynamics
Third Edition

John David Jackson
Professor Emeritus of Physics,
University of California, Berkeley

JOHN WILEY & SONS, INC.
New York • Chichester • Weinheim • Brisbane • Singapore • Toronto

This book was set in 10 on 12 Times Ten by UG and printed and bound by Hamilton Printing Company.

This book is printed on acid-free paper.

The paper in this book was manufactured by a mill whose forest management programs include sustained yield harvesting of its timberlands. Sustained yield harvesting principles ensure that the numbers of trees cut each year does not exceed the amount of new growth.

Library of Congress Cataloging-in-Publication Data

Jackson, John David, 1925–
    Classical electrodynamics / John David Jackson.—3rd ed.
      p. cm.
    Includes index.
    ISBN 0-471-30932-X (cloth : alk. paper)
    1. Electrodynamics.   I. Title.
    QC631.J3   1998
    537.6—dc21

                                     97-46873
                                       CIP

Printed in the United States of America

10 9 8 7 6 5

**To the memory of my father,
Walter David Jackson**

Preface

It has been 36 years since the appearance of the first edition of this book, and 23 years since the second. Such intervals may be appropriate for a subject whose fundamental basis was completely established theoretically 134 years ago by Maxwell and experimentally 110 years ago by Hertz. Still, there are changes in emphasis and applications. This third edition attempts to address both without any significant increase in size. Inevitably, some topics present in the second edition had to be eliminated to make room for new material. One major omission is the chapter on plasma physics, although some pieces appear elsewhere. Readers who miss particular topics may, I hope, be able to avail themselves of the second edition.

The most visible change is the use of SI units in the first 10 chapters. Gaussian units are retained in the later chapters, since such units seem more suited to relativity and relativistic electrodynamics than SI. As a reminder of the system of units being employed, the running head on each left-hand page carries "—SI" or "—G" depending on the chapter.

My tardy adoption of the universally accepted SI system is a recognition that almost all undergraduate physics texts, as well as engineering books at all levels, employ SI units throughout. For many years Ed Purcell and I had a pact to support each other in the use of Gaussian units. Now I have betrayed him! Although this book is formally dedicated to the memory of my father, I dedicate this third edition informally to the memory of Edward Mills Purcell (1912–1997), a marvelous physicist with deep understanding, a great teacher, and a wonderful man.

Because of the increasing use of personal computers to supplement analytical work or to attack problems not amenable to analytic solution, I have included some new sections on the *principles* of some numerical techniques for electrostatics and magnetostatics, as well as some elementary problems. Instructors may use their ingenuity to create more challenging ones. The aim is to provide an understanding of such methods before blindly using canned software or even *Mathematica* or *Maple*.

There has been some rearrangement of topics—Faraday's law and quasi-static fields are now in Chapter 5 with magnetostatics, permitting a more logical discussion of energy and inductances. Another major change is the consolidation of the discussion of radiation by charge-current sources, in both elementary and exact multipole forms, in Chapter 9. All the applications to scattering and diffraction are in Chapter 10.

The principles of optical fibers and dielectric waveguides are discussed in two new sections in Chapter 8. In Chapter 13 the treatment of energy loss has been shortened and strengthened. Because of the increasing importance of synchrotron radiation as a research tool, the discussion in Chapter 14 has been augmented by a detailed section on the physics of wigglers and undulators for synchroton light sources. There is new material in Chapter 16 on radiation reaction and models of classical charged particles, as well as changed emphasis.

There is much tweaking by small amounts throughout. I hope the reader will not notice, or will notice only greater clarity. To mention but a few minor additions: estimating self-inductances, Poynting's theorem in lossy materials, polarization potentials (Hertz vectors), Goos–Hänchen effect, attenuation in optical fibers, London penetration depth in superconductors. And more problems, of course! Over 110 new problems, a 40% increase, all aimed at educating, not discouraging.

In preparing this new edition, I have benefited from questions, suggestions, criticism, and advice from many students, colleagues, and newfound friends. I am in debt to all. Particular thanks for help in various ways go to Myron Bander, David F. Bartlett, Robert N. Cahn, John Cooper, John L. Gammel, David J. Griffiths, Leroy T. Kerth, Kwang J. Kim, Norman M. Kroll, Harry J. Lipkin, William Mendoza, Gerald A. Miller, William A. Newcomb, Ivan Otero, Alan M. Portis, Fritz Rohrlich, Wayne M. Saslow, Chris Schmid, and George H. Trilling.

J. David Jackson
Berkeley, California, 1998

Preface to the Second Edition

In the thirteen years since the appearance of the first edition, my interest in classical electromagnetism has waxed and waned, but never fallen to zero. The subject is ever fresh. There are always important new applications and examples. The present edition reflects two efforts on my part: the refinement and improvement of material already in the first edition; the addition of new topics (and the omission of a few).

The major purposes and emphasis are still the same, but there are extensive changes and additions. A major augmentation is the "Introduction and Survey" at the beginning. Topics such as the present experimental limits on the mass of the photon and the status of linear superposition are treated there. The aim is to provide a survey of those basics that are often assumed to be well known when one writes down the Maxwell equations and begins to solve specific examples. Other major changes in the first half of the book include a new treatment of the derivation of the equations of macroscopic electromagnetism from the microscopic description; a discussion of symmetry properties of mechanical and electromagnetic quantities; sections on magnetic monopoles and the quantization condition of Dirac; Stokes's polarization parameters; a unified discussion of the frequency dispersion characteristics of dielectrics, conductors, and plasmas; a discussion of causality and the Kramers-Kronig dispersion relations; a simplified, but still extensive, version of the classic Sommerfeld–Brillouin problem of the arrival of a signal in a dispersive medium (recently verified experimentally); an unusual example of a resonant cavity; the normal-mode expansion of an arbitrary field in a wave guide; and related discussions of sources in a guide or cavity and the transmission and reflection coefficients of flat obstacles in wave guides.

Chapter 9, on simple radiating systems and diffraction, has been enlarged to include scattering at long wavelengths (the blue sky, for example) and the optical theorem. The sections on scalar and vectorial diffraction have been improved.

Chapters 11 and 12, on special relativity, have been rewritten almost completely. The old pseudo-Euclidean metric with $x_4 = ict$ has been replaced by $g^{\mu\nu}$ (with $g^{00} = +1$, $g^{ii} = -1$, $i = 1, 2, 3$). The change of metric necessitated a complete revision and thus permitted substitution of modern experiments and concerns about the experimental basis of the special theory for the time-honored aberration of starlight and the Michelson–Morley experiment. Other aspects have been modernized, too. The extensive treatment of relativistic kinematics of the first edition has been relegated to the problems. In its stead is a discussion of the Lagrangian for the electromagnetic fields, the canonical and symmetric stress-energy tensor, and the Proca Lagrangian for massive photons.

Significant alterations in the remaining chapters include a new section on transition radiation, a completely revised (and much more satisfactory) semiclassical treatment of radiation emitted in collisions that stresses momentum transfer instead of impact parameter, and a better derivation of the coupling of multipole fields to their sources. The collection of formulas and page references to special functions on the front and back flyleaves is a much requested addition. Of the 278 problems, 117 (more than 40 per cent) are new.

ix

The one area that remains almost completely unchanged is the chapter on magnetohydrodynamics and plasma physics. I regret this. But the book obviously has grown tremendously, and there are available many books devoted exclusively to the subject of plasmas or magnetohydrodynamics.

Of minor note is the change from Maxwell's equations and a Green's function to the Maxwell equations and a Green function. The latter boggles some minds, but is in conformity with other usage (Bessel function, for example). It is still Green's theorem, however, because that's whose theorem it is.

Work on this edition began in earnest during the first half of 1970 on the occasion of a sabbatical leave spent at Clare Hall and the Cavendish Laboratory in Cambridge. I am grateful to the University of California for the leave and indebted to N. F. Mott for welcoming me as a visitor to the Cavendish Laboratory and to R. J. Eden and A. B. Pippard for my appointment as a Visiting Fellow of Clare Hall. Tangible and intangible evidence at the Cavendish of Maxwell, Rayleigh and Thomson provided inspiration for my task; the stimulation of everyday activities there provided necessary diversion.

This new edition has benefited from questions, suggestions, comments and criticism from many students, colleagues, and strangers. Among those to whom I owe some specific debt of gratitude are A. M. Bincer, L. S. Brown, R. W. Brown, E. U. Condon, H. H. Denman, S. Deser, A. J. Dragt, V. L. Fitch, M. B. Halpern, A. Hobson, J. P. Hurley, D. L. Judd, L. T. Kerth, E. Marx, M. Nauenberg, A. B. Pippard, A. M. Portis, R. K. Sachs, W. M. Saslow, R. Schleif, V. L. Telegdi, T. Tredon, E. P. Tryon, V. F. Weisskopf, and Dudley Williams. Especially helpful were D. G. Boulware, R. N. Cahn, Leverett Davis, Jr., K. Gottfried, C. K. Graham, E. M. Purcell, and E. H. Wichmann. I send my thanks and fraternal greetings to all of these people, to the other readers who have written to me, and the countless students who have struggled with the problems (and sometimes written asking for solutions to be dispatched before some deadline!). To my mind, the book is better than ever. May each reader benefit and enjoy!

J. D. Jackson
Berkeley, California, 1974

Preface to the First Edition

Classical electromagnetic theory, together with classical and quantum mechanics, forms the core of present-day theoretical training for undergraduate and graduate physicists. A thorough grounding in these subjects is a requirement for more advanced or specialized training.

Typically the undergraduate program in electricity and magnetism involves two or perhaps three semesters beyond elementary physics, with the emphasis on the fundamental laws, laboratory verification and elaboration of their consequences, circuit analysis, simple wave phenomena, and radiation. The mathematical tools utilized include vector calculus, ordinary differential equations with constant coefficients, Fourier series, and perhaps Fourier or Laplace transforms, partial differential equations, Legendre polynomials, and Bessel functions.

As a general rule, a two-semester course in electromagnetic theory is given to beginning graduate students. It is for such a course that my book is designed. My aim in teaching a graduate course in electromagnetism is at least threefold. The first aim is to present the basic subject matter as a coherent whole, with emphasis on the unity of electric and magnetic phenomena, both in their physical basis and in the mode of mathematical description. The second, concurrent aim is to develop and utilize a number of topics in mathematical physics which are useful in both electromagnetic theory and wave mechanics. These include Green's theorems and Green's functions, orthonormal expansions, spherical harmonics, cylindrical and spherical Bessel functions. A third and perhaps most important purpose is the presentation of new material, especially on the interaction of relativistic charged particles with electromagnetic fields. In this last area personal preferences and prejudices enter strongly. My choice of topics is governed by what I feel is important and useful for students interested in theoretical physics, experimental nuclear and high-energy physics, and that as yet ill-defined field of plasma physics.

The book begins in the traditional manner with electrostatics. The first six chapters are devoted to the development of Maxwell's theory of electromagnetism. Much of the necessary mathematical apparatus is constructed along the way, especially in Chapter 2 and 3, where boundary-value problems are discussed thoroughly. The treatment is initially in terms of the electric field E and the magnetic induction B, with the derived macroscopic quantities, D and H, introduced by suitable averaging over ensembles of atoms or molecules. In the discussion of dielectrics, simple classical models for atomic polarizability are described, but for magnetic materials no such attempt to made. Partly this omission was a question of space, but truly classical models of magnetic susceptibility are not possible. Furthermore, elucidation of the interesting phenomenon of ferromagnetism needs almost a book in itself.

The next three chapters (7–9) illustrate various electromagnetic phenomena, mostly of a macroscopic sort. Plane waves in different media, including plasmas as well as dispersion and the propagation of pulses, are treated in Chapter 7. The discussion of wave guides and cavities in Chapter 8 is developed for systems of arbitrary cross section, and the problems of attenuation in guides and the Q of a cavity are handled in a very general way which emphasizes the physical processes involved. The elementary theory of multipole radiation from a localized source and diffraction occupy Chapter 9. Since the simple scalar theory of diffraction is covered in many optics textbooks, as well as undergraduate books on electricity and magnetism, I have presented an improved, although still approximate, theory of diffraction based on vector rather than scalar Green's theorems.

The subject of magnetohydrodynamics and plasmas receives increasingly more attention from physicists and astrophysicists. Chapter 10 represents a survey of this complex field with an introduction to the main physical ideas involved.

The first nine or ten chapters constitute the basic material of classical electricity and magnetism. A graduate student in physics may be expected to have been exposed to much of this material, perhaps at a somewhat lower level, as an undergraduate. But he obtains a more mature view of it, understands it more deeply, and gains a considerable technical ability in analytic methods of solution when he studies the subject at the level of this book. He is then prepared to go on to more advanced topics. The advanced topics presented here are predominantly those involving the interaction of charged particles with each other and with electromagnetic fields, especially when moving relativistically.

The special theory of relativity had its origins in classical electrodynamics. And even after almost 60 years, classical electrodynamics still impresses and delights as a beautiful example of the covariance of physical laws under Lorentz transformations. The special theory of relativity is discussed in Chapter 11, where all the necessary formal apparatus is developed, various kinematic consequences are explored, and the covariance of electrodynamics is established. The next chapter is devoted to relativistic particle kinematics and dynamics. Although the dynamics of charged particles in electromagnetic fields can properly be considered electrodynamics, the reader may wonder whether such things as kinematic transformations of collision problems can. My reply is that these examples occur naturally once one has established the four-vector character of a particle's momentum and energy, that they serve as useful practice in manipulating Lorentz transformations, and that the end results are valuable and often hard to find elsewhere.

Chapter 13 on collisions between charged particles emphasizes energy loss and scattering and develops concepts of use in later chapters. Here for the first time in the book I use semiclassical arguments based on the uncertainty principle to obtain approximate quantum-mechanical expressions for energy loss, etc., from the classical results. This approach, so fruitful in the hands of Niels Bohr and E. J. Williams, allows one to see clearly how and when quantum-mechanical effects enter to modify classical considerations.

The important subject of emission of radiation by accelerated point charges is discussed in detail in Chapters 14 and 15. Relativistic effects are stressed, and expressions for the frequency and angular dependence of the emitted radiation are developed in sufficient generality for all applications. The examples treated range from synchrotron radiation to bremsstrahlung and radiative beta processes. Cherenkov radiation and the Weizsäcker–Williams method of virtual quanta are also discussed. In the atomic and nuclear collision processes semiclassical arguments are again employed to obtain approximate quantum-mechanical results. I lay considerable stress on this point because I feel that it is important for the student to see that radiative effects such as bremsstrahlung are almost entirely classical in nature, even though involving small-scale collisions. A student who meets bremsstrahlung for the first time as an example of a calculation in quantum field theory will not understand its physical basis.

Multipole fields form the subject matter of Chapter 16. The expansion of scalar and vector fields in spherical waves is developed from first principles with no restrictions as to the relative dimensions of source and wavelength. Then the properties of electric and magnetic multipole radiation fields are considered. Once the connection to the multiple moments of the source has been made, examples of atomic and nuclear multipole radiation are discussed, as well as a macroscopic source whose dimensions are comparable to a wavelength. The scattering of a plane electromagnetic wave by a spherical object is treated in some detail in order to illustrate a boundary-value problem with vector spherical waves.

In the last chapter the difficult problem of radiative reaction is discussed. The treatment is physical, rather than mathematical, with the emphasis on delimiting the areas where approximate radiative corrections are adequate and on finding where and why existing theories fail. The original Abraham–Lorentz theory of the self-force is presented, as well as more recent classical considerations.

The book ends with an appendix on units and dimensions and a bibliography. In the appendix I have attempted to show the logical steps involved in setting up a system of units, without haranguing the reader as to the obvious virtues of *my* choice of units. I have provided two tables which I hope will be useful, one for converting equations and symbols and the other for converting a given quantity of something from so many Gaussian units to so many mks units, and vice versa. The bibliography lists books which I think the reader may find pertinent and useful for reference or additional study. These books are referred to by author's name in the reading lists at the end of each chapter.

This book is the outgrowth of a graduate course in classical electrodynamics which I have taught off and on over the past eleven years, at both the University of Illinois and McGill University. I wish to thank my colleagues and students at both institutions for countless helpful remarks and discussions. Special mention must be made of Professor P. R. Wallace of McGill, who gave me the opportunity and encouragement to teach what was then a rather unorthodox course in electromagnetism, and Professors H. W. Wyld and G. Ascoli of Illinois, who have been particularly free with many helpful suggestions on the treatment of various topics. My thanks are also extended to Dr. A. N. Kaufman for reading and commenting on a preliminary version of the manuscript, and to Mr. G. L. Kane for his zealous help in preparing the index.

J. D. Jackson
Urbana, Illinois, January, 1962

Contents

Chapter 3 / *Boundary-Value Problems in Electrostatics: II*   *95*

Chapter 4 / *Multipoles, Electrostatics of Macroscopic Media, Dielectrics*   *145*

Chapter 5 / *Magnetostatics, Faraday's Law, Quasi-Static Fields*   *174*

Chapter 6 / *Maxwell Equations, Macroscopic Electromagnetism, Conservation Laws* *237*

Chapter 10 / *Scattering and Diffraction*   *456*

Chapter 11 / *Special Theory of Relativity*   *514*

Chapter 15 / *Bremsstrahlung, Method of Virtual Quanta, Radiative Beta Processes*   *708*

Chapter 16 / *Radiation Damping, Classical Models of Charged Particles*   *745*

Appendix on Units and Dimensions   *775*

Introduction and Survey

Although amber and lodestone were known to the ancient Greeks, electro-dynamics developed as a quantitative subject in less than a hundred years. Cavendish's remarkable experiments in electrostatics were done from 1771 to 1773. Coulomb's monumental researches began to be published in 1785. This marked the beginning of quantitative research in electricity and magnetism on a worldwide scale. Fifty years later Faraday was studying the effects of time-varying currents and magnetic fields. By 1864 Maxwell had published his famous paper on a dynamical theory of the electromagnetic field. Twenty-four years later (1888) Hertz published his discovery of transverse electromagnetic waves, which propagated at the same speed as light, and placed Maxwell's theory on a firm experimental footing.

The story of the development of our understanding of electricity and mag-netism and of light is, of course, much longer and richer than the mention of a few names from one century would indicate. For a detailed account of the fas-cinating history, the reader should consult the authoritative volumes by *Whittaker*.* A briefer account, with emphasis on optical phenomena, appears at the beginning of *Born and Wolf*.

Since the 1960s there has been a true revolution in our understanding of the basic forces and constituents of matter. Now (1990s) classical electrodynamics rests in a sector of the unified description of particles and interactions known as the *standard model*. The standard model gives a coherent quantum-mechanical description of electromagnetic, weak, and strong interactions based on funda-mental constituents—quarks and leptons—interacting via force carriers—pho-tons, W and Z bosons, and gluons. The unified theoretical framework is gener-ated through principles of continuous gauge (really phase) invariance of the forces and discrete symmetries of particle properties.

From the point of view of the standard model, classical electrodynamics is a limit of quantum electrodynamics (for small momentum and energy transfers, and large average numbers of virtual or real photons). Quantum electrodynamics, in turn, is a consequence of a spontaneously broken symmetry in a theory in which initially the weak and electromagnetic interactions are unified and the force carriers of both are massless. The symmetry breaking leaves the electro-magnetic force carrier (photon) massless with a Coulomb's law of infinite range, while the weak force carriers acquire masses of the order of 80–90 GeV/c^2 with a weak interaction at low energies of extremely short range (2×10^{-18} meter). Because of the origins in a unified theory, the range and strength of the weak interaction are related to the electromagnetic coupling (the fine structure con-stant $\alpha \approx 1/137$).

*Italicized surnames denote books that are cited fully in the Bibliography.

Despite the presence of a rather large number of quantities that must be taken from experiment, the standard model (together with general relativity at large scales) provides a highly accurate description of nature in all its aspects, from far inside the nucleus, to microelectronics, to tables and chairs, to the most remote galaxy. Many of the phenomena are classical or explicable with nonrelativistic quantum mechanics, of course, but the precision of the agreement of the standard model with experiment in atomic and particle physics where relativistic quantum mechanics rules is truly astounding. Classical mechanics and classical electrodynamics served as progenitors of our current understanding, and still play important roles in practical life and at the research frontier.

This book is self-contained in that, though some mathematical background (vector calculus, differential equations) is assumed, the subject of electrodynamics is developed from its beginnings in electrostatics. Most readers are not coming to the subject for the first time, however. The purpose of this introduction is therefore not to set the stage for a discussion of Coulomb's law and other basics, but rather to present a review and a survey of classical electromagnetism. Questions such as the current accuracy of the inverse square law of force (mass of the photon), the limits of validity of the principle of linear superposition, and the effects of discreteness of charge and of energy differences are discussed. "Bread and butter" topics such as the boundary conditions for macroscopic fields at surfaces between different media and at conductors are also treated. The aim is to set classical electromagnetism in context, to indicate its domain of validity, and to elucidate some of the idealizations that it contains. Some results from later in the book and some nonclassical ideas are used in the course of the discussion. Certainly a reader beginning electromagnetism for the first time will not follow all the arguments or see their significance. For others, however, this introduction will serve as a springboard into the later parts of the book, beyond Chapter 5, and will remind them of how the subject stands as an experimental science.

I.1   *Maxwell Equations in Vacuum, Fields, and Sources*

The equations governing electromagnetic phenomena are the Maxwell equations,

$$\nabla \cdot \mathbf{D} = \rho$$
$$\nabla \times \mathbf{H} - \frac{\partial \mathbf{D}}{\partial t} = \mathbf{J}$$
$$\nabla \times \mathbf{E} + \frac{\partial \mathbf{B}}{\partial t} = 0 \qquad \text{(I.1a)}$$
$$\nabla \cdot \mathbf{B} = 0$$

where for external sources in vacuum, $\mathbf{D} = \epsilon_0 \mathbf{E}$ and $\mathbf{B} = \mu_0 \mathbf{H}$. The first two equations then become

$$\nabla \cdot \mathbf{E} = \rho/\epsilon_0$$
$$\nabla \times \mathbf{B} - \frac{\partial \mathbf{E}}{c^2 \partial t} = \mu_0 \mathbf{J} \qquad \text{(I.1b)}$$

Implicit in the Maxwell equations is the continuity equation for charge density and current density,

$$\frac{\partial \rho}{\partial t} + \mathbf{\nabla} \cdot \mathbf{J} = 0 \tag{I.2}$$

This follows from combining the time derivative of the first equation in (I.1a) with the divergence of the second equation. Also essential for consideration of charged particle motion is the Lorentz force equation,

$$\mathbf{F} = q(\mathbf{E} + \mathbf{v} \times \mathbf{B}) \tag{I.3}$$

which gives the force acting on a point charge q in the presence of electromagnetic fields.

These equations have been written in SI units, the system of electromagnetic units used in the first 10 chapters of this book. (Units and dimensions are discussed in the Appendix.) The Maxwell equations are displayed in the commoner systems of units in Table 2 of the Appendix. Essential to electrodynamics is the speed of light in vacuum, given in SI units by $c = (\mu_0 \epsilon_0)^{-1/2}$. As discussed in the Appendix, the meter is now defined in terms of the second (based on a hyperfine transition in cesium-133) and the speed of light ($c = 299\ 792\ 458$ m/s, exactly). These definitions assume that the speed of light is a universal constant, consistent with evidence (see Section 11.2.C) indicating that to a high accuracy the speed of light in vacuum is independent of frequency from very low frequencies to at least $\nu \simeq 10^{24}$ Hz (4 GeV photons). For most practical purposes we can approximate $c \simeq 3 \times 10^8$ m/s or to be considerably more accurate, $c = 2.998 \times 10^8$ m/s.

The electric and magnetic fields $\mathbf{E}$ and $\mathbf{B}$ in (I.1) were originally introduced by means of the force equation (I.3). In Coulomb's experiments forces acting between localized distributions of charge were observed. There it is found useful (see Section 1.2) to introduce the electric field $\mathbf{E}$ as the force per unit charge. Similarly, in Ampère's experiments the mutual forces of current-carrying loops were studied (see Section 5.2). With the identification of $NAq\mathbf{v}$ as a current in a conductor of cross-sectional area A with N charge carriers per unit volume moving at velocity $\mathbf{v}$, we see that $\mathbf{B}$ in (I.3) is defined in magnitude as a force per unit current. Although $\mathbf{E}$ and $\mathbf{B}$ thus first appear just as convenient replacements for forces produced by distributions of charge and current, they have other important aspects. First, their introduction decouples conceptually the sources from the test bodies experiencing electromagnetic forces. If the fields $\mathbf{E}$ and $\mathbf{B}$ from two source distributions are the same at a given point in space, the force acting on a test charge or current at that point will be the same, regardless of how different the source distributions are. This gives $\mathbf{E}$ and $\mathbf{B}$ in (I.3) meaning in their own right, independent of the sources. Second, electromagnetic fields can exist in regions of space where there are no sources. They can carry energy, momentum, and angular momentum and so have an existence totally independent of charges and currents. In fact, though there are recurring attempts to eliminate explicit reference to the fields in favor of action-at-a-distance descriptions of the interaction of charged particles, the concept of the electromagnetic field is one of the most fruitful ideas of physics, both classically and quantum mechanically.

The concept of $\mathbf{E}$ and $\mathbf{B}$ as ordinary fields is a classical notion. It can be thought of as the classical limit (limit of large quantum numbers) of a quantum-mechanical description in terms of real or virtual photons. In the domain of macroscopic phenomena and even some atomic phenomena, the discrete photon aspect of the electromagnetic field can usually be ignored or at least glossed over. For example, 1 meter from a 100-watt light bulb, the root mean square electric field is of the order of 50 V/m and there are of the order of 10^{15} visible photons/ cm$^2 \cdot$s. Similarly, an isotropic FM antenna with a power of 100 watts at 10^8 Hz produces an rms electric field of only 0.5 mV/m at a distance of 100 kilometers, but this still corresponds to a flux of 10^{12} photons/cm$^2 \cdot$s, or about 10^9 photons in a volume of 1 wavelength cubed (27 m^3) at that distance. Ordinarily an apparatus will not be sensible to the individual photons; the cumulative effect of many photons emitted or absorbed will appear as a continuous, macroscopically observable response. Then a completely classical description in terms of the Maxwell equations is permitted and is appropriate.

How is one to decide a priori when a classical description of the electromagnetic fields is adequate? Some sophistication is occasionally needed, but the following is usually a sufficient criterion: When the number of photons involved can be taken as large but the momentum carried by an individual photon is small compared to the momentum of the material system, then the response of the material system can be determined adequately from a classical description of the electromagnetic fields. For example, each 10^8 Hz photon emitted by our FM antenna gives it an impulse of only 2.2×10^{-34} N$\cdot$s. A classical treatment is surely adequate. Again, the scattering of light by a free electron is governed by the classical Thomson formula (Section 14.8) at low frequencies, but by the laws of the Compton effect as the momentum $\hbar\omega/c$ of the incident photon becomes significant compared to mc. The photoelectric effect is nonclassical for the matter system, since the quasi-free electrons in the metal change their individual energies by amounts equal to those of the absorbed photons, but the photoelectric current can be calculated quantum mechanically for the electrons using a classical description of the electromagnetic fields.

The quantum nature of the electromagnetic fields must, on the other hand, be taken into account in spontaneous emission of radiation by atoms, or by any other system that initially lacks photons and has only a small number of photons present finally. The average behavior may still be describable in essentially classical terms, basically because of conservation of energy and momentum. An example is the classical treatment (Section 16.2) of the cascading of a charged particle down through the orbits of an attractive potential. At high particle quantum numbers, a classical description of particle motion is adequate, and the secular changes in energy and angular momentum can be calculated classically from the radiation reaction because the energies of the successive photons emitted are small compared to the kinetic or potential energy of the orbiting particle.

The sources in (I.1) are $\rho(\mathbf{x}, t)$, the electric charge density, and $\mathbf{J}(\mathbf{x}, t)$, the electric current density. In classical electromagnetism they are assumed to be continuous distributions in $\mathbf{x}$, although we consider from time to time localized distributions that can be approximated by points. The magnitudes of these point charges are assumed to be completely arbitrary, but are known to be restricted in reality to discrete values. The basic unit of charge is the magnitude of the charge on the electron,

$$|q_e| = 4.803\ 206\ 8(15) \times 10^{-10}\ \text{esu}$$
$$= 1.602\ 177\ 33(49) \times 10^{-19}\ \text{C}$$

where the errors in the last two decimal places are shown in parentheses. The charges on the proton and on all presently known particles or systems of particles are integral multiples of this basic unit.* The experimental accuracy with which it is known that the multiples are exactly integers is phenomenal (better than 1 part in 10^{20}). The experiments are discussed in Section 11.9, where the question of the Lorentz invariance of charge is also treated.

The discreteness of electric charge does not need to be considered in most macroscopic applications. A 1-microfarad capacitor at a potential of 150 volts, for example, has a total of 10^{15} elementary charges on each electrode. A few thousand electrons more or less would not be noticed. A current of 1 microampere corresponds to 6.2×10^{12} elementary charges per second. There are, of course, some delicate macroscopic or almost macroscopic experiments in which the discreteness of charge enters. Millikan's famous oil drop experiment is one. His droplets were typically 10^{-4} cm in radius and had a few or few tens of elementary charges on them.

There is a lack of symmetry in the appearance of the source terms in the Maxwell equations (I.1a). The first two equations have sources; the second two do not. This reflects the experimental *absence of magnetic charges* and currents. Actually, as is shown in Section 6.11, particles could have magnetic as well as electric charge. If all particles in nature had the same ratio of magnetic to electric charge, the fields and sources could be redefined in such a way that the usual Maxwell equations (I.1a) emerge. In this sense it is somewhat a matter of convention to say that no magnetic charges or currents exist. Throughout most of this book it is assumed that only electric charges and currents act in the Maxwell equations, but some consequences of the existence of a particle with a different magnetic to electric charge ratio, for example, a magnetic monopole, are described in Chapter 6.

I.2 Inverse Square Law or the Mass of the Photon

The distance dependence of the electrostatic law of force was shown quantitatively by Cavendish and Coulomb to be an inverse square law. Through Gauss's law and the divergence theorem (see Sections 1.3 and 1.4) this leads to the first of the Maxwell equations (I.1b). The original experiments had an accuracy of only a few percent and, furthermore, were at a laboratory length scale. Experiments at higher precision and involving different regimes of size have been performed over the years. It is now customary to quote the tests of the inverse square law in one of two ways:

(a) Assume that the force varies as $1/r^{2+\epsilon}$ and quote a value or limit for ϵ.

(b) Assume that the electrostatic potential has the "Yukawa" form (see Section 12.8), $r^{-1}e^{-\mu r}$ and quote a value or limit for μ or μ^{-1}. Since $\mu = m_\gamma c/\hbar$, where m_γ is the assumed mass of the photon, the test of the inverse square law is sometimes phrased in terms of an upper limit on m_γ. Laboratory experiments usually give ϵ and perhaps μ or m_γ; geomagnetic experiments give μ or m_γ.

---

*Quarks have charges ⅔ and −⅓ in these units, but are never (so far) seen individually.

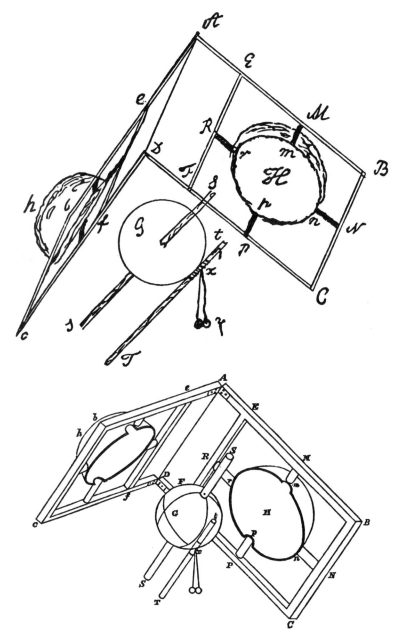

Figure I.1 Cavendish's apparatus for establishing the inverse square law of electrostatics. Top, facsimile of Cavendish's own sketch; bottom, line drawing by a draughtsman. The inner globe is 12.1 inches in diameter, the hollow pasteboard hemispheres slightly larger. Both globe and hemispheres were covered with tinfoil "to make them the more perfect conductors of electricity." (Figures reproduced by permission of the Cambridge University Press.)

The original experiment with concentric spheres by Cavendish* in 1772 gave an upper limit on ϵ of $|\epsilon| \leq 0.02$. His apparatus is shown in Fig. I.1. About 100 years later Maxwell performed a very similar experiment at Cambridge[†] and set an upper limit of $|\epsilon| \leq 5 \times 10^{-5}$. Two other noteworthy laboratory experiments based on Gauss's law are those of Plimpton and Lawton,[‡] which gave $|\epsilon| < 2 \times 10^{-9}$, and the recent one of Williams, Faller, and Hill.[§] A schematic drawing of the apparatus of the latter experiment is shown in Fig. I.2. Though not a static experiment ($\nu = 4 \times 10^6$ Hz), the basic idea is almost the same as Cavendish's. He looked for a charge on the inner sphere after it had been brought into electrical contact with the charged outer sphere and then disconnected; he found none. Williams, Faller, and Hill looked for a voltage difference between two concentric shells when the outer one was subjected to an alternating voltage of ± 10 kV with respect to ground. Their sensitivity was such that a voltage difference of less than 10^{-12} V could have been detected. Their null result, when interpreted by means of the Proca equations (Section 12.8), gives a limit of $\epsilon = (2.7 \pm 3.1) \times 10^{-16}$.

Measurements of the earth's magnetic field, both on the surface and out from the surface by satellite observation, permit the best direct limits to be set on ϵ or equivalently the photon mass m_γ. The geophysical and also the laboratory observations are discussed in the reviews by Kobzarev and Okun' and by Goldhaber and Nieto, listed at the end of this introduction. The surface measurements of the earth's magnetic field give slightly the best value (see Problem 12.15), namely,

$$m_\gamma < 4 \times 10^{-51} \text{ kg}$$

or

$$\mu^{-1} > 10^8 \text{ m}$$

For comparison, the electron mass is $m_e = 9.1 \times 10^{-31}$ kg. The laboratory experiment of Williams, Faller, and Hill can be interpreted as setting a limit $m_\gamma < 1.6 \times 10^{-50}$ kg, only a factor of 4 poorer than the geomagnetic limit.

A *rough* limit on the photon mass can be set quite easily by noting the existence of very low frequency modes in the earth-ionosphere resonant cavity (Schumann resonances, discussed in Section 8.9). The double Einstein relation, $h\nu = m_\gamma c^2$, suggests that the photon mass must satisfy an inequality, $m_\gamma < h\nu_0/c^2$, where ν_0 is any electromagnetic resonant frequency. The lowest Schumann resonance has $\nu_0 \simeq 8$ Hz. From this we calculate $m_\gamma < 6 \times 10^{-50}$ kg, a very small value only one order of magnitude above the best limit. While this argument has crude validity, more careful consideration (see Section 12.8 and the references given there) shows that the limit is roughly $(R/H)^{1/2} \simeq 10$ times larger, $R \simeq 6400$ km being the radius of the earth, and $H \simeq 60$ km being the height of the iono-

*H. Cavendish, *Electrical Researches*, ed. J. C. Maxwell, Cambridge University Press, Cambridge (1879), pp. 104–113.

[†]*Ibid.*, see note 19.

[‡]S. J. Plimpton and W. E. Lawton, *Phys. Rev.* **50**, 1066 (1936).

[§]E. R. Williams, J. E. Faller, and H. A. Hill, *Phys. Rev. Lett.* **26**, 721 (1971).

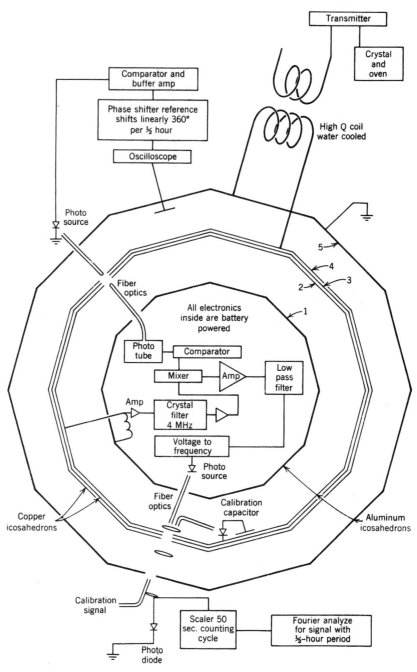

Figure I.2   Schematic diagram of the "Cavendish" experiment of Williams, Faller, and Hill. The concentric icosahedrons are conducting shells. A 4 MHz voltage of 10 kV peak is applied between shells 5 and 4. Shell 4 and its contiguous shells 2 and 3 are roughly 1.5 meters in diameter and contain shell 1 inside. The voltage difference between shells 1 and 2 (if any) appears across the inductor indicated at about 8 o'clock in shell 1. The amplifier and optics system are necessary to extract the voltage information to the outside world. They are equivalent to Cavendish's system of strings that automatically opened the hinged hemispheres and brought up the pith balls to test for charge on the inner sphere. (Figure reproduced with permission of the authors.)

sphere.* In spite of this dilution factor, the limit of 10^{-48} kg set by the mere existence of Schumann resonances is quite respectable.

The laboratory and geophysical tests show that on length scales of order 10^{-2} to 10^7 m, the inverse square law holds with extreme precision. At smaller distances we must turn to less direct evidence often involving additional assumptions. For example, Rutherford's historical analysis of the scattering of alpha particles by thin foils substantiates the Coulomb law of force down to distances of the order of 10^{-13} m, provided the alpha particle and the nucleus can be treated as classical point charges interacting statically and the charge cloud of the electrons can be ignored. All these assumptions can be, and have been, tested, of course, but only within the framework of the validity of quantum mechanics, linear superposition (see below), and other (very reasonable) assumptions. At still smaller distances, relativistic quantum mechanics is necessary, and strong interaction effects enter to obscure the questions as well as the answers. Nevertheless, elastic scattering experiments with positive and negative electrons at center of mass energies of up to 100 GeV have shown that quantum electrodynamics (the relativistic theory of point electrons interacting with massless photons) holds to distances of the order of 10^{-18} m. We conclude that the photon mass can be taken to be zero (the inverse square force law holds) over the whole classical range of distances and deep into the quantum domain as well. The inverse square law is known to hold over at least 25 orders of magnitude in the length scale!

I.3   *Linear Superposition*

The Maxwell equations in vacuum are *linear* in the fields **E** and **B**. This linearity is exploited so often, for example, with hundreds of different telephone conversations on a single microwave link, that it is taken for granted. There are, of course, circumstances where nonlinear effects occur—in magnetic materials, in crystals responding to intense laser beams, even in the devices used to put those telephone conversations on and off the microwave beam. But here we are concerned with fields in vacuum or the microscopic fields inside atoms and nuclei.

What evidence do we have to support the idea of linear superposition? At the macroscopic level, all sorts of experiments test linear superposition at the level of 0.1% accuracy—groups of charges and currents produce electric and magnetic forces calculable by linear superposition, transformers perform as expected, standing waves are observed on transmission lines—the reader can make a list. In optics, slit systems show diffraction patterns; x-ray diffraction tells us about crystal structure; white light is refracted by a prism into the colors of the rainbow and recombined into white light again. At the macroscopic and even at the atomic level, linear superposition is remarkably valid.

It is in the subatomic domain that departures from linear superposition can be legitimately sought. As charged particles approach each other very closely, electric field strengths become enormous. If we think of a charged particle as a

---

*The basic point is that, to the extent that H/R is negligible, the extremely low frequency (ELF) propagation is the same as in a parallel plate transmission line in the fundamental TEM mode. This propagation is unaffected by a finite photon mass, except through changes in the static capacitance and inductance per unit length. Explicit photon mass effects occur in order $(H/R)\,\mu^2$.

localized distribution of charge, we see that its electromagnetic energy grows larger and larger as the charge is localized more and more. In attempting to avoid infinite self-energies of point particles, it is natural to speculate that some sort of saturation occurs, that field strengths have some upper bound. Such classical nonlinear theories have been studied in the past. One well-known example is the theory of Born and Infeld.* The vacuum is given electric and magnetic permeabilities,

$$\frac{\epsilon}{\epsilon_0} = \frac{\mu_0}{\mu} = \left[1 + \frac{1}{b^2} (c^2 B^2 - E^2) \right]^{-1/2} \tag{I.4}$$

where b is a maximum field strength. Equation (I.4) is actually a simplification proposed earlier by Born alone. It suffices to illustrate the general idea. Fields are obviously modified at short distances; all electromagnetic energies are finite. But such theories suffer from arbitrariness in the manner of how the nonlinearity occurs and also from grave problems with a transition to a quantum theory. Furthermore, there is no evidence of this kind of classical nonlinearity. The quantum mechanics of many-electron atoms is described to high precision by normal quantum theory with the interactions between nucleus and electrons and between electrons and electrons given by a *linear superposition* of pairwise potentials (or retarded relativistic interactions for fine effects). Field strengths of the order of 10^{11}–10^{17} V/m exist at the orbits of electrons in atoms, while the electric field at the edge of a heavy nucleus is of the order of 10^{21} V/m. Energy level differences in light atoms like helium, calculated on the basis of linear superposition of electromagnetic interactions, are in agreement with experiment to accuracies that approach 1 part in 10^6. And Coulomb energies of heavy nuclei are consistent with linear superposition of electromagnetic effects. It is possible, of course, that for field strengths greater than 10^{21} V/m nonlinear effects could occur. One place to look for such effects is in superheavy nuclei ($Z > 110$), both in the atomic energy levels and in the nuclear Coulomb energy.† At the present time there is no evidence for any classical nonlinear behavior of vacuum fields at short distances.

There *is* a *quantum-mechanical nonlinearity* of electromagnetic fields that arises because the uncertainty principle permits the momentary creation of an electron-positron pair by two photons and the subsequent disappearance of the pair with the emission of two different photons, as indicated schematically in Fig. I.3. This process is called the scattering of light by light.‡S The two incident plane waves $e^{i\mathbf{k}_1 \cdot \mathbf{x} - i\omega_1 t}$ and $e^{i\mathbf{k}_2 \cdot \mathbf{x} - i\omega_2 t}$ do not merely add coherently, as expected with linear superposition, but interact and (with small probability) transform into two different plane waves with wave vectors $\mathbf{k}_3$ and $\mathbf{k}_4$. This nonlinear feature of

---

*M. Born and L. Infeld, *Proc. R. Soc. London* **A144**, 425 (1934). See M. Born, *Atomic Physics*, Blackie, London (1949), Appendix VI, for an elementary discussion.

†An investigation of the effect of a Born–Infeld type of nonlinearity on the atomic energy levels in superheavy elements has been made by J. Rafelski, W. Greiner, and L. P. Fulcher, *Nuovo Cimento* **13B**, 135 (1973).

‡When two of the photons in Fig. I.3 are virtual photons representing interaction to second order with a static nuclear Coulomb field, the process is known as Delbrück scattering. See Section 15.8 of J. M. Jauch and F. Rohrlich, *The Theory of Photons and Electrons*, Addison-Wesley, Reading, MA (1955).

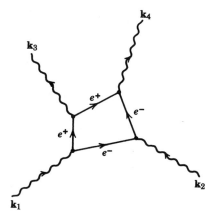

Figure I.3   The scattering of light by light. Schematic diagram of the process by which photon-photon scattering occurs.

quantum electrodynamics can be expressed, at least for slowly varying fields, in terms of electric and magnetic permeability tensors of the vacuum:

$$D_i = \epsilon_0 \sum_k \epsilon_{ik} E_k, \qquad B_i = \mu_0 \sum_k \mu_{ik} H_k$$

where

$$\epsilon_{ik} = \delta_{ik} + \frac{e_G^4 \hbar}{45 \pi m^4 c^7} [2(E^2 - c^2 B^2)\delta_{ik} + 7 c^2 B_i B_k] + \cdots$$

$$\mu_{ik} = \delta_{ik} + \frac{e_G^4 \hbar}{45 \pi m^4 c^7} [2(c^2 B^2 - E^2)\delta_{ik} + 7 E_i E_k] + \cdots$$

(I.5)

Here e_G and m are the charge (in Gaussian units) and mass of the electron. These results were first obtained by Euler and Kockel in 1935.* We observe that in the classical limit ($\hbar \to 0$), these nonlinear effects go to zero. Comparison with the classical Born–Infeld expression (I.4) shows that for small nonlinearities, the quantum-mechanical field strength

$$b_q = \frac{\sqrt{45\pi}}{2} \sqrt{\frac{e_G^2}{\hbar c}} \frac{e_G}{r_0^2} \simeq 0.51 \frac{e_G}{r_0^2}$$

plays a role analogous to the Born–Infeld parameter b. Here $r_0 = e_G^2/mc^2 \simeq 2.8 \times 10^{-15}$ m is the classical electron radius and $e_G/r_0^2 = 1.8 \times 10^{20}$ V/m is the electric field at the surface of such a classical electron. Two comments in passing: (a) the ϵ_{ik} and μ_{ik} in (I.5) are approximations that fail for field strengths approaching b_q or when the fields vary too rapidly in space or time ($\hbar/mc$ setting the critical scale of length and $\hbar/mc^2$ of time); (b) the chance numerical coincidence of b_q and $e_G/2r_0^2$ is suggestive but probably not significant, since b_q involves Planck's constant $\hbar$.

In analogy with the polarization $\mathbf{P} = (\mathbf{D} - \mathbf{E})/4\pi$, we speak of the field-dependent terms in (I.5) as *vacuum polarization* effects. In addition to the scattering of light by light or Delbrück scattering, vacuum polarization causes very small shifts in atomic energy levels. The dominant contribution involves a virtual electron-positron pair, just as in Fig. I.3, but with only two photon lines instead

*H. Euler and B. Kockel, *Naturwissenschaften* **23**, 246 (1935).

of four. If the photons are real, the process contributes to the mass of the photon and is decreed to vanish. If the photons are virtual, however, as in the electromagnetic interaction between a nucleus and an orbiting electron, or indeed for any externally applied field, the creation and annihilation of a virtual electron-positron pair from time to time causes observable effects.

Vacuum polarization is manifest by a modification of the electrostatic interaction between two charges at short distances, described as a screening of the "bare" charges with distance, or in more modern terms as a "running" coupling constant. Since the charge of a particle is defined as the strength of its electromagnetic coupling observed at large distances (equivalent to negligible momentum transfers), the presence of a screening action by electron-positron pairs closer to the charge implies that the "bare" charge observed at short distances is larger than the charge defined at large distances. Quantitatively, the lowest order quantum-electrodynamic result for the Coulomb potential energy between two charges $Z_1 e$ and $Z_2 e$, corrected for vacuum polarization, is

$$V(r) = \hbar c \frac{Z_1 Z_2 \alpha}{r} \left[1 + \frac{2\alpha}{3\pi} \int_{2m}^{\infty} d\kappa \frac{\sqrt{\kappa^2 - 4m^2}}{\kappa^2} \left(1 + \frac{2m^2}{\kappa^2} \right) e^{-\kappa r} \right] \quad (I.6)$$

where α is the fine structure constant ($\approx 1/137$), m is the inverse Compton wavelength (electron mass, multiplied by $c/\hbar$). The integral, a superposition of Yukawa potentials ($e^{-\kappa r}/r$) is the one-loop contribution of all the virtual pairs. It increases the magnitude of the potential energy at distances of separation inside the electron Compton wavelength ($\hbar/mc = \alpha a_0 \approx 3.86 \times 10^{-13}$ m).

Because of its short range, the added vacuum polarization energy is unimportant in light atoms, except for very precise measurements. It is, however, important in high Z atoms and in muonic atoms, where the heavier mass of the muon ($m_\mu \approx 207\, m_e$) means that, even in the lightest muonic atoms, the Bohr radius is well inside the range of the modified potential. X-ray measurements in medium-mass muonic atoms provide a highly accurate verification of the vacuum polarization effect in (I.6).

The idea of a "running" coupling constant, that is, an effective strength of interaction that changes with momentum transfer, is illustrated in electromagnetism by exhibiting the spatial Fourier transform of the interaction energy (I.6):

$$\tilde{V}(Q^2) = \frac{4\pi Z_1 Z_2\, \alpha(Q^2)}{Q^2} \quad (I.7)$$

The $1/Q^2$ dependence is characteristic of the Coulomb potential (familiar in Rutherford scattering), but now the strength is governed by the so-called running coupling constant $\alpha(Q^2)$, the reciprocal of which is

$$[\alpha(Q^2)]^{-1} \approx \frac{1}{\alpha(0)} - \frac{1}{3\pi} \ln\left(\frac{Q^2}{m^2 e^{5/3}} \right) \quad (I.8)$$

Here $\alpha(0) = 1/137.036\ldots$ is the fine structure constant, e is the base of natural logarithms, and Q^2 is the square of the wavenumber (momentum) transfer. The expression (I.8) is an approximation for large Q^2/m^2. The running coupling $\alpha(Q^2)$ increases slowly with increasing Q^2 (shorter distances); the particles are penetrating inside the cloud of screening electron-positron pairs and experiencing a larger effective product of charges.

Since the lowest order vacuum polarization energy is proportional to α times the external charges, we describe it as a linear effect, even though it involves (in α) the square of the internal charge of the electron and positron. Small higher order effects, such as in Fig. I.3 with three of the photons corresponding to the third power of the external field or charge, are truly nonlinear interactions.

The final conclusion about linear superposition of fields *in vacuum* is that in the classical domain of sizes and attainable field strengths there is abundant evidence for the validity of linear superposition and no evidence against it. In the atomic and subatomic domain there are small quantum-mechanical nonlinear effects whose origins are in the coupling between charged particles and the electromagnetic field. They modify the interactions between charged particles and cause interactions between electromagnetic fields even if physical particles are absent.

I.4   *Maxwell Equations in Macroscopic Media*

So far we have considered electromagnetic fields and sources in vacuum. The Maxwell equations (I.1b) for the electric and magnetic fields **E** and **B** can be thought of as equations giving the fields everywhere in space, provided all the sources ρ and **J** are specified. For a small number of definite sources, determination of the fields is a tractable problem; but for macroscopic aggregates of matter, the solution of the equations is almost impossible. There are two aspects here. One is that the number of individual sources, the charged particles in every atom and nucleus, is prohibitively large. The other aspect is that for macroscopic observations the detailed behavior of the fields, with their drastic variations in space over atomic distances, is not relevant. What *is* relevant is the average of a field or a source over a volume large compared to the volume occupied by a single atom or molecule. We call such averaged quantities the *macroscopic* fields and macroscopic sources. It is shown in detail in Section 6.6 that the *macroscopic Maxwell equations* are of the form (I.1a) with **E** and **B** the averaged **E** and **B** of the microscopic or vacuum Maxwell equations, while **D** and **H** are no longer simply multiples of **E** and **B**, respectively. The *macroscopic* field quantities **D** and **H**, called the electric displacement and magnetic field (with **B** called the magnetic induction), have components given by

$$
\begin{aligned}
D_\alpha &= \epsilon_0\, E_\alpha + \left(P_\alpha - \sum_\beta \frac{\partial Q'_{\alpha\beta}}{\partial x_\beta} + \cdots \right) \\
H_\alpha &= \frac{1}{\mu_0}\, B_\alpha - (M_\alpha + \cdots)
\end{aligned}
\tag{I.9}
$$

The quantities **P**, **M**, $Q'_{\alpha\beta}$, and similar higher order objects represent the macroscopically averaged electric dipole, magnetic dipole, and electric quadrupole, and higher moment densities of the material medium in the presence of applied fields. Similarly, the charge and current densities ρ and **J** are macroscopic averages of the "free" charge and current densities in the medium. The bound charges and currents appear in the equations via **P**, **M**, and $Q'_{\alpha\beta}$.

The macroscopic Maxwell equations (I.1a) are a set of eight equations involving the components of the four fields **E**, **B**, **D**, and **H**. The four homogeneous equations can be solved formally by expressing **E** and **B** in terms of the scalar potential Φ and the vector potential **A**, but the inhomogeneous equations cannot be solved until the derived fields **D** and **H** are known in terms of **E** and **B**. These connections, which are implicit in (I.9), are known as *constitutive relations*,

$$\mathbf{D} = \mathbf{D}[\mathbf{E}, \mathbf{B}]$$
$$\mathbf{H} = \mathbf{H}[\mathbf{E}, \mathbf{B}]$$

In addition, for conducting media there is the generalized Ohm's law,

$$\mathbf{J} = \mathbf{J}[\mathbf{E}, \mathbf{B}]$$

The square brackets signify that the connections are not necessarily simple and may depend on past history (hysteresis), may be nonlinear, etc.

In most materials the electric quadrupole and higher terms in (I.9) are completely negligible. Only the *electric and magnetic polarizations* **P** and **M** are significant. This does not mean, however, that the constitutive relations are then simple. There is tremendous diversity in the electric and magnetic properties of matter, especially in crystalline solids, with ferroelectric and ferromagnetic materials having nonzero **P** or **M** in the absence of applied fields, as well as more ordinary dielectric, diamagnetic, and paramagnetic substances. The study of these properties is one of the provinces of solid-state physics. In this book we touch only very briefly and superficially on some more elementary aspects. Solid-state books such as *Kittel* should be consulted for a more systematic and extensive treatment of the electromagnetic properties of bulk matter.

In substances other than ferroelectrics or ferromagnets, for weak enough fields the presence of an applied electric or magnetic field induces an electric or magnetic polarization proportional to the magnitude of the applied field. We then say that the response of the medium is linear and write the Cartesian components of **D** and **H** in the form,*

$$\left. \begin{aligned} D_\alpha &= \sum_\beta \epsilon_{\alpha\beta} E_\beta \\ H_a &= \sum_\beta \mu'_{\alpha\beta} B_\beta \end{aligned} \right\} \tag{I.10}$$

The tensors $\epsilon_{\alpha\beta}$ and $\mu'_{\alpha\beta}$ are called the electric permittivity or dielectric tensor and the inverse magnetic permeability tensor. They summarize the linear response of the medium and are dependent on the molecular and perhaps crystalline structure of the material, as well as bulk properties like density and temperature. For simple materials the linear response is often isotropic in space. Then $\epsilon_{\alpha\beta}$ and $\mu'_{\alpha\beta}$ are diagonal with all three elements equal, and $\mathbf{D} = \epsilon\mathbf{E}$, $\mathbf{H} = \mu'\mathbf{B} = \mathbf{B}/\mu$.

To be generally correct Eqs. (I.10) should be understood as holding for the Fourier transforms in space and time of the field quantities. This is because the basic linear connection between **D** and **E** (or **H** and **B**) can be nonlocal. Thus

$$D_\alpha(\mathbf{x}, t) = \sum_\beta \int d^3x' \int dt'\, \epsilon_{\alpha\beta}(\mathbf{x}', t') E_\beta(\mathbf{x} - \mathbf{x}', t - t')$$

---

*Precedent would require writing $B_\alpha = \Sigma_\beta \mu_{\alpha\beta} H_\beta$, but this reverses the natural roles of **B** as the basic magnetic field and **H** as the derived quantity. In Chapter 5 we revert to the traditional usage.

where $\epsilon_{\alpha\beta}(\mathbf{x}', t')$ may be localized around $\mathbf{x}' = 0$, $t' = 0$, but is nonvanishing for some range away from the origin. If we introduce the Fourier transforms $D_\alpha(\mathbf{k}, \omega)$, $E_\beta(\mathbf{k}, \omega)$, and $\epsilon_{\alpha\beta}(\mathbf{k}, \omega)$ through

$$f(\mathbf{k}, \omega) = \int d^3x \int dt \ f(\mathbf{x}, t)e^{-i\mathbf{k}\cdot\mathbf{x}+i\omega t}$$

Eq. (I.10) can be written in terms of the Fourier transforms as

$$D_\alpha(\mathbf{k}, \omega) = \sum_\beta \ \epsilon_{\alpha\beta}(\mathbf{k}, \omega)E_\beta(\mathbf{k}, \omega) \tag{I.11}$$

A similar equation can be written $H_\alpha(\mathbf{k}, \omega)$ in terms of $B_\beta(\mathbf{k}, \omega)$. The permeability tensors are therefore functions of frequency and wave vector in general. For visible light or electromagnetic radiation of longer wavelength it is often permissible to neglect the nonlocality in space. Then $\epsilon_{\alpha\beta}$ and $\mu'_{\alpha\beta}$ are functions only of frequency. This is the situation discussed in Chapter 7, which gives a simplified treatment of the high frequency properties of matter and explores the consequences of causality. For conductors and superconductors long-range effects can be important. For example, when the electronic collisional mean free path in a conductor becomes large compared to the skin depth, a spatially local form of Ohm's law is no longer adequate. Then the dependence on wave vector also enters. In the understanding of a number of properties of solids the concept of a dielectric constant as a function of wave vector and frequency is fruitful. Some exemplary references are given in the suggested reading at the end of this introduction.

For orientation we mention that at low frequencies ($\nu \lesssim 10^6$ Hz) where all charges, regardless of their inertia, respond to applied fields, solids have dielectric constants typically in the range of $\epsilon_{\alpha\alpha}/\epsilon_0 \sim 2\text{–}20$ with larger values not uncommon. Systems with permanent molecular dipole moments can have much larger and temperature-sensitive dielectric constants. Distilled water, for example, has a static dielectric constant of $\epsilon/\epsilon_0 = 88$ at 0°C and $\epsilon/\epsilon_0 = 56$ at 100°C. At optical frequencies only the electrons can respond significantly. The dielectric constants are in the range, $\epsilon_{\alpha\alpha}/\epsilon_0 \sim 1.7\text{–}10$, with $\epsilon_{\alpha\alpha}/\epsilon_0 \approx 2\text{–}3$ for most solids. Water has $\epsilon/\epsilon_0 = 1.77\text{–}1.80$ over the visible range, essentially independent of temperature from 0 to 100°C.

The type of response of materials to an applied magnetic field depends on the properties of the individual atoms or molecules and also on their interactions. *Diamagnetic* substances consist of atoms or molecules with no net angular momentum. The response to an applied magnetic field is the creation of circulating atomic currents that produce a very small bulk magnetization opposing the applied field. With the definition of $\mu'_{\alpha\beta}$ in (I.10) and the form of (I.9), this means $\mu_0\mu'_{\alpha\alpha} > 1$. Bismuth, the most diamagnetic substance known, has $(\mu_0\mu'_{\alpha\alpha} - 1) \approx 1.8 \times 10^{-4}$. Thus diamagnetism is a very small effect. If the basic atomic unit of the material has a net angular momentum from unpaired electrons, the substance is *paramagnetic*. The magnetic moment of the odd electron is aligned parallel to the applied field. Hence $\mu_0\mu'_{\alpha\alpha} < 1$. Typical values are in the range $(1 - \mu_0\mu'_{\alpha\alpha}) \approx 10^{-2}\text{–}10^{-5}$ at room temperature, but decreasing at higher temperatures because of the randomizing effect of thermal excitations.

Ferromagnetic materials are paramagnetic but, because of interactions between atoms, show drastically different behavior. Below the Curie temperature (1040 K for Fe, 630 K for Ni), ferromagnetic substances show spontaneous magnetization; that is, all the magnetic moments in a microscopically large region called a domain are aligned. The application of an external field tends to cause the domains to change and the moments in different domains to line up together, leading to the saturation of the bulk magnetization. Removal of the field leaves a considerable fraction of the moments still aligned, giving a permanent magnetization that can be as large as $B_r = \mu_0 M_r \gtrsim 1$ tesla.

For data on the dielectric and magnetic properties of materials, the reader can consult some of the basic physics handbooks* from which he or she will be led to more specific and detailed compilations.

Materials that show a linear response to weak fields eventually show *nonlinear behavior* at high enough field strengths as the electronic or ionic oscillators are driven to large amplitudes. The linear relations (I.10) are modified to, for example,

$$D_\alpha = \sum_\beta \epsilon^{(1)}_{\alpha\beta} E_\beta + \sum_{\beta,\gamma} \epsilon^{(2)}_{\alpha\beta\gamma} E_\beta E_\gamma + \cdots \tag{I.12}$$

For static fields the consequences are not particularly dramatic, but for time-varying fields it is another matter. A large amplitude wave of two frequencies ω_1 and ω_2 generates waves in the medium with frequencies 0, $2\omega_1$, $2\omega_2$, $\omega_1 + \omega_2$, $\omega_1 - \omega_2$, as well as the original ω_1 and ω_2. From cubic and higher nonlinear terms an even richer spectrum of frequencies can be generated. With the development of lasers, nonlinear behavior of this sort has become a research area of its own, called *nonlinear optics*, and also a laboratory tool. At present, lasers are capable of generating light pulses with peak electric fields approaching 10^{12} or even 10^{13} V/m. The static electric field experienced by the electron in its orbit in a hydrogen atom is $e_G/a_0^2 \simeq 5 \times 10^{11}$ V/m. Such laser fields are thus seen to be capable of driving atomic oscillators well into their nonlinear regime, capable indeed of destroying the sample under study! References to some of the literature of this specialized field are given in the suggested reading at the end of this introduction. The reader of this book will have to be content with basically linear phenomena.

I.5   *Boundary Conditions at Interfaces Between Different Media*

The Maxwell equations (I.1) are differential equations applying locally at each point in space-time $(\mathbf{x}, t)$. By means of the divergence theorem and Stokes's theorem, they can be cast in integral form. Let V be a finite volume in space, S the closed surface (or surfaces) bounding it, da an element of area on the surface, and $\mathbf{n}$ a unit normal to the surface at da pointing outward from the enclosed volume. Then the divergence theorem applied to the first and last equations of (I.1a) yields the integral statements

$$\oint_S \mathbf{D} \cdot \mathbf{n} \, da = \int_V \rho \, d^3x \tag{I.13}$$

$$\oint_S \mathbf{B} \cdot \mathbf{n} \, da = 0 \tag{I.14}$$

CRC Handbook of Chemistry and Physics, ed. D. R. Lide, 78th ed., CRC Press, Boca Raton, FL (1997–98).

 American Institute of Physics Handbook, ed. D. E. Gray, McGraw Hilll, New York, 3rd edition (1972), Sections 5.d and 5.f.

The first relation is just Gauss's law that the total flux of **D** out through the surface is equal to the charge contained inside. The second is the magnetic analog, with no net flux of **B** through a closed surface because of the nonexistence of magnetic charges.

Similarly, let C be a closed contour in space, S' an open surface spanning the contour, $d\mathbf{l}$ a line element on the contour, da an element of area on S', and $\mathbf{n}'$ a unit normal at da pointing in the direction given by the right-hand rule from the sense of integration around the contour. Then applying Stokes's theorem to the middle two equations in (I.1a) gives the integral statements

$$\oint_C \mathbf{H} \cdot d\mathbf{l} = \int_{S'} \left[\mathbf{J} + \frac{\partial \mathbf{D}}{\partial t} \right] \cdot \mathbf{n}' \, da \tag{I.15}$$

$$\oint_C \mathbf{E} \cdot d\mathbf{l} = -\int_{S'} \frac{\partial \mathbf{B}}{\partial t} \cdot \mathbf{n}' \, da \tag{I.16}$$

Equation (I.15) is the Ampère–Maxwell law of magnetic fields and (I.16) is Faraday's law of electromagnetic induction.

These familiar integral equivalents of the Maxwell equations can be used directly to deduce the relationship of various normal and tangential components of the fields on either side of a surface between different media, perhaps with a surface charge or current density at the interface. An appropriate geometrical arrangement is shown in Fig. I.4. An infinitesimal Gaussian pillbox straddles the boundary surface between two media with different electromagnetic properties. Similarly, the infinitesimal contour C has its long arms on either side of the boundary and is oriented so that the normal to its spanning surface is tangent to the interface. We first apply the integral statements (I.13) and (I.14) to the volume of the pillbox. In the limit of a very shallow pillbox, the side surface does

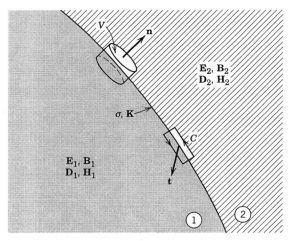

Figure I.4   Schematic diagram of boundary surface (heavy line) between different media. The boundary region is assumed to carry idealized surface charge and current densities σ and **K**. The volume V is a small pillbox, half in one medium and half in the other, with the normal **n** to its top pointing from medium 1 into medium 2. The rectangular contour C is partly in one medium and partly in the other and is oriented with its plane perpendicular to the surface so that its normal **t** is tangent to the surface.

not contribute to the integrals on the left in (I.13) and (I.14). Only the top and bottom contribute. If the top and the bottom are parallel, tangent to the surface, and of area Δa, then the left-hand integral in (I.13) is

$$\oint_S \mathbf{D} \cdot \mathbf{n} \, da = (\mathbf{D}_2 - \mathbf{D}_1) \cdot \mathbf{n} \, \Delta a$$

and similarly for (I.14). If the charge density ρ is singular at the interface so as to produce an idealized surface charge density σ, then the integral on the right in (I.13) is

$$\int_V \rho \, d^3x = \sigma \, \Delta a$$

Thus the *normal components* of $\mathbf{D}$ and $\mathbf{B}$ on either side of the boundary surface are related according to

$$(\mathbf{D}_2 - \mathbf{D}_1) \cdot \mathbf{n} = \sigma \qquad (\text{I.17})$$
$$(\mathbf{B}_2 - \mathbf{B}_1) \cdot \mathbf{n} = 0 \qquad (\text{I.18})$$

In words, we say that the normal component of $\mathbf{B}$ is continuous and the discontinuity of the normal component of $\mathbf{D}$ at any point is equal to the surface charge density at that point.

In an analogous manner the infinitesimal Stokesian loop can be used to determine the discontinuities of the tangential components of $\mathbf{E}$ and $\mathbf{H}$. If the short arms of the contour C in Fig. I.4 are of negligible length and each long arm is parallel to the surface and has length Δl, then the left-hand integral of (I.16) is

$$\oint_C \mathbf{E} \cdot d\mathbf{l} = (\mathbf{t} \times \mathbf{n}) \cdot (\mathbf{E}_2 - \mathbf{E}_1) \, \Delta l$$

and similarly for the left-hand side of (I.15). The right-hand side of (I.16) vanishes because $\partial \mathbf{B}/\partial t$ is finite at the surface and the area of the loop is zero as the length of the short sides goes to zero. The right-hand side of (I.15) does not vanish, however, if there is an idealized surface current density $\mathbf{K}$ flowing exactly on the boundary surface. In such circumstances the integral on the right of (I.15) is

$$\int_{S'} \left[\mathbf{J} + \frac{\partial \mathbf{D}}{\partial t} \right] \cdot \mathbf{t} \, da = \mathbf{K} \cdot \mathbf{t} \, \Delta l$$

The second term in the integral vanishes by the same argument that was just given. The *tangential components* of $\mathbf{E}$ and $\mathbf{H}$ on either side of the boundary are therefore related by

$$\mathbf{n} \times (\mathbf{E}_2 - \mathbf{E}_1) = 0 \qquad (\text{I.19})$$
$$\mathbf{n} \times (\mathbf{H}_2 - \mathbf{H}_1) = \mathbf{K} \qquad (\text{I.20})$$

In (I.20) it is understood that the surface current $\mathbf{K}$ has only components parallel to the surface at every point. The tangential component of $\mathbf{E}$ across an interface is continuous, while the tangential component of $\mathbf{H}$ is discontinuous by an amount whose magnitude is equal to the magnitude of the surface current density and whose direction is parallel to $\mathbf{K} \times \mathbf{n}$.

The discontinuity equations (I.17)–(I.20) are useful in solving the Maxwell equations in different regions and then connecting the solutions to obtain the fields throughout all space.

I.6 Some Remarks on Idealizations in Electromagnetism

In the preceding section we made use of the idea of surface distributions of charge and current. These are obviously mathematical idealizations that do not exist in the physical world. There are other abstractions that occur throughout electromagnetism. In electrostatics, for example, we speak of holding objects at a fixed potential with respect to some zero of potential usually called "ground." The relations of such idealizations to the real world is perhaps worthy of a little discussion, even though to the experienced hand most will seem obvious.

First we consider the question of maintaining some conducting object at a fixed electrostatic potential with respect to some reference value. Implicit is the idea that the means does not significantly disturb the desired configuration of charges and fields. To maintain an object at fixed potential it is necessary, at least from time to time, to have a conducting path or its equivalent from the object to a source of charge far away ("at infinity") so that as other charged or uncharged objects are brought in the vicinity, charge can flow to or from the object, always maintaining its potential at the desired value. Although more sophisticated means are possible, metallic wires are commonly used to make the conducting path. Intuitively we expect small wires to be less perturbing than large ones. The reason is as follows:

> Since the quantity of electricity on any given portion of a wire at a given potential diminishes indefinitely when the diameter of the wire is indefinitely diminished, the distribution of electricity on bodies of considerable dimensions will not be sensibly affected by the introduction of very fine metallic wires into the field, such as are used to form electrical connexions between these bodies and the earth, an electrical machine, or an electrometer.*

The electric field in the immediate neighborhood of the thin wire is very large, of course. However, at distances away of the order of the size of the "bodies of considerable dimensions" the effects can be made small. An important historical illustration of Maxwell's words is given by the work of Henry Cavendish 200 years ago. By experiments done in a converted stable of his father's house, using Leyden jars as his sources of charge, thin wires as conductors, and suspending the objects in the room, Cavendish measured the amounts of charge on cylinders, discs, etc., held at fixed potential and compared them to the charge on a sphere (the same sphere shown in Fig. I.1) at the same potential. His values of capacitance, so measured, are accurate to a few per cent. For example, he found the ratio of the capacitance of a sphere to that of a thin circular disc of the same radius was 1.57. The theoretical value is $\pi/2$.

There is a practical limit to the use of finer and finer wires. The charge per unit length decreases only logarithmically [as the reciprocal of $\ln(d/a)$, where a

---

*J. C. Maxwell, *A Treatise on Electricity and Magnetism*, Dover, New York, 1954 reprint of the 3rd edition (1891), Vol. 1, p. 96.

is the mean radius of the wire and d is a typical distance of the wire from some conducting surface]. To minimize the perturbation of the system below some level, it is necessary to resort to other means to maintain potentials, comparison methods using beams of charged particles intermittently, for example.

When a conducting object is said to be *grounded*, it is assumed to be connected by a very fine conducting filament to a remote reservoir of charge that serves as the common zero of potential. Objects held at fixed potentials are similarly connected to one side of a voltage source, such as a battery, the other side of which is connected to the common "ground." Then, when initially electrified objects are moved relative to one another in such a way that their distributions of electricity are altered, but their potentials remain fixed, the appropriate amounts of charge flow from or to the remote reservoir, assumed to have an inexhaustible supply. The idea of grounding something is a well-defined concept in electrostatics, where time is not a factor, but for oscillating fields the finite speed of propagation blurs the concept. In other words, stray inductive and capacitive effects can enter significantly. Great care is then necessary to ensure a "good ground."

Another idealization in macroscopic electromagnetism is the idea of a surface charge density or a surface current density. The physical reality is that the charge or current is confined to the immediate neighborhood of the surface. If this region has thickness small compared to the length scale of interest, we may approximate the reality by the idealization of a region of infinitesimal thickness and speak of a surface distribution. Two different limits need to be distinguished. One is the limit in which the "surface" distribution is confined to a region near the surface that is *macroscopically small, but microscopically large*. An example is the penetration of time-varying fields into a very good, but not perfect, conductor, described in Section 8.1. It is found that the fields are confined to a thickness δ, called the skin depth, and that for high enough frequencies and good enough conductivities δ can be macroscopically very small. It is then appropriate to integrate the current density $\mathbf{J}$ over the direction perpendicular to the surface to obtain an effective surface current density $\mathbf{K}_{eff}$.

The other limit is *truly microscopic* and is set by quantum-mechanical effects in the atomic structure of materials. Consider, for instance, the distribution of excess charge of a conducting body in electrostatics. It is well known that this charge lies entirely on the surface of a conductor. We then speak of a *surface charge density* σ. There is no electric field inside the conductor, but there is, in accord with (I.17), a normal component of electric field just outside the surface. At the microscopic level the charge is not exactly at the surface and the field does not change discontinuously. The most elementary considerations would indicate that the transition region is a few atomic diameters in extent. The ions in a metal can be thought of as relatively immobile and localized to 1 angstrom or better; the lighter electrons are less constrained. The results of model calculations* are shown in Fig. I.5. They come from a solution of the quantum-mechanical many-electron problem in which the ions of the conductor are approximated by a continuous constant charge density for $x < 0$. The electron density ($r_s = 5$) is roughly appropriate to copper and the heavier alkali metals.

---

*N. D. Lang and W. Kohn, *Phys. Rev.* **B1**, 4555 (1970); **B3**, 1215 (1971); V. E. Kenner, R. E. Allen, and W. M. Saslow, *Phys. Lett.* **38A**, 255 (1972).

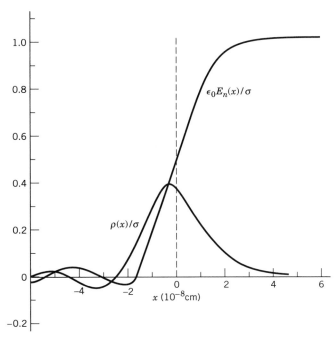

Figure I.5   Distribution of excess charge at the surface of a conductor and of the normal component of the electric field. The ions of the solid are confined to $x < 0$ and are approximated by a constant continuous charge distribution through which the electrons move. The bulk of the excess charge is confined to within ± 2 Å of the "surface."

The excess electronic charge is seen to be confined to a region within ± 2 Å of the "surface" of the ionic distribution. The electric field rises smoothly over this region to its value of σ "outside" the conductor. For macroscopic situations where 10^{-9} m is a negligible distance, we can idealize the charge density and electric field behavior as $\rho(x) = \sigma\delta(x)$ and $E_n(x) = \sigma\theta(x)/\epsilon_0$, corresponding to a truly surface density and a step-function jump of the field.

  We see that the theoretical treatment of classical electromagnetism involves several idealizations, some of them technical and some physical. The subject of electrostatics, discussed in the first chapters of the book, developed as an experimental science of *macroscopic* electrical phenomena, as did virtually all other aspects of electromagnetism. The extension of these macroscopic laws, even for charges and currents in vacuum, to the *microscopic* domain was for the most part an unjustified extrapolation. Earlier in this introduction we discussed some of the limits to this extrapolation. The point to be made here is the following. With hindsight we know that many aspects of the laws of classical electromagnetism apply well into the atomic domain provided the sources are treated quantum mechanically, that the averaging of electromagnetic quantities over volumes containing large numbers of molecules so smooths the rapid fluctuations that static applied fields induce static average responses in matter, and that excess charge is *on* the surface of a conductor in a macroscopic sense. Thus Coulomb's and Ampère's macroscopic observations and our mathematical abstractions from them have a wider applicability than might be supposed by a supercautious physicist. The absence for air of significant electric or magnetic susceptibility certainly simplifies matters!

References and Suggested Reading

The history of electricity and magnetism is in large measure the history of science itself. We have already cited Whittaker's two volumes, the first covering the period up to 1900, as well as the shorter account emphasizing optics in
>   Born and Wolf.

Another readable account, with perceptive discussion of the original experiments, is
>   N. Feather, *Electricity and Matter*, University Press, Edinburgh (1968).

The experimental tests of the inverse square nature of Coulomb's law or, in modern language, the mass of the photon, are reviewed by
>   I. Yu. Kobzarev and L. B. Okun', *Usp. Fiz. Nauk* **95**, 131 (1968) [transl., *Sov. Phys. Usp.* **11**, 338 (1968).]

and
>   A. S. Goldhaber and M. M. Nieto, *Rev. Mod. Phys.* **43**, 277 (1971).

An accessible treatment of the gauge principle in the construction of field theories, building on classical electrodynamics and ordinary quantum mechanics, can be found in
>   I. J. R. Aitchison and A. J. G. Hey, *Gauge Theories in Particle Physics*, 2nd ed., Adam Hilger, Bristol (1989).

Suggested reading on the topic of the macroscopic Maxwell equations and their derivation from the microscopic equations can be found at the end of Chapter 6. The basic physics of dielectrics, ferroelectrics, and magnetic materials can be found in numerous books on solid-state physics, for example,
>   Ashcroft and Mermin
>   Beam
>   Kittel
>   Wert and Thomson
>   Wooten

The second of these is aimed at electrical engineers and stresses practical topics like semiconductors. The last one is mainly on optical properties. The need for spatial non-locality in treating the surface impedance of metals (the anomalous skin effect) is discussed in several places by
>   A. B. Pippard, *Advances in Electronics and Electron Physics*, Vol. VI, ed. L. Marton, Academic Press, New York (1954), pp. 1–45; *Reports on Progress in Physics*, Vol. XXIII, pp. 176–266 (1960); *The Dynamics of Conduction Electrons*, Gordon and Breach, New York (1965).

The concept of a wave-vector and frequency-dependent dielectric constant $\epsilon(\mathbf{k}, \omega)$ is developed by
>   Kittel, *Advanced Topic D*.
>   D. Pines, *Elementary Excitations in Solids*, W. A. Benjamin, New York (1963), Chapters 3 and 4.
>   F. Stern, *Solid State Physics*, Vol. 15, eds. F. Seitz and D. Turnbull, Academic Press, New York, pp. 299–408.

The field of nonlinear optics is now nearly 40 years old. Beginnings and introductions can be found in
>   J. A. Giordmaine, *Phys. Today* **22**(1), 38 (1969).
>   N. Bloembergen, *Am. J. Phys.* **35**, 989 (1967).

Nonlinear optical phenomena and applications are discussed in

R. L. Sutherland, *Handbook of Nonlinear Optics*, Marcel Dekker, New York (1966).

Some texts and monographs on the subject are

R. W. Boyd, *Nonlinear Optics*, Academic Press, New York (1990).

M. Schubert and B. Wilhelmi, *Nonlinear Optics and Quantum Electronics*, Wiley, New York (1986).

Y. R. Shen, *The Principles of Nonlinear Optics*, Wiley, New York (1984).

CHAPTER 1

Introduction to Electrostatics

We begin our discussion of electrodynamics with the subject of *electrostatics*— phenomena involving time-independent distributions of charge and fields. For most readers this material is in the nature of a review. In this chapter especially we do not elaborate significantly. We introduce concepts and definitions that are important for later discussion and present some essential mathematical apparatus. In subsequent chapters the mathematical techniques are developed and applied.

One point of physics should be mentioned. Historically, electrostatics developed as a science of *macroscopic* phenomena. As indicated at the end of the Introduction, such idealizations as point charges or electric fields at a point must be viewed as mathematical constructs that permit a description of the phenomena at the macroscopic level, but that may fail to have meaning microscopically.

1.1   *Coulomb's Law*

All of electrostatics stems from the quantitative statement of Coulomb's law concerning the force acting between charged bodies at rest with respect to each other. Coulomb, in an impressive series of experiments, showed experimentally that the force between two small charged bodies separated in air a distance large compared to their dimensions

varies directly as the magnitude of each charge,
varies inversely as the square of the distance between them,
is directed along the line joining the charges, and
is attractive if the bodies are oppositely charged and repulsive if the bodies have the same type of charge.

Furthermore it was shown experimentally that the total force produced on one small charged body by a number of the other small charged bodies placed around it is the *vector* sum of the individual two-body forces of Coulomb. Strictly speaking, Coulomb's conclusions apply to charges in vacuum or in media of negligible susceptibility. We defer consideration of charges in dielectrics to Chapter 4.

1.2   *Electric Field*

Although the thing that eventually gets measured is a force, it is useful to introduce a concept one step removed from the forces, the concept of an electric field due to some array of charged bodies. At the moment, the electric field can be defined as the force per unit charge acting at a given point. It is a vector function of position, denoted by **E**. One must be careful in its definition, however. It is not necessarily the force that one would observe by placing one unit of charge on a pith ball and placing it in position. The reason is that one unit of charge may be so large that its presence alters appreciably the field configuration of the array. Consequently one must use a limiting process whereby the ratio of the force on the small test body to the charge on it is measured for smaller and smaller amounts of charge.* Experimentally, this ratio and the direction of the force will become constant as the amount of test charge is made smaller and smaller. These limiting values of magnitude and direction define the magnitude and direction of the electric field **E** at the point in question. In symbols we may write

$$\mathbf{F} = q\mathbf{E} \tag{1.1}$$

where **F** is the force, **E** the electric field, and q the charge. In this equation it is assumed that the charge q is located at a point, and the force and the electric field are evaluated at that point.

Coulomb's law can be written down similarly. If **F** is the force on a point charge q_1, located at $\mathbf{x}_1$, due to another point charge q_2, located at $\mathbf{x}_2$, then Coulomb's law is

$$\mathbf{F} = kq_1q_2 \frac{\mathbf{x}_1 - \mathbf{x}_2}{|\mathbf{x}_1 - \mathbf{x}_2|^3} \tag{1.2}$$

Note that q_1 and q_2 are algebraic quantities, which can be positive or negative. The constant of proportionality k depends on the system of units used.

The electric field at the point **x** due to a point charge q_1 at the point $\mathbf{x}_1$ can be obtained directly:

$$\mathbf{E}(\mathbf{x}) = kq_1 \frac{\mathbf{x} - \mathbf{x}_1}{|\mathbf{x} - \mathbf{x}_1|^3} \tag{1.3}$$

as indicated in Fig. 1.1. The constant k differs in different systems of units.[†] In electrostatic units (esu), $k = 1$ and unit charge is chosen as that charge that exerts a force of one dyne on an equal point charge located one centimeter away. The esu unit of charge is called the *statcoulomb*, and the electric field is measured in *statvolts per centimeter*. In the SI system, which we employ here, $k = (4\pi\epsilon_0)^{-1} = 10^{-7}c^2$, where $\epsilon_0 \approx 8.854 \times 10^{-12}$ farad per meter (F/m) is called the permittivity of free space. The SI unit of charge is the *coulomb* (C), and the electric field is measured in *volts per meter* (V/m). One coulomb (1 C) produces an electric field

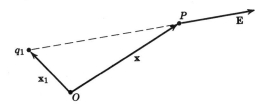

Figure 1.1

*The discreteness of electric charge (see Section I.1) means that this mathematical limit is impossible to realize physically. This is an example of a mathematical idealization in macroscopic electrostatics.

[†]The question of units is discussed in detail in the Appendix.

of approximately 8.9874×10^9 V/m (8.9874 GV/m) at a distance of 1 meter. One electron ($q \approx 1.602 \times 10^{-19}$ C) produces a field of approximately 1.44×10^{-9} V/m (1.44 nV/m) at 1 meter.

The experimentally observed linear superposition of forces due to many charges means that we may write the electric field at $\mathbf{x}$ due to a system of point charges q_i, located at $\mathbf{x}_i$, $i = 1, 2, \ldots, n$, as the vector sum:

$$\mathbf{E}(\mathbf{x}) = \frac{1}{4\pi\epsilon_0} \sum_{i=1}^{n} q_i \frac{\mathbf{x} - \mathbf{x}_i}{|\mathbf{x} - \mathbf{x}_i|^3} \tag{1.4}$$

If the charges are so small and so numerous that they can be described by a charge density $\rho(\mathbf{x}')$ [if Δq is the charge in a small volume $\Delta x\, \Delta y\, \Delta z$ at the point $\mathbf{x}'$, then $\Delta q = \rho(\mathbf{x}')\, \Delta x\, \Delta y\, \Delta z$], the sum is replaced by an integral:

$$\mathbf{E}(\mathbf{x}) = \frac{1}{4\pi\epsilon_0} \int \rho(\mathbf{x}') \frac{\mathbf{x} - \mathbf{x}'}{|\mathbf{x} - \mathbf{x}'|^3} d^3x' \tag{1.5}$$

where $d^3x' = dx'\, dy'\, dz'$ is a three-dimensional volume element at $\mathbf{x}'$.

At this point it is worthwhile to introduce the *Dirac delta function*. In one dimension, the delta function, written $\delta(x - a)$, is a mathematically improper function having the properties:

1. $\delta(x - a) = 0$ for $x \neq a$, and

2. $\int \delta(x - a)\, dx = 1$ if the region of integration includes $x = a$, and is zero otherwise.

The delta function can be given an intuitive, but nonrigorous, meaning as the limit of a peaked curve such as a Gaussian that becomes narrower and narrower, but higher and higher, in such a way that the area under the curve is always constant. L. Schwartz's theory of distributions is a comprehensive rigorous mathematical approach to delta functions and their manipulations.*

From the definitions above it is evident that, for an arbitrary function $f(x)$,

3. $\int f(x)\, \delta(x - a)\, dx = f(a)$.

The integral of $f(x)$ times the derivative of a delta function is simply understood if the delta function is thought of as a well-behaved, but sharply peaked, function. Thus the definition is

4. $\int f(x)\, \delta'(x - a)\, dx = -f'(a)$

where a prime denotes differentiation with respect to the argument.

If the delta function has as argument a function $f(x)$ of the independent variable x, it can be transformed according to the rule,

5. $\delta(f(x)) = \sum_i \dfrac{1}{\left| \dfrac{df}{dx}(x_i) \right|} \delta(x - x_i)$

where $f(x)$ is assumed to have only simple zeros, located at $x = x_i$.

In more than one dimension, we merely take products of delta functions in each dimension. In three dimensions, for example, with Cartesian coordinates,

6. $\delta(\mathbf{x} - \mathbf{X}) = \delta(x_1 - X_1)\, \delta(x_2 - X_2)\, \delta(x_3 - X_3)$

---

*A useful, rigorous account of the Dirac delta function is given by *Lighthill*. See also *Dennery and Krzywicki* (Section III.13). (Full references for items cited in the text or footnotes by italicized author only will be found in the Bibliography.)

is a function that vanishes everywhere except at $\mathbf{x} = \mathbf{X}$, and is such that

7. $\displaystyle\int_{\Delta V} \delta(\mathbf{x} - \mathbf{X})\, d^3x = \begin{cases} 1 & \text{if } \Delta V \text{ contains } \mathbf{x} = \mathbf{X} \\ 0 & \text{if } \Delta V \text{ does not contain } \mathbf{x} = \mathbf{X} \end{cases}$

Note that a delta function has the dimensions of an inverse volume in whatever number of dimensions the space has.

A discrete set of point charges can be described with a charge density by means of delta functions. For example,

$$\rho(\mathbf{x}) = \sum_{i=1}^{n} q_i\, \delta(\mathbf{x} - \mathbf{x}_i) \tag{1.6}$$

represents a distribution of n point charges q_i, located at the points $\mathbf{x}_i$. Substitution of this charge density (1.6) into (1.5) and integration, using the properties of the delta function, yields the discrete sum (1.4).

1.3   *Gauss's Law*

The integral (1.5) is not always the most suitable form for the evaluation of electric fields. There is another integral result, called *Gauss's law*, which is sometimes more useful and furthermore leads to a differential equation for $\mathbf{E}(\mathbf{x})$. To obtain Gauss's law we first consider a point charge q and a *closed* surface S, as shown in Fig. 1.2. Let r be the distance from the charge to a point on the surface, $\mathbf{n}$ be the outwardly directed unit normal to the surface at that point, da be an

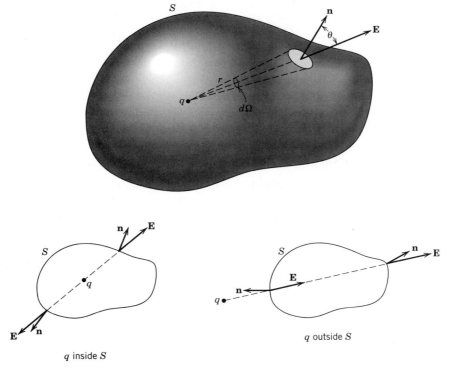

Figure 1.2 Gauss's law. The normal component of electric field is integrated over the closed surface S. If the charge is inside (outside) S, the total solid angle subtended at the charge by the inner side of the surface is 4π (zero).

element of surface area. If the electric field **E** at the point on the surface due to the charge q makes an angle θ with the unit normal, then the normal component of **E** times the area element is:

$$\mathbf{E} \cdot \mathbf{n} \, da = \frac{q}{4\pi\epsilon_0} \frac{\cos\theta}{r^2} \, da \tag{1.7}$$

Since **E** is directed along the line from the surface element to the charge q, $\cos\theta \, da = r^2 \, d\Omega$, where $d\Omega$ is the element of solid angle subtended by da at the position of the charge. Therefore

$$\mathbf{E} \cdot \mathbf{n} \, da = \frac{q}{4\pi\epsilon_0} \, d\Omega \tag{1.8}$$

If we now integrate the normal component of **E** over the whole surface, it is easy to see that

$$\oint_S \mathbf{E} \cdot \mathbf{n} \, da = \begin{cases} q/\epsilon_0 & \text{if } q \text{ lies } inside \text{ } S \\ 0 & \text{if } q \text{ lies } outside \text{ } S \end{cases} \tag{1.9}$$

This result is Gauss's law for a single point charge. For a discrete set of charges, it is immediately apparent that

$$\oint_S \mathbf{E} \cdot \mathbf{n} \, da = \frac{1}{\epsilon_0} \sum_i q_i \tag{1.10}$$

where the sum is over only those charges *inside* the surface S. For a continuous charge density $\rho(\mathbf{x})$, Gauss's law becomes:

$$\oint_S \mathbf{E} \cdot \mathbf{n} \, da = \frac{1}{\epsilon_0} \int_V \rho(\mathbf{x}) \, d^3x \tag{1.11}$$

where V is the volume enclosed by S.

Equation (1.11) is one of the basic equations of electrostatics. Note that it depends upon

the inverse square law for the force between charges,

the central nature of the force, and

the linear superposition of the effects of different charges.

Clearly, then, Gauss's law holds for Newtonian gravitational force fields, with matter density replacing charge density.

It is interesting to note that, even before the experiments of Cavendish and Coulomb, Priestley, taking up an observation of Franklin that charge seemed to reside on the outside, but not the inside, of a metal cup, reasoned by analogy with Newton's law of universal gravitation that the electrostatic force must obey an inverse square law with distance. The present status of the inverse square law is discussed in Section I.2.

1.4   *Differential Form of Gauss's Law*

Gauss's law can be thought of as being an integral formulation of the law of electrostatics. We can obtain a differential form (i.e., a differential equation) by using the divergence theorem. The *divergence theorem* states that for any well-behaved vector field $\mathbf{A}(\mathbf{x})$ defined within a volume V surrounded by the closed surface S the relation

$$\oint_S \mathbf{A} \cdot \mathbf{n} \, da = \int_V \boldsymbol{\nabla} \cdot \mathbf{A} \, d^3x$$

holds between the volume integral of the divergence of $\mathbf{A}$ and the surface integral of the outwardly directed normal component of $\mathbf{A}$. The equation in fact can be used as the definition of the divergence (see *Stratton*, p. 4).

To apply the divergence theorem we consider the integral relation expressed in Gauss's theorem:

$$\oint_S \mathbf{E} \cdot \mathbf{n} \, da = \frac{1}{\epsilon_0} \int_V \rho(\mathbf{x}) \, d^3x$$

Now the divergence theorem allows us to write this as

$$\int_V (\boldsymbol{\nabla} \cdot \mathbf{E} - \rho/\epsilon_0) \, d^3x = 0 \tag{1.12}$$

for an arbitrary volume V. We can, in the usual way, put the integrand equal to zero to obtain

$$\boldsymbol{\nabla} \cdot \mathbf{E} = \rho/\epsilon_0 \tag{1.13}$$

which is the differential form of Gauss's law of electrostatics. This equation can itself be used to solve problems in electrostatics. However, it is often simpler to deal with scalar rather then vector functions of position, and then to derive the vector quantities at the end if necessary (see below).

1.5   *Another Equation of Electrostatics and the Scalar Potential*

The single equation (1.13) is not enough to specify completely the three components of the electric field $\mathbf{E}(\mathbf{x})$. Perhaps some readers know that a vector field can be specified almost* completely if its divergence and curl are given everywhere in space. Thus we look for an equation specifying curl $\mathbf{E}$ as a function of position. Such an equation, namely,

$$\boldsymbol{\nabla} \times \mathbf{E} = 0 \tag{1.14}$$

follows directly from our generalized Coulomb's law (1.5):

$$\mathbf{E}(\mathbf{x}) = \frac{1}{4\pi\epsilon_0} \int \rho(\mathbf{x}') \frac{\mathbf{x} - \mathbf{x}'}{|\mathbf{x} - \mathbf{x}'|^3} \, d^3x'$$

The vector factor in the integrand, viewed as a function of $\mathbf{x}$, is the negative gradient of the scalar $1/|\mathbf{x} - \mathbf{x}'|$:

$$\frac{\mathbf{x} - \mathbf{x}'}{|\mathbf{x} - \mathbf{x}'|^3} = -\boldsymbol{\nabla}\left(\frac{1}{|\mathbf{x} - \mathbf{x}'|}\right)$$

---

*Up to the gradient of a scalar function that satisfies the Laplace equation. See Section 1.9 on uniqueness.

Since the gradient operation involves **x**, but not the integration variable **x′**, it can be taken outside the integral sign. Then the field can be written

$$\mathbf{E}(\mathbf{x}) = \frac{-1}{4\pi\epsilon_0} \boldsymbol{\nabla} \int \frac{\rho(\mathbf{x}')}{|\mathbf{x} - \mathbf{x}'|} \, d^3x' \tag{1.15}$$

Since the curl of the gradient of any well-behaved scalar function of position vanishes ($\boldsymbol{\nabla} \times \boldsymbol{\nabla}\psi = 0$, for all ψ), (1.14) follows immediately from (1.15).

Note that $\boldsymbol{\nabla} \times \mathbf{E} = 0$ depends on the central nature of the force between charges, and on the fact that the force is a function of relative distances only, but does not depend on the inverse square nature.

In (1.15) the electric field (a vector) is derived from a scalar by the gradient operation. Since one function of position is easier to deal with than three, it is worthwhile concentrating on the scalar function and giving it a name. Consequently we define the *scalar potential* $\Phi(\mathbf{x})$ by the equation:

$$\mathbf{E} = -\boldsymbol{\nabla}\Phi \tag{1.16}$$

Then (1.15) shows that the scalar potential is given in terms of the charge density by

$$\Phi(\mathbf{x}) = \frac{1}{4\pi\epsilon_0} \int \frac{\rho(\mathbf{x}')}{|\mathbf{x} - \mathbf{x}'|} \, d^3x' \tag{1.17}$$

where the integration is over all charges in the universe, and Φ is arbitrary only to the extent that a constant can be added to the right-hand side of (1.17).

The scalar potential has a physical interpretation when we consider the work done on a test charge q in transporting it from one point (A) to another point (B) in the presence of an electric field $\mathbf{E}(\mathbf{x})$, as shown in Fig. 1.3. The force acting on the charge at any point is

$$\mathbf{F} = q\mathbf{E}$$

so that the work done in moving the charge from A to B is

$$W = -\int_A^B \mathbf{F} \cdot d\mathbf{l} = -q \int_A^B \mathbf{E} \cdot d\mathbf{l} \tag{1.18}$$

The minus sign appears because we are calculating the work done *on* the charge against the action of the field. With definition (1.16) the work can be written

$$W = q \int_A^B \boldsymbol{\nabla}\Phi \cdot d\mathbf{l} = q \int_A^B d\Phi = q(\Phi_B - \Phi_A) \tag{1.19}$$

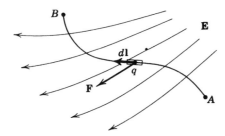

Figure 1.3

which shows that $q\Phi$ can be interpreted as the potential energy of the test charge in the electrostatic field.

From (1.18) and (1.19) it can be seen that the line integral of the electric field between two points is independent of the path and is the negative of the potential difference between the points:

$$\int_A^B \mathbf{E} \cdot d\mathbf{l} = -(\Phi_B - \Phi_A) \tag{1.20}$$

This follows directly, of course, from definition (1.16). If the path is closed, the line integral is zero,

$$\oint \mathbf{E} \cdot d\mathbf{l} = 0 \tag{1.21}$$

a result that can also be obtained directly from Coulomb's law. Then application of *Stokes's theorem* [if $\mathbf{A}(\mathbf{x})$ is a well-behaved vector field, S is an arbitrary open surface, and C is the closed curve bounding S,

$$\oint_C \mathbf{A} \cdot d\mathbf{l} = \int_S (\nabla \times \mathbf{A}) \cdot \mathbf{n} \, da$$

where $d\mathbf{l}$ is a line element of C, $\mathbf{n}$ is the normal to S, and the path C is traversed in a right-hand screw sense relative to $\mathbf{n}$] leads immediately back to $\nabla \times \mathbf{E} = 0$.

1.6 Surface Distributions of Charges and Dipoles and Discontinuities in the Electric Field and Potential

One of the common problems in electrostatics is the determination of electric field or potential due to a given surface distribution of charges. Gauss's law (1.11) allows us to write down a partial result directly. If a surface S, with a unit normal $\mathbf{n}$ directed from side 1 to side 2 of the surface, has a surface-charge density of $\sigma(\mathbf{x})$ (measured in coulombs per square meter) and electric fields $\mathbf{E}_1$ and $\mathbf{E}_2$ on either side of the surface, as shown in Fig. 1.4, then Gauss's law tells us immediately that

$$(\mathbf{E}_2 - \mathbf{E}_1) \cdot \mathbf{n} = \sigma/\epsilon_0 \tag{1.22}$$

This does not determine $\mathbf{E}_1$ and $\mathbf{E}_2$ unless there are no other sources of field and the geometry and form of σ are especially simple. All that (1.22) says is that there

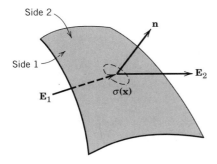

Figure 1.4 Discontinuity in the normal component of electric field across a surface layer of charge.

is a discontinuity of σ/ϵ_0 in the normal component of electric field in crossing a surface with a surface-charge density σ, the crossing being made in the direction of **n**.

The tangential component of electric field can be shown to be continuous across a boundary surface by using (1.21) for the line integral of **E** around a closed path. It is only necessary to take a rectangular path with negligible ends and one side on either side of the boundary.

An expression for the potential (hence the field, by differentiation) at any point in space (not just at the surface) can be obtained from (1.17) by replacing $\rho \, d^3x$ by $\sigma \, da$:

$$\Phi(\mathbf{x}) = \frac{1}{4\pi\epsilon_0} \int_s \frac{\sigma(\mathbf{x}')}{|\mathbf{x} - \mathbf{x}'|} \, da' \tag{1.23}$$

For volume or surface distributions of charge, the potential is everywhere continuous, even within the charge distribution. This can be shown from (1.23) or from the fact that **E** is bounded, even though discontinuous across a surface distribution of charge. With point or line charges, or dipole layers, the potential is no longer continuous, as will be seen immediately.

Another problem of interest is the potential due to a dipole-layer distribution on a surface S. A dipole layer can be imagined as being formed by letting the surface S have a surface-charge density $\sigma(\mathbf{x})$ on it, and another surface S', lying close to S, have an equal and opposite surface-charge density on it at neighboring points, as shown in Fig. 1.5. The dipole-layer distribution of strength $D(\mathbf{x})$ is formed by letting S' approach infinitesimally close to S while the surface-charge density $\sigma(\mathbf{x})$ becomes infinite in such a manner that the product of $\sigma(\mathbf{x})$ and the local separation $d(\mathbf{x})$ of S and S' approaches the limit $D(\mathbf{x})$:

$$\lim_{d(\mathbf{x}) \to 0} \sigma(\mathbf{x}) \, d(\mathbf{x}) = D(\mathbf{x})$$

The direction of the dipole moment of the layer is normal to the surface S and in the direction going from negative to positive charge.

To find the potential due to a dipole layer we can consider a single dipole and then superpose a surface density of them, or we can obtain the same result by performing mathematically the limiting process described in words above on the surface-density expression (1.23). The first way is perhaps simpler, but the second gives useful practice in vector calculus. Consequently we proceed with

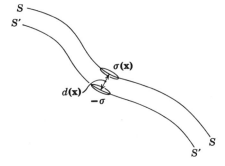

Figure 1.5 Limiting process involved in creating a dipole layer.

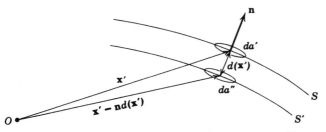

Figure 1.6   Dipole-layer geometry.

the limiting process. With **n**, the unit normal to the surface S, directed away from S', as shown in Fig. 1.6, the potential due to the two close surfaces is

$$\Phi(\mathbf{x}) = \frac{1}{4\pi\epsilon_0} \int_S \frac{\sigma(\mathbf{x}')}{|\mathbf{x} - \mathbf{x}'|}\, da' - \frac{1}{4\pi\epsilon_0} \int_{S'} \frac{\sigma(\mathbf{x}')}{|\mathbf{x} - \mathbf{x}' + \mathbf{n}d|}\, da''$$

For small d we can expand $|\mathbf{x} - \mathbf{x}' + \mathbf{n}d|^{-1}$. Consider the general expression $|\mathbf{x} + \mathbf{a}|^{-1}$, where $|\mathbf{a}| \ll |\mathbf{x}|$. We write a Taylor series expansion in three dimensions:

$$\frac{1}{|\mathbf{x} + \mathbf{a}|} = \frac{1}{x} + \mathbf{a} \cdot \nabla\left(\frac{1}{x}\right) + \cdots$$

In this way we find that as $d \to 0$ the potential becomes

$$\Phi(\mathbf{x}) = \frac{1}{4\pi\epsilon_0} \int_S D(\mathbf{x}')\mathbf{n} \cdot \nabla'\left(\frac{1}{|\mathbf{x} - \mathbf{x}'|}\right) da' \qquad (1.24)$$

In passing we note that the integrand in (1.24) is the potential of a point dipole with dipole moment $\mathbf{p} = \mathbf{n}\, D\, da'$. The potential at $\mathbf{x}$ caused by a dipole $\mathbf{p}$ at $\mathbf{x}'$ is

$$\Phi(\mathbf{x}) = \frac{1}{4\pi\epsilon_0} \frac{\mathbf{p} \cdot (\mathbf{x} - \mathbf{x}')}{|\mathbf{x} - \mathbf{x}'|^3} \qquad (1.25)$$

Equation (1.24) has a simple geometrical interpretation. We note that

$$\mathbf{n} \cdot \nabla'\left(\frac{1}{|\mathbf{x} - \mathbf{x}'|}\right) da' = -\frac{\cos\theta\, da'}{|\mathbf{x} - \mathbf{x}'|^2} = -d\Omega$$

where $d\Omega$ is the element of solid angle subtended at the observation point by the area element da', as indicated in Fig. 1.7. Note that $d\Omega$ has a positive sign if θ is

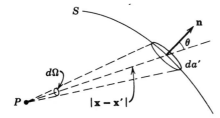

Figure 1.7   The potential at P due to the dipole layer D on the area element da' is just the negative product of D and the solid angle element $d\Omega$ subtended by da' at P.

an acute angle (i.e., when the observation point views the "inner" side of the dipole layer). The potential can be written:

$$\Phi(\mathbf{x}) = -\frac{1}{4\pi\epsilon_0} \int_S D(\mathbf{x}') \, d\Omega \qquad (1.26)$$

For a constant surface-dipole-moment density D, the potential is just the product of the moment divided by $4\pi\epsilon_0$ and the solid angle subtended at the observation point by the surface, regardless of its shape.

There is a discontinuity in potential in crossing a double layer. This can be seen by letting the observation point come infinitesimally close to the double layer. The double layer is now imagined to consist of two parts, one being a small disc directly under the observation point. The disc is sufficiently small that it is sensibly flat and has constant surface-dipole-moment density D. Evidently the total potential can be obtained by linear superposition of the potential of the disc and that of the remainder. From (1.26) it is clear that the potential of the disc alone has a discontinuity of D/ϵ_0 in crossing from the inner to the outer side, being $-D/2\epsilon_0$ on the inner side and $+D/2\epsilon_0$ on the outer. The potential of the remainder alone, with its hole where the disc fits in, is continuous across the plane of the hole. Consequently the total potential jump in crossing the surface is:

$$\Phi_2 - \Phi_1 = D/\epsilon_0 \qquad (1.27)$$

This result is analogous to (1.22) for the discontinuity of electric field in crossing a surface-charge density. Equation (1.27) can be interpreted "physically" as a potential drop occurring "inside" the dipole layer; it can be calculated as the product of the field between the two layers of surface charge times the separation before the limit is taken.

1.7 *Poisson and Laplace Equations*

In Sections 1.4 and 1.5 it was shown that the behavior of an electrostatic field can be described by the two differential equations:

$$\nabla \cdot \mathbf{E} = \rho/\epsilon_0 \qquad (1.13)$$

and

$$\nabla \times \mathbf{E} = 0 \qquad (1.14)$$

the latter equation being equivalent to the statement that $\mathbf{E}$ is the gradient of a scalar function, the scalar potential Φ:

$$\mathbf{E} = -\nabla\Phi \qquad (1.16)$$

Equations (1.13) and (1.16) can be combined into one partial differential equation for the single function $\Phi(\mathbf{x})$:

$$\nabla^2\Phi = -\rho/\epsilon_0 \qquad (1.28)$$

This equation is called the *Poisson equation*. In regions of space that lack a charge density, the scalar potential satisfies the *Laplace equation*:

$$\nabla^2\Phi = 0 \qquad (1.29)$$

We already have a solution for the scalar potential in expression (1.17):

$$\Phi(\mathbf{x}) = \frac{1}{4\pi\epsilon_0} \int \frac{\rho(\mathbf{x}')}{|\mathbf{x} - \mathbf{x}'|} d^3x' \tag{1.17}$$

To verify directly that this does indeed satisfy the Poisson equation (1.28), we operate with the Laplacian on both sides. Because it turns out that the resulting integrand is singular, we invoke a limiting procedure. Define the "a-potential" $\Phi_a(\mathbf{x})$ by

$$\Phi_a(\mathbf{x}) = \frac{1}{4\pi\epsilon_0} \int \frac{\rho(\mathbf{x}')}{\sqrt{(\mathbf{x} - \mathbf{x}')^2 + a^2}} d^3x'$$

The actual potential (1.17) is then the limit of the "a-potential" as $a \to 0$. Taking the Laplacian of the "a-potential" gives

$$\begin{aligned}
\nabla^2 \Phi_a(\mathbf{x}) &= \frac{1}{4\pi\epsilon_0} \int \rho(\mathbf{x}') \nabla^2 \left(\frac{1}{\sqrt{r^2 + a^2}} \right) d^3x' \\
&= -\frac{1}{4\pi\epsilon_0} \int \rho(\mathbf{x}') \left[\frac{3a^2}{(r^2 + a^2)^{5/2}} \right] d^3x'
\end{aligned} \tag{1.30}$$

where $r = |\mathbf{x} - \mathbf{x}'|$. The square-bracketed expression is the negative Laplacian of $1/\sqrt{r^2 + a^2}$. It is well-behaved everywhere for nonvanishing a, but as a tends to zero it becomes infinite at $r = 0$ and vanishes for $r \neq 0$. It has a volume integral equal to 4π for arbitrary a. For the purposes of integration, divide space into two regions by a sphere of fixed radius R centered on $\mathbf{x}$. Choose R such that $\rho(\mathbf{x}')$ changes little over the interior of the sphere, and imagine a much smaller than R and tending toward zero. If $\rho(\mathbf{x}')$ is such that (1.17) exists, the contribution to the integral (1.30) from the exterior of the sphere will vanish like a^2 as $a \to 0$. We thus need consider only the contribution from inside the sphere. With a Taylor series expansion of the well-behaved $\rho(\mathbf{x}')$ around $\mathbf{x}' = \mathbf{x}$, one finds

$$\nabla^2 \Phi_a(\mathbf{x}) = -\frac{1}{\epsilon_0} \int_0^R \frac{3a^2}{(r^2 + a^2)^{5/2}} \left[\rho(\mathbf{x}) + \frac{r^2}{6} \nabla^2 \rho + \cdots \right] r^2 dr + O(a^2)$$

Direct integration yields

$$\nabla^2 \Phi_a(\mathbf{x}) = -\frac{1}{\epsilon_0} \rho(\mathbf{x}) \left(1 + O(a^2/R^2) \right) + O(a^2, a^2 \log a) \nabla^2 \rho + \cdots$$

In the limit $a \to 0$, we obtain the Poisson equation (1.28).

The singular nature of the Laplacian of $1/r$ can be exhibited formally in terms of a Dirac delta function. Since $\nabla^2(1/r) = 0$ for $r \neq 0$ and its volume integral is -4π, we can write the formal equation, $\nabla^2(1/r) = -4\pi\delta(\mathbf{x})$ or, more generally,

$$\nabla^2 \left(\frac{1}{|\mathbf{x} - \mathbf{x}'|} \right) = -4\pi\delta(\mathbf{x} - \mathbf{x}') \tag{1.31}$$

1.8 Green's Theorem

If electrostatic problems always involved localized discrete or continuous distributions of charge with no boundary surfaces, the general solution (1.17) would be the most convenient and straightforward solution to any problem. There would be no need of the Poisson or Laplace equation. In actual fact, of course, many, if not most, of the problems of electrostatics involve finite regions of space, with or without charge inside, and with prescribed boundary conditions on the bounding surfaces. These boundary conditions may be simulated by an appropriate distribution of charges outside the region of interest (perhaps at infinity), but (1.17) becomes inconvenient as a means of calculating the potential, except in simple cases (e.g., method of images).

To handle the boundary conditions it is necessary to develop some new mathematical tools, namely, the identities or theorems due to George Green (1824). These follow as simple applications of the divergence theorem. The divergence theorem:

$$\int_V \boldsymbol{\nabla} \cdot \mathbf{A} \, d^3x = \oint_S \mathbf{A} \cdot \mathbf{n} \, da$$

applies to any well-behaved vector field $\mathbf{A}$ defined in the volume V bounded by the closed surface S. Let $\mathbf{A} = \phi \, \boldsymbol{\nabla}\psi$, where ϕ and ψ are arbitrary scalar fields. Now

$$\boldsymbol{\nabla} \cdot (\phi \, \boldsymbol{\nabla}\psi) = \phi \, \nabla^2\psi + \boldsymbol{\nabla}\phi \cdot \boldsymbol{\nabla}\psi \tag{1.32}$$

and

$$\phi \, \boldsymbol{\nabla}\psi \cdot \mathbf{n} = \phi \, \frac{\partial \psi}{\partial n} \tag{1.33}$$

where $\partial/\partial n$ is the normal derivative at the surface S (directed outward from inside the volume V). When (1.32) and (1.33) are substituted into the divergence theorem, there results *Green's first identity*:

$$\int_V (\phi \, \nabla^2\psi + \boldsymbol{\nabla}\phi \cdot \boldsymbol{\nabla}\psi) \, d^3x = \oint_S \phi \, \frac{\partial \psi}{\partial n} \, da \tag{1.34}$$

If we write down (1.34) again with ϕ and ψ interchanged, and then subtract it from (1.34), the $\boldsymbol{\nabla}\phi \cdot \boldsymbol{\nabla}\psi$ terms cancel, and we obtain Green's second identity or *Green's theorem*:

$$\int_V (\phi \, \nabla^2\psi - \psi \, \nabla^2\phi) \, d^3x = \oint_S \left[\phi \, \frac{\partial \psi}{\partial n} - \psi \, \frac{\partial \phi}{\partial n} \right] da \tag{1.35}$$

The Poisson differential equation for the potential can be converted into an integral equation if we choose a particular ψ, namely $1/R \equiv 1/|\mathbf{x} - \mathbf{x}'|$, where $\mathbf{x}$ is the observation point and $\mathbf{x}'$ is the integration variable. Further, we put $\phi = \Phi$, the scalar potential, and make use of $\nabla^2\Phi = -\rho/\epsilon_0$. From (1.31) we know that $\nabla^2(1/R) = -4\pi\delta(\mathbf{x} - \mathbf{x}')$, so that (1.35) becomes

$$\int_V \left[-4\pi\Phi(\mathbf{x}') \, \delta(\mathbf{x} - \mathbf{x}') + \frac{1}{\epsilon_0 R} \rho(\mathbf{x}') \right] d^3x' = \oint_S \left[\Phi \, \frac{\partial}{\partial n'} \left(\frac{1}{R} \right) - \frac{1}{R} \frac{\partial \Phi}{\partial n'} \right] da'$$

If the point $\mathbf{x}$ lies within the volume V, we obtain:

$$\Phi(\mathbf{x}) = \frac{1}{4\pi\epsilon_0} \int_V \frac{\rho(\mathbf{x}')}{R} \, d^3x' + \frac{1}{4\pi} \oint_S \left[\frac{1}{R} \frac{\partial \Phi}{\partial n'} - \Phi \, \frac{\partial}{\partial n'} \left(\frac{1}{R} \right) \right] da' \tag{1.36}$$

If **x** lies outside the surface S, the left-hand side of (1.36) is zero.* [Note that this is consistent with the interpretation of the surface integral as being the potential due to a surface-charge density $\sigma = \epsilon_0 \, \partial\Phi/\partial n'$ and a dipole layer $D = -\epsilon_0\Phi$. The discontinuities in electric field and potential (1.22) and (1.27) across the surface then lead to zero field and zero potential outside the volume V.]

Two remarks are in order about result (1.36). First, if the surface S goes to infinity and the electric field on S falls off faster than R^{-1}, then the surface integral vanishes and (1.36) reduces to the familiar result (1.17). Second, for a charge-free volume, the potential anywhere inside the volume (a solution of the Laplace equation) is expressed in (1.36) in terms of the potential and its normal derivative only on the surface of the volume. This rather surprising result is not a solution to a boundary-value problem, but only an integral statement, since the arbitrary specification of both Φ and $\partial\Phi/\partial n$ (*Cauchy boundary conditions*) is an overspecification of the problem. This is discussed in detail in the next sections, where techniques yielding solutions for appropriate boundary conditions are developed using Green's theorem (1.35).

1.9   Uniqueness of the Solution with Dirichlet or Neumann Boundary Conditions

What boundary conditions are appropriate for the Poisson (or Laplace) equation to ensure that a unique and well-behaved (i.e., physically reasonable) solution will exist inside the bounded region? Physical experience leads us to believe that specification of the potential on a closed surface (e.g., a system of conductors held at different potentials) defines a unique potential problem. This is called a *Dirichlet problem*, or *Dirichlet boundary conditions*. Similarly it is plausible that specification of the electric field (normal derivative of the potential) everywhere on the surface (corresponding to a given surface-charge density) also defines a unique problem. Specification of the normal derivative is known as the *Neumann boundary condition*. We now proceed to prove these expectations by means of Green's first identity (1.34).

We want to show the uniqueness of the solution of the Poisson equation, $\nabla^2\Phi = -\rho/\epsilon_0$, inside a volume V subject to either Dirichlet or Neumann boundary conditions on the closed bounding surface S. We suppose, to the contrary, that there exist two solutions Φ_1 and Φ_2 satisfying the same boundary conditions. Let

$$U = \Phi_2 - \Phi_1 \tag{1.37}$$

Then $\nabla^2 U = 0$ inside V, and $U = 0$ or $\partial U/\partial n = 0$ on S for Dirichlet and Neumann boundary conditions, respectively. From Green's first identity (1.34), with $\phi = \psi = U$, we find

$$\int_V (U \, \nabla^2 U + \nabla U \cdot \nabla U) \, d^3x = \oint_S U \frac{\partial U}{\partial n} \, da \tag{1.38}$$

---

*The reader may complain that (1.36) has been obtained in an illegal fashion since $1/|\mathbf{x} - \mathbf{x}'|$ is not well-behaved inside the volume V. Rigor can be restored by using a limiting process, as in the preceding section, or by excluding a small sphere around the offending point, $\mathbf{x} = \mathbf{x}'$. The result is still (1.36).

With the specified properties of U, this reduces (for both types of boundary condition) to:

$$\int_V |\boldsymbol{\nabla} U|^2 \, d^3x = 0$$

which implies $\boldsymbol{\nabla} U = 0$. Consequently, inside V, U is constant. For Dirichlet boundary conditions, $U = 0$ on S so that, inside V, $\Phi_1 = \Phi_2$ and the solution is unique. Similarly, for Neumann boundary conditions, the solution is unique, apart from an unimportant arbitrary additive constant.

From the right-hand side of (1.38) it is evident that there is also a unique solution to a problem with mixed boundary conditions (i.e., Dirichlet over part of the surface S, and Neumann over the remaining part).

It should be clear that a solution to the Poisson equation with both Φ and $\partial\Phi/\partial n$ specified arbitrarily on a closed boundary (Cauchy boundary conditions) does not exist, since there are unique solutions for Dirichlet and Neumann conditions separately and these will in general not be consistent. This can be verified with (1.36). With arbitrary values of Φ and $\partial\Phi/\partial n$ inserted on the right-hand side, it can be shown that the values of $\Phi(\mathbf{x})$ and $\boldsymbol{\nabla}\Phi(\mathbf{x})$ as $\mathbf{x}$ approaches the surface are in general inconsistent with the assumed boundary values. The question of whether Cauchy boundary conditions on an *open* surface define a unique electrostatic problem requires more discussion than is warranted here. The reader may refer to *Morse and Feshbach* (Section 6.2, pp. 692–706) or to *Sommerfeld* (*Partial Differential Equations in Physics*, Chapter II) for a detailed discussion of these questions. The conclusion is that electrostatic problems are specified only by Dirichlet *or* Neumann boundary conditions on a closed surface (part or all of which may be at infinity, of course).

1.10 *Formal Solution of Electrostatic Boundary-Value Problem with Green Function*

The solution of the Poisson or Laplace equation in a finite volume V with either Dirichlet or Neumann boundary conditions on the bounding surface S can be obtained by means of Green's theorem (1.35) and so-called Green functions.

In obtaining result (1.36)—not a solution—we chose the function ψ to be $1/|\mathbf{x} - \mathbf{x}'|$, it being the potential of a unit point source, satisfying the equation:

$$\nabla'^2\left(\frac{1}{|\mathbf{x} - \mathbf{x}'|}\right) = -4\pi\delta(\mathbf{x} - \mathbf{x}') \tag{1.31}$$

The function $1/|\mathbf{x} - \mathbf{x}'|$ is only one of a class of functions depending on the variables $\mathbf{x}$ and $\mathbf{x}'$, and called *Green functions*, which satisfy (1.31). In general,

$$\nabla'^2 G(\mathbf{x}, \mathbf{x}') = -4\pi\delta(\mathbf{x} - \mathbf{x}') \tag{1.39}$$

where

$$G(\mathbf{x}, \mathbf{x}') = \frac{1}{|\mathbf{x} - \mathbf{x}'|} + F(\mathbf{x}, \mathbf{x}') \tag{1.40}$$

with the function F satisfying the Laplace equation inside the volume V:

$$\nabla'^2 F(\mathbf{x}, \mathbf{x}') = 0 \tag{1.41}$$

In facing the problem of satisfying the prescribed boundary conditions on Φ *or* $\partial\Phi/\partial n$, we can find the key by considering result (1.36). As has been pointed out already, this is not a solution satisfying the correct type of boundary conditions because both Φ *and* $\partial\Phi/\partial n$ appear in the surface integral. It is at best an integral relation for Φ. With the generalized concept of a Green function and its additional freedom [via the function $F(\mathbf{x}, \mathbf{x}')$], there arises the possibility that we can use Green's theorem with $\psi = G(\mathbf{x}, \mathbf{x}')$ and choose $F(\mathbf{x}, \mathbf{x}')$ to eliminate one or the other of the two surface integrals, obtaining a result that involves only Dirichlet or Neumann boundary conditions. Of course, if the necessary $G(\mathbf{x}, \mathbf{x}')$ depended in detail on the exact form of the boundary conditions, the method would have little generality. As will be seen immediately, this is not required, and $G(\mathbf{x}, \mathbf{x}')$ satisfies rather simple boundary conditions on S.

With Green's theorem (1.35), $\phi = \Phi$, $\psi = G(\mathbf{x}, \mathbf{x}')$, and the specified properties of G (1.39), it is simple to obtain the generalization of (1.36):

$$\Phi(\mathbf{x}) = \frac{1}{4\pi\epsilon_0} \int_V \rho(\mathbf{x}')G(\mathbf{x}, \mathbf{x}') \, d^3x'$$
$$+ \frac{1}{4\pi} \oint_S \left[G(\mathbf{x}, \mathbf{x}') \frac{\partial\Phi}{\partial n'} - \Phi(\mathbf{x}') \frac{\partial G(\mathbf{x}, \mathbf{x}')}{\partial n'} \right] da' \qquad (1.42)$$

The freedom available in the definition of G (1.40) means that we can make the surface integral depend only on the chosen type of boundary conditions. Thus, for *Dirichlet boundary conditions* we demand:

$$G_D(\mathbf{x}, \mathbf{x}') = 0 \qquad \text{for } \mathbf{x}' \text{ on } S \qquad (1.43)$$

Then the first term in the surface integral in (1.42) vanishes and the solution is

$$\Phi(\mathbf{x}) = \frac{1}{4\pi\epsilon_0} \int_V \rho(\mathbf{x}')G_D(\mathbf{x}, \mathbf{x}') \, d^3x' - \frac{1}{4\pi} \oint_S \Phi(\mathbf{x}') \frac{\partial G_D}{\partial n'} \, da' \qquad (1.44)$$

For *Neumann boundary conditions* we must be more careful. The obvious choice of boundary condition on $G(\mathbf{x}, \mathbf{x}')$ seems to be

$$\frac{\partial G_N}{\partial n'} (\mathbf{x}, \mathbf{x}') = 0 \qquad \text{for } \mathbf{x}' \text{ on } S$$

since that makes the second term in the surface integral in (1.42) vanish, as desired. But an application of Gauss's theorem to (1.39) shows that

$$\oint_S \frac{\partial G}{\partial n'} \, da' = -4\pi$$

Consequently the simplest allowable boundary condition on G_N is

$$\frac{\partial G_N}{\partial n'} (\mathbf{x}, \mathbf{x}') = -\frac{4\pi}{S} \qquad \text{for } \mathbf{x}' \text{ on } S \qquad (1.45)$$

where S is the total area of the boundary surface. Then the solution is

$$\Phi(\mathbf{x}) = \langle\Phi\rangle_S + \frac{1}{4\pi\epsilon_0} \int_V \rho(\mathbf{x}')G_N(\mathbf{x}, \mathbf{x}') \, d^3x' + \frac{1}{4\pi} \oint_S \frac{\partial\Phi}{\partial n'} G_N \, da' \qquad (1.46)$$

where $\langle\Phi\rangle_S$ is the average value of the potential over the whole surface. The customary Neumann problem is the so-called exterior problem in which the volume V is bounded by two surfaces, one closed and finite, the other at infinity. Then the surface area S is infinite; the boundary condition (1.45) becomes homogeneous; the average value $\langle\Phi\rangle_S$ vanishes.

We note that the Green functions satisfy simple boundary conditions (1.43) or (1.45) which do not depend on the detailed form of the Dirichlet (or Neumann) boundary values. Even so, it is often rather involved (if not impossible) to determine $G(\mathbf{x}, \mathbf{x}')$ because of its dependence on the shape of the surface S. We will encounter such problems in Chapters 2 and 3.

The mathematical symmetry property $G(\mathbf{x}, \mathbf{x}') = G(\mathbf{x}', \mathbf{x})$ can be proved for the Green functions satisfying the Dirichlet boundary condition (1.43) by means of Green's theorem with $\phi = G(\mathbf{x}, \mathbf{y})$ and $\psi = G(\mathbf{x}', \mathbf{y})$, where $\mathbf{y}$ is the integration variable. Since the Green function, as a function of one of its variables, is a potential due to a unit point source, the symmetry merely represents the physical interchangeability of the source and the observation points. For Neumann boundary conditions the symmetry is not automatic, but can be imposed as a separate requirement.*

As a final, important remark we note the physical meaning of $F(\mathbf{x}, \mathbf{x}')/4\pi\epsilon_0$. It is a solution of the Laplace equation inside V and so represents the potential of a system of charges *external to the volume V*. It can be thought of as the potential due to an external distribution of charges chosen to satisfy the homogeneous boundary conditions of zero potential (or zero normal derivative) on the surface S when combined with the potential of a point charge at the source point $\mathbf{x}'$. Since the potential at a point $\mathbf{x}$ on the surface due to the point charge depends on the position of the source point, the external distribution of charge $F(\mathbf{x}, \mathbf{x}')$ must also depend on the "parameter" $\mathbf{x}'$. From this point of view, we see that the method of images (to be discussed in Chapter 2) is a physical equivalent of the determination of the appropriate $F(\mathbf{x}, \mathbf{x}')$ to satisfy the boundary conditions (1.43) or (1.45). For the Dirichlet problem with conductors, $F(\mathbf{x}, \mathbf{x}')/4\pi\epsilon_0$ can also be interpreted as the potential due to the surface-charge distribution induced on the conductors by the presence of a point charge at the source point $\mathbf{x}'$.

1.11   *Electrostatic Potential Energy and Energy Density; Capacitance*

In Section 1.5 it was shown that the product of the scalar potential and the charge of a point object could be interpreted as potential energy. More precisely, if a point charge q_i is brought from infinity to a point $\mathbf{x}_i$ in a region of localized electric fields described by the scalar potential Φ (which vanishes at infinity), the work done on the charge (and hence its potential energy) is given by

$$W_i = q_i\Phi(\mathbf{x}_i) \tag{1.47}$$

The potential Φ can be viewed as produced by an array of $(n - 1)$ charges $q_j(j = 1, 2, \ldots, n - 1)$ at positions $\mathbf{x}_j$. Then

$$\Phi(\mathbf{x}_i) = \frac{1}{4\pi\epsilon_0} \sum_{j=1}^{n-1} \frac{q_j}{|\mathbf{x}_i - \mathbf{x}_j|} \tag{1.48}$$

*See K.-J. Kim and J. D. Jackson, *Am. J. Phys.* **61**, (12) 1144–1146 (1993).

so that the potential energy of the charge q_i is

$$W_i = \frac{q_i}{4\pi\epsilon_0} \sum_{j=1}^{n-1} \frac{q_j}{|\mathbf{x}_i - \mathbf{x}_j|} \tag{1.49}$$

The *total* potential energy of all the charges due to all the forces acting between them is:

$$W = \frac{1}{4\pi\epsilon_0} \sum_{i=1}^{n} \sum_{j<i} \frac{q_i q_j}{|\mathbf{x}_i - \mathbf{x}_j|} \tag{1.50}$$

as can be seen most easily by adding each charge in succession. A more symmetric form can be written by summing over i and j unrestricted, and then dividing by 2:

$$W = \frac{1}{8\pi\epsilon_0} \sum_{i} \sum_{j} \frac{q_i q_j}{|\mathbf{x}_i - \mathbf{x}_j|} \tag{1.51}$$

It is understood that $i = j$ terms (infinite "self-energy" terms) are omitted in the double sum.

For a continuous charge distribution [or, in general, using the Dirac delta functions (1.6)] the potential energy takes the form:

$$W = \frac{1}{8\pi\epsilon_0} \int \int \frac{\rho(\mathbf{x})\rho(\mathbf{x}')}{|\mathbf{x} - \mathbf{x}'|} d^3x \, d^3x' \tag{1.52}$$

Another expression, equivalent to (1.52), can be obtained by noting that one of the integrals in (1.52) is just the scalar potential (1.17). Therefore

$$W = \frac{1}{2} \int \rho(\mathbf{x})\Phi(\mathbf{x}) \, d^3x \tag{1.53}$$

Equations (1.51), (1.52), and (1.53) express the electrostatic potential energy in terms of the positions of the charges and so emphasize the interactions between charges via Coulomb forces. An alternative, and very fruitful, approach is to emphasize the electric field and to interpret the energy as being stored in the electric field surrounding the charges. To obtain this latter form, we make use of the Poisson equation to eliminate the charge density from (1.53):

$$W = \frac{-\epsilon_0}{2} \int \Phi \, \nabla^2\Phi \, d^3x$$

Integration by parts leads to the result:

$$W = \frac{\epsilon_0}{2} \int |\nabla\Phi|^2 \, d^3x = \frac{\epsilon_0}{2} \int |\mathbf{E}|^2 \, d^3x \tag{1.54}$$

where the integration is over all space. In (1.54) all explicit reference to charges has gone, and the energy is expressed as an integral of the square of the electric field over all space. This leads naturally to the identification of the integrand as an energy density w:

$$w = \frac{\epsilon_0}{2} |\mathbf{E}|^2 \tag{1.55}$$

This expression for energy density is intuitively reasonable, since regions of high fields "must" contain considerable energy.

There is perhaps one puzzling thing about (1.55). The energy density is positive definite. Consequently its volume integral is necessarily nonnegative. This seems to contradict our impression from (1.51) that the potential energy of two charges of opposite sign is negative. The reason for this apparent contradiction is that (1.54) and (1.55) contain "self-energy" contributions to the energy density, whereas the double sum in (1.51) does not. To illustrate this, consider two point charges q_1 and q_2 located at $\mathbf{x}_1$ and $\mathbf{x}_2$, as in Fig. 1.8. The electric field at the point P with coordinate $\mathbf{x}$ is

$$\mathbf{E} = \frac{1}{4\pi\epsilon_0} \frac{q_1(\mathbf{x} - \mathbf{x}_1)}{|\mathbf{x} - \mathbf{x}_1|^3} + \frac{1}{4\pi\epsilon_0} \frac{q_2(\mathbf{x} - \mathbf{x}_2)}{|\mathbf{x} - \mathbf{x}_2|^3}$$

so that the energy density (1.55) is

$$32\pi^2\epsilon_0 w = \frac{q_1^2}{|\mathbf{x} - \mathbf{x}_1|^4} + \frac{q_2^2}{|\mathbf{x} - \mathbf{x}_2|^4} + 2\frac{q_1 q_2(\mathbf{x} - \mathbf{x}_1)\cdot(\mathbf{x} - \mathbf{x}_2)}{|\mathbf{x} - \mathbf{x}_1|^3 |\mathbf{x} - \mathbf{x}_2|^3} \qquad (1.56)$$

Clearly the first two terms are "self-energy" contributions. To show that the third term gives the proper result for the interaction potential energy we integrate over all space:

$$W_{\text{int}} = \frac{q_1 q_2}{16\pi^2\epsilon_0} \int \frac{(\mathbf{x} - \mathbf{x}_1)\cdot(\mathbf{x} - \mathbf{x}_2)}{|\mathbf{x} - \mathbf{x}_1|^3 |\mathbf{x} - \mathbf{x}_2|^3} d^3x \qquad (1.57)$$

A change of integration variable to $\boldsymbol{\rho} = (\mathbf{x} - \mathbf{x}_1)/|\mathbf{x} - \mathbf{x}_2|$ yields

$$W_{\text{int}} = \frac{1}{4\pi\epsilon_0} \frac{q_1 q_2}{|\mathbf{x}_1 - \mathbf{x}_2|} \times \frac{1}{4\pi} \int \frac{\boldsymbol{\rho}\cdot(\boldsymbol{\rho} + \mathbf{n})}{\rho^3 |\boldsymbol{\rho} + \mathbf{n}|^3} d^3\rho \qquad (1.58)$$

where $\mathbf{n}$ is a unit vector in the direction $(\mathbf{x}_1 - \mathbf{x}_2)$. Using the fact that $(\boldsymbol{\rho} + \mathbf{n})/|\boldsymbol{\rho} + \mathbf{n}|^3 = -\nabla_\rho(1/|\boldsymbol{\rho} + \mathbf{n}|)$, the dimensionless integral can easily be shown to have the value 4π, so that the interaction energy reduces to the expected value.

Forces acting between charged bodies can be obtained by calculating the change in the total electrostatic energy of the system under small virtual displacements. Examples of this are discussed in the problems. Care must be taken to exhibit the energy in a form showing clearly the factors that vary with a change in configuration and those that are kept constant.

As a simple illustration we calculate the force per unit area on the surface of a conductor with a surface-charge density $\sigma(\mathbf{x})$. In the immediate neighborhood of the surface the energy density is

$$w = \frac{\epsilon_0}{2} |\mathbf{E}|^2 = \sigma^2/2\epsilon_0 \qquad (1.59)$$

If we now imagine a small outward displacement Δx of an elemental area Δa of the conducting surface, the electrostatic energy decreases by an amount that is the product of energy density w and the excluded volume $\Delta x \, \Delta a$:

$$\Delta W = -\sigma^2 \, \Delta a \, \Delta x/2\epsilon_0 \qquad (1.60)$$

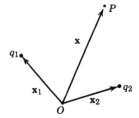

Figure 1.8

This means that there is an outward force per unit area equal to $\sigma^2/2\epsilon_0 = w$ at the surface of the conductor. This result is normally derived by taking the product of the surface-charge density and the electric field, with care taken to eliminate the electric field due to the element of surface-charge density itself.

For a system of n conductors, each with potential V_i and total charge Q_i $(i = 1, 2, \ldots, n)$ in otherwise empty space, the electrostatic potential energy can be expressed in terms of the potentials alone and certain geometrical quantities called coefficients of capacity. For a given configuration of the conductors, the linear functional dependence of the potential on the charge density implies that the potential of the ith conductor can be written as

$$V_i = \sum_{j=1}^{n} p_{ij} Q_j \qquad (i = 1, 2, \ldots, n)$$

where the p_{ij} depend on the geometry of the conductors. These n equations can be inverted to yield the charge on the ith conductor in terms of all the potentials:

$$Q_i = \sum_{j=1}^{n} C_{ij} V_j \qquad (i = 1, 2, \ldots, n) \tag{1.61}$$

The coefficients C_{ii} are called capacities or capacitances while the C_{ij}, $i \neq j$, are called coefficients of induction. The *capacitance of a conductor is therefore the total charge on the conductor when it is maintained at unit potential, all other conductors being held at zero potential.* Sometimes the capacitance of a system of conductors is also defined. For example, the capacitance of two conductors carrying equal and opposite charges in the presence of other grounded conductors is defined as the ratio of the charge on one conductor to the potential difference between them. The equations (1.61) can be used to express this capacitance in terms of the coefficients C_{ij}.

The potential energy (1.53) for the system of conductors is

$$W = \frac{1}{2} \sum_{i=1}^{n} Q_i V_i = \frac{1}{2} \sum_{i=1}^{n} \sum_{j=1}^{n} C_{ij} V_i V_j \tag{1.62}$$

The expression of the energy in terms of the potentials V_i and the C_{ij}, or in terms of the charges Q_i and the coefficients p_{ij}, permits the application of variational methods to obtain approximate values of capacitances. It can be shown, based on the technique of the next section (see Problems 1.17 and 1.18), that there are variational principles giving upper and lower bounds on C_{ii}. The principles permit estimation with known error of the capacitances of relatively involved configurations of conductors. High-speed computational techniques permit the use of elaborate trial functions involving several parameters. It must be remarked, however, that the need for a Green function satisfying Dirichlet boundary conditions in the lower bound makes the error estimate nontrivial. Further consideration of this technique for calculating capacitances is left to the problems at the end of this and subsequent chapters.

1.12 *Variational Approach to the Solution of the Laplace and Poisson Equations*

Variational methods play prominent roles in many areas of classical and quantum physics. They provide formal techniques for the derivation of "equations of motion" and also practical methods for obtaining approximate, but often accurate, solutions to problems not amenable to other approaches. Estimates of resonant frequencies of acoustic resonators and energy eigenvalues of atomic systems come readily to mind.

The far-reaching concept that physical systems in equilibrium have minimal energy content is generalized to the consideration of energy-like functionals. As an example, consider the functional

$$I[\psi] = \frac{1}{2} \int_V \boldsymbol{\nabla}\psi \cdot \boldsymbol{\nabla}\psi \, d^3x - \int_V g\psi \, d^3x \tag{1.63}$$

where the function $\psi(\mathbf{x})$ is well-behaved inside the volume V and on its surface S (which may consist of several separate surfaces), and $g(\mathbf{x})$ is a specified "source" function without singularities within V. We now examine the first-order change in the functional when we change $\psi \to \psi + \delta\psi$, where the modification $\delta\psi(\mathbf{x})$ is infinitesimal within V. The difference $\delta I = I[\psi + \delta\psi] - I[\psi]$ is

$$\delta I = \int_V \boldsymbol{\nabla}\psi \cdot \boldsymbol{\nabla}(\delta\psi) \, d^3x - \int_V g\delta\psi \, d^3x + \cdots \tag{1.64}$$

The neglected term is semipositive definite and is second order in $\delta\psi$. Use of Green's first identity with $\phi = \delta\psi$ and $\psi = \psi$ yields

$$\delta I = \int_V [-\nabla^2\psi - g] \, \delta\psi \, d^3x + \oint_S \delta\psi \frac{\partial\psi}{\partial n} \, da \tag{1.65}$$

Provided $\delta\psi = 0$ on the boundary surface S (so that the surface integral vanishes), the first-order change in $I[\psi]$ vanishes if $\psi(\mathbf{x})$ satisfies

$$\nabla^2\psi = -g \tag{1.66}$$

Recalling that the neglected term in (1.64) is semipositive definite, we see that $I[\psi]$ is a stationary minimum if ψ satisfies a Poisson-like equation within the volume V and the departures $\delta\psi$ vanish on the boundary. With $\psi \to \Phi$ and $g \to \rho/\epsilon_0$, the minimization of the functional yields the "equation of motion" of the electrostatic potential in the presence of a charge density and Dirichlet boundary conditions (Φ given on S and so $\delta\Phi = 0$ there).

The derivation of the Poisson equation from the variational functional is the formal aspect. Equally important, the stationary nature of the extremum of $I[\psi]$ permits a practical approach to an approximate solution for $\psi(\mathbf{x})$. We choose a flexible "trial" function $\psi(\mathbf{x}) = A\Psi(\mathbf{x}, \alpha, \beta, \ldots)$ that depends on a normalization constant A and some number of other parameters, $\alpha, \beta, \ldots$, and is constructed to satisfy the given boundary conditions on the surface S. The function Ψ may be a sum of terms with the parameters as coefficients, or a single function of several parameters; it should be chosen with some eye toward the expected form of the solution. (Intuition plays a role here!) Calculation of $I[\psi]$ gives the function, $I(A, \alpha, \beta, \ldots)$. We now vary the parameters to locate the extremum (actually a minimum) of $I(A, \alpha, \beta, \ldots)$. With the optimum parameters, the trial solution is the best possible approximation to the true solution with the particular functional form chosen. For the Laplace equation, the normalization constant is determined by the Dirichlet boundary values of ψ. For the Poisson equation, it is determined by the source strength $g(\mathbf{x})$, as well as the boundary values on S.

A different functional is necessary for Neumann boundary conditions. Suppose that the boundary conditions on ψ are specified by $\partial\psi/\partial n|_s = f(\mathbf{s})$, where $\mathbf{s}$ locates a point on the surface S. The appropriate functional is

$$I[\psi] = \frac{1}{2} \int_V \nabla\psi \cdot \nabla\psi \, d^3x - \int_V g\psi d^3x - \oint_S f\psi \, da \qquad (1.67)$$

The same steps as before with $\psi \rightarrow \psi + \delta\psi$ lead to the first-order difference in functionals,

$$\delta I = \int_V [-\nabla^2\psi - g] \, \delta\psi \, d^3x + \oint_S \left(\frac{\partial\psi}{\partial n} - f(\mathbf{s})\right) \delta\psi \, da \qquad (1.68)$$

The requirement that δI vanish independent of $\delta\psi$ implies

$$\nabla^2\psi = -g \text{ within } V \qquad \text{and} \qquad \frac{\partial\psi}{\partial n} = f(\mathbf{s}) \text{ on } S \qquad (1.69)$$

Again the functional is a stationary minimum for ψ satisfying (1.69). Approximate solutions can be found by the use of trial functions that satisfy the Neumann boundary conditions, just as described above for Dirichlet boundary conditions.

As a simple application to the Poisson equation, consider the two-dimensional problem of a hollow circular cylinder of unit radius centered on the z-axis, with an interior source density $g(\mathbf{x}) = g(\rho)$, azimuthally symmetric and independent of z. The potential vanishes at $\rho = 1$. The "equation of motion" for ψ (a function of ρ alone) in polar coordinates is

$$\frac{1}{\rho}\frac{\partial}{\partial\rho}\left(\rho\frac{\partial\psi}{\partial\rho}\right) = -g(\rho) \qquad (1.70)$$

For trial functions we consider finite polynomials in powers of $(1 - \rho)$ and ρ. A three-parameter function of the first type is

$$\Psi_1 = \alpha_1(1 - \rho) + \beta_1(1 - \rho)^2 + \gamma_1(1 - \rho)^3 \qquad (1.71)$$

This choice might seem natural because it automatically builds in the boundary condition at $\rho = 1$, but it contains a flaw that makes it a less accurate representation of ψ than the power series in ρ. The reason is that, if the source density g is well behaved and finite at the origin, Gauss's law shows that ψ has a maximum or minimum there with vanishing slope. The requirements at both the origin and $\rho = 1$ are met by a three-parameter trial function in powers of ρ:

$$\Psi_2 = \alpha\rho^2 + \beta\rho^3 + \gamma\rho^4 - (\alpha + \beta + \gamma) \qquad (1.72)$$

We expect this trial function in general to be a better approximation to ψ than Ψ_1 for the same number of variational parameters. [We could, of course, impose the constraint, $\alpha_1 + 2\beta_1 + 3\gamma_1 = 0$ on (1.71) to get the proper behavior at the origin, but that would reduce the number of parameters from three to two.]

The functional integral (1.63) for Ψ_2 is easily shown to be

$$\frac{1}{2\pi} I[\Psi_2] = \left[\frac{1}{2}\alpha^2 + \frac{6}{5}\alpha\beta + \frac{4}{3}\alpha\gamma + \frac{3}{4}\beta^2\right.$$
$$\left. + \frac{12}{7}\beta\gamma + \gamma^2\right] - [e_2\alpha + e_3\beta + e_4\gamma] \qquad (1.73)$$

where $e_n = \int_0^1 g(\rho)(\rho^n - 1) \, \rho \, d\rho$.

The integral for Ψ_1 has the same form as (1.73), but different coefficients. As described above, we seek an extremum of (1.73) by setting the partial derivatives with respect to the parameters α, β, and γ equal to zero. The three coupled algebraic linear equations yield the "best" values,

$$\alpha = 225e_2 - 420e_3 + 210e_4$$

$$\beta = -420e_2 + \frac{2450}{3} e_3 - 420e_4 \tag{1.74}$$

$$\gamma = 210e_2 - 420e_3 + \frac{441}{2} e_4$$

These values can be inserted into (1.73) to give $I[\Psi_2]_{min}$ as a not very illuminating function of the e_n. One would then find that the "kinetic" (first) bracket was equal to half the "potential" (second) bracket and opposite in sign, a characteristic of the extremum.

To go further we must specify $g(\rho)$. The results for the best trial functions Ψ_1 and Ψ_2 are shown in Fig. 1.9 for the source density,

$$g(\rho) = -5(1 - \rho) + 10^4\rho^5(1 - \rho)^5 \tag{1.75}$$

The choice of source is arbitrary and is chosen to give a potential that is not quite featureless. The "best" parameters for Ψ_2 are $\alpha = 2.915$, $\beta = -7.031$, and $\gamma = 3.642$. The variational integral has the value, $I[\Psi_2]_{min} = -1.5817$, compared to $I[\psi]_{exact} = -1.6017$. The fractional error is 1.3%.

Note that the trial function Ψ_1 fails rather badly for $\rho < 0.3$ because it does

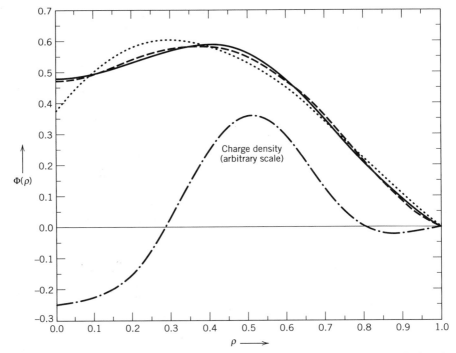

Figure 1.9 Comparison of the exact solution $\psi(\rho)$ (solid curve) with two variational approximations for the potential, Ψ_1 (dotted curve) and Ψ_2 (dashed curve). The charge density (1.75) is indicated by the dash-dot curve (arbitrary scale).

not respect the vanishing slope at $\rho = 0$. Nonetheless, it gives $I[\Psi_1]_{min} = -1.5136$, which is somewhat, but not greatly, worse than Ψ_2 (5.5% error). The insensitivity of $I[\Psi]$ to errors in the trial function illustrates both a strength and a weakness of the variational method. If the principle is used to estimate eigenvalues (related to the value of $I[\Psi]$), it does well. Used as a method of estimating a solution $\psi \approx \Psi$, it can fail badly, at least in parts of the configuration space.

The reader will recognize from (1.70) that a polynomial source density leads to an exact polynomial solution for ψ, but the idea here is to illustrate the variational method, not to demonstrate a class of explicit solutions. Further illustration is left to the problems at the end of this and later chapters.

1.13 *Relaxation Method for Two-Dimensional Electrostatic Problems*

The relaxation method is an iterative numerical scheme (sometimes called iterative finite difference method) for the solution of the Laplace or Poisson equation in two dimensions. Here we present only its basic ideas and its connection with the variational method. First we consider the Laplace equation with Dirichlet boundary conditions within a two-dimensional region S with a boundary contour C. We imagine the region S spanned by a square lattice with lattice spacing h (and the boundary contour C approximated by a step-like boundary linking lattice sites along C). The independent variables are the integers (i, j) specifying the sites; the dependent variables are the trial values of the potential $\psi(i, j)$ at each site. The potential values on the boundary sites are assumed given.

To establish the variational nature of the method and to specify the iterative scheme, we imagine the functional integral $I[\psi]$ over S as a sum over small domains of area h^2, as shown in Fig. 1.10a. We consider the neighboring trial values of the potential as fixed, while the value at the center of the subarea is a variational quantity to be optimized. The spacing is small enough to permit us to approximate the derivatives in, say, the northeast quarter of the subarea by

$$\left(\frac{\partial \psi}{\partial x}\right)_{NE} = \frac{1}{h}(\psi_E - \psi_0); \qquad \left(\frac{\partial \psi}{\partial y}\right)_{NE} = \frac{1}{h}(\psi_N - \psi_0)$$

and similarly for the other three quarters. The functional integral over the northeast quarter is

$$\begin{aligned}
I_{NE} &= \frac{1}{2}\int_0^{h/2} dx \int_0^{h/2} dy \left[\left(\frac{\partial \psi}{\partial x}\right)^2 + \left(\frac{\partial \psi}{\partial y}\right)^2\right] \\
&\approx \frac{1}{8}[(\psi_0 - \psi_N)^2 + (\psi_0 - \psi_E)^2]
\end{aligned} \tag{1.76}$$

The complete integral over the whole (shaded) subarea is evidently

$$I \approx \frac{1}{4}[(\psi_0 - \psi_N)^2 + (\psi_0 - \psi_E)^2 + (\psi_0 - \psi_S)^2 + (\psi_0 - \psi_W)^2] \tag{1.77}$$

Minimizing this integral with respect to ψ_0 gives the optimum value,

$$(\psi_0)_{optimum} = \frac{1}{4}(\psi_N + \psi_E + \psi_S + \psi_W) \tag{1.78}$$

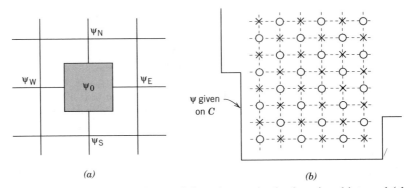

Figure 1.10 (a) Enlargement of one of the subareas in the functional integral (shaded). The trial values of the potential at the neighboring sites are labeled ψ_N, ψ_S, ψ_E, and ψ_W, while the value at the center of the subarea is ψ_0. (b) One possible iteration is to replace the trial values at the lattice sites ($\circ$) with the average of the values at the surrounding sites ($\times$).

The integral is minimized if ψ_0 is equal to the average of the values at the "cross" points.

Now consider the whole functional integral, that is, the sum of the integrals over all the subareas. We guess a set of $\psi(i, j)$ initially and approximate the functional integral $I[\psi]$ by the sum of terms of the form of (1.77). Then we go over the lattice and replace half the values, indicated by the circles in Fig. 1.10b, by the average of the points (crosses) around them. The new set of trial values $\psi(i, j)$ will evidently minimize $I[\psi]$ more than the original set of values; the new set will be closer to the correct solution. Actually, there is no need to do the averaging for only half the points—that was just a replication for half of the subareas of the process for Fig. 1.10a.

There are many improvements that can be made. One significant one concerns the type of averaging. We could have taken the average of the values at the corners of the large square in Fig. 1.10a instead of the "cross" values. Or we could take some linear combination of the two. It can be shown (see Problem 1.21) by Taylor series expansion of any well-behaved function $F(x, y)$ that a particular weighted average,

$$\langle\langle F(x, y)\rangle\rangle \equiv \frac{4}{5} \langle F\rangle_c + \frac{1}{5} \langle F\rangle_s \tag{1.79}$$

where the "cross" and "square" averages are

$$\langle F(x, y)\rangle_c = \frac{1}{4} [F(x+h, y) + F(x, y+h) + F(x-h, y) + F(x, y-h)] \tag{1.80a}$$

$$\langle F(x, y)\rangle_s = \frac{1}{4} [F(x+h, y+h) + F(x+h, y-h) \\ + F(x-h, y+h) + F(x-h, y-h)] \tag{1.80b}$$

yields

$$\langle\langle F(x, y)\rangle\rangle = F(x, y) + \frac{3}{10} h^2 \nabla^2 F + \frac{1}{40} h^4 \nabla^2(\nabla^2 F) + O(h^6) \tag{1.81}$$

In (1.81) the Laplacians of F are evaluated at (x, y). If $F(x, y)$ is a solution of the Laplace equation, the weighted averaging over the eight adjacent lattice sites in (1.79) gives F at the center with corrections only of order h^6. Instead of (1.78), which is the same as (1.80a), a better iteration scheme uses $\psi_{new}(i, j) = \langle\langle\psi(i, j)\rangle\rangle$ $+ O(h^6)$. With either the "cross" or "square" averaging separately, the error is $O(h^4)$. The increase in accuracy with $\langle\langle\psi\rangle\rangle$ is at the expense of twice as much computation for each lattice site, but for the same accuracy, far fewer lattice sites are needed: $\langle\langle N\rangle\rangle = O(\langle N\rangle^{2/3})$, where $\langle\langle N\rangle\rangle$ is the number of sites needed with $\langle\langle\psi\rangle\rangle$ and $\langle N\rangle$ is the corresponding number with the "cross" or "square" average.

Equation (1.81) has an added advantage in application to the Poisson equation, $\nabla^2\psi = -g$. The terms of order h^2 and h^4 can be expressed directly in terms of the specified charge density and the simplest approximation for its Laplacian. It is easy to show that the new value of the trial function at (i, j) is generated by

$$\psi_{new}(i, j) = \langle\langle\psi(i, j)\rangle\rangle + \frac{h^2}{5} g(i, j) + \frac{h^2}{10} \langle g(i, j)\rangle_c + O(h^6) \qquad (1.82)$$

where $\langle g\rangle_c$ is the "cross" average of g, according to (1.80a).

A basic procedure for the iterative numerical solution of the Laplace or Poisson equation in two dimensions with Dirichlet boundary conditions is as follows:

1. A square lattice spacing h is chosen and the lattice sites, including the sites on the boundary, are labeled in some manner [which we denote here as (i, j)].

2. The values of the potential at the boundary sites are entered in a table of the potential at all sites.

3. A guess is made for the values, called $\Phi_{old}(i, j)$, at all interior sites. A constant value everywhere is easiest. These are added to the table or array of "starting" values.

4. The first iteration cycle begins by systematically going over the lattice sites, one by one, and computing $\langle\langle\Phi(i, j)\rangle\rangle$ with (1.79) or one of the averages in (1.80). This quantity (or (1.82) for the Poisson equation) is entered as $\Phi_{new}(i, j)$ in a table of "new" values of the potential at each site. Note that the sites next to the boundary benefit from the known boundary values, and so their $\langle\langle\Phi\rangle\rangle$ values are likely initially to be closer to the ultimate values of the potential than those for sites deep in the interior. With each iteration, the accuracy works its way from the boundaries into the interior.

5. Once all interior sites have been processed, the set of $\Phi_{old}(i, j)$ is replaced by the set of $\Phi_{new}(i, j)$, and the iteration cycle begins again.

6. Iterations continue until some desired level of accuracy is achieved. For example, one might continue iterations until the absolute value of the difference of old and new values is less than some preassigned value at every interior site.

The scheme just outlined is called Jacobian iteration. It requires two arrays of values of the potential at the lattice sites during each iteration. A better scheme, called Gauss–Seidel iteration, employs a trivial change: one replaces $\Phi_{old}(i, j)$ with $\Phi_{new}(i, j)$ as soon as the latter is determined. This means that during an iteration one benefits immediately from the improved values. Typically, at any given site, $\langle\langle\Phi\rangle\rangle$ is made up half of old values and half of new ones, depending on the path over the lattice. There are many other improvements possible—consult *Press et al.*, *Numerical Recipes*, or some of the references cited at the end of the chapter. The relaxation method is also applicable to magnetic field problems, as described briefly in Section 5.14.

References and Suggested Reading

On the mathematical side, the subject of delta functions is treated simply but rigorously by
> Lighthill
> Dennery and Kryzwicki

For a discussion of different types of partial differential equations and the appropriate boundary conditions for each type, see
> Morse and Feshbach, Chapter 6
> Sommerfeld, *Partial Differential Equations in Physics*, Chapter II
> Courant and Hilbert, Vol. II, Chapters III–VI

The general theory of Green functions is treated in detail by
> Friedman, Chapter 3
> Morse and Feshbach, Chapter 7

The general theory of electrostatics is discussed extensively in many of the older books. Notable, in spite of some old-fashioned notation, are
> Maxwell, Vol. 1, Chapters II and IV
> Jeans, Chapters II, VI, VII
> Kellogg

Of more recent books, mention may be made of the treatment of the general theory by Stratton, Chapter III, and parts of Chapter II.

Readers interested in variational methods applied to electromagnetic problems can consult
> Cairo and Kahan
> Collin, Chapter 4
> Sadiku, Chapter 4

and
> Pólya and Szegö

for elegant and powerful mathematical techniques.

The classic references to relaxation methods are the two books by R. V. Southwell:
> *Relaxation Methods in Engineering Science*, Oxford University Press, Oxford (1940).
> *Relaxation Methods in Theoretical Physics*, Oxford University Press, Oxford (1946).

Physicists will be more comfortable with the second volume, but much basic material is in the first. More modern references on relaxation and other numerical methods are
> Sadiku
> Zhou

Problems

1.1 Use Gauss's theorem [and (1.21) if necessary] to prove the following:

(a) Any excess charge placed on a conductor must lie entirely on its surface. (A conductor by definition contains charges capable of moving freely under the action of applied electric fields.)

(b) A closed, hollow conductor shields its interior from fields due to charges outside, but does not shield its exterior from the fields due to charges placed inside it.

(c) The electric field at the surface of a conductor is normal to the surface and has a magnitude σ/ϵ_0, where σ is the charge density per unit area on the surface.

1.2 The Dirac delta function in three dimensions can be taken as the improper limit as $\alpha \to 0$ of the Gaussian function

$$D(\alpha; x, y, z) = (2\pi)^{-3/2}\alpha^{-3} \exp\left[-\frac{1}{2\alpha^2}(x^2 + y^2 + z^2)\right]$$

Consider a general orthogonal coordinate system specified by the surfaces $u =$ constant, $v =$ constant, $w =$ constant, with length elements du/U, dv/V, dw/W in the three perpendicular directions. Show that

$$\delta(\mathbf{x} - \mathbf{x}') = \delta(u - u')\,\delta(v - v')\,\delta(w - w') \cdot UVW$$

by considering the limit of the Gaussian above. Note that as $\alpha \to 0$ only the infinitesimal length element need be used for the distance between the points in the exponent.

1.3 Using Dirac delta functions in the appropriate coordinates, express the following charge distributions as three-dimensional charge densities $\rho(\mathbf{x})$.

(a) In spherical coordinates, a charge Q uniformly distributed over a spherical shell of radius R.

(b) In cylindrical coordinates, a charge λ per unit length uniformly distributed over a cylindrical surface of radius b.

(c) In cylindrical coordinates, a charge Q spread uniformly over a flat circular disc of negligible thickness and radius R.

(d) The same as part (c), but using spherical coordinates.

1.4 Each of three charged spheres of radius a, one conducting, one having a uniform charge density within its volume, and one having a spherically symmetric charge density that varies radially as r^n ($n > -3$), has a total charge Q. Use Gauss's theorem to obtain the electric fields both inside and outside each sphere. Sketch the behavior of the fields as a function of radius for the first two spheres, and for the third with $n = -2, +2$.

1.5 The time-averaged potential of a neutral hydrogen atom is given by

$$\Phi = \frac{q}{4\pi\epsilon_0}\frac{e^{-\alpha r}}{r}\left(1 + \frac{\alpha r}{2}\right)$$

where q is the magnitude of the electronic charge, and $\alpha^{-1} = a_0/2$, a_0 being the Bohr radius. Find the distribution of charge (both continuous and discrete) that will give this potential and interpret your result physically.

1.6 A simple capacitor is a device formed by two insulated conductors adjacent to each other. If equal and opposite charges are placed on the conductors, there will be a certain difference of potential between them. The ratio of the magnitude of the charge on one conductor to the magnitude of the potential difference is called the capacitance (in SI units it is measured in farads). Using Gauss's law, calculate the capacitance of

(a) two large, flat, conducting sheets of area A, separated by a small distance d;

(b) two concentric conducting spheres with radii a, b ($b > a$);

(c) two concentric conducting cylinders of length L, large compared to their radii a, b ($b > a$).

(d) What is the inner diameter of the outer conductor in an air-filled coaxial cable whose center conductor is a cylindrical wire of diameter 1 mm and whose capacitance is 3×10^{-11} F/m? 3×10^{-12} F/m?

1.7 Two long, cylindrical conductors of radii a_1 and a_2 are parallel and separated by a distance d, which is large compared with either radius. Show that the capacitance per unit length is given approximately by

$$C \simeq \pi\epsilon_0 \left(\ln \frac{d}{a} \right)^{-1}$$

where a is the geometrical mean of the two radii.

Approximately what gauge wire (state diameter in millimeters) would be necessary to make a two-wire transmission line with a capacitance of 1.2×10^{-11} F/m if the separation of the wires was 0.5 cm? 1.5 cm? 5.0 cm?

1.8 **(a)** For the three capacitor geometries in Problem 1.6 calculate the total electrostatic energy and express it alternatively in terms of the equal and opposite charges Q and $-Q$ placed on the conductors *and* the potential difference between them.

(b) Sketch the energy density of the electrostatic field in each case as a function of the appropriate linear coordinate.

1.9 Calculate the attractive force between conductors in the parallel plate capacitor (Problem 1.6a) and the parallel cylinder capacitor (Problem 1.7) for

(a) fixed charges on each conductor;

(b) fixed potential difference between conductors.

1.10 Prove the *mean value theorem*: For charge-free space the value of the electrostatic potential at any point is equal to the average of the potential over the surface of *any* sphere centered on that point.

1.11 Use Gauss's theorem to prove that at the surface of a curved charged conductor, the normal derivative of the electric field is given by

$$\frac{1}{E} \frac{\partial E}{\partial n} = -\left(\frac{1}{R_1} + \frac{1}{R_2} \right)$$

where R_1 and R_2 are the principal radii of curvature of the surface.

1.12 Prove *Green's reciprocation theorem*: If Φ is the potential due to a volume-charge density ρ within a volume V and a surface-charge density σ on the conducting surface S bounding the volume V, while Φ' is the potential due to another charge distribution ρ' and σ', then

$$\int_V \rho\Phi' \, d^3x + \int_S \sigma\Phi' \, da = \int_V \rho'\Phi \, d^3x + \int_S \sigma'\Phi \, da$$

1.13 Two infinite grounded parallel conducting planes are separated by a distance d. A point charge q is placed between the planes. Use the reciprocation theorem of Green to prove that the total induced charge on one of the planes is equal to $(-q)$ times the fractional perpendicular distance of the point charge from the other plane. (*Hint*: As your comparison electrostatic problem with the same surfaces choose one whose charge densities and potential are known and simple.)

1.14 Consider the symmetric and antisymmetric combinations of the electrostatic Green function $G(\mathbf{x}, \mathbf{x}')$ of Section 1.10,

$$S(\mathbf{x}, \mathbf{x}') = (\tfrac{1}{2})[G(\mathbf{x}, \mathbf{x}') + G(\mathbf{x}', \mathbf{x})]$$
$$A(\mathbf{x}, \mathbf{x}') = (\tfrac{1}{2})[G(\mathbf{x}, \mathbf{x}') - G(\mathbf{x}', \mathbf{x})]$$

Show that within the volume V, $\nabla'^2 A = 0$. Then apply Green's theorem with $\phi = \Phi$, $\psi = S$ and also $\phi = \Phi$, $\psi = A$ to obtain relations similar to (1.42). Show that the antisymmetric Green function gives no contribution to the potential for either Dirichlet or Neumann boundary conditions. [The antisymmetric form must vanish for Dirichlet boundary conditions—see Section 1.10—but may be nonvanishing for Neumann boundary conditions. Your result here shows, however, that one may symmetrize the Neumann Green function or not, as one wishes. See Problem 3.26 for an explicit example.]

1.15   Prove *Thomson's theorem*: If a number of surfaces are fixed in position and a given total charge is placed on each surface, then the electrostatic energy in the region bounded by the surfaces is an absolute minimum when the charges are placed so that every surface is an equipotential, as happens when they are conductors.

1.16   Prove the following theorem: If a number of conducting surfaces are fixed in position with a given total charge on each, the introduction of an uncharged, insulated conductor into the region bounded by the surfaces lowers the electrostatic energy.

1.17   A volume V in vacuum is bounded by a surface S consisting of several separate conducting surfaces S_i. One conductor is held at *unit* potential and all the other conductors at zero potential.

(a)   Show that the capacitance of the one conductor is given by

$$C = \epsilon_0 \int_V |\nabla \Phi|^2 \, d^3x$$

where $\Phi(\mathbf{x})$ is the solution for the potential.

(b)   Show that the true capacitance C is always less than or equal to the quantity

$$C[\Psi] = \epsilon_0 \int_V |\nabla \Psi|^2 \, d^3x$$

where Ψ is any trial function satisfying the boundary conditions on the conductors. This is a variational principle for the capacitance that yields an *upper bound*.

1.18   Consider the configuration of conductors of Problem 1.17, with all conductors except S_1 held at zero potential.

(a)   Show that the potential $\Phi(\mathbf{x})$ anywhere in the volume V *and* on any of the surfaces S_i can be written

$$\Phi(\mathbf{x}) = \frac{1}{4\pi\epsilon_0} \oint_{S_1} \sigma_1(\mathbf{x}') G(\mathbf{x}, \mathbf{x}') \, d^3x'$$

where $\sigma_1(\mathbf{x}')$ is the surface charge density on S_1 and $G(\mathbf{x}, \mathbf{x}')$ is the Green function potential for a point charge in the presence of all the surfaces that are held at zero potential (but with S_1 absent). Show also that the electrostatic energy is

$$W = \frac{1}{8\pi\epsilon_0} \oint_{S_1} da \oint_{S_1} da' \, \sigma_1(\mathbf{x}) G(\mathbf{x}, \mathbf{x}') \sigma_1(\mathbf{x}')$$

where the integrals are only over the surface S_1.

(b)   Show that the variational expression

$$C^{-1}[\sigma] = \frac{\displaystyle\oint_{S_1} da \oint_{S_1} da' \sigma(\mathbf{x}) G(\mathbf{x}, \mathbf{x}') \sigma(\mathbf{x}')}{\displaystyle 4\pi\epsilon_0 \left[\oint_{S_1} \sigma(\mathbf{x}) \, da \right]^2}$$

with an arbitrary integrable function $\sigma(\mathbf{x})$ defined on S_1, is stationary for small variations of σ away from σ_1. Use Thomson's theorem to prove that the reciprocal of $C^{-1}[\sigma]$ gives a *lower bound* to the true capacitance of the conductor S_1.

1.19   For the cylindrical capacitor of Problem 1.6c, evaluate the variational upper bound of Problem 1.17b with the naive trial function, $\Psi_1(\rho) = (b - \rho)/(b - a)$. Compare the variational result with the exact result for $b/a = 1.5, 2, 3$. Explain the trend of your results in terms of the functional form of Ψ_1. An improved trial function is treated by *Collin* (pp. 151–152).

1.20   In estimating the capacitance of a given configuration of conductors, comparison with known capacitances is often helpful. Consider two configurations of n conductors in which the $(n - 1)$ conductors held at zero potential are the same, but the one conductor whose capacitance we wish to know is different. In particular, let the conductor in one configuration have a closed surface S_1 and in the other configuration have surface S_1', with S_1' totally inside S_1.

(a)   Use the extremum principle of Section 1.12 and the variational principle of Problem 1.17 to prove that the capacitance C' of the conductor with surface S_1' is less than or equal to the capacitance C of the conductor with surface S_1 that encloses S_1'.

(b)   Set upper and lower limits for the capacitance of a conducting cube of side a. Compare your limits and also their average with the numerical value, $C \simeq 0.655(4\pi\epsilon_0 a)$.

(c)   By how much do you estimate the capacitance per unit length of the two-wire system of Problem 1.7 will change (larger? smaller?) if *one* of the wires is replaced by a wire of square cross section whose side is equal to the diameter of the other wire?

1.21   A two-dimensional potential problem consists of a unit square area ($0 \le x \le 1$, $0 \le y \le 1$) bounded by "surfaces" held at zero potential. Over the entire square there is a uniform charge density of unit strength (per unit length in z).

(a)   Apply the variational principle (1.63) for the Poisson equation with the "variational" trial function $\Psi(x, y) = A \cdot x(1 - x) \cdot y(1 - y)$ to determine the best value of the constant A. [I use quotation marks around variational because there are no parameters to vary except the overall scale.]

(b)   The exact (albeit series) solution for this problem is [see Problems 2.15 and 2.16]

$$4\pi\epsilon_0\Phi(x, y) = \frac{16}{\pi^2} \sum_{m=0}^{\infty} \frac{\sin[(2m + 1)\pi x]}{(2m + 1)^3} \left\{ 1 - \frac{\cosh[(2m + 1)\pi(y - \frac{1}{2})]}{\cosh[(2m + 1)\pi/2]} \right\}$$

For $y = 0.25$ and $y = 0.5$, plot and compare the simple variational solution of part a with the exact solution as functions of x.

1.22   Two-dimensional relaxation calculations commonly use sites on a square lattice with spacing $\Delta x = \Delta y = h$, and label the sites by (i, j), where i, j are integers and $x_i = ih + x_0$, $y_j = jh + y_0$. The value of the potential at (i, j) can be approximated by the average of the values at neighboring sites. [Recall the relevant theorem about harmonic functions.] But what average?

(a)   If $F(x, y)$ is a well-behaved function in the neighborhood of the origin, but not necessarily harmonic, by explicit Taylor series expansions, show that the "cross" sum

$$S_c = F(h, 0) + F(0, h) + F(-h, 0) + F(0, -h)$$

can be expressed as

$$S_c = 4F(0, 0) + h^2\nabla^2 F + \frac{h^4}{12} (F_{xxxx} + F_{yyyy}) + O(h^6)$$

(b) Similarly, show that the "square" sum,

$$S_S = F(h, h) + F(-h, h) + F(-h, -h) + F(h, -h)$$

can be expressed as

$$S_S = 4F(0, 0) + 2h^2\nabla^2 F - \frac{h^4}{3}(F_{xxxx} + F_{yyyy}) + \frac{h^4}{2}\nabla^2(\nabla^2 F) + O(h^6)$$

Here F_{xxxx} is the fourth partial derivative of F with respect to x, evaluated at $x = 0, y = 0$, etc. If $\nabla^2 F = 0$, the averages $S_c/4$ and $S_s/4$ each give the value of $F(0, 0)$, correct to order h^3 inclusive. Note that an improvement can be obtained by forming the "improved" average,

$$\tilde{S} = \frac{1}{5}\left[S_c + \frac{1}{4}S_s\right]$$

where

$$\tilde{S} = F(0, 0) + \frac{3}{10}h^2\nabla^2 F + \frac{h^4}{40}\nabla^2(\nabla^2 F) + O(h^6)$$

If $\nabla^2 F = 0$, then $\tilde{S}$ gives $F(0, 0)$, correct to order h^5 inclusive. For Poisson's equation, the charge density and its lowest order Laplacian can be inserted for the same accuracy.

1.23 A transmission line consists of a long straight conductor with a hollow square region in its interior, with a square conductor of one-quarter the area of the hollow region centered in the empty space, with edges parallel to the inner sides of outer conductor. If the conductors are raised to different potentials, the potential and electric field in the space between them exhibit an eightfold symmetry; the basic unit is sketched in the accompanying figure. The efficacy of the relaxation method in determining the properties of the transmission line can be illustrated by a simple calculation.

(a) Using only the four interior points indicated in the figure, write down the relaxation equation for each point for the "cross" and the "improved" averaging schemes (defined in Problem 1.22) if the inner conductor has $\Phi = 100$ V and the outer has $\Phi = 0$. By performing either the relaxation iteration process or solving the set of algebraic equations for each scheme, find estimates for the potential at each of the four points for the two schemes.

(b) From the results of part a make the best estimate (or estimates) you can for the capacitance per unit length of the transmission line.

(c) (Optional) Using your favorite computational tools, repeat the relaxation calculation with half the lattice spacing (21 interior points) and compare.

Answer: $\Phi_1 = 48.87$ V, $\Phi_2 = 47.18$ V, $\Phi_3 = 38.34$ V, $\Phi_4 = 19.81$ V and $C = 10.23$ ϵ_0 F/m [from an accurate numerical calculation].

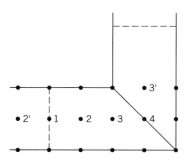

Problem 1.23

1.24   Consider solution of the two-dimensional Poisson equation problem of Problem 1.21, a unit square with zero potential on the boundary and a constant unit charge density in the interior, by the technique of relaxation. Choose $h = 0.25$ so that there are nine interior sites. Use symmetry to reduce the number of needed sites to three, at (0.25, 0.25), (0.5, 0.25), and (0.5, 0.5). With so few sites, it is easy to do the iterations with a block of paper and a pocket calculator, but suit yourself.

(a)   Use the "improved grid" averaging of Problem 1.22 and the simple (Jacobian) iteration scheme, starting with $\Phi = 1.0$ at all three interior sites. Do at least six iterations, preferably eight or ten.

(b)   Repeat the iteration procedure with the same starting values, but using Gauss–Seidel iteration.

(c)   Graph the two sets of results of each iteration versus iteration number and compare with the exact values, $4\pi\epsilon_0\Phi(0.25, 0.25) = 0.5691$, $4\pi\epsilon_0\Phi(0.5, 0.25) = 0.7205$, $4\pi\epsilon_0\Phi(0.5, 0.5) = 0.9258$. Comment on rate of convergence and final accuracy.

CHAPTER 2

Boundary-Value Problems in Electrostatics: I

Many problems in electrostatics involve boundary surfaces on which either the potential or the surface-charge density is specified. The formal solution of such problems was presented in Section 1.10, using the method of Green functions. In practical situations (or even rather idealized approximations to practical situations) the discovery of the correct Green function is sometimes easy and sometimes not. Consequently a number of approaches to electrostatic boundary-value problems have been developed, some of which are only remotely connected to the Green function method. In this chapter we will examine three of these special techniques: (1) the method of images, which is closely related to the use of Green functions; (2) expansion in orthogonal functions, an approach directly through the differential equation and rather remote from the direct construction of a Green function; (3) an introduction to finite element analysis (FEA), a broad class of numerical methods. A major omission is the use of complex-variable techniques, including conformal mapping, for the treatment of two-dimensional problems. The topic is important, but lack of space and the existence of self-contained discussions elsewhere accounts for its absence. The interested reader may consult the references cited at the end of the chapter.

2.1 Method of Images

The method of images concerns itself with the problem of one or more point charges in the presence of boundary surfaces, for example, conductors either grounded or held at fixed potentials. Under favorable conditions it is possible to infer from the geometry of the situation that a small number of suitably placed charges of appropriate magnitudes, external to the region of interest, can simulate the required boundary conditions. These charges are called *image charges*, and the replacement of the actual problem with boundaries by an enlarged region with image charges but not boundaries is called the *method of images*. The image charges must be external to the volume of interest, since their potentials must be solutions of the Laplace equation inside the volume; the "particular integral" (i.e., solution of the Poisson equation) is provided by the sum of the potentials of the charges inside the volume.

A simple example is a point charge located in front of an infinite plane conductor at zero potential, as shown in Fig. 2.1. It is clear that this is equivalent to the problem of the original charge and an equal and opposite charge located at the mirror-image point behind the plane defined by the position of the conductor.

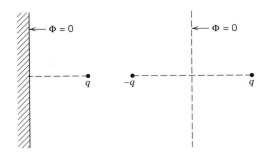

Figure 2.1   Solution by method of images. The original potential problem is on the left, the equivalent-image problem on the right.

2.2   *Point Charge in the Presence of a Grounded Conducting Sphere*

As an illustration of the method of images we consider the problem illustrated in Fig. 2.2 of a point charge q located at $\mathbf{y}$ relative to the origin, around which is centered a grounded conducting sphere of radius a. We seek the potential $\Phi(\mathbf{x})$ such that $\Phi(|\mathbf{x}| = a) = 0$. By symmetry it is evident that the image charge q' (assuming that only one image is needed) will lie on the ray from the origin to the charge q. If we consider the charge q *outside* the sphere, the image position $\mathbf{y}'$ will lie inside the sphere. The potential due to the charges q and q' is:

$$\Phi(\mathbf{x}) = \frac{q/4\pi\epsilon_0}{|\mathbf{x} - \mathbf{y}|} + \frac{q'/4\pi\epsilon_0}{|\mathbf{x} - \mathbf{y}'|} \tag{2.1}$$

We now must try to choose q' and $|\mathbf{y}'|$ such that this potential vanishes at $|\mathbf{x}| = a$. If $\mathbf{n}$ is a unit vector in the direction $\mathbf{x}$, and $\mathbf{n}'$ a unit vector in the direction $\mathbf{y}$, then

$$\Phi(\mathbf{x}) = \frac{q/4\pi\epsilon_0}{|x\mathbf{n} - y\mathbf{n}'|} + \frac{q'/4\pi\epsilon_0}{|x\mathbf{n} - y'\mathbf{n}'|} \tag{2.2}$$

If x is factored out of the first term and y' out of the second, the potential at $x = a$ becomes:

$$\Phi(x = a) = \frac{q/4\pi\epsilon_0}{a\left|\mathbf{n} - \dfrac{y}{a}\mathbf{n}'\right|} + \frac{q'/4\pi\epsilon_0}{y'\left|\mathbf{n}' - \dfrac{a}{y'}\mathbf{n}\right|} \tag{2.3}$$

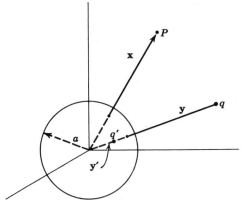

Figure 2.2   Conducting sphere of radius a, with charge q and image charge q'.

From the form of (2.3) it will be seen that the choices:

$$\frac{q}{a} = -\frac{q'}{y'}, \qquad \frac{y}{a} = \frac{a}{y'}$$

make $\Phi(x = a) = 0$, for all possible values of $\mathbf{n} \cdot \mathbf{n}'$. Hence the magnitude and position of the image charge are

$$q' = -\frac{a}{y} q, \qquad y' = \frac{a^2}{y} \tag{2.4}$$

We note that, as the charge q is brought closer to the sphere, the image charge grows in magnitude and moves out from the center of the sphere. When q is just outside the surface of the sphere, the image charge is equal and opposite in magnitude and lies just beneath the surface.

Now that the image charge has been found, we can return to the original problem of a charge q outside a grounded conducting sphere and consider various effects. The actual charge density induced on the surface of the sphere can be calculated from the normal derivative of Φ at the surface:

$$\sigma = -\epsilon_0 \frac{\partial \Phi}{\partial x}\bigg|_{x=a} = -\frac{q}{4\pi a^2} \left(\frac{a}{y}\right) \frac{1 - \dfrac{a^2}{y^2}}{\left(1 + \dfrac{a^2}{y^2} - 2\dfrac{a}{y}\cos\gamma\right)^{3/2}} \tag{2.5}$$

where γ is the angle between $\mathbf{x}$ and $\mathbf{y}$. This charge density in units of $-q/4\pi a^2$ is shown plotted in Fig. 2.3 as a function of γ for two values of y/a. The concentra-

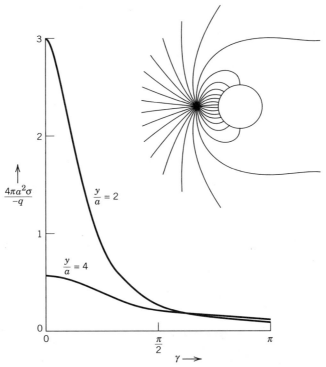

Figure 2.3   Surface-charge density σ induced on the grounded sphere of radius a as a result of the presence of a point charge q located a distance y away from the center of the sphere. σ is plotted in units of $-q/4\pi a^2$ as a function of the angular position γ away from the radius to the charge for $y = 2a$, $4a$. Inset shows lines of force for $y = 2a$.

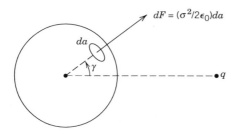

Figure 2.4

tion of charge in the direction of the point charge q is evident, especially for $y/a = 2$. It is easy to show by direct integration that the total induced charge on the sphere is equal to the magnitude of the image charge, as it must be, according to Gauss's law.

The force acting on the charge q can be calculated in different ways. One (the easiest) way is to write down immediately the force between the charge q and the image charge q'. The distance between them is $y - y' = y(1 - a^2/y^2)$. Hence the attractive force, according to Coulomb's law, is:

$$|\mathbf{F}| = \frac{1}{4\pi\epsilon_0} \frac{q^2}{a^2} \left(\frac{a}{y}\right)^3 \left(1 - \frac{a^2}{y^2}\right)^{-2} \qquad (2.6)$$

For large separations the force is an inverse cube law, but close to the sphere it is proportional to the inverse square of the distance away from the surface of the sphere.

The alternative method for obtaining the force is to calculate the total force acting on the surface of the sphere. The force on each element of area da is $(\sigma^2/2\epsilon_0) \, da$, where σ is given by (2.5), as indicated in Fig. 2.4. But from symmetry it is clear that only the component parallel to the radius vector from the center of the sphere to q contributes to the total force. Hence the total force acting on the sphere (equal and opposite to the force acting on q) is given by the integral:

$$|\mathbf{F}| = \frac{q^2}{32\pi^2\epsilon_0 a^2} \left(\frac{a}{y}\right)^2 \left(1 - \frac{a^2}{y^2}\right)^2 \int \frac{\cos\gamma}{\left(1 + \frac{a^2}{y^2} - \frac{2a}{y}\cos\gamma\right)^3} \, d\Omega \qquad (2.7)$$

Integration immediately yields (2.6).

The whole discussion has been based on the understanding that the point charge q is *outside* the sphere. Actually, the results apply equally for the charge q *inside* the sphere. The only change necessary is in the surface-charge density (2.5), where the normal derivative out of the conductor is now radially inward, implying a change in sign. The reader may transcribe all the formulas, remembering that now $y \leq a$. The angular distributions of surface charge are similar to those of Fig. 2.3, but the total induced surface charge is evidently equal to $-q$, independent of y.

2.3   *Point Charge in the Presence of a Charged, Insulated, Conducting Sphere*

In the preceding section we considered the problem of a point charge q near a grounded sphere and saw that a surface-charge density was induced on the sphere. This charge was of total amount $q' = -aq/y$, and was distributed over the surface in such a way as to be in equilibrium under all forces acting.

If we wish to consider the problem of an insulated conducting sphere with total charge Q in the presence of a point charge q, we can build up the solution for the potential by linear superposition. In an operational sense, we can imagine that we start with the grounded conducting sphere (with its charge q' distributed over its surface). We then disconnect the ground wire and add to the sphere an amount of charge $(Q - q')$. This brings the total charge on the sphere up to Q. To find the potential we merely note that the added charge $(Q - q')$ will distribute itself *uniformly* over the surface, since the electrostatic forces due to the point charge q are already balanced by the charge q'. Hence the potential due to the added charge $(Q - q')$ will be the same as if a point charge of that magnitude were at the origin, at least for points outside the sphere.

The potential is the superposition of (2.1) and the potential of a point charge $(Q - q')$ at the origin:

$$\Phi(\mathbf{x}) = \frac{1}{4\pi\epsilon_0}\left[\frac{q}{|\mathbf{x} - \mathbf{y}|} - \frac{aq}{y\left|\mathbf{x} - \dfrac{a^2}{y^2}\mathbf{y}\right|} + \frac{Q + \dfrac{a}{y}q}{|\mathbf{x}|}\right] \tag{2.8}$$

The force acting on the charge q can be written down directly from Coulomb's law. It is directed along the radius vector to q and has the magnitude:

$$\mathbf{F} = \frac{1}{4\pi\epsilon_0}\frac{q}{y^2}\left[Q - \frac{qa^3(2y^2 - a^2)}{y(y^2 - a^2)^2}\right]\frac{\mathbf{y}}{y} \tag{2.9}$$

In the limit of $y \gg a$, the force reduces to the usual Coulomb's law for two small charged bodies. But close to the sphere the force is modified because of the induced charge distribution on the surface of the sphere. Figure 2.5 shows the force as a function of distance for various ratios of Q/q. The force is expressed in units of $q^2/4\pi\epsilon_0 y^2$; positive (negative) values correspond to a repulsion (attraction). If the sphere is charged oppositely to q, or is uncharged, the force is attractive at all distances. Even if the charge Q is the same sign as q, however, the force becomes attractive at very close distances. In the limit of $Q \gg q$, the point of zero force (unstable equilibrium point) is very close to the sphere, namely, at $y \simeq a(1 + \frac{1}{2}\sqrt{q/Q})$. Note that the asymptotic value of the force is attained as soon as the charge q is more than a few radii away from the sphere.

This example exhibits a general property that explains why an excess of charge on the surface does not immediately leave the surface because of mutual repulsion of the individual charges. As soon as an element of charge is removed from the surface, the image force tends to attract it back. If sufficient work is done, of course, charge can be removed from the surface to infinity. The work function of a metal is in large part just the work done against the attractive image force to remove an electron from the surface.

2.4   Point Charge Near a Conducting Sphere at Fixed Potential

Another problem that can be discussed easily is that of a point charge near a conducting sphere held at a fixed potential V. The potential is the same as for

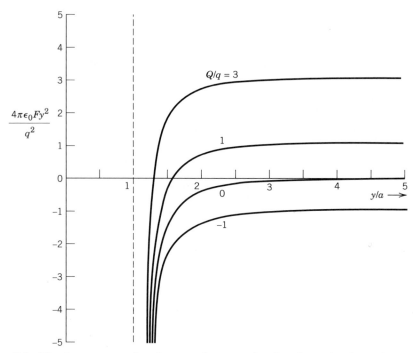

Figure 2.5 The force on a point charge q due to an insulated, conducting sphere of radius a carrying a total charge Q. Positive values mean a repulsion, negative an attraction. The asymptotic dependence of the force has been divided out. $4\pi\epsilon_0 Fy^2/q^2$ is plotted versus y/a for $Q/q = -1, 0, 1, 3$. Regardless of the value of Q, the force is always attractive at close distances because of the induced surface charge.

the charged sphere, except that the charge $(Q - q')$ at the center is replaced by a charge (Va). This can be seen from (2.8), since at $|\mathbf{x}| = a$ the first two terms cancel and the last term will be equal to V as required. Thus the potential is

$$\Phi(\mathbf{x}) = \frac{1}{4\pi\epsilon_0}\left[\frac{q}{|\mathbf{x} - \mathbf{y}|} - \frac{aq}{y\left|\mathbf{x} - \dfrac{a^2}{y^2}\mathbf{y}\right|}\right] + \frac{Va}{|\mathbf{x}|} \tag{2.10}$$

The force on the charge q due to the sphere at fixed potential is

$$\mathbf{F} = \frac{q}{y^2}\left[Va - \frac{1}{4\pi\epsilon_0}\frac{qay^3}{(y^2 - a^2)^2}\right]\frac{\mathbf{y}}{y} \tag{2.11}$$

For corresponding values of $4\pi\epsilon_0 Va/q$ and Q/q this force is very similar to that of the charged sphere, shown in Fig. 2.5, although the approach to the asymptotic value (Vaq/y^2) is more gradual. For $Va \gg q$, the unstable equilibrium point has the equivalent location $y \simeq a(1 + \frac{1}{2}\sqrt{q/4\pi\epsilon_0 Va})$.

2.5 *Conducting Sphere in a Uniform Electric Field by Method of Images*

As a final example of the method of images we consider a conducting sphere of radius a in a uniform electric field E_0. A uniform field can be thought of as being produced by appropriate positive and negative charges at infinity. For example, if there are two charges $\pm Q$, located at positions $z = \mp R$, as shown in Fig. 2.6a, then in a region near the origin whose dimensions are very small compared to R there is an approximately constant electric field $E_0 \simeq 2Q/4\pi\epsilon_0 R^2$ parallel to the z axis. In the limit as $R, Q \to \infty$, with Q/R^2 constant, this approximation becomes exact.

If now a conducting sphere of radius a is placed at the origin, the potential will be that due to the charges $\pm Q$ at $\mp R$ and their images $\mp Qa/R$ at $z = \mp a^2/R$:

$$\Phi = \frac{Q/4\pi\epsilon_0}{(r^2 + R^2 + 2rR\cos\theta)^{1/2}} - \frac{Q/4\pi\epsilon_0}{(r^2 + R^2 - 2rR\cos\theta)^{1/2}}$$

$$- \frac{aQ/4\pi\epsilon_0}{R\left(r^2 + \dfrac{a^4}{R^2} + \dfrac{2a^2 r}{R}\cos\theta\right)^{1/2}} + \frac{aQ/4\pi\epsilon_0}{R\left(r^2 + \dfrac{a^4}{R^2} - \dfrac{2a^2 r}{R}\cos\theta\right)^{1/2}} \qquad (2.12)$$

where Φ has been expressed in terms of the spherical coordinates of the observation point. In the first two terms R is much larger than r by assumption. Hence we can expand the radicals after factoring out R^2. Similarly, in the third and fourth terms, we can factor out r^2 and then expand. The result is:

$$\Phi = \frac{1}{4\pi\epsilon_0}\left[-\frac{2Q}{R^2}r\cos\theta + \frac{2Q}{R^2}\frac{a^3}{r^2}\cos\theta\right] + \cdots \qquad (2.13)$$

where the omitted terms vanish in the limit $R \to \infty$. In that limit $2Q/4\pi\epsilon_0 R^2$ becomes the applied uniform field, so that the potential is

$$\Phi = -E_0\left(r - \frac{a^3}{r^2}\right)\cos\theta \qquad (2.14)$$

The first term $(-E_0 z)$ is, of course, just the potential of a uniform field E_0 which could have been written down directly instead of the first two terms in (2.12).

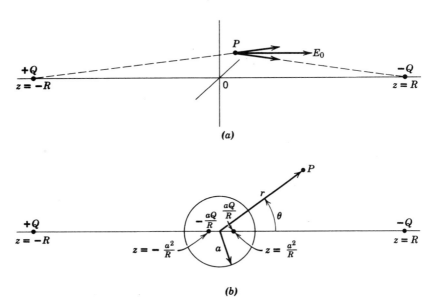

(a)

(b)

Figure 2.6 Conducting sphere in a uniform electric field by the method of images.

The second is the potential due to the induced surface-charge density or, equivalently, the image charges. Note that the image charges form a dipole of strength $D = Qa/R \times 2a^2/R = 4\pi\epsilon_0 E_0 a^3$. The induced surface-charge density is

$$\sigma = -\epsilon_0 \left.\frac{\partial \Phi}{\partial r}\right|_{r=a} = 3\epsilon_0 E_0 \cos\theta \tag{2.15}$$

We note that the surface integral of this charge density vanishes, so that there is no difference between a grounded and an insulated sphere.

2.6   *Green Function for the Sphere; General Solution for the Potential*

In preceding sections the problem of a conducting sphere in the presence of a point charge was discussed by the method of images. As mentioned in Section 1.10, the potential due to a unit source and its image (or images), chosen to satisfy homogeneous boundary conditions, is just the Green function (1.43 or 1.45) appropriate for Dirichlet or Neumann boundary conditions. In $G(\mathbf{x}, \mathbf{x}')$ the variable $\mathbf{x}'$ refers to the location P' of the unit source, while the variable $\mathbf{x}$ is the point P at which the potential is being evaluated. These coordinates and the sphere are shown in Fig. 2.7. For Dirichlet boundary conditions on the sphere of radius a the Green function defined via (1.39) for a unit source and its image is given by (2.1) with $q \to 4\pi\epsilon_0$ and relations (2.4). Transforming variables appropriately, we obtain the Green function:

$$G(\mathbf{x}, \mathbf{x}') = \frac{1}{|\mathbf{x} - \mathbf{x}'|} - \frac{a}{x'\left|\mathbf{x} - \dfrac{a^2}{x'^2}\mathbf{x}'\right|} \tag{2.16}$$

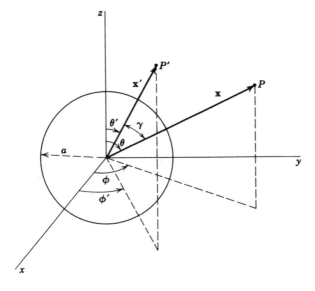

Figure 2.7

In terms of spherical coordinates this can be written:

$$G(\mathbf{x}, \mathbf{x}') = \frac{1}{(x^2 + x'^2 - 2xx' \cos \gamma)^{1/2}} - \frac{1}{\left(\dfrac{x^2 x'^2}{a^2} + a^2 - 2xx' \cos \gamma\right)^{1/2}} \quad (2.17)$$

where γ is the angle between $\mathbf{x}$ and $\mathbf{x}'$. The symmetry in the variables $\mathbf{x}$ and $\mathbf{x}'$ is obvious in the form (2.17), as is the condition that $G = 0$ if either $\mathbf{x}$ or $\mathbf{x}'$ is on the surface of the sphere.

For solution (1.44) of the Poisson equation we need not only G, but also $\partial G/\partial n'$. Remembering that $\mathbf{n}'$ is the unit normal outward from the volume of interest (i.e., inward along $\mathbf{x}'$ toward the origin), we have

$$\left.\frac{\partial G}{\partial n'}\right|_{x'=a} = -\frac{(x^2 - a^2)}{a(x^2 + a^2 - 2ax \cos \gamma)^{3/2}} \quad (2.18)$$

[Note that this is essentially the induced surface-charge density (2.5).] Hence the solution of the Laplace equation *outside* a sphere with the potential specified on its surface is, according to (1.44),

$$\Phi(\mathbf{x}) = \frac{1}{4\pi} \int \Phi(a, \theta', \phi') \frac{a(x^2 - a^2)}{(x^2 + a^2 - 2ax \cos \gamma)^{3/2}} \, d\Omega' \quad (2.19)$$

where $d\Omega'$ is the element of solid angle at the point (a, θ', ϕ') and $\cos \gamma = \cos \theta \cos \theta' + \sin \theta \sin \theta' \cos(\phi - \phi')$. For the *interior* problem, the normal derivative is radially outward, so that the sign of $\partial G/\partial n'$ is opposite to (2.18). This is equivalent to replacing the factor $(x^2 - a^2)$ by $(a^2 - x^2)$ in (2.19). For a problem with a charge distribution, we must add to (2.19) the appropriate integral in (1.44), with the Green function (2.17).

2.7   *Conducting Sphere with Hemispheres at Different Potentials*

As an example of the solution (2.19) for the potential outside a sphere with prescribed values of potential on its surface, we consider the conducting sphere of radius a made up of two hemispherical shells separated by a small insulating ring. The hemispheres are kept at different potentials. It will suffice to consider the potentials as $\pm V$, since arbitrary potentials can be handled by superposition of the solution for a sphere at fixed potential over its whole surface. The insulating ring lies in the $z = 0$ plane, as shown in Fig. 2.8, with the upper (lower) hemisphere at potential $+V$ $(-V)$.

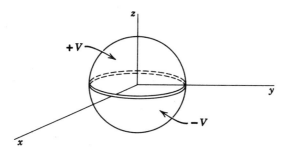

Figure 2.8

From (2.19) the solution for $\Phi(x, \theta, \phi)$ is given by the integral:

$$\Phi(x, \theta, \phi) = \frac{V}{4\pi} \int_0^{2\pi} d\phi' \left\{ \int_0^1 d(\cos\theta') \right.$$
$$\left. - \int_{-1}^0 d(\cos\theta') \right\} \frac{a(x^2 - a^2)}{(a^2 + x^2 - 2ax\cos\gamma)^{3/2}} \tag{2.20}$$

By a suitable change of variables in the second integral $(\theta' \to \pi - \theta', \phi' \to \phi' + \pi)$, this can be cast in the form:

$$\Phi(x, \theta, \phi) = \frac{Va(x^2 - a^2)}{4\pi} \int_0^{2\pi} d\phi' \int_0^1 d(\cos\theta')[(a^2 + x^2 - 2ax\cos\gamma)^{-3/2}$$
$$-(a^2 + x^2 + 2ax\cos\gamma)^{-3/2}] \tag{2.21}$$

Because of the complicated dependence of $\cos\gamma$ on the angles (θ', ϕ') and (θ, ϕ), equation (2.21) cannot in general be integrated in closed form.

As a special case we consider the potential on the positive z axis. Then $\cos\gamma = \cos\theta'$, since $\theta = 0$. The integration is elementary, and the potential can be shown to be

$$\Phi(z) = V\left[1 - \frac{(z^2 - a^2)}{z\sqrt{z^2 + a^2}}\right] \tag{2.22}$$

At $z = a$, this reduces to $\Phi = V$ as required, while at large distances it goes asymptotically as $\Phi \simeq 3Va^2/2z^2$.

In the absence of a closed expression for the integrals in (2.21), we can expand the denominator in power series and integrate term by term. Factoring out $(a^2 + x^2)$ from each denominator, we obtain

$$\Phi(x, \theta, \phi) = \frac{Va(x^2 - a^2)}{4\pi(a^2 + x^2)^{3/2}} \int_0^{2\pi} d\phi' \int_0^1 d(\cos\theta')[(1 - 2\alpha\cos\gamma)^{-3/2}$$
$$- (1 + 2\alpha\cos\gamma)^{-3/2}] \tag{2.23}$$

where $\alpha = ax/(a^2 + x^2)$. We observe that in the expansion of the radicals only odd powers of $\alpha\cos\gamma$ will appear:

$$[(1 - 2\alpha\cos\gamma)^{-3/2} - (1 + 2\alpha\cos\gamma)^{-3/2}] = 6\alpha\cos\gamma + 35\alpha^3\cos^3\gamma + \cdots \tag{2.24}$$

It is now necessary to integrate odd powers of $\cos\gamma$ over $d\phi'\, d(\cos\theta')$:

$$\left.\begin{aligned}
\int_0^{2\pi} d\phi' \int_0^1 d(\cos\theta')\cos\gamma &= \pi\cos\theta \\
\int_0^{2\pi} d\phi' \int_0^1 d(\cos\theta')\cos^3\gamma &= \frac{\pi}{4}\cos\theta(3 - \cos^2\theta)
\end{aligned}\right\} \tag{2.25}$$

If (2.24) and (2.25) are inserted into (2.23), the potential becomes

$$\Phi(x, \theta, \phi) = \frac{3Va^2}{2x^2}\left(\frac{x^3(x^2 - a^2)}{(x^2 + a^2)^{5/2}}\right)\cos\theta\left[1 + \frac{35}{24}\frac{a^2x^2}{(a^2 + x^2)^2}(3 - \cos^2\theta) + \cdots\right] \tag{2.26}$$

We note that only odd powers of $\cos \theta$ appear, as required by the symmetry of the problem. If the expansion parameter is (a^2/x^2), rather than α^2, the series takes on the form:

$$\Phi(x, \theta, \phi) = \frac{3Va^2}{2x^2}\left[\cos \theta - \frac{7a^2}{12x^2}\left(\frac{5}{2}\cos^3\theta - \frac{3}{2}\cos \theta\right) + \cdots\right] \quad (2.27)$$

For large values of x/a this expansion converges rapidly and so is a useful representation for the potential. Even for $x/a = 5$, the second term in the series is only of the order of 2%. It is easily verified that, for $\cos \theta = 1$, expression (2.27) agrees with the expansion of (2.22) for the potential on the axis. [The particular choice of angular factors in (2.27) is dictated by the definitions of the Legendre polynomials. The two factors are, in fact, $P_1(\cos \theta)$ and $P_3(\cos \theta)$, and the expansion of the potential is one in Legendre polynomials of odd order. We establish this in a systematic fashion in Section 3.3.] Further consideration of both the exterior and interior problem of the two hemispheres is found in Problem 2.22.

2.8   *Orthogonal Functions and Expansions*

The representation of solutions of potential problems (or any mathematical physics problem) by expansions in orthogonal functions forms a powerful technique that can be used in a large class of problems. The particular orthogonal set chosen depends on the symmetries or near symmetries involved. To recall the general properties of orthogonal functions and expansions in terms of them, we consider an interval (a, b) in a variable ξ with a set of real or complex functions $U_n(\xi)$, $n = 1, 2, \ldots$, square integrable and orthogonal on the interval (a, b). The orthogonality condition on the functions $U_n(\xi)$ is expressed by

$$\int_a^b U_n^*(\xi)U_m(\xi)\, d\xi = 0, \qquad m \neq n \quad (2.28)$$

If $n = m$, the integral is nonzero. We assume that the functions are normalized so that the integral is unity. Then the functions are said to be *orthonormal*, and they satisfy

$$\int_a^b U_n^*(\xi)U_m(\xi)\, d\xi = \delta_{nm} \quad (2.29)$$

An arbitrary function $f(\xi)$, square integrable on the interval (a, b), can be expanded in a series of the orthonormal functions $U_n(\xi)$. If the number of terms in the series is finite (say N),

$$f(\xi) \leftrightarrow \sum_{n=1}^{N} a_n U_n(\xi) \quad (2.30)$$

then we can ask for the "best" choice of coefficients a_n so that we get the "best" representation of the function $f(\xi)$. If "best" is defined as minimizing the mean square error M_N:

$$M_N = \int_a^b \left| f(\xi) - \sum_{n=1}^{N} a_n U_n(\xi) \right|^2 d\xi \quad (2.31)$$

it is easy to show that the coefficients are given by

$$a_n = \int_a^b U_n^*(\xi) f(\xi) \, d\xi \tag{2.32}$$

where the orthonormality condition (2.29) has been used. This is the standard result for the coefficients in an orthonormal function expansion.

If the number of terms N in series (2.30) is taken larger and larger, we intuitively expect that our series representation of $f(\xi)$ is "better" and "better." Our intuition will be correct provided the set of orthonormal functions is *complete*, completeness being defined by the requirement that there exist a finite number N_0 such that for $N > N_0$ the mean square error M_N can be made smaller than any arbitrarily small positive quantity. Then the series representation

$$\sum_{n=1}^{\infty} a_n U_n(\xi) = f(\xi) \tag{2.33}$$

with a_n given by (2.32) is said to *converge in the mean* to $f(\xi)$. Physicists generally leave the difficult job of proving completeness of a given set of functions to the mathematicians. All orthonormal sets of functions normally occurring in mathematical physics have been proven to be complete.

Series (2.33) can be rewritten with the explicit form (2.32) for the coefficients a_n:

$$f(\xi) = \int_a^b \left\{ \sum_{n=1}^{\infty} U_n^*(\xi') U_n(\xi) \right\} f(\xi') \, d\xi' \tag{2.34}$$

Since this represents any function $f(\xi)$ on the interval (a, b), it is clear that the sum of bilinear terms $U_n^*(\xi') U_n(\xi)$ must exist only in the neighborhood of $\xi' = \xi$. In fact, it must be true that

$$\sum_{n=1}^{\infty} U_n^*(\xi') U_n(\xi) = \delta(\xi' - \xi) \tag{2.35}$$

This is the so-called *completeness* or *closure relation*. It is analogous to the orthonormality condition (2.29), except that the roles of the continuous variable ξ and the discrete index n have been interchanged.

The most famous orthogonal functions are the sines and cosines, an expansion in terms of them being a *Fourier series*. If the interval in x is $(-a/2, a/2)$, the orthonormal functions are

$$\sqrt{\frac{2}{a}} \sin\left(\frac{2\pi mx}{a}\right), \qquad \sqrt{\frac{2}{a}} \cos\left(\frac{2\pi mx}{a}\right)$$

where m is a non-negative integer and for $m = 0$ the cosine function is $1/\sqrt{a}$. The series equivalent to (2.33) is customarily written in the form:

$$f(x) = \tfrac{1}{2}A_0 + \sum_{m=1}^{\infty} \left[A_m \cos\left(\frac{2\pi mx}{a}\right) + B_m \sin\left(\frac{2\pi mx}{a}\right) \right] \tag{2.36}$$

where

$$A_m = \frac{2}{a} \int_{-a/2}^{a/2} f(x) \cos\left(\frac{2\pi mx}{a}\right) dx$$

$$B_m = \frac{2}{a} \int_{-a/2}^{a/2} f(x) \sin\left(\frac{2\pi mx}{a}\right) dx \tag{2.37}$$

If the interval spanned by the orthonormal set has more than one dimension, formulas (2.28)–(2.33) have obvious generalizations. Suppose that the space is two-dimensional, and the variable ξ ranges over the interval (a, b) while the variable η has the interval (c, d). The orthonormal functions in each dimension are $U_n(\xi)$ and $V_m(\eta)$. Then the expansion of an arbitrary function $f(\xi, \eta)$ is

$$f(\xi, \eta) = \sum_n \sum_m a_{nm} U_n(\xi) V_m(\eta) \tag{2.38}$$

where

$$a_{nm} = \int_a^b d\xi \int_c^d d\eta \, U_n^*(\xi) V_m^*(\eta) f(\xi, \eta) \tag{2.39}$$

If the interval (a, b) becomes infinite, the set of orthogonal functions $U_n(\xi)$ may become a continuum of functions, rather than a denumerable set. Then the Kronecker delta symbol in (2.29) becomes a Dirac delta function. An important example is the *Fourier integral*. Start with the orthonormal set of complex exponentials,

$$U_m(x) = \frac{1}{\sqrt{a}} e^{i(2\pi mx/a)} \tag{2.40}$$

$m = 0, \pm 1, \pm 2, \ldots$, on the interval $(-a/2, a/2)$, with the expansion:

$$f(x) = \frac{1}{\sqrt{a}} \sum_{m=-\infty}^{\infty} A_m e^{i(2\pi mx/a)} \tag{2.41}$$

where

$$A_m = \frac{1}{\sqrt{a}} \int_{-a/2}^{a/2} e^{-i(2\pi mx'/a)} f(x') \, dx' \tag{2.42}$$

Then let the interval become infinite ($a \to \infty$), at the same time transforming

$$\left. \begin{array}{c} \dfrac{2\pi m}{a} \to k \\[2ex] \displaystyle\sum_m \to \int_{-\infty}^{\infty} dm = \frac{a}{2\pi} \int_{-\infty}^{\infty} dk \\[2ex] A_m \to \sqrt{\dfrac{2\pi}{a}} A(k) \end{array} \right\} \tag{2.43}$$

The resulting expansion, equivalent to (2.41), is the *Fourier integral*,

$$f(x) = \frac{1}{\sqrt{2\pi}} \int_{-\infty}^{\infty} A(k) e^{ikx} \, dk \tag{2.44}$$

where

$$A(k) = \frac{1}{\sqrt{2\pi}} \int_{-\infty}^{\infty} e^{-ikx} f(x) \, dx \tag{2.45}$$

The orthogonality condition is

$$\frac{1}{2\pi} \int_{-\infty}^{\infty} e^{i(k-k')x} \, dx = \delta(k - k') \tag{2.46}$$

while the completeness relation is

$$\frac{1}{2\pi} \int_{-\infty}^{\infty} e^{ik(x-x')} \, dk = \delta(x - x') \tag{2.47}$$

These last integrals serve as convenient representations of a delta function. We note in (2.44)–(2.47) the complete equivalence of the two continuous variables x and k.

2.9 Separation of Variables; Laplace Equation in Rectangular Coordinates

The partial differential equations of mathematical physics are often solved conveniently by a method called *separation of variables*. In the process, one often generates orthogonal sets of functions that are useful in their own right. Equations involving the three-dimensional Laplacian operator are known to be separable in eleven different coordinate systems (see *Morse and Feshbach*, pp. 509, 655). We discuss only three of these in any detail—rectangular, spherical, and cylindrical—beginning with the simplest, rectangular coordinates.

The Laplace equation in rectangular coordinates is

$$\frac{\partial^2 \Phi}{\partial x^2} + \frac{\partial^2 \Phi}{\partial y^2} + \frac{\partial^2 \Phi}{\partial z^2} = 0 \tag{2.48}$$

A solution of this *partial* differential equation can be found in terms of three *ordinary* differential equations, all of the same form, by the assumption that the potential can be represented by a product of three functions, one for each coordinate:

$$\Phi(x, y, z) = X(x)Y(y)Z(z) \tag{2.49}$$

Substitution into (2.48) and division of the result by (2.49) yields

$$\frac{1}{X(x)} \frac{d^2 X}{dx^2} + \frac{1}{Y(y)} \frac{d^2 Y}{dy^2} + \frac{1}{Z(z)} \frac{d^2 Z}{dz^2} = 0 \tag{2.50}$$

where total derivatives have replaced partial derivatives, since each term involves a function of one variable only. If (2.50) is to hold for arbitrary values of the independent coordinates, each of the three terms must be separately constant:

$$\left. \begin{aligned} \frac{1}{X} \frac{d^2 X}{dx^2} &= -\alpha^2 \\[2mm] \frac{1}{Y} \frac{d^2 Y}{dy^2} &= -\beta^2 \\[2mm] \frac{1}{Z} \frac{d^2 Z}{dz^2} &= \gamma^2 \end{aligned} \right\} \tag{2.51}$$

where

$$\alpha^2 + \beta^2 = \gamma^2$$

If we arbitrarily choose α^2 and β^2 to be positive, then the solutions of the three ordinary differential equations (2.51) are $e^{\pm i\alpha x}$, $e^{\pm i\beta y}$, $e^{\pm \sqrt{\alpha^2+\beta^2}\,z}$. The potential (2.49) can thus be built up from the product solutions:

$$\Phi = e^{\pm i\alpha x} e^{\pm i\beta y} e^{\pm \sqrt{\alpha^2+\beta^2}\,z} \tag{2.52}$$

At this stage α and β are completely arbitrary. Consequently (2.52), by linear superposition, represents a very large class of solutions to the Laplace equation.

To determine α and β it is necessary to impose specific boundary conditions on the potential. As an example, consider a rectangular box, located as shown in Fig. 2.9, with dimensions (a, b, c) in the (x, y, z) directions. All surfaces of the box are kept at zero potential, except the surface $z = c$, which is at a potential $V(x, y)$. It is required to find the potential everywhere inside the box. Starting with the requirement that $\Phi = 0$ for $x = 0$, $y = 0$, $z = 0$, it is easy to see that the required forms of X, Y, Z are

$$\left.\begin{aligned}
X &= \sin \alpha x \\
Y &= \sin \beta y \\
Z &= \sinh(\sqrt{\alpha^2 + \beta^2}z)
\end{aligned}\right\} \tag{2.53}$$

To have $\Phi = 0$ at $x = a$ and $y = b$, we must have $\alpha a = n\pi$ and $\beta b = m\pi$. With the definitions,

$$\left.\begin{aligned}
\alpha_n &= \frac{n\pi}{a} \\[1em]
\beta_m &= \frac{m\pi}{b} \\[1em]
\gamma_{nm} &= \pi\sqrt{\frac{n^2}{a^2} + \frac{m^2}{b^2}}
\end{aligned}\right\} \tag{2.54}$$

we can write the partial potential Φ_{nm}, satisfying all the boundary conditions except one,

$$\Phi_{nm} = \sin(\alpha_n x) \sin(\beta_m y) \sinh(\gamma_{nm} z) \tag{2.55}$$

The potential can be expanded in terms of these Φ_{nm} with initially arbitrary coefficients (to be chosen to satisfy the final boundary condition):

$$\Phi(x, y, z) = \sum_{n,m=1}^{\infty} A_{nm} \sin(\alpha_n x) \sin(\beta_m y) \sinh(\gamma_{nm} z) \tag{2.56}$$

There remains only the boundary condition $\Phi = V(x, y)$ at $z = c$:

$$V(x, y) = \sum_{n,m=1}^{\infty} A_{nm} \sin(\alpha_n x) \sin(\beta_m y) \sinh(\gamma_{nm} c) \tag{2.57}$$

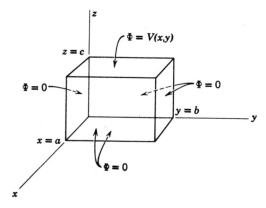

Figure 2.9   Hollow, rectangular box with five sides at zero potential, while the sixth ($z = c$) has the specified potential $\Phi = V(x, y)$.

This is just a double Fourier series for the function $V(x, y)$. Consequently the coefficients A_{nm} are given by:

$$A_{nm} = \frac{4}{ab \sinh(\gamma_{nm}c)} \int_0^a dx \int_0^b dy \, V(x, y) \sin(\alpha_n x) \sin(\beta_m y) \qquad (2.58)$$

If the rectangular box has potentials different from zero on all six sides, the required solution for the potential inside the box can be obtained by a linear superposition of six solutions, one for each side, equivalent to (2.56) and (2.58). The problem of the solution of the Poisson equation, that is, the potential inside the box with a charge distribution inside, as well as prescribed boundary conditions on the surface, requires the construction of the appropriate Green function, according to (1.43) and (1.44). Discussion of this topic will be deferred until we have treated the Laplace equation in spherical and cylindrical coordinates. For the moment, we merely note that the solution given by (2.56) and (2.58) is equivalent to the surface integral in the Green function solution (1.44).

2.10 A Two-Dimensional Potential Problem; Summation of a Fourier Series

We now consider briefly the solution by separation of variables of the two-dimensional Laplace equation in Cartesian coordinates. By two-dimensional problems we mean those in which the potential can be assumed to be independent of one of the coordinates, say, z. This is usually only an approximation, but may hold true to high accuracy, as in a long uniform transmission line. If the potential is independent of z, the basic solutions of the previous section reduce to the products

$$e^{\pm i\alpha x} e^{\pm \alpha y}$$

where α is any real or complex constant. The imposition of boundary conditions on the potential will determine what values of α are permitted and the form of the linear superposition of different solutions required.

A simple problem that can be used to demonstrate the separation of variables technique and also to establish connection with the use of complex variables is indicated in Fig. 2.10. The potential in the region, $0 \le x \le a$, $y \le 0$, is desired, subject to the boundary conditions that $\Phi = 0$ at $x = 0$ and $x = a$, while $\Phi = V$ at $y = 0$ for $0 \le x \le a$ and $\Phi \to 0$ for large y. Inspection of the basic solutions shows that α is real and that, to have the potential vanish at $x = 0$ and $x = a$ for all y and as $y \to \infty$, the proper linear combinations are $e^{-\alpha y} \sin(\alpha x)$ with $\alpha = n\pi/a$. The linear combination of solutions satisfying the boundary conditions on three of the four boundary surfaces is thus

$$\Phi(x, y) = \sum_{n=1}^{\infty} A_n \exp(-n\pi y/a) \sin(n\pi x/a) \qquad (2.59)$$

The coefficients A_n are determined by the requirement that $\Phi = V$ for $y = 0$, $0 \le x \le a$. As discussed in Section 2.8, the Fourier coefficients are

$$A_n = \frac{2}{a} \int_0^a \Phi(x, 0) \sin(n\pi x/a) \, dx \qquad (2.60)$$

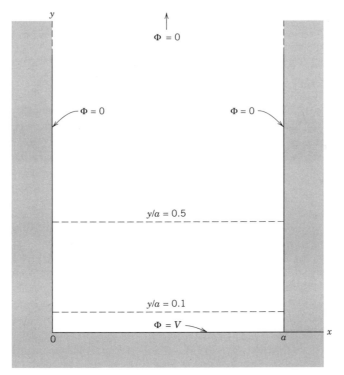

Figure 2.10   Two-dimensional potential problem.

With $\Phi(x, 0) = V$, one finds

$$A_n = \frac{4V}{\pi n} \begin{cases} 1 & \text{for } n \text{ odd} \\ 0 & \text{for } n \text{ even} \end{cases}$$

The potential $\Phi(x, y)$ is therefore determined to be

$$\Phi(x, y) = \frac{4V}{\pi} \sum_{n \text{ odd}} \frac{1}{n} \exp(-n\pi y/a) \sin(n\pi x/a) \qquad (2.61)$$

For small values of y many terms in the series are necessary to give an accurate approximation, but for $y \gtrsim a/\pi$ it is evident that only the first few terms are appreciable. The potential rapidly approaches its asymptotic form given by the first term,

$$\Phi(x, y) \rightarrow \frac{4V}{\pi} \exp(-\pi y/a) \sin(\pi x/a) \qquad (2.62)$$

Parenthetically, we remark that this general behavior is characteristic of all boundary-value problems of this type, independently of whether $\Phi(x, 0)$ is a constant, provided the first term in the series is nonvanishing. The coefficient A_1 (2.60) will be different, but the smooth behavior in x of the asymptotic solution sets in for $y \gtrsim a$, regardless of the complexities of $\Phi(x, 0)$. This is shown quantitatively for the present example in Fig. 2.11 where the potential along the two dashed lines, $y/a = 0.1, 0.5$, of Fig. 2.10 is plotted. The solid curves are the exact potential, the dotted, the first term (2.62). Close to the boundary ($y/a = 0.1$) the curves differ appreciably, but for $y/a = 0.5$ the asymptotic form is already an excellent approximation.

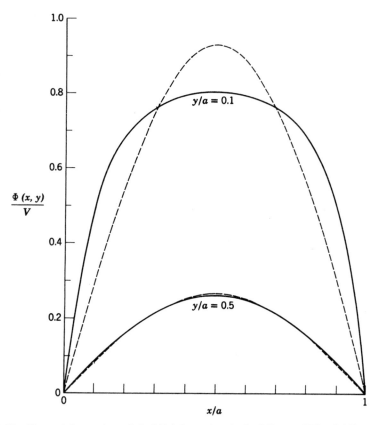

Figure 2.11 Potentials at $y/a = 0.1, 0.5$ (along the dashed lines of Fig. 2.10) as functions of x/a. The solid curves are the exact solution; the dashed curves are the first term in the series solution (2.61).

There are many Fourier series that can be summed to give an answer in closed form. The series in (2.61) is one of them. We proceed as follows. Observing that $\sin\theta = \mathrm{Im}(e^{i\theta})$, where Im stands for the imaginary part, we see that (2.61) can be written as

$$\Phi(x, y) = \frac{4V}{\pi}\, \mathrm{Im} \sum_{n \text{ odd}} \frac{1}{n}\, e^{(in\pi/a)(x+iy)}$$

With the definition,

$$Z = e^{(i\pi/a)(x+iy)} \tag{2.63}$$

this can be put in the suggestive form,

$$\Phi(x, y) = \frac{4V}{\pi}\, \mathrm{Im} \sum_{n \text{ odd}} \frac{Z^n}{n}$$

At this point we can perhaps recall the expansion,*

$$\ln(1 + Z) = Z - \tfrac{1}{2}Z^2 + \tfrac{1}{3}Z^3 - \tfrac{1}{4}Z^4 + \cdots$$

---

*Alternatively, we observe that $(d/dZ)(\sum_{n=1}^{\infty} Z^n/n) = \sum_{n=0}^{\infty} Z^n = 1/(1 - Z)$. Integration then gives $\sum_{n=1}^{\infty} Z^n/n = -\ln(1 - Z)$.

Evidently,

$$\sum_{n \text{ odd}} \frac{Z^n}{n} = \frac{1}{2} \ln\left(\frac{1 + Z}{1 - Z}\right)$$

and

$$\Phi(x, y) = \frac{2V}{\pi} \text{Im}\left[\ln\left(\frac{1 + Z}{1 - Z}\right)\right] \tag{2.64}$$

Since the imaginary part of a logarithm is equal to the phase of its argument, we consider

$$\frac{1 + Z}{1 - Z} = \frac{(1 + Z)(1 - Z^*)}{|1 - Z|^2} = \frac{1 - |Z|^2 + 2i \text{ Im } Z}{|1 - Z|^2}$$

The phase of the argument of the logarithm is thus $\tan^{-1}[2 \text{ Im } Z/(1 - |Z|^2)]$. With the explicit form (2.63) of Z substituted, it is found that the potential becomes

$$\Phi(x, y) = \frac{2V}{\pi} \tan^{-1}\left(\frac{\sin \dfrac{\pi x}{a}}{\sinh \dfrac{\pi y}{a}}\right) \tag{2.65}$$

The branch of the tangent curve corresponds to the angle lying between 0 and $\pi/2$. The infinite series (2.61) has been transformed into the explicit closed form (2.65). The reader may verify that the boundary conditions are satisfied and that the asymptotic form (2.62) emerges in a simple manner.

The potential (2.64) with Z given by (2.63) is obviously related to functions of a complex variable. This connection is a direct consequence of the fact that the real or the imaginary part of an analytic function satisfies the Laplace equation in two dimensions as a result of the Cauchy–Riemann equations. As mentioned at the beginning of the chapter, we omit discussion of the complex-variable technique, not because it is unimportant but for lack of space and because completely adequate discussions exist elsewhere. Some of these sources are listed at the end of the chapter. The methods of summation of Fourier series, with many examples, are described in *Collin* (Appendix A.6).

2.11 *Fields and Charge Densities in Two-Dimensional Corners and Along Edges*

In many practical situations conducting surfaces come together in a way that can be approximated, on the small scale at least, as the intersection of two planes. The edges of the box shown in Fig. 2.9 are one example, the corners at $x = 0$, $y = 0$ and $x = a$, $y = 0$ in Fig. 2.10 another. It is useful therefore to have an understanding of how the potential fields, and the surface-charge densities behave in the neighborhood of such sharp "corners" or edges. To be able to look at them closely enough to have the behavior of the fields determined in functional form solely by the properties of the "corner" and not by the details of the overall configuration, we assume that the "corners" are infinitely sharp.

The general situation in two dimensions is shown in Fig. 2.12. Two conducting planes intersect at an angle β. The planes are assumed to be held at potential V. Remote from the origin and not shown in the figure are other conductors or possibly configurations of charges that specify the potential problem uniquely. Since we are interested in the functional behavior of the fields, etc. near the origin, but not in the absolute magnitudes, we leave the "far away" behavior unspecified as much as possible.

The geometry of Fig. 2.12 suggests use of polar rather than Cartesian coordinates. In terms of the polar coordinates (ρ, ϕ), the Laplace equation in two dimensions is

$$\frac{1}{\rho}\frac{\partial}{\partial \rho}\left(\rho \frac{\partial \Phi}{\partial \rho}\right) + \frac{1}{\rho^2}\frac{\partial^2 \Phi}{\partial \phi^2} = 0 \tag{2.66}$$

Using the separation of variables approach, we substitute

$$\Phi(\rho, \phi) = R(\rho)\Psi(\phi)$$

This leads, upon multiplication by ρ^2/Φ, to

$$\frac{\rho}{R}\frac{d}{d\rho}\left(\rho \frac{dR}{d\rho}\right) + \frac{1}{\Psi}\frac{d^2\Psi}{d\phi^2} = 0 \tag{2.67}$$

Since the two terms are separately functions of ρ and ϕ respectively, each one must be constant:

$$\frac{\rho}{R}\frac{d}{d\rho}\left(\rho \frac{dR}{d\rho}\right) = \nu^2, \qquad \frac{1}{\Psi}\frac{d^2\Psi}{d\phi^2} = -\nu^2 \tag{2.68}$$

The solutions to these equations are

$$\left.\begin{array}{l} R(\rho) = a\rho^\nu + b\rho^{-\nu} \\ \Psi(\phi) = A \cos(\nu\phi) + B \sin(\nu\phi) \end{array}\right\} \tag{2.69}$$

For the special circumstance of $\nu = 0$, the solutions are

$$\left.\begin{array}{l} R(\rho) = a_0 + b_0 \ln \rho \\ \Psi(\phi) = A_0 + B_0\phi \end{array}\right\} \tag{2.70}$$

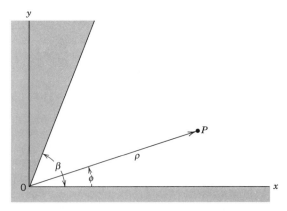

Figure 2.12 Intersection of two conducting planes defining a corner in two dimensions with opening angle β.

These are the building blocks with which we construct the potential by linear superposition.

Although not central to our present purpose, we note the general solution of the Laplace equation in two dimensions when the full azimuthal range is permitted as, for example, for the potential between two cylindrical surfaces, $\rho = a$ and $\rho = b$, on which the potential is given as a function of ϕ. If there is no restriction on ϕ, it is necessary that ν be a positive or negative integer or zero to ensure that the potential is single-valued. Furthermore, for $\nu = 0$, the constant B_0 in (2.70) must vanish for the same reason. The general solution is therefore of the form,

$$\Phi(\rho, \phi) = a_0 + b_0 \ln \rho + \sum_{n=1}^{\infty} a_n \rho^n \sin(n\phi + \alpha_n)$$
$$+ \sum_{n=1}^{\infty} b_n \rho^{-n} \sin(n\phi + \beta_n) \tag{2.71}$$

If the origin is included in the volume in which there is no charge, all the b_n are zero. Only a constant and positive powers of ρ appear. If the origin is excluded, the b_n can be different from zero. In particular, the logarithmic term is equivalent to a line charge on the axis with charge density per unit length $\lambda = -2\pi\epsilon_0 b_0$, as is well known.

For the situation of Fig. 2.12 the azimuthal angle is restricted to the range $0 \leq \phi \leq \beta$. The boundary conditions are that $\Phi = V$ for all $\rho \geq 0$ when $\phi = 0$ and $\phi = \beta$. This requires that $b_0 = B_0 = 0$ in (2.70) and $b = 0$ and $A = 0$ in (2.69). Furthermore, it requires that ν be chosen to make $\sin(\nu\beta) = 0$. Hence

$$\nu = \frac{m\pi}{\beta}, \qquad m = 1, 2, \ldots$$

and the general solution becomes

$$\Phi(\rho, \phi) = V + \sum_{m=1}^{\infty} a_m \rho^{m\pi/\beta} \sin(m\pi\phi/\beta) \tag{2.72}$$

The still undetermined coefficients a_m depend on the potential remote from the corner at $\rho = 0$. Since the series involves positive powers of $\rho^{\pi/\beta}$, for small enough ρ only the first term in the series will be important.* Thus, *near $\rho = 0$, the potential is approximately*

$$\Phi(\rho, \phi) \simeq V + a_1 \rho^{\pi/\beta} \sin(\pi\phi/\beta) \tag{2.73}$$

The electric field components are

$$\left. \begin{array}{l} E_\rho(\rho, \phi) = -\dfrac{\partial \Phi}{\partial \rho} \simeq -\dfrac{\pi a_1}{\beta} \rho^{(\pi/\beta)-1} \sin(\pi\phi/\beta) \\[4mm] E_\phi(\rho, \phi) = -\dfrac{1}{\rho}\dfrac{\partial \Phi}{\partial \phi} \simeq -\dfrac{\pi a_1}{\beta} \rho^{(\pi/\beta)-1} \cos(\pi\phi/\beta) \end{array} \right\} \tag{2.74}$$

---

*Here we make a necessary assumption about the remote boundary conditions, namely, that they are such that the coefficient a_1 is not zero. Ordinarily this is of no concern, but special symmetries might make a_1, or even a_2, etc., vanish. These unusual examples must be treated separately.

The surface-charge densities at $\phi = 0$ and $\phi = \beta$ are equal and are approximately

$$\sigma(\rho) = \epsilon_0 E_\phi(\rho, 0) \simeq -\frac{\epsilon_0 \pi a_1}{\beta} \rho^{(\pi/\beta)-1} \qquad (2.75)$$

The components of the field and the surface-charge density near $\rho = 0$ all vary with distance as $\rho^{(\pi/\beta)-1}$. This dependence on ρ is shown for some special cases in Fig. 2.13. For a very deep corner (small β) the power of ρ becomes very large. Essentially no charge accumulates in such a corner. For $\beta = \pi$ (a flat surface), the field quantities become independent of ρ, as is intuitively obvious. When $\beta > \pi$, the two-dimensional corner becomes an edge and the field and the surface-charge density become singular as $\rho \to 0$. For $\beta = 2\pi$ (the edge of a thin sheet) the singularity is as $\rho^{-1/2}$. This is still integrable so that the charge within a finite distance from the edge is finite, but it implies that field strengths become very large at the edges of conducting sheets (or, in fact, for any configuration where $\beta > \pi$).

The preceding two-dimensional electrostatic considerations apply to many three-dimensional situations, even with time-varying fields. If the edge is a sharp edge of finite length, as the edge of a cube away from a corner, then sufficiently close to the edge the variation of the potential along the edge can be ignored. The two-dimensional considerations apply, although the coefficient a_1 in (2.75) may vary with distance along the edge. Similarly, the electrostatic arguments are valid even for time-varying fields. The point here is that with time dependence another length enters, namely, the wavelength. Provided one is concerned with distances away from the edge that are small compared to a wavelength, as well as other relevant distances, the behavior of the fields reduces to electrostatic or magnetostatic behavior. In the diffraction of microwaves by a hole in a thin conducting sheet, for example, the fields are singular as $\rho^{-1/2}$ as $\rho \to 0$, where ρ is the distance from the boundary of the hole, and this fact must be taken into account in any exact solution of the diffraction problem.

The singular behavior of the fields near sharp edges is the reason for the effectiveness of lightning rods. In the idealized situation discussed here the field strength increases without limit as $\rho \to 0$, but for a thin sheet of thickness d with a smoothly rounded edge it can be inferred that the field strength at the surface will be proportional to $d^{-1/2}$. For small enough d this can be very large. In absolute vacuum such field strengths are possible; in air, however, electrical breakdown and a discharge will occur if the field strength exceeds a certain value (depending on the exact shape of the electrode, its proximity to the other elec-

Figure 2.13 Variation of the surface-charge density (and the electric field) with distance ρ from the "corner" or edge for opening angles $\beta = \pi/4$, $\pi/2$, π, $3\pi/2$, and 2π.

trodes, etc., but greater than about 2.5×10^6 V/m for air at normal temperature and pressure (NTP), sometimes by a factor of 4). In thunderstorms, with large potential differences between the ground and the thunderclouds, a grounded sharp conducting edge, or better, a point (see Section 3.4), will have breakdown occur around it first and will then provide one end of the jagged conducting path through the air along which the lightning discharge travels.

2.12   *Introduction to Finite Element Analysis for Electrostatics*

Finite element analysis (FEA) encompasses a variety of numerical approaches for the solution of boundary-value problems in physics and engineering. Here we sketch only an introduction to the essential ideas, using Galerkin's method for two-dimensional electrostatics as an illustration. The generalization to three dimensions is mentioned briefly at the end. The reader who wishes a deeper introduction may consult *Binns, Lawrenson, and Trowbridge, Ida and Bastos, Sadiku, Strang,* or *Zhou.*

Consider the Poisson equation, $\nabla^2 \psi = -g$ in a two-dimensional region R, with Dirichlet boundary conditions on the boundary curve C. We construct the vanishing integral,

$$\int_R [\phi \, \nabla^2 \psi + g\phi] \, dx \, dy = 0 \qquad (2.76)$$

where $\phi(x, y)$ is a test function specified for the moment only as piecewise continuous in R and vanishing on C. Use of Green's first identity on the first term above leads to

$$\int_R [\boldsymbol{\nabla}\phi \cdot \boldsymbol{\nabla}\psi - g\phi] \, dx \, dy = 0 \qquad (2.77)$$

The surface integral vanishes because ϕ vanishes on C. Galerkin's method consists first of approximating the desired solution $\psi(x, y)$ by a finite expansion in terms of a set of localized, linearly independent functions, $\phi_{ij}(x, y)$, with support only in a finite neighborhood of $x = x_i, y = y_j$. For definiteness, we imagine the region R spanned by a square lattice with lattice spacing h. Then a possible choice for $\phi_{ij}(x, y)$ is,

$$\phi_{ij}(x, y) = (1 - |x - x_i|/h)(1 - |y - y_j|/h) \qquad (2.78)$$

for $|x - x_i| \le h, |y - y_j| \le h$; otherwise, $\phi_{ij}(x, y) = 0$. The sum of all the ϕ_{ij} over the square lattice is unity. Other choices of the localized functions are possible, of course. Whatever the choice, if the number of lattice sites, including the boundary, is N_0, the expansion of $\psi(x, y)$ takes the form

$$\psi(x, y) \approx \sum_{k,l}^{(N_0)} \Psi_{kl}\phi_{kl}(x, y) \qquad (2.79)$$

Apart from the known values at sites on the boundary, the constant coefficients Ψ_{kl} may be thought of as the approximate values of $\psi(x_k, y_l)$. If the lattice spacing h is small enough, the expansion (2.79) will be a reasonable approximate to the true ψ, provided the coefficients are chosen properly.

The second step in Galerkin's method is to choose the test function ϕ in (2.77) to be the $(i, j)^{\text{th}}$ function on the expansion set, with i and j running successively over all N *internal* sites of the lattice. The typical equation derived from (2.77) is

$$\sum_{k,l}^{(N_0)} \Psi_{kl} \int_R \nabla\phi_{ij}(x, y) \cdot \nabla\phi_{kl}(x, y)\, dx\, dy = g(x_i, y_j) \int_R \phi_{ij}(x, y)\, dx\, dy \quad (2.80)$$

While the integrals are indicated as being over the whole region R, ϕ_{ij} has support only in a small region around the site (x_i, y_j). In (2.80) it is assumed that $g(x, y)$ varies slowly enough on the scale of the cell size to be approximated in the integral on the right by its value at the lattice site. Once the integrals have been performed, (2.80) becomes one of N coupled inhomogeneous linear algebraic equations for the N unknowns, Ψ_{kl}. The coupling among the Ψ_{kl} is confined to a small number of sites near (x_i, y_j), as indicated in Fig. 2.14 for the localized function (2.78). It is left as a problem to show that the needed integrals for the functions (2.78) are

$$\int_R \phi_{ij}(x, y)dx\, dy = h^2 \qquad (2.81)$$

$$\int_R \nabla\phi_{ij}(x, y) \cdot \nabla\phi_{kl}(x, y)dx\, dy = \begin{Bmatrix} 8/3 \\ -1/3 \end{Bmatrix} \text{ for } \begin{cases} k = i, & l = j \\ k = i \pm 1, l = j \\ k = i, & l = j \pm 1 \\ k = i \pm 1, l = j \pm 1 \end{cases}$$

When the site (i, j) is adjacent to the boundary, there are three or more terms on the left-hand side of (2.80) that are $(-1/3)$ times known boundary values of

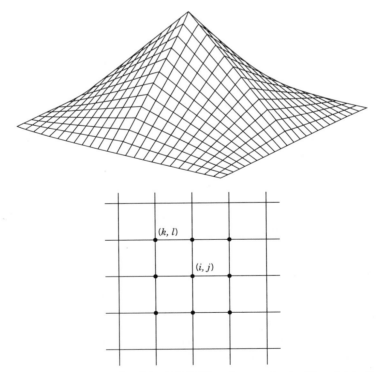

Figure 2.14   Sketch of the $\phi_{ij}(x, y)$ in (2.78). The sites marked with a dot in the lattice (bottom) are those coupled by the integrals on the left in (2.80) for the localized function (2.78).

ψ. These can be moved to the right-hand side as part of the inhomogeneity. If we write (2.80) in matrix form, $\mathbf{K\Psi} = \mathbf{G}$, with $\mathbf{K}$ an $N \times N$ square matrix and $\mathbf{\Psi}$ and $\mathbf{G}$ N-column vectors, the matrix $\mathbf{K}$ is a "sparse" matrix, with only a few nonvanishing elements in any row or column. The solution of the matrix operator equation by inversion of such a sparse matrix can be accomplished rapidly by special numerical techniques (see *Press et al.*). Concrete illustration of this approach is left to the problems at the end of the chapter.

A square lattice is not optimal in many problems because the solution may change more rapidly in some parts of the domain of interest than in other parts. In such regions one wishes to have a finer mesh. An FEA method with a standard generic shape, but permitting different sizes, will be more flexible and therefore superior. We describe the popular triangle as the basic unit in two dimensions.

The triangular element is assumed to be small enough that the field variable changes little over the element and may be approximated by a linear form in each direction. The basic triangular element $e(1, 2, 3)$ is shown in Fig. 2.15. Within this region, we approximate the field variable $\psi(x, y) \approx \psi_e(x, y) = A + Bx + Cy$. The three values (ψ_1, ψ_2, ψ_3) at the nodes or vertices determine the coefficients (A, B, C). It is useful, however, to systematize the procedure for numerical computation by defining three shape functions $N_j^{(e)}(x, y)$, one for each vertex, such that $N_j^{(e)} = 1$ when $x = x_j$, $y = y_j$ and $N_j^{(e)} = 0$ at the other vertices. The shape functions for the element e vanish outside that triangular domain.

Consider $N_1^{(e)} = a_1 + b_1 x + c_1 y$. Demand that

$$a_1 + b_1 x_1 + c_1 y_1 = 1$$
$$a_1 + b_1 x_2 + c_1 y_2 = 0$$
$$a_1 + b_1 x_3 + c_1 y_3 = 0$$

The determinant D of the coefficients on the left is

$$D = \begin{vmatrix} 1 & x_1 & y_1 \\ 1 & x_2 & y_2 \\ 1 & x_3 & y_3 \end{vmatrix} = (x_2 - x_1)(y_3 - y_1) - (x_3 - x_1)(y_2 - y_1)$$

The determinant D is invariant under rotations of the triangle; in fact, $D = 2S_e$, where S_e is the area of the triangle. The coefficients (a_1, b_1, c_1) are

$$a_1 = \frac{1}{2S_e} (x_2 y_3 - x_3 y_2)$$

$$b_1 = \frac{1}{2S_e} (y_2 - y_3)$$

$$c_1 = \frac{-1}{2S_e} (x_2 - x_3)$$

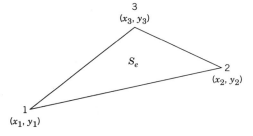

Figure 2.15   Basic triangular element $e(1, 2, 3)$ with area S_e for FEA in two dimensions.

The other $N_j^{(e)}$ can be written down by cyclic permutation of indices. The N_j and their coefficients obey the following relations:

$$\sum_{i=1}^{3} N_i^{(e)}(x, y) = 1; \quad \sum_{i=1}^{3} a_i = 1; \quad \sum_{i=1}^{3} b_i = 0; \quad \sum_{i=1}^{3} c_i = 0;$$

$$a_j + b_j \bar{x}_e + c_j \bar{y}_e = \frac{1}{3} \quad (j = 1, 2, 3)$$

Here $\bar{x}_e = (x_1 + x_2 + x_3)/3$ and $\bar{y}_e = (y_1 + y_2 + y_3)/3$ are the coordinates of the center of gravity of the triangular element e.

The shape functions for the triangular elements spanning the region R can be used in the Galerkin method as the localized linearly independent expansion set. The field variable $\psi(x, y)$ has the expansion,

$$\psi(x, y) \approx \sum_{f,j} \Psi_j^{(f)} N_j^{(f)}(x, y) \tag{2.82}$$

where the sum goes over all the triangles f and over the vertices of each triangle. The constants $\Psi_j^{(f)}$ are the desired values of the field at the vertices. (There is redundant labeling here because adjacent triangles have some vertices in common.) It is worth noting that despite the shift from one set of shape functions to another as the point (x, y) crosses from one triangle to one adjacent to it, the function defined by the right-hand side of (2.82) is continuous. Because of the linearity of the shape functions, the value of (2.82) along the common side of the two triangles from either representation is the same weighted average of the values at each end, with no contributions from the shape functions for the vertices not in common.

We return to the Poisson equation with Dirichlet boundary conditions and the vanishing integral (2.77). With the expansion (2.82) for $\psi(x, y)$, we choose the test function $\phi(x, y) = N_i^{(e)}(x, y)$ for some particular element e and vertex i (only avoiding vertices on the boundary because we require $\phi = 0$ on C). The choice reduces the integral [and the sum in (2.82)] to one over the particular element chosen, just as did the choice of the localized function in (2.80). The integral, with the inhomogeneity transferred to the right-hand side, is

$$\sum_{j=1}^{3} \Psi_j^{(e)} \int_e \nabla N_i^{(e)} \cdot \nabla N_j^{(e)} \, dx \, dy = \int_e g N_i^{(e)} \, dx \, dy \tag{2.83}$$

If $g(x, y)$ changes very little over the element e, it can be approximated by its value $g_e \equiv g(\bar{x}_e, \bar{y}_e)$ at the center of gravity of the triangle and factored out of the right-hand integral. The remaining integral is

$$\int_e N_i^{(e)} \, dx \, dy = S_e(a_i + b_i \bar{x}_e + c_i \bar{y}_e) = \tfrac{1}{3} S_e \tag{2.84}$$

For the left-hand integral in (2.83), the linearity of the shape functions means that the integrand is a constant. We note that $\partial N_i^{(e)}/\partial x = b_i$, $\partial N_i^{(e)}/\partial y = c_i$, and define

$$k_{ij}^{(e)} = S_e(b_i b_j + c_i c_j) \tag{2.85}$$

The coefficients $k_{ij}^{(e)}$ form an array of dimensionless coupling coefficients for the triangle e. It is straightforward to show that they depend on the shape of the

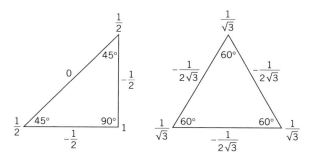

Figure 2.16   Examples of the triangular coupling coefficients. The "diagonal" coefficients are at the corners (vertices) and the "off-diagonal" coefficients along the sides, between vertices.

triangle, but not its orientation or size. Two examples are shown in Fig. 2.16, where the diagonal elements $k_{ii}^{(e)}$ are located at the corresponding vertices (i) and the off-diagonal elements $k_{ij}^{(e)}$ along the line connecting vertex i with vertex j.

With the definition (2.85) of the coupling coefficients, (2.83) becomes

$$\sum_{j=1}^{3} k_{ij}^{(e)} \Psi_j^{(e)} = \frac{S_e}{3} g_e \qquad (i = 1, 2, 3) \qquad (2.86)$$

For each element e there are three algebraic equations, except when the side(s) of the triangle form part of the boundary. The three coupled equations can be written in matrix form, $\mathbf{k}^{(e)} \mathbf{\Psi}^{(e)} = \mathbf{G}^{(e)}$.

The result for one element must now be generalized to include all the triangular elements spanning R. Let the number of interior vertices or nodes be N and the total number of vertices, including the boundary, be N_0. Label the internal nodes with $j = 1, 2, 3, \ldots, N$, and the boundary nodes by $j = N + 1, N + 2, \ldots, N_0$. Now enlarge and rearrange the matrix $\mathbf{k}^{(e)} \rightarrow \mathbf{K}$, where $\mathbf{K}$ is an $N \times N$ matrix with rows and columns labeled by the node index. Similarly, define the N-column vectors, $\mathbf{\Psi}$ and $\mathbf{G}$. For each triangular element in turn, add the elements of $k_{ij}^{(e)}$ and $S_e g_e/3$ to the appropriate rows and columns of $\mathbf{K}$ and $\mathbf{G}$. The end result is the matrix equation

$$\mathbf{K\Psi} = \mathbf{G} \qquad (2.87)$$

where

$$\mathbf{K} = (k_{ij}) \qquad \text{with } k_{ii} = \sum_T k_{ii}^{(e)} \qquad \text{and} \qquad k_{ij} = \sum_E k_{ij}^{(e)}, \qquad i \neq j$$

$$G_i = \frac{1}{3} \sum_T S_e g_e - \sum_{j=N+1}^{N_0} k_{ij}^{(e)} \Psi_j^{(e)} \qquad (2.88)$$

The summation over T means over all the triangles connected to the internal node i; the summation over E means a sum over all the triangles with a side from internal node i to internal node j. The final sum in G_i contains, for nodes connected directly to the boundary nodes, the known boundary values of ψ there and the corresponding $k_{ij}^{(e)}$ values (not present in the matrix $\mathbf{K}$). The reader may ponder Fig. 2.17 to be convinced of the correctness of (2.88). Just as for the square lattice, the $N \times N$ matrix $\mathbf{K}$ is a symmetric sparse matrix, with positive diagonal elements. As mentioned earlier, there are special efficient methods of inverting such matrices, even if very large.

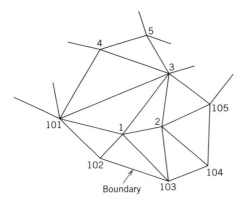

Figure 2.17 A part of the array of triangular elements spanning the region R, assumed to have 100 internal nodes.

The obvious generalization of the triangle to three-dimensional FEA is to add another vertex out of the plane to make a tetrahedron the basic element of volume. Now four shape functions, $N_j^{(e)}(x, y, z)$, are used to give an approximation to the field variable within the tetrahedron. The algebra is more involved, but the concept is the same.

Our discussion is a bare introduction to finite element analysis. Many variants exist in every branch of physics and engineering. National laboratories and commercial companies have "canned" FEA packages: POISSON is one such package, developed at the Lawrence Berkeley National Laboratory jointly with Livermore National Laboratory; TOSCA and CARMEN are two developed at the Rutherford–Appleton Laboratory in Britain.

References and Suggested Reading

The method of images and the related technique of inversion are treated in many books; among the better or more extensive discussions are those by
  Jeans, Chapter VIII
  Maxwell, Vol. 1, Chapter XI
  Smythe, Chapters IV and V
The classic use of inversion by Lord Kelvin in 1847 to obtain the charge distribution on the inside and outside surfaces of a thin, charged, conducting spherical bowl is discussed in
  Kelvin, p. 186
  Jeans, pp. 250–251
A truly encyclopedic source of examples of all sorts with numerous diagrams is the book by Durand, especially Chapters III and IV. Durand discusses inversion on pp. 107–114.

Complex variables and conformal mapping techniques for the solution of two-dimensional potential problems are discussed by
  Durand, Chapter X
  Jeans, Chapter VIII, Sections 306–337
  Maxwell, Vol. I, Chapter XII
  Morse and Feshbach, pp. 443–453, 1215–1252
  Smythe, Chapter IV, Sections 4.09–4.31
  Thomson, Chapter 3
A useful little mathematics book on conformal mapping is
  Bieberbach

There are, in addition, many engineering books devoted to the subject, e.g.,
>Gibbs
>Rothe, Ollendorff, and Polhausen

Elementary, but clear, discussions of the mathematical theory of Fourier series and integrals, and orthogonal expansions, can be found in
>Churchill
>Hildebrand, Chapter 5

A somewhat old-fashioned treatment of Fourier series and integrals, but with many examples and problems, is given by
>Byerly

The literature on numerical methods is vast and growing. A good guidepost to pertinent literature is
>Paul L. DeVries, Resource Letter CP-1: Computational Physics, *Am. J. Phys.* **64**, 364–368 (1996)

In addition to the books cited at the beginning of Section 2.12, two others are
>P. Hammond and J. K. Sykulski, *Engineering Electromagnetism, Physical Processes and Computation*, Oxford University Press, New York (1994).
>C. W. Steele, *Numerical Computation of Electric and Magnetic Fields*, Van Nostrand, New York (1987).

The first of these has a brief but clear discussion of FEA in Chapter 7; the second treats FEA and related topics in greater depth.

Problems

2.1   A point charge q is brought to a position a distance d away from an infinite plane conductor held at zero potential. Using the method of images, find:

  (a)   the surface-charge density induced on the plane, and plot it;

  (b)   the force between the plane and the charge by using Coulomb's law for the force between the charge and its image;

  (c)   the total force acting on the plane by integrating $\sigma^2/2\epsilon_0$ over the whole plane;

  (d)   the work necessary to remove the charge q from its position to infinity;

  (e)   the potential energy between the charge q and its image [compare the answer to part d and discuss].

  (f)   Find the answer to part d in electron volts for an electron originally one angstrom from the surface.

2.2   Using the method of images, discuss the problem of a point charge q *inside* a hollow, grounded, conducting sphere of inner radius a. Find

  (a)   the potential inside the sphere;

  (b)   the induced surface-charge density;

  (c)   the magnitude and direction of the force acting on q.

  (d)   Is there any change in the solution if the sphere is kept at a fixed potential V? If the sphere has a total charge Q on its inner and outer surfaces?

2.3   A straight-line charge with constant linear charge density λ is located perpendicular to the x-y plane in the first quadrant at (x_0, y_0). The intersecting planes $x = 0$, $y \geq 0$ and $y = 0$, $x \geq 0$ are conducting boundary surfaces held at zero potential. Consider the potential, fields, and surface charges in the first quadrant.

(a) The well-known potential for an isolated line charge at (x_0, y_0) is $\Phi(x, y) = (\lambda/4\pi\epsilon_0)\ln(R^2/r^2)$, where $r^2 = (x - x_0)^2 + (y - y_0)^2$ and R is a constant. Determine the expression for the potential of the line charge in the presence of the intersecting planes. Verify explicitly that the potential and the tangential electric field vanish on the boundary surfaces.

(b) Determine the surface charge density σ on the plane $y = 0$, $x \geq 0$. Plot σ/λ versus x for $(x_0 = 2, y_0 = 1)$, $(x_0 = 1, y_0 = 1)$, and $(x_0 = 1, y_0 = 2)$.

(c) Show that the total charge (per unit length in z) on the plane $y = 0$, $x \geq 0$ is

$$Q_x = -\frac{2}{\pi} \lambda \tan^{-1}\left(\frac{x_0}{y_0}\right)$$

What is the total charge on the plane $x = 0$?

(d) Show that far from the origin $[\rho \gg \rho_0$, where $\rho = \sqrt{(x^2 + y^2)}$ and $\rho_0 = \sqrt{(x_0^2 + y_0^2)}]$ the leading term in the potential is

$$\Phi \to \Phi_{\text{asym}} = \frac{4\lambda}{\pi\epsilon_0} \frac{(x_0 y_0)(xy)}{\rho^4}$$

Interpret.

2.4 A point charge is placed a distance $d > R$ from the center of an equally charged, isolated, conducting sphere of radius R.

(a) Inside of what distance from the surface of the sphere is the point charge attracted rather than repelled by the charged sphere?

(b) What is the limiting value of the force of attraction when the point charge is located a distance $a (= d - R)$ from the surface of the sphere, if $a \ll R$?

(c) What are the results for parts a and b if the charge on the sphere is twice (half) as large as the point charge, but still the same sign?

[Answers: (a) $d/R - 1 = 0.6178$, (b) $F = -q^2/(16\pi\epsilon_0 a^2)$, i.e., image force, (c) for $Q = 2q$, $d/R - 1 = 0.4276$; for $Q = q/2$, $d/R - 1 = 0.8823$. The answer for part b is the same.]

2.5 (a) Show that the work done to remove the charge q from a distance $r > a$ to infinity against the force, Eq. (2.6), of a grounded conducting sphere is

$$W = \frac{q^2 a}{8\pi\epsilon_0(r^2 - a^2)}$$

Relate this result to the electrostatic potential, Eq. (2.3), and the energy discussion of Section 1.11.

(b) Repeat the calculation of the work done to remove the charge q against the force, Eq. (2.9), of an isolated charged conducting sphere. Show that the work done is

$$W = \frac{1}{4\pi\epsilon_0}\left[\frac{q^2 a}{2(r^2 - a^2)} - \frac{q^2 a}{2r^2} - \frac{qQ}{r}\right]$$

Relate the work to the electrostatic potential, Eq. (2.8), and the energy discussion of Section 1.11.

2.6 The electrostatic problem of a point charge q outside an isolated, charged conducting sphere is equivalent to that of three charges, the original and two others, one located at the center of the sphere and another ("the image charge") inside the now imaginary sphere, on the line joining the center and the original charge.

If the point charge and sphere are replaced by two conducting spheres of radii r_a and r_b, carrying total charges Q_a and Q_b, respectively, with centers separated by a distance $d > r_a + r_b$, there is an equivalence with an infinite set of charges within each sphere, one at the center and a set of images along the line joining the centers. The charges and their locations can be determined iteratively, starting with a charge $q_a(1)$ at the center of the first sphere and $q_b(1)$ correspondingly for the second sphere. The charge $q_b(1)$ has its image $q_a(2)$ within the first sphere and vice versa. Then the image charge within the first sphere induces another image within the second sphere, and so on. The sum of all the charges within each sphere must be scaled to be equal to Q_a or Q_b.

The electrostatic potential outside the spheres, the force between the spheres, etc. can be found by summing the contributions from all the charges.

(a) Show that the charges and their positions are determined iteratively by the relations,

$$q_a(j) = -r_a q_b(j-1)/d_b(j-1), \qquad x_a(j) = r_a^2/d_b(j-1), \qquad d_a(j) = d - x_a(j)$$
$$q_b(j) = -r_b q_a(j-1)/d_a(j-1), \qquad x_b(j) = r_b^2/d_a(j-1), \qquad d_b(j) = d - x_b(j)$$

for $j = 2, 3, 4, \ldots$, with $d_a(1) = d_b(1) = d$, and $x_a(1) = x_b(1) = 0$.

(b) Find the image charges and their locations as well as the potentials on the spheres and force between them by means of a suitable computer program. [In computing the potential on each sphere, evaluate it in different places: e.g., in the equatorial plane and at the pole opposite the other sphere. This permits a check on the equipotential of the conductor and on the accuracy of computation.]

(c) As an example, show that for two equally charged spheres of the same radius R, the force between them when almost in contact is 0.6189 times the value that would be obtained if all the charge on each sphere were concentrated at its center. Show numerically and by explicit summation of the series that the capacitance of two identical conducting spheres in contact is $C/4\pi\epsilon_0 R = 1.3863 \cdots [= \ln 4]$.

Reference: J. A. Soules, *Am. J. Phys.* **58**, 1195 (1990).

2.7 Consider a potential problem in the half-space defined by $z \geq 0$, with Dirichlet boundary conditions on the plane $z = 0$ (and at infinity).

(a) Write down the appropriate Green function $G(\mathbf{x}, \mathbf{x}')$.

(b) If the potential on the plane $z = 0$ is specified to be $\Phi = V$ inside a circle of radius a centered at the origin, and $\Phi = 0$ outside that circle, find an integral expression for the potential at the point P specified in terms of cylindrical coordinates (ρ, ϕ, z).

(c) Show that, along the axis of the circle ($\rho = 0$), the potential is given by

$$\Phi = V\left(1 - \frac{z}{\sqrt{a^2 + z^2}}\right)$$

(d) Show that at large distances ($\rho^2 + z^2 \gg a^2$) the potential can be expanded in a power series in $(\rho^2 + z^2)^{-1}$, and that the leading terms are

$$\Phi = \frac{Va^2}{2}\frac{z}{(\rho^2 + z^2)^{3/2}}\left[1 - \frac{3a^2}{4(\rho^2 + z^2)} + \frac{5(3\rho^2 a^2 + a^4)}{8(\rho^2 + z^2)^2} + \cdots\right]$$

Verify that the results of parts c and d are consistent with each other in their common range of validity.

2.8   A two-dimensional potential problem is defined by two straight parallel line charges separated by a distance R with equal and opposite linear charge densities λ and $-\lambda$.

(a)   Show by direct construction that the surface of constant potential V is a circular cylinder (circle in the transverse dimensions) and find the coordinates of the axis of the cylinder and its radius in terms of R, λ, and V.

(b)   Use the results of part a to show that the capacitance per unit length C of two right-circular cylindrical conductors, with radii a and b, separated by a distance $d > a + b$, is

$$C = \frac{2\pi\epsilon_0}{\cosh^{-1}\left(\dfrac{d^2 - a^2 - b^2}{2ab}\right)}$$

(c)   Verify that the result for C agrees with the answer in Problem 1.7 in the appropriate limit and determine the next nonvanishing order correction in powers of a/d and b/d.

(d)   Repeat the calculation of the capacitance per unit length for two cylinders inside each other $(d < |b - a|)$. Check the result for concentric cylinders $(d = 0)$.

2.9   An insulated, spherical, conducting shell of radius a is in a uniform electric field E_0. If the sphere is cut into two hemispheres by a plane perpendicular to the field, find the force required to prevent the hemispheres from separating

(a)   if the shell is uncharged;

(b)   if the total charge on the shell is Q.

2.10   A large parallel plate capacitor is made up of two plane conducting sheets with separation D, one of which has a small hemispherical boss of radius a on its inner surface $(D \gg a)$. The conductor with the boss is kept at zero potential, and the other conductor is at a potential such that far from the boss the electric field between the plates is E_0.

(a)   Calculate the surface-charge densities at an arbitrary point on the plane and on the boss, and sketch their behavior as a function of distance (or angle).

(b)   Show that the total charge on the boss has the magnitude $3\pi\epsilon_0 E_0 a^2$.

(c)   If, instead of the other conducting sheet at a different potential, a point charge q is placed directly above the hemispherical boss at a distance d from its center, show that the charge induced on the boss is

$$q' = -q\left[1 - \frac{d^2 - a^2}{d\sqrt{d^2 + a^2}}\right]$$

2.11   A line charge with linear charge density τ is placed parallel to, and a distance R away from, the axis of a conducting cylinder of radius b held at fixed voltage such that the potential vanishes at infinity. Find

(a)   the magnitude and position of the image charge(s);

(b)   the potential at any point (expressed in polar coordinates with the origin at the axis of the cylinder and the direction from the origin to the line charge as the x axis), including the asymptotic form far from the cylinder;

(c)   the induced surface-charge density, and plot it as a function of angle for $R/b = 2$, 4 in units of $\tau/2\pi b$;

(d)   the force on the charge.

2.12 Starting with the series solution (2.71) for the two-dimensional potential problem with the potential specified on the surface of a cylinder of radius b, evaluate the coefficients formally, substitute them into the series, and sum it to obtain the potential *inside* the cylinder in the form of Poisson's integral:

$$\Phi(\rho, \phi) = \frac{1}{2\pi} \int_0^{2\pi} \Phi(b, \phi') \frac{b^2 - \rho^2}{b^2 + \rho^2 - 2b\rho \cos(\phi' - \phi)} \, d\phi'$$

What modification is necessary if the potential is desired in the region of space bounded by the cylinder and infinity?

2.13 **(a)** Two halves of a long hollow conducting cylinder of inner radius b are separated by small lengthwise gaps on each side, and are kept at different potentials V_1 and V_2. Show that the potential inside is given by

$$\Phi(\rho, \phi) = \frac{V_1 + V_2}{2} + \frac{V_1 - V_2}{\pi} \tan^{-1}\left(\frac{2b\rho}{b^2 - \rho^2} \cos \phi\right)$$

where ϕ is measured from a plane perpendicular to the plane through the gap.

(b) Calculate the surface-charge density on each half of the cylinder.

2.14 A variant of the preceding two-dimensional problem is a long hollow conducting cylinder of radius b that is divided into equal quarters, alternate segments being held at potential $+V$ and $-V$.

(a) Solve by means of the series solution (2.71) and show that the potential inside the cylinder is

$$\Phi(\rho, \phi) = \frac{4V}{\pi} \sum_{n=0}^{\infty} \left(\frac{\rho}{b}\right)^{4n+2} \frac{\sin[(4n + 2)\phi]}{2n + 1}$$

(b) Sum the series and show that

$$\Phi(\rho, \phi) = \frac{2V}{\pi} \tan^{-1}\left(\frac{2\rho^2 b^2 \sin 2\phi}{b^4 - \rho^4}\right)$$

(c) Sketch the field lines and equipotentials.

2.15 **(a)** Show that the Green function $G(x, y; x', y')$ appropriate for Dirichlet boundary conditions for a square two-dimensional region, $0 \le x \le 1, 0 \le y \le 1$, has an expansion

$$G(x, y; x', y') = 2 \sum_{n=1}^{\infty} g_n(y, y') \sin(n\pi x) \sin(n\pi x')$$

where $g_n(y, y')$ satisfies

$$\left(\frac{\partial^2}{\partial y'^2} - n^2\pi^2\right) g_n(y, y') = -4\pi\delta(y' - y) \quad \text{and} \quad g_n(y, 0) = g_n(y, 1) = 0$$

(b) Taking for $g_n(y, y')$ appropriate linear combinations of $\sinh(n\pi y')$ and $\cosh(n\pi y')$ in the two regions, $y' < y$ and $y' > y$, in accord with the boundary conditions and the discontinuity in slope required by the source delta function, show that the explicit form of G is

$$G(x, y; x', y')$$
$$= 8 \sum_{n=1}^{\infty} \frac{1}{n \sinh(n\pi)} \sin(n\pi x) \sin(n\pi x') \sinh(n\pi y_<) \sinh[n\pi(1 - y_>)]$$

where $y_<(y_>)$ is the smaller (larger) of y and y'.

2.16   A two-dimensional potential exists on a unit square area $(0 \leq x \leq 1, 0 \leq y \leq 1)$ bounded by "surfaces" held at zero potential. Over the entire square there is a uniform charge density of unit strength (per unit length in z). Using the Green function of Problem 2.15, show that the solution can be written as

$$\Phi(x, y) = \frac{4}{\pi^3 \epsilon_0} \sum_{m=0}^{\infty} \frac{\sin[(2m + 1)\pi x]}{(2m + 1)^3} \left\{ 1 - \frac{\cosh[(2m + 1)\pi(y - \frac{1}{2})]}{\cosh[(2m + 1)\pi/2]} \right\}$$

2.17 (a)   Construct the free-space Green function $G(x, y; x', y')$ for two-dimensional electrostatics by integrating $1/R$ with respect to $(z' - z)$ between the limits $\pm Z$, where Z is taken to be very large. Show that apart from an inessential constant, the Green function can be written alternately as

$$G(x, y; x', y') = -\ln[(x - x')^2 + (y - y')^2]$$
$$= -\ln[\rho^2 + \rho'^2 - 2\rho\rho' \cos(\phi - \phi')]$$

(b)   Show explicitly by separation of variables in polar coordinates that the Green function can be expressed as a Fourier series in the azimuthal coordinate,

$$G = \frac{1}{2\pi} \sum_{-\infty}^{\infty} e^{im(\phi - \phi')} g_m(\rho, \rho')$$

where the radial Green functions satisfy

$$\frac{1}{\rho'} \frac{\partial}{\partial \rho'} \left(\rho' \frac{\partial g_m}{\partial \rho'} \right) - \frac{m^2}{\rho'^2} g_m = -4\pi \frac{\delta(\rho - \rho')}{\rho}$$

Note that $g_m(\rho, \rho')$ for fixed ρ is a different linear combination of the solutions of the homogeneous radial equation (2.68) for $\rho' < \rho$ and for $\rho' > \rho$, with a discontinuity of slope at $\rho' = \rho$ determined by the source delta function.

(c)   Complete the solution and show that the free-space Green function has the expansion

$$G(\rho, \phi; \rho', \phi') = -\ln(\rho_>^2) + 2 \sum_{m=1}^{\infty} \frac{1}{m} \left(\frac{\rho_<}{\rho_>} \right)^m \cdot \cos[m(\phi - \phi')]$$

where $\rho_<(\rho_>)$ is the smaller (larger) of ρ and ρ'.

2.18 (a)   By finding appropriate solutions of the radial equation in part b of Problem 2.17, find the Green function for the interior Dirichlet problem of a cylinder of radius b $[g_m(\rho, \rho' = b) = 0$. See (1.40)]. First find the series expansion akin to the free-space Green function of Problem 2.17. Then show that it can be written in closed form as

$$G = \ln \left[\frac{\rho^2 \rho'^2 + b^4 - 2\rho\rho' b^2 \cos(\phi - \phi')}{b^2(\rho^2 + \rho'^2 - 2\rho\rho' \cos(\phi - \phi'))} \right]$$

or

$$G = \ln \left[\frac{(b^2 - \rho^2)(b^2 - \rho'^2) + b^2 |\mathbf{\rho} - \mathbf{\rho}'|^2}{b^2 |\mathbf{\rho} - \mathbf{\rho}'|^2} \right]$$

(b)   Show that the solution of the Laplace equation with the potential given as $\Phi(b, \phi)$ on the cylinder can be expressed as Poisson's integral of Problem 2.12.

(c)   What changes are necessary for the Green function for the *exterior* problem $(b < \rho < \infty)$, for both the Fourier expansion and the closed form? [Note that the exterior Green function is not rigorously correct because it does not vanish for ρ or $\rho' \to \infty$. For situations in which the potential falls off fast enough as $\rho \to \infty$, no mistake is made in its use.]

2.19 Show that the two-dimensional Green function for Dirichlet boundary conditions for the annular region, $b \leq \rho \leq c$ (concentric cylinders) has the expansion

$$G = \frac{\ln(\rho_<^2/b^2)\,\ln(c^2/\rho_>^2)}{\ln(c^2/b^2)} + 2\sum_{m=1}^{\infty} \frac{\cos[m(\phi - \phi')]}{m[1 - (b/c)^{2m}]}\,(\rho_<^m - b^{2m}/\rho_<^m)(1/\rho_>^m - \rho_>^m/c^{2m})$$

2.20 Two-dimensional electric quadrupole focusing fields for particle accelerators can be modeled by a set of four symmetrically placed line charges, with linear charge densities $\pm\lambda$, as shown in the left-hand figure (the right-hand figure shows the electric field lines).

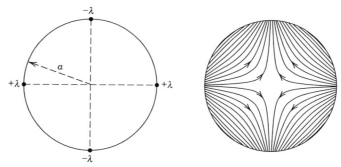

Problem 2.20

The charge density in two dimensions can be expressed as

$$\sigma(\rho, \phi) = \frac{\lambda}{a}\sum_{n=0}^{3} (-1)^n\,\delta(\rho - a)\,\delta(\phi - n\pi/2)$$

(a) Using the Green function expansion from Problem 2.17c, show that the electrostatic potential is

$$\Phi(\rho, \phi) = \frac{\lambda}{\pi\epsilon_0}\sum_{k=0}^{\infty} \frac{1}{2k + 1}\left(\frac{\rho_<}{\rho_>}\right)^{4k+2} \cos[(4k + 2)\,\phi]$$

(b) Relate the solution of part a to the real part of the complex function

$$w(z) = \frac{2\lambda}{4\pi\epsilon_0}\ln\left[\frac{(z - ia)(z + ia)}{(z - a)(z + a)}\right]$$

where $z = x + iy = \rho e^{i\phi}$. Comment on the connection to Problem 2.3.

(c) Find expressions for the Cartesian components of the electric field near the origin, expressed in terms of x and y. Keep the $k = 0$ and $k = 1$ terms in the expansion. For $y = 0$ what is the relative magnitude of the $k = 1$ (2^6-pole) contribution to E_x compared to the $k = 0$ (2^2-pole or quadrupole) term?

2.21 Use Cauchy's theorem to derive the Poisson integral solution. Cauchy's theorem states that if $F(z)$ is analytic in a region R bounded by a closed curve C, then

$$\frac{1}{2\pi i}\oint_C \frac{F(z')\,dz'}{z' - z} = \begin{cases} F(z) \\ 0 \end{cases} \quad \text{if } z \text{ is} \begin{matrix} \text{inside} \\ \text{outside} \end{matrix}\ R$$

Hint: You may wish to add an integral that vanishes (associated with the image point) to the integral for the point inside the circle.

2.22 **(a)** For the example of oppositely charged conducting hemispherical shells separated by a tiny gap, as shown in Figure 2.8, show that the interior potential $(r < a)$ on the z axis is

$$\Phi_{in}(z) = V \frac{a}{z} \left[1 - \frac{(a^2 - z^2)}{a\sqrt{a^2 + z^2}} \right]$$

Find the first few terms of the expansion in powers of z and show that they agree with (2.27) with the appropriate substitutions.

(b) From the result of part a and (2.22), show that the radial electric field on the positive z axis is

$$E_r(z) = \frac{Va^2}{(z^2 + a^2)^{3/2}} \left(3 + \frac{a^2}{z^2} \right)$$

for $z > a$, and

$$E_r(z) = -\frac{V}{a} \left[\frac{3 + (a/z)^2}{(1 + (z/a)^2)^{3/2}} - \frac{a^2}{z^2} \right]$$

for $|z| < a$. Show that the second form is well behaved at the origin, with the value, $E_r(0) = -3V/2a$. Show that at $z = a$ (north pole inside) it has the value $-(\sqrt{2} - 1)V/a$. Show that the radial field at the north pole outside has the value $\sqrt{2} \, V/a$.

(c) Make a sketch of the electric field lines, both inside and outside the conducting hemispheres, with directions indicated. Make a *plot* of the radial electric field along the z axis from $z = -2a$ to $z = +2a$.

2.23 A hollow cube has conducting walls defined by six planes $x = 0$, $y = 0$, $z = 0$, and $x = a$, $y = a$, $z = a$. The walls $z = 0$ and $z = a$ are held at a constant potential V. The other four sides are at zero potential.

(a) Find the potential $\Phi(x, y, z)$ at any point inside the cube.

(b) Evaluate the potential at the center of the cube numerically, accurate to three significant figures. How many terms in the series is it necessary to keep in order to attain this accuracy? Compare your numerical result with the average value of the potential on the walls. See Problem 2.28.

(c) Find the surface-charge density on the surface $z = a$.

2.24 In the two-dimensional region shown in Fig. 2.12, the angular functions appropriate for Dirichlet boundary conditions at $\phi = 0$ and $\phi = \beta$ are $\Phi(\phi) = A_m \sin(m\pi\phi/\beta)$. Show that the completeness relation for these functions is

$$\delta(\phi - \phi') = \frac{2}{\beta} \sum_{m=1}^{\infty} \sin(m\pi\phi/\beta) \sin(m\pi\phi'/\beta) \qquad \text{for } 0 < \phi, \phi' < \beta$$

2.25 Two conducting planes at zero potential meet along the z axis, making an angle β between them, as in Fig. 2.12. A unit line charge parallel to the z axis is located between the planes at position (ρ', ϕ').

(a) Show that $(4\pi\epsilon_0)$ times the potential in the space between the planes, that is, the Dirichlet Green function $G(\rho, \phi; \rho', \phi')$, is given by the infinite series

$$G(\rho, \phi; \rho', \phi') = 4 \sum_{m=1}^{\infty} \frac{1}{m} \rho_<^{m\pi/\beta} \rho_>^{-m\pi/\beta} \sin(m\pi\phi/\beta) \sin(m\pi\phi'/\beta)$$

(b) By means of complex-variable techniques or other means, show that the series can be summed to give a closed form,

$$G(\rho, \phi; \rho', \phi') = \ln\left\{\frac{(\rho)^{2\pi/\beta} + (\rho')^{2\pi/\beta} - 2(\rho\rho')^{\pi/\beta}\cos[\pi(\phi + \phi')/\beta]}{(\rho)^{2\pi/\beta} + (\rho')^{2\pi/\beta} - 2(\rho\rho')^{\pi/\beta}\cos[\pi(\phi - \phi'/\beta]}\right\}$$

(c) Verify that you obtain the familiar results when $\beta = \pi$ and $\beta = \pi/2$.

2.26 The two-dimensional region, $\rho \geq a$, $0 \leq \phi \leq \beta$, is bounded by conducting surfaces at $\phi = 0$, $\rho = a$, and $\phi = \beta$ held at zero potential, as indicated in the sketch. At large ρ the potential is determined by some configuration of charges and/or conductors at fixed potentials.

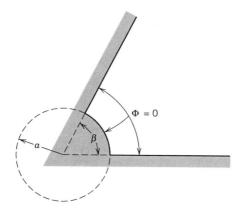

Problem 2.26

(a) Write down a solution for the potential $\Phi(\rho, \phi)$ that satisfies the boundary conditions for finite ρ.

(b) Keeping only the lowest nonvanishing terms, calculate the electric field components E_ρ and E_ϕ and also the surface-charge densities $\sigma(\rho, 0)$, $\sigma(\rho, \beta)$, and $\sigma(a, \phi)$ on the three boundary surfaces.

(c) Consider $\beta = \pi$ (a plane conductor with a half-cylinder of radius a on it). Show that far from the half-cylinder the lowest order terms of part b give a uniform electric field normal to the plane. Sketch the charge density on and in the neighborhood of the half-cylinder. For fixed electric field strength far from the plane, show that the total charge on the half-cylinder (actually charge per unit length in the z direction) is twice as large as would reside on a strip of width $2a$ in its absence. Show that the extra portion is drawn from regions of the plane nearby, so that the total charge on a strip of width large compared to a is the same whether the half-cylinder is there or not.

2.27 Consider the two-dimensional wedge-shaped region of Problem 2.26, with $\beta = 2\pi$. This corresponds to a semi-infinite thin sheet of conductor on the positive x axis from $x = a$ to infinity with a conducting cylinder of radius a fastened to its edge.

(a) Sketch the surface-charge densities on the cylinder and on the top and bottom of the sheet, using the lowest order solution.

(b) Calculate the total charge on the cylinder and compare with the total deficiency of charge on the sheet near the cylinder, that is, the total difference in charge for a finite compared with $a = 0$, assuming that the charge density far from the cylinder is the same.

2.28   A closed volume is bounded by conducting surfaces that are the n sides of a regular polyhedron (n = 4, 6, 8, 12, 20). The n surfaces are at different potentials V_i, $i = 1, 2, \ldots, n$. Prove in the simplest way you can that the potential at the center of the polyhedron is the average of the potential on the n sides. This problem bears on Problem 2.23b, and has an interesting similarity to the result of Problem 1.10.

2.29   For the Galerkin method on a two-dimensional square lattice with lattice spacing h, verify the relations (2.81) for the localized "pyramid" basis functions, $\phi_{ij}(x, y)$ = $(1 - |x|/h)(1 - |y|/h)$, $|x| < h$, $|y| < h$, where x and y are measured from the site (i, j). In particular,

$$\int dx \int dy \, \phi_{i,j}(x, y) = h^2; \qquad \int dx \int dy \, \nabla\phi_{i,j} \cdot \nabla\phi_{i,j} = \frac{8}{3};$$

$$\int dx \int dy \, \nabla\phi_{i+1,j} \cdot \nabla\phi_{i,j} = -\frac{1}{3}; \qquad \int dx \int dy \, \nabla\phi_{i,j+1} \cdot \nabla\phi_{i,j} = -\frac{1}{3};$$

$$\int dx \int dy \, \nabla\phi_{i+1,j+1} \cdot \nabla\phi_{i,j} = -\frac{1}{3}$$

2.30   Using the results of Problem 2.29, apply the Galerkin method to the integral equivalent of the Poisson equation with zero potential on the boundary,

$$\int_V dx \, dy [\nabla\phi_{i,j} \cdot \nabla\psi - 4\pi\rho\phi_{i,j}] = 0 \text{ with } \psi(x, y) = \sum_{i',j'=1}^{N} \psi_{i',j'} \phi_{i',j'}(x, y)$$

for the lattice of Problem 1.24, with its three independent lattice sites. Show that you get three coupled equations for the $\psi_{i,j}$ values (ψ_1, ψ_2, ψ_3) and solve to find the "Galerkin" approximations for the potential at these sites. Compare with the exact values and the results of the various iterations of Problem 1.24c. Comment.

CHAPTER 3

Boundary-Value Problems in Electrostatics: II

In this chapter the discussion of boundary-value problems is continued. Spherical and cylindrical geometries are first considered, and solutions of the Laplace equation are represented by expansions in series of the appropriate orthonormal functions. Only an outline is given of the solution of the various ordinary differential equations obtained from the Laplace equation by separation of variables, but the properties of the different functions are summarized.

The problem of construction of Green functions in terms of orthonormal functions arises naturally in the attempt to solve the Poisson equation in the various geometries. Explicit examples of Green functions are obtained and applied to specific problems, and the equivalence of the various approaches to potential problems is discussed.

3.1   Laplace Equation in Spherical Coordinates

In spherical coordinates (r, θ, ϕ), shown in Fig. 3.1, the Laplace equation can be written in the form:

$$\frac{1}{r}\frac{\partial^2}{\partial r^2}(r\Phi) + \frac{1}{r^2 \sin\theta}\frac{\partial}{\partial\theta}\left(\sin\theta\frac{\partial\Phi}{\partial\theta}\right) + \frac{1}{r^2 \sin^2\theta}\frac{\partial^2\Phi}{\partial\phi^2} = 0 \qquad (3.1)$$

If a product form for the potential is assumed, then it can be written:

$$\Phi = \frac{U(r)}{r}P(\theta)Q(\phi) \qquad (3.2)$$

When this is substituted into (3.1), there results the equation:

$$PQ\frac{d^2U}{dr^2} + \frac{UQ}{r^2 \sin\theta}\frac{d}{d\theta}\left(\sin\theta\frac{dP}{d\theta}\right) + \frac{UP}{r^2 \sin^2\theta}\frac{d^2Q}{d\phi^2} = 0$$

If we multiply by $r^2 \sin^2\theta/UPQ$, we obtain:

$$r^2 \sin^2\theta\left[\frac{1}{U}\frac{d^2U}{dr^2} + \frac{1}{Pr^2 \sin\theta}\frac{d}{d\theta}\left(\sin\theta\frac{dP}{d\theta}\right)\right] + \frac{1}{Q}\frac{d^2Q}{d\phi^2} = 0 \qquad (3.3)$$

The ϕ dependence of the equation has now been isolated in the last term. Consequently that term must be a constant which we call $(-m^2)$:

$$\frac{1}{Q}\frac{d^2Q}{d\phi^2} = -m^2 \qquad (3.4)$$

95

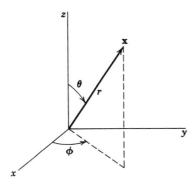

Figure 3.1

This has solutions

$$Q = e^{\pm im\phi} \tag{3.5}$$

For Q to be single valued, m must be an integer if the full azimuthal range is allowed. By similar considerations we find separate equations for $P(\theta)$ and $U(r)$:

$$\frac{1}{\sin\theta}\frac{d}{d\theta}\left(\sin\theta\frac{dP}{d\theta}\right) + \left[l(l+1) - \frac{m^2}{\sin^2\theta}\right]P = 0 \tag{3.6}$$

$$\frac{d^2U}{dr^2} - \frac{l(l+1)}{r^2}U = 0 \tag{3.7}$$

where $l(l+1)$ is another real constant.

From the form of the radial equation it is apparent that a single power of r (rather than a power series) will satisfy it. The solution is found to be:

$$U = Ar^{l+1} + Br^{-l} \tag{3.8}$$

but l is as yet undetermined.

3.2   *Legendre Equation and Legendre Polynomials*

The θ equation for $P(\theta)$ is customarily expressed in terms of $x = \cos\theta$, instead of θ itself. Then it takes the form:

$$\frac{d}{dx}\left[(1-x^2)\frac{dP}{dx}\right] + \left[l(l+1) - \frac{m^2}{1-x^2}\right]P = 0 \tag{3.9}$$

This equation is called the generalized Legendre equation, and its solutions are the associated Legendre functions. Before considering (3.9) we outline the solution by power series of the ordinary Legendre differential equation with $m^2 = 0$;

$$\frac{d}{dx}\left[(1-x^2)\frac{dP}{dx}\right] + l(l+1)P = 0 \tag{3.10}$$

We assume that the whole range of $\cos\theta$, including the north and south poles, is in the region of interest. The desired solution should then be single valued, finite, and continuous on the interval $-1 \le x \le 1$ in order that it represent a physical potential. The solution will be assumed to be represented by a power series of the form:

$$P(x) = x^\alpha \sum_{j=0}^{\infty} a_j x^j \tag{3.11}$$

where α is a parameter to be determined. When this is substituted into (3.10), there results the series:

$$\sum_{j=0}^{\infty} \{(\alpha + j)(\alpha + j - 1)a_j x^{\alpha+j-2} + [(\alpha + j)(\alpha + j + 1) \mp l(l + 1)]a_j x^{\alpha+j}\} = 0 \tag{3.12}$$

In this expansion the coefficient of each power of x must vanish separately. For $j = 0, 1$ we find that

$$\left. \begin{array}{l} \text{if } a_0 \neq 0, \text{ then } \alpha(\alpha - 1) = 0 \\ \text{if } a_1 \neq 0, \text{ then } \alpha(\alpha + 1) = 0 \end{array} \right\} \tag{3.13}$$

while for a general j value

$$a_{j+2} = \left[\frac{(\alpha + j)(\alpha + j + 1) - l(l + 1)}{(\alpha + j + 1)(\alpha + j + 2)} \right] a_j \tag{3.14}$$

A moment's thought shows that the two relations (3.13) are equivalent and that it is sufficient to choose *either* a_0 or a_1 different from zero, but not both. Making the former choice, we have $\alpha = 0$ or $\alpha = 1$. From (3.14) we see that the power series has only even powers of x ($\alpha = 0$) or only odd powers of x ($\alpha = 1$).

For either of the series $\alpha = 0$ or $\alpha = 1$ it is possible to prove the following properties:

the series converges for $x^2 < 1$, regardless of the value of l;

the series diverges at $x = \pm 1$, unless it terminates.

Since we want a solution that is finite at $x = \pm 1$, as well as for $x^2 < 1$, we demand that the series terminate. Since α and j are positive integers or zero, the recurrence relation (3.14) will terminate only if l *is zero or a positive integer*. Even then only one of the two series converges at $x = \pm 1$. If l is even (odd), then only the $\alpha = 0$ ($\alpha = 1$) series terminates.* The polynomials in each case have x^l as their highest power of x, the next highest being x^{l-2}, and so on, down to $x^0(x)$ for l even (odd). By convention these polynomials are normalized to have the value unity at $x = +1$ and are called the *Legendre polynomials* of order l, $P_l(x)$. The first few Legendre polynomials are:

$$\left. \begin{array}{l} P_0(x) = 1 \\ P_1(x) = x \\ P_2(x) = \tfrac{1}{2}(3x^2 - 1) \\ P_3(x) = \tfrac{1}{2}(5x^3 - 3x) \\ P_4(x) = \tfrac{1}{8}(35x^4 - 30x^2 + 3) \end{array} \right\} \tag{3.15}$$

---

*For example, if $l = 0$ the $\alpha = 1$ series has a general coefficient $a_j = a_0/(j + 1)$ for $j = 0, 2, 4, \ldots$. Thus the series is $a_0(x + \tfrac{1}{3}x^3 + \tfrac{1}{5}x^5 + \cdots)$. This is just a_0 times the power series expansion of a function $Q_0(x) = \tfrac{1}{2}\ln(1 + x)/(1 - x)$, which clearly diverges at $x = \pm 1$. For each l value there is a similar function $Q_l(x)$ with logarithms in it as the partner to the well-behaved polynomial solution. See *Magnus et al.* (pp. 151 ff). *Whittaker and Watson* (Chapter XV) give a treatment using analytic functions.

By manipulation of the power series solutions (3.11) and (3.14) it is possible to obtain a compact representation of the Legendre polynomials, known as *Rodrigues' formula*:

$$P_l(x) = \frac{1}{2^l l!} \frac{d^l}{dx^l} (x^2 - 1)^l \tag{3.16}$$

[See, for example, *Arfken*.]

The Legendre polynomials form a complete orthogonal set of functions on the interval $-1 \le x \le 1$. To prove the orthogonality we can appeal directly to the differential equation (3.10). We write down the differential equation for $P_l(x)$, multiply by $P_{l'}(x)$, and then integrate over the interval:

$$\int_{-1}^{1} P_{l'}(x) \left\{ \frac{d}{dx} \left[(1 - x^2) \frac{dP_l}{dx} \right] + l(l + 1) P_l(x) \right\} dx = 0 \tag{3.17}$$

Integrating the first term by parts, we obtain

$$\int_{-1}^{1} \left[(x^2 - 1) \frac{dP_l}{dx} \frac{dP_{l'}}{dx} + l(l + 1)(P_{l'}(x) P_l(x)) \right] dx = 0 \tag{3.18}$$

If we now write down (3.18) with l and l' interchanged and subtract it from (3.18), the result is the orthogonality condition:

$$[l(l + 1) - l'(l' + 1)] \int_{-1}^{1} P_{l'}(x) P_l(x) \, dx = 0 \tag{3.19}$$

For $l \ne l'$, the integral must vanish. For $l = l'$, the integral is finite. To determine its value it is necessary to use an explicit representation of the Legendre polynomials, e.g., Rodrigues' formula. Then the integral is explicitly:

$$N_l \equiv \int_{-1}^{1} [P_l(x)]^2 \, dx = \frac{1}{2^{2l}(l!)^2} \int_{-1}^{1} \frac{d^l}{dx^l} (x^2 - 1)^l \frac{d^l}{dx^l} (x^2 - 1)^l \, dx$$

Integration by parts l times yields the result:

$$N_l = \frac{(-1)^l}{2^{2l}(l!)^2} \int_{-1}^{1} (x^2 - 1)^l \frac{d^{2l}}{dx^{2l}} (x^2 - 1)^l \, dx$$

The differentiation $2l$ times of $(x^2 - 1)^l$ yields the constant $(2l)!$, so that

$$N_l = \frac{(2l)!}{2^{2l}(l!)^2} \int_{-1}^{1} (1 - x^2)^l \, dx$$

The remaining integral can be done by brute force, but also by induction. We write the integrand as

$$(1 - x^2)^l = (1 - x^2)(1 - x^2)^{l-1} = (1 - x^2)^{l-1} + \frac{x}{2l} \frac{d}{dx} (1 - x^2)^l$$

Thus we have

$$N_l = \left(\frac{2l - 1}{2l} \right) N_{l-1} + \frac{(2l - 1)!}{2^{2l}(l!)^2} \int_{-1}^{1} x \, d[1 - x^2)^l]$$

Integration by parts in the last integral yields

$$N_l = \left(\frac{2l - 1}{2l}\right)N_{l-1} - \frac{1}{2l}N_l$$

or

$$(2l + 1)N_l = (2l - 1)N_{l-1} \tag{3.20}$$

This shows that $(2l + 1)N_l$ is independent of l. For $l = 0$, with $P_0(x) = 1$, we have $N_0 = 2$. Thus $N_l = 2/(2l + 1)$ and the orthogonality condition can be written:

$$\int_{-1}^{1} P_{l'}(x)P_l(x)\, dx = \frac{2}{2l + 1}\, \delta_{l'l} \tag{3.21}$$

and the orthonormal functions in the sense of Section 2.8 are

$$U_l(x) = \sqrt{\frac{2l + 1}{2}}\, P_l(x) \tag{3.22}$$

Since the Legendre polynomials form a complete set of orthogonal functions, any function $f(x)$ on the interval $-1 \le x \le 1$ can be expanded in terms of them. The Legendre series representation is:

$$f(x) = \sum_{l=0}^{\infty} A_l P_l(x) \tag{3.23}$$

where

$$A_l = \frac{2l + 1}{2} \int_{-1}^{1} f(x)P_l(x)\, dx \tag{3.24}$$

As an example, consider the function shown in Fig. 3.2:

$$f(x) = +1 \qquad \text{for } x > 0$$
$$ = -1 \qquad \text{for } x < 0$$

Then

$$A_l = \frac{2l + 1}{2}\left[\int_0^1 P_l(x)\, dx - \int_{-1}^0 P_l(x)\, dx\right]$$

Since $P_l(x)$ is odd (even) about $x = 0$ if l is odd (even), only the odd l coefficients are different from zero. Thus, for l odd,

$$A_l = (2l + 1)\int_0^1 P_l(x)\, dx \tag{3.25}$$

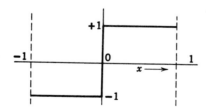

Figure 3.2

By means of Rodrigues' formula the integral can be evaluated, yielding

$$A_l = (-\tfrac{1}{2})^{(l-1)/2} \frac{(2l + 1)(l - 2)!!}{2\left(\dfrac{l + 1}{2}\right)!} \tag{3.26}$$

where $(2n + 1)!! \equiv (2n + 1)(2n - 1)(2n - 3) \cdots \times 5 \times 3 \times 1$. Thus the series for $f(x)$ is:

$$f(x) = \tfrac{3}{2}P_1(x) - \tfrac{7}{8}P_3(x) + \tfrac{11}{16}P_5(x) - \cdots \tag{3.27}$$

Certain recurrence relations among Legendre polynomials of different order are useful in evaluating integrals, generating higher order polynomials from lower order ones, etc. From Rodrigues' formula it is a straightforward matter to show that

$$\frac{dP_{l+1}}{dx} - \frac{dP_{l-1}}{dx} - (2l + 1)P_l = 0 \tag{3.28}$$

This result, combined with differential equation (3.10), can be made to yield various recurrence formulas, some of which are:

$$(l + 1)P_{l+1} - (2l + 1)xP_l + lP_{l-1} = 0$$

$$\frac{dP_{l+1}}{dx} - x\frac{dP_l}{dx} - (l + 1)P_l = 0 \tag{3.29}$$

$$(x^2 - 1)\frac{dP_l}{dx} - lxP_l + lP_{l-1} = 0$$

As an illustration of the use of these recurrence formulas, consider the evaluation of the integral:

$$I_1 = \int_{-1}^{1} xP_l(x)P_{l'}(x)\, dx \tag{3.30}$$

From the first of the recurrence formulas (3.29) we obtain an expression for $xP_l(x)$. Therefore (3.30) becomes

$$I_1 = \frac{1}{2l + 1}\int_{-1}^{1} P_{l'}(x)[l + 1)P_{l+1}(x) + lP_{l-1}(x)]\, dx$$

The orthogonality integral (3.21) can now be employed to show that the integral vanishes unless $l' = l \pm 1$, and that, for those values,

$$\int_{-1}^{1} xP_l(x)P_{l'}(x)\, dx = \begin{cases} \dfrac{2(l + 1)}{(2l + 1)(2l + 3)}, & l' = l + 1 \\[3mm] \dfrac{2l}{(2l - 1)(2l + 1)}, & l' = l - 1 \end{cases} \tag{3.31}$$

These are really the same result with the roles of l and l' interchanged. In a similar manner it is easy to show that

$$\int_{-1}^{1} x^2 P_l(x) P_{l'}(x)\, dx = \begin{cases} \dfrac{2(l+1)(l+2)}{(2l+1)(2l+3)(2l+5)}, & l' = l+2 \\[3mm] \dfrac{2(2l^2+2l-1)}{(2l-1)(2l+1)(2l+3)}, & l' = l \end{cases} \tag{3.32}$$

where it is assumed that $l' \geq l$.

3.3 *Boundary-Value Problems with Azimuthal Symmetry*

From the form of the solution of the Laplace equation in spherical coordinates (3.2), it will be seen that for a problem possessing azimuthal symmetry $m = 0$ in (3.5). This means that the general solution for such a problem is:

$$\Phi(r,\ \theta) = \sum_{l=0}^{\infty} [A_l r^l + B_l r^{-(l+1)}] P_l(\cos\theta) \tag{3.33}$$

The coefficients A_l and B_l can be determined from the boundary conditions. Suppose that the potential is specified to be $V(\theta)$ on the surface of a sphere of radius a, and it is required to find the potential inside the sphere. If there are no charges at the origin, the potential must be finite there. Consequently $B_l = 0$ for all l. The coefficients A_l are found by evaluating (3.33) on the surface of the sphere:

$$V(\theta) = \sum_{l=0}^{\infty} A_l a^l P_l(\cos\theta) \tag{3.34}$$

This is just a Legendre series of the form (3.23), so that the coefficients A_l are:

$$A_l = \frac{2l+1}{2a^l} \int_0^{\pi} V(\theta) P_l(\cos\theta) \sin\theta\, d\theta \tag{3.35}$$

If, for example, $V(\theta)$ is that of Section 2.7, with two hemispheres at equal and opposite potentials,

$$V(\theta) = \begin{cases} +V, & (0 \leq \theta < \pi/2) \\ -V, & (\pi/2 < \theta \leq \pi) \end{cases}$$

then the coefficients are proportional to those in (3.27). Thus the potential inside the sphere is

$$\Phi(r,\ \theta) = V\left[\frac{3}{2}\frac{r}{a}P_1(\cos\theta) - \frac{7}{8}\left(\frac{r}{a}\right)^3 P_3(\cos\theta) + \frac{11}{16}\left(\frac{r}{a}\right)^5 P_5(\cos\theta) \cdots\right] \tag{3.36}$$

To find the potential outside the sphere we merely replace $(r/a)^l$ by $(a/r)^{l+1}$. The resulting potential can be seen to be the same as (2.27), obtained by another means.

Series (3.33), with its coefficients determined by the boundary conditions, is a unique expansion of the potential. This uniqueness provides a means of obtaining the solution of potential problems from a knowledge of the potential in a limited domain, namely on the symmetry axis. On the symmetry axis (3.33) becomes (with $z = r$):

$$\Phi(z = r) = \sum_{l=0}^{\infty} [A_l r^l + B_l r^{-(l+1)}] \tag{3.37}$$

valid for positive z. For negative z each term must be multiplied by $(-1)^l$. Suppose that, by some means, we can evaluate the potential $\Phi(z)$ on the symmetry axis. If this potential function can be expanded in a power series in $z = r$ of the form (3.37), with known coefficients, then the solution for the potential at any point in space is obtained by multiplying each power of r^l and $r^{-(l+1)}$ by $P_l(\cos\theta)$.

At the risk of boring the reader, we return to the problem of the hemispheres at equal and opposite potentials. We have already obtained the series solution in two different ways, (2.27) and (3.36). The method just stated gives a third way. For a point on the axis we have found the closed form (2.22):

$$\Phi(z = r) = V\left[1 - \frac{r^2 - a^2}{r\sqrt{r^2 + a^2}}\right]$$

This can be expanded in powers of a^2/r^2:

$$\Phi(z = r) = \frac{V}{\sqrt{\pi}} \sum_{j=1}^{\infty} (-1)^{j-1} \frac{(2j - \frac{1}{2})\Gamma(j - \frac{1}{2})}{j!} \left(\frac{a}{r}\right)^{2j}$$

Comparison with expansion (3.37) shows that only odd l values ($l = 2j - 1$) enter. The solution, valid for all points outside the sphere, is consequently:

$$\Phi(r, \theta) = \frac{V}{\sqrt{\pi}} \sum_{j=1}^{\infty} (-1)^{j-1} \frac{(2j - \frac{1}{2})\Gamma(j - \frac{1}{2})}{j!} \left(\frac{a}{r}\right)^{2j} P_{2j-1}(\cos\theta)$$

This is the same solution as already obtained, (2.27) and (3.36).

An important expansion is that of the potential at $\mathbf{x}$ due to a unit point charge at $\mathbf{x}'$:

$$\frac{1}{|\mathbf{x} - \mathbf{x}'|} = \sum_{l=0}^{\infty} \frac{r_<^l}{r_>^{l+1}} P_l(\cos\gamma) \tag{3.38}$$

where $r_<$ ($r_>$) is the smaller (larger) of $|\mathbf{x}|$ and $|\mathbf{x}'|$, and γ is the angle between $\mathbf{x}$ and $\mathbf{x}'$, as shown in Fig. 3.3. This can be proved by rotating axes so that $\mathbf{x}'$ lies

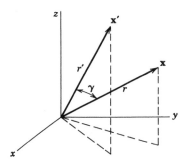

Figure 3.3

along the z axis. Then the potential satisfies the Laplace equation, possesses azimuthal symmetry, and can be expanded according to (3.33), except at the point $\mathbf{x} = \mathbf{x}'$:

$$\frac{1}{|\mathbf{x} - \mathbf{x}'|} = \sum_{l=0}^{\infty} (A_l r^l + B_l r^{-(l+1)}) P_l(\cos \gamma)$$

If the point $\mathbf{x}$ is on the z axis. the right-hand side reduces to (3.37), while the left-hand side becomes:

$$\frac{1}{|\mathbf{x} - \mathbf{x}'|} \equiv \frac{1}{(r^2 + r'^2 - 2rr' \cos \gamma)^{1/2}} \rightarrow \frac{1}{|r - r'|}$$

Expanding, we find, for $\mathbf{x}$ on axis,

$$\frac{1}{|\mathbf{x} - \mathbf{x}'|} = \frac{1}{r_>} \sum_{l=0}^{\infty} \left(\frac{r_<}{r_>}\right)^l$$

For points off the axis it is only necessary, according to (3.33) and (3.37), to multiply each term by $P_l(\cos \gamma)$. This proves the general result (3.38).

Another example is the potential due to a total charge q uniformly distributed around a circular ring of radius a, located as shown in Fig. 3.4, with its axis the z axis and its center at $z = b$. The potential at a point P on the axis of symmetry with $z = r$ is just $q/4\pi\epsilon_0$ divided by the distance AP:

$$\Phi(z = r) = \frac{1}{4\pi\epsilon_0} \frac{q}{(r^2 + c^2 - 2cr \cos \alpha)^{1/2}}$$

where $c^2 = a^2 + b^2$ and $\alpha = \tan^{-1}(a/b)$. The inverse distance AP can be expanded using (3.38). Thus, for $r > c$,

$$\Phi(z = r) = \frac{q}{4\pi\epsilon_0} \sum_{l=0}^{\infty} \frac{c^l}{r^{l+1}} P_l(\cos \alpha)$$

For $r < c$, the corresponding form is:

$$\Phi(z = r) = \frac{q}{4\pi\epsilon_0} \sum_{l=0}^{\infty} \frac{r^l}{c^{l+1}} P_l(\cos \alpha)$$

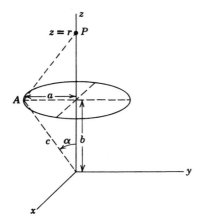

Figure 3.4   Ring of charge of radius a and total charge q located on the z axis with center at $z = b$.

The potential at *any point* in space is now obtained by multiplying each member of these series by $P_l(\cos \theta)$:

$$\Phi(r, \theta) = \frac{q}{4\pi\epsilon_0} \sum_{l=0}^{\infty} \frac{r_<^l}{r_>^{l+1}} P_l(\cos \alpha)P_l(\cos \theta)$$

where $r_<$ ($r_>$) is the smaller (larger) of r and c.

3.4 *Behavior of Fields in a Conical Hole or Near a Sharp Point*

Before turning to more complicated boundary-value problems, we consider one with azimuthal symmetry, but with only a limited range of θ. This is a three-dimensional analog of the situation discussed in Section 2.11. Suppose that the limited angular region, $0 \leq \theta \leq \beta, 0 \leq \phi \leq 2\pi$, is bounded by a conical conducting surface, as indicated in Fig. 3.5. For $\beta < \pi/2$, the region can be thought of as a deep conical hole bored in a conductor. For $\beta > \pi/2$, the region of space is that surrounding a pointed conical conductor.

The treatment of Section 3.2 for the Legendre differential equation needs modification. With the assumption of azimuth symmetry, (3.10) is still applicable, but we now seek solutions finite and single-valued on the range of $x = \cos \theta$ of $\cos \beta \leq x \leq 1$. Furthermore, since the conducting surface $\theta = \beta$ is at fixed potential, which we can take to be zero, the solution in $\cos \theta$ must vanish at $\theta = \beta$ to satisfy the boundary conditions. Since we demand regularity at $x = 1$ it is convenient to make a series expansion around $x = 1$ instead of $x = 0$, as was done with (3.11). With the introduction of the variable

$$\xi = \tfrac{1}{2}(1 - x)$$

the Legendre equation (3.10) becomes

$$\frac{d}{d\xi}\left[\xi(1 - \xi)\frac{dP}{d\xi}\right] + \nu(\nu + 1)P = 0 \tag{3.39}$$

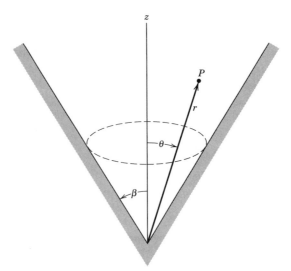

Figure 3.5

where l has been replaced by ν to avoid confusion. The corresponding radial solutions for $U(r)/r$ in (3.2) are r^ν and $r^{-\nu-1}$. With a power series solution,

$$P(\xi) = \xi^\alpha \sum_{j=0}^{\infty} a_j \xi^j$$

substituted into (3.39), the vanishing of the coefficient of the lowest power of ξ requires $\alpha = 0$. The recursion relation between successive coefficients in the series is then

$$\frac{a_{j+1}}{a_j} = \frac{(j - \nu)(j + \nu + 1)}{(j + 1)^2} \tag{3.40}$$

Choosing $a_0 = 1$ to normalize the solution to unity at $\xi = 0$ ($\cos\theta = 1$), we have the series representation

$$P_\nu(\xi) = 1 + \frac{(-\nu)(\nu + 1)}{1!\,1!}\,\xi + \frac{(-\nu)(-\nu + 1)(\nu + 1)(\nu + 2)}{2!\,2!}\,\xi^2 + \cdots \tag{3.41}$$

We first observe that if ν is zero or a positive integer the series terminates. The reader can verify that for $\nu = l = 0, 1, 2, \ldots$, the series (3.41) is exactly the Legendre *polynomials* (3.15). For ν not equal to an integer, (3.41) represents a generalization and is called a *Legendre function of the first kind and order ν*. The series (3.41) is an example of a hypergeometric function $_2F_1(a, b; c; z)$ whose series expansion is

$$_2F_1(a, b; c; z) = 1 + \frac{ab}{c}\frac{z}{1!} + \frac{a(a + 1)b(b + 1)}{c(c + 1)}\frac{z^2}{2!} + \cdots$$

Comparison with (3.41) shows that the Legendre function can be written

$$P_\nu(x) = {}_2F_1\left(-\nu, \nu + 1; 1; \frac{1 - x}{2}\right) \tag{3.42}$$

Here we have returned to our customary variable $x = \cos\theta$. The properties of the hypergeometric functions are well known (see *Morse and Feshbach*, Chapter 5, *Dennery and Krzywicki*, Sections IV.16–18, *Whittaker and Watson*, Chapter XIV). The Legendre function $P_\nu(x)$ is regular at $x = 1$ and for $|x| < 1$, but is singular at $x = -1$ unless ν is an integer. Depending on the value of ν, it has a certain number of zeros on the range $|x| < 1$. Since the polynomial $P_l(x)$ has l zeros for $|x| < 1$, we anticipate that for real ν more and more zeros occur as ν gets larger and larger. Furthermore, the zeros are distributed more or less uniformly on the interval. In particular, the first zero moves closer and closer to $x = 1$ as ν increases.

The basic solution to the Laplace boundary-value problem of Fig. 3.5 is

$$Ar^\nu P_\nu(\cos\theta)$$

where $\nu > 0$ is required for a finite potential at the origin. Since the potential must vanish at $\theta = \beta$ for all r, it is necessary that

$$P_\nu(\cos\beta) = 0 \tag{3.43}$$

This is an eigenvalue condition on ν. From what was just stated about the zeros of P_ν it is evident that (3.43) has an infinite number of solutions, $\nu = \nu_k$

($k = 1, 2, \ldots$), which we arrange in order of increasing magnitude. For $\nu = \nu_1$, $x = \cos \beta$ is the first zero of $P_{\nu_1}(x)$. For $\nu = \nu_2$, $x = \cos \beta$ is the second zero of $P_{\nu_2}(x)$, and so on. The complete solution for the azimuthally symmetric potential in the region $0 \le \theta \le \beta$ is*

$$\Phi(r, \theta) = \sum_{k=1}^{\infty} A_k r^{\nu_k} P_{\nu_k}(\cos \theta) \tag{3.44}$$

In the spirit of Section 2.11 we are interested in the general behavior of the potential and fields in the neighborhood of $r = 0$ and not in the full solution with specific boundary conditions imposed at large r. Thus we approximate the behavior of the potential near $r = 0$ by the first term in (3.44) and write

$$\Phi(r, \theta) \simeq A r^{\nu} P_{\nu}(\cos \theta) \tag{3.45}$$

where now ν is the *smallest root* of (3.43). The components of electric field and the surface-charge density on the conical conductor are

$$\left.\begin{array}{l} E_r = -\dfrac{\partial \Phi}{\partial r} \simeq -\nu A r^{\nu-1} P_{\nu}(\cos \theta) \\[1.2em] E_\theta = -\dfrac{1}{r}\dfrac{\partial \Phi}{\partial \theta} \simeq A r^{\nu-1} \sin \theta\, P_{\nu}'(\cos \theta) \\[1.2em] \sigma(r) = -\dfrac{1}{4\pi} E_\theta\big|_{\theta=\beta} \simeq -\dfrac{A}{4\pi} r^{\nu-1} \sin \beta\, P_{\nu}'(\cos \beta) \end{array}\right\} \tag{3.46}$$

Here the prime on P_ν denotes differential with respect to its argument. The fields and charge density all vary as $r^{\nu-1}$ as $r \to 0$.

The order ν for the first zero of $P_\nu(\cos \beta)$ is plotted as a function of β in Fig. 3.6. Obviously, for $\beta \ll 1$, $\nu \gg 1$. An approximate expression for ν in this domain can be obtained from the Bessel function approximation,[†]

$$P_\nu(\cos \theta) \simeq J_0\left((2\nu + 1) \sin \frac{\theta}{2}\right) \tag{3.47}$$

valid for large ν and $\theta < 1$. The first zero of $J_0(x)$ is at $x = 2.405$. This gives

$$\nu \simeq \frac{2.405}{\beta} - \frac{1}{2} \tag{3.48a}$$

Since $|\mathbf{E}|$ and σ vary as $r^{\nu-1}$ there are evidently very small fields and very little charge deep in a conical hole as $\beta \to 0$. For $\beta = \pi/2$, the conical conductor becomes a plane. There $\nu = 1$ and $\sigma \propto 1$, as expected. For $\beta > \pi/2$, the geometry is that of a conical point. Then $\nu < 1$ and the field is singular at $r = 0$. For $\beta \to \pi$, $\nu \to 0$, but rather slowly. An approximation for $(\pi - \beta)$ small is

$$\nu \simeq \left[2 \ln\left(\frac{2}{\pi - \beta}\right)\right]^{-1} \tag{3.48b}$$

This shows that for $(\pi - \beta) \simeq 10°$, $\nu \simeq 0.2$ and even for $(\pi - \beta) \simeq 1°$, $\nu \simeq 0.1$. In any event, for a narrow conical point the fields near the point vary as $r^{-1+\epsilon}$

---

*The orthogonality of the functions $P_{\nu_k}(\cos \theta)$ on the interval $\cos \beta \le x \le 1$ can be shown in the same way as for $P_l(\cos \theta)$—see (3.17)–(3.19). Completeness can also be shown.

[†]Bessel functions are discussed in Section 3.7.

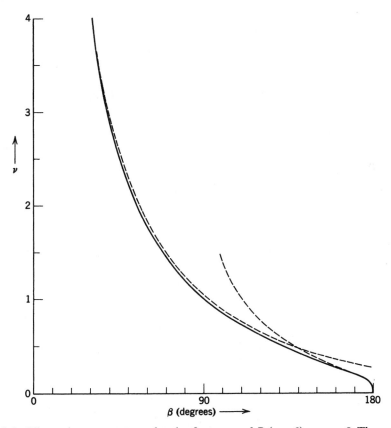

Figure 3.6   The order parameter ν for the first zero of $P_\nu(\cos \beta)$ versus β. The range $0 < \beta < 90°$ corresponds to a conical hole, while $90° < \beta < 180°$ represents a conical point. Near $r = 0$ the fields and surface-charge density are proportional to $r^{\nu-1}$. The dashed curves are the approximate expressions, (3.48a) and (3.48b).

where $\epsilon \ll 1$. Very high fields exist around the point. The efficacy of such points in lightning rods is discussed in Section 2.11.

An extended discussion of potential problems of this general kind by R. N. Hall [*J. Appl. Phys.* **20,** 925 (1949)] includes graphs for a number of the roots ν_k of (3.43) as functions of β.

3.5   *Associated Legendre Functions and the Spherical Harmonics $Y_{lm}(\theta, \phi)$*

So far we have dealt with potential problems possessing azimuthal symmetry with solutions of the form (3.33). Unless the range in θ is restricted, as in Section 3.4, these involve only ordinary Legendre polynomials. The general potential problem can, however, have azimuthal variations so that $m \neq 0$ in (3.5) and (3.9). Then we need the generalization of $P_l(\cos \theta)$, namely, the solution of (3.9) with l and m both arbitrary. In essentially the same manner as for the ordinary Legendre functions it can be shown that to have finite solutions on the interval $-1 \leq x \leq 1$, the parameter l *must be zero or a positive integer* and the *integer m* can take on only the values $-l, -(l - 1), \ldots, 0, \ldots, (l - 1), l$. The solution having these properties is called an associated Legendre function $P_l^m(x)$. For positive m it is defined by the formula*:

$$P_l^m(x) = (-1)^m(1 - x^2)^{m/2} \frac{d^m}{dx^m} P_l(x) \tag{3.49}$$

If Rodrigues' formula is used to represent $P_l(x)$, a definition valid for both positive and negative m is obtained:

$$P_l^m(x) = \frac{(-1)^m}{2^l l!} (1 - x^2)^{m/2} \frac{d^{l+m}}{dx^{l+m}} (x^2 - 1)^l \tag{3.50}$$

$P_l^{-m}(x)$ and $P_l^m(x)$ are proportional, since the differential equation (3.9) depends only on m^2 and m is an integer. It can be shown that

$$P_l^{-m}(x) = (-1)^m \frac{(l - m)!}{(l + m)!} P_l^m(x) \tag{3.51}$$

For fixed m the functions $P_l^m(x)$ form an orthogonal set in the index l on the interval $-1 \le x \le 1$. By the same means as for the Legendre functions the orthogonality relation can be obtained:

$$\int_{-1}^{1} P_{l'}^m(x)P_l^m(x) \, dx = \frac{2}{2l + 1} \frac{(l + m)!}{(l - m)!} \delta_{l'l} \tag{3.52}$$

The solution of the Laplace equation was decomposed into a product of factors for the three variables r, θ, and ϕ. It is convenient to combine the angular factors and construct orthonormal functions over the unit sphere. We will call these functions *spherical harmonics*, although this terminology is often reserved for solutions of the generalized Legendre equation (3.9). Our spherical harmonics are sometimes called "tesseral harmonics" in older books. The functions $Q_m(\phi) = e^{im\phi}$ form a complete set of orthogonal functions in the index m on the interval $0 \le \phi \le 2\pi$. The functions $P_l^m(\cos \theta)$ form a similar set in the index l for each m value on the interval $-1 \le \cos \theta \le 1$. Therefore their product $P_l^m Q_m$ will form a complete orthogonal set on the surface of the unit sphere in the two indices l, m. From the normalization condition (3.52) it is clear that the suitably normalized functions, denoted by $Y_{lm}(\theta, \phi)$, are

$$Y_{lm}(\theta, \phi) = \sqrt{\frac{2l + 1}{4\pi} \frac{(l - m)!}{(l + m)!}} P_l^m(\cos \theta)e^{im\phi} \tag{3.53}$$

From (3.51) it can be seen that

$$Y_{l,-m}(\theta, \phi) = (-1)^m Y_{lm}^*(\theta, \phi) \tag{3.54}$$

The normalization and orthogonality conditions are

$$\int_0^{2\pi} d\phi \int_0^{\pi} \sin \theta \, d\theta \, Y_{l'm'}^*(\theta, \phi)Y_{lm}(\theta, \phi) = \delta_{l'l}\delta_{m'm} \tag{3.55}$$

The completeness relation, equivalent to (2.35), is

$$\sum_{l=0}^{\infty} \sum_{m=-l}^{l} Y_{lm}^*(\theta', \phi')Y_{lm}(\theta, \phi) = \delta(\phi - \phi')\delta(\cos \theta - \cos \theta') \tag{3.56}$$

---

*The choice of phase for $P_l^m(x)$ is that of *Magnus et al.* and E. U. Condon and G. H. Shortley in *Theory of Atomic Spectra*, Cambridge University Press (1953). For explicit expressions and recursion formulas, see *Magnus et al.*, Section 4.3.

For a few small l values and $m \geq 0$ the list below shows the explicit form of the $Y_{lm}(\theta, \phi)$. For negative m values (3.54) can be used.

SPHERICAL HARMONICS $Y_{lm}(\theta, \phi)$

$$l = 0 \qquad Y_{00} = \frac{1}{\sqrt{4\pi}}$$

$$l = 1 \quad \begin{cases} Y_{11} = -\sqrt{\dfrac{3}{8\pi}} \sin\theta e^{i\phi} \\[4mm] Y_{10} = \sqrt{\dfrac{3}{4\pi}} \cos\theta \end{cases}$$

$$l = 2 \quad \begin{cases} Y_{22} = \dfrac{1}{4}\sqrt{\dfrac{15}{2\pi}} \sin^2\theta e^{2i\phi} \\[4mm] Y_{21} = -\sqrt{\dfrac{15}{8\pi}} \sin\theta \cos\theta e^{i\phi} \\[4mm] Y_{20} = \sqrt{\dfrac{5}{4\pi}} \left(\tfrac{3}{2}\cos^2\theta - \tfrac{1}{2}\right) \end{cases}$$

$$l = 3 \quad \begin{cases} Y_{33} = -\dfrac{1}{4}\sqrt{\dfrac{35}{4\pi}} \sin^3\theta e^{3i\phi} \\[4mm] Y_{32} = \dfrac{1}{4}\sqrt{\dfrac{105}{2\pi}} \sin^2\theta \cos\theta e^{2i\phi} \\[4mm] Y_{31} = -\dfrac{1}{4}\sqrt{\dfrac{21}{4\pi}} \sin\theta \, (5\cos^2\theta - 1)e^{i\phi} \\[4mm] Y_{30} = \sqrt{\dfrac{7}{4\pi}} \left(\tfrac{5}{2}\cos^3\theta - \tfrac{3}{2}\cos\theta\right) \end{cases}$$

Note that, for $m = 0$,

$$Y_{l0}(\theta, \phi) = \sqrt{\frac{2l + 1}{4\pi}} \, P_l(\cos\theta) \tag{3.57}$$

An arbitrary function $g(\theta, \phi)$ can be expanded in spherical harmonics:

$$g(\theta, \phi) = \sum_{l=0}^{\infty} \sum_{m=-l}^{l} A_{lm} Y_{lm}(\theta, \phi) \tag{3.58}$$

where the coefficients are

$$A_{lm} = \int d\Omega \, Y^*_{lm}(\theta, \phi) g(\theta, \phi)$$

A point of interest to us in the next section is the form of the expansion for $\theta = 0$. With definition (3.57), we find:

$$[g(\theta, \phi)]_{\theta=0} = \sum_{l=0}^{\infty} \sqrt{\frac{2l + 1}{4\pi}} \, A_{l0} \tag{3.59}$$

where

$$A_{l0} = \sqrt{\frac{2l + 1}{4\pi}} \int d\Omega \, P_l(\cos \theta) g(\theta, \phi) \tag{3.60}$$

All terms in the series with $m \neq 0$ vanish at $\theta = 0$.

The general solution for a boundary-value problem in spherical coordinates can be written in terms of spherical harmonics and powers of r in a generalization of (3.33):

$$\Phi(r, \theta, \phi) = \sum_{l=0}^{\infty} \sum_{m=-l}^{l} [A_{lm}r^l + B_{lm}r^{-(l+1)}]Y_{lm}(\theta, \phi) \tag{3.61}$$

If the potential is specified on a spherical surface, the coefficients can be determined by evaluating (3.61) on the surface and using (3.58).

3.6 *Addition Theorem for Spherical Harmonics*

A mathematical result of considerable interest and use is called the *addition theorem* for spherical harmonics. Two coordinate vectors $\mathbf{x}$ and $\mathbf{x}'$, with spherical coordinates (r, θ, ϕ) and (r', θ', ϕ'), respectively, have an angle γ between them, as shown in Fig. 3.7. The addition theorem expresses a Legendre polynomial of order l in the angle γ in terms of products of the spherical harmonics of the angles θ, ϕ and θ', ϕ':

$$P_l(\cos \gamma) = \frac{4\pi}{2l + 1} \sum_{m=-l}^{l} Y^*_{lm}(\theta', \phi')Y_{lm}(\theta, \phi) \tag{3.62}$$

where $\cos \gamma = \cos \theta \cos \theta' + \sin \theta \sin \theta' \cos(\phi - \phi')$. To prove this theorem we consider the vector $\mathbf{x}'$ as fixed in space. Then $P_l(\cos \gamma)$ is a function of the angles θ, ϕ, with the angles θ', ϕ' as parameters. It may be expanded in a series (3.58):

$$P_l(\cos \gamma) = \sum_{l'=0}^{\infty} \sum_{m=-l'}^{l'} A_{l'm}(\theta', \phi')Y_{l'm}(\theta, \phi) \tag{3.63}$$

Comparison with (3.62) shows that only terms with $l' = l$ appear. To see why this is so, note that if coordinate axes are chosen so that $\mathbf{x}'$ is on the z axis, then γ becomes the usual polar angle and $P_l(\cos \gamma)$ satisfies the equation:

$$\nabla'^2 P_l(\cos \gamma) + \frac{l(l + 1)}{r^2} P_l(\cos \gamma) = 0 \tag{3.64}$$

where ∇'^2 is the Laplacian referred to these new axes. If the axes are now rotated to the position shown in Fig. 3.7, $\nabla'^2 = \nabla^2$ and r is unchanged.* Consequently $P_l(\cos \gamma)$ still satisfies an equation of the form (3.64); i.e., it is a spherical harmonic of order l. This means that it is a linear combination of Y_{lm}'s of that order only:

$$P_l(\cos \gamma) = \sum_{m=-l}^{l} A_m(\theta', \phi')Y_{lm}(\theta, \phi) \tag{3.65}$$

The coefficients $A_m(\theta', \phi')$ are given by

$$A_m(\theta', \phi') = \int Y^*_{lm}(\theta, \phi)P_l(\cos \gamma) \, d\Omega \tag{3.66}$$

---

*The proof that $\nabla'^2 = \nabla^2$ under rotations follows most easily from noting that $\nabla^2\psi = \nabla \cdot \nabla\psi$ is an operator scalar product and that all scalar products are invariant under rotations.

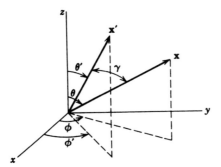

Figure 3.7

To evaluate this coefficient we note that it may be viewed, according to (3.60), as the $m' = 0$ coefficient in an expansion of the function $\sqrt{4\pi/(2l + 1)}\ Y^*_{lm}(\theta, \phi)$ in a series of $Y_{lm'}(\gamma, \beta)$ referred to the primed axis of (3.64). From (3.59) it is then found that, since only one l value is present, coefficient (3.66) is

$$A_m(\theta', \phi') = \frac{4\pi}{2l + 1}\ \{Y^*_{lm}[\theta(\gamma, \beta), \phi(\gamma, \beta)]\}_{\gamma=0} \qquad (3.67)$$

In the limit $\gamma \to 0$, the angles (θ, ϕ), as functions of (γ, β), go over into (θ', ϕ'). Thus addition theorem (3.62) is proved. Sometimes the theorem is written in terms of $P^m_l(\cos\theta)$ rather than Y_{lm}. Then it has the form:

$$P_l(\cos\gamma) = P_l(\cos\theta)P_l(\cos\theta') \qquad (3.68)$$
$$+ 2 \sum_{m=1}^{l} \frac{(l - m)!}{(l + m)!}\ P^m_l(\cos\theta)P^m_l(\cos\theta')\ \cos[m(\phi - \phi')]$$

If the angle γ goes to zero, there results a "sum rule" for the squares of Y_{lm}'s:

$$\sum_{m=-l}^{l} |Y_{lm}(\theta, \phi)|^2 = \frac{2l + 1}{4\pi} \qquad (3.69)$$

The addition theorem can be used to put expansion (3.38) of the potential at **x** due to a unit charge at **x'** into its most explicit form. Substituting (3.62) for $P_l(\cos\gamma)$ into (3.38), we obtain

$$\frac{1}{|\mathbf{x} - \mathbf{x'}|} = 4\pi \sum_{l=0}^{\infty} \sum_{m=-l}^{l} \frac{1}{2l + 1} \frac{r^l_<}{r^{l+1}_>}\ Y^*_{lm}(\theta', \phi')Y_{lm}(\theta, \phi) \qquad (3.70)$$

Equation (3.70) gives the potential in a completely factorized form in the coordinates **x** and **x'**. This is useful in any integrations over charge densities, etc., where one variable is the variable of integration and the other is the coordinate of the observation point. The price paid is that there is a double sum rather than a single term.

3.7   *Laplace Equation in Cylindrical Coordinates; Bessel Functions*

In cylindrical coordinates (ρ, ϕ, z), as shown in Fig. 3.8, the Laplace equation takes the form:

$$\frac{\partial^2\Phi}{\partial\rho^2} + \frac{1}{\rho}\frac{\partial\Phi}{\partial\rho} + \frac{1}{\rho^2}\frac{\partial^2\Phi}{\partial\phi^2} + \frac{\partial^2\Phi}{\partial z^2} = 0 \qquad (3.71)$$

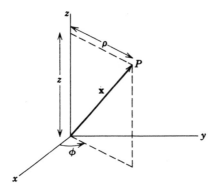

Figure 3.8

The separation of variables is accomplished by the substitution:

$$\Phi(\rho,\ \phi,\ z) = R(\rho)Q(\phi)Z(z) \tag{3.72}$$

In the usual way this leads to the three ordinary differential equations:

$$\frac{d^2Z}{dz^2} - k^2 Z = 0 \tag{3.73}$$

$$\frac{d^2Q}{d\phi^2} + \nu^2 Q = 0 \tag{3.74}$$

$$\frac{d^2R}{d\rho^2} + \frac{1}{\rho}\frac{dR}{d\rho} + \left(k^2 - \frac{\nu^2}{\rho^2}\right)R = 0 \tag{3.75}$$

The solutions of the first two equations are elementary:

$$Z(z) = e^{\pm kz} \tag{3.76}$$
$$Q(\phi) = e^{\pm i\nu\phi}$$

For the potential to be single-valued when the full azimuth is allowed, ν must be an integer. But barring some boundary-condition requirement in the z direction, the parameter k is arbitrary. For the present we assume that k is real and positive.

The radial equation can be put in a standard form by the change of variable $x = k\rho$. Then it becomes

$$\frac{d^2R}{dx^2} + \frac{1}{x}\frac{dR}{dx} + \left(1 - \frac{\nu^2}{x^2}\right)R = 0 \tag{3.77}$$

This is the Bessel equation, and the solutions are called *Bessel functions* of order ν. If a power series solution of the form

$$R(x) = x^\alpha \sum_{j=0}^{\infty} a_j x^j \tag{3.78}$$

is assumed, then it is found that

$$\alpha = \pm\nu \tag{3.79}$$

and

$$a_{2j} = -\frac{1}{4j(j + \alpha)}\,a_{2j-2} \tag{3.80}$$

for $j = 1, 2, 3, \ldots$. All odd powers of x^j have vanishing coefficients. The recursion formula can be iterated to obtain

$$a_{2j} = \frac{(-1)^j \Gamma(\alpha + 1)}{2^{2j} j! \; \Gamma(j + \alpha + 1)} a_0 \tag{3.81}$$

It is conventional to choose the constant $a_0 = [2^\alpha \Gamma(\alpha + 1)]^{-1}$. Then the two solutions are

$$J_\nu(x) = \left(\frac{x}{2}\right)^\nu \sum_{j=0}^\infty \frac{(-1)^j}{j! \; \Gamma(j + \nu + 1)} \left(\frac{x}{2}\right)^{2j} \tag{3.82}$$

$$J_{-\nu}(x) = \left(\frac{x}{2}\right)^{-\nu} \sum_{j=0}^\infty \frac{(-1)^j}{j! \; \Gamma(j - \nu + 1)} \left(\frac{x}{2}\right)^{2j} \tag{3.83}$$

These solutions are called Bessel functions of the first kind of order $\pm \nu$. The series converge for all finite values of x. If ν *is not an integer*, these two solutions $J_{\pm\nu}(x)$ form a pair of linearly independent solutions to the second-order Bessel equation. However, if ν is an integer, it is well known that the solutions are linearly dependent. In fact, for $\nu = m$, an integer, it can be seen from the series representation that

$$J_{-m}(x) = (-1)^m J_m(x) \tag{3.84}$$

Consequently it is necessary to find another linearly independent solution when ν is an integer. It is customary, even if ν is not an integer, to replace the pair $J_{\pm\nu}(x)$ by $J_\nu(x)$ and $N_\nu(x)$, the Neumann function (or Bessel function of the second kind):

$$N_\nu(x) = \frac{J_\nu(x) \cos \nu\pi - J_{-\nu}(x)}{\sin \nu\pi} \tag{3.85}$$

For ν not an integer, $N_\nu(x)$ is clearly linearly independent of $J_\nu(x)$. In the limit $\nu \to$ integer, it can be shown that $N_\nu(x)$ is still linearly independent of $J_\nu(x)$. As expected, it involves $\log x$. Its series representation is given in the reference books.

The Bessel functions of the third kind, called *Hankel functions*, are defined as linear combinations of $J_\nu(x)$ and $N_\nu(x)$:

$$\left. \begin{array}{l} H_\nu^{(1)}(x) = J_\nu(x) + iN_\nu(x) \\ H_\nu^{(2)}(x) = J_\nu(x) - iN_\nu(x) \end{array} \right\} \tag{3.86}$$

The Hankel functions form a fundamental set of solutions to the Bessel equation, just as do $J_\nu(x)$ and $N_\nu(x)$.

The functions J_ν, N_ν, $H_\nu^{(1)}$, $H_\nu^{(2)}$ all satisfy the recursion formulas

$$\Omega_{\nu-1}(x) + \Omega_{\nu+1}(x) = \frac{2\nu}{x} \Omega_\nu(x) \tag{3.87}$$

$$\Omega_{\nu-1}(x) - \Omega_{\nu+1}(x) = 2 \frac{d\Omega_\nu(x)}{dx} \tag{3.88}$$

where $\Omega_\nu(x)$ is any one of the cylinder functions of order ν. These may be verified directly from the series representation (3.82).

For reference purposes, the limiting forms of the various kinds of Bessel function are given for small and large values of their argument. For simplicity, we show only the leading terms:

$$x \ll 1 \qquad J_\nu(x) \to \frac{1}{\Gamma(\nu + 1)} \left(\frac{x}{2}\right)^\nu \tag{3.89}$$

$$N_\nu(x) \to \begin{cases} \dfrac{2}{\pi} \left[\ln\left(\dfrac{x}{2}\right) + 0.5772 \cdots \right], & \nu = 0 \\[2ex] -\dfrac{\Gamma(\nu)}{\pi} \left(\dfrac{2}{x}\right)^\nu & \nu \neq 0 \end{cases} \tag{3.90}$$

In these formulas ν is assumed to be real and nonnegative.

$$x \gg 1, \nu \qquad J_\nu(x) \to \sqrt{\frac{2}{\pi x}} \cos\left(x - \frac{\nu\pi}{2} - \frac{\pi}{4} \right) \tag{3.91}$$

$$N_\nu(x) \to \sqrt{\frac{2}{\pi x}} \sin\left(x - \frac{\nu\pi}{2} - \frac{\pi}{4} \right)$$

The transition from the small x behavior to the large x asymptotic form occurs in the region of $x \sim \nu$.

From the asymptotic forms (3.91) it is clear that each Bessel function has an infinite number of roots. We will be chiefly concerned with the roots of $J_\nu(x)$:

$$J_\nu(x_{\nu n}) = 0 \qquad (n = 1, 2, 3, \ldots) \tag{3.92}$$

$x_{\nu n}$ is the nth root of $J_\nu(x)$. For the first few integer values of ν, the first three roots are:

$$\nu = 0, \qquad x_{0n} = 2.405, \ 5.520, \ 8.654, \ldots$$
$$\nu = 1, \qquad x_{1n} = 3.832, \ 7.016, \ 10.173, \ldots$$
$$\nu = 2, \qquad x_{2n} = 5.136, \ 8.417, \ 11.620, \ldots$$

For higher roots, the asymptotic formula

$$x_{\nu n} \simeq n\pi + (\nu - \tfrac{1}{2}) \frac{\pi}{2}$$

gives adequate accuracy (to at least three figures). Tables of roots are given in *Jahnke, Emde, and Lösch* (p. 194) and *Abramowitz and Stegun* (p. 409).

Having found the solution of the radial part of the Laplace equation in terms of Bessel functions, we can now ask in what sense the Bessel functions form an orthogonal, complete set of functions. We consider only Bessel functions of the first kind, and we show that $\sqrt{\rho}\, J_\nu(x_{\nu n}\rho/a)$, for fixed $\nu \geq 0$, $n = 1, 2, \ldots$, form an orthogonal set on the interval $0 \leq \rho \leq a$. The demonstration starts with the differential equation satisfied by $J_\nu(x_{\nu n}\rho/a)$:

$$\frac{1}{\rho} \frac{d}{d\rho} \left[\rho \frac{dJ_\nu\left(x_{\nu n}\dfrac{\rho}{a}\right)}{d\rho} \right] + \left(\frac{x_{\nu n}^2}{a^2} - \frac{\nu^2}{\rho^2} \right) J_\nu\left(x_{\nu n}\frac{\rho}{a}\right) = 0 \tag{3.93}$$

If we multiply the equation by $\rho J_\nu(x_{\nu n'}\rho/a)$ and integrate from 0 to a, we obtain

$$\int_0^a J_\nu\left(x_{\nu n'}\frac{\rho}{a}\right)\frac{d}{d\rho}\left[\rho\frac{dJ_\nu\left(x_{\nu n}\frac{\rho}{a}\right)}{d\rho}\right]d\rho +$$

$$\int_0^a \left(\frac{x_{\nu n}^2}{a^2}-\frac{\nu^2}{\rho^2}\right)\rho J_\nu\left(x_{\nu n'}\frac{\rho}{a}\right)J_\nu\left(x_{\nu n}\frac{\rho}{a}\right)d\rho = 0$$

Integration by parts, combined with the vanishing of $(\rho J_\nu J_\nu')$ at $\rho = 0$ (for $\nu \geq 0$) and $\rho = a$, leads to the result:

$$-\int_0^a \rho\frac{dJ_\nu\left(x_{\nu n'}\frac{\rho}{a}\right)}{d\rho}\frac{dJ_\nu\left(x_{\nu n}\frac{\rho}{a}\right)}{d\rho}d\rho + \int_0^a \left(\frac{x_{\nu n}^2}{a^2}-\frac{\nu^2}{\rho^2}\right)\rho J_\nu\left(x_{\nu n'}\frac{\rho}{a}\right)J_\nu\left(x_{\nu n}\frac{\rho}{a}\right)d\rho = 0$$

If we now write down the same expression, with n and n' interchanged, and subtract, we obtain the orthogonality condition:

$$(x_{\nu n}^2 - x_{\nu n'}^2)\int_0^a \rho J_\nu\left(x_{\nu n'}\frac{\rho}{a}\right)J_\nu\left(x_{\nu n}\frac{\rho}{a}\right)d\rho = 0 \tag{3.94}$$

Adroit use of the differential equation, and the recursion formulas (3.87) and (3.88) leads to the normalization integral:

$$\int_0^a \rho J_\nu\left(x_{\nu n'}\frac{\rho}{a}\right)J_\nu\left(x_{\nu n}\frac{\rho}{a}\right)d\rho = \frac{a^2}{2}[J_{\nu+1}(x_{\nu n})]^2\delta_{n'n} \tag{3.95}$$

Assuming that the set of Bessel functions is complete, we can expand an arbitrary function of ρ on the interval $0 \leq \rho \leq a$ in a Fourier–Bessel series:

$$f(\rho) = \sum_{n=1}^\infty A_{\nu n}J_\nu\left(x_{\nu n}\frac{\rho}{a}\right) \tag{3.96}$$

where

$$A_{\nu n} = \frac{2}{a^2 J_{\nu+1}^2(x_{\nu n})}\int_0^a \rho f(\rho)J_\nu\left(\frac{x_{\nu n}\rho}{a}\right)d\rho \tag{3.97}$$

Our derivation of (3.96) involved the restriction $\nu \geq 0$. Actually it can be proved to hold for all $\nu \geq -1$.

Expansion (3.96) and (3.97) is the conventional Fourier–Bessel series and is particularly appropriate to functions that vanish at $\rho = a$ (e.g., homogeneous Dirichlet boundary conditions on a cylinder; see the following section). But it will be noted that an alternative expansion is possible in a series of functions $\sqrt{\rho}\,J_\nu(y_{\nu n}\rho/a)$ where $y_{\nu n}$ is the nth root of the equation $[dJ_\nu(x)]/dx = 0$. The reason is that, in proving the orthogonality of the functions, all that is demanded is that the quantity $[\rho J_\nu(k\rho)(d/d\rho)J_\nu(k'\rho) - \rho J_\nu(k'\rho)(d/d\rho)J_\nu(k\rho)]$ vanish at the end points $\rho = 0$ and $\rho = a$. The requirement is met by $\lambda = x_{\nu n}/a$ or $\lambda = y_{\nu n}/a$, where $J_\nu(x_{\nu n}) = 0$ and $J_\nu'(y_{\nu n}) = 0$, or, more generally, by $\rho(d/d\rho)J_\nu(k\rho) + \lambda J_\nu(k\rho) = 0$ at the end points, with λ a constant independent of k. The expansion in terms of the set $\sqrt{\rho}\,J_\nu(y_{\nu n}\rho/a)$ is especially useful for functions with vanishing slope at $\rho = a$. (See Problem 3.11.)

A Fourier–Bessel series is only one type of expansion involving Bessel functions. Some of the other possibilities are:

Neumann series: $\displaystyle\sum_{n=0}^{\infty} a_n J_{\nu+n}(z)$

Kapteyn series: $\displaystyle\sum_{n=0}^{\infty} a_n J_{\nu+n}((\nu + n)z)$

Schlömilch series: $\displaystyle\sum_{n=1}^{\infty} a_n J_{\nu}(nx)$

The reader may refer to *Watson* (Chapters XVI–XIX) for a detailed discussion of the properties of these series. Kapteyn series occur in the discussion of the Kepler motion of planets and of radiation by rapidly moving charges (see Problems 14.14 and 14.15).

Before leaving the properties of Bessel functions, we note that if, in the separation of the Laplace equation, the separation constant k^2 in (3.73) had been taken as $-k^2$, then $Z(z)$ would have been $\sin kz$ or $\cos kz$ and the equation for $R(\rho)$ would have been:

$$\frac{d^2R}{d\rho^2} + \frac{1}{\rho}\frac{dR}{d\rho} - \left(k^2 + \frac{\nu^2}{\rho^2}\right)R = 0 \tag{3.98}$$

With $k\rho = x$, this becomes

$$\frac{d^2R}{dx^2} + \frac{1}{x}\frac{dR}{dx} - \left(1 + \frac{\nu^2}{x^2}\right)R = 0 \tag{3.99}$$

The solutions of this equation are called *modified Bessel functions*. It is evident that they are just Bessel functions of pure imaginary argument. The usual choices of linearly independent solutions are denoted by $I_\nu(x)$ and $K_\nu(x)$. They are defined by

$$I_\nu(x) = i^{-\nu}J_\nu(ix) \tag{3.100}$$

$$K_\nu(x) = \frac{\pi}{2}i^{\nu+1}H_\nu^{(1)}(ix) \tag{3.101}$$

and are real functions for real x and ν. Their limiting forms for small and large x are, assuming real $\nu \geq 0$:

$$x \ll 1 \qquad I_\nu(x) \rightarrow \frac{1}{\Gamma(\nu + 1)}\left(\frac{x}{2}\right)^\nu \tag{3.102}$$

$$K_\nu(x) \rightarrow \begin{cases} -\left[\ln\left(\dfrac{x}{2}\right) + 0.5772 \cdots\right], & \nu = 0 \\[2mm] \dfrac{\Gamma(\nu)}{2}\left(\dfrac{2}{x}\right)^\nu, & \nu \neq 0 \end{cases} \tag{3.103}$$

$$x \gg 1,\ \nu \qquad I_\nu(x) \rightarrow \frac{1}{\sqrt{2\pi x}}e^x\left[1 + 0\left(\frac{1}{x}\right)\right] \tag{3.104}$$

$$K_\nu(x) \rightarrow \sqrt{\frac{\pi}{2x}}\,e^{-x}\left[1 + 0\left(\frac{1}{x}\right)\right]$$

3.8   *Boundary-Value Problems in Cylindrical Coordinates*

The solution of the Laplace equation in cylindrical coordinates is $\Phi = R(\rho)Q(\phi)Z(z)$, where the separate factors are given in the previous section. Consider now the specific boundary-value problem shown in Fig. 3.9. The cylinder has a radius a and a height L, the top and bottom surfaces being at $z = L$ and $z = 0$. The potential on the side and the bottom of the cylinder is zero, while the top has a potential $\Phi = V(\rho, \phi)$. We want to find the potential at any point inside the cylinder. In order that Φ be single valued and vanish at $z = 0$,

$$Q(\phi) = A \sin m\phi + B \cos m\phi$$
$$Z(z) = \sinh kz$$

where $\nu = m$ is an integer and k is a constant to be determined. The radial factor is

$$R(\rho) = CJ_m(k\rho) + DN_m(k\rho)$$

If the potential is finite at $\rho = 0$, $D = 0$. The requirement that the potential vanish at $\rho = a$ means that k can take on only those special values:

$$k_{mn} = \frac{x_{mn}}{a} \qquad (n = 1, 2, 3, \ldots)$$

where x_{mn} are the roots of $J_m(x_{mn}) = 0$.

Combining all these conditions, we find that the general form of the solution is

$$\Phi(\rho, \phi, z) = \sum_{m=0}^{\infty} \sum_{n=1}^{\infty} J_m(k_{mn}\rho) \sinh(k_{mn}z)(A_{mn} \sin m\phi \tag{3.105a}$$
$$+ B_{mn} \cos m\phi)$$

At $z = L$, we are given the potential as $V(\rho, \phi)$. Therefore we have

$$V(\rho, \phi) = \sum_{m,n} \sinh(k_{mn}L)J_m(k_{mn}\rho)(A_{mn} \sin m\phi + B_{mn} \cos m\phi)$$

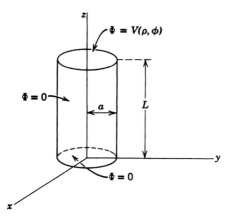

Figure 3.9

This is a Fourier series in ϕ and a Fourier–Bessel series in ρ. The coefficients are, from (2.37) and (3.97),

$$A_{mn} = \frac{2 \operatorname{cosech}(k_{mn}L)}{\pi a^2 J_{m+1}^2(k_{mn}a)} \int_0^{2\pi} d\phi \int_0^a d\rho \, \rho V(\rho, \phi) J_m(k_{mn}\rho) \sin m\phi$$

and $\hspace{9cm}$ (3.105b)

$$B_{mn} = \frac{2 \operatorname{cosech}(k_{mn}L)}{\pi a^2 J_{m+1}^2(k_{mn}a)} \int_0^{2\pi} d\phi \int_0^a d\rho \, \rho V(\rho, \phi) J_m(k_{mn}\rho) \cos m\phi$$

with the proviso that, for $m = 0$, we use $\frac{1}{2}B_{0n}$ in the series.

The particular form of expansion (3.105a) is dictated by the requirement that the potential vanish at $z = 0$ for arbitrary ρ and at $\rho = a$ for arbitrary z. For different boundary conditions the expansion would take a different form. An example where the potential is zero on the end faces and equal to $V(\phi, z)$ on the side surface is left as Problem 3.9 for the reader.

The Fourier–Bessel series (3.105) is appropriate for a finite interval in ρ, $0 \leq \rho \leq a$. If $a \to \infty$, the series goes over into an integral in a manner entirely analogous to the transition from a trigonometric Fourier series to a Fourier integral. Thus, for example, if the potential in charge-free space is finite for $z \geq 0$ and vanishes for $z \to \infty$, the general form of the solution for $z \geq 0$ must be

$$\Phi(\rho, \phi, z) = \sum_{m=0}^{\infty} \int_0^{\infty} dk \, e^{-kz} J_m(k\rho)[A_m(k) \sin m\phi + B_m(k) \cos m\phi] \quad (3.106)$$

If the potential is specified over the whole plane $z = 0$ to be $V(\rho, \phi)$ the coefficients are determined by

$$V(\rho, \phi) = \sum_{m=0}^{\infty} \int_0^{\infty} dk \, J_m(k\rho)[A_m(k) \sin m\phi + B_m(k) \cos m\phi]$$

The variation in ϕ is just a Fourier series. Consequently the coefficients $A_m(k)$ and $B_m(k)$ are separately specified by the integral relations:

$$\frac{1}{\pi} \int_0^{2\pi} V(\rho, \phi) \begin{Bmatrix} \sin m\phi \\ \cos m\phi \end{Bmatrix} d\phi = \int_0^{\infty} J_m(k'\rho) \begin{Bmatrix} A_m(k') \\ B_m(k') \end{Bmatrix} dk' \quad (3.107)$$

These radial integral equations of the first kind can be easily solved, since they are *Hankel transforms*. For our purposes, the integral relation,

$$\int_0^{\infty} x J_m(kx) J_m(k'x) \, dx = \frac{1}{k} \delta(k' - k) \quad (3.108)$$

can be exploited to invert equations (3.107). Multiplying both sides by $\rho J_m(k\rho)$ and integrating over ρ, we find with the help of (3.108) that the coefficients are determined by integrals over the whole area of the plane $z = 0$:

$$\begin{Bmatrix} A_m(k) \\ B_m(k) \end{Bmatrix} = \frac{k}{\pi} \int_0^{\infty} d\rho \, \rho \int_0^{2\pi} d\phi \, V(\rho, \phi) J_m(k\rho) \begin{Bmatrix} \sin m\phi \\ \cos m\phi \end{Bmatrix} \quad (3.109)$$

As usual, for $m = 0$, we must use $\frac{1}{2}B_0(k)$ in series (3.106).

While on the subject of expansions in terms of Bessel functions, we observe that the functions $J_\nu(kx)$ for fixed ν, $\operatorname{Re}(\nu) > -1$, form a complete, orthogonal (in k) set of functions on the interval, $0 < x < \infty$. For each m value (and fixed ϕ and z), the expansion in k in (3.106) is a special case of the expansion,

$$A(x) = \int_0^\infty \tilde{A}(k)J_\nu(kx)\,dk, \text{ where } \tilde{A}(k) = k\int_0^\infty xA(x)J_\nu(kx)\,dx \quad (3.110)$$

An important example of these expansions occurs in spherical coordinates, with spherical Bessel functions, $j_l(kr)$, $l = 0, 1, 2, \ldots$. For present purposes we merely note the definition,

$$j_l(z) = \sqrt{\frac{\pi}{2z}}\,J_{l+1/2}(z) \quad (3.111)$$

[Details of spherical Bessel functions may be found in Chapter 9.] The orthogonality relation (3.108) evidently becomes

$$\int_0^\infty r^2 j_l(kr)j_l(k'r)\,dr = \frac{\pi}{2k^2}\,\delta(k - k') \quad (3.112)$$

The completeness relation has the same form, with $r \to k$, $k \to r$, $k' \to r'$. The Fourier–spherical Bessel expansion for a given l is then

$$A(r) = \int_0^\infty \tilde{A}(k)j_l(kr)\,dk, \text{ where } \tilde{A}(k) = \frac{2k^2}{\pi}\int_0^\infty r^2 A(r)j_l(kr)\,dr \quad (3.113)$$

Such expansions are useful for current decay in conducting media or time-dependent magnetic diffusion for which angular symmetry reduces consideration to one or a few l values. See Problems 5.35 and 5.36.

3.9   *Expansion of Green Functions in Spherical Coordinates*

To handle problems involving distributions of charge as well as boundary values for the potential (i.e., solutions of the Poisson equation), it is necessary to determine the Green function $G(\mathbf{x}, \mathbf{x}')$ that satisfies the appropriate boundary conditions. Often these boundary conditions are specified on surfaces of some separable coordinate system (e.g., spherical or cylindrical boundaries). Then it is convenient to express the Green function as a series of products of the functions appropriate to the coordinates in question. We first illustrate the type of expansion involved by considering spherical coordinates.

For the case of no boundary surfaces, except at infinity, we already have the expansion of the Green function, namely (3.70):

$$\frac{1}{|\mathbf{x} - \mathbf{x}'|} = 4\pi \sum_{l=0}^\infty \sum_{m=-l}^l \frac{1}{2l + 1} \frac{r_<^l}{r_>^{l+1}}\,Y_{lm}^*(\theta', \phi')Y_{lm}(\theta, \phi)$$

Suppose that we wish to obtain a similar expansion for the Green function appropriate for the "exterior" problem with a spherical boundary at $r = a$. The result is readily found from the image form of the Green function (2.16). Using expansion (3.70) for both terms in (2.16), we obtain:

$$G(\mathbf{x}, \mathbf{x}') = 4\pi \sum_{l,m} \frac{1}{2l + 1} \left[\frac{r_<^l}{r_>^{l+1}} - \frac{1}{a}\left(\frac{a^2}{rr'}\right)^{l+1} \right] Y_{lm}^*(\theta', \phi')Y_{lm}(\theta, \phi) \quad (3.114)$$

To see clearly the structure of (3.114) and to verify that it satisfies the boundary conditions, we exhibit the radial factors separately for $r < r'$ and for $r > r'$:

$$\left[\frac{r_<^l}{r_>^{l+1}} - \frac{1}{a}\left(\frac{a^2}{rr'}\right)^{l+1}\right] = \begin{cases} \dfrac{1}{r'^{l+1}}\left[r^l - \dfrac{a^{2l+1}}{r^{l+1}}\right], & r < r' \\[4mm] \left[r'^l - \dfrac{a^{2l+1}}{r'^{l+1}}\right]\dfrac{1}{r^{l+1}}, & r > r' \end{cases} \tag{3.115}$$

First of all, we note that for either r or r' equal to a the radial factor vanishes, as required. Similarly, as r or $r' \to \infty$, the radial factor vanishes. It is symmetric in r and r'. Viewed as a function of r, for fixed r', the radial factor is just a linear combination of the solutions r^l and $r^{-(l+1)}$ of the radial part (3.7) of the Laplace equation. It is admittedly a different linear combination for $r < r'$ and for $r > r'$. The reason for this, which will become apparent below, is connected with the fact that the Green function is a solution of the Poisson equation with a delta function inhomogeneity.

Now that we have seen the general structure of the expansion of a Green function in separable coordinates we turn to the systematic construction of such expansions from first principles. A Green function for a Dirichlet potential problem satisfies the equation

$$\nabla_x^2 G(\mathbf{x}, \mathbf{x}') = -4\pi\delta(\mathbf{x} - \mathbf{x}') \tag{3.116}$$

subject to the boundary conditions $G(\mathbf{x}, \mathbf{x}') = 0$ for either $\mathbf{x}$ or $\mathbf{x}'$ on the boundary surface S. For spherical boundary surfaces we desire an expansion of the general form (3.114). Accordingly we exploit the fact that the delta function can be written*

$$\delta(\mathbf{x} - \mathbf{x}') = \frac{1}{r^2}\,\delta(r - r')\,\delta(\phi - \phi')\,\delta(\cos\theta - \cos\theta')$$

and that the completeness relation (3.56) can be used to represent the angular delta functions:

$$\delta(\mathbf{x} - \mathbf{x}') = \frac{1}{r^2}\,\delta(r - r')\sum_{l=0}^{\infty}\sum_{m=-l}^{l} Y_{lm}^*(\theta', \phi')Y_{lm}(\theta, \phi) \tag{3.117}$$

Then the Green function, considered as a function of $\mathbf{x}$, can be expanded as

$$G(\mathbf{x}, \mathbf{x}') = \sum_{l=0}^{\infty}\sum_{m=-l}^{l} A_{lm}(r|r', \theta', \phi')Y_{lm}(\theta, \phi) \tag{3.118}$$

Substitution of (3.117) and (3.118) into (3.116) leads to the results

$$A_{lm}(r|r', \theta', \phi') = g_l(r, r')Y_{lm}^*(\theta', \phi') \tag{3.119}$$

---

*To express $\delta(\mathbf{x} - \mathbf{x}') = \delta(x_1 - x_1')\delta(x_2 - x_2')\delta(x_3 - x_3')$ in terms of the coordinates (ξ_1, ξ_2, ξ_3), related to (x_1, x_2, x_3) via the Jacobian $J(x_i, \xi_i)$, we note that the meaningful quantity is $\delta(\mathbf{x} - \mathbf{x}')\,d^3x$. Hence

$$\delta(\mathbf{x} - \mathbf{x}') = \frac{1}{|J(x_i, \xi_i)|}\,\delta(\xi_1 - \xi_1')\,\delta(\xi_2 - \xi_2')\,\delta(\xi_3 - \xi_3')$$

See Problem 1.2.

with

$$\frac{1}{r} \frac{d^2}{dr^2} (rg_l(r, r')) - \frac{l(l + 1)}{r^2} g_l(r, r') = -\frac{4\pi}{r^2} \delta(r - r') \qquad (3.120)$$

The radial Green function is seen to satisfy the homogeneous radial equation (3.7) for $r \neq r'$. Thus it can be written as

$$g_l(r, r') = \begin{cases} Ar^l + Br^{-(l+1)} & \text{for } r < r' \\ A'r^l + B'r^{-(l+1)} & \text{for } r > r' \end{cases}$$

The coefficients A, B, A', B' are functions of r' to be determined by the boundary conditions, the requirement implied by $\delta(r - r')$ in (3.120), and the symmetry of $g_l(r, r')$ in r and r'. Suppose that the boundary surfaces are concentric spheres at $r = a$ and $r = b$. The vanishing of $G(\mathbf{x}, \mathbf{x}')$ for $\mathbf{x}$ on the surface implies the vanishing of $g_l(r, r')$ for $r = a$ and $r = b$. Consequently $g_l(r, r')$ becomes

$$g_l(r, r') = \begin{cases} A\left(r^l - \dfrac{a^{2l+1}}{r^{l+1}}\right), & r < r' \\ B'\left(\dfrac{1}{r^{l+1}} - \dfrac{r^l}{b^{2l+1}}\right), & r > r' \end{cases} \qquad (3.121)$$

The symmetry in r and r' requires that the coefficients $A(r')$ and $B'(r')$ be such that $g_l(r, r')$ can be written

$$g_l(r, r') = C\left(r_<^l - \frac{a^{2l+1}}{r_<^{l+1}}\right)\left(\frac{1}{r_>^{l+1}} - \frac{r_>^l}{b^{2l+1}}\right) \qquad (3.122)$$

where $r_<$ ($r_>$) is the smaller (larger) of r and r'. To determine the constant C we must consider the effect of the delta function in (3.120). If we multiply both sides of (3.120) by r and integrate over the interval from $r = r' - \epsilon$ to $r = r' + \epsilon$, where ϵ is very small, we obtain

$$\left\{\frac{d}{dr} [rg_l(r, r')]\right\}_{r' + \epsilon} - \left\{\frac{d}{dr} [rg_l(r, r')]\right\}_{r' - \epsilon} = -\frac{4\pi}{r'} \qquad (3.123)$$

Thus there is a discontinuity in slope at $r = r'$, as indicated in Fig. 3.10.

For $r = r' + \epsilon, r_> = r, r_< = r'$. Hence

$$\left\{\frac{d}{dr} [rg_l(r, r')]\right\}_{r' + \epsilon} = C\left(r'^l - \frac{a^{2l+1}}{r'^{l+1}}\right)\left[\frac{d}{dr}\left(\frac{1}{r^l} - \frac{r^{l+1}}{b^{2l+1}}\right)\right]_{r=r'}$$

$$= -\frac{C}{r'}\left[1 - \left(\frac{a}{r'}\right)^{2l+1}\right]\left[l + (l + 1)\left(\frac{r'}{b}\right)^{2l+1}\right]$$

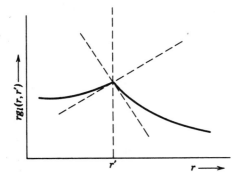

Figure 3.10   Discontinuity in slope of the radial Green function.

Similarly

$$\left\{ \frac{d}{dr} [rg_l(r, r')] \right\}_{r'-\epsilon} = \frac{C}{r'} \left[l + 1 + l \left(\frac{a}{r'} \right)^{2l+1} \right] \left[1 - \left(\frac{r'}{b} \right)^{2l+1} \right]$$

Substituting these derivatives into (3.123), we find:

$$C = \frac{4\pi}{(2l + 1) \left[1 - \left(\dfrac{a}{b} \right)^{2l+1} \right]} \tag{3.124}$$

Combination of (3.124), (3.122), (3.119), and (3.118) yields the expansion of the Green function for a spherical shell bounded by $r = a$ and $r = b$:

$$G(\mathbf{x}, \mathbf{x}') = 4\pi \sum_{l=0}^{\infty} \sum_{m=-l}^{l} \frac{Y^*_{lm}(\theta', \phi') Y_{lm}(\theta, \phi)}{(2l + 1) \left[1 - \left(\dfrac{a}{b} \right)^{2l+1} \right]} \left(r_<^l - \frac{a^{2l+1}}{r_<^{l+1}} \right) \left(\frac{1}{r_>^{l+1}} - \frac{r_>^l}{b^{2l+1}} \right) \tag{3.125}$$

For the special cases $a \to 0$, $b \to \infty$, and $b \to \infty$, we recover the expansions (3.70) and (3.114), respectively. For the "interior" problem with a sphere of radius b, we merely let $a \to 0$. Whereas the expansion for a single sphere is most easily obtained from the image solution, the general result (3.125) for a spherical shell is rather difficult to obtain by the method of images, since it involves an infinite set of images.

3.10 Solution of Potential Problems with the Spherical Green Function Expansion

The general solution to the Poisson equation with specified values of the potential on the boundary surface is (see Section 1.10):

$$\Phi(\mathbf{x}) = \frac{1}{4\pi\epsilon_0} \int_V \rho(\mathbf{x}') G(\mathbf{x}, \mathbf{x}') \, d^3x' - \frac{1}{4\pi} \oint_S \Phi(\mathbf{x}') \frac{\partial G}{\partial n'} \, da' \tag{3.126}$$

For purposes of illustration let us consider the potential *inside* a sphere of radius b. First we will establish the equivalence of the surface integral in (3.126) to the method of Section 3.5, equations (3.61) and (3.58). With $a = 0$ in (3.125), the normal derivative, evaluated at $r' = b$, is:

$$\frac{\partial G}{\partial n'} = \frac{\partial G}{\partial r'} \bigg|_{r'=b} = -\frac{4\pi}{b^2} \sum_{l,m} \left(\frac{r}{b} \right)^l Y^*_{lm}(\theta', \phi') Y_{lm}(\theta, \phi) \tag{3.127}$$

Consequently the solution of the Laplace equation inside $r = b$ with $\Phi = V(\theta', \phi')$ on the surface is, according to (3.126):

$$\Phi(\mathbf{x}) = \sum_{l,m} \left[\int V(\theta', \phi') Y^*_{lm}(\theta', \phi') \, d\Omega' \right] \left(\frac{r}{b} \right)^l Y_{lm}(\theta, \phi) \tag{3.128}$$

For the case considered, this is the same form of solution as (3.61) with (3.58). There is a *third* form of solution for the sphere, the so-called Poisson integral (2.19). The equivalence of this solution to the Green function expansion solution

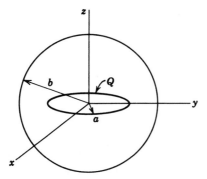

Figure 3.11   Ring of charge of radius a and total charge Q inside a grounded, conducting sphere of radius b.

is implied by the fact that both were derived from the general expression (3.126) and the image Green function. The explicit demonstration of the equivalence of (2.19) and the series solution (3.61) will be left to the problems.

We now turn to the solution of problems with charge distributed in the volume, so that the volume integral in (3.126) is involved. It is sufficient to consider problems in which the potential vanishes on the boundary surfaces. By linear superposition of a solution of the Laplace equation, the general situation can be obtained. The first illustration is that of a hollow grounded sphere of radius b with a concentric ring of charge of radius a and total charge Q. The ring of charge is located in the x-y plane, as shown in Fig. 3.11. The charge density of the ring can be written with the help of delta functions in angle and radius as

$$\rho(\mathbf{x}') = \frac{Q}{2\pi a^2}\, \delta(r' - a)\, \delta(\cos\theta') \tag{3.129}$$

In the volume integral over the Green function only terms in (3.125) with $m = 0$ will survive because of azimuthal symmetry. Then, using (3.57) and remembering that $a \to 0$ in (3.125), we find

$$\Phi(\mathbf{x}) = \frac{1}{4\pi\epsilon_0} \int \rho(\mathbf{x}')G(\mathbf{x}, \mathbf{x}')\, d^3x'$$
$$= \frac{Q}{4\pi\epsilon_0} \sum_{l=0}^{\infty} P_l(0) r_<^l \left(\frac{1}{r_>^{l+1}} - \frac{r_>^l}{b^{2l+1}} \right) P_l(\cos\theta) \tag{3.130}$$

where now $r_<$ $(r_>)$ is the smaller (larger) of r and a. Using the fact that $P_{2n+1}(0) = 0$ and $P_{2n}(0) = [(-1)^n(2n - 1)!!]/2^n n!$, (3.130) can be written as

$$\Phi(\mathbf{x}) = \frac{Q}{4\pi\epsilon_0} \sum_{n=0}^{\infty} \frac{(-1)^n(2n - 1)!!}{2^n n!}\, r_<^{2n} \left(\frac{1}{r_>^{2n+1}} - \frac{r_>^{2n}}{b^{4n+1}} \right) P_{2n}(\cos\theta) \tag{3.131}$$

In the limit $b \to \infty$, it will be seen that (3.130) or (3.131) reduces to the expression at the end of Section 3.3 for a ring of charge in free space. The present result can be obtained alternatively by using that result and the images for a sphere.

A second example of charge densities, illustrated in Fig. 3.12, is that of a hollow grounded sphere with a uniform line charge of total charge Q located on the z axis between the north and south poles of the sphere. Again with the help of delta functions, the volume-charge density can be written:

$$\rho(\mathbf{x}') = \frac{Q}{2b} \frac{1}{2\pi r'^2} [\delta(\cos\theta' - 1) + \delta(\cos\theta' + 1)] \tag{3.132}$$

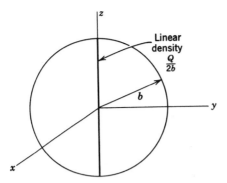

Figure 3.12 Uniform line charge of length $2b$ and total charge Q inside a grounded, conducting sphere of radius b.

The two delta functions in $\cos\theta$ correspond to the two halves of the line charge, above and below the x-y plane. The factor $2\pi r'^2$ in the denominator assures that the charge density has a constant *linear* density $Q/2b$. With this density in (3.126) we obtain

$$\Phi(\mathbf{x}) = \frac{Q}{8\pi\epsilon_0 b}\sum_{l=0}^{\infty}[P_l(1) + P_l(-1)]P_l(\cos\theta)\int_0^b r'^l_<\left(\frac{1}{r'^{l+1}_>} - \frac{r'^l_>}{b^{2l+1}}\right)dr' \quad (3.133)$$

The integral must be broken up into the intervals $0 \le r' < r$ and $r \le r' \le b$. Then we find

$$\int_0^b = \left(\frac{1}{r^{l+1}} - \frac{r^l}{b^{2l+1}}\right)\int_0^r r'^l \, dr' + r^l\int_r^b\left(\frac{1}{r'^{l+1}} - \frac{r'^l}{b^{2l+1}}\right)dr' \quad (3.134)$$

$$= \frac{(2l + 1)}{l(l + 1)}\left[1 - \left(\frac{r}{b}\right)^l\right]$$

For $l = 0$ this result is indeterminate. Applying L'Hospital's rule, we have, for $l = 0$ only,

$$\int_0^b = \lim_{l\to 0}\frac{\dfrac{d}{dl}\left[1 - \left(\dfrac{r}{b}\right)^l\right]}{\dfrac{d}{dl}(l)} = \lim_{l\to 0}\left[-\frac{d}{dl}\,e^{l\,\ln(r/b)}\right] = \ln\left(\frac{b}{r}\right) \quad (3.135)$$

This can be verified by direct integration in (3.133) for $l = 0$. Using the fact that $P_l(-1) = (-1)^l$, the potential (3.133) can be put in the form:

$$\Phi(\mathbf{x}) = \frac{Q}{4\pi\epsilon_0 b}\left\{\ln\left(\frac{b}{r}\right) + \sum_{j=1}^{\infty}\frac{(4j + 1)}{2j(2j + 1)}\left[1 - \left(\frac{r}{b}\right)^{2j}\right]P_{2j}(\cos\theta)\right\} \quad (3.136)$$

The presence of the logarithm for $l = 0$ reminds us that the potential diverges along the z axis. This is borne out by the series in (3.136), which diverges for $\cos\theta = \pm 1$, except at $r = b$ exactly. The peculiarity that the logarithm has argument (b/r) instead of $(b/r\sin\theta)$ is addressed in Problem 3.8.

The surface-charge density on the grounded sphere is readily obtained from (3.136) by differentiation:

$$\sigma(\theta) = \epsilon_0\left.\frac{\partial\Phi}{\partial r}\right|_{r=b} = -\frac{Q}{4\pi b^2}\left[1 + \sum_{j=1}^{\infty}\frac{(4j + 1)}{(2j + 1)}P_{2j}(\cos\theta)\right] \quad (3.137)$$

The leading term shows that the total charge induced on the sphere is $-Q$, the other terms integrating to zero over the surface of the sphere.

3.11 *Expansion of Green Functions in Cylindrical Coordinates*

The expansion of the potential of a unit point charge in cylindrical coordinates affords another useful example of Green function expansions. We present the initial steps in general enough fashion to permit the procedure to be readily adapted to finding Green functions for potential problems with cylindrical boundary surfaces. The starting point is the equation for the Green function:

$$\nabla_x^2 G(\mathbf{x}, \mathbf{x}') = -\frac{4\pi}{\rho} \delta(\rho - \rho') \, \delta(\phi - \phi') \, \delta(z - z') \tag{3.138}$$

where the delta function has been expressed in cylindrical coordinates. The ϕ and z delta functions can be written in terms of orthonormal functions:

$$\left.\begin{aligned}
\delta(z - z') &= \frac{1}{2\pi} \int_{-\infty}^{\infty} dk \, e^{ik(z-z')} = \frac{1}{\pi} \int_{0}^{\infty} dk \, \cos[k(z - z')] \\
\delta(\phi - \phi') &= \frac{1}{2\pi} \sum_{m=-\infty}^{\infty} e^{im(\phi-\phi')}
\end{aligned}\right\} \tag{3.139}$$

We expand the Green function in similar fashion:

$$G(\mathbf{x}, \mathbf{x}') = \frac{1}{2\pi^2} \sum_{m=-\infty}^{\infty} \int_{0}^{\infty} dk \, e^{im(\phi-\phi')} \cos[k(z - z')] g_m(k, \rho, \rho') \tag{3.140}$$

Then substitution into (3.138) leads to an equation for the radial Green function $g_m(k, \rho, \rho')$:

$$\frac{1}{\rho} \frac{d}{d\rho} \left(\rho \frac{dg_m}{d\rho} \right) - \left(k^2 + \frac{m^2}{\rho^2} \right) g_m = -\frac{4\pi}{\rho} \delta(\rho - \rho') \tag{3.141}$$

For $\rho \neq \rho'$ this is just equation (3.98) for the modified Bessel functions, $I_m(k\rho)$ and $K_m(k\rho)$. Suppose that $\psi_1(k\rho)$ is some linear combination of I_m and K_m which satisfies the correct boundary conditions for $\rho < \rho'$, and that $\psi_2(k\rho)$ is a linearly independent combination that satisfies the proper boundary conditions for $\rho > \rho'$. Then the symmetry of the Green function in ρ and ρ' requires that

$$g_m(k, \rho, \rho') = \psi_1(k\rho_<)\psi_2(k\rho_>) \tag{3.142}$$

The normalization of the product $\psi_1\psi_2$ is determined by the discontinuity in slope implied by the delta function in (3.141):

$$\left.\frac{dg_m}{d\rho}\right|_{+} - \left.\frac{dg_m}{d\rho}\right|_{-} = -\frac{4\pi}{\rho'} \tag{3.143}$$

where $|_{\pm}$ means evaluated at $\rho = \rho' \pm \epsilon$. From (3.142) it is evident that

$$\left[\left.\frac{dg_m}{d\rho}\right|_{+} - \left.\frac{dg_m}{d\rho}\right|_{-}\right] = k(\psi_1\psi_2' - \psi_2\psi_1') = kW[\psi_1, \psi_2] \tag{3.144}$$

where primes mean differentiation with respect to the argument, and $W[\psi_1, \psi_2]$ is the Wronskian of ψ_1 and ψ_2. Equation (3.141) is of the Sturm–Liouville type

$$\frac{d}{dx}\left[p(x)\frac{dy}{dx}\right] + g(x)y = 0 \qquad (3.145)$$

and it is well known that the Wronskian of two linearly independent solutions of such an equation is proportional to $[1/p(x)]$. Hence the possibility of satisfying (3.143) for all values of ρ' is assured. Clearly we must demand that the normalization of the product $\psi_1\psi_2$ be such that the Wronskian has the value

$$W[\psi_1(x), \psi_2(x)] = -\frac{4\pi}{x} \qquad (3.146)$$

If there are no boundary surfaces, $g_m(k, \rho, \rho')$ must be finite at $\rho = 0$ and vanish at $\rho \to \infty$. Consequently $\psi_1(k\rho) = AI_m(k\rho)$ and $\psi_2(k\rho) = K_m(k\rho)$. The constant A is to be determined from the Wronskian condition (3.146). Since the Wronskian is proportional to $(1/x)$ for all values of x, it does not matter where we evaluate it. Using the limiting forms (3.102) and (3.103) for small x [or (3.104) for large x], we find

$$W[I_m(x), K_m(x)] = -\frac{1}{x} \qquad (3.147)$$

so that $A = 4\pi$. The expansion of $1/|\mathbf{x} - \mathbf{x}'|$ therefore becomes:

$$\frac{1}{|\mathbf{x} - \mathbf{x}'|} = \frac{2}{\pi}\sum_{m=-\infty}^{\infty}\int_0^{\infty} dk\, e^{im(\phi - \phi')}\cos[k(z - z')]I_m(k\rho_<)K_m(k\rho_>) \qquad (3.148)$$

This can also be written entirely in terms of real functions as:

$$\frac{1}{|\mathbf{x} - \mathbf{x}'|} = \frac{4}{\pi}\int_0^{\infty} dk\, \cos[k(z - z')]$$
$$\times \left\{\tfrac{1}{2}I_0(k\rho_<)K_0(k\rho_>) + \sum_{m=1}^{\infty}\cos[m(\phi - \phi')]I_m(k\rho_<)K_m k\rho_>)\right\} \qquad (3.149)$$

A number of useful mathematical results can be obtained from this expansion. If we let $\mathbf{x}' \to 0$, only the $m = 0$ term survives, and we obtain the integral representation:

$$\frac{1}{\sqrt{\rho^2 + z^2}} = \frac{2}{\pi}\int_0^{\infty}\cos kz\, K_0(k\rho)\, dk \qquad (3.150)$$

If we replace ρ^2 in (3.150) by $R^2 = \rho^2 + \rho'^2 - 2\rho\rho'\cos(\phi - \phi')$, then we have on the left-hand side the inverse distance $|\mathbf{x} - \mathbf{x}'|^{-1}$ with $z' = 0$, i.e., just (3.149) with $z' = 0$. Then comparison of the right-hand sides of (3.149) and (3.150) (which must hold for *all* values of z) leads to the identification:

$$K_0(k\sqrt{\rho^2 + \rho'^2 - 2\rho\rho'\cos(\phi - \phi')}) =$$
$$I_0(k\rho_<)K_0(k\rho_>) + 2\sum_{m=1}^{\infty}\cos[m(\phi - \phi')]I_m(k\rho_<)K_m(k\rho_>) \qquad (3.151)$$

In this last result we can take the limit $k \to 0$ and obtain an expansion for the Green function for (two-dimensional) polar coordinates:

$$
\ln\left(\frac{1}{\rho^2 + \rho'^2 - 2\rho\rho' \cos(\phi - \phi')}\right) =
$$
$$
2\ln\left(\frac{1}{\rho_>}\right) + 2 \sum_{m=1}^{\infty} \frac{1}{m} \left(\frac{\rho_<}{\rho_>}\right)^m \cos[m(\phi - \phi')]
\tag{3.152}
$$

This representation can be verified by a systematic construction of the two-dimensional Green function for the Poisson equation along the lines leading to (3.148). See Problem 2.17.

3.12   Eigenfunction Expansions for Green Functions

Another technique for obtaining expansions of Green functions is the use of eigenfunctions for some related problem. This approach is intimately connected with the methods of Sections 3.9 and 3.11.

To specify what we mean by eigenfunctions, we consider an elliptic differential equation of the form

$$
\nabla^2 \psi(\mathbf{x}) + [f(\mathbf{x}) + \lambda]\psi(\mathbf{x}) = 0
\tag{3.153}
$$

If the solutions $\psi(\mathbf{x})$ are required to satisfy homogeneous boundary conditions on the surface S of the volume of interest V, then (3.153) will not in general have well-behaved (e.g., finite and continuous) solutions, except for certain values of λ. These values of λ, denoted by λ_n, are called *eigenvalues* (or *characteristic values*) and the solutions $\psi_n(\mathbf{x})$ are called *eigenfunctions*.* The eigenvalue differential equation is written:

$$
\nabla^2 \psi_n(\mathbf{x}) + [f(\mathbf{x}) + \lambda_n]\psi_n(\mathbf{x}) = 0
\tag{3.154}
$$

By methods similar to those used to prove the orthogonality of the Legendre or Bessel functions, it can be shown that the eigenfunctions are orthogonal:

$$
\int_V \psi_m^*(\mathbf{x})\psi_n(\mathbf{x}) \, d^3x = \delta_{mn}
\tag{3.155}
$$

where the eigenfunctions are assumed normalized. The spectrum of eigenvalues λ_n may be a discrete set, or a continuum, or both. It will be assumed that the totality of eigenfunctions forms a complete set.

Suppose now that we wish to find the Green function for the equation:

$$
\nabla_x^2 G(\mathbf{x}, \mathbf{x}') + [f(\mathbf{x}) + \lambda]G(\mathbf{x}, \mathbf{x}') = -4\pi\delta(\mathbf{x} - \mathbf{x}')
\tag{3.156}
$$

where λ is *not* equal to one of the eigenvalues λ_n of (3.154). Furthermore, suppose that the Green function is to have the same boundary conditions as the eigenfunctions of (3.154). Then the Green function can be expanded in a series of the eigenfunctions of the form:

$$
G(\mathbf{x}, \mathbf{x}') = \sum_n a_n(\mathbf{x}')\psi_n(\mathbf{x})
\tag{3.157}
$$

---

*The reader familiar with wave mechanics will recognize (3.153) as equivalent to the Schrödinger equation for a particle in a potential.

Substitution into the differential equation for the Green function leads to the result:

$$\sum_m a_m(\mathbf{x}')(\lambda - \lambda_m)\psi_m(\mathbf{x}) = -4\pi\delta(\mathbf{x} - \mathbf{x}') \tag{3.158}$$

If we multiply both sides by $\psi_n^*(\mathbf{x})$ and integrate over the volume V, the orthogonality condition (3.155) reduces the left-hand side to one term, and we find:

$$a_n(\mathbf{x}') = 4\pi \frac{\psi_n^*(\mathbf{x}')}{\lambda_n - \lambda} \tag{3.159}$$

Consequently the eigenfunction expansion of the Green function is:

$$G(\mathbf{x}, \mathbf{x}') = 4\pi \sum_n \frac{\psi_n^*(\mathbf{x}')\psi_n(\mathbf{x})}{\lambda_n - \lambda} \tag{3.160}$$

For a continuous spectrum the sum is replaced by an integral.

Specializing the foregoing considerations to the Poisson equation, we place $f(\mathbf{x}) = 0$ and $\lambda = 0$ in (3.156). As a first, essentially trivial, illustration we let (3.154) be the wave equation over all space:

$$(\nabla^2 + k^2)\psi_{\mathbf{k}}(\mathbf{x}) = 0 \tag{3.161}$$

with the continuum of eigenvalues, k^2, and the eigenfunctions:

$$\psi_{\mathbf{k}}(\mathbf{x}) = \frac{1}{(2\pi)^{3/2}} e^{i\mathbf{k}\cdot\mathbf{x}} \tag{3.162}$$

These eigenfunctions have delta function normalization:

$$\int \psi_{\mathbf{k}'}^*(\mathbf{x})\psi_{\mathbf{k}}(\mathbf{x})\, d^3x = \delta(\mathbf{k} - \mathbf{k}') \tag{3.163}$$

Then, according to (3.160), the infinite space Green function has the expansion:

$$\frac{1}{|\mathbf{x} - \mathbf{x}'|} = \frac{1}{2\pi^2} \int d^3k \frac{e^{i\mathbf{k}\cdot(\mathbf{x}-\mathbf{x}')}}{k^2} \tag{3.164}$$

This is just the three-dimensional Fourier integral representation of $1/|\mathbf{x} - \mathbf{x}'|$.

As a second example, consider the Green function for a Dirichlet problem inside a rectangular box defined by the six planes, $x = 0$, $y = 0$, $z = 0$, $x = a$, $y = b$, $z = c$. The expansion is to be made in terms of eigenfunctions of the wave equation:

$$(\nabla^2 + k_{lmn}^2)\psi_{lmn}(x, y, z) = 0 \tag{3.165}$$

where the eigenfunctions which vanish on all the boundary surfaces are

$$\psi_{lmn}(x, y, z) = \sqrt{\frac{8}{abc}} \sin\left(\frac{l\pi x}{a}\right) \sin\left(\frac{m\pi y}{b}\right) \sin\left(\frac{n\pi z}{c}\right)$$

and

$$k_{lmn}^2 = \pi^2\left(\frac{l^2}{a^2} + \frac{m^2}{b^2} + \frac{n^2}{c^2}\right)$$

(3.166)

The expansion of the Green function is therefore:

$$G(\mathbf{x}, \mathbf{x}') = \frac{32}{\pi abc} \tag{3.167}$$

$$\times \sum_{l,m,n=1}^{\infty} \frac{\sin\left(\dfrac{l\pi x}{a}\right) \sin\left(\dfrac{l\pi x'}{a}\right) \sin\left(\dfrac{m\pi y}{b}\right) \sin\left(\dfrac{m\pi y'}{b}\right) \sin\left(\dfrac{n\pi z}{c}\right) \sin\left(\dfrac{n\pi z'}{c}\right)}{\dfrac{l^2}{a^2} + \dfrac{m^2}{b^2} + \dfrac{n^2}{c^2}}$$

To relate expansion (3.167) to the type of expansions obtained in Sections 3.9 and 3.11, namely, (3.125) for spherical coordinates and (3.148) for cylindrical coordinates, we write down the analogous expansion for the rectangular box. If the x and y coordinates are treated in the manner of (θ, ϕ) or (ϕ, z) in those cases, while the z coordinate is singled out for special treatment, we obtain the Green function:

$$G(\mathbf{x}, \mathbf{x}') = \frac{16\pi}{ab} \sum_{l,m=1}^{\infty} \sin\left(\frac{l\pi x}{a}\right) \sin\left(\frac{l\pi x'}{a}\right) \sin\left(\frac{m\pi y}{b}\right) \sin\left(\frac{m\pi y'}{b}\right)$$
$$\times \frac{\sinh(K_{lm}z_<) \sinh[K_{lm}(c - z_>)]}{K_{lm} \sinh(K_{lm}c)} \tag{3.168}$$

where $K_{lm} = \pi(l^2/a^2 + m^2/b^2)^{1/2}$. If (3.167) and (3.168) are to be equal, it must be that the sum over n in (3.167) is just the Fourier series representation on the interval $(0, c)$ of the one-dimensional Green function in z in (3.168):

$$\frac{\sinh(K_{lm}z_<) \sinh[K_{lm}(c - z_>)]}{K_{lm} \sinh(K_{lm}c)} = \frac{2}{c} \sum_{n=1}^{\infty} \frac{\sin\left(\dfrac{n\pi z'}{c}\right)}{K_{lm}^2 + \left(\dfrac{n\pi}{c}\right)^2} \sin\left(\frac{n\pi z}{c}\right) \tag{3.169}$$

The verification that (3.169) is the correct Fourier representation is left as an exercise for the reader.

Further illustrations of this technique will be found in the problems at the end of the chapter.

3.13 *Mixed Boundary Conditions; Conducting Plane with a Circular Hole*

The potential problems discussed so far in this chapter have been of the orthodox kind in which the boundary conditions are of one type (usually Dirichlet) over the whole boundary surface. In the uniqueness proof for solutions of the Laplace or Poisson equation (Section 1.9) it was pointed out, however, that mixed boundary conditions, where the potential is specified over part of the boundary and its normal derivative is specified over the remainder, also lead to well-defined, unique boundary-value problems. Textbooks tend to mention the possibility of mixed boundary conditions when making the uniqueness proof and to ignore such problems in subsequent discussion. The reason, as we shall see, is that mixed boundary conditions are much more difficult to handle than the normal type.

To illustrate the difficulties encountered with mixed boundary conditions, we consider the problem of an infinitely thin, grounded, conducting plane with a circular hole of radius a cut in it, and with the electric field far from the hole being normal to the plane, constant in magnitude, and having different values on either side of the plane. The geometry is sketched in Fig. 3.13. The plane is at $z = 0$; the hole is centered on the origin of coordinates; the nonvanishing asymptotic electric field components are $E_z = -E_0$ for $z > 0$ and $E_z = -E_1$ for $z < 0$. The problem may seem contrived, but with $E_0 = 0$ or $E_1 = 0$ it has application for radiation from small holes in the walls of wave guides, where "small" is defined as small compared to a wavelength so that electrostatic considerations can apply (see Section 9.5).

Since the electric field is specified far from the hole, we write the potential as

$$\Phi = \begin{cases} E_0 z + \Phi^{(1)} & (z > 0) \\ E_1 z + \Phi^{(1)} & (z < 0) \end{cases} \tag{3.170}$$

If the hole were not there, $\Phi^{(1)}$ would be zero. The top surface of the sheet would have a uniform surface charge density $-\epsilon_0 E_0$ and the bottom surface a charge density $\epsilon_0 E_1$. The potential $\Phi^{(1)}$ can thus be thought of as resulting from a rearrangement of surface charge in the neighborhood of the hole. Since this charge density is located on the plane $z = 0$, the potential $\Phi^{(1)}$ can be represented as

$$\Phi^{(1)}(x, y, z) = \frac{1}{4\pi\epsilon_0} \int \frac{\sigma^{(1)}(x', y')\, dx'\, dy'}{\sqrt{(x - x')^2 + (y - y')^2 + z^2}}$$

This shows that $\Phi^{(1)}$ is *even* in z, so that $E_x^{(1)}$ and $E_y^{(1)}$ are even in z, but $E_z^{(1)}$ is *odd*. We note that $E_x^{(1)}$ and $E_y^{(1)}$ are the x and y components of the total electric field, but that, because of (3.170), $E_z^{(1)}$ is not the total z component. Thus, even though it is odd in z, it does not vanish at $z = 0$. Rather, it is discontinuous there.

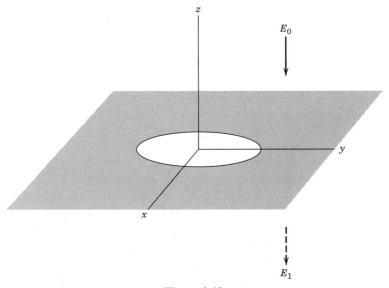

Figure 3.13

Since the *total z component* of electric field must be continuous across $z = 0$ in the hole, we must have (for $\rho < a$)

$$-E_0 + E_z^{(1)}|_{z=0^+} = -E_1 + E_z^{(1)}|_{z=0^-}$$

Because $E_z^{(1)}$ is odd in z, this relation determines the normal component of the electric field to be

$$E_z^{(1)}|_{z=0^+} = -E_z^{(1)}|_{z=0^-} = \tfrac{1}{2}(E_0 - E_1)$$

provided (x, y) lie inside the opening ($0 \le \rho < a$). For points on the conducting surface ($a \le \rho < \infty$), the electric field is not known, but the potential is zero by hypothesis. From (3.170) this means that $\Phi^{(1)} = 0$ there. Note that in the opening we do not know the potential. We therefore have an electrostatic boundary-value problem with the following *mixed boundary conditions*:

$$\frac{\partial \Phi^{(1)}}{\partial z}\bigg|_{z=0^+} = -\tfrac{1}{2}(E_0 - E_1) \qquad \text{for } 0 \le \rho < a$$

and
$$(3.171)$$

$$\Phi^{(1)}|_{z=0} = 0 \qquad \text{for } a \le \rho < \infty$$

Because of the azimuthal symmetry of the geometry, the potential $\Phi^{(1)}$ can be written in terms of cylindrical coordinates [from (3.106)] as

$$\Phi^{(1)}(\rho, z) = \int_0^\infty dk \, A(k) e^{-k|z|} J_0(k\rho) \tag{3.172}$$

Before proceeding to see how $A(k)$ is determined by the boundary conditions, we relate $A(k)$ and its derivatives at $k = 0$ to the asymptotic behavior of the potential. For large ρ or $|z|$ the rapid oscillations of $J_0(k\rho)$ or the rapid decrease of $e^{-k|z|}$ imply that the integral in (3.172) receives its important contributions from the region around $k = 0$. The asymptotic behavior of $\Phi^{(1)}$ is therefore related to the behavior of $A(k)$ at small k. We assume that $A(k)$ can be expanded in a Taylor series around $k = 0$:

$$A(k) = \sum_{l=0}^\infty \frac{k^l}{l!} \frac{d^l A}{dk^l}(0)$$

With this series inserted into (3.172), the potential $\Phi^{(1)}$ becomes

$$\Phi^{(1)}(\rho, z) = \sum_{l=0}^\infty \frac{d^l A}{dk^l}(0) \, B_l(\rho, z) \tag{3.173}$$

where

$$B_l(\rho, z) = \frac{1}{l!} \int_0^\infty dk \, k^l e^{-k|z|} J_0(k\rho) \tag{3.174}$$

The integral (3.174) can evidently be written

$$B_l = \frac{1}{l!} \left(-\frac{d}{d|z|}\right)^l \int_0^\infty dk \, e^{-k|z|} J_0(k\rho)$$

Using a result from Problem 3.16c, we find that B_l is

$$B_l = \frac{1}{l!} \left(-\frac{d}{d|z|}\right)^l \left(\frac{1}{\sqrt{\rho^2 + z^2}}\right) \tag{3.175}$$

The reader should not be surprised to find that explicit calculation yields

$$B_l = \frac{P_l(|\cos\theta|)}{r^{l+1}} \tag{3.176}$$

where $\cos\theta = z/r$ and $r = \sqrt{\rho^2 + z^2}$. The asymptotic expansion (3.173) is thus an expansion of the spherical harmonic form (3.33):

$$\Phi^{(1)} = \sum_{l=0}^{\infty} \frac{d^l A}{dk^l}(0) \cdot \frac{P_l(|\cos\theta|)}{r^{l+1}} \tag{3.177}$$

As is discussed in the next chapter, this expansion in powers of r^{-1} is called a multipole expansion. The $l = 0$ coefficient, $A(0)$, is the *total charge* (divided by $4\pi\epsilon_0$). The $l = 1$ coefficient, $dA(0)/dk$, is the *dipole moment* in the z-direction, and so on. Once the function $A(k)$ is known these quantities that describe the asymptotic behavior of the potential can be evaluated without explicit construction of the potential itself.

We are now ready to discuss the mixed boundary value problem. With the assumed form (3.172) for $\Phi^{(1)}$, the boundary conditions (3.171) become a pair of integral equations of the first kind for $A(k)$:

$$\int_0^{\infty} dk\, kA(k)J_0(k\rho) = \tfrac{1}{2}(E_0 - E_1) \qquad \text{for } 0 \le \rho < a$$
$$\int_0^{\infty} dk\, A(k)J_0(k\rho) = 0 \qquad \text{for } a \le \rho < \infty \tag{3.178}$$

Such pairs of integral equations, with one of the pair holding over one part of the range of the independent variable and the other over the other part of the range, are known as *dual integral equations*. The general theory of such integral equations is complicated and not highly developed.* Just over a hundred years ago H. Weber solved the closely related problem of the potential of a charged circular disc by means of certain discontinuous integrals involving Bessel functions. We appeal to a generalization of Weber's formulas. Consider the dual integral equations,

$$\int_0^{\infty} dy\, yg(y)J_n(yx) = x^n \qquad \text{for } 0 \le x < 1$$
$$\int_0^{\infty} dy\, g(y)J_n(yx) = 0 \qquad \text{for } 1 \le x < \infty \tag{3.179}$$

Examination of the formula of Sonine and Schafheitlin for the integral of $J_\mu(at)J_\nu(bt)t^{-\lambda}$ (see *Watson*, pp. 398 ff, or *Magnus* et al., p. 99) shows that the solution for $g(y)$ is

$$g(y) = \frac{\Gamma(n+1)}{\sqrt{\pi}\,\Gamma(n+\frac{3}{2})} j_{n+1}(y) = \frac{\Gamma(n+1)}{\Gamma(n+\frac{3}{2})} \frac{J_{n+\frac{3}{2}}(y)}{(2y)^{1/2}} \tag{3.180}$$

In this relation $j_n(y)$ is the spherical Bessel function of order n (see Section 9.6).

---

*One monograph, I. N. Sneddon, *Mixed Boundary Value Problems in Potential Theory*, North-Holland, Amsterdam, and Wiley-Interscience, New York (1966), is devoted to our subject. See also *Tranter* (Chapter VIII).

For our pair of equations (3.178) we have $n = 0$, $x = \rho/a$, $y = ka$. Therefore $A(k)$ is

$$A(k) = \frac{(E_0 - E_1)a^2}{\pi} j_1(ka) = \frac{(E_0 - E_1)}{\pi} \left[\frac{\sin ka}{k^2} - \frac{a \cos ka}{k} \right] \quad (3.181)$$

The expansion of $A(k)$ for small k takes the form,

$$A(k) \simeq \frac{(E_0 - E_1)a^2}{3\pi} \left[ka - \frac{(ka)^3}{10} + \cdots \right]$$

This means that total charge associated with $\Phi^{(1)}$ is zero and the leading term in the asymptotic potential (3.177) is the $l = 1$ contribution,

$$\Phi^{(1)} \to \frac{(E_0 - E_1)a^3}{3\pi} \cdot \frac{|z|}{r^3} \quad (3.182)$$

falling off with distance as r^{-2} and having an *effective electric dipole moment*,

$$\mathbf{p} = \mp \frac{4\epsilon_0}{3} (\mathbf{E}_0 - \mathbf{E}_1)a^3 \quad (z \gtrless 0) \quad (3.183)$$

The reversal of the effective dipole moment depending on whether the observation point is above or below the plane is a consequence of the fact that a true dipole potential is odd in z, whereas (3.182) is even. The idea that a small hole in a plane conducting sheet is equivalent far from the opening to a dipole normal to the surface is important in discussing the consequences of such openings in the walls of waveguides and cavities. Figure 9.4 depicts the origin of the dipole-like field as a consequence of the penetration of the field lines through the hole to terminate on the side with the smaller constant field. The picture is given quantitative meaning through (3.182) and (3.183).

The added potential $\Phi^{(1)}$ in the neighborhood of the opening must be calculated from the exact expression,

$$\Phi^{(1)}(\rho, z) = \frac{(E_0 - E_1)}{\pi} a^2 \int_0^\infty dk \, j_1(ka) e^{-k|z|} J_0(k\rho) \quad (3.184)$$

The integral,* after an integration by parts to replace j_1 with j_0, can be expressed as a sum of the imaginary parts of the Laplace transforms (for complex p) of $J_\nu(k\rho)/k$ for $\nu = 0, 1$. The result, after some simplifications, is

$$\Phi^{(1)}(\rho, z) = \frac{(E_0 - E_1)a}{\pi} \left[\sqrt{\frac{R - \lambda}{2}} - |z| \tan^{-1}\left(\sqrt{\frac{2}{R + \lambda}} \right) \right] \quad (3.185)$$

where

$$\lambda = \frac{1}{a^2} (z^2 + \rho^2 - a^2), \qquad R = \sqrt{\lambda^2 + 4z^2/a^2}$$

Some special cases are of interest. The added potential on the axis ($\rho = 0$) is

$$\Phi^{(1)}(0, z) = \frac{(E_0 - E_1)a}{\pi} \left[1 - \frac{|z|}{a} \tan^{-1}\left(\frac{a}{|z|} \right) \right]$$

*For integrals of the kind encountered here, see *Watson* (Chapter 13), *Gradshteyn and Ryzhik*, *Magnus, Oberhettinger, and Soni*, or the *Bateman Manuscript Project*.

For $|z| \gg a$ thus reduces to (3.182) with $r = |z|$, while for $|z| \rightarrow 0$ it is approximated by the first term. In the plane of the opening ($z = 0$) the potential $\Phi^{(1)}$ is

$$\Phi^{(1)}(\rho, 0) = \frac{(E_0 - E_1)}{\pi} \sqrt{a^2 - \rho^2}$$

for $0 \leq \rho < a$ (and zero, of course, for $\rho \geq a$). The *tangential electric field in the opening* is a radial field,

$$\mathbf{E}_{\text{tan}}(\rho, 0) = \frac{(E_0 - E_1)}{\pi} \frac{\rho}{\sqrt{a^2 - \rho^2}} \tag{3.186}$$

The *normal component* of electric field in the opening is, from the first equation in (3.171), just the average of the uniform fields above and below the plane, that is,

$$E_z(\rho, 0) = -\tfrac{1}{2}(E_0 + E_1) \tag{3.187}$$

We note that the magnitude of the electric field has a square root singularity at the edge of the opening, in agreement with the considerations of Section 2.11. The surface-charge densities on the upper and lower sides of the conducting plane in the neighborhood of the hole can be evaluated in a straightforward manner. The explicit calculation is left to the problems.

Equipotential contours near the circular hole for the full potential (3.170) are shown in Fig. 3.14 for the situation where $E_1 = 0$. At distances more than

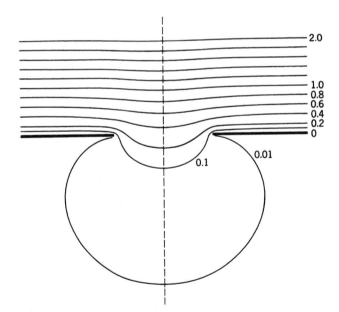

Figure 3.14 Equipotential contours near a circular hole in a conducting plane with a normal electric field E_0 far from the hole on one side and no field asymptotically on the other ($E_1 = 0$). The numbers are the values of the potential Φ in units of aE_0. The distribution is rotationally symmetric about the vertical dashed line through the center of the hole.

two or three times the radius away from the hole, its presence is hardly discernible.

The classic problem of a charged conducting disc is discussed in detail by Sneddon (*op. cit.*). The mixed boundary conditions for the disc or hole can be avoided by separating the Laplace equation in elliptic coordinates. The disc (or hole) is then taken to be the limiting form of an oblate spheroidal surface. For this approach, see, for example, *Smythe* (pp. 124, 171) or *Jeans* (p. 244).

References and Suggested Reading

The subjects of the special functions of mathematical physics, the solution of ordinary differential equations, hypergeometric functions, and Sturm–Liouville theory are covered in many books. For the reader who does not already have his favorite, some of the possibilities are

> Arfken
> Dennery and Kryzwicki
> Morse and Feshbach
> Whittaker and Watson

A more elementary treatment, with well-chosen examples and problems, can be found in
> Hildebrand, Chapters 4, 5, and 8

A somewhat old-fashioned source of the theory and practice of Legendre polynomials and spherical harmonics, with many examples and problems, is
> Byerly

For purely mathematical properties of spherical functions one of the most useful one-volume references is
> Magnus, Oberhettinger, and Soni

For more detailed mathematical properties, see
> Watson, for Bessel functions
> Bateman Manuscript Project books, for all types of special functions

Electrostatic problems in cylindrical, spherical, and other coordinates are discussed extensively in
> Durand, Chapter XI
> Jeans, Chapter VIII
> Smythe, Chapter V
> Stratton, Chapter III

Problems

3.1 Two concentric spheres have radii a, b ($b > a$) and each is divided into two hemispheres by the same horizontal plane. The upper hemisphere of the inner sphere and the lower hemisphere of the outer sphere are maintained at potential V. The other hemispheres are at zero potential

Determine the potential in the region $a \leq r \leq b$ as a series in Legendre polynomials. Include terms at least up to $l = 4$. Check your solution against known results in the limiting cases $b \to \infty$, and $a \to 0$.

3.2 A spherical surface of radius R has charge uniformly distributed over its surface with a density $Q/4\pi R^2$, except for a spherical cap at the north pole, defined by the cone $\theta = \alpha$.

(a) Show that the potential inside the spherical surface can be expressed as

$$\Phi = \frac{Q}{8\pi\epsilon_0} \sum_{l=0}^{\infty} \frac{1}{2l+1} [P_{l+1}(\cos\alpha) - P_{l-1}(\cos\alpha)] \frac{r^l}{R^{l+1}} P_l(\cos\theta)$$

where, for $l = 0$, $P_{l-1}(\cos\alpha) = -1$. What is the potential outside?

(b) Find the magnitude and the direction of the electric field at the origin.

(c) Discuss the limiting forms of the potential (part a) and electric field (part b) as the spherical cap becomes (1) very small, and (2) so large that the area with charge on it becomes a very small cap at the south pole.

3.3   A thin, flat, conducting, circular disc of radius R is located in the x-y plane with its center at the origin, and is maintained at a fixed potential V. With the information that the charge density on a disc at fixed potential is proportional to $(R^2 - \rho^2)^{-1/2}$, where ρ is the distance out from the center of the disc,

(a) show that for $r > R$ the potential is

$$\Phi(r, \theta, \phi) = \frac{2V}{\pi} \frac{R}{r} \sum_{l=0}^{\infty} \frac{(-1)^l}{2l+1} \left(\frac{R}{r}\right)^{2l} P_{2l}(\cos\theta)$$

(b) find the potential for $r < R$.

(c) What is the capacitance of the disc?

3.4   The surface of a hollow conducting sphere of inner radius a is divided into an *even number* of equal segments by a set of planes; their common line of intersection is the z axis and they are distributed uniformly in the angle ϕ. (The segments are like the skin on wedges of an apple, or the earth's surface between successive meridians of longitude.) The segments are kept at fixed potentials $\pm V$, alternately.

(a) Set up a series representation for the potential inside the sphere for the general case of $2n$ segments, and carry the calculation of the coefficients in the series far enough to determine exactly which coefficients are different from zero. For the nonvanishing terms, exhibit the coefficients as an integral over $\cos\theta$.

(b) For the special case of $n = 1$ (two hemispheres) determine explicitly the potential up to and including all terms with $l = 3$. By a coordinate transformation verify that this reduces to result (3.36) of Section 3.3.

3.5   A hollow sphere of inner radius a has the potential specified on its surface to be $\Phi = V(\theta, \phi)$. Prove the equivalence of the two forms of solution for the potential inside the sphere:

(a) $$\Phi(\mathbf{x}) = \frac{a(a^2 - r^2)}{4\pi} \int \frac{V(\theta', \phi')}{(r^2 + a^2 - 2ar\cos\gamma)^{3/2}} d\Omega'$$

where $\cos\gamma = \cos\theta\cos\theta' + \sin\theta\sin\theta'\cos(\phi - \phi')$.

(b) $$\Phi(\mathbf{x}) = \sum_{l=0}^{\infty} \sum_{m=-l}^{l} A_{lm} \left(\frac{r}{a}\right)^l Y_{lm}(\theta, \phi)$$

where $A_{lm} = \int d\Omega' \, Y_{lm}^*(\theta', \phi') V(\theta', \phi')$.

3.6   Two point charges q and $-q$ are located on the z axis at $z = +a$ and $z = -a$, respectively.

(a) Find the electrostatic potential as an expansion in spherical harmonics and powers of r for both $r > a$ and $r < a$.

(b) Keeping the product $qa \equiv p/2$ constant, take the limit of $a \to 0$ and find the potential for $r \neq 0$. This is by definition a dipole along the z axis and its potential.

(c) Suppose now that the dipole of part b is surrounded by a *grounded* spherical shell of radius b concentric with the origin. By linear superposition find the potential everywhere inside the shell.

3.7 Three point charges $(q, -2q, q)$ are located in a straight line with separation a and with the middle charge $(-2q)$ at the origin of a grounded conducting spherical shell of radius b, as indicated in the sketch.

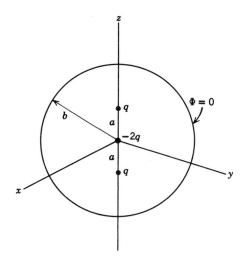

Problem 3.7

(a) Write down the potential of the three charges in the absence of the grounded sphere. Find the limiting form of the potential as $a \to 0$, but the product $qa^2 = Q$ remains finite. Write this latter answer in spherical coordinates.

(b) The presence of the grounded sphere of radius b alters the potential for $r < b$. The added potential can be viewed as caused by the surface-charge density induced on the inner surface at $r = b$ or by image charges located at $r > b$. Use linear superposition to satisfy the boundary conditions and find the potential everywhere inside the sphere for $r < a$ and $r > a$. Show that in the limit $a \to 0$,

$$\Phi(r,\,\theta,\,\phi) \to \frac{Q}{2\pi\epsilon_0 r^3}\left(1 - \frac{r^5}{b^5}\right)P_2(\cos\theta)$$

3.8 There is a puzzling aspect of the solution (3.136) for the potential inside a grounded sphere with a uniformly charged wire along a diameter. Very close to the wire (i.e., for $\rho = r\sin\theta \ll b$), the potential should be that of a uniformly charged wire, namely, $\Phi = (Q/4\pi\epsilon_0 b)\ln(b/\rho) + \Phi_0$. The solution (3.136) does not explicitly have this behavior.

(a) Show by use of the Legendre differential equation (3.10) and some integration by parts, that $\ln(\operatorname{cosec}\theta)$ has the appropriate expansion in spherical harmonics to permit the solution (3.136) to be written in the alternative form,

$$\Phi(\mathbf{x}) = \frac{Q}{4\pi\epsilon_0 b}\left\{\ln\left(\frac{2b}{r\sin\theta}\right) - 1 - \sum_{j=1}^{\infty}\frac{4j+1}{2j(2j+1)}\left(\frac{r}{b}\right)^{2j}P_{2j}(\cos\theta)\right\}$$

in which the expected behavior near the wire is manifest. Give an interpretation of the constant term $\Phi_0 = -Q/4\pi\epsilon_0 b$. Note that in this form, for any $r/b < 1$ the Legendre polynomial series is rapidly convergent at all angles.

(b) Show by use of the expansion (3.38) that

$$\frac{1}{2}\left(\frac{1}{\sin\theta/2} + \frac{1}{\cos\theta/2}\right) = 2\sum_{j=0}^{\infty} P_{2j}(\cos\theta)$$

and that therefore the charge density on the inner surface of the sphere, Eq. (3.137), can be expressed alternatively as

$$\sigma(\theta) = -\frac{Q}{4\pi b^2}\left\{\frac{1}{2}\left(\frac{1}{\sin\theta/2} + \frac{1}{\cos\theta/2}\right) - \sum_{j=0}^{\infty}\frac{1}{2j+1} P_{2j}(\cos\theta)\right\}$$

The (integrable) singular behavior at $\theta = 0$ and $\theta = \pi$ is now exhibited explicitly. The series provides corrections in $\ln(1/\theta)$ as $\theta \to 0$.

3.9   A hollow right circular cylinder of radius b has its axis coincident with the z axis and its ends at $z = 0$ and $z = L$. The potential on the end faces is zero, while the potential on the cylindrical surface is given as $V(\phi, z)$. Using the appropriate separation of variables in cylindrical coordinates, find a series solution for the potential anywhere inside the cylinder.

3.10   For the cylinder in Problem 3.9 the cylindrical surface is made of two equal half-cylinders, one at potential V and the other at potential $-V$, so that

$$V(\phi, z) = \begin{cases} V & \text{for } -\pi/2 < \phi < \pi/2 \\ -V & \text{for } \pi/2 < \phi < 3\pi/2 \end{cases}$$

(a)   Find the potential inside the cylinder.

(b)   Assuming $L \gg b$, consider the potential at $z = L/2$ as a function of ρ and ϕ and compare it with two-dimensional Problem 2.13.

3.11   A modified Bessel–Fourier series on the interval $0 \le \rho \le a$ for an arbitrary function $f(\rho)$ can be based on the "homogeneous" boundary conditions:

$$\text{At } \rho = 0, \qquad \rho\, J_\nu(k\rho)\, \frac{dJ_\nu(k'\rho)}{d\rho} = 0$$

$$\text{At } \rho = a, \qquad \frac{d}{d\rho}\ln[J_\nu(k\rho)] = -\frac{\lambda}{a} \qquad (\lambda \text{ real})$$

The first condition restricts ν. The second condition yields eigenvalues $k = y_{\nu n}/a$, where $y_{\nu n}$ is the nth positive root of $x\, dJ_\nu(x)/dx + \lambda J_\nu(x) = 0$.

(a)   Show that the Bessel functions of different eigenvalues are orthogonal in the usual way.

(b)   Find the normalization integral and show that an arbitrary function $f(\rho)$ can be expanded on the interval in the modified Bessel–Fourier series

$$f(\rho) = \sum_{n=1}^{\infty} A_n J_\nu\left(\frac{y_{\nu n}\rho}{a}\right)$$

with the coefficients A_n given by

$$A_n = \frac{2}{a^2}\left[\left(1 - \frac{\nu^2}{y_{\nu n}^2}\right)J_\nu^2(y_{\nu n}) + \left(\frac{dJ_\nu(y_{\nu n})}{dy_{\nu n}}\right)^2\right]^{-1} \int_0^a f(\rho)\, \rho\, J_\nu\left(\frac{y_{\nu n}\rho}{a}\right) d\rho$$

The dependence on λ is implicit in this form, but the square bracket has alternative forms:

$$\left[\left(1 - \frac{\nu^2}{y_{\nu n}^2}\right)J_\nu^2(y_{\nu n}) + \left(\frac{dJ_\nu(y_{\nu n})}{dy_{\nu n}}\right)^2\right] = \left(1 + \frac{\lambda^2 - \nu^2}{y_{\nu n}^2}\right)J_\nu^2(y_{\nu n})$$

$$= \left(1 + \frac{y_{\nu n}^2 - \nu^2}{\lambda^2}\right)\left[\frac{dJ_\nu(y_{\nu n})}{dy_{\nu n}}\right]^2$$

$$= [J_\nu^2(y_{\nu n}) - J_{\nu-1}(y_{\nu n})J_{\nu+1}(y_{\nu n})]$$

For $\lambda \to \infty$ we recover the result of (3.96) and (3.97). The choice $\lambda = 0$ is another simple alternative.

3.12 An infinite, thin, plane sheet of conducting material has a circular hole of radius a cut in it. A thin, flat disc of the same material and slightly smaller radius lies in the plane, filling the hole, but separated from the sheet by a very narrow insulating ring. The disc is maintained at a fixed potential V, while the infinite sheet is kept at zero potential.

   (a) Using appropriate cylindrical coordinates, find an integral expression involving Bessel functions for the potential at any point above the plane.

   (b) Show that the potential a perpendicular distance z above the *center* of the disc is

$$\Phi_0(z) = V\left(1 - \frac{z}{\sqrt{a^2 + z^2}}\right)$$

   (c) Show that the potential a perpendicular distance z above the *edge* of the disc is

$$\Phi_a(z) = \frac{V}{2}\left[1 - \frac{kz}{\pi a}K(k)\right]$$

   where $k = 2a/(z^2 + 4a^2)^{1/2}$, and $K(k)$ is the complete elliptic integral of the first kind.

3.13 Solve for the potential in Problem 3.1, using the appropriate Green function obtained in the text, and verify that the answer obtained in this way agrees with the direct solution from the differential equation.

3.14 A line charge of length $2d$ with a total charge Q has a linear charge density varying as $(d^2 - z^2)$, where z is the distance from the midpoint. A grounded, conducting, spherical shell of inner radius $b > d$ is centered at the midpoint of the line charge.

   (a) Find the potential everywhere inside the spherical shell as an expansion in Legendre polynomials.

   (b) Calculate the surface-charge density induced on the shell.

   (c) Discuss your answers to parts a and b in the limit that $d \ll b$.

3.15 Consider the following "spherical cow" model of a battery connected to an external circuit. A sphere of radius a and conductivity σ is embedded in a uniform medium of conductivity σ'. Inside the sphere there is a uniform (chemical) force in the z direction acting on the charge carriers; its strength as an effective electric field entering Ohm's law is F. In the steady state, electric fields exist inside and outside the sphere and surface charge resides on its surface.

   (a) Find the electric field (in addition to F) and current density everywhere in space. Determine the surface-charge density and show that the electric dipole moment of the sphere is $p = 4\pi\epsilon_0\sigma a^3 F/(\sigma + 2\sigma')$.

(b) Show that the total current flowing out through the upper hemisphere of the sphere is

$$I = \frac{2\sigma\sigma'}{\sigma + 2\sigma'} \cdot \pi a^2 F$$

Calculate the total power dissipation outside the sphere. Using the lumped circuit relations, $P = I^2 R_e = I V_e$, find the effective external resistance R_e and voltage V_e.

(c) Find the power dissipated within the sphere and deduce the effective internal resistance R_i and voltage V_i.

(d) Define the total voltage through the relation, $V_t = (R_e + R_i)I$ and show that $V_t = 4aF/3$, as well as $V_e + V_i = V_t$. Show that IV_t is the power supplied by the "chemical" force.

Reference: W. M. Saslow, *Am. J. Phys.* **62**, 495–501 (1994).

3.16 **(a)** Starting from the Bessel differential equation and appropriate limiting procedures, verify the generalization of (3.108),

$$\frac{1}{k}\,\delta(k - k') = \int_0^\infty \rho J_\nu(k\rho)J_\nu(k'\rho)\,d\rho$$

or equivalently that

$$\frac{1}{\rho}\,\delta(\rho - \rho') = \int_0^\infty k J_\nu(k\rho)J_\nu(k\rho')\,dk$$

where $\mathrm{Re}(\nu) > -1$.

(b) Obtain the following expansion:

$$\frac{1}{|\mathbf{x} - \mathbf{x}'|} = \sum_{m=-\infty}^{\infty}\int_0^\infty dk\; e^{im(\phi-\phi')}J_m(k\rho)J_m(k\rho')e^{-k(z_> - z_<)}$$

(c) By appropriate limiting procedures prove the following expansions:

$$\frac{1}{\sqrt{\rho^2 + z^2}} = \int_0^\infty e^{-k|z|}J_0(k\rho)\,dk$$

$$J_0(k\sqrt{\rho^2 + \rho'^2 - 2\rho\rho'\cos\phi}) = \sum_{m=-\infty}^{\infty} e^{im\phi}J_m(k\rho)J_m(k\rho')$$

$$e^{ik\rho\cos\phi} = \sum_{m=-\infty}^{\infty} i^m e^{im\phi}J_m(k\rho)$$

(d) From the last result obtain an integral representation of the Bessel function:

$$J_m(x) = \frac{1}{2\pi i^m}\int_0^{2\pi} e^{ix\cos\phi - im\phi}\,d\phi$$

Compare the standard integral representations.

3.17 The Dirichlet Green function for the unbounded space between the planes at $z = 0$ and $z = L$ allows discussion of a point charge or a distribution of charge between parallel conducting planes held at zero potential.

(a) Using cylindrical coordinates show that one form of the Green function is

$$G(\mathbf{x}, \mathbf{x}')$$
$$= \frac{4}{L}\sum_{n=1}^{\infty}\sum_{m=-\infty}^{\infty} e^{im(\phi-\phi')}\sin\left(\frac{n\pi z}{L}\right)\sin\left(\frac{n\pi z'}{L}\right)I_m\left(\frac{n\pi}{L}\rho_<\right)K_m\left(\frac{n\pi}{L}\rho_>\right)$$

(b) Show that an alternative form of the Green function is

$$G(\mathbf{x}, \mathbf{x}') = 2 \sum_{m=-\infty}^{\infty} \int_0^{\infty} dk \, e^{im(\phi - \phi')} J_m(k\rho) J_m(k\rho') \frac{\sinh(kz_<) \sinh[k(L - z_>)]}{\sinh(kL)}$$

3.18 The configuration of Problem 3.12 is modified by placing a conducting plane held at zero potential parallel to and a distance L away from the plane with the disc insert in it. For definiteness put the grounded plane at $z = 0$ and the other plane with the center of the disc on the z axis at $z = L$.

(a) Show that the potential between the planes can be written in cylindrical coordinates (z, ρ, ϕ) as

$$\Phi(z, \rho) = V \int_0^{\infty} d\lambda \, J_1(\lambda) J_0(\lambda \rho / a) \frac{\sinh(\lambda z / a)}{\sinh(\lambda L / a)}$$

(b) Show that in the limit $a \to \infty$ with z, ρ, L fixed the solution of part a reduces to the expected result. Viewing your result as the lowest order answer in an expansion in powers of a^{-1}, consider the question of corrections to the lowest order expression if a is *large* compared to ρ and L, but not infinite. Are there difficulties? Can you obtain an explicit *estimate* of the corrections?

(c) Consider the limit of $L \to \infty$ with $(L - z)$, a and ρ fixed and show that the results of Problem 3.12 are recovered. What about corrections for $L \gg a$, but not $L \to \infty$?

3.19 Consider a point charge q between two infinite parallel conducting planes held at zero potential. Let the planes be located at $z = 0$ and $z = L$ in a cylindrical coordinate system, with the charge on the z axis at $z = z_0$, $0 < z_0 < L$. Use Green's reciprocation theorem of Problem 1.12 with problem 3.18 as the comparison problem.

(a) Show that the amount of induced charge on the plate at $z = L$ inside a circle of radius a whose center is on the z axis is given by

$$Q_L(a) = -\frac{q}{V} \Phi(z_0, 0)$$

where $\Phi(z_0, 0)$ is the potential of Problem 3.18 evaluated at $z = z_0$, $\rho = 0$. Find the *total* charge induced on the upper plate. Compare with the solution (in method and answer) of Problem 1.13.

(b) Show that the induced charge *density* on the upper plate can be written as

$$\sigma(\rho) = -\frac{q}{2\pi} \int_0^{\infty} dk \, \frac{\sinh(kz_0)}{\sinh(kL)} k J_0(k\rho)$$

This integral can be expressed (see, e.g., *Gradshteyn* and *Ryzhik*, p. 728, formula 6.666) as an infinite series involving the modified Bessel functions $K_0(n\pi\rho/L)$, showing that at large radial distances the induced charge density falls off as $(\rho)^{-1/2} e^{-\pi\rho/L}$.

(c) Show that the charge density at $\rho = 0$ is

$$\sigma(0) = -\frac{\pi q}{8L^2} \sec^2\left(\frac{\pi z_0}{2L}\right)$$

3.20 (a) From the results of Problem 3.17 or from first principles show that the potential at a point charge q between two infinite parallel conducting planes held at zero potential can be written as

$$\Phi(z, \rho) = \frac{q}{\pi \epsilon_0 L} \sum_{n=1}^{\infty} \sin\left(\frac{n\pi z_0}{L}\right) \sin\left(\frac{n\pi z}{L}\right) K_0\left(\frac{n\pi\rho}{L}\right)$$

where the planes are at $z = 0$ and $z = L$ and the charge is on the z axis at the point $z = z_0$.

(b) Calculate the induced surface-charge densities $\sigma_0(\rho)$ and $\sigma_L(\rho)$ on the lower and upper plates. The result for $\sigma_L(\rho)$ is

$$\sigma_L(\rho) = \frac{q}{L^2} \sum_{n=1}^{\infty} (-1)^n n \, \sin\left(\frac{n\pi z_0}{L}\right) K_0\left(\frac{n\pi\rho}{L}\right)$$

Discuss the connection of this expression with that of Problem 3.19b and 3.19c.

(c) From the answer in part b, calculate the total charge Q_L on the plate at $z = L$. By summing the Fourier series or by other means of comparison, check your answer against the known expression of Problem 1.13 [C. Y. Fong and C. Kittel, *Am. J. Phys.* **35**, 1091 (1967).]

3.21 (a) By using the Green function of Problem 3.17b in the limit $L \to \infty$, show that the capacitance of a flat, thin, circular, conducting disc of radius R located parallel to, and a distance d above, a grounded conducting plane is given by

$$\frac{4\pi\epsilon_0}{C} = \int_0^{\infty} dk(1 - e^{-2kd}) \frac{\left[\int_0^R \rho J_0(k\rho)\sigma(\rho) \, d\rho\right]^2}{\left[\int_0^R \rho\sigma(\rho) \, d\rho\right]^2}$$

where $\sigma(\rho)$ is the charge density on the disc.

(b) Use the expression in part a as a variational or stationary principle for C^{-1} with the approximation that $\sigma(\rho) =$ constant. Show explicitly that you obtain the correct limiting value for C^{-1} as $d \ll R$. Determine an approximate value of C^{-1} for an isolated disc $(d \gg R)$ and evaluate the ratio of it to the exact result, $4\pi\epsilon_0/C = (\pi/2)R^{-1}$.

(c) As a better trial form for $\sigma(\rho)$ consider a linear combination of a constant and $(R^2 - \rho^2)^{-1/2}$, the latter being the correct form for an isolated disc.

For part b the following integrals may be of use:

$$\int_0^{\infty} dt \left[\frac{J_1(t)}{t}\right]^2 = \frac{4}{3\pi}, \qquad \int_0^{\infty} \frac{dt \, J_1^2(t)}{t} = \frac{1}{2}$$

3.22 The geometry of a two-dimensional potential problem is defined in polar coordinates by the surfaces $\phi = 0$, $\phi = \beta$, and $\rho = a$, as indicated in the sketch.

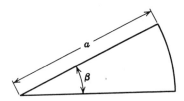

Problem 3.22

Using separation of variables in polar coordinates, show that the Green function can be written as

$$G(\rho, \phi; \rho', \phi') = \sum_{m=1}^{\infty} \frac{4}{m} \rho_<^{m\pi/\beta} \left(\frac{1}{\rho_>^{m\pi/\beta}} - \frac{\rho_>^{m\pi/\beta}}{a^{2m\pi/\beta}}\right) \sin\left(\frac{m\pi\phi}{\beta}\right) \sin\left(\frac{m\pi\phi'}{\beta}\right)$$

Problem 2.25 may be of use.

3.23   A point charge q is located at the point (ρ', ϕ', z') inside a grounded cylindrical box defined by the surfaces $z = 0$, $z = L$, $\rho = a$. Show that the potential inside the box can be expressed in the following alternative forms:

$$\Phi(\mathbf{x}, \mathbf{x}') = \frac{q}{\pi\epsilon_0 a} \sum_{m=-\infty}^{\infty} \sum_{n=1}^{\infty} \frac{e^{im(\phi-\phi')}J_m\left(\frac{x_{mn}\rho}{a}\right)J_m\left(\frac{x_{mn}\rho'}{a}\right)}{x_{mn}J_{m+1}^2(x_{mn})\sinh\left(\frac{x_{mn}L}{a}\right)}$$

$$\times \sinh\left[\frac{x_{mn}}{a}z_<\right]\sinh\left[\frac{x_{mn}}{a}(L-z_>)\right]$$

$$\phi(\mathbf{x}, \mathbf{x}') = \frac{q}{\pi\epsilon_0 L} \sum_{m=-\infty}^{\infty} \sum_{n=1}^{\infty} e^{im(\phi-\phi')}\sin\left(\frac{n\pi z}{L}\right)\sin\left(\frac{n\pi z'}{L}\right)\frac{I_m\left(\frac{n\pi\rho_<}{L}\right)}{I_m\left(\frac{n\pi a}{L}\right)}$$

$$\times \left[I_m\left(\frac{n\pi a}{L}\right)K_m\left(\frac{n\pi\rho_>}{L}\right) - K_m\left(\frac{n\pi a}{L}\right)I_m\left(\frac{n\pi\rho_>}{L}\right)\right]$$

$$\Phi(\mathbf{x}, \mathbf{x}') = \frac{2q}{\pi\epsilon_0 L a^2} \sum_{m=-\infty}^{\infty} \sum_{k=1}^{\infty} \sum_{n=1}^{\infty} \frac{e^{im(\phi-\phi')}\sin\left(\frac{k\pi z}{L}\right)\sin\left(\frac{k\pi z'}{L}\right)J_m\left(\frac{x_{mn}\rho}{a}\right)J_m\left(\frac{x_{mn}\rho'}{a}\right)}{\left[\left(\frac{x_{mn}}{a}\right)^2 + \left(\frac{k\pi}{L}\right)^2\right]J_{m+1}^2(x_{mn})}$$

Discuss the relation of the last expansion (with its extra summation) to the other two.

3.24   The walls of the conducting cylindrical box of Problem 3.23 are all at zero potential, except for a disc in the upper end, defined by $\rho = b < a$, at potential V.

   (a)   Using the various forms of the Green function obtained in Problem 3.23, find three expansions for the potential inside the cylinder.

   (b)   For each series, calculate numerically the ratio of the potential at $\rho = 0$, $z = L/2$ to the potential of the disc, assuming $b = L/4 = a/2$. Try to obtain at least two-significant-figure accuracy. Is one series less rapidly convergent than the others? Why?
   (*Abramowitz and Stegun* have tables; Mathematica has Bessel functions, as does the software of *Press et al.*)

3.25   Consider the surface-charge densities for the problem of Section 3.13 of the conducting plane with a circular hole of radius a.

   (a)   Show that the surface-charge densities on the top and bottom of the plane for $\rho \geq a$ are

$$\sigma_+(\rho) = -\epsilon_0 E_0 + \Delta\sigma(\rho)$$
$$\sigma_-(\rho) = \epsilon_0 E_1 + \Delta\sigma(\rho)$$

where

$$\Delta\sigma(\rho) = -\epsilon_0 \frac{(E_0 - E_1)}{\pi}\left[\frac{a}{\sqrt{\rho^2 - a^2}} - \sin^{-1}\left(\frac{a}{\rho}\right)\right]$$

How does $\Delta\sigma(\rho)$ behave for large ρ? Is $\Delta\sigma(\rho)$, defined in terms of $\Phi^{(1)}$, zero for $\rho < a$? Explain.

(b) Show by direct integration that

$$\lim_{R \to \infty} \left[2\pi \int_a^R d\rho \, \rho(\sigma_+ + \sigma_-) + 2\pi\epsilon_0 \int_0^R d\rho \, \rho(E_0 - E_1) \right] = 0$$

Interpret.

3.26 Consider the Green function appropriate for Neumann boundary conditions for the volume V between the concentric spherical surfaces defined by $r = a$ and $r = b$, $a < b$. To be able to use (1.46) for the potential, impose the simple constraint (1.45). Use an expansion in spherical harmonics of the form,

$$G(\mathbf{x}, \mathbf{x}') = \sum_{l=0}^{\infty} g_l(r, r') P_l(\cos \gamma)$$

where $g_l(r, r') = r_<^l / r_>^{l+1} + f_l(r, r')$.

(a) Show that for $l > 0$, the radial Green function has the symmetric form

$$g_l(r, r') = \frac{r_<^l}{r_>^{l+1}} +$$

$$\frac{1}{(b^{2l+1} - a^{2l+1})} \left[\frac{l+1}{l} (rr')^l + \frac{l}{l+1} \frac{(ab)^{2l+1}}{(rr')^{l+1}} + a^{2l+1} \left(\frac{r^l}{r'^{l+1}} + \frac{r'^l}{r^{l+1}} \right) \right]$$

(b) Show that for $l = 0$

$$g_0(r, r') = \frac{1}{r_>} - \left(\frac{a^2}{a^2 + b^2} \right) \frac{1}{r'} + \lambda + \frac{\mu}{r}$$

where λ and μ are arbitrary. Show explicitly in (1.46) that answers for the potential $\Phi(\mathbf{x})$ are independent of λ and μ.

[The arbitrariness in the Neumann Green function can be removed by symmetrizing g_0 in r and r'.]

3.27 Apply the Neumann Green function of Problem 3.26 to the situation in which the normal electric field is $E_r = -E_0 \cos \theta$ at the outer surface ($r = b$) and is $E_r = 0$ on the inner surface ($r = a$).

(a) Show that the electrostatic potential inside the volume V is

$$\Phi(\mathbf{x}) = E_0 \frac{r \cos \theta}{1 - p^3} \left(1 + \frac{a^3}{2r^3} \right)$$

where $p = a/b$. Find the components of the electric field,

$$E_r(r, \theta) = -E_0 \frac{\cos \theta}{1 - p^3} \left(1 - \frac{a^3}{r^3} \right), \qquad E_\theta(r, \theta) = E_0 \frac{\sin \theta}{1 - p^3} \left(1 + \frac{a^3}{2r^3} \right)$$

(b) Calculate the Cartesian or cylindrical components of the field, E_z and E_ρ, and make a sketch or computer plot of the lines of electric force for a typical case of $p = 0.5$.

CHAPTER 4

Multipoles, Electrostatics of Macroscopic Media, Dielectrics

This chapter is first concerned with the potential due to localized charge distributions and its expansion in multipoles. The development is made in terms of spherical harmonics, but contact is established with the rectangular components for the first few multipoles. The energy of a multipole in an external field is then discussed. An elementary derivation of the macroscopic equations of electrostatics is sketched, but a careful treatment is deferred to Chapter 6. Dielectrics and the appropriate boundary conditions are then described, and some typical boundary-value problems with dielectrics are solved. Simple classical models are used to illustrate the main features of atomic polarizability and susceptibility. Finally the question of electrostatic energy and forces in the presence of dielectrics is discussed.

4.1 Multipole Expansion

A localized distribution of charge is described by the charge density $\rho(\mathbf{x}')$, which is nonvanishing only inside a sphere of radius R around some origin.* The potential outside the sphere can be written as an expansion in spherical harmonics:

$$\Phi(\mathbf{x}) = \frac{1}{4\pi\epsilon_0} \sum_{l=0}^{\infty} \sum_{m=-l}^{l} \frac{4\pi}{2l+1} q_{lm} \frac{Y_{lm}(\theta, \phi)}{r^{l+1}} \tag{4.1}$$

where the particular choice of constant coefficients is made for later convenience. Equation (4.1) is called a multipole expansion; the $l = 0$ term is called the monopole term, $l = 1$ are the dipole terms, etc. The reason for these names becomes clear below. The problem to be solved is the determination of the constants q_{lm} in terms of the properties of the charge density $\rho(\mathbf{x}')$. The solution is very easily obtained from the integral (1.17) for the potential:

$$\Phi(\mathbf{x}) = \frac{1}{4\pi\epsilon_0} \int \frac{\rho(\mathbf{x}')}{|\mathbf{x} - \mathbf{x}'|} d^3x'$$

with expansion (3.70) for $1/|\mathbf{x} - \mathbf{x}'|$. Since we are interested at the moment in the potential outside the charge distribution, $r_< = r'$ and $r_> = r$. Then we find:

$$\Phi(\mathbf{x}) = \frac{1}{\epsilon_0} \sum_{l,m} \frac{1}{2l+1} \left[\int Y_{lm}^*(\theta', \phi') r'^l \rho(\mathbf{x}') \, d^3x' \right] \frac{Y_{lm}(\theta, \phi)}{r^{l+1}} \tag{4.2}$$

---

*The sphere of radius R is an arbitrary conceptual device employed merely to divide space into regions with and without charge. If the charge density falls off with distance faster than any power, the expansion in multipoles is valid at large enough distances.

145

Consequently the coefficients in (4.1) are:

$$q_{lm} = \int Y^*_{lm}(\theta', \phi') r''\rho(\mathbf{x}') \, d^3x' \tag{4.3}$$

These coefficients are called *multipole moments*. To see the physical interpretation of them we exhibit the first few explicitly in terms of Cartesian coordinates:

$$q_{00} = \frac{1}{\sqrt{4\pi}} \int \rho(\mathbf{x}') \, d^3x' = \frac{1}{\sqrt{4\pi}} q \tag{4.4}$$

$$q_{11} = -\sqrt{\frac{3}{8\pi}} \int (x' - iy')\rho(\mathbf{x}') \, d^3x' = -\sqrt{\frac{3}{8\pi}} (p_x - ip_y)$$

$$q_{10} = \sqrt{\frac{3}{4\pi}} \int z'\rho(\mathbf{x}') \, d^3x' = \sqrt{\frac{3}{4\pi}} p_z \tag{4.5}$$

$$q_{22} = \frac{1}{4}\sqrt{\frac{15}{2\pi}} \int (x' - iy')^2\rho(\mathbf{x}') \, d^3x' = \frac{1}{12}\sqrt{\frac{15}{2\pi}} (Q_{11} - 2iQ_{12} - Q_{22})$$

$$q_{21} = -\sqrt{\frac{15}{8\pi}} \int z'(x' - iy')\rho(\mathbf{x}') \, d^3x' = -\frac{1}{3}\sqrt{\frac{15}{8\pi}} (Q_{13} - iQ_{23}) \tag{4.6}$$

$$q_{20} = \frac{1}{2}\sqrt{\frac{5}{4\pi}} \int (3z'^2 - r'^2)\rho(\mathbf{x}') \, d^3x' = \frac{1}{2}\sqrt{\frac{5}{4\pi}} Q_{33}$$

Only the moments with $m \geq 0$ have been given, since (3.54) shows that for a real charge density the moments with $m < 0$ are related through

$$q_{l,-m} = (-1)^m q^*_{lm} \tag{4.7}$$

In equations (4.4)–(4.6), q is the total charge, or monopole moment, $\mathbf{p}$ is the electric dipole moment:

$$\mathbf{p} = \int \mathbf{x}'\rho(\mathbf{x}') \, d^3x' \tag{4.8}$$

and Q_{ij} is the traceless quadrupole moment tensor:

$$Q_{ij} = \int (3x'_i x'_j - r'^2\delta_{ij})\rho(\mathbf{x}') \, d^3x' \tag{4.9}$$

We see that the lth multipole coefficients [$(2l + 1)$ in number] are linear combinations of the corresponding multipoles expressed in rectangular coordinates. The expansion of $\Phi(\mathbf{x})$ in rectangular coordinates

$$\Phi(\mathbf{x}) = \frac{1}{4\pi\epsilon_0} \left[\frac{q}{r} + \frac{\mathbf{p} \cdot \mathbf{x}}{r^3} + \frac{1}{2}\sum_{i,j} Q_{ij} \frac{x_i x_j}{r^5} + \cdots \right] \tag{4.10}$$

by direct Taylor series expansion of $1/|\mathbf{x} - \mathbf{x}'|$ will be left as an exercise for the reader. It becomes increasingly cumbersome to continue the expansion in (4.10) beyond the quadrupole terms.

The electric field components for a given multipole can be expressed most easily in terms of spherical coordinates. The negative gradient of a term in (4.1) with definite l, m has spherical components:

$$E_r = \frac{(l + 1)}{(2l + 1)\epsilon_0} q_{lm} \frac{Y_{lm}(\theta, \phi)}{r^{l+2}}$$

$$E_\theta = -\frac{1}{(2l + 1)\epsilon_0} q_{lm} \frac{1}{r^{l+2}} \frac{\partial}{\partial \theta} Y_{lm}(\theta, \phi) \qquad (4.11)$$

$$E_\phi = -\frac{1}{(2l + 1)\epsilon_0} q_{lm} \frac{1}{r^{l+2}} \frac{im}{\sin \theta} Y_{lm}(\rho, \phi)$$

$\partial Y_{lm}/\partial\theta$ and $Y_{lm}/\sin \theta$ can be expressed as linear combinations of other Y_{lm}'s, but the expressions are not particularly illuminating and so will be omitted. The proper way to describe a vector multipole field is by *vector* spherical harmonics, discussed in Chapter 9.

For a dipole **p** along the z axis, the fields in (4.11) reduce to the familiar form:

$$E_r = \frac{2p \cos \theta}{4\pi\epsilon_0 r^3}$$

$$E_\theta = \frac{p \sin \theta}{4\pi\epsilon_0 r^3} \qquad (4.12)$$

$$E_\phi = 0$$

These dipole fields can be written in vector form by recombining (4.12) or by directly operating with the gradient on the dipole term in (4.10). The result for the field at a point **x** due to a dipole **p** at the point $\mathbf{x}_0$ is:

$$\mathbf{E}(\mathbf{x}) = \frac{3\mathbf{n}(\mathbf{p} \cdot \mathbf{n}) - \mathbf{p}}{4\pi\epsilon_0 |\mathbf{x} - \mathbf{x}_0|^3} \qquad (4.13)$$

where **n** is a unit vector directed from $\mathbf{x}_0$ to **x**.

There are two important remarks to be made. The first concerns the relationship of the Cartesian multipole moments like (4.8) to the spherical multipole moments (4.3). The former are $(l + 1)(l + 2)/2$ in number and for $l > 1$ are more numerous than the $(2l + 1)$ spherical components. There is no contradiction here. The root of the differences lies in the different rotational transformation properties of the two types of multipole moments; the Cartesian tensors are reducible, the spherical, irreducible—see Problem 4.3. Note that for $l = 2$ we have recognized the difference by defining a *traceless* Cartesian quadrupole moment (4.9).

The second remark is that in general the multipole moment coefficients in the expansion (4.1) depend on the choice of origin. As a blatant example, consider a *point charge e* located at $\mathbf{x}_0 = (r_0, \theta_0, \phi_0)$. Its potential has a multipole expansion of the form (4.1) with multipole moments,

$$q_{lm} = er_0^l Y^*_{lm}(\theta_0, \phi_0)$$

These are nonvanishing for all l, m in general. Only the $l = 0$ multipole $q_{00} = e/\sqrt{4\pi}$ is independent of the location of the point charge. For two point charges $+ e$ and $-e$ at $\mathbf{x}_0$ and $\mathbf{x}_1$, respectively, the multipole moments are

$$q_{lm} = e[r_0^l Y^*_{lm}(\theta_0, \phi_0) - r_1^l Y^*_{lm}(\theta_1, \phi_1)]$$

Now the $l = 0$ multipole moment of the system vanishes, and the $l = 1$ moments are

$$q_{10} = \sqrt{\frac{3}{4\pi}} \, e(z_0 - z_1)$$

$$q_{11} = -\sqrt{\frac{3}{8\pi}} \, e[(x_0 - x_1) - i(y_0 - y_1)]$$

These moments are independent of the location of the origin, depending only on the relative position of the two charges, but all higher moments depend on the location of the origin as well. These simple examples are special cases of general theorem (see Problem 4.4). The values of q_{lm} for the *lowest nonvanishing multipole moment* of any charge distribution are *independent of the choice of origin* of the coordinates, but all higher multipole moments do in general depend on the location of the origin.

Before leaving the general formulation of multipoles, we consider a result that is useful in elucidating the basic difference between electric and magnetic dipoles (see Section 5.6) as well as in other contexts. Consider a localized charge distribution $\rho(\mathbf{x})$ that gives rise to an electric field $\mathbf{E}(\mathbf{x})$ throughout space. We wish to calculate the integral of $\mathbf{E}$ over the *volume* of a sphere of radius R. We begin by examining the problem in general, but then specialize to the two extremes shown in Fig. 4.1, one in which the sphere contains all of the charge and the other in which the charge lies external to the sphere. Choosing the origin of coordinates at the center of the sphere, we have the volume integral of the electric field,

$$\int_{r<R} \mathbf{E}(\mathbf{x}) \, d^3x = -\int_{r<R} \boldsymbol{\nabla}\Phi \, d^3x \tag{4.14}$$

This can be converted to an integral over the surface of the sphere:

$$\int_{r<R} \mathbf{E}(\mathbf{x}) \, d^3x = -\int_{r=R} R^2 \, d\Omega \, \Phi(\mathbf{x})\mathbf{n} \tag{4.15}$$

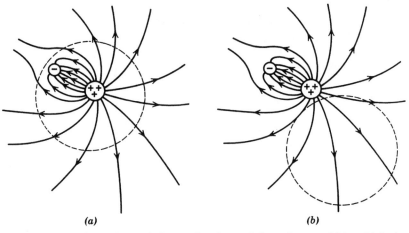

(a)   *(b)*

Figure 4.1   Two configurations of charge density and the spheres within which the volume integral of electric field is to be calculated.

where **n** is the outwardly directed normal ($\mathbf{n} = \mathbf{x}/R$). Substitution of (1.17) for the potential leads to

$$\int_{r<R} \mathbf{E}(\mathbf{x})\, d^3x = -\frac{R^2}{4\pi\epsilon_0} \int d^3x'\, \rho(\mathbf{x}') \int_{r=R} d\Omega\, \frac{\mathbf{n}}{|\mathbf{x} - \mathbf{x}'|} \qquad (4.16)$$

To perform the angular integration we first observe that **n** can be written in terms of the spherical angles (θ, ϕ) as

$$\mathbf{n} = \mathbf{i} \sin\theta \cos\phi + \mathbf{j} \sin\theta \sin\phi + \mathbf{k} \cos\theta$$

Evidently the different components of **n** are linear combinations of Y_{lm} for $l = 1$ *only*. When (3.38) or (3.70) is inserted into (4.16), orthogonality of the Y_{lm} will eliminate all but the $l = 1$ term in the series. Thus we have

$$\int_{r=R} d\Omega\, \frac{\mathbf{n}}{|\mathbf{x} - \mathbf{x}'|} = \frac{r_<}{r_>^2} \int d\Omega\, \mathbf{n} \cos\gamma \qquad (4.16')$$

where $\cos\gamma = \cos\theta \cos\theta' + \sin\theta \sin\theta' \cos(\phi - \phi')$. The angular integral is equal to $4\pi\mathbf{n}'/3$, where $\mathbf{n}' = \mathbf{r}'/r'$. Thus the integral (4.16) is

$$\int_{r<R} \mathbf{E}(\mathbf{x})\, d^3x = -\frac{R^2}{3\epsilon_0} \int d^3x'\, \frac{r_<}{r_>^2}\, \mathbf{n}' \rho(\mathbf{x}') \qquad (4.17)$$

where $(r_<, r_>) = (r', R)$ or (R, r') depending on which of r' and R is larger.

If the sphere of radius R completely encloses the charge density, as indicated in Fig. 4.1a, then $r_< = r'$ and $r_> = R$ in (4.17). The volume integral of the electric field over the sphere then becomes

$$\int_{r<R} \mathbf{E}(\mathbf{x})\, d^3x = -\frac{\mathbf{p}}{3\epsilon_0} \qquad (4.18)$$

where **p** is the electric dipole moment (4.8) of the charge distribution with respect to the center of the sphere. Note that this volume integral is independent of the size of the *spherical* region of integration provided all the charge is inside.

If, on the other hand, the situation is as depicted in Fig. 4.1b, with the charge all exterior to the sphere of interest, $r_< = R$ and $r_> = r'$ in (4.17). Then we have

$$\int_{r<R} \mathbf{E}(\mathbf{x})\, d^3x = -\frac{R^3}{3\epsilon_0} \int d^3x'\, \frac{\mathbf{n}'}{r'^2}\, \rho(\mathbf{x}')$$

From Coulomb's law (1.5) the integral can be recognized to be the negative of $4\pi\epsilon_0$ times the electric field at the center of the sphere. Thus the volume integral of **E** is

$$\int_{r<R} \mathbf{E}(\mathbf{x})\, d^3x = \frac{4\pi}{3} R^3 \mathbf{E}(0) \qquad (4.19)$$

In other words, the average value of the electric field over a spherical volume containing no charge is the value of the field at the center of the sphere.

The result (4.18) implies modification of (4.13) for the electric field of a dipole. To be consistent with (4.18), the dipole field must be written as

$$\mathbf{E}(\mathbf{x}) = \frac{1}{4\pi\epsilon_0} \left[\frac{3\mathbf{n}(\mathbf{p} \cdot \mathbf{n}) - \mathbf{p}}{|\mathbf{x} - \mathbf{x}_0|^3} - \frac{4\pi}{3} \mathbf{p}\delta(\mathbf{x} - \mathbf{x}_0) \right] \qquad (4.20)$$

The added delta function does not contribute to the field away from the site of the dipole. Its purpose is to yield the required volume integral (4.18), with the *convention* that the spherically symmetric (around $\mathbf{x}_0$) volume integral of the first term is zero (from angular integration), the singularity at $\mathbf{x} = \mathbf{x}_0$ causing an otherwise ambiguous result. Equation (4.20) and its magnetic dipole counterpart (5.64), when handled carefully, can be employed as if the dipoles were idealized point dipoles, the delta function terms carrying the essential information about the actually finite distributions of charge and current.

4.2 *Multipole Expansion of the Energy of a Charge Distribution in an External Field*

If a localized charge distribution described by $\rho(\mathbf{x})$ is placed in an *external* potential $\Phi(\mathbf{x})$, the electrostatic energy of the system is:

$$W = \int \rho(\mathbf{x})\Phi(\mathbf{x}) \, d^3x \qquad (4.21)$$

If the potential Φ is slowly varying over the region where $\rho(\mathbf{x})$ is nonnegligible, then it can be expanded in a Taylor series around a suitably chosen origin:

$$\Phi(\mathbf{x}) = \Phi(0) + \mathbf{x} \cdot \nabla\Phi(0) + \frac{1}{2}\sum_i \sum_j x_i x_j \frac{\partial^2 \Phi}{\partial x_i \, \partial x_j}(0) + \cdots \qquad (4.22)$$

Utilizing the definition of the electric field $\mathbf{E} = -\nabla\Phi$, the last two terms can be rewritten. Then (4.22) becomes:

$$\Phi(\mathbf{x}) = \Phi(0) - \mathbf{x} \cdot \mathbf{E}(0) - \frac{1}{2}\sum_i \sum_j x_i x_j \frac{\partial E_j}{\partial x_i}(0) + \cdots$$

Since $\nabla \cdot \mathbf{E} = 0$ for the external field, we can subtract

$$\tfrac{1}{6}r^2\nabla \cdot \mathbf{E}(0)$$

from the last term to obtain finally the expansion:

$$\Phi(\mathbf{x}) = \Phi(0) - \mathbf{x} \cdot \mathbf{E}(0) - \frac{1}{6}\sum_i \sum_j (3x_i x_j - r^2\delta_{ij}) \frac{\partial E_j}{\partial x_i}(0) + \cdots \qquad (4.23)$$

When this is inserted into (4.21) and the definitions of total charge, dipole moment (4.8), and quadrupole moment (4.9) are employed, the energy takes the form:

$$W = q\Phi(0) - \mathbf{p} \cdot \mathbf{E}(0) - \frac{1}{6}\sum_i \sum_j Q_{ij} \frac{\partial E_j}{\partial x_i}(0) + \cdots \qquad (4.24)$$

This expansion shows the characteristic way in which the various multipoles interact with an external field—the charge with the potential, the dipole with the electric field, the quadrupole with the field gradient, and so on.

In nuclear physics the quadrupole interaction is of particular interest. Atomic nuclei can possess electric quadrupole moments, and their magnitudes and signs reflect the nature of the forces between neutrons and protons, as well as the shapes of the nuclei themselves. The energy levels or states of a nucleus are described by the quantum numbers of total angular momentum J and its projection M along the z axis, as well as others, which we will denote by a general index α. A given nuclear state has associated with it a quantum-mechanical charge density* $\rho_{JM\alpha}(\mathbf{x})$, which depends on the quantum numbers (J, M, α) but is cylindrically symmetric about the z axis. Thus the only nonvanishing quadrupole moment is q_{20} in (4.6), or Q_{33} in (4.9).[†] The quadrupole moment of a nuclear state is defined as the value of $(1/e) Q_{33}$ with the charge density $\rho_{JM\alpha}(\mathbf{x})$, where e is the protonic charge:

$$Q_{JM\alpha} = \frac{1}{e} \int (3z^2 - r^2)\rho_{JM\alpha}(\mathbf{x}) \, d^3x \qquad (4.25)$$

The dimensions of $Q_{JM\alpha}$ are consequently (length)^2. Unless the circumstances are exceptional (e.g., nuclei in atoms with completely closed electronic shells), nuclei are subjected to electric fields that possess field gradients in the neighborhood of the nuclei. Consequently, according to (4.24), the energy of the nuclei will have a contribution from the quadrupole interaction. The states of different M value for the same J will have different quadrupole moments $Q_{JM\alpha}$, and so a degeneracy in M value that may have existed will be removed by the quadrupole coupling to the "external" (crystal lattice, or molecular) electric field. Detection of these small energy differences by radiofrequency techniques allows the determination of the quadrupole moment of the nucleus.[‡]

The interaction energy between two dipoles $\mathbf{p}_1$ and $\mathbf{p}_2$ can be obtained directly from (4.24) by using the dipole field (4.20). Thus, the mutual potential energy is

$$W_{12} = \frac{\mathbf{p}_1 \cdot \mathbf{p}_2 - 3(\mathbf{n} \cdot \mathbf{p}_1)(\mathbf{n} \cdot \mathbf{p}_2)}{4\pi\epsilon_0 |\mathbf{x}_1 - \mathbf{x}_2|^3} \qquad (4.26)$$

where $\mathbf{n}$ is a unit vector in the direction $(\mathbf{x}_1 - \mathbf{x}_2)$ and it is assumed that $\mathbf{x}_1 \neq \mathbf{x}_2$. The dipole-dipole interaction is attractive or repulsive, depending on the orientation of the dipoles. For fixed orientation and separation of the dipoles, the value of the interaction, averaged over the relative positions of the dipoles, is zero. If the moments are generally parallel, attraction (repulsion) occurs when the moments are oriented more or less parallel (perpendicular) to the line joining their centers. For antiparallel moments the reverse is true. The extreme values of the potential energy are equal in magnitude.

4.3 *Elementary Treatment of Electrostatics with Ponderable Media*

In Chapters 1, 2, and 3 we considered electrostatic potentials and fields in the presence of charges and conductors, but no other ponderable media. We there-

---

*See *Blatt and Weisskopf* (pp. 23 ff.) for an elementary discussion of the quantum aspects of the problem.

[†]Actually Q_{11} and Q_{22} are different from zero, but are not independent of Q_{33}, being given by $Q_{11} = Q_{22} = -\frac{1}{2}Q_{33}$.

[‡]"The quadrupole moment of a nucleus," denoted by Q, is defined as the value of $Q_{JM\alpha}$ in the state $M = J$. See *Blatt and Weisskopf, loc. cit.*

fore made no distinction between microscopic fields and macroscopic fields, although our treatment of conductors in an idealized fashion with *surface* charge densities implied a macroscopic description. Air is sufficiently tenuous that the neglect of its dielectric properties causes no great error; our results so far are applicable there. But much of electrostatics concerns itself with charges and fields in ponderable media whose respective electric responses must be taken into account. In the Introduction we indicated the need for averaging over macroscopically small, but microscopically large, regions to obtain the Maxwell equations appropriate for macroscopic phenomena. This is done in a careful fashion in Chapter 6, after the Maxwell equations with time variation have been discussed. For the present we merely remind the reader of the outlines of the elementary discussion of polarization in a fashion that glosses over difficult and sometimes subtle aspects of the averaging procedure and the introduction of the macroscopic quantities.

The first observation is that when an averaging is made of the homogeneous equation, $\nabla \times \mathbf{E}_{\text{micro}} = 0$, the same equation, namely,

$$\nabla \times \mathbf{E} = 0 \tag{4.27}$$

holds for the averaged, that is, the macroscopic, electric field $\mathbf{E}$. This means that the electric field is still derivable from a potential $\Phi(\mathbf{x})$ in electrostatics.

If an electric field is applied to a medium made up of a large number of atoms or molecules, the charges bound in each molecule will respond to the applied field and will execute perturbed motions. The molecular charge density will be distorted. The multipole moments of each molecule will be different from what they were in the absence of the field. In simple substances, when there is no applied field the multipole moments are all zero, at least when averaged over many molecules. The dominant molecular multipole with the applied fields is the dipole. There is thus produced in the medium an *electric polarization* $\mathbf{P}$ (dipole moment per unit volume) given by

$$\mathbf{P}(\mathbf{x}) = \sum_i N_i \langle \mathbf{p}_i \rangle \tag{4.28}$$

where $\mathbf{p}_i$ is the dipole moment of the ith type of molecule in the medium, the average is taken over a small volume centered at $\mathbf{x}$ and N_i is the average number per unit volume of the ith type of molecule at the point $\mathbf{x}$. If the molecules have a net charge e_i and, in addition, there is macroscopic excess or free charge, the charge density at the macroscopic level will be

$$\rho(\mathbf{x}) = \sum_i N_i \langle e_i \rangle + \rho_{\text{excess}} \tag{4.29}$$

Usually the average molecular charge is zero. Then the charge density is the excess or free charge (suitably averaged).

If we now look at the medium from a macroscopic point of view, we can build up the potential or field by linear superposition of the contributions from each macroscopically small volume element ΔV at the variable point $\mathbf{x}'$. Thus the charge of ΔV is $\rho(\mathbf{x}') \Delta V$ and the dipole moment of ΔV is $\mathbf{P}(\mathbf{x}') \Delta V$. If there are no higher macroscopic multipole moment densities, the potential $\Delta \Phi(\mathbf{x}, \mathbf{x}')$

caused by the configuration of moments in ΔV can be seen from (4.10) to be given without approximation by

$$\Delta\Phi(\mathbf{x}, \mathbf{x}') = \frac{1}{4\pi\epsilon_0} \left[\frac{\rho(\mathbf{x}')}{|\mathbf{x} - \mathbf{x}'|} \Delta V + \frac{\mathbf{P}(\mathbf{x}') \cdot (\mathbf{x} - \mathbf{x}')}{|\mathbf{x} - \mathbf{x}'|^3} \Delta V \right] \qquad (4.30)$$

provided $\mathbf{x}$ is outside ΔV. We now treat ΔV as (macroscopically) infinitesimal, put it equal to d^3x', and integrate over all space to obtain the potential

$$\Phi(\mathbf{x}) = \frac{1}{4\pi\epsilon_0} \int d^3x' \left[\frac{\rho(\mathbf{x}')}{|\mathbf{x} - \mathbf{x}'|} + \mathbf{P}(\mathbf{x}') \cdot \nabla'\left(\frac{1}{|\mathbf{x} - \mathbf{x}'|}\right) \right] \qquad (4.31)$$

The second term is analogous to the dipole layer potential (1.25), but is for a volume distribution of dipoles. An integration by parts transforms the potential into

$$\Phi(\mathbf{x}) = \frac{1}{4\pi\epsilon_0} \int d^3x' \frac{1}{|\mathbf{x} - \mathbf{x}'|} [\rho(\mathbf{x}') - \nabla' \cdot \mathbf{P}(\mathbf{x}')] \qquad (4.32)$$

This is just the customary expression for the potential caused by a charge distribution $(\rho - \nabla \cdot \mathbf{P})$. With $\mathbf{E} = -\nabla\Phi$, the first Maxwell equation therefore reads

$$\nabla \cdot \mathbf{E} = \frac{1}{\epsilon_0} [\rho - \nabla \cdot \mathbf{P}] \qquad (4.33)$$

The presence of the divergence of $\mathbf{P}$ in the effective charge density can be understood qualitatively. If the polarization is nonuniform there can be a net increase or decrease of charge within any small volume, as indicated schematically in Fig. 4.2.

With the definition of the *electric displacement* $\mathbf{D}$,

$$\mathbf{D} = \epsilon_0\mathbf{E} + \mathbf{P} \qquad (4.34)$$

(4.33) becomes the familiar

$$\nabla \cdot \mathbf{D} = \rho \qquad (4.35)$$

Equations (4.27) and (4.35) are the macroscopic counterparts of (1.13) and (1.14) of Chapter 1.

As discussed in the Introduction, a *constitutive relation* connecting $\mathbf{D}$ and $\mathbf{E}$ is necessary before a solution for the electrostatic potential or fields can be obtained. In the subsequent sections of this chapter we assume that the response of the system to an applied field is linear. This excludes ferroelectricity from discussion, but otherwise is no real restriction provided the field strengths do not

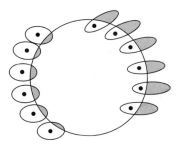

Figure 4.2   Origin of polarization-charge density. Because of spatial variation of polarization, more molecular charge may leave a given small volume than enters it. Only molecules near the boundary are shown.

become extremely large. As a further simplification we suppose that the *medium is isotropic*. Then the induced polarization **P** is parallel to **E** with a coefficient of proportionality that is independent of direction:

$$\mathbf{P} = \epsilon_0 \chi_e \mathbf{E} \tag{4.36}$$

The constant χ_e is called the *electric susceptibility* of the medium. The displacement **D** is therefore proportional to **E**,

$$\mathbf{D} = \epsilon \mathbf{E} \tag{4.37}$$

where

$$\epsilon = \epsilon_0(1 + \chi_e) \tag{4.38}$$

is the electric permittivity; $\epsilon/\epsilon_0 = 1 + \chi_e$ is called the *dielectric constant* or relative electric permittivity.

If the dielectric is not only isotropic, but also uniform, then ϵ is independent of position. The divergence equations (4.35) can then be written

$$\nabla \cdot \mathbf{E} = \rho/\epsilon \tag{4.39}$$

All problems *in that medium* are reduced to those of preceding chapters, except that the electric fields produced by given charges are reduced by a factor ϵ_0/ϵ. The reduction can be understood in terms of a polarization of the atoms that produce fields in opposition to that of the given charge. One immediate consequence is that the capacitance of a capacitor is increased by a factor of ϵ/ϵ_0 if the empty space between the electrodes is filled with a dielectric with dielectric constant ϵ/ϵ_0 (true only to the extent that fringing fields can be neglected).

If the uniform medium does not fill all of the space where there are electric fields or, more generally, if there are different media juxtaposed, not necessarily linear in their responses, we must consider the question of boundary conditions on **D** and **E** at the interfaces between media. These boundary conditions are derived from the full set of Maxwell equations in Section I.5. The results are that the normal components of **D** and the tangential components of **E** on either side of an interface satisfy the *boundary conditions*, valid for time-varying as well as static fields,

$$\left.\begin{array}{c} (\mathbf{D}_2 - \mathbf{D}_1) \cdot \mathbf{n}_{21} = \sigma \\ (\mathbf{E}_2 - \mathbf{E}_1) \times \mathbf{n}_{21} = 0 \end{array}\right\} \tag{4.40}$$

where $\mathbf{n}_{21}$ is a unit normal to the surface, directed from region 1 to region 2, and σ is the macroscopic surface-charge density on the boundary surface (*not* including the polarization charge).

4.4 *Boundary-Value Problems with Dielectrics*

The methods of earlier chapters for the solution of electrostatic boundary-value problems can readily be extended to handle the presence of dielectrics. In this section we treat a few examples of the various techniques applied to dielectric media.

To illustrate the method of images for dielectrics we consider a point charge q embedded in a semi-infinite dielectric ϵ_1 a distance d away from a plane inter-

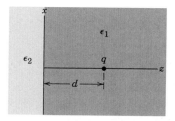

Figure 4.3

face that separates the first medium from another semi-infinite dielectric ϵ_2. The surface may be taken as the plane $z = 0$, as shown in Fig. 4.3. We must find the appropriate solution to the equations:

$$\epsilon_1 \nabla \cdot \mathbf{E} = \rho, \qquad z > 0$$
$$\epsilon_2 \nabla \cdot \mathbf{E} = 0, \qquad z < 0$$

and

$$\nabla \times \mathbf{E} = 0, \qquad \text{everywhere}$$

(4.41)

subject to the boundary conditions at $z = 0$:

$$\lim_{z \to 0^+} \left\{ \begin{array}{c} \epsilon_1 E_z \\ E_x \\ E_y \end{array} \right\} = \lim_{z \to 0^-} \left\{ \begin{array}{c} \epsilon_2 E_z \\ E_x \\ E_y \end{array} \right\}$$

(4.42)

Since $\nabla \times \mathbf{E} = 0$ everywhere, $\mathbf{E}$ is derivable in the usual way from a potential Φ. In attempting to use the image method it is natural to locate an image charge q' at the symmetrical position A' shown in Fig. 4.4. Then for $z > 0$ the potential at a point P described by cylindrical coordinates (ρ, ϕ, z) will be

$$\Phi = \frac{1}{4\pi\epsilon_1} \left(\frac{q}{R_1} + \frac{q'}{R_2} \right), \qquad z > 0$$

(4.43)

where $R_1 = \sqrt{\rho^2 + (d - z)^2}$, $R_2 = \sqrt{\rho^2 + (d + z)^2}$. So far the procedure is completely analogous to the problem with a conducting material in place of the dielectric ϵ_2 for $z < 0$. But we now must specify the potential for $z < 0$. Since there are no charges in the region $z < 0$, it must be a solution of the Laplace equation without singularities in that region. Clearly the simplest assumption is that for $z < 0$ the potential is equivalent to that of a charge q'' at the position A of the actual charge q:

$$\Phi = \frac{1}{4\pi\epsilon_2} \frac{q''}{R_1}, \qquad z < 0$$

(4.44)

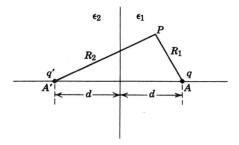

Figure 4.4

Since

$$\frac{\partial}{\partial z}\left(\frac{1}{R_1}\right)\bigg|_{z=0} = -\frac{\partial}{\partial z}\left(\frac{1}{R_2}\right)\bigg|_{z=0} = \frac{d}{(\rho^2 + d^2)^{3/2}}$$

while

$$\frac{\partial}{\partial \rho}\left(\frac{1}{R_1}\right)\bigg|_{z=0} = \frac{\partial}{\partial \rho}\left(\frac{1}{R_2}\right)\bigg|_{z=0} = \frac{-\rho}{(\rho^2 + d^2)^{3/2}}$$

the boundary conditions (4.42) lead to the requirements:

$$q - q' = q''$$
$$\frac{1}{\epsilon_1}(q + q') = \frac{1}{\epsilon_2}q''$$

These can be solved to yield the image charges q' and q'':

$$q' = -\left(\frac{\epsilon_2 - \epsilon_1}{\epsilon_2 + \epsilon_1}\right)q$$

$$q'' = \left(\frac{2\epsilon_2}{\epsilon_2 + \epsilon_1}\right)q \qquad (4.45)$$

For the two cases $\epsilon_2 > \epsilon_1$ and $\epsilon_2 < \epsilon_1$ the lines of force (actually lines of **D**) are shown qualitatively in Fig. 4.5.

The polarization-charge density is given by $-\boldsymbol{\nabla} \cdot \mathbf{P}$. Inside either dielectric, $\mathbf{P} = \epsilon_0\chi_e\mathbf{E}$, so that $-\boldsymbol{\nabla} \cdot \mathbf{P} = -\epsilon_0\chi_e\boldsymbol{\nabla} \cdot \mathbf{E} = 0$, except at the point charge q. At the surface, however, χ_e takes a discontinuous jump, $\nabla\chi_e = (\epsilon_1 - \epsilon_2)/\epsilon_0$ as z passes through $z = 0$. This implies that there is a polarization-surface-charge density on the plane $z = 0$:

$$\sigma_{\text{pol}} = -(\mathbf{P}_2 - \mathbf{P}_1) \cdot \mathbf{n}_{21} \qquad (4.46)$$

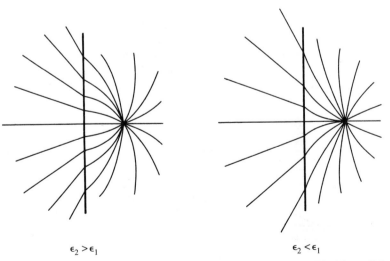

$\epsilon_2 > \epsilon_1$ $\qquad\qquad\qquad\qquad$ $\epsilon_2 < \epsilon_1$

Figure 4.5 Lines of electric displacement for a point charge embedded in a dielectric ϵ_1 near a semi-infinite slab of dielectric ϵ_2.

where $\mathbf{n}_{21}$ is the unit normal from dielectric 1 to dielectric 2, and $\mathbf{P}_i$ is the polarization in the dielectric i at $z = 0$. Since

$$\mathbf{P}_i = (\epsilon_i - \epsilon_0)\mathbf{E}_i = -(\epsilon_i - \epsilon_0)\nabla\Phi(0^{\pm})$$

it is a simple matter to show that the polarization-charge density is

$$\sigma_{\text{pol}} = -\frac{q}{2\pi}\frac{\epsilon_0(\epsilon_2 - \epsilon_1)}{\epsilon_1(\epsilon_2 + \epsilon_1)}\frac{d}{(\rho^2 + d^2)^{3/2}} \tag{4.47}$$

In the limit $\epsilon_2 \gg \epsilon_1$ the dielectric ϵ_2 behaves much like a conductor in that the electric field inside it becomes very small and the surface-charge density (4.47) approaches the value appropriate to a conducting surface, apart from a factor of ϵ_0/ϵ_1.

The second illustration of electrostatic problems involving dielectrics is that of a dielectric sphere of radius a with dielectric constant ϵ/ϵ_0 placed in an initially uniform electric field, which at large distances from the sphere is directed along the z axis and has magnitude E_0, as indicated in Fig. 4.6. Both inside and outside the sphere there are no charges. Consequently the problem is one of solving the Laplace equation with the proper boundary conditions at $r = a$. From the axial symmetry of the geometry we can take the solution to be of the form:

INSIDE:

$$\Phi_{\text{in}} = \sum_{l=0}^{\infty} A_l r^l P_l(\cos\theta) \tag{4.48}$$

OUTSIDE:

$$\Phi_{\text{out}} = \sum_{l=0}^{\infty} [B_l r^l + C_l r^{-(l+1)}]P_l(\cos\theta) \tag{4.49}$$

From the boundary condition at infinity ($\Phi \to -E_0 z = -E_0 r \cos\theta$) we find that the only nonvanishing B_l is $B_1 = -E_0$. The other coefficients are determined from the boundary conditions at $r = a$:

TANGENTIAL E:

NORMAL D:

$$-\frac{1}{a}\frac{\partial\Phi_{\text{in}}}{\partial\theta}\bigg|_{r=a} = -\frac{1}{a}\frac{\partial\Phi_{\text{out}}}{\partial\theta}\bigg|_{r=a}$$

$$-\epsilon\frac{\partial\Phi_{\text{in}}}{\partial r}\bigg|_{r=a} = -\epsilon_0\frac{\partial\Phi_{\text{out}}}{\partial r}\bigg|_{r=a} \tag{4.50}$$

When the series (4.48) and (4.49) are substituted, there result two series of Legendre polynomials equal to zero. Since these must vanish for all θ, the coef-

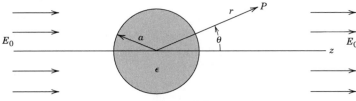

Figure 4.6

ficient of each Legendre polynomial must vanish separately. For the first boundary condition this leads to the relations:

$$A_1 = -E_0 + \frac{C_1}{a^3}$$

$$A_l = \frac{C_l}{a^{2l+1}} \qquad \text{for } l \neq 1 \tag{4.51}$$

while the second gives:

$$(\epsilon/\epsilon_0)A_1 = -E_0 - 2\frac{C_1}{a^3}$$

$$(\epsilon/\epsilon_0)lA_l = -(l+1)\frac{C_l}{a^{2l+1}} \qquad \text{for } l \neq 1 \tag{4.52}$$

The second equations in (4.51) and (4.52) can be satisfied simultaneously only with $A_l = C_l = 0$ for all $l \neq 1$. The remaining coefficients are given in terms of the applied electric field E_0:

$$A_1 = -\left(\frac{3}{2 + \epsilon/\epsilon_0}\right)E_0$$

$$C_1 = \left(\frac{\epsilon/\epsilon_0 - 1}{\epsilon/\epsilon_0 + 2}\right)a^3 E_0 \tag{4.53}$$

The potential is therefore

$$\Phi_{\text{in}} = -\left(\frac{3}{\epsilon/\epsilon_0 + 2}\right)E_0 r \cos\theta$$

$$\Phi_{\text{out}} = -E_0 r \cos\theta + \left(\frac{\epsilon/\epsilon_0 - 1}{\epsilon/\epsilon_0 + 2}\right)E_0 \frac{a^3}{r^2} \cos\theta \tag{4.54}$$

The potential inside the sphere describes a constant electric field parallel to the applied field with magnitude

$$E_{\text{in}} = \frac{3}{\epsilon/\epsilon_0 + 2} E_0 < E_0 \text{ if } \epsilon > \epsilon_0 \tag{4.55}$$

Outside the sphere the potential is equivalent to the applied field E_0 plus the field of an electric dipole at the origin with dipole moment:

$$p = 4\pi\epsilon_0\left(\frac{\epsilon/\epsilon_0 - 1}{\epsilon/\epsilon_0 + 2}\right)a^3 E_0 \tag{4.56}$$

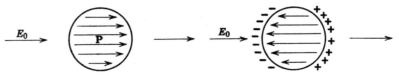

Figure 4.7   Dielectric sphere in a uniform field E_0, showing the polarization on the left and the polarization charge with its associated, opposing, electric field on the right.

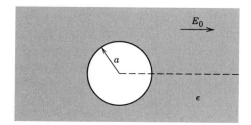

Figure 4.8 Spherical cavity in a dielectric with a uniform field applied.

oriented in the direction of the applied field. The dipole moment can be interpreted as the volume integral of the polarization $\mathbf{P}$. The polarization is

$$\mathbf{P} = (\epsilon - \epsilon_0)\mathbf{E} = 3\epsilon_0\left(\frac{\epsilon/\epsilon_0 - 1}{\epsilon/\epsilon_0 + 2}\right)\mathbf{E}_0 \tag{4.57}$$

It is constant throughout the volume of the sphere and has a volume integral given by (4.56). The polarization-surface-charge density is, according to (4.46), $\sigma_{\text{pol}} = (\mathbf{P} \cdot \mathbf{r})/r$:

$$\sigma_{\text{pol}} = 3\epsilon_0\left(\frac{\epsilon/\epsilon_0 - 1}{\epsilon/\epsilon_0 + 2}\right)E_0 \cos\theta \tag{4.58}$$

This can be thought of as producing an internal field directed oppositely to the applied field, so reducing the field inside the sphere to its value (4.55), as sketched in Fig. 4.7.

The problem of a spherical cavity of radius a in a dielectric medium with dielectric constant ϵ/ϵ_0 and with an applied electric field E_0 parallel to the z axis, as shown in Fig. 4.8, can be handled in exactly the same way as the dielectric sphere. In fact, inspection of boundary conditions (4.50) shows that the results for the cavity can be obtained from those of the sphere by the replacement $\epsilon/\epsilon_0 \rightarrow (\epsilon_0/\epsilon)$. Thus, for example, the field inside the cavity is uniform, parallel to $\mathbf{E}_0$, and of magnitude:

$$E_{\text{in}} = \frac{3\epsilon}{2\epsilon + \epsilon_0} E_0 > E_0 \text{ if } \epsilon > \epsilon_0 \tag{4.59}$$

Similarly, the field outside is the applied field plus that of a dipole at the origin *oriented oppositely* to the applied field and with dipole moment:

$$p = 4\pi\epsilon_0\left(\frac{\epsilon/\epsilon_0 - 1}{2\epsilon/\epsilon_0 + 1}\right)a^3 E_0 \tag{4.60}$$

4.5 *Molecular Polarizability and Electric Susceptibility*

In this section and the next we consider the relation between molecular properties and the macroscopically defined parameter, the electric susceptibility χ_e. Our discussion is in terms of simple classical models of the molecular properties, although a proper treatment necessarily would involve quantum-mechanical considerations. Fortunately, the simpler properties of dielectrics are amenable to classical analysis.

Before examining how the detailed properties of the molecules are related to the susceptibility, we must make a distinction between the fields acting on the molecules in the medium and the applied field. The susceptibility is defined through the relation $\mathbf{P} = \epsilon_0 \chi_e \mathbf{E}$, where $\mathbf{E}$ is the macroscopic electric field. In rarefied media where molecular separations are large there is little difference between the macroscopic field and that acting on any molecule or group of molecules. But in dense media with closely packed molecules the polarization of neighboring molecules gives rise to an internal field $\mathbf{E}_i$ at any given molecule in addition to the average macroscopic field $\mathbf{E}$, so that the total field at the molecule is $\mathbf{E} + \mathbf{E}_i$. The internal field $\mathbf{E}_i$ can be written as the difference of two terms,

$$\mathbf{E}_i = \mathbf{E}_{\text{near}} - \mathbf{E}_P \tag{4.61}$$

where $\mathbf{E}_{\text{near}}$ is the actual contribution of the molecules close to the given molecule and $\mathbf{E}_P$ is the contribution from those molecules treated in an average continuum approximation described by the polarization $\mathbf{P}$. What we are saying here is that close to the molecule in question we must take care to recognize the specific atomic configuration and locations of the nearby molecules. Inside some macroscopically small, but microscopically large, volume V we therefore subtract out the smoothed macroscopic equivalent of the nearby molecular contributions ($\mathbf{E}_P$) and replace it with the correctly evaluated contribution ($\mathbf{E}_{\text{near}}$). This difference is the extra internal field $\mathbf{E}_i$.

The result (4.18) for the integral of the electric field inside a spherical volume of radius R containing a charge distribution can be used to calculate $\mathbf{E}_P$. If the volume V is chosen to be a sphere of radius R containing many molecules, the total dipole moment inside is

$$\mathbf{p} = \frac{4\pi R^3}{3} \mathbf{P}$$

provided V is so small that $\mathbf{P}$ is essentially constant throughout the volume. Then (4.18) shows that the average electric field inside the sphere (just what is desired for $\mathbf{E}_P$) is

$$\mathbf{E}_P = \frac{3}{4\pi R^3} \int_{r<R} \mathbf{E} \, d^3x = -\frac{\mathbf{P}}{3\epsilon_0} \tag{4.62}$$

The internal field can therefore be written

$$\mathbf{E}_i = \frac{1}{3\epsilon_0} \mathbf{P} + \mathbf{E}_{\text{near}} \tag{4.63}$$

The field due to the molecules near by is more difficult to determine. *Lorentz* (p. 138) showed that for atoms in a simple cubic lattice $\mathbf{E}_{\text{near}}$ vanishes at any lattice site. The argument depends on the symmetry of the problem, as can be seen as follows. Suppose that inside the sphere we have a cubic array of dipoles such as are shown in Fig. 4.9, with all their moments constant in magnitude and oriented along the same direction (remember that the sphere is macroscopically small even though it contains very many molecules). The positions of the dipoles are given by the coordinates $\mathbf{x}_{ijk}$ with the components along the coordinate axes (ia, ja, ka), where a is the lattice spacing, and i, j, k each take on positive and

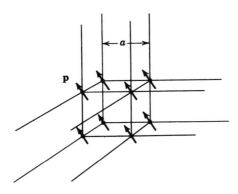

Figure 4.9   Calculation of the internal field: contribution from nearby molecules in a simple cubic lattice.

negative integer values. The field at the origin due to all the dipoles is, according to (4.13),

$$\mathbf{E} = \sum_{i,j,k} \frac{3(\mathbf{p} \cdot \mathbf{x}_{ijk})\mathbf{x}_{ijk} - x_{ijk}^2 \mathbf{p}}{4\pi\epsilon_0 x_{ijk}^5} \tag{4.64}$$

The x component of the field can be written in the form:

$$E_1 = \sum_{ijk} \frac{3(i^2 p_1 + ij p_2 + ik p_3) - (i^2 + j^2 + k^2)p_1}{4\pi\epsilon_0 a^3 (i^2 + j^2 + k^2)^{5/2}} \tag{4.65}$$

Since the indices run equally over positive and negative values, the cross terms involving $(ij p_2 + ik p_3)$ vanish. By symmetry the sums

$$\sum_{ijk} \frac{i^2}{(i^2 + j^2 + k^2)^{5/2}} = \sum_{ijk} \frac{j^2}{(i^2 + j^2 + k^2)^{5/2}} = \sum_{ijk} \frac{k^2}{(i^2 + j^2 + k^2)^{5/2}}$$

are all equal. Consequently

$$E_1 = \sum_{ijk} \frac{[3i^2 - (i^2 + j^2 + k^2)]p_1}{4\pi\epsilon_0 a^3 (i^2 + j^2 + k^2)^{5/2}} = 0 \tag{4.66}$$

Similar arguments show that the y and z components vanish also. Hence $\mathbf{E}_{near} = 0$ for a simple cubic lattice.

If $\mathbf{E}_{near} = 0$ for a highly symmetric situation, it seems plausible that $\mathbf{E}_{near} = 0$ also for completely random situations. Hence we expect amorphous substances to have no internal field due to nearby molecules. For lattices other than simple cubic, the components of $\mathbf{E}_{near}$ are related to the components of $\mathbf{P}$ through a traceless tensor $s_{\alpha\beta}$ that has the symmetry properties of the lattice. Nevertheless, it is a good working assumption that $\mathbf{E}_{near} \simeq 0$ for most materials.

The polarization vector $\mathbf{P}$ was defined in (4.28) as

$$\mathbf{P} = N\langle \mathbf{p}_{mol} \rangle$$

where $\langle \mathbf{p}_{mol} \rangle$ is the average dipole moment of the molecules. This dipole moment is approximately proportional to the electric field acting on the molecule. To exhibit this dependence on electric field we define the *molecular polarizability* γ_{mol} as the ratio of the average molecular dipole moment to ϵ_0 times the applied field at the molecule. Taking account of the internal field (4.63), this gives:

$$\langle \mathbf{p}_{mol} \rangle = \epsilon_0 \gamma_{mol}(\mathbf{E} + \mathbf{E}_i) \tag{4.67}$$

γ_{mol} is, in principle, a function of the electric field, but for a wide range of field strengths is a constant that characterizes the response of the molecules to an applied field. Equation (4.67) can be combined with (4.28) and (4.63) to yield:

$$\mathbf{P} = N\gamma_{\mathrm{mol}}\left(\epsilon_0\mathbf{E} + \frac{1}{3}\,\mathbf{P}\right) \tag{4.68}$$

where we have assumed $\mathbf{E}_{\mathrm{near}} = 0$. Solving for $\mathbf{P}$ in terms of $\mathbf{E}$ and using the fact that $\mathbf{P} = \epsilon_0\chi_e\mathbf{E}$ defines the electric susceptibility of a substance, we find

$$\chi_e = \frac{N\gamma_{\mathrm{mol}}}{1 - \dfrac{1}{3}\,N\gamma_{\mathrm{mol}}} \tag{4.69}$$

as the relation between susceptibility (the macroscopic parameter) and molecular polarizability (the microscopic parameter). Since the dielectric constant is $\epsilon/\epsilon_0 = 1 + \chi_e$, it can be expressed in terms of γ_{mol}, or alternatively the molecular polarizability can be expressed in terms of the dielectric constant:

$$\gamma_{\mathrm{mol}} = \frac{3}{N}\left(\frac{\epsilon/\epsilon_0 - 1}{\epsilon/\epsilon_0 + 2}\right) \tag{4.70}$$

This is called the *Clausius–Mossotti equation*, since Mossotti (in 1850) and Clausius independently (in 1879) established that for any given substance $(\epsilon/\epsilon_0 - 1)(\epsilon/\epsilon_0 + 2)$ should be proportional to the density of the substance.* The relation holds best for dilute substances such as gases. For liquids and solids, (4.70) is only approximately valid, especially if the dielectric constant is large. The interested reader can refer to the books by *Böttcher*, *Debye*, and *Fröhlich* for further details.

4.6   *Models for the Molecular Polarizability*

The polarization of a collection of atoms or molecules can arise in two ways:

the applied field distorts the charge distributions and so produces an induced dipole moment in each molecule;

the applied field tends to line up the initially randomly oriented permanent dipole moments of the molecules.

To estimate the induced moments we consider a simple model of harmonically bound charges (electrons and ions). Each charge e is bound under the action of a restoring force

$$\mathbf{F} = -m\omega_0^2\mathbf{x} \tag{4.71}$$

where m is the mass of the charge, and ω_0 the frequency of oscillation about equilibrium. Under the action of an electric field $\mathbf{E}$ the charge is displaced from its equilibrium by an amount $\mathbf{x}$ given by

$$m\omega_0^2\mathbf{x} = e\mathbf{E}$$

*At optical frequencies, $\epsilon/\epsilon_0 = n^2$, where n is the index of refraction. With n^2 replacing ϵ/ϵ_0 in (4.70), the equation is sometimes called the *Lorentz–Lorenz equation* (1880).

Consequently the induced dipole moment is

$$\mathbf{p}_{\text{mol}} = e\mathbf{x} = \frac{e^2}{m\omega_0^2}\,\mathbf{E} \tag{4.72}$$

This means that the polarizability is $\gamma = e^2/m\omega_0^2\epsilon_0$. If there are a set of charges e_j with masses m_j and oscillation frequencies ω_j in each molecule then the molecular polarizability is

$$\gamma_{\text{mol}} = \frac{1}{\epsilon_0}\sum_j \frac{e_j^2}{m_j\omega_j^2} \tag{4.73}$$

To get a feeling for the order of magnitude of γ we can make two different estimates. Since γ has the dimensions of a volume, its magnitude must be of the order of molecular dimensions or less, namely $\gamma_{\text{el}} \lesssim 10^{-29}$ m^3. Alternatively, we note that the binding frequencies of electrons in atoms must be of the order of light frequencies. Taking a typical wavelength of light as 3000 Å, we find $\omega \approx 6 \times 10^{15}$ s^{-1}. Then the electronic contribution to γ is $\gamma_{\text{el}} \sim (e^2/m\omega^2\epsilon_0) \sim 0.88 \times 10^{-29}$ m^3, consistent with the molecular volume estimate. For gases at NTP the number of molecules per cubic meter is $N = 2.7 \times 10^{25}$, so that their susceptibilities should be of the order of $\chi_e \lesssim 10^{-3}$. This means dielectric constants differing from unity by a few parts in 10^3, or less. Experimentally, typical values of dielectric constant are 1.00054 for air, 1.0072 for ammonia vapor, 1.0057 for methyl alcohol, 1.000068 for helium. For solid or liquid dielectrics, $N \sim 10^{28}$ − 10^{29} molecules/m^3. Consequently, the susceptibility can be of the order of unity (to within a factor $10^{\pm1}$) as is observed.*

The possibility that thermal agitation of the molecules could modify the result (4.73) for the induced dipole polarizability needs consideration. In statistical mechanics the probability distribution of particles in phase space ($\mathbf{p}$, $\mathbf{q}$ space) is some function $f(H)$ of the Hamiltonian. For classical systems,

$$f(H) = e^{-H/kT} \tag{4.74}$$

is the Boltzmann factor. For the simple problem of the harmonically bound charge with an applied field in the z direction, the Hamiltonian is

$$H = \frac{1}{2m}\mathbf{p}^2 + \frac{m}{2}\,\omega_0^2\mathbf{x}^2 - eEz \tag{4.75}$$

where here $\mathbf{p}$ is the momentum of the charged particle. The average value of the dipole moment in the z direction is

$$\langle p_{\text{mol}}\rangle = \frac{\displaystyle\int d^3p \int d^3x\;(ez)f(H)}{\displaystyle\int d^3p \int d^3x\; f(H)} \tag{4.76}$$

If we introduce a displaced coordinate $\mathbf{x}' = \mathbf{x} - eE\hat{\mathbf{z}}/m\omega_0^2$ then

$$H = \frac{1}{2m}\mathbf{p}^2 + \frac{m\omega_0^2}{2}\,(\mathbf{x}')^2 - \frac{e^2E^2}{2m\omega_0^2} \tag{4.77}$$

---

*See, e.g., *CRC Handbook of Chemistry and Physics*, 78th ed., ed. D. R. Lipe, CRC Press, Boca Raton, FL (1997–98).

and

$$\langle p_{\text{mol}} \rangle = \frac{\int d^3p \int d^3x' \left(ez' + \frac{e^2 E}{m\omega_0^2} \right) fH)}{\int d^3p \int d^3x' \, f(H)} \tag{4.78}$$

Since H is even in z' the first integral vanishes. Thus, independent of the form of $f(H)$, we obtain

$$\langle p_{\text{mol}} \rangle = \frac{e^2}{m\omega_0^2} E$$

just as was found in (4.72), ignoring thermal motion.

The second type of polarizability is that caused by the partial orientation of otherwise random permanent dipole moments. This orientation polarization is important in "polar" substances such as HCl and H_2O and was first discussed by Debye (1912). All molecules are assumed to possess a permanent dipole moment $\mathbf{p}_0$, which can be oriented in any direction in space. In the absence of a field, thermal agitation keeps the molecules randomly oriented so that there is no net dipole moment. With an applied field there is a tendency to line up along the field in the configuration of lowest energy. Consequently there will be an average dipole moment. To calculate this we note that the Hamiltonian of the molecule is given by

$$H = H_0 - \mathbf{p}_0 \cdot \mathbf{E} \tag{4.79}$$

where H_0 is a function of only the "internal" coordinates of the molecule. Using the Boltzmann factor (4.74), we can write the average dipole moment as:

$$\langle p_{\text{mol}} \rangle = \frac{\int d\Omega \, p_0 \cos\theta \exp\left(\frac{p_0 E \cos\theta}{kT} \right)}{\int d\Omega \exp\left(\frac{p_0 E \cos\theta}{kT} \right)} \tag{4.80}$$

where we have chosen $\mathbf{E}$ along the z axis, integrated out all the irrelevant variables, and noted that only the component of $\langle \mathbf{p}_0 \rangle$ parallel to the field is different from zero. In general, $(p_0 E/kT)$ is very small compared to unity, except at low temperatures. Hence we can expand the exponentials and obtain the result:

$$\langle p_{\text{mol}} \rangle \simeq \frac{1}{3} \frac{p_0^2}{kT} E \tag{4.81}$$

The orientation polarization depends inversely on the temperature, as might be expected of an effect in which the applied field must overcome the opposition of thermal agitation.

In general both types of polarization, induced (electronic and ionic) and orientation, are present, and the general form of the molecular polarization is

$$\gamma_{\text{mol}} \simeq \gamma_i + \frac{1}{3\epsilon_0} \frac{p_0^2}{kT} \tag{4.82}$$

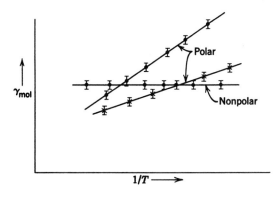

Figure 4.10   Variation of molecular polarizability γ_{mol} with temperature for polar and nonpolar substances: γ_{mol} versus T^{-1}.

This shows a temperature dependence of the form $(a + b/T)$ so that the two types of polarization can be separated experimentally, as indicated in Fig. 4.10. For "polar" molecules, such as HCl and H_2O, the observed permanent dipole moments are of the order of an electronic charge times 10^{-8} cm, in accordance with molecular dimensions.

4.7   *Electrostatic Energy in Dielectric Media*

In Section 1.11 we discussed the energy of a system of charges in free space. The result obtained there,

$$W = \frac{1}{2} \int \rho(\mathbf{x})\Phi(\mathbf{x})\, d^3x \tag{4.83}$$

for the energy due to a charge density $\rho(\mathbf{x})$ and a potential $\Phi(\mathbf{x})$ cannot in general be taken over as it stands in our macroscopic description of dielectric media. The reason becomes clear when we recall how (4.83) was obtained. We thought of the final configuration of charge as being created by assembling bit by bit the elemental charges, bringing each one in from infinitely far away against the action of the then existing electric field. The total work done was given by (4.83). With dielectric media, work is done not only to bring real (macroscopic) charge into position, but also to produce a certain state of polarization in the medium. If ρ and Φ in (4.83) represent macroscopic variables, it is certainly not evident that (4.83) represents the total work, including that done on the dielectric.

To be general in our description of dielectrics, we will not initially make any assumptions about linearity, uniformity, etc., of the response of a dielectric to an applied field. Rather, let us consider a small change in the energy δW due to some sort of change $\delta\rho$ in the macroscopic charge density ρ existing in all space. The work done to accomplish this change is

$$\delta W = \int \delta\rho(\mathbf{x})\Phi(\mathbf{x})\, d^3x \tag{4.84}$$

where $\Phi(\mathbf{x})$ is the potential due to the charge density $\rho(\mathbf{x})$ already present. Since $\boldsymbol{\nabla} \cdot \mathbf{D} = \rho$, we can relate the change $\delta\rho$ to a change in the displacement of $\delta\mathbf{D}$:

$$\delta\rho = \boldsymbol{\nabla} \cdot (\delta\mathbf{D}) \tag{4.85}$$

Then the energy change δW can be cast into the form:

$$\delta W = \int \mathbf{E} \cdot \delta \mathbf{D} \, d^3x \tag{4.86}$$

where we have used $\mathbf{E} = -\nabla\Phi$ and have assumed that $\rho(\mathbf{x})$ was a localized charge distribution. The total electrostatic energy can now be written down formally, at least, by allowing $\mathbf{D}$ to be brought from an initial value $\mathbf{D} = 0$ to its final value $\mathbf{D}$:

$$W = \int d^3x \int_0^D \mathbf{E} \cdot \delta \mathbf{D} \tag{4.87}$$

If the medium is *linear*, then

$$\mathbf{E} \cdot \delta \mathbf{D} = \tfrac{1}{2}\delta(\mathbf{E} \cdot \mathbf{D}) \tag{4.88}$$

and the total electrostatic energy is

$$W = \frac{1}{2} \int \mathbf{E} \cdot \mathbf{D} \, d^3x \tag{4.89}$$

This last result can be transformed into (4.83) by using $\mathbf{E} = -\nabla\Phi$ and $\nabla \cdot \mathbf{D} = \rho$, or by going back to (4.84) and assuming that ρ and Φ are connected linearly. Thus we see that (4.83) *is valid macroscopically only if the behavior is linear.* Otherwise the energy of a final configuration must be calculated from (4.87) and might conceivably depend on the past history of the system (hysteresis effects).

A problem of considerable interest is the change in energy when a dielectric object with a linear response is placed in an electric field whose sources are fixed. Suppose that initially the electric field $\mathbf{E}_0$ due to a certain distribution of charges $\rho_0(\mathbf{x})$ exists in a medium of electric susceptibility ϵ_0, which may be a function of position (for the moment ϵ_0 is not the susceptibility of the vacuum). The initial electrostatic energy is

$$W_0 = \frac{1}{2} \int \mathbf{E}_0 \cdot \mathbf{D}_0 \, d^3x$$

where $\mathbf{D}_0 = \epsilon_0 \mathbf{E}_0$. Then with the sources fixed in position a dielectric object of volume V_1 is introduced into the field, changing the field from $\mathbf{E}_0$ to $\mathbf{E}$. The presence of the object can be described by a susceptibility $\epsilon(\mathbf{x})$, which has the value ϵ_1 inside V_1 and ϵ_0 outside V_1. To avoid mathematical difficulties we can imagine $\epsilon(\mathbf{x})$ to be a smoothly varying function of position that falls rapidly but continuously from ϵ_1 to ϵ_0 at the edge of the volume V_1. The energy now has the value

$$W_1 = \frac{1}{2} \int \mathbf{E} \cdot \mathbf{D} \, d^3x$$

where $\mathbf{D} = \epsilon \mathbf{E}$. The difference in the energy can be written:

$$
\begin{aligned}
W &= \frac{1}{2} \int (\mathbf{E} \cdot \mathbf{D} - \mathbf{E}_0 \cdot \mathbf{D}_0) \, d^3x \\
&= \frac{1}{2} \int (\mathbf{E} \cdot \mathbf{D}_0 - \mathbf{D} \cdot \mathbf{E}_0) \, d^3x + \frac{1}{2} \int (\mathbf{E} + \mathbf{E}_0) \cdot (\mathbf{D} - \mathbf{D}_0) \, d^3x
\end{aligned}
\tag{4.90}
$$

The second integral can be shown to vanish by the following argument. Since $\nabla \times (\mathbf{E} + \mathbf{E}_0) = 0$, we can write

$$\mathbf{E} + \mathbf{E}_0 = -\nabla \Phi$$

Then the second integral becomes:

$$I = -\frac{1}{2} \int \nabla \Phi \cdot (\mathbf{D} - \mathbf{D}_0) \, d^3x$$

Integration by parts transforms this into

$$I = \frac{1}{2} \int \Phi \nabla \cdot (\mathbf{D} - \mathbf{D}_0) \, d^3x = 0$$

since $\nabla \cdot (\mathbf{D} - \mathbf{D}_0) = 0$ because the source charge density $\rho_0(\mathbf{x})$ is assumed unaltered by the insertion of the dielectric object. Consequently the energy change is

$$W = \frac{1}{2} \int (\mathbf{E} \cdot \mathbf{D}_0 - \mathbf{D} \cdot \mathbf{E}_0) \, d^3x \tag{4.91}$$

The integration appears to be over all space, but is actually only over the volume V_1 of the object, since, outside V_1, $\mathbf{D} = \epsilon_0 \mathbf{E}$. Therefore we can write

$$W = -\frac{1}{2} \int_{V_1} (\epsilon_1 - \epsilon_0) \mathbf{E} \cdot \mathbf{E}_0 \, d^3x \tag{4.92}$$

If the medium surrounding the dielectric body is free space, then using the definition of polarization $\mathbf{P}$, (4.92) can then be expressed in the form:

$$W = -\frac{1}{2} \int_{V_1} \mathbf{P} \cdot \mathbf{E}_0 \, d^3x \tag{4.93}$$

where $\mathbf{P}$ is the polarization of the dielectric. This shows that the energy density of a dielectric placed in a field $\mathbf{E}_0$ whose sources are fixed is given by

$$w = -\frac{1}{2} \mathbf{P} \cdot \mathbf{E}_0 \tag{4.94}$$

This result is analogous to the dipole term in the energy (4.24) of a charge distribution in an external field. The factor $\frac{1}{2}$ is due to the fact that (4.94) represents the energy density of a polarizable dielectric in an external field, rather than a permanent dipole. It is the same factor $\frac{1}{2}$ that appears in (4.88).

Equations (4.92) and (4.93) show that a dielectric body will tend to move toward regions of increasing field $\mathbf{E}_0$ provided $\epsilon_1 > \epsilon_0$. To calculate the force acting we can imagine a small generalized displacement of the body $\delta \xi$. Then there will be a change in the energy δW. Since the charges are held fixed, there is no external source of energy and the change in field energy can be interpreted as a change in the potential energy of the body. This means that there is a force acting on the body:

$$F_\xi = -\left(\frac{\partial W}{\partial \xi} \right)_Q \tag{4.95}$$

where the subscript Q has been placed on the partial derivative to indicate that *the sources of the field are kept fixed.*

In practical situations involving the motion of dielectrics the electric fields are often produced by a configuration of electrodes held at *fixed potentials* by connection to an external source such as a battery. To maintain the potentials constant as the distribution of dielectric varies, charge will flow to or from the battery to the electrodes. This means that energy is being supplied from the external source, and it is of interest to compare the energy supplied in that way with the energy change found above *for fixed sources* of the field. We will treat only linear media so that (4.83) is valid. It is sufficient to consider small changes in an existing configuration. From (4.83) it is evident that the change in energy accompanying the changes $\delta\rho(\mathbf{x})$ and $\delta\Phi(\mathbf{x})$ in charge density and potential is

$$\delta W = \frac{1}{2} \int (\rho \, \delta\Phi + \Phi \, \delta\rho) \, d^3x \qquad (4.96)$$

Comparison with (4.84) shows that, if the dielectric properties are not changed, the two terms in (4.96) are equal. If, however, the dielectric properties are altered,

$$\epsilon(\mathbf{x}) \rightarrow \epsilon(\mathbf{x}) + \delta\epsilon(\mathbf{x}) \qquad (4.97)$$

the contributions in (4.96) are not necessarily the same. In fact, we have just calculated the change in energy brought about by introducing a dielectric body into an electric field whose sources were fixed ($\delta\rho = 0$). Equal contributions in (4.96) would imply $\delta W = 0$, but (4.91) or (4.92) are not zero in general. The reason for this difference lies in the existence of the polarization charge. The change in dielectric properties implied by (4.97) can be thought of as a change in the polarization-charge density. If then (4.96) is interpreted as an integral over both free and polarization-charge densities (i.e., a microscopic equation), the two contributions are always equal. However, it is often convenient to deal with macroscopic quantities. Then the equality holds only if the dielectric properties are unchanged.

The process of altering the dielectric properties in some way (by moving the dielectric bodies, by changing their susceptibilities, etc.) in the presence of electrodes at fixed potentials can be viewed as taking place in two steps. In the first step the electrodes are disconnected from the batteries and the charges on them held fixed ($\delta\rho = 0$). With the change (4.97) in dielectric properties, the energy change is

$$\delta W_1 = \frac{1}{2} \int \rho \, \delta\Phi_1 \, d^3x \qquad (4.98)$$

where $\delta\Phi_1$ is the change in potential produced. This can be shown to yield the result (4.92). In the second step the batteries are connected again to the electrodes to restore their potentials to the original values. There will be a flow of charge $\delta\rho_2$ from the batteries accompanying the change in potential* $\delta\Phi_2 = -\delta\Phi_1$. Therefore the energy change in the second step is

$$\delta W_2 = \frac{1}{2} \int (\rho \, \delta\Phi_2 + \Phi \, \delta\rho_2) \, d^3x = -2\delta W_1 \qquad (4.99)$$

---

*Note that it is necessary merely to know that $\delta\Phi_2 = -\delta\Phi_1$ on the electrodes, since that is the only place where free charge resides.

since the two contributions are equal. In the second step we find the external sources changing the energy in the opposite sense and by twice the amount of the initial step. Consequently the net change is

$$\delta W = -\frac{1}{2} \int \rho \, \delta\Phi_1 \, d^3x \qquad (4.100)$$

Symbolically

$$\delta W_V = -\delta W_Q \qquad (4.101)$$

where the subscript denotes the quantity held fixed. If a dielectric with $\epsilon/\epsilon_0 > 1$ moves into a region of greater field strength, the energy increases instead of decreases. For a generalized displacement $d\xi$ the mechanical force acting is now

$$F_\xi = +\left(\frac{\partial W}{\partial \xi}\right)_V \qquad (4.102)$$

References and Suggested Reading

The derivation of the macroscopic equations of electrostatics by averaging over aggregates of atoms is presented in Chapter 6 and by

  Rosenfeld, Chapter II
  Mason and Weaver, Chapter I, Part III
  Van Vleck, Chapter 1

Rosenfeld also treats the classical electron theory of dielectrics. Van Vleck's book is devoted to electric and magnetic susceptibilities. Specific works on electric polarization phenomena are those of

  Böttcher
  Debye
  Fröhlich

Boundary-value problems with dielectrics are discussed in all the references on electrostatics in Chapters 2 and 3.

Our treatment of forces and energy with dielectric media is brief. More extensive discussions, including forces on liquid and solid dielectrics, the electric stress tensor, electrostriction, and thermodynamic effects, may be found in

  Abraham and Becker, Band 1, Chapter V
  Durand, Chapters VI and VII
  Landau and Lifshitz, *Electrodynamics of Continuous Media*
  Maxwell, Vol. 1, Chapter V
  Panofsky and Phillips, Chapter 6
  Stratton, Chapter II

Problems

4.1   Calculate the multipole moments q_{lm} of the charge distributions shown as parts a and b. Try to obtain results for the nonvanishing moments valid for all l, but in each case find the first *two* sets of nonvanishing moments at the very least.

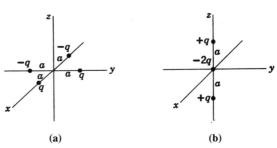

(a)   (b)

Problem 4.1

(c)   For the charge distribution of the second set b write down the multipole expansion for the potential. Keeping only the lowest-order term in the expansion, plot the potential in the x-y plane as a function of distance from the origin for distances greater than a.

(d)   Calculate directly from Coulomb's law the exact potential for b in the x-y plane. Plot it as a function of distance and compare with the result found in part c.

Divide out the asymptotic form in parts c and d to see the behavior at large distances more clearly.

4.2   A point dipole with dipole moment $\mathbf{p}$ is located at the point $\mathbf{x}_0$. From the properties of the derivative of a Dirac delta function, show that for calculation of the potential Φ or the energy of a dipole in an external field, the dipole can be described by an effective charge density

$$\rho_{\text{eff}}(\mathbf{x}) = -\mathbf{p} \cdot \nabla \delta(\mathbf{x} - \mathbf{x}_0)$$

4.3   The lth term in the multipole expansion (4.1) of the potential is specified by the $(2l + 1)$ multipole moments q_{lm}. On the other hand, the Cartesian multipole moments,

$$Q^{(l)}_{\alpha\beta\gamma} = \int \rho(\mathbf{x}) x^\alpha y^\beta z^\gamma \, d^3x$$

with α, β, γ nonnegative integers subject to the constraint $\alpha + \beta + \gamma = l$, are $(l + 1)(l + 2)/2$ in number. Thus, for $l > 1$ there are more Cartesian multipole moments than seem necessary to describe the term in the potential whose radial dependence is r^{-l-1}.

Show that while the q_{lm} transform under rotations as irreducible spherical tensors of rank l, the Cartesian multipole moments correspond to reducible spherical tensors of ranks $l, l - 2, l - 4, \ldots, l_{\min}$, where $l_{\min} = 0$ or 1 for l even or odd, respectively. Check that the number of different tensorial components adds up to the total number of Cartesian tensors. Why are only the q_{lm} needed in the expansion (4.1)?

4.4   (a)   Prove the following theorem: For an arbitrary charge distribution $\rho(\mathbf{x})$ the values of the $(2l + 1)$ moments of the first nonvanishing multipole are independent of the origin of the coordinate axes, but the values of all higher multipole moments do in general depend on the choice of origin. (The different moments q_{lm} for fixed l depend, of course, on the orientation of the axes.)

(b)   A charge distribution has multipole moments q, $\mathbf{p}$, $Q_{ij}, \ldots$ with respect to one set of coordinate axes, and moments q', $\mathbf{p}'$, $Q'_{ij}, \ldots$ with respect to another set whose axes are parallel to the first, but whose origin is located at the point $\mathbf{R} = (X, Y, Z)$ relative to the first. Determine explicitly the connections between the monopole, dipole, and quadrupole moments in the two coordinate frames.

(c) If $q \neq 0$, can $\mathbf{R}$ be found so that $\mathbf{p}' = 0$? If $q \neq 0$, $\mathbf{p} \neq 0$, or at least $\mathbf{p} \neq 0$, can $\mathbf{R}$ be found so that $Q'_{ij} = 0$?

4.5 A localized charge density $\rho(x, y, z)$ is placed in an external electrostatic field described by a potential $\Phi^{(0)}(x, y, z)$. The external potential varies slowly in space over the region where the charge density is different from zero.

(a) From first principles calculate the total *force* acting on the charge distribution as an expansion in multipole moments times derivatives of the electric field, up to and including the quadrupole moments. Show that the force is

$$\mathbf{F} = q\mathbf{E}^{(0)}(0) + \{\nabla[\mathbf{p} \cdot \mathbf{E}^{(0)}(\mathbf{x})]\}_0 + \left\{ \nabla \left[\frac{1}{6} \sum_{j,k} Q_{jk} \frac{\partial E_j^{(0)}}{\partial x_k}(\mathbf{x}) \right] \right\}_0 + \cdots$$

Compare this to the expansion (4.24) of the *energy* W. Note that (4.24) is a number—it is not a function of $\mathbf{x}$ that can be differentiated! What is its connection to $\mathbf{F}$?

(b) Repeat the calculation of part a for the total *torque*. For simplicity, evaluate only one Cartesian component of the torque, say N_1. Show that this component is

$$N_1 = [\mathbf{p} \times \mathbf{E}^{(0)}(0)]_1 + \frac{1}{3} \left[\frac{\partial}{\partial x_3} \left(\sum_j Q_{2j} E_j^{(0)} \right) - \frac{\partial}{\partial x_2} \left(\sum_j Q_{3j} E_j^{(0)} \right) \right]_0 + \cdots$$

4.6 A nucleus with quadrupole moment Q finds itself in a cylindrically symmetric electric field with a gradient $(\partial E_z / \partial z)_0$ along the z axis at the position of the nucleus.

(a) Show that the energy of quadrupole interaction is

$$W = -\frac{e}{4} Q \left(\frac{\partial E_z}{\partial z} \right)_0$$

(b) If it is known that $Q = 2 \times 10^{-28}$ m^2 and that W/h is 10 MHz, where h is Planck's constant, calculate $(\partial E_z / \partial z)_0$ in units of $e/4\pi\epsilon_0 a_0^3$, where $a_0 = 4\pi\epsilon_0 \hbar^2 / me^2 = 0.529 \times 10^{-10}$ m is the Bohr radius in hydrogen.

(c) Nuclear charge distributions can be approximated by a constant charge density throughout a spheroidal volume of semimajor axis a and semiminor axis b. Calculate the quadrupole moment of such a nucleus, assuming that the total charge is Ze. Given that Eu^{153} ($Z = 63$) has a quadrupole moment $Q = 2.5 \times 10^{-28}$ m^2 and a mean radius

$$R = (a + b)/2 = 7 \times 10^{-15} \text{ m}$$

determine the fractional difference in radius $(a - b)/R$.

4.7 A localized distribution of charge has a charge density

$$\rho(\mathbf{r}) = \frac{1}{64\pi} r^2 e^{-r} \sin^2\theta$$

(a) Make a multipole expansion of the potential due to this charge density and determine all the nonvanishing multipole moments. Write down the potential at large distances as a finite expansion in Legendre polynomials.

(b) Determine the potential explicitly at any point in space, and show that near the origin, correct to r^2 inclusive,

$$\Phi(\mathbf{r}) \simeq \frac{1}{4\pi\epsilon_0}\left[\frac{1}{4} - \frac{r^2}{120}\, P_2(\cos\theta)\right]$$

(c) If there exists at the origin a nucleus with a quadrupole moment $Q = 10^{-28}\,\text{m}^2$, determine the magnitude of the interaction energy, assuming that the unit of charge in $\rho(\mathbf{r})$ above is the electronic charge and the unit of length is the hydrogen Bohr radius $a_0 = 4\pi\epsilon_0\hbar^2/me^2 = 0.529 \times 10^{-10}$ m. Express your answer as a frequency by dividing by Planck's constant h.

  The charge density in this problem is that for the $m = \pm 1$ states of the $2p$ level in hydrogen, while the quadrupole interaction is of the same order as found in molecules.

4.8   A very long, right circular, cylindrical shell of dielectric constant ϵ/ϵ_0 and inner and outer radii a and b, respectively, is placed in a previously uniform electric field E_0 with its axis perpendicular to the field. The medium inside and outside the cylinder has a dielectric constant of unity.

(a) Determine the potential and electric field in the three regions, neglecting end effects.

(b) Sketch the lines of force for a typical case of $b \simeq 2a$.

(c) Discuss the limiting forms of your solution appropriate for a solid dielectric cylinder in a uniform field, and a cylindrical cavity in a uniform dielectric.

4.9   A point charge q is located in free space a distance d from the center of a dielectric sphere of radius a ($a < d$) and dielectric constant ϵ/ϵ_0.

(a) Find the potential at all points in space as an expansion in spherical harmonics.

(b) Calculate the rectangular components of the electric field *near* the center of the sphere.

(c) Verify that, in the limit $\epsilon/\epsilon_0 \to \infty$, your result is the same as that for the conducting sphere.

4.10   Two concentric conducting spheres of inner and outer radii a and b, respectively, carry charges $\pm Q$. The empty space between the spheres is half-filled by a hemispherical shell of dielectric (of dielectric constant ϵ/ϵ_0), as shown in the figure.

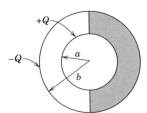

Problem 4.10

(a) Find the electric field everywhere between the spheres.

(b) Calculate the surface-charge distribution on the inner sphere.

(c) Calculate the polarization-charge density induced on the surface of the dielectric at $r = a$.

4.11 The following data on the variation of dielectric constant with pressure are taken from the *Smithsonian Physical Tables*, 9th ed., p. 424:

Air at 292 K		
Pressure (atm)	**ϵ/ϵ_0**	
20	1.0108	Relative density of air as a function of
40	1.0218	pressure is given in *AIP Handbook*,
60	1.0333	[3rd ed., McGraw-Hill, New York
80	1.0439	(1972), p. 4-165].
100	1.0548	

Pentane (C_5H_{12}) at 303 K		
Pressure (atm)	**Density (g/cm³)**	**ϵ/ϵ_0**
1	0.613	1.82
10^3	0.701	1.96
4×10^3	0.796	2.12
8×10^3	0.865	2.24
12×10^3	0.907	2.33

Test the Clausius–Mossotti relation between dielectric constants and density for air and pentane in the ranges tabulated. Does it hold exactly? Approximately? If approximately, discuss fractional variations in density and $(\epsilon/\epsilon_0 - 1)$. For pentane, compare the Clausius–Mossotti relation to the cruder relation, $(\epsilon/\epsilon_0 - 1) \propto$ density.

4.12 Water vapor is a polar gas whose dielectric constant exhibits an appreciable temperature dependence. The following table gives experimental data on this effect. Assuming that water vapor obeys the ideal gas law, calculate the molecular polarizability as a function of inverse temperature and plot it. From the slope of the curve, deduce a value for the permanent dipole moment of the H_2O molecule (express the dipole moment in coulomb-meters).

T(K)	Pressure (cm Hg)	$(\epsilon/\epsilon_0 - 1) \times 10^5$
393	56.49	400.2
423	60.93	371.7
453	65.34	348.8
483	69.75	328.7

4.13 Two long, coaxial, cylindrical conducting surfaces of radii a and b are lowered vertically into a liquid dielectric. If the liquid rises an average height h between the electrodes when a potential difference V is established between them, show that the susceptibility of the liquid is

$$\chi_e = \frac{(b^2 - a^2)\rho g h \ln(b/a)}{\epsilon_0 V^2}$$

where ρ is the density of the liquid, g is the acceleration due to gravity, and the susceptibility of air is neglected.

CHAPTER 5

Magnetostatics, Faraday's Law,
Quasi-Static Fields

5.1 Introduction and Definitions

In the preceding chapters we examined various aspects of electrostatics (i.e., the fields and interactions of stationary charges and boundaries). We now turn to steady-state magnetic phenomena, Faraday's law of induction, and quasi-static fields. From a historical point of view, magnetic phenomena have been known and studied for at least as long as electric phenomena. Lodestones were known in ancient times; the mariner's compass is a very old invention; Gilbert's researches on the earth as a giant magnet date from before 1600. In contrast to electrostatics, the basic laws of magnetic fields did not follow straightforwardly from man's earliest contact with magnetic materials. The reasons are several, but they all stem from the radical difference between magnetostatics and electrostatics: *there are no free magnetic charges* (even though the idea of a magnetic charge density may be a useful mathematical construct in some circumstances). This means that magnetic phenomena are quite different from electric phenomena and that for a long time no connection was established between them. The basic entity in magnetic studies was what we now know as a magnetic dipole. In the presence of magnetic materials the dipole tends to align itself in a certain direction. That direction is by definition the direction of the magnetic-flux density, denoted by **B**, provided the dipole is sufficiently small and weak that it does not perturb the existing field. The magnitude of the flux density can be defined by the mechanical torque **N** exerted on the magnetic dipole:

$$\mathbf{N} = \boldsymbol{\mu} \times \mathbf{B} \tag{5.1}$$

where $\boldsymbol{\mu}$ is the magnetic moment of the dipole, defined in some suitable set of units.

Already, in the definition of the magnetic-flux density **B** (sometimes called the *magnetic induction*), we have a more complicated situation than for the electric field. Further quantitative elucidation of magnetic phenomena did not occur until the connection between currents and magnetic fields was established. A current corresponds to charges in motion and is described by a current density **J**, measured in units of positive charge crossing unit area per unit time, the direction of motion of the charges defining the direction of **J**. In SI units it is measured in coulombs per square meter-second or amperes per square meter. If the current density is confined to wires of small cross section, we usually integrate over the cross-sectional area and speak of a current of so many amperes flowing along the wire.

174

Conservation of charge demands that the charge density at any point in space be related to the current density in that neighborhood by a continuity equation:

$$\frac{\partial \rho}{\partial t} + \nabla \cdot \mathbf{J} = 0 \tag{5.2}$$

This expresses the physical fact that a decrease in charge inside a small volume with time must correspond to a flow of charge out through the surface of the small volume, since the total amount of charge must be conserved. Steady-state magnetic phenomena are characterized by no change in the net charge density anywhere in space. Consequently in magnetostatics

$$\nabla \cdot \mathbf{J} = 0 \tag{5.3}$$

We now proceed to discuss the experimental connection between current and magnetic-flux density and to establish the basic laws of magnetostatics.

5.2 Biot and Savart Law

In 1819 Oersted observed that wires carrying electric currents produced deflections of permanent magnetic dipoles placed in their neighborhood. Thus the currents were sources of magnetic-flux density. Biot and Savart (1820), first, and Ampère (1820–1825), in much more elaborate and thorough experiments, established the basic experimental laws relating the magnetic induction **B** to the currents and established the law of force between one current and another. Although not in the form in which Ampère deduced it, the basic relation is the following. If $d\mathbf{l}$ is an element of length (pointing in the direction of current flow) of a filamentary wire that carries a current I and $\mathbf{x}$ is the coordinate vector from the element of length to an observation point P, as shown in Fig. 5.1, then the elemental flux density $d\mathbf{B}$ at the point P is given in magnitude and direction by

$$d\mathbf{B} = kI \frac{(d\mathbf{l} \times \mathbf{x})}{|\mathbf{x}|^3} \tag{5.4}$$

It should be noted that (5.4) is an inverse square law, just as is Coulomb's law of electrostatics. However, the vector character is very different.

A word of caution about (5.4). There is a temptation to think of (5.4) as the magnetic equivalent of the electric field (1.3) of a point charge and to identify $I\,d\mathbf{l}$ as the analog of q. Strictly speaking this is incorrect. Equation (5.4) has meaning only as one element of a sum over a continuous set, the sum representing the magnetic induction of a current loop or circuit. Obviously the continuity equation (5.3) is not satisfied for the current element $I\,d\mathbf{l}$ standing alone—the

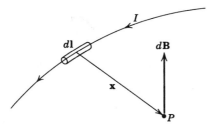

Figure 5.1 Elemental magnetic induction $d\mathbf{B}$ due to current element $I\,d\mathbf{l}$.

current comes from nowhere and disappears after traversing the length $d\mathbf{l}$! One apparent way out of this difficulty is to realize that current is actually charge in motion and to replace $I\,d\mathbf{l}$ by $q\mathbf{v}$ where q is the charge and $\mathbf{v}$ its velocity. The flux density for such a charge in motion would be

$$\mathbf{B} = kq\,\frac{\mathbf{v} \times \mathbf{x}}{|\mathbf{x}|^3} \tag{5.5}$$

in close correspondence with (5.4). But this expression is time dependent and furthermore is valid only for charges whose velocities are small compared to that of light and whose accelerations can be neglected. Since we are considering steady-state magnetic fields in this chapter, we stick with (5.4) and integrate over circuits to obtain physical results.*

In (5.4) and (5.5) the constant k depends in magnitude and dimension on the system of units used, as discussed in detail in the Appendix. In Gaussian units, in which current is measured in esu and magnetic induction in emu, the constant is empirically found to be $k = 1/c$, where c is the speed of light in vacuo. The presence of the speed of light in the equations of magnetostatics is an initial puzzlement resolved within special relativity where v/c has a natural appearance. In Gaussian units, $\mathbf{E}$ and $\mathbf{B}$ have the same dimensions: charge divided by length squared or force per unit charge.

In SI units, $k = \mu_0/4\pi = 10^{-7}$ newton per square ampere (N/A^2) or henry per meter (H/m). Here $\mathbf{B}$ has the dimensions of newtons per ampere-meter (N/A $\cdot$ m) while $\mathbf{E}$ has dimensions of N/C. $\mathbf{B}$ times a speed has the same dimensions as $\mathbf{E}$. Since c is the natural speed in electromagnetism, it is no surprise that in SI units $\mathbf{E}$ and $c\mathbf{B}$ form the field-strength tensor $F^{\mu\nu}$ in a relativistic description (see Chapter 11).

We can linearly superpose the basic magnetic-flux elements (5.4) by integration to determine the magnetic-flux density due to various configurations of current-carrying wires. For example, the magnetic induction $\mathbf{B}$ of the long straight wire shown in Fig. 5.2 carrying a current I can be seen to be directed along the normal to the plane containing the wire and the observation point, so that the lines of magnetic induction are concentric circles around the wire. The magnitude of $\mathbf{B}$ is given by

$$|\mathbf{B}| = \frac{\mu_0}{4\pi}\,IR \int_{-\infty}^{\infty} \frac{dl}{(R^2 + l^2)^{3/2}} = \frac{\mu_0}{2\pi}\frac{I}{R} \tag{5.6}$$

where R is the distance from the observation point to the wire. This is the experimental result first found by Biot and Savart and is known as the Biot–Savart law. Note that the magnitude of the induction $\mathbf{B}$ varies with R in the same way as the electric field due to a long line charge of uniform linear-charge density.

---

*There is an apparent inconsistency here. Currents are, after all, charges in motion. How can (5.4), integrated, yield exact results yet (5.5) be only approximate? The answer is that (5.5) applies to only one charge. If a system of many charges moves in such a way that as the unit of charge goes to zero and the number of charges goes to infinity it produces a steady current flow, then the sum of the exact relativistic fields, *including acceleration effects*, gives a magnetostatic field equal to the field obtained by integrating (5.4) over the circuit. This rather subtle result is discussed for some special situations in Problems 14.23 and 14.24.

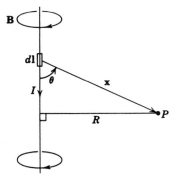

Figure 5.2

This analogy shows that in some circumstances there may be a correspondence between electrostatic and magnetostatic problems, even though the vector character of the fields is different. We see more of that in later sections.

Ampère's experiments did not deal directly with the determination of the relation between currents and magnetic induction, but were concerned rather with the force that one current-carrying wire experiences in the presence of another. Since we have already introduced the idea that a current element produces a magnetic induction, we phrase the force law as the force experienced by a current element $I_1 \, d\mathbf{l}_1$ in the presence of a magnetic induction $\mathbf{B}$. The elemental force is

$$d\mathbf{F} = I_1 \, (d\mathbf{l}_1 \times \mathbf{B}) \tag{5.7}$$

If the external field $\mathbf{B}$ is due to a closed current loop #2 with current I_2, then the total force which a closed current loop #1 with current I_1 experiences is [from (5.4) and (5.7)]:

$$\mathbf{F}_{12} = \frac{\mu_0}{4\pi} I_1 I_2 \oint \oint \frac{d\mathbf{l}_1 \times (d\mathbf{l}_2 \times \mathbf{x}_{12})}{|\mathbf{x}_{12}|^3} \tag{5.8}$$

The line integrals are taken around the two loops; $\mathbf{x}_{12}$ is the vector distance from line element $d\mathbf{l}_2$ to $d\mathbf{l}_1$, as shown in Fig. 5.3. This is the mathematical statement of Ampère's observations about forces between current-carrying loops. By manipulating the integrand it can be put in a form that is symmetric in $d\mathbf{l}_1$ and $d\mathbf{l}_2$ and that explicitly satisfies Newton's third law. Thus

$$\frac{d\mathbf{l}_1 \times (d\mathbf{l}_2 \times \mathbf{x}_{12})}{|\mathbf{x}_{12}|^3} = -(d\mathbf{l}_1 \cdot d\mathbf{l}_2) \frac{\mathbf{x}_{12}}{|\mathbf{x}_{12}|^3} + d\mathbf{l}_2 \left(\frac{d\mathbf{l}_1 \cdot \mathbf{x}_{12}}{|\mathbf{x}_{12}|^3} \right) \tag{5.9}$$

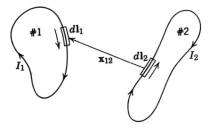

Figure 5.3   Two Ampèrian current loops.

The second term involves a perfect differential in the integral over dl_1. Consequently it gives no contribution to the integral (5.8), provided the paths are closed or extend to infinity. Then Ampère's law of force between current loops becomes

$$\mathbf{F}_{12} = -\frac{\mu_0}{4\pi} I_1 I_2 \oint \oint \frac{(d\mathbf{l}_1 \cdot d\mathbf{l}_2)\mathbf{x}_{12}}{|\mathbf{x}_{12}|^3} \tag{5.10}$$

showing symmetry in the integration, apart from the necessary vectorial dependence on $\mathbf{x}_{12}$.

Each of two long, parallel, straight wires a distance d apart, carrying currents I_1 and I_2, experiences a force per unit length directed perpendicularly toward the other wire and of magnitude,

$$\frac{dF}{dl} = \frac{\mu_0}{2\pi} \frac{I_1 I_2}{d} \tag{5.11}$$

The force is attractive (repulsive) if the currents flow in the same (opposite) directions. The forces that exist between current-carrying wires can be used to define magnetic-flux density in a way that is independent of permanent magnetic dipoles.* We will see later that the torque expression (5.1) and the force result (5.7) are intimately related.

If a current density $\mathbf{J}(\mathbf{x})$ is in an external magnetic-flux density $\mathbf{B}(\mathbf{x})$, the elementary force law implies that the total force on the current distribution is

$$\mathbf{F} = \int \mathbf{J}(\mathbf{x}) \times \mathbf{B}(\mathbf{x}) \, d^3x \tag{5.12}$$

Similarly the total torque is

$$\mathbf{N} = \int \mathbf{x} \times (\mathbf{J} \times \mathbf{B}) \, d^3x \tag{5.13}$$

These general results will be applied to localized current distributions in Section 5.7.

5.3 *Differential Equations of Magnetostatics and Ampère's Law*

The basic law (5.4) for the magnetic induction can be written down in general form for a current density $\mathbf{J}(\mathbf{x})$:

$$\mathbf{B}(\mathbf{x}) = \frac{\mu_0}{4\pi} \int \mathbf{J}(\mathbf{x}') \times \frac{\mathbf{x} - \mathbf{x}'}{|\mathbf{x} - \mathbf{x}'|^3} \, d^3x' \tag{5.14}$$

This expression for $\mathbf{B}(\mathbf{x})$ is the magnetic analog of electric field in terms of the charge density:

$$\mathbf{E}(\mathbf{x}) = \frac{1}{4\pi\epsilon_0} \int \rho(\mathbf{x}') \frac{\mathbf{x} - \mathbf{x}'}{|\mathbf{x} - \mathbf{x}'|^3} \, d^3x' \tag{5.15}$$

Just as this result for $\mathbf{E}$ was not as convenient in some situations as differential equations, so (5.14) is not the most useful form for magnetostatics, even though it contains in principle a description of all the phenomena.

---

*In fact, (5.11) is the basis of the internationally accepted standard of current. See the Appendix.

To obtain the differential equations equivalent to (5.14), we use the relation just above (1.15) to transform (5.14) into the form:

$$\mathbf{B}(\mathbf{x}) = \frac{\mu_0}{4\pi} \boldsymbol{\nabla} \times \int \frac{\mathbf{J}(\mathbf{x}')}{|\mathbf{x} - \mathbf{x}'|} d^3x' \tag{5.16}$$

From (5.16) it follows immediately that the divergence of **B** vanishes:

$$\boldsymbol{\nabla} \cdot \mathbf{B} = 0 \tag{5.17}$$

This is the first equation of magnetostatics and corresponds to $\boldsymbol{\nabla} \times \mathbf{E} = 0$ in electrostatics. By analogy with electrostatics we now calculate the curl of **B**:

$$\boldsymbol{\nabla} \times \mathbf{B} = \frac{\mu_0}{4\pi} \boldsymbol{\nabla} \times \boldsymbol{\nabla} \times \int \frac{\mathbf{J}(\mathbf{x}')}{|\mathbf{x} - \mathbf{x}'|} d^3x' \tag{5.18}$$

With the identity $\boldsymbol{\nabla} \times (\boldsymbol{\nabla} \times \mathbf{A}) = \boldsymbol{\nabla}(\boldsymbol{\nabla} \cdot \mathbf{A}) - \nabla^2\mathbf{A}$ for an arbitrary vector field **A**, expression (5.18) can be transformed into

$$\boldsymbol{\nabla} \times \mathbf{B} = \frac{\mu_0}{4\pi} \boldsymbol{\nabla} \int \mathbf{J}(\mathbf{x}') \cdot \boldsymbol{\nabla}\left(\frac{1}{|\mathbf{x} - \mathbf{x}'|}\right) d^3x' - \frac{\mu_0}{4\pi} \int \mathbf{J}(\mathbf{x}')\nabla^2\left(\frac{1}{|\mathbf{x} - \mathbf{x}'|}\right) d^3x' \tag{5.19}$$

If we use

$$\boldsymbol{\nabla}\left(\frac{1}{|\mathbf{x} - \mathbf{x}'|}\right) = -\boldsymbol{\nabla}'\left(\frac{1}{|\mathbf{x} - \mathbf{x}'|}\right)$$

and

$$\nabla^2\left(\frac{1}{|\mathbf{x} - \mathbf{x}'|}\right) = -4\pi\delta(\mathbf{x} - \mathbf{x}')$$

the integrals in (5.19) can be written:

$$\boldsymbol{\nabla} \times \mathbf{B} = -\frac{\mu_0}{4\pi} \boldsymbol{\nabla} \int \mathbf{J}(\mathbf{x}') \cdot \boldsymbol{\nabla}'\left(\frac{1}{|\mathbf{x} - \mathbf{x}'|}\right) d^3x' + \mu_0\mathbf{J}(\mathbf{x}) \tag{5.20}$$

Integration by parts yields

$$\boldsymbol{\nabla} \times \mathbf{B} = \mu_0\mathbf{J} + \frac{\mu_0}{4\pi} \boldsymbol{\nabla} \int \frac{\boldsymbol{\nabla}' \cdot \mathbf{J}(\mathbf{x}')}{|\mathbf{x} - \mathbf{x}'|} d^3x' \tag{5.21}$$

But for steady-state magnetic phenomena $\boldsymbol{\nabla} \cdot \mathbf{J} = 0$, so that we obtain

$$\boldsymbol{\nabla} \times \mathbf{B} = \mu_0\mathbf{J} \tag{5.22}$$

This is the second equation of magnetostatics, corresponding to $\boldsymbol{\nabla} \cdot \mathbf{E} = \rho/\epsilon_0$ in electrostatics.

In electrostatics Gauss's law (1.11) is the integral form of the equation $\boldsymbol{\nabla} \cdot \mathbf{E} = \rho/\epsilon_0$. The integral equivalent of (5.22) is called *Ampère's law*. It is obtained by applying Stokes's theorem to the integral of the normal component of (5.22) over an open surface S bounded by a closed curve C, as shown in Fig. 5.4. Thus

$$\int_S \boldsymbol{\nabla} \times \mathbf{B} \cdot \mathbf{n} \, da = \mu_0 \int_S \mathbf{J} \cdot \mathbf{n} \, da \tag{5.23}$$

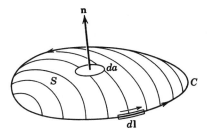

Figure 5.4

is transformed into

$$\oint_C \mathbf{B} \cdot d\mathbf{l} = \mu_0 \int_S \mathbf{J} \cdot \mathbf{n} \, da \qquad (5.24)$$

Since the surface integral of the current density is the total current I passing through the closed curve C, Ampère's law can be written in the form:

$$\oint_C \mathbf{B} \cdot d\mathbf{l} = \mu_0 I \qquad (5.25)$$

Just as Gauss's law can be used for calculation of the electric field in highly symmetric situations, so Ampère's law can be employed in analogous circumstances.

5.4 Vector Potential

The basic differential laws of magnetostatics are

$$\begin{aligned} \boldsymbol{\nabla} \times \mathbf{B} &= \mu_0 \mathbf{J} \\ \boldsymbol{\nabla} \cdot \mathbf{B} &= 0 \end{aligned} \qquad (5.26)$$

The problem is how to solve them. If the current density is zero in the region of interest, $\boldsymbol{\nabla} \times \mathbf{B} = 0$ permits the expression of the vector magnetic induction $\mathbf{B}$ as the gradient of a *magnetic scalar potential*, $\mathbf{B} = -\boldsymbol{\nabla}\Phi_M$. Then (5.26) reduces to the Laplace equation for Φ_M, and all our techniques for handling electrostatic problems can be brought to bear. A large number of problems fall into this class, but we will defer discussion of them until later in the chapter. The reason is that the boundary conditions are different from those encountered in electrostatics, and the problems usually involve macroscopic media with magnetic properties different from free space with charges and currents.

A general method of attack is to exploit the second equation in (5.26). If $\boldsymbol{\nabla} \cdot \mathbf{B} = 0$ everywhere, $\mathbf{B}$ must be the curl of some vector field $\mathbf{A}(\mathbf{x})$, called the *vector potential*,

$$\mathbf{B}(\mathbf{x}) = \boldsymbol{\nabla} \times \mathbf{A}(\mathbf{x}) \qquad (5.27)$$

We have, in fact, already written $\mathbf{B}$ in this form (5.16). Evidently, from (5.16), the general form of $\mathbf{A}$ is

$$\mathbf{A}(\mathbf{x}) = \frac{\mu_0}{4\pi} \int \frac{\mathbf{J}(\mathbf{x}')}{|\mathbf{x} - \mathbf{x}'|} \, d^3x' + \boldsymbol{\nabla}\Psi(\mathbf{x}) \qquad (5.28)$$

The added gradient of an arbitrary scalar function Ψ shows that for a given magnetic induction **B**, the vector potential can be freely transformed according to

$$\mathbf{A} \to \mathbf{A} + \nabla\Psi \qquad (5.29)$$

This transformation is called a *gauge transformation*. Such transformations on **A** are possible because (5.27) specifies only the curl of **A**. The freedom of gauge transformations allows us to make $\nabla \cdot \mathbf{A}$ have any convenient functional form we wish.

If (5.27) is substituted into the first equation in (5.26), we find

$$\nabla \times (\nabla \times \mathbf{A}) = \mu_0 \mathbf{J}$$

or

$$\nabla(\nabla \cdot \mathbf{A}) - \nabla^2 \mathbf{A} = \mu_0 \mathbf{J} \qquad (5.30)$$

If we now exploit the freedom implied by (5.29), we can make the convenient choice of gauge,* $\nabla \cdot \mathbf{A} = 0$. Then each rectangular component of the vector potential satisfies the Poisson equation,

$$\nabla^2 \mathbf{A} = -\mu_0 \mathbf{J} \qquad (5.31)$$

From our discussions of electrostatics it is clear that the solution for **A** in unbounded space is (5.28) with $\Psi = $ constant:

$$\mathbf{A}(\mathbf{x}) = \frac{\mu_0}{4\pi} \int \frac{\mathbf{J}(\mathbf{x}')}{|\mathbf{x} - \mathbf{x}'|} \, d^3x' \qquad (5.32)$$

The condition $\Psi = $ constant can be understood as follows. Our choice of gauge, $\nabla \cdot \mathbf{A} = 0$, reduces to $\nabla^2\Psi = 0$, since the first term in (5.28) has zero divergence because of $\nabla' \cdot \mathbf{J} = 0$. If $\nabla^2\Psi = 0$ holds in all space, Ψ must be at most a constant provided there are no sources at infinity.

5.5 *Vector Potential and Magnetic Induction for a Circular Current Loop*

As an illustration of the calculation of magnetic fields from given current distributions, we consider the problem of a circular loop of radius a, lying in the x-y plane, centered at the origin, and carrying a current I, as shown in Fig. 5.5. The current density **J** has only a component in the ϕ direction,

$$J_\phi = I \sin \theta' \delta(\cos \theta') \frac{\delta(r' - a)}{a} \qquad (5.33)$$

The delta functions restrict current flow to a ring of radius a. The vectorial current density **J** can be written

$$\mathbf{J} = -J_\phi \sin \phi' \mathbf{i} + J_\phi \cos \phi' \mathbf{j} \qquad (5.34)$$

Since the geometry is cylindrically symmetric, we may choose the observation point in the x-z plane ($\phi = 0$) for purposes of calculation. Since the azimuthal

---

*The choice is called the *Coulomb gauge*, for a reason that will become apparent only in Section 6.3.

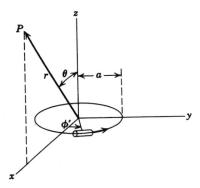

Figure 5.5

integration in (5.32) is symmetric about $\phi' = 0$, the x component of the current does not contribute. This leaves only the y component, which is A_ϕ. Thus

$$A_\phi(r, \theta) = \frac{\mu_0 I}{4\pi a} \int r'^2 \, dr' \, d\Omega' \, \frac{\sin \theta' \cos \phi' \delta(\cos \theta')\delta(r' - a)}{|\mathbf{x} - \mathbf{x}'|} \quad (5.35)$$

where $|\mathbf{x} - \mathbf{x}'| = [r^2 + r'^2 - 2rr'(\cos \theta \cos \theta' + \sin \theta \sin \theta' \cos \phi')]^{1/2}$.

We first consider the straightforward evaluation of (5.35). Integration over the delta functions leaves the result

$$A_\phi(r, \theta) = \frac{\mu_0 I a}{4\pi} \int_0^{2\pi} \frac{\cos \phi' \, d\phi'}{(a^2 + r^2 - 2ar \sin \theta \cos \phi')^{1/2}} \quad (5.36)$$

This integral can be expressed in terms of the complete elliptic integrals K and E:

$$A_\phi(r, \theta) = \frac{\mu_0}{4\pi} \frac{4Ia}{\sqrt{a^2 + r^2 + 2ar \sin \theta}} \left[\frac{(2 - k^2)K(k) - 2E(k)}{k^2} \right] \quad (5.37)$$

where the argument k of the elliptic integrals is defined through

$$k^2 = \frac{4ar \sin \theta}{a^2 + r^2 + 2ar \sin \theta}$$

The components of magnetic induction,

$$\left. \begin{aligned} B_r &= \frac{1}{r \sin \theta} \frac{\partial}{\partial \theta} (\sin \theta A_\phi) \\ B_\theta &= -\frac{1}{r} \frac{\partial}{\partial r} (rA_\phi) \\ B_\phi &= 0 \end{aligned} \right\} \quad (5.38)$$

can also be expressed in terms of elliptic integrals. But the results are not particularly illuminating (useful, however, for computation).

For $a \gg r$, $a \ll r$, or $\theta \ll 1$, an alternative expansion of (5.36) in powers of $a^2 r^2 \sin^2\theta/(a^2 + r^2)^2$ leads to the following approximate expression for the vector potential,

$$A_\phi(r, \theta) = \frac{\mu_0 I a^2 r \sin \theta}{4(a^2 + r^2)^{3/2}} \left[1 + \frac{15 a^2 r^2 \sin^2\theta}{8(a^2 + r^2)^2} + \cdots \right] \quad (5.39)$$

To the same accuracy, the corresponding field components are

$$B_r = \frac{\mu_0 I a^2 \cos\theta}{2(a^2 + r^2)^{3/2}} \left[1 + \frac{15a^2 r^2 \sin^2\theta}{4(a^2 + r^2)^2} + \cdots \right]$$

$$B_\theta = -\frac{\mu_0 I a^2 \sin\theta}{4(a^2 + r^2)^{5/2}} \left[2a^2 - r^2 + \frac{15a^2 r^2 \sin^2\theta(4a^2 - 3r^2)}{8(a^2 + r^2)^2} + \cdots \right]$$

(5.40)

These can easily be specialized to the three regions, near the axis ($\theta \ll 1$), near the center of the loop ($r \ll a$), and far from the loop ($r \gg a$).

Of particular interest are the fields far from the loop:

$$\left. \begin{array}{c} B_r = \dfrac{\mu_0}{2\pi} (I\pi a^2) \dfrac{\cos\theta}{r^3} \\[2ex] B_\theta = \dfrac{\mu_0}{4\pi} (I\pi a^2) \dfrac{\sin\theta}{r^3} \end{array} \right\}$$

(5.41)

Comparison with the electrostatic dipole fields (4.12) shows that the magnetic fields far away from a circular current loop are dipole in character. By analogy with electrostatics we define the magnetic dipole moment of the loop to be

$$m = \pi I a^2$$

(5.42)

We see in the next section that this is a special case of a general result—localized current distributions give dipole fields at large distances; the magnetic moment of a plane current loop is the product of the area of the loop times the current.

Although we have obtained a complete solution to the problem in terms of elliptic integrals, we now illustrate the use of a spherical harmonic expansion to point out similarities and differences between the magnetostatic and electrostatic problems. Thus we return to (5.35) and substitute the spherical expansion (3.70) for $|\mathbf{x} - \mathbf{x}'|^{-1}$:

$$A_\phi = \frac{\mu_0 I}{a} \text{Re} \sum_{l,m} \frac{Y_{lm}(\theta, 0)}{2l + 1} \int r'^2 \, dr' \, d\Omega' \, \delta(\cos\theta') \delta(r' - a) e^{i\phi'} \frac{r_<^l}{r_>^{l+1}} Y^*_{lm}(\theta', \phi')$$

(5.43)

The presence of $e^{i\phi'}$ means that only $m = +1$ will contribute to the sum. Hence

$$A_\phi = 2\pi\mu_0 I a \sum_{l=1}^{\infty} \frac{Y_{l,1}(\theta, 0)}{2l + 1} \frac{r_<^l}{r_>^{l+1}} \left[Y_{l,1}\left(\frac{\pi}{2}, 0\right) \right]$$

(5.44)

where now $r_<$ ($r_>$) is the smaller (larger) of a and r. The square-bracketed quantity is a number depending on l:

$$[\cdots] = \sqrt{\frac{2l + 1}{4\pi l(l + 1)}} \, P_l^1(0)$$

$$= \begin{cases} 0 & \text{for } l \text{ even} \\[2ex] \sqrt{\dfrac{2l + 1}{4\pi l(l + 1)}} \left[\dfrac{(-1)^{n+1}\Gamma(n + \frac{3}{2})}{\Gamma(n + 1)\Gamma(\frac{3}{2})} \right] & \text{for } l = 2n + 1 \end{cases}$$

(5.45)

Then A_ϕ can be written

$$A_\phi = -\frac{\mu_0 I a}{4} \sum_{n=0}^{\infty} \frac{(-1)^n (2n-1)!!}{2^n (n+1)!} \frac{r_<^{2n+1}}{r_>^{2n+2}} P_{2n+1}^1(\cos\theta) \qquad (5.46)$$

where $(2n-1)!! = (2n-1)(2n-3)(\cdots) \times 5 \times 3 \times 1$, and the $n = 0$ coefficient in the sum is unity by definition. To evaluate the radial component of **B** from (5.38) we need

$$\frac{d}{dx}[\sqrt{1-x^2}\, P_l^1(x)] = l(l+1)P_l(x) \qquad (5.47)$$

Then we find

$$B_r = \frac{\mu_0 I a}{2r} \sum_{n=0}^{\infty} \frac{(-1)^n(2n+1)!!}{2^n n!} \frac{r_<^{2n+1}}{r_>^{2n+2}} P_{2n+1}(\cos\theta) \qquad (5.48)$$

The θ component of **B** is similarly

$$B_\theta = -\frac{\mu_0 I a^2}{4} \sum_{n=0}^{\infty} \frac{(-1)^n(2n+1)!!}{2^n(n+1)!} \left\{ \begin{array}{l} -\left(\dfrac{2n+2}{2n+1}\right)\dfrac{1}{a^3}\left(\dfrac{r}{a}\right)^{2n} \\[2ex] \dfrac{1}{r^3}\left(\dfrac{a}{r}\right)^{2n} \end{array} \right\} P_{2n+1}^1(\cos\theta) \qquad (5.49)$$

The upper line holds for $r < a$, and the lower line for $r > a$. For $r \gg a$, only the $n = 0$ term in the series is important. Then, since $P_1^1(\cos\theta) = -\sin\theta$, (5.48) and (5.49) reduce to (5.41). For $r \ll a$, the leading term is again $n = 0$. The fields are then equivalent to a magnetic induction $\mu_0 I/2a$ in the z direction, a result that can be found by elementary means.

We note a characteristic difference between this problem and a corresponding cylindrically symmetric electrostatic problem. Associated Legendre polynomials appear, as well as ordinary Legendre polynomials. This can be traced to the vector character of the current and vector potential, as opposed to the scalar properties of charge and electrostatic potential.

Another mode of attack on the problem of the planar loop is to employ an expansion in cylindrical waves. Instead of (3.70) as a representation of $|\mathbf{x} - \mathbf{x}'|^{-1}$ we may use the cylindrical form (3.148) or (3.149) or that of Problem 3.16b. The application of this technique to the circular loop will be left to the problems.

5.6 Magnetic Fields of a Localized Current Distribution, Magnetic Moment

We now consider the properties of a general current distribution that is localized in a small region of space, "small" being relative to the scale of length of interest to the observer. A complete treatment of this problem, in analogy with the electrostatic multipole expansion, can be made using *vector* spherical harmonics.*

---

*This is not the only way. Scalar potentials can be used. See J. B. Bronzan, *Am. J. Phys.* **39**, 1357 (1971).

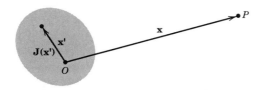

Figure 5.6   Localized current density $\mathbf{J}(\mathbf{x}')$ gives rise to a magnetic induction at the point P with coordinate $\mathbf{x}$.

These are presented in Chapter 9 in connection with multipole radiation. We will be content here with only the lowest order of approximation. Assuming $|\mathbf{x}| \gg |\mathbf{x}'|$, we expand the denominator of (5.32) in powers of $\mathbf{x}'$ measured relative to a suitable origin in the localized current distribution, shown schematically in Fig. 5.6:

$$\frac{1}{|\mathbf{x} - \mathbf{x}'|} = \frac{1}{|\mathbf{x}|} + \frac{\mathbf{x} \cdot \mathbf{x}'}{|\mathbf{x}|^3} + \cdots \tag{5.50}$$

Then a given component of the vector potential will have the expansion,

$$A_i(\mathbf{x}) = \frac{\mu_0}{4\pi} \left[\frac{1}{|\mathbf{x}|} \int J_i(\mathbf{x}') \, d^3x' + \frac{\mathbf{x}}{|\mathbf{x}|^3} \cdot \int J_i(\mathbf{x}')\mathbf{x}' \, d^3x' + \cdots \right] \tag{5.51}$$

The fact that $\mathbf{J}$ is a localized, divergenceless current distribution permits simplification and transformation of the expansion (5.51). Let $f(\mathbf{x}')$ and $g(\mathbf{x}')$ be well-behaved functions of $\mathbf{x}'$ to be chosen below. If $\mathbf{J}(\mathbf{x}')$ is localized but not necessarily divergenceless, we have the identity

$$\int (f\mathbf{J} \cdot \nabla'g + g\mathbf{J} \cdot \nabla'f + fg\nabla' \cdot \mathbf{J}) \, d^3x' = 0 \tag{5.52}$$

This can be established by an integration by parts on the second term, followed by expansion of $f\nabla' \cdot (g\mathbf{J})$. With $f = 1$ and $g = x'_i$, (5.52) with $\nabla' \cdot \mathbf{J} = 0$ establishes that

$$\int J_i(\mathbf{x}') \, d^3x' = 0$$

The first term in (5.51), corresponding to the monopole term in the electrostatic expansion, is therefore absent. With $f = x'_i$, $g = x'_j$ and $\nabla' \cdot \mathbf{J} = 0$, (5.52) yields

$$\int (x'_i J_j + x'_j J_i) \, d^3x' = 0$$

The integral in the second term of (5.51) can therefore be written

$$\mathbf{x} \cdot \int \mathbf{x}' J_i \, d^3x' \equiv \sum_j x_j \int x'_j J_i \, d^3x'$$

$$= -\frac{1}{2} \sum_j x_j \int (x'_i J_j - x'_j J_i) \, d^3x'$$

$$= -\frac{1}{2} \sum_{j,k} \epsilon_{ijk} x_j \int (\mathbf{x}' \times \mathbf{J})_k \, d^3x'$$

$$= -\frac{1}{2} \left[\mathbf{x} \times \int (\mathbf{x}' \times \mathbf{J}) \, d^3x' \right]_i$$

It is customary to define *the magnetic moment density or magnetization* as

$$\mathcal{M}(\mathbf{x}) = \frac{1}{2}\, [\mathbf{x} \times \mathbf{J}(\mathbf{x})] \tag{5.53}$$

and its integral as the *magnetic moment* **m**:

$$\mathbf{m} = \frac{1}{2} \int \mathbf{x}' \times \mathbf{J}(\mathbf{x}')\, d^3x' \tag{5.54}$$

Then the vector potential from the second term in (5.51) is the magnetic dipole vector potential,

$$\mathbf{A}(\mathbf{x}) = \frac{\mu_0}{4\pi}\, \frac{\mathbf{m} \times \mathbf{x}}{|\mathbf{x}|^3} \tag{5.55}$$

This is the lowest nonvanishing term in the expansion of **A** for a localized steady-state current distribution. The magnetic induction **B** outside the localized source can be calculated directly by evaluating the curl of (5.55):

$$\mathbf{B}(\mathbf{x}) = \frac{\mu_0}{4\pi} \left[\frac{3\mathbf{n}(\mathbf{n} \cdot \mathbf{m}) - \mathbf{m}}{|\mathbf{x}|^3} \right] \tag{5.56}$$

Here **n** is a unit vector in the direction **x**. The magnetic induction (5.56) has exactly the form (4.13) of the field of a dipole. This is the generalization of the result found for the circular loop in the last section. Far away from *any* localized current distribution the magnetic induction is that of a magnetic dipole of dipole moment given by (5.54).

If the current is confined to a plane, but otherwise arbitrary, loop, the magnetic moment can be expressed in a simple form. If the current I flows in a closed circuit whose line element is $d\mathbf{l}$, (5.54) becomes

$$\mathbf{m} = \frac{I}{2} \oint \mathbf{x} \times d\mathbf{l}$$

For a plane loop such as that in Fig. 5.7, the magnetic moment is perpendicular to the plane of the loop. Since $\frac{1}{2}|\mathbf{x} \times d\mathbf{l}| = da$, where da is the triangular element of the area defined by the two ends of $d\mathbf{l}$ and the origin, the loop integral gives the total area of the loop. Hence the magnetic moment has magnitude,

$$|\mathbf{m}| = I \times (\text{Area}) \tag{5.57}$$

regardless of the shape of the circuit.

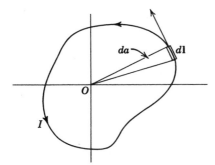

Figure 5.7

If the current distribution is provided by a number of charged particles with charges q_i and masses M_i in motion with velocities $\mathbf{v}_i$, the magnetic moment can be expressed in terms of the orbital angular momentum of the particles. The current density is

$$\mathbf{J} = \sum_i q_i \mathbf{v}_i \delta(\mathbf{x} - \mathbf{x}_i)$$

where $\mathbf{x}_i$ is the position of the ith particle. Then the magnetic moment (5.54) becomes

$$\mathbf{m} = \frac{1}{2} \sum_i q_i (\mathbf{x}_i \times \mathbf{v}_i)$$

The vector product $(\mathbf{x}_i \times \mathbf{v}_i)$ is proportional to the ith particle's orbital angular momentum, $\mathbf{L}_i = M_i(\mathbf{x}_i \times \mathbf{v}_i)$. Thus the moment becomes

$$\mathbf{m} = \sum_i \frac{q_i}{2M_i} \mathbf{L}_i \tag{5.58}$$

If all the particles in motion have the same charge-to-mass ratio $(q_i/M_i = e/M)$, the magnetic moment can be written in terms of the *total* orbital angular momentum $\mathbf{L}$:

$$\mathbf{m} = \frac{e}{2M} \sum_i \mathbf{L}_i = \frac{e}{2M} \mathbf{L} \tag{5.59}$$

This is the well-known classical connection between angular momentum and magnetic moment, which holds for orbital motion even on the atomic scale. But this classical connection fails for the intrinsic moment of electrons and other elementary particles. For electrons, the intrinsic moment is slightly more than twice as large as implied by (5.59), with the spin angular momentum $\mathbf{S}$ replacing $\mathbf{L}$. Thus we speak of the electron having a g factor of $2(1.00116)$. The departure of the magnetic moment from its classical value has its origins in relativistic and quantum-mechanical effects which we cannot consider here.

Before leaving the topic of the fields of a localized current distribution, we consider the spherical volume integral of the magnetic induction $\mathbf{B}$. Just as in the electrostatic case discussed at the end of Section 4.1, there are two limits of interest, one in which the sphere of radius R contains all of the current and the other where the current is completely external to the spherical volume. The volume integral of $\mathbf{B}$ is

$$\int_{r<R} \mathbf{B}(\mathbf{x}) \, d^3x = \int_{r<R} \boldsymbol{\nabla} \times \mathbf{A} \, d^3x \tag{5.60}$$

The volume integral of the curl of $\mathbf{A}$ can be integrated to give a surface integral. Thus

$$\int_{r<R} \mathbf{B} \, d^3x = R^2 \int d\Omega \, \mathbf{n} \times \mathbf{A}$$

where $\mathbf{n}$ is the outwardly directed normal. Substitution of (5.32) for $\mathbf{A}$ and an interchange of the orders of integration permits this to be written as

$$\int_{r<R} \mathbf{B} \, d^3x = -\frac{\mu_0}{4\pi} R^2 \int d^3x' \, \mathbf{J}(\mathbf{x}') \times \int d\Omega \, \frac{\mathbf{n}}{|\mathbf{x} - \mathbf{x}'|}$$

The angular integral is the same one as occurred in the electrostatic situation. Making use of (4.16'), we therefore find for the integral of **B** over a spherical volume,

$$\int_{r<R} \mathbf{B} \, d^3x = \frac{\mu_0}{3} \int \left(\frac{R^2 r_<}{r' r_>^2}\right) \mathbf{x}' \times \mathbf{J}(\mathbf{x}') \, d^3x' \qquad (5.61)$$

where $(r_<, r_>)$ are the smaller and larger of r' and R. If all the current density is contained within the sphere, $r_< = r'$ and $r_> = R$. Then

$$\int_{r<R} \mathbf{B} \, d^3x = \frac{2\mu_0}{3} \mathbf{m} \qquad (5.62)$$

where **m** is the total magnetic moment (5.54). For the opposite extreme of the current all external to the sphere, we have, by virtue of (5.14),

$$\int_{r<R} \mathbf{B} \, d^3x = \frac{4\pi R^3}{3} \mathbf{B}(0) \qquad (5.63)$$

The results (5.62) and (5.63) can be compared with their electrostatic counterparts (4.18) and (4.19). The difference between (5.62) and (4.18) is attributable to the difference in the origins of the fields, one from charges and the other from circulating currents. If we wish to include the information of (5.62) in the magnetic dipole field (5.56), we must add a delta function contribution

$$\mathbf{B}(\mathbf{x}) = \frac{\mu_0}{4\pi} \left[\frac{3\mathbf{n}(\mathbf{n} \cdot \mathbf{m}) - \mathbf{m}}{|\mathbf{x}|^3} + \frac{8\pi}{3} \mathbf{m}\delta(\mathbf{x})\right] \qquad (5.64)$$

The delta function term enters the expression for the hyperfine structure of atomic s states (see the next section).

5.7 *Force and Torque on and Energy of a Localized Current Distribution in an External Magnetic Induction*

If a localized distribution of current is placed in an external magnetic induction **B**(**x**), it experiences forces and torques according to Ampère's laws. The general expressions for the total force and torque are given by (5.12) and (5.13). If the external magnetic induction varies slowly over the region of current, a Taylor series expansion can be utilized to find the dominant terms in the force and torque. A component of **B** can be expanded around a suitable origin,

$$B_k(\mathbf{x}) = B_k(0) + \mathbf{x} \cdot \nabla B_k(0) + \cdots \qquad (5.65)$$

Then the ith component of the force (5.12) becomes

$$F_i = \sum_{jk} \epsilon_{ijk} \left[B_k(0) \int J_j(\mathbf{x}') \, d^3x' + \int J_j(\mathbf{x}')\mathbf{x}' \cdot \nabla B_k(0) \, d^3x' + \cdots\right] \qquad (5.66)$$

Here ϵ_{ijk} is the completely antisymmetric unit tensor ($\epsilon_{ijk} = 1$ for $i = 1$, $j = 2$, $k = 3$, and any cyclic permutation, $\epsilon_{ijk} = -1$ for other permutations, and $\epsilon_{ijk} = 0$ for two or more indices equal). The volume integral of **J** vanishes for steady currents; the lowest order contribution to the force comes from the second term in (5.66). The result above (5.53) can be used [with $\mathbf{x} \to \nabla B_k(0)$] to yield

$$F_i = \sum_{jk} \epsilon_{ijk} (\mathbf{m} \times \nabla)_j B_k(\mathbf{x}) \tag{5.67}$$

After differentiation of $B_k(\mathbf{x})$, $\mathbf{x}$ is to be put to zero. This can be written vectorially as

$$\mathbf{F} = (\mathbf{m} \times \nabla) \times \mathbf{B} = \nabla(\mathbf{m} \cdot \mathbf{B}) - \mathbf{m}(\nabla \cdot \mathbf{B}) \tag{5.68}$$

Since $\nabla \cdot \mathbf{B} = 0$ generally, the lowest order force on a localized current distribution in an external magnetic field $\mathbf{B}$ is

$$\mathbf{F} = \nabla(\mathbf{m} \cdot \mathbf{B}) \tag{5.69}$$

This force represents the rate of change of the total mechanical momentum, including the "hidden mechanical momentum" associated with the presence of electromagnetic momentum. (See Problems 6.5 and 12.8, and the references cited at the end of Chapter 12.) The effective force in Newton's equation of motion of mass times acceleration is (5.69), augmented by $(1/c^2)(d/dt)(\mathbf{E} \times \mathbf{m})$, where $\mathbf{E}$ is the external electric field at the position of the dipole. Apart from angular factors, the relative size of the two contributions is (cB/L) versus $(E/\lambdabar)$, where L is the length scale over which $\mathbf{B}$ changes significantly and $\lambdabar$ is the free-space wavelength of radiation at the typical frequencies present in a Fourier decomposition of the time-varying electric field.

A localized current distribution in a nonuniform magnetic induction experiences a force proportional to its magnetic moment $\mathbf{m}$ and given by (5.69). One simple application of this result is the time-averaged force on a charged particle spiraling in a nonuniform magnetic field. As is well known, a charged particle in a *uniform* magnetic induction moves in a circle at right angles to the field and with constant velocity parallel to the field, tracing out a helical path. The circular motion is, on the time average, equivalent to a circular loop of current that will have a magnetic moment given by (5.57). If the field is not uniform but has a small gradient (so that in one turn around the helix the particle does not feel significantly different field strengths), then the motion of the particle can be discussed in terms of the force on the equivalent magnetic moment. Consideration of the signs of the moment and the force shows that charged particles tend to be repelled by regions of high flux density, independent of the sign of their charge. This is the basis of the "magnetic mirrors," important in the confinement of plasmas.

The total torque on the localized current distribution is found in a similar way by inserting expansion (5.65) into (5.13). Here the zeroth-order term in the expansion contributes. Keeping only this leading term, we have

$$\mathbf{N} = \int \mathbf{x}' \times [\mathbf{J} \times \mathbf{B}(0)] \, d^3x' \tag{5.70}$$

Writing out the triple vector product, we get

$$\mathbf{N} = \int [(\mathbf{x}' \cdot \mathbf{B})\mathbf{J} - (\mathbf{x}' \cdot \mathbf{J})\mathbf{B}] \, d^3x'$$

The first integral has the same form as the one considered in (5.66). Hence we can write down its value immediately. The second integral vanishes for a localized steady-state current distribution, as can be seen from (5.52) with $f = g = r'$. The leading term in the torque is therefore

$$\mathbf{N} = \mathbf{m} \times \mathbf{B}(0) \tag{5.71}$$

This is the familiar expression for the torque on a dipole, discussed in Section 5.1 as one of the ways of defining the magnitude and direction of the magnetic induction.

The potential energy of a permanent magnetic moment (or dipole) in an external magnetic field can be obtained from either the force (5.69) or the torque (5.71). If we interpret the force as the negative gradient of a potential energy U, we find

$$U = -\mathbf{m} \cdot \mathbf{B} \tag{5.72}$$

For a magnetic moment in a uniform field, the torque (5.71) can be interpreted as the negative derivative of U with respect to the angle between $\mathbf{B}$ and $\mathbf{m}$. This well-known result for the potential energy of a dipole shows that the dipole tends to orient itself parallel to the field in the position of lowest potential energy.

We remark in passing that (5.72) is *not* the total energy of the magnetic moment in the external field. In bringing the dipole $\mathbf{m}$ into its final position in the field, work must be done to keep the current $\mathbf{J}$, which produces $\mathbf{m}$, constant. Even though the final situation is a steady state, there is a transient period initially in which the relevant fields are time-dependent. This lies outside our present considerations. Consequently we leave the discussion of the energy of magnetic fields to Section 5.16, following Faraday's law of induction.

The energy expression (5.72) can be employed in the treatment of magnetic effects on atomic energy levels, as in the Zeeman effect or for the fine and hyperfine structure. The fine structure can be viewed as coming from differences in energy of an electron's intrinsic magnetic moment $\boldsymbol{\mu}_e$ in the magnetic field seen in its rest frame. Fine structure, with the subtle complication of Thomas precession, is discussed briefly in Chapter 11. The hyperfine interaction is that of the magnetic moment $\boldsymbol{\mu}_N$ of the nucleus with the magnetic field produced by the electron. The interaction Hamiltonian is (5.72) with $\mathbf{m} = \boldsymbol{\mu}_N$ and $\mathbf{B}$ equal to the magnetic field of the electron, evaluated at the position of the nucleus ($\mathbf{x} = 0$). This field has two parts; one is the dipole field (5.64) and the other is the magnetic field produced by the orbital motion of the electron's charge. The latter is given nonrelativistically by (5.5) and can be expressed as $\mathbf{B}_{\text{orbital}}(0) = \mu_0 e \mathbf{L}/4\pi m r^3$, where $\mathbf{L} = \mathbf{x} \times m\mathbf{v}$ is the orbital angular momentum of the electron about the nucleus. The hyperfine Hamiltonian is therefore

$$\mathcal{H}_{\text{HFS}} = \frac{\mu_0}{4\pi} \left\{ -\frac{8\pi}{3} \boldsymbol{\mu}_e \cdot \boldsymbol{\mu}_N \delta(\mathbf{x}) \right.$$
$$\left. + \frac{1}{r^3} \left[\boldsymbol{\mu}_e \cdot \boldsymbol{\mu}_N - 3 \frac{(\mathbf{x} \cdot \boldsymbol{\mu}_e)(\mathbf{x} \cdot \boldsymbol{\mu}_N)}{r^2} - \frac{e}{m} \mathbf{L} \cdot \boldsymbol{\mu}_N \right] \right\} \tag{5.73}$$

The expectation values of this Hamiltonian in the various atomic (and nuclear spin) states yield the hyperfine energy shifts. For spherically symmetric s states the second term in (5.73) gives a zero expectation value. The hyperfine energy comes solely from the first term:

$$\Delta E = - \frac{\mu_0}{4\pi} \frac{8\pi}{3} |\psi_e(0)|^2 \langle \boldsymbol{\mu}_e \cdot \boldsymbol{\mu}_N \rangle \tag{5.74}$$

For $l \neq 0$, the hyperfine energy comes entirely from the second term in (5.73) because the wave functions for $l \neq 0$ vanish at the origin. These expressions are due to Fermi, who obtained them from the Dirac equation (1930). In applying (5.73) and (5.74) it should be remembered that the charge e is negative and that $\boldsymbol{\mu}_e$ points in the opposite direction to the electron's spin. The energy difference (5.74) between the singlet and triplet states of the $1s$ state of atomic hydrogen is the source of the famous 21 cm line in astrophysics.

The difference of the "contact" term in (5.73) from the *electric* dipole form (4.20) allows us to draw a conclusion concerning the nature of *intrinsic* magnetic moments. While orbital magnetic moments are obviously caused by circulating currents, it is a priori possible that the *intrinsic* magnetic moments of elementary particles such as the electron, positron, muon, proton, and neutron are caused by magnetic *charges*, arranged in magnetically neutral configurations (no net magnetic charge). If the electron and proton magnetic moments were caused by groups of magnetic charges, the coefficient $8\pi/3$ in (5.74) would be replaced by $-4\pi/3$! The astrophysical hyperfine line of atomic hydrogen would be at 42 cm wavelength, and the singlet and triplet states would be reversed. The experimental results on positronium and muonium, as well as the magnetic scattering of neutrons, give strong additional support to the conclusion that *intrinsic* magnetic moments of particles can be attributed to electric currents, not magnetic charges.*

5.8   *Macroscopic Equations, Boundary Conditions on* **B** *and* **H**

So far we have dealt with the basic laws (5.26) of steady-state magnetic fields as microscopic equations in the sense of the Introduction and Chapter 4. We have assumed that the current density **J** was a completely known function of position. In macroscopic problems this is often not true. The atoms in matter have electrons that give rise to effective atomic currents, the current density of which is a rapidly fluctuating quantity. Only its average over a macroscopic volume is known or pertinent. Furthermore, the atomic electrons contribute intrinsic magnetic moments in addition to those from their orbital motion. All these moments can give rise to dipole fields that vary appreciably on the atomic scale of dimensions.

The process of averaging the microscopic equations to obtain a macroscopic description of magnetic fields in ponderable media is discussed in detail in Chapter 6. Here, just as in Chapter 4, we give only a sketch of the elementary

---

*There is a caveat that all particles must have the same origin for their moments. For a pedagogical discussion of the experiments, see J. D. Jackson, The nature of intrinsic magnetic dipole moments, CERN Report No. 77-17, CERN, Geneva (1977), reprinted in *The International Community of Physicists: Essays on Physics and Society in Honor of Victor Frederick Weisskopf*, ed. V. Stefan, AIP Press/ Springer-Verlag, New York (1997).

derivation. The first step is to observe that the averaging of the equation, $\nabla \cdot \mathbf{B}_{\text{micro}} = 0$, leads to the same equation

$$\nabla \cdot \mathbf{B} = 0 \qquad (5.75)$$

for the macroscopic magnetic induction. Thus we can still use the concept of a vector potential $\mathbf{A}(\mathbf{x})$ whose curl gives $\mathbf{B}$. The large number of molecules or atoms per unit volume, each with its molecular magnetic moment $\mathbf{m}_i$, gives rise to an average macroscopic *magnetization* or magnetic moment density,

$$\mathbf{M}(\mathbf{x}) = \sum_i N_i \langle \mathbf{m}_i \rangle \qquad (5.76)$$

where N_i is the average number per unit volume of molecules of type i and $\langle \mathbf{m}_i \rangle$ is the average molecular moment in a small volume at the point $\mathbf{x}$. In addition to the bulk magnetization, we suppose that there is a macroscopic current density $\mathbf{J}(\mathbf{x})$ from the flow of free charge in the medium. Then the vector potential from a small volume ΔV at the point $\mathbf{x}'$ will be

$$\Delta \mathbf{A}(\mathbf{x}) = \frac{\mu_0}{4\pi} \left[\frac{\mathbf{J}(\mathbf{x}') \, \Delta V}{|\mathbf{x} - \mathbf{x}'|} + \frac{\mathbf{M}(\mathbf{x}') \times (\mathbf{x} - \mathbf{x}')}{|\mathbf{x} - \mathbf{x}'|^3} \, \Delta V \right]$$

This is the magnetic analog of (4.30). The second term is the dipole vector potential (5.55). Letting ΔV become the macroscopically infinitesimal d^3x', the total vector potential at $\mathbf{x}$ can be written as the integral over all space,

$$\mathbf{A}(\mathbf{x}) = \frac{\mu_0}{4\pi} \int \left[\frac{\mathbf{J}(\mathbf{x}')}{|\mathbf{x} - \mathbf{x}'|} + \frac{\mathbf{M}(\mathbf{x}') \times (\mathbf{x} - \mathbf{x}')}{|\mathbf{x} - \mathbf{x}'|^3} \right] d^3x' \qquad (5.77)$$

The magnetization term can be rewritten as follows:

$$\int \frac{\mathbf{M}(\mathbf{x}') \times (\mathbf{x} - \mathbf{x}')}{|\mathbf{x} - \mathbf{x}'|^3} \, d^3x' = \int \mathbf{M}(\mathbf{x}') \times \nabla' \left(\frac{1}{|\mathbf{x} - \mathbf{x}'|} \right) d^3x'$$

Now an integration by parts casts the gradient operator over onto the magnetization and also gives a surface integral. If $\mathbf{M}(\mathbf{x}')$ is well behaved and localized, the surface integral vanishes. The vector potential (5.77) then becomes

$$\mathbf{A}(\mathbf{x}) = \frac{\mu_0}{4\pi} \int \frac{[\mathbf{J}(\mathbf{x}') + \nabla' \times \mathbf{M}(\mathbf{x}')]}{|\mathbf{x} - \mathbf{x}'|} \, d^3x' \qquad (5.78)$$

The magnetization is seen to contribute an *effective current density*,

$$\mathbf{J}_M = \nabla \times \mathbf{M} \qquad (5.79)$$

The macroscopic equivalent of the microscopic equation, $\nabla \times \mathbf{B}_{\text{micro}} = \mu_0 \mathbf{J}_{\text{micro}}$, can be read off from (5.78). If the equations (5.26) have (5.32) as a solution, then (5.78) implies that $\mathbf{J} + \mathbf{J}_M$ plays the role of the current in the macroscopic equivalent, that is:

$$\nabla \times \mathbf{B} = \mu_0 [\mathbf{J} + \nabla \times \mathbf{M}] \qquad (5.80)$$

The $\nabla \times \mathbf{M}$ term can be combined with $\mathbf{B}$ to define a new macroscopic field $\mathbf{H}$, called the *magnetic field*,

$$\mathbf{H} = \frac{1}{\mu_0} \mathbf{B} - \mathbf{M} \qquad (5.81)$$

Then the macroscopic equations, replacing (5.26), are

$$\nabla \times \mathbf{H} = \mathbf{J}$$
$$\nabla \cdot \mathbf{B} = 0 \tag{5.82}$$

The introduction of **H** as a macroscopic field is completely analogous to the introduction of **D** for the electrostatic field. The macroscopic equations (5.82) have their electrostatic counterparts,

$$\nabla \cdot \mathbf{D} = \rho$$
$$\nabla \times \mathbf{E} = 0 \tag{5.83}$$

We emphasize that the fundamental fields are **E** and **B**. They satisfy the homogeneous equations in (5.82) and (5.83). The derived fields, **D** and **H**, are introduced as a matter of convenience, to permit us to take into account in an average way the contributions to ρ and **J** of the atomic charges and currents.

To complete the description of macroscopic magnetostatics, there must be a constitutive relation between **H** and **B**. As discussed in the Introduction, for isotropic diamagnetic and paramagnetic substances the simple linear relation

$$\mathbf{B} = \mu\mathbf{H} \tag{5.84}$$

holds, μ being a parameter characteristic of the medium and called the *magnetic permeability*. Typically μ/μ_0 differs from unity by only a few parts in 10^5 ($\mu > \mu_0$ for paramagnetic substances and $\mu < \mu_0$ for diamagnetic). For the ferromagnetic substances, (5.84) must be replaced by a nonlinear functional relationship,

$$\mathbf{B} = \mathbf{F}(\mathbf{H}) \tag{5.85}$$

The phenomenon of hysteresis, shown schematically in Fig. 5.8, implies that **B** is not a single-valued function of **H**. In fact, the function **F**(**H**) depends on the history of preparation of the material. The incremental permeability $\mu(\mathbf{H})$ is defined as the derivative of **B** with respect to **H**, assuming that **B** and **H** are parallel. For high-permeability substances, $\mu(\mathbf{H})/\mu_0$ can be as high as 10^6. Most untreated ferromagnetic materials have a linear relation (5.84) between **B** and **H** for very small fields. Typical values of initial relative permeability range from 10 to 10^4.

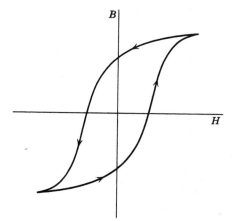

Figure 5.8   Hysteresis loop giving **B** in a ferromagnetic material as a function of **H**.

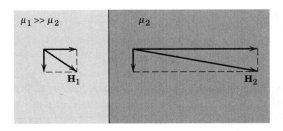

Figure 5.9

The complicated relationship between **B** and **H** in ferromagnetic materials makes analysis of magnetic boundary-value problems inherently more difficult than that of similar electrostatic problems. But the very large values of relative permeability sometimes allow simplifying assumptions on the boundary conditions.

The boundary conditions for **B** and **H** at an interface between two media are derived in Section I.5. There it is shown that the normal components of **B** and the tangential components of **H** on either side of the boundary are related according to

$$(\mathbf{B}_2 - \mathbf{B}_1) \cdot \mathbf{n} = 0 \tag{5.86}$$

$$\mathbf{n} \times (\mathbf{H}_2 - \mathbf{H}_1) = \mathbf{K} \tag{5.87}$$

where **n** is a unit normal pointing from region 1 into region 2 and **K** is the idealized surface current density. For media satisfying linear relations of the form (5.84) the boundary conditions can be expressed alternatively as

$$\mathbf{B}_2 \cdot \mathbf{n} = \mathbf{B}_1 \cdot \mathbf{n}, \qquad \mathbf{B}_2 \times \mathbf{n} = \frac{\mu_2}{\mu_1} \mathbf{B}_1 \times \mathbf{n} \tag{5.88}$$

or

$$\mathbf{H}_2 \cdot \mathbf{n} = \frac{\mu_1}{\mu_2} \mathbf{H}_1 \cdot \mathbf{n}, \qquad \mathbf{H}_2 \times \mathbf{n} = \mathbf{H}_1 \times \mathbf{n} \tag{5.89}$$

If $\mu_1 \gg \mu_2$, the normal component of $\mathbf{H}_2$ is much larger than the normal component of $\mathbf{H}_1$, as shown in Fig. 5.9. In the limit $(\mu_1/\mu_2) \to \infty$, the magnetic field $\mathbf{H}_2$ is normal to the boundary surface, independent of the direction of $\mathbf{H}_1$ (barring the exceptional case of $\mathbf{H}_1$ exactly parallel to the interface). The boundary condition on **H** at the surface of a material of very high permeability is thus the same as for the electric field at the surface of a conductor. We may therefore use electrostatic potential theory for the magnetic field. The surfaces of the high-permeability material are approximately "equipotentials," and the lines of **H** are normal to these equipotentials. This analogy is exploited in many magnet-design problems. The type of field is decided upon, and the pole faces are shaped to be equipotential surfaces. See Section 5.14 for further discussion.

5.9 Methods of Solving Boundary-Value Problems in Magnetostatics

The basic equations of magnetostatics are

$$\nabla \cdot \mathbf{B} = 0, \qquad \nabla \times \mathbf{H} = \mathbf{J} \tag{5.90}$$

with some constitutive relation between **B** and **H**. The variety of situations that can occur in practice is such that a survey of different techniques for solving boundary-value problems in magnetostatics is worthwhile.

A. *Generally Applicable Method of the Vector Potential*

Because of the first equation in (5.90) we can always introduce a vector potential **A(x)** such that

$$\mathbf{B} = \boldsymbol{\nabla} \times \mathbf{A}$$

If we have an explicit constitutive relation, $\mathbf{H} = \mathbf{H}[\mathbf{B}]$, then the second equation in (5.90) can be written

$$\boldsymbol{\nabla} \times \mathbf{H}[\boldsymbol{\nabla} \times \mathbf{A}] = \mathbf{J}$$

This is, in general, a very complicated differential equation, even if the current distribution is simple, unless **H** and **B** are simply related. For linear media with $\mathbf{B} = \mu\mathbf{H}$, the equation becomes

$$\boldsymbol{\nabla} \times \left(\frac{1}{\mu} \boldsymbol{\nabla} \times \mathbf{A} \right) = \mathbf{J} \tag{5.91}$$

If μ is constant over a finite region of space, then in that region (5.91) can be written

$$\boldsymbol{\nabla}(\boldsymbol{\nabla} \cdot \mathbf{A}) - \nabla^2\mathbf{A} = \mu\mathbf{J} \tag{5.92}$$

With the choice of the Coulomb gauge ($\boldsymbol{\nabla} \cdot \mathbf{A} = 0$), this becomes (5.31) with a modified current density, $(\mu/\mu_0)\mathbf{J}$. The situation closely parallels the treatment of uniform isotropic dielectric media where the effective charge density in the Poisson equation is $\epsilon_0\rho/\epsilon$. Solutions of (5.92) in different linear media must be matched across the boundary surfaces using the boundary conditions (5.88) or (5.89).

B. $\mathbf{J} = 0$; *Magnetic Scalar Potential*

If the current density vanishes in some finite region of space, the second equation in (5.90) becomes $\boldsymbol{\nabla} \times \mathbf{H} = 0$. This implies that we can introduce a *magnetic scalar potential* Φ_M such that

$$\mathbf{H} = -\boldsymbol{\nabla}\Phi_M \tag{5.93}$$

just as $\mathbf{E} = -\boldsymbol{\nabla}\Phi$ in electrostatics. With an explicit constitutive relation, this time of $\mathbf{B} = \mathbf{B}[\mathbf{H}]$, the $\boldsymbol{\nabla} \cdot \mathbf{B} = 0$ equation can be written

$$\boldsymbol{\nabla} \cdot \mathbf{B}[-\boldsymbol{\nabla}\Phi_M] = 0$$

Again, this is a very complicated differential equation unless the medium is *linear*, in which case the equation becomes

$$\boldsymbol{\nabla} \cdot (\mu\boldsymbol{\nabla}\Phi_M) = 0 \tag{5.94}$$

If μ is at least *piecewise constant*, in each region the magnetic scalar potential satisfies the Laplace equation,

$$\nabla^2\Phi_M = 0$$

The solutions in the different regions are connected via the boundary conditions (5.89). Note that in this last circumstance of piecewise constancy of μ, we can also write $\mathbf{B} = -\nabla\Psi_M$ with $\nabla^2\Psi_M = 0$. With this alternative scalar potential the boundary conditions (5.88) are appropriate.

The concept of a magnetic scalar potential can be used fruitfully for closed loops of current. It can be shown that Φ_M is proportional to the solid angle subtended by the boundary of the loop at the observation point. See Problem 5.1. Such a potential is evidently multiple-valued.

C. Hard Ferromagnets (M given and J = 0)

A common practical situation concerns "hard" ferromagnets, having a magnetization that is essentially independent of applied fields for moderate field strengths. Such materials can be treated as if they had a fixed, specified magnetization $\mathbf{M}(\mathbf{x})$.

(a) Scalar Potential

Since $\mathbf{J} = 0$, the magnetic scalar potential Φ_M can be employed. The first equation in (5.90) is written as

$$\nabla \cdot \mathbf{B} = \mu_0\nabla \cdot (\mathbf{H} + \mathbf{M}) = 0$$

Then with (5.93) it becomes a magnetostatic Poisson equation,

$$\nabla^2\Phi_M = -\rho_M \tag{5.95}$$

with the *effective magnetic-charge density*,

$$\rho_M = -\nabla \cdot \mathbf{M} \tag{5.96}$$

The solution for the potential Φ_M if there are no boundary surfaces is

$$\Phi_M(\mathbf{x}) = -\frac{1}{4\pi} \int \frac{\nabla' \cdot \mathbf{M}(\mathbf{x}')}{|\mathbf{x} - \mathbf{x}'|} d^3x' \tag{5.97}$$

If $\mathbf{M}$ is well behaved and localized, an integration by parts may be performed to yield

$$\Phi_M(\mathbf{x}) = \frac{1}{4\pi} \int \mathbf{M}(\mathbf{x}') \cdot \nabla'\left(\frac{1}{|\mathbf{x} - \mathbf{x}'|}\right) d^3x'$$

Then

$$\nabla'\left(\frac{1}{|\mathbf{x} - \mathbf{x}'|}\right) = -\nabla\left(\frac{1}{|\mathbf{x} - \mathbf{x}'|}\right)$$

may be used to give

$$\Phi_M(\mathbf{x}) = -\frac{1}{4\pi} \nabla \cdot \int \frac{\mathbf{M}(\mathbf{x}')}{|\mathbf{x} - \mathbf{x}'|} d^3x' \tag{5.98}$$

In passing we observe that far from the region of nonvanishing magnetization the potential may be approximated by

$$\Phi_M(\mathbf{x}) \simeq -\frac{1}{4\pi} \nabla\left(\frac{1}{r}\right) \cdot \int \mathbf{M}(\mathbf{x}') \, d^3x'$$

$$= \frac{\mathbf{m} \cdot \mathbf{x}}{4\pi r^3}$$

where $\mathbf{m} = \int \mathbf{M} \, d^3x$ is the total magnetic moment. This is the scalar potential of a dipole, as can be seen from the electrostatic (4.10). Thus an arbitrary localized distribution of magnetization asymptotically has a dipole field with strength given by the total magnetic moment of the distribution.

While physical distributions of magnetization are mathematically well behaved and without discontinuities, it is sometimes convenient to idealize the reality and treat $\mathbf{M}(\mathbf{x})$ as if it were discontinuous. Thus, if a "hard" ferromagnet has a volume V and surface S, we specify $\mathbf{M}(\mathbf{x})$ inside V and assume that it falls suddenly to zero at the surface S. Application of the divergence theorem to ρ_M (5.96) in a Gaussian pillbox straddling the surface shows that there is an *effective magnetic surface-charge density*,

$$\sigma_M = \mathbf{n} \cdot \mathbf{M} \tag{5.99}$$

where $\mathbf{n}$ is the outwardly directed normal. Then instead of (5.97) the potential is given by

$$\Phi_M(\mathbf{x}) = -\frac{1}{4\pi} \int_V \frac{\boldsymbol{\nabla}' \cdot \mathbf{M}(\mathbf{x}')}{|\mathbf{x} - \mathbf{x}'|} \, d^3x' + \frac{1}{4\pi} \oint_S \frac{\mathbf{n}' \cdot \mathbf{M}(\mathbf{x}') \, da'}{|\mathbf{x} - \mathbf{x}'|} \tag{5.100}$$

An important special case is that of uniform magnetization throughout the volume V. Then the first term vanishes; only the surface integral over σ_M contributes.

It is important to note that (5.98) is generally applicable, even for the limit of discontinuous distributions of $\mathbf{M}$, because we can introduce a limiting procedure *after* transforming (5.97) into (5.98) in order to discuss discontinuities in $\mathbf{M}$. *Never combine the surface integral of σ_M with (5.98)!*

(b) Vector Potential

If we choose to write $\mathbf{B} = \boldsymbol{\nabla} \times \mathbf{A}$ to satisfy $\boldsymbol{\nabla} \cdot \mathbf{B} = 0$ automatically, then we write the second equation of (5.90) as

$$\boldsymbol{\nabla} \times \mathbf{H} = \boldsymbol{\nabla} \times (\mathbf{B}/\mu_0 - \mathbf{M}) = 0$$

This leads to the Poisson equation for $\mathbf{A}$ in the Coulomb gauge,

$$\nabla^2 \mathbf{A} = -\mu_0 \mathbf{J}_M \tag{5.101}$$

where $\mathbf{J}_M$ is the effective magnetic current density (5.79). The solution for the vector potential in the absence of boundary surfaces is

$$\mathbf{A}(\mathbf{x}) = \frac{\mu_0}{4\pi} \int \frac{\boldsymbol{\nabla}' \times \mathbf{M}(\mathbf{x}')}{|\mathbf{x} - \mathbf{x}'|} \, d^3x' \tag{5.102}$$

as was already shown in (5.78). An alternative form is given by the magnetization term in (5.77).

If the distribution of magnetization is discontinuous, it is necessary to add a surface integral to (5.102). Starting from (5.77) it can be shown that for $\mathbf{M}$ discontinuously falling to zero at the surface S bounding the volume V, the generalization of (5.102) is

$$\mathbf{A}(\mathbf{x}) = \frac{\mu_0}{4\pi} \int_V \frac{\boldsymbol{\nabla}' \times \mathbf{M}(\mathbf{x}')}{|\mathbf{x} - \mathbf{x}'|} \, d^3x' + \frac{\mu_0}{4\pi} \oint_S \frac{\mathbf{M}(\mathbf{x}') \times \mathbf{n}'}{|\mathbf{x} - \mathbf{x}'|} \, da' \tag{5.103}$$

The effective surface current $(\mathbf{M} \times \mathbf{n})$ can also be understood by expressing the boundary condition (5.87) for tangential $\mathbf{H}$ in terms of $\mathbf{B}$ and $\mathbf{M}$. Again, if $\mathbf{M}$ is constant throughout the volume, only the surface integral survives.

5.10 *Uniformly Magnetized Sphere*

To illustrate the different methods possible for the solution of a boundary-value problem in magnetostatics, we consider in Fig. 5.10 the simple problem of a sphere of radius a, with a uniform permanent magnetization $\mathbf{M}$ of magnitude M_0 and parallel to the z axis, embedded in a nonpermeable medium.

The simplest method of solution is that of part $C(a)$ of the preceding section, via the magnetic scalar potential in spherical coordinates and a surface magnetic-charge density $\sigma_M(\theta)$. With $\mathbf{M} = M_0\boldsymbol{\epsilon}_3$ and $\sigma_M = \mathbf{n} \cdot \mathbf{M} = M_0 \cos \theta$, the solution (5.100) for the potential is

$$\Phi_M(r, \theta) = \frac{M_0 a^2}{4\pi} \int d\Omega' \frac{\cos \theta'}{|\mathbf{x} - \mathbf{x}'|}$$

With the expansion (3.38) or (3.70) for the inverse distance, only the $l = 1$ term survives. The potential is

$$\Phi_M(r, \theta) = \frac{1}{3} M_0 a^2 \frac{r_<}{r_>^2} \cos \theta \tag{5.104}$$

where $(r_<, r_>)$ are smaller and larger of (r, a). Inside the sphere, $r_< = r$ and $r_> = a$. Then $\Phi_M = (1/3)M_0 r \cos \theta = (1/3)M_0 z$. The magnetic field and magnetic induction inside the sphere are therefore

$$\mathbf{H}_{\text{in}} = -\frac{1}{3}\mathbf{M}, \qquad \mathbf{B}_{\text{in}} = \frac{2\mu_0}{3}\mathbf{M} \tag{5.105}$$

We note that $\mathbf{B}_{\text{in}}$ is parallel to $\mathbf{M}$, while $\mathbf{H}_{\text{in}}$ is antiparallel. Outside the sphere, $r_< = a$ and $r_> = r$. The potential is thus

$$\Phi_M = \frac{1}{3} M_0 a^3 \frac{\cos \theta}{r^2} \tag{5.106}$$

This is the potential of a dipole with dipole moment,

$$\mathbf{m} = \frac{4\pi a^3}{3} \mathbf{M} \tag{5.107}$$

For the *sphere* with uniform magnetization, the fields are not only dipole in character asymptotically, but also close to the sphere. For this special geometry (and this only) there are no higher multipoles.

The lines of $\mathbf{B}$ and $\mathbf{H}$ are shown in Fig. 5.11. The lines of $\mathbf{B}$ are continuous closed paths, but those of $\mathbf{H}$ terminate on the surface because there is an effective surface-charge density σ_M.

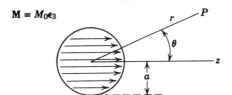

Figure 5.10

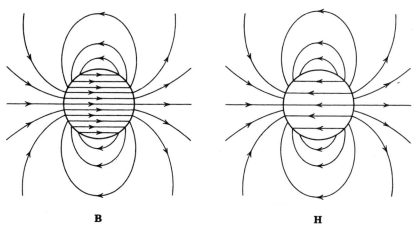

B   **H**

Figure 5.11 Lines of **B** and lines of **H** for a uniformly magnetized sphere. The lines of **B** are closed curves, but the lines of **H** originate on the surface of the sphere where the effective surface magnetic "charge," σ_M, resides.

Brief mention should be made of employing (5.98) instead of (5.100). With $\mathbf{M} = M_0\boldsymbol{\epsilon}_3$ inside the sphere, (5.98) gives

$$\Phi_M(r, \theta) = -\frac{1}{4\pi} M_0 \frac{\partial}{\partial z} \int_0^a r'^2 \, dr' \int d\Omega' \frac{1}{|\mathbf{x} - \mathbf{x}'|} \tag{5.108}$$

Now only the $l = 0$ term in expansion of the inverse separation survives the angular integration and the integral is a function only of r. With $\partial r/\partial z = \cos \theta$, the potential is

$$\Phi_M(r, \theta) = -M_0 \cos \theta \frac{\partial}{\partial r} \int_0^a \frac{r'^2 \, dr'}{r_>}$$

Integration over r' leads directly to the expression (5.104) for Φ_M.

An alternative solution can be accomplished by means of the vector potential and (5.103). Because **M** is uniform inside the sphere the volume current density $\mathbf{J}_M$ vanishes, but there is a surface contribution. With $\mathbf{M} = M_0\boldsymbol{\epsilon}_3$, we have

$$\mathbf{M} \times \mathbf{n}' = M_0 \sin \theta' \boldsymbol{\epsilon}_\phi$$
$$= M_0 \sin \theta'(-\sin \phi'\boldsymbol{\epsilon}_1 + \cos \phi'\boldsymbol{\epsilon}_2)$$

Because of the azimuthal symmetry of the problem we can choose the observation point in the x-z plane ($\phi = 0$), just as in Section 5.5. Then only the y component of $\mathbf{M} \times \mathbf{n}'$ survives integration over the azimuth, giving an azimuthal component of the vector potential,

$$A_\phi(\mathbf{x}) = \frac{\mu_0}{4\pi} M_0 a^2 \int d\Omega' \frac{\sin \theta' \cos \phi'}{|\mathbf{x} - \mathbf{x}'|} \tag{5.109}$$

where $\mathbf{x}'$ has coordinates (a, θ', ϕ'). The angular factor can be written

$$\sin \theta' \cos \phi' = -\sqrt{\frac{8\pi}{3}} \operatorname{Re}[Y_{1,1}(\theta', \phi')] \tag{5.110}$$

Thus with expansion (3.70) for $|\mathbf{x} - \mathbf{x}'|$ only the $l = 1$, $m = 1$ term will survive. Consequently

$$A_\phi(\mathbf{x}) = \frac{\mu_0}{3} M_0 a^2 \left(\frac{r_<}{r_>^2}\right) \sin \theta \qquad (5.111)$$

where $r_<$ $(r_>)$ is the smaller (larger) of r and a. With only a ϕ component of $\mathbf{A}$, the components of the magnetic induction $\mathbf{B}$ are given by (5.38). Equation (5.111) evidently gives the uniform $\mathbf{B}$ inside and the dipole field outside, as found before.

5.11   *Magnetized Sphere in an External Field; Permanent Magnets*

In Section 5.10 we discussed the fields of a uniformly magnetized sphere. Because of the linearity of the field equations we can superpose a uniform magnetic induction $\mathbf{B}_0 = \mu_0 \mathbf{H}_0$ throughout all space. Then we have the problem of a uniformly magnetized sphere in an external field. From (5.105) we find that the magnetic induction and field inside the sphere are now

$$\left. \begin{array}{l} \mathbf{B}_{in} = \mathbf{B}_0 + \dfrac{2\mu_0}{3} \mathbf{M} \\[2ex] \mathbf{H}_{in} = \dfrac{1}{\mu_0} \mathbf{B}_0 - \dfrac{1}{3} \mathbf{M} \end{array} \right\} \qquad (5.112)$$

We now imagine that the sphere is not a permanently magnetized object, but rather a paramagnetic or diamagnetic substance of permeability μ. Then the magnetization $\mathbf{M}$ is a result of the application of the external field. To find the magnitude of $\mathbf{M}$ we use (5.84):

$$\mathbf{B}_{in} = \mu \mathbf{H}_{in} \qquad (5.113)$$

Thus

$$\mathbf{B}_0 + \frac{2\mu_0}{3} \mathbf{M} = \mu \left(\frac{1}{\mu_0} \mathbf{B}_0 - \frac{1}{3} \mathbf{M}\right) \qquad (5.114)$$

This gives a magnetization,

$$\mathbf{M} = \frac{3}{\mu_0} \left(\frac{\mu - \mu_0}{\mu + 2\mu_0}\right) \mathbf{B}_0 \qquad (5.115)$$

We note that this is completely analogous to the polarization $\mathbf{P}$ of a dielectric sphere in a uniform electric field (4.57).

For a ferromagnetic substance, the arguments of the preceding paragraph fail. Equation (5.115) implies that the magnetization vanishes when the external field vanishes. The existence of permanent magnets contradicts this result. The nonlinear relation (5.85) and the phenomenon of hysteresis allow the creation of permanent magnets. We can solve equations (5.112) for one relation between $\mathbf{H}_{in}$ and $\mathbf{B}_{in}$ by eliminating $\mathbf{M}$:

$$\mathbf{B}_{in} + 2\mu_0 \mathbf{H}_{in} = 3\mathbf{B}_0 \qquad (5.116)$$

The hysteresis curve provides the other relation between $\mathbf{B}_{in}$ and $\mathbf{H}_{in}$, so that specific values can be found for any external field. Equation (5.116) corresponds

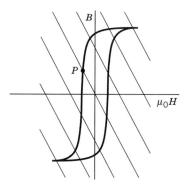

Figure 5.12

to a line with slope -2 on the hysteresis diagram with intercept $3B_0$ *on the y axis*, as in Fig. 5.12. Suppose, for example, that the external field is increased until the ferromagnetic sphere becomes saturated and then decreased to zero. The internal B and H will then be given by the point marked P in Fig. 5.12. The magnetization can be found from (5.112) with $\mathbf{B}_0 = 0$.

The relation (5.116) between $\mathbf{B}_{in}$ and $\mathbf{H}_{in}$ is specific to the sphere. For other geometries other relations pertain. The problem of the ellipsoid can be solved exactly and shows that the slope of the lines (5.116) range from zero for a flat disc to $-\infty$ for a long needle-like object. Thus a larger internal magnetic induction can be obtained with a rod geometry than with spherical or oblate spheroidal shapes.

5.12 Magnetic Shielding, Spherical Shell of Permeable Material in a Uniform Field

Suppose that a certain magnetic induction $\mathbf{B}_0 = \mu_0\mathbf{H}_0$ exists in a region of empty space initially. A permeable body is now placed in the region. The lines of magnetic induction are modified. From our remarks at the end of Section 5.8 concerning media of very high permeability, we would expect the field lines to tend to be normal to the surface of the body. Carrying the analogy with conductors further, if the body is hollow, we would expect the field in the cavity to be smaller than the external field, vanishing in the limit $\mu \to \infty$. Such a reduction in field is said to be due to the *magnetic shielding* provided by the permeable material. It is of considerable practical importance, since essentially field-free regions are often necessary or desirable for experimental purposes or for the reliable working of electronic devices.

As an example of the phenomenon of magnetic shielding we consider a spherical shell of inner (outer) radius a (b), made of material of permeability μ, and placed in a formerly uniform constant magnetic induction $\mathbf{B}_0$, as shown in Fig. 5.13. We wish to find the fields $\mathbf{B}$ and $\mathbf{H}$ everywhere in space, but most particularly in the cavity $(r < a)$, as functions of μ. Since there are no currents present, the magnetic field $\mathbf{H}$ is derivable from a scalar potential, $\mathbf{H} = -\nabla\Phi_M$. Furthermore, since $\mathbf{B} = \mu\mathbf{H}$, the divergence equation $\nabla \cdot \mathbf{B} = 0$ becomes $\nabla \cdot \mathbf{H} = 0$ in the various regions. Thus the potential Φ_M satisfies the Laplace

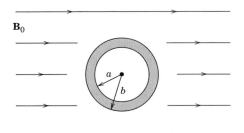

Figure 5.13

equation everywhere. The problem reduces to finding the proper solutions in the different regions to satisfy the boundary conditions (5.89) at $r = a$ and $r = b$.

For $r > b$, the potential must be of the form,

$$\Phi_M = -H_0 r \cos \theta + \sum_{l=0}^{\infty} \frac{\alpha_l}{r^{l+1}} P_l(\cos \theta) \tag{5.117}$$

to give the uniform field, $\mathbf{H} = \mathbf{H}_0$, at large distances. For the inner regions, the potential must be

$$a < r < b \qquad \Phi_M = \sum_{l=0}^{\infty} \left(\beta_l r^l + \gamma_l \frac{1}{r^{l+1}} \right) P_l(\cos \theta)$$

$$r < a \qquad \Phi_M = \sum_{l=0}^{\infty} \delta_l r^l P_l(\cos \theta) \tag{5.118}$$

The boundary conditions at $r = a$ and $r = b$ are that H_θ and B_r be continuous. In terms of the potential Φ_M these conditions become

$$\frac{\partial \Phi_M}{\partial \theta}(b_+) = \frac{\partial \Phi_M}{\partial \theta}(b_-) \qquad \frac{\partial \Phi_M}{\partial \theta}(a_+) = \frac{\partial \Phi_M}{\partial \theta}(a_-)$$

$$\mu_0 \frac{\partial \Phi_M}{\partial r}(b_+) = \mu \frac{\partial \Phi_M}{\partial r}(b_-) \qquad \mu \frac{\partial \Phi_M}{\partial r}(a_+) = \mu_0 \frac{\partial \Phi_M}{\partial r}(a_-) \tag{5.119}$$

The notation $b_\pm$ means the limit $r \to b$ approached from $r \gtrless b$, and similarly for $a_\pm$. These four conditions, which hold for all angles θ, are sufficient to determine the unknown constants in (5.117) and (5.118). All coefficients with $l \ne 1$ vanish. The $l = 1$ coefficients satisfy the four simultaneous equations

$$\begin{aligned}
\alpha_1 - b^3 \beta_1 - \gamma_1 &= b^3 H_0 \\
2\alpha_1 + \mu' b^3 \beta_1 - 2\mu' \gamma_1 &= -b^3 H_0 \\
a^3 \beta_1 + \gamma_1 - a^3 \delta_1 &= 0 \\
\mu' a^3 \beta_1 - 2\mu' \gamma_1 - a^3 \delta_1 &= 0
\end{aligned} \tag{5.120}$$

Here we have used the notation $\mu' = \mu/\mu_0$ to simplify the equations. The solutions for α_1 and δ_1 are

$$\alpha_1 = \left[\frac{(2\mu' + 1)(\mu' - 1)}{(2\mu' + 1)(\mu' + 2) - 2 \dfrac{a^3}{b^3}(\mu' - 1)^2} \right] (b^3 - a^3) H_0 \tag{5.121}$$

$$\delta_1 = -\left[\frac{9\mu'}{(2\mu' + 1)(\mu' + 2) - 2 \dfrac{a^3}{b^3}(\mu' - 1)^2} \right] H_0$$

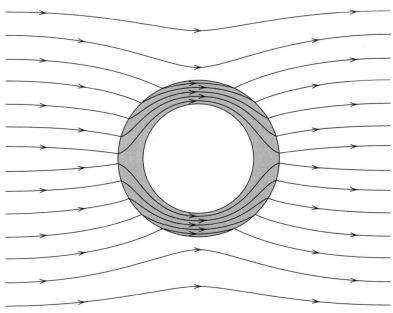

Figure 5.14 Shielding effect of a shell of highly permeable material.

The potential outside the spherical shell corresponds to a uniform field H_0 plus a dipole field (5.41) with dipole moment α_1 oriented parallel to H_0. Inside the cavity, there is a uniform magnetic field parallel to H_0 and equal in magnitude to $-\delta_1$. For $\mu \gg \mu_0$, the dipole moment α_1 and the inner field $-\delta_1$ become

$$\alpha_1 \to b^3 H_0$$

$$-\delta_1 \to \frac{9\mu_0}{2\mu\left(1 - \dfrac{a^3}{b^3}\right)} H_0 \tag{5.122}$$

We see that the inner field is proportional to μ^{-1}. Consequently a shield made of high-permeability material with $\mu/\mu_0 \sim 10^3$ to 10^6 causes a great reduction in the field inside it, even with a relatively thin shell. Figure 5.14 shows the behavior of the lines of $\mathbf{B}$. The lines tend to pass through the permeable medium if possible.

5.13 *Effect of a Circular Hole in a Perfectly Conducting Plane with an Asymptotically Uniform Tangential Magnetic Field on One Side*

Section 3.13 discussed the electrostatic problem of a circular hole in a conducting plane with an asymptotically uniform normal electric field. Its magnetic counterpart has a uniform tangential magnetic field asymptotically. The two examples are useful in the treatment of small holes in wave guides and resonant cavities (see Section 9.5).

Before sketching the solution of the magnetostatic boundary-value problem, we must discuss what we mean by a perfect conductor. Static magnetic fields penetrate conductors, even excellent ones. The conductor modifies the fields only because of its magnetic properties, not its conductivity, unless of course there is current flow inside. With time-varying fields it is often otherwise. It is shown in Section 5.18 that at the interface between conductor and nonconductor, fields with harmonic time dependence penetrate only a distance of the order of $\delta = (2/\mu\omega\sigma)^{1/2}$ into the conductor, where ω is the frequency and σ the conductivity. For any nonvanishing ω, therefore, the skin depth $\delta \to 0$ as $\sigma \to \infty$. Oscillating electric and magnetic fields do not exist inside a perfect conductor. We *define* magnetostatic problems with perfect conductors as the limit of harmonically varying fields as $\omega \to 0$, provided at the same time that $\omega\sigma \to \infty$. Then the magnetic field can exist outside and up to the surface of the conductor, but not inside. The boundary conditions (5.86) and (5.87) show that $\mathbf{B} \cdot \mathbf{n} = 0$, $\mathbf{n} \times \mathbf{H} = \mathbf{K}$ at the surface. These boundary conditions are the magnetostatic counterparts of the electrostatic boundary conditions, $\mathbf{E}_{\text{tan}} = 0$, $\mathbf{D} \cdot \mathbf{n} = \sigma$, at the surface of a conductor, where in this last relation σ is the surface-charge density, not the conductivity!

We consider a perfectly conducting plane at $z = 0$ with a hole of radius a centered at the origin, as shown in Fig. 5.15. For simplicity we assume that the medium surrounding the plane is uniform, isotropic, and linear and that there is a uniform tangential magnetic field $\mathbf{H}_0$ in the y direction in the region $z > 0$ far from the hole, and zero field asymptotically for $z < 0$. Other possibilities can be obtained by linear superposition. Because there are no currents present except on the surface $z = 0$, we can use $\mathbf{H} = -\nabla\Phi_M$, with the magnetic scalar potential $\Phi_M(\mathbf{x})$ satisfying the Laplace equation with suitable mixed boundary conditions. Then we can parallel the solution of Section 3.13.

The potential is written as

$$\Phi_M(x) = \begin{cases} -H_0 y + \Phi^{(1)} & \text{for } z > 0 \\ -\Phi^{(1)} & \text{for } z < 0 \end{cases} \tag{5.123}$$

The reversal of sign for the added potential $\Phi^{(1)}$ below the plane is a consequence of the symmetry properties of the added fields—$H_x^{(1)}$ and $H_y^{(1)}$ are odd in z, while $H_z^{(1)}$ and $\Phi^{(1)}$ are even in z. This can be inferred from (5.14) with the realization that the effective current is only on the surface $z = 0$, as is the effective magnetic-charge density that determines the scalar potential $\Phi^{(1)}$.

From (3.106) the added potential can be written in cylindrical coordinates as

$$\Phi^{(1)}(\mathbf{x}) = \int_0^\infty dk \, A(k)e^{-k|z|}J_1(k\rho) \sin \phi \tag{5.124}$$

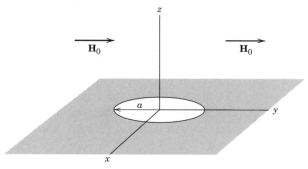

Figure 5.15

Only $m = 1$ enters because the hole is cylindrically symmetric and the asymptotic field varies as $y = \rho \sin \phi$. From the boundary conditions on normal **B** and tangential **H** we find that the boundary conditions on the full potential Φ_M are

Φ_M continuous across $z = 0 \qquad$ for $0 \leq \rho < a$

$\dfrac{\partial \Phi_M}{\partial z} = 0$ at $z = 0 \qquad$ for $a < \rho < \infty$

These requirements imply the dual integral equations,

$$
\begin{aligned}
\int_0^\infty dk\, A(k)J_1(k\rho) &= H_0\rho/2 \qquad \text{for } 0 \leq \rho < a \\
\int_0^\infty dk\, kA(k)J_1(k\rho) &= 0 \qquad\qquad \text{for } a < \rho < \infty
\end{aligned}
\tag{5.125}
$$

These are closely related to, but different from, the electrostatic set (3.178) or (3.179). The necessary pair here are

$$
\begin{aligned}
\int_0^\infty dy\, g(y)J_n(yx) &= x^n \qquad \text{for } 0 \leq x < 1 \\
\int_0^\infty dy\, yg(y)J_n(yx) &= 0 \qquad \text{for } 1 < x < \infty
\end{aligned}
\tag{5.126}
$$

with solution,

$$
g(y) = \frac{2\Gamma(n+1)}{\sqrt{\pi}\,\Gamma(n+\frac{1}{2})}\, j_n(y) = \frac{\Gamma(n+1)}{\Gamma(n+\frac{1}{2})}\left(\frac{2}{y}\right)^{1/2} J_{n+1/2}(y)
\tag{5.127}
$$

In (5.125) we have $g = 2A(k)/H_0a^2$, $n = 1$, $x = \rho/a$, and $y = ka$. Hence

$$
A(k) = \frac{2H_0a^2}{\pi}\, j_1(ka)
\tag{5.128}
$$

The added potential is therefore

$$
\Phi^{(1)}(\mathbf{x}) = \frac{2H_0a^2}{\pi} \int_0^\infty dk\, j_1(ka)e^{-k|z|}J_1(k\rho)\sin\phi
\tag{5.129}
$$

By methods similar to those of Section 3.13 it can be shown that far from the opening the added potential has the asymptotic form

$$
\Phi^{(1)}(\mathbf{x}) \rightarrow \frac{2H_0a^3}{3\pi} \cdot \frac{y}{r^3}
\tag{5.130}
$$

This is the potential of a dipole aligned in the y direction, the direction of $\mathbf{H}_0$. Because of the signs in (5.123), the circular hole is equivalent at large distances to a magnetic dipole with moment

$$
\mathbf{m} = \pm\frac{8a^3}{3}\,\mathbf{H}_0 \qquad \text{for } z \gtrless 0
\tag{5.131}
$$

where $\mathbf{H}_0$ is the tangential magnetic field on the $z = 0^+$ side of the plane in the absence of the hole. Later (Fig. 9.4) we show qualitatively how the magnetic field lines distort to give rise to the dipole field. In the opening itself ($z = 0$, $0 \leq \rho < a$) the tangential and normal components of the magnetic field are

$$\mathbf{H}_{\text{tan}} = \frac{1}{2}\,\mathbf{H}_0$$

$$H_z(\rho,\,0) = \frac{2H_0}{\pi}\,\frac{\rho}{\sqrt{a^2 - \rho^2}}\,\sin\phi \qquad (5.132)$$

Comparison with the corresponding electrostatic problem in Section 3.13 shows similarities and differences. Roughly speaking, the roles of tangential and normal components of fields have been interchanged. The effective dipoles point in the directions of the asymptotic fields, but the magnetic moment (5.131) is a factor of 2 larger than the electrostatic moment (3.183) for the same field strengths. For *arbitrarily shaped holes* the far field in the electrostatic case is still that of a dipole normal to the plane, while the magnetic case has its effective dipole in the plane, but now the direction of the magnetic dipole depends on both the field direction and the orientation of the hole (the hole has an aniso-tropic magnetic susceptibility).

5.14 *Numerical Methods for Two-Dimensional Magnetic Fields*

Magnetic fields in the presence of iron or other highly permeable materials can be evaluated numerically in two dimensions by the relaxation method described in Section 1.13 or, more generally, by the method of finite element analysis of Section 2.12. The problems can be classed as "interior" or "exterior," depending whether the current flow and/or magnetized material and desired field are within the same region.

First consider the boundary conditions for the field components at the smooth interface of a highly permeable medium and a nonpermeable one. Lo-cally, the interface can be approximated by a plane. The boundary conditions are that the tangential component of **H** and the normal component of **B** are continuous across the interface, if there are no surface currents. Figure 5.16 is a

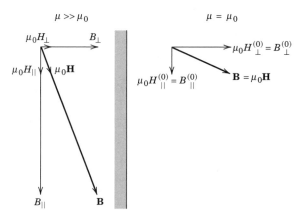

Figure 5.16 Illustration of the effect of large permeability on the components of the magnetic induction and magnetic field on either side of an interface. The sketch has $\mu \approx 5\mu_0$, not a very high permeability!

sketch of the behavior of the field components, similar to Fig. 5.9 but showing both **B** and **H** components. For a given "external" field $\mathbf{B}^{(0)}$ in the nonpermeable region, the components of **B** (and **H**) in the highly permeable medium are more closely parallel to the interface. The magnitude of the magnetic induction just inside the highly permeable medium is

$$|\mathbf{B}|^2 = B_\perp^{(0)2} + \frac{\mu^2}{\mu_0^2} B_\parallel^{(0)2}$$

while the energy per unit volume (see Section 5.16) there is

$$\frac{1}{2\mu} |\mathbf{B}|^2 = \frac{1}{2\mu} B_\perp^{(0)2} + \frac{\mu}{2\mu_0^2} B_\parallel^{(0)2}$$

These two relations are immediately useful in learning the appropriate boundary conditions of "exterior" and "interior" problems in the limit $\mu/\mu_0 \to \infty$.

The most familiar static magnetic fields are those around a permanent magnet of high permeability or an iron core excited by remote current-carrying windings. The region of interest is the nonpermeable region bounded by the highly permeable pole face or faces—the archetypal "exterior" problem. If we suppose that the stored energy within the highly permeable medium is finite, the energy relation shows that, as $\mu/\mu_0 \to \infty$, the parallel component of the magnetic field outside must vanish: the "external" magnetic field at the surface is perpendicular to the interface. These are just the boundary conditions for the electrostatic field at the surface of a conducting boundary, as mentioned at the end of Section 5.8. If there are no currents within the nonpermeable region of interest, then $\nabla \times \mathbf{H} = 0$ there and we can write $\mathbf{H} = -\nabla\Phi_M$. The magnetic scalar potential satisfies the Laplace equation, $\nabla^2\Phi_M = 0$, with the "pole pieces," surfaces of constant potential; the analogy with electrostatics is complete.

For simplicity we restrict our discussion of "interior" problems to two dimensions, with steady current flow only in the third direction in a uniform, highly permeable conducting medium. We are interested in the magnetic induction within the medium—for example, a long iron third rail of a subway system. The current flow produces a magnetic induction both inside and outside the medium. Whatever the magnitudes of the parallel and perpendicular components just outside, the boundary conditions assure that **B** is parallel to the surface of the medium just inside as $\mu/\mu_0 \to \infty$.

If the current density has only a z component, $J_z(x, y)$, the vector potential **A** has only a z component, $A_z(x, y)$, which satisfies the Poisson equation, $\nabla^2 A_z = -\mu J_z$. The field components are $B_x = \partial A_z/\partial y$, $B_y = -\partial A_z/\partial x$, $B_z = 0$. If the internal field **B** is tangential to the boundary C of the region R sketched in Fig. 5.17, we have $\mathbf{n} \cdot (\nabla_\perp \times \mathbf{A}) = (\mathbf{n} \times \nabla_\perp) \cdot \mathbf{A} = 0$ on C. The gradient operator in the x-y plane can be resolved into components parallel to and perpendicular to **n**. The boundary condition thus becomes

$$\frac{\partial A_z}{\partial l} = 0$$

where dl is an element of arc length along C. The vector potential is constant along the boundary curve C. Furthermore, we can infer that in the interior region R the magnetic field lines are parallel to the contours of constant A_z. Because **B** $= \nabla \times \mathbf{A}$, the density of lines of force is given by the derivative of A_z perpen-

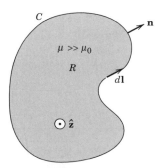

Figure 5.17 Cross section of a long, highly permeable cylindrical conductor with current flow along its length.

dicular to the surfaces of constant value; the spacing of contours of constant A_z with equal increments in A_z will show the intensity of the field as well as its direction.

In implementing numerical methods of solution of the Poisson equation, $\nabla^2 A_z = -\mu J_z$, boundary conditions must be specified. That seems to mean the constant value of A_z on the contour C. But the vector potential is arbitrary to within addition of the gradient of a scalar function χ. With the choice, $\chi = -A_0 \cdot z$, where A_0 is the yet undetermined value of A_z on C, we define $A_z' = A_z(x, y) - A_0$. The Poisson equation problem to be solved then becomes $\nabla^2 A_z' = \mu J_z$ within R with the homogeneous boundary condition $A_z' = 0$ on the boundary C. The value of A_z on C is not physically meaningful and is not needed. With $J_z(x, y)$ specified, the solution by the relaxation technique proceeds as in Section 1.13.

Powerful numerical codes exist to solve more realistic magnetic field problems where, for example, the different permeable materials have large, but not infinite, values of μ_i/μ_0. References are given at the end of the chapter.

5.15 Faraday's Law of Induction

The first quantitative observations relating time-dependent electric and magnetic fields were made by Faraday (1831) in experiments on the behavior of currents in circuits placed in time-varying magnetic fields. Faraday observed that a transient current is induced in a circuit if (a) the steady current flowing in an adjacent circuit is turned on or off, (b) the adjacent circuit with a steady current flowing is moved relative to the first circuit, (c) a permanent magnet is thrust into or out of the circuit. No current flows unless either the adjacent current changes or there is relative motion. Faraday attributed the transient current flow to a changing magnetic flux linked by the circuit. The changing flux induces an electric field around the circuit, the line integral of which is called the *electromotive force*, $\mathscr{E}$. The electromotive force causes a current flow, according to Ohm's law.

We now express Faraday's observations in quantitative mathematical terms. Let the circuit C be bounded by an open surface S with unit normal $\mathbf{n}$, as in Fig. 5.18. The magnetic induction in the neighborhood of the circuit is $\mathbf{B}$. The magnetic flux linking the circuit is defined by

$$F = \int_S \mathbf{B} \cdot \mathbf{n} \, da \qquad (5.133)$$

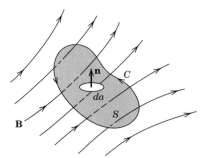

Figure 5.18

The electromotive force around the circuit is

$$\mathscr{E} = \oint_C \mathbf{E}' \cdot d\mathbf{l} \qquad (5.134)$$

where $\mathbf{E}'$ is the electric field at the element $d\mathbf{l}$ of the circuit C. Faraday's observations are summed up in the mathematical law,

$$\mathscr{E} = -k \frac{dF}{dt} \qquad (5.135)$$

The induced electromotive force around the circuit is proportional to the time rate of change of magnetic flux linking the circuit. The sign is specified by Lenz's law, which states that the induced current (and accompanying magnetic flux) is in such a direction as to oppose the change of flux through the circuit.

The constant of proportionality k depends on the choice of units for the electric and magnetic field quantities. It is not, as might at first be supposed, an independent empirical constant to be determined from experiment. As we will see immediately, once the units and dimensions in Ampère's law have been chosen, the magnitude and dimensions of k follow from the assumption of Galilean invariance for Faraday's law. For SI units, $k = 1$; for Gaussian units, $k = c^{-1}$, where c is the velocity of light.

Before the development of special relativity (and even afterward, when investigators were dealing with relative speeds that were small compared with the velocity of light), it was understood, although not often explicitly stated, by all physicists that physical laws should be invariant under Galilean transformations. That is, physical phenomena are the same when viewed by two observers moving with a constant velocity $\mathbf{v}$ relative to one another, provided the coordinates in space and time are related by the Galilean transformation, $\mathbf{x}' = \mathbf{x} - \mathbf{v}t$, $t' = t$. In particular, consider Faraday's observations. It is expected and experimentally verified that the same current is induced in a secondary circuit whether *it is moved* while the primary circuit through which current is flowing is stationary or it is held fixed while *the primary circuit is moved* in the same relative manner.

Let us now consider Faraday's law for a moving circuit and see the consequences of Galilean invariance. Expressing (5.135) in terms of the integrals over $\mathbf{E}'$ and $\mathbf{B}$, we have

$$\oint_C \mathbf{E}' \cdot d\mathbf{l} = -k \frac{d}{dt} \int_s \mathbf{B} \cdot \mathbf{n} \, da \qquad (5.136)$$

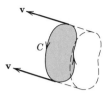

Figure 5.19

The induced electromotive force is proportional to the *total* time derivative of the flux—the flux can be changed by changing the magnetic induction or by changing the shape or orientation or position of the circuit. In form (5.136) we have a far-reaching generalization of Faraday's law. The circuit C can be thought of as any closed geometrical path in space, not necessarily coincident with an electric circuit. Then (5.136) becomes a relation between the fields themselves. It is important to note, however, that the electric field, $\mathbf{E}'$ is the electric field at $d\mathbf{l}$ in the coordinate system or medium in which $d\mathbf{l}$ is at rest, since it is that field that causes current to flow if a circuit is actually present.

If the circuit C is moving with a velocity $\mathbf{v}$ in some direction, as shown in Fig. 5.19, the total time derivative in (5.136) must take into account this motion. The flux through the circuit may change because (a) the flux changes with time at a point, or (b) the translation of the circuit changes the location of the boundary. It is easy to show that the result for the total time derivative of flux through the moving circuit is*

$$\frac{d}{dt}\int_S \mathbf{B}\cdot\mathbf{n}\,da = \int_S \frac{\partial\mathbf{B}}{\partial t}\cdot\mathbf{n}\,da + \oint_C (\mathbf{B}\times\mathbf{v})\cdot d\mathbf{l} \qquad (5.137)$$

Equation (5.136) can now be written in the form,

$$\oint_C [\mathbf{E}' - k(\mathbf{v}\times\mathbf{B})]\cdot d\mathbf{l} = -k\int_S \frac{\partial\mathbf{B}}{\partial t}\cdot\mathbf{n}\,da \qquad (5.138)$$

This is an equivalent statement of Faraday's law applied to the moving circuit C. But we can choose to interpret it differently. We can think of the circuit C and surface S as instantaneously at a certain position in space in the laboratory. Applying Faraday's law (5.136) to that fixed circuit, we find

$$\oint_C \mathbf{E}\cdot d\mathbf{l} = -k\int_S \frac{\partial\mathbf{B}}{\partial t}\cdot\mathbf{n}\,da \qquad (5.139)$$

---

*For a general vector field there is an added term, $\int_s(\mathbf{\nabla}\cdot\mathbf{B})\mathbf{v}\cdot\mathbf{n}\,da$, which gives the contribution of the sources of the vector field swept over by the moving circuit. The general result follows most easily from the use of the convective derivative,

$$\frac{d}{dt} = \frac{\partial}{\partial t} + \mathbf{v}\cdot\mathbf{\nabla}$$

Thus

$$\frac{d\mathbf{B}}{dt} = \frac{\partial\mathbf{B}}{\partial t} + (\mathbf{v}\cdot\mathbf{\nabla})\mathbf{B} = \frac{\partial\mathbf{B}}{\partial t} + \mathbf{\nabla}\times(\mathbf{B}\times\mathbf{v}) + \mathbf{v}(\mathbf{\nabla}\cdot\mathbf{B})$$

where $\mathbf{v}$ is treated as a fixed vector in the differentiation. Use of Stokes's theorem on the second term yields (5.137).

where **E** is now the electric field in the laboratory. The assumption of Galilean invariance implies that the left-hand sides of (5.138) and (5.139) must be equal. This means that the electric field **E**′ in the moving coordinate system of the circuit is

$$\mathbf{E}' = \mathbf{E} + k(\mathbf{v} \times \mathbf{B}) \tag{5.140}$$

To determine the constant k we merely observe the significance of **E**′. A charged particle (e.g., one of the conduction electrons) essentially at rest in a moving circuit will experience a force $q\mathbf{E}'$. When viewed from the laboratory, the charge represents a current $\mathbf{J} = q\mathbf{v}\delta(\mathbf{x} - \mathbf{x}_0)$. From the magnetic force law (5.7) or (5.12) it is evident that this current experiences a force in agreement with (5.140) provided the constant k is equal to unity (SI) or $1/c$ (Gaussian).

Thus we see that, with our choice of units for charge and current, Galilean covariance requires that the present constant k be equal to the constant appearing in the definition of the magnetic field (5.4). Faraday's law (5.136) therefore reads

$$\oint_C \mathbf{E}' \cdot d\mathbf{l} = -\frac{d}{dt} \int_S \mathbf{B} \cdot \mathbf{n} \, da \tag{5.141}$$

where **E**′ is the electric field at $d\mathbf{l}$ in its rest frame of coordinates. The time derivative on the right is a *total* time derivative (5.137). As a by-product we have found that the electric field **E**′ in a coordinate frame moving with a velocity **v** relative to the laboratory is

$$\mathbf{E}' = \mathbf{E} + \mathbf{v} \times \mathbf{B} \tag{5.142}$$

Because we considered a Galilean transformation, the result (5.142) is an approximation valid only for speeds small compared to the speed of light. (The relativistic expressions are derived in Section 11.10.) Faraday's law is no approximation, however. The Galilean transformation was used merely to evaluate the constant k in (5.135), a task for which it was completely adequate.

Faraday's law (5.141) can be put in differential form by use of Stokes's theorem, provided the circuit is held fixed in the chosen reference frame (to have **E** and **B** defined in the *same* frame). The transformation of the electromotive force integral into a surface integral leads to

$$\int_S \left(\boldsymbol{\nabla} \times \mathbf{E} + \frac{\partial \mathbf{B}}{\partial t} \right) \cdot \mathbf{n} \, da = 0$$

Since the circuit C and bounding surface S are arbitrary, the integrand must vanish at all points in space.

Thus the differential form of Faraday's law is

$$\boldsymbol{\nabla} \times \mathbf{E} + \frac{\partial \mathbf{B}}{\partial t} = 0 \tag{5.143}$$

We note that this is the time-dependent generalization of the statement, $\boldsymbol{\nabla} \times \mathbf{E} = 0$, for electrostatic fields.

5.16   *Energy in the Magnetic Field*

In discussing steady-state magnetic fields in the first 14 sections of this chapter we avoided the question of field energy and energy density. The reason was that the creation of a steady-state configuration of currents and associated magnetic fields involves an initial transient period during which the currents and fields are brought from zero to the final values. For such time-varying fields there are induced electromotive forces that cause the sources of current to do work. Since the energy in the field is by definition the total work done to establish it, we must consider these contributions.

Suppose for a moment that we have only a single circuit with a constant current I flowing in it. If the flux through the circuit changes, an electromotive force $\mathscr{E}$ is induced around it. To keep the current constant, the sources of current must do work. To determine the rate, we note that the time rate of change of energy of a particle with velocity $\mathbf{v}$ acted on by a force $\mathbf{F}$ is $dE/dt = \mathbf{v} \cdot \mathbf{F}$. With a changing flux, the added field $\mathbf{E}'$ on each conduction electron of charge q and mean velocity $\mathbf{v}$ gives rise to a change in energy per unit time of $q\mathbf{v} \cdot \mathbf{E}'$ per electron. Summing over all the electrons in the circuit, we find that the sources do work to maintain the current at the rate

$$\frac{dW}{dt} = -I\mathscr{E} = I\frac{dF}{dt}$$

the negative sign following from Lenz's law. This is in addition to ohmic losses in the circuit, which are not to be included in the magnetic-energy content. Thus, if the flux change through a circuit carrying a current I is δF, the work done by the sources is

$$\delta W = I\,\delta F$$

Now we consider the problem of the work done in establishing a general steady-state distribution of currents and fields. We may imagine that the buildup process occurs at an infinitesimal rate so that $\nabla \cdot \mathbf{J} = 0$ holds to any desired degree of accuracy. Then the current distribution can be broken up into a network of elementary current loops, the typical one of which is an elemental tube of current of cross-sectional area $\Delta\sigma$ following a closed path C and spanned by a surface S with normal $\mathbf{n}$, as shown in Fig. 5.20.

We can express the increment of work done against the induced emf in terms of the change in magnetic induction through the loop:

$$\Delta(\delta W) = J\,\Delta\sigma \int_S \mathbf{n} \cdot \delta\mathbf{B}\,da$$

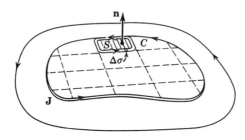

Figure 5.20   Distribution of current density broken up into elemental current loops.

where the extra Δ appears because we are considering only one elemental circuit. If we express **B** in terms of the vector potential **A**, then we have

$$\Delta(\delta W) = J \, \Delta\sigma \int_S (\mathbf{\nabla} \times \delta\mathbf{A}) \cdot \mathbf{n} \, da$$

With application of Stokes's theorem this can be written

$$\Delta(\delta W) = J \, \Delta\sigma \oint_C \delta\mathbf{A} \cdot d\mathbf{l}$$

but $J \, \Delta\sigma \, d\mathbf{l}$ is equal to $\mathbf{J} \, d^3x$, by definition, since $d\mathbf{l}$ is parallel to $\mathbf{J}$. Evidently the sum over all such elemental loops will be the volume integral. Hence the total increment of work done by the external sources due to a change $\delta\mathbf{A}(\mathbf{x})$ in the vector potential is

$$\delta W = \int \delta\mathbf{A} \cdot \mathbf{J} \, d^3x \tag{5.144}$$

An expression involving the magnetic fields rather than **J** and $\delta\mathbf{A}$ can be obtained by using Ampère's law:

$$\mathbf{\nabla} \times \mathbf{H} = \mathbf{J}$$

Then

$$\delta W = \int \delta\mathbf{A} \cdot (\mathbf{\nabla} \times \mathbf{H}) \, d^3x \tag{5.145}$$

The vector identity,

$$\mathbf{\nabla} \cdot (\mathbf{P} \times \mathbf{Q}) = \mathbf{Q} \cdot (\mathbf{\nabla} \times \mathbf{P}) - \mathbf{P} \cdot (\mathbf{\nabla} \times \mathbf{Q})$$

can be used to transform (5.145):

$$\delta W = \int [\mathbf{H} \cdot (\mathbf{\nabla} \times \delta\mathbf{A}) + \mathbf{\nabla} \cdot (\mathbf{H} \times \delta\mathbf{A})] \, d^3x \tag{5.146}$$

If the field distribution is assumed to be localized, the second integral vanishes. With the definition of **B** in terms of **A**, the energy increment can be written:

$$\delta W = \int \mathbf{H} \cdot \delta\mathbf{B} \, d^3x \tag{5.147}$$

This relation is the magnetic equivalent of the electrostatic equation (4.86). In its present form it is applicable to all magnetic media, including ferromagnetic substances. If we assume that the medium is para- or diamagnetic, so that a linear relation exists between **H** and **B**, then

$$\mathbf{H} \cdot \delta\mathbf{B} = \tfrac{1}{2}\delta(\mathbf{H} \cdot \mathbf{B})$$

If we now bring the fields up from zero to their final values, the total magnetic energy will be

$$W = \frac{1}{2} \int \mathbf{H} \cdot \mathbf{B} \, d^3x \tag{5.148}$$

This is the magnetic analog of (4.89).

The magnetic equivalent of (4.83) where the electrostatic energy is expressed in terms of charge density and potential, can be obtained from (5.144) by assuming a linear relation between $\mathbf{J}$ and $\mathbf{A}$. Then we find the magnetic energy to be

$$W = \frac{1}{2} \int \mathbf{J} \cdot \mathbf{A} \, d^3x \qquad (5.149)$$

The magnetic problem of the change in energy when an object of permeability μ_1 is placed in a magnetic field whose current sources are fixed can be treated in close analogy with the electrostatic discussion of Section 4.7. The role of $\mathbf{E}$ is taken by $\mathbf{B}$, that of $\mathbf{D}$ by $\mathbf{H}$. The original medium has permeability μ_0 and existing magnetic induction $\mathbf{B}_0$. After the object is in place the fields are $\mathbf{B}$ and $\mathbf{H}$. It is left as an exercise for the reader to verify that for fixed sources of the field the change in energy is

$$W = \frac{1}{2} \int_{V_1} (\mathbf{B} \cdot \mathbf{H}_0 - \mathbf{H} \cdot \mathbf{B}_0) \, d^3x$$

where the integration is over the volume of the object. This can be written in the alternative forms:

$$W = \frac{1}{2} \int_{V_1} (\mu_1 - \mu_0) \mathbf{H} \cdot \mathbf{H}_0 \, d^3x = \frac{1}{2} \int_{V_1} \left(\frac{1}{\mu_0} - \frac{1}{\mu_1} \right) \mathbf{B} \cdot \mathbf{B}_0 \, d^3x$$

Both μ_1 and μ_0 can be functions of position, but they are assumed independent of field strength.

If the object is in otherwise free space, the change in energy can be expressed in terms of the magnetization as

$$W = \frac{1}{2} \int_{V_1} \mathbf{M} \cdot \mathbf{B}_0 \, d^3x \qquad (5.150)$$

It should be noted that (5.150) is equivalent to the electrostatic result (4.93), except for sign. This sign change arises because the energy W consists of the total energy change occurring when the permeable body is introduced in the field, including the work done by the sources against the induced electromotive forces. In this respect the magnetic problem with fixed currents is analogous to the electrostatic problem with fixed potentials on the surfaces that determine the fields. By an analysis equivalent to that at the end of Section 4.7 we can show that for a small displacement the work done against the induced emf's is twice as large as, and of the opposite sign to, the potential-energy change of the body. Thus, to find the force acting on the body, we consider a generalized displacement ξ and calculate the *positive* derivative of W with respect to the displacement:

$$F_\xi = \left(\frac{\partial W}{\partial \xi} \right)_J \qquad (5.151)$$

The subscript J implies fixed source currents.

The difference between (5.150) and the potential energy (5.72) for a permanent magnetic moment in an external field (apart from the factor $\frac{1}{2}$, which is traced to the linear relation assumed between $\mathbf{M}$ and $\mathbf{B}$) comes from the fact that (5.150) is the total energy required to produce the configuration, whereas (5.72) includes only the work done in establishing the permanent magnetic moment in the field, not the work done in creating the magnetic moment and keeping it permanent.

5.17   Energy and Self- and Mutual Inductances

A.   Coefficients of Self- and Mutual Inductance

Just as the concept of coefficients of capacitance for a system of conductors held at different electrostatic potential is useful (Section 1.11), the concept of self- and mutual inductances are useful for systems of current-carrying circuits. Imagine a system of N distinct current-carrying circuits, the ith one with total current I_i, in otherwise empty space. The circuits are not necessarily thin wires (they can be bus bars, etc.) but are assumed for the present to be nonpermeable. The total energy (5.149) in terms of an integral of $\mathbf{J} \cdot \mathbf{A}/2$ can be expressed as

$$W = \frac{1}{2} \sum_{i=1}^{N} L_i I_i^2 + \sum_{i=1}^{N} \sum_{j>i}^{N} M_{ij} I_i I_j \tag{5.152}$$

where L_i is the self-inductance of the ith circuit and M_{ij} is the mutual inductance between the ith and jth circuits. To establish this result, we first use (5.32) for the vector potential to convert (5.149) to

$$W = \frac{\mu_0}{8\pi} \int d^3x \int d^3x' \, \frac{\mathbf{J}(\mathbf{x}) \cdot \mathbf{J}(\mathbf{x}')}{|\mathbf{x} - \mathbf{x}'|} \tag{5.153}$$

The integrals can now be broken up into sums of separate integrals over each circuit:

$$W = \frac{\mu_0}{8\pi} \sum_{i=1}^{N} \int d^3x_i \sum_{j=1}^{N} \int d^3x_j' \, \frac{\mathbf{J}(\mathbf{x}_i) \cdot \mathbf{J}(\mathbf{x}_j')}{|\mathbf{x}_i - \mathbf{x}_j'|}$$

In the sums there are terms with $i = j$ and terms with $i \neq j$. The former define the first sum in (5.152), the latter, the second. Evidently, the coefficients L_i and M_{ij} are given by

$$L_i = \frac{\mu_0}{4\pi I_i^2} \int_{C_i} d^3x_i \int_{C_i} d^3x_i' \, \frac{\mathbf{J}(\mathbf{x}_i) \cdot \mathbf{J}(\mathbf{x}_i')}{|\mathbf{x}_i - \mathbf{x}_i'|} \tag{5.154}$$

and

$$M_{ij} = \frac{\mu_0}{4\pi I_i I_j} \int_{C_i} d^3x_i \int_{C_j} d^3x_j' \, \frac{\mathbf{J}(\mathbf{x}_i) \cdot \mathbf{J}(\mathbf{x}_j')}{|\mathbf{x}_i - \mathbf{x}_j'|} \tag{5.155}$$

Note that the coefficients of mutual inductance M_{ij} are symmetric in i and j.

These general expressions for self- and mutual inductance are the rigorous versions of the more elementary definitions in terms of flux linkage. To establish the connection, consider the expression for mutual inductance (for which the ambiguities in the definition of flux linkage for self-inductance are absent). The integral over d^3x' times $\mu_0/4\pi$ is just the expression (5.32) for the vector potential $\mathbf{A}(\mathbf{x}_i)$ at position $\mathbf{x}_i$ in the ith circuit caused by the current I_j flowing in the jth circuit. If the ith circuit is imagined to be negligible in cross section compared to the overall scale of both circuits, we can write the integrand $\mathbf{J}(\mathbf{x}_i)\, d^3x$ for the integration over the volume of the ith circuit as $\mathbf{J}\, d^3x = \mathbf{J}_\parallel\, da\, d\mathbf{l}$, where da is a locally defined element of cross-sectional area and $d\mathbf{l}$ is a directed longitudinal differential in the sense of current flow. With the vector potential sensibly constant in the cross-sectional integral at a fixed position along the circuit, the mutual inductance becomes

$$M_{ij} = \frac{1}{I_i I_j} \cdot I_i \oint_{C_i} \mathbf{A}_{ij} \cdot d\mathbf{l} = \frac{1}{I_j} \int_{S_i} (\nabla \times \mathbf{A}_{ij}) \cdot \mathbf{n}\, da$$

where $\mathbf{A}_{ij}$ is the vector potential caused by the jth circuit at the integration point on the ith and the factor I_i comes from the integral over the cross section. Stokes's theorem has been used to obtain the second form. Since the curl of $\mathbf{A}$ is the magnetic induction $\mathbf{B}$, the area integral is just the magnetic-flux linkage (5.133). Thus the mutual inductance is finally

$$M_{ij} = \frac{1}{I_j} F_{ij} \tag{5.156}$$

where F_{ij} is the magnetic flux from circuit j linked within circuit i. For self-inductance, the physical argument is the same, but the ambiguity in the meaning of the self-flux linkage F_{ii} requires a return to the rigorous expression (5.154) based on the magnetic energy.

For both mutual and self-inductance the energy definitions are fundamental. If either the conductors carrying the current are permeable or the medium between the conductors is $(\mu \neq \mu_0)$, (5.152) is valid, but (5.153) is not. It is then best to use the expression (5.148) for the magnetic energy in terms of the fields on the left-hand side of (5.152) in computation of the coefficients of induction.

The presence of terms such as $L\, dI/dt$ or $M_{12}\, dI_2/dt$ in the voltage balance in lumped circuit equations follows immediately from relating the time derivative of the linked flux (dF/dt) to the induced emf $\mathscr{E}$ through (5.135).

B. Estimation of Self-Inductance for Simple Circuits

The self-inductance of simple current-carrying elements can be estimated by consideration of the magnetic energy. Suppose a circular wire of cross-sectional radius a carrying a steady current I forms a loop of circumference C and "area" A (the quotation marks remind us that, since the loop may not be planar, A may stand for a projected area). We imagine that the loop, though relatively arbitrary in shape, does not have kinks in it with radii of curvature as small as the wire size. An example is sketched in Fig. 5.21. There are three length scales here—the wire radius, the dimensions of the loop, represented by $C/2\pi$ or $A^{1/2}$, and the outside region, $r \gg C/2\pi$. From (5.152), the relation between the self-inductance and the magnetic energy, we find that

$$L = \frac{1}{I^2} \int \frac{\mathbf{B} \cdot \mathbf{B}}{\mu}\, d^3x \tag{5.157}$$

Estimation of the magnetic induction will lead to an estimate of the inductance. On the length scale of the wire radius, we may ignore the curvature and consider

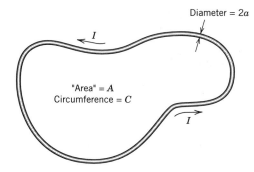

Figure 5.21 Closed current-carrying circuit made of a wire of radius a, length C, and (projected) area A.

the field inside and outside the wire as if it were straight and infinitely long. If the current density is uniform throughout the interior, from symmetry and Ampère's law (5.25) the magnetic induction is azimuthal and equal to

$$B_\phi = \frac{\mu_0 I}{2\pi a} \frac{\rho_<}{\rho_>}$$

where $\rho_<$ ($\rho_>$) is the smaller (larger) of a and ρ. We have assumed that the wire and the medium surrounding it are nonpermeable. The contributions to the inductance per unit length from inside the wire and outside the wire, out to a radius ρ_{max}, are

$$\frac{dL_{in}}{dl} = \frac{\mu_0}{8\pi}; \qquad \frac{dL_{out}(\rho_{max})}{dl} = \frac{\mu_0}{4\pi} \ln\left(\frac{\rho_{max}^2}{a^2}\right)$$

The radial integral outside the wire is limited to $\rho < \rho_{max}$ because the expression for B_ϕ fails to represent the magnetic induction at distances of the order of the middle length scale. If we look to the interior of the loop, it is clear that for $\rho = O(C/2\pi) = O(A^{1/2})$ the isolated straight wire is a very poor representation of the current pattern. Thus we expect* $\rho_{max} = O(A^{1/2})$. There is, of course, a contribution to the inductance from the outside region at distances beyond ρ_{max}. There, at distances large compared to $A^{1/2}$, the slow falloff of the magnetic induction as $1/\rho$ is replaced by a dipole field pattern with $|\mathbf{B}| = O(\mu_0 m/4\pi r^3)$, where $m = O(IA)$ is the magnetic moment of the loop of wire. Because of the rapid decrease of the field beyond ρ_{max}, the contribution per unit length to the inductance from large distances (i.e., $\rho \gtrsim A^{1/2}$) can be estimated to be

$$\frac{dL_{dipole}}{dl} = O\left(\frac{4\pi}{\mu_0 I^2 C} \int_{\rho_{max}}^{\infty} r^2(\mu_0 IA/4\pi r^3)^2 \, dr\right) = O(\mu_0 A^2/4\pi\rho_{max}^2 C)$$

If we set $\rho_{max} = (\xi' A)^{1/2}$, where ξ' is a number of order unity (containing our ignorance),

$$\frac{dL_{dipole}}{dl} = O(\mu_0 A^{1/2}/4\pi C)$$

---

*If the circuit shape is such that $A \ll C^2$, as for an elongated loop, a different estimate of ρ_{max} may be appropriate [e.g., $\rho_{max} = O(A/C)$].

a contribution of order unity compared to the logarithm above. Upon combining the different contributions, the inductance of the loop is estimated to be

$$L \approx \frac{\mu_0}{4\pi} C \left[\ln\left(\frac{\xi A}{a^2}\right) + \frac{1}{2} \right] \tag{5.158}$$

Here we have exhibited the interior contribution explicitly and indicated the uncertainty in the proper value of ρ_{max} and the size of the exterior contribution through the number ξ, of order unity.

Four comments: First, if the wire has a magnetic permeability μ, the interior contribution becomes $\frac{1}{2} \to \mu/2\mu_0$. Second, for a thin wire bent in a circle of radius large compared to the wire radius, a precise calculation (see Problem 5.32) shows that $\xi = 64/\pi e^4 \approx 0.373$. Third, at frequencies high enough to ensure that the skin depth of the wire is small compared to its radius, the interior contribution is absent because the current is confined to near the surface of the wire (see next section). Fourth, if the single turn of wire is replaced by a tight coil of N turns, with the effective cross-sectional radius of the bundle being a, the self-inductance is N^2 times the expression above.

Exercise

Consider a circuit made up of two long, parallel, nonpermeable, circular wires of radii a_1 and a_2, separated by a distance d large compared to the largest radius. Current flows up one wire and back along the other. Ignore the ends. Use the method above to show that the self-inductance per unit length is approximately

$$\frac{dL}{dl} \approx \frac{\mu_0}{\pi} \left[\ln\left(\frac{\xi d}{\sqrt{a_1 a_2}}\right) + \frac{1}{4} \right]$$

where ξ is of order unity. Can you find a reliable value of ξ, within the approximations stated?

5.18 Quasi-Static Magnetic Fields in Conductors; Eddy Currents; Magnetic Diffusion

The magnetostatics of the first 14 sections of this chapter are based on Ampère's law and the absence of magnetic charges. As we saw in Section 5.15, if the magnetic induction varies in time, an electric field is created, according to Faraday's law; the situation is no longer purely magnetic in character. Nevertheless, if the time variation is not too rapid, the magnetic fields dominate and the behavior can be called quasi-static. "Quasi-static" refers to the regime for which the finite speed of light can be neglected and fields treated as if they propagated instantaneously. Said in other, equivalent words, it is the regime where the system is small compared with the electromagnetic wavelength associated with the dominant time scale of the problem. As we learn in subsequent chapters, such a regime permits neglect of the contribution of the Maxwell displacement current to Ampère's law. We consider such fields in conducting media, where Ohm's law relates the electric field to the current density and so back to the magnetic field via the Ampère equation. The relevant equations are

$$\nabla \times \mathbf{H} = \mathbf{J}, \quad \nabla \cdot \mathbf{B} = 0, \quad \nabla \times \mathbf{E} + \frac{\partial \mathbf{B}}{\partial t} = 0, \quad \mathbf{J} = \sigma \mathbf{E} \quad (5.159)$$

With $\mathbf{B} = \nabla \times \mathbf{A}$, Faraday's law shows that the curl of $\mathbf{E} + \partial \mathbf{A}/\partial t$ vanishes. As a result, we can write $\mathbf{E} = -\partial \mathbf{A}/\partial t - \nabla \Phi$. With the assumption of negligible free charge and the time-varying $\mathbf{B}$ as the sole source of the electric field, we may set the scalar potential $\Phi = 0$ and have $\mathbf{E} = -\partial \mathbf{A}/\partial t$. Note that we have the subsidiary conditions, $\nabla \cdot \mathbf{E} = 0$ and $\nabla \cdot \mathbf{A} = 0$. For media of uniform, frequency-independent permeability μ, Ampère's law can be written $\nabla \times \mathbf{B} = \mu \mathbf{J} = \mu \sigma \mathbf{E}$. Elimination of $\mathbf{B}$ and $\mathbf{E}$ in favor of $\mathbf{A}$ and use of the vector identity, $\nabla \times \nabla \times \mathbf{A} = \nabla(\nabla \cdot \mathbf{A}) - \nabla^2 \mathbf{A}$, yields the *diffusion equation* for the vector potential,

$$\nabla^2 \mathbf{A} = \mu \sigma \frac{\partial \mathbf{A}}{\partial t} \quad (5.160)$$

This equation, which obviously also holds for the electric field $\mathbf{E}$, is valid for spatially varying, but frequency-independent σ. If the conductivity is constant in space, it follows that the magnetic induction $\mathbf{B}$ and the current density $\mathbf{J}$ also satisfy the same diffusion equation.

The structure of (5.160) allows us to estimate the time τ for decay of an initial configuration of fields with typical spatial variation defined by the length L. We put $\nabla^2 \mathbf{A} = O(\mathbf{A}/L^2)$ and $\partial \mathbf{A}/\partial t = O(\mathbf{A}/\tau)$. Then

$$\tau = O(\mu \sigma L^2) \quad (5.161)$$

Alternatively, (5.161) can be used to estimate the distance L over which fields exist in a conductor subjected externally to fields with harmonic variation at frequency $\nu = 1/\tau$,

$$L = O\left(\frac{1}{\sqrt{\mu \sigma \nu}}\right) \quad (5.162)$$

For a copper sphere of radius 1 cm, the decay time of some initial $\mathbf{B}$ field inside is of the order of 5–10 milliseconds; for the molten iron core of the earth it is of the order of 10^5 years. This last number is consistent with paleomagnetic data—the last polarity reversal of the earth's field occurred about 10^6 years ago; there is some evidence for a decline to near zero about 5×10^4 years ago and a rise back to its present value.

A. Skin Depth, Eddy Currents, Induction Heating

A simple quantitative illustration of the fields described by (5.160) is afforded by the situation shown in Fig. 5.22: A semi-infinite conductor of uniform conductivity σ and permeability μ occupies the space $z > 0$, with empty space for $z < 0$. The surface at $z = 0^-$ is subjected to a spatially constant, but time-varying, magnetic field in the x direction, $H_x(t) = H_0 \cos \omega t$. We seek a steady-state solution of (5.160) for $z > 0$, subject to appropriate boundary conditions at $z = 0$ and finiteness at $z \to +\infty$. Continuity of the tangential component of $\mathbf{H}$ and the

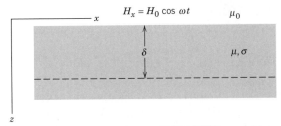

Figure 5.22 At the surface of a semi-infinite conducting permeable medium, a spatially constant magnetic field, $H_x(t) = H_0 \cos \omega t$, is applied parallel to the surface at $z = 0^-$. A localized magnetic field and current flow exists within the medium in the region $z < O(\delta)$.

normal component of **B** across $z = 0$ requires that at $z = 0^+$, the magnetic field have only an x component, $H_x(t) = H_0 \cos \omega t$. The linearity of (5.160) implies that there is only an x component throughout the half-space, $z > 0$ and it is a function of z and t, $H_x(z, t)$.

Because the diffusion equation is second order in the spatial derivatives and first order in the time, it is convenient to use complex notation, with the understanding that the physical fields are found by taking the real parts of the solutions. Thus, the boundary value on H_x is $H_x = H_0 e^{-i\omega t}$, where taking the real part is understood. The steady-state solution for $H_x(z, t)$ can be written

$$H_x(z, t) = h(z)e^{-i\omega t}$$

where, from (5.160), $h(z)$ satisfies

$$\left(\frac{d^2}{dz^2} + i\mu\sigma\omega \right) h(z) = 0 \tag{5.163}$$

A trial solution of the form, $h(z) = e^{ikz}$ leads to the condition

$$k^2 = i\mu\sigma\omega \quad \text{or} \quad k = \pm(1 + i)\sqrt{\frac{\mu\sigma\omega}{2}} \tag{5.164}$$

The square root has the dimensions of an inverse length characteristic of the medium and the frequency [see (5.162)]. The length is called the *skin depth* δ:

$$\delta = \sqrt{\frac{2}{\mu\sigma\omega}} \tag{5.165}$$

For copper at room temperature ($\sigma^{-1} = 1.68 \times 10^{-8} \ \Omega \cdot \text{m}$), $\delta = 6.52 \times 10^{-2}/\sqrt{\nu(\text{Hz})}$ m, where $\nu = \omega/2\pi$. For seawater, $\delta \approx 240/\sqrt{\nu(\text{Hz})}$ m (see Fig. 7.9 and accompanying text).

The solution for H_x is the real part of

$$H_x(z, t) = Ae^{-z/\delta}e^{i(z/\delta - \omega t)} + Be^{z/\delta}e^{-i(z/\delta + \omega t)}$$

with A and B complex numbers. We must choose $B = 0$ to avoid exponentially large fields as $z \to \infty$. Comparison of the solution to the boundary value, $H_x(0^+, t) = H_0 e^{-i\omega t}$, shows that $A = H_0$ and the solution for $z > 0$ is

$$H_x(z, t) = H_0 e^{-z/\delta} \cos(z/\delta - \omega t) \tag{5.166}$$

The magnetic field falls off exponentially in z, with a spatial oscillation of the same scale, being confined mainly to a depth less than the skin depth δ.

Since the field varies in time, there is an accompanying small electric field. From Ampère's and Ohm's laws, together with the existence of only $H_x(z, t)$, we find that there is only a y component of $\mathbf{E}$, given by

$$E_y = \frac{1}{\sigma} \frac{dH_x}{dz} = \frac{-1 + i}{\sigma \delta} H_0 e^{-z/\delta} e^{iz/\delta - i\omega t}$$

Taking the real part and writing $1/\sigma\delta = \mu\delta\omega/2$, we have

$$E_y = \frac{\mu\omega\delta}{\sqrt{2}} H_0 e^{-z/\delta} \cos(z/\delta - \omega t + 3\pi/4) \tag{5.167}$$

To compare the magnitude of the electric field and the magnetic induction, we form the dimensionless ratio,

$$E_y/c\mu H_x = O(\omega\delta/c) \ll 1$$

by the quasi-static assumption. The fields are predominantly magnetic, with a small tangential electric field. The field is associated with a localized current density (for $z > 0$),

$$J_y = \sigma E_y = \frac{\sqrt{2}}{\delta} H_0 e^{-z/\delta} \cos(z/\delta - \omega t + 3\pi/4) \tag{5.168}$$

whose integral in z is an effective surface current,

$$K_y(t) \equiv \int_0^\infty J_y(z, t) \, dz = -H_0 \cos \omega t$$

For very small skin depth, the volume current flow in the region within $O(\delta)$ of the surface acts as a surface current whose magnitude and direction is such as to reduce the magnetic field to zero for $z \gg \delta$. See Section 8.1 for more discussion relevant to waveguides and cavities.

There is resistive heating in the conductor. The time-averaged power input per unit volume is $P_{\text{resistive}} = \langle \mathbf{J} \cdot \mathbf{E} \rangle$ (recall $P = IV = V^2/R$ in a simple lumped resistor circuit). With (5.167) and (5.168), we find

$$P_{\text{resistive}} = \frac{1}{2} \mu\omega H_0^2 e^{-2z/\delta} \tag{5.169}$$

The heating of the conducting medium to a depth of the order of the skin depth is the basis of induction furnaces in steel mills and of microwave cookers in kitchens (where the conductivity of water, or more correctly, the dissipative part of its dielectric susceptibility, causes the losses—see Fig. 7.9). References to more elaborate treatments of eddy currents and induction heating are found at the end of the chapter.

B. Diffusion of Magnetic Fields in Conducting Media

Diffusion of magnetic fields in conducting media can be illustrated with the simple example of two infinite uniform current sheets, parallel to each other and located a distance $2a$ apart, at $z = -a$ and $z = +a$, within an infinite conducting medium of permeability μ and conductivity σ. The currents are such that in the region, $0 < |z| < a$, there is a constant magnetic field H_0 in the x direction and zero field outside. Explicitly, the current density $\mathbf{J}$ is in the y direction, and

$$J_y = H_0[\delta(z + a) - \delta(z - a)]$$

At time $t = 0$, the current is suddenly turned off. The vector potential and magnetic field decay according to (5.160), with variation only in z and t. We use a Laplace transform technique: Separate the space and time dependences by writing

$$H_x(z, t) = \int_0^\infty e^{-pt}\overline{h}(p, z)\, dp$$

Substitution into the diffusion equation (5.160) for H_x leads to the wave equation, $(d^2/dz^2 + k^2)\overline{h}(p, z) = 0$, where $k^2 = \mu\sigma p$. Since the situation is symmetric about $z = 0$, the appropriate solution is $\overline{h} \propto \cos(kz)$. With a change of variable from p to k in the transform integral, $H_x(z, t)$ becomes

$$H_x(z, t) = \int_0^\infty e^{-k^2 t/\mu\sigma} h(k) \cos(kz)\, dk \tag{5.170}$$

The coefficient function $h(k)$ is determined by the initial conditions. At $t = 0^+$, the magnetic field is

$$H_x(z, 0^+) = \int_0^\infty h(k) \cos(kz)\, dk = H_0[\Theta(z + a) - \Theta(z - a)] \tag{5.171}$$

where $\Theta(x)$ is the unit step function, $\Theta(x) = 0$ for $x < 0$ and $\Theta(x) = 1$ for $x > 0$. Exploiting the symmetry in z, we can express the cosine in terms of exponentials and write

$$\frac{1}{2} \int_{-\infty}^\infty h(k) e^{ikz}\, dk = H_0[\Theta(z + a) - \Theta(z - a)] \tag{5.172}$$

where $h(-k) = h(k)$. Inversion of the Fourier integral yields $h(k)$,

$$h(k) = \frac{H_0}{\pi} \int_{-a}^a e^{-ikz}\, dz = \frac{2H_0}{\pi k} \sin(ka) \tag{5.173}$$

The solution for the magnetic field at all times, $t > 0$, is therefore

$$H_x(z, t) = \frac{2H_0}{\pi} \int_0^\infty e^{-\nu t\kappa^2} \frac{\sin \kappa}{\kappa} \cos\left[\left(\frac{z}{a}\right)\kappa\right] d\kappa \tag{5.174}$$

where $\nu = (\mu\sigma a^2)^{-1}$ is a characteristic decay rate [see (5.161)]. The integral in (5.174) can be expressed as the sum of two terms, each identified with a representation of the error function,

$$\Phi(\xi) \equiv \frac{2}{\sqrt{\pi}} \int_0^\xi e^{-x^2}\, dx = \frac{2}{\pi} \int_0^\infty e^{-x^2/4\xi^2} \frac{\sin x}{x}\, dx \tag{5.175}$$

The result is

$$H_x(z, t) = \frac{H_0}{2}\left[\Phi\left(\frac{1 + |z|/a}{2\sqrt{\nu t}}\right) + \Phi\left(\frac{1 - |z|/a}{2\sqrt{\nu t}}\right)\right] \tag{5.176}$$

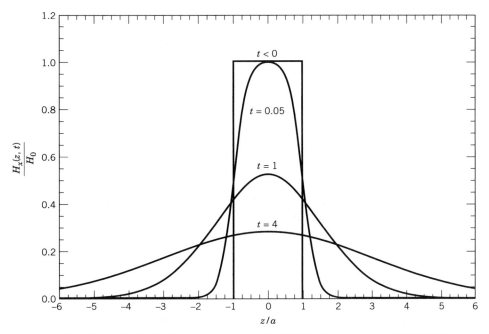

Figure 5.23 Magnetic field distributions given by (5.172) for $\nu t < 0$, and (5.176) for $\nu t = 0.05, 1, 4$ as a function of z/a. The outward diffusion with time for $t > 0$ is manifest; the rise and fall of the field in time at a fixed position can be noted for $1 < |z|/a < 2$.

To understand qualitatively the meaning of the solution we note first that $\Phi(-\xi) = -\Phi(\xi)$, second that $\Phi(\xi) \to 1 - (1/\sqrt{\pi})[1 - 1/2\xi^2 + \cdots]\exp(-\xi^2)$ for $\xi \to \infty$, and third that $\Phi(\xi) \approx (2\xi/\sqrt{\pi})(1 - \xi^2/3 + \cdots)$ for $|\xi| \ll 1$. For $\nu t \to 0$, the arguments in (5.176) are large in magnitude; the solution obviously reduces to the right-hand side of (5.172), as required. For long times ($|\xi| \ll 1$), $H_x(z, t) \to H_0/\sqrt{\pi \nu t}$, independent of $|z|/a$ to leading order in an expansion in $1/\sqrt{\nu t}$. This result is misleading, however, because the coefficients of the higher terms in $1/\nu t$ are z-dependent. A more revealing result is obtained by expanding the error functions in Taylor series in $1/2\sqrt{\nu t}$ to the third order. The result is

$$H_x(z, t) \approx \frac{H_0}{\sqrt{\pi \nu t}} e^{-|z|^2/4\nu t a^2}\left[1 + \frac{1}{12\nu t}\left(|z|^2/2\nu t a^2 - 1\right) + \cdots\right] \quad (5.177)$$

Note that the approximate expression vanishes as $\nu t \to 0$, as it should for any $|z| > a$, and goes to $H_x \approx H_0/\sqrt{\pi \nu t}$ for $\nu t \gg |z|/2a$. For $|z|/a < 5$, it is within a few percent for any $\nu t > 1$. At a given position, the field as a function of time has a maximum at $\nu t \approx |z|^2/2a^2$ [exact for the approximation (5.177)], followed by the very slow decrease as $t^{-1/2}$. Figure 5.23 shows the spatial distributions of the magnetic field at different fixed times.

References and Suggested Reading

Problems of steady-state current flow in an extended resistive medium are analogous to electrostatic potential problems, with the current density replacing the displacement and the conductivity replacing the dielectric constant. But the boundary conditions are generally different. Steady-state current flow is treated in

Jeans, Chapters IX and X

Smythe, Chapter VI

Magnetic fields due to specified current distributions and boundary-value problems in magnetostatics are discussed, with numerous examples, by

Durand, Chapters XIV and XV

Smythe, Chapters VII and XII

The atomic theory of magnetic properties rightly falls in the domain of quantum mechanics. Semiclassical discussions are given by

Abraham and Becker, Band II, Sections 29–34

Durand, pp. 551–573, and Chapter XVII

Landau and Lifshitz, *Electrodynamics of Continuous Media*

Rosenfeld, Chapter IV

More detailed treatments of the subject, including crystallographic aspects and the topic of ferromagnetic domains, can be found in

Chakravarty

Craik

Kittel

Ziman, Chapter 10

An encyclopedic reference to the magnetic properties of alloys and other practical information as well as the theory is

R. M. Bozorth, *Ferromagnetism*, Van Nostrand, New York (1978).

Numerical methods in magnetism are discussed in

Ida and Bastos

Zhou

and in the other references cited in the References and Suggested Reading at the end of Chapter 2.

Faraday's law of induction is somewhat tricky to apply to moving circuits such as Faraday discs or homopolar generators. There are numerous discussions of these questions. A sampling from one journal comprises:

D. R. Corson, *Am. J. Phys.* **24**, 126 (1956)

D. L. Webster, *Am. J. Phys.* **31**, 590 (1963)

E. M. Pugh, *Am. J. Phys.* **32**, 879 (1964)

P. J. Scanlon, R. N. Hendriksen, and J. R. Allen, *Am. J. Phys.* **37**, 698 (1969)

R. D. Eagleton, *Am. J. Phys.* **55**, 621 (1987)

A good treatment of the energy of quasi-stationary currents and forces acting on current-carrying circuits, different from ours, is given by

Panofsky and Phillips, Chapter 10

Stratton, Chapter II

Inductance calculations and forces are handled lucidly by

Abraham and Becker, Band I, Chapters VIII and IX

and in many engineering texts.

The near-neglected topic of eddy currents and induction heating is discussed with many examples in

Smythe, Chapter XX

Magnetic fields in conducting fluids or plasmas play a central role in astrophysics, from the dynamos in the interior of planets to the galaxies. Two references are

E. N. Parker, *Cosmical Magnetic Fields*, Oxford University Press, Oxford (1979).

Ya. B. Zel'dovich, A. A. Ruzmaikin, and D. D. Sokoloff, *Magnetic Fields in Astrophysics*, Gordon & Breach, New York (1983).

Problems

5.1 Starting with the differential expression

$$d\mathbf{B} = \frac{\mu_0 I}{4\pi} \, d\mathbf{l}' \times \frac{\mathbf{x} - \mathbf{x}'}{|\mathbf{x} - \mathbf{x}'|^3}$$

for the magnetic induction at the point P with coordinate $\mathbf{x}$ produced by an increment of current $I \, d\mathbf{l}'$ at $\mathbf{x}'$, show explicitly that for a closed loop carrying a current I the magnetic induction at P is

$$\mathbf{B} = \frac{\mu_0 I}{4\pi} \, \nabla\Omega$$

where Ω is the solid angle subtended by the loop at the point P. This corresponds to a magnetic scalar potential, $\Phi_M = -\mu_0 I \Omega / 4\pi$. The sign convention for the solid angle is that Ω is positive if the point P views the "inner" side of the surface spanning the loop, that is, if a unit normal $\mathbf{n}$ to the surface is defined by the direction of current flow via the right-hand rule, Ω is positive if $\mathbf{n}$ points *away* from the point P, and negative otherwise. This is the same convention as in Section 1.6 for the electric dipole layer.

5.2 A long, right cylindrical, ideal solenoid of arbitrary cross section is created by stacking a large number of identical current-carrying loops one above the other, with N coils per unit length and each loop carrying a current I. [In practice such a solenoid could be wound on a mandrel machined to the arbitrary cross section. After the coil was made rigid (e.g., with epoxy), the mandrel would be withdrawn.]

(a) In the approximation that the solenoidal coil is an ideal current sheet and infinitely long, use Problem 5.1 to establish that at any point inside the coil the magnetic field is axial and equal to

$$H = NI$$

and that $H = 0$ for any point outside the coil.

(b) For a realistic solenoid of circular cross section of radius a $(Na \gg 1)$, but still infinite in length, show that the "smoothed" magnetic field just outside the solenoid (averaged axially over several turns) is not zero, but is the same in magnitude and direction as that of a single wire on the axis carrying a current I, even if $Na \to \infty$. Compare fields inside and out.

5.3 A right-circular solenoid of finite length L and radius a has N turns per unit length and carries a current I. Show that the magnetic induction on the cylinder axis in the limit $NL \to \infty$ is

$$B_z = \frac{\mu_0 NI}{2} \, (\cos \theta_1 + \cos \theta_2)$$

where the angles are defined in the figure.

Problem 5.3

5.4 A magnetic induction $\mathbf{B}$ in a current-free region in a uniform medium is cylindrically symmetric with components $B_z(\rho, z)$ and $B_\rho(\rho, z)$ and with a known $B_z(0, z)$ on the axis of symmetry. The magnitude of the axial field varies slowly in z.

(a) Show that near the axis the axial and radial components of magnetic induction are approximately

$$B_z(\rho, z) \approx B_z(0, z) - \left(\frac{\rho^2}{4}\right)\left[\frac{\partial^2 B_z(0, z)}{\partial z^2}\right] + \cdots$$

$$B_\rho(\rho, z) \approx -\left(\frac{\rho}{2}\right)\left[\frac{\partial B_z(0, z)}{\partial z}\right] + \left(\frac{\rho^3}{16}\right)\left[\frac{\partial^3 B_z(0, z)}{\partial z^3}\right] + \cdots$$

(b) What are the magnitudes of the neglected terms, or equivalently what is the criterion defining "near" the axis?

5.5 **(a)** Use the results of Problems 5.4 and 5.3 to find the axial and radial components of magnetic induction in the *central region* ($|z| \ll L/2$) of a long uniform solenoid of radius a and ends at $z = \pm L/2$, including the value of B_z just inside the coil ($\rho = a^-$).

(b) Use Ampère's law to show that the longitudinal magnetic induction just outside the coil is approximately

$$B_z(\rho = a^+, z) \approx -\left(\frac{2\mu_0 NIa^2}{L^2}\right)\left(1 + \frac{12z^2}{L^2} - \frac{9a^2}{L^2} + \cdots\right)$$

For $L \gg a$, the field outside is negligible compared to inside. How does this axial component compare in size to the azimuthal component of Problem 5.2b?

(c) Show that at the end of the solenoid the magnetic induction near the axis has components

$$B_z \simeq \frac{\mu_0 NI}{2}, \qquad B_\rho \simeq \pm\frac{\mu_0 NI}{4}\left(\frac{\rho}{a}\right)$$

5.6 A cylindrical conductor of radius a has a hole of radius b bored parallel to, and centered a distance d from, the cylinder axis ($d + b < a$). The current density is uniform throughout the remaining metal of the cylinder and is parallel to the axis. Use Ampère's law and principle of linear superposition to find the magnitude and the direction of the magnetic-flux density in the hole.

5.7 A compact circular coil of radius a, carrying a current I (perhaps N turns, each with current I/N), lies in the x-y plane with its center at the origin.

(a) By elementary means [Eq. (5.4)] find the magnetic induction at any point on the z axis.

(b) An identical coil with the same magnitude and sense of the current is located on the same axis, parallel to, and a distance b above, the first coil. With the coordinate origin relocated at the point midway between the centers of the two coils, determine the magnetic induction on the axis near the origin as an expansion in powers of z, up to z^4 inclusive:

$$B_z = \left(\frac{\mu_0 Ia^2}{d^3}\right)\left[1 + \frac{3(b^2 - a^2)z^2}{2d^4} + \frac{15(b^4 - 6b^2a^2 + 2a^4)z^4}{16d^8} + \cdots\right]$$

where $d^2 = a^2 + b^2/4$.

(c) Show that, off-axis near the origin, the axial and radial components, correct to second order in the coordinates, take the form

$$B_z = \sigma_0 + \sigma_2\left(z^2 - \frac{\rho^2}{2}\right); \qquad B_\rho = -\sigma_2 z\rho$$

(d) For the two coils in part b show that the magnetic induction on the z axis for large $|z|$ is given by the expansion in inverse odd powers of $|z|$ obtained from the small z expansion of part b by the formal substitution, $d \to |z|$.

(e) If $b = a$, the two coils are known as a pair of Helmholtz coils. For this choice of geometry the second terms in the expansions of parts b and d are absent ($\sigma_2 = 0$ in part c). The field near the origin is then very uniform. What is the maximum permitted value of $|z|/a$ if the axial field is to be uniform to one part in 10^4, one part in 10^2?

5.8 A localized cylindrically symmetric current distribution is such that the current flows only in the azimuthal direction; the current density is a function only of r and θ (or ρ and z): $\mathbf{J} = \hat{\boldsymbol{\phi}} J(r, \theta)$. The distribution is "hollow" in the sense that there is a current-free region near the origin, as well as outside.

(a) Show that the magnetic field can be derived from the azimuthal component of the vector potential, with a multipole expansion

$$A_\phi(r, \theta) = -\frac{\mu_0}{4\pi} \sum_L m_L r^L P_L^1(\cos \theta)$$

in the interior and

$$A_\phi(r, \theta) = -\frac{\mu_0}{4\pi} \sum_L \mu_L r^{-L-1} P_L^1(\cos \theta)$$

outside the current distribution.

(b) Show that the internal and external multipole moments are

$$m_L = -\frac{1}{L(L+1)} \int d^3x \, r^{-L-1} P_L^1(\cos \theta) \, J(r, \theta)$$

and

$$\mu_L = -\frac{1}{L(L+1)} \int d^3x \, r^L P_L^1(\cos \theta) \, J(r, \theta)$$

5.9 The two circular coils of radius a and separation b of Problem 5.7 can be described in cylindrical coordinates by the current density

$$\mathbf{J} = \hat{\boldsymbol{\phi}} I \delta(\rho - a)[\delta(z - b/2) + \delta(z + b/2)]$$

(a) Using the formalism of Problem 5.8, calculate the internal and external multipole moments for $L = 1, \ldots, 5$.

(b) Using the internal multiple expansion of Problem 5.8, write down explicitly an expression for B_z on the z axis and relate it to the answer of Problem 5.7b.

5.10 A circular current loop of radius a carrying a current I lies in the x-y plane with its center at the origin.

(a) Show that the only nonvanishing component of the vector potential is

$$A_\phi(\rho, z) = \frac{\mu_0 I a}{\pi} \int_0^\infty dk \, \cos kz I_1(k\rho_<)K_1(k\rho_>)$$

where $\rho_<$ ($\rho_>$) is the smaller (larger) of a and ρ.

(b) Show that an alternative expression for A_ϕ is

$$A_\phi(\rho, z) = \frac{\mu_0 I a}{2} \int_0^\infty dk \, e^{-k|z|} J_1(ka)J_1(k\rho)$$

(c) Write down integral expressions for the components of magnetic induction, using the expressions of parts a and b. Evaluate explicitly the components of **B** on the z axis by performing the necessary integrations.

5.11 A circular loop of wire carrying a current I is located with its center at the origin of coordinates and the normal to its plane having spherical angles θ_0, ϕ_0. There is an applied magnetic field, $B_x = B_0(1 + \beta y)$ and $B_y = B_0(1 + \beta x)$.

(a) Calculate the force acting on the loop without making any approximations. Compare your result with the approximate result (5.69). Comment.

(b) Calculate the torque in lowest order. Can you deduce anything about the higher order contributions? Do they vanish for the circular loop? What about for other shapes?

5.12 Two concentric circular loops of radii a, b and currents I, I', respectively ($b < a$), have an angle α between their planes. Show that the torque on one of the loops is about the line of intersection of the two planes containing the loops and has the magnitude.

$$N = \frac{\mu_0 \pi I I' b^2}{2a} \sum_{n=0}^{\infty} \frac{(n+1)}{(2n+1)} \left[\frac{\Gamma(n + \frac{3}{2})}{\Gamma(n+2)\Gamma(\frac{3}{2})} \right]^2 \left(\frac{b}{a} \right)^{2n} P_{2n+1}^1(\cos \alpha)$$

where $P_l^1(\cos \alpha)$ is an associated Legendre polynomial. Determine the sense of the torque for α an acute angle and the currents in the same (opposite) directions.

5.13 A sphere of radius a carries a uniform surface-charge distribution σ. The sphere is rotated about a diameter with constant angular velocity ω. Find the vector potential and magnetic-flux density both inside and outside the sphere.

5.14 A long, hollow, right circular cylinder of inner (outer) radius a (b), and of relative permeability μ_r, is placed in a region of initially uniform magnetic-flux density $\mathbf{B}_0$ at right angles to the field. Find the flux density at all points in space, and sketch the logarithm of the ratio of the magnitudes of **B** on the cylinder axis to $\mathbf{B}_0$ as a function of $\log_{10} \mu_r$ for $a^2/b^2 = 0.5, 0.1$. Neglect end effects.

5.15 Consider two long, straight wires, parallel to the z axis, spaced a distance d apart and carrying currents I in opposite directions. Describe the magnetic field **H** in terms of a magnetic scalar potential Φ_M, with $\mathbf{H} = -\nabla\Phi_M$.

(a) If the wires are parallel to the z axis with positions, $x = \pm d/2$, $y = 0$, show that in the limit of small spacing, the potential is approximately that of a two-dimensional dipole,

$$\Phi_M \approx -\frac{Id \sin \phi}{2\pi\rho} + O(d^2/\rho^2)$$

where ρ and ϕ are the usual polar coordinates.

(b) The closely spaced wires are now centered in a hollow right circular cylinder of steel, of inner (outer) radius a (b) and magnetic permeability $\mu = \mu_r\mu_0$. Determine the magnetic scalar potential in the three regions, $0 < \rho < a$, $a < \rho < b$, and $\rho > b$. Show that the field outside the steel cylinder is a two-dimensional dipole field, as in part a, but with a strength reduced by the factor

$$F = \frac{4\mu_r b^2}{(\mu_r + 1)^2 b^2 - (\mu_r - 1)^2 a^2}$$

Relate your result to Problem 5.14.

(c) Assuming that $\mu_r \gg 1$, and $b = a + t$, where the thickness $t \ll b$, write down an approximate expression for F and determine its numerical value for

$\mu_r = 200$ (typical of steel at 20 G), $b = 1.25$ cm, $t = 3$ mm. The shielding effect is relevant for reduction of stray fields in residential and commercial 60 Hz, 110 or 220 V wiring. The figure illustrates the shielding effect for $a/b = 0.9$, $\mu_r = 100$.

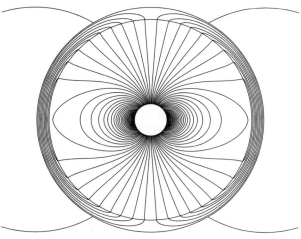

Problem 5.15

5.16   A circular loop of wire of radius a and negligible thickness carries a current I. The loop is centered in a spherical cavity of radius $b > a$ in a large block of soft iron. Assume that the relative permeability of the iron is effectively infinite and that of the medium in the cavity, unity.

(a)   In the approximation of $b \gg a$, show that the magnetic field at the center of the loop is augmented by a factor $(1 + a^3/2b^3)$ by the presence of the iron.

(b)   What is the radius of the "image" current loop (carrying the same current) that simulates the effect of the iron for $r < b$?

5.17   A current distribution $\mathbf{J}(\mathbf{x})$ exists in a medium of unit relative permeability adjacent to a semi-infinite slab of material having relative permeability μ_r and filling the half-space, $z < 0$.

(a)   Show that for $z > 0$ the magnetic induction can be calculated by replacing the medium of permeability μ_r by an image current distribution, $\mathbf{J}^*$, with components,

$$\left(\frac{\mu_r - 1}{\mu_r + 1}\right)J_x(x, y, -z), \qquad \left(\frac{\mu_r - 1}{\mu_r + 1}\right)J_y(x, y, -z), \qquad -\left(\frac{\mu_r - 1}{\mu_r + 1}\right)J_z(x, y, -z)$$

(b)   Show that for $z < 0$ the magnetic induction appears to be due to a current distribution $[2\mu_r/(\mu_r + 1)]\mathbf{J}$ in a medium of unit relative permeability.

5.18   A circular loop of wire having a radius a and carrying a current I is located in vacuum with its center a distance d away from a semi-infinite slab of permeability μ. Find the force acting on the loop when

(a)   the plane of the loop is parallel to the face of the slab,

(b)   the plane of the loop is perpendicular to the face of the slab.

(c)   Determine the limiting form of your answer to parts a and b when $d \gg a$. Can you obtain these limiting values in some simple and direct way?

5.19 A magnetically "hard" material is in the shape of a right circular cylinder of length L and radius a. The cylinder has a permanent magnetization M_0, uniform throughout its volume and parallel to its axis.

    (a) Determine the magnetic field **H** and magnetic induction **B** at all points on the axis of the cylinder, both inside and outside.

    (b) Plot the ratios $\mathbf{B}/\mu_0 M_0$ and $\mathbf{H}/M_0$ on the axis as functions of z for $L/a = 5$.

5.20 **(a)** Starting from the force equation (5.12) and the fact that a magnetization **M** inside a volume V bounded by a surface S is equivalent to a volume current density $\mathbf{J}_M = (\nabla \times \mathbf{M})$ and a surface current density $(\mathbf{M} \times \mathbf{n})$, show that in the absence of macroscopic conduction currents the total magnetic force on the body can be written

$$\mathbf{F} = -\int_V (\nabla \cdot \mathbf{M})\mathbf{B}_e \, d^3x + \int_S (\mathbf{M} \cdot \mathbf{n})\mathbf{B}_e \, da$$

where $\mathbf{B}_e$ is the applied magnetic induction (not including that of the body in question). The force is now expressed in terms of the effective charge densities ρ_M and σ_M. If the distribution of magnetization is not discontinuous, the surface can be at infinity and the force given by just the volume integral.

    (b) A sphere of radius R with uniform magnetization has its center at the origin of coordinates and its direction of magnetization making spherical angles θ_0, ϕ_0. If the external magnetic field is the same as in Problem 5.11, use the expression of part a to evaluate the components of the force acting on the sphere.

5.21 A magnetostatic field is due entirely to a localized distribution of permanent magnetization.

    (a) Show that

$$\int \mathbf{B} \cdot \mathbf{H} \, d^3x = 0$$

provided the integral is taken over all space.

    (b) From the potential energy (5.72) of a dipole in an external field, show that for a continuous distribution of permanent magnetization the magnetostatic energy can be written

$$W = \frac{\mu_0}{2} \int \mathbf{H} \cdot \mathbf{H} \, d^3x = -\frac{\mu_0}{2} \int \mathbf{M} \cdot \mathbf{H} \, d^3x$$

apart from an additive constant, which is independent of the orientation or position of the various constituent magnetized bodies.

5.22 Show that in general a long, straight bar of uniform cross-sectional area A with uniform lengthwise magnetization M, when placed with its flat end against an infinitely permeable flat surface, adheres with a force given approximately by

$$F \simeq \frac{\mu_0}{2} AM^2$$

Relate your discussion to the electrostatic considerations in Section 1.11.

5.23 A right circular cylinder of length L and radius a has a uniform lengthwise magnetization M.

(a) Show that, when it is placed with its flat end against an infinitely permeable plane surface, it adheres with a force

$$F = 2\mu_0 a L M^2 \left[\frac{K(k) - E(k)}{k} - \frac{K(k_1) - E(k_1)}{k_1} \right]$$

where

$$k = \frac{2a}{\sqrt{4a^2 + L^2}}, \qquad k_1 = \frac{a}{\sqrt{a^2 + L^2}}$$

(b) Find the limiting form for the force if $L \gg a$.

5.24 **(a)** For the perfectly conducting plane of Section 5.13 with the circular hole in it and the asymptotically uniform tangential magnetic field $\mathbf{H}_0$ on one side, calculate the added tangential magnetic field $\mathbf{H}^{(1)}$ on the side of the plane with $\mathbf{H}_0$. Show that its components for $\rho > a$ are

$$H_x^{(1)} = \frac{2 H_0 a^3}{\pi} \frac{xy}{\rho^4 \sqrt{\rho^2 - a^2}}$$

$$H_y^{(1)} = \frac{2 H_0 a^3}{\pi} \frac{y^2}{\rho^4 \sqrt{\rho^2 - a^2}} + \frac{H_0}{\pi} \left[\frac{a}{\rho} \sqrt{1 - \frac{a^2}{\rho^2}} - \sin^{-1}\left(\frac{a}{\rho} \right) \right]$$

(b) Sketch the lines of surface current flow in the neighborhood of the hole on both sides of the plane.

5.25 A flat right rectangular loop carrying a constant current I_1 is placed near a long straight wire carrying a current I_2. The loop is oriented so that its center is a perpendicular distance d from the wire; the sides of length a are parallel to the wire and the sides of length b make an angle α with the plane containing the wire and the loop's center. The direction of the current I_1 is the same as that of I_2 in the side of the rectangle nearest the wire.

(a) Show that the interaction magnetic energy

$$W_{12} = \int \mathbf{J}_1 \cdot \mathbf{A}_2 \, d^3x = I_1 F_2$$

(where F_2 is the magnetic flux from I_2 linking the rectangular circuit carrying I_1), is

$$W_{12} = \frac{\mu_0 I_1 I_2 a}{4\pi} \ln\left[\frac{4d^2 + b^2 + 4d\, b\, \cos\alpha}{4d^2 + b^2 - 4d\, b\, \cos\alpha} \right]$$

(b) Calculate the force between the loop and the wire for fixed currents.

(c) Repeat the calculation for a circular loop of radius a, whose plane makes an angle α with respect to the plane containing the center of the loop and the wire. Show that the interaction energy is

$$W_{12} = \mu_0 I_1 I_2 \, d \, \mathrm{Re}\{e^{i\alpha} - \sqrt{e^{2i\alpha} - a^2/d^2}\}$$

Find the force.

(d) For both loops, show that when $d \gg a,b$ the interaction energy reduces to $W_{12} \approx \mathbf{m} \cdot \mathbf{B}$, where $\mathbf{m}$ is the magnetic moment of the loop. Explain the sign.

5.26 A two-wire transmission line consists of a pair of nonpermeable parallel wires of radii a and b separated by a distance $d > a + b$. A current flows down one wire and back the other. It is uniformly distributed over the cross section of each wire. Show that the self-inductance per unit length is

$$L = \frac{\mu_0}{4\pi}\left[1 + 2\ln\left(\frac{d^2}{ab}\right)\right]$$

5.27 A circuit consists of a long thin conducting shell of radius a and a parallel return wire of radius b on axis inside. If the current is assumed distributed uniformly throughout the cross section of the wire, calculate the self-inductance per unit length. What is the self-inductance if the inner conductor is a thin hollow tube?

5.28 Show that the mutual inductance of two circular coaxial loops in a homogeneous medium of permeability μ is

$$M_{12} = \mu\sqrt{ab}\left[\left(\frac{2}{k} - k\right)K(k) - \frac{2}{k}E(k)\right]$$

where

$$k^2 = \frac{4ab}{(a + b^2) + d^2}$$

and a, b are the radii of the loops, d is the distance between their centers, and K and E are the complete elliptic integrals.

Find the limiting value when $d \ll a$, b and $a \simeq b$.

5.29 The figure represents a transmission line consisting of two, parallel *perfect* conductors of arbitrary, but constant, cross section. Current flows down one conductor and returns via the other.

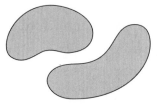

Problem 5.29

Show that the product of the inductance per unit length L and the capacitance per unit length C is

$$LC = \mu\epsilon$$

where μ and ϵ are the permeability and the permittivity of the medium surrounding the conductors. (See the discussion about magnetic fields near perfect conductors at the beginning of Section 5.13.)

5.30 **(a)** Show that a surface current density $K(\phi) = I \cos \phi/2R$ flowing in the axial direction on a right circular cylindrical surface of radius R produces inside the cylinder a uniform magnetic induction $B_0 = \mu_0 I/4R$ in a direction perpendicular to the cylinder axis. Show that the field outside is that of a two-dimensional dipole.

(b) Calculate the total magnetostatic field energy per unit length. How is it divided inside and outside the cylinder?

(c) What is the inductance per unit length of the system, viewed as a long circuit with current flowing up one side of the cylinder and back the other?

Answer: $L = \pi\mu_0/8$.

5.31 An accelerator bending magnet consists of N turns of superconducting cable whose current configuration can be described approximately by the axial current density

$$J_z(\rho, \phi) = \left(\frac{NI}{2R}\right)\cos \phi \; \delta(\rho - R)$$

The right circular current cylinder is centered on the axis of a hollow iron cylinder of inner radius R' ($R' > R$). The relative dimensions (R, R' a few centimeters and a magnet length of several meters) permit the use of a two-dimensional approximation, at least away from the ends of the magnet. Assume that the relative permeability of the iron can be taken as infinite. [Then the outer radius of the iron is irrelevant.]

(a) Show that the magnetic field inside the current sheath is perpendicular to the axis of the cylinder in the direction defined by $\phi = \pm\pi/2$ and has the magnitude

$$B_0 = \left(\frac{\mu_0 NI}{4R}\right)\left[1 + \frac{R^2}{R'^2}\right]$$

(b) Show that the magnetic energy inside $r = R$ is augmented (and that outside diminished) relative to the values in the absence of the iron. (Compare part b of Problem 5.30.)

(c) Show that the inductance per unit length is

$$\frac{dL}{dz} = \left(\frac{\pi\mu_0 N^2}{8}\right)\left[1 + \frac{R^2}{R'^2}\right]$$

5.32 A circular loop of mean radius a is made of wire having a circular cross section of radius b, with $b \ll a$. The sketch shows the relevant dimensions and coordinates for this problem.

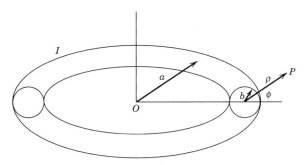

Problem 5.32

(a) Using (5.37), the expression for the vector potential of a filamentary circular loop, and appropriate approximations for the elliptic integrals, show that the vector potential at the point P near the wire is approximately

$$A_\phi = (\mu_0 I/2\pi)[\ln(8a/\rho) - 2]$$

where ρ is the transverse coordinate shown in the figure and corrections are of order $(\rho/a)\cos \phi$ and $(\rho/a)^2$.

(b) Since the vector potential of part a is, apart from a constant, just that outside a straight circular wire carrying a current I, determine the vector potential inside the wire ($\rho < b$) in the same approximation by requiring continuity of

A_ϕ and its radial derivative at $\rho = b$, assuming that the current is uniform in density inside the wire:

$$A_\phi = (\mu_0 I/4\pi)(1 - \rho^2/b^2) + (\mu_0 I/2\pi)[\ln(8a/b) - 2], \qquad \rho < b$$

(c) Use (5.149) to find the magnetic energy, hence the self-inductance,

$$L = \mu_0 a[\ln(8a/b) - 7/4]$$

Are the corrections of order b/a or $(b/a)^2$? What is the change in L if the current is assumed to flow only on the surface of the wire (as occurs at high frequencies when the skin depth is small compared to b)?

5.33 Consider two current loops (as in Fig. 5.3) whose orientation in space is fixed, but whose relative separation can be changed. Let O_1 and O_2 be origins in the two loops, fixed relative to each loop, and $\mathbf{x}_1$ and $\mathbf{x}_2$ be coordinates of elements $d\mathbf{l}_1$ and $d\mathbf{l}_2$, respectively, of the loops referred to the respective origins. Let $\mathbf{R}$ be the relative coordinate of the origins, directed from loop 2 to loop 1.

(a) Starting from (5.10), the expression for the force between the loops, show that it can be written

$$\mathbf{F}_{12} = I_1 I_2 \nabla_R M_{12}(\mathbf{R})$$

where M_{12} is the mutual inductance of the loops,

$$M_{12}(\mathbf{R}) = \frac{\mu_0}{4\pi} \oint \oint \frac{d\mathbf{l}_1 \cdot d\mathbf{l}_2}{|\mathbf{x}_1 - \mathbf{x}_2 + \mathbf{R}|}$$

and it is assumed that the orientation of the loops does not change with $\mathbf{R}$.

(b) Show that the mutual inductance, viewed as a function of $\mathbf{R}$, is a solution of the Laplace equation,

$$\nabla_R^2 M_{12}(\mathbf{R}) = 0$$

The importance of this result is that the uniqueness of solutions of the Laplace equation allows the exploitation of the properties of such solutions, provided a solution can be found for a particular value of $\mathbf{R}$.

5.34 Two identical circular loops of radius a are initially located a distance R apart on a common axis perpendicular to their planes.

(a) From the expression $W_{12} = \int d^3x \, \mathbf{J}_1 \cdot \mathbf{A}_2$ and the result for A_ϕ from Problem 5.10b, show that the mutual inductance of the loops is

$$M_{12} = \mu_0 \pi a^2 \int_0^\infty dk \, e^{-kR} J_1^2(ka)$$

(b) Show that for $R > 2a$, M_{12} has the expansion,

$$M_{12} = \frac{\mu_0 \pi a}{2} \left[\left(\frac{a}{R}\right)^3 - 3\left(\frac{a}{R}\right)^5 + \frac{75}{8}\left(\frac{a}{R}\right)^7 + \cdots \right]$$

(c) Use the techniques of Section 3.3 for solutions of the Laplace equation to show that the mutual inductance for two coplanar identical circular loops of radius a whose centers are separated by a distance $R > 2a$ is

$$M_{12} = -\frac{\mu_0 \pi a}{4} \left[\left(\frac{a}{R}\right)^3 + \frac{9}{4}\left(\frac{a}{R}\right)^5 + \frac{375}{64}\left(\frac{a}{R}\right)^7 + \cdots \right]$$

(d) Calculate the forces between the loops in the common axis and coplanar configurations. Relate the answers to those of Problem 5.18.

5.35 An insulated coil is wound on the surface of a sphere of radius a in such a way as to produce a uniform magnetic induction B_0 in the z direction inside the sphere and dipole field outside the sphere. The medium inside and outside the sphere has a uniform conductivity σ and permeability μ.

(a) Find the necessary surface current density $\mathbf{K}$ and show that the vector potential describing the magnetic field has only an azimuthal component, given by

$$A_\phi = \frac{B_0 a^2}{2} \frac{r_<}{r_>^2} \sin\theta$$

where $r_<$ ($r_>$) is the smaller (larger) of r and a.

(b) At $t = 0$ the current in the coil is cut off. [The coil's presence may be ignored from now on.] With the neglect of Maxwell's displacement current, the decay of the magnetic field is described by the diffusion equation, (5.160). Using a Laplace transform and a spherical Bessel function expansion (3.113), show that the vector potential at times $t > 0$ is given by

$$A_\phi = \frac{3B_0 a}{\pi} \sin\theta \int_0^\infty e^{-\nu t k^2} j_1(k) j_1\left(\frac{kr}{a}\right) dk$$

where $\nu = 1/\mu\sigma a^2$ is a characteristic decay rate and $j_1(x)$ is the spherical Bessel function of order one. Show that the magnetic field at the center of the sphere can be written explicitly in terms of the error function $\Phi(x)$ as

$$B_z(0, t) = B_0\left[\Phi\left(\frac{1}{\sqrt{4\nu t}}\right) - \frac{1}{\sqrt{\pi\nu t}} \exp\left(-\frac{1}{4\nu t}\right)\right]$$

(c) Show that the total magnetic energy at time $t > 0$ can be written

$$W_m = \frac{6B_0^2 a^3}{\mu} \int_0^\infty e^{-2\nu t k^2}[j_1(k)]^2 \, dk$$

Show that at long times ($\nu t \gg 1$) the magnetic energy decays asymptotically as

$$W_m \rightarrow \frac{\sqrt{2\pi}B_0^2 a^3}{24\mu(\nu t)^{3/2}}$$

(d) Find a corresponding expression for the asymptotic form of the vector potential (at fixed r, θ and $\nu t \rightarrow \infty$) and show that it decays as $(\nu t)^{-3/2}$ as well. Since the energy is quadratic in the field strength, there seems to be a puzzle here. Show by numerical or analytic means that the behavior of the magnetic field at time t is such that, for distances small compared to $R = a(\nu t)^{1/2} \gg a$, the field is uniform with strength $(B_0/6\pi^{1/2})(\nu t)^{-3/2}$, and for distances large compared to R, the field is essentially the original dipole field. Explain physically.

5.36 The time-varying magnetic field for $t > 0$ in Problem 5.35 induces an electric field and causes current to flow.

(a) What components of electric field exist? Determine integral expressions for the components of the electric field and find a simple explicit form of the current density $\mathbf{J} = \sigma\mathbf{E}$ at $t = 0^+$. Compare your result with the current density of Problem 5.35a. Find the asymptotic behavior of the electric fields in time.

(b) With Ohm's law and the electric fields found in part a, show that the total power dissipated in the resistive medium can be written

$$P = \frac{12B_0^2 a^3 \nu}{\mu} \int_0^\infty e^{-2\nu t k^2} [k j_1(k)]^2 \, dk$$

Note that the power is the negative time derivative of the *magnetic* energy, W_m.

(c) Because of Ohm's law, the total *electric* energy is $W_e = \epsilon_0 P / 2\sigma$. The total energy is the sum of W_e and W_m; its time derivative should be the negative of the power dissipation. Show that the neglect of the energy in the electric field is the same order of approximation as neglect of the displacement current in the equations governing the magnetic field.

CHAPTER 6

Maxwell Equations, Macroscopic Electromagnetism, Conservation Laws

In the preceding chapters we dealt mostly with steady-state problems in electricity and in magnetism. Similar mathematical techniques were employed, but electric and magnetic phenomena were treated as independent. The only link between them was that the currents that produce magnetic fields are basically electrical in character, being charges in motion. The almost independent nature of electric and magnetic phenomena disappears when we consider time-dependent problems. Faraday's discovery of induction (Section 5.15) destroyed the independence. Time-varying magnetic fields give rise to electric fields and vice versa. We then must speak of *electromagnetic fields*, rather than electric or magnetic fields. The full import of the interconnection between electric and magnetic fields and their essential sameness becomes clear only within the framework of special relativity (Chapter 11). For the present we content ourselves with examining the basic phenomena and deducing the set of equations known as the *Maxwell equations*, which describe the behavior of electromagnetic fields. Vector and scalar potentials, gauge transformations, and Green functions for the wave equation are next discussed, including retarded solutions for the fields, as well as the potentials. There follows a derivation of the macroscopic equations of electromagnetism. Conservation laws for energy and momentum and transformation properties of electromagnetic quantities are treated, as well as the interesting topic of magnetic monopoles.

6.1 *Maxwell's Displacement Current; Maxwell Equations*

The basic laws of electricity and magnetism we have discussed so far can be summarized in differential form by these four (not yet Maxwell) equations:

COULOMB'S LAW	$\mathbf{\nabla} \cdot \mathbf{D} = \rho$
AMPÈRE'S LAW ($\mathbf{\nabla} \cdot \mathbf{J} = 0$)	$\mathbf{\nabla} \times \mathbf{H} = \mathbf{J}$
FARADAY'S LAW	$\mathbf{\nabla} \times \mathbf{E} + \dfrac{\partial \mathbf{B}}{\partial t} = 0$
ABSENCE OF FREE MAGNETIC POLES	$\mathbf{\nabla} \cdot \mathbf{B} = 0$

$$(6.1)$$

Let us recall that all but Faraday's law were derived from steady-state observations. Consequently, from a logical point of view there is no a priori reason to expect that the static equations will hold unchanged for time-dependent fields. In fact, the equations in set (6.1) are inconsistent as they stand.

It required the genius of J. C. Maxwell, spurred on by Faraday's observations, to see the inconsistency in equations (6.1) and to modify them into a consistent set that implied new physical phenomena, at the time unknown but subsequently verified in all details by experiment. For this brilliant stroke in 1865, the modified set of equations is justly known as the *Maxwell equations*.

The faulty equation is Ampère's law. It was derived for steady-state current phenomena with $\nabla \cdot \mathbf{J} = 0$. This requirement on the divergence of $\mathbf{J}$ is contained right in Ampère's law, as can be seen by taking the divergence of both sides:

$$\nabla \cdot \mathbf{J} = \nabla \cdot (\nabla \times \mathbf{H}) \equiv 0 \qquad (6.2)$$

While $\nabla \cdot \mathbf{J} = 0$ is valid for steady-state problems, the general relation is given by the continuity equation for charge and current:

$$\nabla \cdot \mathbf{J} + \frac{\partial \rho}{\partial t} = 0 \qquad (6.3)$$

What Maxwell saw was that the continuity equation could be converted into a vanishing divergence by using Coulomb's law (6.1). Thus

$$\nabla \cdot \mathbf{J} + \frac{\partial \rho}{\partial t} = \nabla \cdot \left(\mathbf{J} + \frac{\partial \mathbf{D}}{\partial t} \right) = 0 \qquad (6.4)$$

Then Maxwell replaced $\mathbf{J}$ in Ampère's law by its generalization

$$\mathbf{J} \to \mathbf{J} + \frac{\partial \mathbf{D}}{\partial t}$$

for time-dependent fields. Thus Ampère's law became

$$\nabla \times \mathbf{H} = \mathbf{J} + \frac{\partial \mathbf{D}}{\partial t} \qquad (6.5)$$

still the same, experimentally verified, law for steady-state phenomena, but now mathematically consistent with the continuity equation (6.3) for time-dependent fields. Maxwell called the added term in (6.5) the *displacement current*. Its presence means that a changing *electric* field causes a magnetic field, even without a current—the converse of Faraday's law. This necessary addition to Ampère's law is of crucial importance for rapidly fluctuating fields. Without it there would be no electromagnetic radiation, and the greatest part of the remainder of this book would have to be omitted. It was Maxwell's prediction that light was an electromagnetic wave phenomenon, and that electromagnetic waves of all frequencies could be produced, that drew the attention of all physicists and stimulated so much theoretical and experimental research into electromagnetism during the last part of the nineteenth century.

The set of four equations,

$$\begin{aligned} \nabla \cdot \mathbf{D} = \rho \qquad & \nabla \times \mathbf{H} = \mathbf{J} + \frac{\partial \mathbf{D}}{\partial t} \\[2mm] \nabla \cdot \mathbf{B} = 0 \qquad & \nabla \times \mathbf{E} + \frac{\partial \mathbf{B}}{\partial t} = 0 \end{aligned} \qquad (6.6)$$

known as the *Maxwell equations*, forms the basis of all classical electromagnetic phenomena. When combined with the Lorentz force equation and Newton's second law of motion, these equations provide a complete description of the classical dynamics of interacting charged particles and electromagnetic fields (see Section 6.7 and Chapters 12 and 16). The range of validity of the Maxwell equations is discussed in the Introduction, as are questions of boundary conditions for the normal and tangential components of fields at interfaces between different media. Constitutive relations connecting **E** and **B** with **D** and **H** were touched on in the Introduction and treated for static phenomena in Chapters 4 and 5. More is said later in this chapter and in Chapter 7.

The units employed in writing the Maxwell equations (6.6) are those of the preceding chapters, namely, SI. For the reader more at home in other units, such as Gaussian, Table 2 of the Appendix summarizes essential equations in the commoner systems. Table 3 of the Appendix allows the conversion of any equation from Gaussian to SI units or vice versa, while Table 4 gives the corresponding conversions for given amounts of any variable.

6.2   *Vector and Scalar Potentials*

The Maxwell equations consist of a set of coupled first-order partial differential equations relating the various components of electric and magnetic fields. They can be solved as they stand in simple situations. But it is often convenient to introduce potentials, obtaining a smaller number of second-order equations, while satisfying some of the Maxwell equations identically. We are already familiar with this concept in electrostatics and magnetostatics, where we used the scalar potential Φ and the vector potential **A**.

Since $\nabla \cdot \mathbf{B} = 0$ still holds, we can define **B** in terms of a vector potential:

$$\mathbf{B} = \nabla \times \mathbf{A} \tag{6.7}$$

Then the other homogeneous equation in (6.6), Faraday's law, can be written

$$\nabla \times \left(\mathbf{E} + \frac{\partial \mathbf{A}}{\partial t} \right) = 0 \tag{6.8}$$

This means that the quantity with vanishing curl in (6.8) can be written as the gradient of some scalar function, namely, a scalar potential Φ:

$$\mathbf{E} + \frac{\partial \mathbf{A}}{\partial t} = -\nabla\Phi$$

or

$$\mathbf{E} = -\nabla\Phi - \frac{\partial \mathbf{A}}{\partial t} \tag{6.9}$$

The definition of **B** and **E** in terms of the potentials **A** and Φ according to (6.7) and (6.9) satisfies identically the two homogeneous Maxwell equations. The dynamic behavior of **A** and Φ will be determined by the two inhomogeneous equations in (6.6).

At this stage it is convenient to restrict our considerations to the vacuum form of the Maxwell equations. Then the inhomogeneous equations in (6.6) can be written in terms of the potentials as

$$\nabla^2 \Phi + \frac{\partial}{\partial t} (\nabla \cdot \mathbf{A}) = -\rho/\epsilon_0 \tag{6.10}$$

$$\nabla^2 \mathbf{A} - \frac{1}{c^2} \frac{\partial^2 \mathbf{A}}{\partial t^2} - \nabla \left(\nabla \cdot \mathbf{A} + \frac{1}{c^2} \frac{\partial \Phi}{\partial t} \right) = -\mu_0 \mathbf{J} \tag{6.11}$$

We have now reduced the set of four Maxwell equations to two equations. But they are still coupled equations. The uncoupling can be accomplished by exploiting the arbitrariness involved in the definition of the potentials. Since **B** is defined through (6.7) in terms of **A**, the vector potential is arbitrary to the extent that the gradient of some scalar function Λ can be added. Thus **B** is left unchanged by the transformation,

$$\mathbf{A} \rightarrow \mathbf{A}' = \mathbf{A} + \nabla\Lambda \tag{6.12}$$

For the electric field (6.9) to be unchanged as well, the scalar potential must be simultaneously transformed,

$$\Phi \rightarrow \Phi' = \Phi - \frac{\partial \Lambda}{\partial t} \tag{6.13}$$

The freedom implied by (6.12) and (6.13) means that we can choose a set of potentials ($\mathbf{A}, \Phi$) to satisfy the *Lorenz condition* (1867),*

$$\nabla \cdot \mathbf{A} + \frac{1}{c^2} \frac{\partial \Phi}{\partial t} = 0 \tag{6.14}$$

This will uncouple the pair of equations (6.10) and (6.11) and leave two inhomogeneous wave equations, one for Φ and one for **A**:

$$\nabla^2 \Phi - \frac{1}{c^2} \frac{\partial^2 \Phi}{\partial t^2} = -\rho/\epsilon_0 \tag{6.15}$$

$$\nabla^2 \mathbf{A} - \frac{1}{c^2} \frac{\partial^2 \mathbf{A}}{\partial t^2} = -\mu_0 \mathbf{J} \tag{6.16}$$

Equations (6.15) and (6.16), plus (6.14), form a set of equations equivalent in all respects to the Maxwell equations in vacuum, as observed by Lorenz and others.

6.3 *Gauge Transformations, Lorenz Gauge, Coulomb Gauge*

The transformation (6.12) and (6.13) is called a *gauge transformation*, and the invariance of the fields under such transformations is called *gauge invariance*. To see that potentials can always be found to satisfy the Lorenz condition, suppose that the potentials **A**, Φ that satisfy (6.10) and (6.11) do not satisfy (6.14). Then let us make a gauge transformation to potentials $\mathbf{A}', \Phi'$ and demand that $\mathbf{A}', \Phi'$ satisfy the Lorenz condition:

$$\nabla \cdot \mathbf{A}' + \frac{1}{c^2} \frac{\partial \Phi'}{\partial t} = 0 = \nabla \cdot \mathbf{A} + \frac{1}{c^2} \frac{\partial \Phi}{\partial t} + \nabla^2 \Lambda - \frac{1}{c^2} \frac{\partial^2 \Lambda}{\partial t^2} \tag{6.17}$$

---

*L. V. Lorenz, Phil. Mag. Ser. 3, **34**, 287 (1867). *See also* p. 294.

Thus, provided a gauge function Λ can be found to satisfy

$$\nabla^2 \Lambda - \frac{1}{c^2} \frac{\partial^2 \Lambda}{\partial t^2} = -\left(\nabla \cdot \mathbf{A} + \frac{1}{c^2} \frac{\partial \Phi}{\partial t} \right) \tag{6.18}$$

the new potentials $\mathbf{A}'$, Φ' will satisfy the Lorenz condition and the wave equations (6.15) and (6.16).

Even for potentials that satisfy the Lorenz condition (6.14) there is arbitrariness. Evidently the *restricted gauge transformation*,

$$\mathbf{A} \rightarrow \mathbf{A} + \nabla \Lambda$$
$$\Phi \rightarrow \Phi - \frac{\partial \Lambda}{\partial t} \tag{6.19}$$

where

$$\nabla^2 \Lambda - \frac{1}{c^2} \frac{\partial^2 \Lambda}{\partial t^2} = 0 \tag{6.20}$$

preserves the Lorenz condition, provided $\mathbf{A}$, Φ satisfy it initially. All potentials in this restricted class are said to belong to the *Lorenz gauge*. The Lorenz gauge is commonly used, first because it leads to the wave equations (6.15) and (6.16), which treat Φ and $\mathbf{A}$ on equivalent footings, and second because it is a concept independent of the coordinate system chosen and so fits naturally into the considerations of special relativity (see Section 11.9).

Another useful gauge for the potentials is the so-called *Coulomb*, *radiation*, or *transverse gauge*. This is the gauge in which

$$\nabla \cdot \mathbf{A} = 0 \tag{6.21}$$

From (6.10) we see that the scalar potential satisfies the Poisson equation,

$$\nabla^2 \Phi = -\rho / \epsilon_0 \tag{6.22}$$

with solution,

$$\Phi(\mathbf{x}, t) = \frac{1}{4\pi\epsilon_0} \int \frac{\rho(\mathbf{x}', t)}{|\mathbf{x} - \mathbf{x}'|} \, d^3x' \tag{6.23}$$

The scalar potential is just the *instantaneous* Coulomb potential due to the charge density $\rho(\mathbf{x}, t)$. This is the origin of the name "Coulomb gauge."

The vector potential satisfies the inhomogeneous wave equation,

$$\nabla^2 \mathbf{A} - \frac{1}{c^2} \frac{\partial^2 \mathbf{A}}{\partial t^2} = -\mu_0 \mathbf{J} + \frac{1}{c^2} \nabla \frac{\partial \Phi}{\partial t} \tag{6.24}$$

The term involving the scalar potential can, in principle, be calculated from (6.23). Since it involves the gradient operator, it is a term that is *irrotational*, that is, has vanishing curl. This suggests that it may cancel a corresponding piece of the current density. The current density (or any vector field) can be written as the sum of two terms,

$$\mathbf{J} = \mathbf{J}_l + \mathbf{J}_t \tag{6.25}$$

where $\mathbf{J}_l$ is called the *longitudinal* or *irrotational* current and has $\nabla \times \mathbf{J}_l = 0$, while $\mathbf{J}_t$ is called the *transverse* or *solenoidal* current and has $\nabla \cdot \mathbf{J}_t = 0$. Starting from the vector identity,

$$\nabla \times (\nabla \times \mathbf{J}) = \nabla(\nabla \cdot \mathbf{J}) - \nabla^2 \mathbf{J} \tag{6.26}$$

together with $\nabla^2(1/|\mathbf{x} - \mathbf{x}'|) = -4\pi\delta(\mathbf{x} - \mathbf{x}')$, it can be shown that $\mathbf{J}_l$ and $\mathbf{J}_t$ can be constructed explicitly from $\mathbf{J}$ as follows:

$$\mathbf{J}_l = -\frac{1}{4\pi} \nabla \int \frac{\nabla' \cdot \mathbf{J}}{|\mathbf{x} - \mathbf{x}'|} d^3x' \tag{6.27}$$

$$\mathbf{J}_t = \frac{1}{4\pi} \nabla \times \nabla \times \int \frac{\mathbf{J}}{|\mathbf{x} - \mathbf{x}'|} d^3x' \tag{6.28}$$

With the help of the continuity equation and (6.23) it is seen that

$$\frac{1}{c^2} \nabla \frac{\partial \Phi}{\partial t} = \mu_0 \mathbf{J}_l \tag{6.29}$$

Therefore the source for the wave equation for $\mathbf{A}$ can be expressed entirely in terms of the *transverse* current (6.28):

$$\nabla^2 \mathbf{A} - \frac{1}{c^2} \frac{\partial^2 \mathbf{A}}{\partial t^2} = -\mu_0 \mathbf{J}_t \tag{6.30}$$

This, of course, is the origin of the name "transverse gauge." The name "radiation gauge" stems from the fact that transverse radiation fields are given by the vector potential alone, the instantaneous Coulomb potential contributing only to the near fields. This gauge is particularly useful in quantum electrodynamics. A quantum-mechanical description of photons necessitates quantization of only the vector potential.

The Coulomb or transverse gauge is often used when no sources are present. Then $\Phi = 0$, and $\mathbf{A}$ satisfies the homogeneous wave equation. The fields are given by

$$\mathbf{E} = -\frac{\partial \mathbf{A}}{\partial t}$$
$$\mathbf{B} = \nabla \times \mathbf{A} \tag{6.31}$$

In passing we note a peculiarity of the Coulomb gauge. It is well known that electromagnetic disturbances propagate with finite speed. Yet (6.23) indicates that the scalar potential "propagates" instantaneously everywhere in space. The vector potential, on the other hand, satisfies the wave equation (6.30), with its implied finite speed of propagation c. At first glance it is puzzling to see how obviously unphysical behavior is avoided. A preliminary remark is that it is the fields, not the potentials, that concern us. A further observation is that the *transverse* current (6.28) extends over all space, even if $\mathbf{J}$ is localized.*

---

*See O. L. Brill and B. Goodman, *Am. J. Phys.* **35**, 832 (1967) for a detailed discussion of causality in the Coulomb gauge. See also Problem 6.20.

6.4   *Green Functions for the Wave Equation*

The wave equations (6.15), (6.16), and (6.30) all have the basic structure

$$\nabla^2\Psi - \frac{1}{c^2}\frac{\partial^2\Psi}{\partial t^2} = -4\pi f(\mathbf{x}, t) \tag{6.32}$$

where $f(\mathbf{x}, t)$ is a known source distribution. The factor c is the velocity of propagation in the medium, assumed here to be without dispersion.

To solve (6.32) it is useful to find a Green function, just as in electrostatics.
We consider the simple situation of no boundary surfaces and proceed to remove
the explicit time dependence by introducing a Fourier transform with respect to
frequency. We suppose that $\Psi(\mathbf{x}, t)$ and $f(\mathbf{x}, t)$ have the Fourier integral
representations,

$$\left.\begin{aligned}\Psi(\mathbf{x}, t) &= \frac{1}{2\pi}\int_{-\infty}^{\infty}\Psi(\mathbf{x}, \omega)e^{-i\omega t}\,d\omega\\[2mm]f(\mathbf{x}, t) &= \frac{1}{2\pi}\int_{-\infty}^{\infty}f(\mathbf{x}, \omega)e^{-i\omega t}\,d\omega\end{aligned}\right\} \tag{6.33}$$

with the inverse transformations,

$$\left.\begin{aligned}\Psi(\mathbf{x}, \omega) &= \int_{-\infty}^{\infty}\Psi(\mathbf{x}, t)e^{i\omega t}\,dt\\[2mm]f(\mathbf{x}, \omega) &= \int_{-\infty}^{\infty}f(\mathbf{x}, t)e^{i\omega t}\,dt\end{aligned}\right\} \tag{6.34}$$

When the representations (6.33) are inserted into (6.32) it is found that the
Fourier transform $\Psi(\mathbf{x}, \omega)$ satisfies the *inhomogeneous Helmholtz wave equation*

$$(\nabla^2 + k^2)\Psi(\mathbf{x}, \omega) = -4\pi f(\mathbf{x}, \omega) \tag{6.35}$$

for each value of ω. Here $k = \omega/c$ is the wave number associated with frequency
ω. In this form, the restriction of no dispersion is unnecessary. A priori, any
connection between k and ω is allowed, although causality imposes some restrictions (see Section 7.10).

Equation (6.35) is an elliptic partial differential equation similar to the
Poisson equation to which it reduces for $k = 0$. The Green function $G(\mathbf{x}, \mathbf{x}')$
appropriate to (6.35) satisfies the inhomogeneous equation

$$(\nabla^2 + k^2)G_k(\mathbf{x}, \mathbf{x}') = -4\pi\delta(\mathbf{x} - \mathbf{x}') \tag{6.36}$$

If there are no boundary surfaces, the Green function can depend only on $\mathbf{R} =
\mathbf{x} - \mathbf{x}'$, and must in fact be spherically symmetric, that is, depend only on
$R = |\mathbf{R}|$. From the form of the Laplacian operator in spherical coordinates [see
(3.1)], it is evident that $G_k(R)$ satisfies

$$\frac{1}{R}\frac{d^2}{dR^2}(RG_k) + k^2 G_k = -4\pi\delta(\mathbf{R}) \tag{6.37}$$

Everywhere except $R = 0$, $RG_k(R)$ satisfies the homogeneous equation

$$\frac{d^2}{dR^2}(RG_k) + k^2(RG_k) = 0$$

with solution,

$$RG_k(R) = Ae^{ikR} + Be^{-ikR}$$

Furthermore, the delta function in (6.37) has influence only at $R \to 0$. In that limit the equation reduces to the Poisson equation, since $kR \ll 1$. We therefore know from electrostatics that the correct normalization is

$$\lim_{kR \to 0} G_k(R) = \frac{1}{R} \tag{6.38}$$

The general solution for the Green function is thus

$$G_k(R) = AG_k^{(+)}(R) + BG_k^{(-)}(R) \tag{6.39}$$

where

$$G_k^{(\pm)}(R) = \frac{e^{\pm ikR}}{R} \tag{6.40}$$

with $A + B = 1$. With the convention of (6.33) for the time dependence, the first term in (6.39) represents a diverging spherical wave propagating from the origin, while the second represents a converging spherical wave.

The choice of A and B in (6.39) depends on the *boundary conditions in time* that specify the physical problem. It is intuitively obvious that, if a source is quiescent until some time $t = 0$ and then begins to function, the appropriate Green function is the first term in (6.39), corresponding to waves radiated outward from the source after it begins to work. Such a description is certainly correct and also convenient, but is not unique or necessary. By suitable specification of the wave amplitude at boundary times, it is possible to employ the second term in (6.39), not the first, to describe the action of the source.

To understand the different time behaviors associated with $G_k^{(+)}$ and $G_k^{(-)}$ we need to construct the corresponding time-dependent Green functions that satisfy

$$\left(\nabla_x^2 - \frac{1}{c^2}\frac{\partial^2}{\partial t^2}\right)G^{(\pm)}(\mathbf{x}, t; \mathbf{x}', t') = -4\pi\delta(\mathbf{x} - \mathbf{x}')\delta(t - t') \tag{6.41}$$

Using (6.34) we see that the source term for (6.35) is

$$-4\pi\delta(\mathbf{x} - \mathbf{x}')e^{i\omega t'}$$

The solutions are therefore $G_k^{(\pm)}(R)e^{i\omega t'}$. From (6.33) the time-dependent Green functions are

$$G^{(\pm)}(R, \tau) = \frac{1}{2\pi}\int_{-\infty}^{\infty}\frac{e^{\pm ikR}}{R} \cdot e^{-i\omega\tau} \, d\omega \tag{6.42}$$

where $\tau = t - t'$ is the relative time appearing in (6.41). The infinite-space Green function is thus a function of only the relative distance R and the relative time

τ between source and observation point. For a nondispersive medium where $k = \omega/c$, the integral in (6.42) is a delta function. The Green functions are

$$G^{(\pm)}(R, \tau) = \frac{1}{R}\, \delta\!\left(\tau \mp \frac{R}{c} \right) \tag{6.43}$$

or, more explicitly,

$$G^{(\pm)}(\mathbf{x}, t; \mathbf{x}', t') = \frac{\delta\!\left(t' - \left[t \mp \frac{|\mathbf{x} - \mathbf{x}'|}{c} \right] \right)}{|\mathbf{x} - \mathbf{x}'|} \tag{6.44}$$

The Green function $G^{(+)}$ is called the *retarded Green function* because it exhibits the causal behavior associated with a wave disturbance. The argument of the delta function shows that an effect observed at the point $\mathbf{x}$ at time t is caused by the action of a source a distance R away at an earlier or retarded time, $t' = t - R/c$. The time difference R/c is just the time of propagation of the disturbance from one point to the other. Similarly, $G^{(-)}$ is called the *advanced Green function*.

Particular integrals of the inhomogeneous wave equation (6.32) are

$$\Psi^{(\pm)}(\mathbf{x}, t) = \int\int G^{(\pm)}(\mathbf{x}, t; \mathbf{x}', t')f(\mathbf{x}', t')\, d^3x'\, dt'$$

To specify a definite physical problem, solutions of the homogeneous equation may be added to either of these. We consider a source distribution $f(\mathbf{x}', t')$ that is localized in time and space. It is different from zero only for a finite interval of time around $t' = 0$. Two limiting situations are envisioned. In the first it is assumed that at time $t \to -\infty$ there exists a wave $\Psi_{\text{in}}(\mathbf{x}, t)$ that satisfies the homogeneous wave equation. This wave propagates in time and space; the source turns on and generates waves of its own. The complete solution for this situation at all times is evidently

$$\Psi(\mathbf{x}, t) = \Psi_{\text{in}}(\mathbf{x}, t) + \int\int G^{(+)}(\mathbf{x}, t; \mathbf{x}', t')f(\mathbf{x}', t')\, d^3x'\, dt' \tag{6.45}$$

The presence of $G^{(+)}$ guarantees that at remotely early times, t, before the source has been activated, there is no contribution from the integral. Only the specified wave Ψ_{in} exists. The second situation is that at remotely late times ($t \to +\infty$) the wave is given as $\Psi_{\text{out}}(\mathbf{x}, t)$, a known solution of the homogeneous wave equation. Then the complete solution for all times is

$$\Psi(\mathbf{x}, t) = \Psi_{\text{out}}(\mathbf{x}, t) + \int\int G^{(-)}(\mathbf{x}, t; \mathbf{x}', t')f(\mathbf{x}', t')\, d^3x'\, dt' \tag{6.46}$$

Now the advanced Green function assures that no signal from the source shall exist explicitly after the source shuts off (all such signals are by assumption included in Ψ_{out}).

The commonest physical situation is described by (6.45) with $\Psi_{\text{in}} = 0$. It is sometimes written with the Green function (6.44) inserted explicitly:

$$\Psi(\mathbf{x}, t) = \int \frac{[f(\mathbf{x}', t')]_{\text{ret}}}{|\mathbf{x} - \mathbf{x}'|}\, d^3x' \tag{6.47}$$

The square bracket [ ]$_{\text{ret}}$ means that the time t' is to be evaluated at the retarded time, $t' = t - |\mathbf{x} - \mathbf{x}'|/c$.

The initial or final value problem at *finite* times has been extensively studied in one, two, and three dimensions. The reader may refer to *Morse and Feshbach* (pp. 843–847) and also to the more mathematical treatment of *Hadamard*.

6.5 *Retarded Solutions for the Fields: Jefimenko's Generalizations of the Coulomb and Biot–Savart Laws; Heaviside–Feynman Expressions for Fields of Point Charge*

Use of the retarded solution (6.47) for the wave equations (6.15) and (6.16) yields

$$\Phi(\mathbf{x}, t) = \frac{1}{4\pi\epsilon_0} \int d^3x' \frac{1}{R} [\rho(\mathbf{x}', t')]_{\text{ret}}$$

$$\mathbf{A}(\mathbf{x}, t) = \frac{\mu_0}{4\pi} \int d^3x' \frac{1}{R} [\mathbf{J}(\mathbf{x}', t')]_{\text{ret}}$$

(6.48)

where we have defined $\mathbf{R} = \mathbf{x} - \mathbf{x}'$, with $R = |\mathbf{x} - \mathbf{x}'|$ and (below) $\hat{\mathbf{R}} = \mathbf{R}/R$. These solutions were first given by Lorenz (op. cit.). In principle, from these two equations the electric and magnetic fields can be computed, but it is often useful to have retarded integral solutions for the fields in terms of the sources.

Either directly from the Maxwell equations or by use of the wave equations for Φ and $\mathbf{A}$, (6.15) and (6.16), and the definitions of the fields in terms of the potentials, (6.7) and (6.9), we can arrive at wave equations for the fields in free space with given charge and current densities,

$$\nabla^2\mathbf{E} - \frac{1}{c^2}\frac{\partial^2\mathbf{E}}{\partial t^2} = -\frac{1}{\epsilon_0}\left(-\nabla\rho - \frac{1}{c^2}\frac{\partial \mathbf{J}}{\partial t}\right)$$

(6.49)

and

$$\nabla^2\mathbf{B} - \frac{1}{c^2}\frac{\partial^2\mathbf{B}}{\partial t^2} = -\mu_0\nabla \times \mathbf{J}$$

(6.50)

The wave equation for each of the Cartesian field components is in the form (6.32). The retarded solutions (6.47) for the fields can immediately be written in the preliminary forms

$$\mathbf{E}(\mathbf{x}, t) = \frac{1}{4\pi\epsilon_0} \int d^3x' \frac{1}{R}\left[-\nabla'\rho - \frac{1}{c^2}\frac{\partial \mathbf{J}}{\partial t'}\right]_{\text{ret}}$$

(6.51)

and

$$\mathbf{B}(\mathbf{x}, t) = \frac{\mu_0}{4\pi} \int d^3x' \frac{1}{R} [\nabla' \times \mathbf{J}]_{\text{ret}}$$

(6.52)

These preliminary expressions can be cast into forms showing explicitly the static limits and the corrections to them by extracting the spatial partial derivatives from the retarded integrands. There is a subtlety here because $\nabla'[f]_{\text{ret}} \neq [\nabla'f]_{\text{ret}}$. The meaning of ∇' under the retarded bracket is a spatial gradient in $\mathbf{x}'$, with t' fixed; the meaning outside the retarded bracket is a spatial gradient with respect to $\mathbf{x}'$, with $\mathbf{x}$ and t fixed. Since $[f(\mathbf{x}', t')]_{\text{ret}} = f(\mathbf{x}', t - R/c)$, it is necessary to correct for the $\mathbf{x}'$ dependence introduced through R when the gradient operator is taken outside. Explicitly, we have

$$[\nabla'\rho]_{\text{ret}} = \nabla'[\rho]_{\text{ret}} - \left[\frac{\partial\rho}{\partial t'}\right]_{\text{ret}} \nabla'(t - R/c) = \nabla'[\rho]_{\text{ret}} - \frac{\hat{\mathbf{R}}}{c}\left[\frac{\partial\rho}{\partial t'}\right]_{\text{ret}} \tag{6.53}$$

and

$$[\nabla' \times \mathbf{J}]_{\text{ret}} = \nabla' \times [\mathbf{J}]_{\text{ret}} + \left[\frac{\partial\mathbf{J}}{\partial t'}\right]_{\text{ret}} \times \nabla'(t - R/c)$$

$$= \nabla' \times [\mathbf{J}]_{\text{ret}} + \frac{1}{c}\left[\frac{\partial\mathbf{J}}{\partial t'}\right]_{\text{ret}} \times \hat{\mathbf{R}} \tag{6.54}$$

If these expressions are substituted into the preliminary forms of the solutions and an integration by parts is performed on the first (gradient or curl) term in each case, we arrive at

$$\mathbf{E}(\mathbf{x}, t) = \frac{1}{4\pi\epsilon_0}\int d^3x' \left\{\frac{\hat{\mathbf{R}}}{R^2}[\rho(\mathbf{x}', t')]_{\text{ret}} + \frac{\hat{\mathbf{R}}}{cR}\left[\frac{\partial\rho(\mathbf{x}', t')}{\partial t'}\right]_{\text{ret}} - \frac{1}{c^2R}\left[\frac{\partial\mathbf{J}(\mathbf{x}', t')}{\partial t'}\right]_{\text{ret}}\right\} \tag{6.55}$$

and

$$\mathbf{B}(\mathbf{x}, t) = \frac{\mu_0}{4\pi}\int d^3x' \left\{[\mathbf{J}(\mathbf{x}', t')]_{\text{ret}} \times \frac{\hat{\mathbf{R}}}{R^2} + \left[\frac{\partial\mathbf{J}(\mathbf{x}', t')}{\partial t'}\right]_{\text{ret}} \times \frac{\hat{\mathbf{R}}}{cR}\right\} \tag{6.56}$$

If the charge and current densities are time independent, the expressions reduce to the familiar static expressions (1.5) and (5.14). The terms involving the time derivatives *and* the retardation provide the generalizations to time-dependent sources. These two results, sometimes known as Jefimenko's generalizations of the Coulomb and Biot–Savart laws, were popularized in this author's text, (*Jefimenko*).

In passing, we note that because the integrands are to be viewed as functions of $\mathbf{x}$, $\mathbf{x}'$, and t, with $t' = t - |\mathbf{x} - \mathbf{x}'|/c$, the time derivatives in the integrands have the property

$$\left[\frac{\partial f(\mathbf{x}', t')}{\partial t'}\right]_{\text{ret}} = \frac{\partial}{\partial t}[f(\mathbf{x}', t')]_{\text{ret}} \tag{6.57}$$

This relation facilitates the specialization of the Jefimenko formulas to the Heaviside–Feynman expressions for the fields of a point charge. With $\rho(\mathbf{x}', t') = q\delta[\mathbf{x}' - \mathbf{r}_0(t')]$ and $\mathbf{J}(\mathbf{x}', t') = \rho\mathbf{v}(t')$, (6.55) and (6.56) specialize to

$$\mathbf{E} = \frac{q}{4\pi\epsilon_0}\left\{\left[\frac{\hat{\mathbf{R}}}{\kappa R^2}\right]_{\text{ret}} + \frac{\partial}{c\partial t}\left[\frac{\hat{\mathbf{R}}}{\kappa R}\right]_{\text{ret}} - \frac{\partial}{c^2\partial t}\left[\frac{\mathbf{v}}{\kappa R}\right]_{\text{ret}}\right\} \tag{6.58}$$

and

$$\mathbf{B} = \frac{\mu_0 q}{4\pi}\left\{\left[\frac{\mathbf{v} \times \hat{\mathbf{R}}}{\kappa R^2}\right]_{\text{ret}} + \frac{\partial}{c\partial t}\left[\frac{\mathbf{v} \times \hat{\mathbf{R}}}{\kappa R}\right]_{\text{ret}}\right\} \tag{6.59}$$

Here R is the distance from the position of the charge to the observation point; $\hat{\mathbf{R}}$ is a unit vector from the charge toward the observation point; $\mathbf{v}$ is the charge's velocity; $\kappa = 1 - \mathbf{v} \cdot \hat{\mathbf{R}}/c$ is a retardation factor. [See Problem 6.2.] It is important to note that now there is a difference between $\partial[\cdots]_{\text{ret}}/\partial t$ and $[\partial \cdots /\partial t]_{\text{ret}}$ because $\mathbf{x}' \to \mathbf{r}_0(t')$, where $\mathbf{r}_0$ is the position of the charge. The fields are functions of $\mathbf{x}$ and t, with $t' = t - |\mathbf{x} - \mathbf{r}_0(t')|/c$. Feynman's expression for the electric field is

$$\mathbf{E} = \frac{q}{4\pi\epsilon_0} \left\{ \left[\frac{\hat{\mathbf{R}}}{R^2} \right]_{\text{ret}} + \frac{[R]_{\text{ret}}}{c} \frac{\partial}{\partial t} \left[\frac{\hat{\mathbf{R}}}{R^2} \right]_{\text{ret}} + \frac{\partial^2}{c^2 \partial t^2} [\hat{\mathbf{R}}]_{\text{ret}} \right\} \quad (6.60)$$

while Heaviside's expression for the magnetic field is

$$\mathbf{B} = \frac{\mu_0 q}{4\pi} \left\{ \left[\frac{\mathbf{v} \times \hat{\mathbf{R}}}{\kappa^2 R^2} \right]_{\text{ret}} + \frac{1}{c[R]_{\text{ret}}} \frac{\partial}{\partial t} \left[\frac{\mathbf{v} \times \hat{\mathbf{R}}}{\kappa} \right]_{\text{ret}} \right\} \quad (6.61)$$

The equivalence of the two sets of expressions for the fields follows from some careful algebra.

6.6 Derivation of the Equations of Macroscopic Electromagnetism

The discussion of electromagnetism in the preceding chapters has been based on the *macroscopic* Maxwell equations,

$$\nabla \cdot \mathbf{B} = 0 \qquad \nabla \times \mathbf{E} + \frac{\partial \mathbf{B}}{\partial t} = 0$$

$$\nabla \cdot \mathbf{D} = \rho \qquad \nabla \times \mathbf{H} - \frac{\partial \mathbf{D}}{\partial t} = \mathbf{J} \quad (6.62)$$

where $\mathbf{E}$ and $\mathbf{B}$ are the macroscopic electric and magnetic field quantities, $\mathbf{D}$ and $\mathbf{H}$ are corresponding derived fields, related to $\mathbf{E}$ and $\mathbf{B}$ through the polarization $\mathbf{P}$ and the magnetization $\mathbf{M}$ of the material medium by

$$\mathbf{D} = \epsilon_0 \mathbf{E} + \mathbf{P}, \qquad \mathbf{H} = \frac{1}{\mu_0} \mathbf{B} - \mathbf{M} \quad (6.63)$$

Similarly, ρ and $\mathbf{J}$ are the macroscopic (free) charge density and current density, respectively. Although these equations are familiar and totally acceptable, we have yet to present a serious derivation of them from a microscopic starting point. This deficiency is remedied in the present section. The derivation remains within a classical framework even though atoms must be described quantum mechanically. The excuse for this apparent inadequacy is that the quantum-mechanical discussion closely parallels the classical one, with quantum-mechanical expectation values replacing the classical quantities in the formulas given below. The reader can examine the statistical mechanical treatments in the literature cited at the end of the chapter.

We consider a microscopic world made up of electrons and nuclei. For dimensions large compared to 10^{-14} m, the nuclei can be treated as point systems, as can the electrons. We assume that the equations governing electromagnetic phenomena for these point charges are the *microscopic* Maxwell equations,

$$\nabla \cdot \mathbf{b} = 0, \qquad \nabla \times \mathbf{e} + \frac{\partial \mathbf{b}}{\partial t} = 0$$

$$\nabla \cdot \mathbf{e} = \eta/\epsilon_0, \qquad \nabla \times \mathbf{b} - \frac{1}{c^2} \frac{\partial \mathbf{e}}{\partial t} = \mu_0 \mathbf{j} \quad (6.64)$$

where $\mathbf{e}$ and $\mathbf{b}$ are the microscopic electric and magnetic fields and η and $\mathbf{j}$ are the microscopic charge and current densities. There are no corresponding fields $\mathbf{d}$ and $\mathbf{h}$ because all the charges are included in η and $\mathbf{j}$. A macroscopic amount of matter at rest contains of the order of $10^{23\pm5}$ electrons and nuclei, all in incessant motion because of thermal agitation, zero point vibration, or orbital motion. The microscopic electromagnetic fields produced by these charges vary extremely rapidly in space and in time. The spatial variations occur over distances of the order of 10^{-10} m or less, and the temporal fluctuations occur with periods ranging from 10^{-13} s for nuclear vibrations to 10^{-17} s for electronic orbital motion. Macroscopic measuring devices generally average over intervals in space and time much larger than these. All the microscopic fluctuations are therefore averaged out, giving relatively smooth and slowly varying macroscopic quantities, such as appear in the macroscopic Maxwell equations.

The question of what type of averaging is appropriate must be examined with some care. At first glance one might think that averages over both space and time are necessary. But this is not true. Only a spatial averaging is necessary. (Parenthetically, we note that a time averaging alone would certainly not be sufficient, as can be seen by considering an ionic crystal whose ions have small zero point vibrations around well-defined and separated lattice sites.) To delimit the domain where we expect a macroscopic description of electromagnetic phenomena to work, we observe that the reflection and refraction of visible light are adequately described by the Maxwell equations with a continuous dielectric constant, whereas x-ray diffraction clearly exposes the atomistic nature of matter. It is plausible therefore to take the length $L_0 = 10^{-8}$ m $= 10^2$ Å as the absolute lower limit to the macroscopic domain. The period of oscillation associated with light of this wavelength is $L_0/c \simeq 3 \times 10^{-17}$ s. In a volume of $L_0^3 = 10^{-24}$ m^3 there are, in ordinary matter, still of the order of 10^6 nuclei and electrons. Thus in any region of macroscopic interest with $L \gg L_0$ there are so many nuclei and electrons that the fluctuations will be completely washed out by a spatial averaging. On the other hand, because the time scale associated with L is actually in the range of atomic and molecular motions, a time-averaging would not be appropriate. There is, nevertheless, no evidence after the spatial averaging of the microscopic time fluctuations of the medium. This is so because, in the absence of special preparation and the establishment of ordering over macroscopic distances, the time variations of the microscopic fields are uncorrelated over distances of order L. All that survive are the frequency components corresponding to oscillators driven at the external, applied frequencies.

The spatial average of a function $F(\mathbf{x}, t)$ with respect to a test function $f(\mathbf{x})$ is defined as

$$\langle F(\mathbf{x}, t) \rangle = \int d^3x' \, f(\mathbf{x}')F(\mathbf{x} - \mathbf{x}', t) \tag{6.65}$$

where $f(\mathbf{x})$ is real, nonzero in some neighborhood of $\mathbf{x} = 0$, and normalized to unity over all space. It is simplest, though not necessary, to imagine $f(\mathbf{x})$ to be nonnegative. To preserve without bias directional characteristics of averaged physical properties, we make $f(\mathbf{x})$ isotropic in space. Two examples are

$$f(\mathbf{x}) = \begin{cases} \dfrac{3}{4\pi R^3}, & r < R \\ 0, & r > R \end{cases}$$

$$f(\mathbf{x}) = (\pi R^2)^{-3/2}e^{-r^2/R^2} \tag{6.66}$$

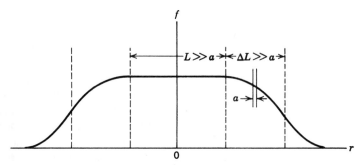

Figure 6.1 Schematic diagram of test function $f(\mathbf{x})$ used in the spatial averaging procedure. The extent L of the plateau region, and also the extent ΔL of the region where f falls to zero, are both large compared to the molecular dimension a.

The first example, a spherical averaging volume of radius R, is a common one in the literature. It has the advantage of conceptual simplicity, but the disadvantage of an abrupt discontinuity at $r = R$. This leads to a fine-scale jitter on the averaged quantities as a single molecule or group of molecules moves in or out of the averaging volume. A smooth test function, exemplified by the Gaussian, eliminates such difficulties provided its scale is large compared to atomic dimensions. Fortunately, the test function $f(\mathbf{x})$ does not need to be specified in detail; all that are needed are general continuity and smoothness properties that permit a rapidly converging Taylor series expansion of $f(\mathbf{x})$ over distances of atomic dimensions, as indicated schematically in Fig. 6.1. This is a great virtue.*

Since space and time derivatives enter the Maxwell equations, we must consider these operations with respect to averaging according to (6.65). Evidently, we have

$$\frac{\partial}{\partial x_i} \langle F(\mathbf{x}, t) \rangle = \int d^3x' \, f(\mathbf{x}') \frac{\partial F}{\partial x_i} (\mathbf{x} - \mathbf{x}', t) = \left\langle \frac{\partial F}{\partial x_i} \right\rangle$$

and (6.67)

$$\frac{\partial}{\partial t} \langle F(\mathbf{x}, t) \rangle = \left\langle \frac{\partial F}{\partial t} \right\rangle$$

The operations of space and time differentiation thus commute with the averaging operation.

We can now consider the averaging of the microscopic Maxwell equations (6.64). The *macroscopic* electric and magnetic field quantities $\mathbf{E}$ and $\mathbf{B}$ are defined as the averages of the microscopic fields $\mathbf{e}$ and $\mathbf{b}$:

$$\mathbf{E}(\mathbf{x}, t) = \langle \mathbf{e}(\mathbf{x}, t) \rangle \tag{6.68}$$
$$\mathbf{B}(\mathbf{x}, t) = \langle \mathbf{b}(\mathbf{x}, t) \rangle$$

Then the averages of the two homogeneous equations in (6.64) become the corresponding macroscopic equations,

$$\langle \boldsymbol{\nabla} \cdot \mathbf{b} \rangle = 0 \rightarrow \boldsymbol{\nabla} \cdot \mathbf{B} = 0 \tag{6.69}$$
$$\left\langle \boldsymbol{\nabla} \times \mathbf{e} + \frac{\partial \mathbf{b}}{\partial t} \right\rangle = 0 \rightarrow \boldsymbol{\nabla} \times \mathbf{E} + \frac{\partial \mathbf{B}}{\partial t} = 0$$

*We are here following the development of G. Russakoff, *Am. J. Physics* **38**, 1188 (1970).

The averaged *in*homogeneous equations from (6.64) become

$$\epsilon_0 \nabla \cdot \mathbf{E} = \langle \eta(\mathbf{x}, t) \rangle \tag{6.70}$$

$$\frac{1}{\mu_0} \nabla \times \mathbf{B} - \epsilon_0 \frac{\partial \mathbf{E}}{\partial t} = \langle \mathbf{j}(\mathbf{x}, t) \rangle$$

Comparison with the inhomogeneous pair of macroscopic equations in (6.62) indicates the already known fact that the derived fields $\mathbf{D}$ and $\mathbf{H}$ are introduced by the extraction from $\langle \eta \rangle$ and $\langle \mathbf{j} \rangle$ of certain contributions that can be identified with the bulk properties of the medium. The examination of $\langle \eta \rangle$ and $\langle \mathbf{j} \rangle$ is therefore the next task.

We consider a medium made up of molecules composed of nuclei and electrons and, in addition, "free" charges that are not localized around any particular molecule. The microscopic charge density can be written as

$$\eta(\mathbf{x}, t) = \sum_j q_j \delta[\mathbf{x} - \mathbf{x}_j(t)] \tag{6.71}$$

where $\mathbf{x}_j(t)$ is the position of the point charge q_j. To distinguish the bound charges from the free ones, we decompose η as

$$\eta = \eta_{\text{free}} + \eta_{\text{bound}} \tag{6.72}$$

and write

$$\eta_{\text{free}} = \sum_{j(\text{free})} q_j \, \delta(\mathbf{x} - \mathbf{x}_j)$$

$$\eta_{\text{bound}} = \sum_{\substack{n \\ (\text{molecules})}} \eta_n(\mathbf{x}, t)$$

where η_n is the charge density of the nth molecule,

$$\eta_n(\mathbf{x}, t) = \sum_{j(n)} q_j \, \delta(\mathbf{x} - \mathbf{x}_j) \tag{6.73}$$

In these and subsequent equations we suppress the explicit time dependence, since the averaging is done at one instant of time. We proceed by averaging the charge density of the nth molecule and then summing up the contributions of all molecules. It is appropriate to express the coordinates of the charges in the nth molecule with respect to an origin at rest in the molecule. Let the coordinate of that fixed point in the molecule (usually chosen as the center of mass) be $\mathbf{x}_n(t)$, and the coordinate of the jth charge in the molecule be $\mathbf{x}_{jn}(t)$ relative to that origin, as indicated in Fig. 6.2. The average of the charge density of the nth molecule is

$$\langle \eta_n(\mathbf{x}, t) \rangle = \int d^3 x' \, f(\mathbf{x}') \, \eta_n(\mathbf{x} - \mathbf{x}', t)$$

$$= \sum_{j(n)} q_j \int d^3 x' \, f(\mathbf{x}') \, \delta(\mathbf{x} - \mathbf{x}' - \mathbf{x}_{jn} - \mathbf{x}_n) \tag{6.74}$$

$$= \sum_{j(n)} q_j \, f(\mathbf{x} - \mathbf{x}_n - \mathbf{x}_{jn})$$

Since $\mathbf{x}_{jn}$ is of order atomic dimensions, the terms in the sum have arguments differing only slightly from $(\mathbf{x} - \mathbf{x}_n)$ on the scale over which $f(\mathbf{x})$ changes appre-

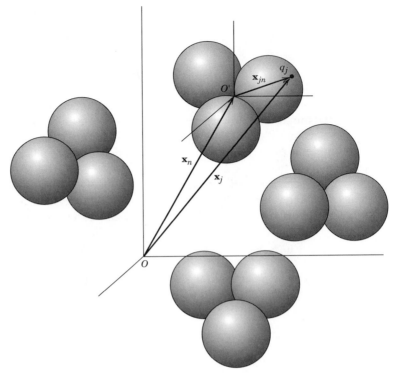

Figure 6.2 Coordinates for the nth molecule. The origin O' is fixed in the molecule (usually it is chosen at the center of mass). The jth charge has coordinate $\mathbf{x}_{jn}$ relative to O', while the molecule is located relative to the fixed (laboratory) axes by the coordinate $\mathbf{x}_n$.

ciably. It is therefore appropriate to make a Taylor series expansion around $(\mathbf{x} - \mathbf{x}_n)$ for each term. This gives

$$\langle \eta_n(\mathbf{x}, t) \rangle = \sum_{j(n)} q_j \left[f(\mathbf{x} - \mathbf{x}_n) - \mathbf{x}_{jn} \cdot \boldsymbol{\nabla} f(\mathbf{x} - \mathbf{x}_n) \right.$$
$$\left. + \frac{1}{2} \sum_{\alpha\beta} (\mathbf{x}_{jn})_\alpha (\mathbf{x}_{jn})_\beta \frac{\partial^2}{\partial x_\alpha \, \partial x_\beta} f(\mathbf{x} - \mathbf{x}_n) + \cdots \right]$$

The various sums over the charges in the molecule are just *molecular multipole moments*:

MOLECULAR CHARGE

$$q_n = \sum_{j(n)} q_j \tag{6.75}$$

MOLECULAR DIPOLE MOMENT

$$\mathbf{p}_n = \sum_{j(n)} q_j \, \mathbf{x}_{jn} \tag{6.76}$$

MOLECULAR QUADRUPOLE MOMENT

$$(Q'_n)_{\alpha\beta} = 3 \sum_{j(n)} q_j (\mathbf{x}_{jn})_\alpha (\mathbf{x}_{jn})_\beta \tag{6.77}$$

In terms of these multipole moments the averaged charge density of the nth molecule is

$$\langle \eta_n(\mathbf{x}, t) \rangle = q_n f(\mathbf{x} - \mathbf{x}_n) - \mathbf{p}_n \cdot \nabla f(\mathbf{x} - \mathbf{x}_n) \tag{6.78}$$
$$+ \frac{1}{6} \sum_{\alpha\beta} (Q'_n)_{\alpha\beta} \frac{\partial^2 f(\mathbf{x} - \mathbf{x}_n)}{\partial x_\alpha \, \partial x_\beta} + \cdots$$

If we attempt to view this equation as the direct result of the definition (6.65) of the spatial averaging, we see that the first term can be thought of as the averaging of a point charge density at $\mathbf{x} = \mathbf{x}_n$, the second as the divergence of the average of a point dipole density at $\mathbf{x} = \mathbf{x}_n$, and so on. Explicitly,

$$\langle \eta_n(\mathbf{x}, t) \rangle = \langle q_n \delta(\mathbf{x} - \mathbf{x}_n) \rangle - \nabla \cdot \langle \mathbf{p}_n \delta(\mathbf{x} - \mathbf{x}_n) \rangle \tag{6.79}$$
$$+ \frac{1}{6} \sum_{\alpha\beta} \frac{\partial^2}{\partial x_\alpha \, \partial x_\beta} \langle (Q'_n)_{\alpha\beta} \delta(\mathbf{x} - \mathbf{x}_n) \rangle + \cdots$$

We thus find that, as far as the result of the averaging process is concerned, *we can view the molecule as a collection of point multipoles* located at one fixed point in the molecule. The detailed extent of the molecular charge distribution is important at the microscopic level, of course, but is replaced in its effect by a sum of multipoles for macroscopic phenomena.

An alternative approach to the spatial averaging of (6.65) via Fourier transforms gives a valuable different perspective. With the spatial Fourier transforms defined by

$$g(\mathbf{x}, t) = \frac{1}{(2\pi)^3} \int d^3 k \, \tilde{g}(\mathbf{k}, t) e^{i\mathbf{k} \cdot \mathbf{x}} \quad \text{and} \quad \tilde{g}(\mathbf{k}, t) = \int d^3 x \, g(\mathbf{x}, t) e^{-i\mathbf{k} \cdot \mathbf{x}} \tag{6.80}$$

straightforward substitution into (6.65) leads to the expression for the average of $F(\mathbf{x}, t)$,

$$\langle F(\mathbf{x}, t) \rangle = \frac{1}{(2\pi)^3} \int d^3 k \, \tilde{f}(\mathbf{k}, t) \tilde{F}(\mathbf{k}, t) e^{i\mathbf{k} \cdot \mathbf{x}} \tag{6.81}$$

an illustration of the "faltung theorem" of Fourier transforms. The convolution of (6.65) has a Fourier transform that is the product of the transforms of the separate functions in the convolution. Thus

$$\text{FT} \, \langle F(\mathbf{x}, t) \rangle = \tilde{f}(\mathbf{k}) \tilde{F}(\mathbf{k}, t) \tag{6.82}$$

The notation FT is introduced to stand for the kernel multiplying the exponential in the first integral above [FT $g(\mathbf{x}, t) \equiv \tilde{g}(\mathbf{k}, t)$] to avoid a clumsy and confusing use of the tilde.

A crucial aspect of $\tilde{f}(\mathbf{k})$ is that $\tilde{f}(0) = 1$, as can be seen from its definition and from the fact that $f(\mathbf{x})$ is normalized to unity. For the Gaussian test function, the Fourier transform is

$$\text{FT} \, f(\mathbf{x}) = \tilde{f}(\mathbf{k}) = e^{-k^2 R^2 / 4} \tag{6.83}$$

Evidently the Fourier transform (6.82) of the averaged quantity contains only low wave numbers, up to but not significantly beyond $k_{\max} = O(1/R)$, the inverse of the length scale of the averaging volume. But because $\tilde{f}(\mathbf{k}) \to 1$ for wave numbers small compared to the cutoff, the FT $\langle F(\mathbf{x}, t) \rangle$ gives a true representation of the long-wavelength aspects of $F(\mathbf{x}, t)$. Only the small-scale (large wave number) aspects are removed, as expected for the averaging.

Consider the averaging of the charge density of the nth molecule shown in Fig. 6.2. The Fourier transform of the averaged quantity is

$$\text{FT} \langle \eta_n(\mathbf{x}, t) \rangle = \tilde{f}(\mathbf{k}) \tilde{\eta}_n(\mathbf{k}, t) \tag{6.84}$$

where

$$\tilde{\eta}_n(\mathbf{k}, t) = \int d^3x' \; \eta_n(\mathbf{x}', t) e^{-i\mathbf{k} \cdot (\mathbf{x}' - \mathbf{x}_n)}$$

Here we have taken the spatial Fourier transform relative to $\mathbf{x}_n$. The qualitative behaviors of the two factors in (6.84) are sketched in Fig. 6.3. Since the support for the product is confined to comparatively small wave numbers, it is appropriate to make a Taylor series expansion of the Fourier transform $\tilde{\eta}_n(\mathbf{k}, t)$ for small $|\mathbf{k}|$,

$$\tilde{\eta}_n(\mathbf{k}, t) \approx \tilde{\eta}_n(0, t) + \mathbf{k} \cdot \nabla_k \tilde{\eta}_n(0, t) + \cdots$$

Explicitly, we have

$$\tilde{\eta}_n(\mathbf{k}, t) = \int d^3x' \; \eta_n(\mathbf{x}', t)[1 - i\mathbf{k} \cdot (\mathbf{x} - \mathbf{x}_n) + \cdots]$$

or

$$\tilde{\eta}_n(\mathbf{k}, t) \approx q_n - i\mathbf{k} \cdot \mathbf{p}_n + \text{quadrupole and higher} \tag{6.85}$$

in terms of the molecule's multipole moments. The averaged molecular charge density can therefore be written as

$$\begin{aligned} \langle \eta_n(\mathbf{x}, t) \rangle &= \frac{1}{(2\pi)^3} \int d^3k \; e^{i\mathbf{k} \cdot (\mathbf{x} - \mathbf{x}_n)} \tilde{f}(\mathbf{k})[q_n - i\mathbf{k} \cdot \mathbf{p}_n + \cdots] \\ &= q_n f(\mathbf{x} - \mathbf{x}_n) - \mathbf{p}_n \cdot \nabla f(\mathbf{x} - \mathbf{x}_n) + \cdots \end{aligned} \tag{6.86}$$

We have arrived at (6.78) by a different and perhaps longer route, but one with the advantage of giving a complementary view of the averaging as a cutoff in wave number space, a point of view stressed by *Robinson*.

The total microscopic charge density (6.72) consists of the free and bound

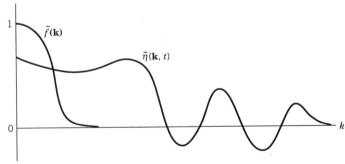

Figure 6.3 Qualitative behavior of the Fourier transforms in (6.84) for the transform of the averaged molecular charge density $\langle \eta_n(\mathbf{x}, t) \rangle$.

charges. Summing up over all the molecules (which may be of different species) and combining with the free charges, we find the *averaged* microscopic charge density to be

$$\langle \eta(\mathbf{x}, t) \rangle = \rho(\mathbf{x}, t) - \nabla \cdot \mathbf{P}(\mathbf{x}, t) + \sum_{\alpha\beta} \frac{\partial^2}{\partial x_\alpha \, \partial x_\beta} Q'_{\alpha\beta}(\mathbf{x}, t) + \cdots \quad (6.87)$$

where ρ is the *macroscopic charge density*,

$$\rho(\mathbf{x}, t) = \left\langle \sum_{j(\text{free})} q_j \delta(\mathbf{x} - \mathbf{x}_j) + \sum_{\substack{n \\ (\text{molecules})}} q_n \delta(\mathbf{x} - \mathbf{x}_n) \right\rangle \quad (6.88)$$

P is the *macroscopic polarization*,

$$\mathbf{P}(\mathbf{x}, t) = \left\langle \sum_{\substack{n \\ (\text{molecules})}} \mathbf{p}_n \delta(\mathbf{x} - \mathbf{x}_n) \right\rangle \quad (6.89)$$

and $Q'_{\alpha\beta}$ is the *macroscopic quadrupole density*,

$$Q'_{\alpha\beta}(\mathbf{x}, t) = \frac{1}{6} \left\langle \sum_{\substack{n \\ (\text{molecules})}} (Q'_n)_{\alpha\beta} \delta(\mathbf{x} - \mathbf{x}_n) \right\rangle \quad (6.90)$$

When (6.87) is inserted in the first equation of (6.70), it gives

$$\sum_\alpha \frac{\partial}{\partial x_\alpha} \left[\epsilon_0 E_\alpha + P_\alpha - \sum_\beta \frac{\partial}{\partial x_\beta} Q'_{\alpha\beta} + \cdots \right] = \rho \quad (6.91)$$

From (6.62) this means that the *macroscopic displacement vector* **D** is defined to have components,

$$D_\alpha = \epsilon_0 E_\alpha + P_\alpha - \sum_\beta \frac{\partial Q'_{\alpha\beta}}{\partial x_\beta} + \cdots \quad (6.92)$$

The first two terms are the familiar result (6.63). The third and higher terms are present in principle, but are almost invariably negligible.

To complete the discussion we must consider $\langle \mathbf{j} \rangle$. Because of its vector nature and the presence of velocities the derivation is considerably more complicated than the earlier treatment of $\langle \eta \rangle$, even though no new principles are involved. We present only the results, leaving the gory details to a problem for those readers who enjoy such challenges. We begin with the microscopic current density,

$$\mathbf{j}(\mathbf{x}, t) = \sum_j q_j \mathbf{v}_j \delta(\mathbf{x} - \mathbf{x}_j(t)) \quad (6.93)$$

where $\mathbf{v}_j = d\mathbf{x}_j/dt$ is the velocity of the jth charge. Again the sum is divided into one over the free charges and one over the molecules. The current density of the nth molecule can be averaged just as in (6.74) to give

$$\langle \mathbf{j}_n(\mathbf{x}, t) \rangle = \sum_{j(n)} q_j (\mathbf{v}_{jn} + \mathbf{v}_n) f(\mathbf{x} - \mathbf{x}_n - \mathbf{x}_{jn}) \quad (6.94)$$

Here we have assumed nonrelativistic motion by writing the velocity of the jth charge as the sum of an internal relative velocity $\mathbf{v}_{jn}$ and the velocity $\mathbf{v}_n = d\mathbf{x}_n/dt$ of the origin O' in the molecule. From this point on the development entails

Taylor series expansions and vector manipulations. A portion of the current involves the *molecular magnetic moment*,

$$\mathbf{m}_n = \sum_{j(n)} \frac{q_j}{2} (\mathbf{x}_{jn} \times \mathbf{v}_{jn}) \tag{6.95}$$

The final result for a component of the *averaged* microscopic current density is

$$\langle j_\alpha(\mathbf{x}, t) \rangle = J_\alpha(\mathbf{x}, t) + \frac{\partial}{\partial t} [D_\alpha(\mathbf{x}, t) - \epsilon_0 E_\alpha(\mathbf{x}, t)] + \sum_{\beta\gamma} \epsilon_{\alpha\beta\gamma} \frac{\partial}{\partial x_\beta} M_\gamma(\mathbf{x}, t)$$

$$+ \sum_\beta \frac{\partial}{\partial x_\beta} \left\langle \sum_{\substack{n \\ \text{(molecules)}}} [(\mathbf{p}_n)_\alpha(\mathbf{v}_n)_\beta - (\mathbf{p}_n)_\beta(\mathbf{v}_n)_\alpha]\delta(\mathbf{x} - \mathbf{x}_n) \right\rangle$$

$$- \frac{1}{6} \sum_{\beta\gamma} \frac{\partial^2}{\partial x_\beta \, \partial x_\gamma} \left\langle \sum_{\substack{n \\ \text{(molecules)}}} [(Q'_n)_{\alpha\beta}(\mathbf{v}_n)_\gamma - (Q'_n)_{\gamma\beta}(\mathbf{v}_n)_\alpha]\delta(\mathbf{x} - \mathbf{x}_n) \right\rangle + \cdots \tag{6.96}$$

The so-far undefined quantities in this rather formidable equation are the *macroscopic current density*

$$\mathbf{J}(\mathbf{x}, t) = \left\langle \sum_{\substack{j \\ \text{(free)}}} q_j\mathbf{v}_j\delta(\mathbf{x} - \mathbf{x}_j) + \sum_{\substack{n \\ \text{(molecules)}}} q_n\mathbf{v}_n\delta(\mathbf{x} - \mathbf{x}_n) \right\rangle \tag{6.97}$$

and the *macroscopic magnetization*

$$\mathbf{M}(\mathbf{x}, t) = \left\langle \sum_{\substack{n \\ \text{(molecules)}}} \mathbf{m}_n\delta(\mathbf{x} - \mathbf{x}_n) \right\rangle \tag{6.98}$$

If the free "charges" also possess intrinsic magnetic moments, these can be included in the definition of **M** in an obvious way. The last terms in (6.96) involve the electric molecular moments and molecular velocities and cannot be given an easy interpretation, except in special cases (see below).

When $\langle \mathbf{j} \rangle$ is inserted in the second equation of (6.70), there results the macroscopic Ampère–Maxwell equation of (6.62) with the derived magnetic field quantity **H** given in terms of **B** and the properties of the medium as

$$\left(\frac{1}{\mu_0}\mathbf{B} - \mathbf{H} \right)_\alpha = M_\alpha + \left\langle \sum_{\substack{n \\ \text{(molecules)}}} (\mathbf{p}_n \times \mathbf{v}_n)_\alpha \, \delta(\mathbf{x} - \mathbf{x}_n) \right\rangle$$

$$- \frac{1}{6} \sum_{\beta\gamma\delta} \epsilon_{\alpha\beta\gamma} \frac{\partial}{\partial x_\delta} \left\langle \sum_{\substack{n \\ \text{(molecules)}}} (Q'_n)_{\delta\beta}(\mathbf{v}_n)_\gamma\delta(\mathbf{x} - \mathbf{x}_n) \right\rangle + \cdots \tag{6.99}$$

The first term of the right-hand side of (6.99) is the familiar result, (6.63). The other terms are generally extremely small; first, because the molecular velocities $\mathbf{v}_n$ are small, typically thermal velocities in a gas or lattice vibrational velocities in a solid and, second, because the velocities fluctuate and tend to average to zero macroscopically. An exception occurs when the medium undergoes bulk motion. For simplicity, suppose that the medium as a whole has a translational velocity **v**. Neglecting any other motion of the molecules, we put $\mathbf{v}_n = \mathbf{v}$ for all n. Then (6.99) becomes, after a little manipulation,

$$\frac{1}{\mu_0}\mathbf{B} - \mathbf{H} = \mathbf{M} + (\mathbf{D} - \epsilon_0\mathbf{E}) \times \mathbf{v} \tag{6.100}$$

where **D** is given by (6.92). This shows that for a medium in motion the electric polarization **P** (and quadrupole density $Q_{\alpha\beta}$) enter the effective magnetization. Equation (6.100) is the nonrelativistic limit of one of the equations of Minkowski's electrodynamics of moving media (see *Pauli*, p. 105).

The reader may consult the book by *de Groot* for a discussion of the relativistic corrections, as well as for a statistical-mechanical treatment of the averaging. From the standpoint of logic and consistency there remains one loose end. In defining the molecular quadrupole moment $(Q'_n)_{\alpha\beta}$ by (6.77) we departed from our convention of Chapter 4, Eq. (4.9), and left $(Q'_n)_{\alpha\beta}$ with a nonvanishing trace. Since we made a point in Chapter 4 of relating the *five* independent components of the traceless quadrupole moment tensor to the $(2l + 1)$ spherical harmonics for $l = 2$, we need to explain why *six* components enter the macroscopic Maxwell equations. If we define a *traceless* molecular quadrupole moment $(Q_n)_{\alpha\beta}$ by means of (4.9), then we have

$$(Q'_n)_{\alpha\beta} = (Q_n)_{\alpha\beta} + \sum_{j(n)} q_j(\mathbf{x}_{jn})^2\delta_{\alpha\beta} \tag{6.101}$$

Introducing a mean square charge radius r_n^2 of the molecular charge distribution by

$$er_n^2 = \sum_{j(n)} q_j(\mathbf{x}_{jn})^2$$

where e is some convenient unit of charge, for example, that of a proton, we can write (6.101) as

$$(Q'_n)_{\alpha\beta} = (Q_n)_{\alpha\beta} + er_n^2\delta_{\alpha\beta}$$

The macroscopic quadrupole density (6.90) thus becomes

$$Q'_{\alpha\beta} = Q_{\alpha\beta} + \frac{1}{6}\left\langle \sum_{\substack{n \\ (\text{molecules})}} er_n^2\delta_{\alpha\beta}\delta(\mathbf{x} - \mathbf{x}_n)\right\rangle$$

where $Q_{\alpha\beta}$ is defined in terms of $(Q_n)_{\alpha\beta}$ just as in (6.90). The net result is that in the averaged microscopic charge density (6.87) the traceless quadrupole density $Q_{\alpha\beta}$ replaces the density $Q'_{\alpha\beta}$ and the charge density ρ is augmented by an additional term,

$$\rho \rightarrow \rho_{\text{free}} + \left\langle \sum_{\substack{n \\ (\text{molecules})}} q_n\delta(\mathbf{x} - \mathbf{x}_n)\right\rangle + \tfrac{1}{6}\nabla^2\left\langle \sum_{\substack{n \\ (\text{molecules})}} er_n^2\delta(\mathbf{x} - \mathbf{x}_n)\right\rangle \tag{6.102}$$

The trace of the tensor $Q'_{\alpha\beta}$ is exhibited with the charge density because it is an $l = 0$ contribution in terms of the multipole expansion. The molecular charge and mean square radius terms together actually represent the first two terms in an expansion of the $l = 0$ molecular multipole as we go beyond the static limit. In the Fourier-transformed wave number space, they correspond to the first two terms in the expansion of the charge form factor in powers of k^2. This can be seen from the definition of the form factor $F(k^2)$ for a charge density $\rho(\mathbf{x})$:

$$
\begin{aligned}
F(k^2) &= \int d^3x \, \rho(\mathbf{x}) \langle e^{i\mathbf{k}\cdot\mathbf{x}} \rangle_{l=0 \text{ part}} \\
&= \int d^3x \, \rho(\mathbf{x}) \frac{\sin kr}{kr} \\
&\simeq \int \rho d^3x - \tfrac{1}{6} k^2 \int r^2 \rho \, d^3x + \cdots
\end{aligned}
$$

With the correspondence $\mathbf{k} \leftrightarrow -i\boldsymbol{\nabla}$, the general equivalence of the form factor expansion and (6.102) is established.

In an interesting monograph alluded to above, *Robinson* gives a discussion of the connection between the microscopic equations and the macroscopic equations similar to ours. However, he makes a distinction between the spatial averaging (6.65) with the test function $f(\mathbf{x})$, called "truncation" (of the wave number spectrum) by him, and the statistical-mechanical averaging over various sorts of ensembles. Robinson holds that each macroscopic problem has its own appropriate lower limit of relevant lengths and that this sets the size of the test function to be used, before any considerations of statistical averaging are made.

6.7 *Poynting's Theorem and Conservation of Energy and Momentum for a System of Charged Particles and Electromagnetic Fields*

The forms of the laws of conservation of energy and momentum are important results to establish for the electromagnetic field. We begin by considering conservation of energy, often called *Poynting's theorem* (1884). For a single charge q the rate of doing work by external electromagnetic fields $\mathbf{E}$ and $\mathbf{B}$ is $q\mathbf{v} \cdot \mathbf{E}$, where $\mathbf{v}$ is the velocity of the charge. The magnetic field does no work, since the magnetic force is perpendicular to the velocity. If there exists a continuous distribution of charge and current, the total rate of doing work by the fields in a finite volume V is

$$
\int_V \mathbf{J} \cdot \mathbf{E} \, d^3x \tag{6.103}
$$

This power represents a conversion of electromagnetic energy into mechanical or thermal energy. It must be balanced by a corresponding rate of decrease of energy in the electromagnetic field within the volume V. To exhibit this conservation law explicitly, we use the Maxwell equations to express (6.103) in other terms. Thus we use the Ampère–Maxwell law to eliminate $\mathbf{J}$:

$$
\int_V \mathbf{J} \cdot \mathbf{E} \, d^3x = \int_V \left[\mathbf{E} \cdot (\boldsymbol{\nabla} \times \mathbf{H}) - \mathbf{E} \cdot \frac{\partial \mathbf{D}}{\partial t} \right] d^3x \tag{6.104}
$$

If we now employ the vector identity,

$$
\boldsymbol{\nabla} \cdot (\mathbf{E} \times \mathbf{H}) = \mathbf{H} \cdot (\boldsymbol{\nabla} \times \mathbf{E}) - \mathbf{E} \cdot (\boldsymbol{\nabla} \times \mathbf{H})
$$

and use Faraday's law, the right-hand side of (6.104) becomes

$$\int_V \mathbf{J} \cdot \mathbf{E} \, d^3x = -\int_V \left[\boldsymbol{\nabla} \cdot (\mathbf{E} \times \mathbf{H}) + \mathbf{E} \cdot \frac{\partial \mathbf{D}}{\partial t} + \mathbf{H} \cdot \frac{\partial \mathbf{B}}{\partial t} \right] d^3x \quad (6.105)$$

To proceed further we make two assumptions: (1) the macroscopic medium is linear in its electric and magnetic properties, with negligible dispersion or losses, and (2) the sum of (4.89) and (5.148) represents the total electromagnetic energy density, even for time-varying fields. With these two assumptions and the total energy density denoted by

$$u = \frac{1}{2} (\mathbf{E} \cdot \mathbf{D} + \mathbf{B} \cdot \mathbf{H}) \quad (6.106)$$

(6.105) can be written

$$-\int_V \mathbf{J} \cdot \mathbf{E} \, d^3x = \int_V \left[\frac{\partial u}{\partial t} + \boldsymbol{\nabla} \cdot (\mathbf{E} \times \mathbf{H}) \right] d^3x \quad (6.107)$$

Since the volume V is arbitrary, this can be cast into the form of a differential continuity equation or conservation law,

$$\frac{\partial u}{\partial t} + \boldsymbol{\nabla} \cdot \mathbf{S} = -\mathbf{J} \cdot \mathbf{E} \quad (6.108)$$

The vector $\mathbf{S}$, representing energy flow, is called the *Poynting vector*. It is given by

$$\mathbf{S} = \mathbf{E} \times \mathbf{H} \quad (6.109)$$

and has the dimensions of (energy/area × time). Since only its divergence appears in the conservation law, the Poynting vector seems arbitrary to the extent that the curl of any vector field can be added to it. Such an added term can, however, have no physical consequences. Relativistic considerations (Section 12.10) show that (6.109) is unique.

The physical meaning of the integral or differential form (6.107) or (6.108) is that the time rate of change of electromagnetic energy within a certain volume, plus the energy flowing out through the boundary surfaces of the volume per unit time, is equal to the negative of the total work done by the fields on the sources within the volume. This is the statement of conservation of energy. The assumptions that follow (6.105) really restrict the applicability of the simple version of Poynting's theorem to vacuum macroscopic or microscopic fields. Even for linear media, there is always dispersion (with accompanying losses). Then the right-hand side of (6.105) does not have the simple interpretation exhibited in (6.107). The more realistic situation of linear dispersive media is discussed in the next section.

The emphasis so far has been on the energy of the electromagnetic fields. The work done per unit time per unit volume by the fields ($\mathbf{J} \cdot \mathbf{E}$) is a conversion of electromagnetic energy into mechanical or heat energy. Since matter is ultimately composed of charged particles (electrons and atomic nuclei), we can think of this rate of conversion as a rate of increase of energy of the charged particles per unit volume. Then we can interpret Poynting's theorem for the *microscopic* fields ($\mathbf{E}, \mathbf{B}$) as a statement of conservation of energy of the combined system of particles and fields. If we denote the total energy of the particles within the volume V as E_{mech} and assume that no particles move out of the volume, we have

$$\frac{dE_{\text{mech}}}{dt} = \int_V \mathbf{J} \cdot \mathbf{E} \, d^3x \qquad (6.110)$$

Then Poynting's theorem expresses the conservation of energy for the combined system as

$$\frac{dE}{dt} = \frac{d}{dt} \left(E_{\text{mech}} + E_{\text{field}} \right) = -\oint_S \mathbf{n} \cdot \mathbf{S} \, da \qquad (6.111)$$

where the total field energy within V is

$$E_{\text{field}} = \int_V u \, d^3x = \frac{\epsilon_0}{2} \int_V (\mathbf{E}^2 + c^2 \mathbf{B}^2) \, d^3x \qquad (6.112)$$

The conservation of linear momentum can be similarly considered. The total electromagnetic force on a charged particle is

$$\mathbf{F} = q(\mathbf{E} + \mathbf{v} \times \mathbf{B}) \qquad (6.113)$$

If the sum of all the momenta of all the particles in the volume V is denoted by $\mathbf{P}_{\text{mech}}$, we can write, from Newton's second law,

$$\frac{d\mathbf{P}_{\text{mech}}}{dt} = \int_V (\rho \mathbf{E} + \mathbf{J} \times \mathbf{B}) \, d^3x \qquad (6.114)$$

where we have converted the sum over particles to an integral over charge and current densities for convenience in manipulation. In the same manner as for Poynting's theorem, we use the Maxwell equations to eliminate ρ and $\mathbf{J}$ from (6.114):

$$\rho = \epsilon_0 \mathbf{\nabla} \cdot \mathbf{E}, \qquad \mathbf{J} = \frac{1}{\mu_0} \mathbf{\nabla} \times \mathbf{B} - \epsilon_0 \frac{\partial \mathbf{E}}{\partial t} \qquad (6.115)$$

With (6.115) substituted into (6.114) the integrand becomes

$$\rho \mathbf{E} + \mathbf{J} \times \mathbf{B} = \epsilon_0 \left[\mathbf{E}(\mathbf{\nabla} \cdot \mathbf{E}) + \mathbf{B} \times \frac{\partial \mathbf{E}}{\partial t} - c^2 \mathbf{B} \times (\mathbf{\nabla} \times \mathbf{B}) \right]$$

Then writing

$$\mathbf{B} \times \frac{\partial \mathbf{E}}{\partial t} = -\frac{\partial}{\partial t} (\mathbf{E} \times \mathbf{B}) + \mathbf{E} \times \frac{\partial \mathbf{B}}{\partial t}$$

and adding $c^2 \mathbf{B}(\mathbf{\nabla} \cdot \mathbf{B}) = 0$ to the square bracket, we obtain

$$\rho \mathbf{E} + \mathbf{J} \times \mathbf{B} = \epsilon_0 [\mathbf{E}(\mathbf{\nabla} \cdot \mathbf{E}) + c^2 \mathbf{B}(\mathbf{\nabla} \cdot \mathbf{B})$$

$$- \mathbf{E} \times (\mathbf{\nabla} \times \mathbf{E}) - c^2 \mathbf{B} \times (\mathbf{\nabla} \times \mathbf{B})] - \epsilon_0 \frac{\partial}{\partial t} (\mathbf{E} \times \mathbf{B})$$

The rate of change of mechanical momentum (6.114) can now be written

$$\frac{d\mathbf{P}_{\text{mech}}}{dt} + \frac{d}{dt} \int_V \epsilon_0 (\mathbf{E} \times \mathbf{B}) \, d^3x \qquad (6.116)$$

$$= \epsilon_0 \int_V [\mathbf{E}(\mathbf{\nabla} \cdot \mathbf{E}) - \mathbf{E} \times (\mathbf{\nabla} \times \mathbf{E}) + c^2 \mathbf{B}(\mathbf{\nabla} \cdot \mathbf{B}) - c^2 \mathbf{B} \times (\mathbf{\nabla} \times \mathbf{B})] \, d^3x$$

We may tentatively identify the volume integral on the left as the total electro-magnetic momentum $\mathbf{P}_{\text{field}}$ in the volume V:

$$\mathbf{P}_{\text{field}} = \epsilon_0 \int_V \mathbf{E} \times \mathbf{B} \, d^3x = \mu_0\epsilon_0 \int_V \mathbf{E} \times \mathbf{H} \, d^3x \qquad (6.117)$$

The integrand can be interpreted as a density of electromagnetic momentum. We note that this momentum density is proportional to the energy-flux density $\mathbf{S}$, with proportionality constant c^{-2}.

To complete the identification of the volume integral of

$$\mathbf{g} = \frac{1}{c^2} (\mathbf{E} \times \mathbf{H}) \qquad (6.118)$$

as electromagnetic momentum, and to establish (6.116) as the conservation law for momentum, we must convert the volume integral on the right into a surface integral of the normal component of something that can be identified as momentum flow. Let the Cartesian coordinates be denoted by x_α, $\alpha = 1, 2, 3$. The $\alpha = 1$ component of the electric part of the integrand in (6.116) is given explicitly by

$$[\mathbf{E}(\boldsymbol{\nabla} \cdot \mathbf{E}) - \mathbf{E} \times (\boldsymbol{\nabla} \times \mathbf{E})]_1$$

$$= E_1\left(\frac{\partial E_1}{\partial x_1} + \frac{\partial E_2}{\partial x_2} + \frac{\partial E_3}{\partial x_3}\right) - E_2\left(\frac{\partial E_2}{\partial x_1} - \frac{\partial E_1}{\partial x_2}\right) + E_3\left(\frac{\partial E_1}{\partial x_3} - \frac{\partial E_3}{\partial x_1}\right)$$

$$= \frac{\partial}{\partial x_1}(E_1^2) + \frac{\partial}{\partial x_2}(E_1E_2) + \frac{\partial}{\partial x_3}(E_1E_3) - \frac{1}{2}\frac{\partial}{\partial x_1}(E_1^2 + E_2^2 + E_3^2)$$

This means that we can write the αth component as

$$[\mathbf{E}(\boldsymbol{\nabla} \cdot \mathbf{E}) - \mathbf{E} \times (\boldsymbol{\nabla} \times \mathbf{E})]_\alpha = \sum_\beta \frac{\partial}{\partial x_\beta}(E_\alpha E_\beta - \tfrac{1}{2}\mathbf{E} \cdot \mathbf{E}\delta_{\alpha\beta}) \qquad (6.119)$$

and have the form of a divergence of a second rank tensor on the right-hand side. With the definition of the *Maxwell stress tensor* $T_{\alpha\beta}$ as

$$T_{\alpha\beta} = \epsilon_0[E_\alpha E_\beta + c^2 B_\alpha B_\beta - \tfrac{1}{2}(\mathbf{E} \cdot \mathbf{E} + c^2\mathbf{B} \cdot \mathbf{B})\delta_{\alpha\beta}] \qquad (6.120)$$

we can therefore write (6.116) in component form as

$$\frac{d}{dt}(\mathbf{P}_{\text{mech}} + \mathbf{P}_{\text{field}})_\alpha = \sum_\beta \int_V \frac{\partial}{\partial x_\beta} T_{\alpha\beta} \, d^3x \qquad (6.121)$$

Application of the divergence theorem to the volume integral gives

$$\frac{d}{dt}(\mathbf{P}_{\text{mech}} + \mathbf{P}_{\text{field}})_\alpha = \oint_S \sum_\beta T_{\alpha\beta}n_\beta \, da \qquad (6.122)$$

where $\mathbf{n}$ is the outward normal to the closed surface S. Evidently, if (6.122) represents a statement of conservation of momentum, $\sum_\beta T_{\alpha\beta}n_\beta$ is the αth component of the flow per unit area of momentum across the surface S into the volume V. In other words, it is the force per unit area transmitted across the surface S and acting on the combined system of particles and fields inside V. Equation (6.122) can therefore be used to calculate the forces acting on material objects in electromagnetic fields by enclosing the objects with a boundary surface S and adding up the total electromagnetic force according to the right-hand side of (6.122).

The conservation of angular momentum of the combined system of particles and fields can be treated in the same way as we have handled energy and linear momentum. This is left as a problem for the student (see Problem 6.10).

The discussion of electromagnetic momentum and the stress tensor in fluids and solids entails analysis of interplay of mechanical, thermodynamic, and electromagnetic properties (e.g., $\partial\epsilon/\partial T$ and $\partial\epsilon/\partial\rho$). We refer the reader to *Landau and Lifshitz, Electrodynamics of Continuous Media* (Sections 10, 15, 16, 31, 35), *Stratton* (Chapter 2), and, for a statistical mechanical approach, to *de Groot* (Section 13). We note only that, although a treatment using the *macroscopic* Maxwell equations leads to an apparent electromagnetic momentum, $\mathbf{g} = \mathbf{D} \times \mathbf{B}$ (Minkowski, 1908), the generally accepted expression for a medium at rest is

$$\mathbf{g} = \frac{1}{c^2}\,\mathbf{E} \times \mathbf{H} = \mu_0\epsilon_0\mathbf{E} \times \mathbf{H} = \frac{1}{c^2}\,\mathbf{S} \tag{6.123}$$

We note that $\mathbf{g}$ is the electromagnetic momentum associated with the fields. There is an additional co-traveling momentum within the medium from the mechanical momentum of the electrons in the molecular dipoles in response to the incident traveling wave.* The Minkowski momentum of a plane wave is the "pseudomomentum" of the wave vector ($k = n\omega/c$ or $\hbar k = n(\hbar\omega)/c$ for a photon).

6.8 *Poynting's Theorem in Linear Dispersive Media with Losses*

In the preceding section Poynting's theorem (6.108) was derived with the restriction to linear media with no dispersion or losses (i.e., $\mathbf{D} = \epsilon\mathbf{E}$ and $\mathbf{B} = \mu\mathbf{H}$), with ϵ and μ real and frequency independent. Actual materials exhibit dispersion and losses. To discuss dispersion it is necessary to make a Fourier decomposition in time of both $\mathbf{E}$ and $\mathbf{D}$ (and $\mathbf{B}$ and $\mathbf{H}$). Thus, with

$$\mathbf{E}(\mathbf{x}, t) = \int_{-\infty}^{\infty} d\omega\,\mathbf{E}(\mathbf{x}, \omega)e^{-i\omega t}$$

$$\mathbf{D}(\mathbf{x}, t) = \int_{-\infty}^{\infty} d\omega\,\mathbf{D}(\mathbf{x}, \omega)e^{-i\omega t}$$

the assumption of linearity (and, for simplicity, isotropy) implies that $\mathbf{D}(\mathbf{x}, \omega) = \epsilon(\omega)\mathbf{E}(\mathbf{x}, \omega)$, where $\epsilon(\omega)$ is the complex and frequency-dependent susceptibility. Similarly, $\mathbf{B}(\mathbf{x}, \omega) = \mu(\omega)\mathbf{H}(\mathbf{x}, \omega)$. The reality of the fields implies that $\mathbf{E}(\mathbf{x}, -\omega) = \mathbf{E}^*(\mathbf{x}, \omega)$, $\mathbf{D}(\mathbf{x}, -\omega) = \mathbf{D}^*(\mathbf{x}, \omega)$, and $\epsilon(-\omega) = \epsilon^*(\omega)$. The presence of dispersion carries with it a temporally nonlocal connection between $\mathbf{D}(\mathbf{x}, t)$ and $\mathbf{E}(\mathbf{x}, t)$, discussed in detail in Section 7.10. As a consequence, the term $\mathbf{E} \cdot (\partial\mathbf{D}/\partial t)$ in (6.105) is not simply the time derivative of $(\mathbf{E} \cdot \mathbf{D}/2)$.

We write out $\mathbf{E} \cdot (\partial\mathbf{D}/\partial t)$ in terms of the Fourier integrals, with the spatial dependence implicit,

$$\mathbf{E} \cdot \frac{\partial\mathbf{D}}{\partial t} = \int d\omega \int d\omega'\mathbf{E}^*(\omega')[-i\omega\epsilon(\omega)] \cdot \mathbf{E}(\omega)e^{-i(\omega-\omega')t}$$

*See R. E. Peierls, *Proc. R. Soc. London* **347**, 475 (1976) for a very accessible discussion, of which Problem 6.25 is a simplified version. See also R. Loudon, L. Allen, and D. F. Nelson, *Phys. Rev. E* **55**, 1071 (1997).

Split the integrand into two equal parts and in one make the substitutions, $\omega \to -\omega'$, $\omega' \to -\omega$, and use the reality constraints to obtain

$$\mathbf{E} \cdot \frac{\partial \mathbf{D}}{\partial t} = \frac{1}{2} \int d\omega \int d\omega' \mathbf{E}^*(\omega')[-i\omega\epsilon(\omega) + i\omega'\epsilon^*(\omega')] \cdot \mathbf{E}(\omega)e^{-i(\omega-\omega')t} \quad (6.124)$$

We now suppose that the electric field is dominated by frequency components in a relatively narrow range compared to the characteristic frequency interval over which $\epsilon(\omega)$ changes appreciably. We may then expand the factor $i\omega'\epsilon^*(\omega')$ in the square brackets around $\omega' = \omega$ to get

$$[\cdots] = 2\omega \,\mathrm{Im}\; \epsilon(\omega) - i(\omega - \omega')\frac{d}{d\omega}(\omega\epsilon^*(\omega)) + \cdots$$

Insertion of this approximation into (6.124) leads to

$$\begin{aligned}
\mathbf{E} \cdot \frac{\partial \mathbf{D}}{\partial t} &= \int d\omega \int d\omega' \; \mathbf{E}^*(\omega') \cdot \mathbf{E}(\omega)\omega \,\mathrm{Im}\; \epsilon(\omega)e^{-i(\omega-\omega')t} \\
&+ \frac{\partial}{\partial t}\frac{1}{2}\int d\omega \int d\omega' \; \mathbf{E}^*(\omega') \cdot \mathbf{E}(\omega)\frac{d}{d\omega}[\omega\epsilon^*(\omega)]e^{-i(\omega-\omega')t}
\end{aligned} \quad (6.125)$$

There is a corresponding expression for $\mathbf{H} \cdot \partial\mathbf{B}/\partial t$ with $\mathbf{E} \to \mathbf{H}$ and $\epsilon \to \mu$ on the right-hand side.

First of all note that if ϵ and μ are real and frequency independent we recover the simple connection between the time derivative terms in (6.105) and $\partial u/\partial t$, with u given by (6.106). Second, the first term in (6.125) evidently represents the conversion of electrical energy into heat (or more generally into different forms of radiation*), while the second term must be an effective energy density. A more transparent expression, consistent with our assumption of the dominance of $\mathbf{E}$ and $\mathbf{H}$ by a relatively narrow range of frequencies can be obtained by supposing that $\mathbf{E} = \tilde{\mathbf{E}}(t)\cos(\omega_0 t + \alpha)$, $\mathbf{H} = \tilde{\mathbf{H}}(t)\cos(\omega_0 t + \beta)$, where $\tilde{\mathbf{E}}(t)$ and $\tilde{\mathbf{H}}(t)$ are slowly varying relative to both $1/\omega_0$ and the inverse of the frequency range over which $\epsilon(\omega)$ changes appreciably. If we substitute for the Fourier transforms $\mathbf{E}(\omega)$ and $\mathbf{H}(\omega)$ and average both sides of the sum of (6.125) and its magnetic counterpart over a period of the "carrier" frequency ω_0, we find (after some straightforward manipulation),

$$\left\langle \mathbf{E} \cdot \frac{\partial \mathbf{D}}{\partial t} + \mathbf{H} \cdot \frac{\partial \mathbf{B}}{\partial t} \right\rangle = 2\omega_0 \,\mathrm{Im}\; \epsilon(\omega_0)\langle \mathbf{E}(\mathbf{x}, t) \cdot \mathbf{E}(\mathbf{x}, t)\rangle \quad (6.126a)$$

$$+ 2\omega_0 \,\mathrm{Im}\; \mu(\omega_0)\langle \mathbf{H}(\mathbf{x}, t) \cdot \mathbf{H}(\mathbf{x}, t)\rangle + \frac{\partial u_{\mathrm{eff}}}{\partial t}$$

where the effective electromagnetic energy density is

$$\begin{aligned}
u_{\mathrm{eff}} &= \mathrm{Re}\left[\frac{d(\omega\epsilon)}{d\omega}(\omega_0)\right]\langle \mathbf{E}(\mathbf{x}, t) \cdot \mathbf{E}(\mathbf{x}, t)\rangle \\
&+ \mathrm{Re}\left[\frac{d(\omega\mu)}{d\omega}(\omega_0)\right]\langle \mathbf{H}(\mathbf{x}, t) \cdot \mathbf{H}(\mathbf{x}, t)\rangle
\end{aligned} \quad (6.126b)$$

---

*For example, if the dominant frequencies are near an atomic resonance of the medium where absorption is important ($\mathrm{Im}\; \epsilon \neq 0$), the re-emission of the radiation absorbed at ω may be at ω', where $\omega' \leq \omega$.

The presence of the factors $d(\omega\epsilon)/d\omega$ and $d(\omega\mu)/d\omega$ was first noted by Brillouin (see *Brillouin*, pp. 88–93). Our treatment is similar to *Landau and Lifshitz, Electrodynamics of Continuous Media* (Section 80).

Poynting's theorem in these circumstances reads

$$\frac{\partial u_{\text{eff}}}{\partial t} + \nabla \cdot \mathbf{S} = -\mathbf{J} \cdot \mathbf{E} - 2\omega_0 \operatorname{Im} \epsilon(\omega_0)\langle \mathbf{E}(\mathbf{x}, t) \cdot \mathbf{E}(\mathbf{x}, t)\rangle$$
$$-2\omega_0 \operatorname{Im} \mu(\omega_0)\langle \mathbf{H}(\mathbf{x}, t) \cdot \mathbf{H}(\mathbf{x}, t)\rangle \tag{6.127}$$

The first term on the right describes the explicit ohmic losses, if any, while the next terms represent the absorptive dissipation in the medium, not counting conduction loss. If the conduction current contribution is viewed as part of the dielectric response (see Section 7.5), the $-\mathbf{J} \cdot \mathbf{E}$ term is absent. Equation (6.127) exhibits the local conservation of electromagnetic energy in realistic situations where, as well as energy flow out of the locality ($\nabla \cdot \mathbf{S} \neq 0$), there may be losses from heating of the medium ($\operatorname{Im} \epsilon \neq 0$, $\operatorname{Im} \mu \neq 0$), leading to a (presumed) slow decay of the energy in the fields.

6.9 *Poynting's Theorem for Harmonic Fields; Field Definitions of Impedance and Admittance**

Lumped circuit concepts such as the resistance and reactance of a two-terminal linear network occur in many applications, even in circumstances where the size of the system is comparable to the free-space wavelength, for example, for a resonant antenna. It is useful therefore to have a general definition based on field concepts. This follows from consideration of Poynting's theorem for harmonic time variation of the fields. We assume that all fields and sources have a time dependence $e^{-i\omega t}$, so that we write

$$\mathbf{E}(\mathbf{x}, t) = \operatorname{Re}[\mathbf{E}(\mathbf{x})e^{-i\omega t}] \equiv \tfrac{1}{2}[\mathbf{E}(\mathbf{x})e^{-i\omega t} + \mathbf{E}^*(\mathbf{x})e^{i\omega t}] \tag{6.128}$$

The field $\mathbf{E}(\mathbf{x})$ is in general complex, with a magnitude and phase that change with position. For product forms, such as $\mathbf{J}(\mathbf{x}, t) \cdot \mathbf{E}(\mathbf{x}, t)$, we have

$$\mathbf{J}(\mathbf{x}, t) \cdot \mathbf{E}(\mathbf{x}, t) = \tfrac{1}{4}[\mathbf{J}(\mathbf{x})e^{-i\omega t} + \mathbf{J}^*(\mathbf{x})e^{i\omega t}] \cdot [\mathbf{E}(\mathbf{x})e^{-i\omega t} + \mathbf{E}^*(\mathbf{x})e^{i\omega t}]$$
$$= \tfrac{1}{2} \operatorname{Re}[\mathbf{J}^*(\mathbf{x}) \cdot \mathbf{E}(\mathbf{x}) + \mathbf{J}(\mathbf{x}) \cdot \mathbf{E}(\mathbf{x})e^{-2i\omega t}] \tag{6.129}$$

For time averages of products, the convention is therefore to take one-half of the real part of the product of one complex quantity with the complex conjugate of the other.

For harmonic fields the Maxwell equations become

$$\nabla \cdot \mathbf{B} = 0, \qquad \nabla \times \mathbf{E} - i\omega\mathbf{B} = 0$$
$$\nabla \cdot \mathbf{D} = \rho, \qquad \nabla \times \mathbf{H} + i\omega\mathbf{D} = \mathbf{J} \tag{6.130}$$

*The treatment of this section parallels that of *Fano, Chu, and Adler* (Sections 8.2 and 8.3). The reader can find in this book considerable further discussion of the connection between lumped circuit and field concepts, examples of stray capacitances in inductors, etc. See also the first two chapters of *Adler, Chu, and Fano*.

where all the quantities are complex functions of $\mathbf{x}$, according to the right-hand side of (6.128). Instead of (6.103) we consider the volume integral

$$\frac{1}{2} \int_V \mathbf{J}^* \cdot \mathbf{E} \, d^3x$$

whose real part gives the time-averaged rate of work done by the fields in the volume V. In a development strictly paralleling the steps from (6.103) to (6.107), we have

$$\frac{1}{2} \int_V \mathbf{J}^* \cdot \mathbf{E} \, d^3x = \frac{1}{2} \int_V \mathbf{E} \cdot [\nabla \times \mathbf{H}^* - i\omega \mathbf{D}^*] \, d^3x \qquad (6.131)$$

$$= \frac{1}{2} \int_V [-\nabla \cdot (\mathbf{E} \times \mathbf{H}^*) - i\omega(\mathbf{E} \cdot \mathbf{D}^* - \mathbf{B} \cdot \mathbf{H}^*)] \, d^3x$$

We now define the complex Poynting vector

$$\mathbf{S} = \tfrac{1}{2}(\mathbf{E} \times \mathbf{H}^*) \qquad (6.132)$$

and the harmonic electric magnetic energy densities,

$$w_e = \tfrac{1}{4}(\mathbf{E} \cdot \mathbf{D}^*), \qquad w_m = \tfrac{1}{4}(\mathbf{B} \cdot \mathbf{H}^*) \qquad (6.133)$$

Then (6.131) can be written as

$$\frac{1}{2} \int_V \mathbf{J}^* \cdot \mathbf{E} \, d^3x + 2i\omega \int_V (w_e - w_m) \, d^3x + \oint_S \mathbf{S} \cdot \mathbf{n} \, da = 0 \quad (6.134)$$

This is the analog of (6.107) for harmonic fields. It is a complex equation whose real part gives the conservation of energy for the time-averaged quantities and whose imaginary part relates to the reactive or stored energy and its alternating flow. If the energy densities w_e and w_m have real volume integrals, as occurs for systems with lossless dielectrics and perfect conductors, the real part of (6.134) is

$$\int_V \tfrac{1}{2} \operatorname{Re}(\mathbf{J}^* \cdot \mathbf{E}) \, d^3x + \oint_S \operatorname{Re}(\mathbf{S} \cdot \mathbf{n}) \, da = 0$$

showing that the steady-state, time-averaged rate of doing work on the sources in V by the fields is equal to the average flow of power *into* the volume V through the boundary surface S, as calculated from the normal component of Re $\mathbf{S}$. This is just what would be calculated from the earlier form of Poynting's theorem (6.107) if we assume that the energy density u has a steady part and a harmonically fluctuating part. With losses in the components of the system, the second term in (6.134) has a real part that accounts for this dissipation.

The complex Poynting theorem (6.134) can be used to define the input impedance of a general, two-terminal, linear, passive electromagnetic system. We imagine the system in the volume V surrounded by the boundary surface S, with only its input terminals protruding, as shown in Fig. 6.4. If the complex harmonic input current and voltage are I_i and V_i, the complex power input is $\tfrac{1}{2} I_i^* V_i$. This

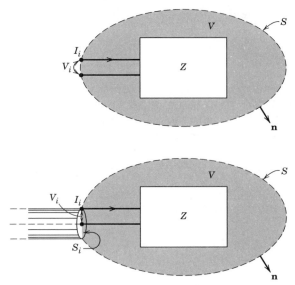

Figure 6.4 Schematic diagrams of arbitrary, two-terminal, linear, passive electromagnetic systems. The surface S completely surrounds the system; only the input terminals protrude. At these terminals, the harmonic input current and voltage are I_i and V_i, with the input impedance Z defined by $V_i = ZI_i$. The upper diagram applies at low frequencies where radiation losses are negligible, while the lower one with its coaxial-line input permits discussion of radiation resistance.

can be written in terms of the Poynting vector by using (6.134) applied to all of space on the *outside* of S as

$$\tfrac{1}{2}I_i^* V_i = -\oint_{S_i} \mathbf{S} \cdot \mathbf{n} \, da \tag{6.135}$$

where the unit normal $\mathbf{n}$ is outwardly directed, as shown in Fig. 6.4, and we have assumed that the input power flow is confined to the area S_i (the cross section of the coaxial line in the lower diagram of Fig. 6.4). By now considering (6.134) for the volume V surrounded by the closed surface S, the right-hand side of (6.135) can be written in terms of integrals over the fields inside the volume V:

$$\tfrac{1}{2}I_i^* V_i = \frac{1}{2} \int_V \mathbf{J}^* \cdot \mathbf{E} \, d^3x + 2i\omega \int_V (w_e - w_m) \, d^3x + \oint_{S-S_i} \mathbf{S} \cdot \mathbf{n} \, da \tag{6.136}$$

The surface integral here represents power flow out of the volume V through the surface S, except for the input surface S_i. If the surface $(S - S_i)$ is taken to infinity, this integral is real and represents escaping radiation (see Chapter 9). At low frequencies it is generally negligible. Then no distinction need be made between S_i and S; the upper diagram in Fig. 6.4 applies.

The input impedance $Z = R - iX$ (electrical engineers please read as $Z = R + jX!$) follows from (6.136) with its definition, $V_i = ZI_i$. Its real and imaginary parts are

$$R = \frac{1}{|I_i|^2} \left\{ \mathrm{Re} \int_V \mathbf{J}^* \cdot \mathbf{E} \, d^3x + 2 \oint_{S-S_i} \mathbf{S} \cdot \mathbf{n} \, da + 4\omega \, \mathrm{Im} \int_V (w_m - w_e) \, d^3x \right\}$$

$$\tag{6.137}$$

$$X = \frac{1}{|I_i|^2} \left\{ 4\omega \, \text{Re} \int_V (w_m - w_e) \, d^3x - \text{Im} \int_V \mathbf{J}^* \cdot \mathbf{E} \, d^3x \right\}$$

(6.138)

In writing (6.137) and (6.138) we have assumed that the power flow out through S is real. The second term in (6.137) is thus the "radiation resistance," important at high frequencies. At low frequencies, in systems where ohmic losses are the only appreciable source of dissipation, these expressions simplify to

$$R \simeq \frac{1}{|I_i|^2} \int_V \sigma \, |\mathbf{E}|^2 \, d^3x$$

(6.139)

$$X \simeq \frac{4\omega}{|I_i|^2} \int_V (w_m - w_e) \, d^3x$$

(6.140)

Here σ is the real conductivity, and the energy densities w_m and w_e (6.133) are also real over essentially the whole volume. The resistance is clearly the value expected from consideration of ohmic heat loss in the circuit. Similarly, the reactance has a plausible form: if magnetic stored energy dominates, as for a lumped inductance, the reactance is positive, etc. The different frequency dependences of the low-frequency reactance for inductances ($X = \omega L$) and capacitances ($X = -1/\omega C$) can be traced to the definition of L in terms of current and voltage ($V = L \, dI/dt$) on the one hand, and of C in terms of charge and voltage ($V = Q/C$) on the other. The treatment of some simple examples is left to the problems at the end of the chapter, as is the derivation of results equivalent to (6.139) and (6.140) for the conductance and susceptance of the complex admittance Y.

6.10 Transformation Properties of Electromagnetic Fields and Sources Under Rotations, Spatial Reflections, and Time Reversal

The fact that related physical quantities have compatible transformation properties under certain types of coordinate transformation is so taken for granted that the significance of such requirements and the limitations that can be thereby placed on the form of the relations is sometimes overlooked. It is useful therefore to discuss explicitly the relatively obvious properties of electromagnetic quantities under rotations, spatial inversions, and time reversal. The notions have direct application for limiting phenomenological constitutive relations, and are applied in the next section where the question of magnetic monopoles is discussed.

It is assumed that the idea of space and time coordinate transformations and their relation to the general conservation laws is familiar to the reader from classical mechanics (see, e.g., *Goldstein*). Only a summary of the main results is given here.

A. Rotations

A rotation in three dimensions is a linear transformation of the coordinates of a point such that the sum of the squares of the coordinates remains invariant.

Such a transformation is called an orthogonal transformation. The transformed coordinates x'_α are given in terms of the original coordinates x_β by

$$x'_\alpha = \sum_\beta a_{\alpha\beta} x_\beta \tag{6.141}$$

The requirement to have $(\mathbf{x}')^2 = (\mathbf{x})^2$ restricts the real transformation coefficients $a_{\alpha\beta}$ to be orthogonal,

$$\sum_\alpha a_{\alpha\beta} a_{\alpha\gamma} = \delta_{\beta\gamma} \tag{6.142}$$

The inverse transformation has $(a^{-1})_{\alpha\beta} = a_{\beta\alpha}$ and the square of the determinant of the matrix (a) is equal to unity. The value $\det(a) = +1$ corresponds to a proper rotation, obtainable from the original configuration by a sequence of infinitesimal steps, whereas $\det(a) = -1$ represents an improper rotation, a reflection plus a rotation.

Physical quantities are classed as rotational tensors of various ranks depending on how they transform under rotations. Coordinates $\mathbf{x}_i$, velocities $\mathbf{v}_i$, momenta $\mathbf{p}_i$ have components that transform according to the basic transformation law (6.141) and are tensors of rank one, or *vectors*. Scalar products of vectors, such as $\mathbf{x}_1 \cdot \mathbf{x}_2$ or $\mathbf{v}_1 \cdot \mathbf{p}_2$, are invariant under rotations and so are tensors of rank zero, or *scalars*. Groups of quantities that transform according to

$$B'_{\alpha\beta} = \sum_{\gamma,\delta} a_{\alpha\gamma} a_{\beta\delta} B_{\gamma\delta} \tag{6.143}$$

are called second-rank tensors or, commonly, *tensors*. The Maxwell stress tensor is one such group of quantities. Higher rank tensor transformations follow obviously.

In considering electromagnetic fields and other physical quantities, we deal with one or more functions of coordinates and perhaps other kinematic variables. There then arises the choice of an "active" or a "passive" view of the rotation. We adopt the active view—the coordinate axes are considered fixed and the physical system is imagined to undergo a rotation. Thus, for example, two charged particles with initial coordinates $\mathbf{x}_1$ and $\mathbf{x}_2$ form a system that under a rotation is transformed so that the coordinates of the particles are now $\mathbf{x}'_1$ and $\mathbf{x}'_2$, as shown in Fig. 6.5. The components of each coordinate vector transform according to (6.141), but electrostatic potential is unchanged because it is a function only of the distance between the two points, $R = |\mathbf{x}_1 - \mathbf{x}_2|$, and R^2 is a sum of scalar products of vectors and so is invariant under the rotation. The electrostatic potential is one example of a scalar under rotations. In general, if a physical quantity ϕ, which is a function of various coordinates denoted collectively by $\mathbf{x}_i$ (possibly including coordinates such as velocities and momenta), is such that when the physical system is rotated with $\mathbf{x}_i \to \mathbf{x}'_i$, the quantity remains unchanged,

$$\phi'(\mathbf{x}'_i) = \phi(\mathbf{x}_i) \tag{6.144}$$

then ϕ is a scalar function under rotations. Similarly, if a set of three physical quantities $V_\alpha(\mathbf{x}_i)$ ($\alpha = 1, 2, 3$) transform under rotation of the system according to

$$V'_\alpha(\mathbf{x}'_i) = \sum_\beta a_{\alpha\beta} V_\beta(\mathbf{x}_i) \tag{6.145}$$

then the V_α form the components of a vector, and so on for higher rank tensors.

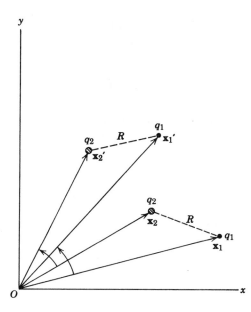

Figure 6.5   Active rotation of a system of two charges.

Differential vector operations have definite transformation properties under rotations. For example, the gradient of a scalar, $\nabla\phi$, transforms as a vector, the divergence of a vector, $\nabla \cdot \mathbf{V}$, is a scalar, and the Laplacian operator ∇^2 is a scalar operator in the sense that its application to a function or set of functions does not alter their rotational transformation properties.

Special mention must be made of the cross product of two vectors:

$$\mathbf{A} = \mathbf{B} \times \mathbf{C} \qquad (6.146)$$

In component form this vector shorthand reads

$$A_\alpha = \sum_{\beta\gamma} \epsilon_{\alpha\beta\gamma} B_\beta C_\gamma$$

where $\epsilon_{\alpha\beta\gamma} = +1$ for $\alpha = 1$, $\beta = 2$, $\gamma = 3$ and cyclic permutations, $\epsilon_{\alpha\beta\gamma} = -1$ for other permutations, and vanishes for two or more indices equal. Because of the presence of two vectors on the right-hand side, the cross product has some attributes of a traceless antisymmetric second-rank tensor. Since such a tensor has only three independent components, we treat it as a vector. This practice is justified, of course, only insofar as it transforms under rotations according to (6.141). In actual fact, the transformation law for the cross product (6.146) is

$$A'_\alpha = \det(a) \sum_{\beta} a_{\alpha\beta} A_\beta \qquad (6.147)$$

For proper rotations, the only kind we have considered so far, $\det(a) = +1$; thus (6.147) is in agreement with the basic coordinate transformation (6.141). Under *proper* rotations, the cross product transforms as a vector.

B. Spatial Reflection or Inversion

Spatial reflection in a plane corresponds to changing the signs of the normal components of the coordinate vectors of all points and to leaving the components parallel to the plane unchanged. Thus, for reflection in the *x*-*y* plane, $\mathbf{x}_i =$

$(x_i, y_i, z_i) \to \mathbf{x}'_i = (x_i, y_i, - z_i)$. Space inversion corresponds to reflection of all three components of every coordinate vector through the origin, $\mathbf{x}_i \to \mathbf{x}'_i = -\mathbf{x}_i$. Spatial inversion or reflection is a discrete transformation that, for more than two coordinates, cannot in general be accomplished by proper rotations. It corresponds to $\det(a) = -1$, and for the straightforward inversion operation is given by (6.141) with $a_{\alpha\beta} = -\delta_{\alpha\beta}$. It follows that vectors change sign under spatial inversion, but cross products, which behave according to (6.147), do not. We are thus forced to distinguish two kinds of vectors (under general rotations):

Polar vectors (or just vectors) that transform according to (6.145) and for $\mathbf{x}_i \to \mathbf{x}'_i = -\mathbf{x}_i$ behave as

$$\mathbf{V} \to \mathbf{V}' = -\mathbf{V}$$

Axial or pseudovectors that transform according to (6.147) and for $\mathbf{x}_i \to \mathbf{x}'_i = -\mathbf{x}_i$ behave as

$$\mathbf{A} \to \mathbf{A}' = \mathbf{A}$$

Similar distinctions must be made for scalars under rotations. We speak of *scalars* or *pseudoscalars*, depending on whether the quantities do not or do change sign under spatial inversion. The triple scalar product $\mathbf{a} \cdot (\mathbf{b} \times \mathbf{c})$ is an example of a pseudoscalar quantity, provided $\mathbf{a}$, $\mathbf{b}$, $\mathbf{c}$ are all polar vectors. (We see here in passing a dangerous aspect of our usual notation. The writing of a vector as $\mathbf{a}$ does not tell us whether it is a polar or an axial vector.) The transformation properties of higher rank tensors under spatial inversion can be deduced directly if they are built up by taking products of components of polar or axial vectors. If a tensor of rank N transforms under spatial inversion with a factor $(-1)^N$, we call it a true tensor or just a *tensor*, while if the factor is $(-1)^{N+1}$ we call it a *pseudotensor* of rank N.

C. Time Reversal

The basic laws of physics are invariant (at least at the classical level) to the sense of direction of time. This does not mean that the equations are even in t, but that, under the time reversal transformation $t \to t' = -t$, the related physical quantities transform in a consistent fashion so that the *form* of the equation is the same as before. Thus, for a particle of momentum $\mathbf{p}$ and position $\mathbf{x}$ moving in an external potential $U(\mathbf{x})$, Newton's equation of motion,

$$\frac{d\mathbf{p}}{dt} = -\boldsymbol{\nabla} U(\mathbf{x})$$

is invariant under time reversal provided $\mathbf{x} \to \mathbf{x}' = \mathbf{x}$ and $\mathbf{p} \to \mathbf{p}' = -\mathbf{p}$. The sign change for the momentum is, of course, intuitively obvious from its relation to the velocity, $\mathbf{v} = d\mathbf{x}/dt$. The consequence of the invariance of Newton's laws under time reversal is that, if a certain initial configuration of a system of particles evolves under the action of various forces into some final configuration, a possible state of motion of the system is that the time-reversed final configuration (all positions the same, but all velocities reversed) will evolve over the reversed path to the time-reversed initial configuration.

The transformation properties of various mechanical quantities under rota-

Table 6.1   Transformation Properties of Various Physical Quantities under Rotations, Spatial Inversion and Time Reversal[a]

Physical Quantity		Rotation (rank of tensor)	Space Inversion (name)	Time Reversal
I. *Mechanical*				
Coordinate	$\mathbf{x}$	1	Odd (vector)	Even
Velocity	$\mathbf{v}$	1	Odd (vector)	Odd
Momentum	$\mathbf{p}$	1	Odd (vector)	Odd
Angular momentum	$\mathbf{L} = \mathbf{x} \times \mathbf{p}$	1	Even (pseudovector)	Odd
Force	$\mathbf{F}$	1	Odd (vector)	Even
Torque	$\mathbf{N} = \mathbf{x} \times \mathbf{F}$	1	Even (pseudovector)	Even
Kinetic energy	$p^2/2m$	0	Even (scalar)	Even
Potential energy	$U(\mathbf{x})$	0	Even (scalar)	Even
II. *Electromagnetic*				
Charge density	ρ	0	Even (scalar)	Even
Current density	$\mathbf{J}$	1	Odd (vector)	Odd
Electric field	$\mathbf{E}$			
Polarization	$\mathbf{P}$	1	Odd (vector)	Even
Displacement	$\mathbf{D}$			
Magnetic induction	$\mathbf{B}$			
Magnetization	$\mathbf{M}$	1	Even (pseudovector)	Odd
Magnetic field	$\mathbf{H}$			
Poynting vector	$\mathbf{S} = \mathbf{E} \times \mathbf{H}$	1	Odd (vector)	Odd
Maxwell stress tensor	$T_{\alpha\beta}$	2	Even (tensor)	Even

[a]For quantities that are functions of $\mathbf{x}$ and t, it is necessary to be very clear what is meant by evenness or oddness under space inversion or time reversal. For example, the magnetic induction is such that under space inversion, $\mathbf{B}(\mathbf{x}, t) \to \mathbf{B}_I(\mathbf{x}, t) = +\mathbf{B}(-\mathbf{x}, t)$, while under time reversal, $\mathbf{B}(\mathbf{x}, t) \to \mathbf{B}_T(\mathbf{x}, t) = -\mathbf{B}(\mathbf{x}, -t)$.

tions, spatial inversion, and time reversal are summarized in the first part of Table 6.1.

D. Electromagnetic Quantities

Just as with the laws of mechanics, it is true (i.e., consistent with all known experimental facts) that the forms of the equations governing electromagnetic phenomena are invariant under rotations, space inversion, and time reversal. This implies that the different electromagnetic quantities have well-defined transformation properties under these operations. It is an experimental fact that electric charge is invariant under Galilean and Lorentz transformations and is a scalar under rotations. It is natural, convenient, and permissible to assume that charge is also a scalar under spatial inversion and even under time reversal. The point here is that physically measurable quantities like force involve the product of charge and field. The transformation properties attributed to fields like $\mathbf{E}$ and $\mathbf{B}$ thus depend on the convention chosen for the charge.

With charge a true scalar under all three transformations, *charge density ρ* is also a *true scalar*. From the fact that the *electric field* is force per unit charge, we see that $\mathbf{E}$ is a *polar vector, even under time reversal*. This also follows from the

Maxwell equation, $\nabla \cdot \mathbf{E} = \rho/\epsilon_0$, since both sides must transform in the same manner under the transformations.

The first term in the Maxwell equation representing Faraday's law,

$$\nabla \times \mathbf{E} + \frac{\partial \mathbf{B}}{\partial t} = 0$$

transforms as a pseudovector under rotations and spatial inversion, and is even under time reversal. To preserve the invariance of form it is therefore necessary that the *magnetic induction* $\mathbf{B}$ *be a pseudovector, odd under time reversal*. The left-hand side of the Ampère–Maxwell equation,

$$\frac{1}{\mu_0} \nabla \times \mathbf{B} - \epsilon_0 \frac{\partial \mathbf{E}}{\partial t} = \mathbf{J}$$

can be seen to transform as a polar vector, odd under time reversal. This implies that the *current density* $\mathbf{J}$ is a *polar vector, odd under time reversal*, as expected from its definition in terms of charge times velocity.

We have just seen that the microscopic fields and sources have well-defined transformation properties under rotations, spatial inversion, and time reversal. From the derivation of the macroscopic Maxwell equations in Section 6.6 and the definitions of $\mathbf{P}, \mathbf{M}$, etc., it can be seen that $\mathbf{E}, \mathbf{P}, \mathbf{D}$ all transform in the same way, as do $\mathbf{B}, \mathbf{M}, \mathbf{H}$. The various transformation properties for electromagnetic quantities are summarized in the second part of Table 6.1.

To illustrate the usefulness of arguments on the symmetry properties listed in Table 6.1, we consider the phenomenological structure of a spatially local constitutive relation specifying the polarization $\mathbf{P}$ for an isotropic, linear, non-dissipative medium in a uniform, constant, external magnetic induction $\mathbf{B}_0$. The relation is first order in the electric field $\mathbf{E}$, by assumption, but we require an expansion in powers of $\mathbf{B}_0$ up to second order. Since $\mathbf{P}$ is a polar vector, and even under time reversal, the various terms to be multiplied by scalar coefficients must transform in the same way. To zeroth order in $\mathbf{B}_0$, only $\mathbf{E}$ is available. To first order in $\mathbf{B}_0$, possible terms involving $\mathbf{E}$ linearly are

$$\mathbf{E} \times \mathbf{B}_0, \qquad \frac{\partial \mathbf{E}}{\partial t} \times \mathbf{B}_0, \qquad \frac{\partial^2 \mathbf{E}}{\partial t^2} \times \mathbf{B}_0, \cdots$$

All these are permitted by rotational and spatial inversion grounds, but only those involving odd time derivatives transform properly under time reversal. For the second order in $\mathbf{B}_0$, the possibilities are

$$(\mathbf{B}_0 \cdot \mathbf{B}_0)\mathbf{E}, \qquad (\mathbf{E} \cdot \mathbf{B}_0)\mathbf{B}_0, \qquad (\mathbf{B}_0 \cdot \mathbf{B}_0) \frac{\partial \mathbf{E}}{\partial t}, \cdots$$

Here only the terms with zero or even time derivatives of $\mathbf{E}$ satisfy all the requirements. The most general spatially local expression for the polarization, correct to second order in the constant magnetic field B_0, is thus

$$\frac{1}{\epsilon_0} \mathbf{P} = \chi_0 \mathbf{E} + \chi_1 \frac{\partial \mathbf{E}}{\partial t} \times \mathbf{B}_0 + \chi_2 (\mathbf{B}_0 \cdot \mathbf{B}_0)\mathbf{E} + \chi_3 (\mathbf{E} \cdot \mathbf{B}_0)\mathbf{B}_0 + \cdots \quad (6.148)$$

where the χ_i are real scalar coefficients and higher time derivatives of $\mathbf{E}$ can occur, odd for the terms linear in $\mathbf{B}_0$ and even for the zeroth and second powers of $\mathbf{B}_0$. At low frequencies, the response of essentially all material systems is via electric forces. This means that at zero frequency there should be no dependence of **P** on $\mathbf{B}_0$, and a more realistic form is

$$\frac{1}{\epsilon_0} \mathbf{P} = \chi_0 \mathbf{E} + \chi_1 \frac{\partial \mathbf{E}}{\partial t} \times \mathbf{B}_0 + \chi_2'(\mathbf{B}_0 \cdot \mathbf{B}_0) \frac{\partial^2 \mathbf{E}}{\partial t^2} + \chi_3'\left(\frac{\partial^2 \mathbf{E}}{\partial t^2} \cdot \mathbf{B}_0\right)\mathbf{B}_0 \quad (6.149)$$

where we have exhibited only the lowest order time derivatives for each power of $\mathbf{B}_0$. At optical frequencies this equation permits an understanding of the gyrotropic behavior of waves in an isotropic medium in a constant magnetic field.*

Another example, the Hall effect, is left to the problems. It, as well as thermogalvanomagnetic effects and the existence of magnetic structure in solids, are discussed in *Landau and Lifshitz* (*op. cit.*).

In certain circumstances the constraints of space-time symmetries must be relaxed in constitutive relations. For example, the optical rotatory power of chiral molecules is described phenomenologically by the constitutive relations, $\mathbf{P} = \epsilon_0 \chi_0 \mathbf{E} + \xi \partial \mathbf{B}/\partial t$ and $\mu_0 \mathbf{M} = \chi_0' \mathbf{B} + \xi' \partial \mathbf{E}/\partial t$. The added terms involve pseudoscalar quantities ξ and ξ' that reflect the underlying lack of parity symmetry for chiral substances. (Quantum mechanically, nonvanishing ξ or ξ' requires both electric and magnetic dipole operators to have nonvanishing matrix elements between the same pair of states, something that cannot occur for states of definite parity.)

6.11 On the Question of Magnetic Monopoles

At the present time (1998) there is no experimental evidence for the existence of magnetic charges or monopoles. But chiefly because of an early, brilliant theoretical argument of Dirac,[†] the search for monopoles is renewed whenever a new energy region is opened up in high-energy physics or a new source of matter, such as rocks from the moon, becomes available. Dirac's argument, outlined below, is that the mere existence of one magnetic monopole in the universe would offer an explanation of the discrete nature of electric charge. Since the quantization of charge is one of the most profound mysteries of the physical world, Dirac's idea has great appeal. The history of the theoretical ideas and experimental searches up to 1990 are described in the resource letter of Goldhaber and Trower.[‡] Some other references appear at the end of the chapter.

There are some necessary preliminaries before examining Dirac's argument. One question that arises is whether it is possible to tell that particles have magnetic as well as electric charge. Let us suppose that there exist magnetic charge and current densities, ρ_m and $\mathbf{J}_m$, in addition to the electric densities, ρ_e and $\mathbf{J}_e$. The Maxwell equations would then be

$$\nabla \cdot \mathbf{D} = \rho_e, \qquad \nabla \times \mathbf{H} = \frac{\partial \mathbf{D}}{\partial t} + \mathbf{J}_e$$

$$\nabla \cdot \mathbf{B} = \rho_m, \qquad -\nabla \times \mathbf{E} = \frac{\partial \mathbf{B}}{\partial t} + \mathbf{J}_m$$

$\qquad (6.150)$

*See Landau and Lifshitz, *Electrodynamics of Continuous Media*, p. 334, Problem 3, p. 337.

[†]P. A. M. Dirac, *Proc. R. Soc. London* **A133**, 60 (1931); *Phys. Rev.* **74**, 817 (1948).

[‡]A. S. Goldhaber and W. P. Trower, Resource Letter MM-1: Magnetic Monopoles, *Am. J. Phys.* **58**, 429–439 (1990).

The magnetic densities are assumed to satisfy the same form of the continuity equation as the electric densities. It appears from these equations that the existence of magnetic charge and current would have observable electromagnetic consequences. Consider, however, the following *duality transformation**:

$$\mathbf{E} = \mathbf{E}' \cos \xi + Z_0 \mathbf{H}' \sin \xi, \qquad Z_0 \mathbf{D} = Z_0 \mathbf{D}' \cos \xi + \mathbf{B}' \sin \xi$$
$$Z_0 \mathbf{H} = -\mathbf{E}' \sin \xi + Z_0 \mathbf{H}' \cos \xi, \qquad \mathbf{B} = -Z_0 \mathbf{D}' \sin \xi + \mathbf{B}' \cos \xi \tag{6.151}$$

For a real (pseudoscalar) angle ξ, such a transformation leaves quadratic forms such as $\mathbf{E} \times \mathbf{H}$, $(\mathbf{E} \cdot \mathbf{D} + \mathbf{B} \cdot \mathbf{H})$, and the components of the Maxwell stress tensor $T_{\alpha\beta}$ invariant. If the sources are transformed in the same way,

$$Z_0 \rho_e = Z_0 \rho_e' \cos \xi + \rho_m' \sin \xi, \qquad Z_0 \mathbf{J}_e = Z_0 \mathbf{J}_e' \cos \xi + \mathbf{J}_m' \sin \xi$$
$$\rho_m = -Z_0 \rho_e' \sin \xi + \rho_m' \cos \xi, \qquad \mathbf{J}_m = -Z_0 \mathbf{J}_e' \sin \xi + \mathbf{J}_m' \cos \xi \tag{6.152}$$

then it is straightforward algebra to show that the generalized Maxwell equations (6.150) are invariant, that is, the equations for the primed fields $(\mathbf{E}', \mathbf{D}', \mathbf{B}', \mathbf{H}')$ are the same as (6.150) with the primed sources present.

The invariance of the equations of electrodynamics under duality transformations shows that it is a matter of convention to speak of a particle possessing an electric charge, but not magnetic charge. The only meaningful question is whether *all* particles have the same ratio of magnetic to electric charge. If they do, then we can make a duality transformation, choosing the angle ξ so that $\rho_m = 0$, $\mathbf{J}_m = 0$. We then have the Maxwell equations as they are usually known.

If, by convention, we choose the electric and magnetic charges of an electron to be $q_e = -e$, $q_m = 0$, then it is known that for a proton, $q_e = +e$ (with the present limits of error being $|q_e(\text{electron}) + q_e(\text{proton})|/e \sim 10^{-20}$) and $|q_m(\text{nucleon})| < 2 \times 10^{-24} Z_0 e$.

This extremely small limit on the magnetic charge of a proton or neutron follows directly from knowing that the average magnetic field at the surface of the earth is not more than 10^{-4} T. The conclusion, to a very high degree of precision, is that the particles of ordinary matter possess only electric charge or, equivalently, they all have the same ratio of magnetic to electric charge. For other, unstable, particles the question of magnetic charge is more open, but no positive evidence exists.

The transformation properties of ρ_m and $\mathbf{J}_m$ under rotations, spatial inversion, and time reversal are important. From the known behavior of $\mathbf{E}$ and $\mathbf{B}$ in the usual formulation we deduce from the second line in (6.150) that

ρ_m *is a pseudoscalar density, odd under time reversal*, and
$\mathbf{J}_m$ *is a pseudovector density, even under time reversal.*

Since the symmetries of ρ_m under both spatial inversion and time reversal are opposite to those of ρ_e, it is a necessary consequence of the existence of a particle with both electric and magnetic charges that space inversion and time reversal are no longer valid symmetries of the laws of physics. It is a fact, of course, that

---

*The presence of the "impedance of free space," $Z_0 = \sqrt{\mu_0/\epsilon_0}$, in the transformation is a consequence of the presence of the dimensionful parameters ϵ_0 and μ_0 in the SI system. Magnetic charge density differs in dimensions from electric charge density in SI units. For users of Gaussian units, put $Z_0 \rightarrow 1$.

these symmetry principles are not exactly valid in the realm of elementary particle physics, but present evidence is that their violation is extremely small and associated somehow with the weak interactions. Future developments linking electromagnetic, weak, and perhaps strong, interactions may utilize particles carrying magnetic charge as the vehicle for violation of space inversion and time reversal symmetries. With no evidence for monopoles, this remains speculation.

In spite of the negative evidence for the existence of magnetic monopoles, let us turn to Dirac's ingenious proposal. By considering the quantum mechanics of an electron in the presence of a magnetic monopole, he showed that consistency required the quantization condition,

$$\frac{eg}{4\pi\hbar} = \frac{\alpha g}{Z_0 e} = \frac{n}{2} \qquad (n = 0, \pm 1, \pm 2, \ldots) \tag{6.153}$$

where e is the electronic charge, $\alpha = e^2/4\pi\epsilon_0\hbar c$ is the fine structure constant ($\alpha \approx 1/137$), and g is the magnetic charge of the monopole. The discrete nature of electric charge thus follows from the existence of a monopole. The magnitude of e is not determined, except in terms of the magnetic charge g. The argument can be reversed. With the known value of the fine structure constant, we infer the existence of magnetic monopoles with charges g whose *magnetic "fine structure"* constant is

$$\frac{g^2}{4\pi\mu_0\hbar c} = \frac{n^2}{4}\left(\frac{4\pi\epsilon_0\hbar c}{e^2}\right) \simeq \frac{137}{4} n^2$$

Such monopoles are known as *Dirac monopoles*. Their coupling strength is enormous, making their extraction from matter with dc magnetic fields and their subsequent detection very simple in principle. For instance, the energy loss in matter by a relativistic Dirac monopole is approximately the same as that of a relativistic heavy nucleus with $Z = 137n/2$. It can presumably be distinguished from such a nucleus if it is brought to rest because it will not show an increase in ionization at the end of its range (see Problem 13.11).

6.12   Discussion of the Dirac Quantization Condition

Semiclassical considerations can illuminate the Dirac quantization condition (6.153). First, we consider the deflection at large impact parameters of a particle of charge e and mass m by the field of a stationary magnetic monopole of magnetic charge g. At sufficiently large impact parameter, the change in the state of motion of the charged particle can be determined by computing the impulse of the force, assuming the particle is undeflected. The geometry is shown in Fig. 6.6. The particle is incident parallel to the z axis with an impact parameter b and a speed v and is acted on by the radially directed magnetic field of the monopole, $\mathbf{B} = g\mathbf{r}/4\pi r^3$, according to the Lorentz force (6.113). In the approximation that the particle is undeflected, the only force acting throughout the collision is a y component,

$$F_y = evB_x = \frac{eg}{4\pi} \frac{vb}{(b^2 + v^2t^2)^{3/2}} \tag{6.154}$$

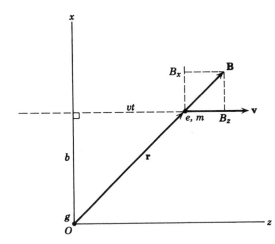

Figure 6.6 Charged particle passing a magnetic monopole at large impact parameter.

The impulse transmitted by this force is

$$\Delta p_y = \frac{egvb}{4\pi} \int_{-\infty}^{\infty} \frac{dt}{(b^2 + v^2t^2)^{3/2}} = \frac{eg}{2\pi b} \qquad (6.155)$$

Since the impulse is in the y direction, the particle is deflected out of the plane of Fig. 6.6, that is, in the azimuthal direction. Evidently the particle's angular momentum is changed by the collision, a result that is not surprising in the light of the noncentral nature of the force. The magnitude of the change in angular momentum *is* somewhat surprising, however. There is no z component of **L** initially, but there is finally. The change in L_z is

$$\Delta L_z = b \,\Delta p_y = \frac{eg}{2\pi} \qquad (6.156)$$

The change in the z component of angular momentum of the particle is independent of the impact parameter b and the speed v of the charged particle. It depends only on the product eg and is a universal value for a charged particle passing a stationary monopole, no matter how far away. If we assume that any change of angular momentum must occur in integral multiples of $\hbar$, we are led immediately to the Dirac quantization condition (6.153).*

The peculiarly universal character of the change in the angular momentum (6.156) of a charged particle in passing a magnetic monopole can be understood by considering the angular momentum contained in the fields of a point electric charge in the presence of a point magnetic monopole. If the monopole g is at $\mathbf{x} = \mathbf{R}$ and the charge e is at $\mathbf{x} = 0$, as indicated in Fig. 6.7, the magnetic and electric fields in all of space are

$$\mathbf{H} = -\frac{g}{4\pi\mu_0} \mathbf{\nabla}\left(\frac{1}{r'}\right) = \frac{g}{4\pi\mu_0} \frac{\mathbf{n'}}{r'^2}, \qquad \mathbf{E} = -\frac{e}{4\pi\epsilon_0} \mathbf{\nabla}\left(\frac{1}{r}\right) = \frac{e}{4\pi\epsilon_0} \frac{\mathbf{n}}{r^2} \qquad (6.157)$$

where $r' = |\mathbf{x} - \mathbf{R}|$, $r = |\mathbf{x}|$, and $\mathbf{n'}$ and $\mathbf{n}$ are unit vectors in the directions of $(\mathbf{x} - \mathbf{R})$ and $\mathbf{x}$, respectively. The angular momentum $\mathbf{L}_{em}$ is given by the volume integral of $\mathbf{x} \times \mathbf{g}$, where $\mathbf{g} = (\mathbf{E} \times \mathbf{H})/c^2$ is the electromagnetic momentum density.

*This argument is essentially due to A. S. Goldhaber, *Phys. Rev.* **140**, B1407 (1965).

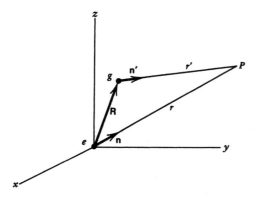

Figure 6.7

The total *momentum* of the fields $\mathbf{P}_{em}$ (volume integral of $\mathbf{g}$) vanishes. This follows from the fact that $\mathbf{P}_{em}$ is a vector and the only vector available is $\mathbf{R}$. Thus $\mathbf{P}_{em} = (\mathbf{R}/R)P$, where P is the volume integral of $\mathbf{g} \cdot (\mathbf{R}/R)$. But $\mathbf{g} \cdot \mathbf{R} \propto \mathbf{R} \cdot (\mathbf{n} \times \mathbf{n}')$. Since $\mathbf{R}$ lies in the plane defined by the vectors $\mathbf{n}$ and $\mathbf{n}'$, the triple scalar product vanishes and so does $\mathbf{P}_{em}$. This vanishing of the total momentum means that the *angular momentum*

$$\mathbf{L}_{em} = \frac{1}{c^2} \int \mathbf{x} \times (\mathbf{E} \times \mathbf{H}) \, d^3x \tag{6.158}$$

is independent of choice of origin. To evaluate $\mathbf{L}_{em}$ one can first substitute from (6.157) for the electric field:

$$\frac{4\pi}{\mu_0} \mathbf{L}_{em} = e \int \frac{1}{r} \mathbf{n} \times (\mathbf{n} \times \mathbf{H}) \, d^3x = -e \int \frac{1}{r} [\mathbf{H} - \mathbf{n}(\mathbf{n} \cdot \mathbf{H})] \, d^3x$$

Using a vector identity from the front flyleaf, this can be expressed as

$$4\pi \mathbf{L}_{em} = -e \int (\mathbf{B} \cdot \nabla)\mathbf{n} \, d^3x$$

where $\mathbf{B} = \mu_0 \mathbf{H}$. Integration by parts gives

$$4\pi \mathbf{L}_{em} = e \int \mathbf{n}(\nabla \cdot \mathbf{B}) \, d^3x - e \int_S \mathbf{n}(\mathbf{B} \cdot \mathbf{n}_S) \, da$$

where the second integral is over a surface S at infinity and $\mathbf{n}_S$ is the outward normal to that surface. With $\mathbf{B}$ from (6.157) this surface integral reduces to $(g/4\pi)\int \mathbf{n} \, d\Omega = 0$, since $\mathbf{n}$ is radially directed and has zero angular average. Since $\mathbf{B}$ is caused by a point monopole at $\mathbf{x} = \mathbf{R}$, its divergence is $\nabla \cdot \mathbf{B} = g \, \delta(\mathbf{x} - \mathbf{R})$. The field angular momentum is therefore*

$$\mathbf{L}_{em} = \frac{eg}{4\pi} \frac{\mathbf{R}}{R} \tag{6.159}$$

---

*This result was first stated by J. J. Thomson, *Elements of the Mathematical Theory of Electricity and Magnetism*, Cambridge University Press, Section 284 of the third (1904) and subsequent editions. The argument of Section 284 is exactly the converse of ours. From the conservation of angular momentum, Thomson deduces the magnetic part $e(\mathbf{v} \times \mathbf{B})$ of the Lorentz force.

It is directed along the line from the electric to the magnetic charge and has magnitude equal to the product of the charges (in SI units) divided by 4π. If we now think of the collision process of Fig. 6.6 and the total angular momentum of the system, that is, the sum of the angular momenta of the particle and the electromagnetic field, we see that the total angular momentum is conserved. The change (6.156) in the angular momentum of the particle is just balanced by the change in the electromagnetic angular momentum (6.159) caused by the reversal of the direction $\mathbf{R}$. A systematic discussion of the classical and quantum-mechanical scattering problem, including the electromagnetic angular momentum, is given by Goldhaber (*loc. cit.*).

The Thomson result (6.159) was used by Saha[*] and independently by Wilson[†] to derive the Dirac condition (6.153) by semiclassical means. To get $n/2$ *instead of n* when only the field angular momentum is considered, it is necessary to postulate half-integral quantization of $\mathbf{L}_{em}$, a somewhat undesirable hypothesis for the electromagnetic field.

Finally, we present a simplified discussion of Dirac's original (1931) argument leading to (6.153). In discussing the quantum mechanics of an electron in the presence of a magnetic monopole it is desirable to change as little as possible of the formalism of electromagnetic interactions, and to keep, for example, the interaction Hamiltonian in the standard form,

$$H_{\text{int}} = e\Phi - \frac{e}{m}\,\mathbf{p}\cdot\mathbf{A} + \frac{e^2}{2m}\,\mathbf{A}\cdot\mathbf{A}$$

where Φ and $\mathbf{A}$ are the scalar and vector potentials of the external sources. To do this with a magnetic charge it is necessary to employ an artifice. The magnetic charge g is imagined to be the end of a line of dipoles or a tightly wound solenoid that stretches off to infinity, as shown in Fig. 6.8. The monopole and its attached string, as the line of dipoles or solenoid is called, can then be treated more or less normally within the framework of conventional electromagnetic interactions where $\mathbf{B} = \nabla \times \mathbf{A}$, etc. From (5.55) we see that the elemental vector potential $d\mathbf{A}$ for a magnetic dipole element $d\mathbf{m}$ at $\mathbf{x}'$ is

$$d\mathbf{A}(\mathbf{x}) = -\frac{\mu_0}{4\pi}\,d\mathbf{m} \times \nabla\!\left(\frac{1}{|\mathbf{x} - \mathbf{x}'|}\right) \tag{6.160}$$

Thus for a string of dipoles or solenoid whose location is given by the string L the vector potential is

$$\mathbf{A}_L(\mathbf{x}) = -\frac{g}{4\pi}\int_L d\mathbf{l} \times \nabla\!\left(\frac{1}{|\mathbf{x} - \mathbf{x}'|}\right) \tag{6.161}$$

For all points except on the string, this vector potential has a curl that is directed radially outward from the end of the string, varies inversely with distance squared, with total outward flux g, as expected for the $\mathbf{B}$ field of monopole g. On the string itself the vector potential is singular. This singular behavior is equivalent to an intense field $\mathbf{B}'$ *inside* the solenoid and bringing a return contribution of flux $(-g)$ in along the string to cancel the pole's outward flow. So far we have

---

[*]M. N. Saha, *Indian J. Phys.* **10**, 141 (1936); *Phys. Rev.* **75**, 1968 (1949).

[†]H. A. Wilson, *Phys. Rev.* **75**, 309 (1949).

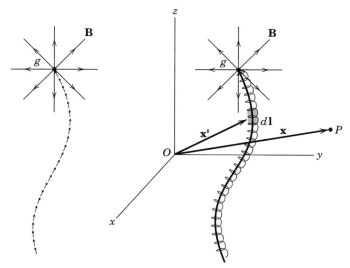

Figure 6.8   Two representations of a magnetic monopole g, one as the termination of a line of dipoles and the other as the end of a tightly wound solenoid, both "strings" stretching off to infinity.

just described a long thin solenoid. To exhibit the field of the monopole alone we write

$$\mathbf{B}_{\text{monopole}} = \mathbf{\nabla} \times \mathbf{A} - \mathbf{B}'$$

where $\mathbf{B}'$ exists only on the string (inside the solenoid). Dirac now argued that to describe the interaction of the electron with a magnetic *monopole*, rather than with a long thin solenoid, it is mandatory that the electron never "see" the singular field $\mathbf{B}'$. He thus required the electronic wave function to vanish along the string. This arbitrary postulate has been criticized, but discussion of such aspects leads us too far afield and is not central to our limited purpose. Dirac's later work (1948) treats the question of the unobservability of the strings in detail.

If (6.161) for $\mathbf{A}_L(\mathbf{x})$ is accepted as the appropriate vector potential for a monopole and its string L, there remains the problem of the arbitrariness of the location of the string. Clearly, the physical observables should not depend on where the string is. We now show that a choice of different string positions is equivalent to different choices of gauge for the vector potential. Indeed, the requirements of gauge invariance of the Schrödinger equation and single-valuedness of the wave function lead to the Dirac quantization condition (6.153). Consider two different strings L and L', as shown in Fig. 6.9. The difference of the two vector potentials is given by (6.161) with the integral taken along the closed path $C = L' - L$ around the area S. By Problem 5.1, this can be written

$$\mathbf{A}_{L'}(\mathbf{x}) = \mathbf{A}_L(\mathbf{x}) + \frac{g}{4\pi} \mathbf{\nabla}\Omega_C(\mathbf{x}) \tag{6.162}$$

where Ω_C is the solid angle subtended by the contour C at the observation point $\mathbf{x}$. Comparison with the gauge transformation equations, $\mathbf{A} \to \mathbf{A}' = \mathbf{A} + \mathbf{\nabla}\chi$, $\Phi \to \Phi' = \Phi - (1/c)(\partial\chi/\partial t)$, shows that a change in string from L to L' is equivalent to a gauge transformation, $\chi = g\Omega_C/4\pi$.

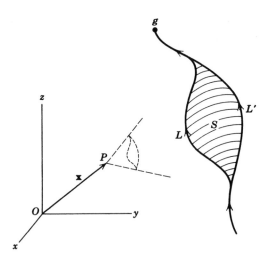

Figure 6.9 Two different strings L and L' give monopole vector potentials differing by a gauge transformation involving the gradient of the solid angle $\Omega_C(\mathbf{x})$ subtended at the observation point P by the surface S spanning the contour $C = L' - L$.

It is well known in quantum mechanics* that a change in the gauge of the electromagnetic potentials leaves the form of the Schrödinger equation invariant, provided the wave function is transformed according to

$$\psi \to \psi' = \psi e^{ie\chi/\hbar}$$

where e is the charge of the particle and χ is the gauge function. A change in the location of the string from L to L' must therefore be accompanied by a modification of the phase of the wave function of the electron,

$$\psi \to \psi' = \psi e^{ieg\Omega_C/4\pi\hbar} \tag{6.163}$$

Since Ω_C changes suddenly by 4π as the electron crosses the surface S, the wave function will be multiple-valued unless we require

$$\frac{eg}{\hbar} = 2\pi n \qquad (n = 0, \pm 1, \pm 2, \dots)$$

This is the Dirac quantization condition (6.153). It follows from the general requirements of gauge invariance and single-valuedness of the wave function, independent of the location of the monopole's string.

The preceding discussion of magnetic monopoles presents only the most basic concepts. An extensive literature exists on modifications of the quantization condition, attempts at a quantum electrodynamics with magnetic monopoles and electric charges, and other aspects. The interested reader can pursue the subject through the article by Goldhaber and Trower (*op. cit.*) and the references at the end of the chapter.

6.13 Polarization Potentials (Hertz Vectors)

It is sometimes useful to utilize potentials other than the standard scalar and vector potentials as auxiliary fields from which to determine the fundamental electromagnetic fields. The most important of these are the *polarization potentials*

---

*The demonstration is very easy. See, for example, H. A. Kramers, *Quantum Mechanics*, North-Holland, Amsterdam, (1957); Dover reprint (1964), Section 62.

or *Hertz vectors*, introduced by Hertz (1889) and Righi (1901). As the name suggests, these potentials put the electric and magnetic polarization densities to the fore. We consider linear, isotopic media with sources of external polarization densities, $\mathbf{P}_{ext}$ and $\mathbf{M}_{ext}$, but no separate macroscopic charge or current. The media are described by electric and magnetic susceptibilities, ϵ and μ. [The realistic situation of frequency-dependent quantities can be abstracted by choosing a unique sinusoidal time dependence and then using Fourier superposition.]

The macroscopic fields are written

$$\mathbf{D} = \epsilon\mathbf{E} + \mathbf{P}_{ext}; \qquad \mathbf{B} = \mu\mathbf{H} + \mu_0\mathbf{M}_{ext} \tag{6.164}$$

Then with the standard definitions (6.7) and (6.9) of the fields in terms of the scalar and vector potentials, the macroscopic Maxwell equations yield the wave equations,

$$\mu\epsilon\frac{\partial^2\mathbf{A}}{\partial t^2} - \nabla^2\mathbf{A} = \mu\frac{\partial\mathbf{P}_{ext}}{\partial t} + \mu_0\nabla \times \mathbf{M}_{ext} \tag{6.165a}$$

$$\mu\epsilon\frac{\partial^2\Phi}{\partial t^2} - \nabla^2\Phi = -\frac{1}{\epsilon}\nabla \cdot \mathbf{P}_{ext} \tag{6.165b}$$

with

$$\nabla \cdot \mathbf{A} + \mu\epsilon\frac{\partial\Phi}{\partial t} = 0$$

as the Lorenz condition. Two vector polarization potentials, $\mathbf{\Pi}_e$ and $\mathbf{\Pi}_m$, are introduced by writing $\mathbf{A}$ and Φ in a form paralleling the structures of the right-hand sides of the wave equations (6.165), namely,

$$\mathbf{A} = \mu\frac{\partial\mathbf{\Pi}_e}{\partial t} + \mu_0\nabla \times \mathbf{\Pi}_m; \qquad \Phi = -\frac{1}{\epsilon}\nabla \cdot \mathbf{\Pi}_e \tag{6.166}$$

When we substitute these definitions into (6.165), we find that the Lorenz condition is automatically satisfied. The wave equations become the following equations for $\mathbf{\Pi}_e$ and $\mathbf{\Pi}_m$:

$$\nabla \cdot \left[\nabla^2\mathbf{\Pi}_e - \mu\epsilon\frac{\partial^2\mathbf{\Pi}_e}{\partial t^2} + \mathbf{P}_{ext} \right] = 0 \tag{6.167a}$$

$$\mu\frac{\partial}{\partial t}\left[\nabla^2\mathbf{\Pi}_e - \mu\epsilon\frac{\partial^2\mathbf{\Pi}_e}{\partial t^2} + \mathbf{P}_{ext} \right] + \mu_0\nabla \times \left[\nabla^2\mathbf{\Pi}_m - \mu\epsilon\frac{\partial^2\mathbf{\Pi}_m}{\partial t^2} + \mathbf{M}_{ext} \right] = 0 \tag{6.167b}$$

From (6.167a) we find that the square-bracketed quantity can at most be equal to the curl of some vector function, call it $(\mu_0/\mu)\mathbf{V}$. When this form is inserted into (6.167b), we have a vanishing curl of a vector quantity that must therefore be equal at most to the gradient of some scalar field, call it $\partial\xi/\partial t$. The result is that the Hertz vectors satisfy the wave equations,

$$\mu\epsilon\frac{\partial^2\mathbf{\Pi}_e}{\partial t^2} - \nabla^2\mathbf{\Pi}_e = \mathbf{P}_{ext} - \frac{\mu_0}{\mu}\nabla \times \mathbf{V} \tag{6.168a}$$

$$\mu\epsilon\frac{\partial^2\mathbf{\Pi}_m}{\partial t^2} - \nabla^2\mathbf{\Pi}_m = \mathbf{M}_{ext} + \frac{\partial\mathbf{V}}{\partial t} + \nabla\frac{\partial\xi}{\partial t} \tag{6.168b}$$

It is left to the problems (Problem 6.23) to show that the arbitrary functions **V** and ξ may be removed by a gauge transformation on the polarization potentials. We may thus set **V** and ξ equal to zero with no loss of generality.

The electric and magnetic fields are given in terms of the Hertz vectors by

$$\mathbf{E} = \frac{1}{\epsilon} \nabla(\nabla \cdot \mathbf{\Pi}_e) - \mu \frac{\partial^2 \mathbf{\Pi}_e}{\partial t^2} - \mu_0 \nabla \times \frac{\partial \mathbf{\Pi}_m}{\partial t} \tag{6.169a}$$

$$\mathbf{B} = \mu \nabla \times \frac{\partial \mathbf{\Pi}_e}{\partial t} + \mu_0 \nabla \times \nabla \times \mathbf{\Pi}_m \tag{6.169b}$$

Outside the source $\mathbf{P}_{ext}$ the wave equation (6.168a) can be used to express **E** in a form analogous to (6.169b) for **B** with the roles of the electric and magnetic Hertz vectors interchanged.

The wave equations for $\mathbf{\Pi}_e$ and $\mathbf{\Pi}_m$ have solutions that are particularly simple if the external polarization densities are simple. For example, a time-dependent magnetic dipole at the point $\mathbf{x}_0$ has a magnetization density,

$$\mathbf{M}_{ext} = \mathbf{m}(t)\delta(\mathbf{x} - \mathbf{x}_0)$$

From the form of the wave equation (6.41) and its solution (6.47), we deduce that the magnetic Hertz vector is

$$\mathbf{\Pi}_m(\mathbf{x}, t) = \frac{\mathbf{m}(t - \sqrt{\mu\epsilon}R)}{4\pi R}$$

where $R = |\mathbf{x} - \mathbf{x}_0|$.

Illustrations of the use of polarization potentials can be found in *Born and Wolf*, in *Stratton*, and in *Panofsky and Phillips*, who discuss elementary multipole radiation in terms of a Hertz vector. We find it adequate to work with the usual potentials **A** and Φ or the fields themselves.

References and Suggested Reading

The conservation laws for the energy and momentum of electromagnetic fields are discussed in almost all text books. For example,

Panofsky and Phillips, Chapter 10

Stratton, Chapter II

Landau and Lifshitz, *Electrodynamics of Continuous Media* (Sections 15, 16, 34) discuss the Maxwell stress tensor in some detail in considering forces in fluids and solids.

The connection of lumped circuit concepts to a description using fields is given by

Adler, Chu, and Fano

Fano, Chu, and Adler

as has already been mentioned. The description of resonant cavities as circuit elements is treated in a classic paper by

W. W. Hansen, *J. Appl. Phys.* **9**, 654 (1938).

A thought-provoking discussion of the derivation of the macroscopic equations of electromagnetism, as well as of the thermodynamics of electric and magnetic systems, is given by

Robinson

The derivation of the macroscopic Maxwell equations from a statistical-mechanical point of view has long been the subject of research for a school of Dutch physicists. Their conclusions are contained in two comprehensive books,

> de Groot
> de Groot and Suttorp

A treatment of the energy, momentum, and Maxwell stress tensor of electromagnetic fields somewhat at variance with these authors is given by

> Penfield and Haus,
> Haus and Melcher

For the reader who wishes to explore the detailed quantum-mechanical treatment of dielectric constants and macroscopic field equations in matter, the following are suggested:

> S. L. Adler, *Phys. Rev*, **126**, 413 (1962).
> B. D. Josephson, *Phys. Rev.* **152**, 21 (1966).
> G. D. Mahan, *Phys. Rev.* **153**, 983 (1967).

Symmetry properties of electromagnetic fields under reflection and rotation are discussed by

> Argence and Kahan

The subject of magnetic monopoles has an extensive literature. We have already cited the paper by Goldhaber and his review with Trower, as well as the fundamental papers of Dirac. The relevance of monopoles to particle physics is discussed by

> J. Schwinger, *Science* **165**, 757 (1969).

The interest in and status of searches for magnetic monopoles up to the 1980s can be found in

> R. A. Carrigan and W. P. Trower, *Magnetic Monopoles*, NATO Adv. Sci. Inst. Series B, Physics, Vol. 102, Plenum Press, New York (1983).

The mathematical topics in this chapter center around the wave equation. The initial-value problem in one, two, three, and more dimensions is discussed by

> Morse and Feshbach (pp. 843–847)

and, in more mathematical detail, by

> Hadamard

Problems

6.1   In three dimensions the solution to the wave equation (6.32) for a point source in space and time (a light flash at $t' = 0$, $\mathbf{x}' = 0$) is a spherical shell disturbance of radius $R = ct$, namely the Green function $G^{(+)}$ (6.44). It may be initially surprising that in one or two dimensions, the disturbance possesses a "wake," even though the source is a "point" in space and time. The solutions for fewer dimensions than three can be found by superposition in the superfluous dimension(s), to eliminate dependence on such variable(s). For example, a flashing line source of uniform amplitude is equivalent to a point source in two dimensions.

(a)   Starting with the retarded solution to the three-dimensional wave equation (6.47), show that the source $f(\mathbf{x}', t') = \delta(x')\delta(y')\delta(t')$, equivalent to a $t = 0$ point source at the origin in two spatial dimensions, produces a two-dimensional wave,

$$\Psi(x, y, t) = \frac{2c\Theta(ct - \rho)}{\sqrt{c^2 t^2 - \rho^2}}$$

where $\rho^2 = x^2 + y^2$ and $\Theta(\xi)$ is the unit step function [$\Theta(\xi) = 0$ (1) if $\xi < (>) 0$.]

(b) Show that a "sheet" source, equivalent to a point pulsed source at the origin in one space dimension, produces a one-dimensional wave proportional to

$$\Psi(x, t) = 2\pi c\Theta(ct - |x|)$$

6.2 The charge and current densities for a single point charge q can be written formally as

$$\rho(\mathbf{x}', t') = q\delta[\mathbf{x}' - \mathbf{r}(t')]; \qquad \mathbf{J}(\mathbf{x}', t') = q\mathbf{v}(t')\delta[\mathbf{x}' - \mathbf{r}(t')]$$

where $\mathbf{r}(t')$ is the charge's position at time t' and $\mathbf{v}(t')$ is its velocity. In evaluating expressions involving the retarded time, one must put $t' = t_{\text{ret}} = t - R(t')/c$, where $\mathbf{R} = \mathbf{x} - \mathbf{r}(t')$.

(a) As a preliminary to deriving the Heaviside–Feynman expressions for the electric and magnetic fields of a point charge, show that

$$\int d^3x' \, \delta[\mathbf{x}' - \mathbf{r}(t_{\text{ret}})] = \frac{1}{\kappa}$$

where $\kappa = 1 - \mathbf{v} \cdot \hat{\mathbf{R}}/c$. Note that κ is evaluated at the retarded time.

(b) Starting with the Jefimenko generalizations of the Coulomb and Biot–Savart laws, use the expressions for the charge and current densities for a point charge and the result of part a to obtain the Heaviside–Feynman expressions for the electric and magnetic fields of a point charge,

$$\mathbf{E} = \frac{q}{4\pi\epsilon_0} \left\{ \left[\frac{\hat{\mathbf{R}}}{\kappa R^2}\right]_{\text{ret}} + \frac{\partial}{c\partial t}\left[\frac{\hat{\mathbf{R}}}{\kappa R}\right]_{\text{ret}} - \frac{\partial}{c^2\partial t}\left[\frac{\mathbf{v}}{\kappa R}\right]_{\text{ret}} \right\}$$

and

$$\mathbf{B} = \frac{\mu_0 q}{4\pi} \left\{ \left[\frac{\mathbf{v} \times \hat{\mathbf{R}}}{\kappa R^2}\right]_{\text{ret}} + \frac{\partial}{c\partial t}\left[\frac{\mathbf{v} \times \hat{\mathbf{R}}}{\kappa R}\right]_{\text{ret}} \right\}$$

(c) In our notation Feynman's expression for the electric field is

$$\mathbf{E} = \frac{q}{4\pi\epsilon_0} \left\{ \left[\frac{\hat{\mathbf{R}}}{R^2}\right]_{\text{ret}} + \frac{[R]_{\text{ret}}}{c}\frac{\partial}{\partial t}\left[\frac{\hat{\mathbf{R}}}{R^2}\right]_{\text{ret}} + \frac{\partial^2}{c^2\partial t^2}[\hat{\mathbf{R}}]_{\text{ret}} \right\}$$

while Heaviside's expression for the magnetic field is

$$\mathbf{B} = \frac{\mu_0 q}{4\pi} \left\{ \left[\frac{\mathbf{v} \times \hat{\mathbf{R}}}{\kappa^2 R^2}\right]_{\text{ret}} + \frac{1}{c[R]_{\text{ret}}}\frac{\partial}{\partial t}\left[\frac{\mathbf{v} \times \hat{\mathbf{R}}}{\kappa}\right]_{\text{ret}} \right\}$$

Show the equivalence of the two sets of expressions for the fields.
References: O. Heaviside, *Electromagnetic Theory*, Vol. 3 (1912), p. 464, Eq. (214). R. P. Feynman, *The Feynman Lectures in Physics*, Vol. 1 (1963), Chapter 28, Eq. (28.3).

6.3 The homogeneous diffusion equation (5.160) for the vector potential for quasi-static fields in unbounded conducting media has a solution to the initial value problem of the form,

$$\mathbf{A}(\mathbf{x}, t) = \int d^3x' \, G(\mathbf{x} - \mathbf{x}', t)\mathbf{A}(\mathbf{x}', 0)$$

where $\mathbf{A}(\mathbf{x}', 0)$ describes the initial field configuration and G is an appropriate kernel.

(a) Solve the initial value problem by use of a three-dimensional Fourier transform in space for $\mathbf{A}(\mathbf{x}, t)$. With the usual assumptions on interchange of orders of integration, show that the Green function has the Fourier representation,

$$G(\mathbf{x} - \mathbf{x}', t) = \frac{1}{(2\pi)^3} \int d^3k \; e^{-k^2 t/\mu\sigma} e^{i\mathbf{k}\cdot(\mathbf{x}-\mathbf{x}')}$$

and it is assumed that $t > 0$.

(b) By introducing a Fourier decomposition in both space and time, and performing the frequency integral in the complex ω plane to recover the result of part a, show that $G(\mathbf{x}-\mathbf{x}', t)$ is the diffusion Green function that satisfies the inhomogeneous equation,

$$\frac{\partial G}{\partial t} - \frac{1}{\mu\sigma} \nabla^2 G = \delta^{(3)}(\mathbf{x} - \mathbf{x}')\delta(t)$$

and vanishes for $t < 0$. [Note that μ must be constant in space here, but σ does not.]

(c) Show that if σ is uniform throughout all space, the Green function is

$$G(\mathbf{x}, t; \mathbf{x}', 0) = \Theta(t)\left(\frac{\mu\sigma}{4\pi t}\right)^{3/2} \exp\left(\frac{-\mu\sigma|\mathbf{x} - \mathbf{x}'|^2}{4t}\right)$$

(d) Suppose that at time $t' = 0$, the initial vector potential $\mathbf{A}(\mathbf{x}', 0)$ is nonvanishing only in a localized region of linear extent d around the origin. The time dependence of the fields is observed at a point P far from the origin, i.e., $|\mathbf{x}| = r \gg d$. Show that there are three regimes of time, $0 < t \leq T_1$, $T_1 \leq t \leq T_2$, and $t \gg T_2$. Give plausible definitions of T_1 and T_2, and describe qualitatively the time dependence at P. Show that in the last regime, the vector potential is proportional to the volume integral of $\mathbf{A}(\mathbf{x}', 0)$ times $t^{-3/2}$, assuming that integral exists. Relate your discussion to those of Section 5.18.B and Problems 5.35 and 5.36.

6.4 A uniformly magnetized and conducting sphere of radius R and total magnetic moment $m = 4\pi MR^3/3$ rotates about its magnetization axis with angular speed ω In the steady state no current flows in the conductor. The motion is nonrelativistic; the sphere has no excess charge on it.

(a) By considering Ohm's law in the moving conductor, show that the motion induces an electric field and a uniform volume charge density in the conductor, $\rho = -m\omega/\pi c^2 R^3$.

(b) Because the sphere is electrically neutral, there is no monopole electric field outside. Use symmetry arguments to show that the lowest possible electric multipolarity is quadrupole. Show that only a quadrupole field exists outside and that the quadrupole moment tensor has nonvanishing components, $Q_{33} = -4m\omega R^2/3c^2$, $Q_{11} = Q_{22} = -Q_{33}/2$.

(c) By considering the radial electric fields inside and outside the sphere, show that the necessary surface-charge density $\sigma(\theta)$ is

$$\sigma(\theta) = \frac{1}{4\pi R^2} \cdot \frac{4m\omega}{3c^2} \cdot \left[1 - \frac{5}{2} P_2(\cos\theta)\right]$$

(d) The rotating sphere serves as a unipolar induction device if a stationary circuit is attached by a slip ring to the pole and a sliding contact to the equator. Show that the line integral of the electric field from the equator contact to the pole contact (by any path) is $\mathscr{E} = \mu_0 m\omega/4\pi R$.

[See Landau and Lifshitz, *Electrodynamics of Continuous Media*, p. 221, for an alternative discussion of this electromotive force.]

6.5 A localized electric charge distribution produces an electrostatic field, $\mathbf{E} = -\nabla\Phi$. Into this field is placed a small localized time-independent current density $\mathbf{J}(\mathbf{x})$, which generates a magnetic field $\mathbf{H}$.

(a) Show that the momentum of these electromagnetic fields, (6.117), can be transformed to

$$\mathbf{P}_{\text{field}} = \frac{1}{c^2} \int \Phi \mathbf{J}\ d^3x$$

provided the product $\Phi\mathbf{H}$ falls off rapidly enough at large distances. How rapidly is "rapidly enough"?

(b) Assuming that the current distribution is localized to a region small compared to the scale of variation of the electric field, expand the electrostatic potential in a Taylor series and show that

$$\mathbf{P}_{\text{field}} = \frac{1}{c^2}\ \mathbf{E}(0) \times \mathbf{m}$$

where $\mathbf{E}(0)$ is the electric field at the current distribution and $\mathbf{m}$ is the magnetic moment, (5.54), caused by the current.

(c) Suppose the current distribution is placed instead in a *uniform* electric field $\mathbf{E}_0$ (filling all space). Show that, no matter how complicated is the localized $\mathbf{J}$, the result in part a is augmented by a surface integral contribution from infinity equal to minus one-third of the result of part b, yielding

$$\mathbf{P}_{\text{field}} = \frac{2}{3c^2}\ \mathbf{E}_0 \times \mathbf{m}$$

Compare this result with that obtained by working directly with (6.117) and the considerations at the end of Section 5.6.

6.6 **(a)** Consider a circular toroidal coil of mean radius a and N turns, with a small uniform cross section of area A (both height and width small compared to a). The toroid has a current I flowing in it and there is a point charge Q located at its center. Calculate all the components of field momentum of the system; show that the component along the axis of the toroid is

$$(\mathbf{P}_{\text{field}})_{\mathbf{z}} \approx \pm\frac{\mu_0 QINA}{4\pi a^2}$$

where the sign depends on the sense of the current flow in the coil. Assume that the electric field of the charge penetrates unimpeded into the region of nonvanishing magnetic field, as would happen for a toroid that is actually a set of N small nonconducting tubes inside which ionized gas moves to provide the current flow.

Check that the answer conforms to the approximation of Problem 6.5b.

(b) If $Q = 10^{-6}$ C ($\approx 6 \times 10^{12}$ electronic charges), $I = 1.0$ A, $N = 2000$, $A = 10^{-4}$ m^2, $a = 0.1$ m, find the electric field at the toroid in volts per meter, the magnetic induction in tesla, and the electromagnetic momentum in newton-seconds. Compare with the momentum of a 10 μg insect flying at a speed of 0.1 m/s.

[Note that the system of charge and toroid is at rest. Its *total* momentum must vanish. There must therefore be a canceling "hidden" mechanical momentum—see Problem 12.8.]

6.7   The microscopic current $\mathbf{j}(\mathbf{x}, t)$ can be written as

$$\mathbf{j}(\mathbf{x}, t) = \sum_j q_j \mathbf{v}_j \delta(\mathbf{x} - \mathbf{x}_j(t))$$

where the point charge q_j is located at the point $\mathbf{x}_j(t)$ and has velocity $\mathbf{v}_j = d\mathbf{x}_j(t)/dt$. Just as for the charge density, this current can be broken up into a "free" (conduction) electron contribution and a bound (molecular) current contribution.

Following the averaging procedures of Section 6.6 *and* assuming *nonrelativistic* addition of velocities, consider the averaged current, $\langle \mathbf{j}(\mathbf{x}, t)\rangle$.

(a)   Show that the averaged current can be written in the form of (6.96) with the definitions (6.92), (6.97), and (6.98).

(b)   Show that for a medium whose internal molecular velocities can be neglected, but which is in bulk motion (i.e., $\mathbf{v}_n = \mathbf{v}$ for all n),

$$\frac{1}{\mu_0} \mathbf{B} - \mathbf{H} = \mathbf{M} + (\mathbf{D} - \epsilon_0 \mathbf{E}) \times \mathbf{v}$$

This shows that a moving polarization ($\mathbf{P}$) produces an effective magnetization density.

Hints for part a: Consider quantities like $(d\mathbf{p}_n/dt)$, $(dQ_{\alpha\beta}^{\prime(n)}/dt)$ and see what they look like. Also note that

$$\frac{df}{dt}(\mathbf{x} - \mathbf{x}_n(t)) = -\mathbf{v}_n \cdot \nabla f(\mathbf{x} - \mathbf{x}_n(t))$$

6.8   A dielectric sphere of dielectric constant ϵ and radius a is located at the origin. There is a uniform applied electric field E_0 in the x direction. The sphere rotates with an angular velocity ω about the z axis. Show that there is a magnetic field $\mathbf{H} = -\nabla\Phi_M$, where

$$\Phi_M = \frac{3}{5}\left(\frac{\epsilon - \epsilon_0}{\epsilon + 2\epsilon_0}\right)\epsilon_0 E_0 \omega \left(\frac{a}{r_>}\right)^5 \cdot xz$$

where $r_>$ is the larger of r and a. The motion is nonrelativistic.

You may use the results of Section 4.4 for the dielectric sphere in an applied field.

6.9   Discuss the conservation of energy and linear momentum for a macroscopic system of sources and electromagnetic fields in a uniform, isotropic medium described by a permittivity ϵ and a permeability μ. Show that in a straightforward calculation the energy density, Poynting vector, field-momentum density, and Maxwell stress tensor are given by the Minkowski expressions,

$$u = \frac{1}{2}\left(\epsilon E^2 + \mu H^2\right)$$

$$\mathbf{S} = \mathbf{E} \times \mathbf{H}$$

$$\mathbf{g} = \mu\epsilon \mathbf{E} \times \mathbf{H}$$

$$T_{ij} = [\epsilon E_i E_j + \mu H_i H_j - \tfrac{1}{2}\delta_{ij}(\epsilon E^2 + \mu H^2)]$$

What modifications arise if ϵ and μ are functions of position?

6.10 With the same assumptions as in Problem 6.9 discuss the conservation of angular momentum. Show that the differential and integral forms of the conservation law are

$$\frac{\partial}{\partial t} (\mathscr{L}_{\text{mech}} + \mathscr{L}_{\text{field}}) + \boldsymbol{\nabla} \cdot \overset{\leftrightarrow}{\mathbf{M}} = 0$$

and

$$\frac{d}{dt} \int_V (\mathscr{L}_{\text{mech}} + \mathscr{L}_{\text{field}}) \, d^3x + \int_S \mathbf{n} \cdot \overset{\leftrightarrow}{\mathbf{M}} \, da = 0$$

where the field angular-momentum density is

$$\mathscr{L}_{\text{field}} = \mathbf{x} \times \mathbf{g} = \mu\epsilon \, \mathbf{x} \times (\mathbf{E} \times \mathbf{H})$$

and the flux of angular momentum is described by the tensor

$$\overset{\leftrightarrow}{\mathbf{M}} = \overset{\leftrightarrow}{\mathbf{T}} \times \mathbf{x}$$

Note: Here we have used the dyadic notation for M_{ij} and T_{ij}. A double-headed arrow conveys a fairly obvious meaning. For example, $\mathbf{n} \cdot \overset{\leftrightarrow}{\mathbf{M}}$ is a vector whose jth component is $\Sigma_i n_i M_{ij}$. The second-rank $\overset{\leftrightarrow}{\mathbf{M}}$ can be written as a third-rank tensor, $M_{ijk} = T_{ij}x_k - T_{ik}x_j$. But in the indices j and k it is antisymmetric and so has only three independent elements. Including the index i, M_{ijk} therefore has nine components and can be written as a pseudotensor of the second rank, as above.

6.11 A transverse plane wave is incident normally in vacuum on a perfectly absorbing flat screen.

(a) From the law of conservation of linear momentum, show that the pressure (called radiation pressure) exerted on the screen is equal to the field energy per unit volume in the wave.

(b) In the neighborhood of the earth the flux of electromagnetic energy from the sun is approximately 1.4 kW/m². If an interplanetary "sailplane" had a sail of mass 1 g/m² of area and negligible other weight, what would be its maximum acceleration in meters per second squared due to the solar radiation pressure? How does this compare with the acceleration due to the solar "wind" (corpuscular radiation)?

6.12 Consider the definition of the admittance $Y = G - iB$ of a two-terminal linear passive network in terms of field quantities by means of the complex Poynting theorem of Section 6.9.

(a) By considering the complex conjugate of (6.134) obtain general expressions for the conductance G and susceptance B for the general case including radiation loss.

(b) Show that at low frequencies the expressions equivalent to (6.139) and (6.140) are

$$G \simeq \frac{1}{|V_i|^2} \int_V \sigma \, |\mathbf{E}|^2 \, d^3x$$

$$B \simeq -\frac{4\omega}{|V_i|^2} \int_V (w_m - w_e) \, d^3x$$

6.13 A parallel plate capacitor is formed of two flat rectangular perfectly conducting sheets of dimensions a and b separated by a distance d small compared to a or b. Current is fed in and taken out uniformly along the adjacent edges of length b.

With the input current and voltage defined at this end of the capacitor, calculate the input impedance or admittance using the field concepts of Section 6.9.

(a) Calculate the electric and magnetic fields in the capacitor correct to second order in powers of the frequency, but neglecting fringing fields.

(b) Show that the expansion of the reactance (6.140) in powers of the frequency to an appropriate order is the same as that obtained for a lumped circuit consisting of a capacitance $C = \epsilon_0 ab/d$ in series with an inductance $L = \mu_0 ad/3b$.

6.14 An ideal circular parallel plate capacitor of radius a and plate separation $d \ll a$ is connected to a current source by axial leads, as shown in the sketch. The current in the wire is $I(t) = I_0 \cos \omega t$.

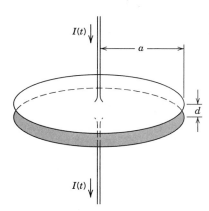

Problem 6.14

(a) Calculate the electric and magnetic fields between the plates to second order in powers of the frequency (or wave number), neglecting the effects of fringing fields.

(b) Calculate the volume integrals of w_e and w_m that enter the definition of the reactance X, (6.140), to second order in ω. Show that in terms of the input current I_i, defined by $I_i = -i\omega Q$, where Q is the *total charge* on one plate, these energies are

$$\int w_e \, d^3x = \frac{1}{4\pi\epsilon_0} \frac{|I_i|^2 \, d}{\omega^2 a^2}, \qquad \int w_m \, d^3x = \frac{\mu_0}{4\pi} \frac{|I_i|^2 \, d}{8} \left(1 + \frac{\omega^2 a^2}{12c^2}\right)$$

(c) Show that the equivalent series circuit has $C \simeq \pi\epsilon_0 a^2/d$, $L \simeq \mu_0 d/8\pi$, and that an estimate for the resonant frequency of the system is $\omega_{\mathrm{res}} \simeq 2\sqrt{2} \, c/a$. Compare with the first root of $J_0(x)$.

6.15 If a conductor or semiconductor has current flowing in it because of an applied electric field, and a transverse magnetic field is applied, there develops a component of electric field in the direction orthogonal to both the applied electric field (direction of current flow) and the magnetic field, resulting in a voltage difference between the sides of the conductor. This phenomenon is known as the *Hall effect*.

(a) Use the known properties of electromagnetic fields under rotations and spatial reflections and the assumption of Taylor series expansions around zero magnetic field strength to show that for an isotropic medium the generalization of Ohm's law, correct to second order in the magnetic field, must have the form

$$\mathbf{E} = \rho_0\mathbf{J} + R(\mathbf{H} \times \mathbf{J}) + \beta_1 H^2\mathbf{J} + \beta_2(\mathbf{H} \cdot \mathbf{J})\mathbf{H}$$

where ρ_0 is the resistivity in the absence of the magnetic field and R is called the Hall coefficient.

(b) What about the requirements of time reversal invariance?

6.16 (a) Calculate the force in newtons acting on a Dirac monopole of the minimum magnetic charge located a distance 0.5 Å from and in the median plane of a magnetic dipole with dipole moment equal to one nuclear magneton $(e\hbar/2m_p)$.

(b) Compare the force in part a with atomic forces such as the direct electrostatic force between charges (at the same separation), the spin-orbit force, the hyperfine interaction. Comment on the question of binding of magnetic monopoles to nuclei with magnetic moments. Assume that the monopole mass is at least that of a proton.

Reference: D. Sivers, *Phys. Rev.* **D2**, 2048 (1970).

6.17 (a) For a particle possessing both electric and magnetic charges, show that the generalization of the Lorentz force is

$$\mathbf{F} = q_e\mathbf{E} + q_m\mathbf{H} + q_e\mathbf{v} \times \mathbf{B} - q_m\mathbf{v} \times \mathbf{D}$$

(b) Show that this expression for the force is invariant under a duality transformation of both fields and charges, (6.151) and (6.152).

(c) Show that the Dirac quantization condition, (6.153), is generalized for two particles possessing electric and magnetic charges e_1, g_1 and e_2, g_2, respectively, to

$$\frac{e_1g_2 - e_2g_1}{\hbar} = 2\pi n$$

and that the relation is invariant under a duality transformation of the charges.

6.18 Consider the Dirac expression

$$\mathbf{A}(\mathbf{x}) = \frac{g}{4\pi} \int_L \frac{d\mathbf{l}' \times (\mathbf{x} - \mathbf{x}')}{|\mathbf{x} - \mathbf{x}'|^3}$$

for the vector potential of a magnetic monopole and its associated string L. Suppose for definiteness that the monopole is located at the origin and the string along the negative z axis.

(a) Calculate $\mathbf{A}$ explicitly and show that in spherical coordinates it has components

$$A_r = 0, \qquad A_\theta = 0, \qquad A_\phi = \frac{g(1 - \cos\theta)}{4\pi r \sin\theta} = \left(\frac{g}{4\pi r}\right)\tan\left(\frac{\theta}{2}\right)$$

(b) Verify that $\mathbf{B} = \nabla \times \mathbf{A}$ is the Coulomb-like field of a point charge, except perhaps at $\theta = \pi$.

(c) With the $\mathbf{B}$ determined in part b, evaluate the total magnetic flux passing through the circular loop of radius $R \sin\theta$ shown in the figure. Consider $\theta < \pi/2$ and $\theta > \pi/2$ separately, but always calculate the upward flux.

(d) From $\oint \mathbf{A} \cdot d\mathbf{l}$ around the loop, determine the total magnetic flux through the loop. Compare the result with that found in part c. Show that they are equal for $0 < \theta < \pi/2$, but have a *constant* difference for $\pi/2 < \theta < \pi$. Interpret this difference.

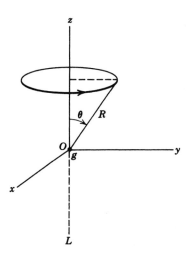

Problem 6.18

6.19 **(a)** Apply space inversion to the monopole vector potential of Problem 6.18 and show that the vector potential becomes

$$A'_\phi = -g \frac{(1 + \cos\theta)}{4\pi r \sin\theta} = -\frac{g}{4\pi r} \cot\left(\frac{\theta}{2}\right)$$

with the other components vanishing. Show explicitly that its curl gives the magnetic field of a magnetic monopole, except perhaps at $\theta = 0$. [Remember the space-inversion properties of the magnetic charge!]

(b) Show that the difference, $\delta\mathbf{A} = \mathbf{A}' - \mathbf{A}$, can be expressed as the gradient of a scalar function, indicating that the original and space-inverted vector potentials differ by a gauge transformation.

(c) Interpret the gauge function in terms of Fig. 6.9. [*Hint:* Choose the contour C to be a rectangle lying in a plane containing the z axis, with three sides at infinity.]

6.20 An example of the preservation of causality and finite speed of propagation in spite of the use of the Coulomb gauge is afforded by a dipole source that is flashed on and off at $t = 0$. The effective charge and current densities are

$$\rho(\mathbf{x}, t) = \delta(x)\delta(y)\delta'(z)\delta(t)$$
$$J_z(\mathbf{x}, t) = -\delta(x)\delta(y)\delta(z)\delta'(t)$$

where a prime means differentiation with respect to the argument. This dipole is of unit strength and it points in the negative z direction.

(a) Show that the instantaneous Coulomb potential (6.23) is

$$\Phi(\mathbf{x}, t) = -\frac{1}{4\pi\epsilon_0} \delta(t) \frac{z}{r^3}$$

(b) Show that the *transverse* current $\mathbf{J}_t$ is

$$\mathbf{J}_t(\mathbf{x}, t) = -\delta'(t)\left[\frac{2}{3}\boldsymbol{\epsilon}_3\delta(\mathbf{x}) - \frac{\boldsymbol{\epsilon}_3}{4\pi r^3} + \frac{3}{4\pi r^3}\mathbf{n}(\boldsymbol{\epsilon}_3 \cdot \mathbf{n})\right]$$

where the factor of 2/3 multiplying the delta function comes from treating the gradient of z/r^3 according to (4.20).

(c) Show that the electric and magnetic fields are causal and that the electric field components are

$$E_x(\mathbf{x}, t) = \frac{1}{4\pi\epsilon_0} \frac{c}{r} \left[-\delta''(r - ct) + \frac{3}{r} \delta'(r - ct) - \frac{3}{r^2} \delta(r - ct) \right] \sin\theta \cos\theta \cos\phi$$

E_y is the same as E_x, with $\cos\phi$ replaced by $\sin\phi$, and

$$E_z(\mathbf{x}, t) = \frac{1}{4\pi\epsilon_0} \frac{c}{r} \left[\sin^2\theta\delta''(r - ct) + (3\cos^2\theta - 1) \cdot \left(\frac{\delta'(r - ct)}{r} - \frac{\delta(r - ct)}{r^2} \right) \right]$$

Hint: While the answer in part b displays the transverse current explicitly, the less explicit form

$$\mathbf{J}_t(\mathbf{x}, t) = -\delta'(t) \left[\epsilon_3 \delta(\mathbf{x}) + \frac{1}{4\pi} \nabla \frac{\partial}{\partial z} \left(\frac{1}{r} \right) \right]$$

can be used with (6.47) to calculate the vector potential and the fields for part c. An alternative method is to use the Fourier transforms in time of $\mathbf{J}_t$ and $\mathbf{A}$, the Green function (6.40) and its spherical wave expansion from Chapter 9.

6.21 An electric dipole of dipole moment $\mathbf{p}$, fixed in direction, is located at a position $\mathbf{r}_0(t)$ with respect to the origin. Its velocity $\mathbf{v} = d\mathbf{r}_0/dt$ is nonrelativistic.

(a) Show that the dipole's charge and current densities can be expressed formally as

$$\rho(\mathbf{x}, t) = -(\mathbf{p} \cdot \nabla)\delta(\mathbf{x} - \mathbf{r}_0(t)); \qquad \mathbf{J}(\mathbf{x}, t) = -\mathbf{v}(\mathbf{p} \cdot \nabla)\delta(\mathbf{x} - \mathbf{r}_0(t))$$

(b) Show that the off-center moving dipole gives rise to a magnetic dipole field and an electric quadrupole field in addition to an electric dipole field, with moments

$$\mathbf{m} = \tfrac{1}{2}\mathbf{p} \times \mathbf{v}$$

and

$$Q_{ij} = 3(x_{0i}p_j + x_{0j}p_i) - 2\mathbf{r}_0 \cdot \mathbf{p}\delta_{ij}$$

[There are, of course, still higher moments.]

(c) Show that the quasi-static electric quadrupole field is

$$\mathbf{E}(\mathbf{x}) = \frac{1}{4\pi\epsilon_0} \frac{1}{r^4} [15\mathbf{n}(\mathbf{n} \cdot \mathbf{r}_0)(\mathbf{n} \cdot \mathbf{p}) - 3\mathbf{r}_0(\mathbf{n} \cdot \mathbf{p}) - 3\mathbf{p}(\mathbf{n} \cdot \mathbf{r}_0) - 3\mathbf{n}(\mathbf{r}_0 \cdot \mathbf{p})]$$

where $\mathbf{n}$ is a unit vector in the radial direction.

6.22 (a) For the off-center, slowly moving, electric dipole of Problem 6.21, show that the quasi-static vector potential produced by the current flow associated with the dipole motion is

$$\mathbf{A}(\mathbf{x}, t) = \frac{\mu_0 \mathbf{v}(\mathbf{n} \cdot \mathbf{p})}{4\pi r^2} = \frac{\mu_0}{4\pi} \left[\frac{1}{2} \frac{(\mathbf{p} \times \mathbf{v}) \times \mathbf{x}}{r^3} + \frac{1}{2} \frac{[\mathbf{p}(\mathbf{x} \cdot \mathbf{v}) + \mathbf{v}(\mathbf{x} \cdot \mathbf{p})]}{r^3} \right]$$

where the first term of the second form (antisymmetric in $\mathbf{v}$ and $\mathbf{p}$) is the vector potential of the magnetic dipole whose moment is given in Problem 6.21. The added term is symmetric in $\mathbf{v}$ and $\mathbf{p}$.

(b) Show that the magnetic field of the symmetric term is

$$\mathbf{B}_{\text{sym}} = -\frac{3\mu_0}{8\pi r^3} \mathbf{n} \times [\mathbf{p}(\mathbf{v} \cdot \mathbf{n}) + \mathbf{v}(\mathbf{p} \cdot \mathbf{n})]$$

(c) By calculating its curl, show that $\mathbf{B}_{\text{sym}}$ is consistent with being the quasi-static magnetic field associated with the electric quadrupole field of Problem 6.21c.

(d) Show that the total magnetic field (computed from the first form of the vector potential, i.e., the sum of $\mathbf{B}_{\text{sym}}$ and the magnetic *dipole* field) is

$$\mathbf{B} = \frac{\mu_0}{4\pi}\,\mathbf{v} \times \frac{[3\mathbf{n}(\mathbf{n} \cdot \mathbf{p}) - \mathbf{p}]}{r^3}$$

Comment.

6.23 The wave equations (6.168) for the Hertz vectors contain arbitrary source terms involving the functions $\mathbf{V}$ and ξ. Consider the gauge transformations

$$\mathbf{\Pi}'_e = \mathbf{\Pi}_e + \mu_0 \nabla \times \mathbf{G} - \nabla g; \qquad \mathbf{\Pi}'_m = \mathbf{\Pi}_m - \mu\,\frac{\partial \mathbf{G}}{\partial t}$$

where $\mathbf{G}$ and g are well-behaved functions of space and time.

(a) Show that, if $\mathbf{G}$ and g satisfy the wave equations

$$\left(\mu\epsilon\,\frac{\partial^2}{\partial t^2} - \nabla^2\right)\left\{\begin{matrix} \mathbf{G} \\ g \end{matrix}\right\} = \left\{\begin{matrix} \dfrac{1}{\mu}\,(\mathbf{V} + \nabla\xi) \\ 0 \end{matrix}\right\}$$

then the new polarization potentials $\mathbf{\Pi}'_e$ and $\mathbf{\Pi}'_m$ satisfy (6.168) with vanishing $\mathbf{V}$ and ξ.

(b) Show that the gauge transformation on the Hertz vectors is equivalent to a gauge transformation on $\mathbf{A}$ and $\mathbf{\Phi}$. What is the gauge function Λ of (6.19) in terms of $\mathbf{G}$ and g?

6.24 A current distribution $\mathbf{J}(\mathbf{x}, t)$ localized near the origin varies slowly in time.

(a) Use the Jefimenko expressions (6.55) and (6.56) for the retarded fields to evaluate the quasi-static fields far from the current distribution. Assuming that there are no electric multipole moments and retaining only the magnetic dipole contributions, show that the magnetic and electric fields at the point $(\mathbf{x} = \hat{\mathbf{r}}r, t)$ to first order in an expansion in successive time derivatives are

$$\mathbf{B} = \frac{\mu_0}{4\pi}\frac{1}{r^3}\left(1 + \frac{r}{c}\frac{\partial}{\partial t}\right)[3(\mathbf{m}(t - r/c) \cdot \hat{\mathbf{r}})\hat{\mathbf{r}} - \mathbf{m}(t - r/c)]$$

$$\mathbf{E} = \frac{\mu_0}{4\pi}\frac{1}{r^2}\,\hat{\mathbf{r}} \times \frac{\partial \mathbf{m}(t - r/c)}{\partial t}$$

(b) The construction and excitation of an infinite, straight, right circular solenoid of radius a, with N turns per unit length, are such that its current $I(t)$ is the same everywhere along its length and is changed very slowly in time. Show that the fields far from the solenoid are approximately

$$\mathbf{B} \approx \frac{\pi\mu_0}{8c}\frac{Na^2}{\rho}\frac{\partial I(t - \rho/c)}{\partial t}\,\hat{\mathbf{z}}$$

$$\mathbf{E} \approx -\frac{\mu_0}{2}\frac{Na^2}{\rho}\frac{\partial I(t - \rho/c)}{\partial t}\,\hat{\boldsymbol{\phi}}$$

where ρ is the perpendicular distance from the axis, provided $\max(|dI/dt/I|) \ll c/\rho$. A long solenoid with a time-varying current has a magnetic and an electric field outside it, in contrast to the static situation. Verify that Faraday's law is satisfied. Does the time-varying "outside" magnetic induction contribute to Faraday's law to this order?

6.25 **(a)** Starting with the Lorentz force expression (6.114), show that in the dipole approximation the force acting on a neutral atom at rest can be expressed as

$$\frac{d\mathbf{P}_{atom}}{dt} = (\mathbf{d} \cdot \boldsymbol{\nabla}) \, \mathbf{E} + \dot{\mathbf{d}} \times \mathbf{B}$$

where $\mathbf{d}$ is the atomic dipole moment and $\mathbf{E}$ and $\mathbf{B}$ are the electric and magnetic fields at the site of the atom.

(b) For a uniform plane wave of frequency ω in a nonmagnetic tenuous dielectric medium with index of refraction $n(\omega)$, show that the time rate of change of mechanical momentum per unit volume $\mathbf{g}_{mech}$ accompanying the electromagnetic momentum $\mathbf{g}_{em}$ (6.118) of the wave is

$$\frac{d\mathbf{g}_{mech}}{dt} = \frac{1}{2} \, (n^2 - 1) \, \frac{d\mathbf{g}_{em}}{dt}$$

[see Peierls (loc. cit.) for corrections for dense media and non-uniform waves.]

Note of explanation:

The reader may be startled to find (in all but the earliest printings) the association of Danish physicist Ludvig V. Lorenz's name instead of Dutch physicist Hendrik A. Lorentz's with the relation (6.14) between the scalar and vector potentials. Yet it is a fact that in 1867 Lorenz, in a paper entitled "On the identity of the vibrations of light with electrical currents," (op. cit.) exploited the retarded solutions for the potentials, derived (6.14) and equations equivalent to wave equations for the electric field, and discussed the characteristics of light propagation in conductors and transparent media, contemporaneously with Maxwell. H. A. Lorentz has ample recognition in physics terminology without the mis-attribution of (6.14) to him (by others, beginning around 1900). As Van Bladel* observes, it is up to textbook authors to accord Lorenz his due.[†]

---

*J. Van Bladel, IEEE Antennas and Propagation Magazine **33**, No. 2, 69 (April 1991).

[†]An earlier author who deplored the lack of recognition of Lorenz's contributions is A. O'Rahilly, *Electromagnetic Theory,* Dover Publications, New York (1965) [originally published as *Electromagnetics,* Longman Green and Cork University Press (1938)], footnote, p. 184.

CHAPTER 7

Plane Electromagnetic Waves and Wave Propagation

This chapter on plane waves in unbounded, or perhaps semi-infinite, media treats first the basic properties of plane electromagnetic waves in nonconducting media—their transverse nature, linear and circular polarization states. Then the important Fresnel formulas for reflection and refraction at a plane interface are derived and applied. This is followed by a survey of the high-frequency dispersion properties of dielectrics, conductors, and plasmas. The richness of nature is illustrated with a panoramic view (Fig. 7.9) of the index of refraction and absorption coefficient of liquid water over 20 decades of frequency. Then comes a simplified discussion of propagation in the ionosphere, and of magnetohydrodynamic waves in a conducting fluid. The ideas of phase and group velocities and the spreading of a pulse or wave packet as it propagates in a dispersive medium come next. The important subject of causality and its consequences for the dispersive properties of a medium are discussed in some detail, including the Kramers–Kronig dispersion relations and various sum rules derived from them. The chapter concludes with a treatment of the classic problem of the arrival of a signal in a dispersive medium, first discussed by Sommerfeld and Brillouin (1914) but only recently subjected to experimental test.

7.1 *Plane Waves in a Nonconducting Medium*

A basic feature of the Maxwell equations for the electromagnetic field is the existence of traveling wave solutions which represent the transport of energy from one point to another. The simplest and most fundamental electromagnetic waves are transverse, plane waves. We proceed to see how such solutions can be obtained in simple nonconducting media described by spatially constant permeability and susceptibility. In the absence of sources, the Maxwell equations in an infinite medium are

$$\nabla \cdot \mathbf{B} = 0, \qquad \nabla \times \mathbf{E} + \frac{\partial \mathbf{B}}{\partial t} = 0$$

$$\nabla \cdot \mathbf{D} = 0, \qquad \nabla \times \mathbf{H} - \frac{\partial \mathbf{D}}{\partial t} = 0 \tag{7.1}$$

Assuming solutions with harmonic time dependence $e^{-i\omega t}$, from which we can build an arbitrary solution by Fourier superposition, the equations for the amplitudes $\mathbf{E}(\omega, \mathbf{x})$, etc. read

$$\nabla \cdot \mathbf{B} = 0, \qquad \nabla \times \mathbf{E} - i\omega\mathbf{B} = 0$$
$$\nabla \cdot \mathbf{D} = 0, \qquad \nabla \times \mathbf{H} + i\omega\mathbf{D} = 0$$

For uniform isotropic linear media we have $\mathbf{D} = \epsilon\mathbf{E}$, $\mathbf{B} = \mu\mathbf{H}$, where ϵ and μ may in general be complex functions of ω. We assume for the present that they are real (no losses). Then the equations for $\mathbf{E}$ and $\mathbf{H}$ are

$$\nabla \times \mathbf{E} - i\omega\mathbf{B} = 0, \qquad \nabla \times \mathbf{B} + i\omega\mu\epsilon\mathbf{E} = 0 \tag{7.2}$$

The zero-divergence equations are not independent, but are obtained by taking divergences in (7.2). By combining the two equations we get the Helmholtz wave equation

$$(\nabla^2 + \mu\epsilon\omega^2)\begin{Bmatrix} \mathbf{E} \\ \mathbf{B} \end{Bmatrix} = 0 \tag{7.3}$$

Consider as a possible solution a plane wave traveling in the x direction, $e^{ikx-i\omega t}$. From (7.3) we find the requirement that the wave number k and the frequency ω are related by

$$k = \sqrt{\mu\epsilon}\,\omega \tag{7.4}$$

The *phase velocity* of the wave is

$$v = \frac{\omega}{k} = \frac{1}{\sqrt{\mu\epsilon}} = \frac{c}{n}, \qquad n = \sqrt{\frac{\mu}{\mu_0}\frac{\epsilon}{\epsilon_0}} \tag{7.5}$$

The quantity n is called the *index of refraction* and is usually a function of frequency. The primordial solution in one dimension is

$$u(x, t) = ae^{ikx-i\omega t} + be^{-ikx-i\omega t} \tag{7.6}$$

Using $k = \omega v$ from (7.5), this can be written

$$u_k(x, t) = ae^{ik(x-vt)} + be^{-ik(x+vt)}$$

If the medium is nondispersive ($\mu\epsilon$ independent of frequency), the Fourier superposition theorem (2.44) and (2.45) can be used to construct a general solution of the form

$$u(x, t) = f(x - vt) + g(x + vt) \tag{7.7}$$

where $f(z)$ and $g(z)$ are arbitrary functions. Equation (7.7) represents waves traveling in the positive and negative x directions with speeds equal to the phase velocity v.

If the medium is dispersive, the basic solution (7.6) still holds, but when we build up a wave as an arbitrary function of x and t, the dispersion produces modifications. Equation (7.7) no longer holds. The wave changes shape as it propagates (see Sections 7.8, 7.9, and 7.11).

We now consider an electromagnetic plane wave of frequency ω and wave vector $\mathbf{k} = k\mathbf{n}$ and require that it satisfy not only the Helmholtz wave equation (7.3) but also all the Maxwell equations. The constraint imposed by (7.3) is essentially kinematic; those imposed by the Maxwell equations, dynamic. With the convention that the physical electric and magnetic fields are obtained by taking the real parts of complex quantities, we write the plane wave fields as

$$\mathbf{E}(\mathbf{x}, t) = \mathscr{E}e^{ik\mathbf{n}\cdot\mathbf{x}-i\omega t} \tag{7.8}$$
$$\mathbf{B}(x, t) = \mathscr{B}e^{ik\mathbf{n}\cdot\mathbf{x}-i\omega t}$$

where $\mathscr{E}$, $\mathscr{B}$, and **n** are constant vectors. Each component of **E** and **B** satisfies (7.3) provided

$$k^2 \, \mathbf{n} \cdot \mathbf{n} = \mu\epsilon\omega^2 \tag{7.9}$$

To recover (7.4) it is necessary that **n** be a unit vector such that $\mathbf{n} \cdot \mathbf{n} = 1$. With the wave equation satisfied, there only remains the fixing of the vectorial properties so that the Maxwell equations (7.1) are valid. The divergence equations in (7.1) demand that

$$\mathbf{n} \cdot \mathscr{E} = 0 \qquad \text{and} \qquad \mathbf{n} \cdot \mathscr{B} = 0 \tag{7.10}$$

This means that **E** and **B** are both perpendicular to the direction of propagation **n**. Such a wave is called a *transverse wave*. The curl equations provide a further restriction, namely

$$\mathscr{B} = \sqrt{\mu\epsilon} \, \mathbf{n} \times \mathscr{E} \tag{7.11}$$

The factor $\sqrt{\mu\epsilon}$ can be written $\sqrt{\mu\epsilon} = n/c$, where n is the index of refraction defined in (7.5). We thus see that $c\mathbf{B}$ and **E**, which have the *same dimensions*, have the *same magnitude* for plane electromagnetic waves in free space and differ by the index of refraction in ponderable media. In engineering literature the magnetic field **H** is often displayed in parallel to **E** instead of **B**. The analog of (7.11) for **H** is

$$\mathscr{H} = \mathbf{n} \times \mathscr{E}/Z \tag{7.11'}$$

where $Z = \sqrt{\mu/\epsilon}$ is an impedance. In vacuum, $Z = Z_0 = \sqrt{\mu_0/\epsilon_0} \approx 376.7$ ohms, the impedance of free space.

If **n** is real, (7.11) implies that $\mathscr{E}$ and $\mathscr{B}$ have the same phase. It is then useful to introduce a set of real mutually orthogonal unit vectors $(\boldsymbol{\epsilon}_1, \boldsymbol{\epsilon}_2, \mathbf{n})$, as shown in Fig. 7.1. In terms of these unit vectors the field strengths $\mathscr{E}$ and $\mathscr{B}$ are

$$\mathscr{E} = \boldsymbol{\epsilon}_1 E_0, \qquad \mathscr{B} = \boldsymbol{\epsilon}_2 \sqrt{\mu\epsilon} \, E_0 \tag{7.12}$$

or

$$\mathscr{E} = \boldsymbol{\epsilon}_2 E_0', \qquad \mathscr{B} = -\boldsymbol{\epsilon}_1 \sqrt{\mu\epsilon} \, E_0' \tag{7.12'}$$

where E_0 and E_0' are constants, possibly complex. The wave described by (7.8) and (7.12) or (7.12') is a transverse wave propagating in the direction **n**. It rep-

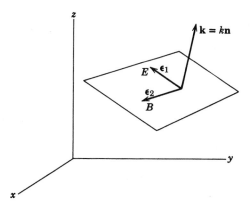

Figure 7.1 Propagation vector **k** and two orthogonal polarization vectors $\boldsymbol{\epsilon}_1$ and $\boldsymbol{\epsilon}_2$.

resents a time-averaged flux of energy given by the real part of the complex Poynting vector:

$$\mathbf{S} = \frac{1}{2}\,\mathbf{E} \times \mathbf{H}^*$$

The energy flow (energy per unit area per unit time) is

$$\mathbf{S} = \frac{1}{2}\,\sqrt{\frac{\epsilon}{\mu}}\,|E_0|^2\,\mathbf{n} \tag{7.13}$$

The time-averaged energy density u is correspondingly

$$u = \frac{1}{4}\left(\epsilon\mathbf{E} \cdot \mathbf{E}^* + \frac{1}{\mu}\,\mathbf{B} \cdot \mathbf{B}^*\right)$$

This gives

$$u = \frac{\epsilon}{2}\,|E_0|^2 \tag{7.14}$$

The ratio of the magnitude of (7.13) to (7.14) shows that the speed of energy flow is $v = 1/\sqrt{\mu\epsilon}$, as expected from (7.5).

In the discussion that follows (7.11) we assumed that $\mathbf{n}$ was a real unit vector. This does not yield the most general possible solution for a plane wave. Suppose that $\mathbf{n}$ is complex, and written as $\mathbf{n} = \mathbf{n}_R + i\mathbf{n}_I$. Then the exponential in (7.8) becomes

$$e^{ik\mathbf{n}\cdot\mathbf{x}-i\omega t} = e^{-k\mathbf{n}_I\cdot\mathbf{x}}e^{ik\mathbf{n}_R\cdot\mathbf{x}-i\omega t}$$

The wave possesses exponential growth or decay in some directions. It is then called an *inhomogeneous plane wave*. The surfaces of constant amplitude and constant phase are still planes, but they are no longer parallel. The relations (7.10) and (7.11) still hold. The requirement $\mathbf{n} \cdot \mathbf{n} = 1$ has real and imaginary parts,*

$$n_R^2 - n_I^2 = 1 \tag{7.15}$$
$$\mathbf{n}_R \cdot \mathbf{n}_I = 0$$

The second of these conditions shows that $\mathbf{n}_R$ and $\mathbf{n}_I$ are orthogonal. The coordinate axes can be oriented so that $\mathbf{n}_R$ is in the x direction and $\mathbf{n}_I$ in the y direction. The first equation in (7.15) can be satisfied generally by writing

$$\mathbf{n} = \mathbf{e}_1 \cosh\theta + i\mathbf{e}_2 \sinh\theta \tag{7.16}$$

where θ is a real constant and $\mathbf{e}_1$ and $\mathbf{e}_2$ are real unit vectors in the x and y directions (not to be confused with $\boldsymbol{\epsilon}_1$ and $\boldsymbol{\epsilon}_2$!). The most general vector $\boldsymbol{\mathscr{E}}$ satisfying $\mathbf{n} \cdot \boldsymbol{\mathscr{E}} = 0$ is then

$$\boldsymbol{\mathscr{E}} = (i\mathbf{e}_1 \sinh\theta - \mathbf{e}_2 \cosh\theta)A + \mathbf{e}_3 A' \tag{7.17}$$

where A and A' are complex constants. For $\theta \neq 0$, $\boldsymbol{\mathscr{E}}$ in general has components in the direction(s) of $\mathbf{n}$. It is easily verified that for $\theta = 0$, the solutions (7.12) and (7.12') are recovered.

We encounter simple examples of inhomogeneous plane waves in the discussion of total internal reflection and refraction in a conducting medium later in the chapter, although in the latter case the inhomogeneity arises from a com-

---

*Note that if $\mathbf{n}$ is complex it does not have unit magnitude, that is, $\mathbf{n} \cdot \mathbf{n} = 1$ does not imply $|\mathbf{n}|^2 = 1$.

plex wave number, not a complex unit vector **n**. Inhomogeneous plane waves form a general basis for the treatment of boundary-value problems for waves and are especially useful in the solution of diffraction in two dimensions. The interested reader can refer to the book by *Clemmow* for an extensive treatment with examples.

7.2 Linear and Circular Polarization; Stokes Parameters

The plane wave (7.8) and (7.12) is a wave with its electric field vector always in the direction $\boldsymbol{\epsilon}_1$. Such a wave is said to be linearly polarized with polarization vector $\boldsymbol{\epsilon}_1$. Evidently the wave described in (7.12′) is linearly polarized with polarization vector $\boldsymbol{\epsilon}_2$ and is linearly independent of the first. Thus the two waves,

$$\mathbf{E}_1 = \boldsymbol{\epsilon}_1 E_1 e^{i\mathbf{k}\cdot\mathbf{x}-i\omega t}$$
$$\mathbf{E}_2 = \boldsymbol{\epsilon}_2 E_2 e^{i\mathbf{k}\cdot\mathbf{x}-i\omega t}$$

with

$$\mathbf{B}_j = \sqrt{\mu\epsilon}\,\frac{\mathbf{k}\times\mathbf{E}_j}{k}, \qquad j = 1, 2$$

(7.18)

can be combined to give the most general homogeneous plane wave propagating in the direction $\mathbf{k} = k\mathbf{n}$,

$$\mathbf{E}(\mathbf{x}, t) = (\boldsymbol{\epsilon}_1 E_1 + \boldsymbol{\epsilon}_2 E_2)e^{i\mathbf{k}\cdot\mathbf{x}-i\omega t} \tag{7.19}$$

The amplitudes E_1 and E_2 are complex numbers, to allow the possibility of a phase difference between waves of different linear polarization.

If E_1 *and* E_2 have the *same phase*, (7.19) represents a *linearly polarized* wave, with its polarization vector making an angle $\theta = \tan^{-1}(E_2/E_1)$ with $\boldsymbol{\epsilon}_1$ and a magnitude $E = \sqrt{E_1^2 + E_2^2}$, as shown in Fig. 7.2.

If E_1 *and* E_2 have *different phases*, the wave (7.19) is *elliptically polarized*. To understand what this means let us consider the simplest case, *circular polarization*. Then E_1 and E_2 have the same magnitude, but differ in phase by 90°. The wave (7.19) becomes:

$$\mathbf{E}(\mathbf{x}, t) = E_0(\boldsymbol{\epsilon}_1 \pm i\boldsymbol{\epsilon}_2)e^{i\mathbf{k}\cdot\mathbf{x}-i\omega t} \tag{7.20}$$

with E_0 the common real amplitude. We imagine axes chosen so that the wave is propagating in the positive z direction, while $\boldsymbol{\epsilon}_1$ and $\boldsymbol{\epsilon}_2$ are in the x and y directions, respectively. Then the components of the actual electric field, obtained by taking the real part of (7.20), are

$$\left.\begin{aligned}
E_x(\mathbf{x}, t) &= E_0 \cos(kz - \omega t) \\
E_y(\mathbf{x}, t) &= \mp E_0 \sin(kz - \omega t)
\end{aligned}\right\} \tag{7.21}$$

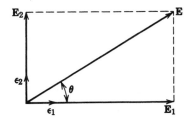

Figure 7.2 Electric field of a linearly polarized wave.

At a *fixed point in space*, the fields (7.21) are such that the electric vector is constant in magnitude, but sweeps around in a circle at a frequency ω, as shown in Fig. 7.3. For the upper sign ($\epsilon_1 + i\epsilon_2$), the rotation is counterclockwise when the observer is facing into the oncoming wave. This wave is called *left circularly polarized* in optics. In the terminology of modern physics, however, such a wave is said to have *positive helicity*. The latter description seems more appropriate because such a wave has a positive projection of angular momentum on the z axis (see Problem 7.29). For the lower sign ($\epsilon_1 - i\epsilon_2$), the rotation of **E** is clockwise when looking into the wave; the wave is *right circularly polarized* (optics); it has *negative helicity*.

The two circularly polarized waves (7.20) form an equally acceptable set of basic fields for description of a general state of polarization. We introduce the complex orthogonal unit vectors:

$$\epsilon_{\pm} = \frac{1}{\sqrt{2}}\,(\epsilon_1 \pm i\epsilon_2) \tag{7.22}$$

with properties

$$\begin{aligned}
\epsilon_{\pm}^* \cdot \epsilon_{\mp} &= 0 \\
\epsilon_{\pm}^* \cdot \epsilon_3 &= 0 \\
\epsilon_{\pm}^* \cdot \epsilon_{\pm} &= 1
\end{aligned} \tag{7.23}$$

Then a general representation, equivalent to (7.19), is

$$\mathbf{E}(\mathbf{x}, t) = (E_+\epsilon_+ + E_-\epsilon_-)e^{i\mathbf{k}\cdot\mathbf{x}-i\omega t} \tag{7.24}$$

where E_+ and E_- are complex amplitudes. If E_+ and E_- have different magnitudes, but the same phase, (7.24) represents an elliptically polarized wave with principal axes of the ellipse in the directions of ϵ_1 and ϵ_2. The ratio of semimajor to semiminor axis is $|(1 + r)/(1 - r)|$, where $E_-/E_+ = r$. If the amplitudes have a phase difference between them, $E_-/E_+ = re^{i\alpha}$, then it is easy to show that the ellipse traced out by the **E** vector has its axes rotated by an angle $(\alpha/2)$. Figure 7.4 shows the general case of elliptical polarization and the ellipses traced out by both **E** and **B** at a given point in space.

For $r = \pm 1$ we get back a linearly polarized wave.

The polarization content of a plane electromagnetic wave is known if it can be written in the form of either (7.19) or (7.24) with known coefficients (E_1, E_2) or (E_+, E_-). In practice, the converse problem arises. Given that the wave is of the form (7.8), how can we determine from observations on the beam the state of polarization in all its particulars? A useful vehicle for this are the four *Stokes*

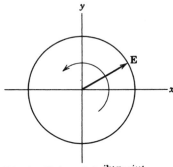

$$\mathbf{E}(\mathbf{x}, t) = E_0\,(\epsilon_1 + i\epsilon_2)e^{i\mathbf{k}\cdot\mathbf{x}-i\omega t}$$

Figure 7.3 Electric field of a circularly polarized wave.

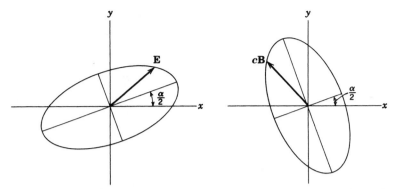

Figure 7.4   Electric field and magnetic induction for an elliptically polarized wave.

parameters, proposed by G. G. Stokes in 1852. These parameters are quadratic in the field strength and can be determined through intensity measurements only, in conjunction with a linear polarizer and a quarter-wave plate or equivalents. Their measurement determines completely the state of polarization of the wave.

The Stokes parameters can be motivated by observing that for a wave propagating in the z direction, the scalar products,

$$\boldsymbol{\epsilon}_1 \cdot \mathbf{E}, \qquad \boldsymbol{\epsilon}_2 \cdot \mathbf{E}, \qquad \boldsymbol{\epsilon}_+^* \cdot \mathbf{E}, \qquad \boldsymbol{\epsilon}_-^* \cdot \mathbf{E} \qquad (7.25)$$

are the amplitudes of radiation, respectively, with linear polarization in the x direction, linear polarization in the y direction, positive helicity, and negative helicity. Note that for circular polarization the *complex conjugate* of the appropriate polarization vector must be used, in accord with (7.23). The squares of these amplitudes give a measure of the intensity of each type of polarization. Phase information is also needed; this is obtained from cross products. We give definitions of the Stokes parameters with respect to both the linear polarization and the circular polarization bases, in terms of the projected amplitudes (7.25) and also explicitly in terms of the magnitudes and relative phases of the components. For the latter purpose we define each of the scalar coefficients in (7.19) and (7.24) as a magnitude times a phase factor:

$$\begin{aligned} E_1 &= a_1 e^{i\delta_1}, & E_2 &= a_2 e^{i\delta_2} \\ E_+ &= a_+ e^{i\delta_+}, & E_- &= a_- e^{i\delta_-} \end{aligned} \qquad (7.26)$$

In terms of the linear polarization basis $(\boldsymbol{\epsilon}_1, \boldsymbol{\epsilon}_2)$, the Stokes parameters are*

$$\begin{aligned} s_0 &= |\boldsymbol{\epsilon}_1 \cdot \mathbf{E}|^2 + |\boldsymbol{\epsilon}_2 \cdot \mathbf{E}|^2 = a_1^2 + a_2^2 \\ s_1 &= |\boldsymbol{\epsilon}_1 \cdot \mathbf{E}|^2 - |\boldsymbol{\epsilon}_2 \cdot \mathbf{E}|^2 = a_1^2 - a_2^2 \\ s_2 &= 2\,\mathrm{Re}[(\boldsymbol{\epsilon}_1 \cdot \mathbf{E})^*(\boldsymbol{\epsilon}_2 \cdot \mathbf{E})] = 2a_1 a_2 \cos(\delta_2 - \delta_1) \\ s_3 &= 2\,\mathrm{Im}[(\boldsymbol{\epsilon}_1 \cdot \mathbf{E})^*(\boldsymbol{\epsilon}_2 \cdot \mathbf{E})] = 2a_1 a_2 \sin(\delta_2 - \delta_1) \end{aligned} \qquad (7.27)$$

If the circular polarization basis $(\boldsymbol{\epsilon}_+, \boldsymbol{\epsilon}_-)$ is used instead, the definitions read

$$\begin{aligned} s_0 &= |\boldsymbol{\epsilon}_+^* \cdot \mathbf{E}|^2 + |\boldsymbol{\epsilon}_-^* \cdot \mathbf{E}|^2 = a_+^2 + a_-^2 \\ s_1 &= 2\,\mathrm{Re}[(\boldsymbol{\epsilon}_+^* \cdot \mathbf{E})^*(\boldsymbol{\epsilon}_-^* \cdot \mathbf{E})] = 2a_+ a_- \cos(\delta_- - \delta_+) \\ s_2 &= 2\,\mathrm{Im}[(\boldsymbol{\epsilon}_+^* \cdot \mathbf{E})^*(\boldsymbol{\epsilon}_-^* \cdot \mathbf{E})] = 2a_+ a_- \sin(\delta_- - \delta_+) \\ s_3 &= |\boldsymbol{\epsilon}_+^* \cdot \mathbf{E}|^2 - |\boldsymbol{\epsilon}_-^* \cdot \mathbf{E}|^2 = a_+^2 - a_-^2 \end{aligned} \qquad (7.28)$$

*The notation for the Stokes parameters is unfortunately not uniform. Stokes himself used (A, B, C, D); other labelings are (I, Q, U, V) and (I, M, C, S). Our notation is that of *Born and Wolf*.

The expressions (7.27) and (7.28) show an interesting rearrangement of roles of the Stokes parameters with respect to the two bases. The parameter s_0 measures the relative intensity of the wave in either case. The parameter s_1 gives the preponderance of x-linear polarization over y-linear polarization, while s_2 and s_3 in the linear basis give phase information. We see from (7.28) that s_3 has the interpretation of the difference in relative intensity of positive and negative helicity, while in this basis s_1 and s_2 concern the phases. The four Stokes parameters are not independent, since they depend on only three quantities, a_1, a_2, and $\delta_2 - \delta_1$. They satisfy the relation

$$s_0^2 = s_1^2 + s_2^2 + s_3^2 \tag{7.29}$$

Discussion of the operational steps needed to measure the Stokes parameters and so determine the state of polarization of a plane wave would take us too far afield. We refer the reader to Section 13.13 of *Stone* for details. Also neglected, except for the barest mention, is the important problem of quasi-monochromatic radiation. Beams of radiation, even if monochromatic enough for the purposes at hand, actually consist of a superposition of finite wave trains. By Fourier's theorem they thus contain a range of frequencies and are not completely monochromatic. One way of viewing this is to say that the magnitudes and phases $(a_i\ \delta_i)$ in (7.26) vary slowly in time, slowly, that is, when compared to the frequency ω. The observable Stokes parameters then become averages over a relatively long time interval, and are written as

$$s_2 = 2\langle a_1 a_2 \cos(\delta_2 - \delta_1) \rangle$$

for example, where the angle brackets indicate the macroscopic time average. One consequence of the averaging process is that the Stokes parameters for a quasi-monochromatic beam satisfy an inequality,

$$s_0^2 \geq s_1^2 + s_2^2 + s_3^2$$

rather than the equality, (7.29). "Natural light," even if monochromatic to a high degree, has $s_1 = s_2 = s_3 = 0$. Further discussion of quasi-monochromatic light and partial coherence can be found in *Born and Wolf*, Chapter 10.

An astrophysical example of the use of Stokes parameters to describe the state of polarization is afforded by the study of optical and radiofrequency radiation from the pulsar in the Crab nebula. The optical light shows some small amount of linear polarization, while the radio emission at $\omega \simeq 2.5 \times 10^9$ s^{-1} has a high degree of linear polarization.* At neither frequency is there evidence for circular polarization. Information of this type obviously helps to elucidate the mechanism of radiation from these fascinating objects.

7.3 *Reflection and Refraction of Electromagnetic Waves at a Plane Interface Between Dielectrics*

The reflection and refraction of light at a plane surface between two media of different dielectric properties are familiar phenomena. The various aspects of the phenomena divide themselves into two classes.

---

*See *The Crab Nebula and Related Supernova Remnants*, eds. M. C. Kafatos and R. B. C. Henry, Cambridge University Press, New York (1985).

1. Kinematic properties:
    (a) Angle of reflection equals angle of incidence.
    (b) Snell's law: $(\sin i)/(\sin r) = n'/n$, where i, r are the angles of incidence and refraction, while n, n' are the corresponding indices of refraction.
2. Dynamic properties:
    (a) Intensities of reflected and refracted radiation.
    (b) Phase changes and polarization.

The kinematic properties follow immediately from the wave nature of the phenomena and from the fact that there are boundary conditions to be satisfied. But they do not depend on the detailed nature of the waves or the boundary conditions. On the other hand, the dynamic properties depend entirely on the specific nature of electromagnetic fields and their boundary conditions.

The coordinate system and symbols appropriate to the problem are shown in Fig. 7.5. The media below and above the plane $z = 0$ have permeabilities and permittivities μ, ϵ and μ', ϵ', respectively. The indices of refraction, defined through (7.5), are $n = \sqrt{\mu\epsilon/\mu_0\epsilon_0}$ and $n' = \sqrt{\mu'\epsilon'/\mu_0\epsilon_0}$. A plane wave with wave vector $\mathbf{k}$ and frequency ω is incident from medium μ, ϵ. The refracted and reflected waves have wave vectors $\mathbf{k}'$ and $\mathbf{k}''$, respectively, and $\mathbf{n}$ is a unit normal directed from medium μ, ϵ into medium μ', ϵ'.

According to (7.18), the three waves are:

INCIDENT

$$\mathbf{E} = \mathbf{E}_0 e^{i\mathbf{k}\cdot\mathbf{x}-i\omega t}$$

$$\mathbf{B} = \sqrt{\mu\epsilon}\,\frac{\mathbf{k}\times\mathbf{E}}{k} \tag{7.30}$$

REFRACTED

$$\mathbf{E}' = \mathbf{E}_0' e^{i\mathbf{k}'\cdot\mathbf{x}-i\omega t}$$

$$\mathbf{B}' = \sqrt{\mu'\epsilon'}\,\frac{\mathbf{k}'\times\mathbf{E}'}{k'} \tag{7.31}$$

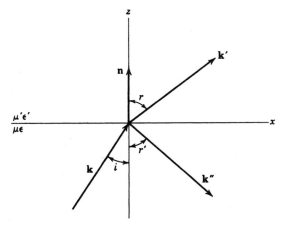

Figure 7.5 Incident wave $\mathbf{k}$ strikes plane interface between different media, giving rise to a reflected wave $\mathbf{k}''$ and a refracted wave $\mathbf{k}'$.

REFLECTED

$$\mathbf{E}'' = \mathbf{E}_0'' e^{i\mathbf{k}'' \cdot \mathbf{x} - i\omega t}$$

$$\mathbf{B}'' = \sqrt{\mu\epsilon}\,\frac{\mathbf{k}'' \times \mathbf{E}''}{k} \tag{7.32}$$

The wave numbers have the magnitudes

$$|\mathbf{k}| = |\mathbf{k}''| = k = \omega\sqrt{\mu\epsilon}$$

$$|\mathbf{k}'| = k' = \omega\sqrt{\mu'\epsilon'} \tag{7.33}$$

The existence of boundary conditions at $z = 0$, which must be satisfied at all points on the plane at all times, implies that the spatial (and time) variation of all fields must be the same at $z = 0$. Consequently, we must have the phase factors all equal at $z = 0$,

$$(\mathbf{k} \cdot \mathbf{x})_{z=0} = (\mathbf{k}' \cdot \mathbf{x})_{z=0} = (\mathbf{k}'' \cdot \mathbf{x})_{z=0} \tag{7.34}$$

independent of the nature of the boundary conditions. Equation (7.34) contains the kinematic aspects of reflection and refraction. We see immediately that all three wave vectors must lie in a plane. Furthermore, in the notation of Fig. 7.5,

$$k \sin i = k' \sin r = k'' \sin r' \tag{7.35}$$

Since $k'' = k$, we find $i = r'$; the angle of incidence equals the angle of reflection. Snell's law is

$$\frac{\sin i}{\sin r} = \frac{k'}{k} = \sqrt{\frac{\mu'\epsilon'}{\mu\epsilon}} = \frac{n'}{n} \tag{7.36}$$

The dynamic properties are contained in the boundary conditions—normal components of $\mathbf{D}$ and $\mathbf{B}$ are continuous; tangential components of $\mathbf{E}$ and $\mathbf{H}$ are continuous. In terms of fields (7.30)–(7.32) these boundary conditions at $z = 0$ are:

$$[\epsilon(\mathbf{E}_0 + \mathbf{E}_0'') - \epsilon'\mathbf{E}_0'] \cdot \mathbf{n} = 0$$

$$[\mathbf{k} \times \mathbf{E}_0 + \mathbf{k}'' \times \mathbf{E}_0'' - \mathbf{k}' \times \mathbf{E}_0'] \cdot \mathbf{n} = 0$$

$$(\mathbf{E}_0 + \mathbf{E}_0'' - \mathbf{E}_0') \times \mathbf{n} = 0 \tag{7.37}$$

$$\left[\frac{1}{\mu}(\mathbf{k} \times \mathbf{E}_0 + \mathbf{k}'' \times \mathbf{E}_0'') - \frac{1}{\mu'}(\mathbf{k}' \times \mathbf{E}_0')\right] \times \mathbf{n} = 0$$

In applying these boundary conditions it is convenient to consider two separate situations, one in which the incident plane wave is linearly polarized with its polarization vector perpendicular to the plane of incidence (the plane defined by $\mathbf{k}$ and $\mathbf{n}$), and the other in which the polarization vector is parallel to the plane of incidence. The general case of arbitrary elliptic polarization can be obtained by appropriate linear combinations of the two results, following the methods of Section 7.2.

We first consider the electric field perpendicular to the plane of incidence, as shown in Fig. 7.6a. All the electric fields are shown directed away from the viewer. The orientations of the $\mathbf{B}$ vectors are chosen to give a positive flow of energy in the direction of the wave vectors. Since the electric fields are all parallel

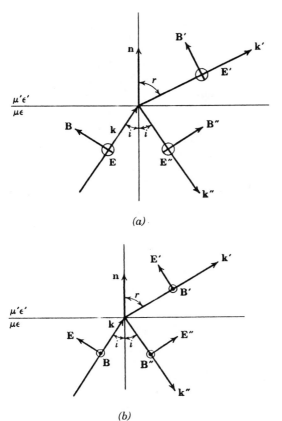

(a)

(b)

Figure 7.6 Reflection and refraction with polarization (*a*) perpendicular and (*b*) parallel to the plane of incidence.

to the surface, the first boundary condition in (7.37) yields nothing. The third and fourth equations in (7.37) give

$$E_0 + E_0'' - E_0' = 0$$

$$\sqrt{\frac{\epsilon}{\mu}}\, (E_0 - E_0'') \cos i - \sqrt{\frac{\epsilon'}{\mu'}}\, E_0' \cos r = 0 \tag{7.38}$$

while the second, using Snell's law, duplicates the third. The relative amplitudes of the refracted and reflected waves can be found from (7.38). These are:

E Perpendicular to Plane of Incidence

$$\frac{E_0'}{E_0} = \frac{2n \cos i}{n \cos i + \dfrac{\mu}{\mu'} \sqrt{n'^2 - n^2 \sin^2 i}}$$

$$\frac{E_0''}{E_0} = \frac{n \cos i - \dfrac{\mu}{\mu'} \sqrt{n'^2 - n^2 \sin^2 i}}{n \cos i + \dfrac{\mu}{\mu'} \sqrt{n'^2 - n^2 \sin^2 i}} \tag{7.39}$$

The square root in these expressions is $n' \cos r$, but Snell's law has been used to express it in terms of the angle of incidence. For optical frequencies it is usually permitted to put $\mu/\mu' = 1$. Equations (7.39), and (7.41) and (7.42) below, are most often employed in optical contexts with real n and n', but they are also valid for complex dielectric constants.

If the electric field is parallel to the plane of incidence, as shown in Fig. 7.6b, the boundary conditions involved are normal D, tangential E, and tangential H [the first, third, and fourth equations in (7.37)]. The tangential E and H continuous demand that

$$\cos i(E_0 - E_0'') - \cos r \, E_0' = 0$$
$$\sqrt{\frac{\epsilon}{\mu}} \, (E_0 + E_0'') - \sqrt{\frac{\epsilon'}{\mu'}} \, E_0' = 0 \tag{7.40}$$

Normal D continuous, plus Snell's law, merely duplicates the second of these équations. The relative amplitudes of refracted and reflected fields are therefore

E PARALLEL TO PLANE OF INCIDENCE

$$\frac{E_0'}{E_0} = \frac{2nn' \cos i}{\dfrac{\mu}{\mu'} \, n'^2 \cos i + n\sqrt{n'^2 - n^2 \sin^2 i}} \tag{7.41}$$

$$\frac{E_0''}{E_0} = \frac{\dfrac{\mu}{\mu'} \, n'^2 \cos i - n\sqrt{n'^2 - n^2 \sin^2 i}}{\dfrac{\mu}{\mu'} \, n'^2 \cos i + n\sqrt{n'^2 - n^2 \sin^2 i}}$$

For normal incidence ($i = 0$), both (7.39) and (7.41) reduce to

$$\left. \begin{aligned} \frac{E_0'}{E_0} &= \frac{2}{\sqrt{\dfrac{\mu\epsilon'}{\mu'\epsilon}} + 1} \rightarrow \frac{2n}{n' + n} \\[2em] \frac{E_0''}{E_0} &= \frac{\sqrt{\dfrac{\mu\epsilon'}{\mu'\epsilon}} - 1}{\sqrt{\dfrac{\mu\epsilon'}{\mu'\epsilon}} + 1} \rightarrow \frac{n' - n}{n' + n} \end{aligned} \right\} \tag{7.42}$$

where the results on the right hold for $\mu' = \mu$. For the reflected wave the sign convention is that for polarization parallel to the plane of incidence. This means that if $n' > n$ there is a phase reversal for the reflected wave.

7.4 *Polarization by Reflection and Total Internal Reflection; Goos–Hänchen Effect*

Two aspects of the dynamical relations on reflection and refraction are worthy of mention. The first is that for polarization parallel to the plane of incidence there is an angle of incidence, called *Brewster's angle*, for which there is no reflected wave. With $\mu' = \mu$ for simplicity, we find that the amplitude of the reflected wave in (7.41) vanishes when the angle of incidence is equal to Brewster's angle,

$$i_B = \tan^{-1}\left(\frac{n'}{n}\right) \tag{7.43}$$

For a typical ratio $n'/n = 1.5$, $i_B \simeq 56°$. If a plane wave of mixed polarization is incident on a plane interface at the Brewster angle, the reflected radiation is *completely* plane-*polarized* with polarization vector *perpendicular* to the plane of incidence. This behavior can be utilized to produce beams of plane-polarized light but is not as efficient as other means employing anisotropic properties of some dielectric media. Even if the unpolarized wave is reflected at angles other than the Brewster angle, there is a tendency for the reflected wave to be predominantly polarized perpendicular to the plane of incidence. The success of dark glasses that selectively transmit only one direction of polarization depends on this fact. In the domain of radiofrequencies, receiving antennas can be oriented to discriminate against surface-reflected waves (and also waves reflected from the ionosphere) in favor of the directly transmitted wave.

The second phenomenon is called *total internal reflection*. The word "internal" implies that the incident and reflected waves are in a medium of larger index of refraction than the refracted wave ($n > n'$). Snell's law (7.36) shows that, if $n > n'$, then $r > i$. Consequently, $r = \pi/2$ when $i = i_0$, where

$$i_0 = \sin^{-1}\left(\frac{n'}{n}\right) \tag{7.44}$$

For waves incident at $i = i_0$, the refracted wave is propagated parallel to the surface. There can be no energy flow across the surface. Hence at that angle of incidence there must be total reflection. What happens if $i > i_0$? To answer this we first note that, for $i > i_0$, $\sin r > 1$. This means that r is a complex angle with a purely imaginary cosine.

$$\cos r = i\sqrt{\left(\frac{\sin i}{\sin i_0}\right)^2 - 1} \tag{7.45}$$

The meaning of these complex quantities becomes clear when we consider the propagation factor for the refracted wave:

$$e^{i\mathbf{k}'\cdot\mathbf{x}} = e^{ik'(x\sin r + z\cos r)} = e^{-k'[(\sin i/\sin i_0)^2 - 1]^{1/2}z}e^{ik'(\sin i/\sin i_0)x} \tag{7.46}$$

This shows that, for $i > i_0$, the refracted wave is propagated only parallel to the surface and is attenuated exponentially beyond the interface. The attenuation occurs within a very few wavelengths of the boundary, except for $i \simeq i_0$.

Even though fields exist on the other side of the surface there is no energy flow through the surface. Hence total internal reflection occurs for $i \geq i_0$. The lack of energy flow can be verified by calculating the time-averaged *normal* component of the Poynting vector just inside the surface:

$$\mathbf{S} \cdot \mathbf{n} = \frac{1}{2}\,\text{Re}[\mathbf{n} \cdot (\mathbf{E}' \times \mathbf{H}'^*)] \tag{7.47}$$

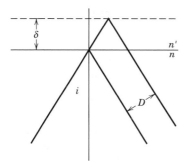

Figure 7.7 Geometrical interpretation of the Goos–Hänchen effect, the lateral displacement of a totally internally-reflected beam of radiation because of the penetration of the evanescent wave into the region of smaller index refraction.

with $\mathbf{H}' = (\mathbf{k}' \times \mathbf{E}')/\mu'\omega$, we find

$$\mathbf{S} \cdot \mathbf{n} = \frac{1}{2\omega\mu'} \, \text{Re}[(\mathbf{n} \cdot \mathbf{k}') \, |\mathbf{E}_0'|^2]$$

But $\mathbf{n} \cdot \mathbf{k}' = k' \cos r$ is purely imaginary, so that $\mathbf{S} \cdot \mathbf{n} = 0$.

The purely imaginary value (7.45) of $\cos r$, times n', is the appropriate quantity to replace the square root appearing in the Fresnel formula, (7.39) and (7.41). Inspection shows that the ratios E_0''/E_0 are now of modulus unity, as is expected physically for *total* internal reflection. The reflected wave does, however, suffer a phase change that is different for the two kinds of incidence and depends on the angle of incidence and on (n/n'). These phase changes can be utilized to convert one kind of polarization into another. Fresnel's rhombus is one such device, whereby linearly polarized light with equal amplitudes in the plane of incidence and perpendicular to it is converted by two successive internal reflections, each involving a relative phase change of 45°, into circularly polarized light (see *Born and Wolf*, p. 50).

The evanescent wave penetrating into the region $z > 0$ has an exponential decay in the perpendicular direction, $e^{-z/\delta}$, where $\delta^{-1} = k\sqrt{\sin^2 i - \sin^2 i_0}$. The penetration of the wave into the "forbidden" region is the physical origin of the *Goos–Hänchen effect*: If a beam of radiation having a finite transverse extent undergoes total internal reflection, the reflected beam emerges displaced laterally with respect to the prediction of a geometrical ray reflected at the boundary.* If we imagine that the beam is reflected from the plane a distance δ beyond the boundary, as indicated in Fig. 7.7, the beam should emerge with a transverse displacement of $D \approx 2\delta \sin i$. More careful calculation (see Problem 7.7) shows that this naive result is modified somewhat, with D dependent on the state of polarization of the radiation. The first-order expressions for D for the two states of linear polarization are

$$D_\perp = \frac{\lambda}{\pi} \frac{\sin i}{\sqrt{\sin^2 i - \sin^2 i_0}} ; \qquad D_\parallel = D_\perp \cdot \frac{\sin^2 i_0}{[\sin^2 i - \cos^2 i \cdot \sin^2 i_0]} \qquad (7.48)$$

where λ is the wavelength in the medium of higher index of refraction.

The phenomenon of internal reflection is exploited in many applications where it is desired to transmit light without loss in intensity. In nuclear and

*F. Goos and H. Hänchen, *Ann. Phys.* (*Leipzig*) (6) **1**, 333–346 (1947). For an extensive discussion of the effect, with many references, see the four-part article, H. K. V. Lotsch, *Optik*, **32**, 116–137, 189–204, 299–319, 553–568 (1970).

particle physics, plastic "light pipes" are used to carry light from scintillators (excited by the passage of a charged particle or energetic photon) to photomultipliers, where the light is converted into useful electrical signals. If the light pipe is large in cross-sectional dimension compared to the wavelength of the light involved, the considerations presented here for a plane interface have approximate validity. In telecommunications, optical fibers exploit total internal reflection for transmission of modulated light signals over long distances. The various transverse dimensions of a multilayered fiber are not always very large compared to a wavelength. Then the precise geometry must be taken into account; the language of modes in a waveguide may be more appropriate—see Chapter 8.

7.5 *Frequency Dispersion Characteristics of Dielectrics, Conductors, and Plasmas*

In Section 7.1 we saw that in the absence of dispersion an arbitrary wave train (7.7) travels without distortion. In reality all media show some dispersion. Only over a limited range of frequencies, or in vacuum, can the velocity of propagation be treated as constant in frequency. Of course, all the results of the preceding sections that involve a single frequency component are valid in the presence of dispersion. The values of μ and ϵ need only be interpreted as those appropriate to the frequency being considered. Where a superposition of a range of frequencies occurs, however, new effects arise as a result of the frequency dependence of ϵ and μ. To examine some of these consequences, we need to develop at least a simple model of dispersion.

A. *Simple Model for $\epsilon(\omega)$*

Almost all of the physics of dispersion is illustrated by an extension to time-varying fields of the classical model described in Section 4.6. For simplicity we neglect the difference between the applied electric field and the local field. The model is therefore appropriate only for substances of relatively low density. [This deficiency can be removed by use of (4.69), if desired.] The relative permeability will be taken equal to unity. The equation of motion for an electron of charge $-e$ bound by a harmonic force (4.71) and acted on by an electric field $\mathbf{E}(\mathbf{x}, t)$ is

$$m[\ddot{\mathbf{x}} + \gamma\dot{\mathbf{x}} + \omega_0^2\mathbf{x}] = -e\mathbf{E}(\mathbf{x}, t) \qquad (7.49)$$

where γ measures the phenomenological damping force. Magnetic force effects are neglected in (7.49). We make the additional approximation that the amplitude of oscillation is small enough to permit evaluation of the electric field at the average position of the electron. If the field varies harmonically in time with frequency ω as $e^{-i\omega t}$, the dipole moment contributed by one electron is

$$\mathbf{p} = -e\mathbf{x} = \frac{e^2}{m}(\omega_0^2 - \omega^2 - i\omega\gamma)^{-1}\mathbf{E} \qquad (7.50)$$

If we suppose that there are N molecules per unit volume with Z electrons per molecule, and that, instead of a single binding frequency for all, there are f_j

electrons per molecule with binding frequency ω_j and damping constant γ_j, then the dielectric constant, $\epsilon/\epsilon_0 = 1 + \chi_e$, is given by

$$\frac{\epsilon(\omega)}{\epsilon_0} = 1 + \frac{Ne^2}{\epsilon_0 m} \sum_j f_j(\omega_j^2 - \omega^2 - i\omega\gamma_j)^{-1} \tag{7.51}$$

where the *oscillator strengths* f_j satisfy the sum rule,

$$\sum_j f_j = Z \tag{7.52}$$

With suitable quantum-mechanical definitions of f_j, γ_j, and ω_j, (7.51) is an accurate description of the atomic contribution to the dielectric constant.

B. Anomolous Dispersion and Resonant Absorption

The damping constants γ_j are generally small compared with the binding or *resonant frequencies* ω_j. This means that $\epsilon(\omega)$ is approximately real for most frequencies. The factor $(\omega_j^2 - \omega^2)^{-1}$ is positive for $\omega < \omega_j$ and negative for $\omega > \omega_j$. Thus, at low frequencies, below the smallest ω_j, all the terms in the sum in (7.51) contribute with the same positive sign and $\epsilon(\omega)$ is greater than unity. As successive ω_j values are passed, more and more negative terms occur in the sum, until finally the whole sum is negative and $\epsilon(\omega)$ is less than one. In the neighborhood of any ω_j, of course, there is rather violent behavior. The real part of the denominator in (7.51) vanishes for that term at $\omega = \omega_j$ and the term is large and purely imaginary. The general features of the real and imaginary parts of $\epsilon(\omega)$ around two successive resonant frequencies are shown in Fig. 7.8. *Normal dispersion* is associated with an increase in Re $\epsilon(\omega)$ with ω, *anomalous dispersion* with the reverse. Normal dispersion is seen to occur everywhere except in the neighborhood of a resonant frequency. And only where there is anomalous dispersion is the imaginary part of ϵ appreciable. Since a positive imaginary part to ϵ represents dissipation of energy from the electromagnetic wave into the medium, the regions where Im ϵ is large are called regions of *resonant absorption*.*

The attenuation of a plane wave is most directly expressed in terms of the real and imaginary parts of the wave number k. If the wave number is written as

$$k = \beta + i\frac{\alpha}{2} \tag{7.53}$$

then the parameter α is known as the attenuation constant or absorption coefficient. The *intensity* of the wave falls off as $e^{-\alpha z}$. Equation (7.5) yields the connection between (α, β) and (Re ϵ, Im ϵ):

$$\beta^2 - \frac{\alpha^2}{4} = \frac{\omega^2}{c^2} \text{ Re } \epsilon/\epsilon_0 \tag{7.54}$$

$$\beta\alpha = \frac{\omega^2}{c^2} \text{ Im } \epsilon/\epsilon_0$$

---

*If Im $\epsilon < 0$, energy is given to the wave by the medium; amplification occurs, as in a maser or laser. See M. Borenstein and W. E. Lamb, *Phys. Rev.* **A5**, 1298 (1972).

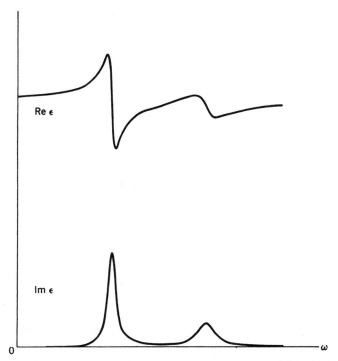

Figure 7.8   Real and imaginary parts of the dielectric constant $\epsilon(\omega)/\epsilon_0$ in the neighborhood of two resonances. The region of anomalous dispersion is also the frequency interval where absorption occurs.

If $\alpha \ll \beta$, as occurs unless the absorption is very strong or Re ϵ is negative, the attenuation constant α can be written approximately as

$$\alpha \simeq \frac{\text{Im } \epsilon(\omega)}{\text{Re } \epsilon(\omega)} \beta \qquad (7.55)$$

where $\beta = \sqrt{\text{Re}(\epsilon/\epsilon_0)} \, \omega/c$. The fractional decrease in intensity per wavelength divided by 2π is thus given by the ratio, Im ϵ/Re ϵ.

C. Low-Frequency Behavior, Electric Conductivity

In the limit $\omega \to 0$ there is a qualitative difference in the response of the medium depending on whether the lowest resonant frequency is zero or nonzero. For insulators the lowest resonant frequency is different from zero. Then at $\omega = 0$ the molecular polarizability is given by (4.73), corresponding to the limit $\omega = 0$ in (7.51). The elementary aspects of dielectrics in the static limit have been discussed in Section 4.6.

If some fraction f_0 of the electrons per molecule are "free" in the sense of having $\omega_0 = 0$, the dielectric constant is singular at $\omega = 0$. If the contribution of the free electrons is exhibited separately, (7.51) times ϵ_0 becomes

$$\epsilon(\omega) = \epsilon_b(\omega) + i \frac{Ne^2 f_0}{m\omega(\gamma_0 - i\omega)} \qquad (7.56)$$

where $\epsilon_b(\omega)$ is the contribution of all the other dipoles. The singular behavior can be understood if we examine the Maxwell–Ampère equation

$$\nabla \times \mathbf{H} = \mathbf{J} + \frac{d\mathbf{D}}{dt}$$

and assume that the medium obeys Ohm's law, $\mathbf{J} = \sigma\mathbf{E}$ and has a "normal" dielectric constant ϵ_b. With harmonic time dependence the equation becomes

$$\nabla \times \mathbf{H} = -i\omega\left(\epsilon_b + i\frac{\sigma}{\omega}\right)\mathbf{E} \tag{7.57}$$

If, on the other hand, we did not insert Ohm's law explicitly but attributed instead all the properties of the medium to the dielectric constant, we would identify the quantity in brackets on the right-hand side of (7.57) with $\epsilon(\omega)$. Comparison with (7.56) yields an expression for the conductivity:

$$\sigma = \frac{f_0 Ne^2}{m(\gamma_0 - i\omega)} \tag{7.58}$$

This is essentially the model of Drude (1900) for the electrical conductivity, with $f_0 N$ being the number of free electrons per unit volume in the medium. The damping constant γ_0/f_0 can be determined empirically from experimental data on the conductivity. For copper, $N \simeq 8 \times 10^{28}$ atoms/m^3 and at normal temperatures the low-frequency conductivity is $\sigma \simeq 5.9 \times 10^7$ $(\Omega \cdot \text{m})^{-1}$. This gives $\gamma_0/f_0 \simeq 4 \times 10^{13}$ s^{-1}. Assuming that $f_0 \sim 1$, this shows that up to frequencies well beyond the microwave region ($\omega \lesssim 10^{11}$ s^{-1}) conductivities of metals are essentially real (i.e., current in phase with the field) and independent of frequency. At higher frequencies (in the infrared and beyond) the conductivity is complex and varies with frequency in a way described qualitatively by the simple result (7.58). The problem of electrical conductivity is really a quantum-mechanical one in which the Pauli principle plays an important role. The free electrons are actually valence electrons of the isolated atoms that become quasi-free and move relatively unimpeded through the lattice (provided their energies lie in certain intervals or bands) when the atoms are brought together to form a solid. The damping effects come from collisions involving appreciable momentum transfer between the electrons and lattice vibrations, lattice imperfections, and impurities.*

The foregoing considerations show that the distinction between dielectrics and conductors is an artificial one, at least away from $\omega = 0$. If the medium possesses free electrons it is a conductor at low frequencies; otherwise, an insulator.† But at nonzero frequencies the "conductivity" contribution to $\epsilon(\omega)$ (7.51) merely appears as a resonant amplitude like the rest. The dispersive properties of the medium can be attributed as well to a complex dielectric constant as to a frequency-dependent conductivity *and* a dielectric constant.

---

*See R. G. Chambers, *Electrons in Metals and Semiconductors*, Chapman & Hall, New York (1990), or G. Lehmann and P. Ziesche, *Electronic Properties of Metals*, Elsevier, New York (1990).

†In terms of the quantum-mechanical band structure of the solid, the conductor has some electrons in a partially filled band, while the insulator has its bands filled to the full extent permitted by the Pauli principle. A "free" electron must have nearby energy-conserving quantum states to which it can move. In a partially filled band there are such states, but a filled band has, by definition, no such states available.

D. High-Frequency Limit, Plasma Frequency

At frequencies far above the highest resonant frequency the dielectric constant (7.51) takes on the simple form

$$\frac{\epsilon(\omega)}{\epsilon_0} \simeq 1 - \frac{\omega_p^2}{\omega^2} \tag{7.59}$$

where

$$\omega_p^2 = \frac{NZe^2}{\epsilon_0 m} \tag{7.60}$$

The frequency ω_p, which depends only on the total number NZ of electrons per unit volume, is called the *plasma frequency* of the medium. The wave number is given in the limit by

$$ck = \sqrt{\omega^2 - \omega_p^2} \tag{7.61}$$

Sometimes (7.61) is expressed as $\omega^2 = \omega_p^2 + c^2k^2$, and is called a dispersion relation or equation for $\omega = \omega(k)$. In dielectric media, (7.59) applies only for $\omega^2 \gg \omega_p^2$. The dielectric constant is then close to unity, although slightly less, and increases with frequency somewhat as the highest frequency part of the curve shown in Fig. 7.8. The wave number is real and varies with frequency as for a mode in a waveguide with cutoff frequency ω_p. (See Fig. 8.4.)

In certain situations, such as in the ionosphere or in a tenuous electronic plasma in the laboratory, the electrons are free and the damping is negligible. Then (7.59) holds over a wide range of frequencies, including $\omega < \omega_p$. For frequencies lower than the plasma frequency, the wave number (7.61) is purely imaginary. Such waves incident on a plasma are reflected and the fields inside fall off exponentially with distance from the surface. At $\omega = 0$ the attenuation constant is

$$\alpha_{\text{plasma}} \simeq \frac{2\omega_p}{c} \tag{7.62}$$

On the laboratory scale, plasma densities are of the order of $10^{18} - 10^{22}$ electrons/ m^3. This means $\omega_p \simeq 6 \times 10^{10} - 6 \times 10^{12}$ s^{-1}, so that typically attenuation lengths (α^{-1}) are of the order of 0.2 cm to 2×10^{-3} cm for static or low-frequency fields. The expulsion of fields from within a plasma is a well-known effect in controlled thermonuclear processes and is exploited in attempts at confinement of hot plasma.

The reflectivity of metals at optical and higher frequencies is caused by essentially the same behavior as for the tenuous plasma. The dielectric constant of a metal is given by (7.56). At high frequencies ($\omega \gg \gamma_0$) this takes the approximate form,

$$\epsilon(\omega) \simeq \epsilon_b(\omega) - \frac{\omega_p^2}{\omega^2} \epsilon_0$$

where $\omega_p^2 = ne^2/m^*\epsilon_0$ is the plasma frequency squared of the conduction electrons, given an effective mass m^* to include partially the effects of binding. For $\omega \ll \omega_p$ the behavior of light incident on the metal is approximately the same as for the plasma described by (7.59). The light penetrates only a very short distance into the metal and is almost entirely reflected. But when the frequency is increased into the domain where $\epsilon(\omega) > 0$, the metal suddenly can transmit light and its reflectivity changes drastically. This occurs typically in the ultraviolet and leads to the terminology "ultraviolet transparency of metals." Determination of the critical frequency gives information on the density or the effective mass of the conduction electrons.*

E. Index of Refraction and Absorption Coefficient of Liquid Water as a Function of Frequency

As an example of the overall frequency behavior of the real part of the index of refraction and the absorption coefficient of a real medium, we take the ubiquitous substance, water. Our intent is to give a broad view and to indicate the tremendous variations that are possible, rather than to discuss specific details. Accordingly, we show in Fig. 7.9, on a log-log plot with 20 decades in frequency and 11 decades in absorption, a compilation of the gross features of $n(\omega) = \mathrm{Re}\,\sqrt{\mu\epsilon/\mu_0\epsilon_0}$ and $\alpha(\omega) = 2\,\mathrm{Im}\,\sqrt{\mu\epsilon}\,\omega$ for liquid water at NTP. The upper part of the graph shows the interesting, but not spectacular, behavior of $n(\omega)$. At very low frequencies, $n(\omega) \simeq 9$, a value arising from the partial orientation of the permanent dipole moments of the water molecules. Above 10^{10} Hz the curve falls relatively smoothly to the structure in the infrared. In the visible region, shown by the vertical dashed lines, $n(\omega) \simeq 1.34$, with little variation. Then in the ultraviolet there is more structure. Above 6×10^{15} Hz ($h\nu \simeq 25$ eV) there are no data on the real part of the index of refraction. The asymptotic approach to unity shown in the figure assumes (7.59).

Much more dramatic is the behavior of the absorption coefficient α. At frequencies below 10^8 Hz the absorption coefficient is extremely small. The data seem unreliable (two different sets are shown), probably because of variations in sample purity. As the frequency increases toward 10^{11} Hz, the absorption coefficient increases rapidly to $\alpha \simeq 10^4$ m^{-1}, corresponding to an attenuation length of 100 μm in liquid water. This is the well-known microwave absorption by water. It is the phenomenon (in moist air) that terminated the trend during World War II toward better and better resolution in radar by going to shorter and shorter wavelengths.

In the infrared region absorption bands associated with vibrational modes of the molecule and possibly oscillations of a molecule in the field of its neighbors cause the absorption to reach peak values of $\alpha \simeq 10^6$ m^{-1}. Then the absorption coefficient falls precipitously over $7\frac{1}{2}$ decades to a value of $\alpha < 3 \times 10^{-1}$ m^{-1} in a narrow frequency range between 4×10^{14} Hz and 8×10^{14} Hz. It then rises again by more than 8 decades by 2×10^{15} Hz. This is a dramatic absorption *window* in what we call the *visible region*. The extreme transparency of water here has its origins in the basic energy level structure of the atoms and molecules. The reader may meditate on the fundamental question of biological evolution on this water-soaked planet, of why animal eyes see the spectrum from red to

---

*See Chapter 4 of D. Pines, *Elementary Excitations in Solids*, W. A. Benjamin, New York (1963), for a discussion of these and other dielectric properties of metals in the optical and ultraviolet region. More generally, see F. Wooten, *Optical Properties of Solids*, Academic Press, New York (1972) and *Handbook of Optical Constants of Solids*, ed. E. D. Palik, Academic Press, Boston (1991).

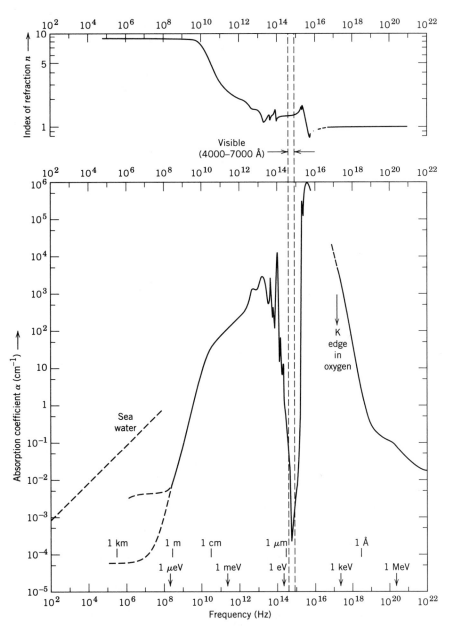

Figure 7.9 The index of refraction (top) and absorption coefficient (bottom) for liquid water as a function of linear frequency. Also shown as abscissas are an energy scale (arrows) and a wavelength scale (vertical lines). The visible region of the frequency spectrum is indicated by the vertical dashed lines. The absorption coefficient for seawater is indicated by the dashed diagonal line at the left. Note that the scales are logarithmic in both directions.

violet and of why the grass is green. Mother Nature has certainly exploited her window! In the very far ultraviolet the absorption has a peak value of $\alpha \simeq 1.1 \times 10^8$ m^{-1} at $\nu \simeq 5 \times 10^{15}$ Hz (21 eV). This is exactly at the plasmon energy $\hbar\omega_p$, corresponding to a collective excitation of all the electrons in the molecule. The attenuation is given in order of magnitude by (7.62). At higher frequencies data are absent until the photoelectric effect, and then Compton scattering and other high-energy processes take over. There the nuclear physicists have studied the absorption in detail. The behavior is basically governed by the atomic properties and the density, not by the fact that the substance is water.

At the low-frequency end of the graph in Fig. 7.9 we have indicated the absorption coefficient of *seawater*. At low frequencies, seawater has an electrical conductivity $\sigma \simeq 4.4 \ \Omega^{-1} \ m^{-1}$. From (7.57) we find that below about 10^8 Hz $\alpha \simeq (2\mu_0\omega\sigma)^{1/2}$. The absorption coefficient is thus proportional to $\sqrt{\omega}$ and becomes very small at low frequencies. The line shown is $\alpha \ (m^{-1}) = 8.4 \times 10^{-3}\sqrt{\nu(\mathrm{Hz})}$. At 10^2 Hz, the attenuation length in seawater is $\alpha^{-1} \simeq 10$ meters. This means that 1% of the intensity at the surface will survive at 50 meters below the surface. If one had a large fleet of submarines scattered throughout the oceans of the world and wished to be able to send messages from a land base to the submerged vessels, one would be led to consider extremely low-frequency (ELF) communications. The existence of prominent resonances of the earth-ionosphere cavity in the range from 8 Hz to a few hundred hertz (see Section 8.9) makes that region of the frequency spectrum specially attractive, as does the reduced attenuation. With wavelengths of the order of 5×10^3 km, very large antennas are needed (still small compared to a wavelength!).*

7.6 Simplified Model of Propagation in the Ionosphere and Magnetosphere

The propagation of electromagnetic waves in the ionosphere is described in zeroth approximation by the dielectric constant (7.59), but the presence of the earth's magnetic field modifies the behavior significantly. The influence of a static external magnetic field is also present for many laboratory plasmas. To illustrate the influence of an external magnetic field, we consider the simple problem of a tenuous electronic plasma of uniform density with a strong, static, uniform, magnetic induction $\mathbf{B}_0$ and transverse waves propagating parallel to the direction of $\mathbf{B}_0$. (The more general problem of an arbitrary direction of propagation is contained in Problem 7.17.) If the amplitude of electronic motion is small and collisions are neglected, the equation of motion is approximately

$$m\ddot{\mathbf{x}} - e\mathbf{B}_0 \times \dot{\mathbf{x}} = -e\mathbf{E}e^{-i\omega t} \qquad (7.63)$$

where the influence of the $\mathbf{B}$ field of the transverse wave has been neglected compared to the static induction $\mathbf{B}_0$ and the electronic charge has been written as $-e$. It is convenient to consider the transverse waves as circularly polarized. Thus we write

$$\mathbf{E} = (\boldsymbol{\epsilon}_1 \pm i\boldsymbol{\epsilon}_2)E \qquad (7.64)$$

and a similar expression for $\mathbf{x}$. Since the direction of $\mathbf{B}_0$ is taken orthogonal to $\boldsymbol{\epsilon}_1$ and $\boldsymbol{\epsilon}_2$, the cross product in (7.63) has components only in the direction $\boldsymbol{\epsilon}_1$ and

---

*For detailed discussion of ELF communications, see the conference proceedings, *ELF/VLF/LF Radio Propagation and Systems Aspects*, (AGARD-CP-529), Brussels, 28 September–2 October, 1992, AGARD, Neuilly sur Seine, France (1993).

ϵ_2 and the transverse components decouple. The steady-state solution of (7.63) is

$$\mathbf{x} = \frac{e}{m\omega(\omega \mp \omega_B)} \mathbf{E} \tag{7.65}$$

where ω_B is the frequency of precession of a charged particle in a magnetic field,

$$\omega_B = \frac{eB_0}{m} \tag{7.66}$$

The frequency dependence of (7.65) can be understood by the transformation of (7.63) to a coordinate system precessing with frequency ω_B about the direction of $\mathbf{B}_0$. The static magnetic field is eliminated; the rate of change of momentum there is caused by a rotating electric field of effective frequency $(\omega \pm \omega_B)$, depending on the sign of the circular polarization.

The amplitude of oscillation (7.65) gives a dipole moment for each electron and yields, for a bulk sample, the dielectric constant

$$\epsilon_\mp/\epsilon_0 = 1 - \frac{\omega_p^2}{\omega(\omega \mp \omega_B)} \tag{7.67}$$

The upper sign corresponds to a positive helicity wave (left-handed circular polarization in the optics terminology), while the lower is for negative helicity. For propagation antiparallel to the magnetic field $\mathbf{B}_0$, the signs are reversed. This is the extension of (7.59) to include a static magnetic induction. It is not completely general, since it applies only to waves propagating along the static field direction. But even in this simple example we see the essential characteristic that waves of right-handed and left-handed circular polarizations propagate differently. The ionosphere is birefringent. For propagation in directions other than parallel to the static field $\mathbf{B}_0$ it is straightforward to show that, if terms of the order of ω_B^2 are neglected compared to ω^2 and $\omega\omega_B$, the dielectric constant is still given by (7.67). But the precession frequency (7.66) is now to be interpreted as that due to only the component of B_0 parallel to the direction of propagation. This means that ω_B in (7.67) is a function of angle—the medium is not only birefringent, but also anisotropic (see Problem 7.17).

For the ionosphere a typical maximum density of free electrons is 10^{10}–10^{12} electrons/m^3, corresponding to a plasma frequency of the order of $\omega_p \simeq 6 \times 10^6$ –6×10^7 s^{-1}. If we take a value of 30 μT as representative of the earth's magnetic field, the precession frequency is $\omega_B \simeq 6 \times 10^6$ s^{-1}.

Figure 7.10 shows $\epsilon_\pm/\epsilon_0$ as a function of frequency for two values of the ratio of (ω_p/ω_B). In both examples there are wide intervals of frequency where one of ϵ_+ or ϵ_- is positive while the other is negative. At such frequencies one state of circular polarization cannot propagate in the plasma. Consequently a wave of that polarization incident on the plasma will be totally reflected. The other state of polarization will be partially transmitted. Thus, when a linearly polarized wave is incident on a plasma, the reflected wave will be elliptically polarized, with its major axis generally rotated away from the direction of the polarization of the incident wave.

The behavior of radio waves reflected from the ionosphere is explicable in terms of these ideas, but the presence of several layers of plasma with densities

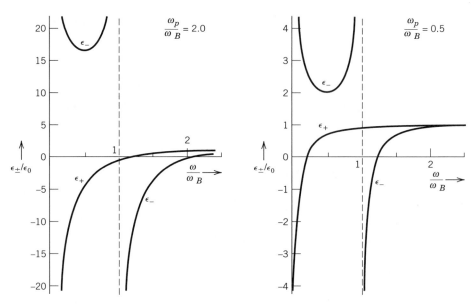

Figure 7.10 Dielectric constants as functions of frequency for model of the ionosphere (tenuous electronic plasma in a static, uniform magnetic induction). $\epsilon_{\pm}(\omega)$ apply to the right and left circularly polarized waves propagating parallel to the magnetic field. ω_B is the gyration frequency; ω_p is the plasma frequency. The two sets of curves correspond to $\omega_p/\omega_B = 2.0, 0.5$.

and relative positions varying with height and time makes the problem considerably more complicated than our simple example. The electron densities at various heights can be inferred by studying the reflection of pulses of radiation transmitted vertically upwards. The number n_0 of free electrons per unit volume increases slowly with height in a given layer of the ionosphere, as shown in Fig. 7.11, reaches a maximum, and then falls with further increase in height. A pulse of a given frequency ω_1 enters the layer without reflection because of the slow change in n_0. When the density n_0 is large enough, however, $\omega_p(h_1) \simeq \omega_1$. Then the dielectric constants (7.67) vanish and the pulse is reflected. The actual density n_0 where the reflection occurs is given by the roots of the right-hand side of (7.67). By observing the time interval between the initial transmission and reception of the reflected signal the height h_1 corresponding to that density can be found. By varying the frequency ω_1 and studying the change in time intervals, the electron

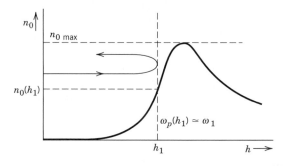

Figure 7.11 Electron density as a function of height in a layer of the ionosphere (schematic).

density as a function of height can be determined. If the frequency ω_1 is too high, the index of refraction does not vanish and very little reflection occurs. The frequency above which reflections disappear determines the maximum electron density in a given layer. A somewhat more quantitative treatment using the Wentzel–Kramers–Brillouin (WKB) approximation is sketched in Problem 7.14.

The behavior of $\epsilon_-(\omega)$ at low frequencies is responsible for a peculiar magnetospheric propagation phenomenon called "whistlers." As $\omega \to 0$, $\epsilon_-(\omega)$ tends to positive infinity as $\epsilon_-/\epsilon_0 \simeq \omega_p^2/\omega\omega_B$. Propagation occurs, but with a wave number (7.5),

$$k_- \simeq \frac{\omega_p}{c} \sqrt{\frac{\omega}{\omega_B}}$$

This corresponds to a highly dispersive medium. Energy transport is governed by the group velocity (7.86)—see Section 7.8—which is

$$v_g(\omega) \simeq 2v_p(\omega) \simeq 2c \frac{\sqrt{\omega_B\omega}}{\omega_p}$$

Pulses of radiation at different frequencies travel at different speeds: the lower the frequency, the slower the speed. A thunderstorm in one hemisphere generates a wide spectrum of radiation, some of which propagates more or less along the dipole field lines of the earth's magnetic field in a fashion described approximately by (7.67). The higher frequency components reach the antipodal point first, the lower frequency ones later. This gives rise at 10^5 Hz and below to *whistlers*, so named because the signal, as detected in an audio receiver, is a whistlelike sound beginning at high audio frequencies and falling rapidly through the audible range. With the estimates given above for ω_p and ω_B and distances of the order of 10^4 km, the reader can verify that the time scale for the whistlers is measured in seconds. Further discussion on whistlers can be found in the reading suggestions at the end of the chapter and in the problems.

7.7 *Magnetohydrodynamic Waves*

In the preceding section we discussed in terms of a dielectric constant the propagation of waves in a dilute plasma in an external magnetic field with negligible collisions. In contrast, in conducting fluids or dense ionized gases, collisions are sufficiently rapid that Ohm's law holds for a wide range of frequencies. Under the action of applied fields the electrons and ions move in such a way that, apart from high-frequency jitter, there is no separation of charge, although there can be current flow. Electric fields arise from external charges, current flow, or time-varying magnetic fields. At low frequencies the Maxwell displacement current is usually neglected. The nonrelativistic mechanical motion is described in terms of a single conducting fluid with the usual hydrodynamic variables of density, velocity, and pressure, with electromagnetic and gravitational forces. The combined system of equations describes *magnetohydrodynamics* (MHD).

The electromagnetic equations are those of Section 5.18, with the Ohm's law in (5.159) generalized for a fluid in motion to $\mathbf{J} = \sigma(\mathbf{E} + \mathbf{v} \times \mathbf{B})$, in accord with the discussion of Section 5.15. The generalization of (5.160), but for the magnetic induction, is

$$\frac{\partial \mathbf{B}}{\partial t} = \boldsymbol{\nabla} \times (\mathbf{v} \times \mathbf{B}) + \frac{1}{\mu\sigma} \nabla^2 \mathbf{B} \tag{7.68}$$

where for simplicity we have assumed that the conductivity and permeability are independent of position.

Consider the idealization of a compressible, nonviscous, "perfectly conducting" fluid in the absence of gravity, but in an external magnetic field. By perfectly conducting we mean that the conductivity is so large that the second term on the right-hand side of (7.68) can be neglected—the diffusion time (5.161) is very long compared to the time scale of interest. The hydrodynamic equations are

$$\frac{\partial \rho}{\partial t} + \boldsymbol{\nabla} \cdot (\rho \mathbf{v}) = 0$$

$$\rho \frac{\partial \mathbf{v}}{\partial t} + \rho(\mathbf{v} \cdot \boldsymbol{\nabla})\mathbf{v} = -\boldsymbol{\nabla}p - \frac{1}{\mu} \mathbf{B} \times (\boldsymbol{\nabla} \times \mathbf{B}) \tag{7.69}$$

The first equation is conservation of matter; the second is the Newton equation of motion with the mechanical pressure force density and the magnetic force density, $\mathbf{J} \times \mathbf{B}$, in which $\mathbf{J}$ has been replaced by $\boldsymbol{\nabla} \times \mathbf{H}$. The magnetic force can be written as

$$-\frac{1}{\mu} \mathbf{B} \times (\boldsymbol{\nabla} \times \mathbf{B}) = -\boldsymbol{\nabla}\left(\frac{1}{2\mu} \mathbf{B}^2\right) + \frac{1}{2} (\mathbf{B} \cdot \boldsymbol{\nabla})\mathbf{B}$$

The first term represents the gradient of a magnetic pressure; the second is an additional tension. Equation (7.69) must be supplemented by an equation of state.

In the absence of a magnetic field, the mechanical equations can describe small-amplitude, longitudinal, compressional (sound) waves with a speed s, the square of which is equal to the derivative of the pressure p with respect to the density ρ at constant entropy. With the adiabatic gas law, $p = K\rho^\gamma$, where γ is the ratio of specific heats, $s^2 = \gamma p_0/\rho_0$. By analogy, we anticipate longitudinal MHD waves in a conducting fluid in an external field B_0, with a speed squared of the order of the magnetic pressure divided by the equilibrium density,

$$v_{\text{MHD}} = O\sqrt{B_0^2/2\mu\rho_0}$$

To exhibit these waves we consider the combined equations of motion (7.68) and (7.69), with the neglect of the $\nabla^2\mathbf{B}/\mu\sigma$ term in (7.68), with an unperturbed configuration consisting of a spatially uniform, time-independent magnetic induction $\mathbf{B}_0$ throughout a stationary fluid of constant equilibrium density ρ_0. We then allow for small-amplitude departures from equilibrium,

$$\mathbf{B} = \mathbf{B}_0 + \mathbf{B}_1(\mathbf{x}, t)$$
$$\rho = \rho_0 + \rho_1(\mathbf{x}, t) \tag{7.70}$$
$$\mathbf{v} = \mathbf{v}_1(\mathbf{x}, t)$$

If equations (7.69) and (7.68) are linearized in the small quantities, they become:

$$\frac{\partial \rho_1}{\partial t} + \rho_0 \mathbf{\nabla} \cdot \mathbf{v}_1 = 0$$

$$\rho_0 \frac{\partial \mathbf{v}_1}{\partial t} + s^2 \mathbf{\nabla} \rho_1 + \frac{\mathbf{B}_0}{\mu} \times (\mathbf{\nabla} \times \mathbf{B}_1) = 0 \tag{7.71}$$

$$\frac{\partial \mathbf{B}_1}{\partial t} - \mathbf{\nabla} \times (\mathbf{v}_1 \times \mathbf{B}_0) = 0$$

where s^2 is the square of the sound velocity. These equations can be combined to yield an equation for $\mathbf{v}_1$ alone:

$$\frac{\partial^2 \mathbf{v}_1}{\partial t^2} - s^2 \mathbf{\nabla}(\mathbf{\nabla} \cdot \mathbf{v}_1) + \mathbf{v}_A \times \mathbf{\nabla} \times [\mathbf{\nabla} \times (\mathbf{v}_1 \times \mathbf{v}_A)] = 0 \tag{7.72}$$

where we have introduced a vectorial *Alfvén velocity*:

$$\mathbf{v}_A = \frac{\mathbf{B}_0}{\sqrt{\mu \rho_0}} \tag{7.73}$$

The wave equation (7.72) for $\mathbf{v}_1$ is somewhat involved, but it allows simple solutions for waves propagating parallel or perpendicular to the magnetic field direction.* With $\mathbf{v}_1(\mathbf{x}, t)$ a plane wave with wave vector $\mathbf{k}$ and frequency ω:

$$\mathbf{v}_1(\mathbf{x}, t) = \mathbf{v}_1 e^{i\mathbf{k} \cdot \mathbf{x} - i\omega t} \tag{7.74}$$

equation (7.72) becomes:

$$-\omega^2 \mathbf{v}_1 + (s^2 + v_A^2)(\mathbf{k} \cdot \mathbf{v}_1)\mathbf{k} + \mathbf{v}_A \cdot \mathbf{k}[(\mathbf{v}_A \cdot \mathbf{k})\mathbf{v}_1 \\ - (\mathbf{v}_A \cdot \mathbf{v}_1)\mathbf{k} - (\mathbf{k} \cdot \mathbf{v}_1)\mathbf{v}_A] = 0 \tag{7.75}$$

If $\mathbf{k}$ *is perpendicular* to $\mathbf{v}_A$ the last term vanishes. Then the solution for $\mathbf{v}_1$ is a *longitudinal* magnetosonic wave with a phase velocity:

$$u_{\text{long}} = \sqrt{s^2 + v_A^2} \tag{7.76}$$

Note that this wave propagates with a velocity that depends on the sum of hydrostatic and magnetic pressures, apart from factors of the order of unity. If $\mathbf{k}$ *is parallel* to $\mathbf{v}_A$, (7.75) reduces to

$$(k^2 v_A^2 - \omega^2)\mathbf{v}_1 + \left(\frac{s^2}{v_A^2} - 1\right)k^2(\mathbf{v}_A \cdot \mathbf{v}_1)\mathbf{v}_A = 0 \tag{7.77}$$

There are two types of wave motion possible in this case. There is an ordinary longitudinal wave ($\mathbf{v}_1$ parallel to $\mathbf{k}$ and $\mathbf{v}_A$) with phase velocity equal to the sound velocity s. But there is also a *transverse* wave ($\mathbf{v}_1 \cdot \mathbf{v}_A = 0$) with a phase velocity equal to the Alfvén velocity v_A. This Alfvén wave is a purely magnetohydrodynamic phenomenon, which depends only on the magnetic field (tension) and the density (inertia).

For mercury at room temperature the Alfvén velocity is 7.67 B_0 (tesla) m/s,

---

*The determination of the characteristics of the waves for arbitrary direction of propagation is left to Problem 7.18.

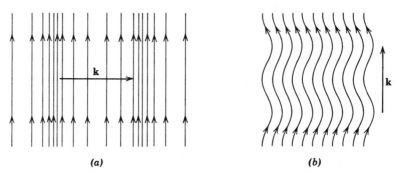

Figure 7.12 Magnetohydrodynamic waves.

compared with the sound speed of 1.45×10^3 m/s. At all laboratory field strengths the Alfvén velocity is much less than the speed of sound. In astrophysical problems, on the other hand, the Alfvén velocity can become very large because of the much smaller densities. In the sun's photosphere, for example, the density is of the order of 10^{-4} kg/m³ ($\sim 6 \times 10^{22}$ hydrogen atoms/m³) so that $v_A \simeq 10^5\, B(\text{T})$ m/s. Solar magnetic fields appear to be of the order of 1 or 2×10^{-4} T at the surface, with much larger values around sunspots. For comparison, the velocity of sound is of the order of 10^4 m/s in both the photosphere and the chromosphere.

The magnetic fields of these different waves can be found from the third equation in (7.71):

$$
\mathbf{B}_1 = \begin{cases}
\dfrac{k}{\omega}\, v_1 \mathbf{B}_0 & \text{for } \mathbf{k} \perp \mathbf{B}_0 \\[2mm]
0 & \text{for the longitudinal } \mathbf{k} \parallel \mathbf{B}_0 \\[2mm]
-\dfrac{k}{\omega}\, B_0 \mathbf{v}_1 & \text{for the transverse } \mathbf{k} \parallel \mathbf{B}_0
\end{cases} \qquad (7.78)
$$

The magnetosonic wave moving perpendicular to $\mathbf{B}_0$ causes compressions and rarefactions in the lines of force without changing their direction, as indicated in Fig. 7.12a. The Alfvén wave parallel to $\mathbf{B}_0$ causes the lines of force to oscillate back and forth laterally (Fig. 7.12b). In either case the lines of force are "frozen in" and move with the fluid.

Inclusion of the effects of fluid viscosity, finite, not infinite, conductivity, and the displacement current add complexity to the analysis. Some of these elaborations are treated in the problems.

7.8 *Superposition of Waves in One Dimension; Group Velocity*

In the preceding sections plane wave solutions to the Maxwell equations were found and their properties discussed. Only monochromatic waves, those with a definite frequency and wave number, were treated. In actual circumstances such idealized solutions do not arise. Even in the most monochromatic light source or the most sharply tuned radio transmitter or receiver, one deals with a finite (although perhaps small) spread of frequencies or wavelengths. This spread may originate in the finite duration of a pulse, in inherent broadening in the source, or in many other ways. Since the basic equations are linear, it is in principle an elementary matter to make the appropriate linear superposition of solutions with different frequencies. In general, however, several new features arise.

1. If the medium is dispersive (i.e., the dielectric constant is a function of the frequency of the fields), the phase velocity is not the same for each frequency component of the wave. Consequently different components of the wave travel with different speeds and tend to change phase with respect to one another.

2. In a dispersive medium the velocity of energy flow may differ greatly from the phase velocity, or may even lack precise meaning.

3. In a dissipative medium, a pulse of radiation will be attenuated as it travels with or without distortion, depending on whether the dissipative effects are or are not sensitive functions of frequency.

The essentials of these dispersive and dissipative effects are implicit in the ideas of Fourier series and integrals (Section 2.8). For simplicity, we consider scalar waves in only one dimension. The scalar amplitude $u(x, t)$ can be thought of as one of the components of the electromagnetic field. The basic solution to the wave equation has been exhibited in (7.6). The relationship between frequency ω and wave number k is given by (7.4) for the electromagnetic field. Either ω or k can be viewed as the independent variable when one considers making a linear superposition. Initially we will find it most convenient to use k as an independent variable. To allow for the possibility of dispersion we will consider ω as a general function of k:

$$\omega = \omega(k) \tag{7.79}$$

Since the dispersive properties cannot depend on whether the wave travels to the left or to the right, ω must be an even function of k, $\omega(-k) = \omega(k)$. For most wavelengths ω is a smoothly varying function of k. But, as we have seen in Section 7.5, at certain frequencies there are regions of "anomalous dispersion" where ω varies rapidly over a narrow interval of wavelengths. With the general form (7.79), our subsequent discussion can apply equally well to electromagnetic waves, sound waves, de Broglie matter waves, etc. For the present we assume that k and $\omega(k)$ are real, and so exclude dissipative effects.

From the basic solutions (7.6) we can build up a general solution of the form

$$u(x, t) = \frac{1}{\sqrt{2\pi}} \int_{-\infty}^{\infty} A(k)e^{ikx - i\omega(k)t} \, dk \tag{7.80}$$

The factor $1\sqrt{2\pi}$ has been inserted to conform with the Fourier integral notation of (2.44) and (2.45). The amplitude $A(k)$ describes the properties of the linear superposition of the different waves. It is given by the transform of the spatial amplitude $u(x, t)$, evaluated at $t = 0$*:

$$A(k) = \frac{1}{\sqrt{2\pi}} \int_{-\infty}^{\infty} u(x, 0)e^{-ikx} \, dx \tag{7.81}$$

If $u(x, 0)$ represents a harmonic wave e^{ik_0x} for all x, the orthogonality relation (2.46) shows that $A(k) = \sqrt{2\pi} \, \delta(k - k_0)$, corresponding to a monochromatic

---

*The following discussion slights somewhat the initial-value problem. For a second-order differential equation we must specify not only $u(x, 0)$ but also $\partial u(x, 0)/\partial t$. This omission is of no consequence for the rest of the material in this section. It is remedied in the following section.

traveling wave $u(x, t) = e^{ik_0x - i\omega(k_0)t}$, as required. If, however, at $t = 0$, $u(x, 0)$ represents a finite wave train with a length of order Δx, as shown in Figure 7.13, then the amplitude $A(k)$ is not a delta function. Rather, it is a peaked function with a breadth of the order of Δk, centered around a wave number k_0, which is the dominant wave number in the modulated wave $u(x, 0)$. If Δx and Δk are defined as the rms deviations from the average values of x and k [defined in terms of the *intensities* $|u(x, 0)|^2$ and $|A(k)|^2$], it is possible to draw the general conclusion:

$$\Delta x \, \Delta k \geq \tfrac{1}{2} \tag{7.82}$$

The reader may readily verify that, for most reasonable pulses or wave packets that do not cut off too violently, Δx times Δk lies near the lower limiting value in (7.82). This means that short wave trains with only a few wavelengths present have a very wide distribution of wave numbers of monochromatic waves, and conversely that long sinusoidal wave trains are almost monochromatic. Relation (7.82) applies equally well to distributions in time and in frequency.

The next question is the behavior of a pulse or finite wave train in time. The pulse shown at $t = 0$ in Fig. 7.13 begins to move as time goes on. The different frequency or wave-number components in it move at different phase velocities. Consequently there is a tendency for the original coherence to be lost and for the pulse to become distorted in shape. At the very least, we might expect it to propagate with a rather different velocity from, say, the average phase velocity of its component waves. The general case of a highly dispersive medium or a very sharp pulse with a great spread of wave numbers present is difficult to treat. But the propagation of a pulse which is not too broad in its wave-number spectrum, or a pulse in a medium for which the frequency depends weakly on wave number, can be handled in the following approximate way. The wave at time t is given by (7.80). If the distribution $A(k)$ is fairly sharply peaked around some value k_0, then the frequency $\omega(k)$ can be expanded around that value of k:

$$\omega(k) = \omega_0 + \frac{d\omega}{dk}\bigg|_0 (k - k_0) + \cdots \tag{7.83}$$

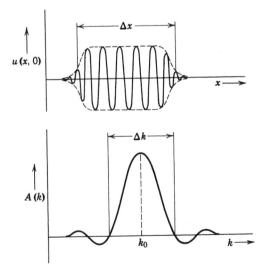

Figure 7.13 A harmonic wave train of finite extent and its Fourier spectrum in wave number.

and the integral performed. Thus

$$u(x, t) \simeq \frac{e^{i[k_0(d\omega/dk)|_0 - \omega_0]t}}{\sqrt{2\pi}} \int_{-\infty}^{\infty} A(k)e^{i[x - (d\omega/dk)|_0 t]k} \, dk \tag{7.84}$$

From (7.81) and its inverse it is apparent that the integral in (7.84) is just $u(x', 0)$, where $x' = x - (d\omega/dk)|_0 t$:

$$u(x, t) \simeq u\left(x - t \frac{d\omega}{dk}\bigg|_0, 0\right) e^{i[k_0(d\omega/dk)|_0 - \omega_0]t} \tag{7.85}$$

This shows that, apart from an overall phase factor, the pulse travels along undistorted in shape with a velocity, called the *group velocity*:

$$v_g = \frac{d\omega}{dk}\bigg|_0 \tag{7.86}$$

If an energy density is associated with the magnitude of the wave (or its absolute square), it is clear that in this approximation the transport of energy occurs with the group velocity, since that is the rate at which the pulse travels along.

For light waves the relation between ω and k is given by

$$\omega(k) = \frac{ck}{n(k)} \tag{7.87}$$

where c is the velocity of light in vacuum, and $n(k)$ is the index of refraction expressed as a function of k. The phase velocity is

$$v_p = \frac{\omega(k)}{k} = \frac{c}{n(k)} \tag{7.88}$$

and is greater or smaller than c depending on whether $n(k)$ is smaller or larger than unity. For most optical wavelengths $n(k)$ is greater than unity in almost all substances. The group velocity (7.86) is

$$v_g = \frac{c}{n(\omega) + \omega(dn/d\omega)} \tag{7.89}$$

In this equation it is more convenient to think of n as a function of ω than of k. For normal dispersion $(dn/d\omega) > 0$, and also $n > 1$; then the velocity of energy flow is less than the phase velocity and also less than c. In regions of anomalous dispersion, however, $dn/d\omega$ can become large and negative as can be inferred from Fig. 7.8. Then the group velocity differs greatly from the phase velocity, often becoming larger than c or even negative. The behavior of group and phase velocities as a function of frequency in the neighborhood of a region of anomalous dispersion is shown in Fig. 7.14. There is no cause for alarm that our ideas of special relativity are violated; group velocity is *generally* not a useful concept in regions of anomalous dispersion. In addition to the existence of significant absorption (see Fig. 7.8), a large $dn/d\omega$ is equivalent to a rapid variation of ω with k. Consequently the approximations made in (7.83) and following equations are no longer valid. Usually a pulse with its dominant frequency components in the neighborhood of a strong absorption line is absorbed and distorted as it travels. As shown by Garret and McCumber,[*] however, there are circumstances

*C. G. B. Garrett and D. E. McCumber, *Phys. Rev. A* **1**, 305 (1970).

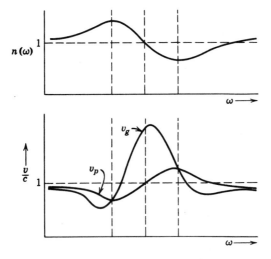

Figure 7.14 Index of refraction $n(\omega)$ as a function of frequency ω at a region of anomalous dispersion; phase velocity v_p and group velocity v_g as functions of ω.

in which "group velocity" can still have meaning, even with anomalous dispersion. Other authors* subsequently verified experimentally what Garrett and McCumber showed theoretically: namely, if absorbers are not too thick, a Gaussian pulse with a central frequency near an absorption line and with support narrow compared to the width of the line (pulse wide in time compared to $1/\gamma$) propagates with appreciable absorption, but more or less retains its shape, the peak of which moves at the group velocity (7.89), even when that quantity is negative. Physically, what occurs is pulse reshaping—the leading edge of the pulse is less attenuated than the trailing edge. Conditions can be such that the *peak* of the greatly attenuated pulse emerges from the absorber before the *peak* of the incident pulse has entered it! (That is the meaning of negative group velocity.) Since a Gaussian pulse does not have a sharply defined front edge, there is no question of violation of causality.

Some experiments are described as showing that photons travel faster than the speed of light through optical "band-gap" devices that reflect almost all of the incident flux over a restricted range of frequencies. While it is true that the centroid of the very small transmitted Gaussian pulse appears slightly in advance of the vacuum transit time, no signal or information travels faster than c. The main results are explicable in conventional classical terms. Some aspects are examined in Problems 7.9–7.11. A review of these and other experiments has been given by Chiao and Steinberg.[†]

7.9 *Illustration of the Spreading of a Pulse as It Propagates in a Dispersive Medium*

To illustrate the ideas of the preceding section and to show the validity of the concept of group velocity, we now consider a specific model for the dependence

*S. Chu and S. Wong, *Phys. Rev. Letters* **48**, 738 (1982); A. Katz, R. R. Alfano, S. Chu, and S. Wong, *Phys. Rev. Letters* **49**, 1292 (1982).

[†]R. Y. Chiao and A. M. Steinberg, in *Progress in Optics*, Vol. 37, ed. E. Wolf, Elsevier, Amsterdam (1997), p. 347–406.

of frequency on wave number and calculate without approximations the propagation of a pulse in this model medium. Before specifying the particular model it is necessary to state the initial-value problem in more detail than was done in (7.80) and (7.81). As noted there, the proper specification of an initial-value problem for the wave equation demands the initial values of both function $u(x, 0)$ and time derivative $\partial u(x, 0)/\partial t$. If we agree to take the real part of (7.80) to obtain $u(x, t)$,

$$u(x, t) = \frac{1}{2} \frac{1}{\sqrt{2\pi}} \int_{-\infty}^{\infty} A(k)e^{ikx - i\omega(k)t} \, dk + \text{c.c.} \tag{7.90}$$

then it is easy to show that $A(k)$ is given in terms of the initial values by:

$$A(k) = \frac{1}{\sqrt{2\pi}} \int_{-\infty}^{\infty} e^{-ikx} \left[u(x, 0) + \frac{i}{\omega(k)} \frac{\partial u}{\partial t} (x, 0) \right] dx \tag{7.91}$$

We take a Gaussian modulated oscillation

$$u(x, 0) = e^{-x^2/2L^2} \cos k_0 x \tag{7.92}$$

as the initial shape of the pulse. For simplicity, we will assume that

$$\frac{\partial u}{\partial t} (x, 0) = 0 \tag{7.93}$$

This means that at times immediately before $t = 0$ the wave consisted of two pulses, both moving toward the origin, such that at $t = 0$ they coalesced into the shape given by (7.92). Clearly at later times we expect each pulse to reemerge on the other side of the origin. Consequently the initial distribution (7.92) may be expected to split into two identical packets, one moving to the left and one to the right. The Fourier amplitude $A(k)$ for the pulse described by (7.92) and (7.93) is

$$\begin{aligned} A(k) &= \frac{1}{\sqrt{2\pi}} \int_{-\infty}^{\infty} e^{-ikx} e^{-x^2/2L^2} \cos k_0 x \, dx \\ &= \frac{L}{2} [e^{-(L^2/2)(k-k_0)^2} + e^{-(L^2/2)(k+k_0)^2}] \end{aligned} \tag{7.94}$$

The symmetry $A(-k) = A(k)$ is a reflection of the presence of two pulses traveling away from the origin, as is seen below.

To calculate the waveform at later times, we must specify $\omega = \omega(k)$. As a model allowing exact calculation and showing the essential dispersive effects, we assume

$$\omega(k) = \nu\left(1 + \frac{a^2 k^2}{2}\right) \tag{7.95}$$

where ν is a constant frequency, and a is a constant length that is a typical wavelength where dispersive effects become important. Equation (7.95) is an approximation to the dispersion equation of the tenuous plasma, (7.59) or (7.61). Since the pulse (7.92) is a modulated wave of wave number $k = k_0$, the approximate arguments of the preceding section imply that the two pulses will travel with the group velocity

$$v_g = \frac{d\omega}{dk}(k_0) = va^2k_0 \tag{7.96}$$

and will be essentially unaltered in shape provided the pulse is not too narrow in space.

The exact behavior of the wave as a function of time is given by (7.90), with (7.94) for $A(k)$:

$$u(x, t) = \frac{L}{2\sqrt{2\pi}} \text{Re} \int_{-\infty}^{\infty} [e^{-(L^2/2)(k-k_0)^2} + e^{-(L^2/2)(k+k_0)^2}] e^{ikx - iv t[1 + (a^2 k^2/2)]} \, dk \tag{7.97}$$

The integrals can be performed by appropriately completing the squares in the exponents. The result is

$$u(x, t) =$$

$$\frac{1}{2}\text{Re} \left\{ \frac{\exp\left[-\dfrac{(x - va^2k_0t)^2}{2L^2\left(1 + \dfrac{ia^2 vt}{L^2}\right)} \right]}{\left(1 + \dfrac{ia^2 vt}{L^2}\right)^{1/2}} \exp\left[ik_0x - iv\left(1 + \dfrac{a^2 k_0^2}{2}\right)t \right] + (k_0 \to -k_0) \right\} \tag{7.98}$$

Equation (7.98) represents two pulses traveling in opposite directions. The peak amplitude of each pulse travels with the group velocity (7.96), while the modulation envelop remains Gaussian in shape. The width of the Gaussian is not constant, however, but increases with time. The width of the envelope is

$$L(t) = \left[L^2 + \left(\frac{a^2 vt}{L} \right)^2 \right]^{1/2} \tag{7.99}$$

Thus the dispersive effects on the pulse are greater (for a given elapsed time), the sharper the envelope. The criterion for a small change in shape is that $L \gg a$. Of course, at long times the width of the Gaussian increases linearly with time

$$L(t) \to \frac{a^2 vt}{L} \tag{7.100}$$

but the time of attainment of this asymptotic form depends on the ratio (L/a). A measure of how rapidly the pulse spreads is provided by a comparison of $L(t)$ given by (7.99), with $v_g t = va^2 k_0 t$. Figure 7.15 shows two examples of curves of the position of peak amplitude $(v_g t)$ and the positions $v_g t \pm L(t)$, which indicate the spread of the pulse, as functions of time. On the left the pulse is not too narrow compared to the wavelength k_0^{-1} and so does not spread too rapidly. The

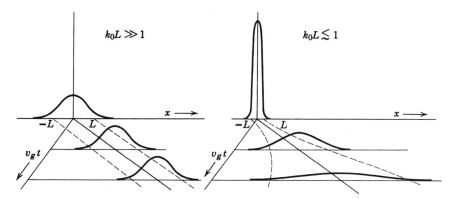

Figure 7.15   Change in shape of a wave packet as it travels along. The broad packet, containing many wavelengths ($k_0 L \gg 1$), is distorted comparatively little, while the narrow packet ($k_0 L \lesssim 1$) broadens rapidly.

pulse on the right, however, is so narrow initially that it is very rapidly spread out and scarcely represents a pulse after a short time.

Although the results above have been derived for a special choice (7.92) of initial pulse shape and dispersion relation (7.95), their implications are of a more general nature. We saw in Section 7.8 that the average velocity of a pulse is the group velocity $v_g = d\omega/dk = \omega'$. The spreading of the pulse can be accounted for by noting that a pulse with an initial spatial width Δx_0 must have inherent in it a spread of wave numbers $\Delta k \sim (1/\Delta x_0)$. This means that the group velocity, when evaluated for various k values within the pulse, has a spread in it of the order

$$\Delta v_g \sim \omega'' \, \Delta k \sim \frac{\omega''}{\Delta x_0} \tag{7.101}$$

At a time t this implies a spread in position of the order of $\Delta v_g t$. If we combine the uncertainties in position by taking the square root of the sum of squares, we obtain the width $\Delta x(t)$ at time t:

$$\Delta x(t) \simeq \sqrt{(\Delta x_0)^2 + \left(\frac{\omega'' t}{\Delta x_0}\right)^2} \tag{7.102}$$

We note that (7.102) agrees exactly with (7.99) if we put $\Delta x_0 = L$. The expression (7.102) for $\Delta x(t)$ shows the general result that, if $\omega'' \neq 0$, a narrow pulse spreads rapidly because of its broad spectrum of wave numbers, and vice versa. All these ideas carry over immediately into wave mechanics. They form the basis of the Heisenberg uncertainty principle. In wave mechanics, the frequency is identified with energy divided by Planck's constant, while wave number is momentum divided by Planck's constant.

The problem of wave packets in a dissipative, as well as dispersive, medium is rather complicated. Certain aspects can be discussed analytically, but the analytical expressions are not readily interpreted physically. Except in special circumstances, wave packets are attenuated and distorted appreciably as they propagate. The reader may refer to *Stratton* (pp. 301–309) for a discussion of the problem, including numerical examples.

7.10   *Causality in the Connection Between D and E; Kramers–Kronig Relations*

A. *Nonlocality in Time*

Another consequence of the frequency dependence of $\epsilon(\omega)$ is a temporally nonlocal connection between the displacement $\mathbf{D}(\mathbf{x}, t)$ and the electric field $\mathbf{E}(\mathbf{x}, t)$. If the monochromatic components of frequency ω are related by

$$\mathbf{D}(\mathbf{x}, \omega) = \epsilon(\omega)\mathbf{E}(\mathbf{x}, \omega) \tag{7.103}$$

the dependence on time can be constructed by Fourier superposition. Treating the spatial coordinate as a parameter, the Fourier integrals in time and frequency can be written

$$\mathbf{D}(\mathbf{x}, t) = \frac{1}{\sqrt{2\pi}} \int_{-\infty}^{\infty} \mathbf{D}(\mathbf{x}, \omega)e^{-i\omega t} \, d\omega$$

and $\hspace{8cm}$ (7.104)

$$\mathbf{D}(\mathbf{x}, \omega) = \frac{1}{\sqrt{2\pi}} \int_{-\infty}^{\infty} \mathbf{D}(\mathbf{x}, t')e^{i\omega t'} \, dt'$$

with corresponding equations for $\mathbf{E}$. The substitution of (7.103) for $\mathbf{D}(\mathbf{x}, \omega)$ gives

$$\mathbf{D}(\mathbf{x}, t) = \frac{1}{\sqrt{2\pi}} \int_{-\infty}^{\infty} \epsilon(\omega)\mathbf{E}(\mathbf{x}, \omega)e^{-i\omega t} \, d\omega$$

We now insert the Fourier representation of $\mathbf{E}(\mathbf{x}, \omega)$ into the integral and obtain

$$\mathbf{D}(\mathbf{x}, t) = \frac{1}{2\pi} \int_{-\infty}^{\infty} d\omega \, \epsilon(\omega)e^{-i\omega t} \int_{-\infty}^{\infty} dt' \, e^{i\omega t'}\mathbf{E}(\mathbf{x}, t')$$

With the assumption that the orders of integration can be interchanged, the last expression can be written as

$$\mathbf{D}(\mathbf{x}, t) = \epsilon_0\left\{\mathbf{E}(\mathbf{x}, t) + \int_{-\infty}^{\infty} G(\tau)\mathbf{E}(\mathbf{x}, t - \tau) \, d\tau\right\} \tag{7.105}$$

where $G(\tau)$ is the Fourier transform of $\chi_e = \epsilon(\omega)/\epsilon_0 - 1$:

$$G(\tau) = \frac{1}{2\pi} \int_{-\infty}^{\infty} [\epsilon(\omega)/\epsilon_0 - 1]e^{-i\omega\tau} \, d\omega \tag{7.106}$$

Equations (7.105) and (7.106) give a nonlocal connection between $\mathbf{D}$ and $\mathbf{E}$, in which $\mathbf{D}$ at time t depends on the electric field at times other than t.* If $\epsilon(\omega)$ is

---

*Equations (7.103) and (7.105) are recognizable as an example of the *faltung* theorem of Fourier integrals: if $A(t)$, $B(t)$, $C(t)$ and $a(\omega)$, $b(\omega)$, $c(\omega)$ are two sets of functions related in pairs by the Fourier inversion formulas (7.104), and

$$c(\omega) = a(\omega)b(\omega)$$

then, under suitable restrictions concerning integrability,

$$C(t) = \frac{1}{\sqrt{2\pi}} \int_{-\infty}^{\infty} A(t')B(t - t') \, dt'$$

independent of ω for all ω, (7.106) yields $G(\tau) \propto \delta(\tau)$ and the instantaneous connection is obtained, but if $\epsilon(\omega)$ varies with ω, $G(\tau)$ is nonvanishing for some values of τ different from zero.

B. Simple Model for G(τ), Limitations

To illustrate the character of the connection implied by (7.105) and (7.106) we consider a one-resonance version of the index of refraction (7.51):

$$\epsilon(\omega)/\epsilon_0 - 1 = \omega_p^2(\omega_0^2 - \omega^2 - i\gamma\omega)^{-1} \tag{7.107}$$

The susceptibility kernel $G(\tau)$ for this model of $\epsilon(\omega)$ is

$$G(\tau) = \frac{\omega_p^2}{2\pi} \int_{-\infty}^{\infty} \frac{e^{-i\omega\tau}}{\omega_0^2 - \omega^2 - i\gamma\omega} d\omega \tag{7.108}$$

The integral can be evaluated by contour integration. The integrand has poles in the lower half-ω-plane at

$$\omega_{1,2} = -\frac{i\gamma}{2} \pm \nu_0, \qquad \text{where } \nu_0^2 = \omega_0^2 - \frac{\gamma^2}{4} \tag{7.109}$$

For $\tau < 0$ the contour can be closed in the upper half-plane without affecting the value of the integral. Since the integrand is regular inside the closed contour, the integral vanishes. For $\tau > 0$, the contour is closed in the lower half-plane and the integral is given by $-2\pi i$ times the residues at the two poles. The kernel (7.108) is therefore

$$G(\tau) = \omega_p^2 e^{-\gamma\tau/2} \frac{\sin \nu_0 \tau}{\nu_0} \theta(\tau) \tag{7.110}$$

where $\theta(\tau)$ is the step function [$\theta(\tau) = 0$ for $\tau < 0$; $\theta(\tau) = 1$ for $\tau > 0$]. For the dielectric constant (7.51) the kernel $G(\tau)$ is just a linear superposition of terms like (7.110). The kernel $G(\tau)$ is oscillatory with the characteristic frequency of the medium and damped in time with the damping constant of the electronic oscillators. The nonlocality in time of the connection between **D** and **E** is thus confined to times of the order of γ^{-1}. Since γ is the width in frequency of spectral lines and these are typically 10^7–10^9 s^{-1}, the departure from simultaneity is of the order of 10^{-7}–10^{-9} s. For frequencies above the microwave region many cycles of the electric field oscillations contribute an average weighed by $G(\tau)$ to the displacement **D** at a given instant of time.

Equation (7.105) is nonlocal in time, but not in space. This approximation is valid provided the spatial variation of the applied fields has a scale that is large compared with the dimensions involved in the creation of the atomic or molecular polarization. For bound charges the latter scale is of the order of atomic dimensions or less, and so the concept of a dielectric constant that is a function only of ω can be expected to hold for frequencies well beyond the visible range. For conductors, however, the presence of free charges with macroscopic mean free paths makes the assumption of a simple $\epsilon(\omega)$ or $\sigma(\omega)$ break down at much lower frequencies. For a good conductor like copper we have seen that the damping constant (corresponding to a collision frequency) is of the order of $\gamma_0 \sim 3 \times 10^{13}$ s^{-1} at room temperature. At liquid helium temperatures, the damping constant may be 10^{-3} times the room temperature value. Taking the Bohr velocity in hydrogen ($c/137$) as typical of electron velocities in metals, we find mean free paths of the order $L \sim c/(137\gamma_0) \sim 10^{-4}$ m at liquid helium temperatures. On the other hand, the conventional skin depth δ (7.77) can be much smaller, of the order of 10^{-7} or 10^{-8} m at microwave frequencies. In such circumstances, Ohm's law must be replaced by a nonlocal expression. The conductivity becomes a tensorial quantity depending on wave number $\mathbf{k}$ and frequency ω. The associated departures from the standard behavior are known collectively as the *anomalous skin effect*. They can be utilized to map out the Fermi surfaces in metals.[*] Similar nonlocal effects occur in superconductors where the electromagnetic properties involve a coherence length of the order of 10^{-6} m.[†] With this brief mention of the limitations of (7.105) and the areas where generalizations have been fruitful we return to the discussion of the physical content of (7.105).

C. Causality and Analyticity Domain of $\epsilon(\omega)$

The most obvious and fundamental feature of the kernel (7.110) is that it vanishes for $\tau < 0$. This means that at time t only values of the electric field *prior* to that time enter in determining the displacement, in accord with our fundamental ideas of causality in physical phenomena. Equation (7.105) can thus be written

$$\mathbf{D}(\mathbf{x}, t) = \epsilon_0 \left\{ \mathbf{E}(\mathbf{x}, t) + \int_0^\infty G(\tau) \mathbf{E}(\mathbf{x}, t - \tau)\, d\tau \right\} \qquad (7.111)$$

This is, in fact, the most general spatially local, linear, and causal relation that can be written between $\mathbf{D}$ and $\mathbf{E}$ in a uniform isotropic medium. Its validity transcends any specific model of $\epsilon(\omega)$. From (7.106) the dielectric constant can be expressed in terms of $G(\tau)$ as

$$\epsilon(\omega)/\epsilon_0 = 1 + \int_0^\infty G(\tau) e^{i\omega\tau}\, d\tau \qquad (7.112)$$

This relation has several interesting consequences. From the reality of $\mathbf{D}$, $\mathbf{E}$, and therefore $G(\tau)$ in (7.111) we can deduce from (7.112) that for complex ω,

$$\epsilon(-\omega)/\epsilon_0 = \epsilon^*(\omega^*)/\epsilon_0 \qquad (7.113)$$

Furthermore, if (7.112) is viewed as a representation of $\epsilon(\omega)/\epsilon_0$ in the complex ω plane, it shows that $\epsilon(\omega)/\epsilon_0$ *is an analytic function of ω in the upper half-plane*, provided $G(\tau)$ is finite for all τ. On the real axis it is necessary to invoke the "physically reasonable" requirement that $G(\tau) \to 0$ as $\tau \to \infty$ to assure that $\epsilon(\omega)/\epsilon_0$ is also analytic there. This is true for dielectrics, but not for conductors, where $G(\tau) \to \sigma/\epsilon_0$ as $\tau \to \infty$ and $\epsilon(\omega)/\epsilon_0$ has a simple pole at $\omega = 0$ ($\epsilon \to i\sigma/\omega$ as $\omega \to 0$). Apart, then, from a possible pole at $\omega = 0$, the dielectric constant $\epsilon(\omega)/\epsilon_0$ is analytic in ω for Im $\omega \geq 0$ as a direct result of the causal relation (7.111)

---

[*]A. B. Pippard, in *Reports on Progress in Physics* **23**, 176 (1960), and the article entitled "The Dynamics of Conduction Electrons," by the same author in *Low-Temperature Physics*, Les Houches Summer School (1961), eds. C. de Witt, B. Dreyfus, and P. G. de Gennes, Gordon and Breach, New York (1962). The latter article has been issued separately by the same publisher.

[†]See, for example, the article "Superconductivity" by M. Tinkham in *Low Temperature Physics, op. cit.*

between **D** and **E**. These properties can be verified, of course, for the models discussed in Sections 7.5.A and 7.5.C.

The behavior of $\epsilon(\omega)/\epsilon_0 - 1$ for large ω can be related to the behavior of $G(\tau)$ at small times. Integration by parts in (7.112) leads to the asymptotic series,

$$\epsilon(\omega)/\epsilon_0 - 1 \simeq \frac{iG(0)}{\omega} - \frac{G'(0)}{\omega^2} + \cdots$$

where the argument of G and its derivatives is $\tau = 0^+$. It is unphysical to have $G(0^-) = 0$, but $G(0^+) \neq 0$. Thus the first term in the series is absent, and $\epsilon(\omega)/\epsilon_0 - 1$ falls off at high frequencies as ω^{-2}, just as was found in (7.59) for the oscillator model. The asymptotic series shows, in fact, that the real and imaginary parts of $\epsilon(\omega)/\epsilon_0 - 1$ behave for large real ω as

$$\mathrm{Re}[\epsilon(\omega)/\epsilon_0 - 1] = O\left(\frac{1}{\omega^2}\right), \qquad \mathrm{Im}\ \epsilon(\omega)/\epsilon_0 = O\left(\frac{1}{\omega^3}\right) \qquad (7.114)$$

These asymptotic forms depend only upon the existence of the derivatives of $G(\tau)$ around $\tau = 0^+$.

D. Kramers–Kronig Relations

The analyticity of $\epsilon(\omega)/\epsilon_0$ in the upper half-ω-plane permits the use of Cauchy's theorem to relate the real and imaginary part of $\epsilon(\omega)/\epsilon_0$ on the real axis. For any point z inside a closed contour C in the upper half-ω-plane, Cauchy's theorem gives

$$\epsilon(z)/\epsilon_0 = 1 + \frac{1}{2\pi i} \oint_C \frac{[\epsilon(\omega')/\epsilon_0 - 1]}{\omega' - z}\, d\omega'$$

The contour C is now chosen to consist of the real ω axis and a great semicircle at infinity in the upper half-plane. From the asymptotic expansion just discussed or the specific results of Section 7.5.D, we see that $\epsilon/\epsilon_0 - 1$ vanishes sufficiently rapidly at infinity so that there is no contribution to the integral from the great semicircle. Thus the Cauchy integral can be written

$$\epsilon(z)/\epsilon_0 = 1 + \frac{1}{2\pi i} \int_{-\infty}^{\infty} \frac{[\epsilon(\omega')/\epsilon_0 - 1]}{\omega' - z}\, d\omega' \qquad (7.115)$$

where z is now any point in the upper half-plane and the integral is taken along the real axis. Taking the limit as the complex frequency approaches the real axis from above, we write $z = \omega + i\delta$ in (7.115):

$$\epsilon(\omega)/\epsilon_0 = 1 + \frac{1}{2\pi i} \int_{-\infty}^{\infty} \frac{[\epsilon(\omega')/\epsilon_0 - 1]}{\omega' - \omega - i\delta}\, d\omega' \qquad (7.116)$$

For real ω the presence of the $i\delta$ in the denominator is a mnemonic for the distortion of the contour along the real axis by giving it an infinitesimal semicircular detour *below* the point $\omega' = \omega$. The denominator can be written formally as

$$\frac{1}{\omega' - \omega - i\delta} = P\left(\frac{1}{\omega' - \omega}\right) + \pi i \delta(\omega' - \omega) \qquad (7.117)$$

where P means principal part. The delta function serves to pick up the contribution from the small semicircle going in a positive sense halfway around the pole at $\omega' = \omega$. Use of (7.117) and a simple rearrangement turns (7.116) into

$$\epsilon(\omega)/\epsilon_0 = 1 + \frac{1}{\pi i} P \int_{-\infty}^{\infty} \frac{[\epsilon(\omega')/\epsilon_0 - 1]}{\omega' - \omega} \, d\omega' \qquad (7.118)$$

The real and imaginary parts of this equation are

$$\text{Re } \epsilon(\omega)/\epsilon_0 = 1 + \frac{1}{\pi} P \int_{-\infty}^{\infty} \frac{\text{Im } \epsilon(\omega')/\epsilon_0}{\omega' - \omega} \, d\omega'$$

$$\text{Im } \epsilon(\omega)/\epsilon_0 = -\frac{1}{\pi} P \int_{-\infty}^{\infty} \frac{[\text{Re } \epsilon(\omega')/\epsilon_0 - 1]}{\omega' - \omega} \, d\omega' \qquad (7.119)$$

These relations, or the ones recorded immediately below, are called *Kramers–Kronig relations* or *dispersion relations*. They were first derived by H. A. Kramers (1927) and R. de L. Kronig (1926) independently. The symmetry property (7.113) shows that Re $\epsilon(\omega)$ is even in ω, while Im $\epsilon(\omega)$ is odd. The integrals in (7.119) can thus be transformed to span only positive frequencies:

$$\text{Re } \epsilon(\omega)/\epsilon_0 = 1 + \frac{2}{\pi} P \int_{0}^{\infty} \frac{\omega' \, \text{Im } \epsilon(\omega')/\epsilon_0}{\omega'^2 - \omega^2} \, d\omega'$$

$$\text{Im } \epsilon(\omega)/\epsilon_0 = -\frac{2\omega}{\pi} P \int_{0}^{\infty} \frac{[\text{Re } \epsilon(\omega')/\epsilon_0 - 1]}{\omega'^2 - \omega^2} \, d\omega' \qquad (7.120)$$

In writing (7.119) and (7.120) we have tacitly assumed that $\epsilon(\omega)/\epsilon_0$ was regular at $\omega = 0$. For conductors the simple pole at $\omega = 0$ can be exhibited separately with little further complication.

The Kramers–Kronig relations are of very general validity, following from little more than the assumption of the causal connection (7.111) between the polarization and the electric field. Empirical knowledge of Im $\epsilon(\omega)$ from absorption studies allows the calculation of Re $\epsilon(\omega)$ from the first equation in (7.120). The connection between absorption and anomalous dispersion, shown in Fig. 7.8, is contained in the relations. The presence of a very narrow absorption line or band at $\omega = \omega_0$ can be approximated by taking

$$\text{Im } \epsilon(\omega') \simeq \frac{\pi K}{2\omega_0} \delta(\omega' - \omega_0) + \cdots$$

where K is a constant and the dots indicate the other (smoothly varying) contributions to Im ϵ. The first equation in (7.120) then yields

$$\text{Re } \epsilon(\omega) \simeq \bar{\epsilon} + \frac{K}{\omega_0^2 - \omega^2} \qquad (7.121)$$

for the behavior of Re $\epsilon(\omega)$ near, but not exactly at, $\omega = \omega_0$. The term $\bar{\epsilon}$ represents the slowly varying part of Re ϵ resulting from the more remote contributions to Im ϵ. The approximation (7.121) exhibits the rapid variation of Re $\epsilon(\omega)$ in the neighborhood of an absorption line, shown in Fig. 7.8 for lines of finite width. A more realistic description for Im ϵ would lead to an expression for Re ϵ in complete accord with the behavior shown in Fig. 7.8. The demonstration of this is left to the problems at the end of the chapter.

Relations of the general type (7.119) or (7.120) connecting the dispersive and absorptive aspects of a process are extremely useful in all areas of physics. Their widespread application stems from the very small number of physically well-founded assumptions necessary for their derivation. References to their application in particle physics, as well as solid-state physics, are given at the end of the chapter. We end with mention of two *sum rules* obtainable from (7.120). It was shown in Section 7.5.D, within the context of a specific model, that the dielectric constant is given at high frequencies by (7.59). The form of (7.59) is, in fact, quite general, as shown above (Section 7.10.C). The plasma frequency can therefore be *defined* by means of (7.59) as

$$\omega_p^2 = \lim_{\omega \to \infty} \{\omega^2 [1 - \epsilon(\omega)/\epsilon_0]\}$$

Provided the falloff of Im $\epsilon(\omega)$ at high frequencies is given by (7.114), the first Kramers–Kronig relation yields a *sum rule for* ω_p^2:

$$\omega_p^2 = \frac{2}{\pi} \int_0^\infty \omega \text{ Im } \epsilon(\omega)/\epsilon_0 \, d\omega \qquad (7.122)$$

This relation is sometimes known as the sum rule for oscillator strengths. It can be shown to be equivalent to (7.52) for the dielectric constant (7.51), but is obviously more general.

The second sum rule concerns the integral over the real part of $\epsilon(\omega)$ and follows from the second relation (7.120). With the assumption that $[\text{Re } \epsilon(\omega')/\epsilon_0 - 1] = -\omega_p^2/\omega'^2 + O(1/\omega'^4)$ for all $\omega' > N$, it is straightforward to show that for $\omega > N$

$$\text{Im } \epsilon(\omega)/\epsilon_0 = \frac{2}{\pi\omega} \left\{ -\frac{\omega_p^2}{N} + \int_0^N [\text{Re } \epsilon(\omega')/\epsilon_0 - 1] \, d\omega' \right\} + O\left(\frac{1}{\omega^3}\right)$$

It was shown in Section 7.10.C that, excluding conductors and barring the unphysical happening that $G(0^+) \neq 0$, Im $\epsilon(\omega)$ behaves at large frequencies as ω^{-3}. It therefore follows that the expression in curly brackets must vanish. We are thus led to a *second sum rule*,

$$\frac{1}{N} \int_0^N \text{Re } \epsilon(\omega)/\epsilon_0 \, d\omega = 1 + \frac{\omega_p^2}{N^2} \qquad (7.123)$$

which, for $N \to \infty$, states that the average value of Re $\epsilon(\omega)/\epsilon_0$ over all frequencies is equal to unity. For conductors, the plasma frequency sum rule (7.122) still holds, but the second sum rule (sometimes called a *superconvergence relation*) has an added term $-\pi\sigma/2\epsilon_0 N$, on the right hand side (see Problem 7.23). These optical sum rules and several others are discussed by Altarelli et al.*

7.11 *Arrival of a Signal After Propagation Through a Dispersive Medium*

Some of the effects of dispersion have been considered in the preceding sections. There remains one important aspect, the actual arrival at a remote point of a

---

*M. Altarelli, D. L. Dexter, H. M. Nussenzveig, and D. Y. Smith, *Phys. Rev.* **B6**, 4502 (1972).

wave train that initially has a well-defined beginning. How does the signal build up? If the phase velocity or group velocity is greater than the velocity of light in vacuum for important frequency components, does the signal propagate faster than allowed by causality and relativity? Can the arrival time of the disturbance be given an unambiguous definition? These questions were examined authoritatively by Sommerfeld and Brillouin in papers published in *Annalen der Physik* in 1914.* The original papers, plus subsequent work by Brillouin, are contained in English translation in the book, *Wave Propagation and Group Velocity*, by Brillouin. A briefer account is given in Sommerfeld's *Optics*, Chapter III. A complete discussion is lengthy and technically complicated.† We treat only the qualitative features. The reader can obtain more detail in the cited literature or the second edition of this book, from which the present account is abbreviated.

For definiteness we consider a plane wave train normally incident from vacuum on a semi-infinite uniform medium of index of refraction $n(\omega)$ filling the region $x > 0$. From the Fresnel equations (7.42) and Problem 7.20, the amplitude of the electric field of the wave for $x > 0$ is given by

$$u(x, t) = \int_{-\infty}^{\infty} \left[\frac{2}{1 + n(\omega)} \right] A(\omega) e^{ik(\omega)x - i\omega t} \, d\omega \tag{7.124}$$

where

$$A(\omega) = \frac{1}{2\pi} \int_{-\infty}^{\infty} u_i(0, t) e^{i\omega t} \, dt \tag{7.125}$$

is the Fourier transform of the real incident electric field $u_i(x, t)$ evaluated just outside the medium, at $x = 0^-$. The wave number $k(\omega)$ is

$$k(\omega) = \frac{\omega}{c} n(\omega) \tag{7.126}$$

and is generally complex, with positive imaginary part corresponding to absorption of energy during propagation. Many media are sufficiently transparent that the wave number can be treated as real for most purposes, but there is always some damping present. [Parenthetically we observe that in (7.124) frequency, not wave number, is used as the independent variable. The change from the practice of Sections 7.8 and 7.9 is dictated by the present emphasis on the *time* development of the wave at a fixed point in space.]

We suppose that the incident wave has a well-defined front edge that reaches $x = 0$ not before $t = 0$. Thus $u(0, t) = 0$ for $t < 0$. With additional physically reasonable mathematical requirements, this condition on $u(0, t)$ assures that $A(\omega)$ is analytic in the upper half-ω-plane [just as condition (7.112) assured the analyticity of $\epsilon(\omega)$ there]. Generally, $A(\omega)$ will have singularities in the lower half-ω-plane determined by the exact form of $u(x, t)$. We assume that $A(\omega)$ is bounded for $|\omega| \to \infty$.

The index of refraction $n(\omega)$ is crucial in determining the detailed nature of the propagation of the wave in the medium. Some general features follow, how-

---

*A Sommerfeld, *Ann. Phys (Leipzig)* **44**, 177 (1914). L. Brillouin, *Ann. Phys. (Leipzig)* **44**, 203 (1914).

†An exhaustive treatment is given in K. E. Oughstun and G. C. Sherman, *Electromagnetic Pulse Propagation in Causal Dielectrics*, Springer-Verlag, Berlin (1994).

ever, from the global properties of $n(\omega)$. Just as $\epsilon(\omega)$ is analytic in the upper half-ω-plane, so is $n(\omega)$. Furthermore, (7.59) shows that for $|\omega| \to \infty$, $n(\omega) \to 1 - \omega_p^2/2\omega^2$. A simple one-resonance model of $n(\omega)$ based on (7.51), with resonant frequency ω_0 and damping constant γ, leads to the singularity structure shown in Fig. 7.16. The poles of $\epsilon(\omega)$ become branch cuts in $n(\omega)$. A multiresonance expression for ϵ leads to a much more complex cut structure, but the upper plane analyticity and the asymptotic behavior for large $|\omega|$ remain.

The proof that no signal can propagate faster than the speed of light in vacuum, whatever the detailed properties of the medium, is now straightforward. We consider evaluating the amplitude (7.124) by contour integration in the complex ω plane. Since $n(\omega) \to 1$ for $|\omega| \to \infty$, the argument of the exponential in (7.124) becomes

$$i\phi(\omega) = i[k(\omega)x - \omega t] \to \frac{i\omega(x - ct)}{c}$$

for large $|\omega|$. Evidently, we obtain a vanishing contribution to the integral by closing the contour with a great semicircle at infinity in the upper half-plane for $x > ct$ and in the lower half-plane for $x < ct$. With $n(\omega)$ and $A(\omega)$ both analytic in the upper half-ω-plane, the whole integrand is analytic there. Cauchy's theorem tells us that if the contour is closed in the upper half-plane $(x > ct)$, the integral vanishes. We have therefore established that

$$u(x, t) = 0 \qquad \text{for } (x - ct) > 0 \qquad (7.127)$$

provided only that $A(\omega)$ and $n(\omega)$ are analytic for $\text{Im } \omega > 0$ and $n(\omega) \to 1$ for $|\omega| \to \infty$. Since the specific form of $n(\omega)$ does not enter, we have a general proof that no signal propagates with a velocity greater than c, whatever the medium.

For $ct > x$, the contour is to be closed in the lower half-plane, enveloping the singularities. The integral is dominated by different singularities at different

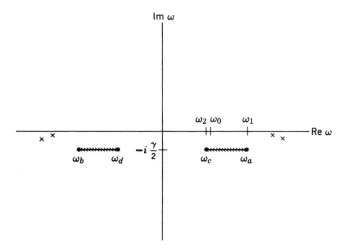

Figure 7.16   Branch cuts defining the singularities of a simple one-resonance model for the index of refraction $n(\omega)$. For transparent media the branch cuts lie much closer to (but still below) the real axis than shown here. More realistic models for $n(\omega)$ have more complicated cut structures, all in the lower half-ω-plane. The crosses mark the possible locations of singularities of $A(\omega)$.

times. Brillouin and Sommerfeld used the method of steepest descent* to evaluate (7.124) in various regimes. We sketch the chief aspects using the concepts
of the less rigorous method of stationary phase. The method of stationary phase
is based on the idea that the phase $\phi(\omega)$ in an integral such as (7.124) is generally
large and rapidly varying. The rapid oscillations of $e^{i\phi}$ over most of the range of
integration mean that the integrand averages almost to zero. Exceptions to the
cancellation occur only when $\phi(\omega)$ is "stationary," that is, when $\phi(\omega)$ has an
extremum. The integral can therefore be estimated by approximating the integral
at each of the points of stationary phase by a Taylor series expansion of $\phi(\omega)$
and summing these contributions.

We use the idea of stationary phase to discuss the qualitative aspects of
the arrival of the signal without explicit use of the integration formulas. With
$\phi(\omega) = k(\omega)x - \omega t$ and $k(\omega)$ given by (7.126), the stationary phase condition
$\partial\phi/\partial\omega = 0$ becomes

$$ c\frac{dk}{d\omega} = n(\omega) + \omega\frac{dn}{d\omega} = \frac{t}{t_0} \qquad \text{for } t > t_0 = x/c \qquad (7.128) $$

The earliest part of the wave occurs when t/t_0 is infinitesimally larger than unity.
From the global properties of $n(\omega)$ we see that the point of stationary phase is
at $|\omega| \to \infty$, where $n \to 1$. Explicitly, we have

$$ c\frac{dk}{d\omega} \approx 1 + \frac{\omega_p^2}{2\omega^2} = \frac{t}{t_0}, \qquad \text{for } t \gtrsim t_0 $$

showing that the frequency of stationary phase $\omega_s \approx \omega_p/\sqrt{2(t/t_0 - 1)}$ depends
only on t/t_0 and ω_p^2, a global property of the index of refraction. The incident
wave's $A(\omega_s)$ is presumably very small. The earliest part of the signal is therefore
extremely small and of very high frequency, bearing no resemblance to the incident wave. This part of the signal is called the *first or Sommerfeld precursor*.
At somewhat later times, the frequency ω_s slowly decreases; the signal grows
very slowly in amplitude, and its structure is complex.

Only when t/t_0 in (7.128) reaches $n(0)$ is there a qualitative change in the
amplitude. Because $\omega = 0$ is now a point of stationary phase, the high frequency
of oscillation is replaced by much lower frequencies. More important is the fact
that $d^2k(\omega)/d\omega^2 = 0$ at $\omega = 0$. In such circumstances the stationary phase
approximation fails, giving an infinite result. One must improve the approximation to include cubic terms in the Taylor series expansion of $\phi(\omega)$ around
$\omega = \omega_s$. The amplitude is expressible in terms of Airy integrals (of rainbow
fame). The wave becomes relatively large in amplitude and of long period for
times $t \gtrsim n(0)t_0$. This phase of development is called the *second or Brillouin
precursor.*

At still later times, there are several points of stationary phase. The wave
depends in detail on the exact form of $n(\omega)$. Eventually, the behavior of $A(\omega)$
begins to dominate the integral. By then the main part of the wave has arrived
at the point x. The amplitude behaves in time as if it were the initial wave propagating with the appropriate phase velocity and attenuation.

The sequence of arrival of the tiny, high-frequency Sommerfeld precursor,
the larger and slower oscillating Brillouin precursor, and then the main signal,

---

*See *Jeffreys and Jeffreys* (Section 17.04) or *Born and Wolf* (Appendix III) for a discussion of this
method, originally developed by P. Debye.

and indeed their detailed appearance, can differ greatly depending upon the specifics of $n(\omega)$, $A(\omega)$, and the position x of observation. A textbook example can be found in *Oughstun and Sherman* (*op. cit.*, Fig. 9.10, p. 383).

References and Suggested Reading

The whole subject of optics as an electromagnetic phenomenon is treated authoritatively by
>Born and Wolf.

Their first chapter covers plane waves, polarization, and reflection and refraction, among other topics. Stokes parameters are covered there and also by
>Stone.

A very complete discussion of plane waves incident on boundaries of dielectrics and conductors is given by
>Stratton, Chapter IX.

Another good treatment of electromagnetic waves in both isotropic and anisotropic media is that of
>Landau and Lifshitz, *Electrodynamics of Continuous Media*, Chapters X and XI.

A more elementary, but clear and thorough, approach to plane waves and their properties appears in
>Adler, Chu, and Fano, Chapters 7 and 8.

The unique optical properties of water (Fig. 7.9) and the use of Kramers–Kronig relations are discussed by
>M. W. Williams and E. T. Arakawa, Optical and dielectric properties of materials relevant for biological research, in *Handbook on Synchrotron Radiation*, Vol. 4, eds. S. Ebashi, M. Koch, and E. Rubenstein, North-Holland, New York (1991), pp. 95–145.

See also Vol. 1 (1972) of
>Franks, Felix, ed., *Water: A Comprehensive Treatise* in 7 volumes, Plenum Press, New York (1972–82).

Propagation of waves in the ionosphere has, because of its practical importance, an extensive literature. The physical and mathematical aspects are covered in
>Budden
>Wait

The special topic of whistlers is discussed in detail in
>R. A. Helliwell, *Whistlers and Related Ionospheric Phenomena*, Stanford University Press, Stanford, CA (1965).

The topic of magnetohydrodynamic and plasma waves has a very large literature. The American Journal of Physics Resource Letters,
>G. Bekefi and S. C. Brown, Plasma Physics: Waves and radiation processes in plasma, *Am. J. Phys.* **34**, 1001 (1966)
>C. L. Grabbe, Plasma waves and instabilities, *Am. J. Phys.* **52**, 970 (1984)

will lead the reader to appropriate references for further study. For discussion of MHD in the context of astrophysics, see
>Alfvén and Fälthammar
>Cowling
>L. F. Burlaga, *Interplanetary Magnetohydrodynamics*, Oxford University Press, New York (1995).

A survey of the basic physics involved in the interaction of light with matter can be found in the semipopular article by
>V. F. Weisskopf, *Sci. Am.* **219**, 3, 60 (September 1968).

The propagation of waves in dispersive media is discussed in detail in the book by Brillouin.

The distortion and attenuation of pulses in dissipative materials are covered by Stratton, pp. 301–309.

The generalization of dispersion to nonlocality in space as well as time via $\epsilon(\omega, \mathbf{k})$ is an important concept for inelastic electromagnetic interactions. Problem 7.26 introduces some of the ideas. The uses of $\epsilon(\omega, \mathbf{k})$ are discussed in

Platzman, P. M., and P. A. Wolff, *Waves and Interactions in Solid State Plasmas*, Academic Press, New York (1972).

Schnatterly, S. E., Inelastic electron scattering spectroscopy, in *Solid State Physics*, Vol. 34, eds. H. Ehrenreich, F. Seitz, and D. Turnbull, Academic Press, New York (1979), pp. 275–358.

Kramers–Kronig dispersion relations and their generalizations find application in many areas of physics. Examples in high-energy physics can be found in

S. Gasiorowicz, *Elementary Particle Physics*, Wiley, New York (1965).

G. Kallen, *Elementary Particle Physics*, Addison-Wesley, Reading, MA (1964).

G. R. Screaton, ed., *Dispersion Relations* (Scottish Universities' Summer School 1960), Oliver and Boyd, Edinburgh and London (1961).

Some uses in solid-state physics are discussed in the article by

F. Stern, in *Solid State Physics*, Vol. 15, eds. F. Seitz and D. Turnbull, Academic Press, New York (1963), pp. 299–408.

Problems

7.1 For each set of Stokes parameters given below deduce the amplitude of the electric field, up to an overall phase, in both linear polarization and circular polarization bases and make an accurate drawing similar to Fig. 7.4 showing the lengths of the axes of one of the ellipses and its orientation.

(a) $s_0 = 3,$ $\quad s_1 = -1,$ $\quad s_2 = 2,$ $\quad s_3 = -2;$

(b) $s_0 = 25,$ $\quad s_1 = 0,$ $\quad s_2 = 24,$ $\quad s_3 = 7.$

7.2 A plane wave is incident on a layered interface as shown in the figure. The indices of refraction of the three nonpermeable media are n_1, n_2, n_3. The thickness of the intermediate layer is d. Each of the other media is semi-infinite.

(a) Calculate the transmission and reflection coefficients (ratios of transmitted and reflected Poynting's flux to the incident flux), and sketch their behavior as a function of frequency for $n_1 = 1, n_2 = 2, n_3 = 3; n_1 = 3, n_2 = 2, n_3 = 1;$ and $n_1 = 2, n_2 = 4, n_3 = 1.$

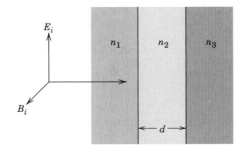

Problem 7.2

(b) The medium n_1 is part of an optical system (e.g., a lens); medium n_3 is air ($n_3 = 1$). It is desired to put an optical coating (medium n_2) on the surface so that there is no reflected wave for a frequency ω_0. What thickness d and index of refraction n_2 are necessary?

7.3   Two plane semi-infinite slabs of the same uniform, isotropic, nonpermeable, lossless dielectric with index of refraction n are parallel and separated by an air gap ($n = 1$) of width d. A plane electromagnetic wave of frequency ω is incident on the gap from one of the slabs with angle of incidence i. For linear polarization *both* parallel to *and* perpendicular to the plane of incidence,

  (a)   calculate the ratio of power transmitted into the second slab to the incident power and the ratio of reflected to incident power;

  (b)   for i greater than the critical angle for total internal reflection, sketch the ratio of transmitted power to incident power as a function of d measured in units of wavelength in the gap.

7.4   A plane-polarized electromagnetic wave of frequency ω in free space is incident normally on the flat surface of a nonpermeable medium of conductivity σ and dielectric constant ϵ.

  (a)   Calculate the amplitude and phase of the reflected wave relative to the incident wave for arbitrary σ and ϵ.

  (b)   Discuss the limiting cases of a very poor and a very good conductor, and show that for a good conductor the reflection coefficient (ratio of reflected to incident intensity) is approximately

$$R \simeq 1 - 2\frac{\omega}{c}\delta$$

where δ is the skin depth.

7.5   A plane polarized electromagnetic wave $\mathbf{E} = \mathbf{E}_i e^{i\mathbf{k}\cdot\mathbf{x} - i\omega t}$ is incident normally on a flat uniform sheet of an *excellent* conductor ($\sigma \gg \omega\epsilon_0$) having a thickness D. Assuming that in space and in the conducting sheet $\mu/\mu_0 = \epsilon/\epsilon_0 = 1$, discuss the reflection and transmission of the incident wave.

  (a)   Show that the amplitudes of the reflected and transmitted waves, correct to the first order in $(\epsilon_0\omega/\sigma)^{1/2}$, are:

$$\frac{E_r}{E_i} = \frac{-(1 - e^{-2\lambda})}{(1 - e^{-2\lambda}) + \gamma(1 + e^{-2\lambda})}$$

$$\frac{E_t}{E_i} = \frac{2\gamma e^{-\lambda}}{(1 - e^{-2\lambda}) + \gamma(1 + e^{-2\lambda})}$$

where

$$\gamma = \sqrt{\frac{2\epsilon_0\omega}{\sigma}}\,(1 - i) = \frac{\omega\delta}{c}(1 - i)$$

$$\lambda = (1 - i)D/\delta$$

and $\delta = \sqrt{2/\omega\mu\sigma}$ is the penetration depth.

  (b)   Verify that for zero thickness and infinite thickness you obtain the proper limiting results.

  (c)   Show that, except for sheets of very small thickness, the transmission coefficient is

$$T = \frac{8(\mathrm{Re}\ \gamma)^2 e^{-2D/\delta}}{1 - 2e^{-2D/\delta}\cos(2D/\delta) + e^{-4D/\delta}}$$

Sketch log T as a function of (D/δ), assuming Re $\gamma = 10^{-2}$. Define "very small thickness."

7.6 A plane wave of frequency ω is incident normally from vacuum on a semi-infinite slab of material with a *complex* index of refraction $n(\omega)$ $[n^2(\omega) = \epsilon(\omega)/\epsilon_0]$.

(a) Show that the ratio of reflected power to incident power is

$$R = \left| \frac{1 - n(\omega)}{1 + n(\omega)} \right|^2$$

while the ratio of power transmitted into the medium to the incident power is

$$T = \frac{4 \operatorname{Re} n(\omega)}{|1 + n(\omega)|^2}$$

(b) Evaluate $\operatorname{Re}[i\omega(\mathbf{E} \cdot \mathbf{D}^* - \mathbf{B} \cdot \mathbf{H}^*)/2]$ as a function of (x, y, z). Show that this rate of change of energy per unit volume accounts for the relative transmitted power T.

(c) For a conductor, with $n^2 = 1 + i(\sigma/\omega\epsilon_0)$, σ real, write out the results of parts a and b in the limit $\epsilon_0\omega \ll \sigma$. Express your answer in terms of δ as much as possible. Calculate $\frac{1}{2} \operatorname{Re}(\mathbf{J}^* \cdot \mathbf{E})$ and compare with the result of part b. Do both enter the complex form of Poynting's theorem?

7.7 A ribbon beam of plane-polarized radiation of wavelength λ is totally reflected internally at a plane boundary between two media with indices of refraction n and n' ($n' < n$). As discussed in Section 7.4, the ratio of the reflected to incident amplitudes is a complex number of modulus unity, $E_0''/E_0 = \exp[i\phi(i, i_0)]$ for the angle of incidence $i > i_0$, where $\sin i_0 = n'/n$.

(a) Show that for a "monochromatic" ribbon beam of radiation in the z direction with an electric field amplitude, $E(x)e^{(ikz-i\omega t)}$, where $E(x)$ is finite in transverse extent (but many wavelengths broad), the lowest order approximation in terms of plane waves is

$$\mathbf{E}(x, z, t) = \boldsymbol{\epsilon} \int d\kappa \, A(\kappa) e^{i\kappa x + ikz - i\omega t}$$

where $\boldsymbol{\epsilon}$ is a polarization vector, and $A(\kappa)$ is the Fourier transform of $E(x)$, with support in κ around $\kappa = 0$ small compared to k. The finite beam consists of plane waves with a small range of angles of incidence, centered around the geometrical optics value.

(b) Consider the reflected beam and show that for $i > i_0$ the electric field can be expressed approximately as

$$\mathbf{E}''(x, z, t) = \boldsymbol{\epsilon}''E(x'' - \delta x) \exp[i\mathbf{k}'' \cdot \mathbf{x} - i\omega t + i\phi(i, i_0)]$$

where $\boldsymbol{\epsilon}''$ is a polarization vector, x'' is the x coordinate perpendicular to $\mathbf{k}''$, the reflected wave vector, and $\delta x = -(1/k)[d\phi(i, i_0)/di]$.

(c) With the Fresnel expressions of Section 7.3 for the phases $\phi(i, i_0)$ for the two states of plane polarization, show that the lateral displacements of reflected beams with respect to the geometric optics position are

$$D_\perp = \frac{\lambda}{\pi} \frac{\sin i}{(\sin^2 i - \sin^2 i_0)^{1/2}} \quad \text{and} \quad D_\parallel = D_\perp \cdot \frac{\sin^2 i_0}{[\sin^2 i - \cos^2 i \cdot \sin^2 i_0]}$$

The displacement is known as the *Goos–Hänchen effect* (*op. cit.*).

7.8 A monochromatic plane wave of frequency ω is incident normally on a stack of layers of various thicknesses t_j and lossless indices of refraction n_j. Inside the stack, the wave has both forward and backward moving components. The change in the wave through any interface and also from one side of a layer to the other can be described by means of 2×2 transfer matrices. If the electric field is written as

$$E = E_+ e^{ikx} + E_- e^{-ikx}$$

in each layer, the transfer matrix equation $E' = TE$ is explicitly

$$\begin{pmatrix} E'_+ \\ E'_- \end{pmatrix} = \begin{pmatrix} t_{11} & t_{12} \\ t_{21} & t_{22} \end{pmatrix} \begin{pmatrix} E_+ \\ E_- \end{pmatrix}$$

(a) Show that the transfer matrix for propagation inside, but across, a layer of index of refraction n_j and thickness t_j is

$$T_{\text{layer}}(n_j, t_j) = \begin{pmatrix} e^{ik_j t_j} & 0 \\ 0 & e^{-ik_j t_j} \end{pmatrix} = I \cos(k_j t_j) + i\sigma_3 \sin(k_j t_j)$$

where $k_j = n_j \omega/c$, I is the unit matrix, and σ_k are the Pauli spin matrices of quantum mechanics. Show that the inverse matrix is T^*.

(b) Show that the transfer matrix to cross an interface from n_1 ($x < x_0$) to n_2 ($x > x_0$) is

$$T_{\text{interface}}(2, 1) = \frac{1}{2} \begin{pmatrix} n + 1 & -(n - 1) \\ -(n - 1) & n + 1 \end{pmatrix} = I \frac{(n + 1)}{2} - \sigma_1 \frac{(n - 1)}{2}$$

where $n = n_1/n_2$.

(c) Show that for a complete stack, the incident, reflected, and transmitted waves are related by

$$E_{\text{trans}} = \frac{\det(T)}{t_{22}} E_{\text{inc}}, \qquad E_{\text{refl}} = -\frac{t_{21}}{t_{22}} E_{\text{inc}}$$

where t_{ij} are the elements of T, the product of the forward-going transfer matrices, including from the material filling space on the incident side into the first layer and from the last layer into the medium filling the space on the transmitted side.

7.9 A stack of optical elements consists of N layers with index of refraction n and thickness t_1, separated by air gaps ($n_2 = 1$) of thickness t_2. A monochromatic plane wave is incident normally. With appropriate thicknesses, a modest number of layers can cause almost total reflection of a given range of frequencies, even for normal n values (e.g., $1.3 < n < 1.8$).

(a) Show the transfer matrix for the stack is $T_{\text{stack}} = T^N (I \cos \alpha_2 - i\sigma_3 \sin \alpha_2)$, where $\alpha_2 = \omega t_2/c$, and the single air gap plus foil transfer matrix is

$$T = (1/4n)\{[(n + 1)^2 \cos(\alpha_1 + \alpha_2) - (n - 1)^2 \cos(\alpha_1 - \alpha_2)]I$$
$$+ i\sigma_3[(n + 1)^2 \sin(\alpha_1 + \alpha_2) + (n - 1)^2 \sin(\alpha_1 - \alpha_2)]$$
$$+ 2\sigma_1(n^2 - 1) \sin \alpha_1 \sin \alpha_2$$
$$- 2\sigma_2(n^2 - 1) \sin \alpha_1 \cos \alpha_2\}$$

with $\alpha_1 = n\omega t_1/c$.

(b) If all the layers (both air gaps and foils) have optical thicknesses of a quarter-wavelength of the incident wave, show that

$$T = -\exp(-\lambda\sigma_1), \qquad \text{where } \lambda = \ln(n)$$

is (roughly) the amplitude "decay constant" per layer. Show that the fractional transmitted intensity is

$$\frac{|E_{\text{trans}}|^2}{|E_{\text{inc}}|^2} = \text{sech}^2[N \ln(n)] = \frac{4n^{2N}}{(n^{2N} + 1)^2} \rightarrow 4 \exp[-N \ln(n^2)]$$

The asymptotic form holds for $n^{2N} \gg 1$.

7.10 An arbitrary optical element of length L is placed in a uniform nonabsorbing medium with index of refraction $n(\omega)$ with its front face at $x = 0$ and its back face at $x = L$. If a monochromatic plane wave of frequency ω with amplitude $\psi_{\text{inc}}(\omega, x, t)$ $= \exp[ik(\omega)x - \omega t]$ is incident on the front face of the element, the transmitted wave amplitude is $\psi_{\text{trans}}(\omega, x, t) = T(\omega) \exp[ik(\omega)(x - L) - \omega t]$, where the relative transmission amplitude $T(\omega) = \tau(\omega) \exp[i\phi(\omega)]$ is a complex quantity of magnitude $\tau(\omega)$ and phase $\phi(\omega)$.

A plane wave of radiation $\psi_{\text{inc}}(x, t)$, consisting of a coherent superposition of different frequencies centered around $\omega = \omega_0$, with support $A(\omega)$ narrow on the scale of variation of $\tau(\omega)$, $\phi(\omega)$ and $k(\omega)$, is incident on the optical element. Show that the transmitted wave for $x > L$ is approximately

$$\psi_{\text{trans}}(x, t) \approx \tau(\omega_0)e^{i\beta}\psi_{\text{inc}}(x', t')$$

where β is a constant phase and $x' = x - L$, $t' = t - T$. The transit or group delay time (sometimes attributed in another context to E. P. Wigner) is $T = [d\phi(\omega)/d\omega]_{\omega=\omega_0}$. If $cT < L$, some authors speak of superluminal propagation through the element. Discuss.

7.11 A simple example of the transit time of the preceding problem is afforded by a slab of lossless dielectric of thickness d and index of refraction n in vacuum.

(a) For a plane wave incident normally, show that the magnitude of the transmitted amplitude is

$$|\tau(\omega)| = \frac{4n}{\sqrt{[(n + 1)^2 - (n - 1)^2 \cos(2z)]^2 + [(n - 1)^2 \sin(2z)]^2}}$$

while its phase is

$$\phi(\omega) = z + \arctan\left[\frac{(n - 1)^2 \sin(2z)}{(n + 1)^2 - (n - 1)^2 \cos(2z)}\right]$$

where $z = n\omega d/c$.

(b) Neglecting dispersion, show that for $z \to 0$ and $z = \pi$, $|\tau| = 1.0$ and $cT/d = (n^2 + 1)/2$, while for $z = \pi/2$ (quarter-wave plate), $|\tau| = 2n/(n^2 + 1)$ and $cT/d = 2n^2/(n^2 + 1)$. Show also that cT/d, averaged over any integer number of quarter-wavelength optical paths, is $\langle cT/d \rangle = n$. Does this result tell you anything about what you might expect for the observed transit time of a long wave train ($\Delta\omega/\omega \ll 1$) through a piece of window glass? Explain.

(c) Calculate numerically and plot the results as functions of z for the magnitude of the transmission amplitude, its phase, and the transit time in units of d/c for $n = 1.5$ and $n = 2.0$.

7.12 The time dependence of electrical disturbances in good conductors is governed by the frequency-dependent conductivity (7.58). Consider longitudinal electric fields in a conductor, using Ohm's law, the continuity equation, and the differential form of Coulomb's law.

(a) Show that the time-Fourier-transformed charge density satisfies the equation

$$[\sigma(\omega) - i\omega\epsilon_0]\rho(\mathbf{x}, \omega) = 0$$

(b) Using the representation $\sigma(\omega) = \sigma_0/(1 - i\omega\tau)$, where $\sigma_0 = \epsilon_0\omega_p^2\tau$ and τ is a damping time, show that in the approximation $\omega_p\tau \gg 1$ any initial disturbance will oscillate with the plasma frequency and decay in amplitude with a decay constant $\lambda = 1/2\tau$. Note that if you use $\sigma(\omega) \approx \sigma(0) = \sigma_0$ in part a, you will find no oscillations and extremely rapid damping with the (wrong) decay constant $\lambda_w = \sigma_0/\epsilon_0$.

Reference: W. M. Saslow and G. Wilkinson, *Am. J. Phys.* **39**, 1244 (1971).

7.13 A stylized model of the ionosphere is a medium described by the dielectric constant (7.59). Consider the earth with such a medium beginning suddenly at a height h and extending to infinity. For waves with polarization both perpendicular to the plane of incidence (from a horizontal antenna) and in the plane of incidence (from a vertical antenna),

(a) show from Fresnel's equations for reflection and refraction that for $\omega > \omega_p$ there is a range of angles of incidence for which reflection is not total, but for larger angles there is total reflection back toward the earth.

(b) A radio amateur operating at a wavelength of 21 meters in the early evening finds that she can receive distant stations located more than 1000 km away, but none closer. Assuming that the signals are being reflected from the F layer of the ionosphere at an effective height of 300 km, calculate the electron density. Compare with the known maximum and minimum F layer densities of $\sim 2 \times 10^{12}$ m^{-3} in the daytime and $\sim (2\text{-}4) \times 10^{11}$ m^{-3} at night.

7.14 A simple model of propagation of radio waves in the earth's atmosphere or ionosphere consists of a flat earth at $z = 0$ and a nonuniform medium with $\epsilon = \epsilon(z)$ for $z > 0$. Consider the Maxwell equations under the assumption that the fields are independent of y and can be written as functions of z times $e^{i(kx - \omega t)}$.

(a) Show that the wave equation governing the propagation for $z > 0$ is

$$\frac{d^2 F}{dz^2} + q^2(z)F = 0$$

where

$$q^2(z) = \omega^2 \mu_0 \epsilon(z) - k^2$$

and $F = E_y$ for *horizontal* polarization, and

$$q^2(z) = \omega^2 \mu_0 \epsilon(z) + \frac{1}{2\epsilon}\frac{d^2\epsilon}{dz^2} - \frac{3}{4\epsilon^2}\left(\frac{d\epsilon}{dz}\right)^2 - k^2$$

with $F = \sqrt{\epsilon/\epsilon_0}\, E_z$ for *vertical* polarization.

(b) Use the WKB approximation to treat the propagation of waves directed vertically into the ionosphere ($k = 0$), assuming that the dielectric constant is given by (7.59) with a plasma frequency $\omega_p(z)$ governed by an electron density like that shown in Fig. 7.11. Verify that the qualitative arguments in Section 7.6 hold, with departures in detail only for $\omega \sim \omega_{p,\text{max}}$.

(c) Using the WKB results of part b and the concepts of the propagation of a pulse from Section 7.8, define an effective height of the ionosphere $h'(\omega)$ by calculating the time T for a pulse of dominant frequency ω to travel up and be reflected back ($h' \equiv cT/2$). [The WKB approximation is discussed in most books on quantum mechanics.]

7.15 The partially ionized interstellar medium (mostly hydrogen) responds to optical frequencies as an electronic plasma in a weak magnetic field. The broad-spectrum pulses from a pulsar allow determination of some average properties of the interstellar medium (e.g., mean electron density and mean magnetic field). The treatment of an electronic plasma in a magnetic field of Section 7.6 is pertinent.

(a) Ignoring the weak magnetic field and assuming that $\max(\omega_p) \ll \omega$, show that c times the transit time of a pulse of mean frequency ω from a pulsar a distance R away is

$$ct(\omega) \approx R + \frac{e^2}{2\epsilon_0 m_e \omega^2} \int n_e(z)\, dz$$

where $n_e(z)$ is the electron density along the path of the light.

(b) The presence of the magnetic field causes a rotation of the plane of linear polarization (Faraday effect). Show that to lowest order in the magnetic field, the polarized light from the pulsar has its polarization rotated through an angle $\delta\theta(\omega)$:

$$\delta\theta(\omega) \approx -\frac{e^3}{2\epsilon_0 cm_e^2 \omega^2} \int n_e(z)B_{\parallel}(z)\, dz$$

where $B_{\parallel}(z)$ is the component of **B** parallel to the path of the light.

(c) Assuming you had an independent measure of the pulsar distance R, what observations would you make in order to infer $\langle n_e \rangle$ and $\langle B_{\parallel} \rangle$? What assumptions, if any, about the polarization are necessary?

7.16 Plane waves propagate in a homogeneous, nonpermeable, but *anisotropic* dielectric. The dielectric is characterized by a tensor ϵ_{ij}, but if coordinate axes are chosen as the principle axes, the components of displacement along these axes are related to the electric-field components by $D_i = \epsilon_i E_i$ ($i = 1, 2, 3$), where ϵ_i are the eigenvalues of the matrix ϵ_{ij}.

(a) Show that plane waves with frequency ω and wave vector **k** must satisfy

$$\mathbf{k} \times (\mathbf{k} \times \mathbf{E}) + \mu_0 \omega^2 \mathbf{D} = 0$$

(b) Show that for a given wave vector $\mathbf{k} = k\mathbf{n}$ there are two distinct modes of propagation with different phase velocities $v = \omega/k$ that satisfy the Fresnel equation

$$\sum_{i=1}^{3} \frac{n_i^2}{v^2 - v_i^2} = 0$$

where $v_i = 1/\sqrt{\mu_0 \epsilon_i}$ is called a principal velocity, and n_i is the component of **n** along the ith principal axis.

(c) Show that $\mathbf{D}_a \cdot \mathbf{D}_b = 0$, where $\mathbf{D}_a$, $\mathbf{D}_b$ are the displacements associated with the two modes of propagation.

7.17 Consider the problem of dispersion and waves in an electronic plasma when a uniform external magnetic induction $\mathbf{B}_0$ is present, as in Section 7.6.

(a) Show that in general the susceptibility tensor $\chi_{jk}(\omega)$ defined through $D_j = \sum_k \epsilon_{jk} E_k$, and $\epsilon_{jk} = \epsilon_0(\delta_{jk} + \chi_{jk})$, is

$$\chi_{jk} = -\frac{\omega_p^2}{\omega^2(\omega^2 - \omega_B^2)} \left[\omega^2 \delta_{jk} - \omega_B^2 b_j b_k - i\omega\omega_B \epsilon_{jkl} b_l \right]$$

where **b** is a unit vector in the direction of $\mathbf{B}_0$.

(b) By straightforward diagonalization of the dielectric tensor ϵ_{jk} or by an airtight argument based on the approach and results of Section 7.6, find the eigenvalues ϵ_j, $j = 1, 2, 3$.

(c)   A plane wave (ω, $\mathbf{k} = k\mathbf{n}$) must satisfy the vector equation of Problem 7.16a. Show that in terms of χ_{jk} the electric field and wave number must satisfy the three homogeneous equations,

$$(1 - \xi)E_j + \xi n_j(\mathbf{n} \cdot \mathbf{E}) + \sum_k \chi_{jk}E_k = 0, \qquad j = 1, 2, 3$$

where $\xi = (ck/\omega)^2$. Keeping only first-order terms in an expansion of χ_{jk} in powers of ω_B/ω, show that the effective dielectric constant for propagation of the plane wave is

$$\epsilon_\pm/\epsilon_0 \approx 1 - \frac{\omega_p^2}{\omega^2} \mp \frac{\omega_p^2\omega_B\mathbf{b} \cdot \mathbf{n}}{\omega^3}$$

for positive and negative helicity waves.

7.18   Magnetohydrodynamic waves can occur in a compressible, nonviscous, perfectly conducting fluid in a uniform static magnetic induction $\mathbf{B}_0$. If the propagation direction is not parallel or perpendicular to $\mathbf{B}_0$, the waves are not separated into purely longitudinal (magnetosonic) or transverse (Alfvén) waves. Let the angle between the propagation direction $\mathbf{k}$ and the field $\mathbf{B}_0$ be θ.

(a)   Show that there are three different waves with phase velocities given by

$$u_1^2 = (v_A \cos\theta)^2$$
$$u_{2,3}^2 = \tfrac{1}{2}(s^2 + v_A^2) \pm \tfrac{1}{2}[(s^2 + v_A^2)^2 - 4s^2v_A^2 \cos^2\theta]^{1/2}$$

where s is the sound velocity in the fluid, and $v_A = (B_0^2/\mu\rho_0)^{1/2}$ is the Alfvén velocity.

(b)   Find the velocity eigenvectors for the three different waves, and prove that the first (Alfvén) wave is always transverse, while the other two are neither longitudinal nor transverse.

(c)   Evaluate the phase velocities and eigenvectors of the mixed waves in the approximation that $v_A \gg s$. Show that for one wave the only appreciable component of velocity is parallel to the magnetic field, while for the other the only component is perpendicular to the field and in the plane containing $\mathbf{k}$ and $\mathbf{B}_0$.

7.19   An approximately monochromatic plane wave packet in one dimension has the instantaneous form, $u(x, 0) = f(x) e^{ik_0 x}$, with $f(x)$ the modulation envelope. For each of the forms $f(x)$ below, calculate the wave-number spectrum $|A(k)|^2$ of the packet, sketch $|u(x, 0)|^2$ and $|A(k)|^2$, evaluate explicitly the rms deviations from the means Δx and Δk (defined in terms of the intensities $|u(x, 0)|^2$ and $|A(k)|^2$), and test inequality (7.82).

(a) $f(x) = Ne^{-\alpha|x|/2}$

(b) $f(x) = Ne^{-\alpha^2 x^2/4}$

(c) $f(x) = \begin{cases} N(1 - \alpha|x|) & \text{for } \alpha|x| < 1 \\ 0 & \text{for } \alpha|x| > 1 \end{cases}$

(d) $f(x) = \begin{cases} N & \text{for } |x| < a \\ 0 & \text{for } |x| > a \end{cases}$

7.20   A homogeneous, isotropic, nonpermeable dielectric is characterized by an index of refraction $n(\omega)$, which is in general complex in order to describe absorptive processes.

(a) Show that the general solution for plane waves in one dimension can be written

$$u(x, t) = \frac{1}{\sqrt{2\pi}} \int_{-\infty}^{\infty} d\omega \, e^{-i\omega t}[A(\omega)e^{i(\omega/c)n(\omega)x} + B(\omega)e^{-i(\omega/c)n(\omega)x}]$$

where $u(x, t)$ is a component of **E** or **B**.

(b) If $u(x, t)$ is real, show that $n(-\omega) = n^*(\omega)$.

(c) Show that, if $u(0, t)$ and $\partial u(0, t)/\partial x$ are the boundary values of u and its derivative at $x = 0$, the coefficients $A(\omega)$ and $B(\omega)$ are

$$\begin{Bmatrix} A(\omega) \\ B(\omega) \end{Bmatrix} = \frac{1}{2} \frac{1}{\sqrt{2\pi}} \int_{-\infty}^{\infty} dt \, e^{i\omega t} \left[u(0, t) \mp \frac{ic}{\omega n(\omega)} \frac{\partial u}{\partial x}(0, t) \right]$$

7.21 Consider the nonlocal (in time) connection between **D** and **E**,

$$\mathbf{D}(\mathbf{x}, t) = \epsilon_0 \left\{ \mathbf{E}(\mathbf{x}, t) + \int d\tau \, G(\tau)\mathbf{E}(\mathbf{x}, t - \tau) \right\}$$

with the $G(\tau)$ appropriate for the single-resonance model,

$$\epsilon(\omega)/\epsilon_0 = 1 + \omega_p^2(\omega_0^2 - \omega^2 - i\gamma\omega)^{-1}$$

(a) Convert the nonlocal connection between **D** and **E** into an instantaneous relation involving derivatives of **E** with respect to time by expanding the electric field in the integral in a Taylor series in τ. Evaluate the integrals over $G(\tau)$ explicitly up to at least $\partial^2\mathbf{E}/\partial t^2$.

(b) Show that the series obtained in part a can be obtained formally by converting the frequency-representation relation, $\mathbf{D}(\mathbf{x}, \omega) = \epsilon(\omega)\mathbf{E}(\mathbf{x}, \omega)$ into a space-time relation,

$$\mathbf{D}(\mathbf{x}, t) = \epsilon\left(i\frac{\partial}{\partial t} \right)\mathbf{E}(\mathbf{x}, t)$$

where the variable ω in $\epsilon(\omega)$ is replaced by $\omega \rightarrow i(\partial/\partial t)$.

7.22 Use the Kramers–Kronig relation (7.120) to calculate the real part of $\epsilon(\omega)$, given the imaginary part of $\epsilon(\omega)$ for positive ω as

(a) $\text{Im } \epsilon/\epsilon_0 = \lambda[\theta(\omega - \omega_1) - \theta(\omega - \omega_2)], \qquad \omega_2 > \omega_1 > 0$

(b) $\text{Im } \epsilon/\epsilon_0 = \dfrac{\lambda\gamma\omega}{(\omega_0^2 - \omega^2)^2 + \gamma^2\omega^2}$

In each case sketch the behavior of $\text{Im } \epsilon(\omega)$ and the result for $\text{Re } \epsilon(\omega)$ as functions of ω. Comment on the reasons for similarities or differences of your results as compared with the curves in Fig. 7.8. The step function is $\theta(x) = 0$, $x < 0$ and $\theta(x) = 1$, $x > 0$.

7.23 Discuss the extension of the Kramers–Kronig relations (7.120) for a medium with a static electrical conductivity σ. Show that the first equation in (7.120) is unchanged, but that the second is changed into

$$\text{Im } \epsilon(\omega) = \frac{\sigma}{\omega} - \frac{2\omega}{\pi} P \int_0^{\infty} \frac{[\text{Re } \epsilon(\omega') - \epsilon_0]}{\omega'^2 - \omega^2} d\omega'$$

[*Hint:* Consider $\epsilon(\omega) - i\sigma/\omega$ as analytic for $\text{Im } \omega \geq 0$.]

7.24 (a) Use the relation (7.113) and the analyticity of $\epsilon(\omega)/\epsilon_0$ for $\text{Im } \omega \geq 0$ to prove that on the positive imaginary axis $\epsilon(\omega)/\epsilon_0$ is real and monotonically decreasing away from the origin toward unity as $\omega \to i\infty$, provided Im $\epsilon \geq 0$ for real positive frequencies.

(b) With the assumption that Im ϵ vanishes for finite real ω only at $\omega = 0$, show that $\epsilon(\omega)$ has no zeros in the upper half-ω-plane.

(c) Write down a Kramers–Kronig relation for $\epsilon_0/\epsilon(\omega)$ and deduce a sum rule similar to (7.122), but as an integral over Im$[\epsilon_0/\epsilon(\omega)]$.

(d) With the one-resonance model (7.107) for $\epsilon(\omega)$ determine Im $\epsilon(\omega)$ and Im$[1/\epsilon(\omega)]$ and verify explicitly that the sum rules (7.122) and part c are satisfied.

7.25 Equation (7.67) is an expression for the square of the index of refraction for waves propagating along field lines through a plasma in a uniform external magnetic field. Using this as a model for propagation in the magnetosphere, consider the arrival of a whistler signal (actually the Brillouin precursor and subsequently of Section 7.11).

(a) Make a reasonably careful sketch of $c\, dk/d\omega$, where $k = \omega n(\omega)/c$, for the positive helicity wave, assuming $\omega_p/\omega_B \gtrsim 1$. Indicate the interval where $c\, dk/d\omega$ is imaginary, but do not try to sketch it there!

(b) Show that on the interval, $0 < \omega < \omega_B$, the minimum of $c\, dk/d\omega$ occurs at $\omega/\omega_B \simeq \frac{1}{4}$, provided $\omega_p/\omega_B \gtrsim 1$. Find approximate expressions for $c\, dk/d\omega$ for ω near zero and for ω near ω_B.

(c) By means of the method of stationary phase and the general structure of the solution to Problem 7.20a, show that the arrival of a whistler is signaled by a rising and falling frequency as a function of time, the falling frequency component being the source of the name.

(d) (Optional) Consider the form of the signal in the Brillouin precursor. Show that it consists of a modulated waveform of frequency $\omega_0 = \omega_B/4$ whose envelope is the Airy integral. This then evolves into a signal beating with the two frequencies of part c.

7.26 A charged particle (charge Ze) moves at constant velocity $\mathbf{v}$ through a medium described by a dielectric function $\epsilon(\mathbf{q}, \omega)/\epsilon_0$ or, equivalently, by a conductivity function $\sigma(\mathbf{q}, \omega) = i\omega[\epsilon_0 - \epsilon(\mathbf{q}, \omega)]$. It is desired to calculate the energy loss per unit time by the moving particle in terms of the dielectric function $\epsilon(\mathbf{q}, \omega)$ in the approximation that the electric field is the negative gradient of the potential and current flow obeys Ohm's law, $\mathbf{J}(\mathbf{q}, \omega) = \sigma(\mathbf{q}, \omega)\mathbf{E}(\mathbf{q}, \omega)$.

(a) Show that with suitable normalization, the Fourier transform of the particle's charge density is

$$\rho(\mathbf{q}, \omega) = \frac{Ze}{(2\pi)^3} \delta(\omega - \mathbf{q} \cdot \mathbf{v})$$

(b) Show that the Fourier components of the electrostatic potential are

$$\phi(\mathbf{q}, \omega) = \frac{\rho(\mathbf{q}, \omega)}{q^2 \, \epsilon(\mathbf{q}, \omega)}$$

(c) Starting from $dW/dt = \int \mathbf{J} \cdot \mathbf{E} \, d^3x$ show that the energy loss per unit time can be written as

$$-\frac{dW}{dt} = \frac{Z^2 e^2}{4\pi^3} \int \frac{d^3q}{q^2} \int_0^\infty d\omega \, \omega \, \text{Im}\left[\frac{1}{\epsilon(\mathbf{q}, \omega)}\right] \delta(\omega - \mathbf{q} \cdot \mathbf{v})$$

[This shows that Im$[\epsilon(\mathbf{q}, \omega)]^{-1}$ is related to energy loss and provides, by studying characteristic energy losses in thin foils, information on $\epsilon(\mathbf{q}, \omega)$ for solids.]

7.27 The angular momentum of a distribution of electromagnetic fields in vacuum is given by

$$\mathbf{L} = \frac{1}{\mu_0 c^2} \int d^3x \, \mathbf{x} \times (\mathbf{E} \times \mathbf{B})$$

where the integration is over all space.

(a) For fields produced a finite time in the past (and so localized to a finite region of space) show that provided the magnetic field is eliminated in favor of the vector potential $\mathbf{A}$, the angular momentum can be written in the form

$$\mathbf{L} = \frac{1}{\mu_0 c^2} \int d^3x \left[\mathbf{E} \times \mathbf{A} + \sum_{j=1}^{3} E_j (\mathbf{x} \times \nabla) A_j \right]$$

The first term is sometimes identified with the "spin" of the photon and the second with its "orbital" angular momentum because of the presence of the angular momentum operator $\mathbf{L}_{op} = -i(\mathbf{x} \times \nabla)$.

(b) Consider an expansion of the vector potential in the radiation gauge in terms of plane waves:

$$\mathbf{A}(\mathbf{x}, t) = \sum_{\lambda} \int \frac{d^3k}{(2\pi)^3} \left[\boldsymbol{\epsilon}_\lambda(\mathbf{k}) a_\lambda(\mathbf{k}) e^{i\mathbf{k}\cdot\mathbf{x} - i\omega t} + \text{c.c.} \right]$$

The polarization vectors $\boldsymbol{\epsilon}_\lambda(\mathbf{k})$ are conveniently chosen as the positive and negative helicity vectors $\boldsymbol{\epsilon}_\pm = (1/\sqrt{2})(\boldsymbol{\epsilon}_1 \pm i\boldsymbol{\epsilon}_2)$ where $\boldsymbol{\epsilon}_1$ and $\boldsymbol{\epsilon}_2$ are real orthogonal vectors in the x-y plane whose positive normal is in the direction of $\mathbf{k}$.

Show that the time average of the first (spin) term of $\mathbf{L}$ can be written as

$$\mathbf{L}_{\text{spin}} = \frac{2}{\mu_0 c} \int \frac{d^3k}{(2\pi)^3} \, \mathbf{k} [|a_+(\mathbf{k})|^2 - |a_-(\mathbf{k})|^2]$$

Can the term "spin" angular momentum be justified from this expression? Calculate the energy of the field in terms of the plane wave expansion of $\mathbf{A}$ and compare.

7.28 A circularly polarized plane wave moving in the z direction has a finite extent in the x and y directions. Assuming that the amplitude modulation is slowly varying (the wave is many wavelengths broad), show that the electric and magnetic fields are given approximately by

$$\mathbf{E}(x, y, z, t) \simeq \left[E_0(x, y)(\mathbf{e}_1 \pm i\mathbf{e}_2) + \frac{i}{k}\left(\frac{\partial E_0}{\partial x} \pm i\frac{\partial E_0}{\partial y}\right)\mathbf{e}_3 \right] e^{ikz - i\omega t}$$

$$\mathbf{B} \simeq \mp i\sqrt{\mu\epsilon}\, \mathbf{E}$$

where $\mathbf{e}_1, \mathbf{e}_2, \mathbf{e}_3$ are unit vectors in the x, y, z directions.

7.29 For the circularly polarized wave of Problem 7.28 with $E_0(x, y)$ a real function of x and y, calculate the time-averaged component of angular momentum parallel to the direction of propagation. Show that the ratio of this component of angular momentum to the energy of the wave is

$$\frac{L_3}{U} = \pm\omega^{-1}$$

Interpret this result in terms of quanta of radiation (photons). Show that for a cylindrically symmetric, finite plane wave, the transverse components of angular momentum vanish.

7.30   Starting with the expression for the total energy of an arbitrary superposition of plane electromagnetic waves (7.11) in otherwise empty space, show that the total number of photons (defined for each plane wave of wave vector **k** and polarization $\boldsymbol{\epsilon}$ as its energy divided by $\hbar c k$) is given by the double integral

$$N = \frac{\epsilon_0}{4\pi^2 \hbar c} \int d^3x \int d^3x' \left[\frac{\mathbf{E}(\mathbf{x}, t) \cdot \mathbf{E}(\mathbf{x}', t) + c^2\, \mathbf{B}(\mathbf{x}, t) \cdot \mathbf{B}(\mathbf{x}', t)}{|\mathbf{x} - \mathbf{x}'|^2} \right]$$

CHAPTER 8

Waveguides, Resonant Cavities, and Optical Fibers

Electromagnetic fields in the presence of metallic boundaries form a practical aspect of the subject of considerable importance. At high frequencies where the wavelengths are of the order of meters or less, the only practical way of generating and transmitting electromagnetic radiation involves metallic structures with dimensions comparable to the wavelengths involved. At much higher (infrared) frequencies, dielectric optical fibers are exploited in the telecommunications industry. In this chapter we consider first the fields in the neighborhood of a conductor and discuss their penetration into the surface and the accompanying resistive losses. Then the problems of waves guided in hollow metal pipes and of resonant cavities are treated from a fairly general viewpoint, with specific illustrations included along the way. Attenuation in waveguides and Q values of cavities are discussed from two different points of view. The earth-ionosphere system as a novel resonant cavity is treated next. Then we discuss multimode and single-mode propagation in optical fibers. The normal mode expansion for an arbitrary field in a waveguide is presented and applied to the fields generated by a localized source, with brief mention of the use of the normal mode expansion in the treatment of obstacles in waveguides by variational methods.

8.1 *Fields at the Surface of and Within a Conductor*

In Section 5.18 the concept of skin depth and effective surface current was introduced by a simple example of a planar interface between conductor and vacuum, with a spatially uniform, time-varying magnetic field at the interface. Here we generalize the circumstances, at least conceptually, even though the mathematics is much the same.

First consider a surface with unit normal **n** directed outward from a *perfect* conductor on one side into a nonconducting medium on the other side. Then, just as in the static case, there is no electric field inside the conductors. The charges inside a perfect conductor are assumed to be so mobile that they move instantly in response to changes in the fields, no matter how rapid, and always produce the correct surface-charge density Σ (capital Σ is used to avoid confusion with the conductivity σ):

$$\mathbf{n} \cdot \mathbf{D} = \Sigma \qquad (8.1)$$

352

to give zero electric field inside the perfect conductor. Similarly, for time-varying magnetic fields, the surface charges move in response to the tangential magnetic field to produce always the correct surface current $\mathbf{K}$:

$$\mathbf{n} \times \mathbf{H} = \mathbf{K} \tag{8.2}$$

to have zero magnetic field inside the perfect conductor. The other two boundary conditions are on normal $\mathbf{B}$ and tangential $\mathbf{E}$:

$$\mathbf{n} \cdot (\mathbf{B} - \mathbf{B}_c) = 0 \tag{8.3}$$
$$\mathbf{n} \times (\mathbf{E} - \mathbf{E}_c) = 0$$

where the subscript c refers to the conductor. From these boundary conditions we see that just outside the surface of a perfect conductor only *normal* $\mathbf{E}$ and *tangential* $\mathbf{H}$ fields can exist, and that the fields drop abruptly to zero inside the perfect conductor. This behavior is indicated schematically in Fig. 8.1.

The fields in the neighborhood of the surface of a good, but not perfect, conductor must behave approximately the same as for a perfect conductor. In Section 5.18 we saw that inside a conductor the fields are attenuated exponentially in a characteristic length δ, called the *skin depth*. For good conductors and moderate frequencies, δ is a small fraction of a centimeter. Consequently, boundary conditions (8.1) and (8.2) are approximately true for a good conductor, aside from a thin transitional layer at the surface.

If we wish to examine that thin transitional region, however, care must be taken. First of all, Ohm's law $\mathbf{J} = \sigma\mathbf{E}$ shows that with a finite conductivity there cannot actually be a surface layer of current, as implied in (8.2). Instead, the boundary condition on the magnetic field is

$$\mathbf{n} \times (\mathbf{H} - \mathbf{H}_c) = 0 \tag{8.4}$$

To explore the changes produced by a finite, rather than an infinite, conductivity, we employ a successive approximation scheme. First we assume that just outside the conductor there exists only a normal electric field $\mathbf{E}_\perp$ and a tangential magnetic field $\mathbf{H}_\parallel$, as for a perfect conductor. The values of these fields are assumed to have been obtained from the solution of an appropriate boundary-value problem. Then we use the boundary conditions and the Maxwell equations in the conductor to find the fields within the transition layer and small corrections to the fields outside. In solving the Maxwell equations within the conductor we make use of the fact that the spatial variation of the fields normal to the surface

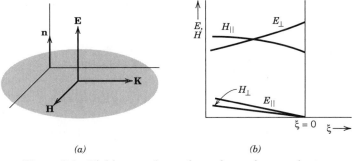

(a) (b)

Figure 8.1 Fields near the surface of a perfect conductor.

is much more rapid than the variations parallel to the surface. This means that we can safely neglect all derivatives with respect to coordinates parallel to the surface compared to the normal derivative.

If there exists a tangential $\mathbf{H}_\parallel$ outside the surface, boundary condition (8.4) implies the same $\mathbf{H}_\parallel$ inside the surface. With the neglect of the displacement current in the conductor, the Maxwell curl equations become

$$\mathbf{E}_c \simeq \frac{1}{\sigma} \nabla \times \mathbf{H}_c$$

$$\mathbf{H}_c = -\frac{i}{\mu_c \omega} \nabla \times \mathbf{E}_c \tag{8.5}$$

where a harmonic variation $e^{-i\omega t}$ has been assumed. If $\mathbf{n}$ is the unit normal *outward* from the conductor and ξ is the normal coordinate *inward* into the conductor, then the gradient operator can be written

$$\nabla \simeq -\mathbf{n} \frac{\partial}{\partial \xi}$$

neglecting the other derivatives when operating on the fields within the conductor. With this approximation the Maxwell curl equations (8.5) become

$$\mathbf{E}_c \simeq -\frac{1}{\sigma} \mathbf{n} \times \frac{\partial \mathbf{H}_c}{\partial \xi}$$

$$\mathbf{H}_c \simeq \frac{i}{\mu_c \omega} \mathbf{n} \times \frac{\partial \mathbf{E}_c}{\partial \xi} \tag{8.6}$$

These can be combined to yield

$$\frac{\partial^2}{\partial \xi^2} (\mathbf{n} \times \mathbf{H}_c) + \frac{2i}{\delta^2} (\mathbf{n} \times \mathbf{H}_c) \simeq 0$$

$$\mathbf{n} \cdot \mathbf{H}_c \simeq 0 \tag{8.7}$$

where δ is the skin depth defined previously:

$$\delta = \left(\frac{2}{\mu_c \omega \sigma} \right)^{1/2} \tag{8.8}$$

The second equation in (8.7) shows that inside the conductor $\mathbf{H}$ is parallel to the surface, consistent with our boundary conditions. The solution for $\mathbf{H}_c$ is

$$\mathbf{H}_c = \mathbf{H}_\parallel e^{-\xi/\delta} e^{i\xi/\delta} \tag{8.9}$$

where $\mathbf{H}_\parallel$ is the tangential magnetic field outside the surface. From (8.6) the electric field in the conductor is approximately

$$\mathbf{E}_c \simeq \sqrt{\frac{\mu_c \omega}{2\sigma}} (1 - i)(\mathbf{n} \times \mathbf{H}_\parallel) e^{-\xi/\delta} e^{i\xi/\delta} \tag{8.10}$$

These solutions for $\mathbf{H}$ and $\mathbf{E}$ inside the conductor exhibit the properties discussed in Section 5.18: rapid exponential decay, phase difference, and magnetic field much larger than the electric field. Furthermore, they show that, for a good con-

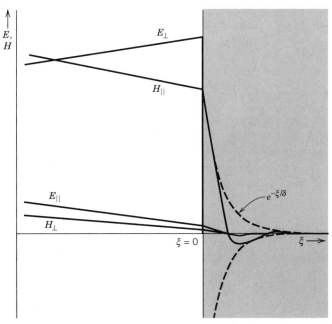

Figure 8.2   Fields near the surface of a good, but not perfect, conductor. For $\xi > 0$, the dashed curves show the envelope of the damped oscillations of $\mathbf{H}_c$ (8.9).

ductor, the fields in the conductor are parallel to the surface* and propagate normal to it, with magnitudes that depend only on the tangential magnetic field $\mathbf{H}_\parallel$ that exists just outside the surface.

From the boundary condition on tangential $\mathbf{E}$ (8.3) we find that *just outside* the surface there exists a small tangential electric field given by (8.10), evaluated at $\xi = 0$:

$$\mathbf{E}_\parallel \simeq \sqrt{\frac{\mu_c \omega}{2\sigma}} \, (1 - i)(\mathbf{n} \times \mathbf{H}_\parallel) \qquad (8.11)$$

In this approximation there is also a small normal component of $\mathbf{B}$ just outside the surface. This can be obtained from Faraday's law of induction and gives $\mathbf{B}_\perp$ of the same order of magnitude as $\mathbf{E}_\parallel$. The amplitudes of the fields both inside and outside the conductor are indicated schematically in Fig. 8.2.

The existence of a small tangential component of $\mathbf{E}$ outside the surface, in addition to the normal $\mathbf{E}$ and tangential $\mathbf{H}$, means that there is a power flow into the conductor. The time-averaged power absorbed per unit area is

$$\frac{dP_{\text{loss}}}{da} = -\frac{1}{2} \, \text{Re}[\mathbf{n} \cdot \mathbf{E} \times \mathbf{H}^*] = \frac{\mu_c \omega \delta}{4} \, |\mathbf{H}_\parallel|^2 \qquad (8.12)$$

---

*From the continuity of the tangential component of $\mathbf{H}$ and the equation connecting $\mathbf{E}$ to $\nabla \times \mathbf{H}$ on either side of the surface, one can show that there exists in the conductor a small normal component of electric field, $\mathbf{E}_c \cdot \mathbf{n} \simeq (i\omega\epsilon/\sigma)E_\perp$, but this is of the next order in small quantities compared with (8.10). Note that our discussion here presupposes a tangential component of $\mathbf{H}$. In situations in which the lowest order approximation is essentially electrostatic, the present treatment is inapplicable. Different approximations must be employed. See T. H. Boyer, *Phys. Rev.* **A9**, 68 (1974).

This result can be given a simple interpretation as ohmic losses in the body of the conductor. According to Ohm's law, there exists a current density $\mathbf{J}$ near the surface of the conductor:

$$\mathbf{J} = \sigma \mathbf{E}_c = \frac{1}{\delta}(1 - i)(\mathbf{n} \times \mathbf{H}_\parallel)e^{-\xi(1-i)/\delta} \tag{8.13}$$

The time-averaged rate of dissipation of energy per unit volume in ohmic losses is $\frac{1}{2}\mathbf{J} \cdot \mathbf{E}^* = (1/2\sigma)|\mathbf{J}|^2$, as written in (5.169). The integral of (5.169) in z leads directly to (8.12).

The current density $\mathbf{J}$ is confined to such a small thickness just below the surface of the conductor that it is equivalent to an effective surface current $\mathbf{K}_{\text{eff}}$:

$$\mathbf{K}_{\text{eff}} = \int_0^\infty \mathbf{J}\, d\xi = \mathbf{n} \times \mathbf{H}_\parallel \tag{8.14}$$

Comparison with (8.2) shows that a good conductor behaves effectively like a perfect conductor, with the idealized surface current replaced by an equivalent surface current, which is actually distributed throughout a very small, but finite, thickness at the surface. The power loss can be written in terms of the effective surface current:

$$\frac{dP_{\text{loss}}}{da} = \frac{1}{2\sigma\delta}|\mathbf{K}_{\text{eff}}|^2 \tag{8.15}$$

This shows that $1/\sigma\delta$ plays the role of a surface resistance of the conductor.* Equation (8.15), with $\mathbf{K}_{\text{eff}}$ given by (8.14), or (8.12) will allow us to calculate approximately the resistive losses for practical cavities, transmission lines, and waveguides, provided we have solved for the fields in the idealized problem of infinite conductivity.

8.2 *Cylindrical Cavities and Waveguides*

A practical situation of great importance is the propagation or excitation of electromagnetic waves in hollow metallic cylinders. If the cylinder has end surfaces, it is called a cavity; otherwise, a waveguide. In our discussion of this problem the boundary surfaces are assumed to be perfect conductors. The losses occurring in practice can be accounted for adequately by the methods of Section 8.1. A cylindrical surface S of general cross-sectional contour is shown in Fig. 8.3. For simplicity, the cross-sectional size and shape are assumed constant along the cylinder axis. With a sinusoidal time dependence $e^{-i\omega t}$ for the fields inside the cylinder, the Maxwell equations take the form

$$\begin{aligned} \nabla \times \mathbf{E} = i\omega\mathbf{B} && \nabla \cdot \mathbf{B} = 0 \\ \nabla \times \mathbf{B} = -i\mu\epsilon\omega\mathbf{E} && \nabla \cdot \mathbf{E} = 0 \end{aligned} \tag{8.16}$$

---

*The coefficient of proportionality linking $\mathbf{E}_\parallel$ and $\mathbf{K}_{\text{eff}}$ is called the *surface impedance* Z_s. For a good conductor (8.11) yields $Z_s = (1 - i)/\sigma\delta$, but the concept of surface impedance obviously has wider applicability.

Figure 8.3 Hollow, cylindrical waveguide of arbitrary cross-sectional shape.

where it is assumed that the cylinder is filled with a uniform nondissipative medium having permittivity ϵ and permeability μ. It follows that both $\mathbf{E}$ and $\mathbf{B}$ satisfy

$$(\nabla^2 + \mu\epsilon\omega^2)\begin{Bmatrix}\mathbf{E}\\\mathbf{B}\end{Bmatrix} = 0 \tag{8.17}$$

Because of the cylindrical geometry it is useful to single out the spatial variation of the fields in the z direction and to assume

$$\begin{matrix}\mathbf{E}(x, y, z, t)\\\mathbf{B}(x, y, z, t)\end{matrix}\Big\} = \begin{cases}\mathbf{E}(x, y)e^{\pm ikz - i\omega t}\\\mathbf{B}(x, y)e^{\pm ikz - i\omega t}\end{cases} \tag{8.18}$$

Appropriate linear combinations can be formed to give traveling or standing waves in the z direction. The wave number k is, at present, an unknown parameter that may be real or complex. With this assumed z dependence of the fields the wave equation reduces to the two-dimensional form

$$[\nabla_t^2 + (\mu\epsilon\omega^2 - k^2)]\begin{Bmatrix}\mathbf{E}\\\mathbf{B}\end{Bmatrix} = 0 \tag{8.19}$$

where ∇_t^2 is the transverse part of the Laplacian operator:

$$\nabla_t^2 = \nabla^2 - \frac{\partial^2}{\partial z^2} \tag{8.20}$$

It is useful to separate the fields into components parallel to and transverse to the z axis:

$$\mathbf{E} = \mathbf{E}_z + \mathbf{E}_t \tag{8.21}$$

where

$$\begin{aligned}\mathbf{E}_z &= \hat{\mathbf{z}}E_z\\\mathbf{E}_t &= (\hat{\mathbf{z}} \times \mathbf{E}) \times \hat{\mathbf{z}}\end{aligned} \tag{8.22}$$

and $\hat{\mathbf{z}}$ is a unit vector in the z direction. Similar definitions hold for the magnetic field $\mathbf{B}$. The Maxwell equations (8.16) can be written out in terms of transverse and parallel components as

$$\frac{\partial \mathbf{E}_t}{\partial z} + i\omega\hat{\mathbf{z}} \times \mathbf{B}_t = \nabla_t E_z, \qquad \hat{\mathbf{z}} \cdot (\nabla_t \times \mathbf{E}_t) = i\omega B_z \tag{8.23}$$

$$\frac{\partial \mathbf{B}_t}{\partial z} - i\mu\epsilon\omega\hat{\mathbf{z}} \times \mathbf{E}_t = \nabla_t B_z, \qquad \hat{\mathbf{z}} \cdot (\nabla_t \times \mathbf{B}_t) = -i\mu\epsilon\omega E_z \tag{8.24}$$

$$\nabla_t \cdot \mathbf{E}_t = -\frac{\partial E_z}{\partial z}, \qquad \nabla_t \cdot \mathbf{B}_t = -\frac{\partial B_z}{\partial z} \tag{8.25}$$

It is evident from the first equations in (8.23) and (8.24) that if E_z and B_z are known the transverse components of $\mathbf{E}$ and $\mathbf{B}$ are determined, assuming the z

dependence is given by (8.18). Explicitly, assuming propagation in the positive z direction and the nonvanishing of at least one of E_z and B_z, the transverse fields are

$$\mathbf{E}_t = \frac{i}{(\mu\epsilon\omega^2 - k^2)} [k\boldsymbol{\nabla}_t E_z - \omega\hat{\mathbf{z}} \times \boldsymbol{\nabla}_t B_z] \tag{8.26a}$$

and

$$\mathbf{B}_t = \frac{i}{(\mu\epsilon\omega^2 - k^2)} [k\boldsymbol{\nabla}_t B_z + \mu\epsilon\omega\,\hat{\mathbf{z}} \times \boldsymbol{\nabla}_t E_z] \tag{8.26b}$$

For waves in the opposite direction, change the sign of k.

Before considering the kinds of field that can exist inside a hollow cylinder, we take note of a degenerate or special type of solution, called the *transverse electromagnetic* (TEM) wave. This solution has only field components transverse to the direction of propagation. From the second equation in (8.23) and the first in (8.25) it is seen that $E_z = 0$ and $B_z = 0$ imply that $\mathbf{E}_t = \mathbf{E}_{\text{TEM}}$ satisfies

$$\boldsymbol{\nabla}_t \times \mathbf{E}_{\text{TEM}} = 0, \qquad \boldsymbol{\nabla}_t \cdot \mathbf{E}_{\text{TEM}} = 0$$

This means that $\mathbf{E}_{\text{TEM}}$ is a solution of an *electrostatic* problem in two dimensions. There are three main consequences. The first is that the axial wave number is given by the infinite-medium value,

$$k = k_0 = \omega\sqrt{\mu\epsilon} \tag{8.27}$$

as can be seen from (8.19). The second consequence is that the magnetic field, deduced from the first equation in (8.24), is

$$\mathbf{B}_{\text{TEM}} = \pm\sqrt{\mu\epsilon}\,\hat{\mathbf{z}} \times \mathbf{E}_{\text{TEM}} \tag{8.28}$$

for waves propagating as $e^{\pm ikz}$. The connection between $\mathbf{B}_{\text{TEM}}$ and $\mathbf{E}_{\text{TEM}}$ is just the same as for plane waves in an infinite medium. The final consequence is that the TEM mode cannot exist inside a single, hollow, cylindrical conductor of infinite conductivity. The surface is an equipotential; the electric field therefore vanishes inside. It is necessary to have two or more cylindrical surfaces to support the TEM mode. The familiar coaxial cable and the parallel-wire transmission line are structures for which this is the dominant mode. (See Problems 8.1 and 8.2.) An important property of the TEM mode is the absence of a cutoff frequency. The wave number (8.27) is real for all ω. This is not true for the modes occurring in hollow cylinders (see below).

In hollow cylinders (and on transmission lines at high frequencies) there occur two types of field configuration. Their existence can be seen from considering the wave equations (8.19) satisfied by the longitudinal components, E_z and B_z, and the boundary conditions to be satisfied. Provided the fields are time-varying, perfect conductivity assures that both $\mathbf{E}$ (and $\mathbf{D}$) and $\mathbf{B}$ (and $\mathbf{H}$) vanish within the conductor. (For the latter, the skin depth is vanishingly small.) The presence of surface charges and currents at the interface allows the existence of a normal component of $\mathbf{D}$ at the boundary, and also a tangential component of $\mathbf{H}$, but the tangential component of $\mathbf{E}$ and the normal component of $\mathbf{B}$ must be continuous across the boundary. Thus, for a perfectly conducting cylinder the boundary conditions are

$$\mathbf{n} \times \mathbf{E} = 0, \qquad \mathbf{n} \cdot \mathbf{B} = 0$$

where $\mathbf{n}$ is a unit normal at the surface S. It is evident that the boundary condition on E_z is

$$E_z|_S = 0 \tag{8.29}$$

From the component of the first equation in (8.24) parallel to $\mathbf{n}$ it can be inferred that the corresponding boundary condition on B_z is

$$\left.\frac{\partial B_z}{\partial n}\right|_S = 0 \tag{8.30}$$

where $\partial/\partial n$ is the normal derivative at a point on the surface. The two-dimensional wave equations (8.19) for E_z and B_z, together with the boundary conditions (8.29) and (8.30), specify eigenvalue problems of the usual sort. For a given frequency ω, only certain values of wave number k can occur (typical waveguide situation), or, for a given k, only certain ω values are allowed (typical resonant cavity situation). *Since the boundary conditions on E_z and B_z are different, the eigenvalues will in general be different.* The fields thus naturally divide themselves into two distinct categories:

TRANSVERSE MAGNETIC (TM) WAVES

$B_z = 0$ everywhere; boundary condition, $E_z|_S = 0$

TRANSVERSE ELECTRIC (TE) WAVES

$E_z = 0$ everywhere; boundary condition, $\left.\dfrac{\partial B_z}{\partial n}\right|_S = 0$

The designations "electric (or E) waves" and "magnetic (or H) waves" are sometimes used instead of TM and TE waves, respectively, corresponding to a specification of the axial component of the fields. The various TM and TE waves, plus the TEM wave if it can exist, constitute a complete set of fields to describe an arbitrary electromagnetic disturbance in a waveguide or cavity.

8.3 Waveguides

For the propagation of waves inside a hollow waveguide of uniform cross section, it is found from (8.26a, b) that the transverse magnetic and electric fields for both TM and TE waves are related by

$$\mathbf{H}_t = \frac{\pm 1}{Z}\, \hat{\mathbf{z}} \times \mathbf{E}_t \tag{8.31}$$

where Z is called the *wave impedance* and is given by

$$Z = \begin{cases} \dfrac{k}{\epsilon\omega} = \dfrac{k}{k_0}\sqrt{\dfrac{\mu}{\epsilon}} & \text{(TM)} \\[2ex] \dfrac{\mu\omega}{k} = \dfrac{k_0}{k}\sqrt{\dfrac{\mu}{\epsilon}} & \text{(TE)} \end{cases} \tag{8.32}$$

where k_0 is given by (8.27). The plus (minus) sign in (8.31) goes with z dependence, e^{ikz} (e^{-ikz}). The transverse fields are determined by the longitudinal fields, according to (8.26):

TM WAVES

$$\mathbf{E}_t = \pm \frac{ik}{\gamma^2} \nabla_t \psi$$

TE WAVES

$$\mathbf{H}_t = \pm \frac{ik}{\gamma^2} \nabla_t \psi \tag{8.33}$$

where $\psi e^{\pm ikz}$ is $E_z(H_z)$ for TM (TE) waves* and γ^2 is defined below. The scalar function ψ satisfies the two-dimensional wave equation (8.19),

$$(\nabla_t^2 + \gamma^2)\psi = 0 \tag{8.34}$$

where

$$\gamma^2 = \mu\epsilon\omega^2 - k^2 \tag{8.35}$$

subject to the boundary condition,

$$\psi|_S = 0 \qquad \text{or} \qquad \left.\frac{\partial\psi}{\partial n}\right|_S = 0 \tag{8.36}$$

for TM (TE) waves.

Equation (8.34) for ψ, together with boundary condition (8.36), specifies an eigenvalue problem. It is easy to see that the constant γ^2 must be nonnegative. Roughly speaking, it is because ψ must be oscillatory to satisfy boundary condition (8.36) on opposite sides of the cylinder. There will be a spectrum of eigenvalues γ_λ^2 and corresponding solutions ψ_λ, $\lambda = 1, 2, 3, \ldots$, which form an orthogonal set. These different solutions are called the *modes of the guide*. For a given frequency ω, the wave number k is determined for each value of λ:

$$k_\lambda^2 = \mu\epsilon\omega^2 - \gamma_\lambda^2 \tag{8.37}$$

If we define a *cutoff frequency* ω_λ,

$$\omega_\lambda = \frac{\gamma_\lambda}{\sqrt{\mu\epsilon}} \tag{8.38}$$

then the wave number can be written:

$$k_\lambda = \sqrt{\mu\epsilon}\sqrt{\omega^2 - \omega_\lambda^2} \tag{8.39}$$

We note that, for $\omega > \omega_\lambda$, the wave number k_λ is real; waves of the λ mode can propagate in the guide. For frequencies less than the cutoff frequency, k_λ is imaginary; such modes cannot propagate and are called *cutoff modes* or *evanescent modes*. The behavior of the axial wave number as a function of frequency is shown qualitatively in Fig. 8.4. We see that at any given frequency only a finite number of modes can propagate. It is often convenient to choose the dimensions

---

*We have changed from $\mathbf{E}$ and $\mathbf{B}$ to $\mathbf{E}$ and $\mathbf{H}$ as our basic fields to eliminate factors of μ when using the wave impedances. (Like ordinary impedance, wave impedance involves voltage and current and so $\mathbf{E}$ and $\mathbf{H}$.)

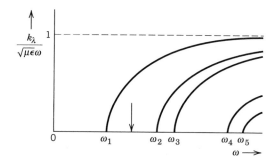

Figure 8.4 Wave number k_λ versus frequency ω for various modes λ. ω_λ is the cutoff frequency.

of the guide so that at the operating frequency only the lowest mode can occur. This is shown by the vertical arrow on the figure.

Since the wave number k_λ is always less than the free-space value $\sqrt{\mu\epsilon}\omega$, the wavelength in the guide is always greater than the free-space wavelength. In turn, the phase velocity v_p is larger than the infinite space value:

$$v_p = \frac{\omega}{k_\lambda} = \frac{1}{\sqrt{\mu\epsilon}} \frac{1}{\sqrt{1 - \left(\dfrac{\omega_\lambda}{\omega}\right)^2}} > \frac{1}{\sqrt{\mu\epsilon}} \tag{8.40}$$

The phase velocity becomes infinite exactly at cutoff.

8.4   Modes in a Rectangular Waveguide

As an important illustration of the general features described in Section 8.3 we consider the propagation of TE waves in a rectangular waveguide with inner dimensions a, b, as shown in Fig. 8.5. The wave equation for $\psi = H_z$ is

$$\left(\frac{\partial^2}{\partial x^2} + \frac{\partial^2}{\partial y^2} + \gamma^2\right)\psi = 0 \tag{8.41}$$

with boundary conditions $\partial\psi/\partial n = 0$ at $x = 0, a$ and $y = 0, b$. The solution for ψ is consequently

$$\psi_{mn}(x, y) = H_0 \cos\left(\frac{m\pi x}{a}\right) \cos\left(\frac{n\pi y}{b}\right) \tag{8.42}$$

where

$$\gamma_{mn}^2 = \pi^2\left(\frac{m^2}{a^2} + \frac{n^2}{b^2}\right) \tag{8.43}$$

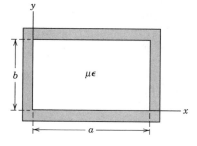

Figure 8.5

The single index λ that specified the modes earlier is replaced by the two positive integers m, n. For there to be nontrivial solutions, m and n cannot both be zero. The cutoff frequency ω_{mn} is given by

$$\omega_{mn} = \frac{\pi}{\sqrt{\mu\epsilon}} \left(\frac{m^2}{a^2} + \frac{n^2}{b^2}\right)^{1/2} \tag{8.44}$$

If $a > b$, the lowest cutoff frequency, that of the dominant TE mode, occurs for $m = 1$, $n = 0$:

$$\omega_{1,0} = \frac{\pi}{\sqrt{\mu\epsilon}\, a} \tag{8.45}$$

This corresponds to half of a free-space wavelength across the guide. The explicit fields for this mode, denoted by $TE_{1,0}$, are:

$$H_z = H_0 \cos\left(\frac{\pi x}{a}\right) e^{ikz - i\omega t}$$

$$H_x = -\frac{ika}{\pi} H_0 \sin\left(\frac{\pi x}{a}\right) e^{ikz - i\omega t} \tag{8.46}$$

$$E_y = i\frac{\omega a \mu}{\pi} H_0 \sin\left(\frac{\pi x}{a}\right) e^{ikz - i\omega t}$$

where $k = k_{1,0}$ is given by (8.39) with $\omega_\lambda = \omega_{1,0}$. The presence of a factor i in H_x (and E_y) means that there is a spatial (or temporal) phase difference of 90° between H_x (and E_y) and H_z in the propagation direction. It happens that the $TE_{1,0}$ mode has the lowest cutoff frequency of both TE and TM modes,* and so is the one used in most practical situations. For a typical choice $a = 2b$ the ratio of cutoff frequencies ω_{mn} for the next few modes to ω_{10} are as follows:

m \ n	0	1	2	3
0		2.00	4.00	6.00
1	1.00	2.24	4.13	
2	2.00	2.84	4.48	
3	3.00	3.61	5.00	
4	4.00	4.48	5.66	
5	5.00	5.39		
6	6.00			

There is a frequency range from cutoff to twice cutoff or to (a/b) times cutoff, whichever is smaller, where the $TE_{1,0}$ mode is the only propagating mode. Beyond that frequency other modes rapidly begin to enter. The field configurations

*This is evident if we note that for the TM modes E_z is of the form

$$E_z = E_0 \sin\left(\frac{m\pi x}{a}\right) \sin\left(\frac{n\pi y}{b}\right)$$

while γ^2 is still given by (8.43). The lowest mode has $m = n = 1$. Its cutoff frequency is greater than that of the TE_{10} mode by the factor $(1 + a^2/b^2)^{1/2}$.

of the $TE_{1,0}$ mode and other modes are shown in many books, for example, *American Institute of Physics Handbook* [ed. D. E. Gray, 3rd edition, McGraw-Hill, New York (1972), p. 5–54].

8.5 *Energy Flow and Attenuation in Waveguides*

The general discussion of Section 8.3 for a cylindrical waveguide of arbitrary cross-sectional shape can be extended to include the flow of energy along the guide and the attenuation of the waves due to losses in the walls having finite conductivity. The treatment is restricted to one mode at a time; degenerate modes are mentioned only briefly. The flow of energy is described by the complex Poynting vector:

$$\mathbf{S} = \tfrac{1}{2}(\mathbf{E} \times \mathbf{H}^*) \tag{8.47}$$

whose real part gives the time-averaged flux of energy. For the two types of field we find, using (8.31) and (8.33):

$$\mathbf{S} = \frac{\omega k}{2\gamma^4} \begin{cases} \epsilon \left[\hat{\mathbf{z}} \, |\nabla_t \psi|^2 + i \, \dfrac{\gamma^2}{k} \, \psi \, \nabla_t \psi^* \right] \\[2mm] \mu \left[\hat{\mathbf{z}} \, |\nabla_t \psi|^2 - i \, \dfrac{\gamma^2}{k} \, \psi^* \, \nabla_t \psi \right] \end{cases} \tag{8.48}$$

where the upper (lower) line is for TM (TE) modes. Since ψ is generally real,* we see that the transverse component of $\mathbf{S}$ represents reactive energy flow and does not contribute to the time-averaged flux of energy. On the other hand, the axial component of $\mathbf{S}$ gives the time-averaged flow of energy along the guide. To evaluate the total power flow P we integrate the axial component of $\mathbf{S}$ over the cross-sectional area A:

$$P = \int_A \mathbf{S} \cdot \hat{\mathbf{z}} \, da = \frac{\omega k}{2\gamma^4} \begin{Bmatrix} \epsilon \\ \mu \end{Bmatrix} \int_A (\nabla_t \psi)^* \cdot (\nabla_t \psi) \, da \tag{8.49}$$

By means of Green's first identity (1.34) applied to two dimensions, (8.49) can be written:

$$P = \frac{\omega k}{2\gamma^4} \begin{Bmatrix} \epsilon \\ \mu \end{Bmatrix} \left[\oint_C \psi^* \frac{\partial \psi}{\partial n} \, dl - \int_A \psi^* \, \nabla_t^2 \psi \, da \right] \tag{8.50}$$

where the first integral is around the curve C, which defines the boundary surface of the cylinder. This integral vanishes for both types of field because of boundary conditions (8.36). By means of the wave equation (8.34) the second integral may be reduced to the normalization integral for ψ. Consequently the transmitted power is

$$P = \frac{1}{2\sqrt{\mu\epsilon}} \left(\frac{\omega}{\omega_\lambda} \right)^2 \left(1 - \frac{\omega_\lambda^2}{\omega^2} \right)^{1/2} \begin{Bmatrix} \epsilon \\ \mu \end{Bmatrix} \int_A \psi^* \psi \, da \tag{8.51}$$

---

*It is possible to excite a guide in such a manner that a given mode or linear combination of modes has a complex ψ. Then a time-averaged transverse energy flow can occur. Since it is a circulatory flow, however, it really represents only stored energy and is not of great practical importance.

where the upper (lower) line is for TM (TE) modes, and we have exhibited all the frequency dependence explicitly.

It is straightforward to calculate the field energy per unit length of the guide in the same way as the power flow. The result is

$$U = \frac{1}{2} \left(\frac{\omega}{\omega_\lambda} \right)^2 \begin{Bmatrix} \epsilon \\ \mu \end{Bmatrix} \int_A \psi^* \psi \, da \tag{8.52}$$

Comparison with the power flow P shows that P and U are proportional. The constant of proportionality has the dimensions of velocity (velocity of energy flow) and is just the group velocity:

$$\frac{P}{U} = \frac{k}{\omega} \frac{1}{\mu\epsilon} = \frac{1}{\sqrt{\mu\epsilon}} \sqrt{1 - \frac{\omega_\lambda^2}{\omega^2}} = v_g \tag{8.53}$$

as can be verified by a direct calculation of $v_g = d\omega/dk$ from (8.39), assuming that the dielectric filling the guide is nondispersive. We note that v_g is always less than the velocity of waves in an infinite medium and falls to zero at cutoff. The product of phase velocity (8.40) and group velocity is constant:

$$v_p v_g = \frac{1}{\mu\epsilon} \tag{8.54}$$

an immediate consequence of the fact that $\omega \, \Delta\omega \propto k \, \Delta k$.

Our considerations so far have applied to waveguides with perfectly conducting walls. The axial wave number k_λ was either real or purely imaginary. If the walls have a finite conductivity, there will be ohmic losses and the power flow along the guide will be attenuated. For walls with large conductivity the wave number will have small additional real and imaginary parts:

$$k_\lambda \simeq k_\lambda^{(0)} + \alpha_\lambda + i\beta_\lambda \tag{8.55}$$

where $k_\lambda^{(0)}$ is the value for perfectly conducting walls. The change α_λ in the real part of the wave number is generally unimportant except near cutoff when $k_\lambda^{(0)} \to 0$. The attenuation constant β_λ can be found either by solving the boundary-value problem over again with boundary conditions appropriate for finite conductivity, or by calculating the ohmic losses by the methods of Section 8.1 and using conservation of energy. We will first use the latter technique. The power flow along the guide will be given by

$$P(z) = P_0 e^{-2\beta_\lambda z} \tag{8.56}$$

Thus the attenuation constant is given by

$$\beta_\lambda = -\frac{1}{2P} \frac{dP}{dz} \tag{8.57}$$

where $-dP/dz$ is the power dissipated in ohmic losses per unit length of the guide. According to the results of Section 8.1, this power loss is

$$-\frac{dP}{dz} = \frac{1}{2\sigma\delta} \oint_C |\mathbf{n} \times \mathbf{H}|^2 \, dl \tag{8.58}$$

where the integral is around the boundary of the guide. With fields (8.31) and (8.33) it is easy to show that for a given mode:

$$-\frac{dP}{dz} = \frac{1}{2\sigma\delta}\left(\frac{\omega}{\omega_\lambda}\right)^2 \oint_C \left\{ \frac{\frac{1}{\mu^2\omega_\lambda^2}\left|\frac{\partial\psi}{\partial n}\right|^2}{\frac{1}{\mu\epsilon\omega_\lambda^2}\left(1 - \frac{\omega_\lambda^2}{\omega^2}\right)|\mathbf{n} \times \nabla_t\psi|^2 + \frac{\omega_\lambda^2}{\omega^2}|\psi|^2} \right\} dl \quad (8.59)$$

where again the upper (lower) line applies to TM (TE) modes.

Since the transverse derivatives of ψ are determined entirely by the size and shape of the waveguide, the frequency dependence of the power loss is explicitly exhibited in (8.59). In fact, the integrals in (8.59) may be simply estimated from the fact that for each mode:

$$(\nabla_t^2 + \mu\epsilon\omega_\lambda^2)\psi = 0 \quad (8.60)$$

This means that, in some average sense, and barring exceptional circumstances, the transverse derivatives of ψ must be of the order of magnitude of $\sqrt{\mu\epsilon}\,\omega_\lambda\psi$:

$$\left\langle\left|\frac{\partial\psi}{\partial n}\right|^2\right\rangle \sim \langle|\mathbf{n} \times \nabla_t\psi|^2\rangle \sim \mu\epsilon\omega_\lambda^2\langle|\psi|^2\rangle \quad (8.61)$$

Consequently, the line integrals in (8.59) can be related to the normalization integral of $|\psi|^2$ over the area. For example,

$$\oint_C \frac{1}{\omega_\lambda^2}\left|\frac{\partial\psi}{\partial n}\right|^2 dl = \xi_\lambda\mu\epsilon\frac{C}{A}\int_A |\psi|^2\, da \quad (8.62)$$

where C is the circumference and A is the area of cross section, while ξ_λ is a dimensionless number of the order of unity. Without further knowledge of the shape of the guide we can obtain the order of magnitude of the attenuation constant β_λ and exhibit completely its frequency dependence. Thus, using (8.59) with (8.62) and (8.51), plus the frequency dependence of the skin depth (8.8), we find

$$\beta_\lambda = \sqrt{\frac{\epsilon}{\mu}}\frac{1}{\sigma\delta_\lambda}\left(\frac{C}{2A}\right)\frac{\left(\frac{\omega}{\omega_\lambda}\right)^{1/2}}{\left(1 - \frac{\omega_\lambda^2}{\omega^2}\right)^{1/2}}\left[\xi_\lambda + \eta_\lambda\left(\frac{\omega_\lambda}{\omega}\right)^2\right] \quad (8.63)$$

where σ is the conductivity (assumed independent of frequency), δ_λ is the skin depth at the cutoff frequency, and ξ_λ, η_λ are dimensionless numbers of the order of unity. For TM modes, $\eta_\lambda = 0$.

For a given cross-sectional geometry it is a straightforward matter to calculate the dimensionless parameters ξ_λ and η_λ in (8.63). For the TE modes with $n = 0$ in a rectangular guide, the values are $\xi_{m,0} = a/(a + b)$ and $\eta_{m,0} = 2b/(a + b)$. For reasonable relative dimensions, these parameters are of order unity, as expected.

The general behavior of β_λ as a function of frequency is shown in Fig. 8.6. Minimum attenuation occurs at a frequency well above cutoff. For TE modes the relative magnitudes of ξ_λ and η_λ depend on the shape of the guide and on λ.

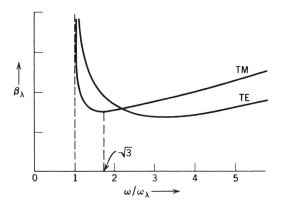

Figure 8.6 Attenuation constant β_λ as a function of frequency for typical TE and TM modes. For TM modes the minimum attenuation occurs at $\omega/\omega_\lambda = \sqrt{3}$, regardless of cross-sectional shape.

Consequently no general statement can be made about the exact frequency for minimum attenuation. But for TM modes the minimum always occurs at $\omega_{\min} = \sqrt{3}\omega_\lambda$. At high frequencies the attenuation increases as $\omega^{1/2}$. In the microwave region typical attenuation constants for copper guides are of the order $\beta_\lambda \sim 10^{-4}\omega_\lambda/c$, giving $1/e$ distances of 200–400 meters.

The approximations employed in obtaining (8.63) break down close to cutoff. Evidence for this is the physically impossible, infinite value of (8.63) at $\omega = \omega_\lambda$.

8.6 *Perturbation of Boundary Conditions*

The use of energy conservation to determine the attenuation constant β_λ is direct and has intuitive appeal, but gives physically meaningless results at cutoff and fails to yield a value for α_λ, the change in the real part of the wave number. Both these defects can be remedied by use of the technique called *perturbation of boundary conditions*. This method is capable, at least in principle, of obtaining answers to any desired degree of accuracy, although we shall apply it only to the lowest order. It also permits the treatment of attenuation for degenerate modes, mentioned briefly at the end of this section and in Problem 8.13. The effect of small distortions of cross section can also be treated. See Problem 8.12.

For definiteness we consider a single TM mode with no other mode (TE or TM) degenerate or nearly degenerate with it. The argument for an isolated TE mode is similar. To reduce the number of sub- and superscripts, we denote the (unperturbed) solution for perfectly conducting walls by a subscript zero and the (perturbed) solution for walls of finite conductivity by no sub- or superscript. Thus the unperturbed problem has a longitudinal electric field $E_z = \psi_0$, where

$$(\nabla_t^2 + \gamma_0^2)\psi_0 = 0, \qquad \psi_0|_s = 0 \tag{8.64}$$

and γ_0^2 is real. For finite, but large, conductivity, $E_z = \psi$ is not zero on the walls, but is given by (8.11). To lowest order, the right-hand side of (8.11) is approximated by the unperturbed fields. By use of the first equation in (8.23) and (8.33), the perturbed boundary condition on ψ can be expressed as

$$\psi|_s \simeq f \left.\frac{\partial\psi_0}{\partial n}\right|_s \tag{8.65}$$

where the small complex parameter f is*

$$f = (1 + i) \frac{\mu_c \delta}{2\mu} \left(\frac{\omega}{\omega_0} \right)^2 \tag{8.66}$$

Here μ_c and μ are the magnetic permeabilities of the conducting walls and the medium in the guide, respectively, δ is the skin depth (8.8), and ω_0 is the cutoff frequency of the unperturbed mode. The perturbed problem, equivalent to (8.64), is thus

$$(\nabla_t^2 + \gamma^2)\psi = 0, \qquad \psi|_S \simeq f \left. \frac{\partial \psi_0}{\partial n} \right|_S \tag{8.67}$$

If only the eigenvalue γ^2 is desired, Green's theorem (1.35) in two dimensions can be employed:

$$\int_A [\phi \, \nabla_t^2 \psi - \psi \, \nabla_t^2 \phi] \, da = \oint_C \left[\psi \frac{\partial \phi}{\partial n} - \phi \frac{\partial \psi}{\partial n} \right] dl$$

where the right-hand side has an *inwardly* directed normal [out of the conductor, in conformity with (8.11) and (8.65)]. With the identifications, $\psi = \psi$ and $\phi = \psi_0^*$, and use of the wave equations (8.64) and (8.67), and their boundary conditions, the statement of Green's theorem becomes

$$(\gamma_0^2 - \gamma^2) \int_A \psi_0^* \psi \, da = f \oint_C \left| \frac{\partial \psi_0}{\partial n} \right|^2 dl \tag{8.68}$$

Since f is assumed to be a small parameter, it is normally consistent to approximate ψ in the integral on the left by its unperturbed value ψ_0. This leads to the final result,

$$\gamma_0^2 - \gamma^2 = k^2 - k^{(0)2} \simeq f \frac{\oint_C \left| \dfrac{\partial \psi_0}{\partial n} \right|^2 dl}{\int_A |\psi_0|^2 \, da} \tag{8.69}$$

From (8.51) and (8.59) of the preceding section one finds that the ratio of integrals on the right-hand side of (8.69) enters a previous result, namely,

$$2k^{(0)} \beta_{\text{TM}}^{(0)} = \frac{\mu_c \delta}{2\mu} \left(\frac{\omega}{\omega_0} \right)^2 \frac{\oint_C \left| \dfrac{\partial \psi_0}{\partial n} \right|^2 dl}{\int_A |\psi_0|^2 \, da} \tag{8.70}$$

where $\beta^{(0)}$ is defined by (8.57) and (8.63). This means that (8.69) can be written as

$$k^2 \simeq k^{(0)2} + 2(1 + i)k^{(0)} \beta^{(0)} \tag{8.71}$$

a result that holds for both TM and TE modes, with the appropriate $\beta^{(0)}$ from Section 8.5. For $k^{(0)} \gg \beta^{(0)}$, (8.71) reduces to the former expression (8.55) with

*More generally, f can be expressed in terms of the surface impedance Z_s as $f = (i\omega/\mu\omega_0^2)Z_s$.

$\alpha = \beta$. At cutoff and below, however, where the earlier results failed, (8.71) yields sensible results because the combination $k^{(0)}\beta^{(0)}$ is finite and well behaved in the neighborhood of $k^{(0)} = 0$. The transition from a propagating mode to a cutoff mode is evidently not a sharp one if the walls are less than perfect conductors, but the attenuation is sufficiently large immediately above and below the cutoff frequency that little error is made in assuming a sharp cutoff.

The discussion of attenuation here and in the preceding section is restricted to one mode at a time. For nondegenerate modes with not too great losses this approximation is adequate. If, however, it happens that a TM and a TE mode are degenerate (as occurs in the rectangular waveguide for $n \neq 0$, $m \neq 0$), then any perturbation, no matter how small, can cause sizable mixing of the two modes. The methods used so far fail in such circumstances. The breakdown of the present method occurs in the perturbed boundary condition (8.65), where there is now on the right-hand side a term involving the tangential derivative of the unperturbed H_z, as well as the normal derivative of E_z. And there is, of course, a corresponding perturbed boundary condition for H_z involving both unperturbed longitudinal fields. The problem is one of degenerate-state perturbation theory, most familiar in the context of quantum mechanics. The perturbed modes are orthogonal linear combinations of the unperturbed TM and TE modes, and the attenuation constants for the two modes have the characteristic expression,

$$\beta = \tfrac{1}{2}(\beta_{\text{TM}} + \beta_{\text{TE}}) \pm \tfrac{1}{2}\sqrt{(\beta_{\text{TM}} - \beta_{\text{TE}})^2 + 4\,|K|^2} \tag{8.72}$$

where β_{TM} and β_{TE} are the values found above, and K is a coupling parameter.

The effects of attenuation and distortion for degenerate modes using perturbation of boundary conditions are addressed in Problem 8.13. See also *Collin*.

8.7 *Resonant Cavities*

Although an electromagnetic cavity resonator can be of any shape whatsoever, an important class of cavities is produced by placing end faces on a length of cylindrical waveguide. We assume that the end surfaces are plane and perpendicular to the axis of the cylinder. As usual, the walls of the cavity are taken to have infinite conductivity, while the cavity is filled with a lossless dielectric with constants μ, ϵ. Because of reflections at the end surfaces, the z dependence of the fields is that appropriate to standing waves:

$$A \sin kz + B \cos kz$$

If the plane boundary surfaces are at $z = 0$ and $z = d$, the boundary conditions can be satisfied at each surface only if

$$k = p\,\frac{\pi}{d} \qquad (p = 0, 1, 2, \ldots) \tag{8.73}$$

For TM fields the vanishing of $\mathbf{E}_t$ at $z = 0$ and $z = d$ requires

$$E_z = \psi(x, y) \cos\left(\frac{p\pi z}{d}\right) \qquad (p = 0, 1, 2, \ldots) \tag{8.74}$$

Similarly for TE fields, the vanishing of H_z at $z = 0$ and $z = d$ requires

$$H_z = \psi(x, y) \sin\left(\frac{p\pi z}{d}\right) \qquad (p = 1, 2, 3, \ldots) \tag{8.75}$$

Then from (8.31) and (8.33) we find the transverse fields:

TM FIELDS

$$\mathbf{E}_t = -\frac{p\pi}{d\gamma^2} \sin\left(\frac{p\pi z}{d}\right) \nabla_t \psi$$
$$\mathbf{H}_t = \frac{i\epsilon\omega}{\gamma^2} \cos\left(\frac{p\pi z}{d}\right) \hat{\mathbf{z}} \times \nabla_t \psi \tag{8.76}$$

TE FIELDS

$$\mathbf{E}_t = -\frac{i\omega\mu}{\gamma^2} \sin\left(\frac{p\pi z}{d}\right) \hat{\mathbf{z}} \times \nabla_t \psi$$
$$\mathbf{H}_t = \frac{p\pi}{d\gamma^2} \cos\left(\frac{p\pi z}{d}\right) \nabla_t \psi \tag{8.77}$$

The boundary conditions at the ends of the cavity are now explicitly satisfied. There remains the eigenvalue problem (8.34)–(8.36), as before. But now the constant γ^2 is:

$$\gamma^2 = \mu\epsilon\omega^2 - \left(\frac{p\pi}{d}\right)^2 \tag{8.78}$$

For each value of p the eigenvalue γ_λ^2 determines an eigenfrequency $\omega_{\lambda p}$:

$$\omega_{\lambda p}^2 = \frac{1}{\mu\epsilon}\left[\gamma_\lambda^2 + \left(\frac{p\pi}{d}\right)^2\right] \tag{8.79}$$

and the corresponding fields of that resonant mode. The resonance frequencies form a discrete set that can be determined graphically on the figure of axial wave number k versus frequency in a waveguide (see Fig. 8.4) by demanding that $k = p\pi/d$. It is usually expedient to choose the various dimensions of the cavity so that the resonant frequency of operation lies well separated from other resonant frequencies. Then the cavity will be relatively stable in operation and insensitive to perturbing effects associated with frequency drifts, changes in loading, etc.

An important practical resonant cavity is the right circular cylinder, perhaps with a piston to allow tuning by varying the height. The cylinder is shown in Fig. 8.7, with inner radius R and length d. For a TM mode the transverse wave equation for $\psi = E_z$, subject to the boundary condition $E_z = 0$ at $\rho = R$, has the solution:

$$\psi(\rho, \phi) = E_0 J_m(\gamma_{mn}\rho)e^{\pm im\phi} \tag{8.80}$$

where

$$\gamma_{mn} = \frac{x_{mn}}{R}$$

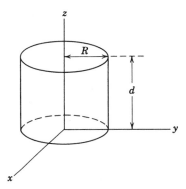

Figure 8.7

x_{mn} is the nth root of the equation, $J_m(x) = 0$. These roots were given earlier, following Eq. (3.92). The integers m and n take on the values $m = 0, 1, 2, \ldots$, and $n = 1, 2, 3, \ldots$. The resonance frequencies are given by

$$\omega_{mnp} = \frac{1}{\sqrt{\mu\epsilon}} \sqrt{\frac{x_{mn}^2}{R^2} + \frac{p^2\pi^2}{d^2}} \qquad (8.81)$$

The lowest TM mode has $m = 0$, $n = 1$, $p = 0$, and so is designated TM$_{0,1,0}$. Its resonance frequency is

$$\omega_{010} = \frac{2.405}{\sqrt{\mu\epsilon}R}$$

The explicit expressions for the fields are

$$\left.\begin{aligned} E_z &= E_0 J_0\left(\frac{2.405\rho}{R}\right)e^{-i\omega t} \\ H_\phi &= -i\sqrt{\frac{\epsilon}{\mu}}\, E_0 J_1\left(\frac{2.405\rho}{R}\right)e^{-i\omega t} \end{aligned}\right\} \qquad (8.82)$$

The resonant frequency for this mode is independent of d. Consequently simple tuning is impossible.

For TE modes, the basic solution (8.80) still applies, but the boundary condition on $H_z[(\partial\psi/\partial\rho)|_R = 0]$ makes

$$\gamma_{mn} = \frac{x_{mn}'}{R}$$

where x_{mn}' is the nth root of $J_m'(x) = 0$. These roots, for a few values of m and n, are tabulated below (for $m \neq 1$, $x = 0$ is a trivial root):

Roots of $J_m'(x) = 0$

$m = 0$: $x_{0n}' = 3.832, 7.016, 10.173, \ldots$

$m = 1$: $x_{1n}' = 1.841, 5.331, \;\; 8.536, \ldots$

$m = 2$: $x_{2n}' = 3.054, 6.706, \;\; 9.970, \ldots$

$m = 3$: $x_{3n}' = 4.201, 8.015, 11.336, \ldots$

The resonance frequencies are given by

$$\omega_{mnp} = \frac{1}{\sqrt{\mu\epsilon}}\left(\frac{x_{mn}'^2}{R^2} + \frac{p^2\pi^2}{d^2}\right)^{1/2} \tag{8.83}$$

where $m = 0, 1, 2, \ldots$, but $n, p = 1, 2, 3, \ldots$. The lowest TE mode has $m = n = p = 1$, and is denoted $TE_{1,1,1}$. Its resonance frequency is

$$\omega_{111} = \frac{1.841}{\sqrt{\mu\epsilon}R}\left(1 + 2.912\frac{R^2}{d^2}\right)^{1/2} \tag{8.84}$$

while the fields are derivable from

$$\psi = H_z = H_0 J_1\left(\frac{1.841\rho}{R}\right)\cos\phi\,\sin\left(\frac{\pi z}{d}\right)e^{-i\omega t} \tag{8.85}$$

by means of (8.77). For d large enough ($d > 2.03R$), the resonance frequency ω_{111} is smaller than that for the lowest TM mode. Then the $TE_{1,1,1}$ mode is the fundamental oscillation of the cavity. Because the frequency depends on the ratio d/R it is possible to provide easy tuning by making the separation of the end faces adjustable.

Variational methods can be exploited to estimate the lowest resonant frequencies of cavities. A variational principle and some examples are presented in the problems (Problems 8.9–8.11).

8.8   *Power Losses in a Cavity; Q of a Cavity*

In the preceding section it was found that resonant cavities have discrete frequencies of oscillation with a definite field configuration for each resonance frequency. This implies that, if one were attempting to excite a particular mode of oscillation in a cavity by some means, no fields of the right sort could be built up unless the exciting frequency were exactly equal to the chosen resonance frequency. In actual fact there will not be a delta function singularity, but rather a narrow band of frequencies around the eigenfrequency over which appreciable excitation can occur. An important source of this smearing out of the sharp frequency of oscillation is the dissipation of energy in the cavity walls and perhaps in the dielectric filling the cavity. A measure of the sharpness of response of the cavity to external excitation is the Q of the cavity, defined as 2π times the ratio of the time-averaged energy stored in the cavity to the energy loss per cycle:

$$Q = \omega_0 \frac{\text{Stored energy}}{\text{Power loss}} \tag{8.86}$$

Here ω_0 is the resonance frequency, assuming no losses. By conservation of energy the power dissipated in ohmic losses is the negative of the time rate of change of stored energy U. Thus from (8.86) we can write an equation for the behavior of U as a function of time:

with solution

$$\left.\begin{aligned}\frac{dU}{dt} &= -\frac{\omega_0}{Q}U \\[2mm] U(t) &= U_0 e^{-\omega_0 t/Q}\end{aligned}\right\} \tag{8.87}$$

If an initial amount of energy U_0 is stored in the cavity, it decays away exponentially with a decay constant inversely proportional to Q. The time dependence in (8.87) implies that the oscillations of the fields in the cavity are damped as follows:

$$E(t) = E_0 e^{-\omega_0 t/2Q} e^{-i(\omega_0 + \Delta\omega)t} \tag{8.88}$$

where we have allowed for a shift $\Delta\omega$ of the resonant frequency as well as the damping. A damped oscillation such as this has not a pure frequency, but a superposition of frequencies around $\omega = \omega_0 + \Delta\omega$. Thus,

$$E(t) = \frac{1}{\sqrt{2\pi}} \int_{-\infty}^{\infty} E(\omega) e^{-i\omega t} \, d\omega$$

where

$$E(\omega) = \frac{1}{\sqrt{2\pi}} \int_{0}^{\infty} E_0 e^{-\omega_0 t/2Q} e^{i(\omega - \omega_0 - \Delta\omega)t} \, dt \tag{8.89}$$

The integral in (8.89) is elementary and leads to a frequency distribution for the energy in the cavity having a resonant line shape:

$$|E(\omega)|^2 \propto \frac{1}{(\omega - \omega_0 - \Delta\omega)^2 + (\omega_0/2Q)^2} \tag{8.90}$$

The resonance shape (8.90), shown in Fig. 8.8, has a full width Γ at half-maximum (confusingly called the half-width) equal to ω_0/Q. For a constant input voltage, the energy of oscillation in the cavity as a function of frequency will follow the resonance curve in the neighborhood of a particular resonant frequency. Thus, the frequency separation $\delta\omega$ between half-power points determines the width Γ and the Q of cavity is

$$Q = \frac{\omega_0}{\delta\omega} = \frac{\omega_0}{\Gamma} \tag{8.91}$$

Q values of several hundreds or thousands are common for microwave cavities.

To determine the Q of a cavity we can calculate the time-averaged energy stored in it and then determine the power loss in the walls. The computations are very similar to those done in Section 8.5 for attenuation in waveguides. We consider here only the cylindrical cavities of Section 8.7, assuming no degener-

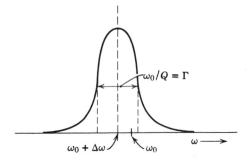

Figure 8.8 Resonance line shape. The full width Γ at half-maximum (of the power) is equal to the unperturbed frequency ω_0 divided by the Q of the cavity.

acies. The energy stored in the cavity for the mode λ, p is, according to (8.74)–(8.77):

$$U = \frac{d}{4} \left\{ \begin{matrix} \epsilon \\ \mu \end{matrix} \right\} \left[1 + \left(\frac{p\pi}{\gamma_\lambda d} \right)^2 \right] \int_A |\psi|^2 \, da \qquad (8.92)$$

where the upper (lower) line applies to TM (TE) modes. For the TM modes with $p = 0$ the result must be multiplied by 2.

The power loss can be calculated by a modification of (8.58):

$$P_{\text{loss}} = \frac{1}{2\sigma\delta} \left[\oint_C dl \int_0^d dz \, |\mathbf{n} \times \mathbf{H}|^2_{\text{sides}} + 2 \int_A da \, |\mathbf{n} \times \mathbf{H}|^2_{\text{ends}} \right] \qquad (8.93)$$

For TM modes with $p \neq 0$ it is easy to show that

$$P_{\text{loss}} = \frac{\epsilon}{\sigma\delta\mu} \left[1 + \left(\frac{p\pi}{\gamma_\lambda d} \right)^2 \right] \left(1 + \xi_\lambda \frac{Cd}{4A} \right) \int_A |\psi|^2 \, da \qquad (8.94)$$

where the dimensionless number ξ_λ is the same one that appears in (8.62), C is the circumference of the cavity, and A is its cross-sectional area. For $p = 0$, ξ_λ must be replaced by $2\xi_\lambda$. Combining (8.92) and (8.94) according to (8.86), and using definition (8.8) for the skin depth δ, we find the Q of the cavity:

$$Q = \frac{\mu}{\mu_c} \frac{d}{\delta} \frac{1}{2 \left(1 + \xi_\lambda \dfrac{Cd}{4A} \right)} \qquad (8.95)$$

where μ_c is the permeability of the metal walls of the cavity. For $p = 0$ modes, (8.95) must be multiplied by 2 and ξ_λ replaced by $2\xi_\lambda$. This expression for Q has an intuitive physical interpretation when written in the form:

$$Q = \frac{\mu}{\mu_c} \left(\frac{V}{S\delta} \right) \times (\text{Geometrical factor}) \qquad (8.96)$$

where V is the volume of the cavity, and S its total surface area. The Q of a cavity is evidently, apart from a geometrical factor, the ratio of the volume occupied by the fields to the volume of the conductor into which the fields penetrate because of the finite conductivity. For the $TE_{1,1,1}$ mode in the right circular cylinder cavity, calculation yields a geometrical factor

$$\left(1 + \frac{d}{R} \right) \frac{\left(1 + 0.344 \dfrac{d^2}{R^2} \right)}{\left(1 + 0.209 \dfrac{d}{R} + 0.242 \dfrac{d^3}{R^3} \right)} \qquad (8.97)$$

that varies from unity for $d/R = 0$ to a maximum of 2.13 at $d/R = 1.91$ and then decreases to 1.42 as $d/R \to \infty$.

Expression (8.96) for Q applies not only to cylindrical cavities but also to cavities of arbitrary shape, with an appropriate geometrical factor of the order of unity.

The use of conservation of energy to discuss losses in a cavity has the same advantages and disadvantages as for waveguides. The Q values can be calculated, but possible shifts in frequency lie outside the scope of the method. The technique of perturbation of boundary conditions, described in Section 8.6, again removes these deficiencies. In fact the analogy is so close to the waveguide situation that the answers can be deduced without performing the calculation explicitly. The unperturbed problem of the resonant frequencies of a cavity with perfectly conducting walls is specified by (8.64) or its equivalent for TE modes. Similarly, the perturbed problem involves solution of (8.67) or equivalent. A result equivalent to (8.69) evidently emerges. The difference $(\gamma_0^2 - \gamma^2)$ is proportional to $(\omega_0^2 - \omega^2)$ where now ω_0 is the unperturbed resonant frequency rather than the cutoff frequency of the waveguide and ω is the perturbed resonant frequency. Thus the analog of (8.69) takes the form,

$$\omega_0^2 - \omega^2 \simeq (1 + i)I \tag{8.98}$$

where I is the ratio of appropriate integrals. In the limit of $I \to 0$, the imaginary part of ω is $-iI/2\omega_0$. From (8.88) this is to be identified with $-i\omega_0/2Q$, and therefore $I = \omega_0^2/Q$. Equation (8.98) can thus be written

$$\omega^2 \simeq \omega_0^2\left[1 - \frac{(1 + i)}{Q}\right] \tag{8.99}$$

where Q is the quantity defined by (8.86) and (8.92), (8.93). Damping is seen to cause equal modifications to the real and imaginary parts of ω^2. For large Q values, the change in the resonant frequency, rather than its square, is

$$\Delta\omega \simeq \text{Im } \omega \simeq -\frac{\omega_0}{2Q}$$

The resonant frequency is always lowered by the presence of resistive losses. The near equality of the real and imaginary parts of the change in ω^2 is a consequence of the boundary condition (8.11) appropriate for relatively good conductors. For very lossy systems or boundaries with different surface impedances, the relative magnitude of the real and imaginary parts of the change in ω^2 can be different from that given by (8.99).

In this section, as in Section 8.6, the discussion has been confined to non-degenerate modes. Generalization to degenerate modes is treated in Problem 8.13.

8.9 Earth and Ionosphere as a Resonant Cavity: Schumann Resonances

A somewhat unusual example of a resonant cavity is provided by the earth itself as one boundary surface and the ionosphere as the other. The lowest resonant modes of such a system are evidently of very low frequency, since the characteristic wavelength must be of the order of magnitude of the earth's radius. In such circumstances the ionosphere and the earth both appear as conductors with real conductivities. Seawater has a conductivity of $\sigma \sim 0.1 \ \Omega^{-1} \ \text{m}^{-1}$, while the ionosphere has $\sigma \sim 10^{-7}\text{–}10^{-4} \ \Omega^{-1} \ \text{m}^{-1}$. The walls of the cavity are thus far from perfectly conducting, especially the outer one. Nevertheless, we idealize the physical reality and consider as a model two perfectly conducting, concentric spheres with radii a and $b = a + h$, where a is the radius of the earth ($a \simeq 6400$ km) and h is the height of the ionosphere above the earth ($h \sim 100$ km). Furthermore, if we are concerned with only the lowest frequencies, we can focus our attention on the TM modes, with only tangential magnetic fields.* The reason for this is that the TM modes, with a radially directed electric field, can satisfy the boundary condition of vanishing tangential electric field at $r = a$ and $r = b$ without appreciable radial variation of the fields. On the other hand, the TE modes, with only tangential electric fields, must have a radial variation of approximately half a wavelength between $r = a$ and $r = b$. The lowest frequencies for the TE modes, are therefore of the order of $\omega_{TE} \sim \pi c/h$, whereas for the lowest TM modes $\omega_{TM} \sim c/a$.

The general problem of modes in a spherical geometry is involved enough that we leave it to Chapter 9. Here we consider only TM modes and assume that the fields are independent of the azimuthal angle ϕ. The last is no real restriction; it is known from consideration of spherical harmonics that the relevant quantity is l, not m. If the radial component of **B** vanishes and the other components do not depend on ϕ, the vanishing of the divergence of **B** requires that only B_ϕ is nonvanishing if the fields are finite at $\theta = 0$. Faraday's law then requires $E_\phi = 0$. Thus the homogeneous Maxwell equations specify that TM modes with no ϕ dependence involve only E_r, E_θ, and B_ϕ. The two curl equations of Maxwell can be combined, after assuming a time-dependence of $e^{-i\omega t}$, into

$$\frac{\omega^2}{c^2} \mathbf{B} - \nabla \times \nabla \times \mathbf{B} = 0 \qquad (8.100)$$

where the relative permeabilities of the medium between the spheres are taken as unity. The ϕ component of (8.100) is

$$\frac{\omega^2}{c^2} (rB_\phi) + \frac{\partial^2}{\partial r^2} (rB_\phi) + \frac{1}{r^2} \frac{\partial}{\partial \theta} \left[\frac{1}{\sin \theta} \frac{\partial}{\partial \theta} (\sin \theta \, rB_\phi) \right] = 0 \qquad (8.101)$$

The angular part of (8.101) can be transformed into

$$\frac{\partial}{\partial \theta} \left[\frac{1}{\sin \theta} \frac{\partial}{\partial \theta} (\sin \theta \, rB_\phi) \right] = \frac{1}{\sin \theta} \frac{\partial}{\partial \theta} \left(\sin \theta \frac{\partial (rB_\phi)}{\partial \theta} \right) - \frac{rB_\phi}{\sin^2 \theta}$$

showing, upon comparison with (3.6) or (3.9), that the θ dependence is given by the associated Legendre polynomials $P_l^m(\cos \theta)$ with $m = \pm 1$. It is natural therefore to write a product solution,

$$B_\phi(r, \theta) = \frac{u_l(r)}{r} P_l^1(\cos \theta) \qquad (8.102)$$

Substitution into (8.101) yields as the differential equation for $u_l(r)$,

$$\frac{d^2 u_l(r)}{dr^2} + \left[\frac{\omega^2}{c^2} - \frac{l(l+1)}{r^2} \right] u_l(r) = 0 \qquad (8.103)$$

with $l = 1, 2, \ldots$ defining the angular dependence of the modes.

The characteristic frequencies emerge from (8.103) when the boundary con-

---

*For a spherical geometry the notation TE (TM) indicates the absence of *radial* electric (magnetic) field components.

ditions appropriate for perfectly conducting walls at $r = a$ and $r = b$ are imposed. The radial and tangential electric fields are

$$E_r = \frac{ic^2}{\omega r \sin \theta} \frac{\partial}{\partial \theta} (\sin \theta B_\phi) = -\frac{ic^2}{\omega r} l(l + 1) \frac{u_l(r)}{r} P_l(\cos \theta)$$

$$E_\theta = -\frac{ic^2}{\omega r} \frac{\partial}{\partial r} (rB_\phi) = -\frac{ic^2}{\omega r} \frac{\partial u_l(r)}{\partial r} P_l^1(\cos \theta)$$

The vanishing of E_θ at $r = a$ and $r = b$ implies that the boundary condition for $u_l(r)$ is

$$\frac{du_l(r)}{dr} = 0 \qquad \text{for } r = a \quad \text{and} \quad r = b \qquad (8.104)$$

The solutions of (8.103) are r times the spherical Bessel functions (see Section 9.6). The boundary conditions (8.104) lead to transcendental equations for the characteristic frequencies. An example is left as a problem; for our present purposes a limiting case suffices. The height h of the ionosphere is sufficiently small compared to the radius a that the limit $h/a \ll 1$ can be assumed. The $l(l + 1)/r^2$ term in (8.103) can be approximated by its value at $r = a$. The solutions of (8.103) are then $\sin(qr)$ and $\cos(qr)$, where q^2 is given by the square bracket in (8.103) evaluated at $r = a$. With the boundary conditions (8.104), the solution is

$$u_l(r) \simeq A \cos [q(r - a)]$$

where $qh = n\pi$, $n = 0, 1, 2, \ldots$. For $n = 1, 2, \ldots$ the frequencies of the modes are evidently larger than $\omega = n\pi c/h$ and are in the domain of frequencies of the TE modes. Only for $n = 0$ are there very-low-frequency modes. The condition $q = 0$ is equivalent to $u_l(r) = $ constant and

$$\omega_l \simeq \sqrt{l(l + 1)} \frac{c}{a} \qquad (8.105)$$

where the equality is exact in the limit $h/a \to 0$. The exact solution shows that to first order in h/a the correct result has a replaced by $(a + \frac{1}{2}h)$. The fields are $E_\theta = 0$, $r^2 E_r \propto P_l(\cos \theta)$, $rB_\phi \propto P_l^1(\cos \theta)$.

The resonant frequencies (8.105) are called *Schumann resonances*.* They are extremely low frequencies: with $a = 6400$ km, the first five resonant frequencies are $\omega_l/2\pi = 10.6, 18.3, 25.8, 33.4, 40.9$ Hz. Schumann resonances manifest themselves as peaks in the noise power spectrum of extremely low frequencies propagating around the earth. Lightning bolts, containing a wide spectrum of frequencies, act as sources of radial electric fields. The frequency components near the Schumann resonances are propagated preferentially because they are normal modes of the earth-ionosphere cavity. The first definitive observations of these peaks in the noise power spectrum were made in 1960,[†] although there is evidence that Nikola Tesla *may* have observed them before 1900.[‡] A typical noise power

*W. O. Schumann, Z. *Naturforsch.* **72**, 149, 250 (1952).

[†]M. Balser and C. A. Wagner, Nature **188**, 638 (1960).

[‡]In U. S. patent 787,412 (April 18, 1905), reprinted in *Nikola Tesla*, Lectures and Patents and Articles, Nikola Tesla Museum, Beograd, Yugoslavia (1956), this remarkable genius clearly outlines the idea of the earth as a resonating circuit (he did not know of the ionosphere), estimates the lowest resonant frequency as 6 Hz (close to the 6.6 Hz for a perfectly conducting sphere), and describes generation and detection of these low-frequency waves. I thank V. L. Fitch for this fascinating piece of history.

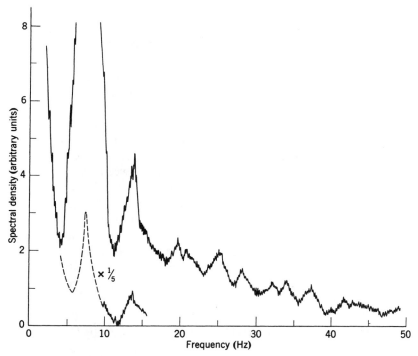

Figure 8.9   Typical noise power spectrum at low frequencies (integrated over 30 s), observed at Lavangsdalen, Norway on June 19, 1965. The prominent Schumann resonances at 8, 14, 20, and 26 Hz, plus peaks at 32, 37, and 43 Hz as well as smaller structure are visible. [After A. Egeland and T. R. Larsen, *Phys. Norv.* **2**, 85 (1967).]

spectrum is shown in Fig. 8.9. The resonances are clearly visible. They shift slightly and change shape from day to day, but have average linear frequencies of 8, 14, 20, 26, 32, 37, and 43 Hz for the first seven peaks. These frequencies are given quite closely by $5.8\sqrt{l(l+1)}$ Hz, the coefficient being 0.78 times $c/2\pi a (= 7.46$ Hz). The lack of precise agreement is not surprising, since, as already noted, the assumption of perfectly conducting walls is rather far from the truth. The Q values are estimated to be of the order of 4 to 10 for the first few resonances, corresponding to rather heavy damping. The effect of the damping on a resonant frequency is in the right direction to account for the differences between the observed values and (8.105), but the simple shift implied by (8.99) is only about half of what is observed. The $\sqrt{l(l+1)}$ variation of the resonant frequencies is, however, quite striking.

The simple picture of a resonant cavity with well-defined, but lossy, walls accounts for the main features of the Schumann resonances, although failing in some quantitative aspects. More realistic and detailed models and discussion of the observations can be found in a review by Galejs,* as well as his monograph, *Galejs*. The use of waveguide and resonant cavity concepts in the treatment of propagation of electromagnetic waves around the earth is discussed in the books by *Budden* and *Wait* listed at the end of this chapter. Two curiosities may be

---

*J. Galejs, *J. Res. Nat. Bur. Stand.* (*U.S.*) **69D**, 1043 (1965). See also T. Madden and W. Thompson, *Rev. Geophys.* **3**, 211 (1965) and F. W. Chapman, D. L. Jones, J. D. W. Todd, and R. A. Challinor, *Radio Sci.* **1**, 1273 (1966).

permitted here. On July 9, 1962, a nuclear explosion was detonated at high altitude over Johnston Island in the Pacific. One consequence of this test was to create observable alterations in the ionosphere and radiation belts on a worldwide scale. Sudden decreases of 3–5% in Schumann resonant frequencies were observed in France and at other stations immediately after the explosion, the changes decaying away over a period of several hours. This is documented in Fig. 17 of the paper by Galejs.

The second curiosity is the proposal* that Schumann resonances can serve as "a global tropical thermometer." The average magnetic field intensity of the fundamental Schumann resonance is expected to be strongly dependent on the frequency of lightning strikes around the world (which are seen from satellite observations to peak strongly in the tropics, $\pm 23°$ latitude). The frequency of lightning strikes at a number of sites in the tropics is known to be dramatically correlated to the average temperature. This lightning-temperature relation provides the physical understanding of the remarkably close correlation of Schumann resonance monthly mean magnetic field strength and monthly mean surface temperature observed at Kingston, Rhode Island, over a 5.5-year period and suggests that the Schumann resonances can serve as a global thermometer!

8.10 Multimode Propagation in Optical Fibers

Optical fibers lie at the heart of high-speed, high-capacity telecommunications. Visible or infrared light, modulated with the signal, is transmitted with little loss through small silica fibers. The very great frequency of the carrier light means that very large bandwidths are available for the signals. The technology has advanced rapidly in the past 25 years; a voluminous technical literature continues to grow. We can discuss only some of the basic principles. The reader wishing more can consult the references given at the end of the chapter.

Transmission via optical fibers falls approximately into two classes—multimode or single-mode propagation. "Cores" (the region where most of the energy flow is located) are typically 50 μm in diameter for multimode propagation, compared to a wavelength of the order of 1 μm, while 5 μm diameters are typical of single-mode fibers. We first consider multimode transmission for which the semigeometrical eikonal or WKB approximation is appropriate. Single-mode propagation is best described in waveguide terms. These concepts are treated in the following section.

Optical fiber cables, of the order of 2 cm in diameter, are actually nests of smaller cables each containing six or eight optical fibers protected by secondary coatings and buffer layers. The operative fiber consists of a cylindrical core of radius a $[2a = O(50 \ \mu\text{m})]$ and index of refraction n_1, surrounded by a cladding of outer radius b $[2b = O(150 \ \mu\text{m})]$ and index of refraction $n_0 < n_1$, as shown in Fig. 8.10a. Since the wavelength of the light is $O(1 \ \mu\text{m})$, the ideas of geometrical optics apply; the interface between core and cladding can be treated as locally flat. If the angle of incidence i of a ray originating within the core is greater than i_0, where $i_0 = \sin^{-1}(n_0/n_1)$ is the critical angle for total internal reflection, the ray will continue to be confined—it will propagate—as shown in Fig. 8.10b and 8.10c.

*E. R. Williams, *Science* **256**, 1184–1187 (1992).

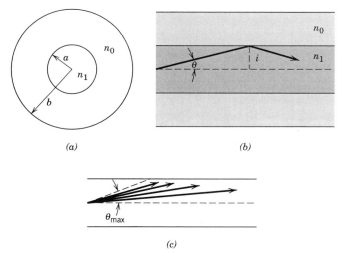

Figure 8.10   Optical fiber core and cladding, with inner cylinder of index of refraction n_1 and cladding of index n_0 ($n_1 > n_0$): (a) cross section of fiber; (b) longitudinal section, showing a total internally reflected ray; (c) longitudinal section of the core, showing meridional propagating rays with complementary angles of incidence $\theta < \theta_{max} = \cos^{-1}(n_0/n_1)$, the critical angle for total internal reflection.

It is convenient to use the complementary angle of incidence θ, measured from the cylinder axis. Propagation occurs for rays with $\theta < \theta_{max} = \cos^{-1}(n_0/n_1)$. It is also convenient to use the parameter

$$\Delta = \frac{n_1^2 - n_0^2}{2n_1^2} \approx 1 - \frac{n_0}{n_1} \tag{8.106}$$

Typical operation has $\Delta \lesssim 1\%$. Then $\theta_{max} \approx \sqrt{2\Delta} \lesssim 0.14$ radian (8 degrees).

The system is, of course, a waveguide with discrete modes, as discussed in Section 8.11. Simple phase-space arguments allow us to estimate the number of propagating modes. The transverse wave number $k_\perp \approx k\theta$ is limited because $\theta < \theta_{max}$. Two-dimensional phase-space number density dN is

$$dN = \pi a^2 \frac{d^2k}{(2\pi)^2} \cdot 2$$

where the first factor is the spatial area, the second the wave-number volume element, and the factor 2 is for two states of polarization. With $d^2k = 2\pi k_\perp\, dk_\perp = 2\pi k^2\theta\, d\theta$, we have

$$N \approx a^2k^2 \int_0^{\theta_{max}} \theta\, d\theta \approx \tfrac{1}{2}(ka\sqrt{2\Delta})^2 = \tfrac{1}{2}V^2 \tag{8.107}$$

Here $V \equiv ka\sqrt{2\Delta}$, called the *fiber parameter* in the literature. Typical numbers are $\lambda = 0.85$ μm, $a = 25$ μm, $n_1 \approx 1.4$ ($ka \approx 260$), and $\Delta = 0.005$, leading to $N \approx 335$. In contrast, single-mode propagation has $a = O(2.7$ μm) and $\Delta = O(0.0025)$. Then $N = O(2)$, one for each state of polarization. Such a phase-space estimate is, of course, is only qualitative.

A core with a single coating is the simplest configuration, but multilayer geometries are possible. Consideration of Snell's law at successive interfaces shows that if the indices of refraction decrease from layer to layer out from the center, a ray leaving the axis at some angle is bent successively more toward the axis until it is totally reflected. In fact, for an arbitrary number of layers outside the core, the critical angle $\theta_{max} = \cos^{-1}(n_{outer}/n_{inner})$, just as for the simple two-index fiber. The limit of many layers is a "graded" index fiber in which the index of refraction varies continuously with radius from the axis. Grading addresses the problem of distortion caused by different optical path lengths for different angles of launch, as we discuss below.

For multimode propagation, especially in graded fibers, the quasi-geometrical description called the eikonal approximation is appropriate. We assume that the medium of propagation is a linear, nonconducting, nonmagnetic material with an index of refraction $n(\mathbf{x}) = \sqrt{\epsilon(\mathbf{x})/\epsilon_0}$ that varies in space slowly on the scale of the local wavelength of the wave. With fields varying in time as $e^{-i\omega t}$, the Maxwell equations for $\mathbf{E}$ and $\mathbf{H}$ can be combined to give Helmholtz wave equations of the form

$$\nabla^2 \mathbf{E} + \mu_0 \omega^2 \epsilon(\mathbf{x}) \mathbf{E} - \nabla\left(\frac{1}{\epsilon} \mathbf{E} \cdot \nabla\epsilon\right) = 0$$

$$\nabla^2 \mathbf{H} + \mu_0 \omega^2 \epsilon(\mathbf{x}) \mathbf{H} - i\omega(\nabla\epsilon) \times \mathbf{E} = 0 \tag{8.108}$$

The assumption that $\epsilon(\mathbf{x})$ changes little over a wavelength allows us to drop the terms involving the gradient of ϵ as the next order of smallness. Then the components of the electric and magnetic fields satisfy

$$\left[\nabla^2 + \frac{\omega^2}{c^2} n^2(\mathbf{x})\right]\psi = 0 \tag{8.109}$$

Locally, the basic solutions are "plane" waves; that is, there is a local wave number $|k(\mathbf{x})| = \omega n(\mathbf{x})/c$. It is suggestive to write, without approximation to (8.109) as yet,

$$\psi = e^{i\omega S(\mathbf{x})/c} \tag{8.110}$$

where the function $S(\mathbf{x})$ is called the *eikonal*. Insertion of (8.110) into (8.109) leads to an equation for S,

$$\frac{\omega^2}{c^2}[n^2(\mathbf{x}) - \nabla S \cdot \nabla S] + i\frac{\omega}{c}\nabla^2 S = 0$$

Consistent with the hypothesis of slow variation of $n(\mathbf{x})$ on the scale of a wavelength (and thus small change in S on the same scale), we neglect the last term as higher order. We then have the *eikonal approximation* of quasi-geometrical optics,

$$\nabla S \cdot \nabla S = n^2(\mathbf{x}) \tag{8.111}$$

To interpret the eikonal S and connect it to geometric ray tracing, we first consider the expansion of $S(\mathbf{x})$ in a Taylor series around some point $\mathbf{x}_0$:

$$S(\mathbf{x}) \approx S(\mathbf{x}_0) + (\mathbf{x} - \mathbf{x}_0) \cdot \nabla S(\mathbf{x}_0) + \cdots$$

The wave amplitude ψ is then

$$\psi(\mathbf{x}) \approx \exp[i\omega S(\mathbf{x}_0)]\exp\left[i(\mathbf{x} - \mathbf{x}_0) \cdot \frac{\omega\nabla S(\mathbf{x}_0)}{c}\right]$$

The form of ψ is that of a plane wave with wave vector $\mathbf{k}(\mathbf{x}_0) = \omega \boldsymbol{\nabla} S(\mathbf{x}_0)/c = n(\mathbf{x}_0)\omega \hat{\mathbf{k}}(\mathbf{x}_0)/c$, where $\hat{\mathbf{k}}(\mathbf{x}_0)$ is a *unit vector* in the direction of $\boldsymbol{\nabla} S(\mathbf{x}_0)$. In general we define $\hat{\mathbf{k}}(\mathbf{x})$ by

$$\boldsymbol{\nabla} S = n(\mathbf{x})\hat{\mathbf{k}}(\mathbf{x}) \tag{8.112}$$

The amplitude $\psi(\mathbf{x})$ describes a wave front that is locally plane and is propagating in the direction defined by $\hat{\mathbf{k}}(\mathbf{x})$. If we imagine advancing incrementally in the direction of $\hat{\mathbf{k}}$, we trace out a path that is the geometrical ray associated with the wave. Figure 8.11a sketches such a path. If the distance along the path is labeled by the variable s, the incremental change $\Delta\mathbf{r}$ has associated with it an incremental distance Δs along the path. In the limit of vanishing increments, the ratio $\Delta\mathbf{r}/\Delta s$ becomes $d\mathbf{r}/ds \equiv \hat{\mathbf{k}}$. We therefore can write a result equivalent to (8.112) to describe the optical ray path $\mathbf{r}(s)$,

$$n(\mathbf{x})\,\frac{d\mathbf{r}}{ds} = \boldsymbol{\nabla} S \tag{8.113}$$

Consider now the change in the left-hand side with s along the path,

$$\frac{d}{ds}\left[n(\mathbf{x})\,\frac{d\mathbf{r}}{ds}\right] = \frac{d}{ds}\,\boldsymbol{\nabla} S = \boldsymbol{\nabla}\,\frac{dS}{ds}$$

But $d/ds = \hat{\mathbf{k}} \cdot \boldsymbol{\nabla}$, so that, from (8.112), $dS/ds = \hat{\mathbf{k}} \cdot \hat{\mathbf{k}} n(\mathbf{r}) = n(\mathbf{r})$. We thus obtain an equation relating the coordinate $\mathbf{r}(s)$ along the ray to the gradient of the index of refraction, a *generalization of Snell's law*,

$$\frac{d}{ds}\left[n(\mathbf{r})\,\frac{d\mathbf{r}}{ds}\right] = \boldsymbol{\nabla} n(\mathbf{r}) \tag{8.114}$$

Rays in a circular fiber fall into two classes:

1. *Meridional rays:* rays that pass through the cylinder axis; they correspond to modes with vanishing azimuthal index m and nonvanishing intensity at $\rho = 0$.

2. *Skew rays:* rays that originate off-axis and whose path is a spiral in space with inner and outer turning points in radius; they correspond to modes with non-vanishing azimuthal index m and vanishing intensity at $\rho = 0$.

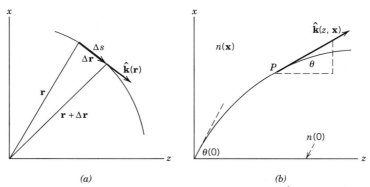

Figure 8.11 (a) Path of wave front defined by ray unit vector $\hat{\mathbf{k}}$. (b) Propagation in the z direction with index of refraction graded in the x direction.

For simplicity, we apply (8.114) only to the transmission of meridional rays in an optical fiber, or equivalently to rays in a "slab" geometry. Let the propagation of radiation be in the x-z plane, generally in the z direction with an index of refraction that is "graded" in the x direction, i.e., $n = n(x)$, as indicated in Fig. 8.11b. Suppose that a ray leaves the origin at an angle $\theta(0)$ with respect to the z axis, as shown. A distance s along the ray (at the point P) the unit vector $\hat{\mathbf{k}}$ makes an angle $\theta(x)$ with the z axis. Note that we write $\theta(x)$, not $\theta(x, z)$, because the coordinate x of the point P on the ray determines the value of z, modulo multiple values if the ray bends back toward the z axis. In terms of $\theta(x)$, the derivatives in (8.114) are $dx/ds = \sin \theta(x)$ and $dz/ds = \cos \theta(x)$. Then the vector equation (8.114) has as its two components,

$$\frac{d}{ds}[n(x) \sin \theta(x)] = \frac{dn(x)}{dx} \quad \text{and} \quad \frac{d}{ds}[n(x) \cos \theta(x)] = 0$$

The second equation has as its integral, $n(x) \cos \theta(x) = n(0) \cos \theta(0)$. If $n(x)$ is a monotonically decreasing function of $|x|$, for any given $\theta(0)$ there is a maximum (and a minimum) value of x attained by the ray, namely, when $\cos \theta(x_{max}) = 1$. The index of refraction at $|x| = x_{max}$ is

$$\bar{n} \equiv n(x_{max}) = n(0) \cos[\theta(0)] \tag{8.115}$$

The parameter $\bar{n}$ is a characteristic of a given ray or trajectory [specified by $\theta(0)$]. From $n(x)$ we can deduce x_{max} and so delimit the lateral extent of that trajectory.

To find the actual path $x(z)$ or $z(x)$ of the ray we must return to the equations for x and z in terms of s. The first integral of the z component of (8.114) is, as we have just seen, $n(dz/ds) = \bar{n}$. This means that we can replace d/ds in the x component of (8.114) by $d/ds = (\bar{n}/n)d/dz$. The equation then reads

$$\frac{\bar{n}}{n(x)} \frac{d}{dz} \left(\bar{n} \frac{dx}{dz} \right) = \frac{dn}{dx}$$

or

$$\bar{n}^2 \frac{d^2x}{dz^2} = \frac{1}{2} \frac{d}{dx}[n^2(x)] \tag{8.116}$$

Equation (8.116) has the structure of Newton's equation of motion of a particle of mass m in a potential $V(x)$, with $t \to z$, $m \to \bar{n}^2$, and $V(x) \to -n^2(x)/2$. Just as in mechanics, use of the "velocity" $x' = dx/dz$ allows one to write $d^2x/dz^2 = d(x'^2/2)/dx$ and find a first integral (conservation of energy in mechanics),

$$\bar{n}^2 x'^2 = n^2(x) - \bar{n}^2 \tag{8.117}$$

the constant of integration being determined by the condition that $x' = 0$ when $n(x) = \bar{n}$. The trajectory $z(x)$ is found from the integral,

$$z(x) = \bar{n} \int_0^x \frac{dx}{\sqrt{n^2(x) - \bar{n}^2}}$$

Here it is assumed that the ray began at the origin with angle $\theta(0)$. For $x \leq x_{max}$, the path represents one-quarter of a cycle of oscillation back and forth across the $x = 0$ line, as shown in Fig. 8.12a. The half-period of the ray (from one crossing of the z axis to the next) is

$$Z = 2\bar{n} \int_0^{x_{\max}} \frac{dx}{\sqrt{n^2(x) - \bar{n}^2}} \tag{8.118}$$

To discuss the transit time of a wave along an optical fiber, we need to examine the physical and optical path lengths along the ray. These path lengths from A to B are

$$L_{\text{phy}} = \int_A^B ds \qquad \text{and} \qquad L_{\text{opt}} = \int_A^B n(x)ds$$

With $ds = (n/\bar{n})\, dz = (n/\bar{n})(dz/dx)\, dx = [n(x)/\sqrt{n^2(x) - \bar{n}^2}]\, dx$, we find the physical and optical path lengths for half a period to be

$$L_{\text{phy}} = 2 \int_0^{x_{\max}} \frac{n(x)\, dx}{\sqrt{n^2(x) - \bar{n}^2}} \qquad \text{and} \qquad L_{\text{opt}} = 2 \int_0^{x_{\max}} \frac{n^2(x)\, dx}{\sqrt{n^2(x) - \bar{n}^2}} \tag{8.119}$$

The transit time of a ray of a given launch angle $\theta(0)$ is given by the optical path divided by c. For a length of fiber $z \gg Z$, the transit time $T(z)$ is

$$T(z) = \frac{L_{\text{opt}}}{Z} \frac{z}{c} \tag{8.120}$$

Different rays, defined by different $\theta(0)$ or $\bar{n}$, have different transit times, a form of dispersion that is geometrical. (Note that cZ/L_{opt} is the ray equivalent of the group velocity within the fiber.) A signal launched with a nonvanishing angular spread will be distorted unless $n(x)$ is chosen to make the transit time largely independent of $\bar{n}$. With a graded profile that decreases monotonically with $|x|$, rays with larger initial angles and so larger $x_{\max}$ will have longer physical paths, but will have larger speeds (phase velocities) $c/n(x)$ in those longer arcs. There is thus an inherent tendency toward equalization of transit times. The grading can in fact be chosen to make all transit times equal (see Problem 8.14). A simple example is shown in Fig. 8.12b. The fractional increase in optical path length L_{opt} [divided by $n(0)$] relative to Z as a function of θ is shown for a simple two-index fiber and a Gaussian-graded fiber with the same values of $n_1 = n(0)$ and n_0 ($\Delta = 0.01$). For $0 < \theta < \theta_{\max} \approx 0.1414$, the graded fiber has a fractional change of less than 10^{-5}; for the simple fiber, the spread is 1%.

    The geometrical dispersion resulting from different launch angles $\theta(0)$ has its counterpart as intermodal and intramodal dispersion when the propagation is described by discrete modes, as in the next section (see Problem 8.16). There is also material dispersion from the optical properties of the dielectrics. The optical path length L_{opt} (8.119) is then modified by having one of the factors of $n(x)$ in the integrand of (8.119) replaced by $d[\omega n(\omega, x)]/d\omega$. For silica, the group velocity in the infinite medium is stationary at $\lambda \approx 1.3$ μm; very large bandwidths and very high information transmission rates are possible there. Absorption is a minimum at $\lambda \approx 1.55$ μm; losses are of the order of 0.2 dB/km (see Section 9.7 for the Rayleigh scattering limit).

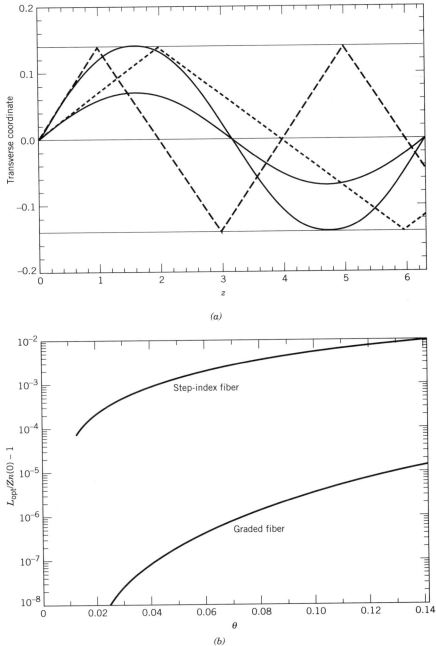

Figure 8.12 (*a*) Rays at critical angles θ_{max} and $\theta_{max}/2$ in a simple fiber with $\Delta = 0.01$ (dashed curves) and a graded fiber of the same radius a, critical angle, and central index, but with a Gaussian profile, $n(x) = n_1 e^{-x^2/b^2}$ for $0 < x < a$, ($b \approx a/\sqrt{\Delta}$) and $n(x) = n_0$ for $x > a$. Note the difference in scales. Units are such that $a = \theta_{max} \approx \sqrt{2}/10$. (*b*) Differences in optical path length (divided by the axial index of refraction) and actual length along a fiber, $[L_{opt}/n(0)Z - 1]$, for a simple two-index fiber with $\Delta = 0.01$ and the Gaussian-graded fiber of (*a*); $\theta_{max} = \sqrt{2}/10$. The compensation from faster phase velocity at larger excursions away from the axis in the graded fiber is striking.

8.11   *Modes in Dielectric Waveguides*

While the geometrical ray description of propagation in optical fibers is appropriate when the wavelength is very short compared to the transverse dimensions of the guiding structure, the wave nature of the fields must be taken into account when these two scales are comparable. Just as in a metallic waveguide, propagation at a given frequency can occur only via certain discrete modes, each with unique transverse field configurations and axial wave numbers. The bound rays $(\theta < \theta_{\max})$ in the geometric description have their counterparts as bound modes, with fields outside the core that decrease exponentially in the radial direction. Unbound rays $(\theta > \theta_{\max})$ correspond to the radiating modes, with oscillatory fields outside the core. Not surprisingly, single-mode propagation is important in optical communication, just as it is in microwave transmission in metallic guides. We now discuss modes in a planar guide and then introduce the circular fiber.

A.   *Modes in a Planar Slab Dielectric Waveguide*

To examine the existence of discrete modes in an optical fiber, we consider the simple situation of a "step-index" planar fiber consisting of a dielectric slab of thickness $2a$ in the x direction and infinite in the other two directions. We look for waves that are traveling in the z direction and are independent of y. The indices of refraction are n_1 and n_2 for the slab and its surrounding medium (cladding), respectively. The surfaces of the slab are at $x = \pm a$, as shown in Fig. 8.13. Geometrically, any ray that makes an angle θ with respect to the z axis less than $\theta_{\max}$ is totally internally reflected; the light is confined and propagates in the z direction, as discussed in the preceding section. The discrete mode structure occurs when we consider the wave nature of the light. Instead of solving the boundary-value problem, as for metallic waveguides, we keep to the optical description (but see Problem 8.15). The path shown in Fig. 8.13 can be thought of as the normal to the wave front of a plane wave, reflected back and forth or alternatively as two plane waves, one with positive x component of wave number, $k_x = k \sin \theta$, and the other with $k_x = -k \sin \theta$. To have a stable transverse field configuration and coherent propagation in the z direction, the accumulation of *transverse* phase on the path from A to just beyond B (with two internal reflections) must be an integer multiple of 2π:

$$4ka \sin \theta + 2\phi = 2p\pi \tag{8.121}$$

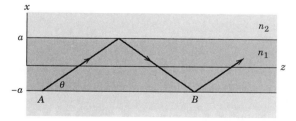

Figure 8.13   Slab dielectric waveguide. Ray or normal to wave front at angle θ shown.

where p is a nonnegative integer, $k = n_1\omega/c$, and ϕ is the phase associated with the total internal reflection, according to the Fresnel formulas (7.39) and (7.42). These phases are easily found to be

$$\phi_{TE} = -2 \arctan \sqrt{\frac{2\Delta}{\sin^2\theta} - 1}$$

$$\phi_{TM} = -2 \arctan\left(\frac{1}{1 - 2\Delta} \sqrt{\frac{2\Delta}{\sin^2\theta} - 1}\right)$$

(8.122)

where $\Delta = (n_1^2 - n_2^2)/2n_1^2$. The subscripts TE and TM in waveguide language correspond to the electric field being perpendicular and parallel to the plane of incidence in the Fresnel equations. Introducing the fiber parameter (frequency variable) $V = ka\sqrt{2\Delta}$ and transverse variable $\xi = \sin\theta/\sqrt{2\Delta}$, (8.121) can be written

$$\tan\left(V\xi - \frac{p\pi}{2}\right) = f\sqrt{\frac{1}{\xi^2} - 1}$$

(8.123)

where $f = 1$ for TE modes and $f = 1/(1 - 2\Delta)$ for TM modes.

The two sides of (8.123) are plotted in Fig. 8.14 for $V = 1$ and $V = 10$. There are seven TE and seven TM modes for $V = 10$. For small Δ the TE and TM modes are almost degenerate. The left-hand side of (8.123) shows that there are roughly $N \approx 4V/\pi$ modes in all, a number that follows from the one-dimensional phase-space estimate,

$$N_{TE} \approx N_{TM} \approx 2a \int_{-k_{max}}^{k_{max}} \frac{dk_x}{2\pi} = \frac{2ka}{\pi} \int_0^{\sqrt{2\Delta}} d(\sin\theta) = \frac{2V}{\pi}$$

An appropriate expression for the roots of (8.123) for TE modes is given in Problem 8.15. The lowest approximation, valid for $V \gg 1$ and small p, is $\xi_p(TE) \approx (p + 1)\pi/2(V + 1)$, showing equal spacing in p, as implied by the phase-space argument.

Although our phase coherence argument relied only on the wave in the interior of the slab, fields exist outside, too. Their influence is expressed in the phases ϕ. From (7.46) we find that the fields outside the slab vary in x as $e^{(-\beta|x|)}$, where

$$\beta = k\sqrt{2\Delta - \sin^2\theta} = \frac{V}{a}\sqrt{1 - \xi^2}$$

(8.124)

For a fixed V, as the mode number p increases ($\xi \to 1$), β gets smaller and smaller; the fields extended farther and farther into the cladding. For angles $\theta > \theta_{max}$ ($\xi > 1$), β becomes imaginary, corresponding to unconfined transverse fields. The slab radiates rather than confines the fields. In the waveguide regime, part of the power propagates within the core (slab) and part outside (see Problem 8.15). For $V = 1$, roughly two-thirds of the power of the TE_0 mode is carried within the core. When $V \gg 1$, the lower modes are almost totally confined. Only for $p \approx p_{max}$ does any appreciable power travel outside the core.

Note that if Δ is very small, $\theta_{max} \approx \sqrt{2\Delta}$ is small. The longitudinal propagation constant $k_z = k \cos\theta \approx k$ for all the waveguide modes, as we saw in the geometrical optics approach. For the TM modes, in which there is a longitudinal

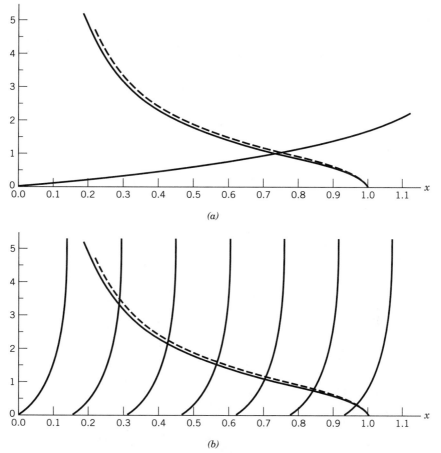

Figure 8.14 Graphic determination of eigenvalues for planar slab optical fiber: $\tan(Vx - p\pi/2) = f\sqrt{1/x^2 - 1}$; $x = \sin\theta/\sqrt{2\Delta} \approx \theta/\theta_{max}$, $V = ka\sqrt{2\Delta}$ ($f = 1$ for TE modes, $f \approx 1 + 2\Delta$ for TM modes); dashed curves have $f = 1.04$. (a) $V = 1$, $x \approx 0.739$ (TE), 0.747 (TM). (b) $V = 10$, seven roots for TE and for TM ($p = 0, \ldots, 6$).

component of electric field E_z, we have $|E_z/E_x| = \tan\theta \le \theta_{max} = \sqrt{2\Delta}$ for small Δ. Thus, to zeroth order in Δ, the TM modes have transverse electric fields and are degenerate with the TE modes. Appropriate linear superposition of two such degenerate modes gives a mode with arbitrary direction of polarization in the x-y plane. Such modes are labeled LP (for linearly polarized, although they can be circularly or elliptically polarized as well). LP modes are approximate descriptions in other geometries, such as circular, provided $\Delta \ll 1$, as is mentioned at the end of this section.

B. Modes in Circular Fibers

Optical fibers come in a wide variety of cross-sectional shapes, many analyzable only by numerical methods. The circular fiber with an index of refraction that is azimuthally symmetric is one of the simplest to discuss, but even it is more complicated than the one-dimensional slab geometry of the preceding section.

We give only an introduction here. The reader wishing to go into more details may consult the references cited at the end of the chapter.

We consider a fiber of uniform cross section with unit relative magnetic permeability and an index of refraction that does not vary along the cylinder axis but may vary in the transverse directions. For the present we do not restrict the problem to a circular cylinder. The Maxwell equations can be combined, as in Section 8.2, with assumed propagation as $e^{ik_z z - i\omega t}$, to yield the Helmholtz wave equations for **H** and **E**,

$$\nabla^2 \mathbf{H} + \frac{n^2 \omega^2}{c^2} \mathbf{H} = i\omega\epsilon_0 (\nabla n^2) \times \mathbf{E}$$

$$\nabla^2 \mathbf{E} + \frac{n^2 \omega^2}{c^2} \mathbf{E} = -\nabla\left[\frac{1}{n^2}(\nabla n^2) \cdot \mathbf{E}\right] \tag{8.125}$$

where we have written $\epsilon = n^2$. Just as in Section 8.2, the transverse components of **E** and **H** can be expressed in terms of the longitudinal fields E_z and H_z. Explicitly, the connections are

$$\mathbf{E}_t = \frac{i}{\gamma^2}[k_z \nabla_t E_z - \omega\mu_0 \hat{\mathbf{z}} \times \nabla_t H_z]$$

and $\tag{8.126}$

$$\mathbf{H}_t = \frac{i}{\gamma^2}[k_z \nabla_t H_z + \omega\epsilon_0 n^2 \hat{\mathbf{z}} \times \nabla_t E_z]$$

where $\gamma^2 = n^2\omega^2/c^2 - k_z^2$ is the radial propagation constant, as for metallic waveguides. If we take the z component of the equations (8.125) and use (8.126) to eliminate the transverse field components (and assume that $\partial n^2/\partial z = 0$), we find generalizations of the two-dimensional scalar wave equation (8.34),

$$\nabla_t^2 H_z + \gamma^2 H_z - \left(\frac{\omega}{\gamma c}\right)^2 (\nabla_t n^2) \cdot \nabla_t H_z = -\frac{\omega k_z \epsilon_0}{\gamma^2} \hat{\mathbf{z}} \cdot [\nabla_t n^2 \times \nabla_t E_z]$$

and $\tag{8.127}$

$$\nabla_t^2 E_z + \gamma^2 H_z - \left(\frac{k_z}{\gamma n}\right)^2 (\nabla_t n^2) \cdot \nabla_t E_z = \frac{\omega k_z \mu_0}{\gamma^2 n^2} \hat{\mathbf{z}} \cdot [\nabla_t n^2 \times \nabla_t H_z]$$

Our first observation is that, in contrast to (8.34) for ideal metallic guides, the equations for E_z and H_z are coupled. In general there is no separation into purely TE or TM modes. We restrict further comments to the simple situation of a core that is a circular cylinder of radius a with an azimuthally symmetric index $n(\rho)$. The cross products on the right-hand sides in (8.127) are proportional to $(\partial n^2/\partial\rho)(\partial/\rho\partial\phi)[E_z, H_z]$. Only if the fields have no azimuthal variation are these right-hand sides zero; only in such circumstances are there separate TE and TM modes. One might think that for a step-index fiber the transverse gradient of n^2 would vanish, at least for $\rho < a$ and for $\rho > a$; but there's the rub. The change from $n = n_1$ inside to $n = n_2$ outside implies a transverse gradient,

$$\nabla_t n^2 = -2n_1^2 \Delta\delta(\rho - a)\hat{\boldsymbol{\rho}}$$

The equations are coupled, unless the fields are independent of azimuth. The modes with both E_z and H_z nonzero are known as *HE* or *EH* hybrid modes. In practice, the solution is found by requiring continuity for normal $\mathbf{D}$ and $\mathbf{B}$ and tangential $\mathbf{E}$ and $\mathbf{H}$ across $\rho = a$. Separation of variables in cylindrical coordinates, assuming variation in azimuth of the form $e^{im\phi}$ leads to solutions for E_z and H_z,

$$\begin{Bmatrix} E_z \\ H_z \end{Bmatrix} = \begin{Bmatrix} A_e \\ A_h \end{Bmatrix} J_m(\gamma\rho)e^{im\phi}, \qquad \rho < a$$

and   (8.128)

$$\begin{Bmatrix} E_z \\ H_z \end{Bmatrix} = \begin{Bmatrix} B_e \\ B_h \end{Bmatrix} K_m(\beta\rho)e^{im\phi}, \qquad \rho > a$$

with the z and t dependences understood. Here $\gamma^2 = n_1^2\omega^2/c^2 - k_z^2$ and $\beta^2 = k_z^2 - n_2^2\omega^2/c^2$. Matching boundary conditions at $\rho = a$, with the transverse components computed from (8.126), leads to a determinantal eigenvalue equation for the various modes (see Problem 8.17). One finds that the TE and TM modes have nonvanishing "cutoff" frequencies, with the lowest corresponding to $V = n_1\omega\sqrt{2\Delta}/c = 2.405$, the first root of $J_0(x)$. In contrast, the lowest HE mode (HE_{11}) has no "cutoff" frequency. For $0 < V < 2.405$, it is the only mode that propagates in the fiber.

The azimuthally symmetric TE or TM modes correspond to meridional rays; the HE or EH modes, which have azimuthal variation, say, as $\sin(m\phi)$ or $\cos(m\phi)$, correspond to skew rays. That "skew ray" modes have longitudinal components of both $\mathbf{E}$ and $\mathbf{H}$ can be understood physically by considering the total internal reflection of such a ray at $\rho = a$. Since the plane containing such a ray and the normal to the surface does not contain the z axis, the electric field vector after reflection will have a different projection on the z axis than before, as will the magnetic field vector. Successive reflections therefore mix TE and TM waves; the eigenmodes have both E_z and H_z nonvanishing.

In fibers with very small Δ, called "weakly guiding waveguides" in the literature, the fields are found to have very small longitudinal components and are closely transverse. The language of plane light waves can be employed. For example, an HE_{11} mode, with azimuthal dependence for E_z of $\cos\phi$, has fields that are approximately linearly polarized and vary as $J_0(\gamma\rho)$ in the radial direction. In the "weakly guided" approximation, this mode is labeled LP_{01}.

The discussion so far (and some further aspects addressed in the problems) provide a brief introduction to the subject of optical waveguides. The literature is extensive and growing. The interested reader may gain entrée by consulting one of the references at the end of the chapter.

8.12   Expansion in Normal Modes; Fields Generated by a Localized Source in a Hollow Metallic Guide

For a given waveguide cross section and frequency ω, the electromagnetic fields in a hollow guide are described by an infinite set of characteristic or normal modes consisting of TE and TM waves, each with its characteristic cutoff frequency. For any given finite frequency, only a finite number of the modes can propagate; the rest are cutoff or evanescent modes. Far away from any source, obstacle, or aperture in the guide, the fields are relatively simple, with only the propagating modes (often just one) present with appreciable amplitude. Near a source or obstacle, however, many modes, both propagating and evanescent, must be superposed in order to describe the fields correctly. The cutoff modes have sizable amplitudes only in the neighborhood of the source or obstacle; their effects decay away over distances measured by the reciprocal of the imaginary part of their wave number. A typical practical problem concerning a source, obstacle, or aperture in a waveguide thus involves as accurate a solution as is possible for the fields in the vicinity of the source, etc., the expansion of those fields in terms of all normal modes of the guide, and a determination of the amplitudes for the one or more propagating modes that will describe the fields far away.

A. Orthonormal Modes

To facilitate the handling of the expansion of fields in the normal modes, it is useful to standardize the notation for the fields of a given mode, treating TE and TM modes on an equal footing and introducing a convenient normalization. Let the subscript λ or μ denote a given mode. One may think of $\lambda = 1, 2, 3, \ldots$ as indicating the modes arranged in some arbitrary order, of increasing cutoff frequency, for example. The subscript λ also conveys whether the mode is a TE or TM wave. The fields for the λ mode propagating in the *positive z* direction are written

$$\mathbf{E}_\lambda^{(+)}(x, y, z) = [\mathbf{E}_\lambda(x, y) + \mathbf{E}_{z\lambda}(x, y)]e^{ik_\lambda z} \tag{8.129}$$
$$\mathbf{H}_\lambda^{(+)}(x, y, z) = [\mathbf{H}_\lambda(x, y) + \mathbf{H}_{z\lambda}(x, y)]e^{ik_\lambda z}$$

where $\mathbf{E}_\lambda$, $\mathbf{H}_\lambda$ are the transverse fields given by (8.31) and (8.33) and $\mathbf{E}_{z\lambda}$, $\mathbf{H}_{z\lambda}$ are the longitudinal fields. The wave number k_λ is given by (8.37) and is taken to be real and positive for propagating modes in lossless guides (and purely imaginary, $k_\lambda = i\kappa_\lambda$, for cutoff modes). A time dependence $e^{-i\omega t}$ is, of course, understood. For a wave propagating in the *negative z* direction the fields are

$$\mathbf{E}_\lambda^{(-)} = [\mathbf{E}_\lambda - \mathbf{E}_{z\lambda}]e^{-ik_\lambda z} \tag{8.130}$$
$$\mathbf{H}_\lambda^{(-)} = [-\mathbf{H}_\lambda + \mathbf{H}_{z\lambda}]e^{-ik_\lambda z}$$

The pattern of signs in (8.130) compared to (8.129) can be understood from the need to satisfy $\boldsymbol{\nabla} \cdot \mathbf{E} = \boldsymbol{\nabla} \cdot \mathbf{H} = 0$ for each direction of propagation and the requirement of positive power flow in the direction of propagation. The overall phase of the fields in (8.130) relative to (8.129) is arbitrary. The choice taken here makes the transverse electric field at $z = 0$ the same for both directions of propagation, just as is done for the voltage waves on transmission lines.

A convenient normalization for the fields in (8.129) and (8.130) is afforded by taking the *transverse electric fields* $\mathbf{E}_\lambda(x, y)$ to be *real*, and requiring that

$$\int \mathbf{E}_\lambda \cdot \mathbf{E}_\mu \, da = \delta_{\lambda\mu} \tag{8.131}$$

where the integral is over the cross-sectional area of the guide. [The orthogonality of the different modes is taken for granted here. The proof is left as a problem

(Problem 8.18), as is the derivation of the other normalization integrals listed below.] From the relation (8.31) between electric and magnetic fields it is evident that (8.131) implies

$$\int \mathbf{H}_\lambda \cdot \mathbf{H}_\mu \, da = \frac{1}{Z_\lambda^2} \delta_{\lambda\mu} \tag{8.132}$$

and that the time-averaged power flow in the λth mode is

$$\frac{1}{2} \int (\mathbf{E}_\lambda \times \mathbf{H}_\mu) \cdot \hat{\mathbf{z}} \, da = \frac{1}{2Z_\lambda} \delta_{\lambda\mu} \tag{8.133}$$

It can also be shown that if (8.131) holds, the longitudinal components are normalized according to

TM Waves

$$\int E_{z\lambda} E_{z\mu} \, da = \frac{-\gamma_\lambda^2}{k_\lambda^2} \delta_{\lambda\mu}$$

TE Waves

$$\int H_{z\lambda} H_{z\mu} \, da = \frac{-\gamma_\lambda^2}{k_\lambda^2 Z_\lambda^2} \delta_{\lambda\mu} \tag{8.134}$$

As an explicit example of these normalized fields we list the transverse electric fields and also H_z and E_z of the TE and TM modes in a rectangular guide. The mode index λ is actually two indices (m, n). The normalized fields are

TM Waves

$$E_{xmn} = \frac{2\pi m}{\gamma_{mn} a \sqrt{ab}} \cos\left(\frac{m\pi x}{a}\right) \sin\left(\frac{n\pi y}{b}\right)$$

$$E_{ymn} = \frac{2\pi n}{\gamma_{mn} b \sqrt{ab}} \sin\left(\frac{m\pi x}{a}\right) \cos\left(\frac{n\pi y}{b}\right) \tag{8.135}$$

$$E_{zmn} = \frac{-2i\gamma_{mn}}{k_\lambda \sqrt{ab}} \sin\left(\frac{m\pi x}{a}\right) \sin\left(\frac{n\pi y}{b}\right)$$

TE Waves

$$E_{xmn} = \frac{-2\pi n}{\gamma_{mn} b \sqrt{ab}} \cos\left(\frac{m\pi x}{a}\right) \sin\left(\frac{n\pi y}{b}\right)$$

$$E_{ymn} = \frac{2\pi m}{\gamma_{mn} a \sqrt{ab}} \sin\left(\frac{m\pi x}{a}\right) \cos\left(\frac{n\pi y}{b}\right) \tag{8.136}$$

$$H_{zmn} = \frac{-2i\gamma_{mn}}{k_\lambda Z_\lambda \sqrt{ab}} \cos\left(\frac{m\pi x}{a}\right) \cos\left(\frac{n\pi y}{b}\right)$$

with γ_{mn} given by (8.43). The transverse magnetic field components can be obtained by means of (8.31). For TM modes, the lowest value of m and n is unity, but for TE modes, $m = 0 \; or \; n = 0$ is allowed. If $m = 0$ or $n = 0$, the normalization must be amended by multiplication of the right-hand sides of (8.136) by $1/\sqrt{2}$.

B. Expansion of Arbitrary Fields

An arbitrary electromagnetic field with time dependence $e^{-i\omega t}$ can be expanded in terms of the normal mode fields (8.129) and (8.130).* It is useful to keep track explicitly of the total fields propagating in the two directions. Thus the arbitrary fields are written in the form

$$\mathbf{E} = \mathbf{E}^{(+)} + \mathbf{E}^{(-)}, \qquad \mathbf{H} = \mathbf{H}^{(+)} + \mathbf{H}^{(-)} \tag{8.137}$$

where

$$\mathbf{E}^{(\pm)} = \sum_{\lambda} A_{\lambda}^{(\pm)} \mathbf{E}_{\lambda}^{(\pm)}, \qquad \mathbf{H}^{(\pm)} = \sum_{\lambda} A_{\lambda}^{(\pm)} \mathbf{H}_{\lambda}^{(\pm)} \tag{8.138}$$

Specification of the expansion coefficients $A_{\lambda}^{(+)}$ and $A_{\lambda}^{(-)}$ determines the fields everywhere in the guide. These may be found from boundary or source conditions in a variety of ways. Here is a useful theorem:

The fields everywhere in the guide are determined uniquely by specification of the transverse components of $\mathbf{E}$ and $\mathbf{H}$ in a plane, z = constant.

Proof: There is no loss in generality in choosing the plane at $z = 0$. Then from (8.137), (8.138), and (8.129), (8.130), the transverse fields are

$$\begin{aligned}
\mathbf{E}_t &= \sum_{\lambda'} (A_{\lambda'}^{(+)} + A_{\lambda'}^{(-)}) \mathbf{E}_{\lambda'} \\
\mathbf{H}_t &= \sum_{\lambda'} (A_{\lambda'}^{(+)} - A_{\lambda'}^{(-)}) \mathbf{H}_{\lambda'}
\end{aligned} \tag{8.139}$$

If the scalar product of both sides of the first equation is formed with $\mathbf{E}_{\lambda}$ and an integration over the cross section of the guide is performed, the orthogonality condition (8.131) implies

$$A_{\lambda}^{(+)} + A_{\lambda}^{(-)} = \int \mathbf{E}_{\lambda} \cdot \mathbf{E}_t \, da$$

Similarly the second equation, with (8.132), yields

$$A_{\lambda}^{(+)} - A_{\lambda}^{(-)} = Z_{\lambda}^2 \int \mathbf{H}_{\lambda} \cdot \mathbf{H}_t \, da$$

The coefficients $A_{\lambda}^{(\pm)}$ are therefore given by

$$A_{\lambda}^{(\pm)} = \frac{1}{2} \int (\mathbf{E}_{\lambda} \cdot \mathbf{E}_t \pm Z_{\lambda}^2 \mathbf{H}_{\lambda} \cdot \mathbf{H}_t) \, da \tag{8.140}$$

This shows that if $\mathbf{E}_t$ and $\mathbf{H}_t$ are given at $z = 0$, the coefficients in the expansion (8.137) and (8.138) are determined. The completeness of the normal mode expansion assures the uniqueness of the representation for all z.

C. Fields Generated by a Localized Source

The fields in a waveguide may be generated by a localized source, as shown schematically in Fig. 8.15. The current density $\mathbf{J}(\mathbf{x}, t)$ is assumed to vary in time as $e^{-i\omega t}$. Because of the oscillating current, fields propagate to the left and to the

---

*We pass over the mathematical problem of the *completeness* of the set of normal modes, and also only remark that more *general* time dependences can be handled by Fourier superposition.

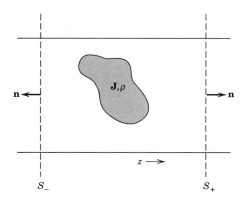

Figure 8.15   Schematic representation of a localized source in a waveguide. The walls of the guide, together with the planes S_+ and S_-, define the volume containing the source.

right. Outside the source, at and to the right of the surface S_+, say, there will be only fields varying as $e^{ik_\lambda z}$ and the electric field can be expressed as

$$\mathbf{E} = \mathbf{E}^{(+)} = \sum_{\lambda'} A_{\lambda'}^{(+)} \mathbf{E}_{\lambda'}^{(+)} \tag{8.141}$$

with a corresponding expression for $\mathbf{H}$. On and to the left of the surface S_- the fields all vary as $e^{-ik_\lambda z}$ and the electric field is

$$\mathbf{E} = \mathbf{E}^{(-)} = \sum_{\lambda'} A_{\lambda'}^{(-)} \mathbf{E}_{\lambda'}^{(-)} \tag{8.142}$$

To determine the coefficients $A_\lambda^{(\pm)}$ in terms of $\mathbf{J}$, we consider a form of the energy flow equation of Poynting's theorem. The identity

$$\boldsymbol{\nabla} \cdot (\mathbf{E} \times \mathbf{H}_\lambda^{(\pm)} - \mathbf{E}_\lambda^{(\pm)} \times \mathbf{H}) = \mathbf{J} \cdot \mathbf{E}_\lambda^{(\pm)} \tag{8.143}$$

follows from the source-free Maxwell equations for $\mathbf{E}_\lambda^{(\pm)}$, $\mathbf{H}_\lambda^{(\pm)}$, and the Maxwell equations with source satisfied by $\mathbf{E}$ and $\mathbf{H}$. Integration of (8.143) over a volume V bounded by a closed surface S leads, via the divergence theorem, to the result,

$$\int_S (\mathbf{E} \times \mathbf{H}_\lambda^{(\pm)} - \mathbf{E}_\lambda^{(\pm)} \times \mathbf{H}) \cdot \mathbf{n} \, da = \int_V \mathbf{J} \cdot \mathbf{E}_\lambda^{(\pm)} \, d^3x \tag{8.144}$$

where $\mathbf{n}$ is an outwardly directly normal. The volume V is now chosen to be the volume bounded by the inner walls of the guide and two surfaces S_+ and S_- of Fig. 8.15. With the assumption of perfectly conducting walls containing no sources or apertures, the part of the surface integral over the walls vanishes. Only the integrals over S_+ and S_- contribute. For definiteness, we choose the *lower* sign in (8.144) and substitute from (8.141) for the integral over S_+:

$$\int_{S_+} = \sum_{\lambda'} A_{\lambda'}^{(+)} \int_{S_+} \hat{\mathbf{z}} \cdot (\mathbf{E}_{\lambda'}^{(+)} \times \mathbf{H}_\lambda^{(-)} - \mathbf{E}_\lambda^{(-)} \times \mathbf{H}_{\lambda'}^{(+)}) \, da$$

With the fields (8.129) and (8.130) and the normalization (8.133), this becomes

$$\int_{S_+} = -\frac{2}{Z_\lambda} A_\lambda^{(+)} \tag{8.145}$$

The part of the surface integral in (8.144) from S_- is

$$\int_{S_-} = -\sum_{\lambda'} A_{\lambda'}^{(-)} \int_{S_-} \hat{\mathbf{z}} \cdot (\mathbf{E}_{\lambda'}^{(-)} \times \mathbf{H}_\lambda^{(-)} - \mathbf{E}_\lambda^{(-)} \times \mathbf{H}_{\lambda'}^{(-)}) \, da$$

which can easily be shown to vanish. For the choice of the lower sign in (8.144), therefore, only the surface S_+ gives a contribution to the left-hand side. Similarly, for the upper sign, only the integral over S_- contributes. It yields (8.145), but with $A_\lambda^{(-)}$ instead of $A_\lambda^{(+)}$. With (8.145) for the left-hand side of (8.144), the coefficients $A_\lambda^{(\pm)}$ are determined to be

$$A_\lambda^{(\pm)} = -\frac{Z_\lambda}{2} \int_V \mathbf{J} \cdot \mathbf{E}_\lambda^{(\mp)} \, d^3x \tag{8.146}$$

where the field $\mathbf{E}_\lambda^{(\mp)}$ of the normal mode λ is normalized according to (8.131). Note that the amplitude for propagation in the *positive z* direction comes from integration of the scalar product of the current with the mode field describing propagation in the *negative z* direction, and vice versa.

It is a simple matter to allow for the presence of apertures (acting as sources or sinks) in the walls of the guide between the two planes S_+ and S_-. Inspection of (8.144) shows that in such circumstances (8.146) is modified to read

$$A_\lambda^{(\pm)} = \frac{Z_\lambda}{2} \int_{\text{apertures}} (\mathbf{E} \times \mathbf{H}_\lambda^{(\mp)}) \cdot \mathbf{n} \, da - \frac{Z_\lambda}{2} \int_V \mathbf{J} \cdot \mathbf{E}_\lambda^{(\mp)} \, d^3x \tag{8.147}$$

where $\mathbf{E}$ is the *exact* tangential electric field in the apertures and $\mathbf{n}$ is *outwardly* directed.

The application of (8.146) to examples of the excitation of waves in guides is left to the problems at the end of the chapter. In the next chapter (Section 9.5) we consider the question of a source that is small compared to a wavelength and derive an approximation to (8.146): the coupling of the electric and magnetic dipole moments of the source to the electric and magnetic fields of the λth mode. The coupling of waveguides by small apertures is also discussed in Section 9.5. The subject of sources and excitation of oscillations in waveguides and cavities is of considerable practical importance in microwave engineering. There is a voluminous literature on the topic. One of the best references is the book by *Collin* (Chapters 5 and 7).

D. *Obstacles in Waveguides*

Discontinuities in the form of obstacles, dielectric slabs, diaphragms, and apertures in walls occur in the practical use of waveguides as carriers of electromagnetic energy and phase information in microwave systems. The expansion of the fields in normal modes is an essential aspect of the analysis. In the second (1975) edition of this book we analyzed the effects of transverse planar obstacles with variational methods (Sections 8.12 and 8.13). Lack of space prevents inclusion of the material here. The reader interested in pursuing these questions can refer to the second edition or the references mentioned below and in the References and Suggested Reading.

Theoretical and experimental study of obstacles, etc. loomed large in the immense radar research effort during the Second World War. The contributions of the United States during 1940–45 are documented in the Massachusetts Institute of Technology Radiation Laboratory Series, published by the McGraw-Hill Book Company, Inc., New York. The general physical principles of microwave circuits are covered in the book by *Montgomery, Dicke, and Purcell*, while a compendium of results on discontinuities in waveguides is provided in the volume by *Marcuvitz*. *Collin*, already cited, is a textbook source.

References and Suggested Reading

Waveguides and resonant cavities are discussed in numerous electrical and communications engineering books, for example,

Ramo, Whinnery, and Van Duzer, Chapters 7, 8, 10, and 11

The two books by Schelkunoff deserve mention for their clarity and physical insight,

Schelkunoff, *Electromagnetic Fields*

Schelkunoff, *Applied Mathematics for Engineers and Scientists*

Among the physics textbooks that treat waveguides, transmission lines, and cavities are

Panofsky and Phillips, Chapter 12

Slater

Smythe, Chapter XIII

Sommerfeld, *Electrodynamics*, Sections 22–25

Stratton, Sections 9.18–9.22

An authoritative discussion appears in

F. E. Borgnis and C. H. Papas, *Electromagnetic Waveguides and Resonators*, Vol. XVI of the *Encyclopaedia of Physics*, ed. S. Flugge, Springer-Verlag, Berlin (1958).

The books by

Collin

Harrington

Johnson

Waldron

are intended for graduate engineers and physicists and are devoted almost completely to guided waves and cavities. The standard theory, plus many specialized topics like discontinuities, are covered in detail. The original work on variational methods for discontinuities is summarized in

J. Schwinger and D. S. Saxon, *Discontinuities in Waveguides, Notes on Lectures by Julian Schwinger*, Gordon & Breach, New York (1968).

Variational principles for eigenfrequencies, etc., as well as discontinuities, are surveyed in

Cairo and Kahan

and also discussed by

Harrington, Chapter 7

Van Bladel, Chapter 13

Waldron, Chapter 8

The definitive compendium of formulas and numerical results on discontinuities, junctions, etc., in waveguides is

Marcuvitz

The mathematical tools for the treatment of these boundary-value problems are presented by

Morse and Feshbach, especially Chapter 13

Perturbation of boundary conditions is discussed by Morse and Feshbach (pp. 1038 ff). Information on special functions may be found in the ever-reliable

Magnus, Oberhettinger, and Soni, and in encyclopedic detail in

Bateman Manuscript Project, *Higher Transcendental Functions*.

Numerical values of special functions, as well as formulas, are given by

Abramowitz and Stegun
Jahnke, Emde, and Lösch

Two books dealing with propagation of electromagnetic waves around the earth and in the ionosphere from the point of view of waveguides and normal modes are

Budden
Wait

See also

Galejs

Schumann resonances are also described in detail in

P. V. Bliokh, A. P. Nicholaenko, and Yu. F. Filtippov, *Schumann Resonances in the Earth-Ionosphere Cavity*, transl. S. Chouet, ed. D. L. Jones, IEE Electromagnetic Wave Series, Vol. 8, Peter Peregrinus, London (1980).

There is a huge literature of the theory and practice of optical fibers for communications. Our discussion in Sections 8.10 and 8.11 has benefited from the comprehensive book

A. W. Snyder and J. D. Love, *Optical Waveguide Theory*, Chapman & Hall, New York (1983).

Books with discussions of the waveguide aspects, as well as much practical detail, are

J. M. Senior, *Optical Fibre Communications*, 2nd ed., Prentice-Hall, New York (1992).
C. Vassallo, *Optical Waveguide Concepts*, Elsevier, New York (1991).

Numerical methods are often required for optical waveguide geometries. A useful reference is

F. A. Fernández and Y. Lu, *Microwave and Optical Waveguide Analysis by the Finite Element Method*, Research Studies Press & Wiley, New York (1996).

Problems

8.1 Consider the electric and magnetic fields in the surface region of an excellent conductor in the approximation of Section 8.1, where the skin depth is very small compared to the radii of curvature of the surface or the scale of significant spatial variation of the fields just outside.

(a) For a single-frequency component, show that the magnetic field $\mathbf{H}$ and the current density $\mathbf{J}$ are such that $\mathbf{f}$, the time-averaged force per unit area at the surface, is given by

$$\mathbf{f} = -\mathbf{n}\,\frac{\mu_c}{4}\,|\mathbf{H}_\parallel|^2$$

where $\mathbf{H}_\parallel$ is the peak parallel component of magnetic field at the surface, μ_c is the magnetic permeability of the conductor, and $\mathbf{n}$ is the outward normal at the surface.

(b) Compare this result with the force calculated with the idealized surface current and magnetic field at the surface, treating the material as a *perfect* conductor. Comment on the comparison.

(c) Assume that the fields are a superposition of different frequencies (all high enough that the approximations still hold). Show that the time-averaged force takes the same form as in part a with $|\mathbf{H}_{\parallel}|^2$ replaced by $2\langle|\mathbf{H}_{\parallel}|^2\rangle$, where the angle brackets $\langle\cdots\rangle$ mean time average.

8.2   A transmission line consisting of two concentric circular cylinders of metal with conductivity σ and skin depth δ, as shown, is filled with a uniform lossless dielectric (μ, ϵ). A TEM mode is propagated along this line.

(a)   Show that the time-averaged power flow along the line is

$$P = \sqrt{\frac{\mu}{\epsilon}}\, \pi a^2 \, |H_0|^2 \, \ln\left(\frac{b}{a}\right)$$

where H_0 is the peak value of the azimuthal magnetic field at the surface of the inner conductor.

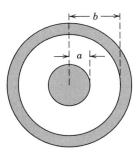

Problem 8.2

(b)   Show that the transmitted power is attenuated along the line as

$$P(z) = P_0 e^{-2\gamma z}$$

where

$$\gamma = \frac{1}{2\sigma\delta} \sqrt{\frac{\epsilon}{\mu}} \, \frac{\left(\dfrac{1}{a} + \dfrac{1}{b}\right)}{\ln\left(\dfrac{b}{a}\right)}$$

(c)   The characteristic impedance Z_0 of the line is defined as the ratio of the voltage between the cylinders to the axial current flowing in one of them at any position z. Show that for this line

$$Z_0 = \frac{1}{2\pi} \sqrt{\frac{\mu}{\epsilon}} \ln\left(\frac{b}{a}\right)$$

(d)   Show that the series resistance and inductance per unit length of the line are

$$R = \frac{1}{2\pi\sigma\delta}\left(\frac{1}{a} + \frac{1}{b}\right)$$

$$L = \left\{\frac{\mu}{2\pi} \ln\left(\frac{b}{a}\right) + \frac{\mu_c \delta}{4\pi}\left(\frac{1}{a} + \frac{1}{b}\right)\right\}$$

where μ_c is the permeability of the conductor. The correction to the inductance comes from the penetration of the flux into the conductors by a distance of order δ.

8.3 **(a)** A transmission line consists of two identical thin strips of metal, shown in cross section in the sketch. Assuming that $b \gg a$, discuss the propagation of a TEM mode on this line, repeating the derivations of Problem 8.2. Show that

$$P = \frac{ab}{2} \sqrt{\frac{\mu}{\epsilon}} |H_0|^2$$

$$\gamma = \frac{1}{a\sigma\delta} \sqrt{\frac{\epsilon}{\mu}}$$

$$Z_0 = \sqrt{\frac{\mu}{\epsilon}} \left(\frac{a}{b}\right)$$

$$R = \frac{2}{\sigma\delta b}$$

$$L = \left(\frac{\mu a + \mu_c \delta}{b}\right)$$

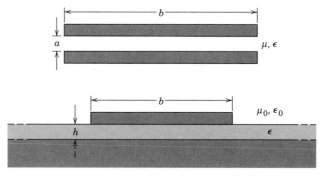

Problem 8.3

where the symbols on the left have the same meanings as in Problem 8.2.

(b) The lower half of the figure shows the cross section of a microstrip line with a strip of width b mounted on a dielectric substrate of thickness h and dielectric constant ϵ, all on a ground plane. What differences occur here compared to part a if $b \gg h$? If $b \ll h$?

8.4 Transverse electric and magnetic waves are propagated along a hollow, right circular cylinder of brass with inner radius R.

(a) Find the cutoff frequencies of the various TE and TM modes. Determine numerically the lowest cutoff frequency (the dominant mode) in terms of the tube radius and the ratio of cutoff frequencies of the next four higher modes to that of the dominant mode. For this part assume that the conductivity of brass is infinite.

(b) Calculate the attenuation constant of the waveguide as a function of frequency for the lowest two modes and plot it as a function of frequency.

8.5 A waveguide is constructed so that the cross section of the guide forms a right triangle with sides of length a, a, $\sqrt{2}a$, as shown. The medium inside has $\mu_r = \epsilon_r = 1$.

(a) Assuming infinite conductivity for the walls, determine the possible modes of propagation and their cutoff frequencies.

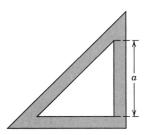

Problem 8.5

(b) For the lowest modes of each type calculate the attenuation constant, assuming that the walls have large, but finite, conductivity. Compare the result with that for a square guide of side a made from the same material.

8.6 A resonant cavity of copper consists of a hollow, right circular cylinder of inner radius R and length L, with flat end faces.

(a) Determine the resonant frequencies of the cavity for all types of waves. With $(1/\sqrt{\mu\epsilon}\, R)$ as a unit of frequency, plot the lowest four resonant frequencies of each type as a function of R/L for $0 < R/L < 2$. Does the same mode have the lowest frequency for all R/L?

(b) If $R = 2$ cm, $L = 3$ cm, and the cavity is made of pure copper, what is the numerical value of Q for the lowest resonant mode?

8.7 A resonant cavity consists of the empty space between two perfectly conducting, concentric spherical shells, the smaller having an outer radius a and the larger an inner radius b. As shown in Section 8.9, the azimuthal magnetic field has a radial dependence given by spherical Bessel functions, $j_l(kr)$ and $n_l(kr)$, where $k = \omega/c$.

(a) Write down the transcendental equation for the characteristic frequencies of the cavity for arbitrary l.

(b) For $l = 1$ use the explicit forms of the spherical Bessel functions to show that the characteristic frequencies are given by

$$\frac{\tan kh}{kh} = \frac{\left(k^2 + \dfrac{1}{ab}\right)}{k^2 + ab\left(k^2 - \dfrac{1}{a^2}\right)\left(k^2 - \dfrac{1}{b^2}\right)}$$

where $h = b - a$.

(c) For $h/a \ll 1$, verify that the result of part b yields the frequency found in Section 8.9, and find the first order correction in h/a. [The result of part b seems to have been derived first by J. J. Thomson and published in his book *Recent Researches in Electricity and Magnetism*, Oxford Clarendon Press, 1893, pp. 373 ff.]

8.8 For the Schumann resonances of Section 8.9 calculate the Q values on the assumption that the earth has a conductivity σ_e and the ionosphere has a conductivity σ_i, with corresponding skin depths δ_e and δ_i.

(a) Show that to lowest order in h/a the Q value is given by $Q = Nh/(\delta_e + \delta_i)$ and determine the numerical factor N for all l.

(b) For the lowest Schumann resonance evaluate the Q value assuming $\sigma_e = 0.1$ $(\Omega m)^{-1}$, $\sigma_i = 10^{-5}$ $(\Omega m)^{-1}$, $h = 10^2$ km.

(c) Discuss the validity of the approximations used in part a for the range of parameters used in part b.

8.9 A hollow volume V containing a uniform isotropic linear medium (ϵ, μ) is bounded by a perfectly conducting closed surface S (which may have more than one disconnected part). A harmonic electric field inside the cavity satisfies the vector Helmholtz equation,

$$\nabla \times (\nabla \times \mathbf{E}) = k^2 \mathbf{E} \qquad \text{with } k^2 = \omega^2 \mu \epsilon$$

The boundary condition is $\mathbf{n} \times \mathbf{E} = 0$ (and $\mathbf{n} \cdot \mathbf{B} = 0$) on S.

(a) Show that

$$k^2 = \frac{\displaystyle\int_V \mathbf{E}^* \cdot [\nabla \times (\nabla \times \mathbf{E})] \, d^3x}{\displaystyle\int_V \mathbf{E}^* \cdot \mathbf{E} \, d^3x}$$

is a variational principle for the eigenvalue k^2 in the sense that a change of $\mathbf{E} \to \mathbf{E} + \delta\mathbf{E}$, where both $\mathbf{E}$ and $\delta\mathbf{E}$ satisfy the boundary conditions on S, leads to only second-order changes in k^2.

(b) Apply the variational principle to the TM_{010} mode of a right cylindrical cavity of radius R and length d, using the trial longitudinal electric field $E_z = E_0 \cos(\pi\rho/2R)$ [no variational parameters]. Show that the estimate of the eigenvalue is

$$kR = \frac{\pi}{2}\sqrt{\frac{\pi^2 + 4}{\pi^2 - 4}}$$

Compare numerically with the known eigenvalue, the root x_{01} of $J_0(x)$.

(c) Repeat the calculation of part b with $E_z = E_0 [1 + \alpha(\rho/R)^2 - (1 + \alpha)(\rho/R)^4]$, where α is a variational parameter. Show that for this trial function the best estimate is

$$kR = \left[80 \cdot \left(\frac{17 - 2\sqrt{34}}{68 + \sqrt{34}}\right) \right]^{1/2}$$

How much better is this truly variational calculation than part b?

8.10 Use the variational principle of Problem 8.9 in terms of the electric field $\mathbf{E}$ to find an estimate of the eigenvalue for k^2 for the TE_{111} mode in a right circular cylinder cavity of radius R and length d with perfectly conducting walls. Use as a trial function $B_z = B_0(\rho/R)(1 - \rho/2R)\cos\phi\sin(\pi z/d)$. [This function satisfies the boundary conditions of $B_z = 0$ at $z = 0$ and $z = d$, and $\partial B_z/\partial\rho = 0$ at $\rho = R$.]

(a) First show that the variational principle can be reexpressed as

$$k^2 = \frac{\displaystyle\int_V (\nabla \times \mathbf{E}^*) \cdot (\nabla \times \mathbf{E}) \, d^3x}{\displaystyle\int_V \mathbf{E}^* \cdot \mathbf{E} \, d^3x}$$

(b) Show that the (transverse) components of the (trial) electric field are

$$E_\rho = B_0(1 - \rho/2R)\sin\phi\sin(\pi z/d); \qquad E_\phi = B_0(1 - \rho/R)\cos\phi\sin(\pi z/d)$$

(c) Calculate the curl of $\mathbf{E}$ and show that the approximation for k^2 is

$$k^2 = \frac{18}{5R^2} + \frac{\pi^2}{d^2}$$

Compare with the exact result. For small enough d/R, this mode has a larger eigenvalue than the TM_{010} mode. Why should the present variational estimate be at all reliable?

(d) The original variational expression in Problem 8.9 has an equivalent integrand in the numerator, $\mathbf{E}^* \cdot [\nabla(\nabla \cdot \mathbf{E}) - \nabla^2 \mathbf{E}]$. Discuss the relative merits of this integrand compared with the square of the curl of $\mathbf{E}$ in part a for the present problem.

8.11 Apply the variational method of Problem 8.9 to estimate the resonant frequency of the lowest TM mode in a "breadbox" cavity with perfectly conducting walls, of length d in the z direction, radius R for the curved quarter-circle "front" of the breadbox, and the "bottom" and "back" of the box defined by the plane segments $(y = 0, 0 < x < R)$ and $(x = 0, 0 < y < R)$, respectively. Use the trial function,

$$E_z = E_0(\rho/R)^\nu(1 - \rho/R)\sin 2\phi$$

for the only component of electric field present. This function gives vanishing tangential component of $\mathbf{E}$ on the boundary surfaces; the index ν is a variational parameter. Show that

$$k^2 R^2 = \frac{(\nu + 2)(2\nu + 3)(\nu^2 + \nu + 4)}{\nu(2\nu + 1)}$$

Minimize with respect to ν to find the best estimate of kR from the given trial function. Compare with the exact answer, $kR = 5.13562$, the first root of $J_2(x)$.

8.12 A waveguide with lossless dielectric inside and perfectly conducting walls has a cross-sectional contour C that departs slightly from a comparison contour C_0 whose fields are known. The difference in boundaries is described by $\delta(x, y)$, the length measured from C_0 to C along the normal to C_0 at the boundary point (x, y). The derivative $d\delta/ds$ along the boundary is higher order in small quantities.

(a) If the eigenvalue parameters and solutions for C and C_0 are (γ^2, ψ) and (γ_0^2, ψ_0), respectively, without degeneracy, show that to first order in δ

$$\gamma^2 - \gamma_0^2 = -\frac{\oint_{C_0} \delta(x, y)\left[\left|\frac{\partial\psi_0}{\partial n}\right|^2 - \psi_0^*\frac{\partial^2\psi_0}{\partial n^2}\right]dl}{\int_{S_0}|\psi_0|^2\,dx\,dy}$$

where only the first (second) term in the numerator occurs for TM (TE) modes. [*Hint:* Follow the same general approach as used in Section 8.6 for the effects of finite conductivity.]

(b) Determine the perturbed value of γ^2 for the lowest TE and TM modes ($TE_{1,0}$ and $TM_{1,1}$) in a rectangular guide if the change in shape is as shown in the accompanying sketch.

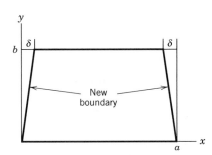

Problem 8.12

8.13 To treat perturbations if there is a degeneracy of modes in guides or cavities under ideal conditions, one must use degenerate-state perturbation theory. Consider the two-dimensional (waveguide) situation in which there is an N-fold degeneracy in the ideal circumstances (or perfect conductivity or chosen shape of cross section), with no other nearby modes. There are N linearly independent solutions $\psi_0^{(i)}$, chosen to be orthogonal, to the transverse wave equation, $(\nabla_t^2 + \gamma_0^2)\psi_0^{(i)} = 0$, $i = 1, 2, \ldots,$ N. In response to the perturbation, the degeneracy is in general lifted. There is a set of perturbed eigenvalues, γ_k^2, with associated eigenmodes, ψ_k, which can be expanded (in lowest order) in terms of the N unperturbed eigenmodes: $\psi_k = \Sigma_i a_i \psi_0^{(i)}$.

(a) Show that the generalization of (8.68) for finite conductivity (and the corresponding expression in Problem 8.12 on distortion of the shape of a waveguide) is the set of algebraic equations,

$$\sum_{i=1}^{N} [(\gamma^2 - \gamma_0^2)N_j\delta_{ji} + \Delta_{ji}]a_i = 0 \qquad (j = 1, 2, \ldots, N)$$

where

$$N_j = \int_A |\psi_0^{(j)}|^2 \, da \quad \text{and} \quad \Delta_{ji} = f \oint_C \frac{\partial \psi_0^{(j)*}}{\partial n} \frac{\partial \psi_0^{(i)}}{\partial n} \, dl$$

for finite conductivity, and

$$\Delta_{ji} = \oint_C \delta(x, y) \left[\frac{\partial \psi_0^{(j)*}}{\partial n} \frac{\partial \psi_0^{(i)}}{\partial n} - \psi_0^{(j)*} \frac{\partial^2 \psi_0^{(i)}}{\partial n^2}\right] dl$$

for distortion of the boundary shape.

(b) The lowest mode in a circular guide of radius R is the twofold degenerate TE_{11} mode, with fields given by

$$\psi^{(\pm)} = B_z = B_0 J_1(\gamma_0\rho) \exp(\pm i\phi) \exp(ikz - i\omega t)$$

The eigenvalue parameter is $\gamma_0 = 1.841/R$, corresponding to the first root of $dJ_1(x)/dx$. Suppose that the circular waveguide is distorted along its length into an elliptical shape with semimajor and semiminor axes, $a = R + \Delta R$, $b = R - \Delta R$, respectively. To first order in $\Delta R/R$, the area and circumference of the guide remain unchanged. Show that the degeneracy is lifted by the distortion and that to first order in $\Delta R/R$, $\gamma_1^2 = \gamma_0^2(1 + \lambda\Delta R/R)$ and $\gamma_2^2 = \gamma_0^2(1 - \lambda\Delta R/R)$. Determine the numerical value of λ and find the eigenmodes as linear combinations of $\psi^{(\pm)}$. Explain physically why the eigenmodes turn out as they do.

8.14 Consider an optical fiber with a graded index of refraction for rays confined to the x-z plane, $n(x) = n(0) \, \mathrm{sech}(\alpha x)$. The fiber has large enough transverse dimensions (x) to contain all rays of interest, which are evidently symmetric about $x = 0$. The invariant $\bar{n} = n(x_{max}) = n(0) \cos\theta(0)$.

(a) Solve the eikonal equations for the transverse coordinate $x(z)$ of the ray and show that

$$\alpha x = \sinh^{-1}[\sinh(\alpha x_{max}) \sin(\alpha z)]$$

where the origin in z is chosen when the ray has $x = 0$. Sketch rays over one half-period for "launch angles" $\theta(0) = \pi/6$, $\pi/4$, and $\pi/3$.

(b) Find the half-period Z of the ray. Does it depend on $\bar{n}$?

(c) Show that the optical path length for a half period, $L_{opt} = \int n(x) \, ds$ is $L_{opt} = n(0)Z$. Comment on the effectiveness of this particular grading of the index.

Hint: In the computation of L_{opt}, a useful change of variables is $\sinh(\alpha x) = \sinh(\alpha x_{max}) \sin t$. The resulting integral can be done by contour integration:

$$\int_0^{\pi/2} \frac{d\theta}{1 + a^2 \sin^2\theta} = \frac{\pi}{2\sqrt{1 + a^2}}$$

8.15 Discuss the TE and TM modes in the dielectric slab waveguide of Section 8.11.A as a boundary-value problem.

    (a) Show that (8.123) emerges as the determining relation for both even and odd modes (in x) and that even or odd p goes with the evenness or oddness of the mode.

    (b) Show that the eigenvalues of ξ for the TE modes are given approximately by

$$\xi \approx \frac{(p + 1)\pi}{2(V + 1)} \left[1 - \frac{(p + 1)^2 \pi^2}{24(V + 1)^3} \right]$$

The lowest order result is accurate for $V \gg 1$ and small p. Check the accuracy of the full expression against solution of (8.123) by Newton's method for $V = 1, 2, 3$.

    (c) Calculate the power flow in the z direction (per unit length in the y direction) within the core ($|x| < a$) and in the cladding ($|x| > a$) for the even TE modes and show that the fractions are

$$F_{core} = \frac{1}{S} \left[1 + \frac{\sin(2V\xi)}{2V\xi} \right] \quad \text{and} \quad F_{clad} = \frac{1}{S} \left[\frac{\cos^2(V\xi)}{V\sqrt{1 - \xi^2}} \right]$$

where

$$S = \left[1 + \frac{\sin(2V\xi)}{2V\xi} \right] + \left[\frac{\cos^2(V\xi)}{V\sqrt{1 - \xi^2}} \right]$$

where ξ is the root of (8.123) for the pth mode. Find corresponding expressions for the odd TE modes.

8.16 The longitudinal phase velocity in the dielectric slab waveguide of the preceding problem is $v_p = \omega/k_z = c/(n_1 \cos\theta_p)$. Intermodal dispersion occurs because the dielectric media have dispersion and also because the group velocity differs intrinsically for different modes.

    (a) Making the approximation that the dielectrics' dispersion can be neglected, show that the group velocity $v_g = d\omega/dk_z$ for the TE$_p$ mode is

$$v_g = \frac{c \cos\theta_p}{n_1} \left[\frac{1 + \beta_p a}{\cos^2\theta_p + \beta_p a} \right]$$

where θ_p is the eigenangle of the pth mode ($\cos\theta_p = \sqrt{1 - 2\Delta\xi_p^2}$) and β_p is given by (8.124). Interpret the departure from $v_p v_g = c^2/n_1^2$ (as in metallic waveguides; $\beta_p \to \infty$) in terms of the Goos–Hänchen effect and ray-like propagation at the simple phase speed c/n_1. [*Hint:* Write the eigenvalue relation (8.121) in terms of the independent variable ω and the dependent variable k_z and differentiate with respect to ω.]

    (b) Write a program to evaluate v_g versus V/V_t, where $V_t = p\pi/2$ is the threshold frequency variable for the pth mode. Make a plot of v_g/c for $n_1 = 1.5, n_2 = 1.0$ for $p = 0, 1, 2, \ldots, 6$ as a function of $V/V_t(p = 1)$ on the range $(0, 10)$.

    (c) Relate the results of part b to the optical path length difference for the step-index fiber shown in Fig. 8.12b. Can you generate a plot from the results for $v_g(p)$ at fixed V for $n_1 = 1.01, n_2 = 1.0$ to compare with the "classical" ray result?

8.17 Consider the propagating modes in a cylindrical optical fiber waveguide of radius a with a step index of refraction, n_1 in the core ($\rho < a$) and $n_2 < n_1$ in the cladding ($\rho > a$). Assume that the fields vary as $e^{im\phi + ik_z z - i\omega t}$. For bound modes, the fields in the core (cladding) are proportional to ordinary (modified) Bessel functions $J_\nu(K_\nu)$ with appropriate values of ν and argument, as in (8.128).

(a) Show that for $m \neq 0$ the eigenvalue relation for the transverse parameter γ_{mn} (and β_{mn}) is

$$\left(\frac{n_1^2}{\gamma}\frac{J_m'}{J_m} + \frac{n_2^2}{\beta}\frac{K_m'}{K_m}\right)\left(\frac{1}{\gamma}\frac{J_m'}{J_m} + \frac{1}{\beta}\frac{K_m'}{K_m}\right) = \frac{m^2}{a^2}\left(\frac{n_1^2}{\gamma^2} + \frac{n_2^2}{\beta^2}\right)\left(\frac{1}{\gamma^2} + \frac{1}{\beta^2}\right)$$

where $\gamma^2 = n_1^2\omega^2/c^2 - k_z^2$ and $\beta^2 = k_z^2 - n_2^2\omega^2/c^2$, while primes indicate derivatives with respect to the argument, and the argument of J_m (K_m) is γa (βa). The first subscript on γ is the azimuthal index m; the second designates the nth root of the eigenvalue equation for fixed m.

(b) Determine the eigenvalue equation for the $m = 0$ modes (TE and TM) and show that the lowest "cutoff" frequency corresponds to $V = 2.405$, the first root of $J_0(x)$, where "cutoff" is the frequency below which the guide radiates rather than confines.

(c) Show that the lowest HE mode (HE_{11}) has no cutoff frequency and that for $V \ll 1$ the decay parameter $\beta a \approx Ae^{-B/V^2}$. Find A and B in terms of n_1 and n_2.

8.18 **(a)** From the use of Green's theorem in two dimensions show that the TM and TE modes in a waveguide defined by the boundary-value problems (8.34) and (8.36) are orthogonal in the sense that

$$\int_A E_{z\lambda}E_{z\mu}\,da = 0 \qquad \text{for } \lambda \neq \mu$$

for TM modes, and a corresponding relation for H_z for TE modes.

(b) Prove that the relations (8.131)–(8.134) form a consistent set of normalization conditions for the fields, including the circumstances when λ is a TM mode and μ is a TE mode.

8.19 The figure shows a cross-sectional view of an infinitely long rectangular waveguide with the center conductor of a coaxial line extending vertically a distance h into its interior at $z = 0$. The current along the probe oscillates sinusoidally in time with frequency ω, and its variation in space can be approximated as $I(y) = I_0 \sin[(\omega/c)(h - y)]$. The thickness of the probe can be neglected. The frequency is such that only the TE_{10} mode can propagate in the guide.

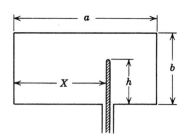

Problem 8.19

(a) Calculate the amplitudes for excitation of both TE and TM modes for all (m, n) and show how the amplitudes depend on m and n for $m, n \gg 1$ for a fixed frequency ω.

(b) For the propagating mode show that the power radiated in the positive z direction is

$$P = \frac{\mu c^2 I_0^2}{\omega kab} \sin^2\left(\frac{\pi X}{a}\right) \sin^4\left(\frac{\omega h}{2c}\right)$$

with an equal amount in the opposite direction. Here k is the wave number for the TE_{10} mode.

(c) Discuss the modifications that occur if the guide, instead of running off to infinity in both directions, is terminated with a perfectly conducting surface at $z = L$. What values of L will maximize the power flow for a fixed current I_0? What is the radiation resistance of the probe (defined as the ratio of power flow to one-half the square of the current at the base of the probe) at maximum?

8.20 An infinitely long rectangular waveguide has a coaxial line terminating in the short side of the guide with the thin central conductor forming a semicircular loop of radius R whose center is a height h above the floor of the guide, as shown in the accompanying cross-sectional view. The half-loop is in the plane $z = 0$ and its radius R is sufficiently small that the current can be taken as having a constant value I_0 everywhere on the loop.

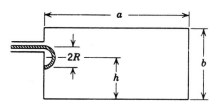

Problem 8.20

(a) Prove that to the extent that the current is constant around the half-loop, the TM modes are not excited. Give a physical explanation of this lack of excitation.

(b) Determine the amplitude for the lowest TE mode in the guide and show that its value is independent of the height h.

(c) Show that the power radiated in either direction in the lowest TE mode is

$$P = \frac{I_0^2}{16} Z \frac{a}{b} \left(\frac{\pi R}{a}\right)^4$$

where Z is the wave impedance of the TE_{10} mode. Here assume $R \ll a, b$.

8.21 A hollow metallic waveguide with a distortion in the form of a localized bend or increase in cross section can support nonpropagating ("bound state") configurations of fields in the vicinity of the distortion. Consider a rectangular guide that has its distortion confined to a plane, as shown in the figure, and TE $_{10}$ as its lowest propagating mode, with perpendicular electric field $E_\perp = \psi$. On either side of the distortion the guide is straight and of width a. Without distortion, $\psi = E_0 \sin(\pi y/a) \exp(\pm ikz)$, where $k^2 = (\omega/c)^2 - (\pi/a)^2$. The distortion is described by a curvature $\kappa(s) = 1/R(s)$ and a width $w(s)$. Locally the element of area in the plane is $dA = h(s, t)\, ds\, dt$, where s is the length along the guide wall and t the transverse coordinate, as shown in the figure, and $h(s, t) = 1 - \kappa(s)t$. In terms of s and t the Laplacian is

$$\nabla^2 \psi = \frac{1}{h}\frac{\partial}{\partial t}\left(h\frac{\partial \psi}{\partial t}\right) + \frac{1}{h}\frac{\partial}{\partial s}\left(\frac{1}{h}\frac{\partial \psi}{\partial s}\right)$$

If the distortions are very small and change slowly in s on the scale of the width a, an ansatz for the solution is

$$\psi(s,\,t) = \frac{u(s)}{\sqrt{h(s,\,t)}} \cdot \sin\left[\frac{\pi t}{w(s)}\right]$$

[The factor in the denominator is equivalent to the factor $\rho^{-1/2}$ familiar from Bessel functions that converts the radial part of the Laplacian in polar coordinates to a simple second partial derivative (plus an additional term without derivatives).]

(a) Show that substitution of the ansatz into the two-dimensional wave equation, $(\nabla^2 + \omega^2/c^2)\psi = 0$, leads to the equation for $u(s)$,

$$\frac{d^2u}{ds^2} + [k^2 - v(s)]u = 0 \text{ with } v(s) = \pi^2\left(\frac{1}{w^2(s)} - \frac{1}{a^2}\right) - \frac{1}{4}\kappa^2(s)$$

if small terms are neglected. Interpret $v(s)$ in analogy with the Schrödinger equation in one dimension.

(b) If the distortion is in the form of a bend through an angle θ with constant radius of curvature $R \gg a$, show that for $\theta a/R \ll 1$ there is a "bound state" at frequency ω_0 where

$$\omega_0^2 \approx \left(\frac{\pi c}{a}\right)^2\left[1 - \left(\frac{\theta a}{8\pi R}\right)^2\right]$$

References: J. Goldstone and R. L. Jaffe, *Phys. Rev. B* **45**, 14100 (1992); J. P. Carini, J. T. Londergan, K. Mullen, and D. P. Murdock, *Phys. Rev. B* **48**, 4503 (1993).

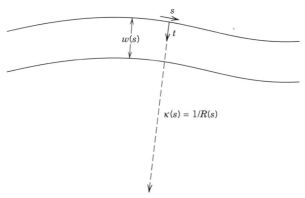

Problem 8.21

CHAPTER 9

Radiating Systems, Multipole Fields and Radiation

In Chapters 7 and 8 we discussed the properties of electromagnetic waves and their propagation in both bounded and unbounded geometries, but very little was said about the generation of such waves. In the present chapter we turn to this question and discuss the emission of radiation by localized systems of oscillating charge and current densities. The initial treatment is straightforward, without elaborate formalism. It addresses simple systems in which electric dipole, magnetic dipole, or electric quadrupole radiation dominates, or the sources are sufficiently simple that direct evaluation of the radiation fields is easy. The simple multipole expansion of a source in a waveguide is also treated, and the effective multipole moments of apertures. These "elementary" discussions are followed by the systematic development of multipole fields of arbitrary order (l, m) and the derivation of exact formulas for multipole radiation of any order by localized harmonic systems. Some comparisons of the simple and systematic approaches are made. Applications to scattering are presented in Chapter 10, along with diffraction and the optical theorem. Considerations of the relativistic Liénard–Wiechert fields and radiation by rapidly moving charged particles are deferred to Chapters 14 and 15.

9.1 Fields and Radiation of a Localized Oscillating Source

For a system of charges and currents varying in time we can make a Fourier analysis of the time dependence and handle each Fourier component separately. We therefore lose no generality by considering the potentials, fields, and radiation from a localized system of charges and currents that vary sinusoidally in time:

$$\rho(\mathbf{x}, t) = \rho(\mathbf{x})e^{-i\omega t}$$
$$\mathbf{J}(\mathbf{x}, t) = \mathbf{J}(\mathbf{x})e^{-i\omega t} \tag{9.1}$$

As usual, the real part of such expressions is to be taken to obtain physical quantities.* The electromagnetic potentials and fields are assumed to have the same time dependence. The sources are located in otherwise empty space.

---

*See Problem 9.1 for some of the subtleties that can arise over factors of 2. There are also factors of 2 in the correspondence between classical and quantum-mechanical quantities. For example, in a one-electron atom our classical dipole moment $\mathbf{p}$ is replaced by $2e\langle f|\mathbf{r}|i\rangle$ for a transition from state i to state f.

It was shown in Chapter 6 that the solution for the vector potential $\mathbf{A}(\mathbf{x}, t)$ in the Lorenz gauge is

$$\mathbf{A}(\mathbf{x}, t) = \frac{\mu_0}{4\pi} \int d^3x' \int dt' \, \frac{\mathbf{J}(\mathbf{x}', t')}{|\mathbf{x} - \mathbf{x}'|} \, \delta\!\left(t' + \frac{|\mathbf{x} - \mathbf{x}'|}{c} - t\right) \qquad (9.2)$$

provided no boundary surfaces are present. The Dirac delta function assures the causal behavior of the fields. With the sinusoidal time dependence (9.1), the solution for $\mathbf{A}$ becomes

$$\mathbf{A}(\mathbf{x}) = \frac{\mu_0}{4\pi} \int \mathbf{J}(\mathbf{x}') \, \frac{e^{ik|\mathbf{x}-\mathbf{x}'|}}{|\mathbf{x} - \mathbf{x}'|} \, d^3x' \qquad (9.3)$$

where $k = \omega/c$ is the wave number, and a sinusoidal time dependence is understood. The magnetic field is given by

$$\mathbf{H} = \frac{1}{\mu_0} \boldsymbol{\nabla} \times \mathbf{A} \qquad (9.4)$$

while, outside the source, the electric field is

$$\mathbf{E} = \frac{iZ_0}{k} \boldsymbol{\nabla} \times \mathbf{H} \qquad (9.5)$$

where $Z_0 = \sqrt{\mu_0/\epsilon_0}$ is the impedance of free space.

Given a current distribution $\mathbf{J}(\mathbf{x}')$, the fields can, in principle at least, be determined by calculating the integral in (9.3). We will consider one or two examples of direct integration of the source integral in Section 9.4. But at present we wish to establish certain simple, but general, properties of the fields in the limit that the source of current is confined to a small region, very small in fact compared to a wavelength. If the source dimensions are of order d and the wavelength is $\lambda = 2\pi c/\omega$, and if $d \ll \lambda$, then there are three spatial regions of interest:

The near (static) zone: $d \ll r \ll \lambda$
The intermediate (induction) zone: $d \ll r \sim \lambda$
The far (radiation) zone: $d \ll \lambda \ll r$

We will see that the fields have very different properties in the different zones. In the near zone the fields have the character of static fields, with radial components and variation with distance that depend in detail on the properties of the source. In the far zone, on the other hand, the fields are transverse to the radius vector and fall off as r^{-1}, typical of radiation fields.

For the near zone where $r \ll \lambda$ (or $kr \ll 1$) the exponential in (9.3) can be replaced by unity. Then the vector potential is of the form already considered in Chapter 5. The inverse distance can be expanded using (3.70), with the result,

$$\lim_{kr \to 0} \mathbf{A}(\mathbf{x}) = \frac{\mu_0}{4\pi} \sum_{l,m} \frac{4\pi}{2l + 1} \frac{Y_{lm}(\theta, \phi)}{r^{l+1}} \int \mathbf{J}(\mathbf{x}') r'^{l} Y_{lm}^*(\theta', \phi') \, d^3x' \qquad (9.6)$$

This shows that the near fields are quasi-stationary, oscillating harmonically as $e^{-i\omega t}$, but otherwise static in character.

In the far zone $(kr \gg 1)$ the exponential in (9.3) oscillates rapidly and determines the behavior of the vector potential. In this region it is sufficient to approximate*

$$|\mathbf{x} - \mathbf{x}'| \simeq r - \mathbf{n} \cdot \mathbf{x}' \tag{9.7}$$

where $\mathbf{n}$ is a unit vector in the direction of $\mathbf{x}$. Furthermore, if only the leading term in kr is desired, the inverse distance in (9.3) can be replaced by r. Then the vector potential is

$$\lim_{kr \to \infty} \mathbf{A}(\mathbf{x}) = \frac{\mu_0}{4\pi} \frac{e^{ikr}}{r} \int \mathbf{J}(\mathbf{x}') e^{-ik\mathbf{n}\cdot\mathbf{x}'} \, d^3x' \tag{9.8}$$

This demonstrates that in the far zone the vector potential behaves as an outgoing spherical wave with an angular dependent coefficient. It is easy to show that the fields calculated from (9.4) and (9.5) are transverse to the radius vector and fall off as r^{-1}. They thus correspond to radiation fields. If the source dimensions are small compared to a wavelength it is appropriate to expand the integral in (9.8) in powers of k:

$$\lim_{kr \to \infty} \mathbf{A}(\mathbf{x}) = \frac{\mu_0}{4\pi} \frac{e^{ikr}}{r} \sum_n \frac{(-ik)^n}{n!} \int \mathbf{J}(\mathbf{x}')(\mathbf{n} \cdot \mathbf{x}')^n \, d^3x' \tag{9.9}$$

The magnitude of the nth term is given by

$$\frac{1}{n!} \int \mathbf{J}(\mathbf{x}')(k\mathbf{n} \cdot \mathbf{x}')^n \, d^3x' \tag{9.10}$$

Since the order of magnitude of $\mathbf{x}'$ is d and kd is small compared to unity by assumption, the successive terms in the expansion of $\mathbf{A}$ evidently fall off rapidly with n. Consequently the radiation emitted from the source will come mainly from the first nonvanishing term in the expansion (9.9). We will examine the first few of these in the following sections.

In the intermediate or induction zone the two alternative approximations leading to (9.6) and (9.8) cannot be made; all powers of kr must be retained. Without marshalling the full apparatus of vector multipole fields, described in Sections 9.6 and beyond, we can abstract enough for our immediate purpose. The key result is the exact expansion (9.98) for the Green function appearing in (9.3). For points outside the source (9.3) then becomes

$$\mathbf{A}(\mathbf{x}) = \mu_0 ik \sum_{l,m} h_l^{(1)}(kr) Y_{lm}(\theta, \phi) \int \mathbf{J}(\mathbf{x}') j_l(kr') Y_{lm}^*(\theta', \phi') \, d^3x' \tag{9.11}$$

If the source dimensions are small compared to a wavelength, $j_l(kr')$ can be approximated by (9.88). Then the result for the vector potential is of the form of (9.6), but with the replacement,

$$\frac{1}{r^{l+1}} \to \frac{e^{ikr}}{r^{l+1}} [1 + a_1(ikr) + a_2(ikr)^2 + \cdots + a_l(ikr)^l] \tag{9.12}$$

---

*Actually (9.7) is valid for $r \gg d$, independent of the value of kr. It is therefore an adequate approximation even in the near zone.

The numerical coefficients a_i come from the explicit expressions for the spherical Hankel functions. The right-hand side of (9.12) shows the transition from the static-zone result (9.6) for $kr \ll 1$ to the radiation-zone form (9.9) for $kr \gg 1$.

Before discussing electric dipole and other types of radiation, we examine the question of electric monopole fields when the sources vary in time. The analog of (9.2) for the scalar potential is

$$\Phi(\mathbf{x}, t) = \frac{1}{4\pi\epsilon_0} \int d^3x' \int dt' \frac{\rho(\mathbf{x}', t')}{|\mathbf{x} - \mathbf{x}'|} \delta\left(t' + \frac{|\mathbf{x} - \mathbf{x}'|}{c} - t\right)$$

The electric monopole contribution is obtained by replacing $|\mathbf{x} - \mathbf{x}'| \to |\mathbf{x}| \equiv r$ under the integral. The result is

$$\Phi_{\text{monopole}}(\mathbf{x}, t) = \frac{q(t' = t - r/c)}{4\pi\epsilon_0}$$

where $q(t)$ is the total charge of the source. Since charge is conserved and a localized source is by definition one that does not have charge flowing into or away from it, the total charge q is independent of time. Thus the *electric monopole part* of the potential (and fields) of a localized source *is* of necessity *static*. The fields with harmonic time dependence $e^{-i\omega t}$, $\omega \neq 0$, have no monopole terms.

We now turn to the lowest order multipole fields for $\omega \neq 0$. Because these fields can be calculated from the vector potential alone via (9.4) and (9.5), we omit explicit reference to the scalar potential in what follows.

9.2 Electric Dipole Fields and Radiation

If only the first term in (9.9) is kept, the vector potential is

$$\mathbf{A}(\mathbf{x}) = \frac{\mu_0}{4\pi} \frac{e^{ikr}}{r} \int \mathbf{J}(\mathbf{x}')d^3x' \tag{9.13}$$

Examination of (9.11) and (9.12) shows that (9.13) is the $l = 0$ part of the series and that it is valid everywhere outside the source, not just in the far zone. The integral can be put in more familiar terms by an integration by parts:

$$\int \mathbf{J} \, d^3x' = -\int \mathbf{x}'(\nabla' \cdot \mathbf{J})d^3x' = -i\omega \int \mathbf{x}'\rho(\mathbf{x}')d^3x' \tag{9.14}$$

since from the continuity equation,

$$i\omega\rho = \nabla \cdot \mathbf{J} \tag{9.15}$$

Thus the vector potential is

$$\mathbf{A}(\mathbf{x}) = -\frac{i\mu_0\omega}{4\pi} \mathbf{p} \frac{e^{ikr}}{r} \tag{9.16}$$

where

$$\mathbf{p} = \int \mathbf{x}'\rho(\mathbf{x}')d^3x' \tag{9.17}$$

is the *electric dipole moment*, as defined in electrostatics by (4.8).

The electric dipole fields from (9.4) and (9.5) are

$$\mathbf{H} = \frac{ck^2}{4\pi} (\mathbf{n} \times \mathbf{p}) \frac{e^{ikr}}{r} \left(1 - \frac{1}{ikr} \right)$$

$$\mathbf{E} = \frac{1}{4\pi\epsilon_0} \left\{ k^2(\mathbf{n} \times \mathbf{p}) \times \mathbf{n} \frac{e^{ikr}}{r} + [3\mathbf{n}(\mathbf{n} \cdot \mathbf{p}) - \mathbf{p}] \left(\frac{1}{r^3} - \frac{ik}{r^2} \right) e^{ikr} \right\}$$

(9.18)

We note that the magnetic field is transverse to the radius vector at all distances, but that the electric field has components parallel and perpendicular to $\mathbf{n}$.

In the radiation zone the fields take on the limiting forms,

$$\mathbf{H} = \frac{ck^2}{4\pi} (\mathbf{n} \times \mathbf{p}) \frac{e^{ikr}}{r}$$

$$\mathbf{E} = Z_0 \mathbf{H} \times \mathbf{n}$$

(9.19)

showing the typical behavior of radiation fields.

In the near zone, on the other hand, the fields approach

$$\mathbf{H} = \frac{i\omega}{4\pi} (\mathbf{n} \times \mathbf{p}) \frac{1}{r^2}$$

$$\mathbf{E} = \frac{1}{4\pi\epsilon_0} [3\mathbf{n}(\mathbf{n} \cdot \mathbf{p}) - \mathbf{p}] \frac{1}{r^3}$$

(9.20)

The electric field, apart from its oscillations in time, is just the static electric dipole field (4.13). The magnetic field times Z_0 is a factor (kr) smaller than the electric field in the region where $kr \ll 1$. Thus the fields in the near zone are dominantly electric in nature. The magnetic field vanishes, of course, in the static limit $k \to 0$. Then the near zone extends to infinity.

The time-averaged power radiated per unit solid angle by the oscillating dipole moment $\mathbf{p}$ is

$$\frac{dP}{d\Omega} = \frac{1}{2} \text{Re}[r^2 \mathbf{n} \cdot \mathbf{E} \times \mathbf{H}^*]$$

(9.21)

where $\mathbf{E}$ and $\mathbf{H}$ are given by (9.19). Thus we find

$$\frac{dP}{d\Omega} = \frac{c^2 Z_0}{32\pi^2} k^4 |(\mathbf{n} \times \mathbf{p}) \times \mathbf{n}|^2$$

(9.22)

The state of polarization of the radiation is given by the vector inside the absolute value signs.* If the components of $\mathbf{p}$ all have the same phase, the angular distribution is a typical dipole pattern,

$$\frac{dP}{d\Omega} = \frac{c^2 Z_0}{32\pi^2} k^4 |\mathbf{p}|^2 \sin^2\theta$$

(9.23)

---

*In writing angular distributions of radiation we will always exhibit the polarization explicitly by writing the absolute square of a vector that is proportional to the electric field. If the angular distribution for some particular polarization is desired, it can then be obtained by taking the scalar product of the vector with the appropriate polarization vector before squaring.

where the angle θ is measured from the direction of $\mathbf{p}$. The total power radiated, independent of the relative phases of the components of $\mathbf{p}$, is

$$P = \frac{c^2 Z_0 k^4}{12\pi} |\mathbf{p}|^2 \tag{9.24}$$

A simple example of an electric dipole radiator is a center-fed, linear antenna whose length d is small compared to a wavelength. The antenna is assumed to be oriented along the z axis, extending from $z = (d/2)$ to $z = -(d/2)$ with a narrow gap at the center for purposes of excitation, as shown in Fig. 9.1. The current is in the same direction in each half of the antenna, having a value I_0 at the gap and falling approximately linearly to zero at the ends:

$$I(z)e^{-i\omega t} = I_0\left(1 - \frac{2|z|}{d}\right)e^{-i\omega t} \tag{9.25}$$

From the continuity equation (9.15) the linear-charge density ρ' (charge per unit length) is constant along each arm of the antenna, with the value,

$$\rho'(z) = \pm\frac{2iI_0}{\omega d} \tag{9.26}$$

the upper (lower) sign being appropriate for positive (negative) values of z. The dipole moment (9.17) is parallel to the z axis and has the magnitude

$$p = \int_{-(d/2)}^{(d/2)} z\rho'(z)\,dz = \frac{iI_0 d}{2\omega} \tag{9.27}$$

The angular distribution of radiated power is

$$\frac{dP}{d\Omega} = \frac{Z_0 I_0^2}{128\pi^2}(kd)^2\sin^2\theta \tag{9.28}$$

while the total power radiated is

$$P = \frac{Z_0 I_0^2 (kd)^2}{48\pi} \tag{9.29}$$

We see that for a fixed input current the power radiated increases as the square of the frequency, at least in the long-wavelength domain where $kd \ll 1$.

The coefficient of $I_0^2/2$ in (9.29) has the dimensions of a resistance and is called the *radiation resistance* R_{rad} of the antenna. It corresponds to the second term in (6.137) and is the total resistance of the antenna if the conductivity is

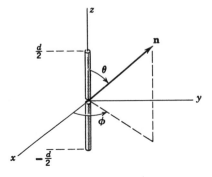

Figure 9.1 Short, center-fed, linear antenna.

perfect. For this short center-fed antenna $R_{rad} \simeq 5(kd)^2$ ohms. In principle the input reactance for the antenna can be calculated by applying (6.138) or (6.140) of Section 6.9. Unfortunately the calculation depends crucially on the strong fields near the gap and thus is sensitive to the exact shape and method of excitation. Since the system is an *electric* dipole and the electrostatic dipole field dominates near the antenna, we can nevertheless say with certainty that the reactance is negative (capacitive) for small kd.

9.3 *Magnetic Dipole and Electric Quadrupole Fields*

The next term in expansion (9.9) leads to a vector potential,

$$\mathbf{A}(\mathbf{x}) = \frac{\mu_0}{4\pi} \frac{e^{ikr}}{r} \left(\frac{1}{r} - ik \right) \int \mathbf{J}(\mathbf{x}')(\mathbf{n} \cdot \mathbf{x}') \, d^3x' \tag{9.30}$$

where we have included the correct terms from (9.12) to make (9.30) valid everywhere outside the source. This vector potential can be written as the sum of two terms: One gives a transverse magnetic induction and the other gives a transverse electric field. These physically distinct contributions can be separated by writing the integrand in (9.30) as the sum of a part symmetric in $\mathbf{J}$ and $\mathbf{x}'$ and a part that is antisymmetric. Thus

$$(\mathbf{n} \cdot \mathbf{x}')\mathbf{J} = \tfrac{1}{2}[(\mathbf{n} \cdot \mathbf{x}')\mathbf{J} + (\mathbf{n} \cdot \mathbf{J})\mathbf{x}'] + \tfrac{1}{2}(\mathbf{x}' \times \mathbf{J}) \times \mathbf{n} \tag{9.31}$$

The second, antisymmetric part is recognizable as the magnetization due to the current $\mathbf{J}$:

$$\mathcal{M} = \tfrac{1}{2}(\mathbf{x} \times \mathbf{J}) \tag{9.32}$$

The first, symmetric term will be shown to be related to the electric quadrupole moment density.

Considering only the magnetization term, we have the vector potential,

$$\mathbf{A}(\mathbf{x}) = \frac{ik\mu_0}{4\pi} (\mathbf{n} \times \mathbf{m}) \frac{e^{ikr}}{r} \left(1 - \frac{1}{ikr} \right) \tag{9.33}$$

where $\mathbf{m}$ is the *magnetic dipole moment*,

$$\mathbf{m} = \int \mathcal{M} d^3x = \tfrac{1}{2} \int (\mathbf{x} \times \mathbf{J}) \, d^3x \tag{9.34}$$

The fields can be determined by noting that the vector potential (9.33) is proportional to the magnetic field (9.18) for an electric dipole. This means that the magnetic field for the present magnetic dipole source will be equal to $1/Z_0$ times the electric field for the electric dipole, with the substitution $\mathbf{p} \to \mathbf{m}/c$. Thus we find

$$\mathbf{H} = \frac{1}{4\pi} \left\{ k^2(\mathbf{n} \times \mathbf{m}) \times \mathbf{n} \frac{e^{ikr}}{r} + [3\mathbf{n}(\mathbf{n} \cdot \mathbf{m}) - \mathbf{m}] \left(\frac{1}{r^3} - \frac{ik}{r^2} \right) e^{ikr} \right\} \tag{9.35}$$

Similarly, the electric field for a magnetic dipole source is the negative of Z_0 times the magnetic field for an electric dipole (with $\mathbf{p} \to \mathbf{m}/c$):

$$\mathbf{E} = -\frac{Z_0}{4\pi} k^2(\mathbf{n} \times \mathbf{m}) \frac{e^{ikr}}{r} \left(1 - \frac{1}{ikr} \right) \tag{9.36}$$

All the arguments concerning the behavior of the fields in the near and far zones are the same as for the electric dipole source, with the interchange $\mathbf{E} \to Z_0\mathbf{H}$, $Z_0\mathbf{H} \to -\mathbf{E}$, $\mathbf{p} \to \mathbf{m}/c$. Similarly the radiation pattern and total power radiated are the same for the two kinds of dipole. The only difference in the radiation fields is in the polarization. For an electric dipole the electric vector lies in the plane defined by $\mathbf{n}$ and $\mathbf{p}$, while for a magnetic dipole it is perpendicular to the plane defined by $\mathbf{n}$ and $\mathbf{m}$.

The integral of the symmetric term in (9.31) can be transformed by an integration by parts and some rearrangement:

$$\frac{1}{2} \int [(\mathbf{n} \cdot \mathbf{x}')\mathbf{J} + (\mathbf{n} \cdot \mathbf{J})\mathbf{x}'] \, d^3x' = -\frac{i\omega}{2} \int \mathbf{x}'(\mathbf{n} \cdot \mathbf{x}')\rho(\mathbf{x}') \, d^3x' \qquad (9.37)$$

The continuity equation (9.15) has been used to replace $\nabla \cdot \mathbf{J}$ by $i\omega\rho$. Since the integral involves second moments of the charge density, this symmetric part corresponds to an electric quadrupole source. The vector potential is

$$\mathbf{A}(\mathbf{x}) = -\frac{\mu_0 ck^2}{8\pi} \frac{e^{ikr}}{r} \left(1 - \frac{1}{ikr}\right) \int \mathbf{x}'(\mathbf{n} \cdot \mathbf{x}')\rho(\mathbf{x}') \, d^3x' \qquad (9.38)$$

The complete fields are somewhat complicated to write down. We content ourselves with the fields in the radiation zone. Then it is easy to see that

$$\left.\begin{array}{l} \mathbf{H} = ik\mathbf{n} \times \mathbf{A}/\mu_0 \\ \mathbf{E} = ikZ_0(\mathbf{n} \times \mathbf{A}) \times \mathbf{n}/\mu_0 \end{array}\right\} \qquad (9.39)$$

Consequently the magnetic field is

$$\mathbf{H} = -\frac{ick^3}{8\pi} \frac{e^{ikr}}{r} \int (\mathbf{n} \times \mathbf{x}')(\mathbf{n} \cdot \mathbf{x}')\rho(\mathbf{x}') \, d^3x' \qquad (9.40)$$

With definition (4.9) for the *quadrupole moment tensor*,

$$Q_{\alpha\beta} = \int (3x_\alpha x_\beta - r^2\delta_{\alpha\beta})\rho(\mathbf{x}) \, d^3x \qquad (9.41)$$

the integral in (9.40) can be written

$$\mathbf{n} \times \int \mathbf{x}'(\mathbf{n} \cdot \mathbf{x}')\rho(\mathbf{x}') \, d^3x' = \tfrac{1}{3}\mathbf{n} \times \mathbf{Q}(\mathbf{n}) \qquad (9.42)$$

The vector $\mathbf{Q}(\mathbf{n})$ is defined as having components,

$$Q_\alpha = \sum_\beta Q_{\alpha\beta}n_\beta \qquad (9.43)$$

We note that it depends in magnitude and direction on the direction of observation as well as on the properties of the source. With these definitions we have the magnetic induction,

$$\mathbf{H} = -\frac{ick^3}{24\pi} \frac{e^{ikr}}{r} \mathbf{n} \times \mathbf{Q}(\mathbf{n}) \qquad (9.44)$$

and the time-averaged power radiated per unit solid angle,

$$\frac{dP}{d\Omega} = \frac{c^2 Z_0}{1152\pi^2} k^6 \, |[\mathbf{n} \times \mathbf{Q}(\mathbf{n})] \times \mathbf{n}|^2 \qquad (9.45)$$

where again the direction of the radiated electric field is given by the vector inside the absolute value signs.

The general angular distribution is complicated. But the total power radiated can be calculated in a straightforward way. With the definition of $\mathbf{Q}(\mathbf{n})$ we can write the angular dependence as

$$|[\mathbf{n} \times \mathbf{Q}(\mathbf{n})] \times \mathbf{n}|^2 = \mathbf{Q}^* \cdot \mathbf{Q} - |\mathbf{n} \cdot \mathbf{Q}|^2$$
$$= \sum_{\alpha,\beta,\gamma} Q^*_{\alpha\beta} Q_{\alpha\gamma} n_\beta n_\gamma - \sum_{\alpha,\beta,\gamma,\delta} Q^*_{\alpha\beta} Q_{\gamma\delta} n_\alpha n_\beta n_\gamma n_\delta \tag{9.46}$$

The necessary angular integrals over products of the rectangular components of $\mathbf{n}$ are readily found to be

$$\left.\begin{array}{l} \displaystyle\int n_\beta n_\gamma \, d\Omega = \frac{4\pi}{3} \delta_{\beta\gamma} \\[3mm] \displaystyle\int n_\alpha n_\beta n_\gamma n_\delta \, d\Omega = \frac{4\pi}{15} (\delta_{\alpha\beta}\delta_{\gamma\delta} + \delta_{\alpha\gamma}\delta_{\beta\delta} + \delta_{\alpha\delta}\delta_{\beta\gamma}) \end{array}\right\} \tag{9.47}$$

Then we find

$$\int |[\mathbf{n} \times \mathbf{Q}(\mathbf{n})] \times \mathbf{n}|^2 \, d\Omega = 4\pi\left\{\frac{1}{3}\sum_{\alpha,\beta}|Q_{\alpha\beta}|^2 \right.$$
$$\left. - \frac{1}{15}\left[\sum_\alpha Q^*_{\alpha\alpha}\sum_\gamma Q_{\gamma\gamma} + 2\sum_{\alpha,\beta}|Q_{\alpha\beta}|^2\right]\right\} \tag{9.48}$$

Since $Q_{\alpha\beta}$ is a tensor whose main diagonal sum is zero, the first term in the square brackets vanishes identically. Thus we obtain the final result for the total power radiated by a quadrupole source:

$$P = \frac{c^2 Z_0 k^6}{1440\pi}\sum_{\alpha,\beta}|Q_{\alpha\beta}|^2 \tag{9.49}$$

The radiated power varies as the sixth power of the frequency for fixed quadrupole moments, compared to the fourth power for dipole radiation.

A simple example of a radiating quadrupole source is an oscillating spheroidal distribution of charge. The off-diagonal elements of $Q_{\alpha\beta}$ vanish. The diagonal elements may be written

$$Q_{33} = Q_0, \qquad Q_{11} = Q_{22} = -\tfrac{1}{2}Q_0 \tag{9.50}$$

Then the angular distribution of radiated power is

$$\frac{dP}{d\Omega} = \frac{c^2 Z_0 k^6}{512\pi^2} Q_0^2 \sin^2\theta \cos^2\theta \tag{9.51}$$

This is a four-lobed pattern, as shown in Fig. 9.2, with maxima at $\theta = \pi/4$ and $3\pi/4$. The total power radiated by this quadrupole is

$$P = \frac{c^2 Z_0 k^6 Q_0^2}{960\pi} \tag{9.52}$$

The labor involved in manipulating higher terms in expansion (9.9) of the vector potential (9.8) becomes increasingly prohibitive as the expansion is extended beyond the electric quadrupole terms. Another disadvantage of the present approach is that physically distinct fields such as those of the magnetic dipole

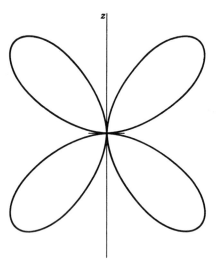

Figure 9.2 A quadrupole radiation pattern.

and the electric quadrupole must be disentangled from the separate terms in (9.9). Finally, the present technique is useful only in the long-wavelength limit. A systematic development of multipole radiation begins in Section 9.6. It involves a fairly elaborate mathematical apparatus, but the price paid is worthwhile. The treatment allows all multipole orders to be handled in the same way; the results are valid for all wavelengths; the physically different electric and magnetic multipoles are clearly separated from the beginning.

9.4 Center-Fed Linear Antenna

A. Approximation of Sinusoidal Current

For certain radiating systems the geometry of current flow is sufficiently simple that integral (9.3) for the vector potential can be found in relatively simple, closed form if the form of the current is assumed known. As an example of such a system we consider a thin, linear antenna of length d which is excited across a small gap at its midpoint. The antenna is assumed to be oriented along the z axis with its gap at the origin, as indicated in Fig. 9.3. If damping due to the emission of radiation is neglected and the antenna is thin enough, the current along the antenna can be taken as sinusoidal in time and space with wave number $k = \omega/c$, and is symmetric on the two arms of the antenna. The current vanishes at the ends of the antenna. Hence the current density can be written

$$\mathbf{J}(\mathbf{x}) = I \sin\left(\frac{kd}{2} - k|z|\right) \delta(x)\, \delta(y)\boldsymbol{\epsilon}_3 \tag{9.53}$$

for $|z| < (d/2)$. The delta functions assure that the current flows only along the z axis. I is the peak value of the current if $kd \geq \pi$. The current at the gap is $I_0 = I \sin(kd/2)$.

With the current density (9.53) the vector potential is in the z direction and in the radiation zone has the form [from (9.8)]:

$$\mathbf{A}(\mathbf{x}) = \hat{\mathbf{z}}\, \frac{\mu_0}{4\pi} \frac{Ie^{ikr}}{r} \int_{-(d/2)}^{(d/2)} \sin\left(\frac{kd}{2} - k|z|\right) e^{-ikz\cos\theta}\, dz \tag{9.54}$$

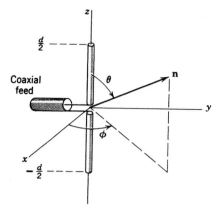

Figure 9.3   Center-fed, linear antenna.

The result of straightforward integration is

$$\mathbf{A}(\mathbf{x}) = \hat{\mathbf{z}}\,\frac{\mu_0}{4\pi}\,\frac{2Ie^{ikr}}{kr}\left[\frac{\cos\left(\dfrac{kd}{2}\cos\theta\right) - \cos\left(\dfrac{kd}{2}\right)}{\sin^2\theta}\right] \tag{9.55}$$

Since the magnetic field in the radiation zone is given by $\mathbf{H} = ik\mathbf{n} \times \mathbf{A}/\mu_0$, its magnitude is $|\mathbf{H}| = k\sin\theta\,|A_3|/\mu_0$. Thus the time-averaged power radiated per unit solid angle is

$$\frac{dP}{d\Omega} = \frac{Z_0I^2}{8\pi^2}\left|\frac{\cos\left(\dfrac{kd}{2}\cos\theta\right) - \cos\left(\dfrac{kd}{2}\right)}{\sin\theta}\right|^2 \tag{9.56}$$

The electric vector is in the direction of the component of $\mathbf{A}$ perpendicular to $\mathbf{n}$. Consequently the polarization of the radiation lies in the plane containing the antenna and the radius vector to the observation point.

The angular distribution (9.56) depends on the value of kd. In the long-wavelength limit ($kd \ll 1$) it is easy to show that it reduces to the dipole result (9.28). For the special values $kd = \pi(2\pi)$, corresponding to a half (two halves) of a wavelength of current oscillation along the antenna, the angular distributions are

$$\frac{dP}{d\Omega} = \frac{Z_0I^2}{8\pi^2}\begin{cases} \dfrac{\cos^2\left(\dfrac{\pi}{2}\cos\theta\right)}{\sin^2\theta}, & kd = \pi \\[4ex] \dfrac{4\cos^4\left(\dfrac{\pi}{2}\cos\theta\right)}{\sin^2\theta}, & kd = 2\pi \end{cases} \tag{9.57}$$

These angular distributions are shown in Fig. 9.7, where they are compared to multipole expansions. The half-wave antenna distribution is seen to be quite similar to a simple dipole pattern, but the full-wave antenna has a considerably sharper distribution.

The full-wave antenna distribution can be thought of as due to the coherent superposition of the fields of two half-wave antennas, one above the other, excited in phase. The intensity at $\theta = \pi/2$, where the waves add algebraically, is four times that of a half-wave antenna. At angles away from $\theta = \pi/2$ the amplitudes tend to interfere, giving the narrower pattern. By suitable arrangement of a set of basic antennas, such as the half-wave antenna, with the phasing of the currents appropriately chosen, arbitrary radiation patterns can be formed by coherent superposition. The interested reader should refer to the electrical engineering literature for detailed treatments of antenna arrays.

B. *The Antenna as a Boundary-Value Problem*

Only for infinitely thin conductors are we justified in assuming that the current along the antenna is sinusoidal, or indeed has any other *known* form. A finite-sized antenna with a given type of excitation is actually a complicated boundary-value problem. Without attempting solution of such problems, we give some preliminary considerations on setting up the boundary-value problem for a straight antenna with circular cross section of radius a and length d, of which the center-fed antenna of Fig. 9.3 is one example. We assume that the conductor is perfectly conducting and has a small enough radius compared to both a wavelength λ and the length d that current flow on the surface has only a longitudinal (z) component, and that the fields have azimuthal symmetry. Then the vector potential $\mathbf{A}$ will have only a z component. With harmonic time dependence of frequency ω and in the Lorentz gauge, the scalar potential and the electric field are given in terms of $\mathbf{A}$ by

$$\Phi(\mathbf{x}) = \frac{-ic}{k} \, \mathbf{\nabla} \cdot \mathbf{A}$$

$$\mathbf{E}(\mathbf{x}) = \frac{ic}{k} \left[\mathbf{\nabla}(\mathbf{\nabla} \cdot \mathbf{A}) + k^2 \mathbf{A} \right]$$

(9.58)

Since $\mathbf{A} = \hat{\mathbf{z}} A_z(\mathbf{x})$, the z component of the electric field is

$$E_z(\mathbf{x}) = \frac{ic}{k} \left(\frac{\partial^2}{\partial z^2} + k^2 \right) A_z(\mathbf{x})$$

But on the surface of the perfectly conducting antenna the tangential component of $\mathbf{E}$ vanishes. We thus establish the important fact that *the vector potential A_z* (*and* also the *scalar potential*) *on the surface of the antenna are strictly sinusoidal*:

$$\left(\frac{\partial^2}{\partial z^2} + k^2 \right) A_z(\rho = a, z) = 0$$

(9.59)

This is an exact statement, in contrast to the much rougher assumption that the current is sinusoidal.

An integral equation for the current can be found from (9.3). If the total current flow in the z direction is $I(z)$, then (9.3) gives for A_z on the surface of the antenna,

$$A_z(\rho = a, z) = \frac{\mu_0}{4\pi} \int_{z_0}^{z_0 + d} I(z') K(z - z') \, dz'$$

where

$$K(z - z') = \frac{1}{\pi} \int_0^\pi \frac{\exp[ik\sqrt{(z - z')^2 + 4a^2 \sin^2 \beta}]}{\sqrt{(z - z')^2 + 4a^2 \sin^2 \beta}} \, d\beta$$

(9.60)

is the azimuthal average of the Green function e^{ikR}/R. The condition (9.59) leads to the integro-differential equation

$$0 = \left(\frac{d^2}{dz^2} + k^2 \right) \int_{z_0}^{z_0+d} I(z')K(z - z') \, dz' \tag{9.61}$$

This can be regarded as a differential equation for the integral, or equivalently one can integrate (9.59) and equate it to $A_z(\rho = a, z)$. The result is the integral equation

$$a_1 \cos kz + a_2 \sin kz = \int_{z_0}^{z_0+d} I(z')K(z - z') \, dz'$$

The constants a_1 and a_2 are determined by the method of excitation and by the boundary conditions that the current vanishes at the ends of the antenna.

The solution of the integral equation is not easy. From the form of (9.60) it is clear that when $z' \simeq z$ care must be taken and the finite radius is important. For $a \to 0$, the current can be shown to be sinusoidal, but the expansion parameter for corrections turns out to be the reciprocal of $\ln(d/a)$. This means that even for $d/a = 10^3$ there can be corrections of the order of 10–15%. When there is a current node near the place of excitation, such corrections can change the antenna's input impedance drastically. Various approximate methods of solution of (9.61) are described by *Jones*. A detailed discussion of his version of the theory and the results of numerical calculations for the current, resistance, and reactance of a linear center-fed antenna are given by *Hallén*. Other references are cited in the suggested reading at the end of the chapter.

9.5 *Multipole Expansion for Localized Source or Aperture in Waveguide*

If a source in the form of a probe or loop or aperture in a waveguide is sufficiently small in dimensions compared to the distances over which the fields vary appreciably, it can be usefully approximated by its lowest order multipole moments, usually electric and magnetic dipoles. Different sources possessing the same lowest order multipole moments will produce sensibly the same excitations in the waveguide. Often the electric dipole or magnetic dipole moments can be calculated from static fields, or even estimated geometrically. Even if the source is not truly small, the multipole expansion gives a qualitative, and often semiquantitative, understanding of its properties.

A. *Current Source Inside Guide*

In Section 8.12 it was shown that the amplitudes $A_\lambda^{(\pm)}$ for excitation of the λth mode are proportional to the integral

$$\int \mathbf{J} \cdot \mathbf{E}_\lambda^{(\mp)} \, d^3x$$

where the integral is extended over the region where $\mathbf{J}$ is different from zero. If the mode fields $\mathbf{E}_\lambda^{(\mp)}$ do not vary appreciably over the source, they can be expanded in Taylor series around some suitably chosen origin. The integral is thus written, dropping the sub- and superscripts on $\mathbf{E}_\lambda^{(\mp)}$:

$$\int \mathbf{J} \cdot \mathbf{E} \, d^3x = \sum_{\alpha=1}^{3} \int J_\alpha(\mathbf{x}) \left[E_\alpha(0) + \sum_{\beta=1}^{3} x_\beta \frac{\partial E_\alpha}{\partial x_\beta}(0) + \cdots \right] d^3x \quad (9.62)$$

From (9.14) and (9.17) we see that the first term is

$$\mathbf{E}(0) \cdot \int \mathbf{J}(\mathbf{x}) \, d^3x = -i\omega\mathbf{p} \cdot \mathbf{E}(0) \quad (9.63)$$

where $\mathbf{p}$ is the electric dipole moment of the source:

$$\mathbf{p} = \frac{i}{\omega} \int \mathbf{J}(\mathbf{x}) \, d^3x$$

This can be transformed into the more familiar form (9.17) by the means of the steps in (9.14), provided the surface integral at the walls of the waveguide can be dropped. This necessitates choosing the origin for the multipole expansion such that $J_\alpha x_\beta$ vanishes at the walls. This remark applies to all the multipole moments. The use of the forms involving the electric and magnetic charge densities ρ and ρ_M requires that $(x_\alpha J_\beta \pm x_\beta J_\alpha)x_\gamma \cdots x_\nu$ vanish at the walls of the guide. The above-mentioned form for the electric dipole and the usual expression (9.34) for the magnetic dipole are correct as they stand, without concern about choice of origin.

The second term in (9.62) is of the same general form as (9.30) and is handled the same way. The product $J_\alpha x_\beta$ is written as the sum of symmetric and antisymmetric terms, just as in (9.31):

$$\sum_{\alpha,\beta} J_\alpha x_\beta \frac{\partial E_\alpha}{\partial x_\beta}(0) = \frac{1}{4} \sum_{\alpha,\beta} (J_\alpha x_\beta - J_\beta x_\alpha) \left[\frac{\partial E_\alpha}{\partial x_\beta}(0) - \frac{\partial E_\beta}{\partial x_\alpha}(0) \right]$$
$$+ \frac{1}{2} \sum_{\alpha,\beta} (J_\alpha x_\beta + J_\beta x_\alpha) \frac{\partial E_\alpha}{\partial x_\beta}(0) \quad (9.64)$$

The first (antisymmetric) part has been written so that the magnetic moment density and the curl of the electric field are clearly visible. With the help of Faraday's law $\nabla \times \mathbf{E} = i\omega\mathbf{B}$, the antisymmetric contribution to the right side of (9.62) can be written

$$\int \left[\sum_{\alpha,\beta} J_\alpha x_\beta \frac{\partial E_\alpha}{\partial x_\beta}(0) \right]_{\text{antisym}} d^3x = i\omega\mathbf{m} \cdot \mathbf{B}(0) \quad (9.65)$$

where $\mathbf{m}$ is the magnetic dipole moment (9.34) of the source. Equations (9.63) and (9.65) give the leading order multipole moment contributions to the source integral (9.62).

Other terms in the expansion in (9.62) give rise to higher order multipoles. The symmetric part of (9.64) can be shown, just as in Section 9.3, to involve the *traceless electric quadrupole moment* (9.41). The first step is to note that if the surface integrals vanish (see above),

$$\int (J_\alpha x_\beta + J_\beta x_\alpha) \, d^3x = -i\omega \int x_\alpha x_\beta \rho(\mathbf{x}) \, d^3x$$

Then the second double sum in (9.64), integrated over the volume of the current distribution, takes the form

$$-\frac{i\omega}{2} \sum_{\alpha,\beta} \frac{\partial E_\alpha}{\partial x_\beta}(0) \int \rho(\mathbf{x}) \, x_\alpha x_\beta \, d^3x$$

The value of the double sum is unchanged by the replacement $x_\alpha x_\beta \to (x_\alpha x_\beta - \frac{1}{3} r^2 \delta_{\alpha\beta})$ because $\nabla \cdot \mathbf{E} = 0$. Thus the symmetric part of the second term in (9.62) is

$$\int \left[\sum_{\alpha,\beta} J_\alpha x_\beta \frac{\partial E_\alpha}{\partial x_\beta}(0) \right]_{\text{sym}} d^3x = -\frac{i\omega}{6} \sum_{\alpha,\beta} Q_{\alpha\beta} \frac{\partial E_\alpha}{\partial x_\beta}(0) \qquad (9.66)$$

Similarly an antisymmetric part of the next terms in (9.62), involving $x_\beta x_\gamma$, gives a contribution

$$\int \left[\frac{1}{2} \sum_{\alpha,\beta,\gamma} J_\alpha x_\beta x_\gamma \frac{\partial^2 E_\alpha}{\partial x_\beta \, \partial x_\gamma}(0) \right]_{\substack{\text{antisym} \\ \text{in}(\alpha,\beta)}} d^3x = \frac{i\omega}{6} \sum_{\alpha,\beta} Q^M_{\alpha\beta} \frac{\partial B_\alpha}{\partial x_\beta}(0) \qquad (9.67)$$

where $Q^M_{\alpha\beta}$ is the *magnetic quadrupole moment* of the source, given by (9.41) with the electric charge density $\rho(\mathbf{x})$ replaced by the magnetic charge density,

$$\rho^M(\mathbf{x}) = -\nabla \cdot \mathbf{\mathcal{M}} = -\frac{1}{2} \nabla \cdot (\mathbf{x} \times \mathbf{J}) \qquad (9.68)$$

If the various contributions are combined, the expression (8.146) for the amplitude $A_\lambda^{(\pm)}$ has as its multipole expansion,

$$A_\lambda^{(\pm)} = i \frac{\omega Z_\lambda}{2} \left\{ \mathbf{p} \cdot \mathbf{E}_\lambda^{(\mp)}(0) - \mathbf{m} \cdot \mathbf{B}_\lambda^{(\mp)}(0) \right.$$
$$\left. + \frac{1}{6} \sum_{\alpha,\beta} \left[Q_{\alpha\beta} \frac{\partial E_{\lambda\alpha}^{(\mp)}}{\partial x_\beta}(0) - Q^M_{\alpha\beta} \frac{\partial B_{\lambda\alpha}^{(\mp)}}{\partial x_\beta}(0) \right] + \cdots \right\} \qquad (9.69)$$

It should be remembered that here the mode fields $E_\lambda^{(\pm)}$ are normalized according to (8.131). The expansion is most useful if the source is such that the series converges rapidly and is adequately approximated by its first terms. The positioning and orientation of probes or antennas to excite preferentially certain modes can be accomplished simply by considering the directions of the electric and magnetic dipole (or higher) moments of the source and the normal mode fields. For example, the excitation of TE modes, with their axial magnetic fields, can be produced by a magnetic dipole antenna whose dipole moment is parallel to the axis of the guide. TM modes cannot be excited by such an antenna, except via higher multipole moments.

B. Aperture in Side Walls of Guide

Apertures in the walls of a waveguide can be considered as sources (or sinks) of energy. In Section 8.12 it was noted that if the guide walls have openings in the volume considered to contain the sources, the amplitudes $A_\lambda^{(\pm)}$ are given by

(8.147) instead of (8.146). With the assumption that there is only one aperture, and no actual current density, the amplitude for excitation of the λth mode is

$$A_\lambda^{(\pm)} = -\frac{Z_\lambda}{2} \int_{\text{aperture}} \mathbf{n} \cdot (\mathbf{E} \times \mathbf{H}_\lambda^{(\mp)}) \, da \qquad (9.70)$$

where $\mathbf{n}$ is an *inwardly* directed normal and the integral is over the aperture in the walls of the guide. If the aperture is small compared to a wavelength or other scale of variation of the fields, the mode field $\mathbf{H}_\lambda^{(\mp)}$ can be expanded just as before. The lowest order term, with $\mathbf{H}_\lambda^{(\mp)}$ treated as constant over the aperture, evidently leads to a coupling of the magnetic dipole type. The next terms, with linear variation of the mode field, give rise to electric dipole and magnetic quadrupole couplings, exactly as for (9.64)–(9.66), but with the roles of electric and magnetic interactions interchanged. The result is an expansion of (9.70) like (9.69):

$$A_\lambda^{(\pm)} = i\frac{\omega Z_\lambda}{4} [\mathbf{p}_{\text{eff}} \cdot \mathbf{E}_\lambda^{(\mp)}(0) - \mathbf{m}_{\text{eff}} \cdot \mathbf{B}_\lambda^{(\mp)}(0) + \cdots] \qquad (9.71)$$

where the effective electric and magnetic dipole moments are

$$\mathbf{p}_{\text{eff}} = \epsilon \mathbf{n} \int (\mathbf{x} \cdot \mathbf{E}_{\text{tan}}) \, da$$

$$\mathbf{m}_{\text{eff}} = \frac{2}{i\mu\omega} \int (\mathbf{n} \times \mathbf{E}_{\text{tan}}) \, da \qquad (9.72)$$

In these expressions the integration is over the aperture, the electric field $\mathbf{E}_{\text{tan}}$ is the exact tangential field in the opening, and in (9.71) the mode fields are evaluated at (the center of) the aperture. The effective moments (9.72) are the equivalent dipoles whose fields (9.18) and (9.35)–(9.36) represent the radiation fields of a small aperture in a flat, perfectly conducting screen (see Problem 10.10). Comparison of (9.71) and (9.69) shows that the dipole moments (9.72) are only half as effective in producing a given amplitude as are the real dipole moments of a source located inside the guide. The effective dipoles of an aperture are in some sense half in and half out of the guide.

C. Effective Dipole Moments of Apertures

On first encounter the effective dipole moments (9.72) are somewhat mysterious. As already mentioned, they have a precise meaning in terms of the electric and magnetic dipole parts of the multipole expansion of the fields radiated through an aperture in a flat perfectly conducting screen (considered later: Problem 10.10). For small apertures they can also be related to the solutions of appropriate static or quasi-static boundary-value problems. Such problems have already been discussed (Sections 3.13 and 5.13), and the results are appropriated below.

If an aperture is very small compared to the distance over which the fields change appreciably, the boundary-value problem can be approximated by one in which the fields "far from the aperture" (measured in units of the aperture dimension) are those that would exist if the aperture were absent. Except for very elongated apertures, it will be sufficiently accurate to take the surface to be flat and the "asymptotic" fields to be the same in all directions away from the aperture. For an opening in a perfectly conducting surface, then, the boundary-value problem is specified by the *normal* electric field $\mathbf{E}_0$ and the *tangential* magnetic field $\mathbf{H}_0$ that would exist in the absence of the opening. The fields $\mathbf{E}_0$ and $\mathbf{H}_0$ are themselves the result of some boundary-value problem, of propagation in a waveguide or reflection of a plane wave from a screen, for example. But for the purpose at hand, they are treated as given. To lowest order their time dependence can be ignored, provided the effective electric dipole moment is related to $\mathbf{E}_0$ and the magnetic moment to $\mathbf{H}_0$. (See, however, Problem 9.20.)

The exact form of the fields around the opening depends on its shape, but some qualitative observations can be made by merely examining the general behavior of the lines of force. Outside a sphere enclosing the aperture the fields may be represented by a multipole expansion. The leading terms will be dipole fields. Figure 9.4 shows the qualitative behavior. The loop of magnetic field protruding above the plane on the left has the appearance of a line of force from a magnetic dipole whose moment is directed oppositely to $\mathbf{H}_0$, as indicated by the direction of the moment $\mathbf{m}^{(+)}$ shown below. The magnetic field below the plane can be viewed as the unperturbed $\mathbf{H}_0$, plus an opposing dipole field (dashed lines in Fig. 9.4) whose moment is oriented parallel to $\mathbf{H}_0$ (denoted by $\mathbf{m}^{(-)}$ below). Similarly, the electric field lines above the plane appear to originate from a vertical dipole moment $\mathbf{p}^{(+)}$ directed along $\mathbf{E}_0$, while below the plane the field has the appearance of the unperturbed normal field $\mathbf{E}_0$, plus the field from a dipole $\mathbf{p}^{(-)}$, directed oppositely to $\mathbf{E}_0$. The use of effective dipole fields is of course restricted to regions some distance from the aperture. Right in the aperture the fields bear no resemblance to dipole fields. Nevertheless, the dipole approximation is useful qualitatively everywhere, and the effective moments are all that are needed to evaluate the couplings of small apertures.

The preceding qualitative discussion has one serious deficiency. While it is correct to state that the *electric* dipole moment is always directed parallel or antiparallel to $\mathbf{E}_0$ and so is normal to the aperture, the *magnetic* dipole moment is not necessarily parallel or antiparallel to $\mathbf{H}_0$. There are two directions in the tangent plane, and the relative orientation of the aperture and the direction of $\mathbf{H}_0$ are relevant in determining the direction of $\mathbf{m}_{\text{eff}}$. Since the effective moments are obviously proportional to the field strength, it is appropriate to speak of the *electric and magnetic polarizabilities* of the aperture. The dipole moments can be written

$$\begin{aligned} \mathbf{p}_{\text{eff}} &= \epsilon_0 \gamma^E \mathbf{E}_0 \\ (\mathbf{m}_{\text{eff}})_\alpha &= \sum_\beta \gamma^M_{\alpha\beta} (\mathbf{H}_0)_\beta \end{aligned} \tag{9.73}$$

where γ^E is the scalar electric polarizability and $\gamma^M_{\alpha\beta}$ is the 2×2 magnetic polarizability tensor. The magnetic tensor can be diagonalized by choosing principal axes for the aperture. There are thus three polarizabilities (one electric and two magnetic) to characterize an arbitrary small aperture. It should be remembered that the signs of the γ's in (9.73) depend on the side of the surface from which the dipole is viewed, as shown in Fig. 9.4. If there are fields on both sides of the surface, the expressions in (9.73) must be modified. For example, if there is a vertically directed electric field $\mathbf{E}_1$ above the surface in Fig. 9.4b, as well as $\mathbf{E}_0$ below, then $\mathbf{E}_0$ in (9.73) is replaced by $(\mathbf{E}_0 - \mathbf{E}_1)$. Other possibilities can be worked out from (9.73) by linear superposition.

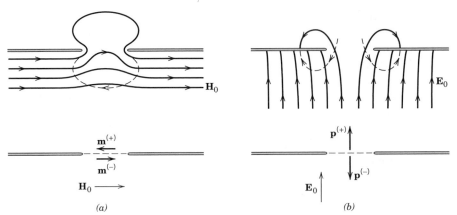

Figure 9.4 Distortion of (*a*) the tangential magnetic field and (*b*) the normal electric field by a small aperture in a perfectly conducting surface. The effective dipole moments, as viewed from above and below the surface, are indicated beneath.

The polarizabilities γ^E and $\gamma^M_{\alpha\beta}$ have the dimensions of length cubed. If a typical dimension of the aperture is d, then it can be expected that the polarizabilities will be d^3 times numerical coefficients of the order of unity, or smaller. The expression (9.72) for $\mathbf{p}_{\text{eff}}$ can be seen to be of the form to yield such a result, since $\mathbf{E}_{\text{tan}}$ is proportional to E_0, and the two-dimensional integral will give E_0 times the cube of a length that is characteristic of the aperture. Furthermore, the vectorial properties of $\mathbf{p}_{\text{eff}}$ in (9.72) correspond to (9.73). On the other hand, the expression in (9.72) for $\mathbf{m}_{\text{eff}}$ is less transparently of the proper form, even though dimensionally correct. Some integrations by parts and use of the Maxwell equations puts it into the equivalent and more satisfying form:

$$\mathbf{m}_{\text{eff}} = 2 \int \mathbf{x}(\mathbf{n} \cdot \mathbf{H}) \, da \qquad (9.74)$$

where $\mathbf{n} \cdot \mathbf{H}$ is the exact normal component of $\mathbf{H}$ in the aperture and the integration is over the plane of the aperture. It is now evident that the connection between $\mathbf{H}_0$ and $\mathbf{m}_{\text{eff}}$ is of the general form shown in (9.73). For a circular opening of radius R the effective dipole moments can be taken from the static solutions of Sections 3.13 and 5.13. The results are

$$\mathbf{p}_{\text{eff}} = -\frac{4\epsilon_0 R^3}{3} \mathbf{E}_0, \qquad \mathbf{m}_{\text{eff}} = \frac{8R^3}{3} \mathbf{H}_0 \qquad (9.75)$$

where the signs are appropriate for the apertures viewed from the side of the surface where $\mathbf{E}$ and $\mathbf{H}$ are nonvanishing, as can be checked from Fig. 9.4. The electric and magnetic polarizabilities are thus

$$\gamma^E = -\frac{4R^3}{3}, \qquad \gamma^M_{\alpha\beta} = \frac{8R^3}{3} \delta_{\alpha\beta} \qquad (9.76)$$

The use of effective dipole moments to describe the electromagnetic properties of small holes can be traced back to Lord Rayleigh.[*] The general theory was developed by H. A. Bethe[†] and has been applied fruitfully to waveguide and

[*]Lord Rayleigh, *Phil. Mag.* **XLIV**, 28 (1897), reprinted in his *Scientific Papers*, Vol. IV, p. 305.
[†]H. A. Bethe, *Phys. Rev.* **66**, 163 (1944).

diffraction problems. It is significant in practical applications that the effective dipole moments of arbitrary apertures can be determined experimentally by electrolytic tank measurements.[‡]

Examples of the use of multipoles to describe excitation and scattering in waveguides and diffraction are left to several problems at the end of the chapter. Other material can be found in the list of suggested reading.

9.6 *Spherical Wave Solutions of the Scalar Wave Equation*

In Chapters 3 and 4 spherical harmonic expansions for the solutions of the Laplace or Poisson equations were used in potential problems with spherical boundaries or to develop multipole expansions of charge densities and their fields. Our approach so far for radiating sources has been "brute force," with creation of the lowest order multipoles more or less by hand. Clearly, treatment of higher multipoles demands a more systematic approach. We therefore turn to the development of vector spherical waves and their relation to time-varying sources.

As a prelude to the vector spherical wave problem, we consider the scalar wave equation. A scalar field $\psi(\mathbf{x}, t)$ satisfying the source-free wave equation,

$$\nabla^2\psi - \frac{1}{c^2}\frac{\partial^2\psi}{\partial t^2} = 0 \tag{9.77}$$

can be Fourier-analyzed in time as

$$\psi(\mathbf{x}, t) = \int_{-\infty}^{\infty} \psi(\mathbf{x}, \omega)e^{-i\omega t}\, d\omega \tag{9.78}$$

with each Fourier component satisfying the Helmholtz wave equation

$$(\nabla^2 + k^2)\psi(\mathbf{x}, \omega) = 0 \tag{9.79}$$

with $k^2 = \omega^2/c^2$. For problems possessing symmetry properties about some origin, it is convenient to have fundamental solutions appropriate to spherical coordinates. The representation of the Laplacian operator in spherical coordinates is given in equation (3.1). The separation of the angular and radial variables follows the well-known expansion

$$\psi(\mathbf{x}, \omega) = \sum_{l,m} f_{lm}(r)Y_{lm}(\theta, \phi) \tag{9.80}$$

where the spherical harmonics Y_{lm} are defined by (3.53). The radial functions $f_{lm}(r)$ satisfy the radial equation, independent of m,

$$\left[\frac{d^2}{dr^2} + \frac{2}{r}\frac{d}{dr} + k^2 - \frac{l(l+1)}{r^2}\right]f_l(r) = 0 \tag{9.81}$$

With the substitution,

$$f_l(r) = \frac{1}{r^{1/2}} u_l(r) \tag{9.82}$$

[‡]S. B. Cohn, *Proc. IRE* **39**, 1416 (1951); **40**, 1069 (1952).

equation (9.81) is transformed into

$$\left[\frac{d^2}{dr^2} + \frac{1}{r}\frac{d}{dr} + k^2 - \frac{(l + \frac{1}{2})^2}{r^2} \right] u_l(r) = 0 \tag{9.83}$$

This equation is just the Bessel equation (3.75) with $\nu = l + \frac{1}{2}$. Thus the solutions for $f_{lm}(r)$ are

$$f_{lm}(r) = \frac{A_{lm}}{r^{1/2}} J_{l+1/2}(kr) + \frac{B_{lm}}{r^{1/2}} N_{l+1/2}(kr) \tag{9.84}$$

It is customary to define *spherical Bessel and Hankel functions*, denoted by $j_l(x), n_l(x), h_l^{(1,2)}(x)$, as follows:

$$j_l(x) = \left(\frac{\pi}{2x} \right)^{1/2} J_{l+1/2}(x)$$

$$n_l(x) = \left(\frac{\pi}{2x} \right)^{1/2} N_{l+1/2}(x) \tag{9.85}$$

$$h_l^{(1,2)}(x) = \left(\frac{\pi}{2x} \right)^{1/2} [J_{l+1/2}(x) \pm iN_{l+1/2}(x)]$$

For real x, $h_l^{(2)}(x)$ is the complex conjugate of $h_l^{(1)}(x)$. From the series expansions (3.82) and (3.83) one can show that

$$j_l(x) = (-x)^l \left(\frac{1}{x}\frac{d}{dx} \right)^l \left(\frac{\sin x}{x} \right)$$

$$n_l(x) = -(-x)^l \left(\frac{1}{x}\frac{d}{dx} \right)^l \left(\frac{\cos x}{x} \right) \tag{9.86}$$

For the first few values of l the explicit forms are:

$$j_0(x) = \frac{\sin x}{x}, \qquad n_0(x) = -\frac{\cos x}{x}, \qquad h_0^{(1)}(x) = \frac{e^{ix}}{ix}$$

$$j_1(x) = \frac{\sin x}{x^2} - \frac{\cos x}{x}, \qquad n_1(x) = -\frac{\cos x}{x^2} - \frac{\sin x}{x}$$

$$h_1^{(1)}(x) = -\frac{e^{ix}}{x}\left(1 + \frac{i}{x} \right)$$

$$j_2(x) = \left(\frac{3}{x^3} - \frac{1}{x} \right)\sin x - \frac{3\cos x}{x^2}, \qquad n_2(x) = -\left(\frac{3}{x^3} - \frac{1}{x} \right)\cos x - 3\frac{\sin x}{x^2}$$

$$h_2^{(1)}(x) = \frac{ie^{ix}}{x}\left(1 + \frac{3i}{x} - \frac{3}{x^2} \right)$$

$$j_3(x) = \left(\frac{15}{x^4} - \frac{6}{x^2} \right)\sin x - \left(\frac{15}{x^3} - \frac{1}{x} \right)\cos x$$

$$n_3(x) = -\left(\frac{15}{x^4} - \frac{6}{x^2} \right)\cos x - \left(\frac{15}{x^3} - \frac{1}{x} \right)\sin x$$

$$h_3^{(1)}(x) = \frac{e^{ix}}{x}\left(1 + \frac{6i}{x} - \frac{15}{x^2} - \frac{15i}{x^3} \right)$$

$$\tag{9.87}$$

From the series (3.82), (3.83), and the definition (3.85) it is possible to calculate the small argument limits ($x \ll 1, l$) to be

$$j_l(x) \rightarrow \frac{x^l}{(2l+1)!!} \left(1 - \frac{x^2}{2(2l+3)} + \cdots \right)$$

$$n_l(x) \rightarrow - \frac{(2l-1)!!}{x^{l+1}} \left(1 - \frac{x^2}{2(1-2l)} + \cdots \right),$$

(9.88)

where $(2l+1)!! = (2l+1)(2l-1)(2l-3) \cdots (5) \cdot (3) \cdot (1)$. Similarly the large argument limits ($x \gg l$) are

$$j_l(x) \rightarrow \frac{1}{x} \sin\left(x - \frac{l\pi}{2}\right)$$

$$n_l(x) \rightarrow -\frac{1}{x} \cos\left(x - \frac{l\pi}{2}\right)$$

(9.89)

$$h_l^{(1)}(x) \rightarrow (-i)^{l+1} \frac{e^{ix}}{x}$$

The spherical Bessel functions satisfy the recursion formulas,

$$\frac{2l+1}{x} z_l(x) = z_{l-1}(x) + z_{l+1}(x)$$

$$z_l'(x) = \frac{1}{2l+1} [lz_{l-1}(x) - (l+1)z_{l+1}(x)]$$

(9.90)

$$\frac{d}{dx} [xz_l(x)] = xz_{l-1}(x) - lz_l(x)$$

where $z_l(x)$ is any one of the functions $j_l(x), n_l(x), h_l^{(1)}(x), h_l^{(2)}(x)$. The Wronskians of the various pairs are

$$W(j_l, n_l) = \frac{1}{i} W(j_l, h_l^{(1)}) = -W(n_l, h_l^{(1)}) = \frac{1}{x^2}$$

(9.91)

The general solution of (9.79) in spherical coordinates can be written

$$\psi(\mathbf{x}) = \sum_{l,m} [A_{lm}^{(1)} h_l^{(1)}(kr) + A_{lm}^{(2)} h_l^{(2)}(kr)] Y_{lm}(\theta, \phi)$$

(9.92)

where the coefficients $A_{lm}^{(1)}$ and $A_{lm}^{(2)}$ will be determined by the boundary conditions.

For reference purposes we present the spherical wave expansion for the outgoing wave Green function $G(\mathbf{x}, \mathbf{x}')$, which is appropriate to the equation,

$$(\nabla^2 + k^2)G(\mathbf{x}, \mathbf{x}') = -\delta(\mathbf{x} - \mathbf{x}')$$

(9.93)

in the infinite domain. This Green function, as was shown in Chapter 6, is

$$G(\mathbf{x}, \mathbf{x}') = \frac{e^{ik|\mathbf{x}-\mathbf{x}'|}}{4\pi |\mathbf{x} - \mathbf{x}'|}$$

(9.94)

The spherical wave expansion for $G(\mathbf{x}, \mathbf{x}')$ can be obtained in exactly the same way as was done in Section 3.9 for the Poisson equation [see especially (3.117) and text following]. An expansion of the form

$$G(\mathbf{x}, \mathbf{x}') = \sum_{l,m} g_l(r, r') Y_{lm}^*(\theta', \phi') Y_{lm}(\theta, \phi)$$

(9.95)

substituted into (9.93) leads to an equation for $g_l(r, r')$:

$$\left[\frac{d^2}{dr^2} + \frac{2}{r} \frac{d}{dr} + k^2 - \frac{l(l+1)}{r^2} \right] g_l = -\frac{1}{r^2} \delta(r - r') \tag{9.96}$$

The solution that satisfies the boundary conditions of finiteness at the origin and outgoing waves at infinity is

$$g_l(r, r') = A j_l(kr_<) h_l^{(1)}(kr_>) \tag{9.97}$$

The correct discontinuity in slope is assured if $A = ik$. Thus the expansion of the Green function is

$$\frac{e^{ik|\mathbf{x} - \mathbf{x}'|}}{4\pi |\mathbf{x} - \mathbf{x}'|} = ik \sum_{l=0}^{\infty} j_l(kr_<) h_l^{(1)}(kr_>) \sum_{m=-l}^{l} Y_{lm}^*(\theta', \phi') Y_{lm}(\theta, \phi) \tag{9.98}$$

Our emphasis so far has been on the radial functions appropriate to the scalar wave equation. We now reexamine the angular functions in order to introduce some concepts of use in considering the vector wave equation. The basic angular functions are the spherical harmonics $Y_{lm}(\theta, \phi)$ (3.53), which are solutions of the equation

$$-\left[\frac{1}{\sin \theta} \frac{\partial}{\partial \theta} \left(\sin \theta \frac{\partial}{\partial \theta} \right) + \frac{1}{\sin^2 \theta} \frac{\partial^2}{\partial \phi^2} \right] Y_{lm} = l(l+1) Y_{lm} \tag{9.99}$$

As is well known in quantum mechanics, this equation can be written in the form:

$$L^2 Y_{lm} = l(l+1) Y_{lm} \tag{9.100}$$

The differential operator $L^2 = L_x^2 + L_y^2 + L_z^2$, where

$$\mathbf{L} = \frac{1}{i} (\mathbf{r} \times \nabla) \tag{9.101}$$

is $\hbar^{-1}$ times the orbital angular-momentum operator of wave mechanics.

The components of $\mathbf{L}$ can be written conveniently in the combinations,

$$L_+ = L_x + iL_y = e^{i\phi} \left(\frac{\partial}{\partial \theta} + i \cot \theta \frac{\partial}{\partial \phi} \right)$$

$$L_- = L_x - iL_y = e^{-i\phi} \left(-\frac{\partial}{\partial \theta} + i \cot \theta \frac{\partial}{\partial \phi} \right) \tag{9.102}$$

$$L_z = -i \frac{\partial}{\partial \phi}$$

We note that $\mathbf{L}$ operates only on angular variables and is independent of r. From definition (9.101) it is evident that

$$\mathbf{r} \cdot \mathbf{L} = 0 \tag{9.103}$$

holds as an operator equation. From the explicit forms (9.102) it is easy to verify that L^2 is equal to the operator on the left side of (9.99).

From the explicit forms (9.102) and recursion relations for Y_{lm} the following useful relations can be established:

$$L_+ Y_{lm} = \sqrt{(l - m)(l + m + 1)} \, Y_{l,m+1}$$
$$L_- Y_{lm} = \sqrt{(l + m)(l - m + 1)} \, Y_{l,m-1} \tag{9.104}$$
$$L_z Y_{lm} = m Y_{lm}$$

Finally we note the following operator equations concerning the commutation properties of $\mathbf{L}$, L^2, and ∇^2:

$$\left.\begin{array}{c} L^2\mathbf{L} = \mathbf{L}L^2 \\ \mathbf{L} \times \mathbf{L} = i\mathbf{L} \\ L_j\nabla^2 = \nabla^2 L_j \end{array}\right\} \tag{9.105}$$

where

$$\nabla^2 = \frac{1}{r}\frac{\partial^2}{\partial r^2}(r) - \frac{L^2}{r^2} \tag{9.106}$$

9.7 Multipole Expansion of the Electromagnetic Fields

With the assumption of a time dependence $e^{-i\omega t}$ the Maxwell equations in a source-free region of empty space may be written

$$\begin{array}{cc} \nabla \times \mathbf{E} = ikZ_0\mathbf{H}, & \nabla \times \mathbf{H} = -ik\mathbf{E}/Z_0 \\ \nabla \cdot \mathbf{E} = 0 & \nabla \cdot \mathbf{H} = 0 \end{array} \tag{9.107}$$

where $k = \omega/c$. If $\mathbf{E}$ is eliminated by combining the two curl equations, we obtain for $\mathbf{H}$,

$$(\nabla^2 + k^2)\mathbf{H} = 0, \qquad \nabla \cdot \mathbf{H} = 0$$

with $\mathbf{E}$ given by

$$\mathbf{E} = \frac{iZ_0}{k}\nabla \times \mathbf{H} \tag{9.108}$$

Alternatively, $\mathbf{H}$ can be eliminated to yield

$$(\nabla^2 + k^2)\mathbf{E} = 0, \qquad \nabla \cdot \mathbf{E} = 0$$

with $\mathbf{H}$ given by

$$\mathbf{H} = -\frac{i}{kZ_0}\nabla \times \mathbf{E} \tag{9.109}$$

Either (9.108) or (9.109) is a set of three equations that is equivalent to the Maxwell equations (9.107).

We wish to find multipole solutions for $\mathbf{E}$ and $\mathbf{H}$. From (9.108) and (9.109) it is evident that each *Cartesian* component of $\mathbf{H}$ and $\mathbf{E}$ satisfies the Helmholtz wave equation (9.79). Hence each such component can be written as an expansion of the general form (9.92). There remains, however, the problem of orchestrating the different components in order to satisfy $\nabla \cdot \mathbf{H} = 0$ and $\nabla \cdot \mathbf{E} = 0$ and to give a pure multipole field of order (l, m). We follow a different and somewhat easier path suggested by Bouwkamp and Casimir.* Consider the scalar quantity $\mathbf{r} \cdot \mathbf{A}$, where $\mathbf{A}$ is a well-behaved vector field. It is straightforward to verify that the Laplacian operator acting on this scalar gives

$$\nabla^2(\mathbf{r} \cdot \mathbf{A}) = \mathbf{r} \cdot (\nabla^2\mathbf{A}) + 2\nabla \cdot \mathbf{A} \tag{9.110}$$

---

*C. J. Bouwkamp and H. B. G. Casimir, *Physica* **20**, 539 (1954). This paper discusses the relationship among a number of different, but equivalent, approaches to multipole radiation.

From (9.108) and (9.109) it therefore follows that the scalars, $\mathbf{r} \cdot \mathbf{E}$ and $\mathbf{r} \cdot \mathbf{H}$, both satisfy the Helmholtz wave equation:

$$(\nabla^2 + k^2)(\mathbf{r} \cdot \mathbf{E}) = 0, \qquad (\nabla^2 + k^2)(\mathbf{r} \cdot \mathbf{H}) = 0 \qquad (9.111)$$

The general solution for $\mathbf{r} \cdot \mathbf{E}$ is given by (9.92), and similarly for $\mathbf{r} \cdot \mathbf{H}$.

We now *define a magnetic multipole field of order* (l, m) by the conditions,

$$\mathbf{r} \cdot \mathbf{H}_{lm}^{(M)} = \frac{l(l + 1)}{k} g_l(kr) Y_{lm}(\theta, \phi)$$

$$\mathbf{r} \cdot \mathbf{E}_{lm}^{(M)} = 0 \qquad (9.112)$$

where

$$g_l(kr) = A_l^{(1)} h_l^{(1)}(kr) + A_l^{(2)} h_l^{(2)}(kr) \qquad (9.113)$$

The presence of the factor of $l(l + 1)/k$ is for later convenience. Using the curl equation in (9.109) we can relate $\mathbf{r} \cdot \mathbf{H}$ to the electric field:

$$Z_0 k\, \mathbf{r} \cdot \mathbf{H} = \frac{1}{i} \mathbf{r} \cdot (\nabla \times \mathbf{E}) = \frac{1}{i} (\mathbf{r} \times \nabla) \cdot \mathbf{E} = \mathbf{L} \cdot \mathbf{E} \qquad (9.114)$$

where $\mathbf{L}$ is given by (9.101). With $\mathbf{r} \cdot \mathbf{H}$ given by (9.112), the electric field of the magnetic multipole must satisfy

$$\mathbf{L} \cdot \mathbf{E}_{lm}^{(M)}(r, \theta, \phi) = l(l + 1) Z_0 g_l(kr) Y_{lm}(\theta, \phi) \qquad (9.115)$$

and $\mathbf{r} \cdot \mathbf{E}_{lm}^{(M)} = 0$. To determine the purely transverse electric field from (9.115), we first observe that the operator $\mathbf{L}$ acts only on the angular variables (θ, ϕ). This means that the radial dependence of $\mathbf{E}_{lm}^{(M)}$ must be given by $g_l(kr)$. Second, the operator $\mathbf{L}$ acting on Y_{lm} transforms the m value according to (9.104), but does not change the l value. Thus the components of $\mathbf{E}_{lm}^{(M)}$ can be at most linear combinations of Y_{lm}'s with different m values and a common l, equal to the l value on the right-hand side of (9.115). A moment's thought shows that for $\mathbf{L} \cdot \mathbf{E}_{lm}^{(M)}$ to yield a *single* Y_{lm}, the components of $\mathbf{E}_{lm}^{(M)}$ must be prepared beforehand to compensate for whatever raising or lowering of m values is done by $\mathbf{L}$. Thus, in the term $L_- E_+$, for example, it must be that E_+ is proportional to $L_+ Y_{lm}$. What this amounts to is that the electric field should be

$$\mathbf{E}_{lm}^{(M)} = Z_0 g_l(kr) \mathbf{L} Y_{lm}(\theta, \phi)$$

together with $\qquad\qquad\qquad\qquad\qquad\qquad\qquad\qquad\qquad\qquad (9.116)$

$$\mathbf{H}_{lm}^{(M)} = -\frac{i}{kZ_0} \nabla \times \mathbf{E}_{lm}^{(M)}$$

Equation (9.116) specifies the electromagnetic fields of a *magnetic* multipole of order (l, m). Because the electric field (9.116) is transverse to the radius vector, these multipole fields are sometimes called *transverse electric (TE)* rather than magnetic.

The fields of an *electric or transverse magnetic (TM) multipole of order* (l, m) are specified similarly by the conditions,

$$\mathbf{r} \cdot \mathbf{E}_{lm}^{(E)} = -Z_0 \frac{l(l + 1)}{k} f_l(kr) Y_{lm}(\theta, \phi)$$

$$\mathbf{r} \cdot \mathbf{H}_{lm}^{(E)} = 0 \qquad (9.117)$$

Then the *electric* multipole fields are

$$\mathbf{H}_{lm}^{(E)} = f_l(kr)\mathbf{L}Y_{lm}(\theta, \phi) \tag{9.118}$$

$$\mathbf{E}_{lm}^{(E)} = \frac{iZ_0}{k} \boldsymbol{\nabla} \times \mathbf{H}_{lm}^{(E)}$$

The radial function $f_l(kr)$ is given by an expression like (9.113).

The fields (9.116) and (9.118) are the spherical wave analogs of the TE and TM cylindrical modes of Chapter 8. Just as in the cylindrical waveguide, the two sets of multipole fields (9.116) and (9.118) can be shown to form a complete set of vector solutions to the Maxwell equations in a source-free region. The terminology electric and magnetic multipole fields will be used, rather than TM and TE, since the sources of each type of field will be seen to be the electric-charge density and the magnetic-moment density, respectively. Since the vector spherical harmonic, $\mathbf{L}Y_{lm}$, plays an important role, it is convenient to introduce the normalized form,*

$$\mathbf{X}_{lm}(\theta, \phi) = \frac{1}{\sqrt{l(l + 1)}} \mathbf{L}Y_{lm}(\theta, \phi) \tag{9.119}$$

with the orthogonality properties,

$$\int \mathbf{X}_{l'm'}^* \cdot \mathbf{X}_{lm} \, d\Omega = \delta_{ll'}\delta_{mm'} \tag{9.120}$$

and

$$\int \mathbf{X}_{l'm'}^* \cdot (\mathbf{r} \times \mathbf{X}_{lm}) \, d\Omega = 0 \tag{9.121}$$

for all l, l', m, m'.

By combining the two types of fields we can write the general solution to the Maxwell equations (9.107):

$$\mathbf{H} = \sum_{l,m} \left[a_E(l, m)f_l(kr)\mathbf{X}_{lm} - \frac{i}{k} a_M(l, m)\boldsymbol{\nabla} \times g_l(kr)\mathbf{X}_{lm} \right]$$

$$\mathbf{E} = Z_0 \sum_{l,m} \left[\frac{i}{k} a_E(l, m)\boldsymbol{\nabla} \times f_l(kr)\mathbf{X}_{lm} + a_M(l, m)g_l(kr)\mathbf{X}_{lm} \right] \tag{9.122}$$

where the coefficients $a_E(l, m)$ and $a_M(l, m)$ specify the amounts of electric (l, m) multipole and magnetic (l, m) multipole fields. The radial functions $f_l(kr)$ and $g_l(kr)$ are of the form (9.113). The coefficients $a_E(l, m)$ and $a_M(l, m)$, as well as the relative proportions in (9.113), are determined by the sources and boundary conditions. To make this explicit, we note that the scalars $\mathbf{r} \cdot \mathbf{H}$ and $\mathbf{r} \cdot \mathbf{E}$ are sufficient to determine the unknowns in (9.122) according to

$$a_M(l, m)g_l(kr) = \frac{k}{\sqrt{l(l + 1)}} \int Y_{lm}^* \mathbf{r} \cdot \mathbf{H} \, d\Omega$$

$$Z_0 a_E(l, m)f_l(kr) = -\frac{k}{\sqrt{l(l + 1)}} \int Y_{lm}^* \mathbf{r} \cdot \mathbf{E} \, d\Omega \tag{9.123}$$

---

*$\mathbf{X}_{lm}$ is defined to be identically zero for $l = 0$. Spherically symmetric solutions to the source-free Maxwell's equations exist only in the static limit $k \to 0$. See Section 9.1.

Knowledge of $\mathbf{r} \cdot \mathbf{H}$ and $\mathbf{r} \cdot \mathbf{E}$ at two different radii, r_1 and r_2, in a source-free region will therefore permit a complete specification of the fields, including determination of the relative proportions of $h_l^{(1)}$ and $h_l^{(2)}$ in f_l and g_l. The use of the scalars $\mathbf{r} \cdot \mathbf{H}$ and $\mathbf{r} \cdot \mathbf{E}$ permits the connection between the sources ρ, $\mathbf{J}$ and the multipole coefficients $a_E(l, m)$ and $a_M(l, m)$ to be established with relative ease (see Section 9.10).

9.8 *Properties of Multipole Fields; Energy and Angular Momentum of Multipole Radiation*

Before considering the connection between the general solution (9.122) and a localized source distribution, we examine the properties of the individual multipole fields (9.116) and (9.118). In the near zone ($kr \ll 1$) the radial function $f_l(kr)$ is proportional to n_l, given by (9.88), unless its coefficient vanishes identically. Excluding this possibility, the limiting behavior of the magnetic field for an electric (l, m) multipole is

$$\mathbf{H}_{lm}^{(E)} \rightarrow -\frac{k}{l} \mathbf{L} \frac{Y_{lm}}{r^{l+1}} \tag{9.124}$$

where the proportionality coefficient is chosen for later convenience. To find the electric field we must take the curl of the right-hand side. A useful operator identity is

$$i\nabla \times \mathbf{L} = \mathbf{r}\nabla^2 - \nabla\left(1 + r\frac{\partial}{\partial r}\right) \tag{9.125}$$

The electric field (9.118) is

$$\mathbf{E}_{lm}^{(E)} \rightarrow \frac{-i}{l} Z_0 \nabla \times \mathbf{L}\left(\frac{Y_{lm}}{r^{l+1}}\right) \tag{9.126}$$

Since (Y_{lm}/r^{l+1}) is a solution of the Laplace equation, the first term in (9.125) vanishes. Consequently the electric field at close distances for an electric (l, m) multipole is

$$\mathbf{E}_{lm}^{(E)} \rightarrow -Z_0\nabla\left(\frac{Y_{lm}}{r^{l+1}}\right) \tag{9.127}$$

This is exactly the electrostatic multipole field of Section 4.1. We note that the magnetic field $\mathbf{H}_{lm}^{(E)}$ is smaller in magnitude than $\mathbf{E}_{lm}^{(E)}/Z_0$ by a factor kr. Hence, in the near zone, the magnetic field of an electric multipole is always much smaller than the electric field. For the magnetic multipole fields (9.116) evidently the roles of $\mathbf{E}$ and $\mathbf{H}$ are interchanged according to the transformation,

$$\mathbf{E}^{(E)} \rightarrow -Z_0\mathbf{H}^{(M)}, \qquad \mathbf{H}^{(E)} \rightarrow \mathbf{E}^{(M)}/Z_0 \tag{9.128}$$

In the far or radiation zone ($kr \gg 1$) the multipole fields depend on the boundary conditions imposed. For definiteness we consider the example of outgoing waves, appropriate to radiation by a localized source. Then the radial function $f_l(kr)$ is proportional to the spherical Hankel function $h_l^{(1)}(kr)$. From the asymptotic form (9.89) we see that in the radiation zone the magnetic induction for an electric (l, m) multipole goes as

$$\mathbf{H}_{lm}^{(E)} \rightarrow (-i)^{l+1} \frac{e^{ikr}}{kr} \mathbf{L} Y_{lm} \tag{9.129}$$

Then the electric field can be written

$$\mathbf{E}_{lm}^{(E)} = Z_0 \frac{(-i)^l}{k^2} \left[\boldsymbol{\nabla} \left(\frac{e^{ikr}}{r} \right) \times \mathbf{L} Y_{lm} + \frac{e^{ikr}}{r} \boldsymbol{\nabla} \times \mathbf{L} Y_{lm} \right] \tag{9.130}$$

Since we have already used the asymptotic form of the spherical Hankel function, we are not justified in keeping powers higher than the first in $(1/r)$. With this restriction and use of the identity (9.125) we find

$$\mathbf{E}_{lm}^{(E)} = -Z_0 (-i)^{l+1} \frac{e^{ikr}}{kr} \left[\mathbf{n} \times \mathbf{L} Y_{lm} - \frac{1}{k} (\mathbf{r} \nabla^2 - \boldsymbol{\nabla}) Y_{lm} \right] \tag{9.131}$$

where $\mathbf{n} = (\mathbf{r}/r)$ is a unit vector in the radial direction. The second term is evidently $1/kr$ times some dimensionless function of angles and can be omitted in the limit $kr \gg 1$. Then we find that the electric field in the radiation zone is

$$\mathbf{E}_{lm}^{(E)} = Z_0 \mathbf{H}_{lm}^{(E)} \times \mathbf{n} \tag{9.132}$$

where $\mathbf{H}_{lm}^{(E)}$ is given by (9.129). These fields are typical radiation fields, transverse to the radius vector and falling off as r^{-1}. For magnetic multipoles the same relation holds because the Poynting vector is directed radially outward for both types of multipole.

The multipole fields of a radiating source can be used to calculate the energy and angular momentum carried off by the radiation. For definiteness we consider a linear superposition of electric (l, m) multipoles with different m values, but all having the same l, and, following (9.122), write the fields as

$$\begin{aligned} \mathbf{H}_l &= \sum_m a_E(l, m) \mathbf{X}_{lm} h_l^{(1)}(kr) e^{-i\omega t} \\ \mathbf{E}_l &= \frac{i}{k} Z_0 \boldsymbol{\nabla} \times \mathbf{H}_l \end{aligned} \tag{9.133}$$

For harmonically varying fields the time-averaged energy density is

$$u = \frac{\epsilon_0}{4} (\mathbf{E} \cdot \mathbf{E}^* + Z_0^2 \mathbf{H} \cdot \mathbf{H}^*) \tag{9.134}$$

In the radiation zone the two terms are equal. Consequently the energy in a spherical shell between r and $(r + dr)$ (for $kr \gg 1$) is

$$dU = \frac{\mu_0 dr}{2k^2} \sum_{m,m'} a_E^*(l, m') a_E(l, m) \int \mathbf{X}_{lm'}^* \cdot \mathbf{X}_{lm} \, d\Omega \tag{9.135}$$

where the asymptotic form (9.89) of the spherical Hankel function has been used. With the orthogonality integral (9.120) this becomes

$$\frac{dU}{dr} = \frac{\mu_0}{2k^2} \sum_m |a_E(l, m)|^2 \tag{9.136}$$

independent of the radius. For a general superpositon of electric and magnetic multipoles the sum over m becomes a sum over l and m and $|a_E|^2$ becomes $|a_E|^2 + |a_M|^2$. The total energy in a spherical shell in the radiation zone is thus an *incoherent sum* over all multipoles.

The time-averaged angular-momentum density is

$$\mathbf{m} = \frac{1}{2c^2} \text{Re}[\mathbf{r} \times (\mathbf{E} \times \mathbf{H}^*)] \tag{9.137}$$

The triple cross product can be expanded and the electric field (9.133) substituted to yield, for a superposition of electric multipoles,

$$\mathbf{m} = \frac{\mu_0}{2\omega} \text{Re}[\mathbf{H}^*(\mathbf{L} \cdot \mathbf{H})] \tag{9.138}$$

Then the angular momentum in a spherical shell between r and $(r + dr)$ in the radiation zone is

$$d\mathbf{M} = \frac{\mu_0 dr}{2\omega k^2} \text{Re} \sum_{m,m'} a_E^*(l, m')a_E(l, m) \int (\mathbf{L} \cdot \mathbf{X}_{lm'})^* \mathbf{X}_{lm} \, d\Omega \tag{9.139}$$

With the explicit form (9.119) for $\mathbf{X}_{lm}$, (9.139) can be written

$$\frac{d\mathbf{M}}{dr} = \frac{\mu_0}{2\omega k^2} \text{Re} \sum_{m,m'} a_E^*(l, m')a_E(l, m) \int Y_{lm'}^* \mathbf{L} Y_{lm} \, d\Omega \tag{9.140}$$

From the properties of $\mathbf{L}Y_{lm}$ listed in (9.104) and the orthogonality of the spherical harmonics we obtain the following expressions for the Cartesian components of $d\mathbf{M}/dr$:

$$\frac{dM_x}{dr} = \frac{\mu_0}{4\omega k^2} \text{Re} \sum_m [\sqrt{(l - m)(l + m + 1)} \, a_E^*(l, m + 1)$$
$$+ \sqrt{(l + m)(l - m + 1)} \, a_E^*(l, m - 1)]a_E(l, m) \tag{9.141}$$

$$\frac{dM_y}{dr} = \frac{\mu_0}{4\omega k^2} \text{Im} \sum_m [\sqrt{(l - m)(l + m + 1)} \, a_E^*(l, m + 1)$$
$$- \sqrt{(l + m)(l - m + 1)} \, a_E^*(l, m - 1)]a_E(l, m) \tag{9.142}$$

$$\frac{dM_z}{dr} = \frac{\mu_0}{2\omega k^2} \sum_m m \, |a_E(l, m)|^2 \tag{9.143}$$

These equations show that for a general lth-order electric multipole that consists of a superposition of different m values only the z component of angular momentum is relatively simple.

For a multipole with a single m value, M_x and M_y vanish, while a comparison of (9.143) and (9.136) shows that

$$\frac{dM_z}{dr} = \frac{m}{\omega} \frac{dU}{dr} \tag{9.144}$$

independent of r. This has the obvious quantum interpretation that the radiation from a multipole of order (l, m) carries off $m\hbar$ units of z component of angular momentum per photon of energy $\hbar\omega$. Even with a superposition of different m values, the same interpretation of (9.143) holds, with each multipole of definite

m contributing *incoherently* its share of the z component of angular momentum. Now, however, the x and y components are in general nonvanishing, with multipoles of adjacent m values contributing in a weighted coherent sum. The behavior contained in (9.140) and exhibited explicitly in (9.141)–(9.143) is familiar in the quantum mechanics of a vector operator and its representation with respect to basis states of J^2 and J_z.* The angular momentum of multipole fields affords a classical example of this behavior, with the z component being diagonal in the (l, m) multipole basis and the x and y components not.

The characteristics of the angular momentum just presented hold true generally, even though our example (9.133) was somewhat specialized. For a superposition of both electric and magnetic multipoles of various (l, m) values, the angular momentum expression (9.139) is generalized to

$$\frac{d\mathbf{M}}{dr} = \frac{\mu_0}{2\omega k^2} \operatorname{Re} \sum_{\substack{l,m \\ l',m'}} \left\{ [a_E^*(l', m')a_E(l, m) + a_M^*(l', m')a_M(l, m)] \int (\mathbf{L} \cdot \mathbf{X}_{l'm'})^* \mathbf{X}_{lm}\, d\Omega \right.$$
$$\left. + i^{l'-l}[a_E^*(l', m')a_M(l, m) - a_M^*(l', m')a_E(l, m)] \int (\mathbf{L} \cdot \mathbf{X}_{l'm'})^* \mathbf{n} \times \mathbf{X}_{lm}\, d\Omega \right\}$$

$$(9.145)$$

The first term in (9.145) is of the same form as (9.139) and represents the sum of the electric and magnetic multipoles separately. The second term is an interference between electric and magnetic multipoles. Examination of the structure of its angular integral shows that the interference is between electric and magnetic multipoles whose l values differ by unity. This is a necessary consequence of the parity properties of the multipole fields (see below). Apart from this complication of interference, the properties of $d\mathbf{M}/dr$ are as before.

The quantum-mechanical interpretation of (9.144) concerned the z component of angular momentum carried off by each photon. In further analogy with quantum mechanics we would expect the ratio of the square of the angular momentum to the square of the energy to have value

$$\frac{M^{(q)^2}}{U^2} = \frac{M_x^2 + M_y^2 + M_z^2)_q}{U^2} = \frac{l(l + 1)}{\omega^2} \qquad (9.146)$$

But from (9.136) and (9.141)–(9.143) the classical result for a pure (l, m) multipole is

$$\frac{M^{(c)^2}}{U^2} = \frac{|M_z|^2}{U^2} = \frac{m^2}{\omega^2} \qquad (9.147)$$

The reason for this difference lies in the quantum nature of the electromagnetic fields for a single photon. If the z component of angular momentum of a single photon is known precisely, the uncertainty principle requires that the other components be uncertain, with mean square values such that (9.146) holds. On the other hand, for a state of the radiation field containing many photons (the classical limit), the mean square values of the transverse components of angular momentum can be made negligible compared to the square of the z component.

---

*See for example, E. U. Condon and G. H. Shortley, *The Theory of Atomic Spectra*, Cambridge University Press, Cambridge (1953), p. 63.

Then the classical limit (9.147) applies. For a (l, m) multipole field containing N photons it can be shown* that

$$\frac{[M^{(q)}(N)]^2}{[U(N)]^2} = \frac{N^2 m^2 + Nl(l + 1) - m^2}{N^2 \omega^2} \tag{9.148}$$

This contains (9.146) and (9.147) as limiting cases.

The quantum-mechanical interpretation of the radiated angular momentum per photon for multipole fields contains the selection rules for multipole transitions between quantum states. A multipole transition of order (l, m) will connect an initial quantum state specified by total angular momentum J and z component M to a final quantum state with J' in the range $|J - l| \leq J' \leq J + l$ and $M' = M - m$. Or, alternatively, with two states (J, M) and (J', M'), possible multipole transitions have (l, m) such that $|J - J'| \leq l \leq J + J'$ and $m = M - M'$.

To complete the quantum-mechanical specification of a multipole transition it is necessary to state whether the parities of the initial and final states are the same or different. The parity of the initial state is equal to the product of the parities of the final state and the multipole field. To determine the parity of a multipole field we merely examine the behavior of the magnetic field $\mathbf{H}_{lm}$ under the parity transformation of inversion through the origin $(\mathbf{r} \rightarrow -\mathbf{r})$. One way of seeing that $\mathbf{H}_{lm}$ specifies the parity of a multipole field is to recall that the interaction of a charged particle and the electromagnetic field is proportional to $(\mathbf{v} \cdot \mathbf{A})$. If $\mathbf{H}_{lm}$ has a certain parity (even or odd) for a multipole transition, then the corresponding $\mathbf{A}_{lm}$ will have the opposite parity, since the curl operation changes parity. Then, because $\mathbf{v}$ is a polar vector with odd parity, the states connected by the interaction operator $(\mathbf{v} \cdot \mathbf{A})$ will differ in parity by the parity of the magnetic field $\mathbf{H}_{lm}$.

For electric multipoles the magnetic field is given by (9.133). The parity transformation $(\mathbf{r} \rightarrow -\mathbf{r})$ is equivalent to $(r \rightarrow r, \theta \rightarrow \pi - \theta, \phi \rightarrow \phi + \pi)$ in spherical coordinates. The operator $\mathbf{L}$ is invariant under inversion. Consequently the parity properties of $\mathbf{H}_{lm}$ for electric multipoles are specified by the transformation of $Y_{lm}(\theta, \phi)$. From (3.53) and (3.50) it is evident that the parity of Y_{lm} is $(-1)^l$. Thus we see that the *parity* of fields of *an electric multipole of order* (l, m) *is* $(-1)^l$. Specifically, the magnetic induction $\mathbf{H}_{lm}$ has parity $(-1)^l$, while the electric field $\mathbf{E}_{lm}$ has parity $(-1)^{l+1}$, since $\mathbf{E}_{lm} = iZ_0 \nabla \times \mathbf{H}_{lm}/k$.

For a *magnetic multipole of order* (l, m) *the parity is* $(-1)^{l+1}$. In this case the electric field $\mathbf{E}_{lm}$ is of the same form as $\mathbf{H}_{lm}$ for electric multipoles. Hence the parities of the fields are just opposite to those of an electric multipole of the same order.

Correlating the parity changes and angular-momentum changes in quantum transitions, we see that only certain combinations of multipole transitions can occur. For example, if the states have $J = \frac{1}{2}$ and $J' = \frac{3}{2}$, the allowed multipole orders are $l = 1, 2$. If the parities of the two states are the same, we see that parity conservation restricts the possibilities, so that only magnetic dipole and electric quadruple transitions occur. If the states differ in parity, then electric dipole and magnetic quadrupole radiation can be emitted or absorbed.

*C. Morette De Witt, and J. H. D. Jensen, *Z. Naturforsch.* **8a**, 267 (1953). Their treatment parallels ours closely, with our classical multipole coefficients $a_E(l, m)$ and $a_M(l, m)$ becoming quantum-mechanical photon annihilation operators (the complex conjugates, a_E^* and a_M^*, become Hermitian conjugate creation operators).

9.9   *Angular Distribution of Multipole Radiation*

For a general localized source distribution, the fields in the radiation zone are given by the superposition

$$\mathbf{H} \rightarrow \frac{e^{ikr-i\omega t}}{kr} \sum_{l,m} (-i)^{l+1}[a_E(l, m)\mathbf{X}_{lm} + a_M(l, m)\mathbf{n} \times \mathbf{X}_{lm}]$$

$$\mathbf{E} \rightarrow Z_0 \mathbf{H} \times \mathbf{n}$$

(9.149)

The coefficients $a_E(l, m)$ and $a_M(l, m)$ will be related to the properties of the source in the next section. The time-averaged power radiated per unit solid angle is

$$\frac{dP}{d\Omega} = \frac{Z_0}{2k^2} \left| \sum_{l,m} (-i)^{l+1}[a_E(l, m)\mathbf{X}_{lm} \times \mathbf{n} + a_M(l, m)\mathbf{X}_{lm}] \right|^2$$

(9.150)

Within the absolute value signs the dimensions are those of magnetic field, but the polarization of the radiation is specified by the directions of the vectors. We note that electric and magnetic multipoles of a given (l, m) have the same angular dependence but have polarizations at right angles to one another. Thus the multipole order may be determined by measurement of the angular distribution of radiated power, but the character of the radiation (electric or magnetic) can be determined only by a polarization measurement.

For a pure multiple of order (l, m) the angular distribution (9.150) reduces to a single term,

$$\frac{dP(l, m)}{d\Omega} = \frac{Z_0}{2k^2} |a(l, m)|^2 |\mathbf{X}_{lm}|^2$$

(9.151)

From definition (9.119) of $\mathbf{X}_{lm}$ and properties (9.104), this can be transformed into the explicit form:

$$\frac{dP(l, m)}{d\Omega} = \frac{Z_0 |a(l, m)|^2}{2k^2 l(l + 1)} \left\{ \begin{array}{l} \frac{1}{2}(l - m)(l + m + 1) |Y_{l,m+1}|^2 \\ + \frac{1}{2}(l + m)(l - m + 1) |Y_{l,m-1}|^2 + m^2 |Y_{lm}|^2 \end{array} \right\}$$

(9.152)

Table 9.1 lists some of the simpler angular distributions.

The dipole distributions are seen to be those of a dipole oscillating parallel to the z axis ($m = 0$) and of two dipoles, one along the x axis and one along the y axis, 90° out of phase ($m = \pm 1$). The dipole and quadrupole angular distributions are plotted as polar intensity diagrams in Fig. 9.5. These are representative of $l = 1$ and $l = 2$ multipole angular distributions, although a general multipole

Table 9.1   Some Angular Distributions: $|\mathbf{X}_{lm}(\theta, \phi)|^2$

		m	
l	0	± 1	± 2
1 Dipole	$\frac{3}{8\pi} \sin^2\theta$	$\frac{3}{16\pi} (1 + \cos^2\theta)$	
2 Quadrupole	$\frac{15}{8\pi} \sin^2\theta \cos^2\theta$	$\frac{5}{16\pi} (1 - 3\cos^2\theta + 4\cos^4\theta)$	$\frac{5}{16\pi} (1 - \cos^4\theta)$

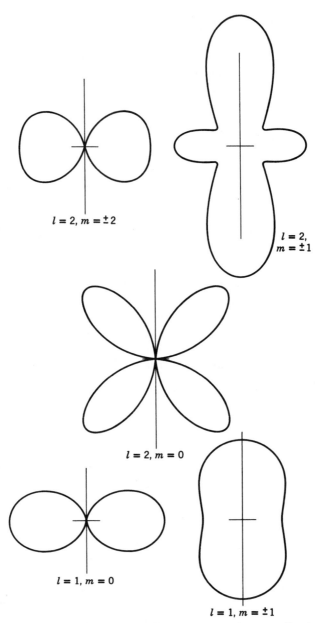

$l = 2, m = \pm 2$

$l = 2,$
$m = \pm 1$

$l = 2, m = 0$

$l = 1, m = 0$

$l = 1, m = \pm 1$

Figure 9.5 Dipole and quadrupole radiation patterns for pure (l, m) multipoles.

distribution of order l will involve a coherent superposition of the $(2l + 1)$ amplitudes for different m, as shown in (9.150).

It can be shown by means of (3.69) that the absolute squares of the vector spherical harmonics obey the sum rule,

$$\sum_{m=-l}^{l} |\mathbf{X}_{lm}(\theta, \phi)|^2 = \frac{2l + 1}{4\pi} \tag{9.153}$$

Hence the radiation distribution will be isotropic from a source that consists of a set of multipoles of order l, with coefficients $a(l, m)$ independent of m, superposed incoherently. This situation usually prevails in atomic and nuclear radiative transitions unless the initial state has been prepared in a special way.

The total power radiated by a pure multipole of order (l, m) is given by the integral of (9.151) over all angles. Since the $\mathbf{X}_{lm}$ are normalized to unity, the power radiated is

$$P(l, m) = \frac{Z_0}{2k^2} |a(l, m)|^2 \tag{9.154}$$

For a general source the angular distribution is given by the coherent sum (9.150). On integration over angles it is easy to show that the interference terms do not contribute. Hence the total power radiated is just an incoherent sum of contributions from the different multipoles:

$$P = \frac{Z_0}{2k^2} \sum_{l,m} [|a_E(l, m)|^2 + |a_M(l, m)|^2] \tag{9.155}$$

9.10   Sources of Multipole Radiation; Multipole Moments

Having discussed the properties of multipole fields, the radiation patterns, and the angular momentum and energy carried off, we now turn to the connection of the fields with the sources that generate them. We assume that there exist localized well-behaved distributions of charge $\rho(\mathbf{x}, t)$, current $\mathbf{J}(\mathbf{x}, t)$, and intrinsic magnetization $\mathcal{M}(\mathbf{x}, t)$. Furthermore, we assume that the time dependence can be analyzed into its Fourier components, and we consider only harmonically varying sources,

$$\rho(\mathbf{x})e^{-i\omega t}, \qquad \mathbf{J}(\mathbf{x})e^{-i\omega t}, \qquad \mathcal{M}(\mathbf{x})e^{-i\omega t} \tag{9.156}$$

where it is understood that we take the real part of such complex quantities. A more general time dependence can be obtained by linear superposition (see also Problem 9.1).

The Maxwell equations for $\mathbf{E}$ and $\mathbf{H}' = \mathbf{B}/\mathcal{M}_0$ are

$$\begin{aligned}
\boldsymbol{\nabla} \cdot \mathbf{H}' &= 0, & \boldsymbol{\nabla} \times \mathbf{E} - ikZ_0\mathbf{H}' &= 0 \\
\boldsymbol{\nabla} \cdot \mathbf{E} &= \rho/\epsilon_0, & \boldsymbol{\nabla} \times \mathbf{H}' + ik\mathbf{E}/Z_0 &= \mathbf{J} + \boldsymbol{\nabla} \times \mathcal{M}
\end{aligned} \tag{9.157}$$

with the continuity equation,

$$i\omega\rho = \boldsymbol{\nabla} \cdot \mathbf{J} \tag{9.158}$$

It is convenient to deal with divergenceless fields. Accordingly, we use as field variables, $\mathbf{H}'$ and

$$\mathbf{E}' = \mathbf{E} + \frac{i}{\omega\epsilon_0} \mathbf{J} \tag{9.159}$$

In the region outside the sources, $\mathbf{E}'$ reduces to $\mathbf{E}$ and $\mathbf{H}'$ to $\mathbf{H}$. In terms of these fields the Maxwell equations read

$$\begin{aligned}
\boldsymbol{\nabla} \cdot \mathbf{H}' &= 0, & \boldsymbol{\nabla} \times \mathbf{E}' - ikZ_0\mathbf{H}' &= \frac{i}{\omega\epsilon_0} \boldsymbol{\nabla} \times \mathbf{J} \\
\boldsymbol{\nabla} \cdot \mathbf{E}' &= 0, & \boldsymbol{\nabla} \times \mathbf{H}' + ik\mathbf{E}'/Z_0 &= \boldsymbol{\nabla} \times \mathcal{M}
\end{aligned} \tag{9.160}$$

The curl equations can be combined to give the inhomogeneous Helmholtz wave equations

$$(\nabla^2 + k^2)\mathbf{H}' = -\nabla \times (\mathbf{J} + \nabla \times \boldsymbol{\mathcal{M}})$$

and   (9.161)

$$(\nabla^2 + k^2)\mathbf{E}' = -iZ_0k\nabla \times \left(\boldsymbol{\mathcal{M}} + \frac{1}{k^2}\nabla \times \mathbf{J}\right)$$

These wave equations, together with $\nabla \cdot \mathbf{H}' = 0$, $\nabla \cdot \mathbf{E}' = 0$, and the curl equations giving $\mathbf{E}'$ in terms of $\mathbf{H}'$ or vice versa, are the counterparts of (9.108) and (9.109) when sources are present.

Since the multipole coefficients in (9.122) are determined according to (9.123) from the scalars $\mathbf{r} \cdot \mathbf{H}'$ and $\mathbf{r} \cdot \mathbf{E}'$, it is sufficient to consider wave equations for them, rather than the vector fields $\mathbf{E}'$ and $\mathbf{H}'$. From (9.110), (9.161) and the vector relation, $\mathbf{r} \cdot (\nabla \times \mathbf{A}) = (\mathbf{r} \times \nabla) \cdot \mathbf{A} = i\mathbf{L} \cdot \mathbf{A}$ for any vector field $\mathbf{A}$, we find the inhomogeneous wave equations

$$(\nabla^2 + k^2)\mathbf{r} \cdot \mathbf{H}' = -i\mathbf{L} \cdot (\mathbf{J} + \nabla \times \boldsymbol{\mathcal{M}}) \qquad (9.162)$$

$$(\nabla^2 + k^2)\mathbf{r} \cdot \mathbf{E}' = Z_0k\mathbf{L} \cdot \left(\boldsymbol{\mathcal{M}} + \frac{1}{k^2}\nabla \times \mathbf{J}\right)$$

The solutions of these scalar wave equations follow directly from the development in Section 6.4. With the boundary condition of outgoing waves at infinity, we have

$$\mathbf{r} \cdot \mathbf{H}'(\mathbf{x}) = \frac{i}{4\pi} \int \frac{e^{ik|\mathbf{x}-\mathbf{x}'|}}{|\mathbf{x}-\mathbf{x}'|} \mathbf{L}' \cdot [\mathbf{J}(\mathbf{x}') + \nabla' \times \boldsymbol{\mathcal{M}}(\mathbf{x}')] \, d^3x'$$

(9.163)

$$\mathbf{r} \cdot \mathbf{E}'(\mathbf{x}) = -\frac{Z_0k}{4\pi} \int \frac{e^{ik|\mathbf{x}-\mathbf{x}'|}}{|\mathbf{x}-\mathbf{x}'|} \mathbf{L}' \cdot \left[\boldsymbol{\mathcal{M}}(\mathbf{x}') + \frac{1}{k^2}\nabla' \times \mathbf{J}(\mathbf{x}')\right] d^3x'$$

To evaluate the multipole coefficients by means of (9.123), we first observe that the requirement of outgoing waves at infinity makes $A_l^{(2)} = 0$ in (9.113). Thus we choose $f_l(kr) = g_l(kr) = h_l^{(1)}(kr)$ in (9.122) as the representation of $\mathbf{E}$ and $\mathbf{H}$ outside the sources. Next we consider the spherical wave representation (9.98) for the Green function in (9.163) and assume that the point $\mathbf{x}$ is outside a spherical surface completely enclosing the sources. Then in the integrations in (9.163), $r_< = r'$, $r_> = r$. The spherical wave projection needed for (9.123) is

$$\frac{1}{4\pi} \int d\Omega \, Y_{lm}^*(\theta, \phi) \frac{e^{ik|\mathbf{x}-\mathbf{x}'|}}{|\mathbf{x}-\mathbf{x}'|} = ik \, h_l^{(1)}(kr)j_l(kr')Y_{lm}^*(\theta', \phi') \quad (9.164)$$

By means of this projection we see that $a_M(l, m)$ and $a_E(l, m)$ are given in terms of the integrands in (9.163) by

$$a_E(l, m) = \frac{ik^3}{\sqrt{l(l+1)}} \int j_l(kr)Y_{lm}^*\mathbf{L} \cdot \left(\boldsymbol{\mathcal{M}} + \frac{1}{k^2}\nabla \times \mathbf{J}\right) d^3x$$

(9.165)

$$a_M(l, m) = \frac{-k^2}{\sqrt{l(l+1)}} \int j_l(kr)Y_{lm}^*\mathbf{L} \cdot (\mathbf{J} + \nabla \times \boldsymbol{\mathcal{M}}) \, d^3x$$

The expressions in (9.165) give the strengths of the various multipole fields outside the source in terms of integrals over the source densities $\mathbf{J}$ and $\boldsymbol{\mathcal{M}}$. They

can be transformed into more useful forms by means of the following identities:
Let $\mathbf{A}(\mathbf{x})$ be any well-behaved vector field. Then

$$\mathbf{L} \cdot \mathbf{A} = i\boldsymbol{\nabla} \cdot (\mathbf{r} \times \mathbf{A}) \tag{9.166}$$

$$\mathbf{L} \cdot (\boldsymbol{\nabla} \times \mathbf{A}) = i\nabla^2(\mathbf{r} \cdot \mathbf{A}) - \frac{i}{r} \frac{\partial}{\partial r} (r^2 \boldsymbol{\nabla} \cdot \mathbf{A})$$

These follow from the definition (9.101) of $\mathbf{L}$ and simple vector identities. With
$\mathbf{A} = \mathcal{M}$ in the first equation and $\mathbf{A} = \mathbf{J}$ in the second, the integral for $a_E(l, m)$ in
(9.165) becomes

$$a_E(l, m) = -\frac{k^3}{\sqrt{l(l + 1)}} \int j_l(kr) Y^*_{lm} \left[\boldsymbol{\nabla} \cdot (\mathbf{r} \times \mathcal{M}) \right.$$

$$\left. + \frac{1}{k^2} \nabla^2(\mathbf{r} \cdot \mathbf{J}) - \frac{ic}{k} \frac{1}{r} \frac{\partial}{\partial r} (r^2 \rho) \right] d^3x$$

where we have used (9.158) to express $\boldsymbol{\nabla} \cdot \mathbf{J}$ in terms of ρ. Use of Green's theorem
on the second term replaces ∇^2 by $-k^2$, while a radial integration by parts on the
third term casts the radial derivative over onto the spherical Bessel function. The
result for the *electric multipole coefficient* is

$$a_E(l, m) = \frac{k^2}{i\sqrt{l(l + 1)}} \int Y^*_{lm} \left\{ c\rho \frac{\partial}{\partial r} [rj_l(kr)] + ik(\mathbf{r} \cdot \mathbf{J})j_l(kr) \right. \left. - ik\boldsymbol{\nabla} \cdot (\mathbf{r} \times \mathcal{M})j_l(kr) \right\} d^3x \tag{9.167}$$

The analogous manipulation with the second equation in (9.165) leads to the
magnetic multipole coefficient,

$$a_M(l, m) = \frac{k^2}{i\sqrt{l(l + 1)}} \int Y^*_{lm} \left\{ \boldsymbol{\nabla} \cdot (\mathbf{r} \times \mathbf{J})j_l(kr) + \boldsymbol{\nabla} \cdot \mathcal{M} \frac{\partial}{\partial r} [rj_l(kr)] \right. \left. - k^2(\mathbf{r} \cdot \mathcal{M})j_l(kr) \right\} d^3x \tag{9.168}$$

These results are exact expressions, valid for arbitrary frequency and source size.
    For many applications in atomic and nuclear physics the source dimensions
are very small compared to a wavelength ($kr_{max} \ll 1$). Then the multipole coefficients can be simplified considerably. The small argument limit (9.88) can be
used for the spherical Bessel functions. Keeping only the lowest powers in kr for
terms involving ρ or $\mathbf{J}$ and $\mathcal{M}$, we find the approximate electric multipole
coefficient,

$$a_E(l, m) \simeq \frac{ck^{l+2}}{i(2l + 1)!!} \left(\frac{l + 1}{l}\right)^{1/2} (Q_{lm} + Q'_{lm}) \tag{9.169}$$

where the multipole moments are

$$Q_{lm} = \int r^l Y^*_{lm} \rho \, d^3x$$

and

$$Q'_{lm} = \frac{-ik}{(l + 1)c} \int r^l Y^*_{lm} \boldsymbol{\nabla} \cdot (\mathbf{r} \times \mathcal{M}) \, d^3x$$

$$\left. \right\} \tag{9.170}$$

The moment Q_{lm} is seen to be the same in form as the electrostatic multipole
moment q_{lm} (4.3). The moment Q'_{lm} is an induced electric multipole moment due

to the magnetization. It is generally at least a factor kr smaller than the normal moment Q_{lm}. For the magnetic multipole coefficient $a_M(l, m)$ the corresponding long-wavelength approximation is

$$a_M(l, m) \simeq \frac{ik^{l+2}}{(2l + 1)!!} \left(\frac{l + 1}{l}\right)^{1/2} (M_{lm} + M'_{lm}) \tag{9.171}$$

where the magnetic multipole moments are

and

$$\left.\begin{array}{l} M_{lm} = -\dfrac{1}{l + 1} \displaystyle\int r^l Y^*_{lm} \, \boldsymbol{\nabla} \cdot (\mathbf{r} \times \mathbf{J}) \, d^3x \\[4mm] M'_{lm} = -\displaystyle\int r^l Y^*_{lm} \, \boldsymbol{\nabla} \cdot \boldsymbol{\mathcal{M}} \, d^3x \end{array}\right\} \tag{9.172}$$

In contrast to the electric multipole moments Q_{lm} and Q'_{lm}, for a system with intrinsic magnetization the magnetic moments M_{lm} and M'_{lm} are generally of the same order of magnitude.

In the long-wavelength limit we see clearly that electric multipole fields are related to the electric-charge density ρ, while the magnetic multipole fields are determined by the magnetic-moment densities, $(\mathbf{r} \times \mathbf{J})/2$ and $\boldsymbol{\mathcal{M}}$.

9.11 Multipole Radiation in Atoms and Nuclei

Although a full discussion of radiative transitions in atoms and nuclei requires a quantum-mechanical treatment, the qualitative aspects can be gleaned from our classical formulas by means of semiclassical arguments and simple estimates of the effective multipole moments. First of all, we note that the transition probability Γ (reciprocal mean life) for emission of a photon of energy $\hbar\omega$ is given by the radiated power divided by $\hbar\omega$. From (9.154) for the power and (9.169) and (9.171) for the amplitudes a_E and a_M in terms of the long-wavelength multipoles, we find the transition probability for an electric multipole (l, m),

$$\Gamma_E(l, m) = \frac{\omega Z_0 k^{2l}}{2\hbar[(2l + 1)!!]^2} \left(\frac{l + 1}{l}\right) |Q_{lm} + Q'_{lm}|^2 \tag{9.173}$$

For a magnetic multipole, $Q_{lm} + Q'_{lm} \to (1/c)[M_{lm} + M'_{lm}]$.

The effective multipole moments can be estimated as to order of magnitude as follows. Suppose that for the system under consideration the effective charge is e, the effective mass of the radiating constituents is m, and the effective size is R. Then the effective magnetization is $|\boldsymbol{\mathcal{M}}| = O(e\hbar/mR^3)$, where $e\hbar/m$ is the effective magnetic moment of the constituents. The most naive estimates of the multipole coefficients are then

$$|Q_{lm}| = O(eR^l); \qquad |Q'_{lm}| = O\!\left(\frac{\hbar\omega}{mc^2} eR^l\right)$$

and

$$\frac{1}{c} |M_{lm} + M'_{lm}| = O\!\left(\frac{e\hbar}{mc} R^{l-1}\right) \tag{9.174}$$

With these order-of-magnitude estimates some qualitative features of atomic and nuclear radiative transitions can be abstracted. In atoms and in nuclei the transition energies $\hbar\omega$ are invariably small compared to the rest energy mc^2 of the constituents. We thus see that $|Q'_{lm}| \ll |Q_{lm}|$ is a universal expectation. Electric multipole transitions of order l (denoted by El) are dominated by the transitional charge density, with negligible contribution from the "magnetization charge." On the other hand, magnetic multipole transitions (Ml) generally have comparable contributions from the orbital and intrinsic magnetizations.

In atoms the electrons are the radiating constituents. The size of the system is $R = O(a_0/Z_{\text{eff}})$, where a_0 is the Bohr radius and Z_{eff} is of order unity for valence electron transitions and of order Z for K- or L-shell x-ray transitions. From (9.174) the relative size of the magnetic multipole moments with respect to the electric of the same order l is $|M|/c|Q| = O(\hbar/mcR) = O(Z_{\text{eff}}/137)$. For the same transition energy, the transition probabilities will be in the ratio

$$\frac{\Gamma_M(l)}{\Gamma_E(l)} = O\left(\frac{Z_{\text{eff}}^2}{(137)^2}\right) \tag{9.175}$$

Only for x-ray transitions in heavy elements are magnetic multipoles even remotely competitive with electric multipoles of the same order. [Note, however, that the Ml transitions have the opposite parity properties to the El for the same l.]

Of interest is the relative size of transition probabilities for multipoles differing by one unit in order. Ignoring factors of order unity, we see from (9.173) and (9.174) that

$$\frac{\Gamma_{E,M}(l+1)}{\Gamma_{E,M}(l)} = O(k^2 R^2) \tag{9.176}$$

In atoms the transition energies are of order $Z_{\text{eff}}^2 mc^2/(137)^2$, while the size is $R = O(137\,\hbar/mcZ_{\text{eff}})$. We thus find $kR = O(Z_{\text{eff}}/137)$ and the ratio for successive El multipoles is of the same order as (9.175). For atomic transitions in which the angular-momentum selection rules permit several multipoles, the lowest multipole generally dominates. For example, if the initial and final angular momenta are $J = \frac{1}{2}$ and $J' = \frac{3}{2}$ and the states have the opposite parity, the allowed multipoles are $E1$ and $M2$. The $E1$ transition will dominate by a factor of order $(Z_{\text{eff}}/137)^4$. If the parities are the same, the allowed transitions are $M1$ and $E2$. Now the two transition mechanisms may be comparable, with transition probabilities much smaller than for opposite parities. In atoms the dominant transitions are $E1$; high angular momentum states de-excite by a cascade of $E1$ transitions, if at all possible.

In nuclei the situation is somewhat different. Successive multipoles of the same type still obey the estimate (9.176), but the transition energies vary significantly. With the nuclear radius $R = 1.4\,A^{1/3} \times 10^{-15}$ m as the effective size, numerically we have $kR \leqq [\hbar\omega(\text{MeV})]\,A^{1/3}/140$. Energies vary from a few keV to several MeV. In heavy nuclei, this corresponds to a range, $kR \leqq 10^{-4}$–10^{-1}. Evidently, for energetic nuclear transitions successive multipoles of the same type are not as suppressed as in atoms. For low energies, however, the suppression of rate with multipole order is dramatic. $M4$ isomeric transitions with energies of the order of 100 keV or less can have mean lives of hours. The nuclear estimates for magnetic relative to electric transition rates of the same order, and for an electric multipole of one higher order relative to a magnetic transition, are

$$\frac{\Gamma_M(l)}{\Gamma_E(l)} = O(0.2\, A^{-2/3}); \qquad \frac{\Gamma_E(l+1)}{\Gamma_M(l)} = O\!\left(\frac{(\hbar\omega[\text{MeV}])^2\, A^{4/3}}{4000}\right) \quad (9.177)$$

In these estimates we have taken the effective magnetization to be roughly $3\, e\hbar/m_\text{N}R^3$, with a g factor of 3 to account for the magnetic moments of nucleons.

Our estimates of the nuclear transition rates are subject to exceptions ascribable to special properties of the nuclear states and interactions. In light to medium mass nuclei, $E1$ transitions are strongly suppressed by the isospin symmetry of nuclear forces, at least at low energies. $M1$ transitions are far commoner than $E1$ transitions and just as intense. In rare earth and transuranic nuclei, $E2$ transitions are often 100 times stronger than our estimate because of significant static and transitional quadrupole moments in these nonspherical nuclei. If allowed by spin-parity, $E2$ transitions then compete favorably with $M1$ transitions.

A proper quantum-mechanical treatment of multipole radiation can be found in *Blatt and Weisskopf*, Chapter XII. Applications to nuclear transitions are cited in the References and Suggested Reading at the end of the chapter.

9.12 Multipole Radiation from a Linear, Center-Fed Antenna

As an illustration of the use of a multipole expansion for a source whose dimensions are comparable to a wavelength, we consider the radiation from a thin, linear, center-fed antenna, as shown in Fig. 9.6. We have already given in Section 9.4 a direct solution for the fields when the current distribution is taken to be sinusoidal. This will serve as a basis of comparison to test the convergence of the multipole expansion. We assume the antenna to lie along the z axis from $-(d/2) \le z \le (d/2)$, and to have a small gap at its center so that it can be suitably

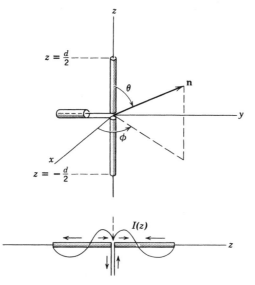

Figure 9.6 Linear, center-fed antenna.

excited. The current along the antenna vanishes at the end points and is an even function of z. For the moment we will not specify it more than to write

$$I(z, t) = I(|z|)e^{-i\omega t}, \qquad I\left(\frac{d}{2}\right) = 0 \tag{9.178}$$

Since the current flows radially, $(\mathbf{r} \times \mathbf{J}) = 0$. Furthermore there is no intrinsic magnetization. Consequently all magnetic multipole coefficients $a_M(l, m)$ vanish. To calculate the electric multipole coefficient $a_E(l, m)$ (9.167) we need expressions for the charge and current densities. The current density $\mathbf{J}$ is a radial current, confined to the z axis. In spherical coordinates this can be written for $r < (d/2)$

$$\mathbf{J}(\mathbf{x}) = \hat{\mathbf{r}} \frac{I(r)}{2\pi r^2} [\delta(\cos\theta - 1) - \delta(\cos\theta + 1)] \tag{9.179}$$

where the delta functions cause the current to flow only upward (or downward) along the z axis. From the continuity equation (9.158) we find the charge density

$$\rho(\mathbf{x}) = \frac{1}{i\omega} \frac{dI(r)}{dr} \left[\frac{\delta(\cos\theta - 1) - \delta(\cos\theta + 1)}{2\pi r^2} \right] \tag{9.180}$$

These expressions for $\mathbf{J}$ and ρ can be inserted into (9.167) to give

$$a_E(l, m) = \frac{k^2}{2\pi\sqrt{l(l + 1)}} \int_0^{d/2} dr \left\{ krj_l(kr)I(r) - \frac{1}{k}\frac{dI}{dr}\frac{d}{dr}[rj_l(kr)] \right\} \tag{9.181}$$

$$\times \int d\Omega \, Y^*_{lm}[\delta(\cos\theta - 1) - \delta(\cos\theta + 1)]$$

The integral over angles is

$$\int d\Omega = 2\pi\delta_{m,0}[Y_{l0}(0) - Y_{l0}(\pi)]$$

showing that only $m = 0$ multipoles occur. This is obvious from the cylindrical symmetry of the antenna. The Legendre polynomials are even (odd) about $\theta = \pi/2$ for l even (odd). Hence, the only nonvanishing multipoles have l odd. The the angular integral has the value,

$$\int d\Omega = \sqrt{4\pi(2l + 1)}, \qquad l \text{ odd}, m = 0$$

With slight manipulation (9.181) can be written

$$a_E(l, 0) = \frac{k}{2\pi} \left[\frac{4\pi(2l + 1)}{l(l + 1)} \right]^{1/2} \int_0^{d/2} \left\{ -\frac{d}{dr}\left[rj_l(kr)\frac{dI}{dr} \right] \right. $$
$$\left. + rj_l(kr)\left(\frac{d^2I}{dr^2} + k^2I \right) \right\} dr \tag{9.182}$$

To evaluate (9.182) we must specify the current $I(z)$ along the antenna. If no radiation occurred, the sinusoidal variation in time at frequency ω would imply a sinusoidal variation in space with wave number $k = \omega/c$. But as discussed in Section 9.4.B, the emission of radiation modifies the current distribution unless the antenna is infinitely thin. The correct current $I(z)$ can be found only by solving a complicated boundary-value problem. Since our purpose here is to compare a multipole expansion with a closed form of solution for a *known* current distribution, we make the same assumption about $I(z)$ as in Section 9.4.A, namely,

$$I(z) = I \sin\left(\frac{kd}{2} - k|z|\right) \tag{9.183}$$

where I is the peak current, and the phase is chosen to ensure that the current vanishes at the ends of the antenna. With a sinusoidal current the second part of the integrand in (9.182) vanishes. The first part is a perfect differential. Consequently we immediately obtain, with $I(z)$ from (9.183),

$$a_E(l, 0) = \frac{I}{\pi d} \left[\frac{4\pi(2l + 1)}{l(l + 1)} \right]^{1/2} \left[\left(\frac{kd}{2}\right)^2 j_l\left(\frac{kd}{2}\right) \right], \qquad l \text{ odd} \tag{9.184}$$

Since we wish to test the multipole expansion when the source dimensions are comparable to a wavelength, we consider the special cases of a half-wave antenna ($kd = \pi$) and a full-wave antenna ($kd = 2\pi$). Table 9.2 shows the $l = 1$ coefficient for these two values of kd, along with the relative values for $l = 3, 5$. From the table it is evident that (a) the coefficients decrease rapidly in magnitude as l increases, and (b) higher l coefficients are more important the larger the source dimensions. But even for the full-wave antenna it is probably adequate to keep only $l = 1$ and $l = 3$ in the angular distribution and certainly adequate for the total power (which involves the squares of the coefficients).

With only dipole and octupole terms in the angular distribution we find that the power radiated per unit solid angle (9.150) is

$$\frac{dP}{d\Omega} = \frac{Z_0 |a_E(1, 0)|^2}{4k^2} \left| \mathbf{L}Y_{1,0} - \frac{a_E(3, 0)}{\sqrt{6}\, a_E(1, 0)} \mathbf{L}Y_{3,0} \right|^2 \tag{9.185}$$

The various factors in the absolute square are

$$|\mathbf{L}Y_{1,0}|^2 = \frac{3}{4\pi} \sin^2\theta$$

$$|\mathbf{L}Y_{3,0}|^2 = \frac{63}{16\pi} \sin^2\theta (5 \cos^2\theta - 1)^2 \tag{9.186}$$

$$(\mathbf{L}Y_{1,0})^* \cdot (\mathbf{L}Y_{3,0}) = \frac{3\sqrt{21}}{8\pi} \sin^2\theta (5 \cos^2\theta - 1)$$

Table 9.2 Multipole Coefficients for Linear Antenna

kd	$a_E(1, 0)$	$a_E(3, 0)/a_E(1, 0)$	$a_E(5, 0)/a_E(1, 0)$
π	$\sqrt{\dfrac{6}{\pi}}\dfrac{I}{d}$	4.95×10^{-2}	1.02×10^{-3}
2π	$\sqrt{6\pi}\,\dfrac{I}{d}$	0.3242	2.39×10^{-2}

With these angular factors (9.185) becomes

$$\frac{dP}{d\Omega} = \lambda \frac{3Z_0 I^2}{\pi^3} \left(\frac{3}{8\pi} \sin^2\theta\right) \left|1 - \sqrt{\frac{7}{8}} \frac{a_E(3, 0)}{a_E(1, 0)} (5\cos^2\theta - 1)\right|^2 \quad (9.187)$$

where the factor λ is equal to 1 for the half-wave antenna and $(\pi^2/4)$ for the full wave. The coefficient of $(5\cos^2\theta - 1)$ in (9.187) is 0.0463 and 0.3033 for the half-wave and full-wave antenna, respectively.

A numerical comparison of the exact and approximate angular distributions, (9.57) and (9.187), is shown in Fig. 9.7. The solid curves are the exact results, the dashed curves the two-term multipole expansions. For the half-wave case (Fig. 9.7a) the simple dipole result [first term in (9.187)] is also shown as a dotted curve. The two-term multipole expansion is almost indistinguishable from the exact result for $kd = \pi$. Even the lowest order approximation is not very far off in this case. For the full-wave antenna (Fig. 9.7b) the dipole approximation is evidently quite poor. But the two-term multipole expansion is reasonably good, differing by less than 5% in the region of appreciable radiation.

The total power radiated is, according to (9.155),

$$P = \frac{Z_0}{2k^2} \sum_{l \text{ odd}} |a_E(l, 0)|^2 \quad (9.188)$$

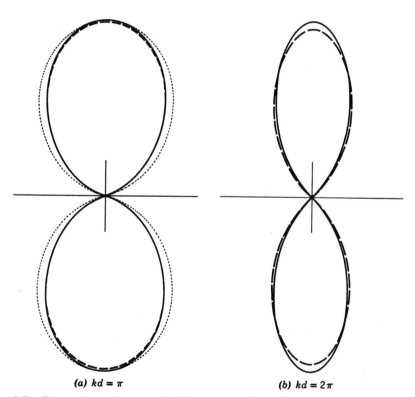

(a) $kd = \pi$ **(b) $kd = 2\pi$**

Figure 9.7   Comparison of exact radiation patterns (solid curves) for half-wave ($kd = \pi$) and full-wave ($kd = 2\pi$) center-fed antennas with two-term multipole expansions (dashed curves). For the half-wave pattern, the dipole approximation (dotted curve) is also shown. The agreement between the exact and two-term multipole results is excellent, especially for $kd = \pi$.

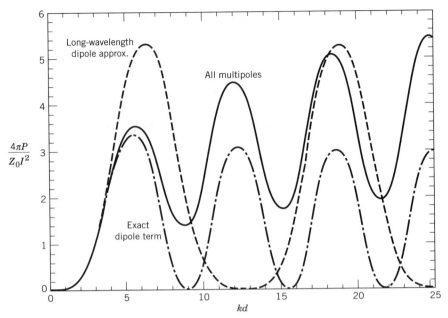

Figure 9.8 Total power radiated by center-fed antenna with sinusoidal current distribution (9.183) versus kd. The ordinate is $4\pi P/Z_0 I^2$, with I the peak current in (9.183). The curve labeled "Long-wavelength dipole approx." employs the long-wavelength dipole moment (9.170) rather than the exact (9.167) used for the curve labeled "Exact dipole term." The curve labeled "All multipoles" is the sum (9.188) [actually up to $E9$].

For the half-wave antenna the coefficients in Table 9.2 show that the power radiated is a factor 1.00244 times larger than the simple dipole result, $(3Z_0 I^2/\pi^3)$. For the full-wave antenna, the power is a factor 1.10565 times larger than the dipole form $(3Z_0 I^2/4\pi)$.

A comparison of the total power (9.188) for the center-fed linear antenna with the lowest multipole power, for both the exact lowest multipole and its long-wavelength approximation, is shown in Fig. 9.8 versus kd. For $kd \lesssim 2\pi$, the power is dominated by the $E1$ multipole, as we have just seen, but for larger kd the higher multipoles contribute more and more. It is noteworthy that the long-wavelength dipole approximation departs significantly from the exact dipole result (and the total power) for $kd > \pi$. The departure, which becomes gross for larger kd, is a consequence of differences between exact multipole moments and the long-wavelength approximations to them when the wavelength becomes comparable to or smaller than source size.

References and Suggested Reading

The simple theory of radiation from a localized source distribution is discussed in all modern textbooks. Treatments analogous to that given here may be found in
Panofsky and Phillips, Chapter 13
Smythe, Chapter 12
Stratton, Chapter 8

More complete discussions of antennas and antenna arrays are given in applied works, such as

Jordan and Balmain
Kraus
Schelkunoff and Friis
Silver

Treatments of antennas as boundary-value problems from various points of view can be found in

Hallén
Jones
Schelkunoff, *Advanced Antenna Theory*

The subject of excitation of waveguides by localized sources and the use of multipole moments is discussed by

Collin

The original literature on the description of small apertures (Bethe holes) in terms of effective dipole moments has been cited in Section 9.5. The basic theory and some applications appear in

Collin
Montgomery, Dicke, and Purcell (pp. 176 ff. and pp. 296 ff.)
Van Bladel

The theory of vector spherical harmonics and multipole vector fields is discussed thoroughly by

Blatt and Weisskopf, Appendix B
Morse and Feshbach, Section 13.3

Applications to nuclear multipole radiation are given in

Blatt and Weisskopf, Chapter XII
Siegbahn, Chapter XIII by S. A. Moszkowski and Chapter XVI (II) by M. Goldhaber and A. W. Sunyar

Problems

9.1 A common textbook example of a radiating system (see Problem 9.2) is a configuration of charges fixed relative to each other but in rotation. The charge density is obviously a function of time, but it is not in the form of (9.1).

(a) Show that for rotating charges one alternative is to calculate *real* time-dependent multipole moments using $\rho(\mathbf{x}, t)$ directly and then compute the multipole moments for a given harmonic frequency with the convention of (9.1) by inspection or Fourier decomposition of the time-dependent moments. Note that care must be taken when calculating $q_{lm}(t)$ to form linear combinations that are real before making the connection.

(b) Consider a charge density $\rho(\mathbf{x}, t)$ that is periodic in time with period $T = 2\pi/\omega_0$. By making a Fourier *series* expansion, show that it can be written as

$$\rho(\mathbf{x}, t) = \rho_0(\mathbf{x}) + \sum_{n=1}^{\infty} \text{Re}[2\rho_n(\mathbf{x})e^{-in\omega_0 t}]$$

where

$$\rho_n(\mathbf{x}) = \frac{1}{T} \int_0^T \rho(\mathbf{x}, t)e^{in\omega_0 t} \, dt$$

This shows explicitly how to establish connection with (9.1).

(c) For a single charge q rotating about the origin in the x-y plane in a circle of radius R at constant angular speed ω_0, calculate the $l = 0$ and $l = 1$ multipole moments by the methods of parts a and b and compare. In method b express the charge density $\rho_n(\mathbf{x})$ in cylindrical coordinates. Are there higher multipoles, for example, quadrupole? At what frequencies?

9.2 A radiating quadrupole consists of a square of side a with charges $\pm q$ at alternate corners. The square rotates with angular velocity ω about an axis normal to the plane of the square and through its center. Calculate the quadrupole moments, the radiation fields, the angular distribution of radiation, and the total radiated power, all in the long-wavelength approximation. What is the frequency of the radiation?

9.3 Two halves of a spherical metallic shell of radius R and infinite conductivity are separated by a very small insulating gap. An alternating potential is applied between the two halves of the sphere so that the potentials are $\pm V \cos \omega t$. In the long-wavelength limit, find the radiation fields, the angular distribution of radiated power, and the total radiated power from the sphere.

9.4 Apply the approach of Problem 9.1b to the current and magnetization densities of the particle of charge q rotating about the origin in the x-y plane in a circle of radius R at constant angular speed ω_0. The motion is such that $\omega_0 R \ll c$.

(a) Find $(J_x)_n$, $(J_y)_n$, and $(J_z)_n$ in terms of cylindrical coordinates for all n. Also determine the components of the orbital "magnetization," $(\mathbf{x} \times \mathbf{J}_n)/2$, and its divergence [which plays the role of a magnetic charge density for magnetic multipoles, as in M_{lm} (9.172)].

(b) What long-wavelength magnetic multipoles (l, m) occur and at what frequencies? [Remember that the multipole order l does not necessarily equal the harmonic number n.]

(c) Use linear superposition to generalize your argument to the four charges rotating in Problem 9.2 at radius $R = a/\sqrt{2}$. What harmonics occur, and what magnetic multipoles at each harmonic? Is there a magnetic multipole contribution at the $E2$ frequency of Problem 9.2? Is it significant relative to the $E2$ radiation?

9.5 **(a)** Show that for harmonic time variation at frequency ω the electric dipole scalar and vector potentials in the Lorenz gauge and the long-wavelength limit are

$$\Phi(\mathbf{x}) = \frac{e^{ikr}}{4\pi\epsilon_0 r^2} \mathbf{n} \cdot \mathbf{p}(1 - ikr)$$

$$\mathbf{A}(\mathbf{x}) = -i \frac{\mu_0 \omega}{4\pi} \frac{e^{ikr}}{r} \mathbf{p} \qquad \text{[this is (9.16)]}$$

where $k = \omega/c$, $\mathbf{n}$ is a unit vector in the radial direction, $\mathbf{p}$ is the dipole moment (9.17), and the time dependence $e^{-i\omega t}$ is understood.

(b) Calculate the electric and magnetic fields *from the potentials* and show that they are given by (9.18).

9.6 **(a)** Starting from the general expression (9.2) for $\mathbf{A}$ and the corresponding expression for Φ, expand both $R = |\mathbf{x} - \mathbf{x}'|$ and $t' = t - R/c$ to first order in $|\mathbf{x}'|/r$ to obtain the electric dipole potentials for arbitrary time variation

$$\Phi(\mathbf{x}, t) = \frac{1}{4\pi\epsilon_0} \left[\frac{1}{r^2} \mathbf{n} \cdot \mathbf{p}_{\text{ret}} + \frac{1}{cr} \mathbf{n} \cdot \frac{\partial \mathbf{p}_{\text{ret}}}{\partial t} \right]$$

$$\mathbf{A}(\mathbf{x}, t) = \frac{\mu_0}{4\pi r} \frac{\partial \mathbf{p}_{\text{ret}}}{\partial t}$$

where $\mathbf{p}_{ret} = \mathbf{p}(t' = t - r/c)$ is the dipole moment evaluated at the retarded time measured from the origin.

(b) Calculate the dipole electric and magnetic fields directly from these potentials and show that

$$\mathbf{B}(\mathbf{x}, t) = \frac{\mu_0}{4\pi}\left[-\frac{1}{cr^2}\,\mathbf{n} \times \frac{\partial \mathbf{p}_{ret}}{\partial t} - \frac{1}{c^2 r}\,\mathbf{n} \times \frac{\partial^2 \mathbf{p}_{ret}}{\partial t^2}\right]$$

$$\mathbf{E}(\mathbf{x}, t) = \frac{1}{4\pi\epsilon_0}\left\{\left(1 + \frac{r}{c}\frac{\partial}{\partial t}\right)\left[\frac{3\mathbf{n}(\mathbf{n} \cdot \mathbf{p}_{ret}) - \mathbf{p}_{ret}}{r^3}\right] + \frac{1}{c^2 r}\,\mathbf{n} \times \left(\mathbf{n} \times \frac{\partial^2 \mathbf{p}_{ret}}{\partial t^2}\right)\right\}$$

(c) Show explicitly how you can go back and forth between these results and the harmonic fields of (9.18) by the substitutions $-i\omega \leftrightarrow \partial/\partial t$ and $\mathbf{p}e^{ikr - i\omega t} \leftrightarrow \mathbf{p}_{ret}(t')$.

9.7 (a) By means of Fourier superposition of different frequencies or equivalent means, show for a real electric dipole $\mathbf{p}(t)$ that the instantaneous radiated power per unit solid angle at a distance r from the dipole in a direction $\mathbf{n}$ is

$$\frac{dP(t)}{d\Omega} = \frac{Z_0}{16\pi^2 c^2}\left|\left[\mathbf{n} \times \frac{d^2\mathbf{p}}{dt'^2}(t')\right] \times \mathbf{n}\right|^2$$

where $t' = t - r/c$ is the retarded time. For a magnetic dipole $\mathbf{m}(t)$, substitute $(1/c)\dot{\mathbf{m}} \times \mathbf{n}$ for $(\mathbf{n} \times \ddot{\mathbf{p}}) \times \mathbf{n}$.

(b) Show similarly for a real quadrupole tensor $Q_{\alpha\beta}(t)$ given by (9.41) with a real charge density $\rho(\mathbf{x}, t)$ that the instantaneous radiated power per unit solid angle is

$$\frac{dP(t)}{d\Omega} = \frac{Z_0}{576\pi^2 c^4}\left|\left[\mathbf{n} \times \frac{d^3\mathbf{Q}}{dt'^3}(\mathbf{n}, t')\right] \times \mathbf{n}\right|^2$$

where $\mathbf{Q}(\mathbf{n}, t)$ is defined by (9.43).

9.8 (a) Show that a classical oscillating electric dipole $\mathbf{p}$ with fields given by (9.18) radiates electromagnetic angular momentum to infinity at the rate

$$\frac{d\mathbf{L}}{dt} = \frac{k^3}{12\pi\epsilon_0}\,\text{Im}[\mathbf{p}^* \times \mathbf{p}]$$

(b) What is the ratio of angular momentum radiated to energy radiated? Interpret.

(c) For a charge e rotating in the x-y plane at radius a and angular speed ω, show that there is only a z component of radiated angular momentum with magnitude $dL_z/dt = e^2 k^3 a^2/6\pi\epsilon_0$. What about a charge oscillating along the z axis?

(d) What are the results corresponding to parts a and b for magnetic dipole radiation?

Hint: The electromagnetic angular momentum density comes from more than the transverse (radiation zone) components of the fields.

9.9 (a) From the electric dipole fields with general time dependence of Problem 9.6, show that the total power and the total rate of radiation of angular momentum through a sphere at large radius r and time t are

$$P(t) = \frac{1}{6\pi\epsilon_0 c^3}\left(\frac{\partial^2 \mathbf{p}_{ret}}{\partial t^2}\right)^2$$

$$\frac{d\mathbf{L}_{em}}{dt} = \frac{1}{6\pi\epsilon_0 c^3}\left(\frac{\partial \mathbf{p}_{ret}}{\partial t} \times \frac{\partial^2 \mathbf{p}_{ret}}{\partial t^2}\right)$$

where the dipole moment $\mathbf{p}$ is evaluated at the retarded time $t' = t - r/c$.

(b) The dipole moment is caused by a particle of mass m and charge e moving nonrelativistically in a fixed central potential $V(r)$. Show that the radiated power and angular momentum for such a particle can be written as

$$P(t) = \frac{\tau}{m} \left(\frac{dV}{dr} \right)^2$$

$$\frac{d\mathbf{L}_{em}}{dt} = \frac{\tau}{m} \left(\frac{dV}{r\,dr} \right) \mathbf{L}$$

where $\tau = e^2/6\pi\epsilon_0 mc^3$ $(= 2e^2/3mc^3$ in Gaussian units) is a characteristic time, $\mathbf{L}$ is the particle's angular momentum, and the right-hand sides are evaluated at the retarded time. Relate these results to those from the Abraham–Lorentz equation for radiation damping [Section 16.2].

(c) Suppose the charged particle is an electron in a hydrogen atom. Show that the inverse time defined by the ratio of the rate of angular momentum radiated to the particle's angular momentum is of the order of $\alpha^4 c/a_0$, where $\alpha = e^2/4\pi\epsilon_0\hbar c \approx 1/137$ is the fine structure constant and a_0 is the Bohr radius. How does this inverse time compare to the observed rate of radiation in hydrogen atoms?

(d) Relate the expressions in parts a and b to those for harmonic time dependence in Problem 9.8.

9.10 The transitional charge and current densities for the radiative transition from the $m = 0$, $2p$ state in hydrogen to the $1s$ ground state are, in the notation of (9.1) and with the neglect of spin,

$$\rho(r, \theta, \phi, t) = \frac{2e}{\sqrt{6}\,a_0^4} \cdot re^{-3r/2a_0} Y_{00} Y_{10} e^{-i\omega_0 t}$$

$$\mathbf{J}(r, \theta, \phi, t) = \frac{-iv_0}{2} \left(\frac{\hat{\mathbf{r}}}{2} + \frac{a_0}{z}\,\hat{\mathbf{z}} \right) \rho(r, \theta, \phi, t)$$

where $a_0 = 4\pi\epsilon_0\hbar^2/me^2 = 0.529 \times 10^{-10}$ m is the Bohr radius, $\omega_0 = 3e^2/32\pi\epsilon_0\hbar a_0$ is the frequency difference of the levels, and $v_0 = e^2/4\pi\epsilon_0\hbar = \alpha c \approx c/137$ is the Bohr orbit speed.

(a) Show that the effective transitional (orbital) "magnetization" is

$$\text{``}\mathcal{M}\text{''}(r, \theta, \phi, t) = -i\frac{\alpha c a_0}{4} \tan\theta(\hat{\mathbf{x}} \sin\phi - \hat{\mathbf{y}} \cos\phi) \cdot \rho(r, \theta, \phi, t)$$

Calculate $\nabla \cdot \text{``}\mathcal{M}\text{''}$ and evaluate all the nonvanishing radiation multipoles in the long-wavelength limit.

(b) In the electric dipole approximation calculate the total time-averaged power radiated. Express your answer in units of $(\hbar\omega_0) \cdot (\alpha^4 c/a_0)$, where $\alpha = e^2/4\pi\epsilon_0\hbar c$ is the fine structure constant.

(c) Interpreting the classically calculated power as the photon energy $(\hbar\omega_0)$ times the transition probability, evaluate numerically the transition probability in units of reciprocal seconds.

(d) If, instead of the semiclassical charge density used above, the electron in the $2p$ state was described by a *circular* Bohr orbit of radius $2a_0$, rotating with the transitional frequency ω_0, what would the radiated power be? Express your answer in the same units as in part b and evaluate the ratio of the two powers numerically.

9.11 Three charges are located along the z axis, a charge $+2q$ at the origin, and charges $-q$ at $z = \pm a \cos \omega t$. Determine the lowest nonvanishing multipole moments, the angular distribution of radiation, and the total power radiated. Assume that $ka \ll 1$.

9.12 An almost spherical surface defined by

$$R(\theta) = R_0[1 + \beta P_2(\cos\theta)]$$

has inside of it a uniform volume distribution of charge totaling Q. The small parameter β varies harmonically in time at frequency ω. This corresponds to surface waves on a sphere. Keeping only lowest order terms in β and making the long-wavelength approximation, calculate the nonvanishing multipole moments, the angular distribution of radiation, and the total power radiated.

9.13 The uniform charge density of Problem 9.12 is replaced by a uniform density of intrinsic magnetization parallel to the z axis and having total magnetic moment M. With the same approximations as above calculate the nonvanishing radiation multipole moments, the angular distribution of radiation, and the total power radiated.

9.14 An antenna consists of a circular loop of wire of radius a located in the x-y plane with its center at the origin. The current in the wire is

$$I = I_0 \cos\omega t = \text{Re } I_0 e^{-i\omega t}$$

  (a) Find the expressions for $\mathbf{E}$, $\mathbf{H}$ in the radiation zone without approximations as to the magnitude of ka. Determine the power radiated per unit solid angle.

  (b) What is the lowest nonvanishing multipole moment ($Q_{l,m}$ or $M_{l,m}$)? Evaluate this moment in the limit $ka \ll 1$.

9.15 Two fixed electric dipoles of dipole moment p are located in the x-y plane a distance $2a$ apart, their axes parallel and perpendicular to the plane, but their moments directed oppositely. The dipoles rotate with constant angular speed ω about a z axis located halfway between them. The motion is nonrelativistic ($\omega a/c \ll 1$).

  (a) Find the lowest nonvanishing multipole moments.

  (b) Show that the magnetic field in the radiation zone is, apart from an overall phase factor,

$$\mathbf{H} = \frac{cpa}{2\pi} k^3[\hat{\mathbf{x}} + i\hat{\mathbf{y}})\cos\theta - \hat{\mathbf{z}}\sin\theta\, e^{i\phi}]\cos\theta\, \frac{e^{ikr}}{r}$$

  (c) Show that the angular distribution of the radiation is proportional to $(\cos^2\theta + \cos^4\theta)$ and the total time-averaged power radiated is

$$P = \frac{4}{15\pi\epsilon_0} ck^6 p^2 a^2$$

Hint: Problem 6.21 is relevant.

9.16 A thin linear antenna of length d is excited in such a way that the sinusoidal current makes a full wavelength of oscillation as shown in the figure.

Problem 9.16

  (a) Calculate exactly the power radiated per unit solid angle and plot the angular distribution of radiation.

  (b) Determine the total power radiated and find a numerical value for the radiation resistance.

9.17 Treat the linear antenna of Problem 9.16 by the multipole expansion method.

   (a) Calculate the multipole moments (electric dipole, magnetic dipole, and electric quadrupole) exactly and in the long-wavelength approximation.

   (b) Compare the shape of the angular distribution of radiated power for the lowest nonvanishing multipole with the exact distribution of Problem 9.16.

   (c) Determine the total power radiated for the lowest multipole and the corresponding radiation resistance using both multipole moments from part a. Compare with Problem 9.16b. Is there a paradox here?

9.18 A qualitative understanding of the result for the reactance of a short antenna whose radiation fields are described by the electric dipole fields of Section 9.2 can be achieved by considering the idealized dipole fields (9.18).

   (a) Show that the integral over all angles at fixed distance r of $\epsilon_0|\mathbf{E}|^2 - \mu_0|\mathbf{H}|^2$ is

$$\int [\epsilon_0|\mathbf{E}|^2 - \mu_0|\mathbf{H}|^2]\, d\Omega = \frac{1}{2\pi\epsilon_0} \frac{|\mathbf{p}|^2}{r^6}\left(1 + \frac{2}{3}k^2 r^2\right)$$

   (b) Using (6.140) for the reactance, show that the contribution X_a to the reactance from fields at distances $r > a$ is

$$X_a = -\frac{\omega|\mathbf{p}|^2}{6\pi\epsilon_0|I_i|^2 a^3}(1 + 2k^2 a^2)$$

   where I_i is the input current.

   (c) For the short center-fed antenna of Section 9.2 show that $X_a \simeq -d^2/24\pi\epsilon_0\omega a^3$, corresponding to an effective capacitance $24\pi\epsilon_0 a^3/d^2$. With $a = d/2$, X_a gives only a small fraction of the total negative reactance of a short antenna. The fields close to the antenna, obviously not dipole in character, contribute heavily. For calculations of reactances of short antennas, see the book by *Schelkunoff and Friis*.

9.19 Consider the excitation of a waveguide in Problem 8.19 from the point of view of multipole moments of the source.

   (a) For the linear probe antenna calculate the multipole moment components of $\mathbf{p}$, $\mathbf{m}$, $Q_{\alpha\beta}$, $Q_{\alpha\beta}^M$ that enter (9.69).

   (b) Calculate the amplitudes for excitation of the $TE_{1,0}$ mode and evaluate the power flow. Compare the multipole expansion result with the answer given in Problem 8.19b. Discuss the reasons for agreement or disagreement. What about the comparison for excitation of other modes?

9.20 **(a)** Verify by direct calculation that the *static* tangential electric field (3.186) in a circular opening in a flat conducting plane, when inserted into the defining equation (9.72) for the electric dipole moment $\mathbf{p}_{eff}$, leads to the expression (9.75).

   (b) Determine the value of $i\mu\omega\mathbf{m}_{eff}$ given by (9.72) with the static electric field in part a.

   (c) Use the *static* normal magnetic field (5.132) for the corresponding magnetic boundary problem with a circular opening to compute via (9.74) the magnetic dipole moment $\mathbf{m}_{eff}$ and compare with (9.75).

   (d) Comment on the differences between the results of parts b and c and the use of the definitions (9.72) in a consistent fashion. [See Section 9 of the article, Diffraction Theory, by C. J. B. Bouwkamp in *Reports on Progress in Physics*, Vol. 17, ed. A. C. Strickland, The Physical Society, London (1954).]

9.21 The fields representing a transverse magnetic wave propagating in a cylindrical waveguide of radius R are:

$$E_z = J_m(\gamma r)e^{im\phi}e^{i\beta z - i\omega t}, \qquad H_z = 0$$

$$E_\phi = \frac{-m\beta}{\gamma^2}\frac{E_z}{r}, \qquad\qquad H_r = -\frac{k}{Z_0\beta}E_\phi$$

$$E_r = \frac{i\beta}{\gamma^2}\frac{\partial E_z}{\partial r}, \qquad\qquad H_\phi = \frac{k}{Z_0\beta}E_r$$

where m is the index specifying the angular dependence, β is the propagation constant, $\gamma^2 = k^2 - \beta^2$ ($k = \omega/c$), where γ is such that $J_m(\gamma R) = 0$. Calculate the ratio of the z component of the electromagnetic angular momentum to the energy in the field. It may be advantageous to perform some integrations by parts, and to use the differential equation satisfied by E_z, to simplify your calculations.

9.22 A spherical hole of radius a in a conducting medium can serve as an electromagnetic resonant cavity.

   (a) Assuming infinite conductivity, determine the transcendental equations for the characteristic frequencies ω_{lm} of the cavity for TE and TM modes.

   (b) Calculate numerical values for the wavelength λ_{lm} in units of the radius a for the four lowest modes for TE and TM waves.

   (c) Calculate explicitly the electric and magnetic fields inside the cavity for the lowest TE and lowest TM mode.

9.23 The spherical resonant cavity of Problem 9.22 has nonpermeable walls of large, but finite, conductivity. In the approximation that the skin depth δ is small compared to the cavity radius a, show that the Q of the cavity, defined by equation (8.86), is given by

$$Q = \frac{a}{\delta} \qquad\qquad \text{for all TE modes}$$

$$Q = \frac{a}{\delta}\left(1 - \frac{l(l+1)}{x_{lm}^2}\right) \qquad \text{for TM modes}$$

where $x_{lm} = (a/c)\omega_{lm}$ for TM modes.

9.24 Discuss the normal modes of oscillation of a perfectly conducting solid sphere of radius a in free space. (This problem was solved by J. J. Thomson in the 1880s.)

   (a) Determine the characteristic equations for the eigenfrequencies for TE and TM modes of oscillation. Show that the roots for ω always have a negative imaginary part, assuming a time dependence of $e^{-i\omega t}$.

   (b) Calculate the eigenfrequencies for the $l = 1$ and $l = 2$ TE and TM modes. Tabulate the wavelength (defined in terms of the real part of the frequency) in units of the radius a and the decay time (defined as the time taken for the *energy* to fall to e^{-1} of its initial value) in units of the transit time (a/c) for each of the modes.

CHAPTER 10

Scattering and Diffraction

The closely related topics of scattering and diffraction are important in many branches of physics. Approaches differ depending on the relative length scales involved—the wavelength of the waves on the one hand, and the size of the target (scatterer or diffractor) on the other. When the wavelength of the radiation is large compared to the dimensions of the target, a simple description in terms of lowest order induced multipoles is appropriate. When the wavelength and size are comparable, a more systematic treatment with multipole fields is required. In the limit of very small wavelength compared to the size of the target, semi-geometric methods can be utilized to obtain the departures from geometrical optics. We begin with the long-wavelength limit of electromagnetic scattering, with some simple examples. Then we develop a perturbation approach to scattering by a medium with small variations in its dielectric properties in order to discuss Rayleigh scattering, the blue sky, and critical opalescence. To introduce the more systematic approach with multipole fields, we first present the multipole expansion of an electromagnetic plane wave and then apply it to the scattering by a conducting sphere.

Diffraction is treated next, first the scalar Huygens–Kirchhoff theory, then a vector generalization that leads naturally to a discussion of Babinet's principle of complementary screens. These tools are applied to diffraction by a circular aperture, with connection to the low-order effective multipoles of Section 9.5 in the long-wavelength limit. Scattering at very short wavelengths and the important optical theorem complete the chapter.

10.1 Scattering at Long Wavelengths

A. Scattering by Dipoles Induced in Small Scatterers

The scattering of electromagnetic waves by systems whose individual dimensions are small compared with a wavelength is a common and important occurrence. In such interactions it is convenient to think of the incident (radiation) fields as inducing electric and magnetic multipoles that oscillate in definite phase relationship with the incident wave and radiate energy in directions other than the direction of incidence. The exact form of the angular distribution of radiated energy is governed by the *coherent superposition* of multipoles induced by the incident fields and in general depends on the state of polarization of the incident wave. If the wavelength of the radiation is long compared to the size of the scatterer, only the lowest multipoles, usually electric and magnetic dipoles, are important. Furthermore, in these circumstances the induced dipoles can be calculated from static or quasi-static boundary-value problems, just as for the small apertures of the preceding chapter (Section 9.5).

456

The customary basic situation is for a plane monochromatic wave to be incident on a scatterer. For simplicity the surrounding medium is taken to have $\mu_r = \epsilon_r = 1$. If the incident direction is defined by the unit vector $\mathbf{n}_0$, and the incident polarization vector is $\boldsymbol{\epsilon}_0$, the incident fields are

$$\mathbf{E}_{\text{inc}} = \boldsymbol{\epsilon}_0 E_0 e^{ik\mathbf{n}_0 \cdot \mathbf{x}}$$
$$\mathbf{H}_{\text{inc}} = \mathbf{n}_0 \times \mathbf{E}_{\text{inc}}/Z_0 \tag{10.1}$$

where $k = \omega/c$ and a time-dependence $e^{-i\omega t}$ is understood. These fields induce dipole moments $\mathbf{p}$ and $\mathbf{m}$ in the small scatterer and these dipoles radiate energy in all directions, as described earlier (Sections 9.2, 9.3). Far away from the scatterer, the scattered (radiated) fields are found from (9.19) and (9.36) to be

$$\mathbf{E}_{\text{sc}} = \frac{1}{4\pi\epsilon_0} k^2 \frac{e^{ikr}}{r} [(\mathbf{n} \times \mathbf{p}) \times \mathbf{n} - \mathbf{n} \times \mathbf{m}/c]$$
$$\mathbf{H}_{\text{sc}} = \mathbf{n} \times \mathbf{E}_{\text{sc}}/Z_0 \tag{10.2}$$

where $\mathbf{n}$ is a unit vector in the direction of observation and r is the distance away from scatterer. The power radiated in the direction $\mathbf{n}$ with polarization $\boldsymbol{\epsilon}$, per unit solid angle, per unit incident flux (power per unit area) in the direction $\mathbf{n}_0$ with polarization $\boldsymbol{\epsilon}_0$, is a quantity with dimensions of area per unit solid angle. It is called the *differential scattering cross section**:

$$\frac{d\sigma}{d\Omega} (\mathbf{n}, \boldsymbol{\epsilon}; \mathbf{n}_0, \boldsymbol{\epsilon}_0) = \frac{r^2 \dfrac{1}{2Z_0} |\boldsymbol{\epsilon}^* \cdot \mathbf{E}_{\text{sc}}|^2}{\dfrac{1}{2Z_0} |\boldsymbol{\epsilon}_0^* \cdot \mathbf{E}_{\text{inc}}|^2} \tag{10.3}$$

The complex conjugation of the polarization vectors in (10.3) is important for the correct handling of circular polarization, as mentioned in Section 7.2. With (10.2) and (10.1), the differential cross section can be written

$$\frac{d\sigma}{d\Omega} (\mathbf{n}, \boldsymbol{\epsilon}; \mathbf{n}_0, \boldsymbol{\epsilon}_0) = \frac{k^4}{(4\pi\epsilon_0 E_0)^2} |\boldsymbol{\epsilon}^* \cdot \mathbf{p} + (\mathbf{n} \times \boldsymbol{\epsilon}^*) \cdot \mathbf{m}/c|^2 \tag{10.4}$$

The dependence of the cross section on $\mathbf{n}_0$ and $\boldsymbol{\epsilon}_0$ is implicitly contained in the dipole moments $\mathbf{p}$ and $\mathbf{m}$. The variation of the differential (and total) scattering cross section with wave number as k^4 (or in wavelength as λ^{-4}) is an almost universal characteristic of the scattering of long-wavelength radiation by any finite system. This dependence on frequency is known as *Rayleigh's law*. Only if both static dipole moments vanish does the scattering fail to obey Rayleigh's law; the scattering is then via quadrupole or higher multipoles (or frequency-dependent dipole moments) and varies as ω^6 or higher. Sometimes the dipole scattering is known as *Rayleigh* scattering, but this term is usually reserved for the *incoherent* scattering by a collection of dipole scatterers.

B. Scattering by a Small Dielectric Sphere

As a first, very simple example of dipole scattering we consider a small dielectric sphere of radius a with $\mu_r = 1$ and a uniform isotropic dielectric constant

---

*In the engineering literature the term *bistatic cross section* is used for $4\pi (d\sigma/d\Omega)$.

$\epsilon_r(\omega)$. From Section 4.4, in particular (4.56), the electric dipole moment is found to be

$$\mathbf{p} = 4\pi\epsilon_0\left(\frac{\epsilon_r - 1}{\epsilon_r + 2}\right)a^3\mathbf{E}_{inc} \tag{10.5}$$

There is no magnetic dipole moment. The differential scattering cross section is

$$\frac{d\sigma}{d\Omega} = k^4 a^6 \left|\frac{\epsilon_r - 1}{\epsilon_r + 2}\right|^2 |\boldsymbol{\epsilon}^* \cdot \boldsymbol{\epsilon}_0|^2 \tag{10.6}$$

The polarization dependence is typical of purely electric dipole scattering. The scattered radiation is linearly polarized in the plane defined by the dipole moment direction ($\boldsymbol{\epsilon}_0$) and the unit vector $\mathbf{n}$.

Typically the incident radiation is unpolarized. It is then of interest to ask for the angular distribution of scattered radiation of a definite state of linear polarization. The cross section (10.6) is averaged over initial polarization $\boldsymbol{\epsilon}_0$ for a fixed choice of $\boldsymbol{\epsilon}$. Figure 10.1 shows a possible set of polarization vectors. The scattering plane is defined by the vectors $\mathbf{n}_0$ and $\mathbf{n}$. The polarization vectors $\boldsymbol{\epsilon}_0^{(1)}$ and $\boldsymbol{\epsilon}^{(1)}$ are in this plane, while $\boldsymbol{\epsilon}_0^{(2)} = \boldsymbol{\epsilon}^{(2)}$ is perpendicular to it. The differential cross sections for scattering with polarizations $\boldsymbol{\epsilon}^{(1)}$ and $\boldsymbol{\epsilon}^{(2)}$, averaged over initial polarizations, are easily shown to be

$$\frac{d\sigma_\parallel}{d\Omega} = \frac{k^4 a^6}{2}\left|\frac{\epsilon_r - 1}{\epsilon_r + 2}\right|^2 \cos^2\theta$$
$$\frac{d\sigma_\perp}{d\Omega} = \frac{k^4 a^6}{2}\left|\frac{\epsilon_r - 1}{\epsilon_r + 2}\right|^2 \tag{10.7}$$

where the subscripts $\parallel$ and $\perp$ indicate polarization parallel to and perpendicular to the scattering plane, respectively. The *polarization* $\Pi(\theta)$ of the scattered radiation is defined by

$$\Pi(\theta) = \frac{\dfrac{d\sigma_\perp}{d\Omega} - \dfrac{d\sigma_\parallel}{d\Omega}}{\dfrac{d\sigma_\perp}{d\Omega} + \dfrac{d\sigma_\parallel}{d\Omega}} \tag{10.8}$$

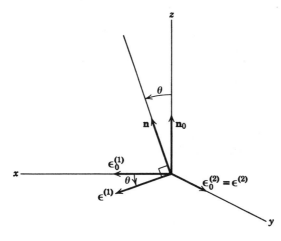

Figure 10.1 Polarization and propagation vectors for the incident and scattered radiation.

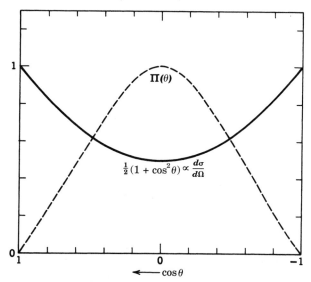

Figure 10.2 Differential scattering cross section (10.10) and the polarization of scattered radiation (10.9) for a small dielectric sphere (dipole approximation).

From (10.7) we find for the (electric dipole) scattering by a small dielectric sphere,

$$\Pi(\theta) = \frac{\sin^2\theta}{1 + \cos^2\theta} \tag{10.9}$$

The differential cross section, summed over scattered polarization, is

$$\frac{d\sigma}{d\Omega} = k^4 a^6 \left| \frac{\epsilon_r - 1}{\epsilon_r + 2} \right|^2 \tfrac{1}{2}(1 + \cos^2\theta) \tag{10.10}$$

and the *total scattering cross section* is

$$\sigma = \int \frac{d\sigma}{d\Omega}\, d\Omega = \frac{8\pi}{3} k^4 a^6 \left| \frac{\epsilon_r - 1}{\epsilon_r + 2} \right|^2 \tag{10.11}$$

The differential cross section (10.10) and the polarization of the scattered radiation (10.9) are shown as functions of $\cos\theta$ in Fig. 10.2. The polarization $\Pi(\theta)$ has its maximum at $\theta = \pi/2$. At this angle the scattered radiation is 100% linearly polarized perpendicular to the scattering plane, and for an appreciable range of angles on either side of $\theta = \pi/2$ is quite significantly polarized. The polarization characteristics of the blue sky are an illustration of this phenomenon, and are, in fact, the motivation that led Rayleigh first to consider the problem. The reader can verify the general behavior on a sunny day with a sheet of linear polarizer or suitable sunglasses.

C. Scattering by a Small Perfectly Conducting Sphere

An example with interesting aspects involving coherence between different multipoles is the scattering by a small perfectly conducting sphere of radius a. The electric dipole moment of such a sphere was shown in Section 2.5 to be

$$\mathbf{p} = 4\pi\epsilon_0 a^3 \mathbf{E}_{\text{inc}} \tag{10.12}$$

The sphere also possesses a magnetic dipole moment. For a perfectly conducting sphere the boundary condition on the magnetic field is that the normal component of **B** vanishes at $r = a$. Either by analogy with the dielectric sphere in a uniform electric field (Section 4.4) with $\epsilon = 0$, or from the magnetically permeable sphere (Section 5.11) with $\mu = 0$, or by a simple direct calculation, it is found that the magnetic moment of the small sphere is

$$\mathbf{m} = -2\pi a^3 \mathbf{H}_{\text{inc}} \tag{10.13}$$

For a linearly polarized incident wave the two dipoles are at right angles to each other and to the incident direction.

The differential cross section (10.4) is

$$\frac{d\sigma}{d\Omega}(\mathbf{n}, \boldsymbol{\epsilon}; \mathbf{n}_0, \boldsymbol{\epsilon}_0) = k^4 a^6 \left|\boldsymbol{\epsilon}^* \cdot \boldsymbol{\epsilon}_0 - \tfrac{1}{2}(\mathbf{n} \times \boldsymbol{\epsilon}^*) \cdot (\mathbf{n}_0 \times \boldsymbol{\epsilon}_0)\right|^2 \tag{10.14}$$

The polarization properties and the angular distribution of scattered radiation are more complicated than for the dielectric sphere. The cross sections analogous to (10.7), for polarization of the scattered radiation parallel to and perpendicular to the plane of scattering, with unpolarized radiation incident, are

$$\frac{d\sigma_{\|}}{d\Omega} = \frac{k^4 a^6}{2}\left|\cos\theta - \tfrac{1}{2}\right|^2$$

$$\frac{d\sigma_{\perp}}{d\Omega} = \frac{k^4 a^6}{2}\left|1 - \tfrac{1}{2}\cos\theta\right|^2 \tag{10.15}$$

The differential cross section summed over both states of scattered polarization can be written

$$\frac{d\sigma}{d\Omega} = k^4 a^6\left[\tfrac{5}{8}(1 + \cos^2\theta) - \cos\theta\right] \tag{10.16}$$

while the polarization (10.8) is

$$\Pi(\theta) = \frac{3\sin^2\theta}{5(1 + \cos^2\theta) - 8\cos\theta} \tag{10.17}$$

The cross section and polarization are plotted versus $\cos\theta$ in Fig. 10.3. The cross section has a *strong backward peaking* caused by electric dipole–magnetic dipole interference. The polarization reaches $\Pi = +1$ at $\theta = 60°$ and is positive through the whole angular range. The polarization thus tends to be similar to that for a small dielectric sphere, as shown in Fig. 10.2, even though the angular distributions are quite different. The total scattering cross section is $\sigma = 10\pi k^4 a^6/3$, of the same order of magnitude as for the dielectric sphere (10.11) if $(\epsilon_r - 1)$ is not small.

Dipole scattering with its ω^4 dependence on frequency can be viewed as the lowest order approximation in an expansion in kd, where d is a length typical of the dimensions of the scatterer. In the domain $kd \sim 1$, more than the lowest order multipoles must be considered. Then the discussion is best accomplished by use of a systematic expansion in spherical multipole fields. In Section 10.4 the scattering by a conducting sphere is examined from this point of view. When $kd \gg 1$, approximation methods of a different sort can be employed, as is illustrated later in this chapter (Section 10.10). Whole books are devoted to the scat-

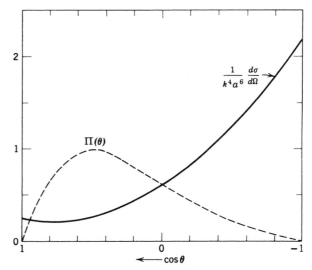

Figure 10.3 Differential scattering cross section (10.16) and polarization of scattered radiation (10.17) for a small perfectly conducting sphere (electric and magnetic dipole approximation).

tering of light by spherical particles possessing arbitrary μ, ϵ, σ. Some references to this literature are given at the end of the chapter.

D. Collection of Scatterers

As a final remark we note that if the scattering system consists of a number of small scatters with fixed spatial separations, each scatterer generates an amplitude of the form (10.2). The scattering cross section results from a coherent superposition of the individual amplitudes. Because the induced dipoles are proportional to the incident fields, evaluated at the position $\mathbf{x}_j$ of the jth scatterer, its moments will possess a phase factor, $e^{ik\mathbf{n}_0 \cdot \mathbf{x}_j}$. Furthermore, if the observation point is far from the whole scattering system, (9.7) shows that the fields (10.2) for the jth scatterer will have a phase factor $e^{-ik\mathbf{n} \cdot \mathbf{x}_j}$. The generalization of (10.4) for such a system is

$$\frac{d\sigma}{d\Omega} = \frac{k^4}{(4\pi\epsilon_0 E_0)^2} \left| \sum_j [\boldsymbol{\epsilon}^* \cdot \mathbf{p}_j + (\mathbf{n} \times \boldsymbol{\epsilon}^*) \cdot \mathbf{m}_j/c] e^{i\mathbf{q} \cdot \mathbf{x}_j} \right|^2 \tag{10.18}$$

where $\mathbf{q} = k\mathbf{n}_0 - k\mathbf{n}$ is the vectorial change in wave vector during the scattering.

The presence of the phase factors $e^{i\mathbf{q} \cdot \mathbf{x}_j}$ in (10.18) means that, apart from the forward direction where $\mathbf{q} = 0$, the scattering depends sensitively on the exact distribution of the scatterers in space. The general behavior can be illustrated by assuming that all the scatterers are identical. Then the cross section is the product of the cross section for one scatterer times a structure factor,*

$$\mathscr{F}(\mathbf{q}) = \left| \sum_j e^{i\mathbf{q} \cdot \mathbf{x}_j} \right|^2 \tag{10.19}$$

---

*We do not consider here the effects of multiple scattering; that is, we assume that the mean free path for scattering is large compared to the dimensions of the scattering array.

Written out as a factor times its complex conjugate, $\mathcal{F}(\mathbf{q})$ is

$$\mathcal{F}(\mathbf{q}) = \sum_j \sum_{j'} e^{i\mathbf{q}\cdot(\mathbf{x}_j - \mathbf{x}_{j'})}$$

If the scatterers are randomly distributed, the terms with $j \neq j'$ can be shown to give a negligible contribution. Only the terms with $j = j'$ are significant. Then $\mathcal{F}(\mathbf{q}) = N$, the total number of scatterers, and the scattering is said to be an *incoherent superposition* of individual contributions. If, on the other hand, the scatterers are very numerous and have a regular distribution in space, the structure factor effectively vanishes everywhere except in the forward direction. There is therefore no scattering by a very large regular array of scatterers, of which single crystals of transparent solids like rock salt or quartz are examples. What small amount of scattering does occur is caused by thermal vibrations away from the perfect lattice, or by impurities, etc. An explicit illustration, also providing evidence for a restriction of the foregoing remarks to the long-wavelength regime, is that of a simple cubic array of scattering centers. The structure factor is well known to be

$$\mathcal{F}(\mathbf{q}) = N^2 \left[\frac{\sin^2\left(\dfrac{N_1 q_1 a}{2}\right)}{N_1^2 \sin^2\left(\dfrac{q_1 a}{2}\right)} \cdot \frac{\sin^2\left(\dfrac{N_2 q_2 a}{2}\right)}{N_2^2 \sin^2\left(\dfrac{q_2 a}{2}\right)} \cdot \frac{\sin^2\left(\dfrac{N_3 q_3 a}{2}\right)}{N_3^2 \sin^2\left(\dfrac{q_3 a}{2}\right)} \right] \quad (10.20)$$

where a is the lattice spacing, N_1, N_2, N_3 are the numbers of lattice sites along the three axes of the array, $N = N_1 N_2 N_3$ is the total number of scatterers and q_1, q_2, q_3 are the components of $\mathbf{q}$ along the axes. At short wavelengths $(ka > \pi)$, (10.20) has peaks when the Bragg scattering condition, $q_i a = 0, 2\pi, 4\pi, \ldots$, is obeyed. This is the situation familiar in x-ray diffraction. But at long wavelengths only the peak at $q_i a = 0$ is relevant because $(q_i a)_{max} = 2ka \ll 1$. In this limit $\mathcal{F}(\mathbf{q})$ is the product of three factors of the form $[(\sin x_i)/x_i]^2$ with $x_i = N_i q_i a/2$. The scattering is thus confined to the region $q_i \lesssim 2\pi/N_i a$, corresponding to angles smaller than λ/L, where λ is the wavelength and L a typical overall dimension of the scattering array.

10.2 *Perturbation Theory of Scattering, Rayleigh's Explanation of the Blue Sky,* Scattering by Gases and Liquids, Attenuation in Optical Fibers*

A. General Theory

If the medium through which an electromagnetic wave is passing is uniform in its properties, the wave propagates undisturbed and undeflected. If, however,

---

*Although Rayleigh's name should undoubtedly be associated with the quantitative explanation of the blue sky, it is of some historical interest that Leonardo da Vinci understood the basic phenomenon around 1500. In particular, his experiments with the scattering of sunlight by wood smoke observed against a dark background (quoted as items 300–302, pp. 237 ff, in Vol. I of Jean Paul Richter, *The Literary Works of Leonardo da Vinci*, 3rd edition, Phaidon, London 1970) (also a Dover reprint entitled *The Notebooks of Leonardo da Vinci*, Vol. 1, pp. 161 ff.) anticipate by 350 years Tyndall's remarkably similar observations [J. Tyndall, *Philos. Trans. R. Soc. London* **160**, 333 (1870)].

there are spatial (or temporal) variations in the electromagnetic properties, the wave is scattered. Some of the energy is deviated from its original course. If the variations in the properties are small in magnitude, the scattering is slight and perturbative methods can be employed. We imagine a comparison situation corresponding to a uniform isotropic medium with electric permittivity ϵ_0 and magnetic permeability μ_0. For the present ϵ_0 and μ_0 are assumed independent of frequency, although when harmonic time dependence is assumed this restriction can be removed in the obvious way. Note that in this section ϵ_0 and μ_0 are not the free-space values! Through the action of some perturbing agent, the medium is supposed to have *small changes* in its response to applied fields, so that $\mathbf{D} \neq \epsilon_0\mathbf{E}$, $\mathbf{B} \neq \mu_0\mathbf{H}$, over certain regions of space. These departures may be functions of time and space variables. Beginning with the Maxwell equations in the absence of sources,

$$\mathbf{\nabla} \cdot \mathbf{B} = 0, \qquad \mathbf{\nabla} \times \mathbf{E} = -\frac{\partial \mathbf{B}}{\partial t}$$
$$\mathbf{\nabla} \cdot \mathbf{D} = 0, \qquad \mathbf{\nabla} \times \mathbf{H} = \frac{\partial \mathbf{D}}{\partial t} \tag{10.21}$$

it is a straightforward matter to arrive at a wave equation for $\mathbf{D}$,

$$\nabla^2\mathbf{D} - \mu_0\epsilon_0\frac{\partial^2\mathbf{D}}{\partial t^2} = -\mathbf{\nabla} \times \mathbf{\nabla} \times (\mathbf{D} - \epsilon_0\mathbf{E}) + \epsilon_0\frac{\partial}{\partial t}\mathbf{\nabla} \times (\mathbf{B} - \mu_0\mathbf{H}) \tag{10.22}$$

This equation is without approximation as yet, although later the right-hand side will be treated as small in some sense.*

If the right-hand side of (10.22) is taken as known, the equation is of the form of (6.32) with the retarded solution (6.47). In general, of course, the right-hand side is unknown and (6.47) must be regarded as an integral relation, rather than a solution. Nevertheless, such an integral formulation of the problem forms a fruitful starting point for approximations. It is convenient to specialize to harmonic time variation with frequency ω for the unperturbed fields and to assume that the departures $(\mathbf{D} - \epsilon_0\mathbf{E})$ and $(\mathbf{B} - \mu_0\mathbf{H})$ also have this time variation. This puts certain limitations on the kind of perturbed problem that can be described by the formalism, but prevents the discussion from becoming too involved. With a time dependence $e^{-i\omega t}$ understood, (10.22) becomes

$$(\nabla^2 + k^2)\mathbf{D} = -\mathbf{\nabla} \times \mathbf{\nabla} \times (\mathbf{D} - \epsilon_0\mathbf{E}) - i\epsilon_0\omega \mathbf{\nabla} \times (\mathbf{B} - \mu_0\mathbf{H}) \tag{10.23}$$

where $k^2 = \mu_0\epsilon_0\omega^2$, and μ_0 and ϵ_0 can be values specific to the frequency ω. The solution of the unperturbed problem, with the right-hand side of (10.23) set equal to zero, will be denoted by $\mathbf{D}^{(0)}(\mathbf{x})$. A formal solution of (10.23) can be obtained from (6.45), if the right-hand side is taken as known. Thus

$$\mathbf{D} = \mathbf{D}^{(0)} + \frac{1}{4\pi}\int d^3x' \frac{e^{ik|\mathbf{x}-\mathbf{x}'|}}{|\mathbf{x} - \mathbf{x}'|}\left\{\begin{array}{l}\mathbf{\nabla}' \times \mathbf{\nabla}' \times (\mathbf{D} - \epsilon_0\mathbf{E}) \\ +i\epsilon_0\omega \mathbf{\nabla}' \times (\mathbf{B} - \mu_0\mathbf{H})\end{array}\right\} \tag{10.24}$$

---

*If prescribed sources $\rho(\mathbf{x}, t)$, $\mathbf{J}(\mathbf{x}, t)$ are present, (10.22) is modified by the addition to the left-hand side of

$$-\left[\mathbf{\nabla}\rho + \mu_0\epsilon_0\frac{\partial\mathbf{J}}{\partial t}\right]$$

If the physical situation is one of scattering, with the integrand in (10.24) confined to some finite region of space and $\mathbf{D}^{(0)}$ describing a wave incident in some direction, the field far away from the scattering region can be written as

$$\mathbf{D} \rightarrow \mathbf{D}^{(0)} + \mathbf{A}_{sc} \frac{e^{ikr}}{r} \tag{10.25}$$

where the *scattering amplitude* $\mathbf{A}_{sc}$ is

$$\mathbf{A}_{sc} = \frac{1}{4\pi} \int d^3x' \, e^{-ik\mathbf{n}\cdot\mathbf{x}'} \left\{ \begin{array}{c} \boldsymbol{\nabla}' \times \boldsymbol{\nabla}' \times (\mathbf{D} - \epsilon_0\mathbf{E}) \\ +i\epsilon_0\omega \, \boldsymbol{\nabla}' \times (\mathbf{B} - \mu_0\mathbf{H}) \end{array} \right\} \tag{10.26}$$

The steps from (10.24) to (10.26) are the same as from (9.3) to (9.8) for the radiation fields. Some integrations by parts in (10.26) allow the scattering amplitude to be expressed as

$$\mathbf{A}_{sc} = \frac{k^2}{4\pi} \int d^3x \, e^{-ik\mathbf{n}\cdot\mathbf{x}} \left\{ \begin{array}{c} [\mathbf{n} \times (\mathbf{D} - \epsilon_0\mathbf{E})] \times \mathbf{n} \\ -\dfrac{\epsilon_0\omega}{k} \, \mathbf{n} \times (\mathbf{B} - \mu_0\mathbf{H}) \end{array} \right\} \tag{10.27}$$

The vectorial structure of the integrand can be compared with the scattered dipole field (10.2). The polarization dependence of the contribution from $(\mathbf{D} - \epsilon_0\mathbf{E})$ is that of an electric dipole, from $(\mathbf{B} - \mu_0\mathbf{H})$ a magnetic dipole. In correspondence with (10.4) the differential scattering cross section is

$$\frac{d\sigma}{d\Omega} = \frac{|\boldsymbol{\epsilon}^* \cdot \mathbf{A}_{sc}|^2}{|\mathbf{D}^{(0)}|^2} \tag{10.28}$$

where $\boldsymbol{\epsilon}$ is the polarization vector of the scattered radiation.

Equations (10.24), (10.27), and (10.28) provide a formal solution to the scattering problem posed at the beginning of the section. The scattering amplitude $\mathbf{A}_{sc}$ is not known, of course, until the fields are known at least approximately. But from (10.24) a systematic scheme of successive approximations can be developed in the same way as the Born approximation series of quantum-mechanical scattering. If the integrand in (10.24) can be approximated to first order, then (10.24) provides a first approximation for $\mathbf{D}$, beyond $\mathbf{D}^{(0)}$. This approximation to $\mathbf{D}$ can be used to give a second approximation for the integrand, and an improved $\mathbf{D}$ can be determined, and so on. Questions of convergence of the series, etc. have been much studied in the quantum-mechanical context. The series is not very useful unless the first few iterations converge rapidly.

B. Born Approximation

We will be content with the lowest order approximation for the scattering amplitude. This is called the *first Born approximation* or just the Born approximation in quantum theory and was actually developed in the present context by Lord Rayleigh in 1881. Furthermore, we shall restrict our discussion to the simple example of spatial variations in the linear response of the medium. Thus we assume that the connections between $\mathbf{D}$ and $\mathbf{E}$ and $\mathbf{B}$ and $\mathbf{H}$ are

$$\begin{aligned} \mathbf{D}(\mathbf{x}) &= [\epsilon_0 + \delta\epsilon(\mathbf{x})]\mathbf{E}(\mathbf{x}) \\ \mathbf{B}(\mathbf{x}) &= [\mu_0 + \delta\mu(\mathbf{x})]\mathbf{H}(\mathbf{x}) \end{aligned} \tag{10.29}$$

where $\delta\epsilon(\mathbf{x})$ and $\delta\mu(\mathbf{x})$ are small in magnitude compared with ϵ_0 and μ_0. The differences appearing in (10.24) and (10.27) are proportional to $\delta\epsilon$ and $\delta\mu$. To lowest order then, the fields in these differences can be approximated by the unperturbed fields:

$$\mathbf{D} - \epsilon_0 \mathbf{E} \simeq \frac{\delta\epsilon(\mathbf{x})}{\epsilon_0} \mathbf{D}^{(0)}(\mathbf{x})$$

$$\mathbf{B} - \mu_0 \mathbf{H} \simeq \frac{\delta\mu(\mathbf{x})}{\mu_0} \mathbf{B}^{(0)}(\mathbf{x})$$

(10.30)

If the unperturbed fields are those of a plane wave propagating in a direction $\mathbf{n}_0$, so that $\mathbf{D}^{(0)}$ and $\mathbf{B}^{(0)}$ are

$$\mathbf{D}^{(0)}(\mathbf{x}) = \epsilon_0 D_0 e^{ik\mathbf{n}_0 \cdot \mathbf{x}}$$

$$\mathbf{B}^{(0)}(\mathbf{x}) = \sqrt{\frac{\mu_0}{\epsilon_0}} \mathbf{n}_0 \times \mathbf{D}^{(0)}(\mathbf{x})$$

the scalar product of the scattering amplitude (10.27) and $\boldsymbol{\epsilon}^*$, divided by D_0, is

$$\frac{\boldsymbol{\epsilon}^* \cdot \mathbf{A}_{sc}^{(1)}}{D_0} = \frac{k^2}{4\pi} \int d^3x \, e^{i\mathbf{q} \cdot \mathbf{x}} \left\{ \begin{array}{l} \boldsymbol{\epsilon}^* \cdot \boldsymbol{\epsilon}_0 \dfrac{\delta\epsilon(\mathbf{x})}{\epsilon_0} \\[2mm] + (\mathbf{n} \times \boldsymbol{\epsilon}^*) \cdot (\mathbf{n}_0 \times \boldsymbol{\epsilon}_0) \dfrac{\delta\mu(\mathbf{x})}{\mu_0} \end{array} \right\}$$

(10.31)

where $\mathbf{q} = k(\mathbf{n}_0 - \mathbf{n})$ is the difference of the incident and scattered wave vectors. The absolute square of (10.31) gives the differential scattering cross section (10.28).

If the wavelength is large compared with the spatial extent of $\delta\epsilon$ and $\delta\mu$, the exponential in (10.31) can be set equal to unity. The amplitude is then a dipole approximation analogous to the preceding section, with the dipole frequency dependence and angular distribution. To establish contact with the results already obtained, suppose that the scattering region is a uniform dielectric sphere of radius a in vacuum. Then $\delta\epsilon$ is constant inside a spherical volume of radius a and vanishes outside. The integral in (10.31) can be performed for arbitrary $|\mathbf{q}|$, with the result,

$$\frac{\boldsymbol{\epsilon}^* \cdot \mathbf{A}_{sc}}{D_0} = k^2 \frac{\delta\epsilon}{\epsilon_0} (\boldsymbol{\epsilon}^* \cdot \boldsymbol{\epsilon}_0) \left[\frac{\sin qa - qa \cos qa}{q^3} \right]$$

In the limit $q \to 0$ the square bracket approaches $a^3/3$. Thus, at very low frequencies or in the forward direction at all frequencies, the Born approximation to the differential cross section for scattering by a dielectric sphere of radius a is

$$\lim_{q \to 0} \left(\frac{d\sigma}{d\Omega} \right)_{Born} = k^4 a^6 \left| \frac{\delta\epsilon}{3\epsilon_0} \right|^2 |\boldsymbol{\epsilon}^* \cdot \boldsymbol{\epsilon}_0|^2$$

(10.32)

Comparison with (10.6) shows that the Born approximation and the exact low frequency result have the expected relationship.

C. Blue Sky: Elementary Argument

The scattering of light by gases, first treated quantitatively by Lord Rayleigh in his celebrated work on the sunset and blue sky,* can be discussed in the present

---

*Lord Rayleigh, *Philos. Mag.* **XLI**, 107, 274, (1871); *ibid.* **XLVII**, 375 (1899); reprinted in his *Scientific Papers*, Vol. I, p. 87, and Vol. 4, p. 397. Rayleigh's papers are well worth reading as examples of a masterful physicist at work.

framework. Since the magnetic moments of most gas molecules are negligible compared to the electric dipole moments, the scattering is purely electric dipole in character. In the preceding section we discussed the angular distribution and polarization of the individual scatterings (see Fig. 10.2). We therefore confine our attention to the total scattering cross section and the attenuation of the incident beam. The treatment is in two parts. The first, elementary argument is adequate for a dilute ideal gas, where the molecules are truly randomly distributed in space relative to each other. The second, based on density fluctuations in the gas, is of more general validity. We now identify ϵ_0 with the electric permittivity of free space.

If the individual molecules, located at $\mathbf{x}_j$, are assumed to possess dipole moments $\mathbf{p}_j = \epsilon_0 \gamma_{\text{mol}} \mathbf{E}(\mathbf{x}_j)$, the effective variation in dielectric constant $\delta\epsilon(\mathbf{x})$ in (10.31) can be written as

$$\delta\epsilon(\mathbf{x}) = \epsilon_0 \sum_j \gamma_{\text{mol}} \, \delta(\mathbf{x} - \mathbf{x}_j) \tag{10.33}$$

The differential scattering cross section obtained from (10.31) and (10.28) is

$$\frac{d\sigma}{d\Omega} = \frac{k^4}{16\pi^2} |\gamma_{\text{mol}}|^2 |\boldsymbol{\epsilon}^* \cdot \boldsymbol{\epsilon}_0|^2 \mathscr{F}(\mathbf{q})$$

where $\mathscr{F}(\mathbf{q})$ is given by (10.19). For a random distribution of scattering centers the structure factor reduces to an incoherent sum, and the cross section is just that for one molecule, times the number of molecules. For a dilute gas the molecular polarizability is related to the dielectric constant by $\epsilon_r \simeq 1 + N\gamma_{\text{mol}}$, where N is the number of molecules per unit volume. The total scattering cross section *per molecule* of the gas is thus

$$\sigma \simeq \frac{k^4}{6\pi N^2} |\epsilon_r - 1|^2 \simeq \frac{2k^4}{3\pi N^2} |n - 1|^2 \tag{10.34}$$

where the last form is written in terms of the index of refraction n, assuming $|n - 1| \ll 1$. The cross section (10.34) represents the power scattered per molecule for a unit incident energy flux. In traversing a thickness dx of the gas, the fractional loss of flux is $N\sigma \, dx$. The incident beam thus has an intensity $I(x) = I_0 e^{-\alpha x}$, where α is the *absorption or attenuation coefficient* (also called the *extinction coefficient*) of (7.53) and is given by

$$\alpha = N\sigma \simeq \frac{2k^4}{3\pi N} |n - 1|^2 \tag{10.35}$$

These results, (10.34) and (10.35), describe what is known as *Rayleigh scattering*, the incoherent scattering by gas molecules or other randomly distributed dipole scatterers, each scattering according to Rayleigh's ω^4 law.

Rayleigh's derivation of (10.35) was in the context of scattering of light by the atmosphere. Evidently the k^4 dependence means that in the visible spectrum the red is scattered least and the violet most. Light received away from the direction of the incident beam is more heavily weighted in high-frequency (blue) components than the spectral distribution of the incident beam, while the transmitted beam becomes increasingly red in its spectral composition, as well as diminishing in overall intensity. The blueness of the sky, the redness of the sunset, the waneness of the winter sun, and the ease of sunburning at midday in summer are all consequences of Rayleigh scattering in the atmosphere. The index of refraction of air in the visible region (4100–6500 Å) and at NTP is $(n - 1) \simeq 2.78 \times 10^{-4}$. With $N = 2.69 \times 10^{19}$ molecules/cm³, typical values of the attenuation length $\Lambda = \alpha^{-1}$ are $\Lambda = 30, 77, 188$ km for violet (4100 Å), green (5200 Å), and red (6500 Å) light, respectively. With an isothermal model of the atmosphere in which the density varies exponentially with height, the following intensities at the earth's surface *relative* to those incident on the top of the atmosphere at each wavelength can be estimated for the sun at zenith and sunrise-sunset:

Color	*Zenith*	*Sunrise-Sunset*
Red (6500 Å)	0.96	0.21
Green (5200 Å)	0.90	0.024
Violet (4100 Å)	0.76	0.000065

These numbers show strikingly the shift to the red of the surviving sunlight at sunrise and sunset.

The actual situation is illustrated in Fig. 10.4. The curve A shows the power spectrum of solar radiation incident on the earth from outside as a function of photon energy. Curve B is a typical spectrum at sea level with the sun directly

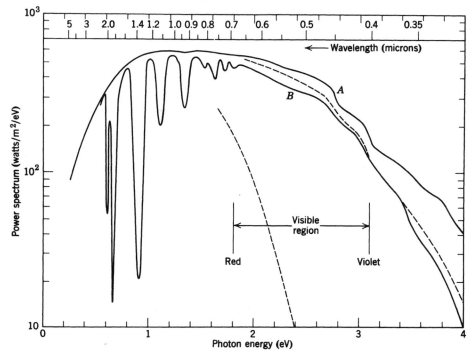

Figure 10.4 Power spectrum of solar radiation (in watts per square meter per electron volt) as a function of photon energy (in electron volts). Curve A is the incident spectrum above the atmosphere. Curve B is a typical sea-level spectrum with the sun at the zenith. The absorption bands below 2 eV are chiefly from water vapor and vary from site to site and day to day. The dashed curves give the expected sea-level spectrum at zenith and at sunrise-sunset if the only attenuation is from Rayleigh scattering by a dry, clean atmosphere.

overhead.* The upper dashed curve is the result expected from curve A if the only attenuation is Rayleigh scattering by a dry, clean, isothermal, exponential atmosphere. In reality the attenuation is greater, mainly because of the presence of water vapor, which has strong absorption bands in the infrared, and ozone, which causes absorption of the ultraviolet, as well as other molecular species and dust. The lower dashed curve indicates roughly the sunrise-sunset spectrum at sea level. Astronauts orbiting the earth see even redder sunsets because the atmospheric path length is doubled.

Detailed observations on the polarization of the scattered light from the sky have been reported.[†] Just as with the attenuation, the reality departs somewhat from the ideal of a dry, clean atmosphere of low density. At 90° the polarization is a function of wavelength and reaches a maximum of approximately 75% at 5500 Å. It is estimated to be less than 100% because of multiple scattering (6%), molecular anisotropy (6%), ground reflection (5%, and especially important in the green when green vegetation is present), and aerosols (8%).

The formula (10.35) for the extinction coefficient is remarkable in its possession of the factor N^{-1} as well as macroscopic quantities such as the index of refraction. If there were no atomicity ($N \to \infty$), there would be no attenuation. Conversely, the observed attenuation can be used to determine N. This point was urged particularly on Rayleigh by Maxwell in private correspondence. If the properties of the atmosphere are assumed to be well enough known, the relative intensity of the light from a definite star as a function of altitude can be used to determine N. Early estimates were made in this way and agree with the results of more conventional methods.

D. Density Fluctuations; Critical Opalescence

An alternative and more general approach to the scattering and attenuation of light in gases and liquids is to consider fluctuations in the density and so the index of refraction. The volume V of fluid is imagined to be divided into cells small compared to a wavelength, but each containing very many molecules. Each cell has volume v with an average number $N_v = vN$ of molecules inside. The actual number of molecules fluctuates around N_v in a manner that depends on the properties of the gas or liquid. Let the departure from the mean of the number of molecules in the jth cell be ΔN_j. The variation in index of refraction $\delta\epsilon$ for the jth cell is

$$\delta\epsilon_j = \frac{\partial\epsilon}{\partial N} \cdot \frac{\Delta N_j}{v}$$

From the Clausius–Mossotti relation (4.70), this can be written

$$\delta\epsilon_j = \frac{(\epsilon_r - 1)(\epsilon_r + 2)}{3Nv} \Delta N_j \tag{10.36}$$

---

*The data in Fig. 10.4 were derived from W. E. Forsythe, *Smithsonian Physical Tables* 9th revised edition, Smithsonian Institution, Washington, DC (1954), Tables 813 and 815, and from K. Ya. Kondratyev, *Radiation in the Atmosphere*, Academic Press, New York (1969), Chapter 5.

[†]T. Gehrels, *J. Opt. Soc. Am.* **52**, 1164 (1962).

With this expression for $\delta\epsilon$ for the jth cell, the integral (10.31), now a sum over cells, becomes

$$\frac{\boldsymbol{\epsilon}^* \cdot \mathbf{A}_{\text{sc}}^{(1)}}{D_0} = \boldsymbol{\epsilon}^* \cdot \boldsymbol{\epsilon}_0 \frac{k^2(\epsilon_r - 1)(\epsilon_r + 2)}{12\pi N \epsilon_r} \sum_j \Delta N_j e^{i\mathbf{q}\cdot\mathbf{x}_j} \qquad (10.37)$$

In forming the absolute square of (10.37) a structure factor similar to (10.19) will occur. If it is assumed that the correlation of fluctuations in different cells (caused indirectly by the intermolecular forces) only extends over a distance small compared to a wavelength, the exponential in (10.37) can be put equal to unity. Then the extinction coefficient α, given by

$$\alpha = \frac{1}{V} \int \left| \frac{\boldsymbol{\epsilon}^* \cdot \mathbf{A}_{\text{sc}}^{(1)}}{D_0} \right|^2 d\Omega$$

is   (10.38)

$$\alpha = \frac{(\omega/c)^4}{6\pi N} \left| \frac{(\epsilon_r - 1)(\epsilon_r + 2)}{3} \right|^2 \cdot \frac{\Delta N_V^2}{NV}$$

where ΔN_V^2 is the mean square number fluctuation in the volume V, defined by

$$\Delta N_V^2 = \sum_{jj'} \Delta N_j \Delta N_{j'}$$

the sum being over all the cells in the volume V. With the use of statistical mechanics* the quantity ΔN_V^2 can be expressed in terms of the *isothermal compressibility* β_T of the medium:

$$\frac{\Delta N_V^2}{NV} = NkT\beta_T, \qquad \beta_T = -\frac{1}{V}\left(\frac{\partial V}{\partial P}\right)_T \qquad (10.39)$$

The attenuation coefficient (10.38) then becomes

$$\alpha = \frac{1}{6\pi N} \left(\frac{\omega}{c}\right)^4 \left| \frac{(\epsilon_r - 1)(\epsilon_r + 2)}{3} \right|^2 \cdot NkT\beta_T \qquad (10.40)$$

This particular expression, first obtained by Einstein in 1910, is called the Einstein–Smoluchowski formula. For a dilute ideal gas, with $|\epsilon - 1| \ll 1$ and $NkT\beta_T = 1$, it reduces to the Rayleigh result (10.35). As the critical point is approached, β_T becomes very large (infinite exactly at the critical point). The scattering and attenuation thus become large there. This is the phenomenon known as *critical opalescence*. The large scattering is directly related to the large fluctuations in density near the critical point, as stressed originally by Smoluchowski (1904). Very near the critical point our treatment so far fails because the correlation length for the density fluctuations becomes greater than a wavelength, as first pointed out by Ornstein and Zernicke (1914).

For large correlation length Λ we must retain the exponential phase factors in (10.37). The absolute square of the scattering amplitude then involves a double sum of $\Delta N_i \Delta N_j e^{i\mathbf{q}\cdot(\mathbf{x}_i - \mathbf{x}_j)}$, which can be expressed as a Fourier transform of the density correlation function. Because there is now additional angular dependence from $\mathbf{q}$, the angular distribution is no longer the simple dipole form. If a corre-

---

*See F. Reif, *Fundamentals of Statistical and Thermal Physics*, McGraw-Hill, New York (1965), pp. 300-1, or L. D. Landau and E. M. Lifshitz, *Statistical Physics*, 3rd edition, Pergamon Press, New York (1980), Chapter XII.

lation function of Yukawa form $e^{-r/\Lambda}/r$ is assumed, it can be shown that the differential attenuation coefficient for unpolarized incident radiation takes the form

$$\frac{d\alpha(\theta)}{d\Omega} = \frac{3}{16\pi} (1 + \cos^2\theta) \; \alpha \left[\frac{1 + \Lambda^2 q^2/NkT\beta_T}{1 + \Lambda^2 q^2} \right] \tag{10.41}$$

where $q^2 = 2(\omega/c)^2(1 - \cos\theta)$ and α is given by (10.40). For $\Lambda q \ll 1$, integration over the normalized angular distribution gives back (10.40), but for $\Lambda \to \infty$, the angular integration yields attenuation proportional to $(c/\Lambda\omega)^2 \ln(\Lambda\omega/c)$ times (10.40). The frequency dependence as ω^4 away from the critical point is altered to roughly ω^2; the scattered light appears "whiter" close to the critical point.

We note that, while our expressions diverge exactly at the critical point and therefore are unphysical, a better treatment yields large but finite attenuation. One consideration is that the correlation length Λ cannot become larger than the dimensions of the fluid container.

References to the early literature can be found in *Fabelinskii*, who discusses the application of light scattering to critical point phenomena and second-order phase transitions. For treatments of the radial density correlation function, see *Rosenfeld* (Chapter V, Section 6), or *Landau and Lifshitz (op. cit.)*.

E. Attenuation in Optical Fibers

It is of interest that the ultimate limiting factor setting the maximum distance between repeater units in optical fiber transmission is the unavoidable attenua-

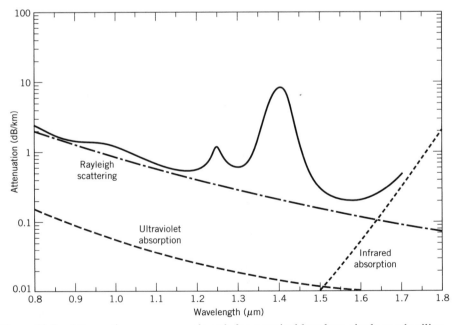

Figure 10.5 Attenuation versus wavelength for a typical low-loss, single-mode silica optical fiber (schematic). Rayleigh scattering sets the lower limit until infrared absorption rises above 1.6 μm. The peaks in the observed attenuation are caused by water (OH ions) dissolved in the glass.

tion caused by Rayleigh scattering, and by infrared absorption at longer wavelengths. The isothermal compressibility of silica glass is $\beta_T \approx 7 \times 10^{-11}$ m^2/N, while the relevant temperature $T \approx 1400$ K (called the fictive temperature) is where the fluctuations are frozen in (approximately the annealing temperature). The effective value of $(\epsilon_r - 1)(\epsilon_r + 2)/3 \approx 1.30$ in (10.40) is somewhat smaller than the 1.51 inferred from an index of refraction of $n = 1.45$ at $\lambda = 1.0$ μm. The net result is that α (km^{-1}) $\approx 0.2/[\lambda$ (μm)]^4. The conversion to decibels per kilometer (a factor of 4.343) gives α (dB/km) $\approx 0.85/[\lambda$ (μm)]^4, shown as the dash-dotted curve in Fig. 10.5, which displays a schematic representation of typical data for a low-loss, single-mode optical fiber. For wavelengths less than 1.5 μm, the attenuation is dominated by Rayleigh scattering, plus the absorption by impurities such as the hydroxyl ions from very small amounts of water dissolved in the glass. At wavelengths longer than 1.6 μm, infrared absorption sets in strongly. The minimum attenuation of about 0.2 dB/km occurs at $\lambda \approx 1.55$ μm. The absorption mean free path at the minimum is 22 km.

10.3   Spherical Wave Expansion of a Vector Plane Wave

In discussing the scattering or absorption of electromagnetic radiation by spherical objects, or localized systems in general, it is useful to have an expansion of a plane electromagnetic wave in spherical waves.

For a scalar field $\psi(\mathbf{x})$ satisfying the wave equation, the necessary expansion can be obtained by using the orthogonality properties of the basic spherical solutions $j_l(kr) Y_{lm}(\theta, \phi)$. An alternative derivation makes use of the spherical wave expansion (9.98) of the Green function $(e^{ikR}/4\pi R)$. We let $|\mathbf{x}'| \to \infty$ on both sides of (9.98). Then we can put $|\mathbf{x} - \mathbf{x}'| \simeq r' - \mathbf{n} \cdot \mathbf{x}$ on the left-hand side, where $\mathbf{n}$ is a unit vector in the direction of $\mathbf{x}'$. On the right side $r_> = r'$ and $r_< = r$. Furthermore we can use the asymptotic form (9.89) for $h_l^{(1)}(kr')$. Then we find

$$\frac{e^{ikr'}}{4\pi r'} e^{-ik\mathbf{n}\cdot\mathbf{x}} = ik \frac{e^{ikr'}}{kr'} \sum_{l,m} (-i)^{l+1} j_l(kr) Y^*_{lm}(\theta', \phi') Y_{lm}(\theta, \phi) \qquad (10.42)$$

Canceling the factor $e^{ikr'}/r'$ on either side and taking the complex conjugate, we have the expansion of a plane wave

$$e^{i\mathbf{k}\cdot\mathbf{x}} = 4\pi \sum_{l=0}^{\infty} i^l j_l(kr) \sum_{m=-l}^{l} Y^*_{lm}(\theta, \phi) Y_{lm}(\theta', \phi') \qquad (10.43)$$

where $\mathbf{k}$ is the wave vector with spherical coordinates k, θ', ϕ'. The addition theorem (3.62) can be used to put this in a more compact form

$$e^{i\mathbf{k}\cdot\mathbf{x}} = \sum_{l=0}^{\infty} i^l (2l + 1) j_l(kr) P_l(\cos\gamma) \qquad (10.44)$$

where γ is the angle between $\mathbf{k}$ and $\mathbf{x}$. With (3.57) for $P_l \cos(\gamma)$, this can also be written as

$$e^{i\mathbf{k}\cdot\mathbf{x}} = \sum_{l=0}^{\infty} i^l \sqrt{4\pi(2l + 1)} \, j_l(kr) Y_{l,0}(\gamma) \qquad (10.45)$$

We now wish to make an equivalent expansion for a circularly polarized plane wave with helicity $\pm$ incident along the z axis,

$$E(x) = (\epsilon_1 \pm i\epsilon_2)e^{ikz}$$
$$cB(x) = \epsilon_3 \times E = \mp iE \tag{10.46}$$

Since the plane wave is finite everywhere, we can write its multipole expansion (9.122) involving only the regular radial functions $j_l(kr)$:

$$E(x) = \sum_{l,m} \left[a_\pm(l, m)j_l(kr)X_{lm} + \frac{i}{k} b_\pm(l, m)\nabla \times j_l(kr)X_{lm} \right]$$
$$cB(x) = \sum_{l,m} \left[\frac{-i}{k} a_\pm(l, m)\nabla \times j_l(kr)X_{lm} + b_\pm(l, m)j_l(kr)X_{lm} \right] \tag{10.47}$$

To determine the coefficients $a_\pm(l, m)$ and $b_\pm(l, m)$ we utilize the orthogonality properties of the vector spherical harmonics X_{lm}. For reference purposes we summarize the basic relation (9.120), as well as some other useful relations:

$$\left.\begin{array}{l} \int [f_l(r)X_{l'm'}]^* \cdot [g_l(r)X_{lm}] \, d\Omega = f_l^* g_l \, \delta_{ll'}\delta_{mm'} \\[2mm] \int [f_l(r)X_{l'm'}]^* \cdot [\nabla \times g_l(r)X_{lm}] \, d\Omega = 0 \\[2mm] \dfrac{1}{k^2} \int [\nabla \times f_l(r)X_{l'm'}]^* \cdot [\nabla \times g_l(r)X_{lm}] \, d\Omega \\[3mm] \qquad = \delta_{ll'}\delta_{mm'}\left\{ f_l^* g_l + \dfrac{1}{k^2 r^2} \dfrac{\partial}{\partial r}\left[rf_l^* \dfrac{\partial}{\partial r}(rg_l) \right] \right\} \end{array}\right\} \tag{10.48}$$

In these relations $f_l(r)$ and $g_l(r)$ are linear combinations of spherical Bessel functions, satisfying (9.81). The second and third relations can be proved using the operator identity (9.125), the representation

$$\nabla = \frac{r}{r} \frac{\partial}{\partial r} - \frac{i}{r^2} r \times L$$

for the gradient operator, and the radial differential equation (9.81).

To determine the coefficients $a_\pm(l, m)$ and $b_\pm(l, m)$ we take the scalar product of both sides of (10.47) with X_{lm}^* and integrate over angles. Then with the first and second orthogonality relations in (10.48) we obtain

$$a_\pm(l, m)j_l(kr) = \int X_{lm}^* \cdot E(x) \, d\Omega \tag{10.49}$$

and

$$b_\pm(l, m)j_l(kr) = c \int X_{lm}^* \cdot B(x) \, d\Omega \tag{10.50}$$

With (10.46) for the electric field, (10.49) becomes

$$a_\pm(l, m)j_l(kr) = \int \frac{(L_\mp Y_{lm})^*}{\sqrt{l(l + 1)}} e^{ikz} \, d\Omega \tag{10.51}$$

where the operators $L_\pm$ are defined by (9.102), and the results of their operating by (9.104). Thus we obtain

$$a_\pm(l, m)j_l(kr) = \frac{\sqrt{(l \pm m)(l \mp m + 1)}}{\sqrt{l(l + 1)}} \int Y^*_{l,m\mp 1}e^{ikz} \, d\Omega \qquad (10.52)$$

If expansion (10.45) for e^{ikz} is inserted, the orthogonality of the Y_{lm}'s evidently leads to the result,

$$a_\pm(l, m) = i^l\sqrt{4\pi(2l + 1)} \, \delta_{m,\pm 1} \qquad (10.53)$$

From (10.50) and (10.46) it is clear that

$$b_\pm(l, m) = \mp ia_\pm(l, m) \qquad (10.54)$$

Then the multipole expansion of the plane wave (10.46) is

$$\mathbf{E}(\mathbf{x}) = \sum_{l=1}^{\infty} i^l\sqrt{4\pi(2l + 1)} \left[j_l(kr)\mathbf{X}_{l,\pm 1} \pm \frac{1}{k} \nabla \times j_l(kr)\mathbf{X}_{l,\pm 1} \right]$$

$$c\mathbf{B}(\mathbf{x}) = \sum_{l=1}^{\infty} i^l\sqrt{4\pi(2l + 1)} \left[\frac{-i}{k} \nabla \times j_l(kr)\mathbf{X}_{l,\pm 1} \mp ij_l(kr)\mathbf{X}_{l,\pm 1} \right] \qquad (10.55)$$

For such a circularly polarized wave the m values of $m = \pm 1$ have the obvious interpretation of ± 1 unit of angular momentum per photon parallel to the propagation direction. This was established in Problems 7.28 and 7.29.

10.4 Scattering of Electromagnetic Waves by a Sphere

If a plane wave of electromagnetic radiation is incident on a spherical obstacle, as indicated schematically in Fig. 10.6, it is scattered, so that far away from the scatterer the fields are represented by a plane wave plus outgoing spherical waves. There may be absorption by the obstacle as well as scattering. Then the total energy flow away from the obstacle will be less than the total energy flow towards it, the difference being absorbed. We will ultimately consider the simple example of scattering by a sphere of radius a and infinite conductivity, but will for a time keep the problem more general.

The fields outside the sphere can be written as a sum of incident and scattered waves:

$$\left.\begin{array}{c} \mathbf{E}(\mathbf{x}) = \mathbf{E}_{\text{inc}} + \mathbf{E}_{\text{sc}} \\ \mathbf{B}(\mathbf{x}) = \mathbf{B}_{\text{inc}} + \mathbf{B}_{\text{sc}} \end{array}\right\} \qquad (10.56)$$

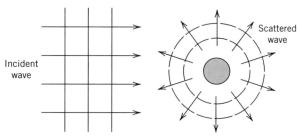

Figure 10.6   Scattering of radiation by a localized object.

where $\mathbf{E}_{inc}$ and $\mathbf{B}_{inc}$ are given by (10.55). Since the scattered fields are outgoing waves at infinity, their expansions must be of the form,

$$
\mathbf{E}_{sc} = \frac{1}{2} \sum_{l=1}^{\infty} i^l \sqrt{4\pi(2l+1)} \left[\alpha_{\pm}(l)h_l^{(1)}(kr)\mathbf{X}_{l,\pm 1} \pm \frac{\beta_{\pm}(l)}{k} \boldsymbol{\nabla} \times h_l^{(1)}(kr)\mathbf{X}_{l,\pm 1} \right]
$$

$$
c\mathbf{B}_{sc} = \frac{1}{2} \sum_{l=1}^{\infty} i^l \sqrt{4\pi(2l+1)} \left[\frac{-i\alpha_{\pm}(l)}{k} \boldsymbol{\nabla} \times h_l^{(1)}(kr)\mathbf{X}_{l,\pm 1} \mp i\beta_{\pm}(l)h_l^{(1)}(kr)\mathbf{X}_{l,\pm 1} \right]
$$

(10.57)

The coefficients $\alpha_{\pm}(l)$ and $\beta_{\pm}(l)$ will be determined by the boundary conditions on the surface of the scatterer. A priori, it is necessary to keep a full sum over m as well as l in (10.57), but for the restricted class of spherically symmetric problems considered here, only $m = \pm 1$ occurs.

Formal expressions for the total scattered and absorbed power in terms of the coefficients of $\alpha(l)$ and $\beta(l)$ can be derived from the scattered and total fields on the surface of a sphere of radius a surrounding the scatterer, with the scattered power being the *outward* component of the Poynting vector formed from the *scattered* fields, integrated over the spherical surface, and the absorbed power being the corresponding *inward* component formed from the *total* fields. With slight rearrangement of the triple scalar products, these can be written

$$
P_{sc} = -\frac{a^2}{2\mu_0} \operatorname{Re} \int \mathbf{E}_{sc} \cdot (\mathbf{n} \times \mathbf{B}_{sc}^*) \, d\Omega \tag{10.58}
$$

$$
P_{abs} = \frac{a^2}{2\mu_0} \operatorname{Re} \int \mathbf{E} \cdot (\mathbf{n} \times \mathbf{B}^*) \, d\Omega \tag{10.59}
$$

Here $\mathbf{n}$ is a radially directed outward normal, $\mathbf{E}_{sc}$ and $\mathbf{B}_{sc}$ are given by (10.57), while $\mathbf{E}$ and $\mathbf{B}$ are the sum of the plane wave fields (10.55) and the scattered fields (10.57). Only the transverse parts of the fields enter these equations. We already know that $\mathbf{X}_{lm}$ is transverse. The other type of term in (10.55) and (10.57) is

$$
\boldsymbol{\nabla} \times f_l(r)\mathbf{X}_{lm} = \frac{i\mathbf{n}\sqrt{l(l+1)}}{r} f_l(r)Y_{lm} + \frac{1}{r} \frac{\partial}{\partial r} [rf_l(r)]\mathbf{n} \times \mathbf{X}_{lm} \tag{10.60}
$$

where f_l is any spherical Bessel function of order l satisfying (9.81). When the multipole expansions of the fields are inserted in (10.58) and (10.59), there results a double sum over l and l' of various scalar products of the form $\mathbf{X}_{lm}^* \cdot \mathbf{X}_{l'm'}$, $\mathbf{X}_{lm}^* \cdot (\mathbf{n} \times \mathbf{X}_{l'm'})$ and $(\mathbf{n} \times \mathbf{X}_{lm}^*) \cdot (\mathbf{n} \times \mathbf{X}_{l'm'})$. On integration over angles, the orthogonality relations (10.48) reduce the double sum to a single sum. Each term in the sum involves products of spherical Bessel functions and derivatives of spherical Bessel functions. Use of the Wronskians (9.91) permits the elimination of all the Bessel functions and yields the following expressions for the total scattering and absorption cross sections (the power scattered or absorbed divided by the incident flux, $1/\mu_0 c$):

$$
\sigma_{sc} = \frac{\pi}{2k^2} \sum_{l} (2l+1)[|\alpha(l)|^2 + |\beta(l)|^2] \tag{10.61}
$$

$$
\sigma_{abs} = \frac{\pi}{2k^2} \sum_{l} (2l+1)[2 - |\alpha(l)+1|^2 - |\beta(l)+1|^2]
$$

The total or extinction cross section is the sum of σ_{sc} and σ_{abs}:

$$\sigma_t = -\frac{\pi}{k^2} \sum_l (2l + 1) \, \text{Re}[\alpha(l) + \beta(l)] \tag{10.62}$$

Not surprisingly, these expressions for the cross sections resemble closely the partial wave expansions of quantum-mechanical scattering.*

The differential scattering cross section is obtained by calculating the scattered power radiated into a given solid angle element $d\Omega$ and dividing by the incident flux. Using the result of Problem 10.6a, we find the scattering cross section for incident polarization $(\boldsymbol{\epsilon}_1 \pm i\boldsymbol{\epsilon}_2)$ to be

$$\frac{d\sigma_{sc}}{d\Omega} = \frac{\pi}{2k^2} \left| \sum_l \sqrt{2l + 1} \, [\alpha_\pm(l)\mathbf{X}_{l,\pm 1} \pm i\beta_\pm(l) \, \mathbf{n} \times \mathbf{X}_{l,\pm 1}] \right|^2 \tag{10.63}$$

The scattered radiation is in general elliptically polarized. Only if $\alpha_\pm(l) = \beta_\pm(l)$ for all l would it be circularly polarized. This means that if the incident radiation is linearly polarized, the scattered radiation will be elliptically polarized; if the incident radiation is unpolarized, the scattered radiation will exhibit partial polarization depending on the angle of observation. Examples of this in the long-wavelength limit were described in Section 10.1 (see Figs. 10.2 and 10.3).

The coefficients $\alpha_\pm(l)$ and $\beta_\pm(l)$ in (10.57) are determined by the boundary conditions on the fields at $r = a$. Normally this would involve the solution of the Maxwell equations inside the sphere and appropriate matching of solutions across $r = a$. If, however, the scatterer is a sphere of radius a whose electromagnetic properties can be described by a *surface impedance Z_s* independent of position (for this the radial variation of the fields just inside the sphere must be rapid compared to the radius), then the boundary conditions take the relatively simple form

$$\mathbf{E}_{\text{tan}} = Z_s \mathbf{n} \times \mathbf{B}/\mu_0 \tag{10.64}$$

where **E** and **B** are evaluated just outside the sphere. From (10.55), (10.57), and (10.60) we have

and

$$
\mathbf{E}_{\text{tan}} = \sum_l i^l \sqrt{4\pi(2l + 1)} \left\{ \left[j_l + \frac{\alpha_\pm(l)}{2} h_l^{(1)} \right] \mathbf{X}_{l,\pm 1} \right. \\
\left. \pm \frac{1}{x} \frac{\partial}{\partial x} \left[x \left(j_l + \frac{\beta_\pm(l)}{2} h_l^{(1)} \right) \right] \mathbf{n} \times \mathbf{X}_{l,\pm 1} \right\}
$$

$$
c\mathbf{n} \times \mathbf{B} = \sum_l i^l \sqrt{4\pi(2l + 1)} \left\{ \frac{i}{x} \frac{\partial}{\partial x} \left[x \left(j_l + \frac{\alpha_\pm(l)}{2} h_l^{(1)} \right) \right] \mathbf{X}_{l,\pm 1} \right. \\
\left. \mp i \left[j_l + \frac{\beta_\pm(l)}{2} h_l^{(1)} \right] \mathbf{n} \times \mathbf{X}_{l,\pm 1} \right\}
$$

*Our results are not completely general. If the sum over m had been included in (10.57), the scattering cross section would have a sum over l and m with the absolute squares of $\alpha(l, m)$ and $\beta(l, m)$. The total cross section would stay as it is, with $\alpha(l) \rightarrow \alpha(l, m = \pm 1)$ and $\beta(l) \rightarrow \beta(l, m = \pm 1)$, depending on the state of polarization of the incident wave (10.46). The absorption cross section can be deduced from taking the difference of σ_t and σ_{sc}.

where $x = ka$ and all the spherical Bessel functions have argument x. The boundary condition (10.64) requires that, for each l value and for each term $\mathbf{X}_{lm}$ and $\mathbf{n} \times \mathbf{X}_{lm}$ separately, the coefficients of $\mathbf{E}_{tan}$ and $\mathbf{n} \times \mathbf{B}$ be proportional, according to

$$
\begin{aligned}
j_l + \frac{\alpha_\pm(l)}{2} h_l^{(1)} &= i\left(\frac{Z_s}{Z_0}\right) \frac{1}{x} \frac{d}{dx}\left[x\left(j_l + \frac{\alpha_\pm(l)}{2} h_l^{(1)}\right)\right] \\
j_l + \frac{\beta_\pm(l)}{2} h_l^{(1)} &= i\left(\frac{Z_0}{Z_s}\right) \frac{1}{x} \frac{d}{dx}\left[x\left(j_l + \frac{\beta_\pm(l)}{2} h_l^{(1)}\right)\right]
\end{aligned}
\tag{10.65}
$$

By means of the relation $2j_l = h_l^{(1)} + h_l^{(2)}$, the coefficients $\alpha_\pm(l)$ and $\beta_\pm(l)$ can be written

$$
\alpha_\pm(l) + 1 = -\left[\frac{h_l^{(2)} - i\left(\dfrac{Z_s}{Z_0}\right) \dfrac{1}{x} \dfrac{d}{dx}(xh_l^{(2)})}{h_l^{(1)} - i\left(\dfrac{Z_s}{Z_0}\right) \dfrac{1}{x} \dfrac{d}{dx}(xh_l^{(1)})}\right]
\tag{10.66}
$$

with $\beta_\pm(l)$ having the same form, but with Z_s/Z_0 replaced by its reciprocal. We note that with the surface impedance boundary condition the coefficients are the same for both states of circular polarization.

For a given Z_s, all the multipole coefficients are determined and the scattering is known in principle. All that remains is to put in numbers. Before proceeding to a specific limit, we make some observations. First, if Z_s is purely imaginary (no dissipation) or if $Z_s = 0$ or $Z_s \to \infty$, $[\alpha_\pm(l) + 1]$ and $[\beta_\pm(l) + 1]$ are numbers of modulus unity. This means that $\alpha_\pm(l)$ and $\beta_\pm(l)$ can be written as

$$
\alpha_\pm(l) = (e^{2i\delta_l} - 1), \qquad \beta_\pm(l) = (e^{2i\delta_l'} - 1)
\tag{10.67}
$$

where the phase angles δ_l and δ_l' are called *scattering phase shifts*. Specifically

$$
\tan \delta_l = j_l(ka)/n_l(ka)
$$

$$
\tan \delta_l' = \left[\frac{\dfrac{d}{dx}(xj_l(x))}{\dfrac{d}{dx}(xn_l(x))}\right]_{x=ka}
\tag{10.68}
$$

if $Z_s = 0$ (perfectly conducting sphere) and $\delta_l \leftrightarrow \delta_l'$ for $Z_s \to \infty$.

The second observation is that (10.66) can be simplified in the low- and high-frequency limits. For $ka \ll l$, the spherical Bessel functions can be approximated according to (9.88). Then we obtain the long-wavelength approximation,

$$
\alpha_\pm(l) \simeq \frac{-2i(ka)^{2l+1}}{(2l+1)[(2l-1)!!]^2}\left[\frac{x - i(l+1)Z_s/Z_0}{x + ilZ_s/Z_0}\right]
\tag{10.69}
$$

and the same form for $\beta_\pm(l)$, with (Z_s/Z_0) replaced by its inverse. For $ka \gg l$, we use (9.89) and obtain

$$
\alpha_\pm(l) \simeq \left(\frac{Z_s/Z_0 - 1}{Z_s/Z_0 + 1}\right)(-1)^{l+1} e^{-2ika} - 1
\tag{10.70}
$$

with $\beta_{\pm}(l) = -\alpha_{\pm}(l)$ via the usual substitution. In the long-wavelength limit, independent of the actual value of Z_s, the scattering coefficients $\alpha_{\pm}(l)$, $\beta_{\pm}(l)$ become small very rapidly as l increases. Usually, only the lowest term ($l = 1$) need be retained for each multipole series. In the opposite limit of $ka \gg 1$, (10.70) shows that for $l \ll ka$, the successive coefficients have comparable magnitudes, but phases that fluctuate widely. For $l \sim l_{max} = ka$, there is a transition region and for $l \gg l_{max}$, (10.69) holds. The use of a partial wave or multipole expansion for such a large number of terms is a delicate matter, necessitating the careful use of digital computers or approximation schemes of the type discussed in Section 10.10.

We specialize now to the long-wavelength limit ($ka \ll 1$) for a perfectly conducting sphere ($Z_s = 0$), and leave examples of slightly more complexity to the problems. Only the $l = 1$ terms in (10.63) are important. From (10.69) we find

$$\alpha_{\pm}(1) = \frac{-1}{2} \beta_{\pm}(1) \simeq -\frac{2i}{3} (ka)^3$$

In this limit the scattering cross section is

$$\frac{d\sigma_{sc}}{d\Omega} \simeq \frac{2\pi}{3} a^2 (ka)^4 \left| \mathbf{X}_{1,\pm1} \mp 2i\mathbf{n} \times \mathbf{X}_{1,\pm1} \right|^2 \tag{10.71}$$

From Table 9.1 we obtain the absolute squared terms,

$$\left| \mathbf{n} \times \mathbf{X}_{1,\pm1} \right|^2 = \left| \mathbf{X}_{1,\pm1} \right|^2 = \frac{3}{16\pi} (1 + \cos^2\theta)$$

The cross terms can be easily worked out:

$$[\pm i(\mathbf{n} \times \mathbf{X}_{1,\pm1})^* \cdot \mathbf{X}_{1,\pm1}] = \frac{-3}{8\pi} \cos\theta$$

Thus the long-wavelength limit of the differential scattering cross section is

$$\frac{d\sigma}{d\Omega} \simeq a^2 (ka)^4 [\tfrac{5}{8}(1 + \cos^2\theta) - \cos\theta] \tag{10.72}$$

Equation (10.72) is the same as (10.16), found by other means and is valid for either state of circular polarization incident, or for an unpolarized incident beam. The generalizations to arbitrary incident polarization and to different surface boundary conditions are left to the problems at the end of the chapter.

The general problem of the scattering of electromagnetic waves by spheres of arbitrary electric and magnetic properties when ka is not small is complicated. It was first systematically attacked by Mie and Debye in 1908–1909. By now, hundreds of papers have been published on the subject. Details of the many aspects of this important problem can be found in the books by *Kerker, King and Wu, Bowman, Senior, and Uslenghi* and other sources cited at the end of the chapter. The book by *Bowman, Senior, and Uslenghi* discusses scattering by other regular shapes besides the sphere.

For scatterers other than spheres, cylinders, etc., there is very little in the way of formal theory. The perturbation theory of Section 10.2 may be used in appropriate circumstances.

10.5 Scalar Diffraction Theory

Although scattering and diffraction are not logically separate, the treatments tend to be separated, with diffraction being associated with departures from geometrical optics caused by the finite wavelength of the waves. Thus diffraction traditionally involves apertures or obstacles whose dimensions are large compared to a wavelength. To lowest approximation the interaction of electromagnetic waves is described by ray tracing (geometrical optics). The next approximation involves the diffraction of the waves around the obstacles or through the apertures with a consequent spreading of the waves. Simple arguments based on Fourier transforms show that the angles of deflection of the waves are confined to the region $\theta \lesssim \lambda/d$, where λ is the wavelength and d is a linear dimension of the aperture or obstacle. The various approximations to be discussed below all work best for $\lambda/d \ll 1$, and fail badly for $\lambda \sim d$ or $\lambda > d$.

The earliest work on diffraction is associated with the names of Huygens, Young, and Fresnel. The first systematic attempt to derive the Fresnel theory from first principles was made by G. Kirchhoff (1882). Kirchhoff's theory, despite its mathematical inconsistency and its physical deficiencies, works remarkably well in the optical domain and has been the basis of most of the work on diffraction. We first derive the basic Kirchhoff integral and its operative approximations, then comment on its mathematical difficulties, and finally describe the modifications of Rayleigh and Sommerfeld that remove the mathematical inconsistencies.

The customary geometry in diffraction involves two spatial regions I and II, separated by a boundary surface S_1, as shown in Fig. 10.7. The surface S_2 is generally taken to be "at infinity," that is, remote from the region of interest. Sources in region I generate fields that propagate outward. The surface S_1 is supposed to be made up of "opaque" portions (the boundary conditions are discussed below) and apertures. The surface S_1 interacts with the fields generated in region I, reflecting some of the energy, absorbing some of it, and allowing some of the fields, modified by their interaction, to pass into region II. The angular distribution of the fields in region II, the diffraction region, is called the *diffraction pattern*. It is the diffracted fields in region II that we wish to express in terms of the fields of the sources and their interaction with the screen and apertures on S_1, or more precisely, in terms of the fields on the surface S_1. It

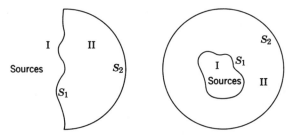

Figure 10.7 Possible diffraction geometries. Region I contains the sources of radiation. Region II is the diffraction region, where the fields satisfy the radiation condition. The right-hand figure is also indicative of scattering, with a finite scatterer in region I instead of an active source, and the surface S_1 an arbitrary mathematical surface enclosing the scatterer rather than a material screen with apertures.

should be obvious that the geometry and mode of description is equally applicable to scattering, with the sources in region I replaced by a scatterer (thought of as a source being driven by the incident wave).

Kirchhoff's method uses Green's theorem (1.35) to express a scalar field (a component of **E** or **B**) inside a closed volume V in terms of the values of the field and its normal derivative on the boundary surface S. Let the scalar field be $\psi(\mathbf{x}, t)$, and let it have harmonic time dependence, $e^{-i\omega t}$. The field ψ is assumed to satisfy the scalar Helmholtz wave equation,

$$(\nabla^2 + k^2)\psi(\mathbf{x}) = 0 \tag{10.73}$$

inside V. We introduce a Green function for the Helmholtz wave equation $G(\mathbf{x}, \mathbf{x}')$, defined by

$$(\nabla^2 + k^2)G(\mathbf{x}, \mathbf{x}') = -\delta(\mathbf{x} - \mathbf{x}') \tag{10.74}$$

In Green's theorem (1.35), we put $\phi = G$, $\psi = \psi$, make use of the wave equations (10.73) and (10.74), and obtain, in analogy to (1.36),

$$\psi(\mathbf{x}) = \oint_S [\psi(\mathbf{x}')\mathbf{n}' \cdot \nabla' G(\mathbf{x}, \mathbf{x}') - G(\mathbf{x}, \mathbf{x}')\mathbf{n}' \cdot \nabla'\psi(\mathbf{x}')] \, da' \tag{10.75}$$

where $\mathbf{n}'$ is an *inwardly* directed normal to the surface S. Equation (10.75) holds if $\mathbf{x}$ is inside V; if it is not, the left-hand side vanishes.

The Kirchhoff diffraction integral is obtained from (10.75) by taking G to be the infinite-space Green function describing outgoing waves,

$$G(\mathbf{x}, \mathbf{x}') = \frac{e^{ikR}}{4\pi R} \tag{10.76}$$

where $\mathbf{R} = \mathbf{x} - \mathbf{x}'$. With this Green function, (10.75) becomes

$$\psi(\mathbf{x}) = -\frac{1}{4\pi} \oint_S \frac{e^{ikR}}{R} \mathbf{n}' \cdot \left[\nabla'\psi + ik\left(1 + \frac{i}{kR}\right)\frac{\mathbf{R}}{R} \psi \right] da' \tag{10.77}$$

This is almost the Kirchhoff integral. To adapt the mathematics to the diffraction context we consider the volume V to be that of region II in Fig. 10.7 and the surface S to consist of $S_1 + S_2$. The integral over S is thus divided into two parts, one over the screen and its apertures (S_1), the other over a surface "at infinity" (S_2). Since the fields in region II are assumed to be transmitted through S_1, they are outgoing waves in the neighborhood of S_2. The fields, hence $\psi(\mathbf{x})$, will satisfy a *radiation condition*,

$$\psi \to f(\theta, \phi) \frac{e^{ikr}}{r}, \qquad \frac{1}{\psi}\frac{\partial\psi}{\partial r} \to \left(ik - \frac{1}{r}\right) \tag{10.78}$$

With this condition on ψ at S_2 it is easily seen that the contribution from S_2 in (10.77) vanishes at least as the inverse of the radius of the hemisphere or sphere as the radius goes to infinity. There remains the integral over S_1. The *Kirchhoff integral* formula reads

$$\psi(\mathbf{x}) = -\frac{1}{4\pi} \int_{S_1} \frac{e^{ikR}}{R} \mathbf{n}' \cdot \left[\nabla'\psi + ik\left(1 + \frac{i}{kR}\right)\frac{\mathbf{R}}{R} \psi \right] da' \tag{10.79}$$

with the integration only over the surface S_1 of the diffracting "screen."

To apply (10.79) it is necessary to know the values of ψ and $\partial\psi/\partial n$ on the surface S_1. Unless the problem has been solved by other means, these values are not known. Kirchhoff's approach was to approximate the values of ψ and $\partial\psi/\partial n$ on S_1 in order to calculate an approximation to the diffracted wave. The *Kirchhoff approximation* consists of the assumptions:

1. ψ and $\partial\psi/\partial n$ vanish everywhere on S_1 except in the openings.
2. The values of ψ and $\partial\psi/\partial n$ in the openings are equal to the values of the incident wave in the absence of any screen or obstacles.

The standard diffraction calculations of classical optics are all based on the Kirchhoff approximation. It is obvious that the recipe can have only limited validity. There is, in fact, a serious mathematical inconsistency in the assumptions of Kirchhoff. It can be shown for the Helmholtz wave equation (10.73), as well as for the Laplace equation, that if ψ and $\partial\psi/\partial n$ are both zero on any finite surface, then $\psi = 0$ everywhere. Thus the only mathematically correct consequence of the first Kirchhoff assumption is that the diffracted field vanishes everywhere. This is, of course, inconsistent with the second assumption. Furthermore, (10.79) does not yield on S_1 the assumed values of ψ and $\partial\psi/\partial n$.

The mathematical inconsistencies in the Kirchhoff approximation can be removed by the choice of a proper Green function in (10.75). Just as in Section 1.10, a Green function appropriate to Dirichlet or Neumann boundary conditions can be constructed. If ψ is known or approximated on the surface S_1, a Dirichlet Green function $G_D(\mathbf{x}, \mathbf{x}')$, satisfying

$$G_D(\mathbf{x}, \mathbf{x}') = 0 \qquad \text{for } \mathbf{x}' \text{ on } S \tag{10.80}$$

is required. Then a *generalized Kirchhoff integral*, equivalent to (10.79), is

$$\psi(\mathbf{x}) = \int_{S_1} \psi(\mathbf{x}') \frac{\partial G_D}{\partial n'} (\mathbf{x}, \mathbf{x}') \, da' \tag{10.81}$$

and a *consistent* approximation is that $\psi = 0$ on S_1 except in the openings and ψ is equal to the incident wave in the openings. If the normal derivative of ψ is to be approximated, a Neumann Green function $G_N(\mathbf{x}, \mathbf{x}')$, satisfying

$$\frac{\partial G_N}{\partial n'} (\mathbf{x}, \mathbf{x}') = 0 \qquad \text{for } \mathbf{x}' \text{ on } S \tag{10.82}$$

is employed. Then the generalized Kirchhoff integral for Neumann boundary conditions reads

$$\psi(\mathbf{x}) = -\int_{S_1} \frac{\partial\psi}{\partial n'} (\mathbf{x}') G_N(\mathbf{x}, \mathbf{x}') \, da' \tag{10.83}$$

Again a consistent approximation scheme can be formulated.

For the important special circumstance in which the surface S_1 is an infinite *plane screen at $z = 0$*, as shown in Fig. 10.8, the method of images can be used to give the Dirichlet and Neumann Green functions explicit form:

$$G_{D,N}(\mathbf{x}, \mathbf{x}') = \frac{1}{4\pi} \left(\frac{e^{ikR}}{R} \mp \frac{e^{ikR'}}{R'} \right) \tag{10.84}$$

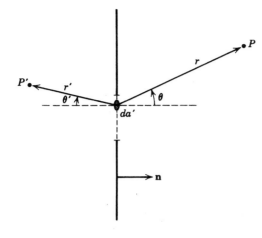

Figure 10.8   Diffraction geometry for a point source at P', a plane screen with apertures, and an observation point at P. The distances from the element of area da' in the aperture to the points P and P' are r and r', respectively. The angles θ and θ' are those between $\mathbf{r}$ and $\mathbf{n}$, and $\mathbf{r}'$ and $-\mathbf{n}$, respectively.

where $\mathbf{R} = \mathbf{x} - \mathbf{x}'$, and $\mathbf{R}' = \mathbf{x} - \mathbf{x}''$, $\mathbf{x}''$ being the mirror image of $\mathbf{x}'$. Explicitly we have

$$R = [(x - x')^2 + (y - y')^2 + (z - z')^2]^{1/2}$$
$$R' = [(x - x')^2 + (y - y')^2 + (z + z')^2]^{1/2}$$

The generalized Kirchhoff integral (10.81) (ψ approximated on S_1) then takes the form,

$$\psi(\mathbf{x}) = \frac{k}{2\pi i} \int_{S_1} \frac{e^{ikR}}{R} \left(1 + \frac{i}{kR}\right) \frac{\mathbf{n}' \cdot \mathbf{R}}{R} \psi(\mathbf{x}') \, da' \tag{10.85}$$

An analogous expression can be written for (10.83), both results attributable to the ubiquitous Rayleigh.*

Comparison of (10.85) with (10.79) shows that (10.85) can be obtained from (10.79) by omitting the first term in the square brackets and doubling the second term. The Neumann result (10.83) specialized to a plane screen is equivalent, on the other hand, to doubling the first term and omitting the second. It might thus appear that the three approximate formulas for the diffracted field are quite different and will lead to very different results. In the domain where they have any reasonable validity they yield, in fact, very similar results. This can be understood by specializing the diffraction problem to a *point source* at position P' on one side of a plane screen and an observation point P on the other side, as shown in Fig. 10.8. The amplitude of the point source is taken to be spherically symmetric and equal to $e^{ikr'}/r'$. Both P and P' are assumed to be many wavelengths away from the screen. With the Kirchhoff approximation in (10.79) and equivalent assumptions in (10.85) and its Neumann boundary condition counterpart, the diffracted fields for all three approximations can be written in the common form,

$$\psi(P) = \frac{k}{2\pi i} \int_{\text{apertures}} \frac{e^{ikr}}{r} \frac{e^{ikr'}}{r'} \mathbb{O}(\theta, \theta') \, da' \tag{10.86}$$

---

*Equation (10.85) was also used by Sommerfeld in his early discussions of diffraction. See Sommerfeld, *Optics*, pp. 197 ff.

where the *obliquity factor* $\mathbb{O}(\theta, \theta')$ is the only point of difference. These factors are

$$\mathbb{O}(\theta, \theta') = \begin{cases} \cos \theta & (\psi \text{ approximated on } S_1) \\ \cos \theta' & \left(\dfrac{\partial \psi}{\partial n} \text{ approximated on } S_1\right) \\ \frac{1}{2}(\cos \theta + \cos \theta') & (\text{Kirchhoff approximation}) \end{cases}$$

where the angles are defined in Fig. 10.8. For apertures whose dimensions are large compared to a wavelength, the diffracted intensity is confined to a narrow range of angles and is governed almost entirely by the interferences between the two exponential factors in (10.86). If the source point P' and the observation point P are far from the screen in terms of the *aperture* dimensions, the obliquity factor in (10.86) can be treated as a constant. Then the relative amplitudes of the different diffracted fields will be the same. For normal incidence all obliquity factors are approximately unity where there is appreciable diffracted intensity. In this case even the absolute magnitudes are the same.

The discussion above explains to some extent why the mathematically inconsistent Kirchhoff approximation has any success at all. The use of Dirichlet or Neumann Green functions gives a better logical structure, but provides little practical improvement without further elaboration of the physics. An important deficiency of the discussion so far is its scalar nature. Electromagnetic fields have vector character. This must be incorporated into any realistic treatment, even if approximate. In the next section we proceed with the task of obtaining the vector equivalent of the Kirchhoff or generalized Kirchhoff integral for a plane screen.

10.6   *Vector Equivalents of the Kirchhoff Integral*

The Kirchhoff integral formula (10.79) is an *exact* formal relation expressing the scattered or diffracted scalar field $\psi(\mathbf{x})$ in region II of Fig. 10.7 in terms of an integral of ψ and $\partial \psi / \partial n$ over the finite surface S_1. Corresponding vectorial relations, expressing $\mathbf{E}$ and $\mathbf{B}$ in terms of surface integrals, are useful as a basis for a vectorial Kirchhoff approximation for diffraction (Section 10.7) and scattering (Section 10.10), and also for formal developments such as the proof of the optical theorem (Section 10.11).

To derive a Kirchhoff integral for the electric field, we begin with (10.75) for each rectangular component of $\mathbf{E}$ and write the obvious vectorial equivalent,

$$\mathbf{E}(\mathbf{x}) = \oint_S [\mathbf{E}(\mathbf{n'} \cdot \nabla'G) - G(\mathbf{n'} \cdot \nabla')\mathbf{E}] \, da' \tag{10.87}$$

provided the point $\mathbf{x}$ is inside the volume V bounded by the surface S. Here, as in (10.75), the unit normal $\mathbf{n'}$ is directed *into* the volume V. Eventually we will specify G to be the infinite-space Green function, (10.76), but for the present we leave it as any solution of (10.74). Because we wish to use certain theorems of vector calculus that apply to well-behaved functions, while G is singular at $\mathbf{x'} = \mathbf{x}$, we must exercise some care. We imagine that the surface S consists of an outer surface S' and an infinitesimally smaller inner surface S'' surrounding the point $\mathbf{x}' = \mathbf{x}$. Then, from Green's theorem, the left-hand side of (10.87) vanishes. Of course, evaluation of the integral over the inner surface S'', in the limit as it shrinks to zero around $\mathbf{x}' = \mathbf{x}$, gives $-\mathbf{E}(\mathbf{x})$. Thus (10.87) is restored in practice, but by excluding the point $\mathbf{x}' = \mathbf{x}$ from the volume V the necessary good mathematical behavior is assured. With this understanding concerning the surface S, we rewrite (10.87) in the form

$$0 = \oint_S [2\mathbf{E}(\mathbf{n}' \cdot \nabla'G) - \mathbf{n}' \cdot \nabla'(G\mathbf{E})] \, da'$$

The divergence theorem can be used to convert the second term into a volume integral, thus yielding

$$0 = \oint_S 2\mathbf{E}(\mathbf{n}' \cdot \nabla'G) \, da' + \int_V \nabla'^2(G\mathbf{E}) \, d^3x'$$

With the use of $\nabla^2\mathbf{A} = \nabla(\nabla \cdot \mathbf{A}) - \nabla \times (\nabla \times \mathbf{A})$ for any vector field $\mathbf{A}$, and the vector calculus theorems,

$$\begin{aligned} \int_V \nabla\phi \, d^3x &= \oint_S \mathbf{n}\phi \, da \\ \int_V \nabla \times \mathbf{A} \, d^3x &= \oint_S (\mathbf{n} \times \mathbf{A}) \, da \end{aligned} \qquad (10.88)$$

where ϕ and $\mathbf{A}$ are any well-behaved scalar and vector functions (and $\mathbf{n}$ is the *outward* normal), we can express the volume integral again as a surface integral. We thus obtain

$$0 = \oint_S [2\mathbf{E}(\mathbf{n}' \cdot \nabla'G) - \mathbf{n}'(\nabla' \cdot (G\mathbf{E})) + \mathbf{n}' \times (\nabla' \times (G\mathbf{E}))] \, da'$$

Carrying out the indicated differentiation of the product $G\mathbf{E}$, and making use of the Maxwell equations, $\nabla' \cdot \mathbf{E} = 0$, $\nabla' \times \mathbf{E} = i\omega\mathbf{B}$, we find

$$0 = \oint_S [i\omega(\mathbf{n}' \times \mathbf{B})G + 2\mathbf{E}(\mathbf{n}' \cdot \nabla'G) - \mathbf{n}'(\mathbf{E} \cdot \nabla'G) + \mathbf{n}' \times (\nabla'G \times \mathbf{E})] \, da'$$

Expansion of the triple cross product and a rearrangement of terms yields the final result,

$$\mathbf{E}(\mathbf{x}) = \oint_S [i\omega(\mathbf{n}' \times \mathbf{B})G + (\mathbf{n}' \times \mathbf{E}) \times \nabla'G + (\mathbf{n}' \cdot \mathbf{E})\nabla'G] \, da' \qquad (10.89)$$

where *now the volume V* bounded by the surface S *contains the point* $\mathbf{x}' = \mathbf{x}$. An analogous expression for $\mathbf{B}$ can be obtained from (10.89) by means of the substitutions, $\mathbf{E} \to c\mathbf{B}$ and $c\mathbf{B} \to -\mathbf{E}$.

Equation (10.89) is the vectorial equivalent of the scalar formula (10.75). To obtain the analog of the Kirchhoff integral (10.79), we consider the geometry of Fig. 10.7 and let the surface S be made up of a finite surface S_1 surrounding the sources or scatterer and a surface S_2 "at infinity." There is no loss of generality in taking S_2 to be a spherical shell of radius $r_0 \to \infty$. The integral in (10.89) can be written as the sum of two integrals, one over S_1 and one over S_2. On the surface S_2 the Green function (10.76) is given, for large enough r_0, by

$$G \to \frac{e^{ikr'}}{4\pi r'} e^{ik\mathbf{n}' \cdot \mathbf{x}}$$

and its gradient by

$$\nabla'G \to -ik\mathbf{n}'G$$

Then the contribution from S_2 to (10.89) is

$$\oint_{S_2} = ik \oint_{S_2} [c(\mathbf{n}' \times \mathbf{B}) - (\mathbf{n}' \times \mathbf{E}) \times \mathbf{n}' - \mathbf{n}'(\mathbf{n}' \cdot \mathbf{E})]G \, da'$$

or

$$\oint_{S_2} = ik \oint_{S_2} [c(\mathbf{n}' \times \mathbf{B}) - \mathbf{E}]G \, da'$$

The fields in region II are diffracted or scattered fields and so satisfy the condition of outgoing waves in the neighborhood of S_2. In particular, the fields $\mathbf{E}$ and $\mathbf{B}$ are mutually perpendicular and transverse to the radius vector. Thus, on S_2, $\mathbf{E} = c\mathbf{n}' \times \mathbf{B} + O(1/r_0^2)$. This shows that

$$\oint_{S_2} \to O\left(\frac{1}{r_0}\right)$$

and the contribution from the integral over S_2 vanishes as $r_0 \to \infty$. For the geometry of Fig. 10.7, then, with S_2 at infinity, the electric field in region II satisfies the *vector Kirchhoff integral relation*,

$$\mathbf{E}(\mathbf{x}) = \oint_{S_1} [ik(\mathbf{n}' \times \mathbf{B})G + (\mathbf{n}' \times \mathbf{E}) \times \nabla'G + (\mathbf{n}' \cdot \mathbf{E})\nabla'G] \, da' \quad (10.90)$$

where G is given by (10.76) and the integral is only over the finite surface S_1.

It is useful to specialize (10.90) to a scattering situation and to exhibit a formal expression for the scattering amplitude as an integral of the scattered fields over S_1. The geometry is shown in Fig. 10.9. On both sides of (10.90) the

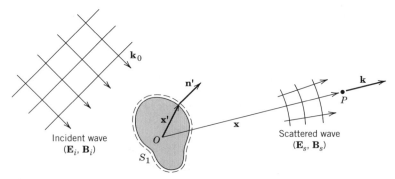

Figure 10.9 Scattering geometry. An incident plane wave with wave vector $\mathbf{k}_0$ and fields $(\mathbf{E}_i, \mathbf{B}_i)$ is scattered by an obstacle (the scatterer), giving rise to scattered fields $(\mathbf{E}_s, \mathbf{B}_s)$ that propagate as spherically diverging waves at large distances. The surface S_1 completely encloses the scatterer.

fields are taken to be the scattered fields $(\mathbf{E}_s, \mathbf{B}_s)$, that is, the total fields $(\mathbf{E}, \mathbf{B})$ minus the incident wave $(\mathbf{E}_i, \mathbf{B}_i)$. If the observation point P is far from the scatterer, then the Green function and the scattered electric field take on their asymptotic forms,

$$G(\mathbf{x}, \mathbf{x}') \rightarrow \frac{1}{4\pi} \frac{e^{ikr}}{r} e^{-i\mathbf{k}\cdot\mathbf{x}'}$$

$$\mathbf{E}_s(\mathbf{x}) \rightarrow \frac{e^{ikr}}{r} \mathbf{F}(\mathbf{k}, \mathbf{k}_0)$$

where $\mathbf{k}$ is the wave vector in the direction of observation, $\mathbf{k}_0$ is the incident wave vector, and $\mathbf{F}(\mathbf{k}, \mathbf{k}_0)$ is the (unnormalized) vectorial scattering amplitude. In this limit, $\nabla'G = -i\mathbf{k}G$. Thus (10.90) can be written as an *integral expression for the scattering amplitude* $\mathbf{F}(\mathbf{k}, \mathbf{k}_0)$:

$$\mathbf{F}(\mathbf{k}, \mathbf{k}_0) = \frac{i}{4\pi} \oint_{S_1} e^{-i\mathbf{k}\cdot\mathbf{x}'}[\omega(\mathbf{n}' \times \mathbf{B}_s) + \mathbf{k} \times (\mathbf{n}' \times \mathbf{E}_s) - \mathbf{k}(\mathbf{n}' \cdot \mathbf{E}_s)] \, da' \quad (10.91)$$

Note carefully how $\mathbf{F}(\mathbf{k}, \mathbf{k}_0)$ depends explicitly on the outgoing direction of $\mathbf{k}$. The dependence on the incident direction specified by $\mathbf{k}_0$ is implicit in the scattered fields $\mathbf{E}_s$ and $\mathbf{B}_s$. Since we know that $\mathbf{k} \cdot \mathbf{F} = 0$, it must be true that in (10.91) the component parallel to $\mathbf{k}$ of the first integral cancels the third integral. It is therefore convenient to resolve the integrand in (10.91) into components parallel and perpendicular to $\mathbf{k}$, and to exhibit the transversality of $\mathbf{F}$ explicitly:

$$\mathbf{F}(\mathbf{k}, \mathbf{k}_0) = \frac{1}{4\pi i} \mathbf{k} \times \oint_{S_1} e^{-i\mathbf{k}\cdot\mathbf{x}'}\left[\frac{c\mathbf{k} \times (\mathbf{n}' \times \mathbf{B}_s)}{k} - \mathbf{n}' \times \mathbf{E}_s \right] da' \quad (10.92)$$

Alternatively, we can ask for the amplitude of scattered radiation with wave vector $\mathbf{k}$ and polarization $\boldsymbol{\epsilon}$. This is given by

$$\boldsymbol{\epsilon}^* \cdot \mathbf{F}(\mathbf{k}, \mathbf{k}_0) = \frac{i}{4\pi} \oint_{S_1} e^{-i\mathbf{k}\cdot\mathbf{x}'}[\omega\boldsymbol{\epsilon}^* \cdot (\mathbf{n}' \times \mathbf{B}_s) + \boldsymbol{\epsilon}^* \cdot (\mathbf{k} \times (\mathbf{n}' \times \mathbf{E}_s))] \, da'$$

$$(10.93)$$

The terms in square brackets can be interpreted as effective electric and magnetic surface currents on S_1 acting as sources for the scattered fields. The various equivalent forms (10.91)–(10.93) are valuable as starting points for the discussion of the scattering of short-wavelength radiation (Section 10.10) and in the derivation of the optical theorem (Section 10.11).

10.7 *Vectorial Diffraction Theory*

The vectorial Kirchhoff integral (10.90) can be used as the basis of an approximate theory of diffraction in exactly the same manner as described below (10.79) for the scalar theory. Unfortunately, the inconsistencies of the scalar Kirchhoff approximation persist.

For the special case of a thin, perfectly conducting, plane screen with apertures, however, it is possible to obtain vectorial relations, akin to the generalized Kirchhoff integral (10.81) or (10.85), in which the boundary conditions *are* satisfied; these relations, moreover, are amenable to consistent approximations. The plane screen is taken at $z = 0$, with the sources supposed to be in the region $z < 0$, and the diffracted fields to be observed in the region $z > 0$. It is convenient to divide the fields into two parts,

$$\mathbf{E} = \mathbf{E}^{(0)} + \mathbf{E}', \qquad \mathbf{B} = \mathbf{B}^{(0)} + \mathbf{B}' \tag{10.94}$$

where $\mathbf{E}^{(0)}$, $\mathbf{B}^{(0)}$ are the fields produced by the sources in the absence of any screen or obstacle (defined for both $z < 0$ and $z > 0$), and $\mathbf{E}'$, $\mathbf{B}'$ are the fields caused by the presence of the plane screen. For $z > 0$, $\mathbf{E}'$, $\mathbf{B}'$ are the diffracted fields, while for $z < 0$, they are the reflected fields. We will call $\mathbf{E}'$, $\mathbf{B}'$ the scattered fields when considering both $z < 0$ and $z > 0$. The scattered fields can be considered as having their origin in the surface-current density and surface-charge density that are necessarily produced on the screen to satisfy the boundary conditions. Certain reflection properties in z of the scattered fields follow from the fact that the surface-current and -charge densities are confined to the $z = 0$ plane. A vector potential $\mathbf{A}'$ and a scalar potential Φ' can be used to construct $\mathbf{E}'$ and $\mathbf{B}'$. Since the surface current flow has no z component, $A'_z = 0$. Furthermore, A'_x, A'_y, and Φ' are evidently even functions of z. The relation of the fields to the potentials shows that the scattered fields have the reflection symmetries,

$$\begin{aligned} E'_x, E'_y, B'_z & \qquad \text{are even in } z \\ E'_z, B'_x, B'_y & \qquad \text{are odd in } z \end{aligned} \tag{10.95}$$

The fields that are odd in z are not necessarily zero over the whole plane $z = 0$. Where the conducting surface exists, $E'_z \neq 0$ implies an associated surface-charge density, equal on the two sides of the surface. Similarly, nonvanishing tangential components of $\mathbf{B}$ imply a surface-current density, equal in magnitude and direction on both sides of the screen. Only in the aperture does continuity require that E'_z, B'_x, B'_y vanish. This leads to the statement that in the apertures of a perfectly conducting plane screen the normal component of $\mathbf{E}$ and the tangential components of $\mathbf{B}$ are the same as in the absence of the screen.

The generalized Kirchhoff integral (10.83) for Neumann boundary conditions can be applied to the components of the vector potential $\mathbf{A}'$. The normal derivatives on the right can be expressed in terms of components of $\mathbf{B}'$. The result, written vectorially is

$$\mathbf{A}'(\mathbf{x}) = \frac{1}{2\pi} \int_{\text{screen}} (\mathbf{n} \times \mathbf{B}') \frac{e^{ikR}}{R} \, da' \tag{10.96}$$

In view of the preceding remarks about the surface current and the tangential components of $\mathbf{B}'$, (10.96) could perhaps have been written down directly. The scattered magnetic field can be obtained by taking the curl of (10.96):

$$\mathbf{B}'(\mathbf{x}) = \frac{1}{2\pi} \nabla \times \int_{\text{screen}} (\mathbf{n} \times \mathbf{B}') \frac{e^{ikR}}{R} \, da' \tag{10.97}$$

In (10.96) and (10.97) the integrand can be evaluated on either side of the screen with $\mathbf{n}$ being normal to the surface. For definiteness, we specify that $\mathbf{n}$ is a unit normal in the positive z direction and the integrand is to be evaluated at $z = 0^+$. The integration extends over the metallic part of the screen; $\mathbf{B}'_{\text{tan}} = 0$ in the apertures. The electric field $\mathbf{E}'$ can be calculated from $\mathbf{E}' = (i/\omega\mu\epsilon)\nabla \times \mathbf{B}'$.

Equation (10.97) can be used for approximations in a consistent way. It is most useful when the diffracting obstacles consist of one or more finite flat segments at $z = 0$, for example, a circular disc. Then the surface current on the obstacles can be approximated in some way—for instance, by using the incident field $\mathbf{B}^{(0)}$ in the integrand. We then have a vectorial version of the generalized Kirchhoff's approximation of the preceding section.

It is useful to construct an expression equivalent to (10.97) for the electric field. From the symmetry of the source-free Maxwell equations with respect to $\mathbf{E}$ and $\mathbf{B}$ it is evident that the electric field $\mathbf{E}'$ can be expressed by analogy with (10.97), as

$$\mathbf{E}'(\mathbf{x}) = \pm\frac{1}{2\pi} \nabla \times \int_{S_1} (\mathbf{n} \times \mathbf{E}') \frac{e^{ikR}}{R} \, da' \tag{10.98}$$

where it is assumed that $\mathbf{E}'$ is known on the whole surface S_1 at $z = 0^+$. The upper (lower) sign applies for $z > 0$ ($z < 0$). It can be verified that (10.98) satisfies the Maxwell equations and yields consistent boundary values at $z = 0$. The reason for the difference in sign for $z \gtrless 0$, as compared to (10.97) for $\mathbf{B}'$, is the opposite reflection properties of $\mathbf{E}'$ compared to $\mathbf{B}'$ [see (10.95)].

There is a practical difficulty with (10.98) as it stands. The integration in (10.98) is over the *whole plane* at $z = 0$. We cannot exploit the vanishing of the tangential components of the electric field on the metallic portions of the screen because it is the *total* electric field whose tangential components vanish, not those of $\mathbf{E}'$. The difficulty can be removed by use of linear superposition. We add $\mathbf{E}^{(0)}$ to the integrand in (10.98) to obtain the full electric field, and subtract the corresponding integral. We thus have, for the diffracted electric field,

$$\mathbf{E}'(\mathbf{x}) = \pm\frac{1}{2\pi} \nabla \times \int_{S_1} (\mathbf{n} \times \mathbf{E}) \frac{e^{ikR}}{R} \, da' - \mathbf{E}^{(1)}(\mathbf{x}) \tag{10.99}$$

where

$$\mathbf{E}^{(1)}(\mathbf{x}) = \pm\frac{1}{2\pi} \nabla \times \int_{S_1} (\mathbf{n} \times \mathbf{E}^{(0)}) \frac{e^{ikR}}{R} \, da' \tag{10.100}$$

The integrand in (10.99) now has support only in the apertures of the screen, as desired. But what is the extra electric field $\mathbf{E}^{(1)}$? Just as (10.98) gives the extra (diffracted) field for $z > 0$ in terms of a surface integral of itself over the whole screen, so (10.100) is equal to the "source" field $\mathbf{E}^{(0)}$ in the region $z > 0$. But because $\mathbf{E}^{(1)}$ is defined by an integral over the surface at $z = 0$, it respects the symmetries of (10.95). A moment's thought will show that this behavior means that for $z < 0$ the sum $\mathbf{E}^{(0)} + \mathbf{E}^{(1)}$ describes the fields of the sources in the presence of a perfectly conducting plane (with no apertures) at $z = 0$: $\mathbf{E}^{(1)}$ (and its partner $\mathbf{B}^{(1)}$) are the *reflected* fields!

If in (10.99) we transfer $\mathbf{E}^{(1)}$ to the left-hand side, we find for $z > 0$ the total electric field, now called the diffracted field, given by

$$\mathbf{E}_{\text{diff}}(\mathbf{x}) = \frac{1}{2\pi} \nabla \times \int_{\text{apertures}} (\mathbf{n} \times \mathbf{E}) \frac{e^{ikR}}{R} \, da' \tag{10.101}$$

where the integration is only over the apertures in the screen and $\mathbf{E}$ is total tangential electric field in the apertures. In the illuminated region ($z < 0$) the total electric field is

$$\mathbf{E}(\mathbf{x}) = \mathbf{E}^{(0)}(\mathbf{x}) + \mathbf{E}^{(1)}(\mathbf{x}) - \mathbf{E}_{\text{diff}}(\mathbf{x}) \tag{10.102}$$

where for both regions $\mathbf{E}_{\text{diff}}(\mathbf{x})$ is given by (10.101). This solution for the diffracted electric field in terms of the tangential electric field in the apertures of a perfectly conducting plane screen was first obtained by Smythe.* It can serve as the basis of a consistent scheme of approximation, with the approximate solutions for $\mathbf{E}_{\text{diff}}$ satisfying the required boundary conditions at $z = 0$ and at infinity. Some examples are discussed in a later section and in the problems.

10.8 Babinet's Principle of Complementary Screens

Before discussing examples of diffraction we wish to establish a useful relation called *Babinet's principle*. Babinet's principle relates the diffraction fields of one diffracting screen to those of the complementary screen. We first discuss the principle in the scalar Kirchhoff approximation. The diffracting screen is assumed to lie in some surface S, which divides space into regions I and II in the sense of Section 10.5. The screen occupies all of the surface S except for certain apertures. The complementary screen is that diffracting screen which is obtained by replacing the apertures by screen and the screen by apertures. If the surface of the original screen is S_a and that of the complementary screen is S_b, then $S_a + S_b = S$, as shown schematically in Fig. 10.10.

If there are sources inside S (in region I) that give rise to a field $\psi(\mathbf{x})$, then in the absence of either screen the field $\psi(\mathbf{x})$ in region II is given by the Kirchhoff integral (10.79) where the surface integral is over the entire surface S. With the screen S_a in position, the field $\psi_a(\mathbf{x})$ in region II is given in the Kirchhoff approximation by (10.79) with the source field ψ in the integrand and the surface integral only over S_b (the apertures). Similarly, for the complementary screen S_b, the field $\psi_b(\mathbf{x})$ is given in the same approximation by a surface integral over S_a. Evidently, then, we have the following relation between the diffraction fields ψ_a and ψ_b:

$$\psi_a + \psi_b = \psi \tag{10.103}$$

This is Babinet's principle as usually formulated in optics. If ψ represents an incident plane wave, for example, Babinet's principle says that the diffraction

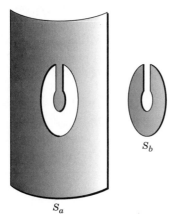

S_b

S_a

Figure 10.10   A diffraction screen S_a and its complementary diffraction screen S_b.

*W. R. Smythe, *Phys. Rev.* **72**, 1066 (1947). See also *Smythe*, Section 12.18.

patterns away from the incident direction are the same for the original screen and its complement.

The result (10.103) also follows from the generalized Kirchhoff integrals (10.81) or (10.83) if the amplitude or its normal derivative is taken equal to that of the incident wave in the apertures and zero elsewhere, in the spirit of the Kirchhoff approximation. All these formulations of Babinet's principle are unsatisfactory in two respects: They are statements about scalar fields, and they are based on a Kirchhoff approximation.

A rigorous statement of Babinet's principle for electromagnetic fields can be made for a thin, perfectly conducting plane screen and its complement. The result follows from the two alternative formulations of this diffraction problem given in the preceding section. The original diffraction problem and its complementary problem are defined by the source fields and screens as follows:

ORIGINAL

$$\mathbf{E}^{(0)}, \qquad \mathbf{B}^{(0)}; \qquad S_a \qquad\qquad (10.104)$$

COMPLEMENT

$$\mathbf{E}_c^{(0)} = c\mathbf{B}^{(0)}, \qquad \mathbf{B}_c^{(0)} = -\mathbf{E}^{(0)}/c; \qquad S_b$$

The complementary situation has a screen that is the complement of the original and has source fields with opposite polarization characteristics. For the original screen S_a the electric field for $z > 0$ is, according to (10.101),

$$\mathbf{E}(\mathbf{x}) = \frac{1}{2\pi} \nabla \times \int_{S_b} (\mathbf{n} \times \mathbf{E}) \frac{e^{ikR}}{R} \, da' \qquad\qquad (10.105)$$

For the complementary screen S_b we choose to use (10.97) instead of (10.101) to express the complementary scattered magnetic field $\mathbf{B}_c'$ for $z > 0$ as

$$\mathbf{B}_c'(\mathbf{x}) = \frac{1}{2\pi} \nabla \times \int_{S_b} (\mathbf{n} \times \mathbf{B}_c') \frac{e^{ikR}}{R} \, da' \qquad\qquad (10.106)$$

In both (10.105) and (10.106) the integration is over the screen S_b because of the boundary conditions on $\mathbf{E}$ and $\mathbf{B}_c'$ in the two cases. Mathematically, (10.105) and (10.106) are of the same form. From the linearity of the Maxwell equations and the relation between the original and complementary source fields, it follows that in the region $z > 0$ the total electric field for the screen S_a is numerically equal to c times the scattered magnetic field for the complementary screen S_b:

$$\mathbf{E}(\mathbf{x}) = c\mathbf{B}_c'(\mathbf{x})$$

The other fields are related by

$$\mathbf{B}(\mathbf{x}) = -\mathbf{E}_c'(\mathbf{x})/c$$

where the minus sign is a consequence of the requirement of outgoing radiation flux at infinity, just as for the source fields. If use is made of (10.94) for the complementary problem to obtain relations between the total fields in the region $z > 0$, *Babinet's principle* for a plane, perfectly conducting thin screen and its complement states that the original fields $(\mathbf{E}, \mathbf{B})$ and the complementary fields $(\mathbf{E}_c, \mathbf{B}_c)$ are related according to

$$\mathbf{E} - c\mathbf{B}_c = \mathbf{E}^{(0)}$$
$$\mathbf{B} + \mathbf{E}_c/c = \mathbf{B}^{(0)} \qquad\qquad (10.107)$$

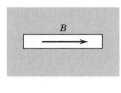

Figure 10.11 Equivalent radiators according to Babinet's principle.

for $z > 0$, provided the complementary diffraction problems are defined by (10.104). These relations are the vectorial analogs of (10.103); they are exact, not approximate, statements for the idealized problem of a perfectly conducting plane screen. For practical situations (finite, but large, conductivity; curved screens whose radii of curvature are large compared to aperture dimensions, etc.), the vectorial Babinet's principle can be expected to hold approximately. It says that the diffracted intensity in directions other than that of the incident field is the same for a screen and its complement. The polarization characteristics are rotated, but this conforms with the altered polarization of the complementary source fields (10.104).

The rigorous vector formulation of Babinet's principle is very useful in microwave problems. For example, consider a narrow slot cut in an infinite, plane, conducting sheet and illuminated with fields that have the magnetic induction along the slot and the electric field perpendicular to it, as shown in Fig. 10.11. The radiation pattern from the slot will be the same as that of a thin linear antenna with its driving electric field along the antenna, as considered in Sections 9.2 and 9.4. The polarization of the radiation will be opposite for the two systems. Elaboration of these ideas makes it possible to design antenna arrays by cutting suitable slots in the sides of waveguides.*

10.9 Diffraction by a Circular Aperture; Remarks on Small Apertures

The subject of diffraction has been extensively studied since Kirchhoff's original work, both in optics, where the scalar theory based on (10.79) generally suffices, and in microwave generation and transmission, where more accurate solutions are needed. Specialized treatises are devoted entirely to the subject of diffraction and scattering. We will content ourselves with a few examples to illustrate the use of the scalar and vector theorems (10.79), (10.85) and (10.101) and to compare the accuracy of the approximation schemes.

Historically, diffraction patterns were classed as Fresnel or Fraunhofer diffraction, depending on the relative geometry involved. There are three length scales to consider, the size d of the diffracting system, the distance r from the system to the observation point, and the wavelength λ. A diffraction pattern only becomes manifest for $r \gg d$. Then in expressions like (10.86) or (10.101) slowly varying factors in the integrands can be treated as constants. Only the phase

*See, for example, *Silver*, Chapter 9.

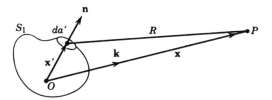

Figure 10.12

factor kR in e^{ikR} needs to be handled with some care. With $r \gg d$, it can be expanded as

$$kR = kr - k\,\mathbf{n} \cdot \mathbf{x}' + \frac{k}{2r}\,[r'^2 - (\mathbf{n} \cdot \mathbf{x}')^2] + \cdots$$

where $\mathbf{n} = \mathbf{x}/r$ is a unit vector in the direction of observation. The successive terms are of order (kr), (kd), $(kd)(d/r),\ldots$. The term *Fraunhofer diffraction* applies if the third and higher terms are negligible compared to unity. For small diffracting systems this always holds, since $kd \ll 1$, and we have supposed $d/r \ll 1$. But for systems that are large compared to a wavelength, (kd^2/r) may be of order unity or larger even though $d/r \ll 1$. Then the term *Fresnel diffraction* applies. In most practical applications the simpler Fraunhofer limit is appropriate. Far enough from any diffracting system it always holds. We consider only the Fraunhofer limit here (except for Problem 10.11).

If the observation point is far from the diffracting system, expansion (9.7) can be used for $R = |\mathbf{x} - \mathbf{x}'|$. Keeping only lowest order terms in $(1/kr)$, the scalar Kirchhoff expression (10.79) becomes

$$\psi(\mathbf{x}) = -\frac{e^{ikr}}{4\pi r} \int_{S_1} e^{-i\mathbf{k}\cdot\mathbf{x}'}[\mathbf{n} \cdot \boldsymbol{\nabla}'\psi(\mathbf{x}') + i\mathbf{k} \cdot \mathbf{n}\psi(\mathbf{x}')]\, da' \qquad (10.108)$$

where $\mathbf{x}'$ is the coordinate of the element of surface area da', r is the length of the vector $\mathbf{x}$ from the origin O to the observation point P, and $\mathbf{k} = k(\mathbf{x}/r)$ is the wave vector in the direction of observation, as indicated in Fig. 10.12. For a plane surface we can use the vector expression (10.101), which reduces in this limit to

$$\mathbf{E}(\mathbf{x}) = \frac{ie^{ikr}}{2\pi r}\,\mathbf{k} \times \int_{S_1} \mathbf{n} \times \mathbf{E}(\mathbf{x}')e^{-i\mathbf{k}\cdot\mathbf{x}'}\, da' \qquad (10.109)$$

As an example of diffraction we consider a plane wave incident at an angle α on a thin, perfectly conducting screen with a circular hole of radius a in it. The polarization vector of the incident wave lies in the plane of incidence. Figure 10.13 shows an appropriate system of coordinates. The screen lies in the x-y plane with the opening centered at the origin. The wave is incident from below, so that the domain $z > 0$ is the region of diffraction fields. The plane of incidence is taken to be the x-z plane. The incident wave's electric field, written out explicitly in rectangular components, is

$$\mathbf{E}_i = E_0(\boldsymbol{\epsilon}_1 \cos \alpha - \boldsymbol{\epsilon}_3 \sin \alpha)e^{ik(z\cos\alpha + x\sin\alpha)} \qquad (10.110)$$

In calculating the diffraction field with (10.108) or (10.109) we will make the customary approximation that the exact field in the surface integral may be replaced by the incident field. For the vector relation (10.109) we need

$$(\mathbf{n} \times \mathbf{E}_i)_{z=0} = E_0\boldsymbol{\epsilon}_2 \cos \alpha\, e^{ik\sin\alpha x'} \qquad (10.111)$$

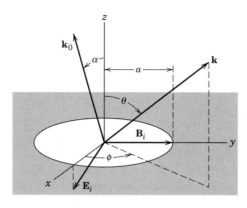

Figure 10.13 Diffraction by a circular hole of radius a.

Then, introducing plane polar coordinates for the integration over the opening, we have

$$\mathbf{E}(\mathbf{x}) = \frac{ie^{ikr}E_0 \cos \alpha}{2\pi r} (\mathbf{k} \times \boldsymbol{\epsilon}_2) \int_0^a \rho \, d\rho \int_0^{2\pi} d\beta e^{ik\rho[\sin \alpha \cos \beta - \sin \theta \cos(\phi - \beta)]} \quad (10.112)$$

where θ, ϕ are the spherical angles of $\mathbf{k}$. If we define the angular function,

$$\xi = (\sin^2\theta + \sin^2\alpha - 2 \sin \theta \sin \alpha \cos \phi)^{1/2}$$

the angular integral can be transformed into

$$\frac{1}{2\pi} \int_0^{2\pi} d\beta' e^{-ik\rho\xi\cos\beta'} = J_0(k\rho\xi)$$

Then the radial integral in (10.112) can be done directly. The resulting electric field in the vector Smythe–Kirchhoff approximation is

$$\mathbf{E}(\mathbf{x}) = \frac{ie^{ikr}}{r} a^2 E_0 \cos \alpha(\mathbf{k} \times \boldsymbol{\epsilon}_2) \frac{J_1(ka\xi)}{ka\xi} \quad (10.113)$$

The time-averaged diffracted power per unit solid angle is

$$\frac{dP}{d\Omega} = P_i \cos \alpha \frac{(ka)^2}{4\pi} (\cos^2\theta + \cos^2\phi \sin^2\theta) \left| \frac{2J_1(ka\xi)}{ka\xi} \right|^2 \quad (10.114)$$

where

$$P_i = \left(\frac{E_0^2}{2Z_0}\right)\pi a^2 \cos \alpha \quad (10.115)$$

is the total power *normally incident* on the aperture. If the opening is large compared to a wavelength ($ka \gg 1$), the factor $[2J_1(ka\xi)/ka\xi]^2$ peaks sharply to a value of unity at $\xi = 0$ and falls rapidly to zero (with small secondary maxima) within a region $\Delta\xi \sim (1/ka)$ away from $\xi = 0$. This means that the main part of the wave passes through the opening in the manner of geometrical optics; only slight diffraction effects occur.* For $ka \sim 1$ the Bessel function varies compara-

---

*To see this explicitly we expand ξ around the geometrical optics direction $\theta = \alpha, \phi = 0$:

$$\xi \simeq \sqrt{(\theta - \alpha)^2 \cos^2\alpha + \phi^2 \sin^2\alpha}$$

For $ka \gg 1$ it is evident that $ka\xi \gg 1$ as soon as θ departs appreciably from α, or ϕ from zero, or both.

tively slowly in angle; the transmitted wave is distributed in directions very different from the incident direction. For $ka \ll 1$, the angular distribution is entirely determined by the factor $(\mathbf{k} \times \boldsymbol{\epsilon}_2)$ in (10.113). But in this limit the assumption of an unperturbed field in the aperture breaks down badly.

The total transmitted power can be obtained by integrating (10.114) over all angles in the forward hemisphere. The ratio of transmitted power to incident power is called the *transmission coefficient T*:

$$T = \frac{\cos \alpha}{\pi} \int_0^{2\pi} d\phi \int_0^{\pi/2} (\cos^2 \theta + \cos^2 \phi \sin^2 \theta) \left| \frac{J_1(ka\xi)}{\xi} \right|^2 \sin \theta \, d\theta \quad (10.116)$$

In the two extreme limits $ka \gg 1$ and $ka \ll 1$, the transmission coefficient approaches the values,

$$T \to \begin{cases} 1, & ka \gg 1 \\ \frac{1}{3}(ka)^2 \cos \alpha, & ka \ll 1 \end{cases}$$

The long-wavelength limit $(ka \ll 1)$ is suspect because of our approximations, but it shows that the transmission is small for very small holes. For normal incidence $(\alpha = 0)$ the transmission coefficient (10.116) can be written

$$T = \int_0^{\pi/2} J_1^2(ka \sin \theta) \left(\frac{2}{\sin \theta} - \sin \theta \right) d\theta$$

With the help of the integral relations,

$$\int_0^{\pi/2} J_n^2(z \sin \theta) \frac{d\theta}{\sin \theta} = \int_0^{2z} \frac{J_{2n}(t)}{t} \, dt$$

$$\int_0^{\pi/2} J_n^2(z \sin \theta) \sin \theta \, d\theta = \frac{1}{2z} \int_0^{2z} J_{2n}(t) \, dt \quad (10.117)$$

and the recurrence formulas (3.87) and (3.88), we can put the transmission coefficient in the alternative forms

$$T = \begin{cases} 1 - \dfrac{1}{ka} \displaystyle\sum_{m=0}^{\infty} J_{2m+1}(2ka) \\ 1 - \dfrac{1}{2ka} \displaystyle\int_0^{2ka} J_0(t) \, dt \end{cases}$$

The transmission coefficient increases more or less monotonically as ka increases, with small oscillations superposed. For $ka \gg 1$, the second form can be used to obtain an asymptotic expression

$$T \simeq 1 - \frac{1}{2ka} - \frac{1}{2\sqrt{\pi}(ka)^{3/2}} \sin\left(2ka - \frac{\pi}{4} \right) + \cdots \quad (10.118)$$

which exhibits the small oscillations explicitly. These approximate expressions for T give the general behavior as a function of ka, but they are not very accurate. Exact calculations, as well as more accurate approximate ones, have been made for the circular opening. These are compared with each other in the book by *King and Wu* (Fig. 41, p. 126). The correct asymptotic expression does not contain the $1/2ka$ term in (10.118), and the coefficient of the term in $(ka)^{-3/2}$ is twice as large.

We now wish to compare our results of the *vector* Smythe–Kirchhoff approximation with the usual scalar theory based on (10.79). For a wave not normally incident, the question immediately arises as to what to choose for the scalar function $\psi(\mathbf{x})$. Perhaps the most consistent assumption is to take the magnitude of the electric or magnetic field. Then the diffracted intensity is treated consistently as proportional to the absolute square of (10.79). If a component of $\mathbf{E}$ or $\mathbf{B}$ is chosen for ψ, we must then decide whether to keep or throw away radial components of the diffracted field in calculating the diffracted power. Choosing the magnitude of $\mathbf{E}$ for ψ, we have, by straightforward calculation with (10.108),

$$\psi(\mathbf{x}) = -ik \, \frac{e^{ikr}}{r} \, a^2 E_0 \left(\frac{\cos\alpha + \cos\theta}{2} \right) \frac{J_1(ka\xi)}{ka\xi}$$

as the scalar equivalent of (10.113). The power radiated per unit solid angle in the *scalar* Kirchhoff approximation is

$$\frac{dP}{d\Omega} \simeq P_i \frac{(ka)^2}{4\pi} \cos\alpha \left(\frac{\cos\alpha + \cos\theta}{2\cos\alpha} \right)^2 \left| \frac{2J_1(ka\xi)}{ka\xi} \right|^2 \qquad (10.119)$$

where P_i is given by (10.115). If the alternative scalar formula (10.85) is used, the obliquity factor $(\cos\alpha + \cos\theta)/2$ in (10.119) is replaced by $\cos\theta$.

If we compare the vector Smythe–Kirchhoff result (10.114) with (10.119), we see similarities and differences. Both formulas contain the same "diffraction" distribution factor $[J_1(ka\xi)/ka\xi]^2$ and the same dependence on wave number. But the scalar result has no azimuthal dependence (apart from that contained in ξ), whereas the vector expression does. The azimuthal variation comes from the polarization properties of the field, and must be absent in a scalar approximation. For normal incidence ($\alpha = 0$) and $ka \gg 1$ the polarization dependence is unimportant. The diffraction is confined to very small angles in the forward direction. Then all scalar and vector approximations reduce to the common expression,

$$\frac{dP}{d\Omega} \simeq P_i \frac{(ka)^2}{\pi} \left| \frac{J_1(ka\sin\theta)}{ka\sin\theta} \right|^2 \qquad (10.120)$$

The vector and scalar approximations are compared in Fig. 10.14 for the angle of incidence equal to 45° and for an aperture one wavelength in diameter ($ka = \pi$). The angular distribution is shown in the plane of incidence (containing the electric field vector of the incident wave) and a plane perpendicular to it. The solid (dashed) curve gives the vector (scalar) approximation in each case. We see that for $ka = \pi$ there is a considerable disagreement between the two approximations. There is reason to believe that the Smythe–Kirchhoff result is close to the correct one, even though the approximation breaks down seriously for $ka \lesssim 1$. The vector approximation and exact calculations for a rectangular opening yield results in surprisingly good agreement, even down to $ka \sim 1$.[*]

---

[*]See J. A. Stratton and L. J. Chu, *Phys. Rev.*, **56**, 99 (1939), for a series of figures comparing the vector Smythe–Kirchhoff approximation with exact calculations by P. M. Morse and P. J. Rubenstein, *Phys. Rev.* **54**, 895 (1938). The alert reader may be puzzled by the apparent discrepancy in the dates of Smythe's publication (*loc. cit.*) and of Stratton and Chu. The two calculations yield the same result, though quite different in appearance and detail of derivation, the earlier one involving a line integral around the boundary of the aperture as well as a surface integral over it.

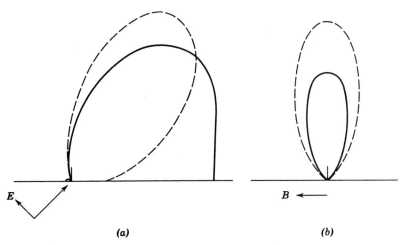

(a)   (b)

Figure 10.14   Fraunhofer diffraction pattern for a circular opening one wavelength in diameter in a thin, plane, conducting sheet. The plane wave is incident on the screen at 45°. The solid curves are the vector Smythe–Kirchhoff approximation, while the dashed curves are the scalar approximation. (*a*) The intensity distribution in the plane of incidence (*E* plane). (*b*) The intensity distribution (enlarged 2.5 times) perpendicular to the plane of incidence (*H* plane).

The diffraction by apertures or obstacles whose dimensions are small compared to a wavelength requires methods different from the Kirchhoff or Kirchhoff-like approximation. The exact formula (10.101) for a plane screen can be used as a starting point. If the radiation fields of (10.101) are expanded in multipoles, as in Sections 9.2–9.3, effective multipole moments (9.72) and (9.74) can be identified in terms of integrals of the exact electric field in the aperture. The derivation of these effective moments is left as Problem 10.10. Once the dipole moments of an aperture are known, the diffraction can be calculated merely by using the dipole fields of Sections 9.2 and 9.3. The example of a circular aperture with effective moments (9.75) is left to the problems. The whole discussion of the physical picture parallels that of Section 9.5.C and is not repeated here.

10.10   Scattering in the Short-Wavelength Limit

Scattering in the long-wavelength limit was discussed in Sections 10.1 and 10.2. The opposite limit, similar to the Kirchhoff domain of diffraction, is a scattering by obstacles large compared to a wavelength. Just as for diffraction by a screen, the zeroth approximation is given by classical ray theory. The wave aspects of the fields give corrections to this, with the scattering confined to angular regions only slightly away from the paths of geometrical optics. For a thin, flat obstacle, the techniques of Section 10.7, perhaps with Babinet's principle, can be used. But for other obstacles we base the calculation on the integral expression (10.93) for the scattering amplitude in terms of the scattered fields $\mathbf{E}_s$, $\mathbf{B}_s$ on a surface S_1 just outside the scatterer.

In the absence of knowledge about the correct fields $\mathbf{E}_s$ and $\mathbf{B}_s$ on the surface, we must make some approximations. If the wavelength is short compared to the dimensions of the obstacle, the surface can be divided approximately into an illuminated region and a shadow region.* The boundary between these regions is sharp only in the limit of geometrical optics. The transition region can be shown to have a width of the order of $(2/kR)^{1/3}R$, where R is a typical radius of curvature of the surface. Since R is of the order of magnitude of the dimensions of the obstacle, the short-wavelength limit will approximately satisfy the geometrical condition. In the shadow region the scattered fields on the surface must be very nearly equal and opposite to the incident fields, regardless of the nature of the scatterer, provided it is "opaque." In the illuminated region, on the other hand, the scattered fields at the surface will depend on the properties of the obstacle. If the wavelength is short compared to the minimum radius of curvature, the Fresnel equations of Section 7.3 can be utilized, treating the surface as locally flat. Eventually we will specialize to a perfectly conducting obstacle, for which the tangential $\mathbf{E}_s$ and the normal $\mathbf{B}_s$ must be equal and opposite to the corresponding incident fields, while the tangential $\mathbf{B}_s$ and normal $\mathbf{E}_s$ will be approximately equal to the incident values [see (10.95)].

Because of the generality of the contribution from the shadow region, it is desirable to consider it separately. We write

$$\boldsymbol{\epsilon}^* \cdot \mathbf{F} = \boldsymbol{\epsilon}^* \cdot \mathbf{F}_{\mathrm{sh}} + \boldsymbol{\epsilon}^* \cdot \mathbf{F}_{\mathrm{ill}} \tag{10.121}$$

If the incident wave is a plane wave with wave vector $\mathbf{k}_0$ and polarization $\boldsymbol{\epsilon}_0$,

$$\begin{aligned} \mathbf{E}_i &= E_0 \boldsymbol{\epsilon}_0 e^{i\mathbf{k}_0 \cdot \mathbf{x}} \\ \mathbf{B}_i &= \mathbf{k}_0 \times \mathbf{E}_i / kc \end{aligned} \tag{10.122}$$

the shadow contribution, from (10.93) with $\mathbf{E}_s \simeq -\mathbf{E}_i$, $\mathbf{B}_s \simeq -\mathbf{B}_i$, is

$$\boldsymbol{\epsilon}^* \cdot \mathbf{F}_{\mathrm{sh}} = \frac{E_0}{4\pi i} \int_{\mathrm{sh}} \boldsymbol{\epsilon}^* \cdot [\mathbf{n}' \times (\mathbf{k}_0 \times \boldsymbol{\epsilon}_0) + \mathbf{k} \times (\mathbf{n}' \times \boldsymbol{\epsilon}_0)] e^{i(\mathbf{k}_0 - \mathbf{k}) \cdot \mathbf{x}'} \, da' \tag{10.123}$$

where the integration is only over the part of S_1 in shadow. A rearrangement of the vector products allows (10.123) to be written

$$\boldsymbol{\epsilon}^* \cdot \mathbf{F}_{\mathrm{sh}} = \frac{E_0}{4\pi i} \int_{\mathrm{sh}} \boldsymbol{\epsilon}^* \cdot [(\mathbf{k} + \mathbf{k}_0) \times (\mathbf{n}' \times \boldsymbol{\epsilon}_0) + (\mathbf{n}' \cdot \boldsymbol{\epsilon}_0)\mathbf{k}_0] e^{i(\mathbf{k}_0 - \mathbf{k}) \cdot \mathbf{x}'} \, da' \tag{10.124}$$

In the short-wavelength limit the magnitudes of $\mathbf{k}_0 \cdot \mathbf{x}'$ and $\mathbf{k} \cdot \mathbf{x}'$ are large compared to unity. The exponential factor in (10.124) will oscillate rapidly and cause the integrand to have a very small average value except in the forward direction where $\mathbf{k} \simeq \mathbf{k}_0$. In that forward region, $\theta \lesssim 1/kR$, the second term in the square bracket is negligible compared to the first because $(\boldsymbol{\epsilon}^* \cdot \mathbf{k}_0)/k$ is of the order of $\sin \theta \ll 1$ (remember $\boldsymbol{\epsilon}^* \cdot \mathbf{k} \equiv 0$ and $\mathbf{k}_0 \simeq \mathbf{k}$). Thus (10.124) can be approximated by

$$\boldsymbol{\epsilon}^* \cdot \mathbf{F}_{\mathrm{sh}} \simeq \frac{iE_0}{2\pi} \boldsymbol{\epsilon}^* \cdot \boldsymbol{\epsilon}_0 \int_{\mathrm{sh}} e^{i(\mathbf{k}_0 - \mathbf{k}) \cdot \mathbf{x}'} (\mathbf{k}_0 \cdot \mathbf{n}') \, da'$$

---

*For a very similar treatment of the scattering of a *scalar* wave by a sphere, see *Morse and Feshbach* (pp. 1551–1555).

The integral over the shadowed side of the obstacle has, in this approximation, the remarkable property of *depending only on the projected area* normal to the incident direction and not at all on the detailed shape of the obstacle. This can be seen from the fact that $(\mathbf{k}_0 \cdot \mathbf{n}') \, da' = k \, dx' \, dy' = k \, d^2x_\perp$ is just k times the projected element of area and $(\mathbf{k}_0 - \mathbf{k}) \cdot \mathbf{x}' = k(1 - \cos\theta)z' - \mathbf{k}_\perp \cdot \mathbf{x}_\perp \simeq -\mathbf{k}_\perp \cdot \mathbf{x}_\perp$. Here we have chosen $\mathbf{k}_0$ along the z axis, introduced two-dimensional vectors, $\mathbf{x}_\perp = x'\mathbf{e}_1 + y'\mathbf{e}_2$, $\mathbf{k}_\perp = k_x\mathbf{e}_1 + k_y\mathbf{e}_2$ in the plane perpendicular to $\mathbf{k}_0$, and approximated to small angles. The final form of the shadow contribution to the scattering when $kR \gg 1$ and $\theta \ll 1$ is therefore

$$\boldsymbol{\epsilon}^* \cdot \mathbf{F}_{\text{sh}} \simeq \frac{ik}{2\pi} E_0(\boldsymbol{\epsilon}^* \cdot \boldsymbol{\epsilon}_0) \int_{\text{sh}} e^{-i\mathbf{k}_\perp \cdot \mathbf{x}_\perp} \, d^2x_\perp \qquad (10.125)$$

In this limit all scatterers of the same projected area give the same shadow-scattering contribution. The polarization character of the scattered radiation is given by the factor $\boldsymbol{\epsilon}^* \cdot \boldsymbol{\epsilon}_0$. Since the scattering is at small angles, the dominant contribution has the same polarization as the incident wave. In quantum-mechanical language we say that the shadow scattering involves *no spin flip*.

For example, consider a scatterer whose projected area is a circular disc of radius a. Then

$$\int_{\text{sh}} e^{-i\mathbf{k}_\perp \cdot \mathbf{x}_\perp} \, d^2x_\perp = 2\pi a^2 \frac{J_1(ka \sin\theta)}{ka \sin\theta} \qquad (10.126)$$

and the shadow-scattering amplitude is

$$\boldsymbol{\epsilon}^* \cdot \mathbf{F}_{\text{sh}} \simeq ika^2 E_0(\boldsymbol{\epsilon}^* \cdot \boldsymbol{\epsilon}_0) \frac{J_1(ka \sin\theta)}{(ka \sin\theta)} \qquad (10.127)$$

The scattering from the illuminated side of the obstacle cannot be calculated without specifying the shape and nature of the surface. We assume, for purposes of illustration, that the illuminated surface is perfectly conducting. In utilizing (10.93) we must know the tangential components of $\mathbf{E}_s$ and $\mathbf{B}_s$ on S_1. As mentioned in the introductory paragraphs of this section, in the short-wavelength limit these are approximately opposite and equal, respectively, to the corresponding components of the incident fields. Thus the contribution from the illuminated side is

$$\boldsymbol{\epsilon}^* \cdot \mathbf{F}_{\text{ill}} = \frac{E_0}{4\pi i} \int_{\text{ill}} \boldsymbol{\epsilon}^* \cdot [-\mathbf{n}' \times (\mathbf{k}_0 \times \boldsymbol{\epsilon}_0) + \mathbf{k} \times (\mathbf{n}' \times \boldsymbol{\epsilon}_0)]e^{i(\mathbf{k}_0 - \mathbf{k}) \cdot \mathbf{x}'} \, da' \qquad (10.128)$$

Comparison with the shadow contribution (10.123) at the same stage shows a sign difference in the first term. This is crucial in giving very different angular behaviors of the two amplitudes. The counterpart of (10.124) is

$$\boldsymbol{\epsilon}^* \cdot \mathbf{F}_{\text{ill}} = \frac{E_0}{4\pi i} \int_{\text{ill}} \boldsymbol{\epsilon}^* \cdot [\mathbf{k} - \mathbf{k}_0) \times (\mathbf{n}' \times \boldsymbol{\epsilon}_0) - (\mathbf{n}' \cdot \boldsymbol{\epsilon}_0)\mathbf{k}_0]e^{i(\mathbf{k}_0 - \mathbf{k}) \cdot \mathbf{x}'} \, da' \qquad (10.129)$$

For $kR \gg 1$, the exponential oscillates rapidly as before, but now, in the forward direction, where we anticipate the major contribution to the integral, the other factor in the integrand goes to zero. This can be traced to the presence of $(\mathbf{k} - \mathbf{k}_0)$ in the first term, rather than the $(\mathbf{k} + \mathbf{k}_0)$ of the shadow amplitude (10.124). The illuminated side of the scatterer thus gives only a modest contribution to the scattering at small angles. This makes perfect sense if we think of the limit of geometrical optics. The illuminated side must give the reflected wave, and the reflection is mainly at angles other than forward.

To proceed further we must specify the *shape* of the illuminated portion of the scatterer, as well as its electromagnetic properties. We assume that the surface is spherical of radius a. Since the contribution is not dominantly forward, we must consider arbitrary scattering angles. The integrand in (10.129) consists of a relatively slowly varying vector function of angles times a rapidly varying exponential. As discussed in Section 7.11, the dominant contribution to such an integral comes from the region of integration where the phase of the exponential is stationary. If (θ, ϕ) are the angular coordinates of $\mathbf{k}$ and (α, β) those of $\mathbf{n}'$, relative to $\mathbf{k}_0$, the phase factor is

$$f(\alpha, \beta) = (\mathbf{k}_0 - \mathbf{k}) \cdot \mathbf{x}' = ka[(1 - \cos\theta)\cos\alpha - \sin\theta\sin\alpha\cos(\beta - \phi)]$$

The stationary point is easily shown to be at angles α_0, β_0, where

$$\alpha_0 = \frac{\pi}{2} + \frac{\theta}{2}$$

$$\beta_0 = \phi$$

These angles are evidently just those appropriate for reflection from the sphere according to geometrical optics. At this point the unit vector $\mathbf{n}'$ points in the direction of $(\mathbf{k} - \mathbf{k}_0)$. If we expand the phase factor around $\alpha = \alpha_0$, $\beta = \beta_0$, we obtain

$$f(\alpha, \beta) = -2ka\sin\frac{\theta}{2}\left[1 - \tfrac{1}{2}\left(x^2 + \cos^2\frac{\theta}{2}y^2\right) + \cdots\right] \tag{10.130}$$

where $x = \alpha - \alpha_0$, $y = \beta - \beta_0$. Then integral (10.129) can be approximated by evaluating the square bracket there at $\alpha = \alpha_0$, $\beta = \beta_0$:

$$\boldsymbol{\epsilon}^* \cdot \mathbf{F}_{\text{ill}} \simeq \frac{ka^2 E_0}{4\pi i}\sin\theta\, e^{-2ika\sin(\theta/2)}\boldsymbol{\epsilon}^* \cdot \boldsymbol{\epsilon}_r\int dx\, e^{i[ka\sin(\theta/2)]x^2}\int dy\, e^{i[ka\sin(\theta/2)\cos^2(\theta/2)]y^2} \tag{10.131}$$

where $\boldsymbol{\epsilon}_r$ is a unit polarization vector defined by

$$\boldsymbol{\epsilon}_r = -\boldsymbol{\epsilon}_0 + 2(\mathbf{n}_r \cdot \boldsymbol{\epsilon}_0)\mathbf{n}_r$$

$\mathbf{n}_r$ being a unit vector in the direction of $(\mathbf{k} - \mathbf{k}_0)$. The vector $\boldsymbol{\epsilon}_r$ is just the polarization expected for reflection, having a component perpendicular to the surface equal to the corresponding component of $\boldsymbol{\epsilon}_0$ and a component parallel to the surface opposite in sign, as shown in Fig. 10.15. The x and y integrals in (10.131) can be approximated using $\int_{-\infty}^{\infty} e^{i\alpha x^2}\, dx = \sqrt{\pi i/\alpha}$ provided $2ka\sin(\theta/2) \gg 1$, giving

$$\boldsymbol{\epsilon}^* \cdot \mathbf{F}_{\text{ill}} \simeq E_0\frac{a}{2}e^{-2ika\sin(\theta/2)}\boldsymbol{\epsilon}^* \cdot \boldsymbol{\epsilon}_r \tag{10.132}$$

For $2ka\sin(\theta/2)$ large, the reflected contribution is constant in magnitude as a function of angle, but it has a rapidly varying phase; as $\theta \to 0$, it vanishes as θ^2.

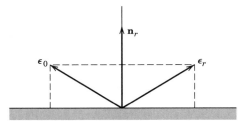

Figure 10.15 Polarization of the reflected wave relative to the incident polarization: $\mathbf{n}_r$ is normal to the surface at the point appropriate for reflection according to geometrical optics. To avoid complexity in the figure, the wave vectors $\mathbf{k}_0$ and $\mathbf{k}$ are not shown, but they are perpendicular to $\boldsymbol{\epsilon}_0$ and $\boldsymbol{\epsilon}_r$, respectively, and so oriented as to make their difference parallel to $\mathbf{n}_r$.

Comparison of the shadow amplitude (10.127) with the reflected amplitude (10.132) shows that in the very forward direction the shadow contribution dominates in magnitude over the reflected amplitude by a factor of $ka \gg 1$, while at angles where $ka \sin \theta \gg 1$, the ratio of the magnitudes is of the order of $1/(ka \sin^3 \theta)^{1/2}$. Thus, the differential scattering cross section (10.3), summed over the outgoing and averaged over the initial polarization states, is given in the two regions by

$$\frac{d\sigma}{d\Omega} \simeq \begin{cases} a^2 (ka)^2 \left| \dfrac{J_1(ka \sin \theta)}{ka \sin \theta} \right|^2, & \theta \lesssim \dfrac{10}{ka} \\[2ex] \dfrac{a^2}{4}, & \theta \gg \dfrac{1}{ka} \end{cases} \qquad (10.133)$$

The scattering in the forward direction is a typical diffraction pattern with a central maximum and smaller secondary maxima, while at larger angles it is isotropic. At intermediate angles there is some interference between the two amplitudes (10.127) and (10.132), causing the cross section to deviate from the sum of the two terms shown in (10.133). Actually, in the present approximation this interference is very small for $ka \gg 1$. There is more interference in the exact solution, as shown in Fig. 10.16, where the dips below unity are indicative of destructive interference.*

The total scattering cross section is obtained by integrating over all angles. Neglecting the interference terms, we find from (10.133) that the shadow diffraction peak gives a contribution of πa^2, and so does the isotropic part. The total scattering cross section is thus $2\pi a^2$, one factor of the geometrical projected area coming from direct reflection and the other from the diffraction scattering that must accompany the formation of a shadow behind the obstacle. The latter part of the total cross section can be shown to be independent of the detailed shape of the scatterer in the short-wavelength limit (Problem 10.16). Similarly, for a general scatterer that is "opaque," the reflected or absorbed part of the total cross section will also be equal to the projected area, although without specifying

---

*For a linearly polarized wave incident, the amount of interference depends on the orientation of the incident polarization vector relative to the plane of observation containing $\mathbf{k}$ and $\mathbf{k}_0$. For $\boldsymbol{\epsilon}_0$ in this plane the interference is much greater than for $\boldsymbol{\epsilon}_0$ perpendicular to it. See *King and Wu* (Appendix) or *Bowman, Senior, and Uslenghi* (pp. 402–405) for numerous graphs with different values of ka.

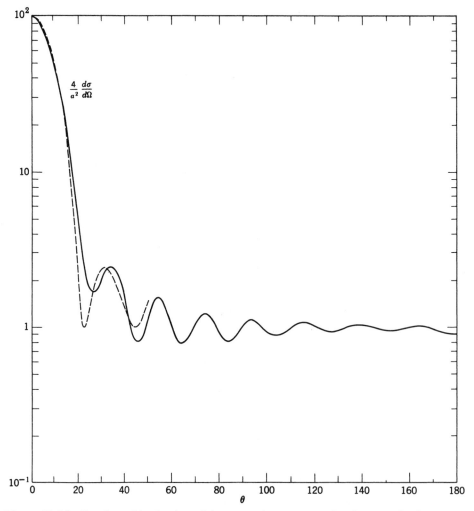

Figure 10.16 Semilogarithmic plot of the scattering cross section for a perfectly conducting sphere as a function of scattering angle, with an unpolarized plane wave incident and $ka = 10$. The solid curve is the exact result (*King and Wu*). The dashed curve is the approximation based on the sum of the amplitudes (10.127) and (10.132).

the properties of the illuminated surface, we cannot say how it is divided between scattering and absorption.

10.11 Optical Theorem and Related Matters

A fundamental relation, called the *optical theorem*, connects the total cross section of a scatterer to the imaginary part of the forward scattering amplitude. The theorem follows from very general considerations of the conservation of energy and power flow, and has its counterpart in the quantum-mechanical scattering of particles through the conservation of probability.

To establish the theorem, we consider the scattering geometry shown in Fig. 10.9. A plane wave with wave vector $\mathbf{k}_0$ and fields $(\mathbf{E}_i, \mathbf{B}_i)$ is incident in vacuum on a finite scatterer that lies inside the surface S_1. The scattered fields $(\mathbf{E}_s, \mathbf{B}_s)$ propagate out from the scatterer and are observed far away in the direction of $\mathbf{k}$. The total fields at all points in space are, by definition,

$$\mathbf{E} = \mathbf{E}_i + \mathbf{E}_s, \qquad \mathbf{B} = \mathbf{B}_i + \mathbf{B}_s$$

The scatterer is, in general, dissipative and absorbs energy from the incident wave. The *absorbed* power can be calculated by integrating the inward-going component of the Poynting vector of the *total* fields over the surface S_1:

$$P_{\text{abs}} = -\frac{1}{2\mu_0} \oint_{S_1} \text{Re}(\mathbf{E} \times \mathbf{B}^*) \cdot \mathbf{n}' \, da' \tag{10.134}$$

The *scattered* power is normally calculated by considering the asymptotic form of the Poynting vector for the scattered fields in the region where these are simple transverse fields falling off as $1/r$. But since there are no sources between S_1 and infinity, the scattered power can equally well be evaluated as an integral over S_1 of the outwardly directed component of the *scattered* Poynting vector:

$$P_{\text{scatt}} = \frac{1}{2\mu_0} \oint_{S_1} \text{Re}(\mathbf{E}_s \times \mathbf{B}_s^*) \cdot \mathbf{n}' \, da' \tag{10.135}$$

The total power P taken from the incident wave, either by scattering or absorption, is the sum of (10.134) and (10.135). With some obvious substitutions and rearrangements, the total power can be written

$$P = -\frac{1}{2\mu_0} \oint_{S_1} \text{Re}[\mathbf{E}_s \times \mathbf{B}_i^* + \mathbf{E}_i^* \times \mathbf{B}_s] \cdot \mathbf{n}' \, da'$$

With the incident wave written explicitly as

$$\mathbf{E}_i = E_0 \boldsymbol{\epsilon}_0 e^{i\mathbf{k}_0 \cdot \mathbf{x}} \tag{10.136}$$
$$c\mathbf{B}_i = \frac{1}{k} \mathbf{k}_0 \times \mathbf{E}_i$$

the total power takes the form,

$$P = \frac{1}{2\mu_0} \text{Re}\left\{ E_0^* \oint_{S_1} e^{-i\mathbf{k}_0 \cdot \mathbf{x}'}\left[\boldsymbol{\epsilon}_0^* \cdot (\mathbf{n}' \times \mathbf{B}_s) + \boldsymbol{\epsilon}_0^* \cdot \frac{\mathbf{k}_0 \times (\mathbf{n}' \times \mathbf{E}_s)}{kc} \right] da' \right\}$$

Comparison with (10.93) for the scattering amplitude shows that the total power is related to the *forward* $(\mathbf{k} = \mathbf{k}_0, \boldsymbol{\epsilon} = \boldsymbol{\epsilon}_0)$ scattering amplitude according to

$$P = \frac{2\pi}{kZ_0} \text{Im}[E_0^* \boldsymbol{\epsilon}_0^* \cdot \mathbf{F}(\mathbf{k} = \mathbf{k}_0)] \tag{10.137}$$

This is the basic result of the optical theorem, although it is customary to express it in a form that is independent of the magnitude of the incident flux. The *total cross section* σ_t (sometimes called the extinction cross section in optics) is defined as the ratio of the total power P to the incident power per unit area, $|E_0|^2/2Z_0$. Similarly, the *normalized scattering amplitude* $\mathbf{f}$ is defined relative to the amplitude of the incident wave at the origin as

$$\mathbf{f}(\mathbf{k}, \mathbf{k}_0) = \frac{\mathbf{F}(\mathbf{k}, \mathbf{k}_0)}{E_0} \tag{10.138}$$

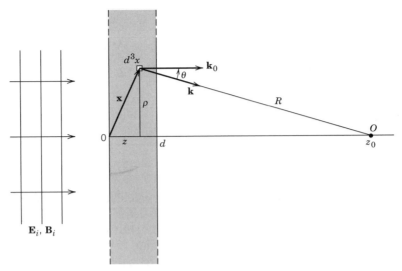

Figure 10.17 A plane wave incident normally on a slab of dielectric of thickness d. The scatterers in the slab give rise to a scattered wave that adds coherently to the incident wave to give a modified wave at the observation point O behind the slab.

In terms of σ_t and **f** the *optical theorem* reads

$$\sigma_t = \frac{4\pi}{k} \, \mathrm{Im}[\boldsymbol{\epsilon}_0^* \cdot \mathbf{f}(\mathbf{k} = \mathbf{k}_0)] \tag{10.139}$$

The notation in (10.139) corresponds to the standard quantum-mechanical conventions. For particles with spin the relevant forward scattering amplitude is the one in which none of the particles change their spin state. For electromagnetic radiation (photons) this is indicated by the presence of the amplitude $\boldsymbol{\epsilon}_0^* \cdot \mathbf{f}$ for scattered radiation with the same polarization finally as it was initially.

The optical theorem relates different aspects of the scattering and absorption of electromagnetic waves for a single scatterer. It is also possible to connect the forward scattering amplitude for a single scatterer to the macroscopic electromagnetic properties, namely the dielectric constant, of a medium composed of a large number of scatterers. We will content ourselves with a brief elementary discussion and refer the reader to the literature for more detailed and rigorous treatments.* Consider a plane wave (10.136) incident normally from the left on a thin slab of uniform material composed of N identical scattering centers per unit volume, as shown in Fig. 10.17). The incident wave impinges on the scattering centers, causing each to generate a scattered wave. The coherent sum of the incident wave and of all the scattered waves gives a modified wave to the right of the slab. Comparison of this modified wave at the observation point O with that expected for a wave transmitted through a slab described by a macroscopic, electric susceptibility $\epsilon(\omega)$ then leads to a relation between ϵ and the scattering amplitude **f**.

*See, for example, the very readable review by M. A. Lax, *Rev. Mod. Phys.* **23**, 287 (1951), or M. L. Goldberger and K. M. Watson, *Collision Theory*, Wiley, New York (1964), Chapter 11, especially pp. 766–775.

The thickness and the density of the slab are assumed to be so small that only single scatterings in the slab need be considered and, as a consequence, the effective exciting field at each scatterer is just the incident field itself. The scattered field produced at the observation point O with cylindrical coordinates $(0, 0, z_0)$ by the $N \, d^3x$ scatterers in the infinitesimal volume element d^3x at the point $\mathbf{x}(\rho, \phi, z)$ in the slab is, in this approximation,

$$d\mathbf{E}_s = \frac{e^{ikR}}{R} \, \mathbf{f}(k, \, \theta, \, \phi) E_0 e^{i\mathbf{k}_0 \cdot \mathbf{x}} N \, d^3x$$

where we have written the scattering amplitude in terms of the scattering angles θ and ϕ, with $\sin \theta = \rho/R$, and have assumed that the observation point is many wavelengths from the slab. The distance from the volume element to O is $R = [\rho^2 + (z_0 - z)^2]^{1/2}$. The presence of the phase factor of the incident wave is necessary to account for the location of the scatterers at $\mathbf{x}$, rather than at the origin of coordinates. The total scattered field is obtained by integration over the volume of the slab:

$$\mathbf{E}_s = NE_0 \int_0^{2\pi} d\phi \int_0^d dz \, e^{ikz} \int_0^\infty \rho \, d\rho \, \frac{e^{ikR}}{R} \, \mathbf{f}(k, \, \theta, \, \phi) \qquad (10.140)$$

Since $\rho \, d\rho = R \, dR$, this expression can be written

$$\mathbf{E}_s = NE_0 \int_0^{2\pi} d\phi \int_0^d dz \, e^{ikz} \int_{|z_0-z|}^\infty dR \, e^{ikR} \, \mathbf{f}(k, \, \theta, \, \phi) \qquad (10.141)$$

where $\cos \theta = (z_0 - z)/R$. We now treat $e^{ikR} \, dR$ as a differential and integrate by parts to obtain for the R integration,

$$\int_{|z_0-z|}^\infty dR \, e^{ikR} \, \mathbf{f}(k, \, \theta, \, \phi) = \frac{1}{ik} \, e^{ikR} \, \mathbf{f}(k, \, \theta, \, \phi) \Bigg|_{R=|z_0-z|}^\infty$$
$$+ \frac{1}{ik} \int_{|z_0-z|}^\infty dR \left(\frac{z_0 - z}{R^2} \right) e^{ikR} \, \frac{d}{d(\cos \theta)} \, \mathbf{f}(k, \, \theta, \, \phi)$$

Provided the indicated derivative of $\mathbf{f}$ is well behaved, the remaining integral is of the order of $1/(k \, |z_0 - z|)$ times the original. Since we have assumed that the observation point is many wavelengths from the slab, this integral can be neglected. Neglecting the oscillating contribution at the upper limit $R \to \infty$ (this can be made to vanish somewhat more plausibly by assuming that the number N of scattering centers per unit volume falls to zero at very large ρ), we have the result

$$\int_{|z_0-z|}^\infty dR \, e^{ikR} \, \mathbf{f}(k, \, \theta, \, \phi) = \frac{i}{k} \, e^{ik|z_0-z|} \, \mathbf{f}(k, 0)$$

The scattered field at O is therefore

$$\mathbf{E}_s = \frac{2\pi i}{k} \, NE_0 \mathbf{f}(k, 0) \int_0^d dz \, e^{ik[z+|z_0-z|]}$$

Since $z_0 > z$ by assumption, we have finally

$$\mathbf{E}_s = \frac{2\pi i}{k} \, NE_0 \mathbf{f}(k, 0) e^{ikz_0} d \qquad (10.142)$$

The *total* electric field at the observation point O is

$$\mathbf{E} = E_0 e^{ikz_0}\left[\boldsymbol{\epsilon}_0 + \frac{2\pi iNd}{k}\mathbf{f}(k, 0)\right] \tag{10.143}$$

correct to first order in the slab thickness d. The amplitude at O for a wave with the same polarization state as the incident wave is

$$\boldsymbol{\epsilon}_0^* \cdot \mathbf{E} = E_0 e^{ikz_0}\left[1 + \frac{2\pi iNd}{k}\boldsymbol{\epsilon}_0^* \cdot \mathbf{f}(k, 0)\right] \tag{10.144}$$

Suppose that we now consider the slab macroscopically, with its electromagnetic properties specified by a dielectric constant $\epsilon(\omega)/\epsilon_0$ appropriate to describe the propagation of the wave of frequency $\omega = ck$ and polarization $\boldsymbol{\epsilon}_0$. A simple calculation using the formulas of Chapter 7 shows that the transmitted wave at $z = z_0$ is given by

$$\boldsymbol{\epsilon}_0^* \cdot \mathbf{E}(\text{macroscopic}) = E_0 e^{ikz_0}\left[1 + ik(\epsilon/\epsilon_0 - 1)\frac{d}{2}\right] \tag{10.145}$$

correct to *first order in d, but with no approximation concerning the smallness of* $|\epsilon/\epsilon_0 - 1|$. Comparison of (10.144) and (10.145) shows that the dielectric constant can be written in terms of the forward scattering amplitude as

$$\epsilon(\omega)/\epsilon_0 = 1 + \frac{4\pi N}{k^2}\boldsymbol{\epsilon}_0^* \cdot \mathbf{f}(k, 0) \tag{10.146}$$

A number of observations are in order. It is obvious that our derivation has been merely indicative, with a number of simplifying assumptions and the notion of a macroscopic description assumed rather than derived. More careful considerations show that the scattering amplitude in (10.146) should be evaluated at the wave number k' in the medium, not at the free-space wave number k, and that there is a multiplier to the second term that gives a measure of the effective exciting field at a scatterer relative to the total coherent field in the medium. The reader can consult the literature cited above for these and other details. Suffice it to say that (10.146) is a reasonable approximation for not too dense substances and provided correlations among neighboring scatterers are not important. It is worthwhile to illustrate (10.146) with the simple electronic oscillator model used in Chapter 7 to describe the dielectric constant. The dipole moment of the atom is given by (7.50), summed over the various oscillators:

$$\mathbf{p} = \frac{e^2}{m}\sum_j f_j(\omega_j^2 - \omega^2 - i\omega\gamma_j)^{-1}E_0\boldsymbol{\epsilon}_0$$

From (10.2) we infer that the atomic scattering amplitude is

$$\mathbf{f}(\mathbf{k}) = \frac{1}{4\pi\epsilon_0}\frac{e^2}{m}\sum_j f_j(\omega_j^2 - \omega^2 - i\omega\gamma_j)^{-1}(\mathbf{k} \times \boldsymbol{\epsilon}_0) \times \mathbf{k}$$

The scalar product of $\boldsymbol{\epsilon}_0^*$ with the forward scattering amplitude is then

$$\boldsymbol{\epsilon}_0^* \cdot \mathbf{f}(\mathbf{k} = \mathbf{k}_0) = \frac{e^2 k^2}{4\pi\epsilon_0 m}\sum_j f_j(\omega_j^2 - \omega^2 - i\omega\gamma_j)^{-1}$$

Substitution into (10.146) yields the dielectric constant

$$\epsilon(\omega)/\epsilon_0 = 1 + \frac{Ne^2}{\epsilon_0 m} \sum_j f_j(\omega_j^2 - \omega^2 - i\omega\gamma_j)^{-1} \qquad (10.147)$$

in agreement with (7.51).

Contact can be established between (10.146) and the optical theorem (10.139) by recalling that the attenuation coefficient α is related to the total cross section of a single scatterer through $\alpha = N\sigma_t$ and to the imaginary part of the wave number in the medium through $\alpha = 2 \operatorname{Im}(k')$. From (10.146) and the relations (7.54) for the real and imaginary parts of k' in terms of $\epsilon(\omega)$ we find

$$\alpha = N\sigma_t = \frac{4\pi N}{\operatorname{Re}(k')} \operatorname{Im}[\boldsymbol{\epsilon}_0^* \cdot \mathbf{f}(\operatorname{Re} k', 0)] \qquad (10.148)$$

where I have improved (10.146) by evaluating $\mathbf{f}$ at the wave number in the medium, as described above. Equation (10.148) indicates that, if we consider scattering by a single scatterer embedded in a medium, the optical theorem and other relations will appear as before, provided we describe the "kinematics" correctly by using the local wave number k' in the medium. The same situation holds in the scattering of electrons in a solid, for example, where the effective mass or other approximation is used to take into account propagation through the lattice.

As a final comment on the optical theorem we note the problem of approximations for $\mathbf{f}$. The optical theorem is an exact relation. If an approximate expression for $\mathbf{f}$ is employed, a manifestly wrong result for the total cross section may be obtained. For example, in the long-wavelength limit we find from (10.2) and (10.5) that the scattering amplitude for a dielectric sphere of radius a is

$$\mathbf{f} = \left(\frac{\epsilon_r - 1}{\epsilon_r + 2}\right) a^3 (\mathbf{k} \times \boldsymbol{\epsilon}_0) \times \mathbf{k}$$

The forward amplitude is

$$\boldsymbol{\epsilon}_0^* \cdot \mathbf{f}(\mathbf{k} = \mathbf{k}_0) = k^2 a^3 \left(\frac{\epsilon_r - 1}{\epsilon_r + 2}\right) \qquad (10.149)$$

For a lossless dielectric, this amplitude is real; the optical theorem (10.139) then yields $\sigma_t = 0$. On the other hand, we know that the total cross section is in this case equal to the scattering cross section (10.11):

$$\sigma_{\mathrm{sc}} = \frac{8\pi}{3} k^4 a^6 \left|\frac{\epsilon_r - 1}{\epsilon_r + 2}\right|^2 \qquad (10.150)$$

Even with a lossy dielectric ($\operatorname{Im} \epsilon \neq 0$), the optical theorem yields a total cross section,

$$\sigma_t = \frac{12\pi k a^3 \operatorname{Im} \epsilon_r}{|\epsilon_r + 2|^2} \qquad (10.151)$$

while the scattering cross section remains (10.150). These seeming contradictions are reflections of the necessity of different orders of approximation required to obtain consistency between the two sides of the optical theorem. In the long-wavelength limit it is necessary to evaluate the forward scattering amplitude to higher order in powers of ω to find the scattering cross section contribution in the total cross section by means of the optical theorem. For lossless or nearly lossless scatterers it is therefore simplest to determine the total cross section directly by integration of the differential scattering cross section over angles. For dissipative scatterers, on the other hand, the optical theorem yields a nonzero answer that has a different (usually a lower power) dependence on ω and other parameters from that of the scattering cross section. This contribution is, of course, the *absorption* cross section to lowest explicit order in ω. It can be calculated from first principles with (10.134), but the optical theorem provides an elegant and convenient method. Examples of these considerations are given in the problems. An analogous situation occurs in quantum-mechanical scattering by a real potential where the first Born approximation yields a real scattering amplitude. The second Born approximation has an imaginary part in the forward direction that gives, via the optical theorem, a total cross section in agreement with the integrated scattering cross section of the first Born approximation.

References and Suggested Reading

Scattering and diffraction are treated in many optics texts. References concerning critical opalescence are cited in Section 10.2. The subject of losses in optical fibers and Rayleigh scattering there can be pursued in

J. M. Senior, *Optical Fiber Communications*, Prentice Hall, New York (1992), Chapter 3.
H. Murata, *Handbook of Optical Fibers and Cables*, Marcel Dekker, New York (1988).

The scattering of radiation by a perfectly conducting sphere is treated briefly in
Morse and Feshbach (pp. 1882–1886)
Panofsky and Phillips, Section 13.9

Much more elaborate discussions, with arbitrary dielectric and conductive properties for the sphere (Mie's problem) are given by
Born and Wolf, Section 13.5
Stratton, Section 9.25

Specialized monographs on the scattering of electromagnetic waves by spheres and other obstacles are
Bowman, Senior, and Uslenghi
H. C. van de Hulst, *Light Scattering by Small Particles*, Wiley, New York (1957)
Kerker
King and Wu

Kerker has a nice historical introduction, including quotations from Leonardo da Vinci and Maxwell in a letter to Rayleigh. See also
Van Bladel

The subject of diffraction has a very extensive literature. A comprehensive treatment of both scalar Kirchhoff and vector theory, with many examples and excellent figures, is given by
Born and Wolf, Chapters VIII, IX, and XI

The review
C. J. Bouwkamp, Diffraction theory, in *Reports on Progress in Physics*, ed. A. C. Stickland, Vol. XVII, pp. 35–100, The Physical Society, London (1954)

has an extensive list of references and treats a number of difficult points. More elementary discussions of the scalar theory are found in

Slater and Frank, Chapters XIII and XIV

Sommerfeld, *Optics*

Stone, Chapters 8, 9, and 10

Sommerfeld also discusses his rigorous calculation of diffraction by a straightedge, one of the few exact solutions in diffraction theory.

The language of diffraction is used increasingly in optical applications, such as Abbé's theory of the resolution of optical instruments and the phase-contrast microscope. An introduction to this area from an electrical engineering viewpoint is provided by

J. W. Goodman, *Introduction to Fourier Optics*, McGraw-Hill, New York (1968).

Born and Wolf and also Stone discuss some aspects of these applications.

Mathematical techniques for diffraction problems are discussed by

Baker and Copson

Jones, Chapters 8 and 9

Morse and Feshbach, Chapter 11

A. Rubinowicz, *Die Beugunswelle in der Kirchhoffschen Theorie der Beugung*, 2nd edition, PWN-Springer, Warsaw-Berlin (1966)

L. A. Weinstein (Vaynshteyn), *Theory of Diffraction and the Factorization Method*, Golem Press, Boulder, CO (1969)

The proof that for short-wavelength scattering or diffraction the transition between the illuminated and shadow regions is of the order of $R/(kR)^{1/3}$ can be found in

V. A. Fock, *Electromagnetic Diffraction and Propagation Problems*, Pergamon Press, Oxford (1965)

Entirely omitted from this chapter is the use of variational principles for diffraction and scattering. The reader may fill this gap by consulting

Bouwkamp, *op. cit.*

Cairo and Kahan

Harrington, Chapter 7

H. Levine and J. Schwinger, *Commun. Pure Appl. Math.* **3**, 355 (1950)

Morse and Feshbach, Section 11.4

Problems

10.1 **(a)** Show that for arbitrary initial polarization, the scattering cross section of a perfectly conducting sphere of radius a, summed over outgoing polarizations, is given in the long-wavelength limit by

$$\frac{d\sigma}{d\Omega}(\boldsymbol{\epsilon}_0, \mathbf{n}_0, \mathbf{n}) = k^4 a^6 \left[\frac{5}{4} - |\boldsymbol{\epsilon}_0 \cdot \mathbf{n}|^2 - \frac{1}{4} |\mathbf{n} \cdot (\mathbf{n}_0 \times \boldsymbol{\epsilon}_0)|^2 - \mathbf{n}_0 \cdot \mathbf{n} \right]$$

where $\mathbf{n}_0$ and $\mathbf{n}$ are the directions of the incident and scattered radiations, respectively, while $\boldsymbol{\epsilon}_0$ is the (perhaps complex) unit polarization vector of the incident radiation ($\boldsymbol{\epsilon}_0^* \cdot \boldsymbol{\epsilon}_0 = 1$; $\mathbf{n}_0 \cdot \boldsymbol{\epsilon}_0 = 0$).

(b) If the incident radiation is linearly polarized, show that the cross section is

$$\frac{d\sigma}{d\Omega}(\boldsymbol{\epsilon}_0, \mathbf{n}_0, \mathbf{n}) = k^4 a^6 \left[\frac{5}{8}(1 + \cos^2\theta) - \cos\theta - \frac{3}{8}\sin^2\theta \cos 2\phi \right]$$

where $\mathbf{n} \cdot \mathbf{n}_0 = \cos\theta$ and the azimuthal angle ϕ is measured from the direction of the linear polarization.

(c) What is the ratio of scattered intensities at $\theta = \pi/2$, $\phi = 0$ and $\theta = \pi/2$, $\phi = \pi/2$? Explain physically in terms of the induced multipoles and their radiation patterns.

10.2 Electromagnetic radiation with elliptic polarization, described (in the notation of Section 7.2) by the polarization vector,

$$\boldsymbol{\epsilon} = \frac{1}{\sqrt{1 + r^2}} (\boldsymbol{\epsilon}_+ + re^{i\alpha}\boldsymbol{\epsilon}_-)$$

is scattered by a perfectly conducting sphere of radius a. Generalize the amplitude in the scattering cross section (10.71), which applies for $r = 0$ or $r = \infty$, and calculate the cross section for scattering in the long-wavelength limit. Show that

$$\frac{d\sigma}{d\Omega} = k^4 a^6 \left[\frac{5}{8} (1 + \cos^2\theta) - \cos\theta - \frac{3}{4} \left(\frac{r}{1 + r^2} \right) \sin^2\theta \cos(2\phi - \alpha) \right]$$

Compare with Problem 10.1.

10.3 A solid uniform sphere of radius R and conductivity σ acts as a scatterer of a plane-wave beam of unpolarized radiation of frequency ω, with $\omega R/c \ll 1$. The conductivity is large enough that the skin depth δ is small compared to R.

(a) Justify and use a magnetostatic scalar potential to determine the magnetic field around the sphere, assuming the conductivity is infinite. (Remember that $\omega \neq 0$.)

(b) Use the technique of Section 8.1 to determine the absorption cross section of the sphere. Show that it varies as $(\omega)^{1/2}$ provided σ is independent of frequency.

10.4 An unpolarized wave of frequency $\omega = ck$ is scattered by a *slightly* lossy uniform isotropic dielectric sphere of radius R much smaller than a wavelength. The sphere is characterized by an ordinary real dielectric constant ϵ_r and a real conductivity σ. The parameters are such that the skin depth δ is very *large* compared to the radius R.

(a) Calculate the differential and total *scattering* cross sections.

(b) Show that the absorption cross section is

$$\sigma_{\text{abs}} = 12\pi R^2 \frac{(RZ_0\sigma)}{(\epsilon_r + 2)^2 + (Z_0\sigma/k)^2}$$

(c) From part a write down the forward scattering amplitude and use the optical theorem to evaluate the total cross section. Compare your answer with the sum of the scattering and absorption cross sections from parts a and b. Comment.

10.5 The scattering by the dielectric sphere of Problem 10.4 was treated as purely electric dipole scattering. This is adequate unless it happens that the real dielectric constant ϵ/ϵ_0 is very large. In these circumstances a magnetic dipole contribution, even though higher order in kR, may be important.

(a) Show that the changing magnetic flux of the incident wave induces an azimuthal current flow in the sphere and produces a magnetic dipole moment,

$$\mathbf{m} = \frac{i4\pi\sigma Z_0}{k\mu_0} (kR)^2 \frac{R^3}{30} \mathbf{B}_{\text{inc}}$$

(b) Show that application of the optical theorem to the coherent sum of the electric and magnetic dipole contributions leads to a total cross section,

$$\sigma_t = 12\pi R^2 (RZ_0\sigma) \left[\frac{1}{(\epsilon_r + 2)^2 + (Z_0\sigma/k)^2} + \frac{1}{90} (kR)^2 \right]$$

(Compare Landau and Lifshitz, *Electrodynamics of Continuous Media*, p. 304).

10.6 **(a)** Show that for the scattered wave (10.57) the *normalized* scattering amplitude (10.138) is

$$\mathbf{f} = \frac{1}{ik} \sqrt{\frac{\pi}{2}} \sum_l \sqrt{2l + 1} \, [\alpha_\pm(l)\mathbf{X}_{l,\pm1} \pm i\beta_\pm(l)\mathbf{n} \times \mathbf{X}_{l,\pm1}]$$

where the polarization vector of the incident wave is $(\boldsymbol{\epsilon}_1 \pm i\boldsymbol{\epsilon}_2)/\sqrt{2}$.

(b) Deduce an expression for the total cross section of σ_t from the optical theorem (10.139) and the above expression for $\mathbf{f}$.

10.7 Discuss the scattering of a plane wave of electromagnetic radiation by a nonpermeable, dielectric sphere of radius a and dielectric constant ϵ_r.

(a) By finding the fields inside the sphere and matching to the incident plus scattered wave outside the sphere, determine without any restriction on ka the multipole coefficients in the scattered wave. Define suitable phase shifts for the problem.

(b) Consider the long-wavelength limit ($ka \ll 1$) and determine explicitly the differential and total scattering cross sections. Compare your results with those of Section 10.1.B.

(c) In the limit $\epsilon \to \infty$ compare your results to those for the perfectly conducting sphere.

10.8 Consider the scattering of a plane wave by a nonpermeable sphere of radius a and very good, but not perfect, conductivity following the spherical multipole field approach of Section 10.4. Assume that $ka \ll 1$ and that the skin depth $\delta < a$.

(a) Show from the analysis of Section 8.1 that

$$\frac{Z_s}{Z_0} = \frac{k\delta}{2} (1 - i)$$

(b) In the long-wavelength limit, show that for $l = 1$ the coefficients $\alpha_\pm(l)$ and $\beta_\pm(l)$ in (10.65) are

$$\alpha_\pm(1) \simeq -\frac{2i}{3} (ka)^3 \frac{\left[\left(1 - \dfrac{\delta}{a}\right) - i\dfrac{\delta}{a} \right]}{\left[\left(1 + \dfrac{\delta}{2a}\right) + i\dfrac{\delta}{2a} \right]}$$

$$\beta_\pm(1) \simeq \frac{4i}{3} (ka)^3$$

(c) Write out explicitly the differential scattering cross section, correct to *first* order in δ/a and lowest order in ka.

(d) Using (10.61), evaluate the absorption cross section. Show that to first order in δ it is $\sigma_{abs} \simeq 3\pi(k\delta)a^2$. How different is the value if $\delta = a$?

10.9 In the scattering of light by a gas very near the critical point the scattered light is observed to be "whiter" (i.e., its spectrum is less predominantly peaked toward the blue) than far from the critical point. Show that this can be understood by the fact that the volumes of the density fluctuations become large enough that Rayleigh's law fails to hold. In particular, consider the lowest order approximation to the scattering by a uniform dielectric sphere of radius a whose dielectric constant ϵ_r differs only slightly from unity.

(a) Show that for $ka \gg 1$, the differential cross section is sharply peaked in the forward direction and the total scattering cross section is approximately

$$\sigma \simeq \frac{\pi}{2} (ka)^2 |\epsilon_r - 1|^2 a^2$$

with a k^2, rather than k^4, dependence on frequency.

(b) Show that for arbitrary ka the total cross section to lowest order in $(\epsilon_r - 1)$ is the expression given in part a, multiplied by the function

$$F(z) = 1 + 5z^{-2} - \tfrac{7}{2}z^{-4}(1 - \cos 2z) - z^{-3} \sin 2z$$
$$- 4(z^{-2} - z^{-4}) \int_0^{2z} \frac{1 - \cos t}{t} \, dt$$

where $z = 2ka$. [This result is due to Lord Rayleigh, 1914.]

10.10 The aperture or apertures in a perfectly conducting plane screen can be viewed as the location of effective sources that produce radiation (the diffracted fields). An aperture whose dimensions are small compared with a wavelength acts as a source of dipole radiation with the contributions of other multipoles being negligible.

(a) Beginning with (10.101) show that the effective electric and magnetic dipole moments can be expressed in terms of integrals of the tangential electric field in the aperture as follows:

$$\mathbf{p} = \epsilon \mathbf{n} \int (\mathbf{x} \cdot \mathbf{E}_{\text{tan}}) \, da \qquad (9.72)$$
$$\mathbf{m} = \frac{2}{i\omega\mu} \int (\mathbf{n} \times \mathbf{E}_{\text{tan}}) \, da$$

where $\mathbf{E}_{\text{tan}}$ is the *exact* tangential electric field in the aperture, $\mathbf{n}$ is the normal to the plane screen, directed into the region of interest, and the integration is over the area of the openings.

(b) Show that the expression for the magnetic moment can be transformed into

$$\mathbf{m} = \frac{2}{\mu} \int \mathbf{x}(\mathbf{n} \cdot \mathbf{B}) \, da \qquad (9.74)$$

Be careful about possible contributions from the edge of the aperture where some components of the fields are singular if the screen is infinitesimally thick.

10.11 A perfectly conducting flat screen occupies half of the x-y plane (i.e., $x < 0$). A plane wave of intensity I_0 and wave number k is incident along the z axis from the region $z < 0$. Discuss the values of the diffracted fields in the plane parallel to the x-y plane defined by $z = Z > 0$. Let the coordinates of the observation point be $(X, 0, Z)$.

(a) Show that, for the usual scalar Kirchhoff approximation and in the limit $Z \gg X$ and $\sqrt{kZ} \gg 1$, the diffracted field is

$$\psi(X, 0, Z, t) \simeq I_0^{1/2} e^{ikZ - i\omega t} \left(\frac{1 + i}{2i}\right) \sqrt{\frac{2}{\pi}} \int_{-\xi}^{\infty} e^{it^2} \, dt$$

where $\xi = (k/2Z)^{1/2} X$.

(b) Show that the intensity can be written

$$I = |\psi|^2 = \frac{I_0}{2} [(C(\xi) + \tfrac{1}{2})^2 + (S(\xi) + \tfrac{1}{2})^2]$$

where $C(\xi)$ and $S(\xi)$ are the so-called Fresnel integrals. Determine the asymptotic behavior of I for ξ large and positive (illuminated region) and ξ large and negative (shadow region). What is the value of I at $X = 0$? Make a sketch of I as a function of X for fixed Z.

(c) Use the vector formula (10.101) to obtain a result equivalent to that of part a. Compare the two expressions.

10.12 A linearly polarized plane wave of amplitude E_0 and wave number k is incident on a circular opening of radius a in an otherwise perfectly conducting flat screen. The incident wave vector makes an angle α with the normal to the screen. The polarization vector is perpendicular to the plane of incidence.

(a) Calculate the diffracted fields and the power per unit solid angle transmitted through the opening, using the vector Smythe–Kirchhoff formula (10.101) with the assumption that the tangential electric field in the opening is the unperturbed incident field.

(b) Compare your result in part a with the standard scalar Kirchhoff approximation and with the result in Section 10.9 for the polarization vector in the plane of incidence.

10.13 Discuss the diffraction of a plane wave by a circular hole of radius a, following Section 10.9, but using a vector Kirchhoff approximation based on (10.90) instead of the Smythe formula (10.101).

(a) Show that the diffracted electric field in this approximation differs from (10.112) in two ways, first, that $\cos \alpha$ is replaced by $(\cos \theta + \cos \alpha)/2$, and second, by the addition of a term proportional to $(\mathbf{k} \times \boldsymbol{\epsilon}_3)$. Compare with the obliquity factors $\mathbb{O}$ of the scalar theory.

(b) Evaluate the ratio of the scattered power for this vector Kirchhoff approximation to that of (10.114) for the conditions shown in Fig. 10.14. Sketch the two angular distributions.

10.14 A rectangular opening with sides of length a and $b \geq a$ defined by $x = \pm(a/2)$, $y = \pm(b/2)$ exists in a flat, perfectly conducting plane sheet filling the x-y plane. A plane wave is normally incident with its polarization vector making an angle β with the long edges of the opening.

(a) Calculate the diffracted fields and power per unit solid angle with the vector Smythe–Kirchhoff relation (10.109), assuming that the tangential electric field in the opening is the incident unperturbed field.

(b) Calculate the corresponding result of the scalar Kirchhoff approximation.

(c) For $b = a$, $\beta = 45°$, $ka = 4\pi$, compute the vector and scalar approximations to the diffracted power per unit solid angle as a function of the angle θ for $\phi = 0$. Plot a graph showing a comparison between the two results.

10.15 A cylindrical coaxial transmission line of inner radius a and outer radius b has its axis along the negative z axis. Both inner and outer conductors end at $z = 0$, and the outer one is connected to an infinite plane copper flange occupying the whole x-y plane (except for the annulus of inner radius a and outer radius b around the origin). The transmission line is excited at frequency ω in its dominant TEM mode, with the peak voltage between the cylinders being V. Use the vector Smythe–Kirchhoff approximation to discuss the radiated fields, the angular distribution of radiation, and the total power radiated.

10.16 (a) Show from (10.125) that the integral of the shadow scattering differential cross section, summed over outgoing polarizations, can be written in the short-wavelength limit as

$$\sigma_{\text{sh}} = \int d^2x_\perp \int d^2x'_\perp \cdot \frac{1}{4\pi^2} \int e^{i(\mathbf{x}_\perp - \mathbf{x}'_\perp) \cdot \mathbf{k}_\perp} \, d^2k_\perp$$

and therefore is equal to the projected area of the scatterer, independent of its detailed shape.

(b) Apply the optical theorem to the "shadow" amplitude (10.125) to obtain the total cross section under the assumption that in the forward direction the contribution from the illuminated side of the scatterer is negligible in comparison.

10.17 (a) Using the approximate amplitudes of Section 10.10, show that, for a linearly polarized plane wave of wave number k incident on a perfectly conducting sphere of radius a in the limit of large ka, the differential scattering cross section in the E plane ($\boldsymbol{\epsilon}_0$, $\mathbf{k}_0$, and $\mathbf{k}$ coplanar) is

$$\frac{d\sigma}{d\Omega} (E \text{ plane}) = \frac{a^2}{4} \left[4 \cot^2\theta \, J_1^2(ka \sin\theta) + 1 \right.$$

$$\left. - 4 \cot\theta \, J_1(ka \sin\theta) \sin\left(2 \, ka \sin\frac{\theta}{2}\right) \right]$$

and in the H plane ($\boldsymbol{\epsilon}_0$ perpendicular to $\mathbf{k}_0$ and $\mathbf{k}$) is

$$\frac{d\sigma}{d\Omega} (H \text{ plane}) = \frac{a^2}{4} \left[4 \csc^2\theta \, J_1^2(ka \sin\theta) + 1 \right.$$

$$\left. + 4 \csc\theta \, J_1(ka \sin\theta) \sin\left(2 \, ka \sin\frac{\theta}{2}\right) \right]$$

(The dashed curve in Fig. 10.16 is the average of these two expressions.)

(b) Look up the exact calculations in *King and Wu* (Appendix) or *Bowman, Senior and Uslenghi* (pp. 402–405). Are the *qualitative* aspects of the interference between the diffractive and reflective amplitudes exhibited in part a in agreement with the exact results? What about quantitative agreement?

10.18 Discuss the diffraction due to a small, circular hole of radius a in a flat, perfectly conducting sheet, assuming that $ka \ll 1$.

(a) If the fields near the screen on the incident side are normal $\mathbf{E}_0 e^{-i\omega t}$ and tangential $\mathbf{B}_0 e^{-i\omega t}$, show that the diffracted electric field in the Fraunhofer zone is

$$\mathbf{E} = \frac{e^{ikr - i\omega t}}{3\pi r} k^2 a^3 \left[2c \, \frac{\mathbf{k}}{k} \times \mathbf{B}_0 + \frac{\mathbf{k}}{k} \times \left(\mathbf{E}_0 \times \frac{\mathbf{k}}{k} \right) \right]$$

where $\mathbf{k}$ is the wave vector in the direction of observation.

(b) Determine the angular distribution of the diffracted radiation and show that the total power transmitted through the hole is

$$P = \frac{2}{27\pi Z_0} k^4 a^6 (4c^2 B_0^2 + E_0^2)$$

10.19 Specialize the discussion of Problem 10.18 to the diffraction of a plane wave by the small, circular hole. Treat the general case of oblique incidence at an angle α to the normal, with polarization in and perpendicular to the plane of incidence.

(a) Calculate the angular distributions of the diffracted radiation and compare them to the vector Smythe–Kirchhoff approximation results of Section 10.9 and Problem 10.12 in the limit $ka \ll 1$.

(b) For the conditions of Fig. 10.14 (but for $ka \ll 1$) compute the diffraction intensity in the plane of incidence and compare the relative values with the solid curve in Fig. 10.14. (Use a protractor and a ruler to read off the values from Fig. 10.14 at several angles.)

(c) Show that the transmission coefficient [defined above (10.116)] for the two states of polarization are

$$T_{\parallel} = \frac{64}{27\pi^2} (ka)^4 \left(\frac{4 + \sin^2\alpha}{4 \cos\alpha} \right)$$

$$T_{\perp} = \frac{64}{27\pi^2} (ka)^4 \cos\alpha$$

Note that these transmission coefficients are a factor $(ka)^2$ smaller than those given by the vector Smythe–Kirchhoff approximation in the same limit.

10.20 A suspension of transparent fibers in a clear liquid is modeled as a collection of scatterers, each being a right circular cylinder of radius a and length L of uniform dielectric material whose electric susceptibility differs from the surrounding medium by a small fractional amount $\delta\epsilon/\epsilon$.

(a) Show that to first order in $\delta\epsilon/\epsilon$ the scattering cross section per scatterer for unpolarized radiation of wave number k is

$$\frac{d\sigma}{d\Omega} = \left| \frac{\delta\epsilon}{\epsilon} \right|^2 \frac{k^4 a^4 L^2}{32} (1 + \cos^2\theta) \left| \frac{2J_1(q_\perp a)}{q_\perp a} \cdot \frac{\sin(q_\parallel L/2)}{q_\parallel L/2} \right|^2$$

where $J_1(z)$ is the Bessel function of order unity and $q_\parallel$ ($q_\perp$) is the component of the wave number transfer parallel (perpendicular) to the cylinder axis.

(b) In the limit of very slender cylinders ($ka \ll 1$), show that the scattering cross section, averaged over the orientation of the cylinder (appropriate for an ensemble of randomly oriented fibers), is

$$\left\langle \frac{d\sigma}{d\Omega} \right\rangle = \left| \frac{\delta\epsilon}{\epsilon} \right|^2 \frac{k^4 a^4 L^2}{32} (1 + \cos^2\theta) \left[\frac{2}{qL} \text{Si}(qL) - \left(\frac{\sin(qL/2)}{qL/2} \right)^2 \right]$$

where $\text{Si}(x) = \int_0^x [(\sin x)/x]\, dx$ is the sine integral (*Abramowitz and Stegun*, p. 231) and $q^2 = 2k^2(1 - \cos\theta)$.

(c) Plot the square-bracketed quantity in part b as a function of $q^2 L^2$ on the range $(0, 100)$. Verify that the cross section is the expected one when $kL \ll 1$ and show that when $kL \gg 1$ (but $ka \ll 1$) the total scattering cross section is approximately

$$\sigma_{\text{scatt}} \approx \frac{11\pi^2}{60} \left| \frac{\delta\epsilon}{\epsilon} \right|^2 k^3 a^4 L \left\{ 1 + O\left[\frac{1}{kL}, \frac{\ln(kL)}{kL} \right] \right\}$$

Comment on the frequency dependence.

CHAPTER 11

Special Theory of Relativity

> Beginning with Chapter 11 we employ Gaussian units instead of SI units for electromagnetic quantities. Explicit factors of c appear in a natural manner in these units, making them more appropriate than SI units for relativistic phenomena. The issue of "rationalization" (suppression of explicit factors of 4π in the Maxwell equations) is another matter. Some workers, especially quantum field theorists, prefer Heaviside–Lorentz units—see the Appendix.

The special theory of relativity has, since its publication by Einstein in 1905, become a commonplace in physics, as taken for granted as Newton's laws of classical mechanics, the Maxwell equations of electromagnetism, or the Schrödinger equation of quantum mechanics. Daily it is employed by scientists in their consideration of precise atomic phenomena, in nuclear physics, and above all in high-energy physics.

The origins of the special theory of relativity lie in electromagnetism. In fact, one can say that the development of the Maxwell equations with the unification of electricity and magnetism and optics forced special relativity on us. Lorentz above all laid the groundwork with his studies of electrodynamics from 1890 onwards. Poincaré made important contributions, but it fell to Einstein to make the crucial generalization to all physical phenomena, not just electrodynamics, and to stress the far-reaching consequences of the second postulate. The special theory of relativity is now believed to apply to all forms of interaction except large-scale gravitational phenomena. It serves as a touchstone in modern physics for the possible forms of interaction between fundamental particles. Only theories consistent with special relativity need to be considered. This often severely limits the possibilities.

The experimental basis and the historical development of the special theory of relativity, as well as many of its elementary consequences, are discussed in many places. A list of books and articles is given at the end of the chapter. We content ourselves with a summary of the key points and some examples of recent definitive experimental confirmations. Then the basic kinematic results are summarized, including coordinate transformations, proper time and time dilatation, the relativistic Doppler shift, and the addition of velocities. The relativistic energy and momentum of a particle are derived from general principles, independent of the force equation. Then the idea of the Lorentz group and its mathematical description is presented and a specific representation in terms of 4×4 matrices

514

is given. The important phenomenon of Thomas precession is then discussed. The experimental basis for the invariance of electric charge, the covariance of electrodynamics, and the explicit transformation properties of electric and magnetic fields follow. The chapter concludes with a treatment of the relativistic equations of motion for spin and a remark on the notation and conventions of relativistic kinematics.

11.1 The Situation Before 1900, Einstein's Two Postulates

In the 40 years before 1900 electromagnetism and optics were correlated and explained in triumphal fashion by the wave theory based on the Maxwell equations. Since previous experience with wave motion had always involved a medium for the propagation of waves, it was natural for physicists to assume that light needed a medium through which to propagate. In view of the known facts about light, it was necessary to assume that this medium, called the ether, permeated all space, was of negligible density, and had negligible interaction with matter. It existed solely as a vehicle for the propagation of electromagnetic waves.

The hypothesis of an ether set electromagnetic phenomena apart from the rest of physics. For a long time it had been known that the laws of mechanics were the same in different coordinate systems moving uniformly relative to one another. We say that the laws of mechanics are invariant under Galilean transformations. To emphasize the distinction between classical mechanics and electromagnetism let us consider explicitly the question of Galilean relativity for each. For two reference frames K and K' with coordinates (x, y, z, t) and (x', y', z', t'), respectively, and moving with relative velocity $\mathbf{v}$, the space and time coordinates in the two frames are related according to Galilean relativity by

$$\mathbf{x}' = \mathbf{x} - \mathbf{v}t$$
$$t' = t \tag{11.1}$$

provided the origins in space and time are chosen suitably. As an example of a mechanical system, consider a group of particles interacting via two-body central potentials. In an obvious notation the equation of motion of the ith particle in the reference frame K' is

$$m_i \frac{d\mathbf{v}_i'}{dt'} = -\nabla_i' \sum_j V_{ij}(|\mathbf{x}_i' - \mathbf{x}_j'|) \tag{11.2}$$

From the connections (11.1) between the coordinates in K and K' it is evident that $\mathbf{v}_i' = \mathbf{v}_i - \mathbf{v}$, $\nabla_i' = \nabla_i$, $d\mathbf{v}_i'/dt' = d\mathbf{v}_i/dt$, and $\mathbf{x}_i' - \mathbf{x}_j' = \mathbf{x}_i - \mathbf{x}_j$. Thus (11.2) can be transformed into

$$m_i \frac{d\mathbf{v}_i}{dt} = -\nabla_i \sum_j V_{ij}(|\mathbf{x}_i - \mathbf{x}_j|) \tag{11.3}$$

namely Newton's equation of motion in the reference frame K.

The preservation of the *form* of the equations of classical mechanics under the transformation (11.1) is in contrast to the change in form of the equations governing wave phenomena. Suppose that a field $\psi(\mathbf{x}', t')$ satisfies the wave equation

$$\left(\sum_i \frac{\partial^2}{\partial x_i'^2} - \frac{1}{c^2} \frac{\partial^2}{\partial t'^2} \right) \psi = 0 \qquad (11.4)$$

in the reference frame K'. By straightforward use of (11.1) it is found that in terms of the coordinates in the reference frame K the wave equation (11.4) becomes

$$\left(\nabla^2 - \frac{1}{c^2} \frac{\partial}{\partial t^2} - \frac{2}{c^2} \mathbf{v} \cdot \nabla \frac{\partial}{\partial t} - \frac{1}{c^2} \mathbf{v} \cdot \nabla \, \mathbf{v} \cdot \nabla \right) \psi = 0 \qquad (11.5)$$

The form of the wave equation is not invariant under Galilean transformations. Furthermore, no kinematic transformation of ψ can restore to (11.5) the appearance of (11.4).* For sound waves the lack of invariance under Galilean transformations is quite acceptable. The wind throws our voices. Sound waves are compressions and rarefactions in the air or in other materials, and the preferred reference frame K' in which (11.4) is valid is obviously the frame in which the transmitting medium is at rest.

So it also appeared for electromagnetism. The vital difference is this. Sound waves and similar wave phenomena are consequences of Galilean classical mechanics. The existence of preferred reference frames where the phenomena are simple is well understood in terms of the bulk motions of the media of propagation. For electromagnetic disturbances, on the other hand, the medium seemed truly ethereal with no manifestation or purpose other than to support the propagation.

When Einstein began to think about these matters there existed several possibilities:

1.   The Maxwell equations were incorrect. The proper theory of electromagnetism was invariant under Galilean transformations.

2.   Galilean relativity applied to classical mechanics, but electromagnetism had a preferred reference frame, the frame in which the luminiferous ether was at rest.

3.   There existed a relativity principle for both classical mechanics and electromagnetism, but it was not Galilean relativity. This would imply that the laws of mechanics were in need of modification.

The first possibility was hardly viable. The amazing successes of the Maxwell theory at the hands of Hertz, Lorentz, and others made it doubtful that the

---

*The reader might wish to ponder the differences between the wave equation and the Schrödinger equation under Galilean transformations. If in K' the Schrödinger equation reads

$$-\frac{\hbar^2}{2m} \nabla'^2 \psi' + V \psi' = i\hbar \frac{\partial \psi'}{\partial t'}$$

then in K the equation has the same form for the wave function ψ provided V is a Galilean invariant and $\psi = \psi' \exp[i(m/\hbar)\mathbf{v} \cdot \mathbf{x} - i(mv^2/2\hbar)t]$. The Schrödinger equation *is* invariant under Galilean transformations.

equations of electromagnetism were in serious error. The second alternative was accepted by most physicists of the time. Efforts to observe motion of the earth and its laboratories relative to the rest frame of the ether, for example, the Michelson–Morley experiment, had failed. But for this important experiment at least, the null result could be explained by the FitzGerald–Lorentz contraction hypothesis (1892) whereby objects moving at a velocity **v** through the ether are contracted in the direction of motion according to the formula

$$L(v) = L_0 \sqrt{1 - \frac{v^2}{c^2}} \tag{11.6}$$

This rather unusual hypothesis apparently lies outside electromagnetism, since it applies to bulk matter, but Lorentz later argued that it was rooted in electrodynamics. He and Poincaré showed that the Maxwell equations are invariant in form under what are known as Lorentz transformations (see Section 11.9) and that the contraction (11.6) held for moving charge densities, etc., in electrodynamics. With the idea that matter is electromagnetic in nature (the discovery of the electron encouraged this hypothesis), it is plausible to assume that (11.6) holds for macroscopic aggregates of electrons and atoms. Lorentz thus saved the ether hypothesis from contradiction with the Michelson–Morley experiment.

Other experiments caused embarrassment to the ether idea. Fizeau's famous experiments (1851, 1853) and later similar experiments by Michelson and Morley (1886) on the velocity of light in moving fluids could be understood only if one supposed that the ether was dragged along partially by the moving fluid, with the effectiveness of the medium in dragging the ether related to its index of refraction!

Apparently it was the implausibility of the explanation of the Fizeau observations, more than anything else, that convinced Einstein of the unacceptability of the hypothesis of an ether. He chose the third alternative above and sought principles of relativity that would encompass classical mechanics, electrodynamics, and indeed all natural phenomena. Einstein's special theory of relativity is based on two postulates:

1. POSTULATE OF RELATIVITY

The laws of nature and the results of all experiments performed in a given frame of reference are independent of the translational motion of the system as a whole. More precisely, there exists a triply infinite set of equivalent Euclidean reference frames moving with constant velocities in rectilinear paths relative to one another in which all physical phenomena occur in an identical manner.

For brevity these equivalent coordinate systems are called *inertial* reference frames. The postulate of relativity, phrased here more or less as by Poincaré, is consistent with all our experience in mechanics where only relative motion between bodies is relevant, and has been an explicit hypothesis in mechanics since the days of Copernicus, if not before. It is also consistent with the Michelson–Morely experiment and makes meaningless the question of detecting motion relative to the ether.

2. POSTULATE OF THE CONSTANCY OF THE SPEED OF LIGHT

The speed of light is finite and independent of the motion of its source.

This postulate, untested when Einstein proposed it (and verified decisively only in recent years—see Section 11.2.B), is simplicity itself. Yet it forces on us such a radical rethinking of our ideas about space and time that it was resisted for many years.

Because special relativity applies to everything, not just light, it is desirable to express the second postulate in terms that convey its generality:

2′. POSTULATE OF A UNIVERSAL LIMITING SPEED

In every inertial frame, there is a finite universal limiting speed C for physical entities.

Experimentally, the limiting speed C is equal to the speed c of light in vacuum. Postulate 2′ (with the first postulate) can be used equally to derive the Lorentz transformation of coordinates (see Problem 11.1). Our own derivation in Section 11.3 is the traditional one, based on Postulates 1 and 2, but, as Mermin has emphasized,* the general structure of the Lorentz transformation can be deduced from the first postulate alone, plus some obvious assumptions, without reference to the speed of light, except as the empirical parameter that distinguishes the transformation from the Galilean (see Problem 11.2).

The history of the special theory of relativity and its gradual establishment through experiments is dealt with in an extensive literature. Some references are given at the end of the chapter. Of particular note is the "Resource letter on relativity" published in the *American Journal of Physics* [**30**, 462 (1962)]. This article contains references to books and journal articles on the history, experimental verification, and laboratory demonstrations on all aspects of special relativity.

In passing we remark that Einstein's postulates require modification of the laws of mechanics for high-speed motions. There was no evidence at the time indicating a failure of Galilean relativity for mechanics. This is basically because relativistic particles and their dynamics were unknown until the discovery of beta rays around 1900. Poincaré had speculated that the speed of light might be a limiting speed for material particles, but Einstein's special theory of relativity originated from his desire to treat all physical phenomena in the same way rather than from any need to "fix up" classical mechanics. The consequences of the special theory for mechanical concepts like momentum and energy are discussed in Section 11.5.

11.2 Some Recent Experiments

Although we omit discussion of the standard material, appealing to the reader's prior knowledge and the existence of many books on the special theory of relativity, there are two experiments worthy of note. One concerns the first postulate, namely the search for an "ether drift" (evidence of motion of the laboratory relative to the ether) and the other the second postulate.

*N. D. Mermin, Relativity without light, *Am. J. Phys.* **52**, 119–124 (1984).

A. Ether Drift

The null result of the Michelson–Morley experiment (1887) established that the velocity of the earth through the presumed ether was less than one-third of its orbital speed of approximately 3×10^4 m/s. The experiment was repeated many times with various modifications, always with no firm evidence of motion relative to the ether. A summary of all available evidence is given by Shankland et al. [*Rev. Mod. Phys.* **27**, 167 (1955)].

As already noted, these null results can be explained without abandoning the concept of an ether by the hypothesis of the FitzGerald–Lorentz contraction. The discovery by Mössbauer (1958) of "recoilless" emission or absorption of gamma rays (called the Mössbauer effect) allows comparison of frequencies to astounding precision and gives the possibility of very accurate ether drift experiments based on the Doppler shift. In the Mössbauer effect the recoil momentum from the emission or absorption of a gamma ray is taken up by the whole solid rather than by the emitting or absorbing nucleus. This means that the *energy* of recoil is totally negligible. A gamma ray is emitted with the full energy E_0 of the nuclear transition, not the reduced energy $E \simeq E_0 - E_0^2/2Mc^2$, where M is the mass of the recoiling nucleus, resulting from the recoil. Furthermore, with such recoilless transitions there are no thermal Doppler shifts. The gamma-ray line thus approaches its natural shape with no broadening or shift in frequency. By employing an absorber containing the same material as the emitter, one can study nuclear resonance absorption or use it as an instrument for the study of extremely small changes of frequency.

To understand the principle of an ether drift experiment based on the Mössbauer effect, we need to recall the classic results of the Doppler shift. The phase of a plane wave is an invariant quantity, the same in all coordinate frames. This is because the elapsed phase of a wave is proportional to the number of wave crests that have passed the observer. Since this is merely a counting operation, it must be independent of coordinate frame. If there is a plane electromagnetic wave in vacuum its phase as observed in the inertial frames K and K', connected by the Galilean coordinate transformation (11.1), is

$$\phi = \omega\left(t - \frac{\mathbf{n} \cdot \mathbf{x}}{c}\right) = \omega'\left(t' - \frac{\mathbf{n}' \cdot \mathbf{x}'}{c'}\right) \tag{11.7}$$

If t and $\mathbf{x}$ are expressed in terms of t' and $\mathbf{x}'$ from (11.1), we obtain

$$\omega\left[t'\left(1 - \frac{\mathbf{n} \cdot \mathbf{v}}{c}\right) - \frac{\mathbf{n} \cdot \mathbf{x}'}{c}\right] = \omega'\left(t' - \frac{\mathbf{n}' \cdot \mathbf{x}'}{c}\right)$$

Since this equality must hold for all t' and $\mathbf{x}'$, it is necessary that the coefficients of t', x_1', x_2', x_3' on both sides be separately equal. We therefore find

$$\left.\begin{aligned} \mathbf{n} &= \mathbf{n}' \\ \omega' &= \omega\left(1 - \frac{\mathbf{n} \cdot \mathbf{v}}{c}\right) \\ c' &= c - \mathbf{n} \cdot \mathbf{v} \end{aligned}\right\} \tag{11.8}$$

These are the standard Doppler shift formulas of Galilean relativity.

The unit wave normal $\mathbf{n}$ is seen from (11.8) to be an invariant, the same in all inertial frames. The direction of energy flow changes, however, from frame to frame. To see this, consider the segments of a plane wave sketched in Fig. 11.1. The segments can be thought of as schematic representations of wave packets. At $t = t' = 0$ the center of the segment is at the point A in both K and K'. If inertial frame K is the preferred reference frame (ether at rest) the wave packet moves in the direction $\mathbf{n}$, arriving after one unit of time at the point B in frame K. The distance AB is equal to c. In frame K' the center of the wave packet arrives at the point B' after one unit of time. Because of the Galilean transformation of coordinates (11.1) the point B' differs from B by a vectorial amount $-\mathbf{v}$, as indicated in the bottom half of Fig. 11.1. The direction of motion of the wave packet, assumed to be the direction of energy flow, is thus not parallel to $\mathbf{n}$ in K', but along a unit vector $\mathbf{m}$ shown in Fig. 11.1 and specified by

$$\mathbf{m} = \frac{c\mathbf{n} - \mathbf{v}}{|c\mathbf{n} - \mathbf{v}|} \tag{11.9}$$

Since the experiments involve photon propagation in the laboratory, it is convenient to have the Doppler formulas (11.8) expressed in terms of the $\mathbf{m}$ appropriate to the laboratory rather than $\mathbf{n}$. It is sufficient to have $\mathbf{n}$ in terms of $\mathbf{m}$ correct to first order in v/c. From (11.9) we find

$$\mathbf{n} \simeq \left(1 - \frac{\mathbf{m} \cdot \mathbf{v}_0}{c}\right)\mathbf{m} + \frac{\mathbf{v}_0}{c} \tag{11.10}$$

where $\mathbf{v}_0$ is the velocity of the laboratory relative to the ether rest frame.

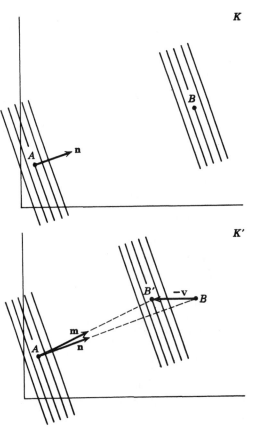

Figure 11.1

Consider now a plane wave whose frequency is ω in the ether rest frame, ω_0 in the laboratory, and ω_1 in an inertial frame K_1 moving with a velocity $\mathbf{v}_1 = \mathbf{u}_1 + \mathbf{v}_0$ relative to the ether rest frame. From (11.8) the observed frequencies are

$$\omega_1 = \omega\left(1 - \frac{\mathbf{n} \cdot \mathbf{v}_1}{c}\right)$$

$$\omega_0 = \omega\left(1 - \frac{\mathbf{n} \cdot \mathbf{v}_0}{c}\right)$$

If ω_1 is expressed in terms of the laboratory frequency ω_0 and the wave normal $\mathbf{n}$ is eliminated by means of (11.10), the result, correct to order v^2/c^2, is easily shown to be

$$\omega_1 \simeq \omega_0\left[1 - \frac{\mathbf{u}_1}{c} \cdot \left(\mathbf{m} + \frac{\mathbf{v}_0}{c}\right)\right] \qquad (11.11)$$

where $\mathbf{u}_1$ is the velocity of the frame K_1 relative to the laboratory, $\mathbf{m}$ is the direction of energy propagation in the laboratory, ω_0 is the frequency of the wave in the laboratory, and $\mathbf{v}_0$ is the velocity of the laboratory with respect to the ether.

Equation (11.11) forms the basis of the analysis of the Mössbauer ether drift experiments. It is a consequence of the validity of both the wave equation in the ether rest frame and Galilean relativity to transform to other inertial frames. Since it involves $\mathbf{v}_0$, it obviously predicts an ether drift effect. Consider two Mössbauer systems, one an emitter and the other an absorber, moving with velocities $\mathbf{u}_1$ and $\mathbf{u}_2$ in the laboratory. From (11.11) the difference in frequency between emitter and absorber is

$$\frac{\omega_1 - \omega_2}{\omega_0} = \frac{1}{c}(\mathbf{u}_2 - \mathbf{u}_1) \cdot \left(\mathbf{m} + \frac{\mathbf{v}_0}{c}\right)$$

If the emitter and absorber are located on the opposite ends of a rod of length $2R$ that is rotated about its center with angular velocity Ω, as indicated in Fig. 11.2, then $(\mathbf{u}_2 - \mathbf{u}_1) \cdot \mathbf{m} = 0$ and the fractional frequency difference is

$$\frac{\omega_1 - \omega_2}{\omega_0} = \frac{2\Omega R}{c^2} \sin \Omega t \, |(\mathbf{v}_0)_\perp| \qquad (11.12)$$

where $(\mathbf{v}_0)_\perp$ is the component of $\mathbf{v}_0$ perpendicular to the axis of rotation.

A resonant absorption experiment of this type was performed in 1963 in Birmingham.* The Mössbauer line was the 14.4 keV gamma ray in ^{57}Fe, following the β^+ decay of ^{57}Co. The isotope ^{57}Fe is stable and occurs with a natural abundance of 2.2%; the absorber was made with iron enriched to 52% in ^{57}Fe. The cobalt source was emplanted in ^{56}Fe. The emitter and absorber foils were located as in Fig. 11.2 with $R \simeq 4$ cm. The observed fractional width of the Mössbauer line was $\Delta\omega/\omega \simeq 2 \times 10^{-12}$. Counters fixed in the laboratory and located symmetrically along a diameter of the circle in the plane of the source and absorber recorded the gamma rays transmitted through the absorber. Two rotational speeds, $\Omega_L = 1257$ s^{-1} and $\Omega_H = 7728$ s^{-1}, were alternated during each 4-hour cycle that data were taken and a diurnal effect connected to the earth's rotation

---

*D. C. Champeney, G. R. Isaak, and A. M. Khan, *Phys. Lett.* **7**, 241 (1963). See also G. R. Isaak, *Phys. Bull.* **21,** 255 (1970).

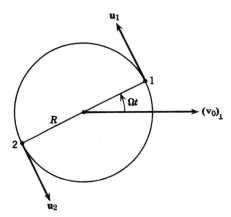

Figure 11.2

was sought. From (11.12) it can be seen that with $\Omega \sim 6000 \text{ s}^{-1}$ and $R = 4$ cm, an ether drift velocity of 200 m/s would produce a total change of frequency of the magnitude of the Mössbauer line width. The data showed no diurnal change in transmission to an accuracy of 1 or 2%. The authors conclude that the magnitude of the component of $\mathbf{v}_0$ past the earth in a plane perpendicular to the earth's axis of rotation is $|(\mathbf{v}_0)_\perp| = 1.6 \pm 2.8$ m/s, a null result. An improved experiment along the same lines in 1970 gave a limit of 5 cm/s (see *Isaak, op. cit.*).

A conceptually similar experiment was performed in 1958 using ammonia masers.[*] The ammonia molecules have a well-defined direction and nonzero speeds when they enter the maser cavity. According to (11.11) there is therefore a shift in the frequency. If the frequencies of two masers whose ammonia molecules travel in opposite directions are compared, there should be an observable beat frequency. Furthermore, if the two masers are rotated together through 180°, the beat frequency should change by $\Delta\omega/\omega_0 = 4\,|\mathbf{u}_{\text{mol}} \cdot \mathbf{v}_0|/c^2$. The null result of this experiment set the component of ether drift velocity at less than 30 m/s.

These two Doppler shift experiments set observable ether drift speed limits 6000 and 1000 times smaller than the speed of the earth in its orbit and make the idea that we can ever detect any motion relative to some "absolute" reference frame quite implausible.

B. Speed of Light from a Moving Source

The second postulate of Einstein, that the speed of light is independent of the motion of the source, destroys the concept of time as a universal variable independent of the spatial coordinates. Because this was a revolutionary and unpalatable idea, many attempts were made to invent theories that would explain all the observed facts without this assumption. The most notable and resilient scheme was Ritz's version of electrodynamics (1908–1911). Ritz kept the two homogeneous Maxwell equations intact, but modified the equations involving the sources in such a way that the speed of light was equal to c only when measured relative to the source. The Ritz theory is in accord with observation for the aberration of star positions, the Fizeau experiments, and the original

---

[*]C. J. Cedarholm, G. F. Bland, B. L. Havens, and C. H. Townes, *Phys. Rev. Lett.* **1**, 342 (1958). See also T. S. Jaseja, A. Javan, J. Murray, and C. H. Townes, *Phys. Rev.* **133**, A1221 (1964).

Michelson–Morley experiment. It is customary, however, to cite Michelson–Morley experiments performed with extraterrestrial light sources (the sun or other stars) and light from binary stars as establishing the second postulate and ruling out Ritz's theory. Unfortunately, it seems clear that most of the early evidence is invalid because of the interaction of the radiation with the matter through which it passes before detection.*

There are, however, some more recent experiments that do not suffer from the criticism of Fox. The most definitive is a beautiful experiment performed at CERN, Geneva, Switzerland, in 1964.† The speed of 6 GeV photons produced in the decay of very energetic neutral pions was measured by time of flight over paths up to 80 meters. The pions were produced by bombardment of a beryllium target by 19.2 GeV protons and had speeds (inferred from measured speeds of charged pions produced in the same bombardment) of $0.99975c$. The timing was done by utilizing the rf structure of the beam. Within experimental error it was found that the speed of the photons emitted by the extremely rapidly moving source was equal to c. If the observed speed is written as $c' = c + kv$, where v is the speed of the source, the experiment showed $k = (0 \pm 1.3) \times 10^{-4}$.

The CERN experiment established conclusively and on a laboratory scale the validity of the second postulate (2) of the special theory of relativity. Other experiments‡ on charged particles and neutrinos independently establish the validity of postulate 2'. See also Section 11.5.

C. Frequency Dependence of the Speed of Light in Vacuum

The speed of light is known to an accuracy of a few parts in 10^9 from measurements at infrared frequencies and lesser accuracy at higher frequencies (or equivalently, the meter is defined to this precision). One can ask whether there is any evidence for a frequency dependence of the speed of electromagnetic waves in vacuum. One possible source of variation is attributable to a photon mass. The group velocity in this case is

$$c(\omega) = c\left(1 - \frac{\omega_0^2}{\omega^2}\right)^{1/2} \tag{11.13}$$

where the photon rest energy is $\hbar\omega_0$. As discussed in the Introduction, the mere existence of normal modes in the earth-ionosphere resonant cavity sets a limit of $\omega_0 < 10c/R$ where R is the radius of the earth. From radiofrequencies ($\omega \sim 10^8 \text{ s}^{-1}$) to $\omega \to \infty$, the change in velocity of propagation from a photon mass is therefore less than $\Delta c/c \simeq 10^{-10}$.

Another source of frequency variation in the speed of light is dispersion of the vacuum, a concept lying outside special relativity but occurring in models with a discrete space-time. The discovery of pulsars make it possible to test this

*See the papers of criticism by J. G. Fox, *Am. J. Phys.* **30**, 297 (1962), **33**, 1 (1965); *J. Opt. Soc.* **57**, 967 (1967). The second paper cited is a detailed discussion of Ritz's emission theory and a critique of the various arguments against it. See also T. Alvager, A. Nilsson, and J. Kjellman, *Ark. Fys.* **26**, 209 (1963).

†T. Alvager, J. M. Bailey, F. J. M. Farley, J. Kjellman, and I. Wallin, *Phys. Lett.* **12**, 260 (1964); *Ark. Fys.* **31**, 145 (1965).

‡G. R. Kalbfleisch, N. Baggett, E. C. Fowler, and J. Alspector, *Phys. Rev. Lett.* **43**, 1361 (1979).

idea with high precision. Pulsar observations cover at least 13 decades of frequency, with any one observing apparatus having a certain "window" in the frequency spectrum. The quite small time duration of the pulse from some pulsars permits a simple estimate for the upper limit of variation on the speed of light for two frequencies ω_1 and ω_2 inside the frequency window of each apparatus:

$$\left| \frac{c(\omega_1) - c(\omega_2)}{c} \right| \leq \frac{c\,\Delta t}{D}$$

where Δt is the pulse duration and D is the distance from the source to observer. For the Crab pulsar Np 0532, $\Delta t \simeq 3 \times 10^{-3}$ s and $D \simeq 6 \times 10^3$ light-years so that $(c\,\Delta t/D) \simeq 1.7 \times 10^{-14}$. Various overlapping observations from $\sim 4 \times 10^8$ Hz through the optical region and up to photon energies of 1 MeV indicate constancy of the speed at the level of $\Delta c/c < 10^{-14}$ by this simple estimation.[*] For higher energies, an experiment at the Stanford Linear Accelerator[†] compared the speed of 7 GeV photons with that of visible light and found $\Delta c/c < 10^{-5}$. Up to very high energies, then, there is no evidence for dispersion of the vacuum. The speed of light is a universal constant, independent of frequency.

11.3 Lorentz Transformations and Basic Kinematic Results of Special Relativity

As is well known, the constancy of the velocity of light, independent of the motion of the source, gives rise to the relations between space and time coordinates in different inertial reference frames known as Lorentz transformations. We derive these results in a more formal manner in Section 11.7, but for the present summarize the elementary derivation and important consequences, omitting the details that can be found in the many textbooks on relativity. The reader who wishes more than a reminder can consult the books listed at the end of the chapter.

A. Simple Lorentz Transformation of Coordinates

Consider two inertial reference frames K and K' with a relative velocity $\mathbf{v}$ between them. The time and space coordinates of a point are (t, x, y, z) and (t', x', y', z') in the frames K and K', respectively. The coordinate axes in the two frames are parallel and oriented so that the frame K' is moving in the positive z direction with speed v, as viewed from K. For simplicity, let the origins of the coordinates in K and K' be coincident at $t = t' = 0$. If a light source at rest at the origin in K (and so moving with a speed v in the negative z direction, as seen from K') is flashed on and off rapidly at $t = t' = 0$, Einstein's second postulate implies that observers in *both* K and K' will see a spherical shell of radiation expanding outward from the respective origins with speed c. The wave front reaches a point (x, y, z) in the frame K at a time t given by the equation,

$$c^2 t^2 - (x^2 + y^2 + z^2) = 0 \tag{11.14}$$

---

[*]J. M. Rawls, *Phys. Rev.* **D5**, 487 (1972).

[†]B. C. Brown et al., *Phys. Rev. Lett.* **30**, 763 (1973).

Similarly, in the frame K' the wave front is specified by

$$c^2 t'^2 - (x'^2 + y'^2 + z'^2) = 0 \qquad (11.14')$$

With the assumption that space-time is homogeneous and isotropic, as implied by the first postulate, the connection between the two sets of coordinates is linear. The quadratic forms (11.14) and (11.14') are then related by

$$c^2 t'^2 - (x'^2 + y'^2 + z'^2) = \lambda^2 [c^2 t^2 - (x^2 + y^2 + z^2)] \qquad (11.15)$$

where $\lambda = \lambda(\mathbf{v})$ is a possible change of scale between frames. With the choice of orientation of axes and considerations of the inverse transformation from K' to K it is straightforward to show that $\lambda(v) = 1$ for all v and that the time and space coordinates in K' are related to those in K by the *Lorentz transformation*

$$\left. \begin{array}{l} x_0' = \gamma(x_0 - \beta x_1) \\ x_1' = \gamma(x_1 - \beta x_0) \\ x_2' = x_2 \\ x_3' = x_3 \end{array} \right\} \qquad (11.16)$$

where we have introduced the suggestive notation $x_0 = ct$, $x_1 = z$, $x_2 = x$, $x_3 = y$ and also the convenient symbols,

$$\boldsymbol{\beta} = \frac{\mathbf{v}}{c}, \qquad \beta = |\boldsymbol{\beta}|$$
$$\gamma = (1 - \beta^2)^{-1/2} \qquad (11.17)$$

The inverse Lorentz transformation is

$$\left. \begin{array}{l} x_0 = \gamma(x_0' + \beta x_1') \\ x_1 = \gamma(x_1' + \beta x_0') \\ x_2 = x_2' \\ x_3 = x_3' \end{array} \right\} \qquad (11.18)$$

It can be found from (11.16) by direct calculation, but we know from the first postulate that it must result from (11.16) by interchange of primed and unprimed variables along with a change in the sign of β. According to (11.16) or (11.18), the coordinates perpendicular to the direction of relative motion are unchanged while the parallel coordinate *and the time* are transformed. This can be contrasted with the Galilean transformation (11.1).

Equations (11.16) and (11.17) describe the special circumstance of a Lorentz transformation from one frame to another moving with velocity $\mathbf{v}$ parallel to the x_1 axis. If the axes in K and K' remain parallel, but the velocity $\mathbf{v}$ of the frame K' in frame K is in an arbitrary direction, the generalization of (11.16) is

$$\left. \begin{array}{l} x_0' = \gamma(x_0 - \boldsymbol{\beta} \cdot \mathbf{x}) \\ \mathbf{x}' = \mathbf{x} + \dfrac{(\gamma - 1)}{\beta^2} (\boldsymbol{\beta} \cdot \mathbf{x})\boldsymbol{\beta} - \gamma \boldsymbol{\beta} x_0 \end{array} \right\} \qquad (11.19)$$

The first equation here follows almost trivially from the first equation in (11.16). The second appears somewhat complicated, but is really only the sorting out of components of $\mathbf{x}$ and $\mathbf{x}'$ parallel and perpendicular to $\mathbf{v}$ for separate treatment in accord with (11.16).

The connection between β and γ given in (11.17) and the ranges $0 \le \beta \le 1$, $1 \le \gamma \le \infty$ allow the alternative parametrization,

and so

$$\left.\begin{array}{l} \beta = \tanh \zeta \\[4pt] \gamma = \cosh \zeta \\[4pt] \gamma\beta = \sinh \zeta \end{array}\right\} \tag{11.20}$$

where ζ is known as the *boost parameter* or *rapidity*. In terms of ζ the first two equations of (11.16) become

$$x_0' = x_0 \cosh \zeta - x_1 \sinh \zeta \tag{11.21}$$
$$x_1' = -x_0 \sinh \zeta + x_1 \cosh \zeta$$

The structure of these equations is reminiscent of a rotation of coordinates, but with hyperbolic functions instead of circular, basically because of the relative negative sign between the space and time terms in (11.14) [see Section 11.7 and (11.95)].

B. 4-Vectors

The Lorentz transformation (11.16), or more generally (11.19), describes the transformation of the coordinates of a point from one inertial frame to another. Just as for rotations in three dimensions, the basic transformation law is defined in terms of the coordinates of a point. In three dimensions we call **x** a vector and speak of x_1, x_2, x_3 as the components of a vector. We designate by the same name any three physical quantities that transform under rotations in the same way as the components of **x**. It is natural therefore to anticipate that there are numerous physical quantities that transform under Lorentz transformations in the same manner as the time and space coordinates of a point. By analogy we speak of *4-vectors*. The coordinate 4-vector is (x_0, x_1, x_2, x_3); we designate the components of an arbitrary 4-vector similarly as (A_0, A_1, A_2, A_3),* where A_1, A_2, A_3 are the components of a 3-vector **A**. The Lorentz transformation law equivalent to (11.16) for an arbitrary 4-vector is

$$\left.\begin{array}{l} A_0' = \gamma(A_0 - \boldsymbol{\beta} \cdot \mathbf{A}) \\[4pt] A_\parallel' = \gamma(A_\parallel - \beta A_0) \\[4pt] \mathbf{A}_\perp' = \mathbf{A}_\perp \end{array}\right\} \tag{11.22}$$

where the parallel and perpendicular signs indicate components relative to the velocity $\mathbf{v} = c\boldsymbol{\beta}$. The invariance from one inertial frame to another embodied through the second postulate in (11.15) has its counterpart for any 4-vector in the invariance,

$$A_0'^2 - |\mathbf{A}'|^2 = A_0^2 - |\mathbf{A}|^2 \tag{11.23}$$

---

*Because we are deferring the explicit algebraic treatment of the Lorentz group to Section 11.7, we do not write a single symbol for this 4-vector. As written, they are the components of the contravariant 4-vector A^α.

where the components $(A_0', \mathbf{A}')$ and $(A_0, \mathbf{A})$ refer to any two inertial reference frames. For two 4-vectors (A_0, A_1, A_2, A_3) and (B_0, B_1, B_2, B_3) the "scalar product" is an invariant, that is,

$$A_0'B_0' - \mathbf{A}' \cdot \mathbf{B}' = A_0 B_0 - \mathbf{A} \cdot \mathbf{B} \tag{11.24}$$

This result can be verified by explicit construction of the left-hand side, using (11.22) for the primed components, or using (11.23) for the sum of two 4-vectors. It is the Lorentz transformation analog of the invariance of $\mathbf{A} \cdot \mathbf{B}$ under rotations in three dimensions.

C. Light Cone, Proper Time, and Time Dilatation

A fruitful concept in special relativity is the idea of the light cone and "space-like" and "timelike" separations between two events. Consider Fig. 11.3, in which the time axis (actually ct) is vertical and the space axes are perpendicular to it. For simplicity only one space dimension is shown. At $t = 0$ a physical system, say a particle, is at the origin. Because the velocity of light is an upper bound on all velocities, the space-time domain can be divided into three regions by a "cone," called the light cone, whose surface is specified by $x^2 + y^2 + z^2 = c^2 t^2$. Light signals emitted at $t = 0$ from the origin would travel out the 45° lines in the figure. But any material system has a velocity less than c. Consequently as time goes on it would trace out a path, called its *world line*, inside the upper half-cone: for example, the curve OB. Since the path of the system lies inside the *upper half-cone* for times $t > 0$, that region is called the *future*. Similarly the *lower half-cone* is called the *past*. The system may have reached O by a path such as AO lying inside the lower half-cone. The shaded region outside the light cone is called *elsewhere*. A system at O can never reach or come from a point in space-time in elsewhere.

The division of space-time into the past-future region (inside the light cone) and elsewhere (outside the light cone) can be emphasized by considering the invariant separation or interval s_{12} between two events $P_1(t_1, \mathbf{x}_1)$ and $P_2(t_2, \mathbf{x}_2)$ in space-time (we are reverting to t and $\mathbf{x}$ temporarily to avoid proliferation of subscripts). The square of the invariant interval is

$$s_{12}^2 = c^2(t_1 - t_2)^2 - |\mathbf{x}_1 - \mathbf{x}_2|^2 \tag{11.25}$$

For any two events P_1 and P_2 there are three possibilities: (1) $s_{12}^2 > 0$, (2) $s_{12}^2 < 0$, (3) $s_{12}^2 = 0$. If $s_{12}^2 > 0$, the events are said to have a *timelike separation*. It is always

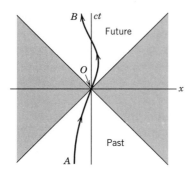

Figure 11.3 World line of a system and the light cone. The unshaded interior of the cone represents the past and the future, while the shaded region outside the cone is called "elsewhere." A point inside (outside) the light cone is said to have a timelike (spacelike) separation from the origin.

possible to find a Lorentz transformation* to a new coordinate frame K' such that $\mathbf{x}_1' = \mathbf{x}_2'$. Then

$$s_{12}^2 = c^2(t_1' - t_2')^2 > 0$$

In the frame K' the two events occur at the same space point, but are separated in time. Referring to Fig. 11.3, one point can be located at the origin and the other lies in the past or future. If $s_{12}^2 < 0$, the events are said to have a *spacelike separation*. Now it is possible to find an inertial frame K'' where $t_1'' = t_2''$. Then

$$s_{12}^2 = - |\mathbf{x}_1'' - \mathbf{x}_2''|^2 < 0$$

In K'' the two events occur at different space points at the same instant of time. In terms of Fig. 11.3, one event is at the origin, while the other lies in the else-where region. The final possibility, $s_{12}^2 = 0$, implies a *lightlike separation*. The events lie on the light cone with respect to each other and can be connected only by light signals.

The division of the separation of two events in space-time into two classes—spacelike separations or timelike separations with the light cone as the boundary surface between—is a Lorentz invariant one. Two events with a space-like separation in one coordinate system have a spacelike separation in all co-ordinate systems. This means that two such events cannot be causally connected. Since physical interactions propagate from one point to another with velocities no greater than that of light, only events with timelike separations can be causally related. An event at the origin in Fig. 11.3 can be influenced causally only by the events that occur in the past region of the light cone.

Another useful concept is *proper time*. Consider a system, which for defi-niteness we will think of as a particle, moving with an instantaneous velocity $\mathbf{u}(t)$ relative to some inertial system K. In a time interval dt its position changes by $d\mathbf{x} = \mathbf{u}\, dt$. From (11.25) the square of the corresponding infinitesimal invariant interval ds is

$$ds^2 = c^2\, dt^2 - |d\mathbf{x}|^2 = c^2\, dt^2(1 - \beta^2)$$

where here $\beta = u/c$. In the coordinate system K' where the system is instanta-neously at rest the space-time increments are $dt' \equiv d\tau, dx' = 0$. Thus the invariant interval is $ds = c\, d\tau$. The increment of time $d\tau$ in the instantaneous rest frame of the system is thus a *Lorentz invariant quantity* that takes the form,

$$d\tau = dt\sqrt{1 - \beta^2(t)} = \frac{dt}{\gamma(t)} \tag{11.26}$$

The time τ is called the *proper time of the particle or system*. It is the time as seen in the rest frame of the system. From (11.26) it follows that a certain proper time interval $\tau_2 - \tau_1$ will be seen in the frame K as a time interval,

$$t_2 - t_1 = \int_{\tau_1}^{\tau_2} \frac{d\tau}{\sqrt{1 - \beta^2(\tau)}} = \int_{\tau_1}^{\tau_2} \gamma(\tau) d\tau \tag{11.27}$$

Equation (11.27) or (11.26) expresses the phenomenon known as *time dila-tation*. A moving clock runs more slowly than a stationary clock. For equal time

---

*By considering equations (11.16), the reader can verify that there exists a Lorentz transformation with $\beta < 1$ provided $s_{12}^2 > 0$. Explicitly, $|\boldsymbol{\beta}| = |\mathbf{x}_1 - \mathbf{x}_2|/c\,|t_1 - t_2|$.

intervals in the clock's rest frame, the time intervals observed in the frame K are greater by a factor of $\gamma > 1$. This paradoxical result is verified daily in high-energy physics laboratories where beams of unstable particles of known lifetimes τ_0 are transported before decay over distances many many times the upper limit on the Galilean decay distance of $c\tau_0$. For example, at the Fermi National Accelerator Laboratory charged pions with energies of 200 GeV are produced and transported 300 meters with less than 3% loss because of decay. With a lifetime of $\tau_0 = 2.56 \times 10^{-8}$ s, the Galilean decay distance is $c\tau_0 = 7.7$ meters. Without time dilatation, only $e^{-300/7.7} \simeq 10^{-17}$ of the pions would survive. But at 200 GeV, $\gamma \simeq 1400$ and the mean free path for pion decay is actually $\gamma c\tau_0 \simeq 11$ km!

A careful test of time dilatation under controlled laboratory conditions is afforded by the study of the decay of mu-mesons orbiting at nearly constant speed in a magnetic field. Such a test, incidental to another experiment, confirms fully the formula (11.27). [See the paper by Bailey et al. cited at the end of Section 11.11.]

A totally different and entertaining experiment on time dilatation has been performed with macroscopic clocks of the type used as official time standards.* The motion of the clocks was relative to the earth in commercial aircraft, the very high precision of the cesium beam atomic clocks compensating for the relatively small speeds of the jet aircraft. The four clocks were flown around the world twice, once in an eastward and once in a westward sense. During the journeys logs were kept of the aircrafts' location and ground speed so that the integral in (11.27) could be calculated. Before and after each journey the clocks were compared with identical clocks at the U.S. Naval Observatory. With allowance for the earth's rotation and the gravitational "red shift" of general relativity, the average observed and calculated time differences in *nanoseconds* are -59 ± 10 and -40 ± 23 for the eastward trip and 273 ± 7 and 275 ± 21 for the westward. The kinematic effect of special relativity is comparable to the general relativistic effect. The agreement between observation and calculation establishes that people who continually fly eastward on jet aircraft age less rapidly than those of us who stay home, but not by much!

D. Relativistic Doppler Shift

As remarked in Section 11.2.A, the phase of a wave is an invariant quantity because the phase can be identified with the mere counting of wave crests in a wave train, an operation that must be the same in all inertial frames. In Section 11.2 the Galilean transformation of coordinates (11.1) was used to obtain the Galilean (nonrelativistic) Doppler shift formulas (11.8). Here we use the Lorentz transformation of coordinates (11.16) to obtain the *relativistic Doppler shift*. Consider a plane wave of frequency ω and wave vector $\mathbf{k}$ in the inertial frame K. In the moving frame K' this wave will have, in general, a different frequency ω' and wave vector $\mathbf{k}'$, but the phase of the wave is an invariant:

$$\phi = \omega t - \mathbf{k} \cdot \mathbf{x} = \omega' t' - \mathbf{k}' \cdot \mathbf{x}' \qquad (11.28)$$

[Parenthetically we remark that because the equations of (11.16) are linear the plane wave in K with phase ϕ indeed remains a plane wave in frame K'.] Using

*J. C. Hafele and R. E. Keating, *Science* **177**, 166, 168 (1972).

(11.16) and the same arguments as we did in going from (11.7) to (11.8), we find that the frequency $\omega' = ck_0'$ and wave vector $\mathbf{k}'$ are given in terms of $\omega = ck_0$ and $\mathbf{k}$ by

$$\left. \begin{array}{l} k_0' = \gamma(k_0 - \boldsymbol{\beta} \cdot \mathbf{k}) \\ k_\parallel' = \gamma(k_\parallel - \beta k_0) \\ \mathbf{k}_\perp' = \mathbf{k}_\perp \end{array} \right\} \tag{11.29}$$

The Lorentz transformation of $(k_0, \mathbf{k})$ has exactly the same form as for $(x_0, \mathbf{x})$. *The frequency and wave number of any plane wave thus form a 4-vector.* The invariance (11.28) of the phase is the invariance of the "scalar product" of two 4-vectors (11.24). This correspondence is, in fact, an alternate path from (11.28) to the transformation law (11.29).

For *light waves*, $|\mathbf{k}| = k_0$, $|\mathbf{k}'| = k_0'$. Then the results (11.29) can be expressed in the more familiar form of the Doppler shift formulas

$$\omega' = \gamma\omega(1 - \beta \cos \theta) \tag{11.30}$$

$$\tan \theta' = \frac{\sin \theta}{\gamma(\cos \theta - \beta)}$$

where θ and θ' are the angles of $\mathbf{k}$ and $\mathbf{k}'$ relative to the direction of $\mathbf{v}$. The inverse equations are obtained by interchanging primed and unprimed quantities and reversing the sign of β.

The first equation in (11.30) is the customary Doppler shift, modified by the factor of γ. Its presence shows that there is a *transverse* Doppler shift, even when $\theta = \pi/2$. This relativistic transverse Doppler shift has been observed spectroscopically with atoms in motion (Ives–Stilwell experiment, 1938). It also has been observed using a precise resonance-absorption Mössbauer experiment, with a nuclear gamma-ray source on the axis of a rapidly rotating cylinder and the absorber attached to the circumference of the cylinder.*

11.4 Addition of Velocities, 4-Velocity

The Lorentz transformation (11.16) or (11.18) for coordinates can be used to obtain the law for addition of velocities. Suppose that there is a moving point P whose velocity vector $\mathbf{u}'$ has spherical coordinates (u', θ', ϕ') in the inertial frame K', as shown in Fig. 11.4. The frame K' is moving with velocity $\mathbf{v} = c\boldsymbol{\beta}$ in the positive x_1 direction with respect to the inertial frame K. We wish to know the components of the velocity $\mathbf{u}$ of the point P as seen from K. From (11.18) the differential expressions for dx_0, dx_1, dx_2, dx_3 are

$$dx_0 = \gamma_v(dx_0' + \beta \, dx_1')$$
$$dx_1 = \gamma_v(dx_1' + \beta \, dx_0')$$
$$dx_2 = dx_2'$$
$$dx_3 = dx_3'$$

*H. J. Hay, J. P. Schiffer, T. E. Cranshaw, and P. A. Egelstaff, *Phys. Rev. Lett.* **4**, 165 (1960). See also T. E. Cranshaw in *Proceedings of the International School of Physics*, Varenna, Course XX, 1961, Academic Press, New York (1962), p. 208.

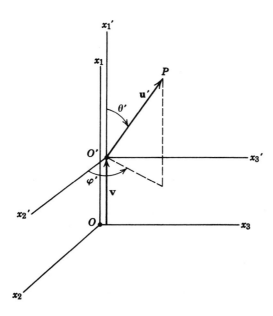

Figure 11.4   Addition of velocities.

where we have put a subscript on γ to distinguish it below from $\gamma_u = (1 - u^2/c^2)^{-1/2}$ and $\gamma_{u'} = (1 - u'^2/c^2)^{-1/2}$. The velocity components in each frame are $u'_i = c\, dx'_i/dx'_0$ and $u_i = c\, dx_i/dx_0$. This means that the components of velocity transform according to

$$u_{\parallel} = \frac{u'_{\parallel} + v}{1 + \dfrac{\mathbf{v} \cdot \mathbf{u}'}{c^2}}$$

$$\mathbf{u}_{\perp} = \frac{\mathbf{u}'_{\perp}}{\gamma_v \left(1 + \dfrac{\mathbf{v} \cdot \mathbf{u}'}{c^2}\right)}$$

$$(11.31)$$

The notation $u_{\parallel}$ and $\mathbf{u}_{\perp}$ refers to components of velocity parallel and perpendicular, respectively, to $\mathbf{v}$. The magnitude of $\mathbf{u}$ and its polar angles θ, ϕ in the frame K are easily found. Since $u_2/u'_2 = u_3/u'_3$, the azimuthal angles in the two frames are equal. Furthermore,

$$\tan \theta = \frac{u' \sin \theta'}{\gamma_v(u' \cos \theta' + v)}$$

and

$$(11.32)$$

$$u = \frac{\sqrt{u'^2 + v^2 + 2u'v \cos \theta' - \left(\dfrac{u'v \sin \theta'}{c}\right)^2}}{1 + \dfrac{u'v}{c^2} \cos \theta'}$$

The inverse results for $\mathbf{u}'$ in terms of $\mathbf{u}$ can be found, as usual, from (11.31) and (11.32), by interchanging primed and unprimed quantities and changing the sign of v.

For speeds u' and v both small compared to c, the velocity addition law (11.31) reduces to the Galilean result, $\mathbf{u} = \mathbf{u}' + \mathbf{v}$, but if either speed is comparable to c modifications appear. It is impossible to obtain a speed greater than that of light by adding two velocities, even if each is very close to c. For the simple case of parallel velocities the addition law is

$$u = \frac{u' + v}{1 + \dfrac{vu'}{c^2}} \tag{11.33}$$

If $u' = c$, then $u = c$ also. This is an explicit example of Einstein's second postulate. The reader can check from the second equation in (11.32) that $u' = c$ implies $u = c$ for nonparallel velocities as well.

The formula for the addition of velocities is in accord with such observational tests as the Fizeau experiments on the speed of light in moving liquids and the aberration of star positions from the motion of the earth in orbit.

The structure of (11.31) makes it obvious that the law of transformation of velocities is not that of 4-vectors, as given by (11.22) and of which (11.16) and (11.29) are examples. There is however, a 4-vector closely related to ordinary velocity. To exhibit this 4-vector we rewrite (11.31). From the second equation in (11.32) it can be shown directly that the factor $(1 + \mathbf{v} \cdot \mathbf{u}'/c^2)$ can be expressed alternatively through

$$\gamma_u = \gamma_v \gamma_{u'} \left(1 + \frac{\mathbf{v} \cdot \mathbf{u}'}{c^2} \right) \tag{11.34}$$

where γ_v, γ_u, $\gamma_{u'}$ are the gammas defined by (11.17) for $\mathbf{v}$, $\mathbf{u}$, and $\mathbf{u}'$, respectively. When (11.34) is substituted into (11.31) those equations become

$$\begin{aligned} \gamma_u u_\parallel &= \gamma_v (\gamma_{u'} u'_\parallel + v \gamma_{u'}) \\ \gamma_u \mathbf{u}_\perp &= \gamma_{u'} \mathbf{u}'_\perp \end{aligned} \tag{11.35}$$

Comparison of (11.34) and (11.35) with the inverse of (11.22) implies that *the four quantities* $(\gamma_u c, \gamma_u \mathbf{u})$ transform in the same way as $(x_0, \mathbf{x})$ and so form a 4-vector under Lorentz transformations. These four quantities are called the *time and space components of the 4-velocity* $(U_0, \mathbf{U})$.

An alternative approach to the 4-velocity is through the concept of proper time τ. Ordinary velocity $\mathbf{u}$ is defined as the time derivative of the coordinate $\mathbf{x}(t)$. The addition law (11.31) for velocities is not a 4-vector transformation law because time is not invariant under Lorentz transformations. But we have seen that the proper time τ *is* a Lorentz invariant. We can thus construct a 4-vector "velocity" by differentiation of the 4-vector $(x_0, \mathbf{x})$ with respect to τ instead of t. Using (11.26) we have

$$\left. \begin{aligned} U_0 &\equiv \frac{dx_0}{d\tau} = \frac{dx_0}{dt} \frac{dt}{d\tau} = \gamma_u c \\ \mathbf{U} &\equiv \frac{d\mathbf{x}}{d\tau} = \frac{d\mathbf{x}}{dt} \frac{dt}{d\tau} = \gamma_u \mathbf{u} \end{aligned} \right\} \tag{11.36}$$

We show in the next section that the components of 4-velocity of a particle are proportional to its total energy and momentum.

11.5   *Relativistic Momentum and Energy of a Particle*

We next consider the relativistic generalizations of the momentum and kinetic energy of a particle. These can be obtained for charged particles from the Lorentz force equation and the transformation properties of electromagnetic fields already established by Lorentz before 1900, but it is useful to give a more general derivation based only on the laws of conservation of energy and momentum and on the kinematics of Lorentz transformations. This approach shows clearly the universality of the relationships, independent of the existence of electromagnetic interactions for the particle in question.

For a particle with speed small compared to the speed of light its momentum and energy are known to be

$$\mathbf{p} = m\mathbf{u}$$
$$E = E(0) + \tfrac{1}{2}mu^2 \tag{11.37}$$

where m is the mass of the particle, $\mathbf{u}$ is its velocity, and $E(0)$ is a constant identified as the rest energy of the particle. In nonrelativistic considerations the rest energies can be ignored; they contribute the same additive constant to both sides of an energy balance equation. In special relativity, however, the rest energy cannot be ignored. We will see below that it is the total energy (the sum of rest energy plus kinetic energy) of a particle that is significant.

We wish to find expressions for the momentum and energy of a particle consistent with the Lorentz transformation law (11.31) of velocities and reducing to (11.37) for nonrelativistic motion. The only possible generalizations consistent with the first postulate are

$$\mathbf{p} = \mathcal{M}(u)\mathbf{u}$$
$$E = \mathcal{E}(u) \tag{11.38}$$

where $\mathcal{M}(u)$ and $\mathcal{E}(u)$ are functions of the magnitude of the velocity $\mathbf{u}$. Comparison with (11.37) yields the limiting values,

$$\mathcal{M}(0) = m$$
$$\frac{\partial \mathcal{E}}{\partial u^2}(0) = \frac{m}{2} \tag{11.39}$$

We make the reasonable assumption that $\mathcal{M}(u)$ and $\mathcal{E}(u)$ are well-behaved monotonic functions of their arguments.

To determine the forms of $\mathcal{M}(u)$ and $\mathcal{E}(u)$ we consider the elastic collision of two *identical* particles and require that conservation of momentum and energy hold in all equivalent inertial frames, as implied by the first postulate. In particular, we consider the collision in two frames K and K' connected by a Lorentz transformation parallel to the z axis. A certain amount of algebra is unavoidable. To keep it to a minimum, two approaches are open. One is to set up the velocities and directions of the particles in such a clever way that the algebra shakes down quickly into an elegant and transparent result. The other is to pick a straightforward kinematic situation and proceed judiciously. The first approach lacks motivation. We adopt the second.

Let the inertial frame K' be the "center of mass" frame with the two identical

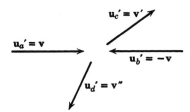

Figure 11.5   Initial and final velocity vectors in the frame K' for the collision of two identical particles.

particles having initial velocities $\mathbf{u}'_a = \mathbf{v}$, $\mathbf{u}'_b = -\mathbf{v}$ along the z axis. The particles collide and scatter, emerging with final velocities, $\mathbf{u}'_c = \mathbf{v}'$, $\mathbf{u}'_d = \mathbf{v}''$. The various velocities are indicated in Fig. 11.5. In K' the conservation equations for momentum and energy read

$$\mathbf{p}'_a + \mathbf{p}'_b = \mathbf{p}'_c + \mathbf{p}'_d$$
$$E'_a + E'_b = E'_c + E'_d$$

or, with the forms (11.38),

$$\mathcal{M}(v)\mathbf{v} - \mathcal{M}(v)\mathbf{v} = \mathcal{M}(v')\mathbf{v}' + \mathcal{M}(v'')\mathbf{v}''$$
$$\mathcal{E}(v) + \mathcal{E}(v) = \mathcal{E}(v') + \mathcal{E}(v'') \tag{11.40}$$

Because the particles are identical it is necessary that $\mathcal{E}(v') = \mathcal{E}(v'')$ and, with the hypothesis of monotonic behavior of $\mathcal{E}(v)$, that $v' = v''$. The second equation in (11.40) then demands $v' = v'' = v$. The first equation requires $\mathbf{v}'' = -\mathbf{v}'$. All four velocities have the same magnitude with the final velocities equal and opposite, just as are the initial velocities. This rather obvious state of affairs is shown in the right-hand diagram of Fig. 11.6 where the scattering angle in K' is denoted by θ'.

We now consider the collision in another inertial frame K moving with a velocity $-\mathbf{v}$ in the z direction with respect to K'. From the transformation equa-

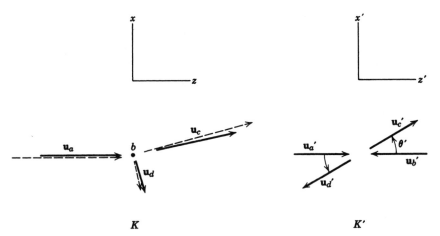

Figure 11.6   Initial and final velocity vectors in frames K and K' for the collision of two identical particles. The lengths and angles of the solid lines representing the velocities correspond to $\theta' = 30°$ and $\beta^2 = \frac{1}{3}$. The dashed lines in K are the results of a Galilean transformation from K' to K.

tions (11.31) for velocity it can be seen that particle b is at rest in K while particle a is incident along the z axis with a velocity

$$\mathbf{u}_a = \frac{2\mathbf{v}}{1 + \dfrac{v^2}{c^2}} = \frac{2c\boldsymbol{\beta}}{1 + \beta^2} \tag{11.41}$$

where $\boldsymbol{\beta} = \mathbf{v}/c$. The velocity components of the final velocities $\mathbf{u}_c$ and $\mathbf{u}_d$ in K are similarly

$$\left.\begin{aligned}
(\mathbf{u}_c)_x &= \frac{c\beta \sin \theta'}{\gamma(1 + \beta^2 \cos \theta')}, & (\mathbf{u}_c)_z &= \frac{c\beta(1 + \cos \theta')}{1 + \beta^2 \cos \theta'} \\
(\mathbf{u}_d)_x &= -\frac{c\beta \sin \theta'}{\gamma(1 - \beta^2 \cos \theta')}, & (\mathbf{u}_d)_z &= \frac{c\beta(1 - \cos \theta')}{1 - \beta^2 \cos \theta'}
\end{aligned}\right\} \tag{11.42}$$

with $\gamma = (1 - \beta^2)^{-1/2}$.

The equations of conservation of momentum and energy in the inertial frame K read

$$\begin{aligned}
\mathcal{M}(u_a)\mathbf{u}_a + \mathcal{M}(u_b)\mathbf{u}_b &= \mathcal{M}(u_c)\mathbf{u}_c + \mathcal{M}(u_d)\mathbf{u}_d \\
\mathcal{E}(u_a) + \mathcal{E}(u_b) &= \mathcal{E}(u_c) + \mathcal{E}(u_d)
\end{aligned} \tag{11.43}$$

It is apparent from (11.41) and (11.42) or the left-hand diagram of Fig. 11.6 that while particle b is at rest the other three velocities are all different in general. Thus the determination of $\mathcal{M}(u)$ and $\mathcal{E}(u)$ from (11.43) seems obscure. We can, however, consider the limiting situation of a glancing collision in which θ' is very small. Then in the frame K, $\mathbf{u}_d$ will be nonrelativistic and $\mathbf{u}_c$ will differ only slightly from $\mathbf{u}_a$. We can therefore make appropriate Taylor series expansions around $\theta' = 0$ and obtain equations involving $\mathcal{M}(u)$, $\mathcal{E}(u)$, and perhaps their first derivatives. Explicitly, the x component of the momentum conservation equation in (11.43) is

$$0 = \mathcal{M}(u_c) \frac{c\beta \sin \theta'}{\gamma(1 + \beta^2 \cos \theta')} - \mathcal{M}(u_d) \frac{c\beta \sin \theta'}{\gamma(1 - \beta^2 \cos \theta')}.$$

Canceling common factors and rearranging terms, we have

$$\mathcal{M}(u_c) = \left(\frac{1 + \beta^2 \cos \theta'}{1 - \beta^2 \cos \theta'}\right)\mathcal{M}(u_d)$$

This relation is valid for all θ' and in particular for $\theta' = 0^+$. Inspection of (11.42) shows that in that limit $\mathbf{u}_c = \mathbf{u}_a$, $\mathbf{u}_d = 0$. Thus we obtain

$$\mathcal{M}(u_a) = \left(\frac{1 + \beta^2}{1 - \beta^2}\right)\mathcal{M}(0) \tag{11.44}$$

From (11.41) it is easy to demonstrate that

$$\frac{1 + \beta^2}{1 - \beta^2} = \frac{1}{\sqrt{1 - \dfrac{u_a^2}{c^2}}} \equiv \gamma_a \tag{11.45}$$

With the value $\mathcal{M}(0) = m$ from (11.39) we thus have

$$\mathcal{M}(u_a) = \gamma_a m$$

or equivalently that the momentum of a particle of mass m and velocity $\mathbf{u}$ is

$$\mathbf{p} = \gamma m \mathbf{u} = \frac{m\mathbf{u}}{\sqrt{1 - \dfrac{u^2}{c^2}}} \tag{11.46}$$

Determining the functional form of $\mathcal{E}(u)$ requires more than the straightforward evaluation of the conservation of energy equation at $\theta' = 0^+$. We must examine the equation for small θ'. From (11.43) we have

$$\mathcal{E}(u_a) + \mathcal{E}(0) = \mathcal{E}(u_c) + \mathcal{E}(u_d) \tag{11.47}$$

where u_c and u_d are functions of θ'. From (11.42) or (11.32) we find, correct to order θ'^2 inclusive,

$$u_c^2 = u_a^2 - \frac{\eta}{\gamma_a^3} + O(\eta^2)$$

$$u_d^2 = \eta + O(\eta^2)$$

where γ_a is given by (11.45) and $\eta = c^2 \beta^2 \theta'^2 / (1 - \beta^2)$ is a convenient expansion parameter. We now expand both sides of (11.47) in Taylor series and equate coefficients of different powers of η:

$$\mathcal{E}(u_a) + \mathcal{E}(0) = \mathcal{E}(u_a) + \eta \cdot \left(\frac{d\mathcal{E}(u_c)}{du_c^2} \cdot \frac{\partial u_c^2}{\partial \eta} \right)_{\eta=0}$$

$$+ \cdots \mathcal{E}(0) + \eta \cdot \left(\frac{d\mathcal{E}(u_d)}{du_d^2} \cdot \frac{\partial u_d^2}{\partial \eta} \right)_{\eta=0} + \cdots$$

The zeroth-order terms give an identity, but the first-order terms yield

$$0 = -\frac{1}{\gamma_a^3} \frac{d\mathcal{E}(u_a)}{du_a^2} + \left(\frac{d\mathcal{E}(u_d)}{du_d^2} \right)_{u_d=0}$$

With the known nonrelativistic value of the second term from (11.39), we find

$$\frac{d\mathcal{E}(u_a)}{du_a^2} = \frac{m}{2} \gamma_a^3 = \frac{m}{2\left(1 - \dfrac{u_a^2}{c^2}\right)^{3/2}}$$

Integration yields the expression,

$$\mathcal{E}(u) = \frac{mc^2}{\left(1 - \dfrac{u^2}{c^2}\right)^{1/2}} + [\mathcal{E}(0) - mc^2] \tag{11.48}$$

for the energy of a particle of mass m and velocity $\mathbf{u}$, up to an arbitrary constant of integration. Parenthetically we remark that in an elastic scattering process the conservation of energy condition can be expressed in terms of kinetic energies alone. Thus the undetermined constant in (11.48) is necessary and is not, as the reader might have conjectured, the result of our Taylor series expansions. Note that the *kinetic* energy $T(u)$ is given unambiguously by

$$T(u) \equiv \mathscr{E}(u) - \mathscr{E}(0) = mc^2 \left[\frac{1}{\left(1 - \dfrac{u^2}{c^2}\right)^{1/2}} - 1 \right] \tag{11.49}$$

Equations (11.46) and (11.48) are the necessary relativistic generalizations for the momentum and energy of a particle, consistent with the conservation laws and the postulates of special relativity. The only remaining question is the value of the rest energy $\mathscr{E}(0)$. We can appeal directly to experiment or we can examine the theoretical framework. First, experiment. Although $\mathscr{E}(0)$ cannot be determined from elastic scattering, it can be found from inelastic processes in which one type of particle is transformed into another or others of different masses. Decay processes are particularly transparent. Consider, for example, the decay of a neutral K-meson into two photons, $K^0 \to \gamma\gamma$. In the rest frame of the K-meson, conservation of energy requires that the sum of the energies of the two photons be equal to $\mathscr{E}_K(0)$. For another decay mode of a neutral K-meson, into two pions, the kinetic energy of each pion in the K-meson's rest frame must be

$$T_\pi = \tfrac{1}{2}\mathscr{E}_K(0) - \mathscr{E}_\pi(0)$$

Measurement of the pion kinetic energy (11.49) and knowledge of $\mathscr{E}_K(0)$ allows determination of $\mathscr{E}_\pi(0)$. In these examples and every other case it is found that the rest energy of a particle (or more complicated system) of mass m is given by the famous Einstein mass-energy relation,

$$\mathscr{E}(0) = mc^2 \tag{11.50}$$

Thus the second, square-bracketed, term in (11.48) is absent. The total energy of a particle of mass m and velocity **u** is

$$E = \gamma mc^2 = \frac{mc^2}{\sqrt{1 - \dfrac{u^2}{c^2}}} \tag{11.51}$$

A second path to the results (11.50) and (11.51) is theoretical. Although the expressions (11.46) and (11.48) for the momentum and energy of a particle were found by applying the principles of special relativity to the conservation of energy and momentum, the properties of **p** and E under Lorentz transformations are not yet explicit. The conservation equations are a set of four equations assumed to be valid in all equivalent inertial frames. Momentum conservation consists of three equations relating the spatial components of vectors. Within the framework of special relativity it is natural to attempt to identify the *four* equations of conservation as relations among 4-vectors. We observe that the momentum (11.46) is proportional to the spatial components of the 4-velocity $(U_0, \mathbf{U})$ defined in (11.36), that is $\mathbf{p} = m\mathbf{U}$. The time component of this 4-vector is $p_0 = mU_0 = m\gamma_u c$. Comparison with (11.48) shows that the energy of a particle differs from cp_0 by an additive constant $[\mathscr{E}(0) - mc^2]$. This means that the four equations of energy and momentum conservation for an arbitrary collision process can be written as

$$\sum_{\substack{a \\ \text{initial}}} (p_0)_a - \sum_{\substack{b \\ \text{final}}} (p_0)_b = \Delta_0$$

$$\sum_{\substack{a \\ \text{initial}}} \mathbf{p}_a - \sum_{\substack{b \\ \text{final}}} \mathbf{p}_b = \Delta \tag{11.52}$$

where (Δ_0, Δ) is a 4-vector with $\Delta = 0$ and

$$c\Delta_0 = \sum_{\substack{b \\ \text{final}}} [\mathscr{E}_b(0) - m_b c^2] - \sum_{\substack{a \\ \text{initial}}} [\mathscr{E}_a(0) - m_a c^2]$$

From the first postulate, (11.52) must be valid in all equivalent inertial frames. But if $\Delta \equiv 0$ in all inertial frames it can be seen from (11.22) that it is necessary that $\Delta_0 \equiv 0$; the 4-vector (Δ_0, Δ) is a null vector. If different types or numbers of particles can occur in the initial and final states of some process, the condition $\Delta_0 = 0$ can only be met by requiring that (11.50) hold for each particle separately. We are thus led to (11.51) as the correct form of the total energy.

The velocity of the particle can evidently be expressed in terms of its momentum and energy from (11.46) and (11.51) as

$$\mathbf{u} = \frac{c^2 \mathbf{p}}{E} \tag{11.53}$$

The invariant "length" of the energy-momentum 4-vector $(p_0 = E/c, \mathbf{p})$ is

$$p_0^2 - \mathbf{p} \cdot \mathbf{p} = (mc)^2 \tag{11.54}$$

We see that the invariant property that characterizes a particle's momentum and energy is its mass, m, sometimes called its rest mass.* Equation (11.54), combined with the conservation equations, forms a powerful and elegant means of treating relativistic kinematics in collision and decay processes (see the problems at the end of the chapter). Note that (11.54) permits the energy E to be expressed in terms of the momentum as

$$E = \sqrt{c^2 p^2 + m^2 c^4} \tag{11.55}$$

The relations (11.46), (11.51), and (11.53) for momentum, energy, and velocity of a particle are so universally accepted that it seems superfluous to speak of experimental tests. It is perhaps worthwhile, nevertheless, to cite some laboratory demonstrations. One is the connection between the kinetic energy (11.49) of a particle and its speed.[†] The speeds of electrons of known kinetic energies from 0.5 to 15 MeV (accelerated through a known voltage in a Van de Graaff generator, verified at the beam catcher by calorimetry) are measured by having bursts of electrons ($\Delta t = 3 \times 10^{-9}$ s) travel a flight path of 8.4 meters. As the energy increases the transit time falls toward a limiting value of 2.8×10^{-8} s, in good agreement with (11.49). Verification of c as a limiting speed for material

---

*Some authors define the mass of a particle to be E/c^2, designating it as m or $m(u)$ and reserving the symbol m_0 for the rest mass. We always use the word "mass" for the Lorentz invariant quantity whose square appears in (11.54).

[†]W. Bertozzi, *Am. J. Phys.* **32**, 551 (1964).

particles has been carried out for 11 GeV electrons ($\gamma \simeq 2 \times 10^7$) in the Stanford experiment cited at the end of Section 11.2, where it was found that the electrons' speed differed fractionally from c by less than 5×10^{-6}. An undergraduate experiment to verify the relation (11.55) between momentum and energy employs a simple magnet with roughly 10 cm radius of curvature for the momentum measurement and a NaI crystal for the energy measurement on beta rays.*

The specification of the kinematic properties of a particle (velocity, momentum, energy) in any inertial frame can be accomplished by giving its mass and either its velocity $\mathbf{u}$ or its momentum $\mathbf{p}$ in that frame. A Lorentz transformation (11.22) of $(p_0, \mathbf{p})$ gives the results in any other frame. It is sometimes convenient to use the two components of $\mathbf{p}$ perpendicular to the z axis and a rapidity ζ (11.20) as kinematic variables. Suppose that a particle has momentum $\mathbf{p}$ in frame K, with transverse momentum $\mathbf{p}_\perp$ and a z component $p_\parallel$. There is a unique Lorentz transformation in the z direction to a frame K' where the particle has no z component of momentum. In K' the particle has momentum and energy,

$$\mathbf{p}' = \mathbf{p}_\perp, \qquad \frac{E'}{c} = \Omega = \sqrt{p_\perp^2 + m^2 c^2} \qquad (11.56)$$

Let the rapidity parameter associated with the Lorentz transformation from K to K' be ζ. Then from the inverse of (11.21) the momentum components and energy of the particle in the original frame K can be written

$$\mathbf{p}_\perp, \qquad p_\parallel = \Omega \sinh \zeta, \qquad \frac{E}{c} = \Omega \cosh \zeta \qquad (11.57)$$

with $\Omega = \sqrt{p_\perp^2 + m^2 c^2}$. The quantity Ω/c is sometimes called the transverse mass (because it depends on $\mathbf{p}_\perp$) or the longitudinal mass (because it is involved in a longitudinal boost). If the particle is at rest in K', that is, $\mathbf{p}_\perp = 0$, then the expressions (11.57) become

$$p = mc \sinh \zeta, \qquad E = mc^2 \cosh \zeta \qquad (11.58)$$

alternatives to (11.46) and (11.51).

One convenience of $\mathbf{p}_\perp^{(i)}$ and $\zeta^{(i)}$ as kinematic variables is that a Lorentz transformation in the z direction shifts all rapidities by a constant amount, $\zeta^{(i)} \rightarrow \zeta^{(i)} - Z$, where Z is the rapidity parameter of the transformation. With these variables, the configuration of particles in a collision process viewed in the laboratory frame differs only by a trivial shift of the origin of rapidity from the same process viewed in the center of mass frame.

11.6 *Mathematical Properties of the Space-Time of Special Relativity*

The kinematics of special relativity presented in the preceding sections can be discussed in a more profound and elegant manner that simultaneously simplifies and illuminates the theory. Three-dimensional rotations in classical and quantum mechanics can be discussed in terms of the group of transformations of the co-

*S. Parker, *Am. J. Phys.* **40**, 241 (1972).

ordinates that leave the norm of the vector **x** invariant. In the special theory of relativity, Lorentz transformations of the four-dimensional coordinates $(x_0, \mathbf{x})$ follow from the invariance of

$$s^2 = x_0^2 - x_1^2 - x_2^2 - x_3^2 \tag{11.59}$$

We can therefore rephrase the kinematics of special relativity as the consideration of the group of all transformations that leave s^2 invariant. Technically, this group is called the *homogeneous Lorentz group*. It contains ordinary rotations as well as the Lorentz transformations of Section 11.3. The group of transformations that leave invariant

$$s^2(x, y) = (x_0 - y_0)^2 - (x_1 - y_1)^2 - (x_2 - y_2)^2 - (x_3 - y_3)^2$$

is called the *inhomogeneous Lorentz group* or the *Poincaré group*. It contains translations and reflections in both space and time, as well as the transformations of the homogeneous Lorentz group. We shall restrict our discussion to the homogeneous transformations and subsequently omit "homogeneous" when referring to the Lorentz group.

From the first postulate it follows that the mathematical equations expressing the laws of nature must be *covariant*, that is, invariant in form, under the transformations of the Lorentz group. They must therefore be relations among Lorentz scalars, 4-vectors, 4-tensors, etc., defined by their transformation properties under the Lorentz group in ways analogous to the familiar specification of tensors of a given rank under three-dimensional rotations. We are thus led to consider briefly the mathematical structure of a space-time whose norm is defined by (11.59).

We begin by summarizing the elements of tensor analysis in a non-Euclidean vector space. The space-time continuum is defined in terms of a four-dimensional space with coordinates x^0, x^1, x^2, x^3. We suppose that there is a well-defined transformation that yields new coordinates x'^0, x'^1, x'^2, x'^3, according to some rule,

$$x'^\alpha = x'^\alpha(x^0, x^1, x^2, x^3) \qquad (\alpha = 0, 1, 2, 3) \tag{11.60}$$

For the moment the transformation law is not specified.

Tensors of rank k associated with the space-time point x are defined by their transformation properties under the transformation $x \to x'$. A *scalar* (tensor of rank zero) is a single quantity whose value is not changed by the transformation. The interval s^2 (11.59) is obviously a Lorentz scalar. For tensors of rank one, called *vectors*, two kinds must be distinguished. The first is called a *contravariant vector* A^α with four components A^0, A^1, A^2, A^3 that are transformed according to the rule

$$A'^\alpha = \frac{\partial x'^\alpha}{\partial x^\beta} A^\beta \tag{11.61}$$

In this equation the derivative is computed from (11.60) and the repeated index β implies a summation over $\beta = 0, 1, 2, 3$. Thus explicitly we have

$$A'^\alpha = \frac{\partial x'^\alpha}{\partial x^0} A^0 + \frac{\partial x'^\alpha}{\partial x^1} A^1 + \frac{\partial x'^\alpha}{\partial x^2} A^2 + \frac{\partial x'^\alpha}{\partial x^3} A^3$$

We will henceforth employ this *summation convention* for repeated indices. A *covariant vector* or tensor of rank one B_α is defined by the rule

$$B'_\alpha = \frac{\partial x^\beta}{\partial x'^\alpha} B_\beta \tag{11.62}$$

or, explicitly, by

$$B'_\alpha = \frac{\partial x^0}{\partial x'^\alpha} B_0 + \frac{\partial x^1}{\partial x'^\alpha} B_1 + \frac{\partial x^2}{\partial x'^\alpha} B_2 + \frac{\partial x^3}{\partial x'^\alpha} B_3$$

The partial derivative in (11.62) is to be calculated from the inverse of (11.60) with x^β expressed as a function of x'^0, x'^1, x'^2, x'^3.

Note that contravariant vectors have superscripts and covariant vectors have subscripts, corresponding to the presence of $\partial x'^\alpha/\partial x^\beta$ and its inverse in the rule of transformation. It can be verified from (11.61) that if the law of transformation (11.60) is linear then the coordinates x^0, x^1, x^2, x^3 form the components of a contravariant vector.

A contravariant tensor of rank two $F^{\alpha\beta}$ consists of 16 quantities that transform according to

$$F'^{\alpha\beta} = \frac{\partial x'^\alpha}{\partial x^\gamma} \frac{\partial x'^\beta}{\partial x^\delta} F^{\gamma\delta} \tag{11.63}$$

A covariant tensor of rank two, $G_{\alpha\beta}$, transforms as

$$G'_{\alpha\beta} = \frac{\partial x^\gamma}{\partial x'^\alpha} \frac{\partial x^\delta}{\partial x'^\beta} G_{\gamma\delta} \tag{11.64}$$

and the mixed second-rank tensor $H^\alpha{}_\beta$ transforms as

$$H'^\alpha{}_\beta = \frac{\partial x'^\alpha}{\partial x^\gamma} \frac{\partial x^\delta}{\partial x'^\beta} H^\gamma{}_\delta \tag{11.65}$$

The generalization to contravariant, covariant, or mixed tensors of arbitrary rank should be obvious from these examples.

The inner or scalar product of two vectors is defined as the product of the components of a covariant and a contravariant vector,

$$B \cdot A \equiv B_\alpha A^\alpha \tag{11.66}$$

With this definition the scalar product is an invariant or scalar under the transformation (11.60). This is established by considering the scalar product $B' \cdot A'$ and employing (11.61) and (11.62):

$$B' \cdot A' = \frac{\partial x^\beta}{\partial x'^\alpha} \frac{\partial x'^\alpha}{\partial x^\gamma} B_\beta A^\gamma = \frac{\partial x^\beta}{\partial x^\gamma} B_\beta A^\gamma = \delta^\beta{}_\gamma B_\beta A^\gamma = B \cdot A$$

The inner product or contraction with respect to any pair of indices, either on the same tensor or one on one tensor and the other on another, is defined in analogy with (11.66). One index is contravariant and the other covariant always.

The results or definitions above are general. The specific geometry of the space-time of special relativity is defined by the invariant interval s^2, (11.59). In differential form, the infinitesimal interval ds that defines the norm of our space is

$$(ds)^2 = (dx^0)^2 - (dx^1)^2 - (dx^2)^2 - (dx^3)^2 \qquad (11.67)$$

Here we have used superscripts on the coordinates because of our present conventions. This *norm* or *metric* is a special case of the general differential length element,

$$(ds)^2 = g_{\alpha\beta} \, dx^\alpha \, dx^\beta \qquad (11.68)$$

where $g_{\alpha\beta} = g_{\beta\alpha}$ is called the *metric tensor*. For the flat space-time of special relativity (in distinction to the curved space-time of general relativity) the metric tensor is diagonal, with elements

$$g_{00} = 1, \qquad g_{11} = g_{22} = g_{33} = -1 \qquad (11.69)$$

The contravariant metric tensor $g^{\alpha\beta}$ is defined as the normalized cofactor of $g_{\alpha\beta}$. For flat space-time it is the same:

$$g^{\alpha\beta} = g_{\alpha\beta} \qquad (11.70)$$

Note that the contraction of the contravariant and covariant metric tensors gives the Kronecker delta in four dimensions:

$$g_{\alpha\gamma}g^{\gamma\beta} = \delta_\alpha{}^\beta \qquad (11.71)$$

where $\delta_\alpha{}^\beta = 0$ for $\alpha \neq \beta$ and $\delta_\alpha{}^\alpha = 1$ for $\alpha = 0, 1, 2, 3$.

Comparison of the invariant length element $(ds)^2$ in (11.68) with the similarly invariant scalar product (11.66) suggests that the covariant coordinate 4-vector x_α can be obtained from the contravariant x^β by contraction with $g_{\alpha\beta}$, that is,

$$x_\alpha = g_{\alpha\beta}x^\beta \qquad (11.72)$$

and its inverse,

$$x^\alpha = g^{\alpha\beta}x_\beta \qquad (11.73)$$

In fact, contraction with $g_{\alpha\beta}$ or $g^{\alpha\beta}$ is the procedure for changing an index on any tensor from being contravariant to covariant, and vice versa. Thus

$$F_{\ldots\ldots}^{\ldots\alpha\ldots} = g^{\alpha\beta}F_{\ldots\beta}^{\ldots\ldots\ldots}$$

and $\qquad\qquad\qquad\qquad\qquad\qquad\qquad\qquad\qquad\qquad\qquad\qquad\qquad$ (11.74)

$$G_{\ldots\ldots\alpha\ldots}^{\ldots\ldots} = g_{\alpha\beta}G_{\ldots\ldots\ldots}^{\ldots\ldots\beta}$$

With the metric tensor (11.69) it follows that if a contravariant 4-vector has components, A^0, A^1, A^2, A^3, its covariant partner has components, $A_0 = A^0, A_1 = -A^1, A_2 = -A^2, A_3 = -A^3$. We write this concisely as

$$A^\alpha = (A^0, \mathbf{A}), \qquad A_\alpha = (A^0, -\mathbf{A}) \qquad (11.75)$$

where the 3-vector $\mathbf{A}$ has components A^1, A^2, A^3. The scalar product (11.66) of two 4-vectors is

$$B \cdot A \equiv B_\alpha A^\alpha = B^0 A^0 - \mathbf{B} \cdot \mathbf{A}$$

in agreement with (11.24).

Consider now the partial derivative operators with respect to x^α and x_α. The transformation properties of these operators can be established directly by using the rules of implicit differentiation. For example, we have

$$\frac{\partial}{\partial x'^\alpha} = \frac{\partial x^\beta}{\partial x'^\alpha} \frac{\partial}{\partial x^\beta}$$

Comparison with (11.62) shows that *differentiation with respect to a contravariant component* of the coordinate vector transforms as the component of a *covariant vector operator*. From (11.72) it follows that differentiation with respect to a covariant component gives a contravariant vector operator. We therefore employ the notation,

$$\partial^\alpha \equiv \frac{\partial}{\partial x_\alpha} = \left(\frac{\partial}{\partial x^0}, \; -\boldsymbol{\nabla} \right)$$

$$\partial_\alpha \equiv \frac{\partial}{\partial x^\alpha} = \left(\frac{\partial}{\partial x^0}, \; \boldsymbol{\nabla} \right)$$

(11.76)

The 4-divergence of a 4-vector A is the invariant,

$$\partial^\alpha A_\alpha = \partial_\alpha A^\alpha = \frac{\partial A^0}{\partial x^0} + \boldsymbol{\nabla} \cdot \mathbf{A}$$

(11.77)

an equation familiar in form from continuity of charge and current density, the Lorentz condition on the scalar and vector potentials, etc. These examples give a first inkling of how the covariance of a physical law emerges provided suitable Lorentz transformation properties are attributed to the quantities entering the equation.

The four-dimensional Laplacian operator is defined to be the invariant contraction,

$$\square \equiv \partial_\alpha \partial^\alpha = \frac{\partial^2}{\partial x^{02}} - \nabla^2$$

(11.78)

This is, of course, just the operator of the wave equation in vacuum.

11.7 *Matrix Representation of Lorentz Transformations, Infinitesimal Generators*

We now turn to the consideration of the Lorentz group of transformations. To make the manipulations explicit and less abstract, it is convenient to use a matrix representation with the components of a contravariant 4-vector forming the elements of a column vector. The coordinates x^0, x^1, x^2, x^3 thus define a coordinate vector whose representative is

$$x = \begin{pmatrix} x^0 \\ x^1 \\ x^2 \\ x^3 \end{pmatrix}$$

(11.79)

Matrix scalar products of 4-vectors (a, b) are defined in the usual way by summing over the products of the elements of a and b, or equivalently by matrix multiplication of the transpose of a on b:

$$(a, b) \equiv \tilde{a}b \tag{11.80}$$

The metric tensor $g_{\alpha\beta}$ has as its representative the square 4×4 matrix

$$g = \begin{pmatrix} 1 & 0 & 0 & 0 \\ 0 & -1 & 0 & 0 \\ 0 & 0 & -1 & 0 \\ 0 & 0 & 0 & -1 \end{pmatrix} \tag{11.81}$$

with $g^2 = I$, the 4×4 unit matrix. The covariant coordinate vector is

$$gx = \begin{pmatrix} x^0 \\ -x^1 \\ -x^2 \\ -x^3 \end{pmatrix} \tag{11.82}$$

obtained by matrix multiplication of g (11.81) on x (11.79). *Note that in the present notation the scalar product* (11.66) *of two 4-vectors reads*

$$a \cdot b = (a, gb) = (ga, b) = \tilde{a}gb \tag{11.83}$$

On the basis of arguments already presented in Section 11.3 we seek a group of linear transformations on the coordinates,

$$x' = Ax \tag{11.84}$$

where A is a square 4×4 matrix, such that the norm (x, gx) is left invariant:

$$\tilde{x}'gx' = \tilde{x}gx \tag{11.85}$$

Substitution of (11.84) into the left-hand side yields the equality,

$$\tilde{x}\tilde{A}gAx = \tilde{x}gx$$

Since this must hold for all coordinate vectors x, A must satisfy the matrix equation,

$$\tilde{A}gA = g \tag{11.86}$$

Certain properties of the transformation matrix A can be deduced immediately from (11.86). The first concerns the determinant of A. Taking the determinant of both sides of (11.86) gives us

$$\det (\tilde{A}gA) = \det g \, (\det A)^2 = \det g$$

Since $\det g = -1 \neq 0$, we obtain

$$\det A = \pm 1$$

There are two classes of transformations: *proper Lorentz transformations*, continuous with the identity transformation and so necessarily having det A = + 1, and *improper Lorentz transformations*. For *improper* transformations it is sufficient, but not necessary, to have det $A = -1$. The fact that det $A = \pm 1$ does not unambiguously sort out the two classes is a consequence of the indefinite metric of space-time. Two examples of improper transformations, $A = g$ (space inversion) with det $A = -1$ and $A = -I$ (space *and* time inversion) with det $A = +1$, illustrate this point.

The second property of A is the number of parameters needed to specify completely a transformation of the group. Since A and g are 4×4 matrices, (11.86) represents 16 equations for the $4^2 = 16$ elements of A. But they are not all independent because of symmetry under transposition. There are thus $16 - (1 + 2 + 3) = 10$ linearly independent equations for the 16 elements of A. This means that there are *six free parameters*—the Lorentz group is a six-parameter group. The six parameters can be conveniently thought of as (a) three parameters (e.g., Euler angles) to specify the relative orientation of the coordinate axes and (b) three parameters (e.g., components of $\boldsymbol{\beta}$) to specify the relative velocity of the two inertial frames. Parenthetically we remark that for every six-parameter A giving a proper Lorentz transformation, there is an improper one represented by $-A$. From now on we consider only proper Lorentz transformations.

The explicit construction of A can proceed as follows. We make the ansatz

$$A = e^L \tag{11.87}$$

where L is a 4×4 matrix. The determinant of A is*

$$\det A = \det (e^L) = e^{\mathrm{Tr}\, L}$$

If L is a real matrix, $\det A = -1$ is excluded. Furthermore, if L is traceless, then $\det A = +1$. Thus, for proper Lorentz transformations, L is a real, traceless 4×4 matrix. Equation (11.86) can be written

$$g\tilde{A}g = A^{-1} \tag{11.88}$$

From the definition (11.87) and the fact that $g^2 = I$ we have

$$\tilde{A} = e^{\tilde{L}}, \qquad g\tilde{A}g = e^{g\tilde{L}g}, \qquad A^{-1} = e^{-L}$$

Therefore (11.88) is equivalent to

$$g\tilde{L}g = -L$$

or

$$\widetilde{gL} = -gL \tag{11.89}$$

The matrix gL is thus antisymmetric. From the properties of g (11.81) it is evident that the general form of L is

$$L = \begin{pmatrix} 0 & L_{01} & L_{02} & L_{03} \\ \hline L_{01} & 0 & L_{12} & L_{13} \\ L_{02} & -L_{12} & 0 & L_{23} \\ L_{03} & -L_{13} & -L_{23} & 0 \end{pmatrix} \tag{11.90}$$

The dashed lines are inserted to set off the 3×3 antisymmetric spatial matrix corresponding to the familiar *rotations* in a fixed inertial frame from the symmetric space-time part of the matrix corresponding to Lorentz transformations or *boosts* from one inertial frame to another.

---

*To prove this, note first that the value of the determinant or the trace of a matrix is unchanged by a similarity transformation. Then make such a transformation to put L in diagonal form. The matrix A will then be diagonal with elements that are the exponentials of the corresponding elements of L. The result follows immediately.

The matrix (11.90), with its six parameters is an explicit construction [through (11.87)] of the transformation matrix A. It is customary, however, to systematize L and its six parameters by introducing a set of six fundamental matrices defined by

$$S_1 = \begin{pmatrix} 0 & 0 & 0 & 0 \\ 0 & 0 & 0 & 0 \\ 0 & 0 & 0 & -1 \\ 0 & 0 & 1 & 0 \end{pmatrix}, \qquad S_2 = \begin{pmatrix} 0 & 0 & 0 & 0 \\ 0 & 0 & 0 & 1 \\ 0 & 0 & 0 & 0 \\ 0 & -1 & 0 & 0 \end{pmatrix},$$

$$S_3 = \begin{pmatrix} 0 & 0 & 0 & 0 \\ 0 & 0 & -1 & 0 \\ 0 & 1 & 0 & 0 \\ 0 & 0 & 0 & 0 \end{pmatrix} \tag{11.91}$$

$$K_1 = \begin{pmatrix} 0 & 1 & 0 & 0 \\ 1 & & & \\ 0 & & \mathbf{0} & \\ 0 & & & \end{pmatrix}, \qquad K_2 = \begin{pmatrix} 0 & 0 & 1 & 0 \\ 0 & & & \\ 1 & & \mathbf{0} & \\ 0 & & & \end{pmatrix}, \qquad K_3 = \begin{pmatrix} 0 & 0 & 0 & 1 \\ 0 & & & \\ 0 & & \mathbf{0} & \\ 1 & & & \end{pmatrix}$$

The matrices S_i evidently generate rotations in three dimensions, while the matrices K_i produce boosts. For reference, we note that the squares of these six matrices are all diagonal and of the form,

$$S_1^2 = \begin{pmatrix} 0 & & & \mathbf{0} \\ & 0 & & \\ & & -1 & \\ \mathbf{0} & & & -1 \end{pmatrix}, \qquad S_2^2 = \begin{pmatrix} 0 & & & \mathbf{0} \\ & -1 & & \\ & & 0 & \\ \mathbf{0} & & & -1 \end{pmatrix},$$

$$S_3^2 = \begin{pmatrix} 0 & & & \mathbf{0} \\ & -1 & & \\ & & -1 & \\ \mathbf{0} & & & 0 \end{pmatrix} \tag{11.92}$$

$$K_1^2 = \begin{pmatrix} 1 & & & \mathbf{0} \\ & 1 & & \\ & & 0 & \\ \mathbf{0} & & & 0 \end{pmatrix}, \qquad K_2^2 = \begin{pmatrix} 1 & & & \mathbf{0} \\ & 0 & & \\ & & 1 & \\ \mathbf{0} & & & 0 \end{pmatrix},$$

$$K_3^2 = \begin{pmatrix} 1 & & & \mathbf{0} \\ & 0 & & \\ & & 0 & \\ \mathbf{0} & & & 1 \end{pmatrix}$$

Furthermore, it can be shown that $(\boldsymbol{\epsilon} \cdot \mathbf{S})^3 = -\boldsymbol{\epsilon} \cdot \mathbf{S}$ and $(\boldsymbol{\epsilon}' \cdot \mathbf{K})^3 = \boldsymbol{\epsilon}' \cdot \mathbf{K}$, where $\boldsymbol{\epsilon}$ and $\boldsymbol{\epsilon}'$ are any real unit 3-vectors. Thus any power of one of the matrices can be expressed as a multiple of the matrix or its square.

The general result (11.90) for L can now be written alternatively as

and

$$\left. \begin{aligned} L &= -\boldsymbol{\omega} \cdot \mathbf{S} - \boldsymbol{\zeta} \cdot \mathbf{K} \\ A &= e^{-\boldsymbol{\omega} \cdot \mathbf{S} - \boldsymbol{\zeta} \cdot \mathbf{K}} \end{aligned} \right\} \tag{11.93}$$

where $\boldsymbol{\omega}$ and $\boldsymbol{\zeta}$ are constant 3-vectors. The three components each of $\boldsymbol{\omega}$ and $\boldsymbol{\zeta}$ correspond to the six parameters of the transformation. To establish contact with earlier results such as (11.16) or (11.21), we consider first a simple situation in which $\boldsymbol{\omega} = 0$ and $\boldsymbol{\zeta} = \zeta\boldsymbol{\epsilon}_1$. Then $L = -\zeta K_1$ and with the help of (11.92) and $K_1^3 = K_1$ we find

$$A = e^L = (I - K_1^2) - K_1 \sinh \zeta + K_1^2 \cosh \zeta \qquad (11.94)$$

Explicitly,

$$A = \begin{pmatrix} \cosh \zeta & -\sinh \zeta & 0 & 0 \\ -\sinh \zeta & \cosh \zeta & 0 & 0 \\ 0 & 0 & 1 & 0 \\ 0 & 0 & 0 & 1 \end{pmatrix} \qquad (11.95)$$

This matrix corresponds exactly to the transformation (11.21).* If $\boldsymbol{\zeta} = 0$ and $\boldsymbol{\omega} = \omega\boldsymbol{\epsilon}_3$, the transformation is similarly found to be

$$A = \begin{pmatrix} 1 & 0 & 0 & 0 \\ 0 & \cos \omega & \sin \omega & 0 \\ 0 & -\sin \omega & \cos \omega & 0 \\ 0 & 0 & 0 & 1 \end{pmatrix} \qquad (11.96)$$

corresponding to a rotation of the coordinate axes in a clockwise sense around the 3-axis.

For a boost (without rotation) in an arbitrary direction,

$$A = e^{-\boldsymbol{\zeta}\cdot\mathbf{K}}$$

The boost vector $\boldsymbol{\zeta}$ can be written in terms of the relative velocity $\boldsymbol{\beta}$ as

$$\boldsymbol{\zeta} = \hat{\boldsymbol{\beta}} \tanh^{-1} \beta$$

where $\hat{\boldsymbol{\beta}}$ is a unit vector in the direction of the relative velocity of the two inertial frames. The pure boost is then

$$A_{\text{boost}}(\boldsymbol{\beta}) = e^{-\hat{\boldsymbol{\beta}}\cdot\mathbf{K}\,\tanh^{-1}\beta} \qquad (11.97)$$

It is left as an exercise to verify that this transformation gives the explicit matrix:

$$A_{\text{boost}}(\boldsymbol{\beta}) = \begin{pmatrix} \gamma & -\gamma\beta_1 & -\gamma\beta_2 & -\gamma\beta_3 \\ -\gamma\beta_1 & 1 + \dfrac{(\gamma-1)\beta_1^2}{\beta^2} & \dfrac{(\gamma-1)\beta_1\beta_2}{\beta^2} & \dfrac{(\gamma-1)\beta_1\beta_3}{\beta^2} \\ -\gamma\beta_2 & \dfrac{(\gamma-1)\beta_1\beta_2}{\beta^2} & 1 + \dfrac{(\gamma-1)\beta_2^2}{\beta^2} & \dfrac{(\gamma-1)\beta_2\beta_3}{\beta^2} \\ -\gamma\beta_3 & \dfrac{(\gamma-1)\beta_1\beta_3}{\beta^2} & \dfrac{(\gamma-1)\beta_2\beta_3}{\beta^2} & 1 + \dfrac{(\gamma-1)\beta_3^2}{\beta^2} \end{pmatrix}$$

$$(11.98)$$

---

*The reader is reminded that in Sections 11.3, 11.4, and 11.5 no distinction is made between subscripts and superscripts. All components of vectors there are to be interpreted as contravariant components, in accordance with (11.75).

The equation $x' = A_{\text{boost}}(\boldsymbol{\beta})x$ is a matrix statement of the four equations of (11.19).

The six matrices (11.91) are a representation of the *infinitesimal generators* of the Lorentz group. Straightforward calculation shows that they satisfy the following commutation relations,

$$[S_i, S_j] = \epsilon_{ijk} S_k$$
$$[S_i, K_j] = \epsilon_{ijk} K_k \qquad (11.99)$$
$$[K_i, K_j] = -\epsilon_{ijk} S_k$$

where the commutator notation is $[A, B] \equiv AB - BA$. The first relation corresponds to the commutation relations for angular momentum, the second relation merely shows that K transforms as a vector under rotations, and the final relation shows that boosts do not in general commute. The commutation relations (11.99), with the characteristic minus sign in the last commutator, specify the algebraic structure of the Lorentz group to be $SL(2, C)$ or $O(3, 1)$.

11.8   *Thomas Precession*

The description of Lorentz transformations in terms of noncommuting matrices demonstrates that in general the result of successive Lorentz transformations depends on the order in which they are performed. The commutation relations (11.99) imply that two successive Lorentz transformations are equivalent to a single Lorentz transformation plus a three-dimensional rotation. An example of the kinematic consequences of the noncommutativity of Lorentz transformations is the phenomenon known as *Thomas precession*.* To motivate the discussion we first describe the physical context.

In 1926 Uhlenbeck and Goudsmit introduced the idea of electron spin and showed that, if the electron had a g factor of 2, the anomalous Zeeman effect could be explained, as well as the existence of multiplet splittings. There was a difficulty, however, in that the observed fine structure intervals were only half the theoretically expected values. If a g factor of unity were chosen, the fine structure intervals were given correctly, but the Zeeman effect was then the normal one. The complete explanation of spin, including correctly the g factor and the proper fine structure interaction, came only with the relativistic electron theory of Dirac. But within the framework of an empirical spin angular momentum and a g factor of 2, Thomas showed in 1927 that the origin of the discrepancy was a relativistic kinematic effect which, when included properly, gave both the anomalous Zeeman effect and the correct fine structure splittings. The *Thomas precession*, as it is called, also gives a qualitative explanation for a spin-orbit interaction in atomic nuclei and shows why the doublets are "inverted" in nuclei.

The Uhlenbeck–Goudsmit hypothesis was that an electron possesses a spin angular momentum **s** (which can take on quantized values of $\pm\hbar/2$ along any axis) and a magnetic moment $\boldsymbol{\mu}$ related to **s** by

$$\boldsymbol{\mu} = \frac{ge}{2mc}\,\mathbf{s} \qquad (11.100)$$

---

*L. H. Thomas, *Phil. Mag.* **3**, 1 (1927).

where the g factor has the value $g = 2$. Suppose that an electron moves with a velocity $\mathbf{v}$ in external fields $\mathbf{E}$ and $\mathbf{B}$. Then the equation of motion for its angular momentum in its rest frame is

$$\left(\frac{d\mathbf{s}}{dt}\right)_{\text{rest frame}} = \boldsymbol{\mu} \times \mathbf{B}' \tag{11.101}$$

where $\mathbf{B}'$ is the magnetic induction in that frame. We will show in Section 11.10 that in a coordinate system moving with the electron the magnetic induction is

$$\mathbf{B}' \simeq \left(\mathbf{B} - \frac{\mathbf{v}}{c} \times \mathbf{E}\right) \tag{11.102}$$

where we have neglected terms of the order of (v^2/c^2). Then (11.101) becomes

$$\left(\frac{d\mathbf{s}}{dt}\right)_{\text{rest frame}} = \boldsymbol{\mu} \times \left(\mathbf{B} - \frac{\mathbf{v}}{c} \times \mathbf{E}\right) \tag{11.103}$$

Equation (11.103) is equivalent to an energy of interaction of the electron spin:

$$U' = -\boldsymbol{\mu} \cdot \left(\mathbf{B} - \frac{\mathbf{v}}{c} \times \mathbf{E}\right) \tag{11.104}$$

In an atom the electric force $e\mathbf{E}$ can be approximated as the negative gradient of a spherically symmetric average potential energy $V(r)$. For one-electron atoms this is, of course, exact. Thus

$$e\mathbf{E} = -\frac{\mathbf{r}}{r}\frac{dV}{dr} \tag{11.105}$$

Then the spin-interaction energy can be written

$$U' = -\frac{ge}{2mc}\mathbf{s} \cdot \mathbf{B} + \frac{g}{2m^2c^2}(\mathbf{s} \cdot \mathbf{L})\frac{1}{r}\frac{dV}{dr} \tag{11.106}$$

where $\mathbf{L} = m(\mathbf{r} \times \mathbf{v})$ is the electron's orbital angular momentum. This interaction energy gives the anomalous Zeeman effect correctly, but has a spin-orbit interaction that is twice too large.

The error in (11.106) can be traced to the incorrectness of (11.101) as an equation of motion for the electron spin. The left-hand side of (11.101) gives the rate of change of spin in the rest frame of the electron. If, as Thomas first pointed out, that coordinate system rotates, then the total time rate of change of the spin, or more generally, any vector $\mathbf{G}$ is given by the well-known result,*

$$\left(\frac{d\mathbf{G}}{dt}\right)_{\text{nonrot}} = \left(\frac{d\mathbf{G}}{dt}\right)_{\text{rest frame}} + \boldsymbol{\omega}_T \times \mathbf{G} \tag{11.107}$$

where $\boldsymbol{\omega}_T$ is the angular velocity of rotation found by Thomas. When applied to the electron spin, (11.107) gives an equation of motion:

$$\left(\frac{d\mathbf{s}}{dt}\right)_{\text{nonrot}} = \mathbf{s} \times \left(\frac{ge\mathbf{B}'}{2mc} - \boldsymbol{\omega}_T\right) \tag{11.108}$$

---

*See, for example, *Goldstein* (pp. 174–177).

The corresponding energy of interaction is

$$U = U' + \mathbf{s} \cdot \boldsymbol{\omega}_T \tag{11.109}$$

where U' is the electromagnetic spin interaction (11.104) or (11.106).

The origin of the Thomas precessional frequency $\boldsymbol{\omega}_T$ is the acceleration experienced by the electron as it moves under the action of external forces. Consider an electron moving with velocity $\mathbf{v}(t)$ with respect to a laboratory inertial frame. The electron's rest frame of coordinates is defined as a co-moving sequence of inertial frames whose successive origins move at each instant with the velocity of the electron. Let the velocity of the rest frame with respect to the laboratory at laboratory time t be $\mathbf{v}(t) = c\boldsymbol{\beta}$, and at laboratory time $t + \delta t$ be $\mathbf{v}(t + \delta t) = c(\boldsymbol{\beta} + \delta\boldsymbol{\beta})$. The connection between the coordinates in the electron's rest frame at time t and the coordinates in the laboratory frame is

$$x' = A_{\text{boost}}(\boldsymbol{\beta})x \tag{11.110}$$

At time $t + \delta t$ the connection is

$$x'' = A_{\text{boost}}(\boldsymbol{\beta} + \delta\boldsymbol{\beta})x \tag{11.111}$$

It is important to note that these transformations of coordinate from the laboratory to the rest frame are defined here in terms of pure Lorentz boosts without rotations. We are interested in the behavior of the coordinate axes of the electron's rest frame as a function of time. Thus we want the connection between the two sets of rest-frame coordinates, x' at time t and x'' at time $t + \delta t$. This relation is

$$x'' = A_T x'$$

where

$$A_T = A_{\text{boost}}(\boldsymbol{\beta} + \delta\boldsymbol{\beta})A_{\text{boost}}^{-1}(\boldsymbol{\beta}) = A_{\text{boost}}(\boldsymbol{\beta} + \delta\boldsymbol{\beta})A_{\text{boost}}(-\boldsymbol{\beta}) \quad (11.112)$$

For purposes of calculating A_T a suitable choice of axes in the laboratory frame is shown in Fig. 11.7. The velocity vector $\boldsymbol{\beta}$ at time t is parallel to the 1 axis and the increment of velocity $\delta\boldsymbol{\beta}$ lies in the 1-2 plane. From (11.98) it follows that

$$A_{\text{boost}}(-\boldsymbol{\beta}) = \begin{pmatrix} \gamma & \gamma\beta & 0 & 0 \\ \gamma\beta & \gamma & 0 & 0 \\ 0 & 0 & 1 & 0 \\ 0 & 0 & 0 & 1 \end{pmatrix} \tag{11.113}$$

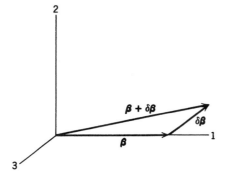

Figure 11.7

Similarly we obtain from (11.98), keeping only first-order terms in $\delta\boldsymbol{\beta}$,

$$A_{\text{boost}}(\boldsymbol{\beta} + \delta\boldsymbol{\beta}) = \begin{pmatrix} \gamma + \gamma^3\beta\,\delta\beta_1 & -(\gamma\beta + \gamma^3\,\delta\beta_1) & -\gamma\,\delta\beta_2 & 0 \\ -(\gamma\beta + \gamma^3\,\delta\beta_1) & \gamma + \gamma^3\beta\,\delta\beta_1 & \left(\dfrac{\gamma-1}{\beta}\right)\delta\beta_2 & 0 \\ -\gamma\,\delta\beta_2 & \left(\dfrac{\gamma-1}{\beta}\right)\delta\beta_2 & 1 & 0 \\ 0 & 0 & 0 & 1 \end{pmatrix} \tag{11.114}$$

Straightforward matrix multiplication according to (11.112) yields

$$A_T = \begin{pmatrix} 1 & -\gamma^2\,\delta\beta_1 & -\gamma\,\delta\beta_2 & 0 \\ -\gamma^2\,\delta\beta_1 & 1 & \left(\dfrac{\gamma-1}{\beta}\right)\delta\beta_2 & 0 \\ -\gamma\,\delta\beta_2 & -\left(\dfrac{\gamma-1}{\beta}\right)\delta\beta_2 & 1 & 0 \\ 0 & 0 & 0 & 1 \end{pmatrix} \tag{11.115}$$

This represents an infinitesimal Lorentz transformation that can be written in terms of the matrices S and K as

$$A_T = I - \left(\frac{\gamma-1}{\beta^2}\right)(\boldsymbol{\beta} \times \delta\boldsymbol{\beta}) \cdot \mathbf{S} - (\gamma^2\,\delta\boldsymbol{\beta}_\parallel + \gamma\,\delta\boldsymbol{\beta}_\perp) \cdot \mathbf{K} \tag{11.116}$$

where $\delta\boldsymbol{\beta}_\parallel$ and $\delta\boldsymbol{\beta}_\perp$ are the components of $\delta\boldsymbol{\beta}$ parallel and perpendicular to $\boldsymbol{\beta}$, respectively. To first order in $\delta\boldsymbol{\beta}$, (11.116) is equivalent to

$$A_T = A_{\text{boost}}(\Delta\boldsymbol{\beta})R(\Delta\boldsymbol{\Omega}) = R(\Delta\boldsymbol{\Omega})A_{\text{boost}}(\delta\boldsymbol{\beta}) \tag{11.117}$$

where

$$A_{\text{boost}}(\Delta\boldsymbol{\beta}) = I - \Delta\boldsymbol{\beta} \cdot \mathbf{K}$$
$$R(\Delta\boldsymbol{\Omega}) = I - \Delta\boldsymbol{\Omega} \cdot \mathbf{S}$$

are commuting infinitesimal boosts and rotations, with velocity,

$$\Delta\boldsymbol{\beta} = \gamma^2\,\delta\boldsymbol{\beta}_\parallel + \gamma\,\delta\boldsymbol{\beta}_\perp$$

and angle of rotation,

$$\Delta\boldsymbol{\Omega} = \left(\frac{\gamma-1}{\beta^2}\right)\boldsymbol{\beta} \times \delta\boldsymbol{\beta} = \frac{\gamma^2}{\gamma+1}\boldsymbol{\beta} \times \delta\boldsymbol{\beta}$$

Thus the pure Lorentz boost (11.111) to the frame with velocity $c(\boldsymbol{\beta} + \delta\boldsymbol{\beta})$ is equivalent to a boost (11.110) to a frame moving with velocity $c\boldsymbol{\beta}$, followed by an infinitesimal Lorentz transformation consisting of a boost with velocity $c\,\Delta\boldsymbol{\beta}$ *and* a rotation $\Delta\boldsymbol{\Omega}$.

In terms of the interpretation of the moving frames as successive rest frames of the electron we do not want rotations as well as boosts. Nonrelativistic equations of motion like (11.101) can be expected to hold provided the evolution of the rest frame is described by infinitesimal boosts *without* rotations. We are thus led to consider the rest-frame coordinates at time $t + \delta t$ that are given from those at time t by the boost $A_{\text{boost}}(\Delta\boldsymbol{\beta})$ instead of A_T. Denoting these coordinates by x''', we have

$$x''' = A_{\text{boost}}(\Delta\boldsymbol{\beta})x'$$

Using (11.117), (11.112), and (11.110) we can express x''' in terms of the laboratory coordinates as

$$x''' = R(-\Delta\boldsymbol{\Omega})A_{\text{boost}}(\boldsymbol{\beta} + \delta\boldsymbol{\beta})x \tag{11.118}$$

The rest system of coordinates defined by x''' is rotated by $-\Delta\boldsymbol{\Omega}$ relative to the boosted laboratory axes (x''). If a physical vector $\mathbf{G}$ has a (proper) time rate of change $(d\mathbf{G}/d\tau)$ in the rest frame, the precession of the rest-frame axes with respect to the laboratory makes the vector have a total time rate of change with respect to the laboratory axes of (11.107), with

$$\boldsymbol{\omega}_T = -\lim_{\delta t \to 0} \frac{\Delta\boldsymbol{\Omega}}{\delta t} = \frac{\gamma^2}{\gamma + 1} \frac{\mathbf{a} \times \mathbf{v}}{c^2} \tag{11.119}$$

where $\mathbf{a}$ is the acceleration in the laboratory frame and, to be precise, $(d\mathbf{G}/dt)_{\text{rest frame}} = \gamma^{-1}(d\mathbf{G}/d\tau)_{\text{rest frame}}$.

The Thomas precession is purely kinematical in origin. If a component of acceleration exists perpendicular to $\mathbf{v}$, for whatever reason, then there is a Thomas precession, independent of other effects such as precession of the magnetic moment in a magnetic field.

For electrons in atoms the acceleration is caused by the screened Coulomb field (11.105). Thus the Thomas angular velocity is

$$\boldsymbol{\omega}_T \simeq \frac{-1}{2c^2} \frac{\mathbf{r} \times \mathbf{v}}{m} \frac{1}{r} \frac{dV}{dr} = \frac{-1}{2m^2c^2} \mathbf{L} \frac{1}{r} \frac{dV}{dr} \tag{11.120}$$

It is evident from (11.109) and (11.106) that the extra contribution to the energy from the Thomas precession reduces the spin-orbit coupling, yielding

$$U = -\frac{ge}{2mc} \mathbf{s} \cdot \mathbf{B} + \frac{(g-1)}{2m^2c^2} \mathbf{s} \cdot \mathbf{L} \frac{1}{r} \frac{dV}{dr} \tag{11.121}$$

With $g = 2$ the spin-orbit interaction of (11.106) is reduced by $\frac{1}{2}$ (sometimes called the *Thomas factor*), as required for the correct spin-orbit interaction energy of an atomic electron.

In atomic nuclei the nucleons experience strong accelerations because of the specifically nuclear forces. The electromagnetic forces are comparatively weak. In an approximate way one can treat the nucleons as moving separately in a short-range, spherically symmetric, attractive, potential well, $V_N(r)$. Then each nucleon will experience in addition a spin-orbit interaction given by (11.109) with the negligible electromagnetic contribution U' omitted:

$$U_N \simeq \mathbf{s} \cdot \boldsymbol{\omega}_T \tag{11.122}$$

where the acceleration in $\boldsymbol{\omega}_T$ is determined by $V_N(r)$. The form of $\boldsymbol{\omega}_T$ is the same as (11.120) with V replaced by V_N. Thus the nuclear spin-orbit interaction is approximately

$$U_N \simeq -\frac{1}{2M^2c^2} \mathbf{s} \cdot \mathbf{L} \frac{1}{r} \frac{dV_N}{dr} \tag{11.123}$$

In comparing (11.123) with atomic formula (11.121) we note that both V and V_N are attractive (although V_N is much larger), so that the signs of the spin-orbit energies are opposite. This means that in nuclei the single particle levels form "inverted" doublets. With a reasonable form for V_N, (11.123) is in qualitative agreement with the observed spin-orbit splittings in nuclei.*

The phenomenon of Thomas precession is presented from a more sophisticated point of view in Section 11.11 where the BMT equation is discussed.

11.9 Invariance of Electric Charge; Covariance of Electrodynamics

The invariance in form of the equations of electrodynamics under Lorentz transformations was shown by Lorentz and Poincaré before the formulation of the special theory of relativity. This invariance of form or *covariance* of the Maxwell and Lorentz force equations implies that the various quantities ρ, $\mathbf{J}$, $\mathbf{E}$, $\mathbf{B}$ that enter these equations transform in well-defined ways under Lorentz transformations. Then the terms of the equations can have consistent behavior under Lorentz transformations.

Consider first the Lorentz force equation for a particle of charge q,

$$\frac{d\mathbf{p}}{dt} = q\left(\mathbf{E} + \frac{\mathbf{v}}{c} \times \mathbf{B}\right) \tag{11.124}$$

We know that $\mathbf{p}$ transforms as the space part of the 4-vector of energy and momentum,

$$p^\alpha = (p_0, \mathbf{p}) = m(U_0, \mathbf{U})$$

where $p_0 = E/c$ and U^α is the 4-velocity (11.36). If we use the proper time τ (11.26) instead of t for differentiation, (11.124) can be written

$$\frac{d\mathbf{p}}{d\tau} = \frac{q}{c}(U_0\mathbf{E} + \mathbf{U} \times \mathbf{B}) \tag{11.125}$$

The left-hand side is the space part of a 4-vector. The corresponding time component equation is the rate of change of energy of the particle (6.110):

$$\frac{dp_0}{d\tau} = \frac{q}{c}\mathbf{U} \cdot \mathbf{E} \tag{11.126}$$

If the force and energy change equations are to be Lorentz covariant, the right-hand sides must form the components of a 4-vector. They involve products of three factors, the charge q, the 4-velocity, and the electromagnetic fields. If the transformation properties of two of the three factors are known and Lorentz covariance is demanded, then the transformation properties of the third factor can be established.

Electric charge is absolutely conserved, as far as we know. Furthermore, the magnitudes of the charges of elementary particles (and therefore of any system

*See, for example, Section. 2.4c of A. Bohr and B. R. Mottelson, *Nuclear Structure*, Vol. 1, W. A. Benjamin, New York (1969).

of charges) are integral multiples of the charge of the proton. In the published literature,* it is experimentally established that the fractional difference between the magnitude of the electron's charge and the proton's charge is less than 10^{-19}, and unpublished results of King push this limit almost two orders of magnitude further.† The results of these experiments can be used to support the *invariance of electric charge* under Lorentz transformations or, more concretely, the independence of the observed charge of a particle on its speed. In his experiments King searched for a residual charge remaining in a container as hydrogen or helium gas is allowed to escape. No effect was observed and a limit of less than $10^{-19}e$ was established for the net charge per molecule for both H_2 and He. Since the electrons in He move at speeds twice as fast as in H_2, the charge of the electron cannot depend significantly on its speed, at least for speeds of the order of $(0.01-0.02)c$. In the experiment of Fraser, Carlson, and Hughes an atomic beam apparatus was used in an attempt to observe electrostatic deflection of beams of "neutral" cesium and potassium atoms. Again, no effect was observed, and a limit of less than 3.5×10^{-19} was set on the fractional difference between the charges of the proton and electron. Cesium and potassium have $Z = 55$ and 19, respectively. Thus the K-shell electrons in cesium at least move with speeds of order $0.4c$. The observed neutrality of the cesium atom at the level of 10^{-18}–10^{-19} is strong evidence for the invariance of electric charge.‡

The *experimental* invariance of electric charge and the requirement of Lorentz covariance of the Lorentz force equation (11.125) and (11.126) determines the Lorentz transformation properties of the electromagnetic field. For example, the requirement from (11.126) that $\mathbf{U} \cdot \mathbf{E}$ be the time component of a 4-vector establishes that the components of $\mathbf{E}$ are the time-space parts of a second rank tensor $F^{\alpha\beta}$, that is, $\mathbf{E} \cdot \mathbf{U} = F^{0\beta}U_\beta$. Although the explicit form of the field strength tensor $F^{\alpha\beta}$ can be found along these lines, we now proceed to examine the Maxwell equations themselves.

For simplicity, we consider the microscopic Maxwell equations, without $\mathbf{D}$ and $\mathbf{H}$. We begin with the charge density $\rho(\mathbf{x}, t)$ and current density $\mathbf{J}(\mathbf{x}, t)$ and the continuity equation

$$\frac{\partial \rho}{\partial t} + \boldsymbol{\nabla} \cdot \mathbf{J} = 0 \tag{11.127}$$

From the discussion at the end of Section 11.6 and especially (11.77) it is natural to postulate that ρ and $\mathbf{J}$ together form a 4-vector J^α:

$$J^\alpha = (c\rho, \mathbf{J}) \tag{11.128}$$

*J. G. King, *Phys. Rev. Lett.* **5**, 562 (1960); V. W. Hughes, L. J. Fraser, and E. R. Carlson, *Z. Phys. D-Atoms, Molecules and Clusters* **10**, 145 (1988). The latter tabulates many of the different methods and results.

†The limits on the measured charge per molecule in units of the electronic charge for H_2, He, and SF_6 were determined as 1.8 ± 5.4, -0.7 ± 4.7, 0 ± 4.3, respectively, all times 10^{-21}. Private communication from J. G. King (1975).

‡Mentioning only the electrons is somewhat misleading. The protons and neutrons inside nuclei move with speeds of the order $(0.2-0.3)c$. Thus the helium results of King already test the invariance of charge at appreciable speeds. Of course, if one is content with invariance at the level of 10^{-10} for $v/c \sim 10^{-3}$ the observed electrical neutrality of bulk matter when heated or cooled will suffice.

Then the continuity equation (11.127) takes the obviously covariant form,

$$\partial_\alpha J^\alpha = 0 \qquad (11.129)$$

where the covariant differential operator ∂_α is given by (11.76). That J^α is a legitimate 4-vector follows from the invariance of electric charge: Consider a large number of elementary charges totaling δq at rest* in a small-volume element d^3x in frame K. They are idealized by a charge density ρ. The total charge $\delta q = \rho\, d^3x$ within the small-volume element is an experimental invariant; it is thus true that $\rho'\, d^3x' = \rho\, d^3x$. But the *four*-dimensional volume element $d^4x = dx^0\, d^3x$ is a Lorentz invariant:

$$d^4x' = \frac{\partial(x'^0, x'^1, x'^2, x'^3)}{\partial(x^0, x^1, x^2, x^3)}\, d^4x = \det A\, d^4x = d^4x$$

The equality $\rho'\, d^3x' = \rho\, d^3x$ then implies that $c\rho$ transforms like x^0, namely, the time component of the 4-vector (11.128).

In the Lorenz family of gauges the wave equations for the vector potential **A** and the scalar potential Φ are

$$\frac{1}{c^2}\frac{\partial^2 \mathbf{A}}{\partial t^2} - \nabla^2 \mathbf{A} = \frac{4\pi}{c}\mathbf{J}$$

$$\frac{1}{c^2}\frac{\partial^2 \Phi}{\partial t^2} - \nabla^2 \Phi = 4\pi\rho \qquad (11.130)$$

with the Lorenz condition,

$$\frac{1}{c}\frac{\partial \Phi}{\partial t} + \nabla \cdot \mathbf{A} = 0 \qquad (11.131)$$

The differential operator form in (11.130) is the invariant four-dimensional Laplacian (11.78), while the right-hand sides are the components of a 4-vector. Obviously, Lorentz covariance requires that the potentials Φ and **A** form a 4-vector potential,

$$A^\alpha = (\Phi, \mathbf{A}) \qquad (11.132)$$

Then the wave equations and the Lorenz condition take on the manifestly covariant forms,

$$\Box A^\alpha = \frac{4\pi}{c}J^\alpha$$

and $\qquad (11.133)$

$$\partial_\alpha A^\alpha = 0$$

The fields **E** and **B** are expressed in terms of the potentials as

$$\mathbf{E} = -\frac{1}{c}\frac{\partial \mathbf{A}}{\partial t} - \nabla\Phi$$

$$\mathbf{B} = \nabla \times \mathbf{A} \qquad (11.134)$$

---

*If there is a conduction current **J** as well as the charge density ρ in K, the total charge within d^3x is not an invariant. See *Møller*, Section 7.5. (His argument assumes the 4-vector character of $c\rho$ and **J**, however.)

The x components of $\mathbf{E}$ and $\mathbf{B}$ are explicitly

$$E_x = -\frac{1}{c}\frac{\partial A_x}{\partial t} - \frac{\partial \Phi}{\partial x} = -(\partial^0 A^1 - \partial^1 A^0) \tag{11.135}$$

$$B_x = \frac{\partial A_z}{\partial y} - \frac{\partial A_y}{\partial z} = -(\partial^2 A^3 - \partial^3 A^2)$$

where the second forms follow from (11.132) and $\partial^\alpha = (\partial/\partial x_0, -\boldsymbol{\nabla})$. These equations imply that the electric and magnetic fields, six components in all, are the elements of a *second-rank, antisymmetric field-strength tensor*,

$$F^{\alpha\beta} = \partial^\alpha A^\beta - \partial^\beta A^\alpha \tag{11.136}$$

Explicitly, the field-strength tensor is, in matrix form,

$$F^{\alpha\beta} = \begin{pmatrix} 0 & -E_x & -E_y & -E_z \\ E_x & 0 & -B_z & B_y \\ E_y & B_z & 0 & -B_x \\ E_z & -B_y & B_x & 0 \end{pmatrix} \tag{11.137}$$

For reference, we record the field-strength tensor with two covariant indices,

$$F_{\alpha\beta} = g_{\alpha\gamma}F^{\gamma\delta}g_{\delta\beta} = \begin{pmatrix} 0 & E_x & E_y & E_z \\ -E_x & 0 & -B_z & B_y \\ -E_y & B_z & 0 & -B_x \\ -E_z & -B_y & B_x & 0 \end{pmatrix} \tag{11.138}$$

The elements of $F_{\alpha\beta}$ are obtained from $F^{\alpha\beta}$ by putting $\mathbf{E} \to -\mathbf{E}$. Another useful quantity is the *dual field-strength tensor* $\mathcal{F}^{\alpha\beta}$. We first define the totally antisymmetric fourth-rank tensor $\epsilon^{\alpha\beta\gamma\delta}$:

$$\epsilon^{\alpha\beta\gamma\delta} = \begin{cases} +1 & \text{for } \alpha = 0, \beta = 1, \gamma = 2, \delta = 3, \text{ and} \\ & \text{any even permutation} \\ -1 & \text{for any odd permutation} \\ 0 & \text{if any two indices are equal} \end{cases} \tag{11.139}$$

Note that the nonvanishing elements all have one time and three (different) space indices and that $\epsilon_{\alpha\beta\gamma\delta} = -\epsilon^{\alpha\beta\gamma\delta}$. The tensor $\epsilon^{\alpha\beta\gamma\delta}$ is a *pseudotensor* under spatial inversions. This can be seen by contracting it with four different 4-vectors and examining the space inversion properties of the resultant rotationally invariant quantity. The dual field-strength tensor is defined by

$$\mathcal{F}^{\alpha\beta} = \tfrac{1}{2}\epsilon^{\alpha\beta\gamma\delta}F_{\gamma\delta} = \begin{pmatrix} 0 & -B_x & -B_y & -B_z \\ B_x & 0 & E_z & -E_y \\ B_y & -E_z & 0 & E_x \\ B_z & E_y & -E_x & 0 \end{pmatrix} \tag{11.140}$$

The elements of the dual tensor $\mathcal{F}^{\alpha\beta}$ are obtained from $F^{\alpha\beta}$ by putting $\mathbf{E} \to \mathbf{B}$ and $\mathbf{B} \to -\mathbf{E}$ in (11.137). This is a special case of the duality transformation (6.151).

To complete the demonstration of the covariance of electrodynamics we must write the Maxwell equations themselves in an explicitly covariant form. The inhomogeneous equations are

$$\nabla \cdot \mathbf{E} = 4\pi\rho$$

$$\nabla \times \mathbf{B} - \frac{1}{c}\frac{\partial \mathbf{E}}{\partial t} = \frac{4\pi}{c}\mathbf{J}$$

In terms of $F^{\alpha\beta}$ and the 4-current J^α these take on the covariant form

$$\partial_\alpha F^{\alpha\beta} = \frac{4\pi}{c} J^\beta \tag{11.141}$$

Similarly, the homogeneous Maxwell equations

$$\nabla \cdot \mathbf{B} = 0, \qquad \nabla \times \mathbf{E} + \frac{1}{c}\frac{\partial \mathbf{B}}{\partial t} = 0$$

can be written in terms of the dual field-strength tensor as

$$\partial_\alpha \mathscr{F}^{\alpha\beta} = 0 \tag{11.142}$$

In terms of $F^{\alpha\beta}$, rather than $\mathscr{F}^{\alpha\beta}$, these homogeneous equations are the four equations

$$\partial_\alpha F_{\beta\gamma} + \partial_\beta F_{\gamma\alpha} + \partial_\gamma F_{\alpha\beta} = 0 \tag{11.143}$$

where α, β, γ are any three of the integers 0, 1, 2, 3.

With the definitions of J^α (11.128), A^α (11.132), and $F^{\alpha\beta}$ (11.136), together with the wave equations (11.133) or the Maxwell equations (11.141) and (11.142), the covariance of the equations of electromagnetism is established. To complete the discussion, we put the Lorentz force and rate of change of energy equations (11.125) and (11.126) in manifestly covariant form,

$$\frac{dp^\alpha}{d\tau} = m\frac{dU^\alpha}{d\tau} = \frac{q}{c} F^{\alpha\beta}U_\beta \tag{11.144}$$

The covariant description of the conservation laws of a combined system of electromagnetic fields and charged particles and a covariant solution for the fields of a moving charge are deferred to Chapter 12, where a Lagrangian formulation is presented.

For the macroscopic Maxwell equations it is necessary to distinguish two field-strength tensors, $F^{\alpha\beta} = (\mathbf{E}, \mathbf{B})$ and $G^{\alpha\beta} = (\mathbf{D}, \mathbf{H})$, where $F^{\alpha\beta}$ is given by (11.137) and $G^{\alpha\beta}$ is obtained from (11.137) by substituting $\mathbf{E} \to \mathbf{D}$ and $\mathbf{B} \to \mathbf{H}$. The covariant form of the Maxwell equations is then

$$\partial_\alpha G^{\alpha\beta} = \frac{4\pi}{c} J^\beta, \qquad \partial_\alpha \mathscr{F}^{\alpha\beta} = 0 \tag{11.145}$$

It is clear that with the fields $(\mathbf{E}, \mathbf{B})$ and $(\mathbf{D}, \mathbf{H})$ transforming as antisymmetric second-rank tensors the polarization $\mathbf{P}$ and the negative magnetization $-\mathbf{M}$ form a similar second-rank tensor. With these quantities given meaning as macroscopic averages of atomic properties *in the rest frame* of the medium, the electrodynamics of macroscopic matter in motion is specified. This is the basis of the electrodynamics of Minkowski and others. For further information on this rather large and important subject, the reader can consult the literature cited at the end of the chapter.

11.10   Transformation of Electromagnetic Fields

Since the fields **E** and **B** are the elements of a second-rank tensor $F^{\alpha\beta}$, their values in one inertial frame K' can be expressed in terms of the values in another inertial frame K according to

$$F'^{\alpha\beta} = \frac{\partial x'^{\alpha}}{\partial x^{\gamma}} \frac{\partial x'^{\beta}}{\partial x^{\delta}} F^{\gamma\delta} \tag{11.146}$$

In the matrix notation of Section 11.7 this can be written

$$F' = AF\tilde{A} \tag{11.147}$$

where F and F' are 4×4 *matrices* (11.137) and A is the Lorentz transformation matrix of (11.93). For the specific Lorentz transformation (11.95), corresponding to a boost along the x_1 axis with speed $c\beta$ from the unprimed frame to the primed frame, the explicit equations of transformation are

$$
\begin{array}{ll}
E_1' = E_1 & B_1' = B_1 \\
E_2' = \gamma(E_2 - \beta B_3) & B_2' = \gamma(B_2 + \beta E_3) \\
E_3' = \gamma(E_3 + \beta B_2) & B_3' = \gamma(B_3 - \beta E_2)
\end{array} \tag{11.148}
$$

Here and below, the subscripts 1, 2, 3 indicate ordinary Cartesian spatial components and are not covariant indices. The inverse of (11.148) is found, as usual, by interchanging primed and unprimed quantities and putting $\beta \to -\beta$. For a general Lorentz transformation from K to a system K' moving with velocity **v** relative to K, the transformation of the fields can be written

$$
\begin{aligned}
\mathbf{E}' &= \gamma(\mathbf{E} + \boldsymbol{\beta} \times \mathbf{B}) - \frac{\gamma^2}{\gamma + 1} \boldsymbol{\beta}(\boldsymbol{\beta} \cdot \mathbf{E}) \\
\mathbf{B}' &= \gamma(\mathbf{B} - \boldsymbol{\beta} \times \mathbf{E}) - \frac{\gamma^2}{\gamma + 1} \boldsymbol{\beta}(\boldsymbol{\beta} \cdot \mathbf{B})
\end{aligned} \tag{11.149}
$$

These are the analogs for the fields of (11.19) for the coordinates. Transformation (11.149) shows that **E** and **B** have no independent existence. A purely electric or magnetic field in one coordinate system will appear as a mixture of electric and magnetic fields in another coordinate frame. Of course certain restrictions apply (see Problem 11.14) so that, for example, a purely electrostatic field in one coordinate system cannot be transformed into a purely magnetostatic field in another. But the fields are completely interrelated, and one should properly speak of the electromagnetic field $F^{\alpha\beta}$, rather than **E** or **B** separately.

If no magnetic field exists in a certain frame K', as for example with one or more point charges at rest in K', the inverse of (11.149) shows that in the frame K the magnetic field **B** and electric field **E** are linked by the simple relation

$$\mathbf{B} = \boldsymbol{\beta} \times \mathbf{E} \tag{11.150}$$

Note that **E** is not the electrostatic field in K', but that field transformed from K' to K.

As an important and illuminating example of the transformation of fields, we consider the fields seen by an observer in the system K when a point charge q moves by in a straight-line path with a velocity **v**. The charge is at rest in the system K', and the transformation of the fields is given by the inverse of (11.148) or (11.149). We suppose that the charge moves in the positive x_1 direction and that its closest distance of approach to the observer is b. Figure 11.8 shows a suitably chosen set of axes. The observer is at the point P. At $t = t' = 0$ the origins of the two coordinate systems coincide and the charge q is at its closest distance to the observer. In the frame K' the observer's point P, where the fields are to be evaluated, has coordinates $x_1' = -vt', x_2' = b, x_3' = 0$, and is a distance $r' = \sqrt{b^2 + (vt')^2}$ away from q. We will need to express r' in terms of the coordinates in K. The only coordinate needing transformation is the time $t' = \gamma[t - (v/c^2)x_1] = \gamma t$, since $x_1 = 0$ for the point P in the frame K. In the rest frame K' of the charge the electric and magnetic fields at the observation point are

$$E_1' = -\frac{qvt'}{r'^3}, \qquad E_2' = \frac{qb}{r'^3}, \qquad E_3' = 0$$

$$B_1' = 0, \qquad B_2' = 0, \qquad B_3' = 0$$

In terms of the coordinates of K the nonvanishing field components are

$$E_1' = -\frac{q\gamma vt}{(b^2 + \gamma^2 v^2 t^2)^{3/2}}, \qquad E_2' = \frac{qb}{(b^2 + \gamma^2 v^2 t^2)^{3/2}} \qquad (11.151)$$

Then, using the inverse of (11.148), we find the transformed fields in the system K:

$$E_1 = E_1' = -\frac{q\gamma vt}{(b^2 + \gamma^2 v^2 t^2)^{3/2}}$$

$$E_2 = \gamma E_2' = \frac{\gamma qb}{(b^2 + \gamma^2 v^2 t^2)^{3/2}} \qquad (11.152)$$

$$B_3 = \gamma\beta E_2' = \beta E_2$$

with the other components vanishing.

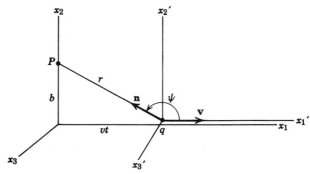

Figure 11.8 Particle of charge q moving at constant velocity **v** passes an observation point P at impact parameter b.

Fields (11.152) exhibit interesting behavior when the velocity of the charge approaches that of light. First of all there is observed a magnetic induction in the x_3 direction already displayed in (11.150). This magnetic field becomes almost equal to the transverse electric field E_2 as $\beta \to 1$. Even at nonrelativistic velocities where $\gamma \simeq 1$, this magnetic induction is equivalent to

$$\mathbf{B} \simeq \frac{q}{c} \frac{\mathbf{v} \times \mathbf{r}}{r^3}$$

which is just the approximate Ampère–Biot–Savart expression for the magnetic field of a moving charge. At high speeds when $\gamma \gg 1$ we see that the peak transverse electric field E_2 $(t = 0)$ becomes equal to γ times its nonrelativistic value. In the same limit, however, the duration of appreciable field strengths at the point P is decreased. A measure of the time interval over which the fields are appreciable is evidently

$$\Delta t \simeq \frac{b}{\gamma v} \tag{11.153}$$

As γ increases, the peak fields increase in proportion, but their duration goes in inverse proportion. The time integral of the fields times v is independent of velocity. Figure 11.9a shows this behavior of the transverse electric and magnetic fields and the longitudinal electric field. For $\beta \to 1$ the observer at P sees nearly equal transverse and mutually perpendicular electric and magnetic fields. These are indistinguishable from the fields of a pulse of plane polarized radiation propagating in the x_1 direction. The extra longitudinal electric field varies rapidly from positive to negative and has zero time integral. If the observer's detecting apparatus has any significant inertia, it will not respond to this longitudinal field. Consequently for practical purposes he will see only the transverse fields. This equivalence of the fields of a relativistic charged particle and those of a pulse of electromagnetic radiation will be exploited in Chapter 15. In Problem 11.18 the fields for $\beta = 1$ are given an explicit realization.

The fields (11.152) and the curves of Fig. 11.9a emphasize the time dependence of the fields at a fixed observation point. An alternative description can be given in terms of the spatial variation of the fields relative to the instantaneous *present position* of the charge in the laboratory. From (11.152) we see that $E_1/E_2 = -vt/b$. Reference to Fig. 11.8 shows that the electric field is thus directed along $\mathbf{n}$, a unit radial vector from the charge's present position to the observation point, just as for a static Coulomb field. By expressing the denominator in (11.152) in terms of r, the radial distance from the present position to the observer, and the angle $\psi = \cos^{-1}(\mathbf{n} \cdot \hat{\mathbf{v}})$ shown in Fig. 11.8, we obtain the electric field in terms of the charge's present position:

$$\mathbf{E} = \frac{q\mathbf{r}}{r^3 \gamma^2 (1 - \beta^2 \sin^2\psi)^{3/2}} \tag{11.154}$$

The magnetic induction is given by (11.150). The electric field is radial, but the lines of force are isotropically distributed only for $\beta = 0$. Along the direction of motion ($\psi = 0, \pi$), the field strength is down by a factor of γ^{-2} relative to isotropy, while in the transverse directions ($\psi = \pi/2$) it is larger by a factor of γ. This whiskbroom pattern of lines of force, shown in Fig. 11.9b, is the spatial "snapshot" equivalent of the temporal behavior sketched in Fig. 11.9a. The compres-

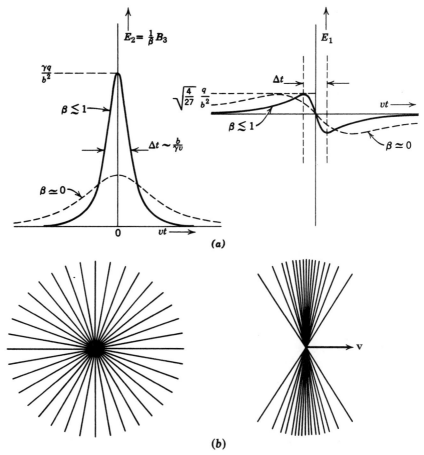

Figure 11.9   Fields of a uniformly moving charged particle. (*a*) Fields at the observation point *P* in Fig. 11.8 as a function of time. (*b*) Lines of electric force for a particle at rest and in motion ($\gamma = 3$). The field lines emanate from the *present* position of the charge.

sion of the lines of force in the transverse direction can be viewed as a consequence of the FitzGerald–Lorentz contraction.

11.11   Relativistic Equation of Motion for Spin in Uniform or Slowly Varying External Fields

The effects of a particle's motion on the precession of its spin have already been discussed in Section 11.8 on Thomas precession. Here we exploit the ideas of Lorentz covariance to give an alternative, more elegant discussion leading to what is known as the BMT equation of motion for the spin.* With the magnetic

---

*Named, not after one of the New York City subway lines, but for V. Bargmann, L. Michel, and V. L. Telegdi, *Phys. Rev. Lett.* **2**, 435 (1959). The equation actually has much earlier origins; Thomas published an equivalent in 1927 (*op. cit.*); Frenkel discussed similar equations contemporaneously; Kramers considered the $g = 2$ equation in the 1930s.

moment given by (11.100), the rest frame equation of motion for the spin, (11.101), is

$$\frac{d\mathbf{s}}{dt'} = \frac{ge}{2mc}\, \mathbf{s} \times \mathbf{B}' \tag{11.155}$$

where primes denote quantities defined in the rest frame and $\mathbf{s}$ is the spin in that frame. This equation applies to a particle of mass m, charge e, spin $\mathbf{s}$ and a magnetic dipole moment with Landé g factor of g. It is a classical equation, but is the same as the quantum-mechanical Heisenberg equation of motion for the spin operator or, equivalently, the equation of motion for the polarization vector of the system.

A. Covariant Equation of Motion

To obtain a relativistic generalization of (11.155) it is first necessary to generalize the spin $\mathbf{s}$ from a 3-vector in the particle's rest frame. There are two avenues open. One is to recall from the end of Section 11.9 that $\mathbf{P}$ and $-\mathbf{M}$ form an antisymmetric second-rank tensor. This suggests that $\boldsymbol{\mu}$, hence $\mathbf{s}$, may be generalized to a second-rank tensor $S^{\alpha\beta}$. A simpler alternative is to define an axial *4-vector* S^α in such a manner that it has only three independent components and reduces to the spin $\mathbf{s}$ in the particle's rest frame.* If S^α denotes the components of the spin 4-vector in the inertial frame K, the time-component in the rest frame K' is, according to (11.22),

$$S'^0 = \gamma(S^0 - \boldsymbol{\beta}\cdot\mathbf{S}) = \frac{1}{c}\,U_\alpha S^\alpha$$

where U^α is the particle's 4-velocity. We see that the vanishing of the time-component in the rest frame is imposed by the covariant constraint,

$$U_\alpha S^\alpha = 0 \tag{11.156}$$

In an inertial frame where the particle's velocity is $c\boldsymbol{\beta}$ the time component of spin is therefore not independent, but is

$$S_0 = \boldsymbol{\beta}\cdot\mathbf{S} \tag{11.157}$$

It is useful to display the explicit connection between S^α and the rest-frame spin $\mathbf{s}$. Use of (11.19) or (11.22) and (11.157) yields

$$\mathbf{s} = \mathbf{S} - \frac{\gamma}{\gamma+1}\,(\boldsymbol{\beta}\cdot\mathbf{S})\boldsymbol{\beta} \tag{11.158}$$

and the inverse expressions

$$\mathbf{S} = \mathbf{s} + \frac{\gamma^2}{\gamma+1}\,(\boldsymbol{\beta}\cdot\mathbf{s})\boldsymbol{\beta} \tag{11.159}$$

$$S_0 = \gamma\boldsymbol{\beta}\cdot\mathbf{s}$$

Specification of the rest-frame 3-vector spin $\mathbf{s}$ determines the components of the 4-vector spin S^α in any inertial frame.

---

*The spin 4-vector S^α is the dual of the tensor $S^{\alpha\beta}$ in the sense that $S^\alpha = (1/2c)\epsilon^{\alpha\beta\gamma\delta}U_\beta S_{\gamma\delta}$, where U^α is the particle's 4-velocity.

The obvious generalization of the left-hand side of (11.155) is $dS^\alpha/d\tau$, where τ is the particle's proper time. The right-hand side must therefore be expressible as a 4-vector. We assume that the equation is linear in the spin S^α and the external fields $F^{\alpha\beta}$. It can also involve U^α and $dU^\alpha/d\tau$, the latter being linear in $F^{\alpha\beta}$ itself. Higher time derivatives are assumed absent. And of course the equation must reduce to (11.155) in the rest frame. With the building blocks S^α, $F^{\alpha\beta}$, U^α, $dU^\alpha/d\tau$ and the requirement of linearity in S^α and $F^{\alpha\beta}$, we can construct the 4-vectors,

$$F^{\alpha\beta}S_\beta, \qquad (S_\lambda F^{\lambda\mu}U_\mu)U^\alpha, \qquad \left(S_\beta \frac{dU^\beta}{d\tau}\right)U^\alpha$$

Other possibilities, such as $F^{\alpha\beta}U_\beta(S_\lambda U^\lambda)$, $(U_\lambda F^{\lambda\mu}U_\mu)S^\alpha$, and $(S_\lambda F^{\lambda\mu}U_\mu)\,dU^\alpha/d\tau$, either vanish, are higher order in $F^{\alpha\beta}$, or reduce to multiples of the three above. The equation of motion must therefore be of the form

$$\frac{dS^\alpha}{d\tau} = A_1 F^{\alpha\beta}S_\beta + \frac{A_2}{c^2}(S_\lambda F^{\lambda\mu}U_\mu)U^\alpha + \frac{A_3}{c^2}\left(S_\beta \frac{dU^\beta}{d\tau}\right)U^\alpha \qquad (11.160)$$

where A_1, A_2, A_3 are constants. The constraint equation (11.156) must hold at all times. This requires

$$\frac{d}{d\tau}(U_\alpha S^\alpha) = S^\alpha \frac{dU_\alpha}{d\tau} + U_\alpha \frac{dS^\alpha}{d\tau} = 0$$

hence

$$(A_1 - A_2)U_\alpha F^{\alpha\beta}S_\beta + (1 + A_3)S_\beta \frac{dU^\beta}{d\tau} = 0 \qquad (11.161)$$

If nonelectromagnetic or field gradient forces are allowed, at least in principle, it is necessary that $A_1 = A_2$ and $A_3 = -1$. Reduction to the rest frame and comparison with (11.155) gives $A_1 = ge/2mc$. Thus (11.160) becomes

$$\frac{dS^\alpha}{d\tau} = \frac{ge}{2mc}\left[F^{\alpha\beta}S_\beta + \frac{1}{c^2}U^\alpha(S_\lambda F^{\lambda\mu}U_\mu)\right] - \frac{1}{c^2}U^\alpha\left(S_\lambda \frac{dU^\lambda}{d\tau}\right) \qquad (11.162)$$

If the electromagnetic fields are uniform in space, or if gradient force terms like $\nabla(\boldsymbol{\mu} \cdot \mathbf{B})$, (5.69), can be neglected, and there are no other appreciable forces on the particle, its translational motion is described by (11.144):

$$\frac{dU^\alpha}{d\tau} = \frac{e}{mc}F^{\alpha\beta}U_\beta \qquad (11.163)$$

Then (11.162) becomes the *BMT equation*:

$$\frac{dS^\alpha}{d\tau} = \frac{e}{mc}\left[\frac{g}{2}F^{\alpha\beta}S_\beta + \frac{1}{c^2}\left(\frac{g}{2} - 1\right)U^\alpha(S_\lambda F^{\lambda\mu}U_\mu)\right] \qquad (11.164)$$

B. Connection to the Thomas Precession

The covariant equation (11.162), or its special case (11.164), contain the Thomas precession of the spin. It occurs in the final term in (11.162), the term that was specified by the requirement (11.156) that the spin 4-vector be orthog-

onal to the 4-velocity. To exhibit the Thomas precession explicitly, we consider
the equation of motion for the rest-frame spin **s**. Using the result

$$S_\lambda \frac{dU^\lambda}{d\tau} = -\gamma \mathbf{S} \cdot \frac{d\mathbf{v}}{d\tau} \tag{11.165}$$

and (11.158) for **s** in terms of **S**, we find that the equations,

$$\frac{d\mathbf{S}}{d\tau} = \mathbf{F} + \gamma^2 \boldsymbol{\beta} \left(\mathbf{S} \cdot \frac{d\boldsymbol{\beta}}{d\tau} \right)$$

and

$$\frac{dS_0}{d\tau} = F_0 + \gamma^2 \left(\mathbf{S} \cdot \frac{d\boldsymbol{\beta}}{d\tau} \right)$$

can be combined to give, after some simplification,

$$\frac{d\mathbf{s}}{d\tau} = \mathbf{F} - \frac{\gamma \boldsymbol{\beta}}{\gamma + 1} F_0 + \frac{\gamma^2}{\gamma + 1} \left[\mathbf{s} \times \left(\boldsymbol{\beta} \times \frac{d\boldsymbol{\beta}}{d\tau} \right) \right] \tag{11.166}$$

In these equations $(F_0, \mathbf{F})$ stand for the time and space components of the terms
with coefficient $(ge/2mc)$ in (11.162). Since $(F_0, \mathbf{F})$ form a 4-vector, with $F_0 = \boldsymbol{\beta} \cdot \mathbf{F}$, the first two terms in (11.166) can be recognized as the torque $\mathbf{F}'$ evaluated
in the rest frame. Dividing both sides by γ and using the definition (11.119) for
the Thomas precession frequency, we find that (11.166) becomes

$$\frac{d\mathbf{s}}{dt} = \frac{1}{\gamma} \mathbf{F}' + \boldsymbol{\omega}_T \times \mathbf{s} \tag{11.167}$$

Since $\mathbf{F}'$ is given by the right-hand side of (11.155), this is just (11.107) of Section
11.8.

For motion in electromagnetic fields where (11.163) holds,

$$\frac{d\boldsymbol{\beta}}{dt} = \frac{e}{\gamma mc} [\mathbf{E} + \boldsymbol{\beta} \times \mathbf{B} - \boldsymbol{\beta}(\boldsymbol{\beta} \cdot \mathbf{E})] \tag{11.168}$$

We also have, from the transformation properties (11.149) of **B**,

$$\frac{1}{\gamma} \mathbf{F}' = \frac{ge}{2mc} \mathbf{s} \times \left[\mathbf{B} - \frac{\gamma}{\gamma + 1} (\boldsymbol{\beta} \cdot \mathbf{B})\boldsymbol{\beta} - \boldsymbol{\beta} \times \mathbf{E} \right] \tag{11.169}$$

When these expressions are inserted into (11.167), it becomes

$$\frac{d\mathbf{s}}{dt} = \frac{e}{mc} \mathbf{s} \times \left[\left(\frac{g}{2} - 1 + \frac{1}{\gamma} \right) \mathbf{B} - \left(\frac{g}{2} - 1 \right) \frac{\gamma}{\gamma + 1} (\boldsymbol{\beta} \cdot \mathbf{B})\boldsymbol{\beta} \right.$$
$$\left. - \left(\frac{g}{2} - \frac{\gamma}{\gamma + 1} \right) \boldsymbol{\beta} \times \mathbf{E} \right] \tag{11.170}$$

This form of the equation of motion of the spin vector is *Thomas's* equation
(4.121) of 1927 (*op. cit.*).

C. *Rate of Change of Longitudinal Polarization*

As an example of the use of (11.170) we consider the rate of change of the
component of spin **s** parallel to the velocity. This is the longitudinal polarization

or net helicity of the particle. If $\hat{\boldsymbol{\beta}}$ is a unit vector in the direction of $\boldsymbol{\beta}$, the longitudinal polarization is $\hat{\boldsymbol{\beta}} \cdot \mathbf{s}$. It changes in time because $\mathbf{s}$ changes and also $\boldsymbol{\beta}$ changes. Explicitly, we have

$$\frac{d}{dt}(\hat{\boldsymbol{\beta}} \cdot \mathbf{s}) = \hat{\boldsymbol{\beta}} \cdot \frac{d\mathbf{s}}{dt} + \frac{1}{\beta}[\mathbf{s} - (\hat{\boldsymbol{\beta}} \cdot \mathbf{s})\hat{\boldsymbol{\beta}}] \cdot \frac{d\boldsymbol{\beta}}{dt}$$

Using (11.168) and (11.170), this can be written, after some algebra, as

$$\frac{d}{dt}(\hat{\boldsymbol{\beta}} \cdot \mathbf{s}) = -\frac{e}{mc}\mathbf{s}_{\perp} \cdot \left[\left(\frac{g}{2} - 1\right)\hat{\boldsymbol{\beta}} \times \mathbf{B} + \left(\frac{g\beta}{2} - \frac{1}{\beta}\right)\mathbf{E}\right] \quad (11.171)$$

where $\mathbf{s}_{\perp}$ is the component of $\mathbf{s}$ perpendicular to the velocity.

Equation (11.171) demonstrates a remarkable property of a particle with $g = 2$. In a purely magnetic field, the spin precesses in such a manner that the longitudinal polarization remains constant, whatever the motion of the particle. If the particle is relativistic ($\beta \to 1$), even the presence of an electric field causes the longitudinal polarization to change only very slowly, at a rate proportional to γ^{-2} times the electric field component perpendicular to $\mathbf{v}$.

The electron and the muon have g factors differing from the Dirac value of 2 by radiative corrections of order $\alpha/\pi = 0.00232$. Because $(g - 2)$ is so small, the longitudinal polarization of a beam of electrons or muons orbiting in a magnetic field changes relatively slowly. This phenomenon permits very precise measurements of the quantity $a \equiv (g - 2)/2$, called the anomaly or the anomalous magnetic moment. The values of a provide accurate tests of the validity of quantum electrodynamics. For muons, 100% longitudinally polarized at birth, the change in polarization is detected by means of the characteristically asymmetric angular distribution of the decay electron from the muon relative to the direction of muon polarization. For electrons from beta decay the initial longitudinal polarization is $\pm\beta$. Its change with time is detected by changes in the asymmetry of Mott scattering (e^-) or the angular distribution of the annihilation photons from positronium formed in an intense magnetic field (e^+). The precision attainable by these techniques is indicated by the recent data:*

$$a(e^-) = 1\ 159\ 652\ 188.4\ (4.3) \times 10^{-12}$$
$$a(e^+) = 1\ 159\ 652\ 187.9\ (4.3) \times 10^{-12}$$
$$a(\mu^{\pm}) = 1\ 165\ 924\ (9) \times 10^{-9}$$

These results are in good agreement with the predictions of quantum electrodynamics, as discussed in detail in the review by Kinoshita.

Further elaboration of spin precession is left to the problems at the end of Chapter 12.

11.12 *Note on Notation and Units in Relativistic Kinematics*

In dealing with Lorentz transformations and relativistic kinematics, it is convenient to adopt a consistent, simple notation and set of units. We have seen that

---

*e^-, e^+: Van Dyck, Schwinberg, and Dehmelt, *Phys. Rev. Lett.* **59**, 26 (1987); $\mu^{\pm}$: J. Bailey et al., *Nucl. Phys. B* **150**, 1 (1979). See also the review, T. Kinoshita, ed., *Quantum Electrodynamics*, World Scientific, Singapore (1990).

various powers of the velocity of light c appear in the formulas of special relativity. These tend to make the formulas cumbersome, although their presence facilitates extracting nonrelativistic limits (by letting $c \to \infty$). In doing relativistic kinematics, it is customary to suppress all factors of c by suitable choice of units. We adopt the convention that all momenta, energies, and masses are measured in energy units, while velocities are measured in units of the velocity of light. In particle kinematics the symbols,

$$\left. \begin{array}{c} p \\ E \\ m \\ \\ v \end{array} \right\} \quad \text{stand for} \quad \left\{ \begin{array}{c} cp \\ E \\ mc^2 \\ \dfrac{v}{c} \end{array} \right.$$

Thus the connection between momentum and total energy is written as $E^2 = p^2 + m^2$, a particle's velocity is $\mathbf{v} = \mathbf{p}/E$, and so on. As energy units, the electron volt (eV), the megaelectron volt ($1 \text{ MeV} = 10^6 \text{ eV}$), and the gigaelectron volt ($1 \text{ GeV} = 10^9 \text{ eV}$) are convenient. One electron volt is the energy gained by a particle with electronic charge when it falls through a potential difference of one volt ($1 \text{ eV} = 1.602 \times 10^{-12} \text{ erg} = 1.602 \times 10^{-19}$ joule).

In addition to eliminating powers of c, it is customary to denote scalar products of 4-vectors by a centered dot between italicized symbols, with scalar products of 3-vectors denoted by a dot between boldface symbols, as usual. Thus we have

$$a \cdot b \equiv a_\alpha b^\alpha = a_0 b_0 - \mathbf{a} \cdot \mathbf{b}$$

Four-vectors may be written with or without an index. Thus conservation of energy and momentum may appear as

$$P = p + q$$

or

$$P^\alpha = p^\alpha + q^\alpha$$

References and Suggested Reading

The theory of relativity has an extensive literature all its own. Perhaps the most lucid, though concise, presentation of special and general relativity is the famous 1921 encyclopedia article (brought up to date in 1956) by
    Pauli
Another authoritative book is
    Møller
Among the older textbooks on special relativity at the graduate level are
    Aharoni
    Bergmann, Chapters I–IX
and, more recently,
    Anderson, Chapters 6 and 7
    Barut, Chapters 1 and 2
    Schwartz
    Tonnelat, Part II

The mathematics (group theory and tensor analysis) of space-time is covered from a physicist's point of view in

> Anderson, Chapters 1–5
> Bergmann, Chapter V
> Schwartz, Chapter 5

The flavor of the original theoretical developments can be obtained by consulting the jointly collected papers of

> Einstein, Lorentz, Minkowski, and Weyl

and in the collection of reprints of

> Kilmister

A very individualistic discussion of special relativity, with extensive references and perceptive commentary, can be found in the two books by

> Arzeliès

The main experiments are summarized in

> Møller, Chapters 1 and 12
> Condon and Odishaw, Part 6, Chapter 8 by E. L. Hill

The barely mentioned topic of the electrodynamics of moving media is treated by

> Cullwick
> Møller, pp. 209 ff.
> Pauli, pp. 99 ff.
> Penfield and Haus

Cullwick discusses many practical situations. Be forewarned, however, about his Chapter 5 by reading his preface to the second edition.

Relativistic kinematics has been relegated to the problems. Useful references in this area are

> Baldin, Gol'danskii, and Rozenthal
> Hagedorn
> Sard, Chapter 4

Another neglected subject is the appearance of rapidly moving objects. This fascinating topic illustrates how careful one must be with concepts such as the FitzGerald–Lorentz contraction. Some pertinent papers are

> R. Penrose, *Proc. Cambridge Philos. Soc.* **55**, 137 (1959).
> J. Terrell, *Phys. Rev.* **116**, 1041 (1959).
> V. F. Weisskopf, *Phys. Today* **13**: 9, 24 (1960).
> M. L. Boas, *Am. J. Phys.* **29**, 283 (1961).
> G. D. Scott and H. J. Van Driel, *Am. J. Phys.* **38**, 971 (1970).

For the reader who wishes a more elementary or leisurely introduction to special relativity, there are a host of undergraduate texts, among them

> Bohm
> Feynman, Vol. 1, Chapters 15–17
> French
> Mermin
> Rindler
> Sard
> Smith
> Taylor and Wheeler

Sard's book is specially recommended. It is at an intermediate level and has a thorough discussion of Thomas precession and the motion of spin. Hagedorn also discusses the BMT equation in detail.

Further references on specific topics can be found in the AAPT Resource Letter on Special Relativity SRT-1, *Am. J. Phys.* **30**, 462 (1962).

Problems

11.1 Two equivalent inertial frames K and K' are such that K' moves in the positive x direction with speed v as seen from K. The spatial coordinate axes in K' are parallel to those in K and the two origins are coincident at times $t = t' = 0$.

(a) Show that the isotropy and homogeneity of space-time and equivalence of different inertial frames (first postulate of relativity) require that the most general transformation between the space-time coordinates (x, y, z, t) and (x', y', z', t') is the linear transformation,

$$x' = f(v^2)x - vf(v^2)t; \qquad t' = g(v^2)t - vh(v^2)x; \qquad y' = y; \qquad z' = z$$

and the inverse,

$$x = f(v^2)x' + vf(v^2)t'; \qquad t = g(v^2)t' + vh(v^2)x'; \qquad y = y'; \qquad z = z'$$

where f, g, and h are functions of v^2, the structures of the x' and x equations are determined by the definition of the inertial frames in relative motion, and the signs in the inverse equation are a reflection of the reversal of roles of the two frames.

(b) Show that consistency of the initial transformation and its inverse require

$$f = g \qquad \text{and} \qquad f^2 - v^2fh = 1$$

(c) If a physical entity has speed u' parallel to the x' axis in K', show that its speed u parallel to the x axis in K is

$$u = \frac{u' + v}{1 + vu'(h/f)}$$

Using the second postulate $2'$ (universal limiting speed C), show that $h = f/C^2$ is required and that the Lorentz transformation of the coordinates results. The universal limiting speed C is to be determined from experiment.

11.2 Consider three inertial frames and coordinates $K(x, t)$, $K'(x', t')$, and $K''(x'', t'')$. Frame K' moves in the x direction with speed v_1 relative to K; frame K'' moves with speed v_2 relative to K', and speed v_3 relative to K. By considering the group property of the transformations of Problem 11.1 (including the results of parts a and b), $(x'', t'') \to (x', t') \to (x, t)$ and $(x'', t'') \to (x, t)$ directly, show that $|h(v^2)/f(v^2)|$ is a universal constant with the dimensions of an inverse speed squared.

This approach obtains the Lorentz transformation without reference to electromagnetism or the second postulate, but requires experiment to show that $h/f > 0$.

Reference: Y. P. Terletskii, *Paradoxes in the Theory of Relativity*, Plenum Press, New York (1968), pp. 17–25.

11.3 Show explicitly that two successive Lorentz transformations in the same direction are equivalent to a single Lorentz transformation with a velocity

$$v = \frac{v_1 + v_2}{1 + (v_1v_2/c^2)}$$

This is an alternative way to derive the parallel-velocity addition law.

11.4 A possible clock is shown in the figure. It consists of a flashtube F and a photocell P shielded so that each views only the mirror M, located a distance d away, and mounted rigidly with respect to the flashtube-photocell assembly. The electronic innards of the box are such that when the photocell responds to a light flash from the mirror, the flashtube is triggered with a negligible delay and emits a short flash toward the mirror. The clock thus "ticks" once every $(2d/c)$ seconds when at rest.

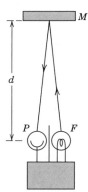

Problem 11.4

(a) Suppose that the clock moves with a uniform velocity v, perpendicular to the line from PF to M, relative to an observer. Using the second postulate of relativity, show by explicit geometrical or algebraic construction that the observer sees the relativistic time dilatation as the clock moves by.

(b) Suppose that the clock moves with a velocity v parallel to the line from PF to M. Verify that here, too, the clock is observed to tick more slowly, by the same time dilatation factor.

11.5 A coordinate system K' moves with a velocity $\mathbf{v}$ relative to another system K. In K' a particle has a velocity $\mathbf{u'}$ and an acceleration $\mathbf{a'}$. Find the Lorentz transformation law for accelerations, and show that in the system K the components of acceleration parallel and perpendicular to $\mathbf{v}$ are

$$\mathbf{a}_{\parallel} = \frac{\left(1 - \dfrac{v^2}{c^2}\right)^{3/2}}{\left(1 + \dfrac{\mathbf{v} \cdot \mathbf{u'}}{c^2}\right)^3} \mathbf{a}'_{\parallel}$$

$$\mathbf{a}_{\perp} = \frac{\left(1 - \dfrac{v^2}{c^2}\right)}{\left(1 + \dfrac{\mathbf{v} \cdot \mathbf{u'}}{c^2}\right)^3} \left(\mathbf{a}'_{\perp} + \frac{\mathbf{v}}{c^2} \times (\mathbf{a'} \times \mathbf{u'})\right)$$

11.6 Assume that a rocket ship leaves the earth in the year 2100. One of a set of twins born in 2080 remains on earth; the other rides in the rocket. The rocket ship is so constructed that it has an acceleration g in its own rest frame (this makes the occupants feel at home). It accelerates in a straight-line path for 5 years (by its own clocks), decelerates at the same rate for 5 more years, turns around, accelerates for 5 years, decelerates for 5 years, and lands on earth. The twin in the rocket is 40 years old.

(a) What year is it on earth?

(b) How far away from the earth did the rocket ship travel?

11.7 In the reference frame K two very evenly matched sprinters are lined up a distance d apart on the y axis for a race parallel to the x axis. Two starters, one beside each man, will fire their starting pistols at slightly different times, giving a handicap to the better of the two runners. The time difference in K is T.

(a) For what range of time differences will there be a reference frame K' in which there is no handicap, and for what range of time differences is there a frame K' in which there is a true (not apparent) handicap?

(b) Determine explicitly the Lorentz transformation to the frame K' appropriate for each of the two possibilities in part a, finding the velocity of K' relative to K and the space-time positions of each sprinter in K'.

11.8 (a) Use the relativistic velocity addition law and the invariance of phase to discuss the Fizeau experiments on the velocity of propagation of light in moving liquids. Show that for liquid flow at a speed v parallel or antiparallel to the path of the light the speed of the light, as observed in the laboratory, is given to first order in v by

$$u = \frac{c}{n(\omega)} \pm v\left(1 - \frac{1}{n^2} + \frac{\omega}{n}\frac{dn(\omega)}{d\omega}\right)$$

where ω is the frequency of the light in the laboratory (in the liquid and outside it) and $n(\omega)$ is the index of refraction of the liquid. Because of the extinction theorem, it is assumed that the light travels with speed $u' = c/n(\omega')$ relative to the moving liquid.

(b) Consult the paper of W. M. Macek, J. R. Schneider, and R. M. Salamon [*J. Appl. Phys.* **35**, 2556 (1964)] and discuss the status of the Fizeau experiments.

11.9 An infinitesimal Lorentz transformation and its inverse can be written as

$$x'^\alpha = (g^{\alpha\beta} + \epsilon^{\alpha\beta})x_\beta$$
$$x^\alpha = (g^{\alpha\beta} + \epsilon'^{\alpha\beta})x'_\beta$$

where $\epsilon^{\alpha\beta}$ and $\epsilon'^{\alpha\beta}$ are infinitesimal.

(a) Show from the definition of the inverse that $\epsilon'^{\alpha\beta} = -\epsilon^{\alpha\beta}$.

(b) Show from the preservation of the norm that $\epsilon^{\alpha\beta} = -\epsilon^{\beta\alpha}$.

(c) By writing the transformation in terms of contravariant components on both sides of the equation, show that $\epsilon^{\alpha\beta}$ is equivalent to the matrix L (11.93).

11.10 (a) For the Lorentz boost and rotation matrices **K** and **S** show that

$$(\boldsymbol{\epsilon} \cdot \mathbf{S})^3 = -\boldsymbol{\epsilon} \cdot \mathbf{S}$$
$$(\boldsymbol{\epsilon}' \cdot \mathbf{K})^3 = \boldsymbol{\epsilon}' \cdot \mathbf{K}$$

where $\boldsymbol{\epsilon}$ and $\boldsymbol{\epsilon}'$ are any real unit 3-vectors.

(b) Use the results of part a to show that

$$\exp(-\zeta\hat{\boldsymbol{\beta}} \cdot \mathbf{K}) = I - \hat{\boldsymbol{\beta}} \cdot \mathbf{K} \sinh \zeta + (\hat{\boldsymbol{\beta}} \cdot \mathbf{K})^2[\cosh \zeta - 1]$$

11.11 Two Lorentz transformations differ by an infinitesimal amount. In the notation of Section 11.7 they are represented by $A_1 = e^L$, $A_2 = e^{L+\delta L}$. Without using explicit matrix representations show that *to first order in* δL the Lorentz transformation $A = A_2 A_1^{-1}$ can be written as

$$A = I + \delta L + \frac{1}{2!}[L, \delta L] + \frac{1}{3!}[L, [L, \delta L]] + \frac{1}{4!}[L, [L, [L, \delta L]]] + \cdots$$

Hint: The early terms can be found by brute force, but alternatively consider the Taylor series expansion in λ of the operator $A(\lambda) = e^{\lambda(L+\delta L)}e^{-\lambda L}$ and then put $\lambda = 1$.

11.12 Apply the result of Problem 11.11 to a purely algebraic deviation of (11.116) on Thomas precession.

(a) With

$$L = - \frac{\boldsymbol{\beta} \cdot \mathbf{K}(\tanh^{-1}\beta)}{\beta}$$

$$L + \delta L = - \frac{(\boldsymbol{\beta} + \delta\boldsymbol{\beta}_{\parallel} + \delta\boldsymbol{\beta}_{\perp}) \cdot \mathbf{K}(\tanh^{-1}\beta')}{\beta'}$$

where $\beta' = \sqrt{(\boldsymbol{\beta} + \delta\boldsymbol{\beta}_{\parallel})^2 + (\delta\boldsymbol{\beta}_{\perp})^2}$, show that

$$\delta L = -\gamma^2 \, \delta\boldsymbol{\beta}_{\parallel} \cdot \mathbf{K} - \frac{\delta\boldsymbol{\beta}_{\perp} \cdot \mathbf{K}(\tanh^{-1}\beta)}{\beta}$$

(b) Using the commutation relations for $\mathbf{K}$ and $\mathbf{S}$, show that

$$C_1 = [L, \delta L] = -\left(\frac{\tanh^{-1}\beta}{\beta}\right)^2 (\boldsymbol{\beta} \times \delta\boldsymbol{\beta}_{\perp}) \cdot \mathbf{S}$$

$$C_2 = [L, C_1] = (\tanh^{-1}\beta)^2 \, \delta L_{\perp}$$

$$C_3 = [L, C_2] = (\tanh^{-1}\beta)^2 C_1$$

$$C_4 = [L, C_3] = (\tanh^{-1}\beta)^4 \, \delta L_{\perp}$$

where $\delta L_{\perp}$ is the term in δL involving $\delta\boldsymbol{\beta}_{\perp}$.

(c) Sum the series of terms for $A_T = A_2 A_1^{-1}$ to obtain

$$A_T = I - (\gamma^2 \, \delta\boldsymbol{\beta}_{\parallel} + \gamma \, \delta\boldsymbol{\beta}_{\perp}) \cdot \mathbf{K} - \frac{\gamma^2}{\gamma + 1} (\boldsymbol{\beta} \times \delta\boldsymbol{\beta}_{\perp}) \cdot \mathbf{S}$$

correct to first order in $\delta\boldsymbol{\beta}$. [See D. Shelupsky, *Am. J. Phys.* **35**, 650 (1967).]

11.13 An infinitely long straight wire of negligible cross-sectional area is at rest and has a uniform linear charge density q_0 in the inertial frame K'. The frame K' (and the wire) move with a velocity $\mathbf{v}$ parallel to the direction of the wire with respect to the laboratory frame K.

(a) Write down the electric and magnetic fields in cylindrical coordinates in the rest frame of the wire. Using the Lorentz transformation properties of the fields, find the components of the electric and magnetic fields in the laboratory.

(b) What are the charge and current densities associated with the wire in its rest frame? In the laboratory?

(c) From the laboratory charge and current densities, calculate directly the electric and magnetic fields in the laboratory. Compare with the results of part a.

11.14 (a) Express the Lorentz scalars $F^{\alpha\beta}F_{\alpha\beta}$, $\mathscr{F}^{\alpha\beta}F_{\alpha\beta}$, and $\mathscr{F}^{\alpha\beta}\mathscr{F}_{\alpha\beta}$ in terms of $\mathbf{E}$ and $\mathbf{B}$. Are there any other invariants quadratic in the field strengths $\mathbf{E}$ and $\mathbf{B}$?

(b) Is it possible to have an electromagnetic field that appears as a purely electric field in one inertial frame and as a purely magnetic field in some other inertial frame? What are the criteria imposed on $\mathbf{E}$ and $\mathbf{B}$ such that there is an inertial frame in which there is no electric field?

(c) For macroscopic media, $\mathbf{E}$, $\mathbf{B}$ form the field tensor $F^{\alpha\beta}$ and $\mathbf{D}$, $\mathbf{H}$ the tensor $G^{\alpha\beta}$. What further invariants can be formed? What are their explicit expressions in terms of the 3-vector fields?

11.15 In a certain reference frame a static, uniform, electric field E_0 is parallel to the x axis, and a static, uniform, magnetic induction $B_0 = 2E_0$ lies in the x-y plane, making an angle θ with the axis. Determine the relative velocity of a reference frame in which the electric and magnetic fields are parallel. What are the fields in that frame for $\theta \ll 1$ and $\theta \rightarrow (\pi/2)$?

11.16 In the rest frame of a conducting medium the current density satisfies Ohm's law, $\mathbf{J}' = \sigma\mathbf{E}'$, where σ is the conductivity and primes denote quantities in the rest frame.

(a) Taking into account the possibility of convection current as well as conduction current, show that the covariant generalization of Ohm's law is

$$J^\alpha - \frac{1}{c^2}(U_\beta J^\beta)U^\alpha = \frac{\sigma}{c}F^{\alpha\beta}U_\beta$$

where U^α is the 4-velocity of the medium.

(b) Show that if the medium has a velocity $\mathbf{v} = c\boldsymbol{\beta}$ with respect to some inertial frame that the 3-vector current in that frame is

$$\mathbf{J} = \gamma\sigma[\mathbf{E} + \boldsymbol{\beta} \times \mathbf{B} - \boldsymbol{\beta}(\boldsymbol{\beta} \cdot \mathbf{E})] + \rho\mathbf{v}$$

where ρ is the charge density observed in that frame.

(c) If the medium is uncharged in its rest frame ($\rho' = 0$), what is the charge density and the expression for $\mathbf{J}$ in the frame of part b? This is the relativistic generalization of the equation above (7.68).

11.17 The electric and magnetic fields (11.152) of a charge in uniform motion can be obtained from Coulomb's law in the charge's rest frame and the fact that the field strength $F^{\alpha\beta}$ is an antisymmetric tensor of rank 2 without considering *explicitly* the Lorentz transformation. The idea is the following. For a charge in uniform motion the only relevant variables are the charge's 4-velocity U^α and the relative coordinate $X^\alpha = x_p^\alpha - x_q^\alpha$, where x_p^α and x_q^α are the 4-vector coordinates of the observation point and the charge, respectively. The only antisymmetric tensor that can be formed is $(X^\alpha U^\beta - X^\beta U^\alpha)$. Thus the electromagnetic field $F^{\alpha\beta}$ must be this tensor multiplied by some scalar function of the possible scalar products, $X_\alpha X^\alpha$, $X_\alpha U^\alpha$, $U_\alpha U^\alpha$.

(a) For the geometry of Fig. 11.8 the coordinates of P and q at a common time in K can be written $x_p^\alpha = (ct, \mathbf{b})$, $x_q^\alpha = (ct, \mathbf{v}t)$, with $\mathbf{b} \cdot \mathbf{v} = 0$. By considering the general form of $F^{\alpha\beta}$ in the rest frame of the charge, show that

$$F^{\alpha\beta} = \frac{q}{c}\frac{(X^\alpha U^\beta - X^\beta U^\alpha)}{\left[\frac{1}{c^2}(U_\alpha X^\alpha)^2 - X_\alpha X^\alpha\right]^{3/2}}$$

Verify that this yields the expressions (11.152) in the inertial frame K.

(b) Repeat the calculation, using as the starting point the common-time coordinates in the rest frame, $x_p'^\alpha = (ct', \mathbf{b} - \mathbf{v}t')$ and $x_q'^\alpha = (ct', 0)$. Show that

$$F^{\alpha\beta} = \frac{q}{c}\frac{(Y^\alpha U^\beta - Y^\beta U^\alpha)}{(-Y_\alpha Y^\alpha)^{3/2}}$$

where $Y'^\alpha = x_p'^\alpha - x_q'^\alpha$. Verify that the fields are the same as in part a. Note that to obtain the results of (11.152) it is necessary to use the time t of the observation point P in K as the time parameter.

(c) Finally, consider the coordinate $x_p^\alpha = (ct, \mathbf{b})$ and the "retarded-time" coordinate $x_q^\alpha = [ct - R, \boldsymbol{\beta}(ct - R)]$ where R is the distance between P and q at the retarded time. Define the difference as $Z^\alpha = [R, \mathbf{b} - \boldsymbol{\beta}(ct - R)]$. Show that in terms of Z^α and U^α the field is

$$F^{\alpha\beta} = \frac{q}{c} \frac{(Z^\alpha U^\beta - Z^\beta U^\alpha)}{\left(\dfrac{1}{c} U_\alpha Z^\alpha\right)^3}$$

11.18   The electric and magnetic fields of a particle of charge q moving in a straight line with speed $v = \beta c$, given by (11.152), become more and more concentrated as $\beta \to 1$, as is indicated in Fig. 11.9. Choose axes so that the charge moves along the z axis in the positive direction, passing the origin at $t = 0$. Let the spatial coordinates of the observation point be (x, y, z) and define the transverse vector $\mathbf{r}_\perp$, with components x and y. Consider the fields and the source in the limit of $\beta = 1$.

(a)   Show that the fields can be written as

$$\mathbf{E} = 2q \frac{\mathbf{r}_\perp}{r_\perp^2} \delta(ct - z); \qquad \mathbf{B} = 2q \frac{\hat{\mathbf{v}} \times \mathbf{r}_\perp}{r_\perp^2} \delta(ct - z)$$

where $\hat{\mathbf{v}}$ is a unit vector in the direction of the particle's velocity.

(b)   Show by substitution into the Maxwell equations that these fields are consistent with a 4-vector source density,

$$J^\alpha = qcv^\alpha \delta^{(2)}(\mathbf{r}_\perp)\delta(ct - z)$$

where the 4-vector $v^\alpha = (1, \hat{\mathbf{v}})$.

(c)   Show that the fields of part a are derivable from either of the following 4-vector potentials,

$$A^0 = A^z = -2q\delta(ct - z) \ln(\lambda r_\perp); \qquad \mathbf{A}_\perp = 0$$

or

$$A^0 = 0 = A^z; \qquad \mathbf{A}_\perp = -2q\Theta(ct - z) \nabla_\perp \ln(\lambda r_\perp)$$

where λ is an irrelevant parameter setting the scale of the logarithm.

Show that the two potentials differ by a gauge transformation and find the gauge function, χ.

Reference: R. Jackiw, D. Kabat, and M. Ortiz, *Phys. Lett. B* **277**, 148 (1992).

11.19   A particle of mass M and 4-momentum P decays into two particles of masses m_1 and m_2.

(a)   Use the conservation of energy and momentum in the form, $p_2 = P - p_1$, and the invariance of scalar products of 4-vectors to show that the total energy of the first particle in the rest frame of the decaying particle is

$$E_1 = \frac{M^2 + m_1^2 - m_2^2}{2M}$$

and that E_2 is obtained by interchanging m_1 and m_2.

(b)   Show that the *kinetic energy* T_i of the ith particle in the same frame is

$$T_i = \Delta M \left(1 - \frac{m_i}{M} - \frac{\Delta M}{2M}\right)$$

where $\Delta M = M - m_1 - m_2$ is the mass excess or Q value of the process.

(c)   The charged pi-meson ($M = 139.6$ MeV) decays into a mu-meson ($m_1 = 105.7$ MeV) and a neutrino ($m_2 = 0$). Calculate the kinetic energies of the mu-meson and the neutrino in the pi-meson's rest frame. The unique kinetic energy of the muon is the signature of a two-body decay. It entered importantly in the discovery of the pi-meson in photographic emulsions by Powell and coworkers in 1947.

11.20 The lambda particle (Λ) is a neutral baryon of mass $M = 1115$ MeV that decays with a lifetime of $\tau = 2.9 \times 10^{-10}$ s into a nucleon of mass $m_1 \simeq 939$ MeV and a pi-meson of mass $m_2 \simeq 140$ MeV. It was first observed in flight by its charged decay mode $\Lambda \rightarrow p + \pi^-$ in cloud chambers. The charged tracks originate from a single point and have the appearance of an inverted vee or lambda. The particles' identities and momenta can be inferred from their ranges and curvature in the magnetic field of the chamber.

(a) Using conservation of momentum and energy and the invariance of scalar products of 4-vectors show that, if the opening angle θ between the two tracks is measured, the mass of the decaying particle can be found from the formula

$$M^2 = m_1^2 + m_2^2 + 2E_1E_2 - 2p_1p_2 \cos \theta$$

where here p_1 and p_2 are the magnitudes of the 3-momenta.

(b) A lambda particle is created with a total energy of 10 GeV in a collision in the top plate of a cloud chamber. How far on the average will it travel in the chamber before decaying? What range of opening angles will occur for a 10 GeV lambda if the decay is more or less isotropic in the lambda's rest frame?

11.21 If a system of mass M decays or transforms at rest into a number of particles, the sum of whose masses is less than M by an amount ΔM,

(a) show that the maximum kinetic energy of the ith particle (mass m_i) is

$$(T_i)_{max} = \Delta M\left(1 - \frac{m_i}{M} - \frac{\Delta M}{2M}\right)$$

(b) determine the maximum kinetic energies in MeV and also the ratios to ΔM for each of the particles in the following decays or transformations of particles at rest:

$$\mu \rightarrow e + \nu + \bar{\nu}$$
$$K^+ \rightarrow \pi^+ + \pi^- + \pi^+$$
$$K^\pm \rightarrow e^\pm + \pi^0 + \nu$$
$$K^\pm \rightarrow \mu^\pm + \pi^0 + \nu$$
$$p + \bar{p} \rightarrow 2\pi^+ + 2\pi^- + \pi^0$$
$$p + \bar{p} \rightarrow K^+ + K^- + 3\pi^0$$

11.22 The presence in the universe of an apparently uniform "sea" of blackbody radiation at a temperature of roughly 3K gives one mechanism for an upper limit on the energies of photons that have traveled an appreciable distance since their creation. Photon-photon collisions can result in the creation of a charged particle and its antiparticle ("pair creation") if there is sufficient energy in the center of "mass" of the two photons. The lowest threshold and also the largest cross section occurs for an electron-positron pair.

(a) Taking the energy of a typical 3K photon to be $E = 2.5 \times 10^{-4}$ eV, calculate the energy for an incident photon such that there is energy just sufficient to make an electron-positron pair. For photons with energies larger than this threshold value, the cross section increases to a maximum of the order of $(e^2/mc^2)^2$ and then decreases slowly at higher energies. This interaction is one mechanism for the disappearance of such photons as they travel cosmological distances.

(b) There is some evidence for a diffuse x-ray background with photons having energies of several hundred electron volts or more. Beyond 1 keV the spectrum falls as E^{-n} with $n \simeq 1.5$. Repeat the calculation of the threshold incident energy, assuming that the energy of the photon in the "sea" is 500 eV.

11.23 In a collision process a particle of mass m_2, at rest in the laboratory, is struck by a particle of mass m_1, momentum $\mathbf{p}_{LAB}$ and total energy E_{LAB}. In the collision the two initial particles are transformed into two others of mass m_3 and m_4. The configurations of the momentum vectors in the center of momentum (cm) frame (traditionally called the center-of-mass frame) and the laboratory frame are shown in the figure.

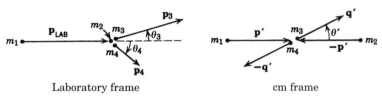

Laboratory frame   cm frame

Problem 11.23

(a) Use invariant scalar products to show that the total energy W in the cm frame has its square given by

$$W^2 = m_1^2 + m_2^2 + 2m_2 E_{LAB}$$

and that the cms 3-momentum $\mathbf{p}'$ is

$$\mathbf{p}' = \frac{m_2 \mathbf{p}_{LAB}}{W}$$

(b) Show that the Lorentz transformation parameters β_{cm} and γ_{cm} describing the velocity of the cm frame in the laboratory are

$$\beta_{cm} = \frac{\mathbf{p}_{LAB}}{m_2 + E_{LAB}}, \qquad \gamma_{cm} = \frac{m_2 + E_{LAB}}{W}$$

(c) Show that the results of parts a and b reduce in the nonrelativistic limit to the familiar expressions,

$$W \simeq m_1 + m_2 + \left(\frac{m_2}{m_1 + m_2}\right) \frac{p_{LAB}^2}{2m_1}$$

$$\mathbf{p}' \simeq \left(\frac{m_2}{m_1 + m_2}\right) \mathbf{p}_{LAB}, \qquad \beta_{cm} \simeq \frac{\mathbf{p}_{LAB}}{m_1 + m_2}$$

11.24 The threshold kinetic energy T_{th} in the laboratory for a given reaction is the kinetic energy of the incident particle on a stationary target just sufficient to make the center of mass energy W equal to the sum of the rest energies of the particles in the final state. Calculate the threshold kinetic energies for the following processes. Express your answers in MeV or GeV and also in units of the rest energy of the incident particle (unless it is a massless particle).

(a) Pi-meson photoproduction, $\gamma p \to \pi^0 p$

$$(m_p = 938.5 \text{ MeV}, \qquad m_{\pi^0} = 135.0 \text{ MeV})$$

(b) Nucleon-antinucleon pair production in nucleon-nucleon collisions, for example, $pp \rightarrow ppp\bar{p}$.

(c) Nucleon-antinucleon pair production in electron-electron collisions, $e^-e^- \rightarrow e^-e^-p\bar{p}$ and $e^+e^- \rightarrow p\bar{p}$ ($m_e = 0.511$ MeV).

11.25 In colliding beam machines such as the Tevatron at Fermilab or the numerous e^+e^- storage rings, counterrotating relativistic beams of particles are stored and made to collide more or less head-on in one or more interaction regions. Let the particles in the two beams have masses m_1 and m_2 and momenta p_1 and p_2, respectively, and let them intersect with an angle θ between the two beams.

(a) Show that, to order $(m/p)^2$ inclusive, the square of the total energy in the cm frame is

$$W^2 = 4p_1p_2 \cos^2 \frac{\theta}{2} + (p_1 + p_2)\left(\frac{m_1^2}{p_1} + \frac{m_2^2}{p_2}\right)$$

(b) Show that the cm inertial frame has a velocity in the laboratory given by

$$\beta_{cm} = \frac{(p_1 + p_2) \sin \theta/2}{(E_1 + E_2) \sin \alpha}$$

where

$$\tan \alpha = \left(\frac{p_1 + p_2}{p_1 - p_2}\right) \tan \frac{\theta}{2}$$

The angle α is defined in the figure.

(c) Check that the results of part b agree with those of Problem 11.23b.

(d) If the crossing angle is $\theta = 20°$ and the colliding protons have $p_1 = p_2 = 100$ GeV/c, is the laboratory frame a reasonable approximation to the cm frame? Consider, for example, a proton-proton inelastic collision involving pion production and examine the collinearity of two pions produced with equal and opposite momenta of 10 GeV/c in the cm frame.

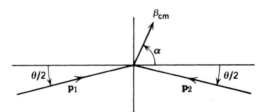

Problem 11.25

11.26 In an elastic scattering process the incident particle imparts energy to the stationary target. The energy ΔE lost by the incident particle appears as recoil kinetic energy of the target. In the notation of Problem 11.23, $m_3 = m_1$ and $m_4 = m_2$, while $\Delta E = T_4 = E_4 - m_4$.

(a) Show that ΔE can be expressed in the following different ways,

$$\Delta E = \frac{m_2}{W^2} p_{LAB}^2 (1 - \cos \theta')$$

$$\Delta E = \frac{2m_2 p_{LAB}^2 \cos^2 \theta_4}{W^2 + p_{LAB}^2 \sin^2 \theta_4}$$

$$\Delta E = \frac{Q^2}{2m_2}$$

where $Q^2 = -(p_1 - p_3)^2 = (\mathbf{p}_1 - \mathbf{p}_3)^2 - (E_1 - E_3)^2$ is the Lorentz invariant momentum transfer (squared).

(b)   Show that for charged particles other than electrons incident on stationary electrons ($m_1 \gg m_2$) the maximum energy loss is approximately

$$\Delta E_{\text{max}} \simeq 2\gamma^2\beta^2 m_e$$

where γ and β are characteristic of the incident particle and $\gamma \ll (m_1/m_e)$. Give this result a simple interpretation by considering the relevant collision in the rest frame of the incident particle and then transforming back to the laboratory.

(c)   For electron-electron collisions, show that the maximum energy transfer is

$$\Delta E_{\text{max}}^{(e)} = (\gamma - 1)m_e$$

11.27   **(a)**   A charge density ρ' of zero total charge, but with a dipole moment $\mathbf{p}$, exists in reference frame K'. There is no current density in K'. The frame K' moves with a velocity $\mathbf{v} = \boldsymbol{\beta} c$ in the frame K. Find the charge and current densities ρ and $\mathbf{J}$ in the frame K and show that there is a magnetic dipole moment, $\mathbf{m} = (\mathbf{p} \times \boldsymbol{\beta})/2$, correct to first order in β. What is the electric dipole moment in K to the same order in β?

(b)   Instead of the charge density, but no current density, in K', consider no charge density, but a current density $\mathbf{J}'$ that has a magnetic dipole moment $\mathbf{m}$. Find the charge and current densities in K and show that to first order in β there is an electric dipole moment $\mathbf{p} = \boldsymbol{\beta} \times \mathbf{m}$ in addition to the magnetic dipole moment.

11.28   Revisit Problems 6.21 and 6.22 from the viewpoint of Lorentz transformations. An electric dipole instantaneously at rest at the origin in the frame K' has potentials, $\Phi' = \mathbf{p} \cdot \mathbf{r}'/r'^3$, and $\mathbf{A}' = 0$ (and thus only an electric field). The frame K' moves with uniform velocity $\mathbf{v} = \boldsymbol{\beta} c$ in the frame K.

(a)   Show that in frame K to first order in β the potentials are

$$\Phi = \frac{\mathbf{p} \cdot \mathbf{R}}{R^3}, \qquad \mathbf{A} = \boldsymbol{\beta} \frac{(\mathbf{p} \cdot \mathbf{R})}{R^3}$$

where $\mathbf{R} = \mathbf{x} - \mathbf{x}_0(t)$, with $\mathbf{v} = d\mathbf{x}_0/dt$ at time t.

(b)   Show explicitly that the potentials in K satisfy the Lorentz condition.

(c)   Show that to first order in β the electric field $\mathbf{E}$ in K is just the electric dipole field (centered at $\mathbf{x}_0$), or a dipole field plus time-dependent higher multipoles, if viewed from a fixed origin, and the magnetic field is $\mathbf{B} = \boldsymbol{\beta} \times \mathbf{E}$. Where is the effective magnetic dipole moment of Problem 6.21 or 11.27a?

11.29   Instead of the electric dipole potential of Problem 11.28, consider a point magnetic moment $\mathbf{m}$ in the moving frame K', with its potentials, $\Phi' = 0$, $\mathbf{A}' = \mathbf{m} \times \mathbf{r}'/r'^3$ (and so only a magnetic field).

(a)   Show that to first order in β the potentials in K are

$$\Phi = \frac{(\boldsymbol{\beta} \times \mathbf{m}) \cdot \mathbf{R}}{R^3}, \qquad \mathbf{A} = \frac{(\mathbf{m} \times \mathbf{R})}{R^3}$$

Note that the scalar potential is the same as the static potential of the electric dipole moment of Problem 11.27b. [But this gives only the irrotational part of the electric field.]

(b) Calculate the electric and magnetic fields in K from the potentials and show that the electric field can be expressed alternatively as

$$\mathbf{E} = \mathbf{E}_{\text{dipole}}(\mathbf{p}_{\text{eff}} = \boldsymbol{\beta} \times \mathbf{m}) - \mathbf{m} \times \frac{[3\mathbf{n}(\mathbf{n} \cdot \boldsymbol{\beta}) - \boldsymbol{\beta}]}{R^3}$$

$$\mathbf{E} = \mathbf{E}_{\text{dipole}}\left(\mathbf{p}_{\text{eff}} = \frac{\boldsymbol{\beta}}{2} \times \mathbf{m}\right) + \frac{3}{2} \mathbf{n} \times \frac{[\mathbf{m}(\mathbf{n} \cdot \boldsymbol{\beta}) + \boldsymbol{\beta}(\mathbf{n} \cdot \mathbf{m})]}{R^3}$$

$$\mathbf{E} = \mathbf{B} \times \boldsymbol{\beta}$$

where $\mathbf{B}$ is the magnetic dipole field. In light of Problem 6.22, comment on the interpretation of the different forms.

11.30 An isotropic linear material medium, characterized by the constitutive relations (in its rest frame K'), $\mathbf{D}' = \epsilon \mathbf{E}'$ and $\mu \mathbf{H}' = \mathbf{B}'$, is in uniform translation with velocity $\mathbf{v}$ in the inertial frame K. By exploiting the fact that $F_{\mu\nu} = (\mathbf{E}, \mathbf{B})$ and $G_{\mu\nu} = (\mathbf{D}, \mathbf{H})$ transform as second rank 4-tensors under Lorentz transformations, show that the macroscopic fields $\mathbf{D}$ and $\mathbf{H}$ are given in terms of $\mathbf{E}$ and $\mathbf{B}$ by

$$\mathbf{D} = \epsilon \mathbf{E} + \gamma^2\left(\epsilon - \frac{1}{\mu}\right)[\beta^2 \mathbf{E}_\perp + \boldsymbol{\beta} \times \mathbf{B}]$$

$$\mathbf{H} = \frac{1}{\mu} \mathbf{B} + \gamma^2\left(\epsilon - \frac{1}{\mu}\right)[-\beta^2 \mathbf{B}_\perp + \boldsymbol{\beta} \times \mathbf{E}]$$

where $\mathbf{E}_\perp$ and $\mathbf{B}_\perp$ are components perpendicular to $\mathbf{v}$.

11.31 Consider a hollow right-circular cylinder of magnetic insulator (relative permeabilities ϵ and μ and inner and outer radii a and b) set in rotation about its axis at angular speed ω in a uniform axial magnetic field B_0. In 1913 the Wilsons measured the voltage difference between its inner and outer surfaces caused by a radial internal electric field. Assuming that locally the relations of Problem 11.30 hold, that the velocity $\mathbf{v} = \omega\rho\hat{\boldsymbol{\phi}}$, and that there are only the field components E_ρ and B_z, which are independent of z and ϕ, solve the equations $\nabla \cdot \mathbf{D} = 0$ and $\nabla \times \mathbf{H} = 0$ within the cylinder and show that the internal fields are

$$E_\rho = -\frac{\mu\omega\rho B_0}{c(1 - \omega^2\rho^2)}\left(1 - \frac{1}{\mu\epsilon}\right); \qquad B_z = \mu B_0\left[\frac{1 - \omega^2\rho^2/c^2\mu\epsilon}{1 - \omega^2\rho^2/c^2}\right]$$

and that for nonrelativistic motion ($\omega b/c \ll 1$) the voltage difference is

$$V = \tfrac{1}{2}\mu\omega B_0(b^2 - a^2)\left(1 - \frac{1}{\mu\epsilon}\right)$$

This experiment was an early validation of special relativity and Minkowski's electrodynamics of material media in motion. If you are curious about how the Wilsons made a magnetic insulator, look up the paper.

Reference: M. Wilson and H. A. Wilson, *Proc. Roy. Soc. London* **A89**, 99–106 (1913).

CHAPTER 12

Dynamics of Relativistic Particles and Electromagnetic Fields

The kinematics of the special theory of relativity was developed in Chapter 11. We now turn to the question of dynamics. In the first part of the chapter we discuss the dynamics of charged particle motion in external electromagnetic fields. The Lagrangian approach to the equations of motion is presented mainly to introduce the concept of a Lorentz invariant action from which covariant dynamical equations can be derived. The transition to a Hamiltonian, with the definition of the canonical momentum, is then discussed. Several sections are devoted to the motion of a charged particle in electric and magnetic fields. Our treatment of motion in a uniform static magnetic field is followed by consideration of motion in a combination of electric and magnetic fields. Then the secular changes (drifts) of a particle's orbit caused by nonuniform magnetic fields and the adiabatic invariance of the linked flux are discussed. The problem of a relativistic Lagrangian for a system of interacting charged particles is addressed, and it is shown that to order v^2/c^2 it is possible to eliminate retardation effects and write a Lagrangian (the Darwin Lagrangian) in terms of the instantaneous positions and velocities of the particles.

In the last five sections of the chapter the emphasis is on fields. First, the Maxwell equations are derived from a suitable Lagrangian. Then, a modified Lagrangian describing a "photon" with mass is presented and its consequences in resonant circuits, transmission lines, and cavities described, as well as its manifestation in superconductors. A covariant generalization of the Hamiltonian for fields is next discussed, along with the conservation laws of energy, momentum, and angular momentum for fields, both source free and in interaction with charged particles. The chapter ends with a derivation of the invariant Green functions that form the basis of the solution of the wave equation with a given 4-vector current density as source.

12.1 Lagrangian and Hamiltonian for a Relativistic Charged Particle in External Electromagnetic Fields

The equations of motion

$$\frac{d\mathbf{p}}{dt} = e\left[\mathbf{E} + \frac{\mathbf{u}}{c} \times \mathbf{B}\right] \tag{12.1}$$

$$\frac{dE}{dt} = e\mathbf{u} \cdot \mathbf{E} \tag{12.2}$$

for a particle of charge e in *external* fields **E** and **B** can be written in the covariant form (11.144):

$$\frac{dU^\alpha}{d\tau} = \frac{e}{mc} F^{\alpha\beta}U_\beta \qquad (12.3)$$

where m is the mass, τ is the proper time, and $U^\alpha = (\gamma c, \gamma\mathbf{u}) = p^\alpha/m$ is the 4-velocity of the particle.

Although the equations of motion (12.1) and (12.2) are sufficient to describe the general motion of a charged particle in external electromagnetic fields (neglecting the emission of radiation), it is useful to consider the formulation of the dynamics from the viewpoint of Lagrangian and Hamiltonian mechanics. The Lagrangian treatment of mechanics is based on the principle at least action or Hamilton's principle. In nonrelativistic mechanics the system is described by generalized coordinates $q_i(t)$ and velocities $\dot{q}_i(t)$. The Lagrangian L is a functional of q_i and $\dot{q}_i$ and perhaps the time explicitly and the action A is defined as the time integral of L along a possible path of the system. The *principle of least action* states that the motion of a mechanical system is such that in going from a configuration a at time t_1 to a configuration b at time t_2, the action

$$A = \int_{t_1}^{t_2} L[q_i(t), \dot{q}_i(t), t]\, dt \qquad (12.4)$$

is an extremum. By considering small variations of the coordinates and velocities away from the actual path and requiring $\delta A = 0$, one obtains (see *Goldstein*, Chapter 2) the Euler–Lagrange equations of motion,

$$\frac{d}{dt}\left(\frac{\partial L}{\partial \dot{q}_i}\right) - \frac{\partial L}{\partial q_i} = 0 \qquad (12.5)$$

We wish to extend the formalism to relativistic particle motion in a manner consistent with the special theory of relativity and leading for charged particles in external fields to (12.1) and (12.2) or (12.3). There are several levels of sophistication possible. The least sophisticated and most familiar treatment continues with ordinary coordinates, velocities, and time and generalizes from the nonrelativistic domain in a straightforward way. More sophisticated is a manifestly covariant discussion. We first present the elementary approach and then indicate the manifestly covariant treatment.

A. Elementary Approach to a Relativistic Lagrangian

To obtain a relativistic Lagrangian for a particle in external fields we first consider the question of the Lorentz transformation properties of the Lagrangian. From the first postulate of special relativity the action integral must be a Lorentz scalar because the equations of motion are determined by the extremum condition, $\delta A = 0$. If we introduce the particle's proper time τ into (12.4) through $dt = \gamma\, d\tau$, the action integral becomes

$$A = \int_{\tau_1}^{\tau_2} \gamma L\, d\tau \qquad (12.6)$$

Since proper time is invariant the condition that A also be invariant requires that γL *be Lorentz invariant.*

The Lagrangian for a free particle can be a function of the velocity of the particle and its mass, but cannot depend on its position. The only Lorentz invariant function of the velocity available is $U_\alpha U^\alpha = c^2$. Thus we conclude that the Lagrangian for a free particle is proportional to $\gamma^{-1} = \sqrt{1 - \beta^2}$. It is easily seen that

$$L_{\text{free}} = -mc^2 \sqrt{1 - \frac{u^2}{c^2}} \tag{12.7}$$

is the proper multiple of γ^{-1} to yield, through (12.5), the free-particle equation of motion,

$$\frac{d}{dt}(\gamma m \mathbf{u}) = 0 \tag{12.8}$$

The action (12.6) is proportional to the integral of the proper time over the path from the initial proper time τ_1 to the final proper time τ_2. This integral is Lorentz invariant, but it depends on the path taken. For purposes of calculation, consider a reference frame in which the particle is initially at rest. From definition (11.26) of proper time it is clear that, if the particle stays at rest in that frame, the integral over proper time will be larger than if it moves with a nonzero velocity along its path. Consequently we see that a straight world line joining the initial and final points of the path gives the maximum integral over proper time or, with the negative sign in (12.7), a minimum for the action integral. This motion at constant velocity is, of course, the solution of the free-particle equation of motion.

The general requirement that γL be Lorentz invariant allows us to determine the Lagrangian for a relativistic charged particle in external electromagnetic fields, provided we know something about the Lagrangian (or equations of motion) for nonrelativistic motion in static fields. A slowly moving charged particle is influenced predominantly by the electric field that is derivable from the scalar potential Φ. The potential energy of interaction is $V = e\Phi$. Since the nonrelativistic Lagrangian is $(T - V)$, the interaction part L_{int} of the relativistic Lagrangian must reduce in the nonrelativistic limit to

$$L_{\text{int}} \to L_{\text{int}}^{NR} = -e\Phi \tag{12.9}$$

Our problem thus becomes that of finding a Lorentz invariant expression for γL_{int} that reduces to (12.9) for nonrelativistic velocities. Since Φ is the time component of the 4-vector potential A^α, we anticipate that γL_{int} will involve the scalar product of A^α with some 4-vector. The only other 4-vectors available are the momentum and position vectors of the particle. Since gamma times the Lagrangian must be translationally invariant as well as Lorentz invariant, it cannot involve the coordinates explicitly. Hence the interaction Lagrangian must be*

$$L_{\text{int}} = -\frac{e}{\gamma c} U_\alpha A^\alpha \tag{12.10}$$

or

$$L_{\text{int}} = -e\Phi + \frac{e}{c} \mathbf{u} \cdot \mathbf{A} \tag{12.11}$$

---

*Without appealing to the nonrelativistic limit, this form of L_{int} can be written down by demanding that γL_{int} be a Lorentz invariant that is (1) linear in the charge of the particle, (2) linear in the electromagnetic potentials, (3) translationally invariant, and (4) a function of no higher than the first time derivative of the particle coordinates. The reader may consider the possibility of an interaction Lagrangian satisfying these conditions, but linear in the field strengths $F^{\alpha\beta}$, rather than the potentials A^α.

The combination of (12.7) and (12.11) yields the complete relativistic Lagrangian for a charged particle:

$$L = -mc^2 \sqrt{1 - \frac{u^2}{c^2}} + \frac{e}{c} \mathbf{u} \cdot \mathbf{A} - e\Phi \tag{12.12}$$

Verification that (12.12) does indeed lead to the Lorentz force equation will be left as an exercise for the reader. Use must be made of the convective derivative $[d/dt = (\partial/\partial t) + \mathbf{u} \cdot \nabla]$ and the standard definitions of the fields in terms of the potentials.

The canonical momentum $\mathbf{P}$ conjugate to the position coordinate $\mathbf{x}$ is obtained by the definition,

$$P_i \equiv \frac{\partial L}{\partial u_i} = \gamma m u_i + \frac{e}{c} A_i \tag{12.13}$$

Thus the conjugate momentum is

$$\mathbf{P} = \mathbf{p} + \frac{e}{c} \mathbf{A} \tag{12.14}$$

where $\mathbf{p} = \gamma m \mathbf{u}$ is the ordinary kinetic momentum. The Hamiltonian H is a function of the coordinate $\mathbf{x}$ and its conjugate momentum $\mathbf{P}$ and is a constant of the motion if the Lagrangian is not an explicit function of time. The Hamiltonian is defined in terms of the Lagrangian as

$$H = \mathbf{P} \cdot \mathbf{u} - L \tag{12.15}$$

The velocity $\mathbf{u}$ must be eliminated from (12.15) in favor of $\mathbf{P}$ and $\mathbf{x}$. From (12.13) or (12.14) we find that

$$\mathbf{u} = \frac{c\mathbf{P} - e\mathbf{A}}{\sqrt{\left(\mathbf{P} - \frac{e\mathbf{A}}{c}\right)^2 + m^2 c^2}} \tag{12.16}$$

When this is substituted into (12.15) and into L (12.12), the Hamiltonian takes on the form:

$$H = \sqrt{(c\mathbf{P} - e\mathbf{A})^2 + m^2 c^4} + e\Phi \tag{12.17}$$

Again the reader may verify that Hamilton's equations of motion can be combined to yield the Lorentz force equation. Equation (12.17) is an expression for the total energy W of the particle. It differs from the free-particle energy by the addition of the potential energy $e\Phi$ and by the replacement $\mathbf{p} \to [\mathbf{P} - (e/c)\mathbf{A}]$. These two modifications are actually only one 4-vector change. This can be seen by transposing $e\Phi$ in (12.17) and squaring both sides. Then

$$(W - e\Phi)^2 - (c\mathbf{P} - e\mathbf{A})^2 = (mc^2)^2 \tag{12.18}$$

This is just the 4-vector scalar product,

$$p_\alpha p^\alpha = (mc)^2 \tag{12.19}$$

where

$$p^\alpha \equiv \left(\frac{E}{c}, \mathbf{p}\right) = \left(\frac{1}{c}(W - e\Phi), \mathbf{P} - \frac{e}{c}\mathbf{A}\right) \tag{12.20}$$

We see that the total energy W/c acts as the time component of a canonically conjugate 4-momentum P^α of which $\mathbf{P}$ given by (12.14) is the space part. A manifestly covariant approach, discussed in the following paragraphs and also in Problem 12.1 leads naturally to this 4-momentum.

In passing we remark on the question of gauge transformations. Obviously the equations of motion (12.1) and (12.2) are invariant under a gauge transformation of the potentials. Since the Lagrangian (12.10) involves the potentials explicitly, it is not invariant. In spite of this lack of invariance of L under gauge transformations it can be shown (Problem 12.2) that the change in the Lagrangian is of such a form (a total time derivative) that it does not alter the action integral or the equations of motion.

B. Manifestly Covariant Treatment of the Relativistic Lagrangian

To make a manifestly covariant description, the customary variables $\mathbf{x}$ and $\mathbf{u}$ are replaced by the 4-vectors x^α and U^α. The free-particle Lagrangian (12.7) can be written in terms of U^α as

$$L_{\text{free}} = -\frac{mc}{\gamma} \sqrt{U_\alpha U^\alpha} \tag{12.21}$$

Then the action integral (12.6) would be

$$A = -mc \int_{\tau_1}^{\tau_2} \sqrt{U_\alpha U^\alpha} \, d\tau \tag{12.22}$$

This manifestly invariant form might be thought to provide the starting point for a variational calculation leading to the equation of motion, $dU^\alpha/d\tau = 0$. There is, however, the equation of constraint,

$$U_\alpha U^\alpha = c^2 \tag{12.23}$$

or the equivalent constraint,

$$U_\alpha \frac{dU^\alpha}{d\tau} = 0 \tag{12.24}$$

on the equations of motion. This can be incorporated by the Lagrange multiplier technique, but we pursue a different, equivalent procedure. The integrand in (12.22) is

$$\sqrt{U_\alpha U^\alpha} \, d\tau = \sqrt{\frac{dx_\alpha}{d\tau} \frac{dx^\alpha}{d\tau}} \, d\tau = \sqrt{g^{\alpha\beta} \, dx_\alpha \, dx_\beta}$$

that is, the infinitesimal length element in 4-space. This suggests that the action integral (12.22) be replaced by

$$A = -mc \int_{s_1}^{s_2} \sqrt{g^{\alpha\beta} \frac{dx_\alpha}{ds} \frac{dx_\beta}{ds}} \, ds \tag{12.25}$$

where the 4-vector coordinate of the particle is $x^\alpha(s)$, with s a parameter that is a monotonically increasing function of τ, but otherwise arbitrary. The action integral is an integral along the world line of the particle, and the principle of least action is the statement that the actual path is the longest path, namely the goedesic.* The Lagrangian variables are now x^α and "the velocity" dx^α/ds, but s is considered as arbitrary. Only after the calculus of variations has been completed do we identify

$$\sqrt{g^{\alpha\beta}\frac{dx_\alpha}{ds}\frac{dx_\beta}{ds}}\, ds = c\, d\tau \tag{12.26}$$

and so impose the constraint (12.23). A straightforward variational calculation with (12.25) yields the Euler–Lagrange equations,

$$mc\frac{d}{ds}\left[\frac{dx^\alpha/ds}{\left(\dfrac{dx_\beta}{ds}\dfrac{dx^\beta}{ds}\right)^{1/2}}\right] = 0 \tag{12.27}$$

or

$$m\frac{d^2x^\alpha}{d\tau^2} = 0 \tag{12.28}$$

as expected for free-particle motion.

For a charged particle in an external field the form of the Lagrangian (12.11) suggests that the manifestly covariant form of the action integral is

$$A = -\int_{s_1}^{s_2}\left[mc\sqrt{g^{\alpha\beta}\frac{dx_\alpha}{ds}\frac{dx_\beta}{ds}} + \frac{e}{c}\frac{dx_\alpha}{ds}A^\alpha(x)\right]ds \tag{12.29}$$

Hamilton's principle yields the Euler–Lagrange equations,

$$\frac{d}{ds}\left[\frac{\partial\tilde{L}}{\partial\left(\dfrac{dx_\alpha}{ds}\right)}\right] - \partial^\alpha\tilde{L} = 0 \tag{12.30}$$

where the Lagrangian is

$$\tilde{L} = -\left[mc\sqrt{g^{\alpha\beta}\frac{dx_\alpha}{ds}\frac{dx_\beta}{ds}} + \frac{e}{c}\frac{dx_\alpha}{ds}A^\alpha(x)\right] \tag{12.31}$$

Explicitly, (12.30), upon division by the square root and use of (12.26), becomes

$$m\frac{d^2x^\alpha}{d\tau^2} + \frac{e}{c}\frac{dA^\alpha(x)}{d\tau} - \frac{e}{c}\frac{dx_\beta}{d\tau}\partial^\alpha A^\beta(x) = 0$$

Since $dA^\alpha/d\tau = (dx_\beta/d\tau)\,\partial^\beta A^\alpha$, this equation can be written as

$$m\frac{d^2x^\alpha}{d\tau^2} = \frac{e}{c}(\partial^\alpha A^\beta - \partial^\beta A^\alpha)\frac{dx_\beta}{d\tau} \tag{12.32}$$

which is the covariant equation of motion (12.3) in different notation.

The transition to the conjugate momenta and a Hamiltonian is simple enough, but has problems of interpretation. The conjugate momentum 4-vector is defined by

$$P^\alpha = -\frac{\partial\tilde{L}}{\partial\left(\dfrac{dx_\alpha}{ds}\right)} = mU^\alpha + \frac{e}{c}A^\alpha \tag{12.33}$$

---

*The geodesic is the longest path or longest proper time for timelike separation of events. See *Rohrlich*, pp. 277–278.

The minus sign is introduced so that (12.33) conforms with (12.14); its origin can be traced to the properties of the Lorentz space-time. A Hamiltonian can be defined by

$$\tilde{H} = P_\alpha U^\alpha + \tilde{L} \tag{12.34}$$

Elimination of U^α by means of (12.33) leads to the expression,

$$\tilde{H} = \frac{1}{m}\left(P_\alpha - \frac{eA_\alpha}{c}\right)\left(P^\alpha - \frac{eA^\alpha}{c}\right) - c\sqrt{\left(P_\alpha - \frac{e}{c}A_\alpha\right)\left(P^\alpha - \frac{e}{c}A^\alpha\right)} \tag{12.35}$$

Hamilton's equations are

$$\frac{dx^\alpha}{d\tau} = \frac{\partial \tilde{H}}{\partial P_\alpha} = \frac{1}{m}\left(P^\alpha - \frac{e}{c}A^\alpha\right)$$

and

$$\frac{dP^\alpha}{d\tau} = -\frac{\partial \tilde{H}}{\partial x_\alpha} = \frac{e}{mc}\left(P_\beta - \frac{eA_\beta}{c}\right)\partial^\alpha A^\beta \tag{12.36}$$

where we have made use of the constraint $\left(P_\alpha - \frac{e}{c}A_\alpha\right)\left(P^\alpha - \frac{e}{c}A^\alpha\right) = m^2 c^2$

after differentiation. These two equations can be immediately shown to be equivalent to the Euler–Langrange equation (12.32).

While the Hamiltonian above is formally satisfactory, it has several problems. The first is that it is by definition a Lorentz scalar, not an energylike quantity. Second, use of (12.23) and (12.33) shows that $\tilde{H} \equiv 0$. Clearly, such a Hamiltonian formulation differs considerably from the familiar nonrelativistic version. The reader can refer to *Barut* (pp. 68 ff.) for a discussion of this and other alternative Hamiltonians.

12.2 *Motion in a Uniform, Static Magnetic Field*

As a first important example of the dynamics of charged particles in electromagnetic fields we consider the motion in a uniform, static, magnetic induction **B**. The equations of motion (12.1) and (12.2) are

$$\frac{d\mathbf{p}}{dt} = \frac{e}{c}\mathbf{v} \times \mathbf{B}, \qquad \frac{dE}{dt} = 0 \tag{12.37}$$

where here the particle's velocity is denoted by **v**. Since the energy is constant in time, the magnitude of the velocity is constant and so is γ. Then the first equation can be written

$$\frac{d\mathbf{v}}{dt} = \mathbf{v} \times \boldsymbol{\omega}_B \tag{12.38}$$

where

$$\boldsymbol{\omega}_B = \frac{e\mathbf{B}}{\gamma m c} = \frac{ec\mathbf{B}}{E} \tag{12.39}$$

is the gyration or precession frequency. The motion described by (12.38) is a circular motion perpendicular to **B** and a uniform translation parallel to **B**. The solution for the velocity is easily shown to be

$$\mathbf{v}(t) = v_\parallel \boldsymbol{\epsilon}_3 + \omega_B a(\boldsymbol{\epsilon}_1 - i\boldsymbol{\epsilon}_2)e^{-i\omega_B t} \tag{12.40}$$

where $\boldsymbol{\epsilon}_3$ is a unit vector parallel to the field, $\boldsymbol{\epsilon}_1$ and $\boldsymbol{\epsilon}_2$ are the other orthogonal unit vectors, $v_\parallel$ is the velocity component along the field, and a is the gyration radius. The convention is that the real part of the equation is to be taken. Then one can see that (12.40) represents a *counterclockwise rotation* (for positive charge e) when viewed in the direction of **B**. Another integration yields the displacement of the particle,

$$\mathbf{x}(t) = \mathbf{X}_0 + v_\parallel t \boldsymbol{\epsilon}_3 + ia(\boldsymbol{\epsilon}_1 - i\boldsymbol{\epsilon}_2)e^{-i\omega_B t} \tag{12.41}$$

The path is a helix of radius a and pitch angle $\alpha = \tan^{-1}(v_\parallel/\omega_B a)$. The magnitude of the gyration radius a depends on the magnetic induction **B** and the transverse momentum $\mathbf{p}_\perp$ of the particle. From (12.39) and (12.40) it is evident that

$$cp_\perp = eBa$$

This form is convenient for the determination of particle momenta. The radius of curvature of the path of a charged particle in a known B allows the determination of its momentum. For particles with charge the same in magnitude as the electronic charge, the momentum can be written numerically as

$$p_\perp(\text{MeV}/c) = 3.00 \times 10^{-4} Ba \text{ (gauss-cm)} = 300\, Ba \text{ (tesla-m)} \tag{12.42}$$

12.3 *Motion in Combined, Uniform, Static Electric and Magnetic Fields*

We now consider a charged particle moving in a combination of electric and magnetic fields **E** and **B**, both uniform and static, but in general not parallel. As an important special case, *perpendicular fields* will be treated first. The energy equation (12.2) shows that the particle's energy is not constant in time. Consequently we cannot obtain a simple equation for the velocity, as was done for a static magnetic field. But an appropriate Lorentz transformation simplifies the equations of motion. Consider a Lorentz transformation to a coordinate frame K' moving with a velocity **u** with respect to the original frame. Then the Lorentz force equation for the particle in K' is

$$\frac{d\mathbf{p}'}{dt'} = e\left(\mathbf{E}' + \frac{\mathbf{v}' \times \mathbf{B}'}{c}\right)$$

where the primed variables are referred to the system K'. The fields **E**' and **B**' are given by relations (11.149) with **v** replaced by **u**. Let us first suppose that $|\mathbf{E}| < |\mathbf{B}|$. If **u** is now chosen perpendicular to the orthogonal vectors **E** and **B**,

$$\mathbf{u} = c\,\frac{\mathbf{E} \times \mathbf{B}}{B^2} \tag{12.43}$$

we find the fields in K' to be

$$
\begin{aligned}
&\mathbf{E}'_\parallel = 0, \quad \mathbf{E}'_\perp = \gamma\left(\mathbf{E} + \frac{\mathbf{u}}{c} \times \mathbf{B}\right) = 0 \\
&\mathbf{B}'_\parallel = 0, \quad \mathbf{B}'_\perp = \frac{1}{\gamma}\mathbf{B} = \left(\frac{B^2 - E^2}{B^2}\right)^{1/2}\mathbf{B}
\end{aligned}
\tag{12.44}
$$

where $\parallel$ and $\perp$ refer to the direction of **u**. In the frame K' the only field acting is a static magnetic field **B**′ which points in the same direction as **B**, but is weaker than **B** by a factor γ^{-1}. Thus the motion in K' is the same as that considered in Section 12.2, namely a spiraling around the lines of force. As viewed from the original coordinate system, this gyration is accompanied by a uniform "drift" **u** perpendicular to **E** and **B** given by (12.43). This drift is sometimes called the $E \times B$ *drift*. The drift can be understood qualitatively by noting that a particle that starts gyrating around **B** is accelerated by the electric field, gains energy, and so moves in a path with a larger radius for roughly half of its cycle. On the other half, the electric field decelerates it, causing it to lose energy and so move in a tighter arc. The combination of arcs produces a translation perpendicular to **E** and **B** as shown in Fig. 12.1. The direction of drift is independent of the sign of the charge of the particle.

The drift velocity **u** (12.43) has physical meaning only if it is less than the velocity of light, i.e., only if $|\mathbf{E}| < |\mathbf{B}|$. If $|\mathbf{E}| > |\mathbf{B}|$, the electric field is so strong that the particle is continually accelerated in the direction of **E** and its average energy continues to increase with time. To see this we consider a Lorentz transformation from the original frame to a system K'' moving with a velocity

$$\mathbf{u}' = c\, \frac{\mathbf{E} \times \mathbf{B}}{E^2} \tag{12.45}$$

relative to the first. In this frame the electric and magnetic fields are

$$\mathbf{E}''_\parallel = 0, \qquad \mathbf{E}''_\perp = \frac{1}{\gamma'}\,\mathbf{E} = \left(\frac{E^2 - B^2}{E^2}\right)^{1/2}\mathbf{E}$$

$$\mathbf{B}''_\parallel = 0, \qquad \mathbf{B}''_\perp = \gamma'\left(\mathbf{B} - \frac{\mathbf{u}' \times \mathbf{E}}{c}\right) = 0 \tag{12.46}$$

Thus in the system K'' the particle is acted on by a purely electrostatic field which causes hyperbolic motion with ever-increasing velocity (see Problem 12.3).

The fact that a particle can move through crossed **E** and **B** fields with the uniform velocity $u = cE/B$ provides the possibility of selecting charged particles according to velocity. If a beam of particles having a spread in velocities is normally incident on a region containing uniform crossed electric and magnetic fields, only those particles with velocities equal to cE/B will travel without deflection. Suitable entrance and exit slits will then allow only a very narrow band of velocities around cE/B to be transmitted, the resolution depending on the geometry, the velocities desired, and the field strengths. When combined with momentum selectors, such as a deflecting magnet, these $\mathbf{E} \times \mathbf{B}$ velocity selectors

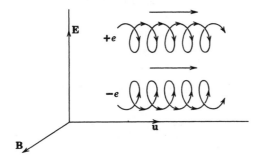

Figure 12.1   E × **B** drift of charged particles in crossed fields.

can extract a very pure and monoenergetic beam of particles of a definite mass from a mixed beam of particles with different masses and momenta. Large-scale devices of this sort are commonly used to provide experimental beams of particles produced in high-energy accelerators.

If **E** has a component parallel to **B**, the behavior of the particle cannot be understood in such simple terms as above. The scalar produce **E · B** is a Lorentz invariant quantity (see Problem 11.14), as is $(B^2 - E^2)$. When the fields were perpendicular (**E · B** = 0), it was possible to find a Lorentz frame where **E** = 0 if $|\mathbf{B}| > |\mathbf{E}|$, or **B** = 0 if $|\mathbf{E}| > |\mathbf{B}|$. In those coordinate frames the motion was relatively simple. If **E · B** ≠ 0, electric and magnetic fields will exist simultaneously in all Lorentz frames, the angle between the fields remaining acute or obtuse depending on its value in the original coordinate frame. Consequently motion in combined fields must be considered. When the fields are static and uniform, it is a straightforward matter to obtain a solution for the motion in Cartesian components. This will be left for Problem 12.6.

12.4 Particle Drifts in Nonuniform, Static Magnetic Fields

In astrophysical and thermonuclear applications it is of considerable interest to know how particles behave in magnetic fields that vary in space. Often the variations are gentle enough that a perturbation solution to the motion, first given by Alfvén, is an adequate approximation. "Gentle enough" generally means that the distance over which **B** changes appreciably in magnitude or direction is large compared to the gyration radius a of the particle. Then the lowest order approximation to the motion is a spiraling around the lines of force at a frequency given by the local value of the magnetic induction. In the next approximation, the orbit undergoes slow changes that can be described as a drifting of the guiding center.

The first type of spatial variation of the field to be considered is a gradient perpendicular to the direction of **B**. Let the gradient at the point of interest be in the direction of the unit vector **n**, with **n · B** = 0. Then, to first order, the gyration frequency can be written

$$\boldsymbol{\omega}_B(\mathbf{x}) = \frac{e}{\gamma mc} \mathbf{B}(\mathbf{x}) \simeq \boldsymbol{\omega}_0 \left[1 + \frac{1}{B_0} \left(\frac{\partial B}{\partial \xi} \right)_0 \mathbf{n} \cdot \mathbf{x} \right] \qquad (12.47)$$

In (12.47) ξ is the coordinate in the direction **n**, and the expansion is about the origin of coordinates where $\omega_B = \omega_0$. Since the direction of **B** is unchanged, the motion parallel to **B** remains a uniform translation. Consequently we consider only modifications in the transverse motion. Writing $\mathbf{v}_\perp = \mathbf{v}_0 + \mathbf{v}_1$, where $\mathbf{v}_0$ is the uniform-field transverse velocity and $\mathbf{v}_1$ is a small correction term, we can substitute (12.47) into the force equation

$$\frac{d\mathbf{v}_\perp}{dt} = \mathbf{v}_\perp \times \boldsymbol{\omega}_B(\mathbf{x}) \qquad (12.48)$$

and, keeping only first-order terms, obtain the approximate result

$$\frac{d\mathbf{v}_1}{dt} \simeq \left[\mathbf{v}_1 + \mathbf{v}_0(\mathbf{n} \cdot \mathbf{x}_0) \frac{1}{B_0} \left(\frac{\partial B}{\partial \xi} \right)_0 \right] \times \boldsymbol{\omega}_0 \qquad (12.49)$$

From (12.40) and (12.41) it is easy to see that for a uniform field the transverse velocity $\mathbf{v}_0$ and coordinate $\mathbf{x}_0$ are related by

$$\left.\begin{array}{c} \mathbf{v}_0 = -\boldsymbol{\omega}_0 \times (\mathbf{x}_0 - \mathbf{X}) \\[2mm] (\mathbf{x}_0 - \mathbf{X}) = \dfrac{1}{\omega_0^2}\,(\boldsymbol{\omega}_0 \times \mathbf{v}_0) \end{array}\right\} \tag{12.50}$$

where $\mathbf{X}$ is the center of gyration of the unperturbed circular motion ($\mathbf{X} = 0$ here). If $(\boldsymbol{\omega}_0 \times \mathbf{v}_0)$ is eliminated in (12.49) in favor of $\mathbf{x}_0$, we obtain

$$\frac{d\mathbf{v}_1}{dt} \simeq \left[\mathbf{v}_1 - \frac{1}{B_0}\left(\frac{\partial B}{\partial \xi}\right)_0 \boldsymbol{\omega}_0 \times \mathbf{x}_0(\mathbf{n} \cdot \mathbf{x}_0)\right] \times \boldsymbol{\omega}_0 \tag{12.51}$$

This shows that apart from oscillatory terms, $\mathbf{v}_1$ has a nonzero average value.

$$\mathbf{v}_G \equiv \langle \mathbf{v}_1 \rangle = \frac{1}{B_0}\left(\frac{\partial B}{\partial \xi}\right)_0 \boldsymbol{\omega}_0 \times \langle (\mathbf{x}_0)_\perp(\mathbf{n} \cdot \mathbf{x}_0)\rangle \tag{12.52}$$

To determine the average value of $(\mathbf{x}_0)_\perp(\mathbf{n} \cdot \mathbf{x}_0)$, it is necessary only to observe that the rectangular components of $(\mathbf{x}_0)_\perp$ oscillate sinusoidally with peak amplitude a and a phase difference of 90°. Hence only the component of $(\mathbf{x}_0)_\perp$ parallel to $\mathbf{n}$ contributes to the average, and

$$\langle (\mathbf{x}_0)_\perp(\mathbf{n} \cdot \mathbf{x}_0)\rangle = \frac{a^2}{2}\,\mathbf{n} \tag{12.53}$$

Thus the gradient drift velocity is given by

$$\mathbf{v}_G = \frac{a^2}{2}\frac{1}{B_0}\left(\frac{\partial B}{\partial \xi}\right)_0(\boldsymbol{\omega}_0 \times \mathbf{n}) \tag{12.54}$$

An alternative form, independent of coordinates, is

$$\frac{\mathbf{v}_G}{\omega_B a} = \frac{a}{2B^2}\,(\mathbf{B} \times \boldsymbol{\nabla}_\perp B) \tag{12.55}$$

From (12.55) it is evident that, if the gradient of the field is such that $a\,|\boldsymbol{\nabla}B/B|$ $\ll 1$, the drift velocity is small compared to the orbital velocity ($\omega_B a$). The particle spirals rapidly while its center of rotation moves slowly perpendicular to both $\mathbf{B}$ and $\boldsymbol{\nabla}B$. The sense of the drift for positive particles is given by (12.55). For negatively charged particles the sign of the drift velocity is opposite; the sign change comes from the definition of ω_B. The gradient drift can be understood qualitatively from consideration of the variation of gyration radius as the particle moves in and out of regions of larger than average and smaller than average field strength. Figure 12.2 shows this qualitative behavior for both signs of charge.

Another type of field variation that causes a drifting of the particle's guiding center is curvature of the lines of force. Consider the two-dimensional field shown in Fig. 12.3. It is locally independent of z. Figure 12.3a shows a constant, uniform magnetic induction $\mathbf{B}_0$, parallel to the x axis. A particle spirals around the field lines with a gyration radius a and a velocity $\omega_B a$, while moving with a uniform velocity $v_\parallel$ along the lines of force. We wish to treat that motion as a zeroth-order approximation to the motion of the particle in the field shown in Fig. 12.3b, where the lines of force are curved with a local radius of curvature R that is large compared to a.

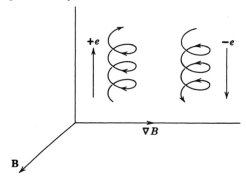

Figure 12.2 Drift of charged particles due to transverse gradient of magnetic field.

The first-order motion can be understood as follows. The particle tends to spiral around a field line, but the field line curves off to the side. As far as the motion of the guiding center is concerned, this is equivalent to a centrifugal acceleration of magnitude $v_\parallel^2/R$. This acceleration can be viewed as arising from an effective electric field

$$\mathbf{E}_{\text{eff}} = \frac{\gamma m}{e} \frac{\mathbf{R}}{R^2} v_\parallel^2 \tag{12.56}$$

in addition to the magnetic induction $\mathbf{B}_0$. From (12.43) we see that the combined effective electric field and the magnetic induction cause a *curvature drift velocity*,

$$\mathbf{v}_C \simeq c \frac{\gamma m}{e} v_\parallel^2 \frac{\mathbf{R} \times \mathbf{B}_0}{R^2 B_0^2} \tag{12.57}$$

With the definition of $\omega_B = eB_0/\gamma mc$, the curvature drift can be written

$$\mathbf{v}_C = \frac{v_\parallel^2}{\omega_B R} \left(\frac{\mathbf{R} \times \mathbf{B}_0}{R B_0} \right) \tag{12.58}$$

The direction of drift is specified by the vector product, in which $\mathbf{R}$ is the radius vector *from* the effective center of curvature *to* the position of the charge. The sign in (12.58) is appropriate for positive charges and is independent of the sign of $v_\parallel$. For negative particles the opposite sign arises from ω_B.

A more straightforward, although pedestrian, derivation of (12.58) can be given by solving the Lorentz force equation directly. If we use cylindrical coor-

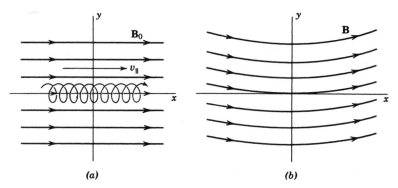

Figure 12.3 (*a*) Particle moving in helical path along lines of uniform, constant magnetic induction. (*b*) Curvature of lines of magnetic induction will cause drift perpendicular to the *x-y* plane.

dinates (ρ, ϕ, z) appropriate to Fig. 12.3b with origin at the center of curvature, the magnetic induction has only a ϕ component, $B_\phi = B_0(R/\rho)$. Then the force equation can be easily shown to give the three equations:

$$\ddot{\rho} - \rho\dot{\phi}^2 = -\omega_B \frac{R}{\rho} \dot{z}$$

$$\rho\ddot{\phi} + 2\dot{\rho}\dot{\phi} = 0 \tag{12.59}$$

$$\ddot{z} = \omega_B \frac{R}{\rho} \dot{\rho}$$

The second equation has a first integral, $\rho^2\dot{\phi} = Rv_\parallel$, a constant. The third equation has a first integral, $\dot{z} = \omega_B \ln(\rho/R) + v_0$, where v_0 is a constant of integration. With the zeroth-order trajectory a helix with radius small compared to R, it is natural to write $\rho = R + x$ and expand $(\rho/R)^n$ and $\ln(\rho/R)$ in powers of x/R. Then $\dot{z} \approx \omega_B x + v_0$, and the radial equation of motion can be approximated by

$$\ddot{x} + \left(\omega_B^2 + 3\frac{v_\parallel^2}{R^2}\right)x \approx \frac{v_\parallel^2}{R} - \omega_B v_0$$

which describes simple harmonic oscillations in x around a displaced equilibrium

$$\langle x \rangle \approx \frac{v_\parallel^2}{\omega_B^2 R} - \frac{v_0}{\omega_B}$$

where we have assumed $v_\parallel \ll \omega_B R$. The mean value of $\dot{z}$ is then

$$\langle \dot{z} \rangle \approx v_0 + \omega_B \langle x \rangle \approx \frac{v_\parallel^2}{\omega_B R} \tag{12.60}$$

This is just the curvature drift given by (12.58).

For regions of space in which there are no currents the gradient drift $\mathbf{v}_G$ (12.55) and the curvature drift $\mathbf{v}_C$ (12.58) can be combined into one simple form. This follows from the fact that for a two-dimensional field such as shown in Fig. 12.3b $\nabla \times \mathbf{B} = 0$ implies

$$\frac{\nabla_\perp B}{B} = -\frac{\mathbf{R}}{R^2}$$

Evidently then, for a two-dimensional field, the sum of $\mathbf{v}_G$ and $\mathbf{v}_C$ is a total drift velocity,

$$\mathbf{v}_D = \frac{1}{\omega_B R} (v_\parallel^2 + \tfrac{1}{2}v_\perp^2)\left(\frac{\mathbf{R} \times \mathbf{B}}{RB}\right) \tag{12.61}$$

where $v_\perp = \omega_B a$ is the transverse velocity of gyration. For singly charged non-relativistic particles in thermal equilibrium, the magnitude of the drift velocity is

$$v_D(\text{cm/s}) = \frac{172\ T(\text{K})}{R(\text{m})\ B(\text{gauss})} \tag{12.62}$$

The particle drifts implied by (12.61) are troublesome in certain types of thermonuclear machine designed to contain hot plasma. A possible configuration is a toroidal tube with a strong field supplied by solenoidal windings around the torus. With typical parameters of $R = 1$ meter, $B = 10^3$ gauss, particles in a 1 eV plasma ($T \simeq 10^4$ K) will have drift velocities $v_D \sim 1.8 \times 10^3$ cm/s. This means that they will drift out to the walls in a small fraction of a second. For hotter plasmas the drift rate is correspondingly greater. One way to prevent this first-order drift in toroidal geometries is to twist the torus into a figure eight. Since the particles generally make many circuits around the closed path before drifting across the tube, they feel no net curvature or gradient of the field. Consequently they experience no net drift, at least to first order in $1/R$. This method of eliminating drifts due to spatial variations of the magnetic field is used in the Stellarator type of thermonuclear machine, in which containment is attempted with a strong, externally produced, axial magnetic field.

12.5 *Adiabatic Invariance of Flux Through Orbit of Particle*

The various motions discussed in the preceding sections have been perpendicular to the lines of magnetic force. These motions, caused by electric fields or by the gradient or curvature of the magnetic field, arise because of the peculiarities of the magnetic-force term in the Lorentz force equation. To complete our general survey of particle motion in magnetic fields, we must consider motion parallel to the lines of force. It turns out that for slowly varying fields a powerful tool is the concept of adiabatic invariants. In celestial mechanics and in the old quantum theory, adiabatic invariants were useful in discussing perturbations on the one hand, and in deciding what quantities were to be quantized on the other. Our discussion will resemble most closely the celestial-mechanical problem, since we are interested in the behavior of a charged particle in slowly varying fields, which can be viewed as small departures from the simple, uniform, static field considered in Section 12.2.

The concept of adiabatic invariance is introduced by considering the action integrals of a mechanical system. If q_i and p_i are the generalized canonical coordinates and momenta, then, for each coordinate which is periodic, the action integral J_i is defined by

$$J_i = \oint p_i \, dq_i \tag{12.63}$$

The integration is over a complete cycle of the coordinate q_i. For a given mechanical system with specified initial conditions the action integrals J_i are constants. If now the properties of the system are changed in some way (e.g., a change in spring constant or mass of some particle), the question arises as to how the action integrals change. It can be proved* that, if the change in property is slow compared to the relevant periods of motion and is not related to the periods (such a change is called an *adiabatic change*), the action integrals are invariant. This means that, if we have a certain mechanical system in some state of motion and we make an adiabatic change in some property so that after a long time we end up with a different mechanical system, the final motion of that different system will be such that the action integrals have the same values as in the initial

*See, for example, M. Born, *The Mechanics of the Atom*, Bell, London (1927), or I. Percival and D. Richards, *Introduction to Dynamics*, Cambridge University Press, Cambridge (1982), Section 9.4.

system. Clearly this provides a powerful tool in examining the effects of slow changes in properties.

For a charged particle in a uniform, static, magnetic induction **B**, the transverse motion is periodic. The action integral for this transverse motion is

$$J = \oint \mathbf{P}_\perp \cdot d\mathbf{l} \tag{12.64}$$

where $\mathbf{P}_\perp$ is the transverse component of the canonical momentum (12.14) and $d\mathbf{l}$ is a directed line element along the circular path of a particle. From (12.14) we find that

$$J = \oint \gamma m \mathbf{v}_\perp \cdot d\mathbf{l} + \frac{e}{c} \oint \mathbf{A} \cdot d\mathbf{l} \tag{12.65}$$

Since $\mathbf{v}_\perp$ is parallel to $d\mathbf{l}$, we find

$$J = \oint \gamma m \omega_B a^2 \, d\theta + \frac{e}{c} \oint \mathbf{A} \cdot d\mathbf{l} \tag{12.66}$$

Applying Stokes's theorem to the second integral and integrating over θ in the first integral, we obtain

$$J = 2\pi \gamma m \omega_B a^2 + \frac{e}{c} \int_S \mathbf{B} \cdot \mathbf{n} \, da \tag{12.67}$$

Since the line element $d\mathbf{l}$ in (12.64) is in a counterclockwise sense relative to **B**, the unit vector **n** is antiparallel to **B**. Hence the integral over the circular orbit subtracts from the first term. This gives

$$J = \gamma m \omega_B \pi a^2 = \frac{e}{c} (B\pi a^2) \tag{12.68}$$

making use of $\omega_B = eB/\gamma mc$. The quantity $B\pi a^2$ is the flux through the particle's orbit.

If the particle moves through regions where the magnetic field strength varies slowly in space or time, the adiabatic invariance of J means that the flux linked by the particle's orbit remains constant. If B increases, the radius a will decrease so that $B\pi a^2$ remains unchanged. This constancy of flux linked can be phrased in several ways involving the particle's orbit radius, its transverse momentum, its magnetic moment. These different statements take the forms:

$$\left. \begin{array}{c} Ba^2 \\ p_\perp^2/B \\ \gamma\mu \end{array} \right\} \quad \text{are adiabatic invariants} \tag{12.69}$$

where $\mu = (e\omega_B a^2/2c)$ is the magnetic moment of the current loop of the particle in orbit. If there are only static magnetic fields present, the speed of the particle is constant and its total energy does not change. Then the magnetic moment μ is itself an adiabatic invariant. In time-varying fields or with static electric fields, μ is an adiabatic invariant only in the nonrelativistic limit.

Let us now consider a simple situation in which a static magnetic field **B** acts

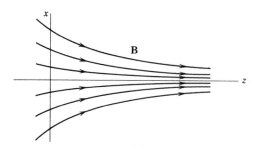

Figure 12.4

mainly in the z direction, but has a small positive gradient in that direction. Figure 12.4 shows the general behavior of the lines of force. In addition to the z component of field there is a small radial component due to the curvature of the lines of force. For simplicity we assume cylindrical symmetry. Suppose that a particle is spiraling around the z axis in an orbit of small radius with a transverse velocity $\mathbf{v}_{\perp0}$ and a component of velocity $v_{\|0}$ parallel to $\mathbf{B}$ at $z = 0$, where the axial field strength is B_0. The speed of the particle is constant so that any position along the z axis

$$v_{\|}^2 + v_{\perp}^2 = v_0^2 \tag{12.70}$$

where $v_0^2 = v_{\perp0}^2 + v_{\|0}^2$ is the square of the speed at $z = 0$. If we assume that the flux linked is a constant of motion, then (12.69) allows us to write

$$\frac{v_{\perp}^2}{B} = \frac{v_{\perp0}^2}{B_0} \tag{12.71}$$

where B is the axial magnetic induction. Then we find the parallel velocity at any position along the z axis given by

$$v_{\|}^2 = v_0^2 - v_{\perp0}^2 \frac{B(z)}{B_0} \tag{12.72}$$

Equation (12.72) for the velocity of the particle in the z direction is equivalent to the first integral of Newton's equation of motion for a particle in a one-dimensional potential*

$$V(z) = \tfrac{1}{2}m \frac{v_{\perp0}^2}{B_0} B(z)$$

If $B(z)$ increases enough, eventually the right-hand side of (12.72) will vanish at some point $z = z_0$. This means that the particle spirals in an ever-tighter orbit along the lines of force, converting more and more translational energy into energy of rotation, until its axial velocity vanishes. Then it turns around, still spiraling in the same sense, and moves back in the negative z direction. The particle is reflected by the magnetic field, as is shown schematically in Fig. 12.5.

Equation (12.72) is a consequence of the assumption that $p_{\perp}^2/B$ is invariant. To show that at least to first order this invariance follows directly from the Lorentz force equation, we consider an explicit solution of the equations of mo-

---

*Note, however, that our discussion is fully relativistic. The analogy with one-dimensional nonrelativistic mechanics is only a formal one.

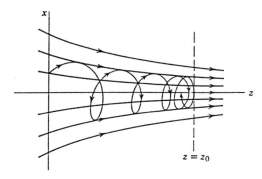

Figure 12.5 Reflection of charged particle out of region of high field strength.

tion. If the magnetic induction along the axis is $B(z)$, there will be a radial component of the field near the axis given by the divergence equation as

$$B_\rho(\rho, z) \simeq -\tfrac{1}{2}\rho \, \frac{\partial B(z)}{\partial z} \tag{12.73}$$

where ρ is the radius out from the axis. The z component of the force equation is

$$\ddot{z} = \frac{e}{\gamma mc} \, (-\rho\dot{\phi}B_\rho) \simeq \frac{e}{2\gamma mc} \, \rho^2 \dot{\phi} \, \frac{\partial B(z)}{\partial z} \tag{12.74}$$

where $\dot{\phi}$ is the angular velocity around the z axis. This can be written, correct to first order in the small variation of $B(z)$, as

$$\ddot{z} \simeq -\frac{v_{\perp 0}^2}{2B_0} \, \frac{\partial B(z)}{\partial z} \tag{12.75}$$

where we have used $\rho^2\dot{\phi} \simeq -(a^2\omega_B)_0 = -(v_{\perp 0}^2/\omega_{B0})$. Equation (12.75) has as its first integral (12.72), showing that to first order in small quantities the constancy of flux linking the orbit follows directly from the equations of motion.

The adiabatic invariance of the flux linking an orbit is useful in discussing particle motions in all types of spatially varying magnetic fields. The simple example described above illustrates the principle of the "magnetic mirror": Charged particles are reflected by regions of strong magnetic field. This mirror property formed the basis of a theory of Fermi for the acceleration of cosmic-ray particles to very high energies in interstellar space by collisions with moving magnetic clouds. The mirror principle can be applied to the containment of a hot plasma for thermonuclear energy production. A magnetic bottle can be constructed with an axial field produced by solenoidal windings over some region of space, and additional coils at each end to provide a much higher field toward the ends. The lines of force might appear as shown in Fig. 12.6. Particles created or injected into the field in the central region will spiral along the axis, but will be reflected by the magnetic mirrors at each end. If the ratio of maximum field B_m in the mirror to the field B in the central region is very large, only particles with a very large component of velocity parallel to the axis can penetrate through the ends. From (12.72) it is evident that the criterion for trapping is

$$\left|\frac{v_{\parallel 0}}{v_{\perp 0}}\right| < \left(\frac{B_m}{B} - 1\right)^{1/2} \tag{12.76}$$

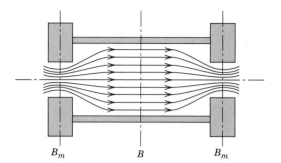

Figure 12.6 Schematic diagram of "mirror" machine for the containment of a hot plasma.

If the particles are injected into the apparatus, it is easy to satisfy requirement (12.76). Then the escape of particles is governed by the rate at which they are scattered by residual gas atoms, etc., in such a way that their velocity components violate (12.76).

Another area of application of these principles is to terrestrial and stellar magnetic fields. The motion of charged particles in the magnetic dipole fields of the sun or earth can be understood in terms of the adiabatic invariant discussed here and the drift velocities of Section 12.4. Some aspects of this topic are left to Problems 12.9 and 12.10 on the trapped particles around the earth (the Van Allen belts).

12.6 Lowest Order Relativistic Corrections to the Lagrangian for Interacting Charge Particles: The Darwin Lagrangian

In Section 12.1 we discussed the general Lagrangian formalism for a relativistic particle in external electromagnetic fields described by the vector and scalar potentials, **A** and Φ. The appropriate interaction Lagrangian was given by (12.11). If we now consider the problem of a conventional Lagrangian description of the interaction of two or more charged particles with each other, we see that it is possible only at nonrelativistic velocities. The Lagrangian is supposed to be a function of the instantaneous velocities and coordinates of all the particles. When the finite velocity of propagation of electromagnetic fields is taken into account, this is no longer possible, since the values of the potentials at one particle due to the other particles depend on their state of motion at "retarded" times. Only when the retardation effects can be neglected is a Lagrangian description in terms of instantaneous positions and velocities possible. In view of this one might think that a Lagrangian could be formulated only in the static limit, i.e., to zeroth order in (v/c). We will now show, however, that lowest order relativistic corrections can be included, giving an approximate Lagrangian for interacting particles, correct to the order of $(v/c)^2$ inclusive.

It is sufficient to consider two interacting particles with charges q_1 and q_2, masses m_1 and m_2, and coordinates $\mathbf{x}_1$ and $\mathbf{x}_2$. The relative separation is $\mathbf{r} = \mathbf{x}_1 - \mathbf{x}_2$. The interaction Lagrangian in the static limit is just the negative of the electrostatic potential energy,

$$L_{\text{int}}^{\text{NR}} = -\frac{q_1 q_2}{r} \qquad (12.77)$$

If attention is directed to the first particle, this can be viewed as the negative of the product of q_1 and the scalar potential Φ_{12} due to the second particle at the position of the first. This is of the same form as (12.9). If we wish to generalize beyond the static limit, we must, according to (12.11), determine both Φ_{12} and $\mathbf{A}_{12}$, at least to some degree of approximation. In general there will be relativistic corrections to both Φ_{12} and $\mathbf{A}_{12}$. But in the *Coulomb gauge*, the scalar potential is given correctly to all orders in v/c by the instantaneous Coulomb potential. Thus, if we calculate in that gauge, the scalar-potential contribution Φ_{12} is already known. All that needs to be considered is the vector potential $\mathbf{A}_{12}$.

If only the lowest order relativistic corrections are desired, retardation effects can be neglected in computing $\mathbf{A}_{12}$. The reason is that the vector potential enters the Lagrangian (12.11) in the combination $q_1(\mathbf{v}_1/c) \cdot \mathbf{A}_{12}$. Since $\mathbf{A}_{12}$ itself is of the order of v_2/c, greater accuracy in calculating $\mathbf{A}_{12}$ is unnecessary. Consequently, we have the magnetostatic expression

$$\mathbf{A}_{12} \simeq \frac{1}{c} \int \frac{\mathbf{J}_t(\mathbf{x}') \, d^3x'}{|\mathbf{x}_1 - \mathbf{x}'|} \tag{12.78}$$

where $\mathbf{J}_t$ is the transverse part of the current due to the second particle, as discussed in Section 6.3. From equations (6.24)–(6.28) it can be shown that the transverse current is

$$\mathbf{J}_t(\mathbf{x}') = q_2 \mathbf{v}_2 \, \delta(\mathbf{x}' - \mathbf{x}_2) - \frac{q_2}{4\pi} \nabla' \left[\frac{\mathbf{v}_2 \cdot (\mathbf{x}' - \mathbf{x}_2)}{|\mathbf{x}' - \mathbf{x}_2|^3} \right] \tag{12.79}$$

When this is inserted in (12.78), the first term can be integrated immediately. Thus

$$\mathbf{A}_{12} \simeq \frac{q_2 \mathbf{v}_2}{cr} - \frac{q_2}{4\pi c} \int \frac{1}{|\mathbf{x}' - \mathbf{x}_1|} \nabla' \left[\frac{\mathbf{v}_2 \cdot (\mathbf{x}' - \mathbf{x}_2)}{|\mathbf{x}' - \mathbf{x}_2|^3} \right] d^3x'$$

By changing variables to $\mathbf{y} = \mathbf{x}' - \mathbf{x}_2$ and integrating by parts, the integral can be put in the form,

$$\mathbf{A}_{12} \simeq \frac{q_2 \mathbf{v}_2}{cr} - \frac{q_2}{4\pi c} \nabla_r \int \frac{\mathbf{v}_2 \cdot \mathbf{y}}{y^3} \frac{1}{|\mathbf{y} - \mathbf{r}|} d^3y$$

The integral can now be done in a straightforward manner to yield

$$\mathbf{A}_{12} \simeq \frac{q_2}{c} \left[\frac{\mathbf{v}_2}{r} - \tfrac{1}{2} \nabla_r \left(\frac{\mathbf{v}_2 \cdot \mathbf{r}}{r} \right) \right]$$

The differentiation of the second term leads to the final result

$$\mathbf{A}_{12} \simeq \frac{q_2}{2cr} \left[\mathbf{v}_2 + \frac{\mathbf{r}(\mathbf{v}_2 \cdot \mathbf{r})}{r^2} \right] \tag{12.80}$$

With expression (12.80) for the vector potential due to the second particle at the position of the first, the interaction Lagrangian for two charged particles, including lowest order relativistic effects, is

$$L_{\text{int}} = \frac{q_1 q_2}{r} \left\{ -1 + \frac{1}{2c^2} \left[\mathbf{v}_1 \cdot \mathbf{v}_2 + \frac{(\mathbf{v}_1 \cdot \mathbf{r})(\mathbf{v}_2 \cdot \mathbf{r})}{r^2} \right] \right\} \tag{12.81}$$

This interaction form was first obtained by Darwin in 1920. It is of importance in a quantum-mechanical discussion of relativistic corrections in two-electron atoms. In the quantum-mechanical problem the velocity vectors are replaced by their corresponding quantum-mechanical operators (Dirac α's). Then the interaction is known as the Breit interaction (1930).*

For a system of interacting charged particles the complete Darwin Lagrangian, correct to order $1/c^2$ inclusive, can be written down by expanding the free-particle Lagrangian (12.7) for each particle and summing up all the interaction terms of the form (12.81). The result is

$$
L_{\text{Darwin}} = \frac{1}{2} \sum_i m_i v_i^2 + \frac{1}{8c^2} \sum_i m_i v_i^4 - \frac{1}{2} \sum_{i,j}' \frac{q_i q_i}{r_{ij}}
$$

$$
+ \frac{1}{4c^2} \sum_{i,j}' \frac{q_i q_j}{r_{ij}} [\mathbf{v}_i \cdot \mathbf{v}_j + (\mathbf{v}_i \cdot \hat{\mathbf{r}}_{ij})(\mathbf{v}_j \cdot \hat{\mathbf{r}}_{ij})]
$$

(12.82)

where $r_{ij} = |\mathbf{x}_i - \mathbf{x}_j|$, $\hat{\mathbf{r}}_{ij}$ is a unit vector in the direction $\mathbf{x}_i - \mathbf{x}_j$, and the prime on the double summation indicates the omission of the (self-energy) terms, $i = j$. Although the Darwin Lagrangian has had its most celebrated application in the quantum-mechanical context of the Breit interaction, it has uses in the purely classical domain. Two examples are cited in the suggested reading at the end of the chapter. See also the problems.

12.7 *Lagrangian for the Electromagnetic Field*

In Section 12.1 we considered the Lagrangian formulation of the equations of motion of a charged particle in an external electromagnetic field. We now examine a Lagrangian description of the electromagnetic field in interaction with specified external sources of charge and current. The Lagrangian approach to continuous fields closely parallels the techniques used for discrete point particles.[†] The finite number of coordinates $q_i(t)$ and $\dot{q}_i(t)$, $i = 1, 2, \ldots, n$, are replaced by an infinite number of degrees of freedom. Each point in space-time x^α corresponds to a finite number of values of the discrete index i. The generalized coordinate q_i is replaced by a continuous field $\phi_k(x)$, with a discrete index ($k = 1, 2, \ldots, n$) and a continuous index (x^α). The generalized velocity $\dot{q}_i$ is replaced by the 4-vector gradient, $\partial^\beta \phi_k$. The Euler–Lagrange equations follow from the stationary property of the action integral with respect to variations $\delta\phi_k$ and $\delta(\partial^\beta \phi_k)$ around the physical values. We thus have the following correspondences:

$$
i \to x^\alpha, k
$$

$$
q_i \to \phi_k(x)
$$

$$
\dot{q}_i \to \partial^\alpha \phi_k(x)
$$

(12.83)

$$
L = \sum_i L_i(q_i, \dot{q}_i) \to \int \mathcal{L}(\phi_k, \partial^\alpha \phi_k) \, d^3x
$$

$$
\frac{d}{dt}\left(\frac{\partial L}{\partial \dot{q}_i}\right) = \frac{\partial L}{\partial q_i} \to \partial^\beta \frac{\partial \mathcal{L}}{\partial(\partial^\beta \phi_k)} = \frac{\partial \mathcal{L}}{\partial \phi_k}
$$

---

*See H. A. Bethe and E. E. Salpeter, *Quantum Mechanics of One- and Two-Electron Atoms*, Springer-Verlag, Berlin; Academic Press, New York (1957), pp. 170 ff.

[†]For more detail and or background than given in our abbreviated account, see *Goldstein* (Chapter 12) or other references cited at the end of the chapter.

where $\mathcal{L}$ is a Lagrangian *density*, corresponding to a definite point in space-time and equivalent to the individual terms in a discrete particle Lagrangian like (12.82). For the electromagnetic field the "coordinates" and "velocities" are A^α and $\partial^\beta A^\alpha$.

The action integral takes the form,

$$A = \iint \mathcal{L} \, d^3x \, dt = \int \mathcal{L} \, d^4x \tag{12.84}$$

The Lorentz-invariant nature of the action is preserved provided the *Lagrangian density $\mathcal{L}$ is a Lorentz scalar* (because the four-dimensional volume element is invariant). In analogy with the situation with discrete particles, we expect the free-field Lagrangian at least to be quadratic in the velocities, that is, $\partial^\beta A^\alpha$ or $F^{\alpha\beta}$. The only Lorentz-invariant quadratic forms are $F_{\alpha\beta}F^{\alpha\beta}$ and $F_{\alpha\beta}\mathcal{F}^{\alpha\beta}$ (see Problem 11.14). The latter is a scalar under proper Lorentz transformations, but a pseudoscalar under inversion. If we demand a scalar $\mathcal{L}$ under inversions as well as proper Lorentz transformations, we must have $\mathcal{L}_{\text{free}}$ as some multiple of $F_{\alpha\beta}F^{\alpha\beta}$. The interaction term in $\mathcal{L}$ involves the source densities. These are described by the current density 4-vector, $J^\alpha(x)$. From the form of the electrostatic and magnetostatic energies, or from the charged-particle interaction Lagrangian (12.10), we anticipate that $\mathcal{L}_{\text{int}}$ is a multiple of $J_\alpha A^\alpha$. With this motivation we postulate the electromagnetic Lagrangian density:

$$\mathcal{L} = -\frac{1}{16\pi} F_{\alpha\beta}F^{\alpha\beta} - \frac{1}{c} J_\alpha A^\alpha \tag{12.85}$$

The coefficient and sign of the interaction terms is chosen to agree with (12.10); the sign and scale of the free Lagrangian is set by the definitions of the field strengths and the Maxwell equations.

In order to use the Euler–Lagrange equation in the form given in (12.83), we substitute the definition of the fields and write

$$\mathcal{L} = -\frac{1}{16\pi} g_{\lambda\mu} g_{\nu\sigma} (\partial^\mu A^\sigma - \partial^\sigma A^\mu)(\partial^\lambda A^\nu - \partial^\nu A^\lambda) - \frac{1}{c} J_\alpha A^\alpha \tag{12.86}$$

In calculating $\partial\mathcal{L}/\partial(\partial^\beta A^\alpha)$ care must be taken to pick up all the terms. There are four different terms, as can be seen from the following explicit calculation:

$$\frac{\partial\mathcal{L}}{\partial(\partial^\beta A^\alpha)} = -\frac{1}{16\pi} g_{\lambda\mu} g_{\nu\sigma} \left\{ \begin{array}{l} \delta_\beta{}^\mu \, \delta_\alpha{}^\sigma F^{\lambda\nu} - \delta_\beta{}^\sigma \, \delta_\alpha{}^\mu F^{\lambda\nu} \\ + \, \delta_\beta{}^\lambda \, \delta_\alpha{}^\nu F^{\mu\sigma} - \delta_\beta{}^\nu \, \delta_\alpha{}^\lambda F^{\mu\sigma} \end{array} \right\}$$

Because of the symmetry of $g_{\alpha\beta}$ and the antisymmetry of $F^{\alpha\beta}$, all four terms are equal and the derivative becomes

$$\frac{\partial\mathcal{L}}{\partial(\partial^\beta A^\alpha)} = -\frac{1}{4\pi} F_{\beta\alpha} = \frac{1}{4\pi} F_{\alpha\beta} \tag{12.87}$$

The other part of the Euler–Lagrange equation is

$$\frac{\partial\mathcal{L}}{\partial A^\alpha} = -\frac{1}{c} J_\alpha \tag{12.88}$$

Thus the equations of motion of the electromagnetic field are

$$\frac{1}{4\pi} \partial^\beta F_{\beta\alpha} = \frac{1}{c} J_\alpha \tag{12.89}$$

These are recognized as a covariant form of the inhomogeneous Maxwell equations (11.141).

The Lagrangian (12.85) yields the inhomogeneous Maxwell equations, but not the homogeneous ones. This is because the definition of the field strength tensor $F^{\alpha\beta}$ in terms of the 4-vector potential A^λ was chosen so that the homogeneous equations were satisfied automatically (see Section 6.2). To see this in our present 4-tensor notation, consider the left-hand side of the homogeneous equations (11.142):

$$\partial_\alpha \mathscr{F}^{\alpha\beta} = \tfrac{1}{2} \partial_\alpha \epsilon^{\alpha\beta\lambda\mu} F_{\lambda\mu}$$
$$= \partial_\alpha \epsilon^{\alpha\beta\lambda\mu} \partial_\lambda A_\mu$$
$$= \epsilon^{\alpha\beta\lambda\mu} \partial_\alpha \partial_\lambda A_\mu$$

But the differential operator $\partial_\alpha \partial_\lambda$ is symmetric in α and λ (assuming A_μ is well behaved), while $\epsilon^{\alpha\beta\lambda\mu}$ is antisymmetric in α and λ. Thus the contraction on α and λ vanishes. The homogeneous Maxwell equations are satisfied trivially.

The conservation of the source current density can be obtained from (12.89) by taking the 4-divergence of both sides:

$$\frac{1}{4\pi} \partial^\alpha \partial^\beta F_{\beta\alpha} = \frac{1}{c} \partial^\alpha J_\alpha$$

The left-hand side has a differential operator that is symmetric in α and β, while $F_{\beta\alpha}$ is antisymmetric. Again the contraction vanishes and we have

$$\partial^\alpha J_\alpha = 0 \tag{12.90}$$

12.8   *Proca Lagrangian; Photon Mass Effects*

The conventional Maxwell equations and the Lagrangian (12.85) are based on the hypothesis that the photon has zero mass. As discussed in the Introduction, it can always be asked what evidence there is for the masslessness of the photon or equivalently for the inverse square law of the Coulomb force and what consequences would result from a nonvanishing mass. A systematic technique for such considerations is the Lagrangian formulation. We modify the Lagrangian density (12.85) by adding a "mass" term. The resulting Lagrangian is known as the Proca Lagrangian, Proca having been the first to consider it (1930, 1936). The *Proca Lagrangian* is

$$\mathscr{L}_{\text{Proca}} = -\frac{1}{16\pi} F_{\alpha\beta} F^{\alpha\beta} + \frac{\mu^2}{8\pi} A_\alpha A^\alpha - \frac{1}{c} J_\alpha A^\alpha \tag{12.91}$$

The parameter μ has dimensions of inverse length and is the reciprocal Compton wavelength of the photon ($\mu = m_\gamma c/\hbar$). Instead of (12.89), the Proca equations of motion are

$$\partial^\beta F_{\beta\alpha} + \mu^2 A_\alpha = \frac{4\pi}{c} J_\alpha \tag{12.92}$$

with the same homogeneous equations, $\partial_\alpha \mathscr{F}^{\alpha\beta} = 0$, as in the Maxwell theory. We observe that in the Proca equations the potentials as well as the fields enter. In contrast to the Maxwell equations, the potentials acquire real physical (observable) significance through the mass term. In the Lorenz gauge, now required by current conservation, (12.92) can be written

$$\Box A_\alpha + \mu^2 A_\alpha = \frac{4\pi}{c} J_\alpha \tag{12.93}$$

and in the static limit takes the form

$$\nabla^2 A_\alpha - \mu^2 A_\alpha = -\frac{4\pi}{c} J_\alpha$$

If the source is a point charge q at rest at the origin, only the time component $A_0 = \Phi$ is nonvanishing. It takes the spherically symmetric Yukawa form

$$\Phi(x) = q \frac{e^{-\mu r}}{r} \tag{12.94}$$

This shows the characteristic feature of the photon mass. There is an exponential falloff of the static potentials and fields, with the $1/e$ distance equal to μ^{-1}. As discussed in the Introduction and also in Problem 12.15, the exponential factor alters the character of the earth's magnetic field sufficiently to permit us to set quite stringent limits on the photon mass from geomagnetic data. It was at one time suggested* that relatively simple laboratory experiments using lumped LC circuits could improve on even these limits, but the idea was conceptually flawed. There is enough subtlety involved that the subject is worth a brief discussion.[†]

The starting point of the argument is (12.93) in the absence of sources. If we assume harmonic time and space variation, the constraint equation on the frequency and wave number is

$$\omega^2 = c^2 k^2 + \mu^2 c^2 \tag{12.95}$$

This is the standard expression for the square of the energy (divided by $\hbar$) for a particle of momentum $\hbar k$ and mass $\mu \hbar / c$. Now consider some resonant system (cavity or lumped circuit). Suppose that when $\mu = 0$ its resonant frequency is ω_0, while for $\mu \neq 0$ the resonant frequency is ω. From the structure of (12.95) it is tempting to write the relation,

$$\omega^2 = \omega_0^2 + \mu^2 c^2 \tag{12.96}$$

Evidently, the smaller the frequency, the larger the fractional difference between ω and ω_0 for a given photon mass. This suggests an experiment with lumped LC circuits. The scheme would be to measure the resonant frequencies of a sequence of circuits whose ω_0^2 values are in known ratios. If the observed resonant frequencies are not in the same proportion, evidence for $\mu \neq 0$ in (12.96) would be found. Franken and Ampulski compared two circuits, one with a certain inductance L and a capacitance C, hence with $\omega_0^2 = (LC)^{-1}$, and another with the same inductor, but two capacitances C in parallel. The squares of the observed fre-

*P. A. Franken and G. W. Ampulski, *Phys. Rev. Lett.* **26**, 115 (1971).

[†]Shortly after the idea was proposed, several analyses based on the Proca equations appeared. Some of these are A. S. Goldhaber and M. M. Nieto, *Phys. Rev. Lett.* **26**, 1390 (1971); D. Park and E. R. Williams, *Phys. Rev. Lett.* **26**, 1393 (1971); N. M. Kroll, *Phys. Rev. Lett.* **26**, 1395 (1971); D. G. Boulware, *Phys. Rev. Lett.* **27**, 55 (1971): N. M. Kroll, *Phys. Rev. Lett.* **27**, 340 (1971).

quencies, corrected for resistive effects, were in the ratio 2:1 within errors. They thus inferred an upper limit on the photon mass, pointing out that in principle improvement of the accuracy by several orders of magnitude was possible if the idea was sound.

What is wrong with the idea? The first observation is that *lumped* circuits are by definition incapable of setting *any* limit on the photon mass.* The lumped circuit concept of a capacitance is a two-terminal box with the property that the current flow I at one terminal and the voltage V between the terminals are related by $I = C\, dV/dt$. Similarly a lumped inductance is a two-terminal box with the governing equation $V = -L\, dI/dt$. When two such boxes are connected, the currents and voltages are necessarily equal, and the combined system is described by the equation, $V + LC\, d^2V/dt^2 = 0$. The resonant frequency of a lumped LC circuit is $\omega_0 = (LC)^{-1/2}$, period.

It is true, of course, that a given set of conducting surfaces or a given coil of wire will have different static properties of capacitance or inductance depending on whether $\mu = 0$. The potentials and fields are all modified by exponential factors of the general form of (12.94). The question then arises as to whether one can set a meaningful limit on μ by means of a "tabletop" experiment, that is, an experiment not with lumped-circuit elements but with ones whose sizes are modest. The reader can verify, for example, that for a solid conducting sphere of radius a at the center of a hollow conducting shell of inner radius b held at zero potential, the capitance is increased by an amount $\mu^2 a^2 b/3$, provided $\mu b \ll 1$. It then turns out that instead of the fractional difference,

$$\frac{\Delta\omega}{\omega_0} \simeq \frac{\mu^2 c^2}{2\omega_0^2} \tag{12.97}$$

that follows from (12.96) with $\omega_0^2 = (LC)^{-1}$, the actual effect of the finite photon mass is

$$\frac{\Delta\omega}{\omega_0} = O(\mu^2 d^2) \tag{12.98}$$

where d is a dimension characteristic of the circuit and ω_0 is the resonant frequency for $\mu = 0$. This makes a "tabletop" experiment possible in principle, but very insensitive in practice to a possible photon mass.

Although the estimate (12.98) says it all, it is of interest to consider the effect of a finite photon mass for transmission lines, waveguides, or resonant cavities. For transmission lines, the effect of the photon mass is the same as for static lumped-circuit parameters. We recall from Chapter 8 that for $\mu = 0$ the TEM modes of a transmission line are degenerate modes, with propagation at a phase velocity equal to the velocity of light. The situation does not alter if $\mu \neq 0$. The only difference is that the *transverse* behavior of the fields is governed by $(\nabla_t^2 - \mu^2)\psi = 0$ instead of the Laplace equation. The capacitance and inductance per unit length of the transmission line are altered by fractional amounts of order $\mu^2 d^2$, but nothing else. (The result of Problem 5.29 still holds.)

For TE and TM modes in a waveguide the situation is more complicated.

---

*I am indebted to E. M. Purcell for emphasizing that this is the point almost universally missed or at least glossed over in discussions of the Franken–Ampulski proposal.

The boundary conditions on fields and potentials must be considered with care. Analysis shows (see Kroll, *op. cit.*) that TM modes have propagation governed by the naive equation (12.96), but that TE modes generally propagate differently. In any event, since the cutoff frequency of a guide is determined by its lateral dimensions, the generally incorrect estimate (12.97) becomes the same as the proper estimate (12.98).

For resonant cavities a rigorous solution is complicated, but for small mass some simple results emerge. For example, for a rectangular cavity, (12.96) holds to a good approximation for modes with l, m, n all different from zero, but fails if any mode number is zero. This is because the fields behave in the direction associated with vanishing l, m, or n as static fields and the arguments already made apply. The low-frequency modes (Schumann resonances) of the earth-ionosphere cavity, discussed in Section 8.9, are of particular interest. These modes have a radial electric field and to the *zeroth order* in h/R, where h is the height of the ionosphere and R the radius of the earth, are TEM modes in a parallel plate geometry. Thus their propagation, hence resonant frequencies, are unaltered from their $\mu = 0$ values. To *first order* in h/R there *is* a mass-dependent change in resonant frequency. The result (see Kroll's second paper cited above) is that (12.97) is modified on its right-hand side by a multiplicative factor $g \simeq 0.44 \, h/R$ for the lowest Schumann mode. With $h \simeq 70$ km, $g \simeq 5 \times 10^{-3}$. This means that the resonant frequency of $\omega_0 \simeq 50 \text{ s}^{-1}$ is a factor of $(1/g)^{1/2} \simeq 14$ less effective in setting a limit on the photon mass than naive considerations imply.

12.9 Effective "Photon" Mass in Superconductivity; London Penetration Depth

A counterpart of Proca electrodynamics is found in the London theory of the electromagnetic behavior of superconductors, formulated to explain the Meissner effect. The Meissner effect (1933) is the expulsion of a magnetic field from the interior of a superconducting material as it makes a transition from the normal state ($T > T_c$) to the superconducting state ($T < T_c$). If the field is applied after the material is superconducting, it does not penetrate into the sample, or rather, it penetrates a very small distance called the London penetration depth λ_L (typically a few tens of nanometers). Rather than being a perfect conductor, a superconductor is perfectly diamagnetic. It is this phenomenon, which is a consequence of an effective "photon" mass for fields within a superconductor, that we explore briefly.

We begin a simple phenomenological discussion by assuming that the current flow within a superconductor is caused by the nonrelativistic motion of charge carriers of charge Q, effective mass m_Q, and density n_Q. If the average local velocity of these carriers is $\mathbf{v}$, the current density is

$$\mathbf{J} = Q n_Q \mathbf{v}$$

In the presence of electromagnetic fields the current can be expressed in terms of the canonical momentum $\mathbf{P}$ through (12.14), $\mathbf{P} = m_Q \mathbf{v} + Q\mathbf{A}/c$:

$$\mathbf{J} = \frac{Q}{m_Q} n_Q \mathbf{P} - \frac{Q^2}{m_Q c} n_Q \mathbf{A}$$

The superconducting state is a coherent state of the charge carriers with vanishing canonical momentum. ($\mathbf{P} = 0$ was an assumption by the Londons, but now has a firm quantum-mechanical foundation—see *Kittel*, Chapter 12.) The effective current density within a superconductor is therefore

$$\mathbf{J} = -\frac{Q^2}{m_Q c} n_Q \mathbf{A} \tag{12.99}$$

With this current density inserted in the Lorenz-gauge wave equation for $\mathbf{A}$ [(6.16), but in Gaussian units], the wave equation takes the Proca form (12.93), but with no source term:

$$\nabla^2 \mathbf{A} - \partial_0^2 \mathbf{A} - \mu^2 \mathbf{A} = 0$$

where $\mu^2 = 4\pi Q^2 n_Q / m_Q c^2$. It follows from (12.99) that the boundary condition on $\mathbf{A}$ at an interface between normal and superconducting media across which no current is flowing is that the normal component of $\mathbf{A}$ vanishes. In the static limit and planar geometry, the solution of the London equation akin to (12.94) is $\mathbf{A} \propto e^{\pm \mu x}$, showing that the *London penetration depth* is $\lambda_L = \mu^{-1}$:

$$\lambda_L = \sqrt{\frac{m_Q c^2}{4\pi Q^2 n_Q}} \tag{12.100}$$

The effective "photon" mass is $(m_\gamma)_{\text{eff}} = \hbar/\lambda_L c$. Since the charge carriers are surely related to electrons in the material, we express the charge Q in units of e, the protonic charge, the mass m_Q in units of m_e, the electronic mass, and write the density of carriers in units of the inverse Bohr radius cubed. Then the rest energy of the "photon" can be written

$$(m_\gamma)_{\text{eff}} c^2 = \left[\left| \frac{Q}{e} \right| \sqrt{\frac{4\pi n_Q a_0^3}{m_Q/m_e}} \right] \cdot \frac{e^2}{a_0}$$

The dimensionless quantity in square brackets is presumably of order unity. The rest energy of the "photon" is thus of the order of the Rydberg energy, that is, a few electron volts.

Experimentally and theoretically, it is known that the charge carriers in low-temperature superconductors are pairs of electrons loosely bound by a second-order interaction through lattice phonons. Thus $Q = -2e$, $m_Q = 2m_e$, and $n_Q = n_{\text{eff}}/2$, where n_{eff} is the effective number of electrons participating in the current flow. A useful formula is $\mu^2 = 8\pi r_0 n_Q$, where $r_0 = 2.818 \times 10^{-15}$ m is the classical electron radius. With $n_Q = O(10^{22}$ cm$^{-3})$ we find

$$\lambda_L = \mu^{-1} = O(4 \times 10^{-6} \text{ cm})$$

The BCS quantum-mechanical theory[*] shows that at zero temperature, $n_Q = n_{\text{eff}}/2 = 2E_F N(0)/3$, where E_F is the Fermi energy of the valence band and $N(0)$ is the density of states (number of states per unit energy of one electronic spin state) at the Fermi surface. For a degenerate free Fermi gas, n_{eff} is equal to the total density of electrons, but in a superconductor the density of states is modified by the interactions and resulting energy gap. Using half the total number of

*J. Bardeen, L. N. Cooper, J. R. Schrieffer, *Phys. Rev.* **108**, 1175 (1957).

valence electrons per unit volume for n_Q in (12.100) yields only order-of-magnitude estimates for λ_L. In passing we note that in high-temperature (cupric oxide) superconductors penetration depths are found to be an order of magnitude smaller than in conventional superconductors.

Measurements of λ_L, especially its temperature dependence, can be accomplished by incorporating the superconducting specimen into a resonant circuit and studying the shift in resonant frequency with change in temperature. In circumstances in which λ_L is small compared to both the wavelength λ associated with the resonant circuit and the sample size, a simple calculation (Problem 12.20) paralleling Section 8.1 leads to a purely reactive surface impedance,

$$Z_s \approx -i \frac{8\pi^2}{c} \frac{\lambda_L}{\lambda} \text{ (Gaussian units)} \qquad \text{or} \qquad Z_s \approx -i \frac{2\pi\lambda_L}{\lambda} Z_0 \text{ (SI units)}$$

With our convention about time dependence ($e^{-i\omega t}$), the impedance is inductive, corresponding to an inductance per unit area, $L = \mu_0\lambda_L$ (SI units).

Our sketch of the simple London theory addresses only the Meissner effect, and not all of it. The magnetic and thermodynamic properties of superconductors, the physical size of the coherent state (coherence length ξ), and many other features are fully addressed only by the microscopic quantum-mechanical theory. The reader wishing to learn more about superconductivity may consult *Ashcroft and Mermin* or *Kittel* and the numerous references cited there. An alternative, perhaps more physical, approach (also by F. London) to the London equations is addressed in Problem 12.21.

12.10   *Canonical and Symmetric Stress Tensors; Conservation Laws*

A. *Generalization of the Hamiltonian: Canonical Stress Tensor*

In particle mechanics the transition to the Hamiltonian formulation and conservation of energy is made by first defining the canonical momentum variables

$$p_i = \frac{\partial L}{\partial \dot{q}_i}$$

and then introducing the Hamiltonian

$$H = \sum_i p_i \dot{q}_i - L \tag{12.101}$$

It can then be shown that $dH/dt = 0$ provided $\partial L/\partial t = 0$. For fields we anticipate having a Hamiltonian density whose volume integral over three-dimensional space is the Hamiltonian. The Lorentz transformation properties of $\mathcal{H}$ can be guessed as follows. Since the energy of a particle is the time component of a 4-vector, the Hamiltonian should transform in the same way. Since $H = \int \mathcal{H} \, d^3x$, and the invariant 4-volume element is $d^4x = d^3x \, dx_0$, it is necessary that the Hamiltonian density $\mathcal{H}$ transform as the time-time component of a second-rank tensor. If the Lagrangian density for some fields is a function of the field variables

$\phi_k(x)$, $\partial^\alpha \phi_k(x)$, $k = 1, 2, \ldots, n$, the Hamiltonian density is defined in analogy with (12.101) as

$$\mathcal{H} = \sum_k \frac{\partial \mathcal{L}}{\partial \left(\dfrac{\partial \phi_k}{\partial t} \right)} \frac{\partial \phi_k}{\partial t} - \mathcal{L} \qquad (12.102)$$

The first factor in the sum is the field momentum canonically conjugate to $\phi_k(x)$ and $\partial \phi_k / \partial t$ is equivalent to the velocity $\dot{q}_i$. The inferred Lorentz transformation properties of $\mathcal{H}$ suggest that the *covariant generalization of the Hamiltonian* density is the *canonical stress tensor*:

$$T^{\alpha\beta} = \sum_k \frac{\partial \mathcal{L}}{\partial (\partial_\alpha \phi_k)} \partial^\beta \phi_k - g^{\alpha\beta} \mathcal{L} \qquad (12.103)$$

For the free electromagnetic field Lagrangian

$$\mathcal{L}_{em} = -\frac{1}{16\pi} F_{\mu\nu} F^{\mu\nu}$$

the canonical stress tensor is

$$T^{\alpha\beta} = \frac{\partial \mathcal{L}_{em}}{\partial (\partial_\alpha A^\lambda)} \partial^\beta A^\lambda - g^{\alpha\beta} \mathcal{L}_{em}$$

where a summation over λ is implied by the repeated index. With the help of (12.87) (but notice the placing of the indices!) we find

$$T^{\alpha\beta} = -\frac{1}{4\pi} g^{\alpha\mu} F_{\mu\lambda} \partial^\beta A^\lambda - g^{\alpha\beta} \mathcal{L}_{em} \qquad (12.104)$$

To elucidate the meaning of the tensor we exhibit some components. With $\mathcal{L} = (\mathbf{E}^2 - \mathbf{B}^2)/8\pi$ and (11.138) we find

$$T^{00} = \frac{1}{8\pi} (\mathbf{E}^2 + \mathbf{B}^2) + \frac{1}{4\pi} \nabla \cdot (\Phi \mathbf{E})$$

$$T^{0i} = \frac{1}{4\pi} (\mathbf{E} \times \mathbf{B})_i + \frac{1}{4\pi} \nabla \cdot (A_i \mathbf{E}) \qquad (12.105)$$

$$T^{i0} = \frac{1}{4\pi} (\mathbf{E} \times \mathbf{B})_i + \frac{1}{4\pi} \left[(\nabla \times \Phi \mathbf{B})_i - \frac{\partial}{\partial x_0} (\Phi E_i) \right]$$

In writing the second terms here we have made use of the free-field equations $\nabla \cdot \mathbf{E} = 0$ and $\nabla \times \mathbf{B} - \partial \mathbf{E}/\partial x_0 = 0$. If we suppose that the fields are localized in some finite region of space (and, because of the finite velocity of propagation, they always are), the integrals over all 3-space at fixed time in some inertial frame of the components T^{00} and T^{0i} can be interpreted, as in Chapter 6, as the total energy and c times the total momentum of the electromagnetic fields in that frame:

$$\int T^{00} \, d^3x = \frac{1}{8\pi} \int (\mathbf{E}^2 + \mathbf{B}^2) \, d^3x = E_{\text{field}}$$

$$\int T^{0i} \, d^3x = \frac{1}{4\pi} \int (\mathbf{E} \times \mathbf{B})_i \, d^3x = cP^i_{\text{field}} \qquad (12.106)$$

These are the usual (Gaussian units) expressions for the total energy and momentum of the fields, discussed in Section 6.7. We note that the components T^{00} and T^{0i} themselves differ from the standard definitions of energy density and momentum density by added divergences. Upon integration over all space, however, the added terms give no contribution, being transformed into surface integrals at infinity where all the fields and potentials are identically zero.

The connection of the time-time and time-space components of $T^{\alpha\beta}$ with the field energy and momentum densities suggests that there is a covariant generalization of the differential conservation law (6.108) of Poynting's theorem. This differential conservation statement is

$$\partial_\alpha T^{\alpha\beta} = 0 \tag{12.107}$$

In proving (12.107) we treat the general situation described by the tensor (12.103) and the Euler–Lagrange equations (12.83). Consider

$$\partial_\alpha T^{\alpha\beta} = \sum_k \partial_\alpha \left[\frac{\partial \mathscr{L}}{\partial(\partial_\alpha \phi_k)} \partial^\beta \phi_k \right] - \partial^\beta \mathscr{L}$$

$$= \sum_k \left[\partial_\alpha \frac{\partial \mathscr{L}}{\partial(\partial_\alpha \phi_k)} \cdot \partial^\beta \phi_k + \frac{\partial \mathscr{L}}{\partial(\partial_\alpha \phi_k)} \partial_\alpha \partial^\beta \phi_k \right] - \partial^\beta \mathscr{L}$$

By means of the equations of motion (12.83) the first term can be transformed so that

$$\partial_\alpha T^{\alpha\beta} = \sum_k \left[\frac{\partial \mathscr{L}}{\partial \phi_k} \partial^\beta \phi_k + \frac{\partial \mathscr{L}}{\partial(\partial_\alpha \phi_k)} \partial^\beta(\partial_\alpha \phi_k) \right] - \partial^\beta \mathscr{L}$$

Since $\mathscr{L} = \mathscr{L}(\phi_k, \partial^\alpha \phi_k)$, the square bracket, summed, is the expression for an implicit differentiation (chain rule). Hence

$$\partial_\alpha T^{\alpha\beta} = \partial^\beta \mathscr{L}(\phi_k, \partial^\alpha \phi_k) - \partial^\beta \mathscr{L} = 0$$

The conservation law or continuity equation (12.107) yields the conservation of total energy and momentum upon integration over all of 3-space at fixed time. Explicitly, we have

$$0 = \int \partial_\alpha T^{\alpha\beta} \, d^3x = \partial_0 \int T^{0\beta} \, d^3x + \int \partial_i T^{i\beta} \, d^3x$$

If the fields are localized the second integral (a divergence) gives no contribution. Then with the identifications (12.106) we find

$$\frac{d}{dt} E_{\text{field}} = 0, \qquad \frac{d}{dt} \mathbf{P}_{\text{field}} = 0 \tag{12.108}$$

In this derivation of the conservation of energy (Poynting's theorem) and momentum and in the definitions (12.106) we have not exhibited manifest covariance. The results are valid for an observer at rest in the frame in which the fields are specified. But the question of transforming from one frame to another has not been addressed. With a covariant differential conservation law, $\partial_\alpha T^{\alpha\beta} = 0$, one expects that a covariant integral statement is also possible. The integrals in (12.106) do not appear to have the transformation properties of the components of a 4-vector. For source-free fields they do in fact transform properly (see Problem 12.18 and *Rohrlich*, Appendix A1-5), but in general do not. To avoid having electromagnetic energy and momentum defined separately in each inertial frame, without the customary connection between frames, one may construct explicitly covariant integral expressions for the electromagnetic energy and momentum, of which the forms (12.106) are special cases, valid in only one reference frame. This is discussed further in Chapter 16 in the context of the classical electromagnetic self-energy problem. (See *Rohrlich*, Section 4-9, for an explicitly covariant treatment of the conservation laws in integral form.)

B. Symmetric Stress Tensor

The canonical stress tensor $T^{\alpha\beta}$, while adequate so far, has a certain number of deficiencies. We have already seen that T^{00} and T^{0i} differ from the usual expressions for energy and momentum densities. Another drawback is its lack of symmetry—see T^{0i} and T^{i0} in (12.105). The question of symmetry arises when we consider the angular momentum of the field,

$$\mathbf{L}_{\text{field}} = \frac{1}{4\pi c} \int \mathbf{x} \times (\mathbf{E} \times \mathbf{B}) \, d^3x$$

The angular momentum density has a covariant generalization in terms of the third-rank tensor,

$$M^{\alpha\beta\gamma} = T^{\alpha\beta}x^\gamma - T^{\alpha\gamma}x^\beta \tag{12.109}$$

Then, just as (12.107) implies (12.108), so the vanishing of the 4-divergence

$$\partial_\alpha M^{\alpha\beta\gamma} = 0 \tag{12.110}$$

implies conservation of the total angular momentum of the field. Direct calculation of (12.110) gives

$$0 = (\partial_\alpha T^{\alpha\beta})x^\gamma + T^{\gamma\beta} - (\partial_\alpha T^{\alpha\gamma})x^\beta - T^{\beta\gamma}$$

With (12.107) eliminating the first and third terms, we see that conservation of angular momentum requires that $T^{\alpha\beta}$ be symmetric. Two final criticisms of $T^{\alpha\beta}$, (12.104), are that it involves the potentials explicitly, and so is not gauge invariant, and that its trace (T^α_α) is not zero, as required for zero-mass photons.

There is a general procedure for constructing a symmetric, traceless, gauge-invariant stress tensor $\Theta^{\alpha\beta}$ from the canonical stress tensor $T^{\alpha\beta}$ (see the references at the end of the chapter). For the electromagnetic $T^{\alpha\beta}$ of (12.104) we proceed directly. We substitute $\partial^\beta A^\lambda = -F^{\lambda\beta} + \partial^\lambda A^\beta$ and obtain

$$T^{\alpha\beta} = \frac{1}{4\pi}\left[g^{\alpha\mu}F_{\mu\lambda}F^{\lambda\beta} + \frac{1}{4}g^{\alpha\beta}F_{\mu\nu}F^{\mu\nu}\right] - \frac{1}{4\pi}g^{\alpha\mu}F_{\mu\lambda}\,\partial^\lambda A^\beta \tag{12.111}$$

The first terms in (12.111) are symmetric in α and β and gauge invariant. With the help of the source-free Maxwell equations, the last term can be written

$$T^{\alpha\beta}_D \equiv -\frac{1}{4\pi}g^{\alpha\mu}F_{\mu\lambda}\,\partial^\lambda A^\beta = \frac{1}{4\pi}F^{\lambda\alpha}\,\partial_\lambda A^\beta$$

$$= \frac{1}{4\pi}(F^{\lambda\alpha}\partial_\lambda A^\beta + A^\beta\partial_\lambda F^{\lambda\alpha}) \tag{12.112}$$

$$= \frac{1}{4\pi}\partial_\lambda(F^{\lambda\alpha}A^\beta)$$

The tensor $T_D^{\alpha\beta}$ has the following easily verified properties:

$$\text{(i)} \qquad \partial_\alpha T_D^{\alpha\beta} = 0$$
$$\text{(ii)} \qquad \int T_D^{0\beta}\, d^3x = 0$$

Thus the differential conservation law (12.107) will hold for the difference $(T^{\alpha\beta} - T_D^{\alpha\beta})$ if it holds for $T^{\alpha\beta}$. Furthermore, the integral relations (12.106) for the total energy and momentum of the fields will also be valid in terms of the difference tensor. We are therefore free to *define the symmetric stress tensor* $\Theta^{\alpha\beta}$:

$$\Theta^{\alpha\beta} = T^{\alpha\beta} - T_D^{\alpha\beta}$$

or

$$\Theta^{\alpha\beta} = \frac{1}{4\pi}\left(g^{\alpha\mu} F_{\mu\lambda} F^{\lambda\beta} + \frac{1}{4} g^{\alpha\beta} F_{\mu\lambda} F^{\mu\lambda} \right) \tag{12.113}$$

Explicit calculation gives the following components,

$$\Theta^{00} = \frac{1}{8\pi}(E^2 + B^2)$$

$$\Theta^{0i} = \frac{1}{4\pi}(\mathbf{E} \times \mathbf{B})_i \tag{12.114}$$

$$\Theta^{ij} = \frac{-1}{4\pi}\left[E_i E_j + B_i B_j - \frac{1}{2}\delta_{ij}(E^2 + B^2) \right]$$

The indices i and j refer to Cartesian components in 3-space. The tensor $\Theta^{\alpha\beta}$ can be written in schematic matrix form as

$$\Theta^{\alpha\beta} = \left(\begin{array}{c:c} u & c\mathbf{g} \\ \hdashline c\mathbf{g} & -T_{ij}^{(M)} \end{array} \right) \tag{12.115}$$

In (12.115) the time-time and time-space components are expressed as the energy and momentum densities (6.106) and (6.118), now in Gaussian units, while the space-space components (12.114) are seen to be just the negative of the Maxwell stress tensor (6.120) in Gaussian units, denoted here by $T_{ij}^{(M)}$ to avoid confusion with the canonical tensor $T^{\alpha\beta}$. The various other, covariant and mixed, forms of the stress tensor are

$$\Theta_{\alpha\beta} = \left(\begin{array}{c:c} u & -c\mathbf{g} \\ \hdashline -c\mathbf{g} & -T_{ij}^{(M)} \end{array} \right) \qquad \Theta^\alpha{}_\beta = \left(\begin{array}{c:c} u & -c\mathbf{g} \\ \hdashline c\mathbf{g} & T_{ij}^{(M)} \end{array} \right)$$

$$\Theta_\alpha{}^\beta = \left(\begin{array}{c:c} u & c\mathbf{g} \\ \hdashline -c\mathbf{g} & T_{ij}^{(M)} \end{array} \right)$$

The differential conservation law

$$\partial_\alpha \Theta^{\alpha\beta} = 0 \tag{12.116}$$

embodies Poynting's theorem and conservation of momentum for free fields. For example, with $\beta = 0$ we have

$$0 = \partial_\alpha \Theta^{\alpha 0} = \frac{1}{c} \left(\frac{\partial u}{\partial t} + \boldsymbol{\nabla} \cdot \mathbf{S} \right)$$

where $\mathbf{S} = c^2 \mathbf{g}$ is the Poynting vector. This is the source-free form of (6.108). Similarly, for $\beta = i$,

$$0 = \partial_\alpha \Theta^{\alpha i} = \frac{\partial g_i}{\partial t} - \sum_{j=1}^{3} \frac{\partial}{\partial x_j} T_{ij}^{(M)}$$

a result equivalent to (6.121) in the absence of sources. The conservation of field angular momentum, defined through the tensor

$$M^{\alpha \beta \gamma} = \Theta^{\alpha \beta} x^\gamma - \Theta^{\alpha \gamma} x^\beta \tag{12.117}$$

is assured by (12.116) and the symmetry of $\Theta^{\alpha \beta}$, as already discussed. There are evidently other conserved quantities in addition to energy, momentum, and angular momentum. The tensor $M^{0\beta\gamma}$ has three time-space components in addition to the space-space components that give the angular momentum density. These three components are a necessary adjunct of the covariant generalization of angular momentum. Their conservation is a statement on the center of mass motion (see Problem 12.19).

C. Conservation Laws for Electromagnetic Fields Interacting with Charged Particles

In the presence of external sources the Lagrangian for the Maxwell equations is (12.85). The symmetric stress tensor for the electromagnetic field retains its form (12.113), but the coupling to the source current makes its divergence non-vanishing. The calculation of the divergence is straightforward:

$$
\begin{aligned}
\partial_\alpha \Theta^{\alpha \beta} &= \frac{1}{4\pi} \left[\partial^\mu (F_{\mu\lambda} F^{\lambda\beta}) + \frac{1}{4} \partial^\beta (F_{\mu\lambda} F^{\mu\lambda}) \right] \\
&= \frac{1}{4\pi} \left[(\partial^\mu F_{\mu\lambda}) F^{\lambda\beta} + F_{\mu\lambda} \partial^\mu F^{\lambda\beta} + \frac{1}{2} F_{\mu\lambda} \partial^\beta F^{\mu\lambda} \right]
\end{aligned}
$$

The first term can be transformed by means of the inhomogeneous Maxwell equations (12.89). Transferring this term to the left-hand side, we have

$$\partial_\alpha \Theta^{\alpha \beta} + \frac{1}{c} F^{\beta\lambda} J_\lambda = \frac{1}{8\pi} F_{\mu\lambda} (\partial^\mu F^{\lambda\beta} + \underline{\partial^\mu F^{\lambda\beta} + \partial^\beta F^{\mu\lambda}})$$

The reason for the peculiar grouping of terms is that the underlined sum can be replaced, by virtue of the homogeneous Maxwell equation ($\partial^\mu F^{\lambda\beta} + \partial^\beta F^{\mu\lambda} + \partial^\lambda F^{\beta\mu} = 0$), by $-\partial^\lambda F^{\beta\mu} = +\partial^\lambda F^{\mu\beta}$. Thus we obtain

$$\partial_\alpha \Theta^{\alpha \beta} + \frac{1}{c} F^{\beta\lambda} J_\lambda = \frac{1}{8\pi} F_{\mu\lambda} (\partial^\mu F^{\lambda\beta} + \partial^\lambda F^{\mu\beta})$$

But the right-hand side is now the contraction (in μ and λ) of one symmetric and one antisymmetric factor. The result is therefore zero. The divergence of the stress tensor is thus

$$\partial_\alpha \Theta^{\alpha\beta} = \frac{-1}{c} F^{\beta\lambda} J_\lambda \qquad (12.118)$$

The time and space components of this equation are

$$\frac{1}{c} \left(\frac{\partial u}{\partial t} + \nabla \cdot \mathbf{S} \right) = -\frac{1}{c} \mathbf{J} \cdot \mathbf{E} \qquad (12.119)$$

and

$$\frac{\partial g_i}{\partial t} - \sum_{j=1}^{3} \frac{\partial}{\partial x_j} T_{ij}^{(M)} = -\left[\rho E_i + \frac{1}{c} (\mathbf{J} \times \mathbf{B})_i \right] \qquad (12.120)$$

These are just the conservation of energy and momentum equations of Chapter 6 for electromagnetic fields interacting with sources described by $J^\alpha = (c\rho, \mathbf{J})$. The negative of the 4-vector on the right-hand side of (12.118) is called the *Lorentz force density*,

$$f^\beta \equiv \frac{1}{c} F^{\beta\lambda} J_\lambda = \left(\frac{1}{c} \mathbf{J} \cdot \mathbf{E}, \rho \mathbf{E} + \frac{1}{c} \mathbf{J} \times \mathbf{B} \right) \qquad (12.121)$$

If the sources are a number of charged particles, the volume integral of f^β leads through the Lorentz force equation (12.1) to the time rate of change of the sum of the energies or the momenta of all particles:

$$\int f^\beta \, d^3x = \frac{dP^\beta_{\text{particles}}}{dt}$$

With the qualification expressed at the end of Section 12.10.A concerning co-variance, the integral over 3-space at fixed time of the left-hand side of (12.118) is the time rate of change of the total energy or momentum of the field. We therefore have the conservation of 4-momentum for the combined system of particles and fields:

$$\int d^3x (\partial_\alpha \Theta^{\alpha\beta} + f^\beta) = \frac{d}{dt} (P^\beta_{\text{field}} + P^\beta_{\text{particles}}) = 0 \qquad (12.122)$$

The discussion above focused on the electromagnetic field, with charged particles only mentioned as the sources of the 4-current density. A more equitable treatment of a combined system of particles and fields involves a Lagrangian having three terms, a free-field Lagrangian, a free-particle Lagrangian, and an interaction Lagrangian that involves both field and particle degrees of freedom. Variation of the action integral with respect to the particle coordinates leads to the Lorentz force equation, just as in Section 12.1, while variation of the field "coordinates" gives the Maxwell equations, as in Section 12.7. However, when self-energy and radiation reaction effects are included, the treatment is not quite so straightforward. References to these aspects are given at the end of the chapter.

Mention should also be made of the action-at-a-distance approach associated with the names of Schwarzschild, Tetrode, and Fokker. The emphasis is on the charged particles and an invariant action principle is postulated with the interaction term involving integrals over the world lines of all the particles. The idea of electromagnetic fields and the Maxwell equations is secondary. This approach is the basis of the Wheeler–Feynman absorber theory of radiation.*

12.11 Solution of the Wave Equation in Covariant Form; Invariant Green Functions

The electromagnetic fields $F^{\alpha\beta}$ arising from an external source $J^{\alpha}(x)$ satisfy the inhomogeneous Maxwell equations

$$\partial_{\alpha} F^{\alpha\beta} = \frac{4\pi}{c} J^{\beta}$$

With the definition of the fields in terms of the potentials this becomes

$$\Box A^{\beta} - \partial^{\beta}(\partial_{\alpha} A^{\alpha}) = \frac{4\pi}{c} J^{\beta}$$

If the potentials satisfy the Lorenz condition, $\partial_{\alpha} A^{\alpha} = 0$, they are then solutions of the four-dimensional wave equation,

$$\Box A^{\beta} = \frac{4\pi}{c} J^{\beta}(x) \tag{12.123}$$

The solution of (12.123) can be accomplished by finding a Green function $D(x, x')$ for the equation

$$\Box_x D(x, x') = \delta^{(4)}(x - x') \tag{12.124}$$

where $\delta^{(4)}(x - x') = \delta(x_0 - x_0') \, \delta(\mathbf{x} - \mathbf{x}')$ is a four-dimensional delta function. In the absence of boundary surfaces, the Green function can depend only on the 4-vector difference $z^{\alpha} = x^{\alpha} - x'^{\alpha}$. Thus $D(x, x') = D(x - x') = D(z)$ and (12.124) becomes

$$\Box_z D(z) = \delta^{(4)}(z)$$

We use Fourier integrals to transform from coordinate to wave number space. The Fourier transform $\tilde{D}(k)$ of the Green function is defined by

$$D(z) = \frac{1}{(2\pi)^4} \int d^4k \, \tilde{D}(k) e^{-ik \cdot z} \tag{12.125}$$

where $k \cdot z = k_0 z_0 - \mathbf{k} \cdot \mathbf{z}$. With the representation of the delta function being

$$\delta^{(4)}(z) = \frac{1}{(2\pi)^4} \int d^4k \, e^{-ik \cdot z} \tag{12.126}$$

one finds that the k-space Green function is

$$\tilde{D}(k) = -\frac{1}{k \cdot k} \tag{12.127}$$

*J. A. Wheeler and R. P. Feynman, *Rev. Mod. Phys.* **21**, 425 (1949).

The Green function $D(z)$ is therefore

$$D(z) = \frac{-1}{(2\pi)^4} \int d^4k \, \frac{e^{-ik\cdot z}}{k \cdot k} \tag{12.128}$$

Because the integrand in (12.128) is singular, the expression as it stands is ambiguous and is given definite meaning only by the handling of the singularities. We proceed by performing the integration over dk_0 first. Thus

$$D(z) = -\frac{1}{(2\pi)^4} \int d^3k \, e^{i\mathbf{k}\cdot\mathbf{z}} \int_{-\infty}^{\infty} dk_0 \, \frac{e^{-ik_0 z_0}}{k_0^2 - \kappa^2} \tag{12.129}$$

where we have introduced the notation, $\kappa = |\mathbf{k}|$. The k_0 integral is given meaning by considering k_0 as a complex variable and treating the integral as a contour integral in the k_0 plane. The integrand has two simple poles, at $k_0 = \pm\kappa$ as shown in Fig. 12.7. Green functions that differ in their behavior are obtained by choosing different contours of integration relative to the poles. Two possible contours are labeled r and a in Fig. 12.7. These open contours may be closed at infinity with a semicircle in the upper or lower half-plane, depending on the sign of z_0 in the exponential. For $z_0 > 0$, the exponential, $e^{-ik_0 z_0}$, increases without limit in the upper half-plane. To use the residue theorem, we must therefore close the contour in the lower half-plane. The opposite holds for $z_0 < 0$.

Consider now the contour r. For $z_0 < 0$, the resulting integral vanishes because the contour is closed in the upper half-plane and encircles no singularities. For $z_0 > 0$, the integral over k_0 is

$$\oint_r dk_0 \, \frac{e^{-ik_0 z_0}}{k_0^2 - \kappa^2} = -2\pi i \, \text{Res}\left(\frac{e^{-ik_0 z_0}}{k_0^2 - \kappa^2}\right)$$

$$= -\frac{2\pi}{\kappa} \sin(\kappa z_0)$$

The Green function (12.129) is then

$$D_r(z) = \frac{\theta(z_0)}{(2\pi)^3} \int d^3k \, e^{i\mathbf{k}\cdot\mathbf{z}} \frac{\sin(\kappa z_0)}{\kappa}$$

The integration over the angles of $\mathbf{k}$ leads to

$$D_r(z) = \frac{\theta(z_0)}{2\pi^2 R} \int_0^{\infty} d\kappa \, \sin(\kappa R) \, \sin(\kappa z_0) \tag{12.130}$$

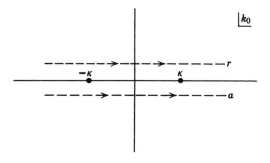

Figure 12.7

where $R = |\mathbf{z}| = |\mathbf{x} - \mathbf{x}'|$ is the spatial distance between x^α and x'^α Using some simple trigonometry and a change of variable, we can write (12.130) as

$$D_r(z) = \frac{\theta(z_0)}{8\pi^2 R} \int_{-\infty}^{\infty} d\kappa [e^{i(z_0 - R)\kappa} - e^{i(z_0 + R)\kappa}]$$

The remaining integrals are just Dirac delta functions. Because $z_0 > 0$ and $R > 0$ the second integral is always zero. The Green function for contour r is therefore

$$D_r(x - x') = \frac{\theta(x_0 - x_0')}{4\pi R} \delta(x_0 - x_0' - R) \tag{12.131}$$

Here we have reintroduced the original variables x and x'. This Green function is called the *retarded or causal Green function* because the source-point time x_0' is always earlier than the observation-point time x_0. Equation (12.131), or its Fourier transform with respect to x_0, $(4\pi R)^{-1} e^{i\omega R/c}$, is the familiar Green function of outgoing waves of Chapter 6.

With the choice of the contour a in Fig. 12.7, an exactly parallel calculation yields the *advanced Green function*,

$$D_a(x - x') = \frac{\theta[-(x_0 - x_0')]}{4\pi R} \delta(x_0 - x_0' + R) \tag{12.132}$$

These Green functions can be put in covariant form by use of the following identity:

$$\begin{aligned}
\delta[(x - x')^2] &= \delta[(x_0 - x_0')^2 - |\mathbf{x} - \mathbf{x}'|^2] \\
&= \delta[(x_0 - x_0' - R)(x_0 - x_0' + R)] \\
&= \frac{1}{2R} [\delta(x_0 - x_0' - R) + \delta(x_0 - x_0' + R)]
\end{aligned}$$

Then, since the theta functions select one or the other of the two terms, we have

$$\begin{aligned}
D_r(x - x') &= \frac{1}{2\pi} \theta(x_0 - x_0') \delta[(x - x')^2] \\
D_a(x - x') &= \frac{1}{2\pi} \theta(x_0' - x_0) \delta[(x - x')^2]
\end{aligned} \tag{12.133}$$

The theta functions, apparently noninvariant, are actually invariant under proper Lorentz transformations when constrained by the delta functions. Thus (12.133) gives the Green functions an explicitly invariant expression. The theta and delta functions in (12.133) show that the retarded (advanced) Green function is different from zero only on the forward (backward) light cone of the source point.

The solution of the wave equation (12.123) can now be written down in terms of the Green functions:

$$A^\alpha(x) = A^\alpha_{\text{in}}(x) + \frac{4\pi}{c} \int d^4x' \, D_r(x - x') J^\alpha(x') \tag{12.134}$$

or

$$A^\alpha(x) = A^\alpha_{\text{out}}(x) + \frac{4\pi}{c} \int d^4x' \, D_a(x - x') J^\alpha(x') \tag{12.135}$$

where A_{in}^α and A_{out}^α are solutions of the homogeneous wave equation. In (12.134) the retarded Green function is used. In the limit $x_0 \to -\infty$, the integral over the sources vanishes, assuming the sources are localized in space and time, because of the retarded nature of the Green function. We see that the free-field potential $A_{in}^\alpha(x)$ has the interpretation of the "incident" or "incoming" potential, specified at $x_0 \to -\infty$. Similarly, in (12.135) with the advanced Green function, the homogeneous solution $A_{out}^\alpha(x)$ is the asymptotic "outgoing" potential, specified at $x_0 \to +\infty$. The *radiation fields* are defined as the difference between the "outgoing" and the "incoming" fields. Their 4-vector potential is

$$A_{rad}^\alpha(x) = A_{out}^\alpha - A_{in}^\alpha = \frac{4\pi}{c} \int d^4x' \, D(x - x') J^\alpha(x') \qquad (12.136)$$

where

$$D(z) = D_r(z) - D_a(z) \qquad (12.137)$$

is the difference between the retarded and advanced Green functions.

The fields of a charged particle moving in a prescribed path will be of interest in Chapter 14. If the particle is a point charge e whose position in the inertial frame K is $\mathbf{r}(t)$, its charge density and current density in that frame are

$$\begin{aligned} \rho(\mathbf{x}, t) &= e \, \delta[\mathbf{x} - \mathbf{r}(t)] \\ \mathbf{J}(\mathbf{x}, t) &= e \, \mathbf{v}(t) \, \delta[\mathbf{x} - \mathbf{r}(t)] \end{aligned} \qquad (12.138)$$

where $\mathbf{v}(t) = d\mathbf{r}(t)/dt$ is the charge's velocity in K. The charge and current densities can be written as a 4-vector current in manifestly covariant form by introducing the charge's 4-vector coordinate $r^\alpha(\tau)$ as a function of the charge's proper time τ and integrating over the proper time with an appropriate additional delta function. Thus

$$J^\alpha(x) = ec \int d\tau \, U^\alpha(\tau) \, \delta^{(4)}[x - r(\tau)] \qquad (12.139)$$

where U^α is the charge's 4-velocity. In the inertial frame K, $r^\alpha = [ct, \mathbf{r}(t)]$ and $U^\alpha = (\gamma c, \gamma \mathbf{v})$. The use of (12.139) in (12.134) to yield the potentials and fields of a moving charge is presented in Section 14.1.

References and Suggested Reading

The Lagrangian and Hamiltonian formalism for relativistic charged particles is treated in every advanced mechanics textbook, as well as in books on electrodynamics. Some useful references are

> Barut, Chapter II
> Corben and Stehle, Chapter 16
> Goldstein, Chapter 7
> Landau and Lifshitz, *Classical Theory of Fields*, Chapters 2 and 3

The motion of charged particles in external electromagnetic fields, especially inhomogeneous magnetic fields, is an increasingly important topic in geophysics, solar physics, and thermonuclear research. The classic reference for these problems is

> Alfvén and Fälthammar

but the basic results are also presented by
> Chandrasekhar, Chapters 2 and 3
> Clemmow and Dougherty, Chapter 4
> Linhart, Chapter 1
> Spitzer, Chapter 1

The book by
> Rossi and Olbert

covers the motion of charged particles in the fields of the earth and sun. The topic of adiabatic invariants is treated thoroughly by
> Northrop

Another important application of relativistic charged-particle dynamics is to high-energy accelerators. An introduction to the physics problems of this field will be found in
> Corben and Stehle, Chapter 17

Some of the practical aspects as well as the basic physics of accelerators can be found in
> Livingston and Blewett
> Persico, Ferrari, and Segre

The Darwin Lagrangian (12.82) has been advocated as a useful tool in the description of low-frequency oscillations in plasmas by
> A. N. Kaufman and P. S. Rostler, *Phys. Fluids* **14**, 446 (1971).

It has also been used in an interesting paper by
> S. Coleman and J. H. Van Vleck, *Phys. Rev.* **171**, 1370 (1968)

to resolve a paradox in the conservation of momentum and Newton's third law.

The issue of "hidden momentum" addressed by Coleman and Van Vleck is treated in a number of pedagogical papers. A valuable early one is
> W. H. Furry, *Am. J. Phys.* **37**, 621 (1969)

More recent works include:
> L. Vaidman, *Am. J. Phys.* **58**, 978 (1990).
> V. Hnizdo, *Am. J. Phys.* **65**, 55 (1997).
> V. Hnizdo, *Am. J. Phys.* **65**, 92 (1997).

The subject of a Lagrangian description of fields is treated in
> Barut, Chapter III
> Corben and Stehle, Chapter 15
> Goldstein, Chapter 11
> Low
> Schwartz, Chapter 6
> Soper

The general construction of the symmetric, gauge-invariant stress tensor $\Theta^{\alpha\beta}$ from $T^{\alpha\beta}$ is described in
> Landau and Lifshitz, *Classical Theory of Fields*, Sections 32 and 94

The particular questions of the interaction of charged particles and electromagnetic fields and the associated conservation laws are covered in
> Barut, Chapters IV and V
> Landau and Lifshitz, *Classical Theory of Fields*, Chapter 4
> Schwartz, Chapter 7

and in an especially thorough fashion by
> Rohrlich

who treats the self-energy and radiation reaction aspects in detail. For a careful discussion of the energy, momentum and mass of electromagnetic fields, consult *Rohrlich's* book, just cited, and also
> F. Rohrlich, *Am. J. Phys.* **38**, 1310 (1970)

The invariant Green functions for the wave equation are derived in almost any book on quantum field theory. One such book, with a concise covariant treatment of classical electrodynamics at its beginning, is

Thirring, Section I.2 and Appendix II

Two reviews on the subject of the photon mass, already cited in the Introduction, are

A. S. Goldhaber and M. M. Nieto, *Rev. Mod. Phys.* **43**, 277 (1971).

I. Yu. Kobzarev and L. B. Okun', *Uspek. Fiz. Nauk.* **95**, 131 (1968) [English transl., *Sov. Phys. Uspek.* **11**, 338 (1968)].

Problems

12.1   **(a)**   Show that the Lorentz invariant Lagrangian

$$L = -\frac{mU_\alpha U^\alpha}{2} - \frac{q}{c}\, U_\alpha A^\alpha$$

gives the correct relativistic equations of motion for a particle of mass m and charge q interacting with an external field described by the 4-vector potential $A^\alpha(x)$.

(b)   Define the canonical momenta and write out the effective Hamiltonian in both covariant and space-time form. The effective Hamiltonian is a Lorentz invariant. What is its value?

12.2   **(a)**   Show from Hamilton's principle that Lagrangians that differ only by a total time derivative of some function of the coordinates and time are equivalent in the sense that they yield the same Euler–Lagrange equations of motion.

(b)   Show explicitly that the gauge transformation $A^\alpha \rightarrow A^\alpha + \partial^\alpha \Lambda$ of the potentials in the charged-particle Lagrangian (12.12) merely generates another equivalent Lagrangian.

12.3   A particle with mass m and charge e moves in a uniform, static, electric field $\mathbf{E}_0$.

(a)   Solve for the velocity and position of the particle as explicit functions of time, assuming that the initial velocity $\mathbf{v}_0$ was perpendicular to the electric field.

(b)   Eliminate the time to obtain the trajectory of the particle in space. Discuss the shape of the path for short and long times (define "short" and "long" times).

12.4   It is desired to make an $E \times B$ velocity selector with uniform, static, crossed, electric and magnetic fields over a length L. If the entrance and exit slit widths are Δx, discuss the interval Δu of velocities, around the mean value $u = cE/B$, that is transmitted by the device as a function of the mass, the momentum or energy of the incident particles, the field strengths, the length of the selector, and any other relevant variables. Neglect fringing effects at the ends. Base your discussion on the practical facts that $L \sim$ few meters, $E_{\max} \sim 3 \times 10^6$ V/m, $\Delta x \sim 10^{-3}$–10^{-4} m, $u \sim 0.5$–$0.995c$. (It is instructive to consider the equation of motion in a frame moving at the mean speed u along the beam direction, as well as in the laboratory.)

References: C. A. Coombes et al., *Phys. Rev.* **112**, 1303 (1958); P. Eberhard, M. L. Good, and H. K. Ticho, *Rev. Sci. Instrum.* **31**, 1054 (1960).

12.5   A particle of mass m and charge e moves in the laboratory in crossed, static, uniform, electric and magnetic fields. $\mathbf{E}$ is parallel to the x axis; $\mathbf{B}$ is parallel to the y axis.

(a)   For $|\mathbf{E}| < |\mathbf{B}|$ make the necessary Lorentz transformation described in Section 12.3 to obtain explicitly parametric equations for the particle's trajectory.

(b)   Repeat the calculation of part a for $|\mathbf{E}| > |\mathbf{B}|$.

12.6 Static, uniform electric and magnetic fields, **E** and **B**, make an angle of θ with respect to each other.

    (a) By a suitable choice of axes, solve the force equation for the motion of a particle of charge e and mass m in rectangular coordinates.

    (b) For **E** and **B** parallel, show that with appropriate constants of integration, etc., the parametric solution can be written

$$x = AR \sin \phi, \qquad y = AR \cos \phi, \qquad z = \frac{R}{\rho} \sqrt{1 + A^2} \cosh(\rho\phi)$$

$$ct = \frac{R}{\rho} \sqrt{1 + A^2} \sinh(\rho\phi)$$

    where $R = (mc^2/eB)$, $\rho = (E/B)$, A is an arbitrary constant, and ϕ is the parameter [actually c/R times the proper time].

12.7 A constant uniform magnetic induction B in the negative z direction exists in a region limited by the planes $x = 0$ and $x = a$. For $x < 0$ and $x > a$, there is no magnetic induction.

    (a) Determine the total electromagnetic momentum **G** in magnitude and direction of the combination of a particle with point charge q at (x_0, y_0, z_0) in the presence of this magnetic induction. Find **G** for the charge located on either side of and within the region occupied by the magnetic field. Assume the particle is at rest or in nonrelativistic motion.

    (b) The particle is normally incident on the field region from $x < 0$ with nonrelativistic momentum p. Assuming that $p > qBa/c$, determine the components of momentum after the particle has emerged into the field-free region, $x > a$. Compare the components of the sum of mechanical (particle) and electromagnetic momenta initially and finally. Why are some components of the sum conserved and some not?

    (c) Assume that $p < qBa/2c$ and that the initial conditions are such that the particle's motion is confined within the region of the magnetic induction at fixed z. Discuss the conservation, or lack of it, of the components of the sum of mechanical and electromagnetic momentum as the particle moves in its path. Comment.

12.8 In Problems 6.5 and 6.6 a nonvanishing momentum of the electromagnetic fields was found for a charge and a current-carrying toroid at rest. This paradox is among situations involving "hidden momentum." Since the field momentum is proportional to $1/c^2$, you may infer that relativistic effects may enter the considerations.

    (a) Consider the charge carriers in the toroid (or other current-carrying systems) of mass m, charge e, and individual mechanical momentum $\mathbf{p} = \gamma m\mathbf{v}$, and the current density $\mathbf{J} = en\mathbf{v}$, where n is the number density of carriers and γ is the relativistic Lorentz factor. Use conservation of energy for each charge carrier to show that, for the "static" field situation of Problem 6.5, the total mechanical momentum of the charge carriers,

$$\mathbf{P}_{mech} = \int d^3x \ \gamma nm\mathbf{v} = -\frac{1}{c^2} \int d^3x \ \Phi\mathbf{J}$$

    just opposite to the field momentum of Problem 6.5a.

    (b) Consider the toroid of Problem 6.6 to be of rectangular cross section, with width w and height h both small compared to a, and hollow tubes of uniform cross section A_w. Show that the electrostatic potential energy difference between the inner and outer vertical segments of each tube yields the change in γmc^2 necessary to generate a net vertical mechanical momentum equal and opposite to the result of Problem 6.6a, with due regard to differences in units.

Reference: Vaidman, *op. cit.*

12.9   The magnetic field of the earth can be represented approximately by a magnetic dipole of magnetic moment $M = 8.1 \times 10^{25}$ gauss-cm^3. Consider the motion of energetic electrons in the neighborhood of the earth under the action of this dipole field (Van Allen electron belts). [Note that **M** points south.]

(a)   Show that the equation for a line of magnetic force is $r = r_0 \sin^2\theta$, where θ is the usual polar angle (colatitude) measured from the axis of the dipole, and find an expression for the magnitude of B along any line of force as a function of θ.

(b)   A positively charged particle circles around a line of force in the equatorial plane with a gyration radius a and a mean radius R ($a \ll R$). Show that the particle's azimuthal position (east longitude) changes approximately linearly in time according to

$$\phi(t) = \phi_0 - \frac{3}{2}\left(\frac{a}{R}\right)^2 \omega_B(t - t_0)$$

where ω_B is the frequency of gyration at radius R.

(c)   If, in addition to its circular motion of part b, the particle has a small component of velocity parallel to the lines of force, show that it undergoes small oscillations in θ around $\theta = \pi/2$ with a frequency $\Omega = (3/\sqrt{2})(a/R)\omega_B$. Find the change in longitude per cycle of oscillation in latitude.

(d)   For an electron of 10 MeV kinetic energy at a mean radius $R = 3 \times 10^7$ m, find ω_B and a, and so determine how long it takes to drift once around the earth and how long it takes to execute one cycle of oscillation in latitude. Calculate the same quantities for an electron of 10 keV at the same radius.

12.10   A charged particle finds itself instantaneously in the equatorial plane of the earth's magnetic field (assumed to be a dipole field) at a distance R from the center of the earth. Its velocity vector at that instant makes an angle α with the equatorial plane ($v_{\parallel}/v_{\perp} = \tan\alpha$). Assuming that the particle spirals along the lines of force with a gyration radius $a \ll R$, and that the flux linked by the orbit is a constant of the motion, find an equation for the maximum magnetic latitude λ reached by the particle as a function of the angle α. Plot a graph (*not a sketch*) of λ versus α. Mark parametrically along the curve the values of α for which a particle at radius R in the equatorial plane will hit the earth (radius R_0) for $R/R_0 = 1.2, 1.5, 2.0, 2.5, 3, 4, 5$.

12.11   Consider the precession of the spin of a muon, initially longitudinally polarized, as the muon moves in a circular orbit in a plane perpendicular to a uniform magnetic field **B**.

(a)   Show that the difference Ω of the spin precession frequency and the orbital gyration frequency is

$$\Omega = \frac{eBa}{m_\mu c}$$

independent of the muon's energy, where $a = (g - 2)/2$ is the magnetic moment anomaly. (Find equations of motion for the components of spin along the mutually perpendicular directions defined by the particle's velocity, the radius vector from the center of the circle to the particle, and the magnetic field.)

(b) For the CERN Muon Storage Ring, the orbit radius is $R = 2.5$ meters and $B = 17 \times 10^3$ gauss. What is the momentum of the muon? What is the time dilatation factor γ? How many periods of precession $T = 2\pi/\Omega$ occur per observed laboratory mean lifetime of the muons? [$m_\mu = 105.66$ MeV, $\tau_0 = 2.2 \times 10^{-6}$ s, $a \simeq \alpha/2\pi$].

(c) Express the difference frequency Ω in units of the orbital rotation frequency and compute how many precessional periods (at the difference frequency) occur per rotation for a 300 MeV muon, a 300 MeV electron, a 5 GeV electron (this last typical of the e^+e^- storage ring at Cornell).

12.12 In Section 11.11 the BMT equation of motion for the spin of a particle of charge e and a magnetic moment with an arbitrary g factor was obtained.

(a) Verify that (11.171) is the correct equation for the time derivative of the longitudinal component of the rest-frame spin vector **s**.

(b) Let $\hat{\mathbf{n}}$ be a unit 3-vector perpendicular to $\hat{\boldsymbol{\beta}}$ and coplanar with $\hat{\boldsymbol{\beta}}$ and **s** ($\hat{\mathbf{n}}$ is generally time dependent). Let θ be the angle between $\hat{\boldsymbol{\beta}}$ and **s**. Show that the time rate of change of θ can be written as

$$\frac{d\theta}{dt} = \frac{e}{mc}\left[\left(\frac{g}{2} - 1\right)\hat{\mathbf{n}} \cdot (\hat{\boldsymbol{\beta}} \times \mathbf{B}) + \left(\frac{g\beta}{2} - \frac{1}{\beta}\right)\hat{\mathbf{n}} \cdot \mathbf{E}\right]$$

where **E** and **B** are the fields in the laboratory and $c\boldsymbol{\beta} = c\beta\hat{\boldsymbol{\beta}}$ is the particle's instantaneous velocity in the laboratory.

(c) For a particle moving undeflected through an $E \times B$ velocity selector and with $(\hat{\mathbf{n}} \times \hat{\boldsymbol{\beta}}) \cdot \mathbf{B} = B$, find $d\theta/dt$ in terms of the gyration frequency $eB/\gamma mc$.

(d) By defining the two 4-vectors, $L^\alpha = (\gamma\beta, \gamma\hat{\boldsymbol{\beta}})$ and $N^\alpha = (0, \hat{\mathbf{n}})$, show that $d\theta/d\tau$ can be written in the quasi-covariant form

$$\frac{d\theta}{d\tau} = \frac{e}{mc}\left[\frac{g}{2}L_\alpha - \frac{1}{v}U_\alpha\right]F^{\alpha\beta}N_\beta$$

where U^α is the particle's 4-velocity.

12.13 **(a)** Specialize the Darwin Lagrangian (12.82) to the interaction of two charged particles (m_1, q_1) and (m_2, q_2). Introduce reduced particle coordinates, $\mathbf{r} = \mathbf{x}_1 - \mathbf{x}_2$, $\mathbf{v} = \mathbf{v}_1 - \mathbf{v}_2$ and also center of mass coordinates. Write out the Lagrangian in the reference frame in which the velocity of the center of mass vanishes and evaluate the canonical momentum components, $p_x = \partial L/\partial v_x$, etc.

(b) Calculate the Hamiltonian to first order in $1/c^2$ and show that it is

$$H = \frac{p^2}{2}\left(\frac{1}{m_1} + \frac{1}{m_2}\right) + \frac{q_1q_2}{r} - \frac{p^4}{8c^2}\left(\frac{1}{m_1^3} + \frac{1}{m_2^3}\right) + \frac{q_1q_2}{2m_1m_2c^2}\left(\frac{p^2 + (\mathbf{p} \cdot \hat{\mathbf{r}})^2}{r}\right)$$

Compare with the various terms in (42.1) of Bethe and Salpeter [*op. cit.* (Section 12.6), p. 193]. Discuss the agreements and disagreements.

12.14 An alternative Lagrangian density for the electromagnetic field is

$$\mathcal{L} = -\frac{1}{8\pi}\partial_\alpha A_\beta \partial^\alpha A^\beta - \frac{1}{c}J_\alpha A^\alpha$$

(a) Derive the Euler–Lagrange equations of motion. Are they the Maxwell equations? Under what assumptions?

(b) Show explicitly, and with what assumptions, that this Lagrangian density differs from (12.85) by a 4-divergence. Does this added 4-divergence affect the action or the equations of motion?

12.15 Consider the Proca equations for a localized steady-state distribution of current that has only a static magnetic moment. This model can be used to study the observable effects of a finite photon mass on the earth's magnetic field. Note that if the magnetization is $\mathcal{M}(\mathbf{x})$ the current density can be written as $\mathbf{J} = c(\mathbf{\nabla} \times \mathcal{M})$.

(a) Show that if $\mathcal{M} = \mathbf{m}f(\mathbf{x})$, where $\mathbf{m}$ is a fixed vector and $f(\mathbf{x})$ is a localized scalar function, the vector potential is

$$\mathbf{A}(\mathbf{x}) = -\mathbf{m} \times \mathbf{\nabla} \int f(\mathbf{x}') \frac{e^{-\mu|\mathbf{x}-\mathbf{x}'|}}{|\mathbf{x}-\mathbf{x}'|} \, d^3x'$$

(b) If the magnetic dipole is a point dipole at the origin $[f(\mathbf{x}) = \delta(\mathbf{x})]$, show that the magnetic field is

$$\mathbf{B}(\mathbf{x}) = [3\hat{\mathbf{r}}(\hat{\mathbf{r}} \cdot \mathbf{m}) - \mathbf{m}]\left(1 + \mu r + \frac{\mu^2 r^2}{3}\right)\frac{e^{-\mu r}}{r^3} - \frac{2}{3}\mu^2 \mathbf{m} \frac{e^{-\mu r}}{r}$$

(c) The result of part b shows that at fixed $r = R$ (on the surface of the earth), the earth's magnetic field will appear as a dipole angular distribution, plus an added constant magnetic field (an apparently external field) antiparallel to $\mathbf{m}$. Satellite and surface observations lead to the conclusion that this "external" field is less than 4×10^{-3} times the dipole field at the magnetic equator. Estimate a lower limit on μ^{-1} in earth radii and an upper limit on the photon mass in grams from this datum.

This method of estimating μ is due to E. Schrödinger, *Proc. R. Irish Acad.* **A49**, 135 (1943). See A. S. Goldhaber and M. M. Nieto, *Phys. Rev. Lett.* **21**, 567 (1968).

12.16 **(a)** Starting with the Proca Lagrangian density (12.91) and following the same procedure as for the electromagnetic fields, show that the symmetric stress-energy-momentum tensor for the Proca *fields* is

$$\Theta^{\alpha\beta} = \frac{1}{4\pi}\left[g^{\alpha\gamma}F_{\gamma\lambda}F^{\lambda\beta} + \frac{1}{4}g^{\alpha\beta}F_{\lambda\nu}F^{\lambda\nu} + \mu^2\left(A^\alpha A^\beta - \frac{1}{2}g^{\alpha\beta}A_\lambda A^\lambda\right)\right]$$

(b) For these fields in interaction with the external source J^β, as in (12.91), show that the differential conservation laws take the same form as for the electromagnetic fields, namely,

$$\partial_\alpha \Theta^{\alpha\beta} = \frac{J_\lambda F^{\lambda\beta}}{c}$$

(c) Show explicitly that the time-time and space-time components of $\Theta^{\alpha\beta}$ are

$$\Theta^{00} = \frac{1}{8\pi}[E^2 + B^2 + \mu^2(A^0 A^0 + \mathbf{A} \cdot \mathbf{A})]$$

$$\Theta^{i0} = \frac{1}{4\pi}[(\mathbf{E} \times \mathbf{B})_i + \mu^2 A^i A^0]$$

12.17 Consider the "Thomson" scattering of Proca waves (photons with mass) by a free electron.

(a) As a preliminary, show that for an incident plane wave of unit amplitude, $\mathbf{A} = \boldsymbol{\epsilon}_0 \cos(kz - \omega t)$, where $\boldsymbol{\epsilon}_0$ is a polarization vector of unit magnitude describing either longitudinal (l) or transverse (t) fields, the time-averaged energy fluxes (measured by Θ^{30}) are $F_t = \omega k/8\pi$ and $F_l = (\mu/\omega)^2 F_t$. Show also for arbitrary polarization that the ratio of time-averaged energy flux to energy density is

$$\frac{\langle \Theta^{30} \rangle}{\langle \Theta^{00} \rangle} = \beta = \frac{k}{\omega} = \sqrt{1 - \frac{\mu^2}{\omega^2}}$$

as expected for particles of mass μ.

(b) For polarization ϵ_0 initially and polarization ϵ finally, show that the "Thomson" cross section for scattering is

$$\frac{d\sigma}{d\Omega}(\epsilon, \epsilon_0) = r_0^2 E_0 \, |\epsilon^* \cdot \epsilon_0|^2 \frac{F_{out}}{F_{in}}$$

where r_0 is the classical electron radius, E_0 is a factor for the efficiency of the incident Proca field in exciting the electron, and the final factor is a ratio of the outgoing to incident fluxes. What is the value of E_0?

(c) For an unpolarized *transverse* wave incident, show that the scattering cross section is

$$\left(\frac{d\sigma}{d\Omega} \right)_t = \frac{r_0^2}{2} \left[1 + \cos^2\theta + \left(\frac{\mu}{\omega} \right)^2 \sin^2\theta \right]$$

where the first term is the familiar transverse to transverse scattering and the second is transverse to longitudinal.

(d) For a *longitudinally* polarized wave incident, show that the cross section, summed over outgoing polarizations, is

$$\left(\frac{d\sigma}{d\Omega} \right)_l = \left(\frac{\mu}{\omega} \right)^2 r_0^2 \left[\sin^2\theta + \left(\frac{\mu}{\omega} \right)^2 \cos^2\theta \right]$$

where the first term is the longitudinal to transverse scattering and the second is longitudinal to longitudinal.

Note that in the limit $\mu/\omega \to 0$, the longitudinal fields decouple and we recover the standard Thomson cross section.

12.18 Prove, by means of the divergence theorem in four dimensions or otherwise, that for source-free electromagnetic fields confined to a finite region of space, the 3-space integrals of Θ^{00} and Θ^{0i} transform as the components of a constant 4-vector, as implied by (12.106).

12.19 Source-free electromagnetic fields exist in a localized region of space. Consider the various conservation laws that are contained in the integral of $\partial_\alpha M^{\alpha\beta\gamma} = 0$ over all space, where $M^{\alpha\beta\gamma}$ is defined by (12.117).

(a) Show that when β and γ are both space indices conservation of the total field angular momentum follows.

(b) Show that when $\beta = 0$ the conservation law is

$$\frac{d\mathbf{X}}{dt} = \frac{c^2 \mathbf{P}_{em}}{E_{em}}$$

where $\mathbf{X}$ is the coordinate of the center of mass of the electromagnetic fields, defined by

$$\mathbf{X} \int u \, d^3x = \int \mathbf{x} u \, d^3x$$

where u is the electromagnetic energy density and E_{em} and $\mathbf{P}_{em}$ are the total energy and momentum of the fields.

12.20 A uniform superconductor with London penetration depth λ_L fills the half-space $x > 0$. The vector potential is tangential and for $x < 0$ is given by

$$A_y = (ae^{ikx} + be^{-ikx})e^{-i\omega t}$$

Find the vector potential inside the superconductor. Determine expressions for the electric and magnetic fields at the surface. Evaluate the surface impedance Z_s (in Gaussian units, $4\pi/c$ times the ratio of tangential electric field to tangential magnetic field). Show that in the appropriate limit your result for Z_s reduces to that given in Section 12.9.

12.21 A two-fluid model for the electrodynamics of superconductors posits two types of electron, normal and superconducting, with number densities, charges, masses, and collisional damping constants, n_j, e_j, m_j, and γ_j, respectively ($j = $ N, S). The electrical conductivity consists of the sum of two terms of the Drude form (7.58) with $f_0 N \to n_j$, $e \to e_j$, $m \to m_j$, $\gamma_0 \to \gamma_j$. The normal (superconducting) electrons are distinguished by $\gamma_N \neq 0$ ($\gamma_S = 0$).

(a) Show that the conductivity of the superconductor at very low frequencies is largely imaginary (inductive) with a small resistive component from the normal electrons.

(b) Show that use of Ohm's law with the conductivity of part a in the Maxwell equations results in the static London equation for the electric field in the limit $\omega \to 0$, with the penetration depth (12.100), provided the carriers are identified with the superconducting component of the electric fluid.

CHAPTER 13

Collisions, Energy Loss, and Scattering of Charged Particles; Cherenkov and Transition Radiation

In this chapter we consider collisions between swiftly moving, charged particles, with special emphasis on the exchange of energy between collision partners and on the accompanying deflections from the incident direction. We also treat Cherenkov radiation and transition radiation, phenomena associated with charged particles in uniform motion through material media.

A fast charged particle incident on matter makes collisions with the atomic electrons and nuclei. If the particle is heavier than an electron (mu or pi meson, K meson, proton, etc.), the collisions with electrons and with nuclei have different consequences. The light electrons can take up appreciable amounts of energy from the incident particle without causing significant deflections, whereas the massive nuclei absorb very little energy but because of their greater charge cause scattering of the incident particle. Thus loss of energy by the incident particle occurs almost entirely in collisions with electrons. The deflection of the particle from its incident direction results, on the other hand, from essentially elastic collisions with the atomic nuclei. The scattering is confined to rather small angles, so that a heavy particle keeps a more or less straight-line path while losing energy until it nears the end of its range. For incident electrons both energy loss and scattering occur in collisions with the atomic electrons. Consequently the path is much less straight. After a short distance, electrons tend to diffuse into the material, rather than go in a rectilinear path.

The subject of energy loss and scattering is an important one and is discussed in several books (see references at the end of the chapter) where numerical tables and graphs are presented. Consequently our discussion emphasizes the physical ideas involved, rather than the exact numerical formulas. Indeed, a full quantum-mechanical treatment is needed to obtain exact results, even though all the essential features are classical or semiclassical in origin. All the orders of magnitude of the quantum effects are easily derivable from the uncertainty principle, as will be seen.

We begin by considering the simple problem of energy transfer to a free electron by a fast heavy particle. Then the effects of a binding force on the electron are explored, and the classical Bohr formula for energy loss is obtained. A description of quantum modifications and the effect of the polarization of the medium is followed by a discussion of the closely related phenomenon of Cherenkov radiation in transparent materials. Then the elastic scattering of incident particles by nuclei and multiple scattering are presented. Finally, we treat

624

transition radiation by a particle passing from one medium to another of different optical properties.

13.1 Energy Transfer in a Coulomb Collision Between Heavy Incident Particle and Stationary Free Electron; Energy Loss in Hard Collisions

A swift particle of charge ze and mass M (energy $E = \gamma Mc^2$, momentum $P = \gamma \beta Mc$) collides with an atomic electron of charge $-e$ and mass m. For energetic collisions the binding of the electron in the atom can be neglected; the electron can be considered free and initially at rest in the laboratory. For all incident particles except electrons and positrons, $M \gg m$. Then the collision is best viewed as elastic Coulomb scattering in the rest frame of the incident particle. The well-known Rutherford scattering formula is

$$\frac{d\sigma}{d\Omega} = \left(\frac{ze^2}{2pv}\right)^2 \operatorname{cosec}^4 \frac{\theta}{2} \tag{13.1}$$

where $p = \gamma \beta mc$ and $v = \beta c$ are the momentum and speed of the electron in the rest frame of the heavy particle (exact in the limit $M/m \to \infty$). The cross section can be given a Lorentz-invariant form by relating the scattering angle to the 4-momentum transfer squared, $Q^2 = -(p - p')^2$. For elastic scattering, $Q^2 = 4p^2 \sin^2(\theta/2)$. The result is

$$\frac{d\sigma}{dQ^2} = 4\pi \left(\frac{ze^2}{\beta c Q^2}\right)^2 \tag{13.2}$$

where βc, the relative speed in each particle's rest frame, is found from $\beta^2 = 1 - (Mmc^2/P \cdot p)^2$.

The cross section for a given energy loss T by the incident particle, that is, the kinetic energy imparted to the initially stationary electron, is proportional to (13.2). If we evaluate the invariant Q^2 in the electron's rest frame, we find $Q^2 = 2mT$. With Q^2 replaced by $2mT$, (13.2) becomes

$$\frac{d\sigma}{dT} = \frac{2\pi z^2 e^4}{mc^2 \beta^2 T^2} \tag{13.3}$$

Equation (13.3) is the cross section per unit energy interval for energy loss T by the massive incident particle in a Coulomb collision with a free stationary electron. Its range of validity for actual collisions in matter is

$$T_{min} < T < T_{max}$$

with T_{min} set by our neglect of binding ($T_{min} \gtrsim \hbar\langle\omega\rangle$ where $\hbar\langle\omega\rangle$ is an estimate of the mean effective atomic binding energy) and T_{max} governed by kinematics. We can find T_{max} by recognizing that the most energetic collision in the rest frame of the incident particle occurs when the electron reverses its direction. After such a collision, the electron has energy $E' = \gamma mc^2$ and momentum $p' = \gamma \beta mc$ in the direction of the incident particle's velocity in the laboratory. The boost to the laboratory gives

$$T_{max} = E - mc^2 = \gamma(E' + \beta c p') - mc^2 = 2\gamma^2 \beta^2 mc^2 \tag{13.4}$$

We note in passing that (13.4) is not correct if the incident particle has too high an energy. The exact answer for T_{max} has a factor in the denominator, $D = 1 + 2mE/M^2c^2 + m^2/M^2$. For muons ($M/m \approx 207$), the denominator must be taken into account if the energy is comparable to 44 GeV or greater. For protons that energy is roughly 340 GeV. For equal masses, it is easy to see that $T_{max} = (\gamma - 1)mc^2$.

When the spin of the electron is taken into account, there is a quantum-mechanical correction to the energy loss cross section, namely, a factor of $1 - \beta^2 \sin^2(\theta/2) = (1 - \beta^2 T/T_{max})$:

$$\left(\frac{d\sigma}{dT}\right)_{qm} = \frac{2\pi z^2 e^4}{mc^2\beta^2 T^2}\left(1 - \beta^2 \frac{T}{T_{max}}\right) \tag{13.5}$$

The energy loss per unit distance in collisions with energy transfer greater than ε for a heavy particle passing through matter with N atoms per unit volume, each with Z electrons, is given by the integral,

$$\frac{dE}{dx}(T > \varepsilon) = NZ \int_\varepsilon^{T_{max}} T \frac{d\sigma}{dT} dT$$

$$= 2\pi NZ \frac{z^2 e^4}{mc^2\beta^2}\left[\ln\left(\frac{2\gamma^2\beta^2 mc^2}{\varepsilon}\right) - \beta^2\right] \tag{13.6}$$

In the result (13.6) we assumed $\varepsilon \ll T_{max}$ and used (13.5) for the energy-transfer cross section. The small term, $-\beta^2$, in the square brackets is the relativistic spin contribution. Equation (13.6) represents the energy loss in close collisions. It is only valid provided $\varepsilon \gg \hbar\langle\omega\rangle$ because binding has been ignored.

An alternative, classical or semiclassical approach throws a different light on the physics of energy loss. In the rest frame of the heavy particle the incident electron approaches at impact parameter b. There is a one-to-one correspondence between b and the scattering angle θ (see Problem 13.1). The energy transfer T can be written as

$$T(b) = \frac{2z^2 e^4}{mv^2} \cdot \frac{1}{b^2 + b_{min}^{(c)2}}$$

with $b_{min}^{(c)} = ze^2/pv$. For $b \gg b_{min}^{(c)}$ the energy transfer varies as b^{-2}, implying that, if the energy transfer is greater than ε, the impact parameter must be less than the maximum,

$$b_{max}^{(c)}(\varepsilon) \approx \left(\frac{2z^2 e^4}{mv^2\varepsilon}\right)^{1/2}$$

When the heavy particle passes through matter it "sees" electrons at all possible impact parameters, with weighting according to the area of an annulus, $2\pi b\, db$. The classical energy loss per unit distance for collisions with transfer greater than ε is therefore

$$\frac{dE}{dx}(T > \varepsilon) = 2\pi NZ \int_0^{b_{max}^{(c)}(\varepsilon)} T(b)b\, db = 2\pi NZ \frac{z^2 e^4}{mc^2\beta^2}\ln\left[\left(\frac{b_{max}^{(c)}(\varepsilon)}{b_{min}^{(c)}}\right)^2\right] \tag{13.7}$$

Substitution of b_{max} and b_{min} leads directly to (13.6), apart from the relativistic spin correction. That we obtain the same result (for a spinless particle) quantum mechanically and classically is a consequence of the validity of the Rutherford cross section in both regimes.

If we wish to find a classical result for the *total* energy loss per unit distance, we must address the influence of atomic binding. Electronic binding can be characterized by the frequency of motion $\langle\omega\rangle$ or its reciprocal, the period. The incident heavy particle produces appreciable time-varying electromagnetic fields at the atom for a time $\Delta t \approx b/\gamma v$ [see (11.153)]. If the characteristic time Δt is long compared to the atomic period, the atom responds adiabatically—it stretches slowly during the encounter and returns to normal, without appreciable energy being transferred. On the other hand, if Δt is very short compared to the characteristic period, the electron can be treated as almost free. The dividing line is $\langle\omega\rangle\Delta t \approx 1$, implying a maximum effective impact parameter

$$b_{\text{max}}^{(c)} \approx \frac{\gamma v}{\langle\omega\rangle} \tag{13.8}$$

beyond which no significant energy transfer is possible. Explicit illustration of this cutoff for a charge bound harmonically is found in Problems 13.2 and 13.3.

If (13.8) is used in (13.7) instead of $b_{\text{max}}^{(c)}(\varepsilon)$, the total classical energy loss per unit distance is approximately

$$\left(\frac{dE}{dx}\right)_{\text{classical}} = 2\pi NZ \frac{z^2 e^4}{mc^2\beta^2} \ln(B_c^2) \tag{13.9}$$

where

$$B_c = \lambda \frac{\gamma^2\beta^3 mc^3}{ze^2\langle\omega\rangle} = \lambda \frac{\gamma^2\beta^2 mc^2}{\eta\hbar\langle\omega\rangle} \tag{13.10}$$

In (13.10) we have inserted a numerical constant λ of the order of unity to allow for our uncertainty in $b_{\text{max}}^{(c)}$. The parameter $\eta = ze^2/\hbar v$ is a characteristic of quantum-mechanical Coulomb scattering: $\eta \ll 1$ is the strongly quantum limit; $\eta \gg 1$ is the classical limit.

Equation (13.9) with (13.10) contains the essentials of the classical energy loss formula derived by Niels Bohr (1915). With many different electronic frequencies, $\langle\omega\rangle$ is the geometric mean of all the frequencies ω_j, weighted with the oscillator strength f_j:

$$Z \ln\langle\omega\rangle = \sum_j f_j \ln \omega_j \tag{13.11}$$

Equation (13.10) is valid for $\eta > 1$ (relatively slow alpha particles, heavy nuclei) but overestimates the energy loss when $\eta < 1$ (muons, protons, even fast alpha particles). We see below that when $\eta < 1$ the correct result sets $\eta = 1$ in (13.10).

13.2   Energy Loss from Soft Collisions; Total Energy Loss

The energy loss in collisions with energy transfers less than ε, including those small compared to electronic binding energies, really can be treated properly only by quantum mechanics, although after the fact we can "explain" the result in semiclassical language. The result, first obtained by Bethe (1930), is

$$\frac{dE}{dx}(T < \varepsilon) = 2\pi NZ \frac{z^2 e^4}{mc^2\beta^2} \{\ln[B_q^2(\varepsilon)] - \beta^2\} \tag{13.12}$$

where

$$B_q(\varepsilon) = \frac{\gamma v(2m\varepsilon)^{1/2}}{\hbar\langle\omega\rangle} \tag{13.13}$$

The effective excitation energy $\hbar\langle\omega\rangle$ is given by (13.11), but now with the proper quantum-mechanical oscillator strengths and frequency differences for the atom, including the contribution from the continuum. The upper limit ε on the energy transfers is assumed to be beyond the limit of appreciable oscillator strength. Such a limit is a consonant with the lower limit ε in Section 13.1, chosen to make the electron essentially free.

The total energy loss per unit length is given by the sum of (13.6) and (13.12):

$$\frac{dE}{dx} = 4\pi NZ \frac{z^2 e^4}{mc^2 \beta^2} \{\ln(B_q) - \beta^2\} \tag{13.14}$$

where

$$B_q = \frac{2\gamma^2 \beta^2 mc^2}{\hbar\langle\omega\rangle} \tag{13.15}$$

The general behavior of both the classical and quantum-mechanical energy loss formulas is illustrated in Fig. 13.1. They are functions only of the speed of the incident heavy particle, the mass and charge of the electron, and the mean excitation energy $\hbar\langle\omega\rangle$. For low energies ($\gamma\beta < 1$) the main dependence is as $1/\beta^2$, while at high energies the slow variation is proportional to $\ln(\gamma)$. The minimum value of dE/dx occurs at $\gamma\beta \approx 3$. The coefficient in (13.12) and (13.14) is numerically equal to $0.150\ z^2(2Z/A)\rho$ MeV/cm, where Z is the atomic number and A the mass number of the material, while ρ (g/cm^3) is its density. Since $2Z/A \approx 1$, the energy loss in MeV·(cm^2/g) for a singly charged particle in aluminum is approximately what is shown in Fig. 13.1. For aluminum the minimum energy loss is roughly 1.7 MeV·(cm^2/g); for lead, it is 1.2 MeV·(cm^2/g). At high energies corrections to the behavior in Fig. 13.1 occur. The energy loss becomes heavy-particle specific, through the mass-dependent denominator D in T_{max}, and

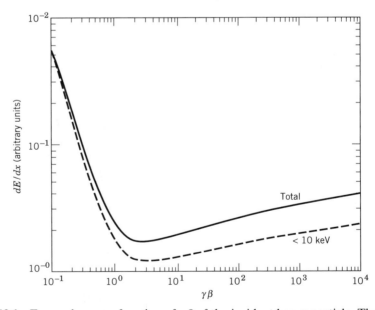

Figure 13.1 Energy loss as a function of $\gamma\beta$ of the incident heavy particle. The solid curve is the total energy loss (13.14) with $\hbar\langle\omega\rangle = 160$ eV (aluminum). The dashed curve is the energy loss in soft collisions (13.12) with $\varepsilon = 10$ keV. The ordinate scale corresponds to the curly-bracketed quantities in (13.12) and (13.14), multiplied by 0.15.

has a different energy variation and dependence on the material, because of the density effect discussed in Section 13.3.

The restricted energy loss shown in Fig. 13.1 is applicable to the energy loss inferred from tracks in photographic emulsions. Electrons with energies greater than about 10 keV have sufficient range to escape from silver bromide grains. The density of blackening along a track is therefore related to the restricted energy loss. Note that it increases more slowly than the total for large $\gamma\beta$—as $\ln(\gamma)$ rather than $\ln(\gamma^2)$. A semiclassical explanation is given below.

Comparison of B_q with the classical B_c (13.10) shows that their ratio is $\eta = ze^2/\hbar v$. To understand how this factor arises, we turn to semiclassical arguments. B_c is the ratio of $b_{max}^{(c)}$ (13.8) to $b_{min}^{(c)} = ze^2/\gamma mv^2$. The uncertainty principle dictates a different b_{min} for $\eta < 1$. In the rest frame of the heavy particle the electron has momentum $p = \gamma mv$. If it is described by a transversely localized wave packet (to define its impact parameter as well as possible), the spread in transverse momenta Δp around zero must satisfy $\Delta p \ll p$; otherwise, its longitudinal direction would be ill-defined. This limit on Δp translates into an uncertainty Δb in impact parameter, $\Delta b \gg \hbar/p$, or in other words, an effective quantum-mechanical lower limit,

$$b_{min}^{(q)} = \frac{\hbar}{\gamma mv} \qquad (13.16)$$

Evidently, in calculating energy loss as an integral over impact parameters, the larger of the two minimum impact parameters should be used. The ratio $b_{min}^{(c)}/b_{min}^{(q)} = \eta$. When $\eta > 1$, the classical lower limit applies; for $\eta < 1$, (13.16) applies and (13.15) is the correct expression for B.

The value of $B_q(\varepsilon)$ in (13.12) can also be understood in terms of impact parameters. The soft collisions contributing to (13.12) come semiclassically from the more distant collisions. The momentum transfer δp to the struck electron in such collisions is related to the energy transfer T according to $\delta p = (2mT)^{1/2}$. On the other hand, the localized electron wave packet has a spread Δp in transverse momenta. To be certain that the collision produces an energy transfer less than ε, we must have $\Delta p < \delta p_{max} = (2m\varepsilon)^{1/2}$, hence $\Delta b > \hbar/(2m\varepsilon)^{1/2}$. The effective minimum impact parameter for soft collisions with energy transfer less than ε is therefore

$$b_{min}^{(q)}(\varepsilon) \approx \frac{\hbar}{(2m\varepsilon)^{1/2}} \qquad (13.17)$$

For collisions so limited in impact parameter between (13.17) and $b_{max} = \gamma v/\langle\omega\rangle$, we find

$$B_q(\varepsilon) \approx \frac{\gamma v(2m\varepsilon)^{1/2}}{\hbar\langle\omega\rangle}$$

in agreement with Bethe's result.

The semiclassical discussion of the minimum and maximum impact parameters elucidates the reason for the difference in the logarithmic growth between the restricted and total energy losses. At high energies the dominant energy dependence is through $dE/dx \propto \ln(B) \approx \ln(b_{max}/b_{min})$. For the total energy loss, the maximum impact parameter is proportional to γ, while the quantum-mechanical minimum impact parameter (13.16) is inversely proportional to γ. The ratio varies as γ^2. For energy loss restricted to energy transfers less than ε, the minimum impact parameter (13.17) is independent of γ, leading to $B_q(\varepsilon) \propto \gamma$.

Despite its attractiveness in making clear the physics, the semiclassical description in terms of impact parameters contains a conceptual difficulty that warrants discussion. Classically, the energy transfer T in each collision is related directly to the impact parameter b. When $b \gg b_{\min}^{(c)}$, $T(b) \approx 2z^2e^4/mv^2b^2$ (Problem 13.1). With increasing b the energy transfer decreases rapidly until at $b = b_{\max} \approx \gamma v/\langle\omega\rangle$ it becomes

$$T(b_{\max}) \approx \frac{z^2}{\gamma^2}\left(\frac{v_0}{v}\right)^4 \left(\frac{\hbar\langle\omega\rangle}{I_H}\right) \hbar\langle\omega\rangle \tag{13.18}$$

Here $v_0 = c/137$ is the orbital speed of an electron in the ground state of hydrogen and $I_H = 13.6$ eV its ionization potential. Since empirically $\hbar\langle\omega\rangle \lesssim ZI_H$, we see that for a fast particle ($v \gg v_0$) the classical energy transfer (13.18) is much smaller than the ionization potential, indeed, smaller than the minimum possible atomic excitation.

We know, however, that energy must be transferred to the atom in discrete quantum jumps. A tiny amount of energy such as (13.18) simply cannot be absorbed by the atom. We might argue that the classical expression for $T(b)$ should be employed only if it is large compared to some typical excitation energy of the atom. This requirement would set quite a different upper limit on the impact parameters from $b_{\max} \approx \gamma v/\langle\omega\rangle$ and lead to wrong results. Could $b_{\max}$ nevertheless be wrong? After all, it came from consideration of the time dependence of the electric and magnetic fields (11.152), without consideration of the system being affected. No, time-dependent perturbations of a quantum system cause significant excitation only if they possess appreciable Fourier components with frequencies comparable to $1/\hbar$ times the lowest energy difference. That was the "adiabatic" argument that led to $b_{\max}$ in the first place. The solution to this conundrum lies in another direction. The classical expressions must be interpreted in a statistical sense.

The classical concept of the transfer of a small amount of energy in every collision is incorrect quantum-mechanically. Instead, while *on the average* over many collisions, a small energy is transferred, the small average results from appreciable amounts of energy transferred in a very small fraction of those collisions. In most collisions no energy is transferred. It is only in a statistical sense that the quantum-mechanical mechanism of discrete energy transfers and the classical process with a continuum of possible energy transfers can be reconciled. The detailed numerical agreement for the averages (but not for the individual amounts) stems from the quantum-mechanical definitions of the oscillator strengths f_j and resonant frequencies ω_j entering $\langle\omega\rangle$. A meaningful semiclassical description requires (a) the statistical interpretation and (b) the use of the uncertainty principle to set appropriate minimum impact parameters.

The discussion so far has been about energy loss by a heavy particle of mass $M \gg m$. For electrons ($M = m$), kinematic modifications occur in the energy loss in hard collisions. The maximum energy loss is $T_{\max} = (\gamma - 1)mc^2$. The argument of the logarithm in (13.6) becomes $(\gamma - 1)mc^2/\varepsilon$. The Bethe expression (13.12) for soft collisions remains the same. The total energy loss for electrons therefore has B_q (13.15) replaced by

$$B_q(\text{electrons}) = \frac{\sqrt{2}\,\gamma\beta\sqrt{\gamma - 1}\,mc^2}{\hbar\langle\omega\rangle} \approx \frac{\sqrt{2}\,\gamma^{3/2}\,mc^2}{\hbar\langle\omega\rangle} \tag{13.19}$$

the last form applicable for relativistic energies. There are spin and exchange effects in addition to the kinematic change, but the dominant effect is in the argument of the logarithm; the other effects only contribute to the added constant.

The expressions for dE/dx represent the *average* collisional energy loss per unit distance by a particle traversing matter. Because the number of collisions per unit distance is finite, even though large, and the spectrum of possible energy transfers in individual collisions is wide, there are fluctuations around the average. These fluctuations produce straggling in energy or range for a particle traversing a certain thickness of matter. If the number of collisions is large enough and the mean energy loss not too great, the final energies of a beam of initially monoenergetic particles of energy E_0 are distributed in Gaussian fashion about the mean $\overline{E}$. With Poisson statistics for the number of collisions producing a given energy transfer T, it can be shown (see, e.g., *Bohr*, Section 2.3, or *Rossi*, Section 2.7) that the mean square deviation in energy from the mean is

$$\Omega^2 = 2\pi N Z z^2 e^4 (\gamma^2 + 1) t \tag{13.20}$$

where t is the thickness traversed. This result holds provided $\Omega \ll \overline{E}$ *and* $\Omega \ll (E_0 - \overline{E})$, and also $\Omega \gg T_{max} \approx 2\gamma^2\beta^2 mc^2$. For ultrarelativistic particles the last condition ultimately fails. Then the distribution in energies is not Gaussian, but is described by the Landau curve. The interested reader may consult the references at the end of the chapter for further details.

13.3 *Density Effect in Collisional Energy Loss*

For particles that are not too relativistic, the observed energy loss is given accurately by (13.14) [or by (13.9) if $\eta > 1$] for particles of all kinds in media of all types. For ultrarelativistic particles, however, the observed energy loss is less than predicted by (13.14), especially for dense substances. In terms of Fig. 13.1 of (dE/dx), the observed energy loss increases beyond the minimum with a slope of roughly one-half that of the theoretical curve, corresponding to only one power of γ in the argument of the logarithm in (13.14) instead of two. In photographic emulsions the energy loss, as measured from grain densities, barely increases above the minimum to a plateau extending to the highest known energies. This again corresponds to a reduction of one power of γ, this time in $B_q(\epsilon)$ (13.13).

This reduction in energy loss, known as the density effect, was first treated theoretically by Fermi (1940). In our discussion so far we have tacitly made one assumption that is not valid in dense substances. We have assumed that it is legitimate to calculate the effect of the incident particle's fields on one electron in one atom at a time, and then sum up incoherently the energy transfers to all the electrons in all the atoms with $b_{min} < b < b_{max}$. Now b_{max} is very large compared to atomic dimensions, especially for large γ. Consequently in dense media there are many atoms lying between the incident particle's trajectory and the typical atom in question if b is comparable to b_{max}. These atoms, influenced themselves by the fast particle's fields, will produce perturbing fields at the chosen atom's position, modifying its response to the fields of the fast particle. Said in another way, in dense media the dielectric polarization of the material alters the particle's fields from their free-space values to those characteristic of macroscopic fields in a dielectric. This modification of the fields due to polarization of the medium must be taken into account in calculating the energy transferred in distant collisions. For close collisions the incident particle interacts with only one atom at a time. Then the free-particle calculation without polarization effects will apply. The dividing impact parameter between close and distant collisions is of the order of atomic dimensions. Since the joining of two logarithms is involved in calculating the sum, the dividing value of b need not be specified with great precision.

We will determine the energy loss in distant collisions ($b \geq a$), assuming that the fields in the medium can be calculated in the continuum approximation of a macroscopic dielectric constant $\epsilon(\omega)$. If a is of the order of atomic dimensions, this approximation will not be good for the closest of the distant collisions, but will be valid for the great bulk of the collisions.

The problem of finding the electric field in the medium due to the incident fast particle moving with constant velocity can be solved most readily by Fourier transforms. If the potentials $A_\mu(x)$ and source density $J_\mu(x)$ are transformed in space and time according to the general rule,

$$F(\mathbf{x}, t) = \frac{1}{(2\pi)^2} \int d^3k \int d\omega \, F(\mathbf{k}, \omega) e^{i\mathbf{k}\cdot\mathbf{x} - i\omega t} \qquad (13.21)$$

then the transformed wave equations become

$$\left[k^2 - \frac{\omega^2}{c^2} \epsilon(\omega) \right] \Phi(\mathbf{k}, \omega) = \frac{4\pi}{\epsilon(\omega)} \rho(\mathbf{k}, \omega)$$

$$\left[k^2 - \frac{\omega^2}{c^2} \epsilon(\omega) \right] \mathbf{A}(\mathbf{k}, \omega) = \frac{4\pi}{c} \mathbf{J}(\mathbf{k}, \omega) \qquad (13.22)$$

The dielectric constant $\epsilon(\omega)$ appears characteristically in positions dictated by the presence of $\mathbf{D}$ in the Maxwell equations. The Fourier transforms of

$$\rho(\mathbf{x}, t) = ze \, \delta(\mathbf{x} - \mathbf{v}t)$$

and $\qquad\qquad\qquad\qquad\qquad\qquad\qquad\qquad\qquad\qquad\qquad\qquad$ (13.23)

$$\mathbf{J}(\mathbf{x}, t) = \mathbf{v}\rho(\mathbf{x}, t)$$

are readily found to be

$$\rho(\mathbf{k}, \omega) = \frac{ze}{2\pi} \delta(\omega - \mathbf{k} \cdot \mathbf{v})$$

$$\mathbf{J}(\mathbf{k}, \omega) = \mathbf{v}\rho(\mathbf{k}, \omega) \qquad (13.24)$$

From (13.22) we see that the Fourier transforms of the potentials are

$$\Phi(\mathbf{k}, \omega) = \frac{2ze}{\epsilon(\omega)} \cdot \frac{\delta(\omega - \mathbf{k} \cdot \mathbf{v})}{k^2 - \frac{\omega^2}{c^2} \epsilon(\omega)}$$

and $\qquad\qquad\qquad\qquad\qquad\qquad\qquad\qquad\qquad\qquad\qquad\qquad\qquad$ (13.25)

$$\mathbf{A}(\mathbf{k}, \omega) = \epsilon(\omega) \frac{\mathbf{v}}{c} \Phi(\mathbf{k}, \omega)$$

From the definitions of the electromagnetic fields in terms of the potentials we obtain their Fourier transforms:

$$\left.\begin{aligned}
\mathbf{E}(\mathbf{k}, \omega) &= i\left[\frac{\omega\epsilon(\omega)}{c}\frac{\mathbf{v}}{c} - \mathbf{k}\right]\Phi(\mathbf{k}, \omega) \\
\mathbf{B}(\mathbf{k}, \omega) &= i\epsilon(\omega)\mathbf{k} \times \frac{\mathbf{v}}{c}\,\Phi(\mathbf{k}, \omega)
\end{aligned}\right\} \tag{13.26}$$

In calculating the energy loss to an electron in an atom at impact parameter b, we evaluate

$$\Delta E = -e\int_{-\infty}^{\infty} \mathbf{v} \cdot \mathbf{E}\, dt = 2e\,\mathrm{Re}\int_{0}^{\infty} i\omega\mathbf{x}(\omega) \cdot \mathbf{E}^*(\omega)\, d\omega \tag{13.27}$$

where $\mathbf{x}(\omega)$ is the Fourier transform in time of the electron's coordinate and $\mathbf{E}(\omega)$ is the Fourier transform in time of the electromagnetic fields at a perpendicular distance b from the path of the particle moving along the x axis. Thus the required electric field is

$$\mathbf{E}(\omega) = \frac{1}{(2\pi)^{3/2}}\int d^3k\,\mathbf{E}(\mathbf{k}, \omega)e^{ibk_2} \tag{13.28}$$

where the observation point has coordinates $(0, b, 0)$. To illustrate the determination of $\mathbf{E}(\omega)$ we consider the calculation of $E_1(\omega)$, the component of $\mathbf{E}$ parallel to $\mathbf{v}$. Inserting the explicit forms from (13.25) and (13.26), we obtain

$$E_1(\omega) = \frac{2ize}{\epsilon(\omega)(2\pi)^{3/2}}\int d^3k\, e^{ibk_2}\left[\frac{\omega\epsilon(\omega)v}{c^2} - k_1\right]\frac{\delta(\omega - vk_1)}{k^2 - \frac{\omega^2}{c^2}\epsilon(\omega)} \tag{13.29}$$

The integral over dk_1 can be done immediately. Then

$$E_1(\omega) = -\frac{2ize\omega}{(2\pi)^{3/2}v^2}\left[\frac{1}{\epsilon(\omega)} - \beta^2\right]\int_{-\infty}^{\infty} dk_2\, e^{ibk_2}\int_{-\infty}^{\infty}\frac{dk_3}{k_2^2 + k_3^2 + \lambda^2}$$

where

$$\lambda^2 = \frac{\omega^2}{v^2} - \frac{\omega^2}{c^2}\epsilon(\omega) = \frac{\omega^2}{v^2}[1 - \beta^2\epsilon(\omega)] \tag{13.30}$$

The integral over dk_3 has the value $\pi/(\lambda^2 + k_2^2)^{1/2}$, so that $E_1(\omega)$ can be written

$$E_1(\omega) = -\frac{ize\omega}{\sqrt{2\pi}\, v^2}\left[\frac{1}{\epsilon(\omega)} - \beta^2\right]\int_{-\infty}^{\infty}\frac{e^{ibk_2}}{(\lambda^2 + k_2^2)^{1/2}}\, dk_2 \tag{13.31}$$

The remaining integral is a representation of a modified Bessel function.* The result is

$$E_1(\omega) = -\frac{ize\omega}{v^2}\left(\frac{2}{\pi}\right)^{1/2}\left[\frac{1}{\epsilon(\omega)} - \beta^2\right]K_0(\lambda b) \tag{13.32}$$

---

*See, for example, *Abramowitz and Stegun* (p. 376, formula 9.6.25); *Magnus, Oberhettinger, and Soni* (Chapter XI), or Bateman Manuscript Project, *Table of Integral Transforms*, Vol. 1 (Chapters I–III).

where the square root of (13.30) is chosen so that λ lies in the fourth quadrant. A similar calculation yields the other fields:

$$
\left.
\begin{aligned}
E_2(\omega) &= \frac{ze}{v}\left(\frac{2}{\pi}\right)^{1/2} \frac{\lambda}{\epsilon(\omega)} K_1(\lambda b) \\
B_3(\omega) &= \epsilon(\omega)\beta E_2(\omega)
\end{aligned}
\right\}
\tag{13.33}
$$

In the limit $\epsilon(\omega) \to 1$ it is easily seen that fields (13.32) and (13.33) reduce to the results of Problem 13.3.

To find the energy transferred to the atom at impact parameter b we merely write down the generalization of (13.27):

$$
\Delta E(b) = 2e \sum_j f_j \, \mathrm{Re} \int_0^\infty i\omega \mathbf{x}_j(\omega) \cdot \mathbf{E}^*(\omega) \, d\omega
$$

where $\mathbf{x}_j(\omega)$ is the amplitude of the jth type of electron in the atom. Rather than use (7.50) for $\mathbf{x}_j(\omega)$ we express the sum of dipole moments in terms of the molecular polarizability and so the dielectric constant. Thus

$$
-e \sum_j f_j \mathbf{x}_j(\omega) = \frac{1}{4\pi N} [\epsilon(\omega) - 1]\mathbf{E}(\omega)
$$

where N is the number of atoms per unit volume. Then the energy transfer can be written

$$
\Delta E(b) = \frac{1}{2\pi N} \, \mathrm{Re} \int_0^\infty -i\omega\epsilon(\omega) \, |\mathbf{E}(\omega)|^2 \, d\omega
\tag{13.34}
$$

The energy loss per unit distance in collisions with impact parameter $b \geq a$ is evidently

$$
\left(\frac{dE}{dx}\right)_{b>a} = 2\pi N \int_0^\infty \Delta E(b) b \, db
\tag{13.35}
$$

If fields (13.32) and (13.33) are inserted into (13.34) and (13.35), we find, after some calculation, the expression due to Fermi,

$$
\left(\frac{dE}{dx}\right)_{b>a} = \frac{2}{\pi} \frac{(ze)^2}{v^2} \, \mathrm{Re} \int_0^\infty i\omega\lambda^* a K_1(\lambda^* a) K_0(\lambda a)\left(\frac{1}{\epsilon(\omega)} - \beta^2\right) d\omega
\tag{13.36}
$$

where λ is given by (13.30). This result can be obtained more elegantly by calculating the electromagnetic energy flow through a cylinder of radius a around the path of the incident particle. By conservation of energy this is the energy lost per unit time by the incident particle. Thus

$$
\left(\frac{dE}{dx}\right)_{b>a} = \frac{1}{v}\frac{dE}{dt} = -\frac{c}{4\pi v} \int_{-\infty}^\infty 2\pi a B_3 E_1 \, dx
$$

The integral over dx at one instant of time is equivalent to an integral at one point on the cylinder over all time. Using $dx = v \, dt$, we have

$$
\left(\frac{dE}{dx}\right)_{b>a} = -\frac{ca}{2} \int_{-\infty}^\infty B_3(t) E_1(t) \, dt
$$

In the standard way this can be converted into a frequency integral,

$$\left(\frac{dE}{dx}\right)_{b>a} = -ca \ \text{Re} \int_0^\infty B_3^*(\omega)E_1(\omega) \, d\omega \tag{13.37}$$

With fields (13.32) and (13.33) this gives the Fermi result (13.36).

The Fermi expression (13.36) bears little resemblance to our earlier results for energy loss. But under conditions where polarization effects are unimportant it yields the same results as before. For example, for nonrelativistic particles ($\beta \ll 1$) it is clear from (13.30) that $\lambda \simeq \omega/v$, independent of $\epsilon(\omega)$. Then in (13.36) the modified Bessel functions are real. Only the imaginary part of $1/\epsilon(\omega)$ contributes to the integral. If we neglect the polarization correction of Section 4.5 to the internal field at an atom, the dielectric constant can be written

$$\epsilon(\omega) \simeq 1 + \frac{4\pi Ne^2}{m} \sum_j \frac{f_j}{\omega_j^2 - \omega^2 - i\omega\Gamma_j} \tag{13.38}$$

where we have used the dipole moment expression (7.50). Assuming that the second term is small, the imaginary part of $1/\epsilon(\omega)$ can be readily calculated and substituted into (13.36). Then the integral over $d\omega$ can be performed in the narrow-resonance approximation. If the small-argument limits of the Bessel functions are used, the nonrelativistic form of (13.9) emerges, with $B_c = v/a\langle\omega\rangle$. If the departure of λ from $\omega/\gamma v$ in (13.30) is neglected, (13.9) emerges with $B_c = \gamma v/a\langle\omega\rangle$.

The density effect evidently comes from the presence of complex arguments in the modified Bessel functions, corresponding to taking into account $\epsilon(\omega)$ in (13.30). Since $\epsilon(\omega)$ there is multiplied by β^2, it is clear that the density effect can be really important only at high energies. The detailed calculations for all energies with some explicit expression such as (13.38) for $\epsilon(\omega)$ are quite complicated and not particularly informative. We content ourselves with the extreme relativistic limit ($\beta \simeq 1$). Furthermore, since the important frequencies in the integral over $d\omega$ are optical frequencies and the radius a is of the order of atomic dimensions, $|\lambda a| \sim (\omega a/c) \ll 1$. Consequently we can approximate the Bessel functions by their small-argument limits (3.103). Then in the relativistic limit the Fermi expression (13.36) is

$$\left(\frac{dE}{dx}\right)_{b>a} \simeq \frac{2}{\pi} \frac{(ze)^2}{c^2} \ \text{Re} \int_0^\infty i\omega\left(\frac{1}{\epsilon(\omega)} - 1\right)\left\{\ln\left(\frac{1.123c}{\omega a}\right) - \frac{1}{2}\ln[1 - \epsilon(\omega)]\right\} d\omega \tag{13.39}$$

It is worthwhile right here to point out that the argument of the second logarithm is actually $[1 - \beta^2\epsilon(\omega)]$. In the limit $\epsilon = 1$, this log term gives a factor γ in the combined logarithm, corresponding to the old result (13.9). Provided $\epsilon(\omega) \neq 1$, we can write this factor as $[1 - \epsilon(\omega)]$, thereby removing one power of γ from the logarithm, in agreement with experiment.

The integral in (13.39) with $\epsilon(\omega)$ given by (13.38) can be performed most easily by using Cauchy's theorem to change the integral over positive real ω to one over positive imaginary ω, minus one over a quarter-circle at infinity. The integral along the imaginary axis gives no contribution. Provided the Γ_j in (13.38) are assumed constant, the result of the integration over the quarter-circle can be written in the simple form:

$$\left(\frac{dE}{dx}\right)_{b>a} = \frac{(ze)^2\omega_p^2}{c^2}\ln\left(\frac{1.123c}{a\omega_p}\right) \tag{13.40}$$

where ω_p is the electronic plasma frequency

$$\omega_p^2 = \frac{4\pi NZe^2}{m} \tag{13.41}$$

The corresponding relativistic expression without the density effect is

$$\left(\frac{dE}{dx}\right)_{b>a} \simeq \frac{(ze)^2\omega_p^2}{c^2}\ln\left(\frac{1.123\gamma c}{a\langle\omega\rangle}\right) \tag{13.42}$$

We see that the density effect produces a simplification in that the asymptotic energy loss no longer depends on the details of atomic structure through $\langle\omega\rangle$ (13.11), but only on the number of electrons per unit volume through ω_p. Two substances having very different atomic structures will produce the same energy loss for ultrarelativistic particles provided their densities are such that the density of electrons is the same in each.

Since there are numerous calculated curves of energy loss based on Bethe's formula (13.14), it is often convenient to tabulate the decrease in energy loss due to the density effect. This is just the difference between (13.40) and (13.42):

$$\lim_{\beta\to 1}\Delta\left(\frac{dE}{dx}\right) = -\frac{(ze)^2\omega_p^2}{c^2}\ln\left(\frac{\gamma\omega_p}{\langle\omega\rangle}\right) \tag{13.43}$$

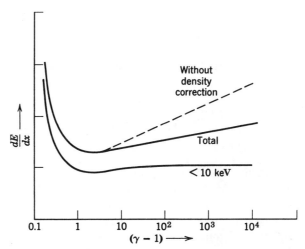

Figure 13.2 Energy loss, including the density effect. The dashed curve is the total energy loss without density correction. The solid curves have the density effect incorporated, the upper one being the total energy loss and the lower one the energy loss due to individual energy transfers of less than 10 keV.

For photographic emulsions, the relevant energy loss is given by (13.12) and (13.13) with $\varepsilon \simeq 10$ keV. With the density correction applied, this becomes constant at high energies with the value,

$$\frac{dE(\epsilon)}{dx} \rightarrow \frac{(ze)^2 \omega_p^2}{2c^2} \ln\left(\frac{2mc^2\varepsilon}{\hbar^2\omega_p^2}\right) \tag{13.44}$$

For silver bromide, $\hbar\omega_p \simeq 48$ eV. Then for singly charged particles (13.44), divided by the density, has the value of approximately 1.02 MeV · (cm²/g). This energy loss is in good agreement with experiment, and corresponds to an increase above the minimum value of less than 10%. Figure 13.2 shows total energy loss and loss from transfers of less than 10 keV for a typical substance. The dashed curve is the Bethe curve for total energy loss without correction for density effect.

13.4 Cherenkov Radiation

The density effect in energy loss is intimately connected to the coherent response of a medium to the passage of a relativistic particle that causes the emission of Cherenkov radiation. They are, in fact, the same phenomenon in different limiting circumstances. The expression (13.36), or better, (13.37), represents the energy lost by the particle into regions a distance greater than $b = a$ away from its path. By varying a we can examine how the energy is deposited throughout the medium. In (13.39) we have considered a to be atomic dimensions and assumed $|\lambda a| \ll 1$. Now we take the opposite limit. If $|\lambda a| \gg 1$, the modified Bessel functions can be approximated by their asymptotic forms. Then the fields (13.32) and (13.33) become

$$E_1(\omega, b) \rightarrow i\frac{ze\omega}{c^2}\left[1 - \frac{1}{\beta^2\epsilon(\omega)}\right]\frac{e^{-\lambda b}}{\sqrt{\lambda b}}$$

$$E_2(\omega, b) \rightarrow \frac{ze}{v\epsilon(\omega)}\sqrt{\frac{\lambda}{b}}\,e^{-\lambda b} \tag{13.45}$$

$$B_3(\omega, b) \rightarrow \beta\epsilon(\omega)E_2(\omega, b)$$

The integrand in (13.37) in this limit is

$$(-caB_3^*E_1) \rightarrow \frac{z^2e^2}{c^2}\left(-i\sqrt{\frac{\lambda^*}{\lambda}}\right)\omega\left[1 - \frac{1}{\beta^2\epsilon(\omega)}\right]e^{-(\lambda+\lambda^*)a} \tag{13.46}$$

The real part of this expression, integrated over frequencies, gives the energy deposited far from the path of the particle. If λ has a positive real part, as is generally true, the exponential factor in (13.46) will cause the expression to vanish rapidly at large distances. All the energy is deposited near the path. This is not true only when λ *is purely imaginary*. Then the exponential is unity; the expression is independent of a; *some of the energy escapes* to infinity *as radiation*. From (13.30) it can be seen that λ can be purely imaginary if $\epsilon(\omega)$ is real (no absorption) and $\beta^2\epsilon(\omega) > 1$. Actually, mild absorption can be allowed for, but in the interests of simplicity we will assume that $\epsilon(\omega)$ is essentially real from now on. The condition $\beta^2 \epsilon(\omega) > 1$ can be written in the more transparent form,

$$v > \frac{c}{\sqrt{\epsilon(\omega)}} \qquad (13.47)$$

This shows that *the speed of the particle must be larger than the phase velocity* of the electromagnetic fields at frequency ω in order *to have emission of Cherenkov radiation* of that frequency.

Consideration of the phase of λ as $\beta^2 \epsilon$ changes from less than unity to greater than unity, assuming that $\epsilon(\omega)$ has an infinitesimal positive imaginary part when $\omega > 0$, shows that

$$\lambda = -i \, |\lambda| \qquad \text{for } \beta^2 \epsilon > 1$$

This means that $(\lambda^*/\lambda)^{1/2} = i$ and (13.46) is real and independent of a. Equation (13.37) then represents the energy radiated as Cherenkov radiation per unit distance along the path of the particle:

$$\left(\frac{dE}{dx} \right)_{\text{rad}} = \frac{(ze)^2}{c^2} \int_{\epsilon(\omega) > (1/\beta^2)} \omega \left(1 - \frac{1}{\beta^2 \epsilon(\omega)} \right) d\omega \qquad (13.48)$$

The integrand obviously gives the differential spectrum in frequency. This is the Frank–Tamm result, first published in 1937 in an explanation of the radiation observed by Cherenkov in 1934. The radiation is evidently not emitted uniformly in frequency. It tends to be emitted in bands situated somewhat below regions of anomalous dispersion, where $\epsilon(\omega) > \beta^{-2}$, as indicated in Fig. 13.3. Of course, if $\beta \simeq 1$ the regions where $\epsilon(\omega) > \beta^{-2}$ may be quite extensive.

Another characteristic feature of Cherenkov radiation is its angle of emission. At large distances from the path the fields become transverse radiation fields. The direction of propagation is given by $\mathbf{E} \times \mathbf{B}$. As shown in Fig. 13.4, the angle θ_C of emission of Cherenkov radiation relative to the velocity of the particle is given by

$$\tan \theta_C = -\frac{E_1}{E_2} \qquad (13.49)$$

From the far fields (13.45) we find

$$\cos \theta_C = \frac{1}{\beta \sqrt{\epsilon(\omega)}} \qquad (13.50)$$

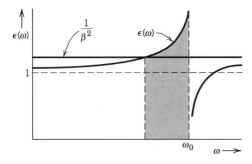

Figure 13.3 Cherenkov band. Radiation is emitted only in shaded frequency range, where $\epsilon(\omega) > \beta^{-2}$.

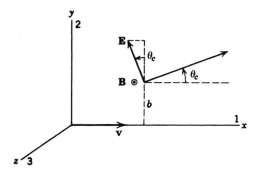

Figure 13.4

The criterion $\beta^2 \epsilon > 1$ can now be rephrased as the requirement that the emission angle θ_C be a physical angle with cosine less than unity. In passing we note from Fig. 13.4 that Cherenkov radiation is completely linearly polarized in the plane containing the direction of observation and the path of the particle.

The emission angle θ_C can be interpreted qualitatively in terms of a "shock" wavefront akin to the bow shock of a boat in water or the shock front accompanying supersonic flight. In Figure 13.5 are sketched two sets of successive spherical wavelets moving out with speed $c/\sqrt{\epsilon}$ from successive instantaneous positions of a particle moving with constant velocity v. On the left v is assumed to be less than, and on the right greater than, $c/\sqrt{\epsilon}$. For $v > c/\sqrt{\epsilon}$ the wavelets interfere so as to produce a "shock" front or wake behind the particle, the angle of which is readily seen to be the complement of θ_C. An observer at rest sees a wavefront moving in the direction of θ_C.

The qualitative behavior shown in Fig. 13.5 can be given quantitative treat-

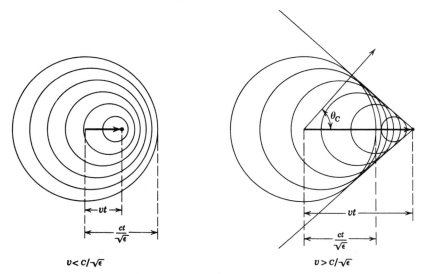

$$v < c/\sqrt{\epsilon} \qquad\qquad v > c/\sqrt{\epsilon}$$

Figure 13.5 Cherenkov radiation. Spherical wavelets of fields of a particle traveling less than and greater than the velocity of light in the medium. For $v > c/\sqrt{\epsilon}$, an electromagnetic "shock" wave appears, moving in the direction given by the Cherenkov angle θ_C.

ment by examining the potentials $\Phi(\mathbf{x}, t)$ or $\mathbf{A}(\mathbf{x}, t)$ constructed from (13.25) with (13.21). For example, the vector potential takes the form,

$$\mathbf{A}(\mathbf{x}, t) = \frac{2ze}{(2\pi)^2} \boldsymbol{\beta} \int d^3k \, \frac{e^{ik_1(x-vt)} e^{i\mathbf{k}_\perp \cdot \boldsymbol{\rho}}}{k_1^2(1 - \beta^2 \epsilon) + k_\perp^2}$$

where $\epsilon = \epsilon(k_1 v)$, while $\boldsymbol{\rho}$ and $\mathbf{k}_\perp$ are transverse coordinates. With the unrealistic, but tractable, approximation that ϵ is a constant the integral can be done in closed form. In the Cherenkov regime ($\beta^2 \epsilon > 1$) the denominator has poles on the path of integration. Choosing the contour for the k_1 integration so that the potential vanishes for points ahead of the particle ($x - vt > 0$), the result is found to be

$$\mathbf{A}(\mathbf{x}, t) = \boldsymbol{\beta} \frac{2ze}{\sqrt{(x - vt)^2 - (\beta^2 \epsilon - 1)\rho^2}} \tag{13.51}$$

inside the Cherenkov cone and zero outside. Note that $\mathbf{A}$ is singular along the shock front, as suggested by the wavelets in Fig. 13.5. The expression (13.51) can be taken as indicative only. The dielectric constant does vary with $\omega = k_1 v$. This functional dependence will remove the mathematical singularity in (13.51).

The properties of Cherenkov radiation can be utilized to measure velocities of fast particles. If the particles of a given velocity pass through a medium of known dielectric constant ϵ, the light is emitted at the Cherenkov angle (13.50). Thus a measurement of the angle allows determination of the velocity. Since the dielectric constant of a medium in general varies with frequency, light of different colors is emitted at somewhat different angles. Narrow-band filters may be employed to select a small interval of frequency and so improve the precision of velocity measurement. For very fast particles ($\beta \lesssim 1$) a gas may be used to provide a dielectric constant differing only slightly from unity and having ($\epsilon - 1$) variable over wide limits by varying the gas pressure. Counting devices using Cherenkov radiation are employed extensively in high-energy physics, as instruments for velocity measurements, as mass analyzers when combined with momentum analysis, and as discriminators against unwanted slow particles.

13.5 *Elastic Scattering of Fast Charged Particles by Atoms*

In Section 13.1 we considered the scattering of electrons by an incident heavy particle in that particle's rest frame in order to treat energy transfers to the electrons. We now turn to the elastic scattering that accompanies passage of swift particles, whether heavy or light, through matter because of interaction with the atoms. Charged particles are elastically scattered by the time-averaged potential created by the atomic nucleus and its associated electrons. The potential is roughly Coulombic in character but is modified at large distances by the screening effect of the electrons and at short distances by the finite size of the nucleus.

For a pure Coulomb field, the scattering cross section is given by the Rutherford formula (13.1), modified at large angles by spin-dependent corrections [see above (13.5)]. At small angles, all particles, regardless of spin, scatter according to the small-angle Rutherford expression

$$\frac{d\sigma}{d\Omega} \approx \left(\frac{2zZe^2}{pv}\right)^2 \cdot \frac{1}{\theta^4} \tag{13.52}$$

Even at $\theta = \pi/2$, the small-angle result is within 30% of the exact Rutherford formula. Such accuracy is sufficient for present purposes.

The singular nature of (13.52) as $\theta \rightarrow 0$ is a consequence of the infinite range of the Coulomb potential. Because of electronic screening, the differential scattering cross section is finite at $\theta = 0$. A simple classical impact parameter calculation (following Problem 13.1b) with a Coulomb force cutoff sharply at $r = a$ gives a small-angle cross section

$$\frac{d\sigma}{d\Omega} \approx \left(\frac{2zZe^2}{pv}\right)^2 \cdot \frac{1}{(\theta^2 + \theta_{min}^2)^2} \tag{13.53}$$

where θ_{min} is the classical cutoff angle,

$$\theta_{min}^{(c)} = \frac{2zZe^2}{pva} \tag{13.54}$$

A better form of screened Coulomb interaction is $V(r) = (zZe^2/r)e^{-r/a}$, with $a \approx 1.4\, a_0 Z^{-1/3}$ (from a rough fit to the Thomas–Fermi atomic potential). A classical calculation with such a potential gives a small-angle cross section for $\theta \rightarrow 0$ that rises less rapidly than θ^{-4}, but still is singular at $\theta = 0$. Quantum mechanically, either the Born approximation or a WKB eikonal approach yields a small-angle cross section of the form (13.53) with θ_{min} the quantum-mechanical cutoff angle

$$\theta_{min}^{(q)} = \frac{\hbar}{pa} \approx \frac{Z^{1/3}}{192} \cdot \frac{mc}{p} \tag{13.55}$$

where p is the incident momentum ($p = \gamma Mv$), and m is the electron's mass. In passing, we note that the ratio of classical to quantum-mechanical angles θ_{min} is $\eta = zZe^2/\hbar v$, in agreement with the corresponding ratio of minimum impact parameters [see below (13.16)]. For fast particles in all but the highest Z substances, $\eta < 1$; the quantum-mechanical expression (13.55) should be used for θ_{min}.

At comparatively large angles (but still small in actual magnitude) the scattering cross section departs from (13.53) because of the finite size of the nucleus. For charged leptons (e, μ, τ) the influence of the finite size is a purely electromagnetic effect, but for hadrons (π, K, p, α, etc.) specifically strong-interaction effects also arise. Since the gross overall effect is to lower the cross section below (13.53) at larger angles for whatever reason, we examine only the electromagnetic aspect. The charge distribution of the atomic nucleus can be approximated crudely by a uniform volume distribution inside a sphere of radius R, falling sharply to zero outside. The electrostatic potential inside the nucleus is parabolic in shape with a finite value at $r = 0$:

$$V(r) = \begin{cases} \dfrac{3zZe^2}{2R}\left(1 - \dfrac{r^2}{3R^2}\right) & r < R \\[2ex] \dfrac{zZe^2}{r} & r > R \end{cases} \quad \text{for} \tag{13.56}$$

The classical scattering cross section from such a potential exhibits singular behavior at a maximum angle given approximately by the classical formula (13.54), but with $a \to R$. This phenomenon is a consequence of the scattering angle $\theta(b) = \Delta p(b)/p$ vanishing at $b = 0$, rising to a maximum at just less than $b = R$, and falling again for larger b. The maximum translates into a vanishing derivative $d\theta/db$ and so an infinite differential cross section. The bizarre classical behavior is the vestige of what occurs quantum mechanically. The wave nature of the incident particle makes the nuclear scattering very much like the scattering of electromagnetic waves by localized scatterers, discussed in Chapter 10. At short wavelengths, the scattering is diffractive, confined to an angular range $\Delta\theta \approx 1/kR$ where $k = p/\hbar$. Depending on the radial dependence of the localized interaction, the scattering cross section may exhibit wiggles or secondary maxima and minima, but it will fall rapidly below the point Coulomb result at larger angles. Said another way, in perturbation theory the scattering amplitude is the product of the Coulomb amplitude for a point charge and a form factor $F(Q^2)$ that is the spatial Fourier transform of the charge distribution. The form factor is defined to be unity at $Q^2 = 0$, but becomes rapidly smaller for $(QR) > 1$. Whatever the viewpoint, the finite nuclear size sets an effective upper limit of the scattering,

$$\theta_{\max} \approx \frac{\hbar}{pR} \approx \frac{274}{A^{1/3}} \cdot \frac{mc}{p} \qquad (13.57)$$

The final expression is based on the estimate, $R = 1.4\,A^{1/3} \times 10^{-15}$ m. We note that $\theta_{\max} \gg \theta_{\min}$ for all physical values of Z and A. If the incident momentum is so small that $\theta_{\max} \gtrsim 1$, the nuclear size has no appreciable effect on the scattering. For an aluminum target, $\theta_{\max} = 1$ when $p \approx 50$ MeV/c, corresponding to 50 MeV kinetic energy for electrons and 1.3 MeV for protons. Only at higher energies are nuclear-sized effects important. At $p \approx 50$ MeV/c, $\theta_{\min} \approx 10^{-4}$ radian in aluminum.

The general behavior of the scattering cross section is shown in Fig. 13.6. The dot-dash curve is the small-angle Rutherford formula (13.52); the solid curve shows the qualitative behavior of the cross section including screening and finite nuclear size. The total scattering cross section can be obtained by integrating (13.53) over the total solid angle,

$$\sigma = \int \frac{d\sigma}{d\Omega} \sin\theta\, d\theta\, d\phi \approx 2\pi \left(\frac{2zZe^2}{pv}\right)^2 \cdot \int_0^\infty \frac{\theta\, d\theta}{(\theta^2 + \theta_{\min}^2)^2} \qquad (13.58)$$

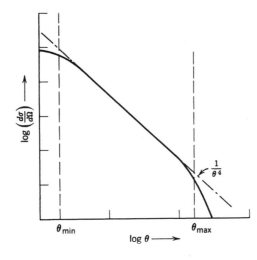

Figure 13.6 Atomic scattering, including effects of electronic screening at small angles and finite nuclear size at large angles.

The result is

$$\sigma \approx \pi\left(\frac{2zZe^2}{pv}\right)^2 \cdot \theta_{min}^{-2} = \pi a^2\left(\frac{2zZe^2}{\hbar v}\right)^2 \tag{13.59}$$

The final expression is obtained by use of (13.55) for θ_{min}. It shows that at high velocities the total scattering cross section can be far smaller than the classical geometrical area πa^2 of the atom.

13.6 Mean Square Angle of Scattering; Angular Distribution of Multiple Scattering

Rutherford scattering is confined to very small angles even for a point Coulomb field, and for fast particles θ_{max} is small compared to unity. Thus there is a very large probability for small-angle scattering. A particle traversing a finite thickness of matter will undergo very many small-angle deflections and will generally emerge at a small angle that is the cumulative statistical superposition of a large number of deflections. Only rarely will the particle be deflected through a large angle; since these events are infrequent, such a particle will have made only one such collision. This circumstance allows us to divide the angular range into two regions—one region at comparatively large angles, which contains only the single scatterings, and one region at very small angles, which contains the multiple or compound scatterings. The complete distribution in angle can be approximated by considering the two regions separately. The intermediate region of so-called plural scattering must allow a smooth transition from small to large angles.

The important quantity in the multiple-scattering region, where there is a large succession of small-angle deflections symmetrically distributed about the incident direction, is the mean square angle for a single scattering. This is defined by

$$\langle\theta^2\rangle = \frac{\int \theta^2 \dfrac{d\sigma}{d\Omega}\, d\Omega}{\int \dfrac{d\sigma}{d\Omega}\, d\Omega} \tag{13.60}$$

With the approximations of Section 13.5 we obtain

$$\langle\theta^2\rangle = 2\theta_{min}^2 \ln\left(\frac{\theta_{max}}{\theta_{min}}\right) \tag{13.61}$$

If the quantum value (13.55) of θ_{min} is used along with θ_{max} (13.57), then with $A \approx 2Z$, (13.61) has the numerical form:

$$\langle\theta^2\rangle \simeq 4\theta_{min}^2 \ln(204 Z^{-1/3}) \tag{13.62}$$

If nuclear size is unimportant (generally only of interest for electrons, and perhaps other particles at very low energies), θ_{max} should be put equal to unity in (13.61). Then instead of $(204Z^{-1/3})$, the argument of the logarithm in (13.62) becomes $\left(\dfrac{192}{Z^{1/3}}\dfrac{p}{mc}\right)^{1/2}$.

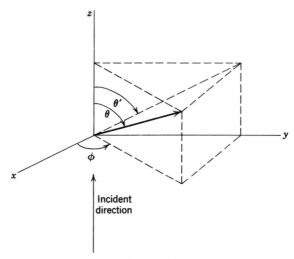

Figure 13.7

It is often desirable to use the projected angle of scattering θ', the projection being made on some convenient plane such as the plane of a photographic emulsion or a bubble chamber, as shown in Fig. 13.7. For small angles it is easy to show that

$$\langle \theta'^2 \rangle = \tfrac{1}{2} \langle \theta^2 \rangle \tag{13.63}$$

In each collision the angular deflections obey the Rutherford formula (13.52) suitably cut off at θ_{min} and θ_{max}, with average value zero (when viewed relative to the forward direction, or as a projected angle) and mean square angle $\langle \theta^2 \rangle$ given by (13.61). Since the successive collisions are independent events, the central-limit theorem of statistics can be used to show that for a large number n of such collisions the distribution in angle will be approximately Gaussian around the forward direction with a mean square angle $\langle \Theta^2 \rangle = n\langle \theta^2 \rangle$. The number of collisions occurring as the particle traverses a thickness t of material containing N atoms per unit volume is

$$n = N\sigma t \simeq \pi N \left(\frac{2zZe^2}{pv} \right)^2 \frac{t}{\theta_{min}^2} \tag{13.64}$$

This means that the mean square angle of the Gaussian is

$$\langle \Theta^2 \rangle \simeq 2\pi N \left(\frac{2zZe^2}{pv} \right)^2 \ln\left(\frac{\theta_{max}}{\theta_{min}} \right) t \tag{13.65}$$

Or, using (13.62) for $\langle \theta^2 \rangle$,

$$\langle \Theta^2 \rangle \simeq 4\pi N \left(\frac{2zZe^2}{pv} \right)^2 \ln(204 Z^{-1/3}) t \tag{13.66}$$

The mean square angle increases linearly with the thickness t. But for reasonable thicknesses such that the particle does not lose appreciable energy, the Gaussian will still be peaked at very small forward angles. Parenthetically, we remark that the numerical coefficient in the logarithm can differ from author to author—for example, *Rossi* has 175 instead of 204. We also note that in practice the Gaussian approximation holds only for large n—see the last paragraph of this section for some elaboration on this point.

The multiple-scattering distribution for the projected angle of scattering is

$$P_M(\theta') \, d\theta' = \frac{1}{\sqrt{\pi \langle \Theta^2 \rangle}} \exp\left(-\frac{\theta'^2}{\langle \Theta^2 \rangle}\right) d\theta' \tag{13.67}$$

where both positive and negative values of θ' are considered. The small-angle Rutherford formula (13.52) can be expressed in terms of the projected angle as

$$\frac{d\sigma}{d\theta'} = \frac{\pi}{2}\left(\frac{2zZe^2}{pv}\right)^2 \frac{1}{\theta'^3} \tag{13.68}$$

This gives a single-scattering distribution for the projected angle:

$$P_S(\theta') \, d\theta' = Nt \frac{d\sigma}{d\theta'} d\theta' = \frac{\pi}{2} Nt \left(\frac{2zZe^2}{pv}\right)^2 \frac{d\theta'}{\theta'^3} \tag{13.69}$$

The single-scattering distribution is valid only for angles large compared to $\langle \Theta^2 \rangle^{1/2}$ and contributes a tail to the Gaussian distribution.

If we express angles in terms of the relative projected angle,

$$\alpha = \frac{\theta'}{\langle \Theta^2 \rangle^{1/2}} \tag{13.70}$$

the multiple- and single-scattering distributions can be written

$$P_M(\alpha) \, d\alpha = \frac{1}{\sqrt{\pi}} e^{-\alpha^2} \, d\alpha \tag{13.71}$$

$$P_S(\alpha) \, d\alpha = \frac{1}{8 \ln(204 Z^{-1/3})} \frac{d\alpha}{\alpha^3}$$

where (13.66) has been used for $\langle \Theta^2 \rangle$. We note that the relative amounts of multiple and single scatterings are independent of thickness in these units, and depend only on Z. Even this Z dependence is not marked. The factor $8 \ln(204 Z^{-1/3})$ has the value 36 for $Z = 13$ (aluminum) and the value 31 for $Z = 82$ (lead). Figure 13.8 shows the general behavior of the scattering distribution as a function of α. The transition from multiple to single scattering occurs in the neighborhood of $\alpha \simeq 2.5$. At this point the Gaussian has a value of 1/600 times its peak value. Thus the single-scattering distribution gives only a very small tail on the multiple-scattering curve.

There are two things that cause departures from the simple behavior shown in Fig. 13.8. The Gaussian shape is the limiting form of the angular distribution for very large n. If the thickness t is such that n (13.64) is not very large (i.e., $n \lesssim 200$), the distribution follows the single-scattering curve to smaller angles than $\alpha \simeq 2.5$, and is more sharply peaked at zero angle than a Gaussian.* On the other hand, if the thickness is great enough, the mean square angle $\langle \Theta^2 \rangle$ becomes comparable with the angle θ_{max} (13.57) which limits the angular width of the single-scattering distribution. For greater thicknesses the multiple-scatter-

---

*For numerical evaluation for very thin samples (e.g., gases), see P. Sigmund and K. B. Winterbon, *Nucl. Instrum. Methods* **119**, 541–557 (1974).

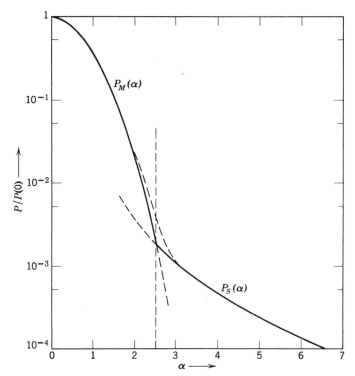

Figure 13.8 Multiple- and single-scattering distributions of projected angle. In the region of plural scattering ($\alpha \sim 2$–3) the dashed curve indicates the smooth transition from the small-angle multiple scattering (approximately Gaussian in shape) to the wide-angle single scattering (proportional to α^{-3}).

ing curve extends in angle beyond the single-scattering region, so that there is no single-scattering tail on the distribution (see Problem 13.8).

13.7 Transition Radiation

A charged particle in uniform motion in a straight line in free space does not radiate. It was shown in Section 13.4, however, that a particle moving at constant velocity can radiate if it is in a material medium and is moving with a speed greater than the phase velocity of light in that medium. This radiation, with its characteristic angle of emission, $\theta_C = \sec^{-1}(\beta\epsilon^{1/2})$, is Cherenkov radiation. There is another type of radiation, *transition radiation*, first noted by Ginsburg and Frank in 1946, that is emitted whenever a charged particle passes suddenly from one medium into another. Far from the boundary in the first medium, the particle has certain fields characteristic of its motion and of that medium. Later, when it is deep in the second medium, it has fields appropriate to its motion and that medium. Even if the motion is uniform throughout, the initial and final fields will be different if the two media have different electromagnetic properties. Evidently the fields must reorganize themselves as the particle approaches and passes through the interface. In this process of reorganization some pieces of the fields are shaken off as transition radiation.

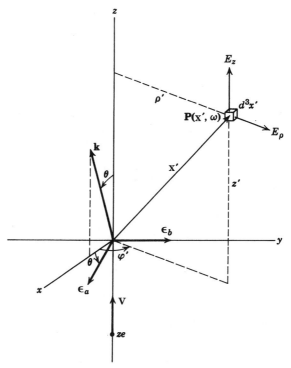

Figure 13.9 A charged particle of charge ze and velocity $\mathbf{v}$ is normally incident along the z axis on a uniform semi-infinite dielectric medium occupying the half-space $z > 0$. The transition radiation is observed at angle θ with respect to the direction of motion of the particle, as specified by the wave vector $\mathbf{k}$ and associated polarization vectors $\boldsymbol{\epsilon}_a$ and $\boldsymbol{\epsilon}_b$.

Important features of transition radiation can be understood without elaborate calculation.* We consider a relativistic particle with charge ze and speed $v = \beta c$ normally incident along the z axis from vacuum ($z < 0$) on a uniform semi-infinite medium ($z > 0$) with index of refraction $n(\omega)$, as indicated in Fig. 13.9. The moving fields of the charged particle induce a time-dependent polarization $\mathbf{P}(\mathbf{x}', t)$ in the medium. The polarization emits radiation. The radiated fields from different points in space combine coherently in the neighborhood of the path and for a certain depth in the medium, giving rise to transition radiation with a characteristic angular distribution and intensity.

The angular distribution and the *formation length D* are a direct consequence of the requirement of coherence for appreciable radiated intensity. The exciting fields of the incident particle are given by (11.152). The dependence at a point $\mathbf{x}' = (z', \rho', \phi')$ on inverse powers of $[\rho'^2 + \gamma^2(z' - vt)^2]$ implies that a Fourier component of frequency ω (a) will move in the z direction with velocity $\mathbf{v}$ and so have an amplitude proportional to $e^{i\omega z'/v}$, and (b) will have significant magnitude radially from the path only out to distances of the order of $\rho'_{max} \simeq \gamma v/\omega$. On the

*The need for a qualitative discussion has been impressed on me by numerous questions from colleagues near and far and by V. F. Weisskopf on the occasion of a seminar by him where he presented a similar discussion.

other hand, the time-dependent polarization at $\mathbf{x}'$ generates a wave whose form in the radiation zone is

$$A = \frac{e^{ikr}}{r} \exp[-ik(z' \cos \theta + \rho' \sin \theta \cos \phi')]$$

where A is proportional to the driving field of the incident particle, $k = n(\omega)\omega/c$ and it is assumed that the radiation is observed in the x-z plane and in the forward hemisphere. Appreciable coherent superposition from different points in the medium will occur provided the product of the driving fields of the particle and the generated wave does not change its phase significantly over the region. The relevant factor in the amplitude is

$$\exp\left(i \frac{\omega}{v} z'\right) \exp\left[-i \frac{\omega}{c} n(\omega) \cos \theta \, z'\right] \exp\left[-i \frac{\omega}{c} n(\omega)\rho' \sin \theta \cos \phi'\right]$$

$$= \exp\left\{i \frac{\omega}{c}\left[\frac{1}{\beta} - n(\omega) \cos \theta\right]z'\right\} \exp\left[-i \frac{\omega}{c} n(\omega)\rho' \sin \theta \cos \phi'\right]$$

In the radial direction coherence will be maintained only if the phase involving ρ' is unity or less in the region $0 < \rho' \lesssim \rho'_{\max}$ where the exciting field is appreciable. Thus radiation will not be appreciable unless

$$\frac{\omega}{c} n(\omega) \frac{\gamma v}{\omega} \sin \theta \lesssim 1$$

or

$$n(\omega)\gamma\theta \lesssim 1 \tag{13.72}$$

for $\gamma \gg 1$. The angular distribution is therefore confined to the forward cone, $\gamma\theta \lesssim 1$, as in all relativistic emission processes.

The z'-dependent factor in the amplitude is

$$\exp\left\{i \frac{\omega}{c}\left[\frac{1}{\beta} - n(\omega) \cos \theta\right]z'\right\}$$

The depth $d(\omega)$ up to which coherence is maintained is therefore

$$\frac{\omega}{c}\left[\frac{1}{\beta} - n(\omega) \cos \theta\right] d(\omega) \simeq 1$$

We approximate $n(\omega) \simeq 1 - (\omega_p^2/2\omega^2)$ for frequencies above the optical region where Cherenkov radiation does not occur, $\beta^{-1} \simeq 1 + 1/2\gamma^2$ for a relativistic particle, and $\cos \theta \simeq 1$, to obtain

$$d(\nu) \simeq \frac{2\gamma c/\omega_p}{\nu + \nu^{-1}} \tag{13.73}$$

where we have introduced a dimensionless frequency variable,

$$\nu = \frac{\omega}{\gamma\omega_p} \tag{13.74}$$

We define the *formation length* D as the largest value of $d(\nu)$ as a function of ν:

$$D = d(1) = \frac{\gamma c}{\omega_p} \tag{13.75}$$

For substances with densities of order of unity, the plasma frequency is $\omega_p \simeq 3 \times 10^{16}\ \text{s}^{-1}$, corresponding to an energy $\hbar\omega_p \simeq 20\ \text{eV}$. Thus $c/\omega_p \simeq 10^{-6}\ \text{cm}$ and even for $\gamma \gtrsim 10^3$ the formation length D is only tens of micrometers. In air at NTP it is a factor of 30 larger because of the reduced density.

The coherence volume adjacent to the particle's path and the surface from which transition radiation of frequency ω comes is evidently

$$V(\omega) \sim \pi\rho_{\max}^2(\omega)\ d(\omega) \sim 2\pi\gamma\left(\frac{c}{\omega_p}\right)^3 \frac{1}{\nu(1 + \nu^2)}$$

This volume decreases in size rapidly for $\nu > 1$. We can therefore expect that in the absence of compensating factors, the spectrum of transition radiation will extend up to, but not appreciably beyond, $\nu \simeq 1$.

We have obtained some insight into the mechanism of transition radiation and its main features. It is confined to small angles in the forward direction ($\gamma\theta \lesssim 1$). It is produced by coherent radiation of the time-varying polarization in a small volume adjacent to the particle's path and at depths into the medium up to the formation length D. Its spectrum extends up to frequencies of the order of $\omega \sim \gamma\omega_p$. It is possible to continue these qualitative arguments and obtain an estimate of the total energy radiated, but the exercise begins to have the appearance of virtuosity based on hindsight. Instead, we turn to an actual calculation of the phenomenon.

An exact calculation of transition radiation is complicated. Some references are given at the end of the chapter. We content ourselves with an approximate calculation that is adequate for most applications and is physically transparent. It is based on the observation that for frequencies above the optical resonance region, the index of refraction is not far from unity. The incident particle's fields at such frequencies are not significantly different in the medium and in vacuum. This means that the Fourier component of the induced polarization $\mathbf{P}(\mathbf{x}', \omega)$ can be evaluated approximately by

$$\mathbf{P}(\mathbf{x}', \omega) \simeq \left[\frac{\epsilon(\omega) - 1}{4\pi}\right]\mathbf{E}_i(\mathbf{x}', \omega) \tag{13.76}$$

where $\mathbf{E}_i$ is the Fourier transform of the electric field of the incident particle in vacuum. The *propagation* of the wave radiated by the polarization must be described properly, however, with the wave number $k = \omega n(\omega)/c$ appropriate to the medium. This is because phase differences are important, as already seen in the qualitative discussion.

The dipole radiation field from the polarization $\mathbf{P}(\mathbf{x}', \omega)\ d^3x'$ in the volume element d^3x' at $\mathbf{x}'$ is, according to (9.18),

$$d\mathbf{E}_{\text{rad}} = \frac{e^{ikR}}{R}\ (\mathbf{k} \times \mathbf{P}) \times \mathbf{k}\ d^3x'$$

where $\mathbf{k}$ is the wave vector in the direction of observation and $R \simeq r - \hat{\mathbf{k}} \cdot \mathbf{x}'$. With the substitution of (13.76) and an integration over the half-space $z' > 0$, the total radiated field at frequency ω is

$$\mathbf{E}_{\text{rad}} = \frac{e^{ikr}}{r} \left[\frac{\epsilon(\omega) - 1}{4\pi} \right] k^2 \int_{z'>0} (\hat{\mathbf{k}} \times \mathbf{E}_i) \times \hat{\mathbf{k}} e^{-i\mathbf{k}\cdot\mathbf{x}'} d^3x'$$

With the approximation,

$$\epsilon(\omega) \simeq 1 - \frac{\omega_p^2}{\omega^2} \tag{13.77}$$

the radiated field for $\omega > \omega_p$ becomes

$$\mathbf{E}_{\text{rad}} \simeq \frac{e^{ikr}}{r} \left(\frac{-\omega_p^2}{4\pi c^2} \right) \int_{z'>0} (\hat{\mathbf{k}} \times \mathbf{E}_i) \times \hat{\mathbf{k}} e^{-i\mathbf{k}\cdot\mathbf{x}'} d^3x' \tag{13.78}$$

From equations given later [see (14.52) and (14.60)], this means that the energy radiated has the differential spectrum in an angle and energy,

$$\frac{d^2 I}{d\omega \, d\Omega} = \frac{c}{32\pi^3} \left(\frac{\omega_p}{c} \right)^4 \left| \int_{z>0} [\hat{\mathbf{k}} \times \mathbf{E}_i(\mathbf{x}, \omega)] \times \hat{\mathbf{k}} e^{-i\mathbf{k}\cdot\mathbf{x}} d^3x \right|^2 \tag{13.79}$$

Note that the driving fields $\mathbf{E}_i$ are defined by the Fourier transform of the fields of Section 11.10. In our approximation it is not necessary to use the more elaborate fields of Sections 13.3 and 13.4. In the notation of Fig. 13.9 the incident fields are (see Problems 13.2 and 13.3)

$$E_\rho(\mathbf{x}, \omega) = \sqrt{\frac{2}{\pi}} \frac{ze\omega}{\gamma v^2} e^{i\omega z/v} K_1 \left(\frac{\omega\rho}{\gamma v} \right)$$

$$E_z(\mathbf{x}, \omega) = -i\sqrt{\frac{2}{\pi}} \frac{ze\omega}{\gamma^2 v^2} e^{i\omega z/v} K_0 \left(\frac{\omega\rho}{\gamma v} \right) \tag{13.80}$$

The integral in (13.79) can be evaluated as follows. We first exploit the fact that the z dependence of $\mathbf{E}_i$ is only via the factor $e^{i\omega z/v}$, and write

$$\mathbf{F} \equiv \int_{z>0} [\hat{\mathbf{k}} \times \mathbf{E}_i(\mathbf{x}, \omega)] \times \hat{\mathbf{k}} e^{-i\mathbf{k}\cdot\mathbf{x}} d^3x$$

$$= \int dx \int dy [\hat{\mathbf{k}} \times \mathbf{E}_i]_{z=0} \times \hat{\mathbf{k}} \, e^{-ikx\sin\theta} \int_0^\infty dz \, \exp\left[i\left(\frac{\omega}{v} - k\cos\theta \right) z \right]$$

$$= \frac{i\left\{ 1 - \exp\left[i\left(\frac{\omega}{v} - k\cos\theta \right) Z \right] \right\}}{\frac{\omega}{v} - k\cos\theta} \int dx \int dy [\hat{\mathbf{k}} \times \mathbf{E}_i]_{z=0} \times \hat{\mathbf{k}} \, e^{-ikx\sin\theta}$$

The upper limit Z on the z integration is a formal device to show that the contributions from different z values add constructively and cause the amplitude to grow until $Z \gtrsim D$. Beyond the depth D the rapidly rotating phase prevents further enhancement. For effectively semi-infinite media (slabs of thickness large compared with D) we drop the oscillating exponential in Z on physical grounds* and obtain, for a *single interface*,

$$\mathbf{F} = \frac{i}{\left(\dfrac{\omega}{v} - k \cos \theta\right)} \iint dx \, dy \, [\hat{\mathbf{k}} \times \mathbf{E}_i]_{z=0} \times \hat{\mathbf{k}} \, e^{-ikx\sin\theta}$$

The electric field transverse to $\hat{\mathbf{k}}$ can be expressed in terms of the components E_ρ, E_z and the polarization vectors $\boldsymbol{\epsilon}_a$ and $\boldsymbol{\epsilon}_b$ shown in Fig. 13.9 as

$$[\hat{\mathbf{k}} \times \mathbf{E}_i] \times \hat{\mathbf{k}} = (E_\rho \cos \theta \cos \phi - E_z \sin \theta)\boldsymbol{\epsilon}_a + E_\rho \sin \phi \boldsymbol{\epsilon}_b$$

where θ is the polar angle of $\hat{\mathbf{k}}$ and the prime has been dropped from the azimuthal angle of integration. The component parallel to $\boldsymbol{\epsilon}_b$ integrates to zero because it is odd in y. Thus, substituting from (13.80), we have

$$\mathbf{F} = \frac{i\boldsymbol{\epsilon}_a}{\left(\dfrac{\omega}{v} - k \cos \theta\right)} \iint dx \, e^{-ikx\sin\theta} \left[\cos \theta \frac{x}{\sqrt{x^2 + y^2}} E_\rho - \sin \theta E_z\right]_{z=0}$$

$$= \frac{i\boldsymbol{\epsilon}_a}{\left(\dfrac{\omega}{v} - k \cos \theta\right)} \sqrt{\frac{2}{\pi}} \frac{ze\omega}{\gamma v^2} \iint dx \, dy \, e^{-ikx\sin\theta}$$

$$\times \left[\cos \theta \frac{x}{\sqrt{x^2 + y^2}} K_1\left(\frac{\omega}{\gamma v} \sqrt{x^2 + y^2}\right) + i \frac{\sin \theta}{\gamma} K_0\left(\frac{\omega}{\gamma v} \sqrt{x^2 + y^2}\right)\right]$$

The first term can be transformed by an integration by parts in x, using

$$\frac{x}{\sqrt{x^2 + y^2}} K_1\left(\frac{\omega}{\gamma v} \sqrt{x^2 + y^2}\right) = -\frac{\gamma v}{\omega} \frac{\partial}{\partial x} K_0\left(\frac{\omega}{\gamma v} \sqrt{x^2 + y^2}\right)$$

so that

$$\mathbf{F} = \boldsymbol{\epsilon}_a \sqrt{\frac{2}{\pi}} \frac{ze \sin \theta \left(k \cos \theta - \dfrac{\omega}{v\gamma^2}\right)}{v\left(\dfrac{\omega}{v} - k \cos \theta\right)} \iint dx \, dy \, e^{-ikx\sin\theta} K_0\left(\frac{\omega}{\gamma v} \sqrt{x^2 + y^2}\right)$$

The remaining integral can be evaluated from the cosine transform,

$$\int_0^\infty K_0(\beta\sqrt{z^2 + t^2}) \cos(\alpha z) dz = \frac{\pi}{2\sqrt{\alpha^2 + \beta^2}} \exp(-|t|\sqrt{\alpha^2 + \beta^2}) \quad (13.81)$$

The result for $\mathbf{F}$ is

$$\mathbf{F} = \boldsymbol{\epsilon}_a \frac{2\sqrt{2\pi} \, ze \sin \theta \left(k \cos \theta - \dfrac{\omega}{v\gamma^2}\right)}{v\left(\dfrac{\omega}{v} - k \cos \theta\right)\left(\dfrac{\omega^2}{\gamma^2 v^2} + k^2 \sin^2\theta\right)} \quad (13.82)$$

---

*A less cavalier treatment of the dependence on thickness is necessary for foils that are not thick compared to D, or when a stack of foils is employed. See Problems 13.13 and 13.14.

In the approximation of relativistic motion ($\gamma \gg 1$), small angles ($\theta \ll 1$), and high frequencies ($\omega \gg \omega_p$), this becomes

$$\mathbf{F} \simeq \epsilon_a 4\sqrt{2\pi}\, \frac{ze}{c}\left(\frac{c}{\omega_p}\right)^2 \frac{\gamma}{\nu^2}\, \frac{\sqrt{\eta}}{\left(1 + \dfrac{1}{\nu^2} + \eta\right)(1 + \eta)} \tag{13.83}$$

where ν is the dimensionless frequency variable (13.74) and $\eta = (\gamma\theta)^2$ is an appropriate angular variable. With $d\Omega = d\phi\, d(\cos\theta) \simeq d\phi\, d\eta/2\gamma^2$, the energy distribution in ν and η is

$$\frac{d^2 I}{d\nu\, d\eta} = \frac{\pi}{\gamma^2} \cdot \gamma\omega_p \cdot \frac{d^2 I}{d\omega\, d\Omega} \tag{13.84}$$

$$\simeq \frac{z^2 e^2 \gamma\omega_p}{\pi c} \cdot \left[\frac{\eta}{\nu^4\left(1 + \dfrac{1}{\nu^2} + \eta\right)^2 (1 + \eta)^2}\right]$$

Angular distributions for fixed ν values are shown in Fig. 13.10. At low frequencies the spectrum peaks at $\eta \simeq 1$ and then falls relatively slowly as η^{-1} until the value $\eta = \nu^{-2}$ is reached. Then it falls off as η^{-3}. For $\nu \gtrsim 1$, the spectrum peaks at $\eta \gtrsim \frac{1}{3}$ and falls at η^{-3} for $\eta \gg 1$. At $\eta = 0$ the denominator in (13.84) is

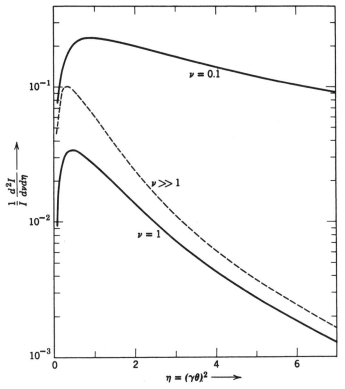

Figure 13.10   Angular distributions of transition radiation at $\nu = 0.1$, $\nu = 1$ and $\nu \gg 1$. The solid curves are the normalized angular distributions, that is, the ratio of (13.84) to (13.87). The dashed curve is ν^4 times that ratio in the limit $\nu \to \infty$.

$(1 + \nu^2)^2$, showing that for $\nu \gg 1$ there is negligible intensity at any angle [cf. coherence volume $V(\omega)$, above].

The energy spectrum, integrated over the angular variable η, is

$$\frac{dI}{d\nu} = \frac{z^2 e^2 \gamma \omega_p}{\pi c} \left[(1 + 2\nu^2) \ln\left(1 + \frac{1}{\nu^2}\right) - 2 \right] \tag{13.85}$$

It has the small and large ν limits,

$$\frac{dI}{d\nu} \simeq \frac{z^2 e^2 \gamma \omega_p}{\pi c} \begin{cases} 2 \ln(1/e\nu), & \nu \ll 1 \\ \dfrac{1}{6\nu^4}, & \nu \gg 1 \end{cases} \tag{13.86}$$

The energy spectrum is shown on a log-log plot in Fig. 13.11. The spectrum diverges logarithmically at low frequencies, where our approximate treatment fails in any event, but it has a finite integral. The total energy emitted in transition radiation per interface is

$$I = \int_0^\infty \frac{dI}{d\nu} \, d\nu = \frac{z^2 e^2 \gamma \omega_p}{3c} = \frac{z^2}{3(137)} \gamma \hbar \omega_p \tag{13.87}$$

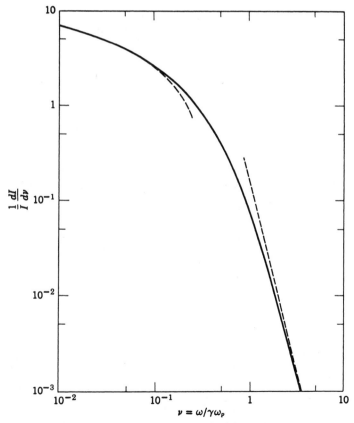

Figure 13.11   Normalized frequency distribution $(1/I)(dI/d\nu)$ of transition radiation as a function of $\nu = \omega/\gamma\omega_p$. The dashed curves are the two approximate expressions in (13.86).

From Fig. 13.11 we can estimate that about half the energy is emitted in the range $0.1 \leq \nu \leq 1$. In quantum language, we say that an appreciable fraction of the energy appears as comparatively energetic photons. For example, with $\gamma = 10^3$ and $\hbar\omega_p = 20$ eV, these quanta are in the soft x-ray region of 2 to 20 keV.

The presence of the factor of γ in (13.87) makes transition radiation attractive as a mechanism for the identification of particles, and perhaps even measurement of their energies, at very high energies where other means are unavailable. The presence of the numerical factor $1/(3 \times 137)$ means that the probability of energetic photon emission per transit of an interface is very small. It is necessary to utilize a stack of many foils with gaps between. The foils can be quite thin, needing to be thick only compared to a formation length D (13.75). Then a particle traversing each foil will emit *twice* (13.87) in transition radiation (see Problem 13.13). A typical set-up might involve 200 Mylar foils of thickness 20 μm, with spacings 150–300 μm.* The coherent superposition of the fields from the different interfaces, two for each foil, causes a modulation of the energy and angular distributions (see Problem 13.14).

References and Suggested Reading

The problems of the penetration of particles through matter interested Niels Bohr all his life. A lovely presentation of the whole subject, with characteristic emphasis on the interplay of classical and quantum-mechanical effects, appears in his comprehensive review article of 1948:

Bohr

Numerical tables and graphs of energy-loss data, as well as key formulas, are given by

Rossi, Chapter 2
Segre, article by H. A. Bethe and J. Ashkin

See also

U. Fano, *Annu. Rev. Nucl. Sci.* **13**, 1 (1963).

Rossi gives a semiclassical treatment of energy loss and scattering similar to ours. He also considers the question of fluctuations in energy loss, including the Landau–Symon theory.

The density effect on the energy loss by extremely relativistic particles is discussed, with numerous results for different substances in graphical form, by

R. M. Sternheimer, in *Methods of Experimental Physics*, Vol. 5A, *Nuclear Physics, Part A*, eds. L. C. L. Yuan and C. S. Wu, Academic Press, New York (1961), pp. 4–55.

Cherenkov radiation is discussed in many places. Its application to particle detectors is described in the book by Yuan and Wu, just mentioned, and also in

D. M. Ritson, ed., *Techniques in High Energy Physics*, Interscience, New York (1961).

Transition radiation is reviewed with extensive bibliographies by

I. M. Frank, *Usp. Fiz. Nauk* **87**, 189 (1965) [transl. *Sov. Phys. Usp.* **8**, 729 (1966)].
F. G. Bass and V. M. Yakovenko, *Usp. Fiz. Nauk* **86**, 189 (1965) [transl. *Sov. Phys. Usp.* **8**, 420 (1965)].

---

*Some examples of practical devices can be found in H. Pieharz, *Nucl. Instrum. Methods A* **367**, 220 (1995), W. Brückner et al., *Nucl. Instrum. Methods A* **378**, 451 (1996), and J. Ruzicka, L. Krupa, and V. A. Fadejev, *Nucl. Instrum. Methods A* **384**, 387 (1997).

The calculation of transition radiation from the traversal of interstellar dust grains by energetic particles, done in the same approximation as in Section 13.7, is given by
L. Durand, *Astrophys. J.* **182**, 417 (1973).

A review of both Cherenkov radiation and transition radiation with much history, is given by
V. L. Ginsburg, *Usp. Fiz. Nauk* **166**, 1033 (1996) [transl. *Phys. Usp.* **39**, 973 (1996)].
For current applications of both Cherenkov and transition radiation, however, the reader must turn to specialized journals such as *Nuclear Instruments and Methods A*. Volume 367 of that journal (1995), a conference proceedings, contains descriptions of several particle physics detectors based on these and other principles.

Problems

13.1   If the light particle (electron) in the Coulomb scattering of Section 13.1 is treated classically, scattering through an angle θ is correlated uniquely to an incident trajectory of impact parameter b according to

$$b = \frac{ze^2}{pv} \cot\frac{\theta}{2}$$

where $p = \gamma m v$ and the differential scattering cross section is $\frac{d\sigma}{d\Omega} = \frac{b}{\sin\theta}\left|\frac{db}{d\theta}\right|$.

(a)   Express the invariant momentum transfer squared in terms of impact parameter and show that the energy transfer $T(b)$ is

$$T(b) = \frac{2z^2e^4}{mv^2} \cdot \frac{1}{b^2 + b^{(c)2}_{\min}}$$

where $b^{(c)}_{\min} = ze^2/pv$ and $T(0) = T_{\max} = 2\gamma^2\beta^2 mc^2$.

(b)   Calculate the small transverse impulse Δp given to the (nearly stationary) light particle by the transverse electric field (11.152) of the heavy particle $q = ze$ as it passes by at large impact parameter b in a (nearly) straight line path at speed v. Find the energy transfer $T \approx (\Delta p)^2/2m$ in terms of b. Compare with the exact classical result of part a. Comment.

13.2   Time-varying electromagnetic fields $\mathbf{E}(\mathbf{x}, t)$ and $\mathbf{B}(\mathbf{x}, t)$ of finite duration act on a charged particle of charge e and mass m bound harmonically to the origin with natural frequency ω_0 and small damping constant Γ. The fields may be caused by a passing charged particle or some other external source. The charge's motion in response to the fields is nonrelativistic and small in amplitude compared to the scale of spatial variation of the fields (dipole approximation). Show that the energy transferred to the oscillator in the limit of very small damping is

$$\Delta E = \frac{\pi e^2}{m}|\mathbf{E}(\omega_0)|^2$$

where $\mathbf{E}(\omega)$ is the symmetric Fourier transform of $\mathbf{E}(0, t)$:

$$\mathbf{E}(0, t) = \frac{1}{\sqrt{2\pi}}\int_{-\infty}^{\infty} \mathbf{E}(\omega)e^{-i\omega t}\, d\omega; \qquad \mathbf{E}(\omega) = \frac{1}{\sqrt{2\pi}}\int_{-\infty}^{\infty} \mathbf{E}(0, t)e^{i\omega t}\, dt$$

13.3 The external fields of Problem 13.2 are caused by a charge ze passing the origin in a straight-line path at speed v and impact parameter b. The fields are given by (11.152).

(a) Evaluate the Fourier transforms for the perpendicular and parallel components of the electric field at the origin and show that

$$E_\perp(\omega) = \frac{ze}{bv}\left(\frac{2}{\pi}\right)^{1/2} \xi K_1(\xi); \qquad E_\|(\omega) = -i\,\frac{ze}{\gamma bv}\left(\frac{2}{\pi}\right)^{1/2} \xi K_0(\xi)$$

where $\xi = \omega b/\gamma v$, and $K_\nu(\xi)$ is the modified Bessel function of the second kind and order ν. [See references to tables of Fourier transforms in Section 13.3.]

(b) Using the result of Problem 13.2, write down the energy transfer ΔE to a harmonically bound charged particle. From the limiting forms of the modified Bessel functions for small and large argument, show that your result agrees with the appropriate limit of $T(b)$ in Problem 13.1 on the one hand and the arguments at the end of Section 13.1 on the adiabatic behavior for $b \gg \gamma v/\omega_0$ on the other.

13.4 (a) Taking $\hbar\langle\omega\rangle = 12Z$ eV in the quantum-mechanical energy-loss formula, calculate the rate of energy loss (in MeV/cm) in air at NTP, aluminum, copper, and lead for a proton and a mu meson, each with kinetic energies of 10, 100, 1000 MeV.

(b) Convert your results to energy loss in units of MeV · (cm²/g) and compare the values obtained in different materials. Explain why all the energy losses in MeV-(cm²/g) are within a factor of 2 of each other, whereas the values in MeV/cm differ greatly.

13.5 Consider the energy loss by close collisions of a fast, but nonrelativistic, heavy particle of charge ze passing through an electronic plasma. Assume that the screened Coulomb interaction $V(r) = ze^2 \exp(-k_D r)/r$, where k_D is the Debye screening parameter, acts between the electrons and the incident particle.

(a) Show that the energy transfer in a collision at impact parameter b is given approximately by

$$\Delta E(b) \simeq \frac{2(ze^2)^2}{mv^2}\, k_D^2 K_1^2(k_D b)$$

where m is the electron mass and v is the velocity of the incident particle.

(b) Determine the energy loss per unit distance traveled for collisions with impact parameter greater than b_{min}. Assuming $k_D b_{min} \ll 1$, show that

$$\left(\frac{dE}{dx}\right)_{k_D b<1} \simeq \frac{(ze)^2}{v^2}\, \omega_p^2 \ln\left(\frac{1}{1.47 k_D b_{min}}\right)$$

where b_{min} is given by the larger of the classical and quantum minimum impact parameters [(13.16) and above].

13.6 The energy loss in a plasma from distant collisions can be found with Fermi's method for the density effect. Consider the nonrelativistic limit of (13.36) with the relative dielectric constant of a plasma given by (7.59) augmented by some damping,

$$\varepsilon(\omega) = 1 - \frac{\omega_p^2}{\omega^2 + i\omega\Gamma}$$

Assume that the arguments of the Bessel functions are small (corresponding to a speed of the incident particle large compared to thermal speeds in the plasma).

(a) Show that the energy loss (13.36) for $k_D b > 1$ becomes

$$\left(\frac{dE}{dx}\right)_{k_D b > 1} \approx \frac{2z^2 e^2}{\pi v^2} \int_0^\infty \mathrm{Re}\left(\frac{i\omega}{\varepsilon(\omega)}\right) \ln\left(\frac{1.123 k_D v}{\omega}\right) d\omega$$

(b) With the assumption that $\Gamma \ll \omega_p$ in $\varepsilon(\omega)$, show that the formula of part a yields

$$\left(\frac{dE}{dx}\right)_{k_D b > 1} \approx \frac{z^2 e^2}{v^2} \omega_p^2 \ln\left(\frac{1.123 k_D v}{\omega_p}\right)$$

Combine with the close-collision result of Problem 13.5 to find the total energy loss of a nonrelativistic particle passing through a plasma,

$$\left(\frac{dE}{dx}\right) \approx \frac{z^2 e^2}{v^2} \omega_p^2 \ln\left(\frac{\Lambda v}{\omega_p b_{min}}\right)$$

where Λ is a number of order unity. The presence of ω_p in the logarithm suggests that the energy loss may be quantized in units of $\hbar \omega_p$. In fact, electrons passing through thin metallic foils do show this discreteness in energy loss, allowing determination of effective plasma frequencies in metals. [See H. Raether, *Springer Tracts in Modern Physics*, Vol. 38, ed. G. Höhler, Springer-Verlag, Berlin (1965), pp. 84–157.]

13.7 With the same approximations as were used to discuss multiple scattering, show that the *projected* transverse displacement y (see Fig. 13.7) of an incident particle is described approximately by a Gaussian distribution,

$$P(y)\, dy = A\, \exp\left[\frac{-y^2}{2\langle y^2 \rangle}\right] dy$$

where the mean square displacement is $\langle y^2 \rangle = (x^2/6)\langle \Theta^2 \rangle$, x being the thickness of the material traversed and $\langle \Theta^2 \rangle$ the mean square angle of scattering.

13.8 If the finite size of the nucleus is taken into account in the "single-scattering" tail of the multiple-scattering distribution, there is a critical thickness x_c beyond which the single-scattering tail is absent.

(a) Define x_c in a reasonable way and calculate its value (in cm) for aluminum and lead, assuming that the incident particle is relativistic.

(b) For these thicknesses calculate the number of collisions that occur and determine whether the Gaussian approximation is valid.

13.9 Assuming that Plexiglas or Lucite has an index of retraction of 1.50 in the visible region, compute the angle of emission of visible Cherenkov radiation for electrons and protons as a function of their kinetic energies in MeV. Determine how many quanta with wavelengths between 4000 and 6000 Å are emitted per centimeter of path in Lucite by a 1 MeV electron, a 500 MeV proton, and a 5 GeV proton.

13.10 A particle of charge ze moves along the z axis with constant speed v, passing $z = 0$ at $t = 0$. The medium through which the particle moves is described by a dielectric constant $\epsilon(\omega)$.

(a) Beginning with the potential $\Phi(\mathbf{k}, \omega)$ of (13.25), show that the potential of frequency ω is given as a function of spatial coordinate $\mathbf{x}$ by

$$\Phi(\omega, \mathbf{x}) = \frac{ze}{v\epsilon(\omega)} \sqrt{\frac{2}{\pi}} K_0\left(\frac{|\omega|\,\rho}{v}\sqrt{1 - \beta^2 \epsilon}\right) e^{i\omega z/v}$$

where z and $\rho = \sqrt{x^2 + y^2}$ are the cylindrical coordinates of the observation point.

(b) Assuming that ϵ is independent of frequency and that $\beta^2\epsilon < 1$, take the Fourier transform with respect to ω of the expression in part a and obtain $\Phi(\mathbf{x}, t)$. Calculate the electric and magnetic fields and compare them to the vacuum fields (11.152). Show that, among other things, the vacuum factor γ is replaced by $\Gamma = (1 - \beta^2\epsilon)^{-1/2}$.

(c) Repeat the calculations of parts a and b with $\beta^2\epsilon > 1$. Show that now

$$\Phi(\omega, \mathbf{x}) = \frac{ze}{v\epsilon(\omega)} \sqrt{\frac{\pi}{2}} e^{i\omega z/v} \left[-N_0\left(\frac{|\omega|\,\rho}{v}\sqrt{\beta^2\epsilon - 1}\right) \pm iJ_0\left(\frac{|\omega|\,\rho}{v}\sqrt{\beta^2\epsilon - 1}\right) \right]$$

for $\omega \gtrless 0$. Calculate the remaining Fourier transform to obtain $\Phi(\mathbf{x}, t)$. Relate your answer to the result given in Section 13.4 for $\mathbf{A}(\mathbf{x}, t)$.

13.11 A magnetic monopole with magnetic charge g passes through matter and loses energy by collisions with electrons, just as does a particle with electric charge ze.

(a) In the same approximation as presented in Section 13.1, show that the energy loss per unit distance is given approximately by (13.14), but with $ze \rightarrow \beta g$, yielding

$$\left(\frac{dE}{dx}\right)_{\substack{\text{magnetic} \\ \text{monopole}}} \simeq 4\pi NZ \frac{g^2e^2}{mc^2} \ln\left(\frac{2\gamma^2 mv^2}{\hbar\langle\omega\rangle}\right)$$

(b) With the Dirac quantization condition (6.153) determining the magnetic charge, what z value is necessary for an ordinary charged particle in order that it lose energy at relativistic speeds at the same rate as a monopole? Sketch for the magnetic monopole a curve of dE/dx equivalent to Fig. 13.1 and comment on the differences.

13.12 A relativistic particle of charge ze moves along the z axis with a constant speed βc. The half-space $z \leq 0$ is filled with a uniform isotropic dielectric medium with plasma frequency ω_1, and the space $z > 0$ with a similar medium whose plasma frequency is ω_2. Discuss the emission of transition radiation as the particle traverses the interface, using the approximation of Section 13.7.

(a) Show that the radiation intensity per unit circular frequency interval and per unit solid angle is given approximately by

$$\frac{d^2I}{d\omega\,d\Omega} \simeq \frac{z^2e^2\theta^2}{\pi^2 c} \left| \frac{1}{\dfrac{1}{\gamma^2} + \dfrac{\omega_1^2}{\omega^2} + \theta^2} - \frac{1}{\dfrac{1}{\gamma^2} + \dfrac{\omega_2^2}{\omega^2} + \theta^2} \right|^2$$

where θ is the angle of emission relative to the velocity of the particle and $\gamma = (1 - \beta^2)^{-1/2}$.

(b) Show that the total energy radiated is

$$I \simeq \frac{z^2e^2}{3c} \cdot \frac{(\omega_1 - \omega_2)^2}{(\omega_1 + \omega_2)} \cdot \gamma$$

13.13 Consider the transition radiation emitted by a relativistic particle traversing a dielectric foil of thickness a perpendicular to its path. Assuming that reflections can be ignored because

$$|[n(\omega) - 1]/[n(\omega) + 1]|$$

is very small, show that the differential angular and frequency spectrum is given by the single-interface result (13.84) times the factor,

$$\mathcal{F} = 4\sin^2\Theta, \quad \text{with} \quad \Theta = \nu\left(1 + \frac{1}{\nu^2} + \eta\right)\frac{a}{4D}$$

Here $D = \gamma c / \omega_p$ is the formation length, $\nu = \omega / \gamma \omega_p$, and $\eta = (\gamma \theta)^2$. Provided $a \gg D$, the factor $\mathcal{F}$ oscillates extremely rapidly in angle or frequency, averaging to $\langle \mathcal{F} \rangle = 2$. For such foils the smoothed intensity distribution is just twice that for a single interface. Frequency distributions for different values of $\Gamma = 2D/a$ are displayed in Fig. 1 of G. B. Yodh, X. Artru, and R. Ramaty, *Astrophys. J.* **181**, 725 (1973).

13.14 Transition radiation is emitted by a relativistic particle traversing normally a uniform array of N dielectric foils, each of thickness a, separated by air gaps (effectively vacuum), each of length b. Assume that multiple reflections can be neglected for the whole stack. This requires

$$\left| \frac{n(\omega) - 1}{n(\omega) + 1} \right| \simeq \frac{\omega_p^2}{4\omega^2} \ll \frac{1}{N}$$

(a) Show that if the dielectric constant of the medium varies in the z direction as $\epsilon(\omega, z) = 1 - (\omega_p^2 / \omega^2) \rho(z)$, the differential spectrum of transition radiation is given approximately by the single-interface result (13.84) times

$$\mathcal{F} = \left| \mu \int dz \, \rho(z) e^{i\omega z / v} \exp\left(-i \cos \theta \int^z k(z') \, dz' \right) \right|^2$$

where $\rho(0) = 1$ by convention, $\mu = \omega / v - k(0) \cos \theta$, and $k(z) = (\omega/c) \sqrt{\epsilon(\omega, z)}$.

(b) Show that for the stack of N foils

$$\mathcal{F} = 4 \sin^2 \Theta \, \frac{\sin^2[N(\Theta + \Psi)]}{\sin^2[\Theta + \Psi]}$$

where Θ is defined in Problem 13.13 and $\Psi = \nu(1 + \eta)(b/4D)$. Compare G. M. Garibyan, *Zh. Eksp. Teor. Fiz.* **60**, 39 (1970) [transl. *Sov. Phys. JETP* **33**, 23 (1971)].

The practical theory of multilayered transition radiation detectors is treated in great detail by X. Artru, G. B. Yodh, and G. Mennessier, *Phys. Rev. D* **12**, 1289 (1975).

13.15 **(a)** Find the number N_γ of transition radiation quanta with frequencies greater than ω_p emitted per interface, starting from the energy spectrum (13.85). Show that for $\gamma \gg 1$,

$$N_\gamma = \frac{z^2 e^2}{\pi \hbar c} \left[(\ln \gamma - 1)^2 + \frac{\pi^2}{12} \right]$$

where terms of order $1/\gamma^2$ have been neglected.

(b) Using the result from part a for the number of photons and the value $\hbar \omega_p = 20$ eV, find the mean energy of the photons (in keV) for $\gamma = 10^3$, 10^4, 10^5.

13.16 A highly relativistic neutral particle of mass m possessing a magnetic moment μ parallel to its direction of motion emits transition radiation as it crosses at right angles a plane interface from vacuum into a dielectric medium characterized at high frequencies by a plasma frequency ω_p. The magnetic moment μ is defined in the particle's rest frame. [The particle could be a neutron or, of more potential interest, a neutrino with a small mass.]

(a) Show that the intensity of transition radiation is given by (13.79), provided the electric field of the incident particle $\mathbf{E}_i$, is given by $(\beta \mu / \gamma z e)$ times the partial derivative in the z direction of E_p in (13.80). Note that the electric field actually points azimuthally, but this affects only the polarization of the radiation, not its intensity.

(b) Show that in the combined limit of $\gamma \gg 1$ and $\omega \gg \omega_p$, the intensity distributions in angle and frequency are given by (13.84) and (13.85), each multiplied by $(\mu\omega/ze\gamma c)^2$.

(c) By expressing μ in units of the Bohr magneton $\mu_B = e\hbar/2m_ec$ and the plasma frequency in atomic units ($\hbar\omega_0 = e^2/a_0 = 27.2$ eV), show that the ratio of frequency distributions of transition radiation emitted by the magnetic moment to that emitted by an electron with the same speed is

$$\frac{dI_\mu(\nu)}{dI_e(\nu)} = \frac{\alpha^4}{4} \left(\frac{\mu}{\mu_B}\right)^2 \left(\frac{\hbar\omega_p}{\hbar\omega_0}\right)^2 \cdot \nu^2$$

where $\alpha = 1/137$ is the fine structure constant and $\nu = \omega/\gamma\omega_p$ is the dimensionless frequency variable.

(d) Calculate the total energy of transition radiation, imposing conservation of energy, that is, $\nu \leq \nu_{max} = mc^2/\hbar\omega_p$. [This constraint will give only a crude estimate of the energy in the quantum regime where $\nu_{max} < 1$ because the derivation is classical throughout.] Show that the ratio of total energies for the magnetic moment and an electron of the same speed can be written as

$$\frac{I_\mu}{I_e} = \frac{\alpha^4}{20} \left(\frac{\mu}{\mu_B}\right)^2 \left(\frac{\hbar\omega_p}{\hbar\omega_0}\right)^2 \cdot G(\nu_{max})$$

where $G \approx 1$ for $\nu_{max} \gg 1$ and $G \approx (10\,\nu_{max}^3/\pi) \cdot [\ln(1/\nu_{max}) - 2/3]$ for $\nu_{max} \ll 1$. For fixed particle energy and magnetic moment, how does the actual amount of radiated energy vary with the particle's mass for very small mass?

Hint: the integrals of Section 2.7 of Gradshteyn and Ryzhik may be of use, although integration by parts is effective.

CHAPTER 14

Radiation by Moving Charges

It is well known that accelerated charges emit electromagnetic radiation. In Chapter 9 we discussed examples of radiation by macroscopic time-varying charge and current densities, which are fundamentally charges in motion. But in one class of radiation phenomena the source is a moving point charge or a small number of such charges. In such problems it is useful to develop the formalism in a way that relates the radiation intensity and polarization directly to properties of the charge's trajectory and motion. Of particular interest are the total radiation emitted, the angular distribution of radiation, and its frequency spectrum. For nonrelativistic motion the radiation is described by the well-known Larmor result (see Section 14.2). But for relativistic particles a number of unusual and interesting effects appear. It is these relativistic aspects that we wish to emphasize. In the present chapter a number of general results are derived and applied to examples of charges undergoing prescribed motions, especially in external force fields.

Deflection of ultrarelativistic electrons in magnetic fields found in accelerators, but also in plasmas and astrophysical contexts, leads to copious emission of radiation called "synchrotron radiation." The basic properties of synchrotron radiation are derived in Sections 14.5 and 14.6. The broad frequency spectrum, often corresponding to millions of harmonics of the basic frequency of particle motion, finds uses in solid-state physics and biology wherever intense beams of x-rays are desirable. These applications have led to the creation of dedicated "light sources" with special "insertion devices" called wigglers and undulators. The physics of these magnetic structures, designed to produce spectral "lines" (actually narrow peaks) of very high brightness and adjustable photon energy, is discussed in Section 14.7.

14.1 *Liénard–Wiechert Potentials and Fields for a Point Charge*

In Section 12.11 it was shown that if there are no incoming fields the 4-vector potential caused by a charged particle in motion is

$$A^\alpha(x) = \frac{4\pi}{c} \int d^4x' \, D_r(x - x') J^\alpha(x') \tag{14.1}$$

where $D_r(x - x')$ is the retarded Green function (12.133) and

$$J^\alpha(x') = ec \int d\tau V^\alpha(\tau) \, \delta^{(4)}[x' - r(\tau)] \tag{14.2}$$

is the charge's 4-vector current, $V^\alpha(\tau)$ its 4-velocity and $r^\alpha(\tau)$ its position. Insertion of the Green function and the current into (14.1) gives, upon integration over d^4x',

$$A^\alpha(x) = 2e \int d\tau V^\alpha(\tau)\, \theta[x_0 - r_0(\tau)]\, \delta\{[x - r(\tau)]^2\} \tag{14.3}$$

The remaining integral over the charge's proper time gives a contribution only at $\tau = \tau_0$, where τ_0 is defined by the light-cone condition,

$$[x - r(\tau_0)]^2 = 0 \tag{14.4}$$

and the retardation requirement $x_0 > r_0(\tau_0)$. The significance of these conditions is shown diagrammatically in Fig. 14.1. The Green function is different from zero only on the backward light cone of the observation point. The world line of the particle $r(\tau)$ intersects the light cone at only two points, one earlier and one later than x_0. The earlier point, $r^\alpha(\tau_0)$, is the only part of the path that contributes to the fields at x^α. To evaluate (14.3) we use the rule,

$$\delta[f(x)] = \sum_i \frac{\delta(x - x_i)}{\left| \left(\dfrac{df}{dx} \right)_{x=x_i} \right|}$$

where the points $x = x_i$ are the zeros of $f(x)$, assumed to be linear. We need

$$\frac{d}{d\tau}[x - r(\tau)]^2 = -2[x - r(\tau)]_\beta V^\beta(\tau) \tag{14.5}$$

evaluated at the one point, $\tau = \tau_0$. The 4-vector potential is therefore

$$A^\alpha(x) = \frac{eV^\alpha(\tau)}{V \cdot [x - r(\tau)]}\bigg|_{\tau=\tau_0} \tag{14.6}$$

where τ_0 is defined by (14.4) and the retardation requirement.

The potentials (14.6) are known as the *Liénard–Wiechert potentials*. They are often written in noncovariant, but perhaps more familiar, form as follows. The light-cone constraint (14.4) implies $x_0 - r_0(\tau_0) = |\mathbf{x} - \mathbf{r}(\tau_0)| \equiv R$. Then

$$
\begin{aligned}
V \cdot (x - r) &= V_0[x_0 - r_0(\tau_0)] - \mathbf{V} \cdot [\mathbf{x} - \mathbf{r}(\tau_0)] \\
&= \gamma cR - \gamma \mathbf{v} \cdot \mathbf{n} R \\
&= \gamma cR(1 - \boldsymbol{\beta} \cdot \mathbf{n})
\end{aligned} \tag{14.7}
$$

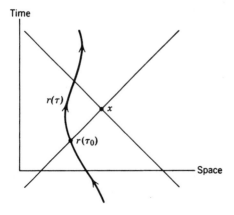

Figure 14.1

where **n** is a unit vector in the direction of $\mathbf{x} - \mathbf{r}(\tau)$ and $\boldsymbol{\beta} = \mathbf{v}(\tau)/c$. The potentials (14.6) can thus be written

$$
\Phi(\mathbf{x}, t) = \left[\frac{e}{(1 - \boldsymbol{\beta} \cdot \mathbf{n})R} \right]_{\text{ret}}, \qquad \mathbf{A}(\mathbf{x}, t) = \left[\frac{e\boldsymbol{\beta}}{(1 - \boldsymbol{\beta} \cdot \mathbf{n})R} \right]_{\text{ret}} \qquad (14.8)
$$

The subscript "ret" means that the quantity in the square brackets is to be evaluated at the retarded time τ_0, given by, $r_0(\tau_0) = x_0 - R$. It is evident that for nonrelativistic motion the potentials reduce to the well-known results.

The electromagnetic fields $F^{\alpha\beta}(x)$ can be calculated directly from (14.6) or (14.8), but it is simpler to return to the integral over $d\tau$, (14.3). In computing $F^{\alpha\beta}$ the differentiation with respect to the observation point x will act on the theta and delta functions. Differentiation of the theta function will give $\delta[x_0 - r_0(\tau)]$ and so constrain the delta function to be $\delta(-R^2)$. There will be no contribution from this differentiation except at $R = 0$. Excluding $R = 0$ from consideration, the derivative $\partial^\alpha A^\beta$ is

$$
\partial^\alpha A^\beta = 2e \int d\tau\, V^\beta(\tau)\, \theta[x_0 - r_0(\tau)]\, \partial^\alpha \delta\{[x - r(\tau)]^2\} \qquad (14.9)
$$

The partial derivative can be written

$$
\partial^\alpha \delta[f] = \partial^\alpha f \cdot \frac{d}{df}\, \delta[f] = \partial^\alpha f \cdot \frac{d\tau}{df} \cdot \frac{d}{d\tau}\, \delta[f]
$$

where $f = [x - r(\tau)]^2$. The indicated differentiation gives

$$
\partial^\alpha \delta[f] = -\frac{(x - r)^\alpha}{V \cdot (x - r)}\, \frac{d}{d\tau}\, \delta[f]
$$

When this is inserted into (14.9) and an integration by parts performed, the result is

$$
\partial^\alpha A^\beta = 2e \int d\tau\, \frac{d}{d\tau} \left[\frac{(x - r)^\alpha V^\beta}{V \cdot (x - r)} \right] \theta[x_0 - r_0(\tau)]\, \delta\{[x - r(\tau)]^2\} \qquad (14.10)
$$

In the integration by parts the differentiation of the theta function gives no contribution, as already indicated. The form of (14.10) is the same as (14.3), with $V^\alpha(\tau)$ replaced by the derivative term. The result can thus be read off by substitution from (14.6). The field strength tensor is

$$
F^{\alpha\beta} = \frac{e}{V \cdot (x - r)}\, \frac{d}{d\tau} \left[\frac{(x - r)^\alpha V^\beta - (x - r)^\beta V^\alpha}{V \cdot (x - r)} \right] \qquad (14.11)
$$

Here r^α and V^α are functions of τ. After differentiation the whole expression is to be evaluated at the retarded proper time τ_0.

The field-strength tensor $F^{\alpha\beta}$ (14.11) is manifestly covariant, but not overly explicit. It is sometimes useful to have the fields **E** and **B** exhibited as explicit functions of the charge's velocity and acceleration. Some of the ingredients needed to carry out the differentiation in (14.11) are

$$
(x - r)^\alpha = (R, R\mathbf{n}), \qquad V^\alpha = (\gamma c, \gamma c \boldsymbol{\beta})
$$

$$
\frac{dV^\alpha}{d\tau} = [c\gamma^4 \boldsymbol{\beta} \cdot \dot{\boldsymbol{\beta}},\, c\gamma^2 \dot{\boldsymbol{\beta}} + c\gamma^4 \boldsymbol{\beta}(\boldsymbol{\beta} \cdot \dot{\boldsymbol{\beta}})] \qquad (14.12)
$$

$$
\frac{d}{d\tau} [V \cdot (x - r)] = -c^2 + (x - r)_\alpha \frac{dV^\alpha}{d\tau}
$$

where $\dot{\boldsymbol{\beta}} = d\boldsymbol{\beta}/dt$ is the ordinary acceleration, divided by c. When these and (14.7) are employed the fields (14.11) can be written in the inelegant, but perhaps more intuitive, forms,

$$\mathbf{B} = [\mathbf{n} \times \mathbf{E}]_{\text{ret}} \qquad (14.13)$$

$$\mathbf{E}(\mathbf{x}, t) = e\left[\frac{\mathbf{n} - \boldsymbol{\beta}}{\gamma^2(1 - \boldsymbol{\beta} \cdot \mathbf{n})^3 R^2}\right]_{\text{ret}} + \frac{e}{c}\left[\frac{\mathbf{n} \times \{(\mathbf{n} - \boldsymbol{\beta}) \times \dot{\boldsymbol{\beta}}\}}{(1 - \boldsymbol{\beta} \cdot \mathbf{n})^3 R}\right]_{\text{ret}} \qquad (14.14)$$

Fields (14.13) and (14.14) divide themselves naturally into "velocity fields," which are independent of acceleration, and "acceleration fields," which depend linearly on $\dot{\boldsymbol{\beta}}$. The velocity fields are essentially static fields falling off as R^{-2}, whereas the acceleration fields are typical radiation fields, both $\mathbf{E}$ and $\mathbf{B}$ being transverse to the radius vector and varying as R^{-1}.

For the special circumstance of a particle in uniform motion the second term in (14.14) is absent. The first term, the velocity field, must be the same as that obtained in Section 11.10 by means of a Lorentz transformation of the static Coulomb field. One way to establish this is to note from (14.11) for $F^{\alpha\beta}$ that if V^α is constant, the field is

$$F^{\alpha\beta} = \frac{ec^2}{[V \cdot (x - r)]^3} \cdot [(x - r)^\alpha V^\beta - (x - r)^\beta V^\alpha] \qquad (14.15)$$

in agreement with the third covariant form in Problem 11.17. It may be worthwhile, nevertheless, to make a transformation of the charge's coordinates from its present position (used in Section 11.10) to the retarded position used here in order to demonstrate explicitly how the different appearing expressions, (11.152) and (14.14), are actually the same. The two positions of the charge are shown in Fig. 14.2 as the points P and P', while O is the observation point. The distance $P'Q$ is $\beta R \cos\theta = \boldsymbol{\beta} \cdot \mathbf{n}R$. Therefore the distance OQ is $(1 - \boldsymbol{\beta} \cdot \mathbf{n})R$. But from the triangles OPQ and $PP'Q$ we have $[(1 - \boldsymbol{\beta} \cdot \mathbf{n})R]^2 = r^2 - (PQ)^2 = r^2 - \beta^2 R^2 \sin^2\theta$. Then from the triangle OMP' we have $R \sin\theta = b$, so that

$$[(1 - \boldsymbol{\beta} \cdot \mathbf{n})R]^2 = b^2 + v^2 t^2 - \beta^2 b^2 = \frac{1}{\gamma^2}(b^2 + \gamma^2 v^2 t^2) \qquad (14.16)$$

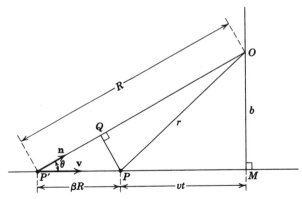

Figure 14.2 Present and retarded positions of a charge in uniform motion.

The transverse component E_2 from (11.152),

$$E_2 = \frac{e\gamma b}{(b^2 + \gamma^2 v^2 t^2)^{3/2}} \tag{14.17a}$$

can thus be written in terms of the retarded position as

$$E_2 = e\left[\frac{b}{\gamma^2(1 - \boldsymbol{\beta} \cdot \mathbf{n})^3 R^3}\right]_{\text{ret}} \tag{14.17b}$$

This is just the transverse component of the velocity field in (14.14). The other components of **E** and **B** come out similarly.

14.2 *Total Power Radiated by an Accelerated Charge: Larmor's Formula and Its Relativistic Generalization*

If a charge is accelerated but is observed in a reference frame where its velocity is small compared to that of light, then in that coordinate frame the acceleration field in (14.14) reduces to

$$\mathbf{E}_a = \frac{e}{c}\left[\frac{\mathbf{n} \times (\mathbf{n} \times \dot{\boldsymbol{\beta}})}{R}\right]_{\text{ret}} \tag{14.18}$$

The instantaneous energy flux is given by the Poynting vector,

$$\mathbf{S} = \frac{c}{4\pi} \mathbf{E} \times \mathbf{B} = \frac{c}{4\pi} |\mathbf{E}_a|^2 \mathbf{n} \tag{14.19}$$

This means that the power radiated per unit solid angle is*

$$\frac{dP}{d\Omega} = \frac{c}{4\pi} |R\mathbf{E}_a|^2 = \frac{e^2}{4\pi c} |\mathbf{n} \times (\mathbf{n} \times \dot{\boldsymbol{\beta}})|^2 \tag{14.20}$$

If Θ is the angle between the acceleration $\dot{\mathbf{v}}$ and **n**, as shown in Fig. 14.3, then the power radiated can be written

$$\frac{dP}{d\Omega} = \frac{e^2}{4\pi c^3} |\dot{\mathbf{v}}|^2 \sin^2\Theta \tag{14.21}$$

This exhibits the characteristic $\sin^2\Theta$ angular dependence, which is a well-known result. We note from (14.18) that the radiation is polarized in the plane containing $\dot{\mathbf{v}}$ and **n**. The total instantaneous power radiated is obtained by integrating (14.21) over all solid angle. Thus

$$P = \frac{2}{3}\frac{e^2}{c^3} |\dot{\mathbf{v}}|^2 \tag{14.22}$$

This is the familiar Larmor result for a nonrelativistic, accelerated charge.

---

*As noted in Chapter 9, in writing angular distributions of radiation we always exhibit the polarization explicitly by writing the absolute square of a vector that is proportional to the electric field. If the angular distribution for some particular polarization is desired, it can be obtained by taking the scalar product of the vector with the appropriate polarization vector before squaring.

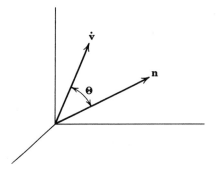

Figure 14.3

Larmor's formula (14.22) can be generalized by arguments about covariance under Lorentz transformations to yield a result that is valid for arbitrary velocities of the charge. Radiated electromagnetic energy behaves under Lorentz transformation like the zeroth component of a 4-vector (see Problem 12.18). This can be used (see *Rohrlich*, p. 109ff.) to show that the power P is a Lorentz invariant. If we can find a Lorentz invariant that reduces to the Larmor formula (14.22) for $\beta \ll 1$, then we have the desired generalization. There are, of course, many Lorentz invariants that reduce to the desired form when $\beta \to 0$. But from (14.14) it is evident that the general result must involve only $\boldsymbol{\beta}$ and $\dot{\boldsymbol{\beta}}$. With this restriction on the order of derivatives that can appear, the result is unique. To find the appropriate generalization we write Larmor's formula in the suggestive form:

$$P = \frac{2}{3}\frac{e^2}{m^2c^3}\left(\frac{d\mathbf{p}}{dt}\cdot\frac{d\mathbf{p}}{dt}\right) \tag{14.23}$$

where m is the mass of the charged particle, and $\mathbf{p}$ its momentum. The Lorentz invariant generalization is

$$P = -\frac{2}{3}\frac{e^2}{m^2c^3}\left(\frac{dp_\mu}{d\tau}\frac{dp^\mu}{d\tau}\right) \tag{14.24}$$

where $d\tau = dt/\gamma$ is the proper time element, and p^μ is the charged particle's momentum-energy 4-vector.* To check that (14.24) reduces properly to (14.23) as $\beta \to 0$ we evaluate the 4-vector scalar product,

$$-\frac{dp_\mu}{d\tau}\frac{dp^\mu}{d\tau} = \left(\frac{d\mathbf{p}}{d\tau}\right)^2 - \frac{1}{c^2}\left(\frac{dE}{d\tau}\right)^2 = \left(\frac{d\mathbf{p}}{d\tau}\right)^2 - \beta^2\left(\frac{dp}{d\tau}\right)^2 \tag{14.25}$$

If (14.24) is expressed in terms of the velocity and acceleration by means of $E = \gamma mc^2$ and $\mathbf{p} = \gamma m\mathbf{v}$, we obtain the Liénard result (1898):

$$P = \frac{2}{3}\frac{e^2}{c}\gamma^6[(\dot{\boldsymbol{\beta}})^2 - (\boldsymbol{\beta} \times \dot{\boldsymbol{\beta}})^2] \tag{14.26}$$

One area of application of the relativistic expression for radiated power is that of charged-particle accelerators. Radiation losses are sometimes the limiting factor in the maximum practical energy attainable. For a given applied force (i.e.,

---

*That (14.24) is unique can be seen by noting that a Lorentz invariant is formed by taking scalar products of 4-vectors or tensors of higher rank. The available 4-vectors are p^μ and $dp^\mu/d\tau$. Only form (14.24) reduces to the Larmor formula for $\beta \to 0$. Contraction of higher rank tensors such as $p^\mu(dp^\nu/d\tau)$ can be shown to vanish, or to give results proportional to (14.24) or m^2.

a given rate of change of momentum), the radiated power (14.24) depends inversely on the square of the mass of the particle involved. Consequently these radiative effects are largest for electrons.

In a linear accelerator the motion is one-dimensional. From (14.25) it is evident that in that case the radiated power is

$$P = \frac{2}{3} \frac{e^2}{m^2 c^3} \left(\frac{dp}{dt} \right)^2 \qquad (14.27)$$

The rate of change of momentum is equal to the change in energy of the particle per unit distance. Consequently

$$P = \frac{2}{3} \frac{e^2}{m^2 c^3} \left(\frac{dE}{dx} \right)^2 \qquad (14.28)$$

showing that for linear motion the power radiated depends only on the external forces that determine the rate of change of particle energy with distance, not on the actual energy or momentum of the particle. The ratio of power radiated to power supplied by the external sources is

$$\frac{P}{(dE/dt)} = \frac{2}{3} \frac{e^2}{m^2 c^3} \frac{1}{v} \frac{dE}{dx} \rightarrow \frac{2}{3} \frac{(e^2/mc^2)}{mc^2} \frac{dE}{dx} \qquad (14.29)$$

where the last form holds for relativistic particles ($\beta \rightarrow 1$). Equation (14.29) shows that the radiation loss in an electron linear accelerator will be unimportant unless the gain in energy is of the order of $mc^2 = 0.511$ MeV in a distance of $e^2/mc^2 = 2.82 \times 10^{-13}$ cm, or of the order of 2×10^{14} MeV/m! Typical energy gains are less than 50 MeV/m. Radiation losses are completely negligible in linear accelerators, whether for electrons or heavier particles.

Circumstances change drastically in circular accelerators like the synchrotron or betatron. In such machines the momentum **p** changes rapidly in direction as the particle rotates, but the change in energy per revolution is small. This means that

$$\left| \frac{d\mathbf{p}}{d\tau} \right| = \gamma \omega \, |\mathbf{p}| \gg \frac{1}{c} \frac{dE}{d\tau} \qquad (14.30)$$

Then the radiated power (14.24) can be written approximately

$$P = \frac{2}{3} \frac{e^2}{m^2 c^3} \gamma^2 \omega^2 \, |\mathbf{p}|^2 = \frac{2}{3} \frac{e^2 c}{\rho^2} \beta^4 \gamma^4 \qquad (14.31)$$

where we have used $\omega = (c\beta/\rho)$, ρ being the orbit radius. This result was first obtained by Liénard in 1898. The radiative-energy loss per revolution is

$$\delta E = \frac{2\pi\rho}{c\beta} P = \frac{4\pi}{3} \frac{e^2}{\rho} \beta^3 \gamma^4 \qquad (14.32)$$

where $1/\rho$ is actually $1/2\pi$ times the path integral around the ring of $[1/\rho(s)]^2$. For high-energy electrons ($\beta \simeq 1$) this has the numerical value,

$$\delta E(\text{MeV}) = 8.85 \times 10^{-2} \frac{[E(\text{GeV})]^4}{\rho(\text{meters})} \qquad (14.33)$$

In the first electron synchrotrons, $\rho \simeq 1$ meter, $E_{\text{max}} \simeq 0.3$ GeV. Hence $\delta E_{\text{max}} \simeq 1$ keV per revolution. This was less than, but not negligible compared to, the energy gain of a few kilovolts per turn. At higher energies the limitation on available radiofrequency power to overcome the radiation loss becomes a dominant consideration. In the 10 GeV Cornell electron synchrotron, for example, the orbit radius is $\rho \sim 100$ meters, the maximum magnetic field is ~ 3.3 kG, and the rf voltage per turn is 10.5 MV at 10 GeV. According to (14.33) the loss per turn is 8.85 MeV. These same general considerations apply to electron-positron storage rings, where rf power must be supplied just to maintain the beams at a constant energy as they circulate. At the LEP ring in Geneva, Switzerland, for beams at 50 GeV the loss per turn is about 300 MeV per electron.

The power radiated in circular electron accelerators can be expressed numerically as

$$P \text{ (watts)} = 10^6 \, \delta E \text{ (MeV)} \, J \text{ (amp)} \qquad (14.34)$$

where J is the circulating beam current. This equation is valid if the emission of radiation from the different electrons in the circulating beam is incoherent. In the largest electron storage rings the radiated power amounts to tens of watts per microampere of beam. While this power dissipation is a waste to high-energy physicists, the radiation has unique properties that make it a valuable research tool. These properties are discussed in Section 14.6, and in greater detail for dedicated "light sources" in Section 14.7.

14.3 *Angular Distribution of Radiation Emitted by an Accelerated Charge*

For an accelerated charge in nonrelativistic motion the angular distribution shows a simple $\sin^2\Theta$ behavior, as given by (14.21), where Θ is measured relative to the direction of acceleration. For relativistic motion the acceleration fields depend on the velocity as well as the acceleration. Consequently the angular distribution is more complicated. From (14.14) the radial component of Poynting's vector can be calculated to be

$$[\mathbf{S} \cdot \mathbf{n}]_{\text{ret}} = \frac{e^2}{4\pi c} \left\{ \frac{1}{R^2} \left| \frac{\mathbf{n} \times [(\mathbf{n} - \boldsymbol{\beta}) \times \dot{\boldsymbol{\beta}}]}{(1 - \boldsymbol{\beta} \cdot \mathbf{n})^3} \right|^2 \right\}_{\text{ret}} \qquad (14.35)$$

It is evident that there are two types of relativistic effect present. One is the effect of the specific spatial relationship between $\boldsymbol{\beta}$ and $\dot{\boldsymbol{\beta}}$, which will determine the detailed angular distribution. The other is a general, relativistic effect arising from the transformation from the rest frame of the particle to the observer's frame and manifesting itself by the presence of the factors $(1 - \boldsymbol{\beta} \cdot \mathbf{n})$ in the denominator of (14.35). For ultrarelativistic particles the latter effect dominates the whole angular distribution.

In (14.35) $\mathbf{S} \cdot \mathbf{n}$ is the energy per unit area per unit time detected at an observation point at time t of radiation emitted by the charge at time $t' = t - R(t')/c$. If we wanted to calculate the energy radiated during a finite period of acceleration, say from $t' = T_1$ to $t' = T_2$, we would write

$$E = \int_{t=T_1+[R(T_1)/c]}^{t=T_2+[R(T_2)/c]} [\mathbf{S} \cdot \mathbf{n}]_{\text{ret}} \, dt = \int_{t'=T_1}^{t'=T_2} (\mathbf{S} \cdot \mathbf{n}) \frac{dt}{dt'} \, dt' \qquad (14.36)$$

Thus we see that the useful and meaningful quantity is $(\mathbf{S} \cdot \mathbf{n})\,(dt/dt')$, the power radiated per unit area in terms of the charge's own time. We therefore define the power radiated per unit solid angle to be

$$\frac{dP(t')}{d\Omega} = R^2(\mathbf{S} \cdot \mathbf{n})\,\frac{dt}{dt'} = R^2 \mathbf{S} \cdot \mathbf{n}(1 - \boldsymbol{\beta} \cdot \mathbf{n}) \tag{14.37}$$

If we imagine the charge to be accelerated only for a short time during which $\boldsymbol{\beta}$ and $\dot{\boldsymbol{\beta}}$ are essentially constant in direction and magnitude, and we observe the radiation far enough away from the charge that $\mathbf{n}$ and R change negligibly during the acceleration interval, then (14.37) is proportional to the angular distribution of the energy radiated. With (14.35) for the Poynting vector, the angular distribution is

$$\frac{dP(t')}{d\Omega} = \frac{e^2}{4\pi c}\,\frac{|\mathbf{n} \times \{(\mathbf{n} - \boldsymbol{\beta}) \times \dot{\boldsymbol{\beta}}\}|^2}{(1 - \mathbf{n} \cdot \boldsymbol{\beta})^5} \tag{14.38}$$

The simplest example of (14.38) is linear motion in which $\boldsymbol{\beta}$ and $\dot{\boldsymbol{\beta}}$ are parallel. If θ is the angle of observation measured from the common direction of $\boldsymbol{\beta}$ and $\dot{\boldsymbol{\beta}}$, then (14.38) reduces to

$$\frac{dP(t')}{d\Omega} = \frac{e^2 \dot{v}^2}{4\pi c^3}\,\frac{\sin^2\theta}{(1 - \beta\cos\theta)^5} \tag{14.39}$$

For $\beta \ll 1$, this is the Larmor result (14.21). But as $\beta \to 1$, the angular distribution is tipped forward more and more and increases in magnitude, as indicated schematically in Fig. 14.4. The angle $\theta_{\max}$ for which the intensity is a maximum is

$$\theta_{\max} = \cos^{-1}\left[\frac{1}{3\beta}\left(\sqrt{1 + 15\beta^2} - 1\right)\right] \to \frac{1}{2\gamma} \tag{14.40}$$

where the last form is the limiting value for $\beta \to 1$. In this same limit the peak intensity is proportional to γ^8. Even for $\beta = 0.5$, corresponding to electrons of $\sim$80 keV kinetic energy, $\theta_{\max} = 38.2°$. For relativistic particles, $\theta_{\max}$ is very small, being of the order of the ratio of the rest energy of the particle to its total energy. Thus the angular distribution is confined to a very narrow cone in the direction of motion. For such small angles the angular distribution (14.39) can be written approximately

$$\frac{dP(t')}{d\Omega} \simeq \frac{8}{\pi}\,\frac{e^2 \dot{v}^2}{c^3}\,\gamma^8\,\frac{(\gamma\theta)^2}{(1 + \gamma^2\theta^2)^5} \tag{14.41}$$

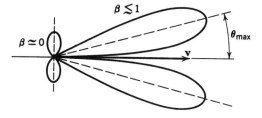

Figure 14.4 Radiation pattern for charge accelerated in its direction of motion. The two patterns are not to scale, the relativistic one (appropriate for $\gamma \sim 2$) having been reduced by a factor $\sim 10^2$ for the same acceleration.

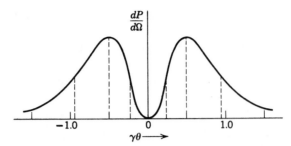

Figure 14.5 Angular distribution of radiation for relativistic particle.

The natural angular unit is evidently γ^{-1}. The angular distribution is shown in Fig. 14.5 with angles measured in these units. The peak occurs at $\gamma\theta = \frac{1}{2}$, and the half-power points at $\gamma\theta = 0.23$ and $\gamma\theta = 0.91$. The root mean square angle of emission of radiation in the relativistic limit is

$$\langle\theta^2\rangle^{1/2} = \frac{1}{\gamma} = \frac{mc^2}{E} \tag{14.42}$$

This is typical of the relativistic radiation patterns, regardless of the vectorial relation between $\boldsymbol{\beta}$ and $\dot{\boldsymbol{\beta}}$. The total power radiated can be obtained by integrating (14.39) over all angles. Thus

$$P(t') = \frac{2}{3}\frac{e^2}{c^3}\dot{v}^2\gamma^6 \tag{14.43}$$

in agreement with (14.26) and (14.27).

Another example of angular distribution of radiation is that for a charge in instantaneously circular motion with its acceleration $\dot{\boldsymbol{\beta}}$ perpendicular to its velocity $\boldsymbol{\beta}$. We choose a coordinate system such that instantaneously $\boldsymbol{\beta}$ is in the z direction and $\dot{\boldsymbol{\beta}}$ is in the x direction. With the customary polar angles θ, ϕ defining the direction of observation, as shown in Fig. 14.6, the general formula (14.38) reduces to

$$\frac{dP(t')}{d\Omega} = \frac{e^2}{4\pi c^3}\frac{|\dot{\mathbf{v}}|^2}{(1 - \beta\cos\theta)^3}\left[1 - \frac{\sin^2\theta\cos^2\phi}{\gamma^2(1 - \beta\cos\theta)^2}\right] \tag{14.44}$$

We note that, although the detailed angular distribution is different from the linear acceleration case, the same characteristic relativistic peaking at forward

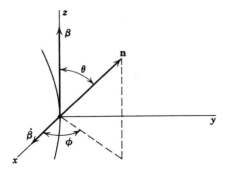

Figure 14.6

angles is present. In the relativistic limit ($\gamma \gg 1$), the angular distribution can be written approximately

$$\frac{dP(t')}{d\Omega} \simeq \frac{2}{\pi} \frac{e^2}{c^3} \gamma^6 \frac{|\dot{\mathbf{v}}|^2}{(1 + \gamma^2 \theta^2)^3} \left[1 - \frac{4\gamma^2 \theta^2 \cos^2 \phi}{(1 + \gamma^2 \theta^2)^2} \right] \tag{14.45}$$

The root mean square angle of emission in this approximation is given by (14.42), just as for one-dimensional motion. The total power radiated can be found by integrating (14.44) over all angles or from (14.26):

$$P(t') = \frac{2}{3} \frac{e^2}{c^3} |\dot{\mathbf{v}}|^2 \gamma^4 \tag{14.46}$$

It is instructive to compare the power radiated for acceleration parallel to the velocity (14.43) or (14.27) with the power radiated for acceleration perpendicular to the velocity (14.46) for the same magnitude of applied force. For circular motion, the magnitude of the rate of change of momentum (which is equal to the applied force) is $\gamma m \dot{\mathbf{v}}$. Consequently, (14.46) can be written

$$P_{\text{circular}}(t') = \frac{2}{3} \frac{e^2}{m^2 c^3} \gamma^2 \left(\frac{d\mathbf{p}}{dt} \right)^2 \tag{14.47}$$

When this is compared to the corresponding result (14.27) for rectilinear motion, we find that for a given magnitude of applied force the radiation emitted with a transverse acceleration is a factor of γ^2 larger than with a parallel acceleration.

14.4   Radiation Emitted by a Charge in Arbitrary, Extremely Relativistic Motion

For a charged particle undergoing arbitrary, extremely relativistic motion the radiation emitted at any instant can be thought of as a coherent superposition of contributions coming from the components of acceleration parallel to and perpendicular to the velocity. But we have just seen that for comparable parallel and perpendicular forces the radiation from the parallel component is negligible (of order $1/\gamma^2$) compared to that from the perpendicular component. Consequently we may neglect the parallel component of acceleration and approximate the radiation intensity by that from the perpendicular component alone. In other words, the radiation emitted by a charged particle in arbitrary, extreme relativistic motion is approximately the same as that emitted by a particle moving instantaneously along the arc of a circular path whose radius of curvature ρ is given by

$$\rho = \frac{v^2}{\dot{v}_\perp} \simeq \frac{c^2}{\dot{v}_\perp} \tag{14.48}$$

where $\dot{v}_\perp$ is the perpendicular component of acceleration. The form of the angular distribution of radiation is (14.44) or (14.45). It corresponds to a narrow cone or searchlight beam of radiation directed along the instantaneous velocity vector of the charge.

For an observer with a frequency-sensitive detector the confinement of the radiation to a narrow pencil parallel to the velocity has important consequences.

The radiation will be visible only when the particle's velocity is directed toward the observer. For a particle in arbitrary motion the observer will detect a pulse or burst of radiation of very short time duration (or a succession of such bursts if the particle is in periodic motion), as sketched in Fig. 14.7. Since the angular width of the beam is of the order of γ^{-1}, the particle will travel only a distance of the order of

$$d = \frac{\rho}{\gamma}$$

corresponding to a time,

$$\Delta t = \frac{\rho}{\gamma v}$$

while illuminating the observer. To make the argument conceptually simple, neglect the curvature of the path during this time and suppose that a sharp rectangular pulse of radiation is emitted. In the time Δt the front edge of the pulse travels a distance,

$$D = c \, \Delta t = \frac{\rho}{\gamma \beta}$$

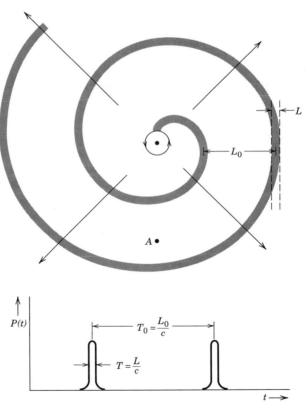

Figure 14.7 A relativistic particle in periodic motion emits a spiral radiation pattern that an observer at the point A detects as short bursts of radiation of time duration $T = L/c$, occurring at regular intervals $T_0 = L_0/c$. The pulse length is given by (14.49), while the interval $T_0 = 2\pi\rho/v \simeq 2\pi\rho/c$. For beautiful diagrams of field lines of radiating particles, see R. Y. Tsien, *Am. J. Phys.* **40**, 46 (1972).

Since the particle is moving in the same direction with speed v and moves a distance d in the time Δt, the rear edge of the pulse will be only a distance

$$L = D - d = \left(\frac{1}{\beta} - 1\right)\frac{\rho}{\gamma} \simeq \frac{\rho}{2\gamma^3} \qquad (14.49)$$

behind the front edge as the pulse moves off. The pulse length is thus L in space, or L/c in time. From general arguments about the Fourier decomposition of finite wave trains this implies that the spectrum of the radiation will contain appreciable frequency components up to a critical frequency,

$$\omega_c \sim \frac{c}{L} \sim \left(\frac{c}{\rho}\right)\gamma^3 \qquad (14.50)$$

For circular motion c/ρ is the angular frequency of rotation ω_0 and even for arbitrary motion it plays the role of a fundamental frequency. Equation (14.50) shows that a relativistic particle emits a broad spectrum of frequencies, up to γ^3 times the fundamental frequency. In a 200 MeV synchrotron, $\gamma_{max} \simeq 400$, while $\omega_0 \simeq 3 \times 10^8 \, \text{s}^{-1}$. The frequency spectrum of emitted radiation extends up to $\sim 2 \times 10^{16} \, \text{s}^{-1}$, or down to a wavelength of 1000 Å, even though the fundamental frequency is in the 100 MHz range. For the 10 GeV machine at Cornell, $\gamma_{max} \simeq 2 \times 10^4$ and $\omega_0 \simeq 3 \times 10^6 \, \text{s}^{-1}$. This means that $\omega_c \simeq 2.4 \times 10^{19} \, \text{s}^{-1}$, corresponding to 16 keV x-rays. In Section 14.6 we discuss in detail the angular distribution of the different frequency components, as well as the total energy radiated as a function of frequency. In Section 14.7 we show how to modify the spectrum with magnetic insertion devices.

14.5 Distribution in Frequency and Angle of Energy Radiated by Accelerated Charges: Basic Results

The qualitative arguments of Section 14.4 show that for relativistic motion the radiated energy is spread over a wide range of frequencies. The range of the frequency spectrum was estimated by appealing to properties of Fourier integrals. The argument can be made precise and quantitative by the use of Parseval's theorem of Fourier analysis.

The general form of the power radiated per unit solid angle is

$$\frac{dP(t)}{d\Omega} = |\mathbf{A}(t)|^2 \qquad (14.51)$$

where

$$\mathbf{A}(t) = \left(\frac{c}{4\pi}\right)^{1/2}[R\mathbf{E}]_{\text{ret}} \qquad (14.52)$$

$\mathbf{E}$ being the electric field (14.14). In (14.51) the instantaneous power is expressed in the observer's time (contrary to the definition in Section 14.3), since we wish to consider a frequency spectrum in terms of the observer's frequencies. For definiteness we think of the acceleration occurring for some finite interval of time, or at least falling off for remote past and future times, so that the total energy radiated is finite. Furthermore, the observation point is considered far enough away from the charge that the spatial region spanned by the charge while accelerated subtends a small solid-angle element at the observation point.

The total energy radiated per unit solid angle is the time integral of (14.51):

$$\frac{dW}{d\Omega} = \int_{-\infty}^{\infty} |\mathbf{A}(t)|^2 \, dt \qquad (14.53)$$

This can be expressed alternatively as an integral over a frequency spectrum by use of Fourier transforms. We introduce the Fourier transform $\mathbf{A}(\omega)$ of $\mathbf{A}(t)$,

$$\mathbf{A}(\omega) = \frac{1}{\sqrt{2\pi}} \int_{-\infty}^{\infty} \mathbf{A}(t) e^{i\omega t} \, dt \qquad (14.54)$$

and its inverse,

$$\mathbf{A}(t) = \frac{1}{\sqrt{2\pi}} \int_{-\infty}^{\infty} \mathbf{A}(\omega) e^{-i\omega t} \, d\omega \qquad (14.55)$$

Then (14.53) can be written

$$\frac{dW}{d\Omega} = \frac{1}{2\pi} \int_{-\infty}^{\infty} dt \int_{-\infty}^{\infty} d\omega \int_{-\infty}^{\infty} d\omega' \, \mathbf{A}^*(\omega') \cdot \mathbf{A}(\omega) e^{i(\omega' - \omega)t} \qquad (14.56)$$

Interchanging the orders of time and frequency integration, we see that the time integral is just a Fourier representation of the delta function $\delta(\omega' - \omega)$. Consequently the energy radiated per unit solid angle becomes

$$\frac{dW}{d\Omega} = \int_{-\infty}^{\infty} |\mathbf{A}(\omega)|^2 \, d\omega \qquad (14.57)$$

The equality of (14.57) and (14.53), with suitable mathematical restrictions on the function $\mathbf{A}(t)$, is a special case of Parseval's theorem. It is customary to integrate only over positive frequencies, since the sign of the frequency has no physical meaning. Then the relation,

$$\frac{dW}{d\Omega} = \int_{0}^{\infty} \frac{d^2 I(\omega, \mathbf{n})}{d\omega \, d\Omega} \, d\omega \qquad (14.58)$$

defines a quantity that is the energy radiated per unit solid angle per unit frequency interval:

$$\frac{d^2 I}{d\omega \, d\Omega} = |\mathbf{A}(\omega)|^2 + |\mathbf{A}(-\omega)|^2 \qquad (14.59)$$

If $\mathbf{A}(t)$ is real, from (14.55) it is evident that $\mathbf{A}(-\omega) = \mathbf{A}^*(\omega)$. Then

$$\frac{d^2 I}{d\omega \, d\Omega} = 2 \, |\mathbf{A}(\omega)|^2 \qquad (14.60)$$

This result relates in a quantitative way the behavior of the power radiated as a function of time to the frequency spectrum of the energy radiated.

By using (14.14) for the electric field of an accelerated charge we can obtain a general expression for the energy radiated per unit solid angle per unit frequency interval in terms of an integral over the trajectory of the particle. We must calculate the Fourier transform (14.54) of $\mathbf{A}(t)$ given by (14.52). Using (14.14), we find

$$\mathbf{A}(\omega) = \left(\frac{e^2}{8\pi^2 c}\right)^{1/2} \int_{-\infty}^{\infty} e^{i\omega t} \left[\frac{\mathbf{n} \times [(\mathbf{n} - \boldsymbol{\beta}) \times \dot{\boldsymbol{\beta}}]}{(1 - \boldsymbol{\beta} \cdot \mathbf{n})^3}\right]_{\text{ret}} dt \qquad (14.61)$$

where ret means evaluated at $t' + [R(t')/c] = t$. We change the variable of integration from t to t', thereby obtaining the result:

$$\mathbf{A}(\omega) = \left(\frac{e^2}{8\pi^2 c}\right)^{1/2} \int_{-\infty}^{\infty} e^{i\omega(t' + [R(t')/c])} \frac{\mathbf{n} \times [(\mathbf{n} - \boldsymbol{\beta}) \times \dot{\boldsymbol{\beta}}]}{(1 - \boldsymbol{\beta} \cdot \mathbf{n})^2} dt' \qquad (14.62)$$

Since the observation point is assumed to be far away from the region of space where the acceleration occurs, the unit vector $\mathbf{n}$ is sensibly constant in time. Furthermore the distance $R(t')$ can be approximated as

$$R(t') \simeq x - \mathbf{n} \cdot \mathbf{r}(t') \qquad (14.63)$$

where x is the distance from an origin O to the observation point P, and $\mathbf{r}(t')$ is the position of the particle relative to O, as shown in Fig. 14.8. Then, apart from an overall phase factor, (14.62) becomes

$$\mathbf{A}(\omega) = \left(\frac{e^2}{8\pi^2 c}\right)^{1/2} \int_{-\infty}^{\infty} e^{i\omega(t - \mathbf{n}\cdot\mathbf{r}(t)/c)} \frac{\mathbf{n} \times [(\mathbf{n} - \boldsymbol{\beta}) \times \dot{\boldsymbol{\beta}}]}{(1 - \boldsymbol{\beta} \cdot \mathbf{n})^2} dt \qquad (14.64)$$

The primes on the time variable have been omitted for brevity. The energy radiated per unit solid angle per unit frequency interval (14.60) is accordingly

$$\frac{d^2 I}{d\omega\, d\Omega} = \frac{e^2}{4\pi^2 c} \left| \int_{-\infty}^{\infty} \frac{\mathbf{n} \times [(\mathbf{n} - \boldsymbol{\beta}) \times \dot{\boldsymbol{\beta}}]}{(1 - \boldsymbol{\beta} \cdot \mathbf{n})^2} e^{i\omega(t - \mathbf{n}\cdot\mathbf{r}(t)/c)}\, dt \right|^2 \qquad (14.65)$$

For a specified motion $\mathbf{r}(t)$ is known, $\boldsymbol{\beta}(t)$ and $\dot{\boldsymbol{\beta}}(t)$ can be computed, and the integral can be evaluated as a function of ω and the direction of $\mathbf{n}$. If accelerated motion of more than one charge is involved, a coherent sum of amplitudes $\mathbf{A}_j(\omega)$, one for each charge, must replace the single amplitude in (14.65) (see Problems 14.23, 15.1, 15.4–15.8).

Even though (14.65) has the virtue of explicitly showing the time interval of integration to be confined to times for which the acceleration is different from zero, a simpler expression for some purposes can be obtained by an integration by parts in (14.64). It is easy to demonstrate that the integrand in (14.64), excluding the exponential, is a perfect differential:

$$\frac{\mathbf{n} \times [(\mathbf{n} - \boldsymbol{\beta}) \times \dot{\boldsymbol{\beta}}]}{(1 - \boldsymbol{\beta} \cdot \mathbf{n})^2} = \frac{d}{dt}\left[\frac{\mathbf{n} \times (\mathbf{n} \times \boldsymbol{\beta})}{1 - \boldsymbol{\beta} \cdot \mathbf{n}}\right] \qquad (14.66)$$

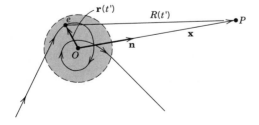

Figure 14.8

Then an integration by parts leads to the intensity distribution:

$$\frac{d^2 I}{d\omega \, d\Omega} = \frac{e^2 \omega^2}{4\pi^2 c} \left| \int_{-\infty}^{\infty} \mathbf{n} \times (\mathbf{n} \times \boldsymbol{\beta}) e^{i\omega(t - \mathbf{n} \cdot \mathbf{r}(t)/c)} \, dt \right|^2 \tag{14.67}$$

The reader may rightly ask whether (14.67) is correct in all circumstances as it stands. Suppose that the acceleration is different from zero only for $T_1 \leq t \leq T_2$. Why then is the integration in (14.67) over all time? The precise answer is that (14.67) can be shown, by adding and subtracting the integrals over the times when the velocity is constant, to follow from (14.65) provided ambiguities at $t = \pm\infty$ are resolved by inserting a convergence factor $e^{-\epsilon|t|}$ in the integrand and taking the limit $\epsilon \to 0$ after evaluating the integral. In processes like beta decay, where the classical description involves the almost instantaneous halting or setting in motion of charges, extra care must be taken to specify each particle's velocity as a physically sensible function of time.

We remind the reader that in (14.67) and (14.65) the polarization of the emitted radiation is specified by the direction of the vector integral in each. The intensity of radiation of a certain fixed polarization can be obtained by taking the scalar product of the appropriate unit polarization vector with the vector integral before forming the absolute square.

For a number of charges e_j in accelerated motion the integrand in (14.67) involves the replacement,

$$e\boldsymbol{\beta} e^{-i(\omega/c)\mathbf{n} \cdot \mathbf{r}(t)} \to \sum_{j=1}^{N} e_j \boldsymbol{\beta}_j e^{-i(\omega/c)\mathbf{n} \cdot \mathbf{r}_j(t)} \tag{14.68}$$

In the limit of a continuous distribution of charge in motion the sum over j becomes an integral over the current density $\mathbf{J}(\mathbf{x}, t)$:

$$e\boldsymbol{\beta} e^{-i(\omega/c)\mathbf{n} \cdot \mathbf{r}(t)} \to \frac{1}{c} \int d^3 x \, \mathbf{J}(\mathbf{x}, t) e^{-i(\omega/c)\mathbf{n} \cdot \mathbf{x}} \tag{14.69}$$

Then the intensity distribution becomes

$$\frac{d^2 I}{d\omega \, d\Omega} = \frac{\omega^2}{4\pi^2 c^3} \left| \int dt \int d^3 x \, \mathbf{n} \times [\mathbf{n} \times \mathbf{J}(\mathbf{x}, t)] e^{i\omega[t - (\mathbf{n} \cdot \mathbf{x})/c]} \right|^2 \tag{14.70}$$

a result that can be obtained from the direct solution of the inhomogeneous wave equation for the vector potential.

14.6 *Frequency Spectrum of Radiation Emitted by a Relativistic Charged Particle in Instantaneously Circular Motion*

In Section 14.4 we saw that the radiation emitted by an extremely relativistic particle subject to arbitrary accelerations is equivalent to that emitted by a particle moving instantaneously at constant speed on an appropriate circular path. The radiation is beamed in a narrow cone in the direction of the velocity vector and is seen by the observer as a short pulse of radiation as the searchlight beam sweeps across the observation point.

To find the distribution of energy in frequency and angle it is necessary to calculate the integral in (14.67). Because the duration of the pulse is very short, it is necessary to know the velocity $\boldsymbol{\beta}$ and position $\mathbf{r}(t)$ over only a small arc of the trajectory whose tangent points in the general direction of the observation point. Figure 14.9 shows an appropriate coordinate system. The segment of trajectory lies in the x-y plane with instantaneous radius of curvature ρ. Since an integral will be taken over the path, the unit vector $\mathbf{n}$ can be chosen without loss of generality to lie in the x-z plane, making an angle θ (the latitude) with the x axis. Only for very small θ will there be appreciable radiation intensity. The origin of time is chosen so that at $t = 0$ the particle is at the origin of coordinates.

The vector part of the integrand in (14.67) can be written

$$\mathbf{n} \times (\mathbf{n} \times \boldsymbol{\beta}) = \beta \left[-\boldsymbol{\epsilon}_\parallel \sin\left(\frac{vt}{\rho}\right) + \boldsymbol{\epsilon}_\perp \cos\left(\frac{vt}{\rho}\right) \sin\theta \right] \tag{14.71}$$

where $\boldsymbol{\epsilon}_\parallel = \boldsymbol{\epsilon}_2$ is a unit vector in the y direction, corresponding to polarization in the plane of the orbit; $\boldsymbol{\epsilon}_\perp = \mathbf{n} \times \boldsymbol{\epsilon}_2$ is the orthogonal polarization vector corresponding approximately to polarization perpendicular to the orbit plane (for θ small). The argument of the exponential is

$$\omega\left(t - \frac{\mathbf{n} \cdot \mathbf{r}(t)}{c}\right) = \omega\left[t - \frac{\rho}{c} \sin\left(\frac{vt}{\rho}\right) \cos\theta \right] \tag{14.72}$$

Since we are concerned with small angles θ and comparatively short times around $t = 0$, we can expand both trigonometric functions in (14.72) to obtain

$$\omega\left(t - \frac{\mathbf{n} \cdot \mathbf{r}(t)}{c}\right) \simeq \frac{\omega}{2} \left[\left(\frac{1}{\gamma^2} + \theta^2\right)t + \frac{c^2}{3\rho^2} t^3 \right] \tag{14.73}$$

where β has been put equal to unity wherever possible. Using the time estimate $\rho/c\gamma$ for t and the estimate $\langle\theta^2\rangle^{1/2}$ (14.42) for θ, it is easy to see that neglected terms in (14.73) are of the order of γ^{-2} times those kept.

With the same type of approximations in (14.71) as led to (14.73), the radiated energy distribution (14.67) can be written

$$\frac{d^2I}{d\omega \, d\Omega} = \frac{e^2\omega^2}{4\pi^2 c} \left| -\boldsymbol{\epsilon}_\parallel A_\parallel(\omega) + \boldsymbol{\epsilon}_\perp A_\perp(\omega) \right|^2 \tag{14.74}$$

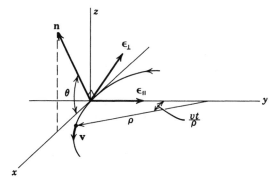

Figure 14.9

where the amplitudes are*

$$A_{\parallel}(\omega) \simeq \frac{c}{\rho} \int_{-\infty}^{\infty} t \exp\left\{ i \frac{\omega}{2} \left[\left(\frac{1}{\gamma^2} + \theta^2 \right) t + \frac{c^2 t^3}{3\rho^2} \right] \right\} dt$$

$$A_{\perp}(\omega) \simeq \theta \int_{-\infty}^{\infty} \exp\left\{ i \frac{\omega}{2} \left[\left(\frac{1}{\gamma^2} + \theta^2 \right) t + \frac{c^2 t^3}{3\rho^2} \right] \right\} dt \tag{14.75}$$

A change of variable to $x = \left[ct \big/ \rho \left(\frac{1}{\gamma^2} + \theta^2 \right)^{1/2} \right]$ and introduction of the parameter ξ,

$$\xi = \frac{\omega\rho}{3c} \left(\frac{1}{\gamma^2} + \theta^2 \right)^{3/2} \tag{14.76}$$

allows us to transform the integrals in $A_{\parallel}(\omega)$ and $A_{\perp}(\omega)$ into the form:

$$A_{\parallel}(\omega) = \frac{\rho}{c} \left(\frac{1}{\gamma^2} + \theta^2 \right) \int_{-\infty}^{\infty} x \exp[i\tfrac{3}{2}\xi(x + \tfrac{1}{3}x^3)] \, dx$$

$$A_{\perp}(\omega) = \frac{\rho}{c} \theta \left(\frac{1}{\gamma^2} + \theta^2 \right)^{1/2} \int_{-\infty}^{\infty} \exp[i\tfrac{3}{2}\xi(x + \tfrac{1}{3}x^3)] \, dx \tag{14.77}$$

The integrals in (14.77) are identifiable as Airy integrals, or alternatively as modified Bessel functions:

$$\int_0^{\infty} x \sin[\tfrac{3}{2}\xi(x + \tfrac{1}{3}x^3)] \, dx = \frac{1}{\sqrt{3}} K_{2/3}(\xi)$$

$$\int_0^{\infty} \cos[\tfrac{3}{2}\xi(x + \tfrac{1}{3}x^3)] \, dx = \frac{1}{\sqrt{3}} K_{1/3}(\xi) \tag{14.78}$$

Consequently the energy radiated per unit frequency interval per unit solid angle is

$$\frac{d^2 I}{d\omega \, d\Omega} = \frac{e^2}{3\pi^2 c} \left(\frac{\omega\rho}{c} \right)^2 \left(\frac{1}{\gamma^2} + \theta^2 \right)^2 \left[K_{2/3}^2(\xi) + \frac{\theta^2}{(1/\gamma^2) + \theta^2} K_{1/3}^2(\xi) \right] \tag{14.79}$$

The first term in the square bracket corresponds to radiation polarized in the plane of the orbit, and the second to radiation polarized perpendicular to that plane.

We now proceed to examine this somewhat complex result. First we integrate over all frequencies and find that the distribution of energy in angle is

$$\frac{dI}{d\Omega} = \int_0^{\infty} \frac{d^2 I}{d\omega \, d\Omega} d\omega = \frac{7}{16} \frac{e^2}{\rho} \frac{1}{(1/\gamma^2 + \theta^2)^{5/2}} \left[1 + \frac{5}{7} \frac{\theta^2}{(1/\gamma^2) + \theta^2} \right] \tag{14.80}$$

---

*The fact that the limits of integration in (14.75) are $t = \pm\infty$ may seem to contradict the approximations made in going from (14.72) to (14.73). The point is that for most frequencies the phase of the integrands in (14.75) oscillates very rapidly and makes the integrands effectively zero for times much smaller than those necessary to maintain the validity of (14.73). Hence the upper and lower limits on the integrals can be taken as infinite without error. Only for frequencies of the order of $\omega \sim (c/\rho) \sim \omega_0$ does the approximation fail. But we have seen in Section 14.4 that for relativistic particles essentially all the frequency spectrum is at much higher frequencies.

This shows the characteristic behavior seen in Section 14.3. Equation (14.80) can be obtained directly, of course, by integrating a slight generalization of the circular-motion power formula (14.44) over all times. As in (14.79), the first term in (14.80) corresponds to polarization parallel to the orbital plane, and the second to perpendicular polarization. Integrating over all angles, we find that seven times as much energy is radiated with parallel polarization as with perpendicular polarization. The radiation from a relativistically moving charge is very strongly, but not completely, polarized in the plane of motion.

The properties of the modified Bessel functions summarized in (3.103) and (3.104) show that the intensity of radiation is negligible for $\xi \gg 1$. From (14.76) we see that this will occur at large angles; the greater the frequency, the smaller the critical angle beyond which there will be negligible radiation. This shows that the radiation is largely confined to the plane containing the motion, as shown by (14.80), being more confined the higher the frequency relative to c/ρ. If ω gets too large, however, we see that ξ will be large at *all* angles. Then there will be negligible total energy emitted at that frequency. The critical frequency ω_c beyond which there is negligible radiation at any angle can be defined by $\xi = 1/2$ for $\theta = 0$. Then we find*

$$\omega_c = \frac{3}{2}\gamma^3\left(\frac{c}{\rho}\right) = \frac{3}{2}\left(\frac{E}{mc^2}\right)^3\frac{c}{\rho} \tag{14.81}$$

This critical frequency is seen to agree with our qualitative estimate (14.50) of Section 14.4. If the motion of the charge is truly circular, then c/ρ is the fundamental frequency of rotation, ω_0. Then we can define a critical harmonic frequency $\omega_c = n_c\omega_0$, with harmonic number,

$$n_c = \frac{3}{2}\left(\frac{E}{mc^2}\right)^3 \tag{14.82}$$

Since the radiation is predominantly in the orbital plane for $\gamma \gg 1$, it is instructive to evaluate the angular distribution (14.79) at $\theta = 0$. For frequencies well below the critical frequency ($\omega \ll \omega_c$), we find

$$\left.\frac{d^2I}{d\omega\,d\Omega}\right|_{\theta=0} \simeq \frac{e^2}{c}\left[\frac{\Gamma(\frac{2}{3})}{\pi}\right]^2\left(\frac{3}{4}\right)^{1/3}\left(\frac{\omega\rho}{c}\right)^{2/3} \tag{14.83}$$

For the opposite limit of $\omega \gg \omega_c$, the result is

$$\left.\frac{d^2I}{d\omega\,d\Omega}\right|_{\theta=0} \simeq \frac{3}{4\pi}\frac{e^2}{c}\gamma^2\frac{\omega}{\omega_c}e^{-\omega/\omega_c} \tag{14.84}$$

These limiting forms show that the spectrum at $\theta = 0$ increases with frequency roughly as $\omega^{2/3}$ well below the critical frequency, reaches a maximum in the neighborhood of ω_c, and then drops exponentially to zero above that frequency.

The spread in angle at a fixed frequency can be estimated by determining

---

*Our present definition of ω_c differs from earlier editions. The present one, defined originally by Schwinger (1949), is in general use.

the angle θ_c at which $\xi(\theta_c) \simeq \xi(0) + 1$. In the low-frequency range ($\omega \ll \omega_c$), $\xi(0)$ is very small, so that $\xi(\theta_c) \simeq 1$. This gives

$$\theta_c \simeq \left(\frac{3c}{\omega\rho}\right)^{1/3} = \frac{1}{\gamma}\left(\frac{2\omega_c}{\omega}\right)^{1/3} \tag{14.85}$$

We note that the low-frequency components are emitted at much wider angles than the average, $\langle\theta^2\rangle^{1/2} \sim \gamma^{-1}$. In the high-frequency limit ($\omega > \omega_c$), $\xi(0)$ is large compared to unity. Then the intensity falls off in angle approximately as

$$\frac{d^2I}{d\omega\, d\Omega} \simeq \left.\frac{d^2I}{d\omega\, d\Omega}\right|_{\theta=0} \cdot e^{-3\omega\gamma^2\theta^2/2\omega_c} \tag{14.86}$$

Thus the critical angle, defined by the $1/e$ point, is

$$\theta_c \simeq \frac{1}{\gamma}\left(\frac{2\omega_c}{3\omega}\right)^{1/2} \tag{14.87}$$

This shows that the high-frequency components are confined to an angular range much smaller than average. Figure 14.10 shows qualitatively the angular distribution for frequencies small compared with, of the order of, and much larger than ω_c. The natural unit of angle $\gamma\theta$ is used.

The frequency distribution of the total energy emitted as the particle passes by can be found by integrating (14.79) over angles:

$$\frac{dI}{d\omega} = 2\pi \int_{-\pi/2}^{\pi/2} \frac{d^2I}{d\omega\, d\Omega} \cos\theta\, d\theta \simeq 2\pi \int_{-\infty}^{\infty} \frac{d^2I}{d\omega\, d\Omega}\, d\theta \tag{14.88}$$

(remember that θ is the latitude). We can estimate the integral for the low-frequency range by using the value of the angular distribution (14.83) at $\theta = 0$ and the critical angle θ_c (14.85). Then we obtain

$$\frac{dI}{d\omega} \sim 2\pi\theta_c \left.\frac{d^2I}{d\omega\, d\Omega}\right|_{\theta=0} \sim \frac{e^2}{c}\left(\frac{\omega\rho}{c}\right)^{1/3} \tag{14.89}$$

showing that the spectrum increases as $\omega^{1/3}$ for $\omega \ll \omega_c$. This gives a very broad, flat spectrum at frequencies below ω_c. For the high-frequency limit where $\omega \gg \omega_c$ we can integrate (14.86) over angles to obtain the reasonably accurate result,

$$\frac{dI}{d\omega} \simeq \sqrt{3\pi/2}\,\frac{e^2}{c}\,\gamma\left(\frac{\omega}{\omega_c}\right)^{1/2} e^{-\omega/\omega_c} \tag{14.90}$$

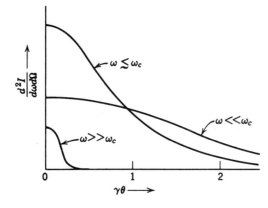

Figure 14.10 Differential frequency spectrum as a function of angle. For frequencies comparable to the critical frequency ω_c, the radiation is confined to angles of the order of γ^{-1}. For much smaller (larger) frequencies, the angular spread is larger (smaller).

A proper integration of (14.79) over angles yields the expression,*

$$\frac{dI}{d\omega} = \sqrt{3} \frac{e^2}{c} \gamma \frac{\omega}{\omega_c} \int_{\omega/\omega_c}^{\infty} K_{5/3}(x) \, dx \qquad (14.91)$$

In the limit $\omega \ll \omega_c$ this reduces to the form (14.89) with a numerical coefficient 3.25, while for $\omega \gg \omega_c$ it is equal to (14.90). The behavior of $dI/d\omega$ as a function of frequency is shown in Fig. 14.11. The peak intensity is of the order of $e^2\gamma/c$, and the total energy is of the order of $e^2\gamma\omega_c/c = 3e^2\gamma^4/\rho$. This is in agreement with the value of $4\pi e^2\gamma^4/3\rho$ for the radiative loss per revolution (14.32) in circular accelerators.

The radiation represented by (14.79) and (14.91) is called *synchrotron radiation* because it was first observed in electron synchrotrons (1948). The theoretical results are much older, however, having been obtained for circular motion by *Schott* (1912) although their expression in the present amenable form is due to Schwinger. For periodic circular motion the spectrum is actually discrete, being composed of frequencies that are integral multiples of the fundamental frequency $\omega_0 = c/\rho$. Since the charged particle repeats its motion at a rate of $c/2\pi\rho$ revolutions per second, it is convenient to talk about the angular distribution of power radiated into the nth multiple of ω_0 instead of the energy radiated per unit frequency interval per passage of the particle. To obtain the harmonic power expressions, we merely multiply $dI/d\omega$ (14.91) or $d^2I/d\omega \, d\Omega$ (14.79) by the repetition rate $c/2\pi\rho$ to convert energy to power, and by $\omega_0 = c/\rho$ to convert per unit frequency interval to per harmonic. Thus

$$\left.\begin{aligned}
\frac{dP_n}{d\Omega} &= \frac{1}{2\pi} \left(\frac{c}{\rho}\right)^2 \left.\frac{d^2I}{d\omega \, d\Omega}\right|_{\omega=n\omega_0} \\
P_n &= \frac{1}{2\pi} \left(\frac{c}{\rho}\right)^2 \left.\frac{dI}{d\omega}\right|_{\omega=n\omega_0}
\end{aligned}\right\} \qquad (14.92)$$

These results have been compared with experiment at various energy synchrotrons.[†] The angular, polarization, and frequency distributions are all in good agreement with theory. Because of the broad frequency distribution shown in Fig. 14.11, covering the visible, ultraviolet, and x-ray regions, synchrotron radiation is a useful tool for studies in condensed matter and biology. We examine synchrotron light sources and some of the insertion devices used to tailor the spectrum for special purposes in the next section.

Synchrotron radiation has been observed in the astronomical realm associated with sunspots, the Crab nebula, and from the particle radiation belts of Jupiter. For the Crab nebula the radiation spectrum extends over a frequency range from radiofrequencies into the extreme ultraviolet, and shows very strong polarization. From detailed observations it can be concluded that electrons with

---

*This result and the differential distribution (14.79) are derived in a somewhat different way by J. Schwinger, *Phys. Rev.* **75**, 1912 (1949). Schwinger later showed that the first-order quantum-mechanical corrections to the classical results involve the replacement of $\omega \to \omega(1 + \hbar\omega/E)$ in $\omega^{-1} \, d^2I/d\omega \, d\Omega$ or $\omega^{-1} \, dI/d\omega$ [*Proc. Natl. Acad. Sci.* **40**, 132 (1954)] and are thus negligible provided $\hbar\omega_c \ll E$, or equivalently, $\gamma \ll (\rho mc/\hbar)^{1/2}$.

†F. R. Elder, R. V. Langmuir, and H. C. Pollock, *Phys. Rev.*, **74**, 52 (1948); D. H. Tomboulain and P. L. Hartman, *Phys. Rev.*, **102**, 1423 (1956); G. Bathow, E. Freytag, and R. Haensel, *J. Appl. Phys.*, **37**, 3449 (1966).

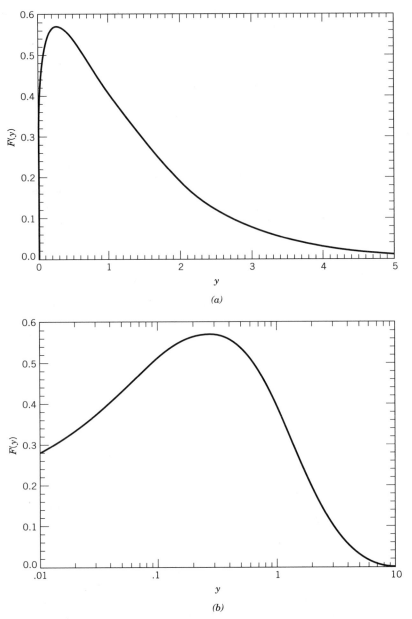

Figure 14.11 Normalized synchrotron radiation spectrum $(1/I)(dI/dy) =$ $(9\sqrt{3}/8\pi)y \int_y^\infty K_{5/3}(x)\,dx$, where $y = \omega/\omega_c$ and $I = 4\pi e^2\gamma^4/3\rho$: (a) linear abscissa scale and (b) logarithmic abscissa scale.

energies ranging up to 10^{13} eV are emitting synchrotron radiation while moving in circular or helical orbits in a magnetic induction of the order of 10^{-4} gauss (see Problem 14.26). The radio emission at $\sim 10^3$ MHz from Jupiter comes from energetic electrons trapped in Van Allen belts at distances from a few to 30–100 radii (R_J) from Jupiter's surface. Data from a space vehicle (Pioneer 10, December 4, 1973, encounter with Jupiter) passing within $2.8R_J$ showed a roughly dipole magnetic field with a dipole moment of $4R_J^3$ gauss. Appreciable fluxes of trapped electrons with energies greater than 3 MeV and a few percent with energies greater than 50 MeV were observed. Taking 1 gauss as a typical field and 5 MeV as a typical energy, Eqs. (12.42) and (14.81) show that the spiraling radius is of the order of 100–200 meters, $\omega_0 \sim 2 \times 10^6$ s^{-1}, and that about 10^3 significant harmonics are radiated.

The treatment of synchrotron radiation presented here is completely classical, but the language of photons can be used, if desired. The number of photons per unit frequency interval is obtained by dividing the intensity distribution (14.91) [or (14.79)] by $\hbar\omega$. Then the photon frequency distribution is

$$\frac{dN}{dy} = \frac{I}{\hbar\omega_c} \cdot \frac{9\sqrt{3}}{8\pi} \int_y^\infty K_{5/3}(x) \, dx \qquad (14.93)$$

where $y = \omega/\omega_c$ and $I = 4\pi e^2 \gamma^4/3\rho$ is the total energy radiated per revolution. Integration over frequency gives the mean number of photons emitted per revolution per particle,

$$N = \frac{5\pi}{\sqrt{3}} \gamma\alpha \qquad (14.94)$$

where α is the fine structure constant. The mean energy per photon is I/N:

$$\langle \hbar\omega \rangle = \frac{8}{15\sqrt{3}} \hbar\omega_c \qquad (14.95)$$

As already remarked, because ω_c is proportional to γ^3 and $\gamma = O(10^4)$ for GeV energies, fundamental wavelengths ($2\pi\rho$) of the order of hundreds of meters give rise to synchrotron photons of wavelengths down to 10^{-10} meter (1 angstrom) or less, corresponding to keV x-rays.

14.7 Undulators and Wigglers for Synchrotron Light Sources

The broad spectrum of radiation emitted by relativistic electrons bent by the magnetic fields of synchrotron storage rings provides a useful source of energetic photons for research and was utilized initially in a "parasitic" mode by biologists and condensed matter physicists. Curved crystals or other devices were used to select specific frequencies from the continuum. As applications grew, the need for brighter sources with the radiation more concentrated in frequency led to the development of magnetic "insertion devices" called wigglers and undulators to be placed in the synchrotron ring. The magnetic properties of these devices cause the electrons to undergo special motion that results in the concentration of the radiation into a much more monochromatic spectrum or series of separated peaks. The basic formula for the radiation is still (14.67), although here we use invariance arguments and Lorentz transformations to make the results more physically understandable.

The essential idea of undulators and wigglers is that a charged particle, usually an electron and usually moving relativistically, ($\gamma \gg 1$), is caused to move transversely to its general forward motion by magnetic fields that alternate periodically. The external magnetic fields induce small transverse oscillations in the motion; the associated accelerations cause radiation to be emitted. A typical configuration of magnets, with an alternating vertical magnetic field at the path

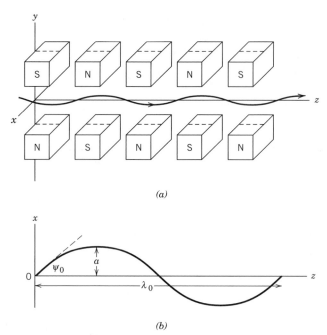

Figure 14.12 (*a*) Schematic diagram of alternating-polarity bending magnets for a wiggler or undulator. (*b*) Sketch of approximately sinusoidal path of electron in the *x-z* plane. The magnet period is λ_0, the maximum transverse amplitude is a, and the maximum angle is ψ_0.

of the particle, is sketched in Fig. 14.12*a*. The path of the particle is in the horizontal (*x-z*) plane.

A. *Qualitative Features*

If the periodicity of the magnetic field structure is λ_0, the particle's path will be approximately sinusoidal in the transverse direction with the same period, as sketched in Fig. 14.12*b*. We have $x \approx a \sin(2\pi z/\lambda_0)$, with the maximum amplitude a dependent on the strength of the wiggler's magnetic field and the particle's energy. The maximum angular deviation ψ_0 away from the forward direction is proportional to a; it is an important parameter, which distinguishes undulators from wigglers. We have

$$\psi_0 = \left(\frac{dx}{dz}\right)_{z=0} = \frac{2\pi a}{\lambda_0} = k_0 a, \qquad \text{where } k_0 = 2\pi/\lambda_0 \qquad (14.96)$$

is the fundamental wave number of the system. [Actually, the time taken for the particle to traverse one period of the magnet structure is $T = \lambda_0/\beta c$ and so the real fundamental wave number of the radiation is βk_0. For $\gamma \gg 1$ the difference is insignificant.]

For $\gamma \gg 1$, the radiation emitted by the charged particle is confined to a narrow angular region of angular width $\Delta\theta = O(1/\gamma)$ about the actual path. As the particle moves in its oscillatory path sketched in Fig. 14.12*b*, the "searchlight" beam of radiation will flick back and forth about the forward direction. Qualitatively different radiation spectra will result, depending on whether ψ_0 is larger or smaller than $\Delta\theta$.

(a) Wiggler ($\psi_0 \gg \Delta\theta$)

For $\psi_0 \gg \Delta\theta$, an observer detects a series of flicks of the searchlight beam, with a repetition rate given by the relation, $\nu_0 = \omega_0/2\pi = ck_0/2\pi$. With λ_0 of the order of a few centimeters, $\nu_0 = O(10 \text{ GHz})$. The phenomenon is very much as in an ordinary synchrotron with bunches spaced a few centimeters apart. The spectrum of radiation extends to frequencies that are γ^3 times the basic frequency $\Omega = c/R$, where R is the effective radius of curvature of the path. The minimum value of R is generally the one of interest. It occurs at the maximum amplitude of the transverse motion and is

$$R = \frac{1}{k_0^2 a} = \frac{\lambda_0}{2\pi\psi_0} \tag{14.97}$$

The wiggler radiation spectrum is a smooth, featureless spectrum very much like the synchrotron radiation spectrum of Fig. 14.11, with a fundamental frequency, $\Omega = 2\pi c\psi_0/\lambda_0$, and a critical frequency γ^3 times this value. If the wiggler magnet structure has N periods, the intensity of radiation will be N times that for a single pass of a particle in the equivalent circular machine.

It is useful to introduce the parameter K, a scaled angle, by

$$K = \gamma\psi_0$$

A wiggler is characterized by $K \gg 1$. In terms of K, its critical frequency is

$$\omega_c = O\left(\gamma^2 K \frac{2\pi c}{\lambda_0}\right) \tag{14.98}$$

Users of synchrotron light sources tend to speak of wavelength rather than frequency. The critical wavelength is

$$\lambda_c = O\left(\frac{\lambda_0}{\gamma^2 K}\right) \tag{14.99}$$

(b) Undulators ($\psi_0 \ll \Delta\theta$ or $K \ll 1$)

If $\psi_0 \ll \Delta\theta$, the searchlight beam of radiation moves negligibly compared to its own angular width. This means that the radiation detected by an observer is an almost *coherent superposition* of the contributions from all the oscillations of the trajectory. For perfect coherence and an infinite number of magnet periods (and infinitesimal angular resolution of the detector), the radiation would be monochromatic. For finite N the spread in frequency is $\Delta\omega/\omega = O(1/N)$; finite angular acceptance also causes a spread because of the Doppler shift. Nevertheless, the frequency spectrum from an undulator is sharply peaked (actually a series of peaks in practice, but with a most intense "fundamental").

The frequency of the "line" from an undulator can be estimated by considering the particle in its rest frame. The FitzGerald–Lorentz contraction means that in that frame the magnet structure is rushing by the particle with a spatial period λ_0/γ. The frequency of simple dipole radiation in that frame is thus $\omega' \approx \gamma(2\pi c/\lambda_0)$. In the laboratory frame the relativistic Doppler shift, $\omega' =$

$\gamma\omega(1 - \beta \cos \theta) \approx \omega(1 + \gamma^2\theta^2)/2\gamma$, leads to a spectral line at an angle θ with frequency

$$\omega \approx \frac{2\gamma^2}{1 + \gamma^2\theta^2} \left(\frac{2\pi c}{\lambda_0}\right) \tag{14.100}$$

Note that at small angles ($\gamma\theta \ll 1$) this frequency has the same γ-dependence as the wiggler's critical frequency, (14.98), for a fixed K.

B. Some Details of the Kinematics and Particle Dynamics

We wish to consider the particle in its average rest frame, in which it executes oscillations both transversely and longitudinally. If its initial Lorentz parameters are γ and β, they remain unchanged because the magnetic field does no work on the particle. But because of the transverse motion, the particle's average speed in the z direction, $\bar{\beta}c$, and its associated $\bar{\gamma}$, are less than the instantaneous parameters. The average rest frame moves with speed $\bar{\beta}c$ with respect to the laboratory.

One way to find $\bar{\beta}$ and $\bar{\gamma}$ is to consider the path shown in Fig. 14.12*b* and compute its length for one cycle:

$$s = \int_0^{\lambda_0} \sqrt{1 + (dx/dz)^2} \, dz \approx \int_0^{\lambda_0} [1 + \tfrac{1}{2}(dx/dz)^2 + \cdots] \, dz \tag{14.101}$$

or

$$s \approx \lambda_0(1 + \tfrac{1}{4}\psi_0^2) \tag{14.102}$$

Here we have assumed that $\psi_0 \ll 1$, and we assume below that $\gamma \gg 1$. Since the particle travels this path as speed βc, we infer that

$$\bar{\beta} = \frac{\beta}{1 + \psi_0^2/4} \approx \beta(1 - \tfrac{1}{4}\psi_0^2) \tag{14.103}$$

Even though $\beta \approx 1$ and $\psi_0 \ll 1$, so that $\bar{\beta} \approx 1$, the difference between $\bar{\beta}$ and β produces a finite (not infinitesimal) difference between $\bar{\gamma}$ and γ:

$$\frac{1}{\bar{\gamma}^2} = 1 - \bar{\beta}^2 \approx 1 - \beta^2(1 - \tfrac{1}{2}\psi_0^2)$$

$$\approx \gamma^{-2} + \tfrac{1}{2}\psi_0^2 = \gamma^{-2}(1 + \tfrac{1}{2}K^2)$$

We therefore find

$$\bar{\gamma} = \frac{\gamma}{\sqrt{1 + \tfrac{1}{2}K^2}} \tag{14.104}$$

Since $K \gg 1$, is possible even if $\psi_0 \ll 1$, $\bar{\gamma}$ can differ significantly from γ, at least for wigglers.

The transverse motion has been assumed to be sinusoidal. How is that connected to the structure of the magnet that causes the motion? With β and γ constant, the x component of the Lorentz force equation can be written $\ddot{x} = -e\beta_z B_y/\gamma m$, where $\beta_y B_z$ is assumed to be negligible or zero. Approximating $z = ct$ and $\beta_z = 1$, we have

$$B_y(z) = -\frac{\gamma mc^2}{e} \frac{d^2x}{dz^2} = \frac{\gamma mc^2}{e} k_0^2 a \sin k_0 z \tag{14.105}$$

The requisite magnetic structure is $B_y = B_0 \sin k_0 z$, where $B_0 = \gamma mc^2 k_0^2 a/e$. Since $K = \gamma k_0 a$, the important parameter K can be expressed in terms of the known field of the magnet and its period,

$$K = \frac{eB_0}{k_0 mc^2} = \frac{eB_0\lambda_0}{2\pi mc^2} \qquad (14.106)$$

An actual magnet structure will be periodic, but not sinusoidal. We can, however, make a Fourier decomposition of the actual B_y in multiples of k_0. Each component will contribute to the motion. The fundamental will dominate. For simplicity, we keep only that contribution.

The longitudinal oscillations can be found, at least approximately, from the constancy of β. We have $\beta_z^2 = \beta^2 - \beta_x^2$. Since $|\beta_x| \ll \beta$, we can write

$$\beta_z \approx \beta - \frac{\beta_x^2}{2\beta} \approx \beta - \frac{\beta_x^2}{2}$$

But $x = a \sin k_0 z \approx a \sin(k_0 ct)$. Thus $\beta_x \approx k_0 a \cos(k_0 ct)$. We then have the component of $\boldsymbol{\beta}$ in the z direction as

$$\begin{aligned}
\beta_z(t) &\approx \beta - \tfrac{1}{2}k_0^2 a^2 \cos^2(k_0 ct) \\
&= \beta - \tfrac{1}{4}k_0^2 a^2 [1 + \cos(2k_0 ct)] \\
&= \overline{\beta} - \frac{K^2}{4\gamma^2} \cos(2k_0 ct)
\end{aligned}$$

Integrating $c\beta_z(t)$ once with respect to t, we find the longitudinal and transverse motions to be

$$z(t) = \overline{\beta}ct - \frac{\lambda_0 K^2}{16\pi\gamma^2} \sin(2k_0 ct) \quad \text{and} \quad x(t) = \frac{\lambda_0 K}{2\pi\gamma} \sin(k_0 ct) \quad (14.107)$$

C. Particle Motion in the Average Rest Frame

It is informative to examine the particle's motion in the frame K', moving with speed $\overline{\beta}$ in the positive z direction. The Lorentz transformations equations are

$$x' = x, \qquad z' = \overline{\gamma}(z - \overline{\beta}ct), \qquad ct' = \overline{\gamma}(ct - \overline{\beta}z)$$

Substituting $z(t)$ from (14.107) into the last equation, we have

$$ct' = \overline{\gamma}\left[ct(1 - \overline{\beta}^2) + \frac{\overline{\beta}K^2}{8k_0\gamma^2} \sin 2\theta\right]$$

where $\theta = k_0 ct$. Neglect of the last term gives the first approximation, $t = \overline{\gamma}t'$. Then with this result inserted into θ, we find a better approximation,

$$t = \overline{\gamma}t' - \frac{1}{4k_0 c}\left(\frac{K^2}{2 + K^2}\right) \sin(2\overline{\gamma}k_0 ct')$$

or $\qquad\qquad\qquad\qquad\qquad\qquad\qquad\qquad\qquad\qquad\qquad\qquad\qquad\qquad$ (14.108)

$$\theta = \overline{\gamma}k_0 ct' - \frac{1}{4}\left(\frac{K^2}{2 + K^2}\right) \sin(2\overline{\gamma}k_0 ct')$$

Usually the first term is adequate, but in computing time derivatives in the moving frame, the second term is necessary when differentiating $\theta(t')$.

The particle's coordinates in the moving frame are

$$x'(t') = \frac{K}{\gamma k_0} \sin \theta(t') = a \sin \theta(t')$$

$$z'(t') = -\frac{\overline{\gamma} K^2}{8\gamma^2 k_0} \sin 2\theta(t') = -\frac{Ka}{8\sqrt{1 + K^2/2}} \sin 2\theta(t')$$

The motion is a figure-eight pattern of the form

$$z' = \mp 2(z')_{max} \cdot \frac{x'}{a} \sqrt{1 - \frac{x'^2}{a^2}} \qquad \text{where} \qquad (z')_{max} = \frac{Ka}{8\sqrt{1 + K^2/2}}$$

Figure 14.13 shows the shape of the particle's orbit in the moving frame for the regime $K \gg 1$. For $K = 1$, the z' amplitude is 0.576 times as large as is shown. For $K \ll 1$, the z' oscillations are negligible; the motion is simple harmonic in the transverse direction.

An important feature of the motion in the moving frame is the maximum speed of the particle. A straightforward calculation yields the square of the particle's speed in the moving frame to be

$$\beta'^2 = \left[\frac{2K^2}{2 + K^2} \cos^2\theta + \frac{K^4}{4(2 + K^2)^2} \cos^2 2\theta\right] \left[1 - \frac{K^2}{2(2 + K^2)} \cos 2\theta\right]^2$$

$$(14.109)$$

where it is now safe to put $\theta(t') = \overline{\gamma} k_0 ct'$. The last factor comes from the form of $d\theta(t')/dt'$. The two limits of K are instructive. For $K \ll 1$, the leading term gives

$$\beta' \approx K \cos \theta, \qquad K \ll 1 \qquad (14.110a)$$

corresponding to *nonrelativistic* simple harmonic motion. This limit is for an undulator. In the opposite limit, $K \to \infty$, the leading behavior is

$$\beta' \approx 1 - (\cos^2\theta - \tfrac{1}{2})^2, \qquad K \to \infty \qquad (14.110b)$$

In this (strong wiggler) limit, the particle's speed varies between $3c/4$ and c in the course of the motion, quite relativistic. From Problems 14.12, 14.14, and 14.15, one can infer that the radiation in the *moving* frame consists of many harmonics of the basic frequency, with an angular distribution that is far from a simple dipole pattern. The laboratory radiation pattern from a strong wiggler is better described by the contributions from the successive segments of the path whose tangents point in the direction of observation.

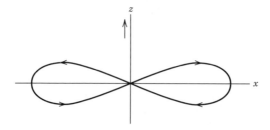

Figure 14.13   Orbit of the particle in the moving (average rest) frame for $K \gg 1$. The arrow indicates the direction of motion in the laboratory frame.

D. Radiation Spectrum from an Undulator

When $K \ll 1$, the motion in the average rest frame is very simple. The particle moves in nonrelativistic simple harmonic motion along the x axis. It emits monochromatic dipole radiation whose power differential distribution is

$$\frac{dP'}{d\Omega'} = \frac{e^2 c}{8\pi} k'^4 a^2 \sin^2\Theta$$

where $k' = \overline{\gamma} k_0$ is the wave number in the moving frame. The coordinates are shown in Fig. 14.14. Now $k'^2 \sin^2\Theta$ can be written as $k'^2 \sin^2\Theta = k'^2 - k'^2 \cos^2\Theta = k_z'^2 + k_y'^2$. With $K = \gamma k_0 a \approx \overline{\gamma} k_0 a$ for $K \ll 1$, the power angular distribution becomes

$$\frac{dP'}{d\Omega'} = \frac{e^2 c}{8\pi} K^2 (k_z'^2 + k_y'^2) \tag{14.111}$$

To find the laboratory spectrum in angle and frequency (actually, either angle *or* frequency), we exploit certain invariances. Since the phase-space density d^3k/ω is a Lorentz invariant, it is useful to consider $\omega' \, d^3P'/d^3k'$, rather than $dP'/d\Omega'$. Inserting a delta function $\delta(k' - \overline{\gamma} k_0)$ to assure the monochromatic nature of the radiation in the moving frame, we have

$$d^3P' = \left[\frac{e^2 c^2 K^2}{8\pi} (k_z'^2 + k_y'^2) \frac{\delta(k' - \overline{\gamma} k_0)}{\overline{\gamma} k_0} \right] \frac{d^3k'}{\omega'} \tag{14.112}$$

where $d^3k' = k'^2 \, dk' \, d\Omega'$. Consider now d^3P'. If we multiply by the time $\Delta t'$ it takes for one period of the magnet structure to pass by the particle in the moving frame ($\Delta t' = \lambda_0/\overline{\gamma}\beta c \approx \lambda_0/\overline{\gamma} c$), we obtain the energy radiated per period into the invariant element of phase space. If we divide by $\hbar\omega' = \hbar c k'$, we obtain the differential *number* d^3N' of photons emitted into d^3k'/ω' per passage of a magnet period. But the *number* of photons is an invariant quantity. We can therefore write the connection between the laboratory differential radiation spectrum and the spectrum in the moving frame as

$$\frac{d^3P}{(d^3k/\omega)} = \frac{\Delta t'}{\Delta t} \cdot \frac{\omega}{\omega'} \cdot \frac{d^3P'}{(d^3k'/\omega')}$$

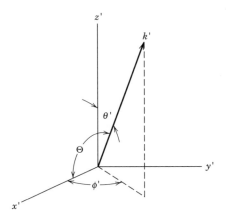

Figure 14.14 Radiation direction and angles in average rest frame. Particle motion is in the x-z plane.

With $\Delta t'/\Delta t = 1/\overline{\gamma}$ and $d^3k/\omega = k\,dk\,d\Omega/c$, we have

$$\frac{d^3P}{dk\,d\Omega} = \frac{e^2cK^2}{8\pi\overline{\gamma}^3} \cdot \frac{k^2}{k_0^2} \cdot (k_z'^2 + k_y'^2) \cdot \delta(k' - \overline{\gamma}k_0) \qquad (14.113)$$

All that remains is to express the primed quantities in terms of the laboratory variables. The Lorentz transformations are

$$k_y' = k_y = k\sin\theta\sin\phi, \qquad \phi' = \phi$$
$$k_z' = \overline{\gamma}k(\cos\theta - \overline{\beta})$$
$$k' = \overline{\gamma}k(1 - \overline{\beta}\cos\theta)$$

Using the constraint of the delta function, we have

$$k = \frac{k_0}{1 - \overline{\beta}\cos\theta}$$

If we make the appropriate approximations for $\overline{\gamma} \gg 1$ (i.e., $\theta \ll 1$, $\overline{\beta} \approx 1 - 1/2\overline{\gamma}^2$, etc.), (14.113) can be written

$$\frac{d^3P}{d\eta\,dk\,d\phi} = \frac{e^2c\overline{\gamma}^2K^2k_0^2}{2\pi}\left[\frac{(1-\eta)^2 + 4\eta\sin^2\phi}{(1+\eta)^4}\right]\delta[k(1+\eta) - 2\overline{\gamma}^2k_0] \quad (14.114)$$

where $\eta = (\overline{\gamma}\theta)^2$ is the natural angle variable to replace $\cos\theta$. Note that, because of the delta function, the frequency and angular distributions are not independent.

(a) Angular Distribution

If we choose to integrate over the frequency spectrum dk, we find the angular distribution of power to be

$$\frac{d^2P}{d\eta\,d\phi} = \frac{e^2c\overline{\gamma}^2K^2k_0^2}{2\pi}\left[\frac{(1-\eta)^2 + 4\eta\sin^2\phi}{(1+\eta)^5}\right] \qquad (14.115)$$

After integration over azimuth, the polar angle spectrum is

$$\frac{dP}{d\eta} = 3P\left[\frac{1+\eta^2}{(1+\eta)^5}\right] \qquad (14.116)$$

where

$$P = \frac{e^2c\overline{\gamma}^2K^2k_0^2}{3} \qquad (14.117)$$

is the total power radiated. It is easy to verify that the average value of η is $\langle\eta\rangle = 1$.

(b) Frequency Distribution

To obtain the frequency distribution emitted into an angular range, $\eta_1 < \eta < \eta_2$, we integrate (14.114) over $d\phi\,d\eta$. The result is

$$\frac{dP}{d\nu} = 3P[\nu(1 - 2\nu + 2\nu^2)] \qquad \text{for } \nu_{min} < \nu < \nu_{max} \qquad (14.118)$$

where $\nu = k/2\overline{\gamma}^2k_0$ and $\nu_{min} = 1/(1 + \eta_2)$, $\nu_{max} = 1/(1 + \eta_1)$. The complete normalized frequency spectrum is plotted in Fig. 14.15a: the sharply peaked spec-

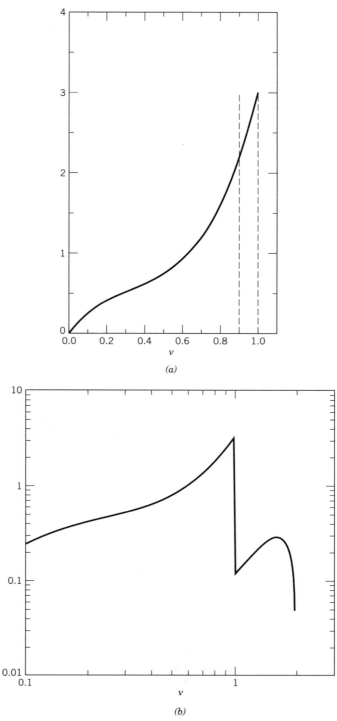

(a)

(b)

Figure 14.15   (*a*) Normalized frequency spectrum for $K \ll 1$ and sinusoidal motion. The dashed lines indicate the frequency interval visible if the angular acceptance is $0 < \gamma\theta < \frac{1}{3}$. (*b*) Log-log plot of intensity of fundamental and first harmonic for $K = 0.5$ with a sinusoidal magnetic field. In real undulators, the spectrum shape depends on details of the undulator structure.

trum between the dashed lines corresponds to an angular acceptance $0 < \eta < 1/9$ ($\theta < 1/3\bar{\gamma}$). Note that this spectrum is for perfectly sinusoidal motion of the particle at all times. If the number N of magnet periods is finite, the duration of the oscillatory motion is finite; the radiated wave train will have a fractional spread in frequency of the order of $1/N$. For large N this spread is generally small compared to the spread from finite acceptance.

For small, but not negligible, K, there are higher harmonics. These can be thought of as coming from higher multipoles caused by the figure-eight motion shown in Fig. 14.13. The first harmonic comes from a coherent superposition of the fields of a dipole in the z direction [$z' \propto \sin 2\theta(t')$] and a quadrupole caused by the x' motion. See Problem 14.27. The resulting frequency spectrum is shown in Fig. 14.15b for $K = 0.5$, with higher harmonics decreasing in intensity, at least for $K < 1$.

(c) Energy of Photons and Number Emitted per Magnet Period

The radiated power is given by (14.117) and the maximum energy of photons in the fundamental is $\hbar\omega_{max} = 2\bar{\gamma}^2 k_0 \hbar c$ (at $\eta = 0$). The amount of energy radiated per passage of one magnet period is $\Delta E = P\Delta t$, where $\Delta t = \lambda_0/c$. The number of photons N_γ emitted per magnet period can thus be estimated to be $N_\gamma \gtrsim P\Delta t/\hbar\omega_{max} = O(\alpha K^2)$, where α is the fine structure constant. A calculation based on (14.118) divided by $\hbar\omega$ gives

$$N_\gamma = \frac{2\pi}{3} \alpha K^2 \tag{14.119}$$

E. Numerical Values and Representative Spectra and Facilities

The parameters K and $\hbar\omega_{max}$ are given for electrons in practical (accelerator) units by

$$K = \frac{eB_0}{k_0 mc^2} = \frac{eB_0\lambda_0}{2\pi mc^2} = 93.4 \, B_0(T)\lambda_0(m)$$

and

$$\hbar\omega_{max}(eV) = \frac{9.496[E \, (GeV)]^2}{(1 + K^2/2)\lambda_0(m)}$$

Typical undulators have $B_0 = O(0.5 \, T)$, $\lambda_0 = O(4 \, cm)$, $E = O(1–7 \, GeV)$. Hence $K = O(2)$ and $\hbar\omega_{max} = O(80 \, eV–4 \, keV)$. Wigglers have $B_0 = O(1 \, T)$ and $\lambda_0 = O(20 \, cm)$. Then $K = O(20)$.

There are dozens of synchrotron light facilities around the world. Typical of the modern dedicated facilities (as of 1998) are

Advanced Light Source (ALS), Lawrence Berkeley National Laboratory, $E = 1.5 \, GeV$

National Synchrotron Light Source (NSLS), Brookhaven National Laboratory, $E = 0.75, 2.5 \, GeV$

European Synchrotron Radiation Facility (ESRF), Grenoble, France, $E = 6 \, GeV$

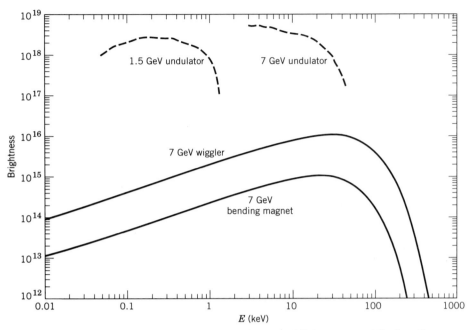

Figure 14.16 Representative photon spectra for actual light sources. The bending magnet and wiggler spectra are continuous and are closely proportional to (14.79), evaluated at $\theta = 0$. The undulator curves are the envelopes of a series of sharp peaks at multiples of the fundamental. See text for definition of brightness.

Tristan Light Source, KEK National Laboratory, Tsukuba, Japan, $E = 6.5$ GeV

Advanced Photon Source (APS), Argonne National Laboratory, $E = 7$ GeV

The lower energy facilities provide photons in the tens of eV to several keV range; the high-energy facilities extend to 10–75 keV, and even higher at reduced flux. Figure 14.16 shows some representative spectra of actual light sources. The spectral brightnesses indicate the typical capabilities available at relatively low-energy rings such as the ALS and the higher energy rings such as the APS. For undulators the smooth curves represent the envelope of the narrow "lines." Brightness or brilliance is defined as the number of photons per second per milliradian in the vertical and horizontal directions per 0.1% fractional bandwidth in photon energy, divided by 2π times the effective source area in square millimeters. High brilliance rather than high flux is generally desired.

F. Additional Comments

There is a vast amount of detail about synchrotron light sources, the design of beams, the transport of photons to experiments, and so on. We make only a few comments here.

1. An undulator's fundamental frequency ω_{max} can be tuned by varying the undulator parameter K by changing the gap in the magnet structure and so changing B_0 [see (14.106)].

2. The simple undulator with beam oscillations in the horizontal plane provides linearly polarized light. Circular polarization can be provided by use of a

helical undulator designed to make the transverse trajectory an ellipse. Alternately, two undulators at right angles with an adjustable longitudinal spacing between them can be used to produce circular polarization or any other state because of the coherent superposition of the radiation from all the magnet periods.

3. Free electron lasers are closely related to wigglers and undulators. An undulator can be thought of as radiating in the forward direction at frequency ω_{max} by spontaneous emission. Addition of a co-traveling electromagnetic wave of almost the same frequency provides the possibility of interaction and stimulated emission and growth of the wave.

Further details about the sources and about their uses in research can be found in the references cited at the end of the chapter.

14.8 *Thomson Scattering of Radiation*

If a plane wave of monochromatic electromagnetic radiation is incident on a free particle of charge e and mass m, the particle will be accelerated and so emit radiation. This radiation will be emitted in directions other than that of the incident plane wave, but for nonrelativistic motion of the particle it will have the same frequency as the incident radiation. The whole process may be described as scattering of the incident radiation.

According to (14.20) the instantaneous power radiated into polarization state $\boldsymbol{\epsilon}$ by a particle of charge e in nonrelativistic motion is

$$\frac{dP}{d\Omega} = \frac{e^2}{4\pi c^3} |\boldsymbol{\epsilon}^* \cdot \dot{\mathbf{v}}|^2 \tag{14.120}$$

The acceleration is provided by the incident plane wave. If its propagation vector is $\mathbf{k}_0$, and its polarization vector $\boldsymbol{\epsilon}_0$, the electric field can be written

$$\mathbf{E}(\mathbf{x}, t) = \boldsymbol{\epsilon}_0 E_0 e^{i\mathbf{k}_0 \cdot \mathbf{x} - i\omega t}$$

Then, from the force equation for nonrelativistic motion, we have the acceleration,

$$\dot{\mathbf{v}}(t) = \boldsymbol{\epsilon}_0 \frac{e}{m} E_0 e^{i\mathbf{k}_0 \cdot \mathbf{x} - i\omega t} \tag{14.121}$$

If we assume that the charge moves a negligible part of a wavelength during one cycle of oscillation, the time average of $|\dot{\mathbf{v}}|^2$ is $\frac{1}{2}\text{Re}(\dot{\mathbf{v}} \cdot \dot{\mathbf{v}}^*)$. Then the average power per unit solid angle can be expressed as

$$\left\langle \frac{dP}{d\Omega} \right\rangle = \frac{c}{8\pi} |E_0|^2 \left(\frac{e^2}{mc^2}\right)^2 |\boldsymbol{\epsilon}^* \cdot \boldsymbol{\epsilon}_0|^2 \tag{14.122}$$

Since the process is most simply viewed as a scattering, it is convenient to introduce a scattering cross section, as in Chapter 10 defined by

$$\frac{d\sigma}{d\Omega} = \frac{\text{Energy radiated/unit time/unit solid angle}}{\text{Incident energy flux in energy/unit area/unit time}} \tag{14.123}$$

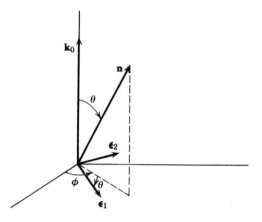

Figure 14.17

The incident energy flux is just the time-averaged Poynting vector for the plane wave, namely, $c|E_0|^2/8\pi$. Thus from (14.122) we obtain the differential scattering cross section,

$$\frac{d\sigma}{d\Omega} = \left(\frac{e^2}{mc^2}\right)^2 |\boldsymbol{\epsilon}^* \cdot \boldsymbol{\epsilon}_0|^2 \tag{14.124}$$

The scattering geometry with a choice of polarization vectors for the outgoing wave is shown in Fig. 14.17. The polarization vector $\boldsymbol{\epsilon}_1$ is in the plane containing $\mathbf{n}$ and $\mathbf{k}_0$; $\boldsymbol{\epsilon}_2$ is perpendicular to it. In terms of unit vectors parallel to the coordinate axes, $\boldsymbol{\epsilon}_1$ and $\boldsymbol{\epsilon}_2$ are

$$\boldsymbol{\epsilon}_1 = \cos\theta(\mathbf{e}_x \cos\phi + \mathbf{e}_y \sin\phi) - \mathbf{e}_z \sin\theta$$
$$\boldsymbol{\epsilon}_2 = -\mathbf{e}_x \sin\phi + \mathbf{e}_y \cos\phi$$

For an incident linearly polarized wave with polarization parallel to the x axis, the angular distribution summed over final polarizations is $(\cos^2\theta \cos^2\phi + \sin^2\phi)$, while for polarization parallel to the y axis it is $(\cos^2\theta \sin^2\phi + \cos^2\phi)$. For unpolarized incident radiation the scattering cross section is therefore

$$\frac{d\sigma}{d\Omega} = \left(\frac{e^2}{mc^2}\right)^2 \cdot \tfrac{1}{2}(1 + \cos^2\theta) \tag{14.125}$$

This is called the *Thomson formula* for scattering of radiation by a free charge, and is appropriate for the scattering of x-rays by electrons or gamma rays by protons. The angular distribution is as shown in Fig. 14.18 by the solid curve. The total scattering cross section, called the *Thomson cross section*, is

$$\sigma_T = \frac{8\pi}{3}\left(\frac{e^2}{mc^2}\right)^2 \tag{14.126}$$

The Thomson cross section is equal to 0.665×10^{-24} cm^2 for electrons. The unit of length $e^2/mc^2 = 2.82 \times 10^{-13}$ cm, is called the *classical electron radius*, since a classical distribution of charge totaling the electronic charge must have a radius of this order if its electrostatic self-energy is to equal the electron mass.

The classical Thomson formula is valid only at low frequencies where the momentum of the incident photon can be ignored. When the photon's momentum $\hbar\omega/c$ becomes comparable to or larger than mc, modifications occur. These

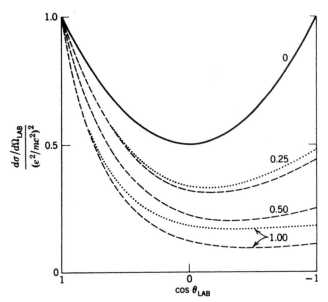

Figure 14.18 Differential scattering cross section of unpolarized radiation by a point charged particle initially at rest in the laboratory. The solid curve is the classical Thomson result. The dashed curves are the quantum-mechanical results for a spinless particle, with the numbers giving the values of $\hbar\omega/mc^2$. For $\hbar\omega/mc^2 = 0.25$, 1.0 the dotted curves show the results for spin $\frac{1}{2}$ point particles (electrons).

can be called quantum-mechanical effects, since the concept of photons as massless particles with momentum and energy is certainly quantum mechanical (*pace*, Newton!), but granting that, most of the modifications are purely kinematical. The most important change is the one observed experimentally by Compton. The energy or momentum of the scattered photon is less than the incident energy because the charged particle recoils during the collision. Applying two-body relativistic kinematics to the process, we find that the ratio of the outgoing to the incident wave number is given by the Compton formula,

$$\frac{k'}{k} = \frac{1}{1 + \dfrac{\hbar\omega}{mc^2}\,(1 - \cos\theta)}$$

where θ is the scattering angle in the laboratory (the rest frame of the target). A quantum-mechanical calculation of the scattering of photons by *spinless* point particles of charge e and mass m yields the cross section,

$$\frac{d\sigma}{d\Omega} = \left(\frac{e^2}{mc^2}\right)^2 \left(\frac{k'}{k}\right)^2 |\boldsymbol{\epsilon}^* \cdot \boldsymbol{\epsilon}_0|^2 \tag{14.127}$$

to be compared with the classical expression (14.124). In the radiation gauge the quantum-mechanical matrix element is the same as the classical amplitude. The factor $(k'/k)^2$ comes entirely from the phase space. Its presence causes the differential cross section to decrease relative to the Thomson result at large angles, as shown by the dashed curves in Fig. 14.18. Also shown in the figure by the dotted curves are the quantum-mechanical results for photon-electron scattering, that is, the scattering by a point spin $\frac{1}{2}$ particle described by the Dirac equation. The curves are generally similar to those for spinless particles, but are somewhat larger at large angles because of scattering by the electron's magnetic moment.* The integral over angles of (14.127) is elementary but slightly involved. We quote only the limiting forms for $\hbar\omega \ll mc^2$ and $\hbar\omega \gg mc^2$:

$$\frac{\sigma}{\sigma_T} = \begin{cases} 1 - 2\,\dfrac{\hbar\omega}{mc^2} + \cdots, & \hbar\omega \ll mc^2 \\[2ex] \dfrac{3}{4}\,\dfrac{mc^2}{\hbar\omega}, & \hbar\omega \gg mc^2 \end{cases} \tag{14.128}$$

For scattering by electrons the low-frequency limit is the same, but at high frequencies there is an additional multiplicative factor, $[\frac{1}{4} + \frac{1}{2}\ln(2\hbar\omega/mc^2)]$.

For protons the departures from the Thomson formula occur at photon energies above about 100 MeV. This is far below the critical energy $\hbar\omega \sim Mc^2 \sim$ 1 GeV, which would be expected in analogy with the electron Compton effect. The reason is that a proton is not a point particle like the electron with nothing but electromagnetic interactions, but is a complex entity having a spread-out charge distribution with a radius of the order of 0.8×10^{-13} cm caused by the strong interactions. The departure (a rapid increase in cross section) from Thomson scattering occurs at photon energies of the order of the rest energy of the pi meson (140 MeV).

References and Suggested Reading

The radiation by accelerated charges is at least touched on in all electrodynamics textbooks, although the emphasis varies considerably. The relativistic aspects are treated in more or less detail in

> Iwanenko and Sokolow, Sections 39–43
> Landau and Lifshitz, *Classical Theory of Fields*, Chapters 8 and 9
> Panofsky and Phillips, Chapters 18 and 19
> Sommerfeld, *Electrodynamics*, Sections 29 and 30

Extensive calculations of the radiation emitted by relativistic particles, anticipating many results rederived in the period 1940–1950, are presented in the interesting monograph by

> Schott

Synchrotron radiation has applications in astrophysics, plasma physics, condensed matter physics, material science, and biology. Synchrotrons and electron storage rings as such are discussed in detail in a classic reference,

> M. Sands, "The physics of electron storage rings," in *Proceedings of the International School of Physics Enrico Fermi, Course No. 46*, ed., B. Touschek, Academic Press, New York (1971), pp. 257–411.

---

*For electrons the cross section equivalent to (14.127) has $|\boldsymbol{\epsilon}^* \cdot \boldsymbol{\epsilon}_0|^2$ replaced by

$$|\boldsymbol{\epsilon}^* \cdot \boldsymbol{\epsilon}_0|^2 + \frac{(k - k')^2}{4kk'}\,[1 + (\boldsymbol{\epsilon}^* \times \boldsymbol{\epsilon}) \cdot (\boldsymbol{\epsilon}_0 \times \boldsymbol{\epsilon}_0^*)]$$

It is known as the *Klein–Nishina formula* for Compton scattering.

Astrophysical applications are treated in detail in

A. G. Pacholczyk, *Radio Astrophysics*, Freeman, San Francisco (1970); *Radio Galaxies*, Pergamon Press, Oxford (1977).

Plasma physics applications are discussed by

Bekefi

The classic reference on the subject of wigglers and undulators is

H. Motz, *J. Appl. Phys.* **22**, 527 (1951).

The production and characteristics of synchrotron radiation from bending magnets, undulators, and wigglers and the many uses are covered exhaustively in the five-volume work,

Handbook on Synchrotron Radiation, eds., E. E. Koch and others, Vols. 1A, 1B, 2, 3, 4, North-Holland, Amsterdam (1983–1991).

In Volume 1A, Chapter 2, S. Krinsky, M. L. Perlman, and R. E. Watson cover all of the theory and comparison with experiment.

An unpublished 1972 treatment of undulators and wigglers by E. M. Purcell, very like that of Section 14.7, appears in

Proceedings, Wiggler Workshop, SSRP Report 70/05, eds., H. Winick and T. Knight, Stanford Linear Accelerator Center (1977), p. IV-18.

The scattering of radiation by charged particles is presented clearly by

Landau and Lifshitz, *Classical Theory of Fields*, Sections 9.11–9.13, and *Electrodynamics of Continuous Media*, Chapters XIV and XV

Problems

14.1 Verify by explicit calculation that the Liénard–Wiechert expressions for *all* components of $\mathbf{E}$ and $\mathbf{B}$ for a particle moving with constant velocity agree with the ones obtained in the text by means of a Lorentz transformation. Follow the general method at the end of Section 14.1.

14.2 A particle of charge e is moving in nearly uniform nonrelativistic motion. For times near $t = t_0$, its vectorial position can be expanded in a Taylor series with fixed vector coefficients multiplying powers of $(t - t_0)$.

(a) Show that, in an inertial frame where the particle is instantaneously at rest at the origin but has a small acceleration $\mathbf{a}$, the Liénard–Wiechert electric field, correct to order $1/c^2$ inclusive, at that instant is $\mathbf{E} = \mathbf{E}_v + \mathbf{E}_a$, where the velocity and acceleration fields are

$$\mathbf{E}_v = e\frac{\hat{\mathbf{r}}}{r^2} + \frac{e}{2c^2 r}\left[\mathbf{a} - 3\hat{\mathbf{r}}(\hat{\mathbf{r}} \cdot \mathbf{a})\right]; \qquad \mathbf{E}_a = -\frac{e}{c^2 r}\left[\mathbf{a} - \hat{\mathbf{r}}(\hat{\mathbf{r}} \cdot \mathbf{a})\right]$$

and that the total electric field to this order is

$$\mathbf{E} = e\frac{\hat{\mathbf{r}}}{r^2} - \frac{e}{2c^2 r}\left[\mathbf{a} + \hat{\mathbf{r}}(\hat{\mathbf{r}} \cdot \mathbf{a})\right]$$

The unit vector $\hat{\mathbf{r}}$ points from the origin to the observation point and r is the magnitude of the distance. Comment on the r dependences of the velocity and acceleration fields. Where is the expansion likely to be valid?

(b) What is the result for the instantaneous magnetic induction $\mathbf{B}$ to the same order? Comment.

(c) Show that the $1/c^2$ term in the electric field has zero divergence and that the curl of the electric field is $\nabla \times \mathbf{E} = e(\hat{\mathbf{r}} \times \mathbf{a})/c^2 r^2$. From Faraday's law, find the magnetic induction $\mathbf{B}$ at times near $t = 0$. Compare with the familiar elementary expression.

14.3   The Heaviside–Feynman expression for the electric field of a particle of charge e in arbitrary motion, an alternative to the Liénard–Wiechert expression (14.14), is

$$\mathbf{E} = e\left[\frac{\mathbf{n}}{R^2}\right]_{\text{ret}} + e\left[\frac{R}{c}\right]_{\text{ret}}\frac{d}{dt}\left[\frac{\mathbf{n}}{R^2}\right]_{\text{ret}} + e\frac{d^2}{c^2\,dt^2}[\mathbf{n}]_{\text{ret}}$$

where the time derivatives are with respect to the time at the observation point. The magnetic field is given by (14.13).

Using the fact that the retarded time is $t' = t - R(t')/c$ and that, as a result,

$$\frac{dt}{dt'} = 1 - \boldsymbol{\beta}(t')\mathbf{n}(t')$$

show that the form above yields (14.14) when the time differentiations are performed.

14.4   Using the Liénard-Wiechert fields, discuss the time-averaged power radiated per unit solid angle in nonrelativistic motion of a particle with charge e, moving

(a)   along the z axis with instantaneous position $z(t) = a\cos\omega_0 t$,

(b)   in a circle of radius R in the x-y plane with constant angular frequency ω_0.

Sketch the angular distribution of the radiation and determine the total power radiated in each case.

14.5   A *nonrelativistic* particle of charge ze, mass m, and kinetic energy E makes a *head-on* collision with a fixed central force field of finite range. The interaction is repulsive and described by a potential $V(r)$, which becomes greater than E at close distances.

(a)   Show that the total energy radiated is given by

$$\Delta W = \frac{4}{3}\frac{z^2 e^2}{m^2 c^3}\sqrt{\frac{m}{2}}\int_{r_{\min}}^{\infty}\left|\frac{dV}{dr}\right|^2\frac{dr}{\sqrt{V(r_{\min}) - V(r)}}$$

where $r_{\min}$ is the closest distance of approach in the collision.

(b)   If the interaction is a Coulomb potential $V(r) = zZe^2/r$, show that the total energy radiated is

$$\Delta W = \frac{8}{45}\frac{zmv_0^5}{Zc^3}$$

where v_0 is the velocity of the charge at infinity.

14.6   **(a)**   Generalize the circumstances of the collision of Problem 14.5 to nonzero angular momentum (impact parameter) and show that the total energy radiated is given by

$$\Delta W = \frac{4z^2 e^2}{3m^2 c^3}\left(\frac{m}{2}\right)^{1/2}\int_{r_{\min}}^{\infty}\left(\frac{dV}{dr}\right)^2\left(E - V(r) - \frac{L^2}{2mr^2}\right)^{-1/2}dr$$

where $r_{\min}$ is the closest distance of approach (root of $E - V - L^2/2mr^2$), $L = mbv_0$, where b is the impact parameter, and v_0 is the incident speed ($E = mv_0^2/2$).

(b)   Specialize to a repulsive Coulomb potential $V(r) = zZe^2/r$. Show that ΔW can be written in terms of impact parameter as

$$\Delta W = \frac{2zmv_0^5}{Zc^3}\left[-t^{-4} + t^{-5}\left(1 + \frac{t^2}{3}\right)\tan^{-1}t\right]$$

where $t = bmv_0^2/zZe^2$ is the ratio of twice the impact parameter to the distance of closest approach in a head-on collision.

Show that in the limit of t going to zero the result of Problem 14.5b is recovered, while for $t \gg 1$ one obtains the approximate result of Problem 14.7a.

(c) Using the relation between the scattering angle θ and $t (= \cot \theta/2)$, show that ΔW can be expressed as

$$\Delta W = \frac{2zmv_0^5}{Zc^3} \tan^3 \frac{\theta}{2} \left[\frac{1}{6} (\pi - \theta)\left(1 + 3 \tan^2 \frac{\theta}{2}\right) - \tan \frac{\theta}{2} \right]$$

(d) What changes occur for an *attractive* Coulomb potential?

14.7 A nonrelativistic particle of charge ze, mass m, and initial speed v_0 is incident on a fixed charge Ze at an impact parameter b that is large enough to ensure that the particle's deflection in the course of the collision is very small.

(a) Using the Larmor power formula and Newton's second law, calculate the total energy radiated, assuming (after you have computed the acceleration) that the particle's trajectory is a straight line at constant speed:

$$\Delta W = \frac{\pi z^4 Z^2 e^6}{3m^2 c^3 v_0} \frac{1}{b^3}$$

(b) The expression found in part a is an approximation that fails at small enough impact parameter. For a repulsive potential the closest distance of approach at zero impact parameter, $r_c = 2zZe^2/mv_0^2$, serves as a length against which to measure b. The approximation will be valid for $b \gg r_c$. Compare the result of replacing b by r_c in part a with the answer of Problem 14.5 for a *head-on* collision.

(c) A radiation cross section χ (with dimensions of energy times area) can be defined classically by multiplying $\Delta W(b)$ by $2\pi b \, db$ and integrating over all impact parameters. Because of the divergence of the expression at small b, one must cut off the integration at some $b = b_{min}$. If, as in Chapter 13, the uncertainty principle is used to specify the minimum impact parameter, one may expect to obtain an approximation to the quantum-mechanical result. Compute such a cross section with the expression from part a. Compare your result with the Bethe–Heitler formula [N^{-1} times (15.30)].

14.8 A swiftly moving particle of charge ze and mass m passes a fixed point charge Ze in an approximately straight-line path at impact parameter b and nearly constant speed v. Show that the total energy radiated in the encounter is

$$\Delta W = \frac{\pi z^4 Z^2 e^6}{4m^2 c^4 \beta} \left(\gamma^2 + \frac{1}{3}\right) \frac{1}{b^3}$$

This is the relativistic generalization of the result of Problem 14.7.

14.9 A particle of mass m, charge q, moves in a plane perpendicular to a uniform, static, magnetic induction B.

(a) Calculate the total energy radiated per unit time, expressing it in terms of the constants already defined and the ratio γ of the particles's total energy to its rest energy.

(b) If at time $t = 0$ the particle has a total energy $E_0 = \gamma_0 mc^2$, show that it will have energy $E = \gamma mc^2 < E_0$ at a time t, where

$$t \approx \frac{3m^3 c^5}{2q^4 B^2} \left(\frac{1}{\gamma} - \frac{1}{\gamma_0}\right)$$

provided $\gamma \gg 1$.

(c)   If the particle is initially nonrelativistic and has a *kinetic* energy T_0 at $t = 0$, what is its kinetic energy at time t?

(d)   If the particle is actually trapped in the magnetic dipole field of the earth and is spiraling back and forth along a line of force, does it radiate more energy while near the equator, or while near its turning points? Why? Make quantitative statements if you can.

14.10   A particle of charge e moves at constant velocity βc for $t < 0$. During the short time interval, $0 < t < \Delta t$, its velocity remains in the same direction, but its speed decreases linearly in time to zero. For $t > \Delta t$, the particle remains at rest.

(a)   Show that the radiant energy emitted per unit solid angle is

$$\frac{dE}{d\Omega} = \frac{e^2\beta^2}{16\pi c \,\Delta t} \frac{(2 - \beta \cos\theta)\,[1 + (1 - \beta \cos\theta)^2]\,\sin^2\theta}{(1 - \beta \cos\theta)^4}$$

where θ is the polar angle relative to the direction of the initial velocity.

(b)   In the limit of $\gamma \gg 1$, show that the angular distribution can be expressed as

$$\frac{dE}{d\xi} \approx \frac{e^2\beta^2\gamma^4}{c\,\Delta t} \frac{\xi}{(1 + \xi)^4}$$

where $\xi = (\gamma\theta)^2$. Show that $\langle \theta^2 \rangle^{1/2} \approx \sqrt{2}/\gamma$ and that the expression for the total energy radiated is in agreement with the result from (14.43) in the same limit.

14.11   A particle of charge ze and mass m moves in external electric and magnetic fields **E** and **B**.

(a)   Show that the classical relativistic result for the instantaneous energy radiated per unit time can be written

$$P = \frac{2}{3} \frac{z^4 e^4}{m^2 c^3}\, \gamma^2 [(\mathbf{E} + \boldsymbol{\beta} \times \mathbf{B})^2 - (\boldsymbol{\beta} \cdot \mathbf{E})^2]$$

where **E** and **B** are evaluated at the position of the particle and γ is the particle's instantaneous Lorentz factor.

(b)   Show that the expression in part a can be put into the manifestly Lorentz-invariant form,

$$P = \frac{2z^4 r_0^2}{3m^2 c} \cdot F^{\mu\nu} p_\nu p^\lambda F_{\lambda\mu}$$

where $r_0 = e^2/mc^2$ is the classical charged particle radius.

14.12   As in Problem 14.4a a charge e moves in simple harmonic motion along the z axis, $z(t') = a \cos(\omega_0 t')$.

(a)   Show that the instantaneous power radiated per unit solid angle is

$$\frac{dP(t')}{d\Omega} = \frac{e^2 c \beta^4}{4\pi a^2} \frac{\sin^2\theta \cos^2(\omega_0 t')}{(1 + \beta \cos\theta \sin \omega_0 t')^5}$$

where $\beta = a\omega_0/c$.

(b)   By performing a time averaging, show that the average power per unit solid angle is

$$\frac{dP}{d\Omega} = \frac{e^2 c \beta^4}{32\pi a^2} \left[\frac{4 + \beta^2 \cos^2\theta}{(1 - \beta^2 \cos^2\theta)^{7/2}}\right] \sin^2\theta$$

(c) Make rough sketches of the angular distribution for nonrelativistic and relativistic motion.

14.13 Show explicitly by use of the Poisson sum formula or other means that, if the motion of a radiating particle repeats itself with periodicity T, the continuous frequency spectrum becomes a discrete spectrum containing frequencies that are integral multiples of the fundamental. Show that a general expression for the time-averaged power radiated per unit solid angle in each multiple m of the fundamental frequency $\omega_0 = 2\pi/T$ is:

$$\frac{dP_m}{d\Omega} = \frac{e^2\omega_0^4 m^2}{(2\pi c)^3} \left| \int_0^{2\pi/\omega_0} \mathbf{v}(t) \times \mathbf{n} \exp\left[im\omega_0\left(t - \frac{\mathbf{n} \cdot \mathbf{x}(t)}{c} \right) \right] dt \right|^2$$

14.14 **(a)** Show that for the simple harmonic motion of a charge discussed in Problem 14.12 the average power radiated per unit solid angle in the mth harmonic is

$$\frac{dP_m}{d\Omega} = \frac{e^2 c \beta^2}{2\pi a^2} m^2 \tan^2\theta\, J_m^2(m\beta \cos\theta)$$

(b) Show that in the nonrelativistic limit the total power radiated is all in the fundamental and has the value

$$P \simeq \frac{2}{3} \frac{e^2}{c^3} \omega_0^4 \overline{a^2}$$

where $\overline{a^2}$ is the mean square amplitude of oscillation.

14.15 A particle of charge e moves in a circular path of radius R in the x-y plane with a constant angular velocity ω_0.

(a) Show that the exact expression for the angular distribution of power radiated into the mth multiple of ω_0 is

$$\frac{dP_m}{d\Omega} = \frac{e^2\omega_0^4 R^2}{2\pi c^3} m^2 \left\{ \left[\frac{dJ_m(m\beta \sin\theta)}{d(m\beta \sin\theta)} \right]^2 + \frac{\cot^2\theta}{\beta^2} J_m^2(m\beta \sin\theta) \right\}$$

where $\beta = \omega_0 R/c$, and $J_m(x)$ is the Bessel function of order m.

(b) Assume nonrelativistic motion and obtain an approximate result for $dP_m/d\Omega$. Show that the results of Problem 14.4b are obtained in this limit.

(c) Assume extreme relativistic motion and obtain the results found in the text for a relativistic particle in instantaneously circular motion. [*Watson* (pp. 79, 249) may be of assistance to you.]

14.16 Exploiting the fact that $k_0 d^3N/d^3k$, the number of quanta per invariant phase-space element d^3k/k_0, is a Lorentz-invariant quantity, show that the energy radiated per unit frequency interval per unit solid angle, (14.79), can be written in the invariant and coordinate-free form

$$\hbar\omega \frac{d^3N}{d^3k} = \frac{4e^2}{3\pi^2 m^2} \left[\frac{(p\cdot k)^2}{[d^2(p\cdot k)/d\tau^2]^2} \left(\frac{d(\epsilon_1 \cdot p)}{d\tau} \right)^2 K_{2/3}^2(\xi) + \frac{(p\cdot k)(\epsilon_2 \cdot p)^2}{2[d^2(p\cdot k)/d\tau^2]} K_{1/3}^2(\xi) \right]$$

where $d\tau$ is the proper time interval of the particle of mass m, p^μ is the 4-momentum of the particle, k^μ is the 4-wave vector of the radiation, and ϵ_1, ϵ_2 are polarization vectors parallel to the acceleration and in the direction $\epsilon_1 \times \mathbf{k}$, respectively. The parameter is

$$\xi = \frac{2\sqrt{2}}{3m} \cdot \frac{(p \cdot k)^{3/2}}{(|d^2(p \cdot k)/d\tau^2|)^{1/2}}$$

This expression can be used to obtain the results of Problem 14.17 in an alternative manner. *Hint:* In proceeding with a solution, it is useful to expand $\mathbf{k} \cdot \mathbf{r}(t)$ around $t = 0$ in terms of the velocity, acceleration, etc. and compare with (14.72). One finds, for example, that $\omega c^2/\rho^2 = -\mathbf{k} \cdot d^2\mathbf{v}(0)/dt^2$, and, because the energy is constant,

$$\rho^2 = \frac{\omega E^3}{m^2 c^4 \, |d^2(p \cdot k)/dt^2|}$$

14.17   A particle of charge e and mass m moves relativistically in a helical path in a uniform magnetic field **B**. The pitch angle of the helix is α ($\alpha = 0$ corresponds to circular motion).

(a)   By arguments similar to those of Section 14.4, show that an observer far from the helix would detect radiation with a fundamental frequency

$$\omega_0 = \frac{\omega_B}{\cos^2 \alpha}$$

and that the spectrum would extend up to frequencies in the order of

$$\omega_c = \frac{3}{2} \gamma^3 \omega_B \cos \alpha$$

where $\omega_B = eB/\gamma mc$. (Take care in determining the radius of curvature ρ of the helical path.)

(b)   From part a and the results of Section 14.6, show that the *power* received by the observer per unit solid angle and per unit circular frequency interval is

$$\frac{d^2 P}{d\omega \, d\Omega} \simeq \frac{3 e^2 \gamma^2}{8 \pi^3 c} \frac{\omega_B}{\cos^2 \alpha} \left(\frac{\omega}{\omega_c}\right)^2 (1 + \gamma^2 \psi^2)^2 \left[K_{2/3}^2(\xi) + \frac{\gamma^2 \psi^2}{1 + \gamma^2 \psi^2} K_{1/3}^2(\xi) \right]$$

where ω_B and ω_c are defined above, $\xi = (\omega/2\omega_c)(1 + \gamma^2 \psi^2)^{3/2}$, and ψ is the angle of observation measured relative to the particle's velocity vector, as in Fig. 14.9.

14.18   **(a)**   By comparison of (14.91) with (14.79), show that the frequency spectrum of the received power for the situation in Problem 14.17 is

$$\frac{dP}{d\omega} = \left(\frac{\sqrt{3} e^2 \gamma \omega_B}{2 \pi c \cos \alpha}\right) G\left(\frac{\omega}{\omega_c}\right)$$

where $G(x) = x \int_x^\infty K_{5/3}(t) \, dt$ and the other symbols are as in Problem 14.17.

This expression shows that the *shape* of the power spectrum in units of ω/ω_c is unchanged by the spiraling.

(b)   Show that the integral over frequencies yields

$$P = \frac{2 e^2 \omega_B^2 \gamma^4}{3c}$$

Comparison with (14.31) shows that the total received power is independent of the pitch angle of motion.

[In doing the integration over solid angles in part a, note that $\psi = 0$ corresponds to $\theta = \pi/2 - \alpha$.]

14.19   Consider the angular and frequency spectrum of radiation produced by a magnetic moment in nonrelativistic motion, using (14.70) and the fact that a magnetization density $\mathcal{M}$ produces an effective current density $\mathbf{J}_{\text{eff}} = c\nabla \times \mathcal{M}$.

(a) Show that a magnetic moment $\boldsymbol{\mu}$ with magnetization, $\mathcal{M} = \boldsymbol{\mu}(t)\,\delta[\mathbf{x} - \mathbf{r}(t)]$, in nonrelativistic motion gives a radiation intensity (energy radiated per unit solid angle per unit frequency interval),

$$\frac{d^2 I_{\text{mag}}}{d\omega\,d\Omega} = \frac{\omega^4}{4\pi^2 c^3}\left|\int dt\,\mathbf{n} \times \boldsymbol{\mu}(t)e^{i\omega(t - \mathbf{n}\cdot\mathbf{r}(t)/c)}\right|^2$$

(b) The magnetic moment is located at the origin and is caused to precess by an external torque such that $\mu_x = \mu_0 \sin\omega_0 t$ and $\mu_z = \mu_0 \cos\omega_0 t$ for the time interval $t = -T/2$ to $t = T/2$, where $\omega_0 T/2\pi \gg 1$. Show that the frequency distribution of the radiation is very strongly peaked at $\omega = \omega_0$, that the angular distribution of radiation is proportional to $(1 + \sin^2\theta \sin^2\phi)$, and that for $T \to \infty$, the total time-averaged power radiated is

$$\langle P \rangle = \frac{2\omega_0^4}{3c^3}\mu_0^2$$

Compare the result with the power calculated by the method of Section 9.3.

14.20 Apply part a of Problem 14.19 to the radiation emitted by a magnetic moment at the origin flipping from pointing down to pointing up, with components,

$$\mu_z = \mu_0 \tanh(\nu t), \qquad \mu_x = \mu_0 \operatorname{sech}(\nu t), \qquad \mu_y = 0$$

where ν^{-1} is characteristic of the time taken to flip.

(a) Find the angular distribution of radiation and show that the intensity per unit frequency interval is

$$\frac{dI_{\text{mag}}}{dx} = \frac{4}{3}\left(\frac{\nu}{c}\right)^3 \mu_0^2 \{16(x/\pi)^4[\operatorname{cosech}^2 x + \operatorname{sech}^2 x]\}$$

where $x = \pi\omega/2\nu$ is a dimensionless frequency variable and the quantity in curly brackets is the normalized frequency distribution in x. Make a plot of this distribution and find the mean value of ω in units of ν.

(b) Apply the method of Problem 9.7 to calculate the instantaneous power and total energy radiated by the flipping dipole. Compare with the answer in part a.

14.21 Bohr's correspondence principle states that in the limit of large quantum numbers the classical power radiated in the fundamental is equal to the product of the quantum energy $(\hbar\omega_0)$ and the reciprocal mean lifetime of the transition from principal quantum number n to $(n-1)$.

(a) Using nonrelativistic approximations, show that in a hydrogen-like atom the transition probability (reciprocal mean lifetime) for a transition from a circular orbit of principal quantum number n to $(n-1)$ is given classically by

$$\frac{1}{\tau} = \frac{2}{3}\frac{e^2}{\hbar c}\left(\frac{Ze^2}{\hbar c}\right)^4 \frac{mc^2}{\hbar}\frac{1}{n^5}$$

(b) For hydrogen compare the classical value from part a with the correct quantum-mechanical results for the mean lives of the transitions $2p \to 1s$ $(1.6 \times 10^{-9}$ s$)$, $4f \to 3d$ $(7.3 \times 10^{-8}$ s$)$, $6h \to 5g$ $(6.1 \times 10^{-7}$ s$)$.

14.22 Periodic motion of charges gives rise to a discrete frequency spectrum in multiples of the basic frequency of the motion. Appreciable radiation in multiples of the fundamental can occur because of relativistic effects (Problems 14.14 and 14.15) even though the components of velocity are truly *sinusoidal*, or it can occur if the components of velocity are not sinusoidal, even though periodic. An example of this latter motion is an electron undergoing nonrelativistic elliptic motion in a hydrogen atom.

The orbit can be specified by the parametric equations

$$x = a(\cos u - \epsilon)$$
$$y = a\sqrt{1 - \epsilon^2} \sin u$$

where

$$\omega_0 t = u - \epsilon \sin u$$

a is the semimajor axis, ϵ is the eccentricity, ω_0 is the orbital frequency, and u is an angle related to the polar angle θ of the particle by $\tan(u/2) = \sqrt{(1 - \epsilon)/(1 + \epsilon)}\,\tan(\theta/2)$. In terms of the binding energy B and the angular momentum L, the various constants are

$$a = \frac{e^2}{2B}, \qquad \epsilon = \sqrt{1 - \frac{2BL^2}{me^4}}, \qquad \omega_0^2 = \frac{8B^3}{me^4}$$

(a) Show that the power radiated in the kth multiple of ω_0 is

$$P_k = \frac{4e^2}{3c^3}(k\omega_0)^4 a^2 \left\{ \frac{1}{k^2}\left[(J_k'(k\epsilon))^2 + \left(\frac{1 - \epsilon^2}{\epsilon^2}\right)J_k^2(k\epsilon) \right] \right\}$$

where $J_k(x)$ is a Bessel function of order k.

(b) Verify that for circular orbits the general result above agrees with part a of Problem 14.21.

14.23 Instead of a single charge e moving with constant velocity $\omega_0 R$ in a circular path of radius R, as in Problem 14.15, N charges q_j move with fixed relative positions θ_j around the same circle.

(a) Show that the power radiated into the mth multiple of ω_0 is

$$\frac{dP_m(N)}{d\Omega} = \frac{dP_m(1)}{d\Omega} F_m(N)$$

where $dP_m(1)/d\Omega$ is the result of part a in Problem 14.15 with $e \to 1$, and

$$F_m(N) = \left| \sum_{j=1}^{N} q_j e^{im\theta_j} \right|^2$$

(b) Show that, if the charges are all equal in magnitude and uniformly spaced around the circle, energy is radiated only into multiples of $N\omega_0$, but with an intensity N^2 times that for a single charge. Give a qualitative explanation of these facts.

(c) For the situation of part b, without detailed calculations show that for non-relativistic motion the dependence on N of the total power radiated is dominantly as β^{2N}, so that in the limit $N \to \infty$ no radiation is emitted.

(d) By arguments like those of part c show that for N relativistic particles of equal charge and symmetrically arrayed, the radiated power varies with N mainly as $e^{-2N/3\gamma^3}$ for $N \gg \gamma^3$, so that again in the limit $N \to \infty$ no radiation is emitted.

(e) What relevance have the results of parts c and d to the radiation properties of a steady current in a loop?

14.24 As an idealization of steady-state currents flowing in a circuit, consider a system of N identical charges q moving with constant *speed* v (but subject to accelerations) in an arbitrary closed path. Successive charges are separated by a constant small interval Δ.

Starting with the Liénard–Wiechert potentials for each particle, and making no assumptions concerning the speed v relative to the velocity of light show that, in the limit $N \to \infty$, $q \to 0$, and $\Delta \to 0$, but $Nq =$ constant and $q/\Delta =$ constant, no radiation is emitted by the system and the electric and magnetic fields of the system are the usual static values.

(Note that for a real circuit the stationary positive ions in the conductors neutralize the bulk charge density of the moving charges.)

14.25 **(a)** Within the framework of approximations of Section 14.6, show that, for a relativistic particle moving in a path with instantaneous radius of curvature ρ, the frequency-angle spectra of radiations with positive and negative helicity are

$$\frac{d^2 I_{\pm}}{d\omega\, d\Omega} = \frac{e^2}{6\pi^2 c} \left(\frac{\omega\rho}{c}\right)^2 \left(\frac{1}{\gamma^2} + \theta^2\right)^2 \left| K_{2/3}(\xi) \pm \frac{\theta}{\left(\dfrac{1}{\gamma^2} + \theta^2\right)^{1/2}} K_{1/3}(\xi) \right|^2$$

(b) From the formulas of Section 14.6 and part a above, discuss the polarization of the total radiation emitted as a function of frequency and angle. In particular, determine the state of polarization at (1) high frequencies ($\omega > \omega_c$) for all angles, (2) intermediate and low frequencies ($\omega < \omega_c$) for large angles, (3) intermediate and low frequencies at very small angles.

(c) See the paper by P. Joos, *Phys. Rev. Letters*, **4**, 558 (1960), for experimental comparison. See also *Handbook on Synchrotron Radiation*, (*op. cit.*), Vol. 1A, p. 139.

14.26 Consider the synchrotron radiation from the Crab nebula. Electrons with energies up to 10^{13} eV move in a magnetic field of the order of 10^{-4} gauss.

(a) For $E = 10^{13}$ eV, $B = 3 \times 10^{-4}$ gauss, calculate the orbit radius ρ, the fundamental frequency $\omega_0 = c/\rho$, and the critical frequency ω_c. What is the energy $\hbar\omega_c$ in keV?

(b) Show that for a relativistic electron of energy E in a constant magnetic field the power spectrum of synchrotron radiation can be written

$$P(E, \omega) = \text{const}\left(\frac{\omega}{E^2}\right)^{1/3} f\left(\frac{\omega}{\omega_c}\right)$$

where $f(x)$ is a cutoff function having the value unity at $x = 0$ and vanishing rapidly for $x \gg 1$ [e.g., $f \simeq \exp(-\omega/\omega_c)$], and $\omega_c = (3/2)(eB/mc)(E/mc^2)^2 \cos\theta$, where θ is the pitch angle of the helical path. Cf. Problem 14.17a.

(c) If electrons are distributed in energy according to the spectrum $N(E)\, dE \propto E^{-n}\, dE$, show that the synchrotron radiation has the power spectrum

$$\langle P(\omega)\rangle\, d\omega \propto \omega^{-\alpha}\, d\omega$$

where $\alpha = (n - 1)/2$.

(d) Observations on the radiofrequency and optical continuous spectrum from the Crab nebula show that on the frequency interval from $\omega \sim 10^8$ s^{-1} to $\omega \sim 6 \times 10^{15}$ s^{-1} the constant $\alpha \simeq 0.35$. At frequencies above 10^{18} s^{-1} the spectrum of radiation falls steeply with $\alpha \gtrsim 1.5$. Determine the index n for the electron-energy spectrum, as well as an upper cutoff for that spectrum. Is this cutoff consistent with the numbers of part a?

(e) The half-life of a particle emitting synchrotron radiation is defined as the time taken for it to lose one half of its initial energy. From the result of

Problem 14.9b, find a formula for the half-life of an electron in years when B is given in milligauss and E in GeV. What is the half-life using the numbers from part a? How does this compare with the known lifetime of the Crab nebula? Must the energetic electrons be continually replenished? From what source?

14.27   Consider the radiation emitted at twice the fundamental frequency in the average rest frame of an electron in the sinusoidal undulator of Sections 14.7.C and 14.7.D. The radiation is a coherent sum of $E1$ radiation from the $z'(t')$ motion and $E2$ radiation from the $x'(t')$ motion.

(a)   Using the techniques and notation of Chapter 9, show that the radiation-zone magnetic induction is given to sufficient accuracy by

$$\mathbf{B} = \frac{-iek'^2 a}{8} \frac{K}{\sqrt{1 + K^2/2}} \, \mathbf{n} \times [\hat{\mathbf{z}} - 4\hat{\mathbf{x}}(\mathbf{n} \cdot \hat{\mathbf{x}})]$$

where $k' = 2\bar{\gamma}k_0$, $\mathbf{n}$ is a unit vector in the direction of $\mathbf{k}'$, and a factor of $\exp[ik'(r' - ct')]/r'$ is understood.

(b)   Show that the time-averaged radiated power in the average rest frame, summed over outgoing polarizations, can be written

$$\frac{dP'}{d\Omega'} = \frac{e^2 c}{8\pi} \frac{K^2}{(1 + K^2/2)} \frac{a^2}{64} \cdot S'$$

where

$$S' = k_x'^4 + k_y'^4 + 18k_x'^2 k_y'^2 + 17k_x'^2 k_z'^2 + k_y'^2 k_z'^2 + 8k' k_x'^2 k_z'$$

(c)   Using the invariance arguments in the text in going from (14.111) to (14.118), show that the laboratory frequency spectrum of the first harmonic is

$$\frac{dP_2}{d\nu} = \frac{3}{16} P_1 \frac{K^2}{(1 + K^2/2)^2} \cdot \nu^2 (10 - 21\nu + 20\nu^2 - 6\nu^3)$$

where $\nu = k/2\bar{\gamma}^2 k_0$ and P_1 is the power in the fundamental, (14.117). For the angular range $\eta_1 < \eta < \eta_2$, the minimum and maximum ν values are $\nu_{min} = 2/(1 + \eta_2)$ and $\nu_{max} = 2/(1 + \eta_1)$. What is the total power radiated in the first harmonic?

CHAPTER 15

Bremsstrahlung, Method of Virtual Quanta, Radiative Beta Processes

In Chapter 14 we discussed radiation by accelerated charges in a general way, deriving formulas for frequency and angular distributions, and presenting examples of radiation by both nonrelativistic and relativistic charged particles in external fields. This chapter is devoted to problems of emission of electromagnetic radiation by charged particles in atomic and nuclear processes.

Particles passing through matter are scattered and lose energy by collisions, as described in detail in Chapter 13. In these collisions the particles undergo acceleration; hence they emit electromagnetic radiation. The radiation emitted during atomic collisions is customarily called *bremsstrahlung* (braking radiation) because it was first observed when high-energy electrons were stopped in a thick metallic target. For nonrelativistic particles the loss of energy by radiation is negligible compared with the collisional energy loss, but for ultrarelativistic particles radiation can be the dominant mode of energy loss.

Our discussion begins with consideration of the radiation spectrum at very low frequencies where a general expression can be derived, valid quantum mechanically as well as classically. The angular distribution, the polarization, and the integrated intensity of radiation emitted in collisions of a general sort are treated before turning to the specific phenomenon of bremsstrahlung in Coulomb collisions. When appropriate, quantum-mechanical modifications are incorporated by treating the kinematics correctly (including the energy and momentum of the photon). All important quantum effects are included in this way, sometimes leading to the exact quantum-mechanical result. Relativistic effects, which can cause significant changes in the results, are discussed in detail.

The creation or annihilation of charged particles is another process in which radiation is emitted. Such processes are purely quantum mechanical in origin. There can be no attempt at a classical explanation of the basic phenomena. But given that the process does occur, we may legitimately ask about the spectrum and intensity of electromagnetic radiation accompanying it. The sudden creation of a fast electron in nuclear beta decay, for example, can be viewed for our purposes as the violent acceleration of a charged particle initially at rest to some final velocity in a very short time interval, or, alternatively, as the sudden switching on of the charge of the moving particle in the same short interval. We discuss nuclear beta decay and orbital-electron capture in these terms in Sections 15.6 and 15.7.

In some radiative processes like bremsstrahlung it is possible to account for the major quantum-mechanical effects merely by treating the conservation of energy and momentum properly in determining the maximum and minimum effective momentum transfers. In other processes like radiative beta decay the quantum effects are more serious. Phase-space modifications occur that have no classical basis. Radiation is emitted in ways that are obscure and not easily related to the acceleration of a charge. Generally, our results are limited to the region of "soft" photons, that is, photons whose energies are small compared to the total energy available. At the upper end of the frequency spectrum our semi-classical expressions can be expected to have only qualitative validity.

15.1 *Radiation Emitted During Collisions*

If a charged particle makes a collision, it undergoes acceleration and emits radiation. If its collision partner is also a charged particle, they both emit radiation, and a coherent superposition of the radiation fields must be made. Since the amplitude of the radiation fields depends on the charge times the acceleration, the lighter particle will radiate more, provided the charges are not too dissimilar. In many applications the mass of one collision partner is much greater than the mass of the other. Then for the emission of radiation it is sufficient to treat the collision as the interaction of the lighter of the two particles with a fixed field of force. We will consider only this situation, leaving more involved cases to the problems at the end of the chapter.

A. *Low-Frequency Limit*

From (14.65) and (14.66) we see that the intensity of radiation emitted by a particle of charge ze during the collision can be expressed as

$$\frac{d^2I}{d\omega \, d\Omega} = \frac{z^2e^2}{4\pi^2c} \left| \int \frac{d}{dt} \left[\frac{\mathbf{n} \times (\mathbf{n} \times \boldsymbol{\beta})}{1 - \mathbf{n} \cdot \boldsymbol{\beta}} \right] e^{i\omega(t - \mathbf{n} \cdot \mathbf{r}(t)/c)} \, dt \right|^2 \tag{15.1}$$

Let us suppose that the collision has a duration τ during which significant acceleration occurs and that the collision changes the particle's velocity from an initial value $c\boldsymbol{\beta}$ to a final value $c\boldsymbol{\beta}'$. The spectrum of radiation at finite frequencies will depend on the details of the collision, but its form at low frequencies depends only on the initial and final velocities. In the limit $\omega \to 0$ the exponential factor in (15.1) is equal to unity. Then the integrand is a perfect differential. The spectrum of radiation with polarization $\boldsymbol{\epsilon}$ is therefore

$$\lim_{\omega \to 0} \frac{d^2I}{d\omega \, d\Omega} = \frac{z^2e^2}{4\pi^2c} \left| \boldsymbol{\epsilon}^* \cdot \left(\frac{\boldsymbol{\beta}'}{1 - \mathbf{n} \cdot \boldsymbol{\beta}'} - \frac{\boldsymbol{\beta}}{1 - \mathbf{n} \cdot \boldsymbol{\beta}} \right) \right|^2 \tag{15.2}$$

The result (15.2) is very general and holds quantum mechanically as well as classically. To establish the connection to the quantum-mechanical form, we first convert (15.2) into a spectrum of photons. The energy of a photon of frequency ω is $\hbar\omega$. By dividing (15.2) by $\hbar^2\omega$ we therefore obtain the differential *number* spectrum per unit *energy* interval and per unit solid angle of "*soft*" *photons* ($\hbar\omega \to 0$) of polarization $\boldsymbol{\epsilon}$:

$$\lim_{\hbar\omega \to 0} \frac{d^2N}{d(\hbar\omega) \, d\Omega_\gamma} = \frac{z^2\alpha}{4\pi^2\hbar\omega} \left| \boldsymbol{\epsilon}^* \cdot \left(\frac{\boldsymbol{\beta}'}{1 - \mathbf{n} \cdot \boldsymbol{\beta}'} - \frac{\boldsymbol{\beta}}{1 - \mathbf{n} \cdot \boldsymbol{\beta}} \right) \right|^2 \tag{15.3}$$

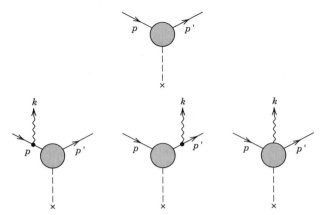

Figure 15.1 Quantum-mechanical diagrams describing the scattering of a particle without photon emission (top) and with the emission of a photon (bottom).

where $\alpha = e^2/\hbar c \simeq 1/137$ is the fine structure constant if e is the proton's charge. The subscript γ on the solid-angle element serves to remind us that it is the solid angle into which the photon goes. The spectrum (15.3) is to be interpreted as follows. Suppose that the collision is caused by an external potential or other interaction. Let the cross section for scattering that causes a change in velocity $c\boldsymbol{\beta} \to c\boldsymbol{\beta}'$ be denoted by $d\sigma/d\Omega_p$, where p stands for particle. Then the cross section for scattering and at the same time for producing a soft photon of energy $\hbar\omega$, per unit energy interval and per unit solid angle, is

$$\frac{d^3\sigma}{d\Omega_p\, d(\hbar\omega)\, d\Omega_\gamma} = \left[\lim_{\hbar\omega\to 0} \frac{d^2N}{d(\hbar\omega)\, d\Omega_\gamma}\right] \cdot \frac{d\sigma}{d\Omega_p} \tag{15.4}$$

The expression (15.3) can be made to appear more relativistically covariant by introducing the energy-momentum 4-vectors of the photon, $k^\mu = (\hbar/c)(\omega, \omega\mathbf{n})$, and of the particle, $p^\mu = Mc(\gamma, \gamma\boldsymbol{\beta})$. It is also valuable to make use of the Lorentz-invariant phase space d^3k/k_0 to write a manifestly invariant expression,*

$$\frac{d^3N}{(d^3k/k_0)} \equiv \frac{c^2}{\hbar\omega} \frac{d^2N}{d(\hbar\omega)\, d\Omega_\gamma} = \frac{c^2}{\hbar(\hbar\omega)^2} \frac{d^2I}{d\omega\, d\Omega_\gamma} \tag{15.5}$$

Then we find from (15.3),

$$\lim_{\hbar\omega\to 0} \frac{d^3N}{(d^3k/k_0)} = \frac{z^2\alpha}{4\pi^2} \left|\frac{\epsilon^* \cdot p'}{k \cdot p'} - \frac{\epsilon^* \cdot p}{k \cdot p}\right|^2 \tag{15.6}$$

where the various scalar products are 4-vector scalar products [in the radiation gauge, $\epsilon^\mu = (0, \boldsymbol{\epsilon})$]. That (15.6) emerges from a quantum-mechanical calculation can be made plausible by considering Fig. 15.1. The upper diagram indicates the scattering process without emission of radiation. The lower three diagrams have scattering and also photon emission. Their contributions add coherently. The two diagrams on the left have the photon emitted by the external lines, that is, before

---

*The fact that ω^{-2} times $d^2I/d\omega\, d\Omega$ is a Lorentz invariant is not restricted to the limit of $\omega \to 0$. We find this result useful in some of our later discussions.

or after the collision; both involve propagators for the particle between the scattering vertex and the photon vertex of the form

$$\frac{1}{(p \pm k)^2 - M^2} = \frac{1}{\pm 2p \cdot k}$$

In the limit $\omega \to 0$ these propagators make the contributions from these two diagrams singular and provide the $(\hbar\omega)^{-1}$ in (15.3). On the other hand, the diagram on the right has the photon emitted from the interior of the scattering vertex. Its contribution is finite as $\omega \to 0$ and so is negligible compared to the first two. The explicit calculation yields (15.4) with (15.6) in the limit that the energy and the momentum of the photon can be neglected in the kinematics. Soft photon emission occurs only from the external lines in any process and is given by the classical result.

B. Polarization and Spectrum Integrated over Angles

Some limiting forms of (15.2) are of interest. If the particle moves *non-relativistically* before and after the collision, then the factors in the denominators can be put equal to unity. The radiated intensity becomes

$$\lim_{\omega \to 0} \frac{d^2 I_{NR}}{d\omega \, d\Omega} = \frac{z^2 e^2}{4\pi^2 c} |\boldsymbol{\epsilon}^* \cdot \Delta\boldsymbol{\beta}|^2 \tag{15.7}$$

where $\Delta\boldsymbol{\beta} = \boldsymbol{\beta}' - \boldsymbol{\beta}$ is the change in velocity in the collision. This is just a dipole radiation pattern and gives, when summed over polarizations, and integrated over angles, the total energy radiated per unit frequency interval per nonrelativistic collision,

$$\lim_{\omega \to 0} \frac{d I_{NR}}{d\omega} = \frac{2 z^2 e^2}{3\pi c} |\Delta\boldsymbol{\beta}|^2 \tag{15.8}$$

For *relativistic* motion in which the change in velocity $\Delta\boldsymbol{\beta}$ is small, (15.2) can be approximated to lowest order in $\Delta\boldsymbol{\beta}$ as

$$\lim_{\omega \to 0} \frac{d^2 I}{d\omega \, d\Omega} \simeq \frac{z^2 e^2}{4\pi^2 c} \left| \boldsymbol{\epsilon}^* \cdot \left(\frac{\Delta\boldsymbol{\beta} + \mathbf{n} \times (\boldsymbol{\beta} \times \Delta\boldsymbol{\beta})}{(1 - \mathbf{n} \cdot \boldsymbol{\beta})^2} \right) \right|^2 \tag{15.9}$$

where $c\boldsymbol{\beta}$ is the initial (or average) velocity.

We now consider the explicit forms of the angular distribution of radiation emitted with a definite state of polarization. In collision problems it is usual that the direction of the incident particle is known and the direction of the radiation is known, but the deflected particle's direction, and consequently that of $\Delta\boldsymbol{\beta}$, are not known. Consequently the plane containing the incident beam direction and the direction of the radiation is a natural one with respect to which one specifies the state of polarization of the radiation.

For simplicity we consider a small angle deflection so that $\Delta\boldsymbol{\beta}$ is approximately perpendicular to the incident direction. Figure 15.2 shows the vectorial relationships. Without loss of generality $\mathbf{n}$, the observation direction, is chosen in the x-z plane, making an angle θ with the incident beam. The change in velocity $\Delta\boldsymbol{\beta}$ lies in the x-y plane, making an angle ϕ with the x axis. Since the direction

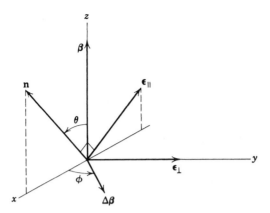

Figure 15.2

of the scattered particle is not observed, we will average over ϕ. The unit vectors $\boldsymbol{\epsilon}_\|$ and $\boldsymbol{\epsilon}_\perp$ are polarization vectors parallel and perpendicular to the plane containing $\boldsymbol{\beta}$ and $\mathbf{n}$.

We leave to Problem 15.6 the demonstration that (15.9) gives the expressions (averaged over ϕ)

$$\lim_{\omega \to 0} \frac{d^2 I_\|}{d\omega \, d\Omega} = \frac{z^2 e^2}{8\pi^2 c} |\Delta\boldsymbol{\beta}|^2 \frac{(\beta - \cos\theta)^2}{(1 - \beta\cos\theta)^4}$$
$$\lim_{\omega \to 0} \frac{d^2 I_\perp}{d\omega \, d\Omega} = \frac{z^2 e^2}{8\pi^2 c} |\Delta\boldsymbol{\beta}|^2 \frac{1}{(1 - \beta\cos\theta)^2}$$

(15.10)

for the low-frequency limits for the two states of linear polarization. These angular distributions are valid for small-angle collisions of all types. The polarization $P(\theta)$, defined as $(d^2 I_\perp - d^2 I_\|)/(d^2 I_\perp + d^2 I_\|)$, vanishes at $\theta = 0$, has a maximum value of $+1$ at $\cos\theta = \beta$, and decreases monotonically for larger angles. For $\gamma \gg 1$, it has the form, $P(\theta) \simeq 2\gamma^2\theta^2/(1 + \gamma^4\theta^4)$. This qualitative behavior is observed experimentally,* but departures from the $\omega \to 0$ limit are significant even for $\omega/\omega_{max} = 0.1$.

The sum of the two terms in (15.10) gives the angular distribution of soft radiation emitted in an arbitrary small-angle collision ($\Delta\boldsymbol{\beta}$ small in magnitude and perpendicular to the incident direction). For relativistic motion the distribution is strongly peaked in the forward direction in the by-now familiar fashion, with a mean angle of emission of the order of $\gamma^{-1} = Mc^2/E$. Explicitly, in the limit $\gamma \gg 1$ we have

$$\lim_{\omega \to 0} \frac{d^2 I}{d\omega \, d\Omega} \simeq \frac{z^2 e^2 \gamma^4 |\Delta\boldsymbol{\beta}|^2}{\pi^2 c} \frac{(1 + \gamma^4\theta^4)}{(1 + \gamma^2\theta^2)^4}$$

(15.11)

The total intensity per unit frequency interval for arbitrary velocity is found by elementary integration from (15.10) to be

$$\lim_{\omega \to 0} \frac{dI}{d\omega} = \frac{2}{3\pi} \frac{z^2 e^2}{c} \gamma^2 |\Delta\boldsymbol{\beta}|^2$$

---

*Some data for electron bremsstrahlung are given by W. Lichtenberg, A. Przybylski, and M. Scheer, *Phys. Rev. A* **11**, 480 (1975).

For nonrelativistic motion this reduces to (15.8). Since the particle's momentum is $\mathbf{p} = \gamma Mc\boldsymbol{\beta}$, this result can be written as

$$\lim_{\omega \to 0} \frac{dI}{d\omega} = \frac{2}{3\pi} \frac{z^2 e^2}{M^2 c^3} Q^2 \tag{15.12}$$

where $Q = |\mathbf{p}' - \mathbf{p}|$ is the magnitude of the momentum transfer in the collision.

Equations (15.10) and (15.12) are valid relativistically, as well as nonrelativistically, provided the change in velocity is not too large. For relativistic motion the criterion is

$$|\Delta\boldsymbol{\beta}| < \frac{2}{\gamma} \quad \text{or} \quad Q < 2Mc \tag{15.13}$$

This can be seen from (15.2). If the two velocities $\boldsymbol{\beta}$ and $\boldsymbol{\beta}'$ have an angle $|\Delta\boldsymbol{\beta}|/\beta$ between them of more than $2/\gamma$, the two terms in the amplitude will not interfere. When the direction of emission $\mathbf{n}$ is such that one of them is large, the other is negligible. The angular distribution will be two searchlight beams, one centered along $\boldsymbol{\beta}$ and the other along $\boldsymbol{\beta}'$, each given by the absolute square of one term. The radiated intensity integrated over angles is then approximately

$$\lim_{\substack{\omega \to 0 \\ Q > 2Mc}} \frac{dI}{d\omega} \simeq \frac{4z^2 e^2}{\pi c} \ln (Q/Mc) \tag{15.14}$$

For $Q > 2Mc$ the radiated intensity of soft photons is logarithmically dependent on Q^2, in contrast to the linear increase with Q^2 shown by (15.12) for smaller momentum transfers. For nonrelativistic motion the momentum transfers are always less than the limit of (15.13). The intensity is therefore given by (15.12) for all momentum transfers.

C. Qualitative Behavior at Finite Frequencies

So far we have concentrated on the very-low-frequency limit of (15.1). It is time to consider the qualitative behavior of the spectrum at finite frequencies. The phase factor in (15.1) controls the behavior at finite frequencies. Appreciable radiation occurs only when the phase changes relatively little during the collision. If the coordinate $\mathbf{r}(t)$ of the particle is written as

$$\mathbf{r}(t) = \mathbf{r}(0) + \int_0^t c\boldsymbol{\beta}(t') \, dt'$$

then, apart from a constant, the phase of the integrand in (15.1) is

$$\Phi(t) = \omega\left(t - \mathbf{n} \cdot \int_0^t \boldsymbol{\beta}(t') \, dt'\right)$$

If we imagine that the collision occurs during a time τ and that $\boldsymbol{\beta}$ changes relatively smoothly from its initial to final value, the criterion for appreciable radiation is

$$\omega\tau(1 - \mathbf{n} \cdot \langle\boldsymbol{\beta}\rangle) < 1 \tag{15.15}$$

where $\langle \boldsymbol{\beta} \rangle = (1/\tau) \int_0^\tau \boldsymbol{\beta}(t) \, dt$ is the average value of $\boldsymbol{\beta}$ during the collision. For nonrelativistic collisions this reduces to

$$\omega\tau < 1$$

At low frequencies the radiated intensity is given by (15.7), but for $\omega\tau > 1$ the oscillating phase factor will cause the integral to be much smaller than when $\omega = 0$. The intensity will thus fall rapidly to zero for $\omega > 1/\tau$. For relativistic motion the situation is more complex. For small $|\Delta\boldsymbol{\beta}|$ but with $\gamma \gg 1$ the criterion (15.15) is approximately

$$\frac{\omega\tau}{2\gamma^2} (1 + \gamma^2\theta^2) < 1 \tag{15.16}$$

Now there is angular dependence. For $\omega\tau < 1$, there is significant radiation at all angles that matter. For $\omega\tau$ on the range, $1 < \omega\tau < \gamma^2$, there is appreciable radiation only out to angles of the order of θ_{max}, where $\theta_{max}^2 = 1/\omega\tau$. For $\omega\tau > \gamma^2$, (15.16) is not satisfied at any angle. Hence the spectrum of radiation in relativistic collisions is given approximately by (15.11) and (15.12) provided $\omega\tau \ll \gamma^2$, but modifications occur in the angular distribution as $\omega\tau$ approaches γ^2, and the intensity at all angles decreases rapidly for $\omega \gtrsim \gamma^2/\tau$.

15.2 Bremsstrahlung in Coulomb Collisions

The most common situation in which a continuum of radiation is emitted is in the collision of a fast particle with an atom. Because of its greater charge, the nucleus is more effective at producing deflections of the incident particle than the electrons. Consequently we ignore the effects of the electrons for the present and consider the radiation emitted in the collision of a particle of charge ze, mass M, and initial velocity $c\boldsymbol{\beta}$ with the Coulomb field of a fixed point charge Ze.

The elastic scattering of a charged particle by a static Coulomb field is given by the Rutherford formula (see Section 13.1):

$$\frac{d\sigma_s}{d\Omega'} = \left(\frac{2zZe^2}{pv}\right)^2 \cdot \frac{1}{(2 \sin \theta'/2)^4} \tag{15.17}$$

where θ' is the scattering angle of the particle. This cross section is correct non-relativistically at all angles, and is true quantum mechanically for the relativistic small-angle scattering of any particle. It is convenient to express (15.17) as a cross section for scattering per unit interval in momentum transfer Q. For elastic scattering,

$$Q^2 = 4p^2\sin^2(\theta'/2) = 2p^2(1 - \cos \theta') \tag{15.18}$$

With $d\Omega' = d\phi' \, d\cos \theta' = -Q \, d\phi' \, dQ/p^2$, integration over azimuth of (15.17) gives

$$\frac{d\sigma_s}{dQ} = 8\pi \left(\frac{zZe^2}{\beta c}\right)^2 \cdot \frac{1}{Q^3} \tag{15.19}$$

In a Coulomb collision with momentum transfer Q the incident particle is accelerated and emits radiation. From Section 15.1 we know that the angular distribution is given by (15.10), at least for small deflections, and the integrated intensity by (15.12). Since the angular distributions have already been discussed, we focus on the frequency spectrum, integrated over angles. In analogy with (15.4) we define the *differential radiation cross section*,

$$\frac{d^2\chi}{d\omega \, dQ} = \frac{dI(\omega, Q)}{d\omega} \cdot \frac{d\sigma_s}{dQ}(Q) \tag{15.20}$$

where $dI(\omega, Q)/d\omega$ is the energy radiated per unit frequency interval in a collision with momentum transfer Q. The differential radiation cross section has dimensions of (area × energy/frequency × momentum). The cross section for photon emission per unit energy interval is obtained by dividing by $\hbar^2\omega$.

The low-frequency radiation spectrum is given by (15.12), provided Q is not too large. Inserting both (15.12) and (15.19) into (15.20) we obtain

$$\frac{d^2\chi}{d\omega \, dQ} = \frac{16}{3} \frac{Z^2 e^2}{c} \left(\frac{z^2 e^2}{Mc^2}\right)^2 \frac{1}{\beta^2} \cdot \frac{1}{Q} \tag{15.21}$$

This result is valid at frequencies and momentum transfers low enough to ensure that the criteria of Section 15.1 are satisfied. The radiation cross section integrated over momentum transfers is

$$\frac{d\chi}{d\omega} \simeq \frac{16}{3} \frac{Z^2 e^2}{c} \left(\frac{z^2 e^2}{Mc^2}\right)^2 \cdot \frac{1}{\beta^2} \int_{Q_{\min}}^{Q_{\max}} \frac{dQ}{Q}$$

or $\tag{15.22}$

$$\frac{d\chi}{d\omega} \simeq \frac{16}{3} \frac{Z^2 e^2}{c} \left(\frac{z^2 e^2}{Mc^2}\right)^2 \cdot \frac{1}{\beta^2} \ln\left(\frac{Q_{\max}}{Q_{\min}}\right)$$

In summing over momentum transfers we have incorporated the limitations on the range of validity of (15.21) by means of maximum and minimum values of Q. At any given frequency (15.21) describes approximately the differential radiation cross section for only a limited range of Q. Outside that range the cross section falls below the estimate (15.21) because one or the other of the factors in (15.20) is much smaller than (15.12) or (15.19) (or zero). This effectively limits the range of Q and leads to (15.22). Determination of the values of $Q_{\max}$ and $Q_{\min}$ for different physical circumstances is our next task.

A. Classical Bremsstrahlung

In our discussion of energy loss in Chapter 13 we saw that classical considerations were applicable provided

$$\eta = \frac{zZe^2}{\hbar v} > 1$$

For particles of modest charges this means $\beta \ll 1$. In this nonrelativistic limit the maximum effective momentum transfer is not restricted by failure of (15.12). The only limit is kinematic. From (15.18) we see that

$$Q_{\max} = 2p = 2Mv \tag{15.23}$$

The lower limit on Q is determined classically by the relation between frequency and collision time that must be satisfied if there is to be significant radiation. From Section 11.10 and Problem 13.1 we have

$$\frac{1}{\tau} \simeq \frac{v}{b}, \qquad Q \simeq \frac{2zZe^2}{bv}$$

so that the condition $\omega < 1/\tau$ can be written in terms of Q as

$$Q > Q^{(c)}_{\min} = \frac{2zZe^2\omega}{v^2} \tag{15.24}$$

The *classical* radiation cross section is therefore

$$\frac{d\chi_c}{d\omega} \simeq \frac{16}{3}\frac{Z^2e^2}{c}\left(\frac{z^2e^2}{Mc^2}\right)^2 \cdot \frac{1}{\beta^2} \cdot \ln\left(\frac{\lambda Mv^3}{zZe^2\omega}\right) \tag{15.25}$$

where λ is a number of order unity that takes into account our ignorance of exactly how the intensity falls to zero around $\omega\tau = 1$. This cross section is meaningful only provided the argument of the logarithm is greater than unity. There is thus an upper limit $\omega^{(c)}_{\max}$ on the frequency spectrum. Phrased in terms of a photon energy it is

$$\hbar\omega^{(c)}_{\max} = \frac{2\lambda}{\eta}\left(\frac{Mv^2}{2}\right) \tag{15.26}$$

Since η is large compared to unity in this classical situation, we find that the range of photon energies is limited to very soft quanta whose energies are all very small compared to the kinetic energy of the incident particle. For $\eta = 10$ the classical spectrum is shown in Fig. 15.3, with $\lambda = 2$ (chosen so that for $\eta = 1$ and $\omega = 0$ the classical and quantum-mechanical cross sections agree).

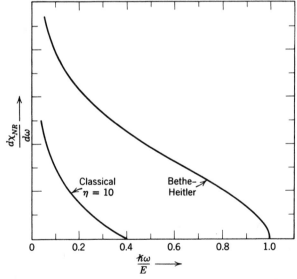

Figure 15.3 Radiation cross section (energy × area/unit frequency) for nonrelativistic Coulomb collisions as a function of frequency in units of the maximum frequency ($E/\hbar$). The classical spectrum is confined to very low frequencies. The curve marked "Bethe–Heitler" is the quantum-mechanical Born approximation result, i.e., (15.29) with $\lambda' = 1$.

B. Nonrelativistic Bremsstrahlung

In the classical limit the energy and the momentum of the photon were not considered. A posteriori such neglect was justified because (15.26) shows that the spectrum is confined to very low-energy photons. But for fast, though still non-relativistic, particles with $\eta < 1$, it is necessary to consider conservation of energy and momentum including the photon. For scattering by a fixed (or massive) center of force, the conservation requirements are

$$E = E' + \hbar\omega \tag{15.27}$$
$$Q^2 = (\mathbf{p} - \mathbf{p}' - \mathbf{k})^2 \simeq (\mathbf{p} - \mathbf{p}')^2$$

where $E = p^2/2M$ and $E' = p'^2/2M$ are the *kinetic* energies of the particle before and after the collision, $\hbar\omega$ and $\mathbf{k} = \hbar\omega\mathbf{n}/c$ are the energy and momentum of the photon, and Q is the momentum transfered to the scattering center, as before. The reader can verify that the neglect of the photon's momentum $\mathbf{k}$ in the second expression for Q^2 is justified independently of the directions of the momenta provided the particles are nonrelativistic.

The maximum momentum transfer effective for radiation is again that allowed by the kinematics. Similarly the minimum effective Q is determined by the kinematics and not by the collision time.* From the second equation in (15.27) we obtain

$$\frac{Q_{max}}{Q_{min}} = \frac{p + p'}{p - p'} = \frac{(\sqrt{E} + \sqrt{E - \hbar\omega})^2}{\hbar\omega} \tag{15.28}$$

The second form is obtained by using conservation of energy. The radiation cross section (15.22) is therefore

$$\frac{d\chi_{NR}}{d\omega} \simeq \frac{16}{3} \frac{Z^2 e^2}{c} \left(\frac{z^2 e^2}{Mc^2}\right)^2 \cdot \frac{1}{\beta^2} \cdot \ln \left[\frac{\lambda'(\sqrt{E} + \sqrt{E - \hbar\omega})^2}{\hbar\omega}\right] \tag{15.29}$$

where again λ' is a number expected to be of order unity. Actually, with $\lambda' = 1$, (15.29) is exactly the quantum-mechanical result in the Born approximation, first calculated by Bethe and Heitler (1934). The shape of the radiation cross section as a function of frequency is shown in Fig. 15.3.

The fact that we have obtained the correct quantum-mechanical Born approximation cross section by semiclassical arguments in which the quantum aspects were included only in the kinematics can be understood from the considerations of Section 15.1, especially Fig. 15.1. In the Born approximation the scattering vertex, drawn as a blob there to indicate complicated things going on, reduces to a point vertex like the photon-particle vertices. The third diagram at the bottom is absent. Only the external lines radiate; the amplitude is given by (15.6); the exact kinematics and phase space conspire to yield (15.29).

The radiation cross section $d\chi/d\omega$ depends on the properties of the particles involved in the collision as $Z^2 z^4/M^2$, showing that the emission of radiation is most important for electrons in materials of high atomic number. The total energy

---

*For soft photons $Q_{min} = p - p'$ can be approximated by $Q_{min} \simeq 2\hbar\omega/v$, while the classical expression (15.24) is $Q_{min}^{(c)} = 2\eta\hbar\omega/v$. With $\eta < 1$, the quantum-mechanical Q_{min} is larger than the classical and so governs the lower cutoff in Q. For more energetic photons $(p - p')$ is even larger. In relativistic collisions $Q_{min}^{(c)}$ is γ^{-3} times its nonrelativistic value and so is much smaller than the quantum minimum [see (15.33)].

lost in radiation by a particle traversing unit thickness of matter containing N fixed charges Ze (atomic nuclei) per unit volume is

$$\frac{dE_{\text{rad}}}{dx} = N \int_0^{\omega_{\text{max}}} \frac{d\chi(\omega)}{d\omega} \, d\omega$$

Using (15.29) for $d\chi/d\omega$ and converting to the variable of integration $x = (\hbar\omega/E)$, we can write the radiative energy loss as

$$\frac{dE_{\text{rad}}}{dx} = \frac{16}{3} NZ \left(\frac{Ze^2}{\hbar c}\right) \frac{z^4 e^4}{Mc^2} \int_0^1 \ln\left(\frac{1 + \sqrt{1-x}}{\sqrt{x}}\right) dx \qquad (15.30)$$

The dimensionless integral has the value unity. For comparison we write the ratio of radiative energy loss to collision energy loss (13.14):

$$\frac{dE_{\text{rad}}}{dE_{\text{coll}}} \simeq \frac{4}{3\pi} z^2 \frac{Z}{137} \frac{m}{M} \left(\frac{v}{c}\right)^2 \frac{1}{\ln B_q} \qquad (15.31)$$

For nonrelativistic particles ($v \ll c$) the radiative loss is completely negligible compared to the collision loss. The fine structure constant ($e^2/\hbar c = 1/137$) enters characteristically whenever there is emission of radiation as an additional step beyond the basic process (here the deflection of the particle in the nuclear Coulomb field). The factor m/M appears because the radiative loss involves the acceleration of the incident particle, while the collision loss involves the acceleration of an electron.

C. Relativistic Bremsstrahlung

For relativistic particles the limits obtained from conservation of energy must be modified. The changes are of two sorts. The first is that the maximum effective Q value is no longer determined by kinematics. It was shown in Section 15.1 that (15.12) is valid only for $Q < 2Mc$. For larger Q the radiated intensity is logarithmic in Q and given by (15.14). Because of the Q^{-3} behavior of (15.19) this means that Q_{max} in (15.22) is

$$Q_{\text{max}} \simeq 2Mc \qquad (15.32)$$

The second modification is that the photon's *momentum* can no longer be ignored in determining the minimum momentum transfer from (15.27). The minimum clearly occurs when all three momenta are parallel:

$$Q_{\text{min}} = p - p' - k$$

For relativistic motion of the particle both initially and finally (even though the photon may carry off appreciable energy), we can approximate $cp \simeq E - M^2c^4/2E$, $cp' \simeq E' - M^2c^4/2E'$, where now E and E' are the total energies. Then we obtain

$$Q_{\text{min}} \simeq \frac{M^2 c^3 \hbar \omega}{2EE'} \qquad (15.33)$$

With (15.32) and (15.33), the radiation cross section (15.22) becomes

$$\frac{d\chi_R}{d\omega} \simeq \frac{16}{3} \frac{Z^2 e^2}{c} \left(\frac{z^2 e^2}{Mc^2}\right)^2 \ln\left(\frac{\lambda'' EE'}{Mc^2 \hbar \omega}\right) \qquad (15.34)$$

with the customary λ'' of order unity. This result is the same as is obtained quantum mechanically in the relativistic limit, provided the photon energy satisfies $\hbar\omega \ll E$. In the limit of $E, E' \gg Mc^2$, the quantum formula is

$$\left(\frac{d\chi_R}{d\omega}\right)_{\text{Born}} \simeq \frac{16}{3} \frac{Z^2 e^2}{c} \left(\frac{z^2 e^2}{Mc^2}\right)^2 \left(1 - \frac{\hbar\omega}{E} + \frac{3\hbar^2\omega^2}{4E^2}\right) \left[\ln\left(\frac{2EE'}{Mc^2\hbar\omega}\right) - \frac{1}{2}\right] \quad (15.35)$$

We note in passing that since $Q_{\max} \simeq 2Mc$, the small change in velocity $\Delta\boldsymbol{\beta}$ always lies in the plane perpendicular to the incident direction in a relativistic collision. The angular distribution of the radiation is thus given by (15.11). The *doubly differential* radiation cross section for energy radiated per unit frequency interval and per unit solid angle for $\hbar\omega \ll E$ is then

$$\frac{d^2\chi_R}{d\omega\, d\Omega_\gamma} \simeq \left[\frac{3}{2\pi} \gamma^2 \frac{(1 + \gamma^4\theta^4)}{(1 + \gamma^2\theta^2)^4}\right] \cdot \frac{d\chi_R}{d\omega} \quad (15.36)$$

where θ is the angle of emission of the photon and $d\chi_R/d\omega$ is given by (15.34). The smallness of $Q_{\max}/p$ justifies the use of the relativistic Rutherford formula (15.19) without quantum-mechanical corrections for spin.

D. Relativistic Bremsstrahlung by a Lorentz Transformation

It is instructive to consider the calculation of relativistic bremsstrahlung from a somewhat different point of view. Suppose that instead of using the laboratory frame where the force center is at rest we view the process as taking place in the rest frame K' of the initial particle. The emission process as it appears in the two frames is indicated schematically in Fig. 15.4. A small-angle deflection in the laboratory corresponds to nonrelativistic motion during the whole collision in the frame K'. The differential radiated intensity in K' is thus given by the sum of the two terms in (15.10) with $\beta = 0$:

$$\frac{d^2 I'}{d\omega'\, d\Omega'} = \frac{z^2 e^2}{8\pi^2 c} |\Delta\boldsymbol{\beta}'|^2 (1 + \cos^2\theta')$$

where primes denote quantities evaluated in the frame K'. The change in velocity can be written for nonrelativistic motion as $\Delta\boldsymbol{\beta}' = \Delta\mathbf{p}'/Mc$, where $\Delta\mathbf{p}'$ is the change of momentum in K'. For a small deflection in the laboratory, $\Delta\mathbf{p}'$ is perpendicular to the direction of motion and so is the same in the laboratory as in

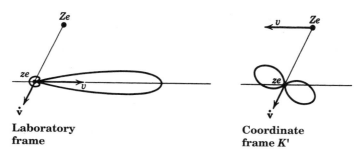

Laboratory frame   **Coordinate frame K'**

Figure 15.4   Radiation emitted during relativistic collisions viewed from the laboratory (nucleus at rest) and the frame K' (incident particle essentially at rest).

K'. Its magnitude is the momentum transfer Q. The radiated energy spectrum can therefore be written as

$$\frac{d^2 I'}{d\omega' \, d\Omega'} = \frac{z^2 e^2}{8\pi^2 c} \left(\frac{Q}{Mc}\right)^2 (1 + \cos^2 \theta')$$

The triply differential radiation cross section for emission of radiation per unit frequency interval, per unit solid angle, and per unit interval in momentum transfer is, in analogy to (15.20),

$$\frac{d^3 \chi'}{d\omega' \, d\Omega' \, dQ} = \frac{z^2 e^2}{8\pi^2 c} \left(\frac{Q}{Mc}\right)^2 \cdot \frac{d\sigma_s}{dQ} \cdot (1 + \cos^2 \theta') \tag{15.37}$$

This is the cross section in frame K'. No primes appear on $d\sigma_s/dQ$ or on Q^2 because to the extent that $\mathbf{Q}$ is transverse, these quantities are obviously invariant under Lorentz transformations.*

The emission of radiation in the frame K' appears as simple dipole radiation in (15.37). To obtain the cross section in the laboratory we must make a Lorentz transformation. In Section 15.1 we saw that (15.5) is a Lorentz-invariant quantity. With what has just been said about $d\sigma_s/dQ$, it is clear that the equation relating the differential cross sections in the two frames is

$$\frac{1}{\omega^2} \frac{d^3 \chi}{d\omega \, d\Omega \, dQ} = \frac{1}{\omega'^2} \frac{d^3 \chi'}{d\omega' \, d\Omega' \, dQ} \tag{15.38}$$

Thus the triply differential cross section in the laboratory is

$$\frac{d^3 \chi}{d\omega \, d\Omega \, dQ} = \frac{2}{3\pi} \cdot \frac{z^2 e^2}{c} \left(\frac{Q}{Mc}\right)^2 \cdot \frac{d\sigma_s}{dQ} \cdot \left[\frac{3}{16\pi} \left(\frac{\omega}{\omega'}\right)^2 (1 + \cos^2 \theta')\right] \tag{15.39}$$

The quantities in the square brackets must, of course, be expressed in terms of (unprimed) laboratory quantities. The relativistic Doppler shift formulas are

$$\omega = \gamma \omega' (1 + \beta \cos \theta')$$

and

$$\omega' = \gamma \omega (1 - \beta \cos \theta)$$

Combining the two equations we obtain

$$\frac{\omega}{\omega'} = \frac{1}{\gamma(1 - \beta \cos \theta)} \simeq \frac{2\gamma}{1 + \gamma^2 \theta^2}$$

and

$$\cos \theta' = \frac{\cos \theta - \beta}{1 - \beta \cos \theta} \simeq \frac{1 - \gamma^2 \theta^2}{1 + \gamma^2 \theta^2} \tag{15.40}$$

---

*Actually, we can use the manifestly invariant 4-momentum transfer whose square is given by $Q^2 = -(p_1 - p_2)^2 = (\mathbf{p}_1 - \mathbf{p}_2)^2 - (E_1 - E_2)^2/c^2$. For elastic scattering by a massive center of force, $E_1 = E_2$, and for small angles and very high energies, the energy difference term can be neglected even for inelastic collisions.

The approximations on the right are appropriate for small angles around the incident direction in the laboratory. With these approximate forms, the square-bracketed quantity in (15.39) becomes

$$\left[\frac{3}{16\pi}\left(\frac{\omega}{\omega'}\right)^2(1 + \cos^2\theta')\right] \simeq \left[\frac{3\gamma^2}{2\pi}\frac{(1 + \gamma^4\theta^4)}{(1 + \gamma^2\theta^2)^4}\right]$$

which is exactly the normalized expression in (15.36). [Use of the exact forms from (15.40) leads to the sum of the two terms in (15.10).] With the Rutherford cross section (15.19), or some other collision cross section for $d\sigma_s/dQ$, if appropriate, we obtain from (15.39) the relativistic bremsstrahlung results as before.

The Doppler shift formulas illustrate an important point. Photons of energies $\hbar\omega'$ in K', emitted at essentially any angle in that frame, appear in the laboratory within the forward cone and with energies of the order of $\hbar\omega \sim \gamma\hbar\omega'$. Thus energetic photons in the laboratory energy range $Mc^2 \ll \hbar\omega \ll \gamma Mc^2$ come from soft quanta with $\hbar\omega' \ll Mc^2$ in the rest frame of the incident particle.

15.3   Screening Effects; Relativistic Radiative Energy Loss

In the treatment of bremsstrahlung so far we have ignored the effects of the atomic electrons. As direct contributors to the acceleration of the incident particle they can be safely ignored, since their contribution per atom is of the order of Z^{-1} times the nuclear one. But they have an indirect effect through their screening of the nuclear charge. The potential energy of the incident particle in the field of the atom can be approximated by the Yukawa form, $V(r) = (zZe^2/r)\exp(-r/a)$, with $a \simeq 1.4a_0Z^{-1/3}$. Instead of (15.17) the scattering cross section is (13.53) with $\theta_{\min}$ given by (13.55). In terms of momentum transfer (15.19) is replaced by

$$\frac{d\sigma_s}{dQ} = 8\pi\left(\frac{zZe^2}{\beta c}\right)^2 \cdot \frac{Q}{(Q^2 + Q_s^2)^2} \tag{15.41}$$

where

$$Q_s = p\theta_{\min}^{(q)} = \frac{\hbar}{a} \simeq \frac{Z^{1/3}}{192}mc \tag{15.42}$$

is the momentum transfer associated with the screening radius a. Note that m is the electronic mass.

The calculation of bremsstrahlung proceeds as at the beginning of Section 15.2, but with the replacement in (15.22),

$$\int_{Q_{\min}}^{Q_{\max}}\frac{dQ}{Q} \rightarrow \int_{Q_{\min}}^{Q_{\max}}\frac{Q^3\,dQ}{(Q^2 + Q_s^2)^2}$$

With the assumption that $Q_{\max}$ is very large compared with both $Q_{\min}$ and Q_s, we find that the logarithm in (15.22) is replaced by

$$\ln\left(\frac{Q_{\max}}{Q_{\min}}\right) \rightarrow \ln\left(\frac{Q_{\max}}{\sqrt{Q_{\min}^2 + Q_s^2}}\right) - \frac{Q_s^2}{2(Q_{\min}^2 + Q_s^2)} \tag{15.43}$$

For $Q_{\min} \gg Q_s$ the effects of screening are unimportant and the results of the preceding section are unaffected. But for $Q_{\min} \lesssim Q_s$, important modifications occur.

From (15.23), (15.28), and (15.32) we see that $Q_{\max}$ can be written in all circumstances as

$$Q_{\max} \simeq 2Mv \tag{15.44}$$

while from (15.28) and (15.33) we find $Q_{\min}$ values,

$$\left. \begin{aligned} Q_{\min}^{(NR)} &= p - p' \simeq \frac{2\hbar\omega}{v} \\ Q_{\min}^{(R)} &\simeq \frac{\hbar\omega}{2\gamma\gamma'c} \simeq \frac{\hbar\omega}{2\gamma^2 c} \end{aligned} \right\} \tag{15.45}$$

The approximations on the right are applicable for soft photons. (Note that, up to factors of 2 in the logarithms, a universal formula for $Q_{\min}$ for soft photons is $Q_{\min} \simeq \hbar\omega/\gamma^2 v$.) Since both values of $Q_{\min}$ are proportional to ω for soft photons, it is clear that *there will always be a frequency below which screening effects are important*. With Q_s given by (15.42), the ratio of $Q_{\min}$ to Q_s for nonrelativistic bremsstrahlung is

$$\frac{Q_{\min}^{(NR)}}{Q_s} \simeq \frac{384}{Z^{1/3}} \cdot \frac{\hbar\omega}{mvc} = \frac{192M\beta}{mZ^{1/3}} \cdot \frac{\hbar\omega}{(\hbar\omega)_{\max}}$$

where $(\hbar\omega)_{\max} = Mv^2/2$. Except for extremely slow speeds, the frequency at which $Q_{\min}^{(NR)} \lesssim Q_s$ is a tiny fraction of the maximum. For example, with 100 keV electrons on a gold target ($Z = 79$), only for $\omega/\omega_{\max} < 0.04$ is screening important. For particles heavier than electrons the factor M/m makes screening totally insignificant in nonrelativistic bremsstrahlung.

For relativistic bremsstrahlung, however, screening effects can be important. The ratio of $Q_{\min}$ to Q_s is now

$$\frac{Q_{\min}^{(R)}}{Q_s} \simeq \frac{96\hbar\omega}{\gamma\gamma'mc^2 Z^{1/3}} = \frac{96M}{\gamma'mZ^{1/3}} \cdot \frac{\hbar\omega}{(\hbar\omega)_{\max}}$$

where $(\hbar\omega)_{\max} = \gamma Mc^2$. The presence of the factor γ' in the denominator implies that at sufficiently high energies $Q_{\min}^{(R)}$ can be less than Q_s for essentially the whole range of frequencies [if $\omega/\omega_{\max} = x$, then $\gamma' = (1 - x)\gamma$]. Then the *screening is said to be complete*. The incident energies for complete screening are defined as $E \gg E_s$, where the critical energy E_s is

$$E_s = \left(\frac{192M}{mZ^{1/3}} \right) Mc^2 \tag{15.46}$$

For energies large compared to E_s, $Q_{\min}$ can be neglected compared to Q_s in (15.43) at all frequencies except the very tip of the spectrum. The radiation cross section in the complete screening limit is thus the constant value,

$$\frac{d\chi}{d\omega} \simeq \frac{16}{3} \frac{Z^2 e^2}{c} \left(\frac{z^2 e^2}{Mc^2} \right)^2 \ln\left(\frac{233M}{mZ^{1/3}} \right) \tag{15.47}$$

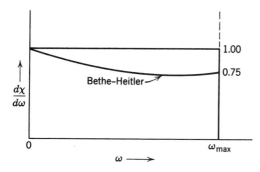

Figure 15.5 Radiation cross section in the complete screening limit. The constant value is the semiclassical result. The curve marked "Bethe–Heitler" is the quantum-mechanical Born approximation.

The numerical coefficient in the logarithm is subject to some uncertainty, of course. Bethe and Heitler found a result with 183 instead of 233 in the logarithm and with the polynomial $(1 - \hbar\omega/E + 3\hbar^2\omega^2/4E^2)$ of (15.35) multiplying it.

For electrons, $E_s \simeq 42$ MeV in aluminum $(Z = 13)$ and 23 MeV in lead $(Z = 82)$. The corresponding values for mu mesons are 2×10^6 MeV and 10^6 MeV. Because of the factor M/m, screening is important only for electrons. When $E > E_s$, the radiation cross section is given by the constant value (15.47) for all frequencies. Figure 15.5 shows the radiation cross section (15.47) in the limit of complete screening, as well as the corresponding Bethe–Heitler result. Their proper quantum treatment involves the slowly varying polynomial, which changes from unity at $\omega = 0$ to 0.75 at $\omega = \omega_{\max}$. For cosmic-ray electrons and for electrons from most high-energy electron accelerators, the bremsstrahlung is in the complete screening limit. Thus the photon spectrum shows a typical $(\hbar\omega)^{-1}$ behavior.

The radiative energy loss was considered in the nonrelativistic limit in Section 15.2.B and was found to be negligible compared to the energy loss by collisions. For ultrarelativistic particles, especially electrons, this is no longer true. The radiative energy loss is given approximately in the limit $\gamma \gg 1$ by

$$\frac{dE_{\rm rad}}{dx} \simeq \frac{16}{3} N \frac{Z^2 e^2}{c} \left(\frac{z^2 e^2}{Mc^2}\right)^2 \int_0^{\gamma Mc^2/\hbar} \ln\left(\frac{Q_{\max}}{\sqrt{Q_{\min}^2 + Q_s^2}}\right) d\omega$$

For negligible screening we find approximately

$$\frac{dE_{\rm rad}}{dx} \simeq \frac{16}{3} N \frac{Z^2 e^2}{\hbar c} \left(\frac{z^2 e^2}{Mc^2}\right)^2 \ln(\lambda\gamma)\gamma Mc^2$$

For higher energies where complete screening occurs this is modified to

$$\frac{dE_{\rm rad}}{dx} \simeq \left[\frac{16}{3} N \frac{Z^2 e^2}{\hbar c} \left(\frac{z^2 e^2}{Mc^2}\right)^2 \ln\left(\frac{233 M}{Z^{1/3} m}\right)\right]\gamma Mc^2 \qquad (15.48)$$

showing that eventually the radiative loss is proportional to the particle's energy.*

The comparison of radiative loss to collision loss now becomes

$$\frac{dE_{\rm rad}}{dE_{\rm coll}} \simeq \frac{4}{3\pi} \left(\frac{Zz^2}{137}\right) \frac{m}{M} \frac{\ln\left(\dfrac{233 M}{Z^{1/3} m}\right)}{\ln B_q} \gamma$$

*With the Bethe–Heitler energy dependence shown in Fig. 15.5, the coefficient 16/3 is replaced by 4; if atomic electrons are counted, the factor of Z^2 is replaced by $Z(Z + 1)$.

The value of γ for which this ratio is unity depends on the particle and on Z. For electrons it is $\gamma \sim 200$ for air and $\gamma \sim 20$ for lead. At higher energies, the radiative energy loss is larger than the collision loss and for ultrarelativistic particles is the dominant loss mechanism.

At energies where the radiative energy loss is dominant, the complete screening result (15.48) holds. Then it is useful to introduce a unit of length X_0, called the *radiation length*, which is the distance a particle travels while its energy falls to e^{-1} of its initial value. By conservation of energy, we may rewrite (15.48) as

$$\frac{dE}{dx} = -\frac{E}{X_0}$$

with solution

$$E(x) = E_0 e^{-x/X_0}$$

where the radiation length X_0 (including quantum corrections, *loc. cit.*) is

$$X_0 = \left[4N \frac{Z(Z+1)e^2}{\hbar c} \left(\frac{z^2 e^2}{Mc^2} \right)^2 \ln \left(\frac{233M}{Z^{1/3}m} \right) \right]^{-1} \tag{15.49}$$

For electrons, some representative values of X_0 are 37 g/cm^2 (310 m) in air at NTP, 24 g/cm^2 (8.9 cm) in aluminum, and 5.8 g/cm^2 (0.51 cm) in lead. In studying the passage of cosmic-ray or man-made high-energy particles through matter, the radiation length X_0 is a convenient unit to employ, since not only the radiative energy loss is governed by it, but also the production of electron-positron pairs by the radiated photons, and so the whole development of the electronic cascade shower.

15.4 Weizsäcker–Williams Method of Virtual Quanta

The emission of bremsstrahlung and other processes involving the electromagnetic interaction of relativistic particles can be viewed in a way that is very helpful in providing physical insight into the processes. This point of view is called the *method of virtual quanta*. It exploits the similarity between the fields of a rapidly moving charged particle and the fields of a pulse of radiation (see Section 11.10) and correlates the effects of the collision of the relativistic charged particle with some system with the corresponding effects produced by the interaction of radiation (the virtual quanta) with the same system. The method was developed independently by C. F. Weizsäcker and E. J. Williams in 1934. Ten years earlier Enrico Fermi had used essentially the same idea to relate the energy loss by ionization to the absorption of x-rays by atoms (see Problem 15.12).

In any given collision we define an "incident particle" and a "struck system." The perturbing fields of the incident particle are replaced by an equivalent pulse of radiation that is analyzed into a frequency spectrum of virtual quanta. Then the effects of the quanta (either scattering or absorption) on the struck system are calculated. In this way the charged-particle interaction is correlated with the photon interaction. Table 15.1 lists a few typical correspondences and specifies the incident particle and struck system. From the table we see that the struck system is not always the target in the laboratory. For bremsstrahlung the struck

Table 15.1   Correspondences Between Charged Particle Interactions
and Photon Interactions

Particle Process	**Incident Particle**	**Struck System**	**Radiative Process**	b_{min}
Bremsstrahlung in electron (light particle)-nucleus collision	Nucleus	Electron (light particle mass M)	Scattering of virtual photons of nuclear Coulomb field by the electron (light particle)	$\hbar/2Mv$
Collisional ionization of atoms (in distant collisions)	Incident particle	Atom	Photoejection of atomic electrons by virtual quanta	a
Electron disintegration of nuclei	Electron (mass m)	Nucleus	Photodisintegration of nuclei by virtual quanta	Larger of $\hbar/\gamma mv$ and R
Production of pions in electron-nuclear collisions	Electron (mass m)	Nucleus	Photoproduction of pions by virtual quanta interactions with nucleus	

system is the lighter of the two collision partners, since its radiation scattering power is greater. For bremsstrahlung in electron-electron collision it is necessary from symmetry to take the sum of two contributions where each electron in turn is the struck system at rest initially in some reference frame.

The chief assumption in the method of virtual quanta is that the effects of the various frequency components of equivalent radiation add incoherently. This will be true provided the perturbing effect of the fields is small, and is consistent with our assumption in Section 15.2.D that the motion of the particle in the frame K' was nonrelativistic throughout the collision.

It is convenient in the discussion of the Weizsäcker–Williams method to use the language of impact parameters rather than momentum transfers in order to make use of results on the Fourier transforms of fields obtained in previous chapters. The connection between the two approaches is via the uncertainty-principle relation,

$$b \sim \frac{\hbar}{Q}$$

With the expression (15.44) for Q_{max} in bremsstrahlung, we see that the minimum impact parameter effective in producing radiation is

$$b_{min} \simeq \frac{\hbar}{Q_{max}} \simeq \frac{\hbar}{2Mv} \tag{15.50}$$

as listed in Table 15.1. The maximum impact parameters corresponding to the Q_{min} values of (15.45) do not need to be itemized. The spectrum of virtual quanta automatically incorporates the cutoff equivalent to Q_{min}.

The spectrum of equivalent radiation for an independent particle of charge

q, velocity $v \simeq c$, passing a struck system S at impact parameter b, can be found from the fields of Section 11.10:

$$E_2(t) = q \frac{\gamma b}{(b^2 + \gamma^2 v^2 t^2)^{3/2}}$$

$$B_3(t) = \beta E_2(t)$$

$$E_1(t) = -q \frac{\gamma v t}{(b^2 + \gamma^2 v^2 t^2)^{3/2}}$$

For $\beta \simeq 1$ the fields $E_2(t)$ and $B_3(t)$ are completely equivalent to a pulse of plane-polarized radiation P_1 incident on S in the x_1 direction, as shown in Fig. 15.6. There is no magnetic field to accompany $E_1(t)$ and so form a proper pulse of radiation P_2 incident along the x_2 direction, as shown. Nevertheless, if the motion of the charged particles in S is nonrelativistic in this coordinate frame, we can add the necessary magnetic field to create the pulse P_2 without affecting the physical problem because the particles in S respond only to electric forces. Even if the particles in S are influenced by magnetic forces, the additional magnetic field implied by replacing $E_1(t)$ by the radiation pulse P_2 is not important, since the pulse P_2 will be seen to be of minor importance anyway.

From the discussion Section 14.5, especially equations (14.51), (14.52), and (14.60), it is evident that the equivalent pulse P_1 has a frequency spectrum (energy per unit area per unit frequency interval) $dI_1(\omega, b)/d\omega$ given by

$$\frac{dI_1}{d\omega}(\omega, b) = \frac{c}{2\pi} |E_2(\omega)|^2 \tag{15.51a}$$

where $E_2(\omega)$ is the Fourier transform (14.54) of $E_2(t)$. Similarly the pulse P_2 has the frequency spectrum

$$\frac{dI_2}{d\omega}(\omega, b) = \frac{c}{2\pi} |E_1(\omega)|^2 \tag{15.51b}$$

The Fourier integrals, calculated in Chapter 13, are given by (13.80). The two frequency spectra are

$$\left. \begin{array}{c} \dfrac{dI_1(\omega, b)}{d\omega} \\[2mm] \dfrac{dI_2(\omega, b)}{d\omega} \end{array} \right\} = \frac{1}{\pi^2} \frac{q^2}{c} \left(\frac{c}{v}\right)^2 \frac{1}{b^2} \left\{ \begin{array}{c} \left(\dfrac{\omega b}{\gamma v}\right)^2 K_1^2\!\left(\dfrac{\omega b}{\gamma v}\right) \\[3mm] \dfrac{1}{\gamma^2} \left(\dfrac{\omega b}{\gamma v}\right)^2 K_0^2\!\left(\dfrac{\omega b}{\gamma v}\right) \end{array} \right. \tag{15.52}$$

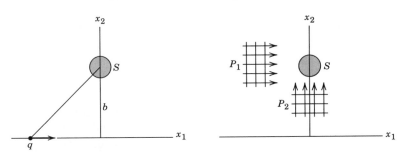

Figure 15.6 Relativistic charged particle passing the struck system S and the equivalent pulses of radiation.

We note that the intensity of the pulse P_2 involves a factor γ^{-2} and so is of little importance for ultrarelativistic particles. The shapes of these spectra are shown in Fig. 15.7. The behavior is easily understood if one recalls that the fields of pulse P_1 are bell-shaped in time with a width $\Delta t \sim b/\gamma v$. Thus the frequency spectrum will contain all frequencies up to a maximum of order $\omega_{max} \sim 1/\Delta t$. On the other hand, the fields of pulse P_2 are similar to one cycle of a sine wave of frequency $\omega \sim \gamma v/b$. Consequently its spectrum will contain only a modest range of frequencies centered around $\gamma v/b$.

In collision problems we must sum the frequency spectra (15.52) over the various possible impact parameters. This gives the energy per unit frequency interval present in the equivalent radiation field. As always in such problems we must specify a minimum impact parameter b_{min}. The method of virtual quanta will be useful only if b_{min} can be so chosen that for impact parameters greater than b_{min} the effects of the incident particle's fields can be represented accurately by the effects of equivalent pulses of radiation, while for small impact parameters the effects of the particle's fields can be neglected or taken into account by other means. Setting aside for the moment how we choose the proper value of b_{min} in general [(15.50) is valid for bremsstrahlung], we can write down the frequency spectrum integrated over possible impact parameters,

$$\frac{dI}{d\omega}(\omega) = 2\pi \int_{b_{min}}^{\infty} \left[\frac{dI_1}{d\omega}(\omega, b) + \frac{dI_2}{d\omega}(\omega, b)\right] b \, db \qquad (15.53)$$

where we have combined the contributions of pulses P_1 and P_2. The result is

$$\frac{dI}{d\omega}(\omega) = \frac{2}{\pi}\frac{q^2}{c}\left(\frac{c}{v}\right)^2\left\{xK_0(x)K_1(x) - \frac{v^2}{2c^2}x^2[K_1^2(x) - K_0^2(x)]\right\} \qquad (15.54)$$

where

$$x = \frac{\omega b_{min}}{\gamma v} \qquad (15.55)$$

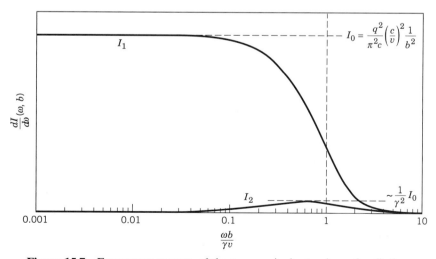

Figure 15.7 Frequency spectra of the two equivalent pulses of radiation.

For low frequencies ($\omega \ll \gamma v/b_{min}$) the energy per unit frequency interval reduces to

$$\frac{dI}{d\omega}(\omega) \simeq \frac{2}{\pi}\frac{q^2}{c}\left(\frac{c}{v}\right)^2\left[\ln\left(\frac{1.123\gamma v}{\omega b_{min}}\right) - \frac{v^2}{2c^2}\right] \tag{15.56}$$

whereas for high frequencies ($\omega \gg \gamma v/b_{min}$) the spectrum falls off exponentially as

$$\frac{dI}{d\omega}(\omega) \simeq \frac{q^2}{c}\left(\frac{c}{v}\right)^2\left(1 - \frac{v^2}{2c^2}\right)\exp\left(-\frac{2\omega b_{min}}{\gamma v}\right) \tag{15.57}$$

Figure 15.8 shows an accurate plot of $I(\omega)$ (15.54) for $v \simeq c$, as well as the low-frequency approximation (15.56). We see that the energy spectrum consists predominantly of low-frequency quanta with a tail extending up to frequencies of the order of $2\gamma v/b_{min}$.

The number spectrum of virtual quanta $N(\hbar\omega)$ is obtained by using the relation

$$\frac{dI}{d\omega}(\omega)\, d\omega = \hbar\omega N(\hbar\omega)\, d(\hbar\omega)$$

Thus the number of virtual quanta per unit energy interval in the low-frequency limit is

$$N(\hbar\omega) \simeq \frac{2}{\pi}\left(\frac{q^2}{\hbar c}\right)\left(\frac{c}{v}\right)^2\frac{1}{\hbar\omega}\left[\ln\left(\frac{1.123\gamma v}{\omega b_{min}}\right) - \frac{v^2}{2c^2}\right] \tag{15.58}$$

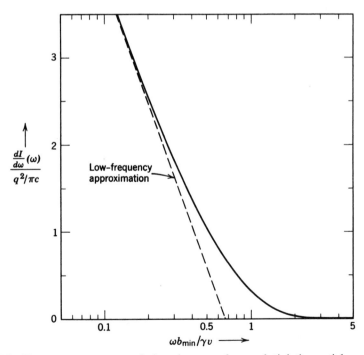

Figure 15.8 Frequency spectrum of virtual quanta for a relativistic particle, with the energy per unit frequency $dI(\omega)/d\omega$ in units of $q^2/\pi c$ and the frequency in units of $\gamma v/b_{min}$. The number of virtual quanta per unit energy interval is obtained by dividing by $\hbar^2\omega$.

The choice of minimum impact parameter b_{min} must be considered. In bremsstrahlung, $b_{min} = \hbar/2Mv$, where M is the mass of the lighter particle, as already discussed. For collisional ionization of atoms, $b_{min} \simeq a$, the atomic radius, with closer impacts treated as collisions between the incident particle and free electrons. In electron disintegration of nuclei or electron production of mesons from nuclei, the wave nature of the particle whose fields provide the virtual quanta sets the effective minimum impact parameter. In these circumstances, $b_{min} = \hbar/\gamma Mv$ or $b_{min} = R$, the nuclear radius, whichever is larger. The values are summarized in Table 15.1.

The quantum-mechanical version* of the Weizsäcker–Williams method of virtual quanta for ultrarelativistic spin $\frac{1}{2}$ electrons ($\beta \to 1$) replaces the square-bracketed quantity in (15.58)—"the logarithm"—by

$$L = \left(\frac{E^2 + E'^2}{2E^2}\right) \ln\left(\frac{2EE'}{mc^2\hbar\omega}\right) - \frac{(E + E')^2}{4E^2} \ln\left(\frac{E + E'}{\hbar\omega}\right) - \frac{E'}{2E} \quad (15.59)$$

where E and $E' = E - \hbar\omega$ are the initial and final energies of the electron. In the limit $\hbar\omega \ll E$, "the logarithm" reduces to

$$L \approx (1 - \hbar\omega/E) \ln (E/mc^2) - \frac{1}{2} + O\left[\frac{(\hbar\omega)^2}{E^2}, \frac{(\hbar\omega)^2}{E^2} \ln\left(\frac{E^2}{mc^2\hbar\omega}\right)\right]$$

which is consistent with (15.58) with $b_{min} \approx c/\omega = \lambdabar$, the wavelength (divided by 2π) of the virtual photon. The quantum-mechanical version finds extensive application in the so-called two-photon processes in electron-positron collisions.[†]

15.5   Bremsstrahlung as the Scattering of Virtual Quanta

The emission of bremsstrahlung in a collision between an incident relativistic particle of charge ze and mass M and an atomic nucleus of charge Ze can be viewed as the scattering of the virtual quanta in the nuclear Coulomb field by the incident particle in the coordinate system K', where the incident particle is at rest. The spectrum of virtual quanta $dI'(\omega')/d\omega'$ is given by (15.54) with $q = Ze$. The minimum impact parameter is $\hbar/2Mv$, so that the frequency spectrum extends up to $\omega' \sim \gamma Mc^2/\hbar$.

The virtual quanta are scattered by the incident particle (the struck system in K') according to the Thomson cross section (14.125) at low frequencies. Thus, in the frame K' and for frequencies small compared to $Mc^2/\hbar$, the differential radiation cross section is approximately

$$\frac{d\chi'}{d\omega'\, d\Omega'} \simeq \left(\frac{z^2e^2}{Mc^2}\right)^2 \cdot \tfrac{1}{2}(1 + \cos^2\theta') \cdot \frac{dI'}{d\omega'}$$

Since the spectrum of virtual quanta extends up to $\gamma Mc^2/\hbar$, the approximation (15.56) can be used for $dI'(\omega')/d\omega'$ in the region $\omega' \ll Mc^2/\hbar$. Thus the radiation cross section in K' becomes

$$\frac{d\chi'}{d\omega'\, d\Omega'} \simeq \frac{1}{\pi} \frac{Z^2e^2}{c} \left(\frac{z^2e^2}{Mc^2}\right)^2 (1 + \cos^2\theta') \cdot \ln\left(\frac{\lambdabar\gamma Mc^2}{\hbar\omega'}\right) \quad (15.60)$$

*R. H. Dalitz and D. R. Yennie, *Phys. Rev.* **105**, 1598 (1957).
†H. Terazawa, *Rev. Mod. Phys.* **45**, 615 (1973).

The cross section in the laboratory can be obtained in the same fashion as in Section 15.2.D. Using (15.38) and the Doppler formulas (15.40) we find

$$\frac{d\chi}{d\omega\,d\Omega} \simeq \frac{16}{3}\frac{Z^2e^2}{c}\left(\frac{z^2e^2}{Mc^2}\right)^2 \ln\left(\frac{2\lambda\gamma^2Mc^2}{\hbar\omega(1+\gamma^2\theta^2)}\right)\cdot\left[\frac{3\gamma^2(1+\gamma^4\theta^4)}{2\pi(1+\gamma^2\theta^2)^4}\right] \quad (15.61)$$

This is essentially the same cross section as (15.36). Upon integration over angles of emission, it yields an expression equal to the soft-photon limit of (15.34).

Equations (15.60) and (15.61) are based on the Thomson scattering cross section and so are restricted to $\omega' < Mc^2/\hbar$ in the rest frame K'. Of course, as observed in Section 15.2.D, such soft photons transform into energetic photons in the laboratory. But the spectrum of virtual quanta contains frequencies up to $\omega' \simeq \gamma Mc^2/\hbar$. For such frequencies the scattering of radiation is not given by the Thomson cross section, but rather by (14.127) for spinless struck particles or the Klein–Nishina formula for particles of spin $\frac{1}{2}$. The angular distribution of scattering of such photons is altered from the dipole form of (15.60), as is shown in Fig. 14.18. More important, the total cross section for scattering decreases rapidly for frequencies larger than $Mc^2/\hbar$, as can be seen from (14.128). This shows that in the frame K' the bremsstrahlung quanta are confined to a frequency range $0 < \omega' \lesssim Mc^2/\hbar$, even though the spectrum of virtual quanta in the nuclear Coulomb field extends to much higher frequencies. The restricted spectrum in K' is required physically by conservation of energy, since in the laboratory system where $\omega \simeq \gamma\omega'$ the frequency spectrum is limited to $0 < \omega < (\gamma Mc^2/\hbar)$. A detailed treatment using the angular distribution of scattering from the Klein–Nishina formula yields a bremsstrahlung cross section in complete agreement with the Bethe–Heitler formulas (Weizsäcker, 1934).

The effects of screening on the bremsstrahlung spectrum can be discussed in terms of the Weizsäcker–Williams method. For a screened Coulomb potential the spectrum of virtual quanta is modified from (15.56). The argument of the logarithm is changed to a constant, as discussed in Section 15.3.

Further applications of the method of virtual quanta to such problems as collisional ionization of atoms and electron disintegration of nuclei are deferred to the problems at the end of the chapter.

15.6 *Radiation Emitted During Beta Decay*

In the process of beta decay an unstable nucleus with atomic number Z transforms spontaneously into another nucleus of atomic number $(Z \pm 1)$ while emitting an electron $(\mp e)$ and a neutrino. The process is written symbolically as

$$Z \to (Z \pm 1) + e^{\mp} + \nu \quad (15.62)$$

The energy released in the decay is shared almost entirely by the electron and the neutrino, with the recoiling nucleus getting a completely negligible share because of its very large mass. Even without knowledge of why or how beta decay takes place, we can anticipate that the sudden creation of a rapidly moving charged particle will be accompanied by the emission of radiation. As mentioned in the introduction, either we can think of the electron initially at rest and being accelerated violently during a short time interval to its final velocity, or we can imagine that its charge is suddenly turned on in the same short time interval. The heavy nucleus receives a negligible acceleration and so does not contribute to the radiation.

For purposes of calculation we can assume that at $t = 0$ an electron is created at the origin with a constant velocity $\mathbf{v} = c\boldsymbol{\beta}$. Then from (15.1) or (15.2) the intensity distribution of radiation is given by

$$\frac{dI}{d\omega\, d\Omega} = \frac{e^2}{4\pi^2 c} \left| \frac{\boldsymbol{\epsilon}^* \cdot \boldsymbol{\beta}}{1 - \mathbf{n} \cdot \boldsymbol{\beta}} \right|^2 \tag{15.63}$$

This is the low-frequency limit of the energy spectrum. The intensity will decrease from this value at frequencies that violate the condition (15.15). Although it is difficult to be precise about the value of $\langle\boldsymbol{\beta}\rangle$ that appears there, if the formation process is imagined to involve a velocity-versus-time curve, such as is sketched in Fig. 15.9, the value of $\langle\beta\rangle$ should not be greater than $\frac{1}{2}$. In that case, the criterion (15.15) is equivalent to $\omega\tau < 1$. The formation time τ can be estimated from the uncertainty principle to be

$$\tau \sim \frac{\hbar}{\Delta E} \sim \frac{\hbar}{E} \tag{15.64}$$

since in the act of beta decay an electron of total energy E is suddenly created. This estimate of τ implies the frequencies for appreciable radiation are limited to $\omega < E/\hbar$. This is just the limit imposed by conservation of energy. The radiation is seen from (15.63) to be linearly polarized in the plane containing the velocity vector of the electron and the direction of observation. The differential distribution in spherical coordinates is

$$\frac{d^2 I}{d\omega\, d\Omega} = \frac{e^2}{4\pi^2 c} \beta^2 \frac{\sin^2\theta}{(1 - \beta\cos\theta)^2} \tag{15.65}$$

while the total intensity per unit frequency interval is

$$\frac{dI}{d\omega}(\omega) = \frac{e^2}{\pi c} \left[\frac{1}{\beta} \ln\left(\frac{1 + \beta}{1 - \beta}\right) - 2 \right] \tag{15.66}$$

For $\beta \ll 1$, (15.66) reduces to $dI/d\omega \simeq 2e^2\beta^2/3\pi c$, showing that for low-energy beta particles the radiated intensity is negligible.

The intensity distribution (15.66) is a typical bremsstrahlung spectrum with number of photons per unit energy range given by

$$N(\hbar\omega) = \frac{e^2}{\pi\hbar c}\left(\frac{1}{\hbar\omega}\right)\left[\frac{1}{\beta}\ln\left(\frac{1 + \beta}{1 - \beta}\right) - 2\right] \tag{15.67}$$

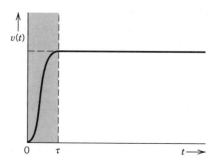

Figure 15.9

It sometimes bears the name *inner bremsstrahlung* to distinguish it from bremsstrahlung emitted by the same beta particle in passing through matter.

The total energy radiated is approximately

$$E_{\text{rad}} = \int_0^{\omega_{\text{max}}} \frac{dI}{d\omega}(\omega)\, d\omega \simeq \frac{e^2}{\pi \hbar c}\left[\frac{1}{\beta}\ln\left(\frac{1+\beta}{1-\beta}\right) - 2\right]E \tag{15.68}$$

For very fast beta particles, the ratio of energy going into radiation to the particle energy is

$$\frac{E_{\text{rad}}}{E} \simeq \frac{2}{\pi}\frac{e^2}{\hbar c}\left[\ln\left(\frac{2E}{mc^2}\right) - 1\right] \tag{15.69}$$

This shows that the radiated energy is a very small fraction of the total energy released in beta decay, even for the most energetic beta processes ($E_{\text{max}} \sim 30mc^2$). Nevertheless, the radiation can be observed and provides useful information for nuclear physicists.

In the actual beta process the energy release is shared by the electron and the neutrino so that the electron has a whole spectrum of energies up to some maximum. Then the radiation spectrum (15.66) must be averaged over the energy distribution of the beta particles. Furthermore, a quantum-mechanical treatment leads to modifications near the upper end of the photon spectrum. These are important details for quantitative comparison with experiment. But the origins of the radiation and its semiquantitative description are given adequately by our classical calculation.

15.7 *Radiation Emitted During Orbital-Electron Capture: Disappearance of Charge and Magnetic Moment*

In beta emission the sudden creation of a fast electron gives rise to radiation. In orbital-electron capture the sudden disappearance of an electron does likewise. Orbital-electron capture is the process whereby an orbital electron around an unstable nucleus of atomic number Z is captured by the nucleus, which is transformed into another nucleus with atomic number $(Z - 1)$, with the simultaneous emission of a neutrino that carries off the excess energy. The process can be written symbolically as

$$Z + e^- \to (Z - 1) + \nu \tag{15.70}$$

Since a virtually undetectable neutrino carries away the decay energy if there is no radiation, the spectrum of photons accompanying orbital-electron capture is of great importance in yielding information about the energy release.

As a simplified model we consider an electron moving in a circular atomic orbit of radius a with a constant angular velocity ω_0. The orbit lies in the x-y plane, as shown in Fig. 15.10, with the nucleus at the center. The observation direction $\mathbf{n}$ is defined by the polar angle θ and lies in the x-z plane. The velocity of the electron is

$$\mathbf{v}(t) = -\boldsymbol{\epsilon}_1 \omega_0 a \sin(\omega_0 t + \alpha) + \boldsymbol{\epsilon}_2 \omega_0 a \cos(\omega_0 t + \alpha) \tag{15.71}$$

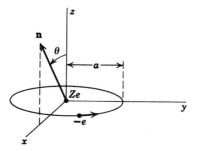

Figure 15.10

where α is an arbitrary phase angle. If the electron vanishes at $t = 0$, the frequency spectrum of emitted radiation (14.67) is approximately*

$$\frac{d^2I}{d\omega\, d\Omega} = \frac{e^2\omega^2}{4\pi^2 c^3} \left| \int_{-\infty}^{0} \mathbf{n} \times [\mathbf{n} \times \mathbf{v}(t)] e^{i\omega t}\, dt \right|^2 \tag{15.72}$$

where we have assumed that $(\omega a/c) \ll 1$ (dipole approximation) and put the retardation factor equal to unity. The integral in (15.72) can be written

$$\int_{-\infty}^{0} dt = -\omega_0 a(\boldsymbol{\epsilon}_\perp I_1 + \boldsymbol{\epsilon}_\parallel \cos\theta I_2) \tag{15.73}$$

where

$$\left.\begin{aligned} I_1 &= \int_{-\infty}^{0} \cos(\omega_0 t + \alpha) e^{i\omega t}\, dt \\ I_2 &= \int_{-\infty}^{0} \sin(\omega_0 t + \alpha) e^{i\omega t}\, dt \end{aligned}\right\} \tag{15.74}$$

and $\boldsymbol{\epsilon}_\perp$, $\boldsymbol{\epsilon}_\parallel$ are unit polarization vectors perpendicular and parallel to the plane containing $\mathbf{n}$ and the z axis. The integrals are elementary and lead to an intensity distribution,

$$\frac{d^2I}{d\omega\, d\Omega} = \frac{e^2\omega^2}{4\pi^2 c^3} \frac{\omega_0^2 a^2}{(\omega^2 - \omega_0^2)^2} [(\omega^2 \cos^2\alpha + \omega_0^2 \sin^2\alpha) \\ + \cos^2\theta(\omega^2 \sin^2\alpha + \omega_0^2 \cos^2\alpha)] \tag{15.75}$$

Since the electron can be captured from any position around the orbit, we average over the phase angle α. Then the intensity distribution is

$$\frac{d^2I}{d\omega\, d\Omega} = \frac{e^2}{4\pi^2 c} \left(\frac{\omega_0 a}{c}\right)^2 \frac{\omega^2(\omega^2 + \omega_0^2)}{(\omega^2 - \omega_0^2)^2} \cdot \frac{1}{2}(1 + \cos^2\theta) \tag{15.76}$$

The total energy radiated per unit frequency interval is

$$\frac{dI(\omega)}{d\omega} = \frac{2}{3\pi} \frac{e^2}{c} \left(\frac{\omega_0 a}{c}\right)^2 \left[\frac{\omega^2(\omega_0^2 + \omega^2)}{(\omega^2 - \omega_0^2)^2}\right] \tag{15.77}$$

---

*To conform to the admonition following (14.67), we should multiply the velocity (15.71) by a factor such as $(1 - e^{t/\tau})\theta(-t)$ in order to bring the velocity to zero continuously in a short time τ near $t = 0$. The reader may verify that in the limit $\omega_0\tau \ll 1$ and $\omega\tau \ll 1$ the results given below emerge.

while the number of photons per unit energy interval is

$$N(\hbar\omega) = \frac{2}{3\pi} \left(\frac{e^2}{\hbar c}\right) \left(\frac{\omega_0 a}{c}\right)^2 \left[\frac{\omega^2(\omega_0^2 + \omega^2)}{(\omega^2 - \omega_0^2)^2}\right] \frac{1}{\hbar\omega} \qquad (15.78)$$

For $\omega \gg \omega_0$ the square-bracketed quantity approaches unity. Then the spectrum is a typical bremsstrahlung spectrum. But for $\omega \simeq \omega_0$ the intensity is very large (infinite in our approximation). The behavior of the photon spectrum is shown in Fig. 15.11. The singularity at $\omega = \omega_0$ may seem alarming, but it is really quite natural and expected. If the electron were to keep orbiting forever, the radiation spectrum would be a sharp line at $\omega = \omega_0$. The sudden termination of the periodic motion produces a broadening of the spectrum in the neighborhood of the characteristic frequency.

Quantum mechanically, the radiation arises when an $l = 1$ electron (mainly from the $2p$ orbit) makes a virtual radiative transition to an $l = 0$ state, from which it can be absorbed by the nucleus. Thus the frequency ω_0 must be identified with the frequency of the characteristic $2p \to 1s$ x-ray, $\hbar\omega_0 \simeq (3Z^2e^2/8a_0)$. Similarly the orbit radius is actually a transitional dipole moment. With the estimate $a \simeq a_0/Z$, where a_0 is the Bohr radius, the photon spectrum (15.78) is

$$N(\hbar\omega) \simeq \frac{3}{32\pi} Z^2 \left(\frac{e^2}{\hbar c}\right)^3 \frac{1}{\hbar\omega} \left[\frac{\omega^2(\omega^2 + \omega_0^2)}{(\omega^2 - \omega_0^2)^2}\right] \qquad (15.79)$$

The essential characteristics of this spectrum are its strong peaking at the x-ray energy and its dependence on atomic number as Z^2.

So far we have considered the radiation that accompanies the disappearance of the charge of an orbital electron in the electron-capture process. An electron possesses a magnetic moment as well as a charge. The disappearance of the magnetic moment also gives rise to radiation, but with a spectrum of quite different character. The intensity distribution in angle and frequency for a point magnetic moment in nonrelativistic motion is given in Problem 14.19a. The electronic magnetic moment can be treated as a constant vector in space until its

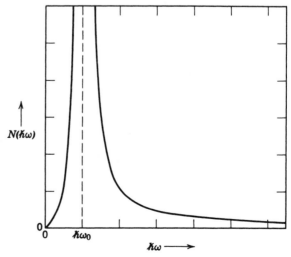

Figure 15.11 Spectrum of photons emitted in orbital-electron capture because of disappearance of the charge of the electron.

disappearance at $t = 0$. Then, in the dipole approximation, the appropriate intensity distribution is

$$\frac{d^2I}{d\omega\,d\Omega} = \frac{\omega^4}{4\pi^2c^3} \left| \int_{-\infty}^{0} \mathbf{n} \times \boldsymbol{\mu}\, e^{i\omega t}\, dt \right|^2 \tag{15.80}$$

which gives

$$\frac{d^2I}{d\omega\,d\Omega} = \frac{\omega^2}{4\pi^2c^3} \mu^2 \sin^2\Theta \tag{15.81}$$

where Θ is the angle between $\boldsymbol{\mu}$ and the observation direction $\mathbf{n}$.

In a semiclassical sense the electronic magnetic moment can be thought of as having a magnitude $\mu = \sqrt{3}(e\hbar/2mc)$, but being observed only through its projection $\mu_z = \pm(e\hbar/2mc)$ on an arbitrary axis. The moment can be thought of as precessing around the axis at an angle $\alpha = \tan^{-1}\sqrt{2}$, so that on the average only the component of the moment along the axis survives. It is easy to show that on averaging over this precession $\sin^2\Theta$ in (15.81) becomes equal to its average value of $\frac{2}{3}$, independent of observation direction. Thus the angular and frequency spectrum becomes

$$\frac{d^2I}{d\omega\,d\Omega} = \frac{e^2}{8\pi^2c} \left(\frac{\hbar\omega}{mc^2}\right)^2 \tag{15.82}$$

The total energy radiated per unit frequency interval is

$$\frac{dI}{d\omega} = \frac{e^2}{2\pi c} \left(\frac{\hbar\omega}{mc^2}\right)^2 \tag{15.83}$$

while the corresponding number of photons per unit energy interval is

$$N(\hbar\omega) = \frac{e^2}{2\pi\hbar c} \frac{\hbar\omega}{(mc^2)^2} \tag{15.84}$$

These spectra are very different in their frequency dependence from a bremsstrahlung spectrum. They increase with increasing frequency, apparently without limit. Of course, we have been forewarned that our classical results are valid only in the low-frequency limit. We can imagine that some sort of uncertainty-principle argument such as was used in Section 15.6 for radiative beta decay holds here and that conservation of energy, at least, is guaranteed. Actually, modifications arise because a neutrino is always emitted in the electron-capture process. The probability of emission of the neutrino can be shown to depend on the square of its energy E_ν. When no photon is emitted, the neutrino has the full decay energy $E_\nu = E_0$. But when a photon of energy $\hbar\omega$ accompanies it, the neutrino's energy is reduced to $E'_\nu = E_0 - \hbar\omega$. Then the probability of neutrino emission is reduced by a factor,

$$\left(\frac{E'_\nu}{E_\nu}\right)^2 = \left(1 - \frac{\hbar\omega}{E_0}\right)^2 \tag{15.85}$$

This means that our classical spectra (15.83) and (15.84) must be corrected by multiplication with (15.85) to take into account the kinematics of the neutrino emission. The modified classical photon spectrum is

$$N(\hbar\omega) = \frac{e^2}{2\pi\hbar c} \frac{\hbar\omega}{(mc^2)^2} \left(1 - \frac{\hbar\omega}{E_0}\right)^2 \tag{15.86}$$

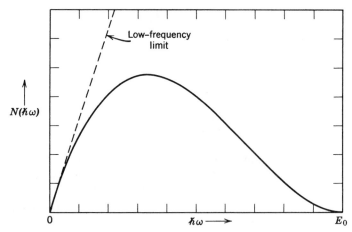

Figure 15.12 Spectrum of photons emitted in orbital-electron capture because of disappearance of the magnetic moment of the electron.

This is essentially the correct quantum-mechanical result. A comparison of the corrected distribution (15.86) and the classical one (15.84) is shown in Fig. 15.12. Evidently the neutrino-emission probability is crucial in obtaining the proper behavior of the photon energy spectrum. For the customary bremsstrahlung spectra such correction factors are less important because the bulk of the radiation is emitted in photons with energies much smaller than the maximum allowable value.

The total radiation emitted in orbital-electron capture is the sum of the contributions from the disappearance of the electric charge and of the magnetic moment. From the different behaviors of (15.79) and (15.86) we see that the upper end of the spectrum will be dominated by the magnetic-moment contri-

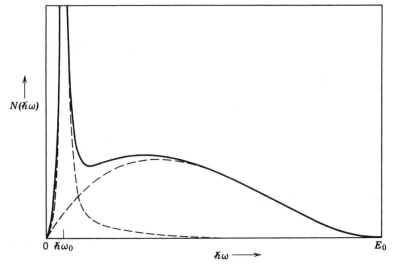

Figure 15.13 Typical photon spectrum for radiative orbital-electron capture with energy release E_0, showing the contributions from the disappearance of the electronic charge and magnetic moment.

bution unless the energy release is very small, whereas the lower end of the spectrum will be dominated by the electric-charge term, especially for high Z. Figure 15.13 shows a typical combined spectrum for $Z \sim 20$–30. Observations on a number of nuclei confirm the general features of the spectra and allow determination of the energy release E_0.

References and Suggested Reading

Classical bremsstrahlung is discussed briefly by
    Landau and Lifshitz, *Classical Theory of Fields*, Section 9.4
    Panofsky and Phillips, Section 19.6
A semiclassical discussion analogous to ours, but using impact parameters throughout and much briefer, appears in
    Rossi, Section 2.12

Bremsstrahlung can be described accurately only by a proper quantum-mechanical treatment. The standard reference is
    Heitler

The method of virtual quanta (Weizsäcker–Williams method) has only one proper reference, the classic article by
    Williams
Short discussions appear in
    Heitler, Appendix 6
    Panofsky and Phillips, Section 18.5
The quantum-mechanical form of the method of virtual quanta is described in detail in Terazawa (*op. cit.*)
The work of Fermi that predates by 10 years the Weizsäcker–Williams treatment is
    E. Fermi, *Z. Phys.* **29**, 315 (1924).
The limitations of Fermi's approach to energy loss were clarified through the use of quantum mechanics by
    E. J. Williams, *Proc. R. Soc. London* **A139**, 163 (1933).

Among the quantum-mechanical treatments of radiative beta processes having comparisons with experiment in some cases, are those by
    C. S. W. Chang and D. L. Falkoff, *Phys. Rev.*, **76**, 365 (1949).
    P. C. Martin and R. J. Glauber, *Phys. Rev.*, **109**, 1307 (1958).
    Siegbahn, Chapter XX (III) by C. S. Wu.

An important topic not mentioned so far is the production of coherent bremsstrahlung beams from high-energy electrons on crystals. Under proper conditions roughly monochromatic, highly polarized photon beams can be produced. A review of the theory and experiment is given by
    G. Diambrini Palazzi, *Rev. Mod. Phys.* **40**, 611 (1968).

Problems

15.1 In radiative collision problems it is useful to have the radiation amplitude expressed explicitly as an integral involving the accelerations of the particles, as in (14.65), for example. In the nonrelativistic limit, particles do not move rapidly or far during the period of acceleration; only the lowest order velocity and retardation effects need be kept in an approximate description.

(a) Show that the integral (times e) in (14.65), which is, apart from an inessential phase, cR times the Fourier transform of the Liénard–Wiechert electric field at distance R, can be expanded in inverse powers of c (remembering that $\beta = v/c$ and $k = \omega/c$) as follows:

$$\mathbf{J} = e \int dt\, e^{i\omega t}(\dot{\boldsymbol{\beta}}_\perp + \beta_\parallel \dot{\boldsymbol{\beta}}_\perp + \dot{\beta}_\parallel \boldsymbol{\beta}_\perp - ikr_\parallel \dot{\boldsymbol{\beta}}_\perp + \cdots)$$

or

$$\mathbf{J} = e \int dt\, e^{i\omega t}\left(\dot{\boldsymbol{\beta}}_\perp + \frac{d}{dt}(\beta_\parallel \boldsymbol{\beta}_\perp) - ikr_\parallel \dot{\boldsymbol{\beta}}_\perp + \cdots \right)$$

where $\beta_\parallel = \boldsymbol{\beta} \cdot \mathbf{n}$, $r_\parallel = \mathbf{r} \cdot \mathbf{n}$, and $\boldsymbol{\beta}_\perp = (\mathbf{n} \times \boldsymbol{\beta}) \times \mathbf{n}$. Neglected terms are of order $1/c^3$ and higher.

(b) Show that the first term in part a corresponds to the electric dipole approximation, while the next terms are the magnetic dipole and the electric quadrupole contributions. [Some integrations by parts are required, with a convergence factor $e^{-\epsilon|t|}$ to give meaning to the integrals, as discussed following (14.67).] For a group of charges, show that the generalization of part a can be written as

$$\mathbf{J} = \frac{1}{c} \int dt\, e^{i\omega t}\left(\frac{d^2\mathbf{p}_\perp(t)}{dt^2} + \frac{d^2\mathbf{m}(t)}{dt^2} \times \mathbf{n} + \frac{1}{6c}\frac{d^3\mathbf{Q}_\perp}{dt^3}(\mathbf{n}, t) + \cdots \right)$$

and the radiated intensity per unit solid angle and per unit frequency as

$$\frac{d^2I}{d\omega\, d\Omega} = \frac{1}{4\pi^2 c^3}\left| \int dt\, e^{i\omega t}\left(\frac{d^2\mathbf{p}_\perp(t)}{dt^2} + \frac{d^2\mathbf{m}(t)}{dt^2} \times \mathbf{n} + \frac{1}{6c}\frac{d^3\mathbf{Q}_\perp}{dt^3}(\mathbf{n}, t) + \cdots \right) \right|^2$$

where

$$\mathbf{p} = \sum_j q_j \mathbf{r}_j, \qquad \mathbf{m} = \sum_j q_j \frac{\mathbf{r}_j \times \boldsymbol{\beta}_j}{2}, \qquad Q_{\alpha\beta} = \sum_j q_j(3x_{j\alpha}x_{j\beta} - r_j^2\,\delta_{\alpha\beta})$$

and the vector $\mathbf{Q}(\mathbf{n})$ has components, $Q_\alpha = \Sigma_\beta Q_{\alpha\beta}n_\beta$. Relate to the treatment of multipole radiation in Sections 9.2 and 9.3 and Problem 9.7.

15.2 A nonrelativistic particle of charge e and mass m collides with a fixed, smooth, hard sphere of radius R. Assuming that the collision is elastic, show that in the dipole approximation (neglecting retardation effects) the classical differential cross section for the emission of photons per unit solid angle per unit energy interval is

$$\frac{d^2\sigma}{d\Omega\, d(\hbar\omega)} = \frac{R^2}{12\pi}\frac{e^2}{\hbar c}\left(\frac{v}{c}\right)^2 \frac{1}{\hbar\omega}(2 + 3\sin^2\theta)$$

where θ is measured relative to the incident direction. Sketch the angular distribution. Integrate over angles to get the total bremsstrahlung cross section. Qualitatively, what factor (or factors) govern the upper limit to the frequency spectrum?

15.3 Treat Problem 15.2 without the assumption of nonrelativistic motion, using (15.2) and assuming the elastic impact is of negligible duration. Show that the cross section for photon emission is now

$$\frac{d^2\sigma}{d\Omega\, d(\hbar\omega)} = \frac{R^2}{4\pi}\frac{e^2}{\hbar c}\frac{\beta^2}{\hbar\omega}\left[\frac{\sin^2\theta}{(1 - \beta\cos\theta)^2} + \frac{1}{\beta^3}\ln\left(\frac{1 + \beta}{1 - \beta}\right) - \frac{2}{\beta^2} \right]$$

15.4   A group of charged particles with charges e_j and coordinates $\mathbf{r}_j(t)$ undergo interactions and are accelerated only during a time $-\tau/2 < t < \tau/2$, during which their velocities change from $c\boldsymbol{\beta}_j$ to $c\boldsymbol{\beta}'_j$.

(a)   Show that for $\omega\tau \ll 1$ the intensity of radiation emitted with polarization $\boldsymbol{\epsilon}$ per unit solid angle and unit frequency interval is

$$\frac{d^2I}{d\omega\,d\Omega} = \frac{1}{4\pi^2 c}|\boldsymbol{\epsilon}^* \cdot \mathbf{E}|^2$$

where

$$\mathbf{E} = \sum_j e_j\left(\frac{\boldsymbol{\beta}'_j}{1 - \mathbf{n}\cdot\boldsymbol{\beta}'_j} - \frac{\boldsymbol{\beta}_j}{1 - \mathbf{n}\cdot\boldsymbol{\beta}_j}\right)e^{-i\omega\mathbf{n}\cdot\mathbf{r}_j(0)/c}$$

(b)   An ω^0 meson of mass 784 MeV decays into $\pi^+\pi^-$ and e^+e^- with branching ratios of 1.3×10^{-2} and 8×10^{-5}, respectively. Show that for both decay modes the frequency spectrum of radiated energy at low frequencies is

$$\frac{dI}{d\omega} = \frac{e^2}{\pi c}\left[\left(\frac{1 + \beta^2}{\beta}\right)\ln\left(\frac{1 + \beta}{1 - \beta}\right) - 2\right] \approx \frac{4e^2}{\pi c}\left[\ln\left(\frac{M_\omega}{m}\right) - \frac{1}{2}\right]$$

where M_ω is the mass of the ω^0 meson and m is the mass of one of the decay products. Evaluate approximately the *total* energy radiated in each decay by integrating the spectrum up to the maximum frequency allowed kinematically. What fraction of the rest energy of the ω^0 is it in each decay?

15.5   A situation closely related to that of Problem 15.4b is the emission of radiation caused by the disappearance of charges and magnetic moments in the annihilation of electrons and positrons to form hadrons in high-energy storage ring experiments. If the differential cross section for the process $e^+e^- \to$ hadrons is $d\sigma_0$, *without* the emission of photons, calculate the cross section for the same process accompanied by a *soft* photon ($\hbar\omega \to 0$) in the energy interval $d(\hbar\omega)$ around $\hbar\omega$. Compare your results with the quantum-mechanical expressions:

$$\frac{d^2\sigma}{d\Omega\,d(\hbar\omega)} = \frac{\alpha}{\pi^2}\frac{d\sigma_0(s')}{\hbar\omega}\sqrt{\frac{s'^2 - 4m^2s'}{s^2 - 4m^2s}}\cdot\frac{\left[\beta^2\sin^2\theta + \dfrac{\hbar^2\omega^2}{s' + 2m^2}(1 - \beta^4\cos^4\theta)\right]}{(1 - \beta^2\cos^2\theta)^2}$$

$$\frac{d\sigma}{d(\hbar\omega)} = \frac{4\alpha}{\pi}\frac{d\sigma_0(s')}{\hbar\omega}\sqrt{\frac{s'^2 - 4m^2s'}{s^2 - 4m^2s}}\left\{\frac{1}{2}\left[\frac{1 + \beta^2}{2\beta}\ln\left(\frac{1 + \beta}{1 - \beta}\right) - 1\right]\right.$$
$$\left. + \frac{\hbar^2\omega^2}{s' + 2m^2}\left[\frac{1}{\beta}\ln\left(\frac{1 + \beta}{1 - \beta}\right) - 1\right]\right\}$$

where $s = (p_1 + p_2)^2$, $s' = (p_1 + p_2 - k)^2$, $\beta =$ electron v/c in c.m. frame. Neglect the emission of radiation by any of the hadrons, all assumed to be much heavier than the electrons.

The factors proportional to ω^2 in the numerators of these expressions can be attributed to the disappearance of the magnetic moments. If you have not included such contributions in your semiclassical calculation, you may consider doing so.

15.6   For the soft-photon limit of radiation emitted when there is a small change $\boldsymbol{\Delta\beta}$ of velocity, (15.9) applies, with convenient polarization vectors shown in Fig. 15.2.

(a)   Show that

$$\boldsymbol{\epsilon}_\parallel \cdot [\boldsymbol{\Delta\beta} + \mathbf{n}\times(\boldsymbol{\beta}\times\boldsymbol{\Delta\beta})] = |\boldsymbol{\Delta\beta}|(\beta - \cos\theta)\cos\phi$$
$$\boldsymbol{\epsilon}_\perp \cdot [\boldsymbol{\Delta\beta} + \mathbf{n}\times(\boldsymbol{\beta}\times\boldsymbol{\Delta\beta})] = |\boldsymbol{\Delta\beta}|(1 - \beta\cos\theta)\sin\phi$$

leading to (15.10) after averaging over ϕ.

(b) Show that in the limit $\gamma \gg 1$ and $\theta \ll 1$

$$\lim_{\omega \to 0} \frac{d^2 I}{d\omega \, d\Omega} \approx \frac{z^2 e^2 \gamma^4 \, |\Delta\boldsymbol{\beta}|^2}{2\pi^2 c} \cdot \left[\frac{(\gamma^2 \theta^2 - 1)^2}{(1 + \gamma^2 \theta^2)^4} + \frac{1}{(1 + \gamma^2 \theta^2)^2} \right]$$

where the first (second) term in square brackets corresponds to the parallel (perpendicular) polarization. This expression leads immediately to (15.11).

(c) Show that the result of part b gives the expression for $P(\theta)$ given following (15.10).

(d) Show that the angular integral of (15.11) or the answer in part b can be written

$$\lim_{\omega \to 0} \frac{dI}{d\omega} \approx \frac{z^2 e^2 \gamma^2 \, |\Delta\boldsymbol{\beta}|^2}{\pi c} \cdot \int_1^\infty \frac{dy}{y^4} (2 - 2y + y^2) = \frac{2}{3} \frac{z^2 e^2 \gamma^2 \, |\Delta\boldsymbol{\beta}|^2}{\pi c}$$

15.7 Consider the radiation emitted in nuclear fission by the sudden creation of two fragments of charge and mass $(Z_1 e, A_1 m)$ and $(Z_2 e, A_2 m)$ recoiling in opposite directions with total c.m.s. kinetic energy E. Treat the nuclei as point charges and their motion after the very short initial period of acceleration is nonrelativistic, but keep terms up to second order in $1/c$, as in Problem 15.1. For simplicity, assume that the fragments move with constant speeds in opposite directions away from the origin for $t > 0$. The relative speed is $c\beta$.

(a) Using the appropriate generalization of Problem 15.1a, show that the intensity of radiation per unit solid angle and per unit photon energy in the c.m. system is

$$\frac{d^2 I}{d(\hbar\omega) \, d\Omega} = \frac{\alpha\beta^2 \sin^2\theta}{4\pi^2 c} |p + q\beta \cos\theta|^2$$

where θ is the angle between the line of recoil and the direction of observation, and

$$\alpha = \frac{e^2}{\hbar c}; \qquad p = \frac{Z_1 A_2 - Z_2 A_1}{A_1 + A_2}; \qquad q = \frac{Z_1 A_2^2 + Z_2 A_1^2}{(A_1 + A_2)^2}$$

Show that the radiated energy per unit photon energy is

$$\frac{dI}{d(\hbar\omega)} = \frac{2\alpha\beta^2}{3\pi} \left(p^2 + \frac{\beta^2 q^2}{5} \right)$$

where the first term is the electric dipole and the second the quadrupole radiation.

(b) As an example of the asymmetric fission of ^{235}U by thermal neutrons, take $Z_1 = 36$, $A_1 = 95$ (krypton), $Z_2 = 56$, $A_2 = 138$ (barium) (three neutrons are emitted during fission), with $E = 170$ MeV and $mc^2 = 931.5$ MeV. What are the values of p^2 and q^2? Determine the total amount of energy (in MeV) radiated by this "inner bremsstrahlung" process, with the substitution, $\beta^2 \to \beta^2(1 - \hbar\omega/E)$, as a crude way to incorporate conservation of energy. What are the relative amounts of energy radiated in the dipole and quadrupole modes? In actual fission, roughly 7 MeV of electromagnetic energy is radiated within 10^{-8} s. How does your estimate compare? If it is much smaller or larger, attempt to explain.

15.8 Two identical point particles of charge q and mass m interact by means of a short-range repulsive interaction that is equivalent to a hard sphere of radius R in their relative separation. Neglecting the electromagnetic *interaction* between the two particles, determine the radiation cross section in the center-of-mass system for a

collision between these identical particles to the lowest nonvanishing approximation. Show that the differential cross section for emission of photons per unit solid
angle per unit energy interval is

$$\frac{d^2\sigma}{d(\hbar\omega)\,d\Omega} = \left(\frac{q^2}{\hbar c}\right) \cdot \frac{\beta^2 R^2}{48\pi} \cdot \frac{1}{\hbar\omega} \left[\beta^2 \left(\frac{2}{5} + 3\sin^2\theta\cos^2\theta\right) + \left(\frac{\omega R}{c}\right)^2 \left(1 - \frac{3}{8}\sin^4\theta\right)\right]$$

where θ is measured relative to the incident direction and $c\beta$ is the relative speed.
By integration over the angles of emission, show that the total cross section for
radiation per unit photon energy $\hbar\omega$ is

$$\frac{d\sigma}{d(\hbar\omega)} = \left(\frac{q^2}{\hbar c}\right) \cdot \frac{\beta^2 R^2}{15} \cdot \frac{1}{\hbar\omega} \left[\beta^2 + \left(\frac{\omega R}{c}\right)^2\right]$$

Compare these results with that of Problem 15.2 as to frequency dependence,
relative magnitude, etc.

15.9   A particle of charge ze, mass m, and nonrelativistic velocity v is deflected in a
screened Coulomb field, $V(r) = Zze^2 e^{-\alpha r}/r$, and consequently emits radiation. Discuss the radiation with the approximation that the particle moves in an almost
straight-line trajectory past the force center.

(a)   Show that, if the impact parameter is b, the energy radiated per unit frequency interval is

$$\frac{dI}{d\omega}(\omega, b) = \frac{8}{3\pi} \frac{Z^2 e^2}{c} \left(\frac{z^2 e^2}{mc^2}\right)^2 \left(\frac{c}{v}\right)^2 \alpha^2 K_1^2(\alpha b)$$

for $\omega \ll v/b$, and negligible for $\omega \gg v/b$.

(b)   Show that the radiation cross section is

$$\frac{d\chi(\omega)}{d\omega} \simeq -\frac{16}{3} \frac{Z^2 e^2}{c} \left(\frac{z^2 e^2}{mc^2}\right)^2 \left(\frac{c}{v}\right)^2 \left\{\frac{x^2}{2} \left[K_0^2(x) - K_1^2(x) + \frac{2K_0(x)K_1(x)}{x}\right]\right\}_{x_1}^{x_2}$$

where $x_1 = \alpha b_{min}$, $x_2 = \alpha b_{max}$.

(c)   With $b_{min} = \hbar/mv$, $b_{max} = v/\omega$, and $\alpha^{-1} = 1.4a_0 Z^{-1/3}$, determine the radiation
cross section in the two limits, $x_2 \ll 1$ and $x_2 \gg 1$. Compare your results
with the "screening" and "no screening" limits of the text.

15.10   A particle of charge ze, mass m, and velocity v is deflected in a hyperbolic path
by a fixed repulsive Coulomb potential, $V(r) = Zze^2/r$. Assume the nonrelativistic
dipole approximation (but no further approximations).

(a)   Show that the energy radiated per unit frequency interval by the particle
when initially incident at impact parameter b is

$$\frac{d}{d\omega} I(\omega, b) = \frac{8}{3\pi} \frac{(zea\omega)^2}{c^3} e^{-(\pi\omega/\omega_0)} \left\{\left[K'_{i\omega/\omega_0}\left(\frac{\omega\epsilon}{\omega_0}\right)\right]^2 + \frac{\epsilon^2 - 1}{\epsilon^2} \left[K_{i\omega/\omega_0}\left(\frac{\omega\epsilon}{\omega_0}\right)\right]^2\right\}$$

(b)   Show that the radiation cross section is

$$\frac{d}{d\omega} \chi(\omega) = \frac{16}{3} \frac{(zeav)^2}{c^3} e^{-(\pi\omega/\omega_0)} \frac{\omega}{\omega_0} K_{i\omega/\omega_0}\left(\frac{\omega}{\omega_0}\right) \left[-K'_{i\omega/\omega_0}\left(\frac{\omega}{\omega_0}\right)\right]$$

(c)   Prove that the radiation cross section reduces to that obtained in the text for
classical bremsstrahlung for $\omega \ll \omega_0$. What is the limiting form for $\omega \gg \omega_0$?

(d)   What modifications occur for an attractive Coulomb interaction?

The hyperbolic path may be described by

$$x = a(\epsilon + \cosh\xi), \qquad y = -b\sinh\xi, \qquad \omega_0 t = (\xi + \epsilon\sinh\xi)$$

where $a = Zze^2/mv^2$, $\epsilon = \sqrt{1 + (b/a)^2}$, $\omega_0 = v/a$.

15.11   Using the method of virtual quanta, discuss the relationship between the cross section for photodisintegration of a nucleus and electrodisintegration of a nucleus.

(a)   Show that, for electrons of energy $E = \gamma mc^2 \gg mc^2$, the electron disintegration cross section is approximately:

$$\sigma_{\text{el}}(E) \simeq \frac{2}{\pi} \frac{e^2}{\hbar c} \int_{\omega_T}^{E/\hbar} \sigma_{\text{photo}}(\omega) \ln\left(\frac{k\gamma^2 mc^2}{\hbar\omega}\right) \frac{d\omega}{\omega}$$

where $\hbar\omega_T$ is the threshold energy for the process and k is a constant of order unity.

(b)   Assuming that $\sigma_{\text{photo}}(\omega)$ has the resonance shape:

$$\sigma_{\text{photo}}(\omega) \simeq \frac{A}{2\pi} \frac{e^2}{Mc} \frac{\Gamma}{(\omega - \omega_0)^2 + (\Gamma/2)^2}$$

where the width Γ is small compared to $(\omega_0 - \omega_T)$, sketch the behavior of $\sigma_{\text{el}}(E)$ as a function of E and show that for $E \gg \hbar\omega_0$,

$$\sigma_{\text{el}}(E) \simeq \frac{2}{\pi} \left(\frac{e^2}{\hbar c}\right) \frac{Ae^2}{Mc} \frac{1}{\omega_0} \ln\left(\frac{kE^2}{mc^2\hbar\omega_0}\right)$$

(c)   In the limit of a very narrow resonance, the photonuclear cross section can be written as $\sigma_{\text{photo}}(\omega) = (Ae^2/Mc)\,\delta(\omega - \omega_0)$. Then the result of part b would represent the electrodisintegration cross section for $\omega > \omega_0$. The corresponding bremsstrahlung-induced cross section is given in the same approximation by (15.47), multiplied by $(Ae^2/Mc\hbar\omega_0)$, where Z is the atomic number of the radiator. Comparisons of the electron- and bremsstrahlung-induced disintegration cross sections of a number of nuclei are given by E. Wolynec et al. *Phys. Rev. C* **11**, 1083 (1975). Calculate the quantity called F as a function of E (with a giant dipole resonance energy $\hbar\omega_0 \approx 20$ MeV) and compare its magnitude and energy dependence (at the high energy end) with the data in Figures 1–5 of Wolynec et al. The comparison is only qualitative at $E \approx \hbar\omega_0$ because of the breadth of the dipole resonance. [F is the ratio of the bremsstrahlung-induced cross section in units of $Z^2 r_0^2$ to the electrodisintegration cross section.]

15.12   A fast particle of charge ze, mass M, and velocity v, collides with a hydrogen-like atom with one electron of charge $-e$, mass m, bound to a nuclear center of charge Ze. The collisions can be divided into two kinds: close collisions where the particle passes through the atom ($b < d$), and distant collisions where the particle passes by outside the atom ($b > d$). The atomic "radius" d can be taken as a_0/Z. For the close collisions the interaction of the incident particle and the electron can be treated as a two-body collision and the energy transfer calculated from the Rutherford cross section. For the distant collisions the excitation and ionization of the atom can be considered the result of the photoelectric effect by the virtual quanta of the incident particle's fields.

For simplicity assume that for photon energies Q greater than the ionization potential I the photoelectric cross section is

$$\sigma_\gamma(Q) = \frac{8\pi^2}{137} \left(\frac{a_0}{Z}\right)^2 \left(\frac{I}{Q}\right)^3$$

(This obeys the empirical $Z^4\lambda^3$ law for x-ray absorption and has a coefficient adjusted to satisfy the dipole sum rule, $\int \sigma_\gamma(Q)\, dQ = 2\pi^2 e^2\hbar/mc$.)

(a) Calculate the differential cross sections $d\sigma/dQ$ for energy transfer Q for close and distant collisions (write them as functions of Q/I as far as possible and in units of $2\pi z^2 e^4/mv^2 I^2$). *Plot* the two distributions for $Q/I > 1$ for non-relativistic motion of the incident particle and $\frac{1}{2}mv^2 = 10^3 I$.

(b) Show that the number of distant collisions measured by the integrated cross section is much larger than the number of close collisions, but that the energy transfer per collision is much smaller. Show that the energy loss is divided approximately equally between the two kinds of collisions, and verify that your total energy loss is in essential agreement with Bethe's result (13.14).

15.13 In the decay of a pi meson at rest a mu meson and a neutrino are created. The total kinetic energy available is $(m_\pi - m_\mu)c^2 = 34$ MeV. The mu meson has a kinetic energy of 4.1 MeV. Determine the number of quanta emitted per unit energy interval because of the sudden creation of the moving mu meson. Assuming that the photons are emitted perpendicular to the direction of motion of the mu meson (actually it is a $\sin^2\theta$ distribution), show that the maximum photon energy is 17 MeV. Find how many quanta are emitted with energies greater than one-tenth of the maximum, and compare your result with the observed ratio of radiative pi-mu decays. [$1.24 \pm 0.25 \times 10^{-4}$ for muons with kinetic energy less than 3.4 MeV. See also H. Primakoff, *Phys. Rev.*, **84**, 1255 (1951).]

15.14 In internal conversion, the nucleus makes a transition from one state to another and an orbital electron is ejected. The electron has a kinetic energy equal to the transition energy minus its binding energy. For a conversion line of 1 MeV determine the number of quanta emitted per unit energy because of the sudden ejection of the electron. What fraction of the electrons will have energies less than 99% of the total energy? Will this low-energy tail on the conversion line be experimentally observable?

15.15 One of the decay modes of a K^+ meson is the three-pion decay, $K^+ \rightarrow \pi^+\pi^+\pi^-$. The energy release is 75 MeV, small enough that the pions can be treated non-relativistically in rough approximation.

(a) Show that the differential spectrum of radiated intensity at low frequencies in the K meson rest frame is approximately

$$\frac{d^2 I}{d\omega \, d\Omega} \simeq \frac{2e^2}{\pi^2 c} \cdot \frac{T_-}{m_\pi c^2} \cdot \sin^2\theta$$

where T_- is the kinetic energy of the negative pion and θ is the angle of emission of the photon relative to the momentum of the negative pion.

(b) Estimate the branching ratio for emission of a photon of energy greater than Δ relative to the nonradiative three-pion decay. What is its numerical value for $\Delta = 1$ MeV? 10 MeV? Compare with experiment ($\sim 2 \times 10^{-3}$ for $\Delta = 11$ MeV).

15.16 One of the decay modes of the charged K meson ($M_K = 493.7$ MeV) is $K^+ \rightarrow \pi^+\pi^0$ ($M_{\pi^+} = 139.6$ MeV, $M_{\pi^0} = 135.0$ MeV). Inner bremsstrahlung is emitted by the creation of the positive pion. A study of this radiative decay mode was made by Edwards et al. [*Phys. Rev.* **D5**, 2720 (1972)].

(a) Calculate the classical distribution in angle and frequency of soft photons and compare with the data of Fig. 6 of Edwards et al. Compute the classical distribution also for $\beta = 0.71$, corresponding to a charged pion of kinetic energy 58 MeV, and compare.

(b) Estimate the number of radiative decays for charged pion kinetic energies on the interval, 55 MeV $\leq T_\pi \leq$ 90 MeV, as a fraction of *all* K^+ decays (the $\pi^+ \pi^0$ decay mode is 21% of all decays). You can treat the kinematics, including the photon, correctly, or you can approximate reality with an idealization that has the neutral pion always with the same momentum and the photon and the charged pion with parallel momenta (see part a for justification of this assumption). This idealization permits you to correlate directly the limits on the charged pion's kinetic energy with that of the photon. Compare your estimate with the experimental value for the branching ratio for $\pi^+ \pi^0 \gamma$ (with the limited range of π^+ energies) of $(2.75 \pm 0.15) \times 10^{-4}$.

CHAPTER 16

Radiation Damping, Classical Models of Charged Particles

16.1 Introductory Considerations

In the preceding chapters the problems of electrodynamics have been divided into two classes: one in which the sources of charge and current are specified and the resulting electromagnetic fields are calculated, and the other in which the external electromagnetic fields are specified and the motions of charged particles or currents are calculated. Waveguides, cavities, and radiation from prescribed multipole sources are examples of the first type of problem, while motion of charges in electric and magnetic fields and energy-loss phenomena are examples of the second type. Occasionally, as in the discussion of bremsstrahlung, the two problems are combined. But the treatment is a stepwise one—first the motion of the charged particle in an external field is determined, neglecting the emission of radiation; then the radiation is calculated from the trajectory as a given source distribution.

It is evident that this manner of handling problems in electrodynamics can be of only approximate validity. The motion of charged particles in external force fields necessarily involves the emission of radiation whenever the charges are accelerated. The emitted radiation carries off energy, momentum, and angular momentum and so must influence the subsequent motion of the charged particles. Consequently the motion of the sources of radiation is determined, in part, by the manner of emission of the radiation. A correct treatment must include the reaction of the radiation on the motion of the sources.

Why is it that we have taken so long in our discussion of electrodynamics to face this fact? Why is it that many answers calculated in an apparently erroneous way agree so well with experiment? A partial answer to the first question lies in the second. There *are* very many problems in electrodynamics that can be put with negligible error into one of the two categories described in the first paragraph. Hence it is worthwhile discussing them without the added and unnecessary complication of including reaction effects. The remaining answer to the first question is that a completely satisfactory classical treatment of the reactive effects of radiation does not exist. The difficulties presented by this problem touch one of the most fundamental aspects of physics, the nature of an elementary particle. Although partial solutions, workable within limited areas, can be given, the basic problem remains unsolved.

In quantum mechanics, the situation at first appeared worse, but development of the renormalization program of quantum field theory in the 1950s led to a consistent theoretical description of electrodynamics (called quantum electrodynamics or QED, the interaction of electrons and positrons with electromagnetic fields) in terms of observed quantities such as mass and static charge. A weak-coupling theory ($\alpha \approx 1/137$), QED has proven remarkably successful in explaining to amazing accuracy the tiny radiative corrections observed in precision atomic experiments (Lamb shift, anomalous magnetic moments, etc.) by calculating to higher and higher orders in perturbation theory. More recently, the success has been extended to weak and strong interactions as well within the standard model, sketched briefly at the beginning of the Introduction. Unfortunately, the strong interactions are not really amenable to accurate calculations via perturbation theory.

In this chapter we address only some of the classical aspects of radiation reaction.

The question of why many problems can apparently be handled neglecting reactive effects of the radiation has the obvious answer that such effects must be of negligible importance. To see qualitatively when this is so, and to obtain semiquantitative estimates of the ranges of parameters where radiative effects are or are not important, we need a simple criterion. One such criterion can be obtained from energy considerations. If an external force field causes a particle of charge e to have an acceleration of typical magnitude a for a period of time T, the energy radiated is of the order of

$$E_{\text{rad}} \sim \frac{2e^2a^2T}{3c^3} \tag{16.1}$$

from the Larmor formula (14.22). If this energy lost in radiation is negligible compared to the relevant energy E_0 of the problem, we can expect that radiative effects will be unimportant. But If $E_{\text{rad}} \gtrsim E_0$, the effects of radiation reaction will be appreciable. The criterion for the regime where radiative effects are unimportant can thus be expressed by

$$E_{\text{rad}} \ll E_0 \tag{16.2}$$

The specification of the relevant energy E_0 demands a little care. We distinguish two apparently different situations, one in which the particle is initially at rest and is acted on by the applied force only for the finite interval T, and one where the particle undergoes continual acceleration, e.g., in quasiperiodic motion at some characteristic frequency ω_0. For the particle at rest initially, a typical energy is evidently its kinetic energy after the period of acceleration. Thus

$$E_0 \sim m(aT)^2$$

The criterion (16.2) for the unimportance of radiative effects then becomes

$$\frac{2}{3}\frac{e^2a^2T}{c^3} \ll ma^2T^2$$

or

$$T \gg \frac{2}{3}\frac{e^2}{mc^3}$$

It is useful to define the *characteristic time* in this relation as

$$\tau = \frac{2}{3}\frac{e^2}{mc^3} \tag{16.3}$$

Then the conclusion is that for time T long compared to τ radiative effects are unimportant. Only when the force is applied so suddenly and for such a short time that $T \sim \tau$ will radiative effects modify the motion appreciably. It is useful to note that the longest characteristic time τ for charged particles is for electrons and that its value is $\tau = 6.26 \times 10^{-24}$ s. This is of the order of the time taken for light to travel 10^{-15} m. Only for phenomena involving such distances or times will we expect radiative effects to play a *crucial* role.

If the motion of the charged particle is quasiperiodic with a typical amplitude d and characteristic frequency ω_0, the mechanical energy of motion can be identified with E_0 and is of the order of

$$E_0 \sim m\omega_0^2 d^2$$

The accelerations are typically $a \sim \omega_0^2 d$, and the time interval $T \sim (1/\omega_0)$. Consequently criterion (16.2) is

$$\frac{2e^2 \omega_0^4 d^2}{3c^3 \omega_0} \ll m\omega_0^2 d^2$$

or   (16.4)

$$\omega_0 \tau \ll 1$$

where τ is given by (16.3). Since ω_0^{-1} is a time appropriate to the mechanical motion, again we see that, if the relevant mechanical time interval is long compared to the characteristic time τ (16.3), radiative reaction effects on the motion will be unimportant.

The examples of the last two paragraphs show that the reactive effects of radiation on the motion of a charged particle can be expected to be important if the external forces are such that the motion changes appreciably in times of the order of τ or over distances of the order of $c\tau$. This is a general criterion within the framework of classical electrodynamics. For motions less violent, the reactive effects are small enough to have a negligible effect on the short-term motion. Their long-term, cumulative effects can be taken into account in an approximate way, as we see immediately.

16.2   *Radiative Reaction Force from Conservation of Energy*

The question now arises as to how to include the reactive effects of radiation in the equations of motion for a charged particle. We begin with a simple plausibility argument based on conservation of energy for a nonrelativistic charged particle. A more fundamental derivation and the incorporation of relativistic effects are deferred to later sections.

If the emission of radiation is neglected, a charged particle of mass m and charge e acted on by an external force $\mathbf{F}_{\text{ext}}$ moves according to the Newton equation of motion:

$$m\dot{\mathbf{v}} = \mathbf{F}_{\text{ext}} \tag{16.5}$$

Since the particle is accelerated, it emits radiation at a rate given by the Larmor power formula (14.22),

$$P(t) = \frac{2}{3}\frac{e^2}{c^3}(\dot{\mathbf{v}})^2 \tag{16.6}$$

To account for this radiative energy loss and its effect on the motion of the particle we modify the Newton equation (16.5) by adding a *radiative reaction force* $\mathbf{F}_{rad}$:

$$m\dot{\mathbf{v}} = \mathbf{F}_{ext} + \mathbf{F}_{rad} \tag{16.7}$$

While $\mathbf{F}_{rad}$ is not determined at this stage, we can see some of the requirements it "must" satisfy:

$\mathbf{F}_{rad}$ "must"   (1) vanish if $\dot{\mathbf{v}} = 0$, since then there is no radiation;
 (2) be proportional to e^2, since (a) the radiated power is proportional to e^2, and (b) the sign of the charge cannot enter in radiative effects;
 (3) in fact involve the characteristic time τ (16.3), since that is apparently the only parameter of significance available.

We determine the form of $\mathbf{F}_{rad}$ by demanding that the work done by this force on the particle in the time interval $t_1 < t < t_2$ be equal to the negative of the energy radiated in that time. Then energy will be conserved, at least over the interval (t_1, t_2). With the Larmor result (16.6), this requirement is

$$\int_{t_1}^{t_2} \mathbf{F}_{rad} \cdot \mathbf{v} \, dt = -\int_{t_1}^{t_2} \frac{2}{3}\frac{e^2}{c^3} \dot{\mathbf{v}} \cdot \dot{\mathbf{v}} \, dt$$

The second integral can be integrated by parts to yield

$$\int_{t_1}^{t_2} \mathbf{F}_{rad} \cdot \mathbf{v} \, dt = \frac{2}{3}\frac{e^2}{c^3} \int_{t_1}^{t_2} \ddot{\mathbf{v}} \cdot \mathbf{v} \, dt - \frac{2}{3}\frac{e^2}{c^3} (\dot{\mathbf{v}} \cdot \mathbf{v}) \Big|_{t_1}^{t_2}$$

If the motion is periodic or such that $(\dot{\mathbf{v}} \cdot \mathbf{v}) = 0$ at $t = t_1$ and $t = t_2$, we may write

$$\int_{t_1}^{t_2} \left(\mathbf{F}_{rad} - \frac{2}{3}\frac{e^2}{c^3} \ddot{\mathbf{v}} \right) \cdot \mathbf{v} \, dt = 0$$

Then it is permissible to identify the radiative reaction force as

$$\mathbf{F}_{rad} = \frac{2}{3}\frac{e^2}{c^3} \ddot{\mathbf{v}} = m\tau\ddot{\mathbf{v}} \tag{16.8}$$

The modified equation of motion then reads

$$m(\dot{\mathbf{v}} - \tau\ddot{\mathbf{v}}) = \mathbf{F}_{ext} \tag{16.9}$$

Equation (16.9) is sometimes called the *Abraham–Lorentz equation of motion*. It can be considered as an equation that includes in some approximate and time-averaged way the reactive effects of the emission of radiation. The equation can be criticized on the grounds that it is second order in time, rather than first, and therefore runs counter to the well-known requirements for a dynamical equation of motion. This difficulty manifests itself immediately in "runaway" solutions. If the external force is zero, it is obvious that (16.9) has two possible solutions,

$$\dot{\mathbf{v}}(t) = \begin{cases} 0 \\ \mathbf{a}e^{t/\tau} \end{cases}$$

where $\mathbf{a}$ is the acceleration at $t = 0$. Only the first solution is reasonable. The method of derivation shows that the second solution is unacceptable, since

$(\dot{\mathbf{v}} \cdot \mathbf{v}) \neq 0$ at t_1 and t_2. It is clear that the equation is useful only in the domain where the reactive term is a small correction. Then the radiative reaction can be treated as a perturbation producing slow or small changes in the state of motion of the particle.

An alternative to (16.9) can be obtained by using the zeroth-order equation of motion, $m\dot{\mathbf{v}} = \mathbf{F}_{ext}$, to evaluate the radiation reaction term. The resulting equation,

$$ m\dot{\mathbf{v}} = \mathbf{F}_{ext} + \tau \frac{d\mathbf{F}_{ext}}{dt} = \mathbf{F}_{ext} + \tau \left[\frac{\partial \mathbf{F}_{ext}}{\partial t} + (\mathbf{v} \cdot \boldsymbol{\nabla})\mathbf{F}_{ext} \right] \qquad (16.10) $$

is a valid equation of motion without runaway solutions or acausal behavior. It is a sensible alternative to the Abraham–Lorentz equation for the classical regime of small radiative effects. It also emerges from a different starting point— see G. W. Ford and R. F. O'Connell [*Phys. Lett. A*, **157**, 217 (1991)]. Relativistic generalizations of (16.9) and (16.10) can be constructed—see Problems 16.7 and 16.9.

To illustrate the use of (16.10) to account for small radiative effects, we consider a particle moving in an attractive, conservative, central force field. In the absence of radiation reaction, the particle's energy and angular momentum are conserved and determine the motion. The emission of radiation causes changes in these quantities. Provided the accelerations are not too violent, the energy and angular momentum will change appreciably only in a time interval that is long compared to the characteristic period of the motion. Thus the motion will instantaneously be essentially the same as in the absence of radiative reaction. The long-term changes can be described by averages over the particle's unperturbed orbit.

If the conservative central force field is described by a potential $V(r)$, the acceleration, neglecting reactive effects, is

$$ \dot{\mathbf{v}} = \frac{-1}{m} \left(\frac{dV}{dr} \right) \frac{\mathbf{r}}{r} \qquad (16.11) $$

By conservation of energy, the rate of change of the particle's total energy is given by the negative of the Larmor power:

$$ \frac{dE}{dt} = -\frac{2}{3} \frac{e^2}{c^3} (\dot{\mathbf{v}})^2 = -\frac{2e^2}{3m^2c^3} \left(\frac{dV}{dr} \right)^2 $$

With the definition of τ (16.3) this can be written

$$ \frac{dE}{dt} = -\frac{\tau}{m} \left(\frac{dV}{dr} \right)^2 \qquad (16.12) $$

Since the change in energy is assumed to be small in one cycle of the orbit, the right-hand side may be replaced by its time-averaged value in terms of the Newtonian orbit. Then we obtain

$$ \left\langle \frac{dE}{dt} \right\rangle \simeq -\frac{\tau}{m} \left\langle \left(\frac{dV}{dr} \right)^2 \right\rangle \qquad (16.13) $$

The secular change in angular momentum can be found by considering the vector product of (16.10) with the radius vector $\mathbf{r}$. Since the angular momentum is $\mathbf{L} = m\mathbf{r} \times \mathbf{v}$, we find

$$\frac{d\mathbf{L}}{dt} = \mathbf{r} \times \mathbf{F}_{\text{ext}} + \tau\mathbf{r} \times \frac{d\mathbf{F}_{\text{ext}}}{dt} = \tau\mathbf{r} \times (\mathbf{v} \cdot \boldsymbol{\nabla})\mathbf{F}_{\text{ext}} \qquad (16.14)$$

where the second form results because the force is central and time independent. With (16.11), the right-hand side of (16.14) is found to be

$$\tau\mathbf{r} \times (\mathbf{v} \cdot \boldsymbol{\nabla})\mathbf{F}_{\text{ext}} = \tau\mathbf{r} \times \left(-\frac{dV}{r\,dr}\right)\mathbf{v}_{\perp} = -\frac{\tau}{m}\frac{dV}{r\,dr}\mathbf{L} \qquad (16.15)$$

With the average of this torque over the slowly changing orbit, the secular rate of change of angular momentum can be written as

$$\left\langle\frac{d\mathbf{L}}{dt}\right\rangle \simeq -\frac{\tau}{m}\left\langle\frac{1}{r}\frac{dV}{dr}\right\rangle\mathbf{L} \qquad (16.16)$$

Note that this result for the decay of the *particle's* angular momentum is exactly the negative of the rate one calculates for the angular momentum *radiated* in electric dipole radiation (Problem 9.9).

Equations (16.13) and (16.16) determine how the particle orbit changes as a function of time because of radiative reaction. Although the detailed behavior depends on the specific law of force, some qualitative statements can be made. If the characteristic frequency of motion is ω_0, the average value in (16.16) can be written

$$\frac{\tau}{m}\left\langle\frac{1}{r}\frac{dV}{dr}\right\rangle \sim \frac{\tau}{m}m\omega_0^2 = \omega_0^2\tau$$

with some dimensionless numerical coefficient of the order of unity. This shows that the characteristic time over which the angular momentum changes is of the order of $1/(\omega_0\tau)\omega_0$. This time is very long compared to the orbital period $2\pi/\omega_0$, provided $\omega_0\tau \ll 1$. Similar arguments can be made with the energy equations.

These equations including radiative effects can be used to discuss practical problems such as the moderation time of a mu or pi meson in cascading from an orbit of very large quantum number around a nucleus down to the low-lying orbits. Over most of the time interval the quantum numbers are sufficiently large that the classical description of continuous motion is an adequate approximation. Discussion of examples of this kind is left to the problems.

16.3 *Abraham–Lorentz Evaluation of the Self-Force*

The derivation of the radiation reaction force in Section 16.2, while plausible, is certainly not rigorous or fundamental. The problem is to give a satisfactory account of the reaction back on the charged particle of its own radiation fields. Thus any systematic discussion must consider the charge structure of the particle and its self-fields. Abraham (1903) and Lorentz (1904) made the first attempt at such a treatment by trying to make a purely electromagnetic model of a charged particle. In the beginning, our discussion is patterned after that given by Lorentz in his book, *Theory of Electrons* (note 18, p. 252).

Let us consider a single charged particle of total charge e with a sharply localized charge density $\rho(\mathbf{x})$ in the particle's rest frame. The particle is in external electromagnetic fields, $\mathbf{E}_{ext}(\mathbf{x}, t)$, $\mathbf{B}_{ext}(\mathbf{x}, t)$. We have seen in Sections 6.7 and 12.10 that the rate of change of mechanical momentum plus electromagnetic momentum in a given volume vanishes, provided there is no flow of momentum out of or into the volume. Abraham and Lorentz proposed that the apparently mechanical momentum of a charged particle is totally electromagnetic in origin. Here we take the more conservative position that the particle's momentum is partly mechanical, but with an electromagnetic contribution. Then the momentum-conservation law can be phrased,

$$\left(\frac{d\mathbf{p}}{dt}\right)_{mech} + \frac{d\mathbf{G}}{dt} = 0$$

or equivalently in terms of the Lorentz force density (12.121),

$$\left(\frac{d\mathbf{p}}{dt}\right)_{mech} = \int \left(\rho\mathbf{E} + \frac{1}{c}\mathbf{J} \times \mathbf{B}\right) d^3x \tag{16.17}$$

In this equation the fields are the *total* fields, and the integration is over the volume of the particle.

In order that (16.17) take on the form of the Newton equation of motion

$$\frac{d\mathbf{p}}{dt} \equiv \left(\frac{d\mathbf{p}}{dt}\right)_{mech} + \left(\frac{d\mathbf{p}}{dt}\right)_{em} = \mathbf{F}_{ext} \tag{16.18}$$

we decompose the total fields into the external fields and the self-fields $\mathbf{E}_s$, $\mathbf{B}_s$ due to the particle's own charge and current densities, ρ and $\mathbf{J}$:

$$\left.\begin{aligned} \mathbf{E} &= \mathbf{E}_{ext} + \mathbf{E}_s \\ \mathbf{B} &= \mathbf{B}_{ext} + \mathbf{B}_s \end{aligned}\right\}$$

Then (16.17) can be written as the Newton equations of motion, with the external force as

$$\mathbf{F}_{ext} = \int \left(\rho\mathbf{E}_{ext} + \frac{1}{c}\mathbf{J} \times \mathbf{B}_{ext}\right) d^3x \tag{16.19}$$

and the electromagnetic contribution to the rate of change of momentum of the particle as

$$\left(\frac{d\mathbf{p}}{dt}\right)_{em} = -\int \left(\rho\mathbf{E}_s + \frac{1}{c}\mathbf{J} \times \mathbf{B}_s\right) d^3x \tag{16.20}$$

Provided the external fields vary only slightly over the extent of the particle, the external force (16.19) becomes just the ordinary Lorentz force on a particle of charge e and velocity $\mathbf{v}$.

To calculate the self-force [the integral on the right-hand side of (16.20)] it is necessary to have a model of the charged particle. We will assume for simplicity that:

the particle is instantaneously at rest;

the charge distribution is rigid and spherically symmetric.

Our results will then necessarily be restricted to nonrelativistic motions and will lack some Lorentz transformation properties.

For a particle instantaneously at rest (16.20) becomes

$$\left(\frac{d\mathbf{p}}{dt}\right)_{\text{em}} = -\int \rho(\mathbf{x}, t)\mathbf{E}_s(x, t)\, d^3x \tag{16.21}$$

The self-field can be expressed in terms of the self-potentials, $\mathbf{A}$ and Φ, so that

$$\left(\frac{d\mathbf{p}}{dt}\right)_{\text{em}} = \int \rho(\mathbf{x}, t)\left[\nabla\Phi(\mathbf{x}, t) + \frac{1}{c}\frac{\partial \mathbf{A}}{\partial t}(\mathbf{x}, t)\right] d^3x \tag{16.22}$$

The potentials are given by $A^\alpha = (\Phi, \mathbf{A})$:

$$A^\alpha(\mathbf{x}, t) = \frac{1}{c}\int \frac{[J^\alpha(\mathbf{x}', t')]_{\text{ret}}}{R}\, d^3x' \tag{16.23}$$

with $J^\alpha = (c\rho, \mathbf{J})$ and $\mathbf{R} = \mathbf{x} - \mathbf{x}'$.

In (16.23) the 4-current must be evaluated at the retarded time t'. This differs from the time t by a time of the order of $\Delta t \sim (a/c)$, where a is the dimension of the particle. For a highly localized charge distribution this time interval is extremely short. During such a short time the motion of the particle can be assumed to change only slightly. Consequently it is natural to make a Taylor series expansion in (16.23) around the time $t' = t$. Since $[\cdots]_{\text{ret}}$ means evaluated at $t' = t - (R/c)$, any retarded quantity has the expansion

$$[\cdots]_{\text{ret}} = \sum_{n=0}^{\infty} \frac{(-1)^n}{n!}\left(\frac{R}{c}\right)^n \frac{\partial^n}{\partial t^n}[\cdots]_{t'=t} \tag{16.24}$$

With this expansion for the retarded 4-current in (16.23), expression (16.22) becomes

$$\left(\frac{d\mathbf{p}}{dt}\right)_{\text{em}} = \sum_{n=0}^{\infty} \frac{(-1)^n}{n!\, c^n}\int d^3x \int d^3x'\, \rho(\mathbf{x}, t)\frac{\partial^n}{\partial t^n}\left[\rho(\mathbf{x}', t)\nabla R^{n-1} + \frac{R^{n-1}}{c^2}\frac{\partial \mathbf{J}(\mathbf{x}', t)}{\partial t}\right]$$

Consider the $n = 0$ and $n = 1$ terms in the scalar potential part (the first term in the square bracket) of the right-hand side. For $n = 0$ the term is proportional to

$$\int d^3x \int d^3x'\, \rho(\mathbf{x}, t)\rho(\mathbf{x}', t)\nabla\left(\frac{1}{R}\right)$$

This is just the electrostatic self-force. For spherically symmetric charge distributions it vanishes. The $n = 1$ term is identically zero, since it involves ∇R^{n-1}. Thus the first nonvanishing contribution from the scalar potential part comes from $n = 2$. This means that we can change the summation indices so that the sum now reads

$$\left(\frac{d\mathbf{p}}{dt}\right)_{\text{em}} = \sum_{n=0}^{\infty} \frac{(-1)^n}{n!\, c^{n+2}}\int d^3x \int d^3x'\, \rho(\mathbf{x}, t)R^{n-1}\frac{\partial^{n+1}}{\partial t^{n+1}}\{\cdots\}$$

where $\hphantom{xxxxxxxxxxxxxxxxxxxxxxxxxxxxxxxxxxxxxxxxxxxxxxxxxx}$ (16.25)

$$\{\cdots\} = \mathbf{J}(\mathbf{x}', t) + \frac{\partial\rho}{\partial t}(\mathbf{x}', t)\frac{\nabla R^{n+1}}{(n+1)(n+2)R^{n-1}}$$

With the continuity equation for charge and current densities, the curly bracket in (16.25) can be written

$$\{\cdots\} = \mathbf{J}(\mathbf{x}', t) - \frac{\mathbf{R}}{n+2}\nabla' \cdot \mathbf{J}(\mathbf{x}', t)$$

In the integral over d^3x' we can integrate the second term by parts. We then have

$$-\int d^3x'\ R^{n-1}\frac{\mathbf{R}}{n+2}\boldsymbol{\nabla}'\cdot\mathbf{J} = +\frac{1}{n+2}\int d^3x'(\mathbf{J}\cdot\boldsymbol{\nabla}')R^{n-1}\mathbf{R}$$

$$=\frac{-1}{n+2}\int d^3x'\ R^{n-1}\left(\mathbf{J} + (n-1)\frac{\mathbf{J}\cdot\mathbf{R}}{R^2}\mathbf{R}\right)$$

This means that the curly bracket in (16.25) is effectively equal to

$$\{\cdots\} = \left(\frac{n+1}{n+2}\right)\mathbf{J}(\mathbf{x}',t) - \left(\frac{n-1}{n+2}\right)\frac{(\mathbf{J}\cdot\mathbf{R})\mathbf{R}}{R^2} \tag{16.26}$$

For a rigid charge distribution the current is

$$\mathbf{J}(\mathbf{x}',t) = \rho(\mathbf{x}',t)\mathbf{v}(t)$$

If the charge distribution is spherically symmetric, the only relevant direction in the problem is that of $\mathbf{v}(t)$. Consequently in the integration over d^3x and d^3x' only the component of (16.26) along the direction of $\mathbf{v}(t)$ survives. Hence (16.26) is equivalent to

$$\{\cdots\} = \rho(\mathbf{x}',t)\mathbf{v}(t)\left[\frac{n+1}{n+2} - \frac{n-1}{n+2}\left(\frac{\mathbf{R}\cdot\mathbf{v}}{Rv}\right)^2\right]$$

Furthermore all directions of $\mathbf{R}$ are equally probable. This means that the second term above can be replaced by its average value of $\frac{1}{3}$. This leads to the final simple form of our curly bracket in (16.25):

$$\{\cdots\} = \tfrac{2}{3}\rho(\mathbf{x}',t)\mathbf{v}(t) \tag{16.27}$$

With (16.27) in (16.25) the self-force becomes, apart from neglected nonlinear terms in time derivatives of $\mathbf{v}$ (which appear for $n \geq 4$),

$$\left(\frac{d\mathbf{p}}{dt}\right)_{\text{em}} = \sum_{n=0}^{\infty}\frac{(-1)^n}{c^{n+2}}\frac{2}{3n!}\frac{\partial^{n+1}\mathbf{v}}{\partial t^{n+1}}\int d^3x'\int d^3x\ \rho(\mathbf{x}')R^{n-1}\rho(\mathbf{x}) \tag{16.28}$$

To proceed further it is convenient to introduce Fourier transforms in time for the external force, the velocity, and the self-force.* The Fourier transform of the velocity $\mathbf{v}(\omega)$ is defined by

$$\mathbf{v}(t) = \frac{1}{\sqrt{2\pi}}\int \mathbf{v}(\omega)e^{-i\omega t}\ d\omega$$

and its inverse, and similarly for the others. If $(d\mathbf{p}/dt)_{\text{mech}} = m_0(d\mathbf{v}/dt)$, the Fourier transform of the force equation (16.18) is

$$-i\omega M(\omega)\mathbf{v}(\omega) = \mathbf{F}_{\text{ext}}(\omega) \tag{16.29}$$

where the "effective mass" $M(\omega)$ is

$$M(\omega) = m_0 + \frac{2}{3c^2}\sum_{n=0}^{\infty}\frac{(i\omega)^n}{n!\ c^n}\int d^3x\int d^3x'\ \rho(\mathbf{x})R^{n-1}\rho(\mathbf{x}')$$

*Here we parallel quantum-mechanical discussions of radiation reaction in the correspondence limit: Nonrelativistic theory, E. J. Moniz and D. E. Sharp, *Phys. Rev. D* **10**, 1133–1136 (1974); fully relativistic quantum theory (QED) of electrons and positrons, F. E. Low, *Ann Phy.* (N.Y.). **265**, No. 2 (1998).

The sum over n can be recognized as $e^{i\omega R/c}/R$, the outgoing wave Green function. Hence $M(\omega)$ can be written

$$M(\omega) = m_0 + \frac{2}{3c^2} \int d^3x \int d^3x' \, \rho(\mathbf{x}) \frac{e^{i\omega R/c}}{R} \rho(\mathbf{x}') \qquad (16.30)$$

The spherically symmetric average of $e^{i\omega R/c}/R$ is

$$\left\langle \frac{e^{i\omega R/c}}{R} \right\rangle = i \frac{\omega}{c} j_0\!\left(\frac{\omega r_<}{c}\right) h_0^{(1)}\!\left(\frac{\omega r_>}{c}\right)$$

For some specific spherically symmetric charge distributions, the spatial integrals in (16.30) may be performed to give an explicit closed form for $M(\omega)$. [See Problem 16.4.]

Alternatively, we can introduce the spatial Fourier transform (form factor) of the charge density to obtain a different expression for $M(\omega)$, a "spectral representation" familiar in quantum mechanics. We define the form factor $f(\mathbf{k})$ through the three-dimensional transform

$$\rho(\mathbf{x}) = \frac{e}{(2\pi)^3} \int d^3k \, f(\mathbf{k}) e^{i\mathbf{k}\cdot\mathbf{x}} \qquad (16.31)$$

where e is the total charge. For a point charge, $f(\mathbf{k}) = 1$. By straightforward substitution and integration, (16.30) is transformed to

$$M(\omega) = m_0 + \frac{e^2}{3\pi^2 c^2} \int d^3k \, \frac{|f(\mathbf{k})|^2}{k^2 - (\omega/c)^2} \qquad (16.32)$$

where ω has a small positive imaginary part.

Equations (16.29) and (16.32) are an almost complete solution for the classical nonrelativistic motion of an extended charged particle, including radiation reaction. ["Almost," because we neglected small nonlinear terms in higher powers of the velocity and we assumed spherical symmetry.] In the limit $\omega \to 0$, (16.32) is $M(0) = m$, the physical mass of the particle, including the contribution of the self-fields:

$$m = m_0 + \frac{e^2}{3\pi^2 c^2} \int d^3k \, \frac{|f(\mathbf{k})|^2}{k^2} \qquad (16.33)$$

In terms of m, the effective mass $M(\omega)$ can be written

$$M(\omega) = m + \frac{e^2 \omega^2}{3\pi^2 c^4} \int d^3k \, \frac{|f(\mathbf{k})|^2}{k^2[k^2 - (\omega/c)^2]} \qquad (16.34)$$

We now comment on the solution we have obtained for the motion of an extended charged particle, including radiation reaction:

1. The self-field contribution to the mass in (16.33) diverges linearly at large k without the form factor, reflecting the fact that the self-fields have an electrostatic energy of the order of e^2/a, where a is a scale parameter determining the size of the charge distribution.

2. The frequency-dependent integral in (16.34) is more convergent by a factor of k^2 than the integral in (16.33) and converges at large k, even if $f(\mathbf{k}) = 1$ (point charge).

3. For a point charge, the integral in (16.34) can be performed easily by contour integration to yield

$$[M(\omega)]_{\text{point}} = m(1 + i\omega\tau) \tag{16.35}$$

Insertion of this expression into (16.29), followed by an inverse Fourier transform, leads back to the Abraham–Lorentz equation, (16.9). The zero in the upper half-complex-ω plane at $\omega\tau = i$ in (16.35) signals the runaway solutions of that equation.

4. For a sufficiently convergent form factor, the integrals in (16.33) and (16.34) are well behaved, with zeros of $M(\omega)$, if any, only in the lower half-ω-plane. [See Problem 16.4.] The particle's response to external forces is causal and without peculiar behavior such as runaway solutions. The particle's extent must be of the order of $c\tau$ or greater, corresponding roughly to the classical charged particle (electron) radius, $r_0 = e^2/mc^2$.

5. While the nonrelativistic approximation causes conceptual difficulties—the self-force contribution in (16.33) is actually $4/3c^2$ times the electrostatic self-energy, rather than $1/c^2$ times it—these are removable by more careful arguments. [An early relativistic treatment was given by Fermi[*]; a covariant description of the electromagnetic parts of the self-energy and momentum is presented in Section 16.5.]

6. A quantum-mechanical treatment of a nonrelativistic extended charged particle in interaction with electromagnetic fields gives essentially the same results, (16.29) and (16.32), for the expectation value of the appropriate operator (Moniz and Sharp, *op. cit.*). The particle's Compton wavelength, $\hbar/mc \approx 137r_0$ plays the formal role of the scale parameter a. The self-field contribution to the mass is then small (or zero, depending on how limits are taken); the particle's motion is causal; no preacceleration or runaway solutions occur. Moniz and Sharp endorse (16.10) as the most sensible form of a classical equation of motion with radiation reaction, to be considered approximately valid when the reactive effects are small.

16.4 *Relativistic Covariance; Stability and Poincaré Stresses*

So far our discussion of the Abraham–Lorentz model of a classical charged particle has been nonrelativistic, with apologies for the paradox of different electromagnetic "masses" from electrostatic and Lorentz force (dynamic) considerations—the infamous 4/3 problem, first noted by J. J. Thomson (1881). The root of the difficulty lies in the nonvanishing of the 4-divergence of the electromagnetic stress tensor (12.113). In contrast to source-free fields, the stress tensor $\Theta^{\alpha\beta}$ of any charged particle model has the divergence (12.118),

$$\partial_\alpha \Theta^{\alpha\beta} = -F^{\beta\lambda}J_\lambda/c = -f^\beta \tag{16.36}$$

where f^α is the Lorentz force density (12.121). As stated in Problem 12.18, only if the 4-divergence of a stress tensor vanishes everywhere do the spatial integrals

---

[*]E. Fermi, *Z. Phy.* **24**, 340 (1922), or *Atti. Accad. Nazl. Lincei Rend.* **31**, 184, 306 (1922).

of $\Theta^{\alpha 0}$ transform as a 4-vector. Thus the usual spatial integrals at a fixed time of the energy and momentum densities,

$$u = \frac{1}{8\pi} (\mathbf{E}^2 + \mathbf{B}^2), \qquad \mathbf{g} = \frac{1}{4\pi c} (\mathbf{E} \times \mathbf{B}) \qquad (16.37)$$

may be used to discuss conservation of electromagnetic energy or momentum in a given inertial frame, but they do not transform as components of a 4-vector unless the fields are source-free.

As Poincaré observed in 1905–1906,* a deficiency of the purely electromagnetic classical models is their lack of stability. Nonelectromagnetic forces are necessary to hold the electric charge in place. Poincaré therefore proposed such forces, described by a stress tensor $P^{\alpha\beta}$ to be added to the electromagnetic $\Theta^{\alpha\beta}$ to give a total stress tensor $S^{\alpha\beta}$,

$$S^{\alpha\beta} = \Theta^{\alpha\beta} + P^{\alpha\beta}$$

The particle's total 4-momentum is then defined to be

$$cP^{\alpha} = \int S^{\alpha 0} \, d^3x \qquad (16.38)$$

where the integral is over all 3-space at a fixed time. The right-hand side of (16.38) transforms as a 4-vector provided

$$\partial_{\alpha} S^{\alpha\beta} = 0 \qquad (16.39)$$

or equivalently, provided

$$\int S^{(0)ij} \, d^3x^{(0)} = 0 \qquad (16.40)$$

with $i, j = 1, 2, 3$, and the superscript (0) denoting the rest frame ($\mathbf{P} = 0$). Condition (16.40) is just the statement that the total self-stress (in the three-dimensional sense) must vanish—the condition for mechanical stability.

Poincaré's solution provides stability and also, because of the generality of the postulates of special relativity, guarantees the proper Lorentz transformation properties for the now stable charged particle. A criticism might be that Poincaré stresses are not known a priori in the way that $\Theta^{\alpha\beta}$ is known for the fields. If we think, however, of macroscopic charged objects, for example a dielectric sphere with charge on its surface, we know that there are "nonelectromagnetic" forces—polarization and quantum-mechanical exchange forces (actually electromagnetic at the fundamental level)—that bind the charge and give the whole system stability. It is not unreasonable then to include Poincaré stresses in our classical models of charged particles, or at least to remember that care must be taken in discussion of the purely electromagnetic aspects of such models.

It is of interest to note that for strongly interacting elementary particles one has a concrete realization of the Poincaré stresses through the gluon field. Con-

---

*H. Poincaré, *Comptes Rendue* **140**, 1504 (1905); *Rendiconti del Circolo Matematico di Palermo* **21**, 129 (1906). The second reference is translated, with modern notation, in H. M. Schwartz, *Am. J. Phys.* **39**, 1287 (1971), **40**, 862, 1282 (1972).

sider the proton, for example. Its three charged quarks are bound together by the gluon field in a stable entity with an extended charge distribution. Setting aside the internal structure and stability of the quarks themselves, the electromagnetic stress tensor $\Theta^{\alpha\beta}$ must be combined with the "Poincaré stress" tensor $\Theta_g^{\alpha\beta}$ of the gluon field to give a divergence-free total stress tensor. The main part of the mass of the proton comes from the strong interactions, not from the electromagnetic contribution to the self-energy—the neutron and proton have the same internal strong interactions, but different electromagnetic; their masses differ by only 0.14% (and in the opposite from expected sense).

In the next section we examine covariant definitions of the total energy and momentum of electromagnetic fields, even in the presence of sources. These definitions have some advantages when purely electromagnetic issues are considered, but in general the nonelectromagnetic forces or stresses must not be forgotten.

16.5 Covariant Definitions of Electromagnetic Energy and Momentum

As emphasized by *Rohrlich*, even if the electromagnetic stress tensor $\Theta^{\alpha\beta}$ is not divergenceless, it is possible to give covariant definitions of the total electromagnetic energy and momentum of a system of fields. The expressions

$$E'_e = \frac{1}{8\pi} \int (\mathbf{E}'^2 + \mathbf{B}'^2) \, d^3x'$$

$$\mathbf{P}'_e = \frac{1}{4\pi c} \int \mathbf{E}' \times \mathbf{B}' \, d^3x'$$

(16.41)

can be considered to define the energy and momentum at a fixed time t' in some particular inertial frame K', to be specified shortly. The integrands in (16.41) are elements of the second-rank tensor $\Theta^{\alpha\beta}$. Evidently we must contract one of the tensor indices with a 4-vector, and the 4-vector must be such as to reduce to d^3x' in the inertial frame K'. We define the timelike 4-vector,

$$d\sigma^\beta = n^\beta \, d^3\sigma$$

where $d^3\sigma$ is an invariant element of three-dimensional "area" on a spacelike hyperplane in four dimensions. The normal to the hyperplane n^β has components $(1, 0, 0, 0)$ in K'. The invariant $d^3\sigma$ is evidently $d^3\sigma = n_\beta \, d\sigma^\beta = d^3x'$. If the inertial frame K' moves with velocity $c\boldsymbol{\beta}$ with respect to an inertial frame K, then in K the 4-vector n^β is

$$n^\beta = (\gamma, \gamma\boldsymbol{\beta})$$

(16.42)

A general definition of the electromagnetic 4-momentum in any frame is therefore

$$cP_e^\alpha = \int \Theta^{\alpha\beta} \, d\sigma_\beta = \int \Theta^{\alpha\beta} n_\beta \, d^3\sigma$$

(16.43)

In K', n_β has only a time component. With $d^3\sigma = d^3x'$, this covariant expression reduces to (16.41). But in the frame K, $n_\beta = (\gamma, -\gamma\boldsymbol{\beta})$ and the covariant definition has time and space components,

$$cP_e^0 = \gamma \int (u - \mathbf{v} \cdot \mathbf{g}) \, d^3\sigma$$

$$cP_e^i = \gamma \int (cg^i + T_{ij}^{(M)}\beta^j) \, d^3\sigma$$
(16.44)

where $T_{ij}^{(M)}$ is the 3×3 Maxwell stress tensor (6.120). If desired, the invariant volume element $d^3\sigma = d^3x'$ can be suppressed in favor of the volume element d^3x in the frame K by means of $d^3x' = \gamma \, d^3x$ (integration at fixed time t).

The definitions (16.43) or (16.44) of the electromagnetic 4-momentum afford a covariant definition starting from the naive expressions (16.41) in any frame K'. Different choices of the frame K' lead to different 4-vectors, of course, but that is no cause for alarm.* There is a natural choice of the frame K' if the electromagnetic mass of the fields is nonvanishing, namely, the rest frame in which

$$\frac{1}{4\pi c} \int \mathbf{E}^{(0)} \times \mathbf{B}^{(0)} \, d^3x^{(0)} = 0$$

We denote this frame where the total electromagnetic momentum $\mathbf{P}_e'$ is zero as $K^{(0)}$ and attach superscripts zero on quantities in that frame to make it clear that it is a special choice of the frame K'. According to (16.41) the electromagnetic rest energy is then

$$E_e^{(0)} = m_e c^2 = \frac{1}{8\pi} \int [\mathbf{E}^{(0)2} + \mathbf{B}^{(0)2}] \, d^3x^{(0)}$$
(16.45)

In the frame K the electromagnetic energy and momentum are given by (16.44) where now $\mathbf{v}$ is the velocity of the rest frame $K^{(0)}$ in K.

For electromagnetic configurations in which all the charges are at rest in some frame (the Abraham–Lorentz model of a charged particle is one example), the general formulas can be reduced to more attractive and transparent forms. Clearly the frame where all the charges are at rest is $K^{(0)}$, since there all is electrostatic and the magnetic field vanishes everywhere in 3-space. For such electrostatic configurations, the magnetic field is given without approximation in the frame K by (11.150):

$$\mathbf{B} = \boldsymbol{\beta} \times \mathbf{E}$$

The integrand in the first equation of (16.44) is thus

$$(u - \mathbf{v} \cdot \mathbf{g}) = \frac{1}{8\pi} (\mathbf{E}^2 + \mathbf{B}^2) - \frac{1}{4\pi} \boldsymbol{\beta} \cdot (\mathbf{E} \times \mathbf{B})$$

$$= \frac{1}{8\pi} (\mathbf{E}^2 + \mathbf{B}^2) - \frac{1}{4\pi} (\boldsymbol{\beta} \times \mathbf{E}) \cdot \mathbf{B}$$

$$= \frac{1}{8\pi} (\mathbf{E}^2 - \mathbf{B}^2)$$

---

*One possible choice for K' is the "laboratory" where the observer is at rest. The discussion of the conservation laws in Chapter 6 may be interpreted in this way.

a Lorentz invariant. Thus the energy in K is given by

$$cP_e^0 = \gamma \int \frac{(\mathbf{E}^2 - \mathbf{B}^2)}{8\pi} d^3\sigma = \gamma^2 \int \frac{(\mathbf{E}^2 - \mathbf{B}^2)}{8\pi} d^3x \qquad (16.46)$$

Similarly, the second equation in (16.44) becomes

$$c\mathbf{P}_e = \gamma\boldsymbol{\beta} \int \frac{(\mathbf{E}^2 - \mathbf{B}^2)}{8\pi} d^3\sigma = \gamma^2\boldsymbol{\beta} \int \frac{(\mathbf{E}^2 - \mathbf{B}^2)}{8\pi} d^3x \qquad (16.47)$$

With the invariant integrand $(\mathbf{E}^2 - \mathbf{B}^2)$ it is clear that we have a 4-vector $P_e^\alpha = (\gamma m_e c, \gamma m_e \mathbf{v})$, where the electromagnetic mass is

$$m_e = \frac{1}{8\pi c^2} \int (\mathbf{E}^2 - \mathbf{B}^2) \, d^3\sigma = \frac{1}{8\pi c^2} \int \mathbf{E}^{(0)2} \, d^3x^{(0)} \qquad (16.48)$$

in agreement with (16.45).

The equation (16.46) for the energy has been used by Butler* to discuss the Trouton–Noble experiment, a test of special relativity involving the question of a torque on a charged suspended capacitor moving with respect to the ether. *Pauli* (Section 44) gives a clear discussion of the Trouton–Noble paradox with emphasis on the early analyses of Lorentz (1904) and von Laue (1911). In a paper that includes as a preamble the proof of the assertion of Problem 12.18, Teukolsky† has revisited the explanation of the Trouton–Noble experiment. He stresses that the removal of the paradox requires consideration of the nonelectromagnetic forces for stability, but that it is a matter of choice whether the balancing of electromagnetic and nonelectromagnetic forces is done in a manifestly covariant way or not. All that matters is that the total stress tensor $S^{\alpha\beta}$ be divergenceless.

16.6   *Covariant Stable Charged Particle*

A. *The Model*

An illuminating example of the considerations of Sections 16.4 and 16.5 is provided by a model of Schwinger‡ for a classical stable spinless charged particle. With its consideration of the Poincaré stresses needed for stability, it may also be viewed as a prototype for the discussion of macroscopic charged mechanical systems. The model is, in fact, a modern generalization of Poincaré's work 77 years earlier [see the middle paper of Schwartz's translation (*op. cit.*)]. In the rest frame K' of the particle, the 4-vector potential is defined as

$$\Phi' = ef(r^2), \qquad \mathbf{A}' = 0$$

with $f(r^2)$ an arbitrary well-behaved function but with the limiting form $f(r^2) \to 1/r$ to define the total charge of the particle as e. We now consider a laboratory frame K in which the particle moves with velocity $\mathbf{v}$ and define the 4-velocity

*J. W. Butler, *Am. J. Phys.* **36**, 936 (1968).

†S. A. Teukolsky, *Am. J. Phys.* **64**, 1104 (1996).

‡J. Schwinger, *Found. Phys.* **13**, 373 (1983).

(divided by c), $v^\alpha = (\gamma, \gamma\boldsymbol{\beta}) = U^\alpha/c$ (11.36), with $v^\alpha v_\alpha = 1$. We introduce a 4-vector coordinate ξ^α perpendicular to v^α,

$$\xi^\alpha = x^\alpha - (v^\beta x_\beta)v^\alpha, \qquad v \cdot \xi \equiv v^\beta \xi_\beta = 0 \qquad (16.49)$$

Then we define the invariant coordinate variable z by

$$z \equiv -\xi^2 = -\xi \cdot \xi = -x \cdot x + (v \cdot x)^2 \qquad (16.50)$$

In the rest frame K', $\xi^0 = 0$, $\boldsymbol{\xi} = \mathbf{x}$, and z becomes $z = r^2$.

The covariant generalization of the rest-frame potentials is

$$A^\alpha = ev^\alpha f(z) \qquad (16.51)$$

To evaluate the fields we need

$$\partial^\mu(z) = \partial^\mu(-\xi \cdot \xi) = -2\xi^\mu$$

Then we have $\partial^\alpha A^\beta = -2e\xi^\alpha v^\beta f'$, where $f' = df(z)/dz$. [Parenthetically, we note that with $\alpha = \beta$ (and summed) we obtain the Lorenz condition on the potentials because $\xi \cdot v = 0$.] The field-strength tensor is

$$F^{\alpha\beta} = -2e(\xi^\alpha v^\beta - \xi^\beta v^\alpha)f' \qquad (16.52)$$

The current density is obtained from the Maxwell equations,

$$J^\beta = \frac{c}{4\pi} \partial_\alpha F^{\alpha\beta} = -\frac{ec}{2\pi} [3f' + 2zf'']v^\beta \qquad (16.53)$$

B. *The Electromagnetic and Poincaré Stress Tensors; Arbitrariness*

The symmetric stress-energy-momentum tensor (12.113) is easily found to be

$$\Theta^{\alpha\beta} = \frac{e^2}{\pi} (f')^2[-\xi^\alpha \xi^\beta + zv^\alpha v^\beta - \frac{z}{2} g^{\alpha\beta}] \qquad (16.54)$$

and its divergence (16.36) is

$$\partial_\alpha \Theta^{\alpha\beta} = -\frac{1}{c} F^{\beta\lambda} J_\lambda = -\frac{e^2}{\pi} \xi^\beta[f'(3f' + 2zf'')] \qquad (16.55)$$

The Lorentz force density [negative of the right-hand side of (16.55)] must be balanced by Poincaré stresses for stability. Schwinger, noting the derivative relation $\partial^\alpha G(z) = -2\xi^\alpha G'$, defines a function $t(z)$ whose derivative is

$$t'(z) \equiv \frac{dt(z)}{dz} = -\frac{e^2}{2\pi} [3(f')^2 + 2zf'f''] \qquad (16.56)$$

He then defines the Poincaré stress tensor to be

$$P^{\alpha\beta} = g^{\alpha\beta} t(z) \qquad (16.57)$$

with its divergence, $\partial_\alpha P^{\alpha\beta} = \partial_\alpha g^{\alpha\beta} t(z) = -2\xi^\beta t'$. But this is just the negative of the right-hand side of (16.55)! We thus have

$$\partial_\alpha(\Theta^{\alpha\beta} + P^{\alpha\beta}) = 0$$

The total stress tensor $S^{\alpha\beta} = \Theta^{\alpha\beta} + P^{\alpha\beta}$ is divergenceless; the spatial integrals of $S^{\alpha 0}$ transform as a 4-vector. The model is covariant and stable.

Before proceeding, we note an arbitrariness in the Poincaré stresses. Any nonelectromagnetic stress tensor with a vanishing divergence may be added to $P^{\alpha\beta}$. Because $v \cdot \xi = 0$, it follows that $\partial_\alpha[v^\alpha v^\beta s(z)] = 0$. This means that we may add $\Delta P^{\alpha\beta} = v^\alpha v^\beta s(z)$, with $s(z)$ arbitrary, without changing the stability or covariance of the model. We will, of course, change the energy and momentum of the particle, as is illustrated below for our special choice of the additional term,

$$P^{\alpha\beta} \to \Pi^{\alpha\beta} = (g^{\alpha\beta} + hv^\alpha v^\beta)\, t(z) \tag{16.58}$$

with h constant. Schwinger discusses the two cases, $h = 0$ and $h = -1$. The components of the total stress tensor $S^{\alpha 0}$ are explicitly

$$S^{00} = \Theta^{00} + (1 + h\gamma^2)t$$
$$S^{i0} = \Theta^{i0} + h\gamma^2\beta^i t$$

showing that when $h = 0$, $S^{00} = (\Theta^{00} + t)$ and $S^{i0} = \Theta^{i0}$ in all frames. When $h = -1$, $S^{00} = \Theta^{00}$ in the particle's rest frame. Schwinger's original choice of Poincaré stresses (16.57) is in some sense the minimal and natural choice, tied directly to the electromagnegic field configuration. Note that the terms proportional to $v^\alpha v^\beta$ contribute to the energy in the rest frame, but not to the stabilizing forces (from the space parts of $\Pi^{\alpha\beta}$). Poincaré had a spherical shell of charge with an arbitrary "pressure" inside, equivalent to our arbitrary $s(z)$ above.

C. The Poincaré Function t(z) and Contributions to the Mass

From the first-order differential equation (16.56) and the physical requirement that $t(z)$ vanish at infinity, an integration by parts leads to

$$t(z) = \frac{e^2}{\pi} \int_z^\infty (f')^2 \, dz' - \frac{e^2}{2\pi} z(f')^2 \tag{16.59}$$

For specific forms of the potential function $f(z)$ it is a straightforward matter of integration to find $t(z)$. It is left as an exercise to show for a spherical shell of charge of radius a and a uniform volume distribution of charge of the same radius that

$$t(z) = \frac{e^2}{8\pi a^4} \left\{ \begin{matrix} \Theta(a^2 - z) \\ 3\Theta(a^2 - z)(1 - z/a^2) \end{matrix} \right\} \quad \text{for} \left\{ \begin{matrix} \text{shell of charge} \\ \text{uniform density} \end{matrix} \right\}$$

The shell of charge provides the most dramatic illustration of the stabilizing effect of the Poincaré stresses. They exist only inside the sphere. Because there are no fields inside the sphere, the electromagnetic stress exists only outside the sphere and gives a destabilizing outward force per unit area at $r = a^+$ equal to $e^2/8\pi a^4$ in the rest frame. At $r = a^-$, the Pioncaré stress provides the stabilizing inward force—the surface layer of charge feels no net force. Continuity across an interface of the total stress tensor contracted with the unit normal is the more general criterion for no net force at the interface.

The electromagnetic contribution to the mass of the particle can be found from (16.48) directly or from either the rest-frame integral of Θ^{00} or $J^0 A^0/2c$. In the first way, we need $E^{(0)2} = 4e^2 \cdot z(f')^2$. Then we find

$$m_e c^2 = e^2 \int_0^\infty z^{3/2}(f')^2 \, dz \tag{16.60}$$

The contribution to the rest mass from the Poincaré stresses is

$$m_p c^2 = (1 + h) \int t(z) \, d^3x'$$

Integration over angles, then substitution of (16.59) for $t(z)$ and an integration by parts leads to the result,

$$m_p c^2 = \tfrac{1}{3}(1 + h) m_e c^2 \tag{16.61}$$

The total mass is therefore

$$M = m_e + m_p = \tfrac{1}{3}(4 + h) m_e \tag{16.62}$$

Note that when $h = 0$ the mass is $4m_e/3$, the "dynamic" result, while if $h = -1$, $M = m_e$, the electrostatic result. On the other hand, if $h \gg 1$, most of the mass is of nonelectromagnetic origin. Neither the 4/3 nor the unity proves anything about the covariance of the energy and momentum of the particle. This property is guaranteed by the divergence-free $S^{\alpha\beta}$, as we now demonstrate.

D. Demonstration of the Covariance of the Particle's Energy and Momentum

The evaluation of the spatial integrals of Θ^{00}, Θ^{i0} and Π^{00}, Π^{i0} and their sums at fixed time x^0 in the laboratory frame illustrates the conspiracy between the electromagnetic and Poincaré stresses to assure the proper Lorentz transformation properties. We begin with Θ^{00}:

$$\Theta^{00} = -\frac{e^2}{\pi} \left[-(\xi^0)^2 + \gamma^2 z - \frac{1}{2} z \right] (f')^2$$

Since we are to integrate Θ^{00} over 3-space at fixed time in K, we need $(\xi^0)^2$ and z evaluated explicitly in K. If we take the 3-axis parallel to $\boldsymbol{\beta}$, from the definition (16.49) and $v \cdot \xi = 0$, we find $\xi^0 = \beta \xi^3$ and $\xi^3 = \gamma^2 (x^3 - \beta x^0)$. With $(\xi^1)^2 + (\xi^2)^2 = (x^1)^2 + (x^2)^2 \equiv \rho^2$, we have

$$z = \rho^2 + \gamma^2 (x^3 - \beta x^0)^2$$

If we define $x'_3 = \gamma(x^3 - \beta x^0)$, which is just the 3-coordinate in K', the volume element d^3x can be written $d^3x = d^3x'/\gamma$. Putting the pieces together, we have the electromagnetic part of the energy as

$$E_e = \frac{e^2}{\gamma\pi} \int d^3x' \, [f'(r'^2)]^2 [(\gamma\beta\rho)^2 + \tfrac{1}{2}r'^2]$$

Averaging over angles introduces a factor of $[(\tfrac{2}{3})\gamma^2\beta^2 + \tfrac{1}{2}]r'^2$ instead of the square bracket. With the definition of m_e through (16.60), we obtain

$$E_e = \left(\frac{4}{3} \gamma - \frac{1}{3\gamma} \right) m_e c^2 \tag{16.63}$$

A corresponding computation of the integral of Θ^{30} gives the electromagnetic momentum

$$cP_e = \int \Theta^{30} \, d^3x = \frac{4}{3} \gamma\beta m_e c^2 \tag{16.64}$$

Clearly the electromagnetic contributions alone do not transform properly.

The nonelectromagnetic contributions to the energy and momentum are

$$E_{\mathrm{p}} = \int \Pi^{00}\, d^3x = (1 + h\gamma^2) \int t(z)\, \frac{d^3x'}{\gamma} = \frac{1}{3}\left(\frac{1}{\gamma} + \gamma h\right) m_e c^2 \tag{16.65}$$

$$cP_{\mathrm{p}} = \int \Pi^{30}\, d^3x = h\gamma^2\beta \int t(z)\, \frac{d^3x'}{\gamma} = \frac{1}{3}\, h\gamma\beta m_e c^2$$

Neither do the Poincaré contributions transform properly. The total stress tensor contributions, the sums of the separate contributions, do, however, yield a proper relativistic energy and momentum:

$$E = \gamma Mc^2, \qquad cP = \gamma\beta Mc^2 \qquad \text{with } M = \tfrac{1}{3}(4 + h)m_e \tag{16.66}$$

the same rest mass as found above. Schwinger's choices of $h = 0$ and $h = -1$ were made to illustrate that either the electrostatic mass or the "dynamic" mass can serve as "the mass" when the charge is stabilized by the Poincaré stresses. Other choices of h are possible and, as noted, above, other totally arbitrary contributions to the mass can be introduced without affecting the question of the covariance of the model.

Although we established the 4-vector nature of energy and momentum using the conventional definitions of the total energy and momentum by taking 3-space integrals at fixed time x^0 in the laboratory frame K, it is of interest to see how the derivation changes if we use the definitions of Section 16.5, which yield covariant expressions for the separate contributions. The appropriate quantities, according to (16.43), 16.54) and (16.58), are

$$cP_e^\alpha = \int \Theta^{\alpha\beta} v_\beta\, d^3\sigma = \frac{e^2}{2\pi}\, v^\alpha \int z(f')^2\, d^3\sigma \tag{16.67}$$

$$cP_{\mathrm{p}}^\alpha = \int \Pi^{\alpha\beta} v_\beta\, d^3\sigma = (1 + h)v^\alpha \int t(z)\, d^3\sigma \tag{16.68}$$

Since the integrands and integration are Lorentz invariants, we may evaluate the integrals in the rest frame. From (16.60) and below, we see that

$$cP_e^\alpha = (m_e c^2)v^\alpha \qquad \text{and} \qquad cP_{\mathrm{p}}^\alpha = \tfrac{1}{3}(1 + h)(m_e c^2)v^\alpha$$

are separately 4-vectors by construction, with a sum equal to (16.66). The simplicity and elegance of the use of the manifestly covariant (16.43) is apparent. The results are, of course, the same either way.

The Poincaré–Schwinger model of a stable charged particle addresses the issue of the Lorentz transformation properties of the particle's energy and momentum, but does not attack the question of radiation reaction. For the spherical shell model, this problem has been treated in detail by Yaghjian[*] who also treats the Poincaré stresses and stability. See also Rohrlich.[†]

16.7   *Line Breadth and Level Shift of a Radiating Oscillator*

The effects of radiative reaction are of great importance in the detailed behavior of atomic systems. Although a complete discussion involves the rather elaborate

[*]A. D. Yaghjian, *Relativistic Dynamics of a Charged Sphere*, Lecture Notes in Physics m11, Springer-Verlag, Berlin, New York (1992).
[†]F. Rohrlich, *Am. J. Phys.* **65**, 1051–1057 (1997).

formalism of quantum electrodynamics, the qualitative features are apparent from a classical treatment. As a typical example we consider a charged particle bound by a one-dimensional linear restoring force with force constant $k = m\omega_0^2$. In the absence of radiation damping, the particle oscillates with constant amplitude at the characteristic frequency ω_0. When the reactive effects are included, the amplitude of oscillation gradually decreases, since energy of motion is being converted into radiant energy. This is the classical analog of spontaneous emission in which an atom makes a transition from an excited state to a state of lower energy by emission of a photon.

If the displacement of the charged particle from equilibrium is $x(t)$ and $\mathbf{F}_{\text{ext}} = -m\omega_0^2 x$, (16.10) becomes

$$m\ddot{x} = -m\omega_0^2 x - m\omega_0^2 \tau \dot{x} \tag{16.69}$$

Because of the expected decay of the amplitude, we assume a solution of the form

$$x(t) = x_0 e^{-\alpha t} \tag{16.70}$$

where α should have a positive real part and an imaginary part close to ω_0 if the radiative damping effects are small. The ansatz leads to a quadratic equation for α,

$$\alpha^2 - \tau\omega_0^2 \alpha + \omega_0^2 = 0$$

with roots

$$\alpha = \tfrac{1}{2}\omega_0^2 \tau \pm i\omega_0 \sqrt{1 - (\omega_0\tau/2)^2} \approx \tfrac{1}{2}\omega_0^2 \tau \pm i(\omega_0 - \tfrac{1}{8}\omega_0^3\tau^2)$$

In the last form we have expanded to order τ^2 in the real part. The real part of α is $\Gamma/2$, where Γ is known as the *decay constant* and the change $\Delta\omega$ in the imaginary part from ω_0 is known as the *level shift**:

$$\Gamma = \omega_0^2 \tau, \qquad \Delta\omega \approx -\tfrac{1}{8}\omega_0^3\tau^2 \tag{16.71a}$$

The alert reader will rightly question the legitimacy of keeping terms of order τ^2 in the solution of an equation that is an approximation valid only for small τ (see Problem 16.10b). In fact, if the Abraham–Lorentz equation (16.9) is used instead of (16.10), the resulting cubic equation in α yields, to order τ^2, the same Γ, but

$$[\Delta\omega]_{\text{A-L}} \approx -\tfrac{5}{8}\omega_0^3\tau^2 \tag{16.71b}$$

The important message here is that the classical level shift $\Delta\omega$ is one power higher order in $\omega_0\tau$ than the decay constant Γ.

The energy of the oscillator decays exponentially as $e^{-\Gamma t}$ because of radiation damping. This means that the emitted radiation appears as a wave train with effective length of the order of c/Γ. Such a finite pulse of radiation is not exactly monochromatic but has a frequency spectrum covering an interval of order Γ. The exact shape of the frequency spectrum is given by the square of the Fourier

---

*The reader is invited to pause at this point and consider the decay constant Γ from various points of view. One is to use the Larmor power formula (16.6) and conservation of energy directly to relate the time-averaged radiated power $\bar{P}(t)$ to the total energy of the oscillator $E(t)$. Another is to ask for the initial energy and amplitude x_0 of the oscillator such that $\Gamma = \bar{P}/\hbar\omega_0$, corresponding to the emission of a single photon of energy $\hbar\omega_0$. These can then be compared to the values for a quantum-mechanical oscillator in its nth quantum state.

transform of the electric field or the acceleration. Neglecting an initial transient (of duration τ), the amplitude of the spectrum is thus proportional to

$$E(\omega) \propto \int_0^\infty e^{-\alpha t} e^{i\omega t} \, dt = \frac{1}{\alpha - i\omega}$$

The energy radiated per unit frequency interval is therefore

$$\frac{dI(\omega)}{d\omega} = I_0 \frac{\Gamma}{2\pi} \frac{1}{(\omega - \omega_0 - \Delta\omega)^2 + (\Gamma/2)^2} \tag{16.72}$$

where I_0 is the total energy radiated. This spectral distribution is called a *resonant line shape*. The width of the distribution at half-maximum intensity is called the *half-width* or *line breadth* and is equal to Γ. Shown in Fig. 16.1 is such a spectral line. Because of the reactive effects of radiation the line is *broadened and shifted* in frequency.

The classical line breadth for electronic oscillators is a universal constant when expressed in terms of wavelength:

$$\Delta\lambda = 2\pi \frac{c}{\omega_0^2} \Gamma = 2\pi c\tau = 1.2 \times 10^{-4} \text{ Å}$$

Quantum mechanically the natural widths of spectral lines vary. To establish a connection with the classical treatment, the quantum-mechanical line width is sometimes written as

$$\Gamma_q = f_{ij} \Gamma$$

where f_{ij} is the "oscillator strength" of the transition $(i \rightarrow j)$. Oscillator strengths vary considerably, sometimes being nearly unity for strong single-electron transitions and sometimes much smaller. For optical transitions, $\lambda \sim 4\text{–}8 \times 10^3$ Å. Thus $\Delta\lambda/\lambda \lesssim 3.5\text{–}1.5 \times 10^{-8}$ and $\omega_0\tau = O(10^{-8})$.

The classical level shift $\Delta\omega$ is smaller than the line width Γ by a factor $\omega_0\tau \ll 1$. Quantum mechanically (and experimentally) this is not so. The reason is that in the quantum theory there is a different mechanism for the level shift, although still involving the electromagnetic field. Even in the absence of photons, the quantized radiation field has nonvanishing expectation values of the *squares* of the electromagnetic field strengths (vacuum fluctuations). These fluctuating

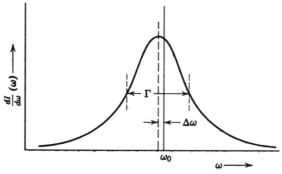

Figure 16.1   Broadening and shifting of spectral line because of radiative reaction. The resonant line shape has width Γ. The level shift is $\Delta\omega$.

fields (along with vacuum fluctuations in the electron-positron field) act on the charged particle to cause a shift in its energy. The quantum-mechanical level shift for an oscillator is of the order of

$$\frac{\Delta \omega_q}{\omega_0} \sim \omega_0 \tau \log \left(\frac{mc^2}{\hbar \omega_0} \right)$$

as compared to the classical shift due to emission of radiation,

$$\frac{|\Delta \omega_c|}{\omega_0} \sim (\omega_0 \tau)^2$$

The quantum-mechanical level shift is seen to be comparable to, or greater than, the line width. The small radiative shift of energy levels of atoms was first observed by Lamb in 1947* and is called the *Lamb shift* in his honor.

16.8 Scattering and Absorption of Radiation by an Oscillator

The scattering of radiation by free charged particles is discussed in Section 14.8. We now consider the scattering and absorption of radiation by bound charges, in particular the scattering of radiation of frequency ω by a single nonrelativistic particle of mass m and charge e bound by a spherically symmetric linear restoring force $m\omega_0^2 \mathbf{x}$. The total force acting on the particle is (neglecting the magnetic field term because of the assumption of nonrelativistic motion)

$$\mathbf{F} = -m\omega_0^2 \mathbf{x} + e\boldsymbol{\epsilon} E_0 e^{i\mathbf{k}\cdot\mathbf{x} - i\omega t}$$

where E_0 is the magnitude and $\boldsymbol{\epsilon}$ the polarization vector of the incident electric field. We introduce a resistive term $m\Gamma'\mathbf{v}$ in the equation of motion to allow for other dissipative processes, corresponding quantum mechanically to other modes of decay besides photon re-emission. With this addition, substitution into (16.10) leads in the electric dipole approximation to the equation of motion,

$$\ddot{\mathbf{x}} + (\Gamma + \Gamma')\dot{\mathbf{x}} + \omega_0^2 \mathbf{x} = \frac{eE_0}{m} \boldsymbol{\epsilon}(1 - i\omega\tau)e^{-i\omega t} \tag{16.73}$$

Here we have neglected the $(\mathbf{v} \cdot \boldsymbol{\nabla})$ term for the incident field because it leads to a v/c correction. The steady-state solution is

$$\mathbf{x} = \frac{eE_0}{m} \boldsymbol{\epsilon} \frac{(1 - i\omega\tau)e^{-i\omega t}}{\omega_0^2 - \omega^2 - i\omega\Gamma_t} \tag{16.74}$$

where $\Gamma_t = \Gamma + \Gamma'$ is the *total decay constant* or *total width* at resonance.

The accelerated motion gives rise to radiation fields given by (14.18),

$$\mathbf{E}_{\text{rad}} = \frac{e}{c^2} \frac{1}{r} [\mathbf{n} \times (\mathbf{n} \times \ddot{\mathbf{x}})]_{\text{ret}}$$

*W. E. Lamb and R. C. Retherford, *Phys. Rev.* **72**, 241 (1947).

The scattering amplitude for scattered radiation of polarization $\boldsymbol{\epsilon}'$ is

$$f = \frac{r}{E_0}\, (e^{ikr - i\omega t}\boldsymbol{\epsilon}')^* \cdot \mathbf{E}_{\mathrm{rad}}$$

or

$$f = \frac{e^2}{mc^2}\, \frac{\omega^2 (1 - i\omega\tau)}{\omega_0^2 - \omega^2 - i\omega\Gamma_t}\, \boldsymbol{\epsilon}'^* \cdot \boldsymbol{\epsilon} \qquad (16.75)$$

The differential scattering cross section is the absolute square of f:

$$\frac{d\sigma}{d\Omega} = \left(\frac{e^2}{mc^2}\right)^2 \left[\frac{\omega^4}{(\omega_0^2 - \omega^2)^2 + \omega^2\Gamma_t^2}\right]|\boldsymbol{\epsilon}'^* \cdot \boldsymbol{\epsilon}|^2 \qquad (16.76)$$

We have omitted the factor of $(1 + \omega^2\tau^2) \approx 1$ in the numerator because the cross section is already proportional to $(c\tau)^2$. The total scattering cross section can be written

$$\sigma_{\mathrm{scatt}} = 6\pi\lambdabar_0^2 \left[\frac{\omega^4\Gamma^2/\omega_0^2}{(\omega_0^2 - \omega^2)^2 + \omega^2\Gamma_t^2}\right] \qquad (16.77)$$

Here $\lambdabar = c/\omega_0$ is the wavelength divided by 2π at resonance and $\Gamma = \omega_0^2\tau$ is the resonant scattering width or radiative decay constant.

The scattering cross section exhibits a resonance at $\omega = \omega_0$ with a peak value of $\sigma_{\mathrm{scatt}}^{\max} = 6\pi\lambdabar_0^2(\Gamma/\Gamma_t)^2$. It is proportional to ω^4 at very low frequencies—Rayleigh's law of scattering, discussed in Chapter 10. At very high frequencies ($\omega \gg \omega_0$, Γ_t), it approaches the Thomson scattering cross section for a free particle. Figure 16.2 shows the scattering cross section over the whole classical range of frequencies.

The sharply resonant scattering at $\omega = \omega_0$ is called *resonance fluorescence*. Quantum mechanically it corresponds to the absorption of radiation by an atom, molecule, or nucleus in a transition from its ground state to an excited state with the subsequent re-emission of the radiation in other directions in the process of de-excitation. The factor $6\pi\lambdabar_0^2$ in the peak cross section is replaced quantum mechanically by the statistical factor,

$$6\pi\lambdabar_0^2 \to 4\pi\lambdabar_0^2 \frac{2J_{ex} + 1}{2(2J_g + 1)}$$

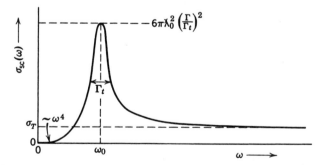

Figure 16.2   Total cross section for the scattering of radiation by an oscillator as a function of frequency. σ_T is the Thomson free-particle scattering cross section.

where J_g and J_{ex} are the angular momenta of the ground and excited states, and $4\pi\lambda_0^2$ is the maximum allowable scattering for any single quantum state. The remaining factors represent a sum over all final magnetic substates and an average over initial ones, the factor 2 being the statistical weight associated with the incident radiation's polarizations. The classical result corresponds to $J_g = 0$ and $J_{ex} = 1$.

The total cross section, scattering plus absorption, is obtained from the scattering amplitude (16.75), including the numerator factor $(1 - i\omega\tau)$ neglected in (16.76), by means of the optical theorem (10.139):

$$\sigma_t = \frac{4\pi}{k} \, \mathrm{Im}[f(\boldsymbol{\epsilon}' = \boldsymbol{\epsilon}, \mathbf{k}' = \mathbf{k})] = 6\pi\lambda_0^2 \left[\frac{\omega^2\Gamma(\Gamma' + \omega^2\Gamma/\omega_0^2)}{(\omega_0^2 - \omega^2)^2 + \omega^2\Gamma_t^2} \right] \quad (16.78)$$

The structure of the numerators in the scattering and total cross sections has a simple interpretation. In (16.78) there is one factor of Γ corresponding to the incident radiation being absorbed. This is multiplied by the sum of widths for all possibilities in the final state—the elastic scattering and the absorptive processes—because it is the total cross section. For the elastic scattering cross section (16.77) there are two factors of Γ, one for the initial and one for the final state. Note that, while the elastic scattering and total cross sections approach the Thomson limiting form at high energies, the inelastic or absorptive cross section has only the resonant shape, vanishing as $1/\omega^2$ at high energies provided Γ' is energy independent.

Just as was done in Section 7.5 in the discussion of the atomic contributions to the polarization and dielectric constant, we can generalize the one-oscillator model to something closer to reality by assuming that there are a number of oscillators with resonant frequencies ω_j, radiative decay constants $\Gamma_j = f_j\omega_j^2\tau$ and absorptive widths Γ_j'. Then the total cross section, for example, becomes

$$\sigma_{\text{total}} = 6\pi \sum_j \lambda_j^2 \left[\frac{\omega^2\Gamma_j(\Gamma_j' + \omega^2\Gamma_j^2/\omega_0^2)}{(\omega_j^2 - \omega^2)^2 + \omega^2\Gamma_{j,t}^2} \right]$$

With the appropriate definitions of f_j, Γ_j, and ω_j, this result is almost the correct quantum-mechanical expression. Lacking are the interference terms from overlapping resonances. The quantum-mechanical scattering amplitude is a coherent superposition of the contributions of all the intermediate states allowed by the selection rules. Usually the states are narrow and separated by energy differences large compared to their widths. Then the interference terms can be ignored. In special situations they must be included, however.

References and Suggested Reading

The history of the attempts at classical models of charged particles and associated questions is treated in interesting detail by
>   Whittaker
The ideas of Abraham, Lorentz, Poincaré, and others are presented by
>   Lorentz, Sections 26–37, 179–184, Note 18
Clear, if brief, treatments of self-energy effects and radiative reaction are given by
>   Abraham and Becker, Band II, Sections 13, 14, 66
>   Landau and Lifshitz, *Classical Theory of Fields*, Section 9.9

Panofsky and Phillips, Chapters 20 and 21
Sommerfeld, *Electrodynamics*, Section 36

A relativistic classical point electron theory was first developed by
P. A. M. Dirac, *Proc. R. Soc. London*, **A167**, 148 (1938).
The Lorentz–Dirac theory is discussed in
Barut
Rohrlich
where other aspects of relativistic classical field theory are treated in detail.

A succinct discussion of "classical electron theory from a modern standpoint" is given by
S. Coleman, in *Electromagnetism: Paths to Research*, ed. D. M. Teplitz, Plenum Press, New York (1982), pp. 183–210.
A useful review of all aspects of classical electron models is
P. Pearle, in *Electromagnetism: Paths to Research*, ed. D. M. Teplitz, Plenum Press, New York (1982), pp. 211–295.

The most detailed and explicit discussion of radiation reaction for a charged spherical shell is that of
Yaghjian (*op. cit.*)

Problems

16.1 A nonrelativistic particle of charge e and mass m is bound by a linear, isotropic, restoring force with force constant $m\omega_0^2$.

Using (16.13) and (16.16) of Section 16.2, show that the energy and angular momentum of the particle both decrease exponentially from their initial values as $e^{-\Gamma t}$, where $\Gamma = \omega_0^2 \tau$. Quantum mechanically, the mean excitation energy of an oscillator decays in exactly the same way because the total radiative transition probability for a state with quantum numbers n_0, l_0 is $\Gamma(n_0, l_0) = n_0 \Gamma$. The decay of the angular momentum approaches the classical law only for $l_0 \gg 1$.

16.2 A nonrelativistic electron of charge $-e$ and mass m bound in an attractive Coulomb potential $(-Ze^2/r)$ moves in a circular orbit in the absence of radiation reaction.

(a) Show that both the energy and angular-momentum equations (16.13) and (16.16) lead to the solution for the slowly changing orbit radius,

$$r^3(t) = r_0^3 - 9Z(c\tau)^3 \frac{t}{\tau}$$

where r_0 is the value of $r(t)$ at $t = 0$.

(b) For circular orbits in a Bohr atom the orbit radius and the principal quantum number n are related by $r = n^2 a_0/Z$. If the transition probability for transitions from $n \to (n-1)$ is defined as $-dn/dt$, show that the result of part a agrees with that found in Problem 14.21.

(c) From part a calculate the numerical value of the times taken for a mu meson of mass $m = 207 m_e$ to fall from a circular orbit with principal quantum number $n_1 = 10$ to one with $n_2 = 4$, and $n_2 = 1$. These are reasonable estimates of the time taken for a mu meson to cascade down to its lowest orbit after capture by an isolated atom.

16.3   An electron moving in an attractive Coulomb field $(-Ze^2/r)$ with binding energy ϵ and angular momentum L has an elliptic orbit,

$$\frac{1}{r} = \frac{Ze^2 m}{L^2}\left[1 + \sqrt{1 - \frac{2\epsilon L^2}{Z^2 e^4 m}}\cos(\theta - \theta_0)\right]$$

The eccentricity of the ellipse is given by the square root multiplying the cosine.

(a)   By performing the appropriate time averages over the orbit, show that the secular changes in energy and angular momentum are

$$\frac{d\epsilon}{dt} = \frac{2^{3/2}}{3}\frac{Z^3 e^8 m^{1/2}}{c^3}\frac{\epsilon^{3/2}}{L^5}\left(3 - \frac{2\epsilon L^2}{Z^2 e^4 m}\right)$$

$$\frac{dL}{dt} = -\frac{2^{5/2}}{3}\frac{Ze^4}{m^{1/2}c^3}\frac{\epsilon^{3/2}}{L^2}$$

(b)   If the initial values of ϵ and L are ϵ_0 and L_0, show that

$$\epsilon(L) = \frac{Z^2 e^4 m}{2L^2}\left[1 - \left(\frac{L}{L_0}\right)^3\right] + \frac{\epsilon_0}{L_0}L$$

Calculate the eccentricity of the ellipse, and show that it decreases from its initial value as $(L/L_0)^{3/2}$, showing that the orbit tends to become circular as time goes on.

(c)   Compare your results here to the special case of a circular orbit of Problem 16.2.

Hint: In performing the time averages make use of Kepler's law of equal areas $(dt = mr^2\,d\theta/L)$ to convert time integrals to angular integrals.

16.4   A classical model of an electron is a spherical shell of charge of radius a and total charge e.

(a)   Using (16.30) for the "mass" $M(\omega)$ and the angular average of $e^{i\omega R/c}/R$, show that

$$M(\omega) = m + \frac{2}{3}\frac{e^2}{ac^2}\cdot\left(\frac{e^{i\xi} - 1 - i\xi}{i\xi}\right)$$

where $\xi = 2\omega a/c$, and $m = m_0 + 2e^2/3ac^2$ is the physical mass of the electron.

(b)   Expand in powers of the frequency (ξ) and show that, to lowest nontrivial order, $M(\omega)$ has a zero in the upper half-plane at $\omega\tau = i$, where $\tau = 2e^2/3mc^3$. What is the physical significance of such a zero?

(c)   For the exact result of part a, show that the zeros of $\omega M(\omega)$, if any, are defined by the two simultaneous equations, proportional to the real and imaginary parts of $i\omega M(\omega)$,

$$e^{-y}\cos x - 1 + y(1 - a/c\tau) = 0$$
$$e^{-y}\sin x - x(1 - a/c\tau) = 0$$

where $x = \text{Re }\xi$ and $y = \text{Im }\xi$. Find the condition on the radius a such that $\omega M(\omega)$ has no zeros in the upper half-ω-plane. Express the condition also in terms of the mechanical mass m_0 for fixed physical mass. What about zeros and/or singularities in the lower half-plane?

16.5   The particle of Problem 16.4 is initially at rest in a spatially uniform, but time-varying electric field $E(t) = E_0\Theta(t)$.

(a) Show that its speed in the direction of the field is given by the integral

$$v(t) = -\frac{eE_0 a}{\pi c} \int_{-\infty}^{\infty} d\xi \, \frac{e^{-iT\xi}}{\xi^2 M(\xi)}$$

where $T = ct/2a$.

(b) From the analytic properties of $M(\xi)$ established in part c of Problem 16.4, show that $v(t) = 0$ for $t \leq 0$ (no preacceleration).

16.6 A particle of bare mass m_0 and charge e has a charge density, $\rho(\mathbf{x}) = e \, e^{-r/a}/4\pi a^2 r$.

(a) Show that the charge form factor is $f(\mathbf{k}) = (1 + k^2 a^2)^{-1}$.

(b) Show that the mass, (16.33), is

$$m = m_0 + \frac{mc\tau}{2a}$$

(c) Show that the zeros of $M(\omega)$, (16.34), in the complex ω plane, are given by

$$\omega\tau = -i(c\tau/a)[1 \pm (1 - 2a/c\tau)^{1/2}]$$

(d) Find the trajectories of the roots in the complex ω plane for $m_0 > 0$ and $m_0 < 0$. Find the limiting form for the roots when $a/c\tau \ll 1$ and $a/c\tau \gg 1$. Discuss.

16.7 The Dirac (1938) relativistic theory of classical point electrons has as its equation of motion,

$$\frac{dp_\mu}{d\tau} = F_\mu^{\text{ext}} + F_\mu^{\text{rad}}$$

where p_μ is the particle's 4-momentum, τ is the particle's proper time, and F_μ^{rad} is the covariant generalization of the radiative reaction force (16.8).

Using the requirement that any force must satisfy $F_\mu p^\mu = 0$, show that

$$F_\mu^{\text{rad}} = \frac{2e^2}{3mc^3} \left[\frac{d^2 p_\mu}{d\tau^2} + \frac{p_\mu}{m^2 c^2} \left(\frac{dp_\nu}{d\tau} \frac{dp^\nu}{d\tau} \right) \right]$$

16.8 (a) Show that for relativistic motion in one dimension the equation of motion of Problem 16.7 can be written in the form,

$$\dot{p} - \frac{2e^2}{3mc^3} \left(\ddot{p} - \frac{p\dot{p}^2}{p^2 + m^2 c^2} \right) = \sqrt{1 + \frac{p^2}{m^2 c^2}} \, f(\tau)$$

where p is the momentum in the direction of motion, a dot means differentiation with respect to proper time, and $f(\tau)$ is the ordinary Newtonian force as a function of proper time.

(b) Show that the substitution of $p = mc \sinh y$ reduces the relativistic equation to the Abraham–Lorentz form (16.9) in y and τ. Write down the general solution for $p(\tau)$, with the initial condition that

$$p(\tau) = p_0 \quad \text{at} \quad \tau = 0$$

16.9 (a) Show that the radiation reaction force in the Lorentz–Dirac equation of Problem 16.7 can be expressed alternatively as

$$F_\mu^{\text{rad}} = \tau \left[\left(g_{\mu\nu} - \frac{p_\mu p_\nu}{m^2 c^2} \right) \frac{d^2 p^\nu}{d\tau^2} \right]$$

(b) The relativistic generalization of (16.10) can be obtained by replacing $d^2 p^\nu/d\tau^2$ by $g^{\nu\lambda} dF_\lambda^{\text{ext}}/d\tau$ in the expression for F_μ^{rad}. Show that the spatial part of the generalization of (16.10) becomes

$$\frac{d\mathbf{p}}{dt} = \mathbf{F} + \tau\left[\gamma\frac{d\mathbf{F}}{dt} - \frac{\gamma^3}{c^2}\frac{d\mathbf{v}}{dt} \times (\mathbf{v} \times \mathbf{F})\right]$$

where $\mathbf{F}$ is the spatial part of $F_{\text{ext}}^\mu/\gamma$. For a charged particle in external electric and magnetic fields $\mathbf{F}$ is the Lorentz force.

Reference: G. W. Ford and R. F. O'Connell, *Phys. Lett. A* **174**, 182 (1993).

16.10 The Abraham–Lorentz equation of motion (16.9) can be replaced by an integro-differential equation if the external force is considered a function of time.

(a) Show that a first integral of (16.9) that eliminates the possibility of "runaway" solutions is

$$m\dot{\mathbf{v}}(t) = \int_0^\infty e^{-s}\mathbf{F}(t + \tau s)\, ds$$

(b) Show that a Taylor series expansion of the force for small τ leads to

$$m\dot{\mathbf{v}}(t) = \sum_{n=0}^\infty \tau^n \frac{d^n\mathbf{F}(t)}{dt^n}$$

The approximate equation (16.10) contains the first two terms of the infinite series.

(c) For a step-function force in one dimension, $F(t) = F_0\Theta(t)$, solve the integro-differential equation of part a for the acceleration and velocity for $t < 0$ and $t > 0$ for a particle at rest at $t = -\infty$. Plot ma/F_0 and $mv/F_0\tau$ in units of t/τ. Compare with the solution from (16.10). Comment.

16.11 A nonrelativistic particle of charge e and mass m is accelerated in one-dimensional motion across a gap of width d by a constant electric field. The mathematical idealization is that the particle has applied to it an external force $m\alpha$ while its coordinate lies in the interval $(0, d)$. Without radiation damping the particle, having initial velocity v_0, is accelerated uniformly for a time $T = (-v_0/\alpha) + \sqrt{(v_0^2/\alpha^2) + (2d/\alpha)}$, emerging at $x = d$ with a final velocity $v_1 = \sqrt{v_0^2 + 2\alpha d}$.

With radiation damping the motion is altered so that the particle takes a time T' to cross the gap and emerges with a velocity v_1'.

(a) Solve the integro-differential equation of motion, including damping, assuming T and T' large compared to τ. Sketch a velocity-versus-time diagram for the motion with and without damping.

(b) Show that to lowest order in τ,

$$T' = T - \tau\left(1 - \frac{v_0}{v_1}\right)$$

$$v_1' = v_1 - \frac{\alpha^2\tau}{v_1}T$$

(c) Verify that the sum of the energy radiated and the change in the particle's kinetic energy is equal to the work done by the applied field.

16.12 A classical model for the description of collision broadening of spectral lines is that the oscillator is interrupted by a collision after oscillating for a time T so that the coherence of the wave train is lost.

(a) Taking the oscillator used in Section 16.7 and assuming that the probability that a collision will occur between time T and $(T + dT)$ is $(ve^{-vT}\, dT)$, where

ν is the mean collision frequency, show that the averaged spectral distribution is

$$\frac{dI(\omega)}{d\omega} = \frac{I_0}{2\pi} \frac{\Gamma + 2\nu}{(\omega - \omega_0)^2 + \left(\dfrac{\Gamma}{2} + \nu\right)^2}$$

so that the breadth of the line is $(2\nu + \Gamma)$.

(b) For the sodium doublet at 5893 Å the oscillator strength is $f = 0.975$, so that the natural width is essentially the classical value, $\Delta\lambda = 1.2 \times 10^{-4}$ Å. Estimate the Doppler width of the line, assuming the sodium atoms are in thermal equilibrium at a temperature of 500K, and compare it with the natural width. Assuming a collision cross section of 10^{-16} cm^2, determine the collision breadth of the sodium doublet as a function of the pressure of the sodium vapor. For what pressure is the collision breadth equal to the natural breadth? The Doppler breadth?

16.13 A single particle oscillator under the action of an applied electric field $\mathbf{E}_0 e^{-i\omega t}$ has a dipole moment given by

$$\mathbf{p} = \alpha(\omega)\mathbf{E}_0 e^{-i\omega t}$$

(a) Show that the total dipole cross section can be written as

$$\sigma_t(\omega) = \frac{2\pi}{c}[-i\omega\alpha(\omega) + \text{c.c.}]$$

(b) Using only the facts that all the normal modes of oscillation must have some damping and that the polarizability $\alpha(\omega)$ must approach the free-particle value $(-e^2/m\omega^2)$ at high frequencies, show that the cross section satisfies the dipole sum rule,

$$\int_0^\infty \sigma_t(\omega)\, d\omega = \frac{2\pi^2 e^2}{mc}$$

(The discussion of Kramers–Kronig dispersion relations in Chapter 7 is clearly relevant.)

Appendix on Units and Dimensions

The question of units and dimensions in electricity and magnetism has exercised a great number of physicists and engineers over the years. This situation is in marked contrast with the almost universal agreement on the basic units of length (centimeter or meter), mass (gram or kilogram), and time (mean solar second). The reason perhaps is that the mechanical units were defined when the idea of "absolute" standards was a novel concept (just before 1800), and they were urged on the professional and commercial world by a group of scientific giants (Borda, Laplace, and others). By the time the problem of electromagnetic units arose there were (and still are) many experts. The purpose of this appendix is to add as little heat and as much light as possible without belaboring the issue.

1 Units and Dimensions; Basic Units and Derived Units

The *arbitrariness* in the *number* of fundamental units and in the *dimensions* of any physical quantity in terms of those units has been emphasized by Abraham, Planck, Bridgman,* Birge,[†] and others. The reader interested in units as such will do well to become familiar with the excellent series of articles by Birge.

The desirable features of a system of units in any field are convenience and clarity. For example, theoretical physicists active in relativistic quantum field theory and the theory of elementary particles find it convenient to *choose* the universal constants such as Planck's quantum of action and the velocity of light in vacuum to be *dimensionless* and of *unit magnitude*. The resulting system of units (called "natural" units) has only *one* basic unit, customarily chosen to be mass. All quantities, whether length or time or force or energy, etc., are expressed in terms of this one unit and have dimensions that are powers of its dimension. There is nothing contrived or less fundamental about such a system than one involving the meter, the kilogram, and the second as basic units. It is merely a matter of convenience.[‡]

A word needs to be said about basic units or standards, considered as independent quantities, and derived units or standards, which are defined in both magnitude and dimension through theory and experiment in terms of the basic units. Tradition requires that mass (m), length (l), and time (t) be treated as basic. But for electrical quantities there has been no compelling tradition. Consider, for example, the unit of current. The "international" ampere (for a long

---

*P. W. Bridgman, *Dimensional Analysis*, Yale University Press, New Haven, CT (1931).

[†]R. T. Birge, *Am. Phys. Teacher* (now *Am. J. Phys.*), **2**, 41 (1934); **3**, 102, 171 (1935).

[‡]In quantum field theory, powers of the coupling constant play the role of other basic units in doing dimensional analysis.

period the accepted practical unit of current) is defined in terms of the mass of silver deposited per unit time by electrolysis in a standard silver voltameter. Such a unit of current is properly considered a basic unit, independent of the mass, length, and time units, since the amount of current serving as the unit is found from a supposedly reproducible experiment in electrolysis.

On the other hand, the presently accepted standard of current, the "absolute" ampere "is that constant current which, if maintained in two straight parallel conductors of infinite length, of negligible circular cross section, and placed one metre apart in vacuum, would produce between these conductors a force equal to $2 \cdot 10^{-7}$ newton per metre of length." This means that the "absolute" ampere is a derived unit, since its definition is in terms of the mechanical force between two wires through equation (A.4) below.* The "absolute" ampere is, by this definition, exactly one-tenth of the em unit of current, the abampere.

Since 1948 the internationally accepted system of electromagnetic standards has been based on the meter, the kilogram, the second, and the above definition of the absolute ampere plus other derived units for resistance, voltage, etc. This seems to be a desirable state of affairs. It avoids such difficulties as arose when, in 1894, by act of Congress (based on recommendations of an international commission of engineers and scientists), independent basic units of current, voltage, and resistance were defined in terms of three independent experiments (silver voltameter, Clark standard cell, specified column of mercury).[†] Soon afterward, because of systematic errors in the experiments outside the claimed accuracy, Ohm's law was no longer valid, by act of Congress!

The Système International d'Unités (SI) has the unit of mass defined since 1889 by a platinum-iridium *kilogram* prototype kept in Sèvres, France. In 1967 the SI *second* was defined to be "the duration of 9 192 631 770 periods of the radiation corresponding to the transition between the two hyperfine levels of the ground state of the cesium-133 atom." The General Conference on Weights and Measures in 1983 adopted a definition of the *meter* based on the speed of light, namely, the *meter* is "the length of the distance traveled in vacuum by light during a time 1/299 792 458 of a second." The speed of light is therefore no longer an experimental number; it is, by definition of the meter, exactly $c = 299\ 792\ 458$ m/s. For electricity and magnetism, the Système International adds the absolute ampere as an additional unit, as already noted. In practice, metrology laboratories around the world define the ampere through the units of electromotive force, the volt, and resistance, the ohm, as determined experimentally from the Josephson effect $(2e/h)$ and the quantum Hall effect (h/e^2), respectively.[‡]

---

*The proportionality constant k_2 in (A.4) is thereby given the magnitude $k_2 = 10^{-7}$ in the SI system. The *dimensions* of the "absolute" ampere, as distinct from its magnitude, depend on the dimensions assigned k_2. In the conventional SI system of electromagnetic units, electric current (I) is arbitrarily chosen as a *fourth* basic dimension. Consequently charge has dimensions It, and k_2 has dimensions of $mlI^{-2}t^{-2}$. If k_2 is taken to be dimensionless, then current has the dimensions $m^{1/2}l^{1/2}t^{-1}$. The question of whether a fourth basic dimension like current is introduced or whether electromagnetic quantities have dimensions given by powers (sometimes fractional) of the three basic mechanical dimensions is a purely subjective matter and has no fundamental significance.

[†]See, for example, F. A. Laws, *Electrical Measurements*, McGraw-Hill, New York (1917), pp. 705–706.

[‡]For a general discussion of SI units in electricity and magnetism and the use of quantum phenomena to define standards, see B. W. Petley, in *Metrology at the Frontiers of Physics and Technology*, eds. L. Corvini and T. J. Quinn, Proceedings of the International School of Physics "Enrico Fermi," Course CX, 27 June–7 July 1989, North-Holland, Amsterdam (1992), pp. 33–61.

2   *Electromagnetic Units and Equations*

In discussing the units and dimensions of electromagnetism we take as our starting point the traditional choice of length (l), mass (m), and time (t) as independent, basic dimensions. Furthermore, we make the commonly accepted definition of current as the time rate of change of charge ($I = dq/dt$). This means that the dimension of the ratio of charge and current is that of time.* The continuity equation for charge and current densities then takes the form:

$$\mathbf{\nabla} \cdot \mathbf{J} + \frac{\partial \rho}{\partial t} = 0 \tag{A.1}$$

To simplify matters we initially consider only electromagnetic phenomena in free space, apart from the presence of charges and currents.

The basic physical law governing electrostatics is Coulomb's law on the force between two point charges q and q', separated by a distance r. In symbols this law is

$$F_1 = k_1 \frac{qq'}{r^2} \tag{A.2}$$

The constant k_1 is a proportionality constant whose magnitude and dimensions *either* are determined by the equation (if the magnitude and dimensions of the unit of charge have been specified independently) *or* are chosen arbitrarily in order to define the unit of charge. Within our present framework all that is determined at the moment is that the product ($k_1 qq'$) has the dimensions (ml^3t^{-2}).

The electric field $\mathbf{E}$ is a derived quantity, customarily defined to be the force per unit charge. A more general definition would be that the electric field be numerically proportional to the force per unit charge, with a proportionality constant that is a universal constant perhaps having dimensions such that the electric field is dimensionally different from force per unit charge. There is, however, nothing to be gained by this extra freedom in the definition of $\mathbf{E}$, since $\mathbf{E}$ is the first derived field quantity to be defined. Only when we define other field quantities may it be convenient to insert dimensional proportionality constants in the definitions in order to adjust the dimensions and magnitude of these fields relative to the electric field. Consequently, with no significant loss of generality the electric field of a point charge q may be defined from (A.2) as the force per unit charge,

$$E = k_1 \frac{q}{r^2} \tag{A.3}$$

All systems of units known to the author use this definition of electric field.

For steady-state magnetic phenomena Ampère's observations form a basis for specifying the interaction and defining the magnetic induction. According to Ampère, the force per unit length between two infinitely long, parallel wires separated by a distance d and carrying currents I and I' is

$$\frac{dF_2}{dl} = 2k_2 \frac{II'}{d} \tag{A.4}$$

---

*From the point of view of special relativity it would be more natural to give current the dimensions of charge divided by length. Then current density J and charge density ρ would have the same dimensions and would form a "natural" 4-vector. This is the choice made in a modified Gaussian system (see the footnote to Table 4, below).

The constant k_2 is a proportionality constant akin to k_1 in (A.2). The dimensionless number 2 is inserted in (A.4) for later convenience in specifying k_2. Because of our choice of the dimensions of current and charge embodied in (A.1), the dimensions of k_2 relative to k_1 are determined. From (A.2) and (A.4) it is easily found that the ratio k_1/k_2 has the dimension of a velocity squared (l^2t^{-2}). Furthermore, by comparison of the magnitude of the two mechanical forces (A.2) and (A.4) for known charges and currents, the magnitude of the ratio k_1/k_2 in free space can be found. The numerical value is closely given by the square of the velocity of light in vacuum. Therefore in symbols we can write

$$\frac{k_1}{k_2} = c^2 \tag{A.5}$$

where c stands for the velocity of light in magnitude and dimension.

The magnetic induction **B** is derived from the force laws of Ampère as being numerically proportional to the force per unit current with a proportionality constant α that may have certain dimensions chosen for convenience. Thus for a long straight wire carrying a current I, the magnetic induction **B** at a distance d has the magnitude (and dimensions)

$$B = 2k_2\alpha \frac{I}{d} \tag{A.6}$$

The dimensions of the ratio of electric field to magnetic induction can be found from (A.1), (A.3), (A.5), and (A.6). The result is that (E/B) has the dimensions $(l/t\alpha)$.

The third and final relation in the specification of electromagnetic units and dimensions is Faraday's law of induction, which connects electric and magnetic phenomena. The observed law that the electromotive force induced around a circuit is proportional to the rate of change of magnetic flux through it takes on the differential form

$$\nabla \times \mathbf{E} + k_3 \frac{\partial \mathbf{B}}{\partial t} = 0 \tag{A.7}$$

where k_3 is a constant of proportionality. Since the dimensions of **E** relative to **B** are established, the dimensions of k_3 can be expressed in terms of previously defined quantities merely by demanding that both terms in (A.7) have the same dimensions. Then it is found that k_3 has the dimensions of α^{-1}. Actually, k_3 is *equal* to α^{-1}. This is established on the basis of Galilean invariance in Section 5.15. But the easiest way to prove the equality is to write all the Maxwell equations in terms of the fields defined here:

$$\left. \begin{aligned} \nabla \cdot \mathbf{E} &= 4\pi k_1 \rho \\[4pt] \nabla \times \mathbf{B} &= 4\pi k_2\alpha \mathbf{J} + \frac{k_2\alpha}{k_1}\frac{\partial \mathbf{E}}{\partial t} \\[4pt] \nabla \times \mathbf{E} + k_3 \frac{\partial \mathbf{B}}{\partial t} &= 0 \\[4pt] \nabla \cdot \mathbf{B} &= 0 \end{aligned} \right\} \tag{A.8}$$

Then for source-free regions the two curl equations can be combined into the wave equation,

$$\nabla^2 \mathbf{B} - k_3 \frac{k_2 \alpha}{k_1} \frac{\partial^2 \mathbf{B}}{\partial t^2} = 0 \tag{A.9}$$

The velocity of propagation of the waves described by (A.9) is related to the combination of constants appearing there. Since this velocity is known to be that of light, we may write

$$\frac{k_1}{k_3 k_2 \alpha} = c^2 \tag{A.10}$$

Combining (A.5) with (A.10), we find

$$k_3 = \frac{1}{\alpha} \tag{A.11}$$

an equality holding for both magnitude and dimensions.

3   *Various Systems of Electromagnetic Units*

The various systems of electromagnetic units differ in their choices of the magnitudes and dimensions of the various constants above. Because of relations (A.5) and (A.11) there are only two constants (e.g., k_2, k_3) that can (and must) be chosen arbitrarily. It is convenient, however, to tabulate all four constants (k_1, k_2, α, k_3) for the commoner systems of units. These are given in Table 1. We note that, apart from dimensions, the em units and SI units are very similar, differing only in various powers of 10 in their mechanical and electromagnetic units. The Gaussian and Heaviside–Lorentz systems differ only by factors of 4π.

Table 1   Magnitudes and Dimensions of the Electromagnetic Constants for Various Systems of Units

The dimensions are given after the numerical values. The symbol c stands for the velocity of light in vacuum ($c \simeq 2.998 \times 10^{10}$ cm/s $\simeq 2.998 \times 10^8$ m/s). The first four systems of units use the centimeter, gram, and second as their fundamental units of length, mass, and time (l, m, t). The SI system uses the meter, kilogram, and second, plus current (I) as a fourth dimension, with the ampere as unit.

System	k_1	k_2	α	k_3
Electrostatic (esu)	1	$c^{-2}(t^2 l^{-2})$	1	1
Electromagnetic (emu)	$c^2(l^2 t^{-2})$	1	1	1
Gaussian	1	$c^{-2}(t^2 l^{-2})$	$c(lt^{-1})$	$c^{-1}(tl^{-1})$
Heaviside–Lorentz	$\dfrac{1}{4\pi}$	$\dfrac{1}{4\pi c^2}(t^2 l^{-2})$	$c(lt^{-1})$	$c^{-1}(tl^{-1})$
SI	$\dfrac{1}{4\pi\epsilon_0} = 10^{-7}c^2$ $(ml^3 t^{-4} I^{-2})$	$\dfrac{\mu_0}{4\pi} \equiv 10^{-7}$ $(mlt^{-2} I^{-2})$	1	1

Only in the Gaussian (and Heaviside–Lorentz) system does k_3 have dimensions. It is evident from (A.7) that, with k_3 having dimensions of a reciprocal velocity, **E** and **B** have the same dimensions. Furthermore, with $k_3 = c^{-1}$, (A.7) shows that for electromagnetic waves in free space **E** and **B** are equal in magnitude as well.

For SI units, (A.10) reads $1/(\mu_0\epsilon_0) = c^2$. With c now defined as a nine-digit number and $k_2 \equiv \mu_0/4\pi = 10^{-7}$ F/m, also by definition, 10^7 times the constant k_1 in Coulomb's law is

$$\frac{10^7}{4\pi\epsilon_0} = c^2 = 89\ 875\ 517\ 873\ 681\ 764$$

an exact 17-digit number (approximately 8.9876×10^{16}). Use of the speed of light without error to define the meter in terms of the second removes the anomaly in SI units of having one of the fundamental proportionality constants ϵ_0 with experimental errors. Note that, although the right-hand side above is the square of the speed of light, the *dimensions* of ϵ_0 (as distinct from its magnitude) are not seconds squared per meter squared because the numerical factor on the left has the dimensions of μ_0^{-1}. The dimensions of $1/\epsilon_0$ and μ_0 are given in Table 1. It is conventional to express the dimensions of ϵ_0 as farads per meter and those of μ_0 as henrys per meter. With $k_3 = 1$ and dimensionless, **E** and c**B** have the same dimensions in SI units; for a plane wave in vacuum they are equal in magnitude.

Only electromagnetic fields in free space have been discussed so far. Consequently only the two fundamental fields **E** and **B** have appeared. There remains the task of defining the macroscopic field variables **D** and **H**. If the averaged electromagnetic properties of a material medium are described by a macroscopic polarization **P** and a magnetization **M**, the general form of the definitions of **D** and **H** are

$$\left. \begin{aligned} \mathbf{D} &= \epsilon_0\mathbf{E} + \lambda\mathbf{P} \\ \mathbf{H} &= \frac{1}{\mu_0}\mathbf{B} - \lambda'\mathbf{M} \end{aligned} \right\} \tag{A.12}$$

where $\epsilon_0, \mu_0, \lambda, \lambda'$ are proportionality constants. Nothing is gained by making **D** and **P** or **H** and **M** have different dimensions. Consequently λ and λ' are chosen as pure numbers ($\lambda = \lambda' = 1$ in rationalized systems, $\lambda = \lambda' = 4\pi$ in unrationalized systems). But there is the choice as to whether **D** and **P** will differ in dimensions from **E**, and **H** and **M** differ from **B**. This choice is made for convenience and simplicity, usually to make the macroscopic Maxwell equations have a relatively simple, neat form. Before tabulating the choices made for different systems, we note that for linear, isotropic media the constitutive relations are always written

$$\left. \begin{aligned} \mathbf{D} &= \epsilon\mathbf{E} \\ \mathbf{B} &= \mu\mathbf{H} \end{aligned} \right\} \tag{A.13}$$

Thus in (A.12) the constants ϵ_0 and μ_0 are the vacuum values of ϵ and μ. The relative permittivity of a substance (often called the *dielectric constant*) is defined as the dimensionless ratio (ϵ/ϵ_0), while the relative permeability (often called the *permeability*) is defined as (μ/μ_0).

Table 2 displays the values of ϵ_0 and μ_0, the defining equations for **D** and **H**, the macroscopic forms of the Maxwell equations, and the Lorentz force equation

Table 2 Definitions of ϵ_0, μ_0, **D**, **H**, Macroscopic Maxwell Equations, and Lorentz Force Equation in Various Systems of Units

Where necessary the dimensions of quantities are given in parentheses. The symbol c stands for the velocity of light in vacuum with dimensions (lt^{-1}).

System	ϵ_0	μ_0	D, H	Macroscopic Maxwell Equations				Lorentz Force per Unit Charge
Electrostatic (esu)	1	c^{-2} $(t^2 l^{-2})$	$\mathbf{D} = \mathbf{E} + 4\pi\mathbf{P}$ $\mathbf{H} = c^2\mathbf{B} - 4\pi\mathbf{M}$	$\nabla \cdot \mathbf{D} = 4\pi\rho$	$\nabla \times \mathbf{H} = 4\pi\mathbf{J} + \dfrac{\partial \mathbf{D}}{\partial t}$	$\nabla \times \mathbf{E} + \dfrac{\partial \mathbf{B}}{\partial t} = 0$	$\nabla \cdot \mathbf{B} = 0$	$\mathbf{E} + \mathbf{v} \times \mathbf{B}$
Electromagnetic (emu)	c^{-2} $(t^2 l^{-2})$	1	$\mathbf{D} = \dfrac{1}{c^2}\mathbf{E} + 4\pi\mathbf{P}$ $\mathbf{H} = \mathbf{B} - 4\pi\mathbf{M}$	$\nabla \cdot \mathbf{D} = 4\pi\rho$	$\nabla \times \mathbf{H} = 4\pi\mathbf{J} + \dfrac{\partial \mathbf{D}}{\partial t}$	$\nabla \times \mathbf{E} + \dfrac{\partial \mathbf{B}}{\partial t} = 0$	$\nabla \cdot \mathbf{B} = 0$	$\mathbf{E} + \mathbf{v} \times \mathbf{B}$
Gaussian	1	1	$\mathbf{D} = \mathbf{E} + 4\pi\mathbf{P}$ $\mathbf{H} = \mathbf{B} - 4\pi\mathbf{M}$	$\nabla \cdot \mathbf{D} = 4\pi\rho$	$\nabla \times \mathbf{H} = \dfrac{4\pi}{c}\mathbf{J} + \dfrac{1}{c}\dfrac{\partial \mathbf{D}}{\partial t}$	$\nabla \times \mathbf{E} + \dfrac{1}{c}\dfrac{\partial \mathbf{B}}{\partial t} = 0$	$\nabla \cdot \mathbf{B} = 0$	$\mathbf{E} + \dfrac{\mathbf{v}}{c} \times \mathbf{B}$
Heaviside–Lorentz	1	1	$\mathbf{D} = \mathbf{E} + \mathbf{P}$ $\mathbf{H} = \mathbf{B} - \mathbf{M}$	$\nabla \cdot \mathbf{D} = \rho$	$\nabla \times \mathbf{H} = \dfrac{1}{c}\left(\mathbf{J} + \dfrac{\partial \mathbf{D}}{\partial t}\right)$	$\nabla \times \mathbf{E} + \dfrac{1}{c}\dfrac{\partial \mathbf{B}}{\partial t} = 0$	$\nabla \cdot \mathbf{B} = 0$	$\mathbf{E} + \dfrac{\mathbf{v}}{c} \times \mathbf{B}$
SI	$\dfrac{10^7}{4\pi c^2}$ $(I^2 t^4 m^{-1} l^{-3})$	$4\pi \times 10^{-7}$ $(ml I^{-2} t^{-2})$	$\mathbf{D} = \epsilon_0 \mathbf{E} + \mathbf{P}$ $\mathbf{H} = \dfrac{1}{\mu_0}\mathbf{B} - \mathbf{M}$	$\nabla \cdot \mathbf{D} = \rho$	$\nabla \times \mathbf{H} = \mathbf{J} + \dfrac{\partial \mathbf{D}}{\partial t}$	$\nabla \times \mathbf{E} + \dfrac{\partial \mathbf{B}}{\partial t} = 0$	$\nabla \cdot \mathbf{B} = 0$	$\mathbf{E} + \mathbf{v} \times \mathbf{B}$

in the five common systems of units of Table 1. For each system of units the continuity equation for charge and current is given by (A.1), as can be verified from the first pair of the Maxwell equations in the table in each case.* Similarly, in all systems the statement of Ohm's law is $\mathbf{J} = \sigma\mathbf{E}$, where σ is the conductivity.

4 Conversion of Equations and Amounts Between SI Units and Gaussian Units

The two systems of electromagnetic units in most common use today are the SI and Gaussian systems. The SI system has the virtue of overall convenience in

Table 3 Conversion Table for Symbols and Formulas

The symbols for mass, length, time, force, and other not specifically electromagnetic quantities are unchanged. To convert any equation in SI variables to the corresponding equation in Gaussian quantities, on both sides of the equation replace the relevant symbols listed below under "SI" by the corresponding "Gaussian" symbols listed on the left. The reverse transformation is also allowed. Residual powers of $\mu_0\epsilon_0$ should be eliminated in favor of the speed of light ($c^2\mu_0\epsilon_0 = 1$). Since the length and time symbols are unchanged, quantities that differ dimensionally from one another only by powers of length and/or time are grouped together where possible.

Quantity	Gaussian	SI
Velocity of light	c	$(\mu_0\epsilon_0)^{-1/2}$
Electric field (potential, voltage)	$\mathbf{E}(\Phi, V)/\sqrt{4\pi\epsilon_0}$	$\mathbf{E}(\Phi, V)$
Displacement	$\sqrt{\epsilon_0/4\pi}\,\mathbf{D}$	$\mathbf{D}$
Charge density (charge, current density, current, polarization)	$\sqrt{4\pi\epsilon_0}\,\rho(q, \mathbf{J}, I, \mathbf{P})$	$\rho(q, \mathbf{J}, I, \mathbf{P})$
Magnetic induction	$\sqrt{\mu_0/4\pi}\,\mathbf{B}$	$\mathbf{B}$
Magnetic field	$\mathbf{H}/\sqrt{4\pi\mu_0}$	$\mathbf{H}$
Magnetization	$\sqrt{4\pi/\mu_0}\,\mathbf{M}$	$\mathbf{M}$
Conductivity	$4\pi\epsilon_0\sigma$	σ
Dielectric constant	$\epsilon_0\epsilon$	ϵ
Magnetic permeability	$\mu_0\mu$	μ
Resistance (impedance)	$R(Z)/4\pi\epsilon_0$	$R(Z)$
Inductance	$L/4\pi\epsilon_0$	L
Capacitance	$4\pi\epsilon_0 C$	C

$$c = 2.997\ 924\ 58 \times 10^8 \text{ m/s}$$

$$\epsilon_0 = 8.854\ 187\ 8 \ldots \times 10^{-12} \text{ F/m}$$

$$\mu_0 = 1.256\ 637\ 0 \ldots \times 10^{-6} \text{ H/m}$$

$$\sqrt{\frac{\mu_0}{\epsilon_0}} = 376.730\ 3 \ldots \ \Omega$$

*Some workers employ a modified Gaussian system of units in which current is defined by $I = (1/c)(dq/dt)$. Then the current density $\mathbf{J}$ in Table 2 must be replaced by $c\mathbf{J}$, and the continuity equation is $\nabla \cdot \mathbf{J} + (1/c)(\partial\rho/\partial t) = 0$. See also the footnote to Table 4.

Table 4   Conversion Table for Given Amounts of a Physical Quantity

The table is arranged so that a given amount of some physical quantity, expressed as so many SI or Gaussian units of that quantity, can be expressed as an equivalent number of units in the other system. Thus the entries in each row stand for the same amount, expressed in different units. All factors of 3 (apart from exponents) should, for accurate work, be replaced by (2.997 924 58), arising from the numerical value of the velocity of light. For example, in the row for displacement (D), the entry ($12\pi \times 10^5$) is actually ($2.997\ 924\ 58 \times 4\pi \times 10^5$) and "9" is actually $10^{-16} c^2 = 8.987\ 55\ldots$. Where a name for a unit has been agreed on or is in common usage, that name is given. Otherwise, one merely reads so many Gaussian units, or SI units.

Physical Quantity	Symbol	SI		Gaussian
Length	l	1 meter (m)	10^2	centimeters (cm)
Mass	m	1 kilogram (kg)	10^3	grams (g)
Time	t	1 second (s)	1	second (s)
Frequency	ν	1 hertz (Hz)	1	hertz (Hz)
Force	F	1 newton (N)	10^5	dynes
Work Energy	$\left.\begin{array}{c}W\\U\end{array}\right\}$	1 joule (J)	10^7	ergs
Power	P	1 watt (W)	10^7	ergs s^{-1}
Charge	q	1 coulomb (C)	3×10^9	statcoulombs
Charge density	ρ	1 C m^{-3}	3×10^3	statcoul cm^{-3}
Current	I	1 ampere (A)	3×10^9	statamperes
Current density	J	1 A m^{-2}	3×10^5	statamp cm^{-2}
Electric field	E	1 volt m^{-1} (Vm^{-1})	$\frac{1}{3} \times 10^{-4}$	statvolt cm^{-1}
Potential	Φ, V	1 volt (V)	$\frac{1}{300}$	statvolt
Polarization	P	1 C m^{-2}	3×10^5	dipole moment cm^{-3}
Displacement	D	1 C m^{-2}	$12\pi \times 10^5$	statvolt cm^{-1} (statcoul cm^{-2})
Conductivity	σ	1 mho m^{-1}	9×10^9	s^{-1}
Resistance	R	1 ohm (Ω)	$\frac{1}{9} \times 10^{-11}$	s cm^{-1}
Capacitance	C	1 farad (F)	9×10^{11}	cm
Magnetic flux	ϕ, F	1 weber (Wb)	10^8	gauss cm^2 or maxwells
Magnetic induction	B	1 tesla (T)	10^4	gauss (G)
Magnetic field	H	1 A m^{-1}	$4\pi \times 10^{-3}$	oersted (Oe)
Magnetization	M	1 A m^{-1}	10^{-3}	magnetic moment cm^{-3}
Inductance*	L	1 henry (H)	$\frac{1}{9} \times 10^{-11}$	

*There is some confusion about the unit of inductance in Gaussian units. This stems from the use by some authors of a modified system of Gaussian units in which current is measured in electromagnetic units, so that the connection between charge and current is $I_m = (1/c)(dq/dt)$. Since inductance is defined through the induced voltage $V = L(dI/dt)$ or the energy $U = \frac{1}{2}LI^2$, the choice of current defined in Section 2 means that our Gaussian unit of inductance is equal in magnitude and dimensions (t^2l^{-1}) to the electrostatic unit of inductance. The electromagnetic current I_m is related to our Gaussian current I by the relation $I_m = (1/c)I$. From the energy definition of inductance, we see that the electromagnetic inductance L_m is related to our Gaussian inductance L through $L_m = c^2L$. Thus L_m has the dimensions of length. The modified Gaussian system generally uses the electromagnetic unit of inductance, as well as current. Then the voltage relation reads $V = (L_m/c)(dI_m/dt)$. The numerical connection between units of inductance is

$$1 \text{ henry} = \tfrac{1}{9} \times 10^{-11} \text{ Gaussian (es) unit} = 10^9 \text{ emu}$$

practical, large-scale phenomena, especially in engineering applications. The Gaussian system is more suitable for microscopic problems involving the electrodynamics of individual charged particles, etc. Previous editions have used Gaussian units throughout, apart from Chapter 8, where factors in square brackets could be omitted for the reader wishing SI units. In this edition, SI units are employed exclusively in the first 10 chapters. For the relativistic electrodynamics of the latter part of the book, we retain Gaussian units as a matter of convenience. A reminder of the units being used appears at the top of every left-hand page, with the designation, **Chapter Heading—SI** or **Chapter Heading—G**. Some may feel it awkward to have two systems of units in use, but the reality is that scientists must be conversant in many languages—SI units are rarely used for electromagnetic interactions in quantum mechanics, but atomic or Hartree units are, and similarly in other fields.

Tables 3 and 4 are designed for general use in conversion from one system to the other. Table 3 is a conversion scheme for *symbols and equations* that allows the reader to convert any equation from the Gaussian system to the SI system and vice versa. Simpler schemes are available for conversion only *from* the SI system *to* the Gaussian system, and other general schemes are possible. But by keeping all mechanical quantities unchanged, the recipe in Table 3 allows the straightforward conversion of quantities that arise from an interplay of electromagnetic and mechanical forces (e.g., the fine structure constant $e^2/\hbar c$ and the plasma frequency $\omega_p^2 = 4\pi n e^2/m$) without additional considerations. Table 4 is a conversion table for units to allow the reader to express a given amount of any physical entity as a certain number of SI units or cgs-Gaussian units.

Bibliography

ABRAHAM, M., AND R. BECKER, *Electricity and Magnetism*, Blackie, London (1937), translation from 8th German edition of *Theorie der Elektrizität*, Band I.

——, *Theorie der Elektrizität*, Band II, *Elektronentheorie*, Teubner, Leipzig (1933).

ABRAMOWITZ, M., AND I. A. STEGUN, eds., *Handbook of Mathematical Functions*, U.S. National Bureau of Standards (1964); Dover, New York (1965).

ADLER, R. B., L. J. CHU, AND R. M. FANO, *Electromagnetic Energy, Transmission and Radiation*, Wiley, New York (1960).

AHARONI, J., *The Special Theory of Relativity*, 2nd edition, Oxford University Press, Oxford (1965).

ALFVÉN, H., AND C.-G. FÄLTHAMMAR, *Cosmical Electrodynamics*, 2nd edition, Oxford University Press, Oxford (1963).

ANDERSON, J. L., *Principles of Relativity Physics*, Academic Press, New York (1967).

ARFKEN, G., AND H. J. WEBER, *Mathematical Methods for Physicists*, 4th edition, Academic Press, New York (1995).

ARGENCE, E., AND T. KAHAN, *Theory of Waveguides and Cavity Resonators*, Blackie, London (1967).

ARZELIÈS, H., *Relativistic Kinematics*, Pergamon Press, Oxford (1966).

——, *Relativistic Point Dynamics*, Pergamon Press, Oxford (1972).

ASHCROFT, N. W., AND N. D. MERMIN, *Solid State Physics*, Holt, Rinehart, and Winston, New York (1976).

BAKER, B. B., AND E. T. COPSON, *Mathematical Theory of Huygens' Principle*, 2nd edition, Oxford University Press, Oxford (1950).

BALDIN, A. M., V. I. GOL'DANSKII, AND I. L. ROZENTHAL, *Kinematics of Nuclear Reactions*, Pergamon Press, New York (1961).

BARUT, A. O., *Electrodynamics and Classical Theory of Fields and Particles*, Macmillan, New York (1964); Dover reprint (1980).

BATEMAN MANUSCRIPT PROJECT, *Higher Transcendental Functions*, 3 vols., edited by A. Erdélyi, McGraw-Hill, New York (1953).

——, *Tables of Integral Transforms*, 2 vols., edited by A. Erdélyi, McGraw-Hill, New York (1954).

BEAM, W. R., *Electronics of Solids*, McGraw-Hill, New York (1965).

BEKEFI, G., *Radiation Processes in Plasma*, Wiley, New York (1966).

BERGMANN, P. G., *Introduction to the Theory of Relativity*, Prentice-Hall, Englewood Cliffs, NJ (1942); Dover reprint (1976).

BIEBERBACH, L., *Conformal Mapping*, Chelsea, New York (1964).

BINNS, K. J., P. J. LAWRENSON, AND C. W. TROWBRIDGE, *The Analytic and Numerical Solution of Electric and Magnetic Fields*, Wiley, New York (1992).

BLATT, J. M., AND V. F. WEISSKOPF, *Theoretical Nuclear Physics*, Wiley, New York (1952).

BOHM, D., *The Special Theory of Relativity*, Benjamin, New York (1965); Addison-Wesley, Reading, MA (1989).

BOHR, N., Penetration of atomic particles through matter, *Kgl. Danske Videnskab. Selskab Mat.-fys. Medd.*, **XVIII**, No. 8 (1948).

BORN, M., AND E. WOLF, *Principles of Optics*, 6th corr. edition, Pergamon Press, New York (1989).

BÖTTCHER, C. J. F., *Theory of Electric Polarization*, Elsevier, New York (1952).

BOWMAN, J. J., T. B. A. SENIOR, AND P. L. E. USLENGHI, *Electromagnetic and Acoustic Scattering by Simple Shapes*, North-Holland, Amsterdam (1969); reprinted (1987).

BRILLOUIN, L., *Wave Propagation and Group Velocity*, Academic Press, New York (1960).

BUDDEN, K. G., *The Propagation of Radio Waves*, Cambridge University Press, New York (1985); paperback, corrected (1988).

BYERLY, W. E., *Fourier Series and Spherical Harmonics*, Ginn, Boston (1893); also Dover reprint.

CAIRO, L., AND T. KAHAN, *Variational Techniques in Electromagnetism*, Blackie, London (1965).

CHANDRASEKHAR, S., *Plasma Physics*, University of Chicago Press, Chicago (1960).

CHAKRAVARTY, A. S., *Introduction to the Magnetic Properties of Solids*, Wiley, New York (1980).

CHURCHILL, R. V., AND J. W. BROWN, *Fourier Series and Boundary Value Problems*, 5th edition, McGraw-Hill, New York (1993).

CLEMMOW, P. C., *The Plane Wave Spectrum Representation of Electromagnetic Fields*, Pergamon Press, Oxford (1966); reissued in IEEE/OUP series (1996).

CLEMMOW, P. C., AND J. P. DOUGHERTY, *Electrodynamics of Particles and Plasmas*, Addison-Wesley, Reading, MA (1969); reprinted (1990).

COLLIN, R. E., *Field Theory of Guided Waves*, 2nd edition, IEEE Press, Piscataway, NJ (1991).

CONDON, E. U., AND H. ODISHAW, eds., *Handbook of Physics*, 2nd edition, McGraw-Hill, New York (1967).

CORBEN, H. C., AND P. STEHLE, *Classical Mechanics*, 2nd edition, Wiley, New York (1960); reprinted (1974).

COURANT, R., AND D. HILBERT, *Methods of Mathematical Physics*, 2 vols., Wiley-Interscience, New York (1962).

COWLING, T. G., *Magnetohydrodynamics*, 2nd edition, Hilger, London (1976).

CRAIK, D. J., *Magnetism: Principles and Applications*, Wiley, New York (1995).

CULLWICK, E. G., *Electromagnetism and Relativity*, 2nd edition, Longmans, London (1959).

DEBYE, P., *Polar Molecules*, Dover, New York (1945).

DENNERY, P., AND A. KRZYWICKI, *Mathematics for Physicists*, Harper & Row, New York (1967).

DURAND, E., *Electrostatique et Magnétostatique*, Masson, Paris (1953).

EINSTEIN, A., H. A. LORENTZ, H. MINKOWSKI, AND H. WEYL, *The Principle of Relativity*. Collected papers, with notes by A. Sommerfeld, Dover, New York (1952).

FABELINSKII, I. L., *Molecular Scattering of Light*, Plenum Press, New York (1968).

FANO, R. M., L. J. CHU, AND R. B. ADLER, *Electromagnetic Fields, Energy, and Forces*, Wiley, New York (1960).

FEYNMAN, R. P., R. B. LEIGHTON, AND M. SANDS, *The Feynman Lectures on Physics*, 3 vols., Addison-Wesley, Reading, MA (1963).

FRENCH, A. P., *Special Relativity*, Norton, New York (1968).

FRIEDMAN, B., *Principles and Techniques of Applied Mathematics*, Wiley, New York (1956); Dover reprint (1990).

FRÖHLICH, H., *Theory of Dielectrics*, Oxford University Press, Oxford (1949).

GALEJS, J., *Terrestrial Propagation of Long Electromagnetic Waves*, Pergamon Press, Oxford (1972).

GIBBS, W. J., *Conformal Transformations in Electrical Engineering*, Chapman & Hall, London (1958).

GOLDSTEIN, H., *Classical Mechanics*, 2nd edition, Addison-Wesley, Reading, MA (1980).

GRADSHTEYN, I. S., AND I. M. RYZHIK, *Tables of Integrals, Series, and Products*, 4th edition, prepared by Yu. V. Geronimus and M. Yu. Tseytlin, corr. enlarged edition by A. Jeffrey, Academic Press, New York (1980).

DEGROOT, S. R., *The Maxwell Equations*, Studies in Statistical Mechanics, Vol. IV, North-Holland, Amsterdam (1969).

DEGROOT, S. R., AND L. G. SUTTORP, *Foundations of Electrodynamics*, North-Holland, Amsterdam (1972).

HADAMARD, J., *Lectures on Cauchy's Problem*, Yale University Press, New Haven, CT (1923); Dover reprint (1952).

HAGEDORN, R., *Relativistic Kinematics*, Benjamin, New York (1963, 1973).

HALLÉN, E., *Electromagnetic Theory*, Chapman & Hall, London (1962).

HARRINGTON, R. F., *Time-Harmonic Electromagnetic Fields*, McGraw-Hill, New York (1961).

HAUS, H. A., AND J. R. MELCHER, *Electromagnetic Fields and Energy*, Prentice-Hall, Englewood Cliffs, NJ (1989).

HEITLER, W., *Quantum Theory of Radiation*, 3rd edition, Oxford University Press, Oxford (1954).

HILDEBRAND, F. B., *Advanced Calculus for Applications*, 2nd edition, Prentice-Hall, Englewood Cliffs, NJ (1976).

IDA, N., AND J. P. A. BASTOS, *Electromagnetics and Calculation of Fields*, Springer-Verlag, New York (1992).

IWANENKO, D., AND A. SOKOLOW, *Klassische Feldtheorie*, Akademie-Verlag, Berlin (1953), translated from the Russian edition (1949).

JAHNKE-EMDE-LÖSCH, *Tables of Higher Functions*, 6th edition, revised by F. Lösch, Teubner, Stuttgart; McGraw-Hill, New York (1960).

JEANS, J. H., *Mathematical Theory of Electricity and Magnetism*, 5th edition, Cambridge University Press, Cambridge (1925); reprinted (1958).

JEFFREYS, H., AND B. S. JEFFREYS, *Methods of Mathematical Physics*, 3rd edition, Cambridge University Press, Cambridge (1956); reprinted (1972).

JEFIMENKO, O. D., *Electricity and Magnetism*, Appleton-Century Crofts, New York (1966); 2nd edition, Electret Scientific, Star City, WV (1989).

JOHNSON, C. C., *Field and Wave Electrodynamics*, McGraw-Hill, New York (1965).

JONES, D. S., *The Theory of Electromagnetism*, Pergamon Press, Oxford (1964).

JORDAN, E. C., AND K. G. BALMAIN, *Electromagnetic Waves and Radiating Systems*, 2nd edition, Prentice-Hall, Englewood Cliffs, NJ (1968).

KELVIN, LORD (SIR W. THOMSON), *Reprints of Papers on Electrostatics and Magnetism*, 2nd edition, Macmillan, London (1884).

KERKER, M., *The Scattering of Light and Other Electromagnetic Radiation*, Academic Press, New York (1969).

KELLOGG, O. D., *Foundations of Potential Theory*, Springer-Verlag, Berlin (1929); reprinted by Ungar, New York and by Springer-Verlag (1967).

KILMISTER, C. W., *Special Theory of Relativity*, Selected Readings in Physics, Pergamon Press, Oxford (1970).

KING, R. W. P., AND T. T. WU, *Scattering and Diffraction of Waves*, Harvard University Press, Cambridge, MA (1959).

KITTEL, C., *Introduction to Solid State Physics*, 7th edition, Wiley, New York (1996).

KRAUS, J. D., *Antennas*, 2nd edition, McGraw-Hill, New York (1988).

LANDAU, L. D., AND E. M. LIFSHITZ, *The Classical Theory of Fields*, 4th revised English edition, translated by M. Hamermesh, Pergamon Press, Oxford, and Addison-Wesley, Reading, MA (1987).

———, *Electrodynamics of Continuous Media*, 2nd edition, Addison-Wesley, Reading, MA (1984).

LIGHTHILL, M. J., *Introduction to Fourier Analysis and Generalised Functions*, Cambridge University Press, Cambridge (1958).

LINHART, J. G., *Plasma Physics*, 2nd edition, North-Holland, Amsterdam (1961).

LIVINGSTON, M. S., AND J. P. BLEWETT, *Particle Accelerators*, McGraw-Hill, New York (1962).

LORENTZ, H. A., *Theory of Electrons*, 2nd edition (1915); reprint, Dover, New York (1952).

LOW, F. E., *Classical Field Theory: Electromagnetism and Gravitation*, Wiley, New York (1997).

MAGNUS, W., F. OBERHETTINGER, AND R. P. SONI, *Formulas and Theorems for the Special Functions of Mathematical Physics*, Springer-Verlag, New York (1966).

MARCUVITZ, N., *Waveguide Handbook*, M.I.T. Radiation Laboratory Series, Vol. 10, McGraw-Hill, New York (1951).

MASON, M., AND W. WEAVER, *The Electromagnetic Field*, University of Chicago Press, Chicago (1929); Dover reprint (1952).

MAXWELL, J. C., *Treatise on Electricity and Magnetism*, 3rd edition (1891), 2 vols., reprint, Dover, New York (1954).

MERMIN, N. D., *Space and Time in Special Relativity*, McGraw-Hill, New York (1968); Waveland Press, Prospect Heights, IL (1989).

MØLLER, C., *The Theory of Relativity*, 2nd edition, Clarendon Press, Oxford (1972).

MONTGOMERY, C. G., R. H. DICKE, AND E. M. PURCELL, *Principles of Microwave Circuits*, M.I.T. Radiation Laboratory Series, Vol. 8, McGraw-Hill, New York (1948). Also available as a Dover reprint.

MORSE, P. M., AND H. FESHBACH, *Methods of Theoretical Physics*, 2 Pts., McGraw-Hill, New York (1953).

NORTHROP, T. G., *The Adiabatic Motion of Charged Particles*, Wiley-Interscience, New York (1963).

PANOFSKY, W. K. H., AND M. PHILLIPS, *Classical Electricity and Magnetism*, 2nd edition, Addison-Wesley, Reading, MA (1962).

PAULI, W., *Theory of Relativity*, Pergamon Press, New York (1958), translated from an article in the *Encyklopedie der mathematischen Wissenschaften*, Vol. V19, Teubner, Leipzig (1921), with supplementary notes by the author (1956).

PENFIELD, P., AND H. A. HAUS, *Electrodynamics of Moving Media*, M.I.T. Press, Cambridge, MA (1967).

PERSICO, E., E. FERRARI, AND S. E. SEGRE, *Principles of Particle Accelerators*, Benjamin, New York (1968).

PÓLYA, G., AND G. SZEGÖ, *Isoperimetric Inequalities in Mathematical Physics*, Annals of Mathematics Series No. 27, Princeton University Press, Princeton, NJ (1951).

PRESS, W. H., B. P. FLANNERY, S. A. TEUKOLSKY, AND W. T. VETTERLING, *Numerical Recipes, The Art of Scientific Computing*, Cambridge University Press, Cambridge (1986).

RAMO, S., J. R. WHINNERY, AND T. VAN DUZER, *Fields and Waves in Communication Electronics*, 3rd edition, Wiley, New York (1994).

RINDLER, W., *Essential Relativity: Special, General, and Cosmological*, rev. 2nd edition, Springer-Verlag, New York (1979).

ROBINSON, F. N. H., *Macroscopic Electromagnetism*, Pergamon Press, Oxford (1973).

ROHRLICH, F., *Classical Charged Particles*, Addison-Wesley, Reading, MA (1965, 1990).

ROSENFELD, L., *Theory of Electrons*, North-Holland, Amsterdam (1951).

ROSSI, B., *High-Energy Particles*, Prentice-Hall, Englewood Cliffs, NJ (1952).

ROSSI, B., AND S. OLBERT, *Introduction to the Physics of Space*, McGraw-Hill, New York (1970).

ROTHE, R., F. OLLENDORFF, AND K. POLHAUSEN, *Theory of Functions as Applied to Engineering Problems*, Technology Press, Cambridge, MA (1933), Dover reprint (1961).

SADIKU, M. N. O., *Numerical Techniques in Electromagnetics*, CRC Press, Boca Raton, FL (1992).

SARD, R. D., *Relativistic Mechanics*, Benjamin, New York (1970).

SCHELKUNOFF, S. A., *Advanced Antenna Theory*, Wiley, New York (1952).

———, *Applied Mathematics for Engineers and Scientists*, 2nd edition, Van Nostrand, New York (1965).

———, *Electromagnetic Fields*, Blaisdell, New York (1963).

SCHELKUNOFF, S. A., AND H. T. FRIIS, *Antennas, Theory and Practice*, Wiley, New York (1952).

SCHOTT, G. A., *Electromagnetic Radiation*, Cambridge University Press, Cambridge (1912).

SCHWARTZ, H. M., *Introduction to Special Relativity*, McGraw-Hill, New York (1968).

SEGRÈ, E., ed., *Experimental Nuclear Physics*, Vol. 1, Wiley, New York (1953).

SIEGBAHN, K., ed., *Beta- and Gamma-Ray Spectroscopy*, North-Holland, Amsterdam; Interscience, New York (1955).

SILVER, S., ed., *Microwave Antenna Theory and Design*, M.I.T. Radiation Laboratory Series, Vol. 12, McGraw-Hill, New York (1949).

SLATER, J. C., *Microwave Electronics*, McGraw-Hill, New York (1950).

SLATER, J. C., AND N. H. FRANK, *Electromagnetism*, McGraw-Hill, New York (1947).

SMITH, J. H., *Introduction to Special Relativity*, Benjamin, New York (1965).

SMYTHE, W. R., *Static and Dynamic Electricity*, 3rd edition, McGraw-Hill, New York (1969); reprinted (1989).

SOMMERFELD, A., *Electrodynamics*, Academic Press, New York (1952).

———, *Optics*, Academic Press, New York (1964).

———, *Partial Differential Equations in Physics*, Academic Press, New York (1964).

SOPER, D. E., *Classical Field Theory*, Wiley, New York (1976).

SPITZER, L., *Physics of Fully Ionized Gases*, 2nd edition, Interscience, New York (1962).

STONE, J. M., *Radiation and Optics*, McGraw-Hill, New York (1963).

STRANG, G., *Introduction to Applied Mathematics*, Wellesley-Cambridge Press, Wellesley, MA (1986).

STRATTON, J. A., *Electromagnetic Theory*, McGraw-Hill, New York (1941).

TAYLOR, E. F., AND J. A. WHEELER, *Spacetime Physics*, 2nd edition, Freeman, San Francisco (1992).

THIRRING, W. E., *Principles of Quantum Electrodynamics*, translated by J. Bernstein, Academic Press, New York (1958).

THOMSON, J. J., *Recent Researches in Electricity and Magnetism*, Clarendon Press, Oxford (1893).

TITCHMARSH, E. C., *Introduction to the Theory of Fourier Integrals*, 3rd edition, Oxford University Press, Oxford (1986).

TONNELAT, M.-A., *Electromagnetic Theory and Relativity*, Reidel, Holland (1966).

TRANTER, C. J., *Integral Transforms in Mathematical Physics*, 3rd edition, Methuen, London (1966).

VAN BLADEL, J., *Electromagnetic Fields*, McGraw-Hill, New York (1964); reprinted (1985).

VAN VLECK, J. H., *Theory of Electric and Magnetic Susceptibilities*, Oxford University Press, Oxford (1932).

WAIT, J. R., *Electromagnetic Waves in Stratified Media*, Pergamon Press, Oxford (1962); reissued in IEEE/OUP series (1996).

WALDRON, R. A., *Theory of Guided Electromagnetic Waves*, Van Nostrand-Reinhold, London (1970).

WATSON, G. N., *Theory of Bessel Functions*, 2nd edition, Cambridge University Press, Cambridge (1952); reprinted (1995).

WERT, C. A., AND R. M. THOMSON, *Physics of Solids*, 2nd edition, McGraw-Hill, New York (1970).

WHITTAKER, E. T., *A History of the Theories of Aether and Electricity*, 2 vols., Nelson,

London, Vol. 1, *The Classical Theories* (1910, revised and enlarged 1951), Vol. 2, *The Modern Theories 1900–1926* (1953); reprinted by Harper Torchbooks, New York (1960); Dover reprint (1989).

WHITTAKER, E. T., AND G. N. WATSON, *A Course in Modern Analysis*, 4th edition, Cambridge University Press, Cambridge (1950).

WILLIAMS, E. J., Correlation of certain collision problems with radiation theory, *Kgl. Danske Videnskab. Selskab Mat.-fys. Medd.*, **XIII**, No. 4 (1935).

WOOTEN, F., *Optical Properties of Solids*, Academic Press, New York (1972).

ZHOU, P.-B., *Numerical Analysis of Electromagnetic Fields*, Springer-Verlag, New York (1993).

ZIMAN, J. M., *Principles of the Theory of Solids*, 2nd edition, Cambridge University Press, Cambridge (1972).

Index

Where to Find Key Material on Special Functions